Recommended Dietary Allowances (RDA) and Adequate Intakes (AI) for Vitamins

Age (yr)	Thiamin RDA (mg/day)	Riboflavin RDA (mg/day)	Niacin RDA (mg/day)[a]	Biotin AI (µg/day)	Pantothenic acid AI (mg/day)	Vitamin B$_6$ RDA (mg/day)	Folate RDA (µg/day)[b]	Vitamin B$_{12}$ RDA (µg/day)	Choline AI (mg/day)	Vitamin C RDA (mg/day)	Vitamin A RDA (µg/day)[c]	Vitamin D AI (µg/day)[d]	Vitamin E RDA (mg/day)[e]	Vitamin K AI (µg/day)
Infants														
0–0.5	0.2	0.3	2	5	1.7	0.1	65	0.4	125	40	400	5	4	2.0
0.5–1	0.3	0.4	4	6	1.8	0.3	80	0.5	150	50	500	5	5	2.5
Children														
1–3	0.5	0.5	6	8	2	0.5	150	0.9	200	15	300	5	6	30
4–8	0.6	0.6	8	12	3	0.6	200	1.2	250	25	400	5	7	55
Males														
9–13	0.9	0.9	12	20	4	1.0	300	1.8	375	45	600	5	11	60
14–18	1.2	1.3	16	25	5	1.3	400	2.4	550	75	900	5	15	75
19–30	1.2	1.3	16	30	5	1.3	400	2.4	550	90	900	5	15	120
31–50	1.2	1.3	16	30	5	1.3	400	2.4	550	90	900	5	15	120
51–70	1.2	1.3	16	30	5	1.7	400	2.4	550	90	900	10	15	120
>70	1.2	1.3	16	30	5	1.7	400	2.4	550	90	900	15	15	120
Females														
9–13	0.9	0.9	12	20	4	1.0	300	1.8	375	45	600	5	11	60
14–18	1.0	1.0	14	25	5	1.2	400	2.4	400	65	700	5	15	75
19–30	1.1	1.1	14	30	5	1.3	400	2.4	425	75	700	5	15	90
31–50	1.1	1.1	14	30	5	1.3	400	2.4	425	75	700	5	15	90
51–70	1.1	1.1	14	30	5	1.5	400	2.4	425	75	700	10	15	90
>70	1.1	1.1	14	30	5	1.5	400	2.4	425	75	700	15	15	90
Pregnancy														
≤18	1.4	1.4	18	30	6	1.9	600	2.6	450	80	750	5	15	75
19–30	1.4	1.4	18	30	6	1.9	600	2.6	450	85	770	5	15	90
31–50	1.4	1.4	18	30	6	1.9	600	2.6	450	85	770	5	15	90
Lactation														
≤18	1.4	1.6	17	35	7	2.0	500	2.8	550	115	1200	5	19	75
19–30	1.4	1.6	17	35	7	2.0	500	2.8	550	120	1300	5	19	90
31–50	1.4	1.6	17	35	7	2.0	500	2.8	550	120	1300	5	19	90

NOTE: For all nutrients, values for infants are AI.

[a] Niacin recommendations are expressed as niacin equivalents (NE), except for recommendations for infants younger than 6 months, which are expressed as preformed niacin.

[b] Folate recommendations are expressed as dietary folate equivalents (DFE).

[c] Vitamin A recommendations are expressed as retinol activity equivalents (RAE).

[d] Vitamin D recommendations are expressed as cholecalciferol and assume an absence of adequate exposure to sunlight.

[e] Vitamin E recommendations are expressed as α-tocopherol.

Recommended Dietary Allowances (RDA) and Adequate Intakes (AI) for Minerals

Age (yr)	Sodium AI (mg/day)	Chloride AI (mg/day)	Potassium AI (mg/day)	Calcium AI (mg/day)	Phosphorus RDA (mg/day)	Magnesium RDA (mg/day)	Iron RDA (mg/day)	Zinc RDA (mg/day)	Iodine RDA (µg/day)	Selenium RDA (µg/day)	Copper RDA (µg/day)	Manganese AI (mg/day)	Fluoride AI (mg/day)	Chromium AI (µg/day)	Molybdenum RDA (µg/day)
Infants															
0–0.5	120	180	400	210	100	30	0.27	2	110	15	200	0.003	0.01	0.2	2
0.5–1	370	570	700	270	275	75	11	3	130	20	220	0.6	0.5	5.5	3
Children															
1–3	1000	1500	3000	500	460	80	7	3	90	20	340	1.2	0.7	11	17
4–8	1200	1900	3800	800	500	130	10	5	90	30	440	1.5	1.0	15	22
Males															
9–13	1500	2300	4500	1300	1250	240	8	8	120	40	700	1.9	2	25	34
14–18	1500	2300	4700	1300	1250	410	11	11	150	55	890	2.2	3	35	43
19–30	1500	2300	4700	1000	700	400	8	11	150	55	900	2.3	4	35	45
31–50	1500	2300	4700	1000	700	420	8	11	150	55	900	2.3	4	35	45
51–70	1300	2000	4700	1200	700	420	8	11	150	55	900	2.3	4	30	45
>70	1200	1800	4700	1200	700	420	8	11	150	55	900	2.3	4	30	45
Females															
9–13	1500	2300	4500	1300	1250	240	8	8	120	40	700	1.6	2	21	34
14–18	1500	2300	4700	1300	1250	360	15	9	150	55	890	1.6	3	24	43
19–30	1500	2300	4700	1000	700	310	18	8	150	55	900	1.8	3	25	45
31–50	1500	2300	4700	1000	700	320	18	8	150	55	900	1.8	3	25	45
51–70	1300	2000	4700	1200	700	320	8	8	150	55	900	1.8	3	20	45
>70	1200	1800	4700	1200	700	320	8	8	150	55	900	1.8	3	20	45
Pregnancy															
≤18	1500	2300	4700	1300	1250	400	27	12	220	60	1000	2.0	3	29	50
19–30	1500	2300	4700	1000	700	350	27	11	220	60	1000	2.0	3	30	50
31–50	1500	2300	4700	1000	700	360	27	11	220	60	1000	2.0	3	30	50
Lactation															
≤18	1500	2300	5100	1300	1250	360	10	14	290	70	1300	2.6	3	44	50
19–30	1500	2300	5100	1000	700	310	9	12	290	70	1300	2.6	3	45	50
31–50	1500	2300	5100	1000	700	320	9	12	290	70	1300	2.6	3	45	50

P9-CFU-810

Tolerable Upper Intake Levels (UL) for Vitamins

Age (yr)	Niacin (mg/day)[a]	Vitamin B_6 (mg/day)	Folate (µg/day)[a]	Choline (mg/day)	Vitamin C (mg/day)	Vitamin A (µg/day)[b]	Vitamin D (µg/day)	Vitamin E (mg/day)[c]
Infants								
0–0.5	—	—	—	—	—	600	25	—
0.5–1	—	—	—	—	—	600	25	—
Children								
1–3	10	30	300	1000	400	600	50	200
4–8	15	40	400	1000	650	900	50	300
9–13	20	60	600	2000	1200	1700	50	600
Adolescents								
14–18	30	80	800	3000	1800	2800	50	800
Adults								
19–70	35	100	1000	3500	2000	3000	50	1000
>70	35	100	1000	3500	2000	3000	50	1000
Pregnancy								
≤18	30	80	800	3000	1800	2800	50	800
19–50	35	100	1000	3500	2000	3000	50	1000
Lactation								
≤18	30	80	800	3000	1800	2800	50	800
19–50	35	100	1000	3500	2000	3000	50	1000

[a]The UL for niacin and folate apply to synthetic forms obtained from supplements, fortified foods, or a combination of the two.

[b]The UL for vitamin A applies to the preformed vitamin only.
[c]The UL for vitamin E applies to any form of supplemental α-tocopherol, fortified foods, or a combination of the two.

Tolerable Upper Intake Levels (UL) for Minerals

Age (yr)	Sodium (mg/day)	Chloride (mg/day)	Calcium (mg/day)	Phosphorus (mg/day)	Magnesium (mg/day)[d]	Iron (mg/day)[b]	Zinc (mg/day)	Iodine (µg/day)	Selenium (µg/day)	Copper (µg/day)	Manganese (mg/day)	Fluoride (mg/day)	Molybdenum (µg/day)	Boron (mg/day)	Nickel (mg/day)	Vanadium (mg/day)
Infants																
0–0.5	—[e]	—[e]	—	—	—	40	4	—	45	—	—	0.7	—	—	—	—
0.5–1	—[e]	—[e]	—	—	—	40	5	—	60	—	—	0.9	—	—	—	—
Children																
1–3	1500	2300	2500	3000	65	40	7	200	90	1000	2	1.3	300	3	0.2	—
4–8	1900	2900	2500	3000	110	40	12	300	150	3000	3	2.2	600	6	0.3	—
9–13	2200	3400	2500	4000	350	40	23	600	280	5000	6	10	1100	11	0.6	—
Adolescents																
14–18	2300	3600	2500	4000	350	45	34	900	400	8000	9	10	1700	17	1.0	—
Adults																
19–70	2300	3600	2500	4000	350	45	40	1100	400	10,000	11	10	2000	20	1.0	1.8
>70	2300	3600	2500	3000	350	45	40	1100	400	10,000	11	10	2000	20	1.0	1.8
Pregnancy																
≤18	2300	3600	2500	3500	350	45	34	900	400	8000	9	10	1700	17	1.0	—
19–50	2300	3600	2500	3500	350	45	40	1100	400	10,000	11	10	2000	20	1.0	—
Lactation																
≤18	2300	3600	2500	4000	350	45	34	900	400	8000	9	10	1700	17	1.0	—
19–50	2300	3600	2500	4000	350	45	40	1100	400	10,000	11	10	2000	20	1.0	—

[d]The UL for magnesium applies to synthetic forms obtained from supplements or drugs only.
[e]Source of intake should be from human milk (or formula) and food only.

NOTE: An Upper Limit was not established for vitamins and minerals not listed and for those age groups listed with a dash (—) because of a lack of data, not because these nutrients are safe to consume at any level of intake. All nutrients can have adverse effects when intakes are excessive.

SOURCE: Adapted with permission from the *Dietary Reference Intakes* series, National Academy Press. Copyright 1997, 1998, 2000, 2001, 2005 by the National Academy of Sciences. Courtesy of the National Academy Press, Washington, D.C.

Nutrition for Health and Health Care

THIRD EDITION

Nutrition for Health and Health Care

Ellie Whitney

Linda Kelly DeBruyne

Kathryn Pinna

Sharon Rady Rolfes

THOMSON
WADSWORTH

Australia • Brazil • Canada • Mexico • Singapore • Spain
United Kingdom • United States

Nutrition for Health and Health Care, Third Edition
Ellie Whitney, Linda Kelly DeBruyne, Kathryn Pinna, Sharon Rady Rolfes

Publisher: Peter Marshall

Development Editor: Elizabeth Howe

Assistant Editor: Elesha Feldman

Editorial Assistant: Lauren Vogelbaum

Technology Project Manager: Donna Kelley

Marketing Manager: Jennifer Somerville

Marketing Assistant: Catie Ronquillo

Marketing Communications Manager: Jessica Perry

Project Manager, Editorial Production: Sandra Craig

Creative Director: Rob Hugel

Art Director: Lee Friedman

Print Buyer: Rebecca Cross

Permissions Editor: Joohee Lee

Production: Dusty Friedman, The Book Company

Text Designer: Ellen Pettengell

Photo Researcher: Roman Barnes

Copy Editor: Patricia Lewis

Illustrations: Imagineering

Cover Designer: Norman Baugher

Cover Images: Fruit background: © G. Biss/Masterfile; nurse portrait: © JLP/Sylvia Torres/Corbis

Compositor: Lachina Publishing Services

Printer: Courier Corporation/Kendallville

Library of Congress Control Number: 2006901296

ISBN 978-0-495-12515-0
ISBN 0-495-12515-6

Thomson Higher Education
10 Davis Drive
Belmont, CA 94002-3098
USA

For more information about our products, contact us at:
Thomson Learning Academic Resource Center
1-800-423-0563

For permission to use material from this text or product, submit a request online at http://www.thomsonrights.com.
Any additional questions about permissions can be submitted by e-mail to thomsonrights@thomson.com.

To the memory of my parents, Edith Tyler Noss, and
Henry H.B. Noss, who supported me with love, discipline,
and pride.

ELLIE WHITNEY

To my son Tyler, whose talents, ambition, and love of life
and family fill me with love and pride. Stay happy and
always look for the good.

LINDA KELLY DeBRUYNE

In memory of my Dad, who charmed my soul
and still lives within my heart.

KATHRYN PINNA

To my son Lyle, whose smile makes my heart smile.

SHARON ROLFES

About the Authors

ELEANOR NOSS WHITNEY, PH.D., received her B.A. in biology from Radcliffe College in 1960 and her Ph.D. in biology from Washington University in St. Louis in 1970. Formerly on the faculty at Florida State University, she now devotes full time to research, writing, and consulting. Her earlier publications include articles in *Science, Genetics,* and other journals. Her textbooks include *Understanding Nutrition, Understanding Normal and Clinical Nutrition, Nutrition Concepts and Controversies,* and *Nutrition and Diet Therapy,* for college students and her most intense interests currently include energy conservation, solar energy uses, alternatively fueled vehicles, and ecosystem restoration.

LINDA KELLY DeBRUYNE, M.S., R.D., received her B.S. in 1980 and her M.S. in 1982 in nutrition and food science at the Florida State University. She is a founding member of Nutrition and Health Associates, an information resource center in Tallahassee, Florida, where her specialty areas are life cycle nutrition and fitness. Her other publications include *Nutrition and Diet Therapy* and a multimedia CD-ROM called *Nutrition Interactive.* As a consultant for a group of Tallahassee pediatricians, she teaches infant nutrition classes to parents. She is a Registered Dietitian and maintains a professional membership in the American Dietetic Association.

KATHRYN PINNA, received her M.S. and Ph.D. degrees in nutrition from the University of California at Berkeley. She has taught nutrition and food science courses in the San Francisco Bay area for over 15 years and currently teaches nutrition classes at City College of San Francisco. She has worked as an outpatient dietitian, Internet consultant, and freelance writer, and is coauthor of the textbook *Understanding Normal and Clinical Nutrition.* She is a Registered Dietitian and a member of the American Dietetic Association and the American Society for Nutrition.

SHARON RADY ROLFES, M.S., R.D., received her B.S. in psychology and criminology in 1974 and her M.S. in nutrition and food science in 1982 from Florida State University. She is a founding member of Nutrition and Health Associates, an information resource center that maintains an ongoing bibliographic database that tracks research in over 1000 nutrition-related topics. Her other publications include the textbooks *Understanding Nutrition, Understanding Normal and Clinical Nutrition,* and a multimedia CD-ROM called *Nutrition Interactive.* In addition to writing, she also teaches at Florida State University and serves as a consultant for various educational projects. She maintains her registration as a dietitian and is a member of the American Dietetic Association.

Brief Contents

Contents

Preface

Health care professionals recognize that optimal health care depends on a thorough understanding of their clients' physical, emotional, intellectual, social, and spiritual needs. Nurses have a broad understanding of medical care and the most frequent contact with clients, so they have a unique opportunity to identify these needs. This third edition of *Nutrition for Health and Health Care* aims to provide future nurses with a basic understanding of nutrition and its role in disease prevention and treatment. Each chapter includes practical information for addressing nutrition concerns and incorporating nutrition into care plans. Although primarily written for nursing students, this book can serve as a valuable resource for students of all health-related professions, including nursing assistants, dietary technicians, health educators, health science majors, dietitians, and physicians.

Chapter Overview

The book begins by introducing basic nutrition concepts (Chapters 1–9) and shows how nutrition supports health. Chapters 10 through 12 describe how nutrient needs change throughout the life cycle. Chapters 13 and 14 explore how nurses can use information from nursing assessments to identify and address a client's nutrient needs. The remaining chapters (Chapters 15–23) introduce nutrition therapy and its role in the prevention and treatment of medical conditions.

New to This Edition

Nutrition is a rapidly expanding science, with many questions remaining to be answered and new "facts" surfacing every day. Each chapter of this book has been substantially revised since the second edition to reflect these many changes. The latest Dietary Reference Intakes and the recently developed *Dietary Guidelines for Americans 2005*, USDA Food Guide, and MyPyramid.gov tool are described in these pages. The clinical chapters include the recently revised nutrition therapies for diabetes, coronary heart disease, and hypertension. To help the student nurse identify appropriate nutrition interventions, the updated NANDA nursing diagnoses (2005–2006 edition) accompany the descriptions of each illness that requires nutrition care.

Pedagogical Features

The text contains abundant pedagogy to help students master "normal" and "clinical" nutrition topics. Within each chapter, definitions and notes in the margins clarify nutrition information, remind readers of previously defined terms, and provide cross-references. How To skill boxes help readers work through calculations or give practical suggestions for applying nutrition advice. The Self Check at the end of each chapter provides questions to help review chapter information. New to this edition are Review Notes, which summarize the information following each major heading. These Review Notes can be used to preview or review key chapter concepts. Also new to the edition are sample NCLEX review questions available on the book's website **www.thomson.edu/nutrition/whitney**. Each chapter concludes with a Nutrition on the Net list of annotated websites.

In the clinical chapters the Nursing Diagnoses feature can help nursing students correlate nutrition care with nursing care. Case Studies guide readers in applying nutrition information to everyday situations. Clinical Applications encourage readers to practice mathematical calculations, synthesize information from previous chapters, or understand how dietary modifications affect patients. Nutrition Assessment Checklists remind readers of assessment parameters particularly relevant to specific stages of the life cycle or groups of disorders. Diet-Drug Interaction boxes identify important nutrient-drug and food-drug interactions.

The Nutrition in Practice sections that follow each chapter explore current topics, advanced subjects, or specialty areas such as ethical issues in nutrition care. This edition includes new Nutrition in Practice topics, such as the health effects of different types of fats, fad diets, nutritional genomics, multiple organ failure, metabolic syndrome, and childhood obesity and chronic diseases.

Appendixes

The appendixes support the book with a wealth of information on the nutrient contents of foods, enteral formulas, Canadian nutrient recommendations and food choices, U.S. nutrient intake recommendations and the exchange system, supplemental information about nutrition assessments, aids to calculations, physical activity and energy requirements, and answers to Self Check questions.

Instructor Tools

A number of helpful ancillary materials are available for instructors including an Instructor's Resource CD-ROM that includes figures and photos from the text as well as PowerPoint lecture presentations, a test bank, and an instructor's manual. A printed version of the instructor's manual and test bank is also available.

Student Ancillaries

A printed study guide for students provides chapter outlines, nutrition applications exercises, multiple choice and discussion questions, math exercises, and case study problems to help students master chapter concepts. The book's website **www.thomson.edu/nutrition/whitney** provides additional quizzing, NCLEX review questions, crossword puzzles, and interactive glossary.

We hope that as you discover the many fascinating aspects of nutrition science, you will enthusiastically apply the concepts in both your personal and professional life.

Ellie Whitney
Linda DeBruyne
Kathryn Pinna
Sharon Rolfes

Acknowledgments

Among the most difficult words to write are those that express the depth of our gratitude to the many dedicated people whose efforts have made this book possible. We extend special thanks to Fran Webb for her expertise and valuable contributions throughout this text. We appreciate the work of Beth Magana in preparing the manuscript files. Thanks also to Joohee Lee for her help in securing permissions, Pat Lewis for her copyediting expertise, Roman Barnes for selecting photographs, and Donna Kelley, Star MacKenzie Burruto, and the folks at First DataBank for compiling the food composition appendix. We are indebted to our editorial team, Peter Marshall and Beth Howe, and to our production team, Sandra Craig and Dusty Friedman, for seeing this project through from start to finish. We would also like to thank those who helped to prepare the many supplements available with the text: Kathleen M. Buckley of the Catholic University of America, Carolyn M. Burger of Miami University, Beth Fontenot of McNeese State University, Vicki Moss of the University of Wisconsin–Oshkosh, Jillann Neely of Onondaga Community College, and Elesha Feldman who managed the preparation of the supplements. We would also like to acknowledge Jennifer Somerville for her marketing efforts. Thanks also to Lauren Vogelbaum for taking care of many details involved in producing this text. To the many others involved in designing, indexing, typesetting, dummying, and marketing, we offer our thanks. We are especially grateful to our associates, family, and friends for their continued encouragement and support, and to the following reviewers who offered excellent suggestions for improving the text.

Reviewers

Sally Aadland *Bismarck State College*

Dorothy S. Akerson *University of Missouri*

Dea H. Baxter *Georgia State University*

Jack R. Brook *Mt. Hood Community College*

Carmen Bruni *Texas A&M International University*

Kathleen M. Buckley *Catholic University of America*

Carolyn M. Burger *Miami University*

Beth Fontenot *McNeese State University*

Jeanine Grillo *Ohlone College*

Susan Grimes *Mercer University*

Nelda W. Malm *Seminole Community College*

Barbara Maxwell *State University of New York Ulster*

Mohey Mowafy *Northern Michigan University*

Vicki A. Moss *University of Wisconsin–Oshkosh*

Jillann Neely *Onondaga Community College*

Kathleen M. Potts *Hocking College*

Tammy Sakanashi *Santa Rosa Junior College*

Karen J. Schmitz *Madonna University*

Kathleen M. Potts *Hocking College*

Tammy Sakanashi *Santa Rosa Junior College*

Karen J. Schmitz *Madonna University*

Nutrition for Health and Health Care

THIRD EDITION

Overview of Nutrition and Health

CHAPTER 1

FIGURE 1-1 The Health Line
No matter how well you maintain your health today, you may still be able to improve tomorrow. Likewise, a person who is well today can slip by failing to maintain health-promoting habits.

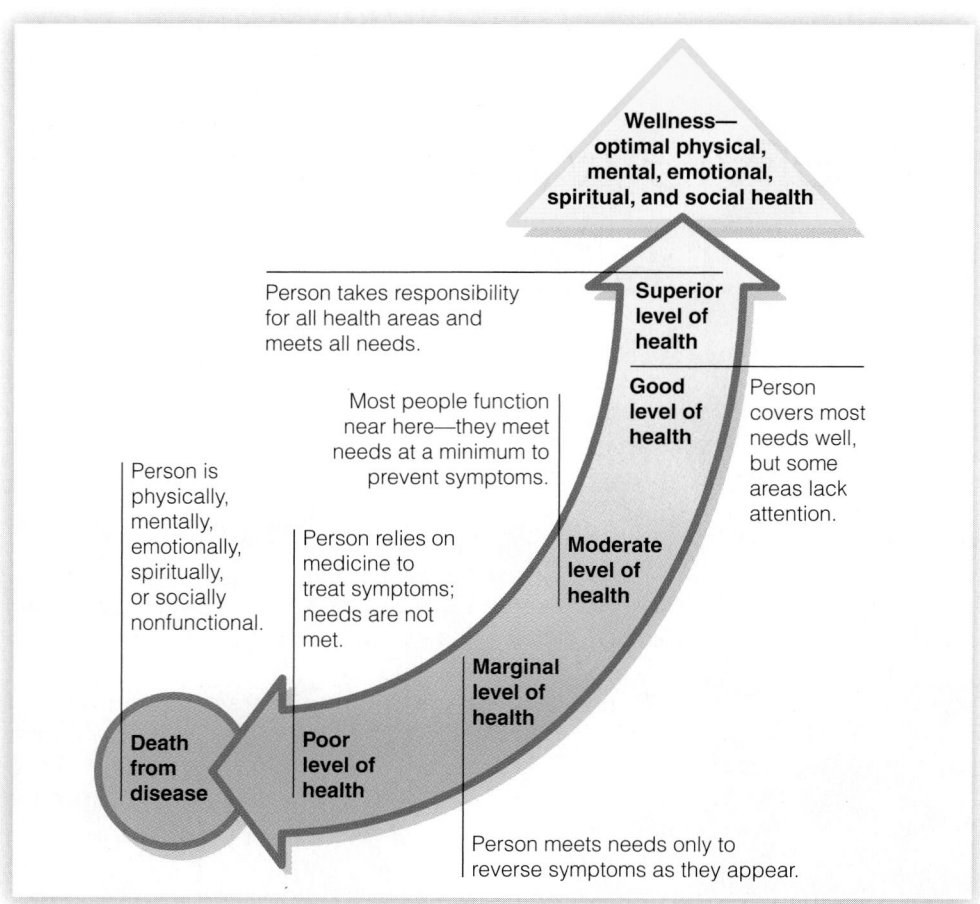

Nurses generally use either *client* or *patient* when referring to an individual under their care. The first 12 chapters of this text emphasize the nutrition concerns of people in good health; therefore, the term *client* is used in these chapters.

health: a range of states with physical, mental, emotional, spiritual, and social components. At a minimum, health means freedom from physical disease, mental disturbances, emotional distress, spiritual discontent, social maladjustment, and other negative states. At a maximum, health means "wellness."

wellness: maximum well-being; the top range of health states; the goal of the person who strives toward realizing his or her full potential physically, mentally, emotionally, spiritually, and socially.

nutrition: the science of foods and the nutrients and other substances they contain, and of their ingestion, digestion, absorption, transport, metabolism, interaction, storage, and excretion. A broader definition includes the study of the environment and of human behavior as it relates to these processes.

Every day, several times a day, you make choices that will either improve your **health** or harm it. Each choice may influence your health only a little, but when these choices are repeated over years and decades, their effects become significant.

The choices people make each day affect not only their physical health but also their **wellness**—all the characteristics that make a person strong, confident, and able to function well with family, friends, and others. People who consistently make poor lifestyle choices, on a daily basis, increase their risks of developing diseases. Figure 1-1 shows how a person's health can fall anywhere along a continuum, from maximum wellness on the one end to total failure to function (death) on the other.

As nurses or other health care professionals, when you take responsibility for your own health by making daily choices and practicing behaviors that enhance your well-being, you prepare yourself physically, mentally, and emotionally to meet the demands of your profession. As health care professionals, however, you have a responsibility to your clients as well as to yourselves. You have unique opportunities to make your clients aware of the benefits of positive health choices and behaviors, to show them how to change their behaviors and make daily choices to enhance their own health, and to serve as role models for those behaviors.

This text focuses on how nutrition choices affect health and disease. The early chapters introduce the basics of nutrition to support good health. The later chapters emphasize diet therapy and its role in supporting health and treating diseases and symptoms.

Food Choices

Sound **nutrition** throughout life does not ensure good health and long life, but it can certainly help to tip the balance in their favor. Nevertheless, most people choose foods for reasons other than their nourishing value. Even people who claim to choose foods primarily for the sake of health or nutrition will admit that other factors also influence their food choices. Because food choices become an integral part of people's lifestyles, they sometimes find it difficult to change their eating habits. Health care professionals who help clients make diet changes must understand the dynamics of food choices because people will alter their eating habits only if their preferences are honored.

Preference Why do people like certain foods? One reason, of course, is their preference for certain tastes. Some tastes are widely liked, such as the sweetness of sugar and the zest of salt. Research suggests that genetics may influence people's taste preferences, a finding that may eventually have implications for clinical nutrition.[1] For example, sensitivity to bitter taste is an inheritable trait. People born with great sensitivity to bitter tastes tend to avoid foods with bitter flavors such as broccoli, cabbage, brussels sprouts, spinach, and grapefruit juice. These foods, as well as many other fruits and vegetables, contain compounds called **phytochemicals** that may reduce the risk of cancer. Thus the role that genetics may play in food selection is gaining importance in cancer research.

Habit Sometimes habit dictates people's food choices. People eat a sandwich for lunch or drink orange juice at breakfast simply because they have always done so.

Associations People also like foods with happy associations—foods eaten in the midst of warm family gatherings on traditional holidays or given to them as children by someone who loved them. By the same token, people can attach intense and unalterable dislikes to foods that they ate when they were sick or that were forced on them when they weren't hungry.

Ethnic Heritage and Tradition Every country, and every region of a country, has its own typical foods and ways of combining them into meals. The foodways of North America reflect the many different cultural and ethnic backgrounds of its inhabitants. Many foods with foreign origins are familiar items on North American menus: tacos, egg rolls, lasagna, and gyros, to name a few. Still others, such as spaghetti and croissants, are almost staples in the "American diet." North American regional cuisines like Cajun and TexMex blend the traditions of several cultures. Table 1-1 (p. 4) presents selected **ethnic diets** and food choices.

Values People's values, environmental ethics, religious beliefs, and political views also influence their food choices. By choosing to eat some foods or avoid others, people make statements that reflect their values. For example, people may select only foods that come in containers that can be reused or recycled. A political activist may boycott fruit or vegetables picked by migrant workers who have been exploited. Some people choose only brands of canned tuna fish that state "dolphin safe" on the label, meaning that dolphins were not killed when the tuna were netted. Labels on other foods carry similar statements or symbols to imply that the foods have been produced in ways that are considered environmentally favorable.[2] These are known as ecolabels. One of the most successful ecolabels is "Certified Organic," meaning the food was grown and processed according to regulations defining the use of synthetic fertilizers, herbicides, pesticides, and other chemical ingredients.

Religion also influences many people's food choices. Jewish law sets forth an extensive set of dietary rules. Many Christians forgo meat on Fridays during Lent, the period prior to Easter. In Islamic dietary laws, permitted or lawful foods are called *halal*. Other faiths prohibit some dietary practices and promote others. Diet planners can foster sound nutrition practices only if they respect and honor each person's values.

Social Interaction Social interaction is another powerful influence on people's food choices. Meals are social events, and the sharing of food is part of hospitality.

© 2001 PhotoDisc, Inc.

Nutrition is only one of the many factors that influence people's food choices.

Nutrition in Practice 8 addresses phytochemicals and their role in disease prevention.

phytochemicals: nonnutrient compounds in plant-derived foods that have biological activity in the body.

ethnic diets: foodways and cuisines typical of national origins, races, cultural heritages, or geographic locations.

TABLE 1-1	Selected Ethnic Cuisines and Food Choices

	Grains	Vegetables	Fruits	Meats, Poultry, Fish, Legumes, Eggs, and Nuts	Milk
Asian[a]	Barley, buckwheat, millet, rice, wheat, rice or wheat noodles	Amaranth, baby corn, bamboo shoots, bok choy, cabbages, mung bean sprouts, scallions, seaweed, snow peas, straw mushrooms, water chestnuts, wild yam	Kumquats, loquats, lychee, mandarin oranges, melons, pears, persimmon, plums	Beef, pork, poultry, fish and other seafood, squid, soybeans, tofu, duck eggs, cashews, peanuts	Soy milk
Mediterranean	Bulgur, couscous, focaccia, Italian bread, pastas, pita pocket bread, polenta, rice	Cucumbers, eggplant, grape leaves, onions, peppers, tomatoes	Dates, figs, grapes, lemons, melons, olives, raisins	Beef, gyros, lamb, pork, sausage, chicken, fish and other seafood, fava beans, lentils, almonds, walnuts	Feta, goat, mozzarella, parmesan, provolone, and ricotta cheeses; yogurt
Mexican	Taco shells, tortillas (corn or flour), rice	Cactus, cassava, chayote, chilies, corn, jicama, onions, tomatoes, tomato salsa, yams	Avocado, bananas, guava, lemons, limes, mango, oranges, papaya, plantain	Beef, chorizo, chicken, fish, refried beans, eggs	Cheese, flan (caramel custard)
U.S. Deep South (West African Influence)[a]	Biscuits, cornbread, grits, macaroni, rice	Black-eyed peas, collards (other leafy greens), corn, hominy, okra, onions, pole beans, potatoes, snap beans, summer squash, sweet potatoes, tomatoes	Apples, bananas, berries, melons, peaches, pears	Beef, pork, sausages, poultry, fish, beans, peanuts	Buttermilk, milk, American cheese, cheddar cheese

[a] Traditional cuisines of China and of West African influence exclude fluid milk as a beverage for adults and use few or no milk products in cooking. Calcium and certain other nutrients of milk are supplied by other foods, such as small fish eaten with the bones or large servings of leafy green vegetables.

Ethnic meals and family gatherings nourish the spirit as well as the body.

It is often considered rude to refuse food or drink being shared by a group or offered by a host. Food brings people together for many different reasons: to celebrate a holiday or special event, to renew an old friendship, to make new friends, to conduct business, and many more. Sometimes food defines status. Lobster and caviar are often associated with wealth; peanut butter and jelly, with children. Sometimes food is used to influence or impress someone. For example, a business executive invites a prospective new client out to dinner in hopes of edging out the competition. In each case, for whatever the purpose, food plays an integral part of the social interaction.

Emotional State People may eat in response to emotional stimuli—for example, to relieve boredom or depression or to calm anxiety. A lonely person may choose to eat rather than to call a friend and risk rejection. A person who has returned home from an exciting evening out may unwind with a late-night snack. Eating in response to emotions can easily lead to overeating and obesity but may be appropriate at times. For example, sharing food at times of bereavement serves both the givers' need to provide comfort and the receiver's need to be cared for and to interact with others.

In contrast, some people do not eat at all, or eat very little, in response to emotions. A depressed person may simply have no appetite for food. A person overcome with grief can easily lose all interest in food. As long as a person's appetite or interest in food returns quickly, little harm is done.

Availability, Convenience, and Economy The influence of these factors on people's food selections is clear. You cannot eat foods if they are not available, if you cannot get to the grocery store, if you do not have the time or skill to prepare them, or if you cannot afford them. Convenience plays a major role in many people's food selections today, but along with convenience, consumers want great taste and quality—foods that are fast, delicious, and nutritious.[3] The demand for fully prepared ready-to-eat foods or foods that can be easily prepared in a microwave oven demonstrates this influence. Consumers today spend more than half their food budget on meals that require no preparation.[4] They frequently eat out, bring ready-to-eat meals home, or have food delivered. Consumer demand for microwavable foods that taste and smell as though they were prepared in a conventional oven has prompted manufacturers to expand the variety of microwavable foods available and to change the formulation and packaging of the foods.[5] Food processors are developing many more microwavable meals and snacks for use at home and office—rising crust pizzas, chicken and fish dishes, precooked entrées, and side dishes such as french fries and rice. Many of the new products are packaged in single-serve containers.

Age Age influences people's food choices. Infants, for example, depend on others to choose foods for them. Older children also rely on others but become more active in selecting foods that taste sweet and are familiar to them and rejecting those whose taste or texture they dislike. In contrast, the links between taste preferences and food choices in adults are less direct than in children. Adults often choose foods based on health concerns such as body weight. Indeed, adults may avoid sweet or familiar foods because of such concerns.

Occupation Some people have jobs that keep them away from home for days at a time, or require them to conduct business in restaurants or at conventions, or involve hectic schedules that allow little or no time for meals at home. For these people, the kinds of restaurants available to them, and the cost of eating out so often, may limit food choices.

Body Image Sometimes people select foods that they associate with ideals of body image. The fashion and movie industries, not the medical community, have defined what people believe to be the ideal body—sometimes an excessively thin body for women or an excessively muscular body for men. Both men and women seek "beautiful bodies," and in doing so, they select or avoid foods that they believe will improve or impair their physical appearance. Such intentions are rational when based on sound nutrition and fitness knowledge, but when based on faddism or carried to extremes, they undermine good health.

Medical Conditions   Sometimes medical conditions and their treatments (including medications) limit the foods a person can select. For example, a person with heart disease might need to adopt a diet low in certain types of fats. The chemotherapy needed to treat cancer can interfere with a person's appetite or limit food choices by causing vomiting. Allergy to certain foods can also limit choices. The second half of this text discusses how diet can be modified to accommodate different medical conditions.

Health and Nutrition   Consumers today cite nutrition as a primary concern in making food choices, yet the foods they choose do not always reflect this concern. Food manufacturers and restaurant chefs have responded to scientific findings linking health with nutrition by offering an abundant selection of health-promoting foods and beverages. In some cases, the health-promoting foods are as simple and familiar as oatmeal or tomatoes. In other cases, the foods have been processed or prepared in a way that provides health benefits, perhaps by reducing the sodium content. In still other cases, manufacturers have developed **functional foods**—products that may contain physiologically active ingredients that provide health benefits. Examples of functional foods include margarine made with plant sterol and stanol esters that lower blood cholesterol and orange juice fortified with calcium to help build strong bones. More and more functional foods are being developed—a trend that will no doubt continue as consumer demand for such products grows.[6]

Consumers may be led to believe that functional foods are a new category of foods. All foods contain thousands of nonnutrient compounds that are biologically active in the body, however, so virtually all of them have some special value in supporting health.

Nutrition in Practice 8 offers more discussion of functional foods.

> **REVIEW NOTES**
>
> A person selects foods for many different reasons.
>
> Food choices influence health—both positively and negatively. Individual food selections neither make nor break a diet's healthfulness, but the balance of foods selected over time can make an important difference to health.
>
> In the interest of health, people are wise to think "nutrition" when making their food choices.

The Nutrients

functional foods: foods that may provide health benefits beyond their nutrient contributions. Functional foods may include whole foods, fortified foods, and modified foods.

nutrients: substances obtained from food and used in the body to provide energy and structural materials and to serve as regulating agents to promote growth, maintenance, and repair. Nutrients may also reduce the risks of some diseases.

You are a collection of molecules that move. All these moving parts are arranged in patterns of extraordinary complexity and order—cells, tissues, and organs. Although the arrangement remains constant, the parts are continually changing, using **nutrients** and energy derived from nutrients. Almost any food you eat is composed of dozens or even hundreds of different kinds of materials. Spinach, for example, is composed mostly of water (95 percent), and most of its solid materials are the compounds carbohydrates, fats (properly called lipids), and proteins. If you could remove these materials, you would find a tiny residue of minerals, vitamins, and other compounds. Water, carbohydrates, fats, proteins, vitamins, and minerals are the six classes of nutrients commonly found in spinach and other foods. Some of the other materials in foods, such as the pigments and other phytochemicals, are not nutrients but may still be important to health. The body can make some nutrients for itself, at least in limited quantities, but it cannot make them all, and it makes some in insufficient quantities to meet its needs. Therefore, the body must obtain many nutrients from foods. The nutrients that foods must supply are called **essential nutrients.**

Carbohydrates, Fats, and Proteins Four of the six classes of nutrients (carbohydrates, fats, proteins, and vitamins) contain carbon, which is found in all living things. They are therefore **organic** (meaning, literally, "alive"). During metabolism, three of these four (carbohydrates, fats, and proteins) provide energy the body can use. These **energy-yielding nutrients** continually replenish the energy you spend daily. Carbohydrates and fats meet most of the body's energy needs; proteins make a significant contribution only when other fuels are unavailable.

Vitamins, Minerals, and Water Vitamins are organic but do not provide energy to the body. They facilitate the release of energy from the three energy-yielding nutrients. In contrast, minerals and water are **inorganic** nutrients. Minerals yield no energy in the human body, but like vitamins, they help to regulate the release of energy, among their many other roles. As for water, it is the medium in which all of the body's processes take place.

kCalories: A Measure of Energy The amount of energy that carbohydrates, fats, and proteins release can be measured in **calories**—tiny units of energy so small that a single apple provides tens of thousands of them. To ease calculations, energy is expressed in 1000-calorie metric units known as kilocalories (shortened to kcalories, but commonly called "calories"). When you read in popular books or magazines that an apple provides "100 calories," understand that it means 100 kcalories. This books uses the term *kcalorie* and its abbreviation *kcal* throughout, as do other scientific books and journals. kCalories are not constituents of foods; they are a measure of the energy foods provide. The energy a food provides depends on how much carbohydrate, fat, and protein the food contains.

Carbohydrate yields 4 kcalories of energy from each gram, and so does protein. Fat yields 9 kcalories per gram. Thus fat has a greater **energy density** than either carbohydrate or protein. If you know how many grams of each nutrient a food contains, you can derive the number of kcalories potentially available from the food (see the accompanying "How to"). Simply multiply the carbohydrate grams

The six classes of nutrients are water, carbohydrates, fats, proteins, vitamins, and minerals.

Metabolism is the set of processes by which nutrients are rearranged into body structures or broken down to yield energy.

Food energy can also be measured in kilojoules (kJ). The kilojoule is the international unit of energy. One kcalorie equals 4.2 kJ.

essential nutrients: nutrients a person must obtain from food because the body cannot make them for itself in sufficient quantities to meet physiological needs.

organic: carbon containing. The four organic nutrients are carbohydrate, fat, protein, and vitamins.

energy-yielding nutrients: the nutrients that break down to yield energy the body can use. The three energy-yielding nutrients are carbohydrate, protein, and fat.

inorganic: not containing carbon or pertaining to living things.

calories: units in which energy is measured. Food energy is measured in **kilocalories** (1000 calories equal 1 kilocalorie), abbreviated **kcalories** or **kcal.** One kcalorie is the amount of heat necessary to raise the temperature of 1 kilogram (kg) of water 1°C. The scientific use of the term **kcalorie** is the same as the popular use of the term *calorie.*

energy density: a measure of the energy a food provides relative to the amount of food (kcalories per gram).

HOW TO   Calculate the Energy Available from Foods

To calculate the energy available from a food, multiply the number of grams of carbohydrate, protein, and fat by 4, 4, and 9, respectively. Then add the results together. For example, 1 slice of bread with 1 tablespoon of peanut butter on it contains 16 grams of carbohydrate, 7 grams of protein, and 9 grams of fat:

16 g carbohydrate × 4 kcal/g = 64 kcal.

7 g protein × 4 kcal/g = 28 kcal.

9 g fat × 9 kcal/g = 81 kcal.

Total = 173 kcal.

From this information, you can calculate the percentage of kcalories each of the energy nutrients contributes to the total.

To determine the percentage of kcalories from fat, for example, divide the 81 fat kcalories by the total 173 kcalories:

81 fat kcal ÷ 173 total kcal = 0.468 (rounded to 0.47).

Then multiply by 100 to get the percentage:

0.47 × 100 = 47%.

Dietary recommendations that urge people to limit fat intake to 20 to 35 percent of kcalories refer to the day's total energy intake, not to individual foods. Still, if the proportion of fat in each food choice throughout a day exceeds 35 percent of kcalories, then the day's total surely will, too. Knowing that this snack provides 47 percent of its kcalories from fat alerts a person to the need to make lower-fat selections at other times that day.

times 4, the protein grams times 4, and the fat grams times 9, and add the results together.

Energy Nutrients in Foods Practically all foods contain mixtures of the energy-yielding nutrients, although foods are sometimes classified by their predominant nutrient. To speak of meat as "a protein" or of bread as "a carbohydrate," however, is inaccurate. Each is rich in a particular nutrient, but a protein-rich food such as beef contains a lot of fat along with the protein, and a carbohydrate-rich food such as cornbread also contains fat (corn oil) and protein. Only a few foods are exceptions to this rule, the common ones being sugar (which is pure carbohydrate) and oil (which is pure fat).

Energy Storage in the Body The body first uses the energy-yielding nutrients to build new compounds and fuel metabolic and physical activities. Excesses are then rearranged into storage compounds, primarily body fat, and put away for later use. Thus, if you take in more energy than you expend, whether from carbohydrate, fat, or protein, the result is usually a gain of body fat. Too much meat (a protein-rich food) is just as fattening as too many potatoes (a carbohydrate-rich food).

Alcohol, Not a Nutrient One other substance contributes energy: alcohol. The body derives energy from alcohol at the rate of 7 kcalories per gram. Alcohol is not a nutrient, however, because it cannot support the body's growth, maintenance, or repair. Nutrition in Practice 19 discusses alcohol's effects on nutrition.

REVIEW NOTES

Foods provide nutrients—substances that support the growth, maintenance, and repair of the body's tissues.

The six classes of nutrients are water, carbohydrates, fats, proteins, vitamins, and minerals.

Vitamins, minerals, and water facilitate a variety of activities in the body.

Foods rich in the energy-yielding nutrients (carbohydrates, fats, and proteins) provide the major materials for building the body's tissues and yield energy the body can use or store.

Energy is measured in kcalories.

Nutrient Recommendations

Appendix C presents nutrient recommendations developed by two international groups, the Food and Agriculture Organization and the World Health Organization (the FAO/WHO recommendations).

Nutrient recommendations are used as standards to measure healthy people's energy and nutrient intakes. Nutrition experts use the recommendations to assess nutrient intakes and to guide people on amounts to consume. Individuals can use them to decide how much of a nutrient they need to consume.

DIETARY REFERENCE INTAKES

Dietary Reference Intakes (DRI): a set of values for the dietary nutrient intakes of healthy people in the United States and Canada. These values are used for planning and assessing diets.

Defining the amounts of energy, nutrients, and other dietary components that best support health is a huge task. Nutrition experts have produced a set of standards that define the amounts of energy, nutrients, other dietary components, and physical activity that best support health. These recommendations are called **Dietary Reference Intakes (DRI)** and reflect the collaborative efforts of scientists in both the United States and Canada.[7] The DRI values are presented on the inside front covers.

---

*The DRI reports are produced by the Food and Nutrition Board, Institute of Medicine of the National Academies, with active involvement of scientists from Canada.

Setting Nutrient Recommendations: RDA and AI One advantage of the DRI is that they apply to the diets of individuals. The DRI committee offers two sets of values to be used as nutrient intake goals by individuals: a set called the **Recommended Dietary Allowances (RDA)** and a set called **Adequate Intakes (AI)**.

Based on solid experimental evidence and other reliable observations, the RDA are the foundation of the DRI. The AI values are also based on scientific findings as much as possible, but estimating their values requires some educated guesswork. The committee establishes an AI value whenever scientific evidence is insufficient to generate an RDA.[8] To see which nutrients have an AI and which have an RDA, turn to the inside front cover.

In the last decade, abundant new research has linked nutrients in the diet with the promotion of health and the prevention of chronic diseases. An advantage of the DRI is that they take into account disease prevention where appropriate, as well as an adequate nutrient intake. For example, the AI for calcium is based on intakes thought to reduce the likelihood of osteoporosis-related fractures later in life.

To ensure that the vitamin and mineral recommendations meet the needs of as many people as possible, the recommendations are set near the top end of the range of the population's estimated average requirements (see Figure 1-2). Small amounts above the daily **requirement** do no harm, whereas amounts below the requirement may lead to health problems. When people's intakes are consistently **deficient**, their nutrient stores decline, and over time this decline leads to deficiency symptoms and poor health.

Facilitating Nutrition Research and Policy: EAR In addition to the RDA and AI, the DRI committee has established another set of values: **Estimated Average Requirements (EAR)**. These values establish average requirements for given life stage and gender groups that researchers and nutrition policymakers use in their work. Nutrition scientists may use the EAR as standards in research. Public health officials may use them to assess nutrient intakes of populations and make recommendations. The EAR values form the scientific basis on which the RDA are set.

Establishing Safety Guidelines: UL The DRI committee also establishes upper limits of intake for nutrients posing a hazard when consumed in excess. These values, the **Tolerable Upper Intake Levels (UL)**, are indispensable to consumers who take supplements. Consumers need to know how much of a nutrient is too much. The UL are also of value to public health officials who set allowances for nutrients that are added to foods and water. The UL values are listed on the inside front cover.

Using Nutrient Recommendations Each of the four DRI categories serves a unique purpose. For example, the EAR are most appropriately used to develop and evaluate nutrition programs for *groups* such as schoolchildren or military personnel. The RDA (or AI if an RDA is not available) can be used to set goals for *individuals*. The UL help to keep nutrient intakes below the amounts that increase the risk of toxicity. With these understandings, professionals can use the DRI for a variety of purposes.

In addition to understanding the unique purposes of the DRI, it is important to keep their uses in perspective. Consider the following:

- The values are recommendations for safe intakes, not minimum requirements; except for energy, they include a generous margin of safety. Figure 1-3 (p. 10) presents an accurate view of how a person's nutrient needs fall within a range, with marginal and danger zones both below and above the range.
- The values reflect daily intakes to be achieved on average, over time. They assume that intakes will vary from day to day, and they are set high enough to ensure that body nutrient stores will meet nutrient needs during periods of inadequate intakes lasting a day or two for some nutrients and up to a month or two for others.

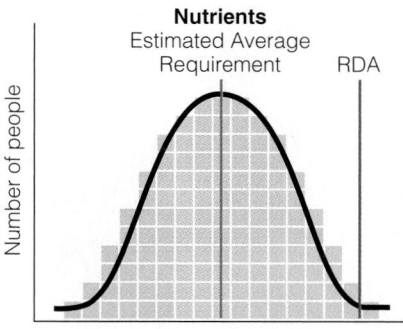

FIGURE 1-2 Nutrient Intake Recommendations

The nutrient intake recommendations are set high enough to cover nearly everyone's requirements (the boxes represent people).

Recommended Dietary Allowances (RDA): a set of values reflecting the average daily amounts of nutrients considered adequate to meet the known nutrient needs of practically all healthy people in a particular life stage and gender group; a goal for dietary intake by individuals.

Adequate Intakes (AI): a set of values that are used as guides for nutrient intakes when scientific evidence is insufficient to determine RDA.

requirement: the lowest continuing intake of a nutrient that will maintain a specified criterion of adequacy.

deficient: in regard to nutrient intake, the amount below which almost all healthy people can be expected, over time, to experience deficiency symptoms.

Estimated Average Requirements (EAR): the average daily nutrient intake levels estimated to meet the requirements of half of the healthy individuals in a given age and gender group; used in nutrition research and policymaking and the basis on which RDA values are set.

Tolerable Upper Intake Levels (UL): A set of values reflecting the highest average daily nutrient intake levels that are likely to pose no risk of toxicity to almost all healthy individuals in a particular life stage and gender group. As intake increases above the UL, the potential risk of adverse health effects increases.

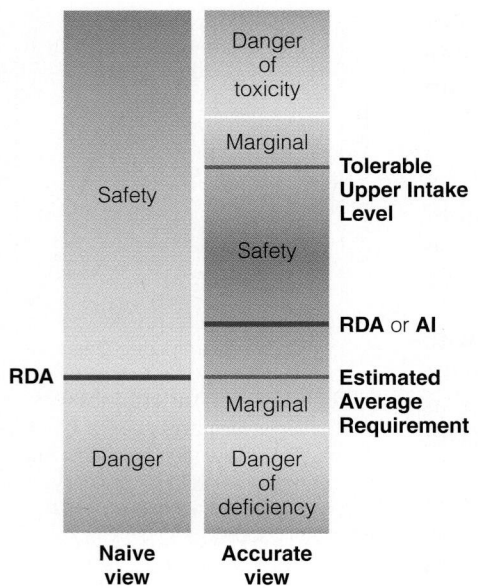

FIGURE 1-3 Naive versus Accurate View of Nutrient Intakes

The RDA or AI for a given nutrient represents a point that lies within a range of appropriate and reasonable intakes between toxicity and deficiency. Both of these recommendations are high enough to provide reserves in times of short-term dietary inadequacies, but not so high as to approach toxicity. Nutrient intakes above or below this range may be equally harmful.

- The values are chosen in reference to specific indicators of nutrient adequacy, such as blood nutrient concentrations, normal growth, and reduction of certain chronic diseases or other disorders when appropriate, rather than prevention of deficiency symptoms alone.
- The recommendations are designed to meet the needs of most healthy people. Medical problems alter nutrient needs as later chapters describe.
- The recommendations are specific for people of both genders as well as various ages and stages of life: infants, children, adolescents, men, women, pregnant women, and lactating women.

Setting Energy Recommendations In contrast to the vitamin and mineral recommendations, the recommendation for energy, called the **Estimated Energy Requirement (EER)**, is not generous because excess energy cannot be excreted and is eventually stored as body fat. Rather, the key to the energy recommendation is balance. For a person who has a body weight, body composition, and physical activity level consistent with good health, energy intake from food should match energy expenditure, so the person achieves energy balance. Enough energy is needed to sustain a healthy, active life, but too much energy leads to obesity. The EER is therefore set at a level of energy intake predicted to maintain energy balance in a healthy adult of a defined age, gender, weight, height, and physical activity level.* Another difference between the requirements for other nutrients and those for energy is that each person has an obvious indicator of whether energy intake is inadequate, adequate, or excessive: body weight. Because any amount of energy in excess of need leads to weight gain, the DRI committee did not set a Tolerable Upper Intake Level.

ACCEPTABLE MACRONUTRIENT DISTRIBUTION RANGES (AMDR)

As noted earlier, the DRI committee considers prevention of chronic disease as well as nutrient adequacy when establishing recommendations. To that end, the committee established healthy ranges of intakes for the energy-yielding nutrients—carbohydrate, fat, and protein—known as **Acceptable Macronutrient Distribution Ranges (AMDR).** Each of these three energy-yielding nutrients contributes to a person's total energy (kcalorie) intake, and those contributions vary in relation to each other. The DRI committee has determined that a diet that provides the energy-yielding nutrients in the following proportions provides adequate energy and nutrients and reduces the risk of chronic disease:

- 45 to 65 percent from carbohydrate.
- 20 to 35 percent from fat.
- 10 to 35 percent from protein.

NUTRITION SURVEYS

How do nutrition experts know whether people are meeting nutrient recommendations? The Dietary Reference Intakes and other major reports that examine the relationships between diet and health depend on information collected from nutrition surveys. Researchers use nutrition surveys to learn which foods people are eating and which supplements they are taking, to assess people's nutritional health, and to determine people's knowledge, attitudes, and behaviors about nutrition and how these relate to health.[9] The resulting wealth of information can be used for a variety of purposes. For example, Congress uses this information to

Estimated Energy Requirement (EER): the dietary energy intake level that is predicted to maintain energy-yielding balance in a healthy adult of a defined age, gender, weight, and physical activity level consistent with good health.

Acceptable Macronutrient Distribution Ranges (AMDR): ranges of intakes for the energy-yielding nutrients that provide adequate energy and nutrients and reduce the risk of chronic disease.

*The EER for children, pregnant women, and lactating women includes energy needs associated with the deposition of tissue or the secretion of milk at rates consistent with good health.

establish public policy on nutrition education, assess food assistance programs, and regulate the food supply. Scientists use the information to establish research priorities. One of the first nutrition surveys, taken before World War II, suggested that up to a third of the U.S. population might be eating poorly. Programs to correct **malnutrition** have been evolving ever since.

Coordinating Nutrition Survey Data The National Nutrition Monitoring program coordinates the many nutrition-related activities of various federal agencies. All major reports that examine the contribution of diet and nutrition status to the health of the people of the United States depend on information collected and coordinated by this national program. One of its most recent projects is the integration of two major national surveys to provide comprehensive data efficiently.[10] One portion of the survey, previously known as the Continuing Survey of Food Intakes by Individuals (CSFII), collects data on the kinds and amounts of foods people eat. Researchers then calculate the energy and nutrients in the foods and compare the amounts consumed with standards such as the DRI. The other portion of the survey, the National Health and Nutrition Examination Survey (NHANES), examines the people themselves, using nutrition assessment methods. The new integrated survey is called What We Eat in America–NHANES and is conducted by the U.S. Department of Agriculture (USDA) and the U.S. Department of Health and Human Services (HHS). Data will be collected for What We Eat in America on a continuous yearly basis. Two days of data will be collected for all respondents. The data provide information on several nutrition-related conditions, including growth retardation, heart disease, and nutrient deficiencies. These data also provide the basis for developing and monitoring national health goals.

Healthy People Reports Healthy People, a program that sets goals for improving the nation's health, was initiated more than 20 years ago. At the start of each decade, the program sets goals for improving the nation's health during the following 10 years. Healthy People 2010 focuses on two overall goals:

- To help people of all ages increase life expectancy and improve their quality of life.
- To eliminate disparities in health that occur by gender, race, ethnicity, disability, income, education, and sexual orientation.

In short, Healthy People 2010 challenges individuals, communities, and professionals—indeed, all of us—to take specific steps to ensure that everyone will enjoy good health and long life.

Nutrition is one of many focus areas in Healthy People 2010; each has its own specific objectives. Table 1-2 (p. 12) lists the nutrition and overweight objectives for 2010.

REVIEW NOTES

The Dietary Reference Intakes (DRI) are a set of nutrient intake values that can be used to plan and evaluate dietary intakes for healthy people.

The Estimated Energy Requirement (EER) defines the energy intake level needed to maintain energy balance in a healthy adult of a defined age, gender, weight, height, and physical activity level.

The Acceptable Macronutrient Distribution Ranges (AMDR) define the proportions contributed by carbohydrate, fat, and protein to a healthy diet.

Nutrition surveys measure people's food consumption and evaluate the nutrition status of populations.

Information gathered from nutrition surveys serves as the basis for many major diet and nutrition reports, including Healthy People 2010.

malnutrition: any condition caused by deficient or excess energy or nutrient intake or by an imbalance of nutrients.

TABLE 1-2	Healthy People 2010 Nutrition and Overweight Objectives

Increase the proportion of adults who are at a *healthy weight*.

Reduce the proportion of adults who are *obese*.

Reduce the proportion of children and adolescents who are *overweight* or *obese*.

Reduce *growth retardation* among low-income children under age 5 years.

Increase the proportion of persons aged 2 years and older who consume at least two daily servings of *fruit*.

Increase the proportion of persons aged 2 years and older who consume at least three daily servings of *vegetables*, with at least one-third being dark green or orange vegetables.

Increase the proportion of persons aged 2 years and older who consume at least six daily servings of *grain products*, with at least three being whole grains.

Increase the proportion of persons aged 2 years and older who consume less than 10 percent of kcalories from *saturated fat*.

Increase the proportion of persons aged 2 years and older who consume no more than 30 percent of kcalories from *total fat*.

Increase the proportion of persons aged 2 years and older who consume 2400 milligrams or less of *sodium*.

Increase the proportion of persons aged 2 years and older who meet dietary recommendations for *calcium*.

Reduce *iron deficiency* among young children, females of childbearing age, and pregnant females.

Reduce *anemia* among low-income pregnant females in their third trimester.

Increase the proportion of children and adolescents aged 6 to 19 years whose intake of *meals and snacks at school* contributes to good overall dietary quality.

Increase the proportion of schools that teach all essential *nutrition education* topics in one course.

Increase the proportion of worksites that offer *nutrition or weight management classes or counseling*.

Increase the proportion of primary care providers who provide nutrition *assessment* when appropriate and who formulate a diet plan for those who need *intervention*.

Increase the proportion of physician office visits made by patients with a diagnosis of cardiovascular disease, diabetes, or hyperlipidemia that include *counseling or education related to diet and nutrition*.

Increase *food security* among U.S. households and in so doing reduce hunger.

Note: "Nutrition and Overweight" is one of 28 focus areas, each with numerous objectives. Several of the other focus areas have nutrition-related objectives.

Source: Health People 2010, **www.healthypeople.gov**

overnutrition: overconsumption of food energy or nutrients sufficient to cause disease or increased susceptibility to disease; a form of malnutrition.

undernutrition: underconsumption of food energy or nutrients severe enough to cause disease or increased susceptibility to disease; a form of malnutrition.

chronic diseases: degenerative diseases characterized by deterioration of the body organs. Examples include heart disease, cancer, and diabetes.

Dietary Guidelines and Food Guides

Today, government authorities are as much concerned about **overnutrition** as they once were about **undernutrition**. Research confirms that dietary excesses, especially of energy, certain fats, and alcohol, contribute to many **chronic diseases,** including heart disease, cancer, stroke, diabetes, and liver disease.[11] Only two common lifestyle habits have more influence on health than a person's choice of diet: smoking and other tobacco use, and excessive drinking of alcohol. Table 1-3 lists the leading causes of death in the United States and shows that the top three are nutrition related (and related to tobacco use), while accidents are alcohol related. Note, however, that although diet is a powerful influence on these diseases, they cannot be prevented by a healthy diet alone; genetics, physical activity, and lifestyle also play a role. Within the range set by genetic inheritance, however, disease development is strongly influenced by the foods a person chooses to eat.

So how can health care professionals help people select foods to create a diet that supplies all the needed nutrients in amounts consistent with good health? The principle is simple enough: encourage clients to eat a variety of foods that supply all the nutrients the body needs. In practice, how do people do this? It helps to keep in mind that a nutritious diet achieves six basic ideals.

DIETARY IDEALS

A nutritious diet has six characteristics. The first is **adequacy.** The earlier discussion on the DRI already addressed the ideal of nutrient adequacy. An adequate diet has enough energy and enough of every nutrient (as well as fiber) to meet the needs of healthy people. Second is **balance:** the food choices do not overemphasize one nutrient or food type at the expense of another. The essential minerals calcium and iron illustrate the importance of dietary balance. Meats, fish, and poultry are rich in iron but poor in calcium. Conversely, milk and milk products are rich in calcium but poor in iron. Use some meat or meat alternates for iron; use some milk and milk products for calcium; and save some space for other foods, too, because a diet consisting of milk and meat alone would not be adequate. For other nutrients, people need grains, vegetables, and fruit.

The third characteristic is **kcalorie control:** the foods provide the amount of energy needed to maintain a healthy body weight—not more, not less. The key to kcalorie control is to select foods that deliver the most nutrients for the least food energy. This fourth characteristic is known as **nutrient density.** Consider calcium sources, for example. Ice cream and fat-free milk both supply calcium, but the milk is "denser" in calcium per kcalorie. A cup of rich ice cream contributes more than 250 kcalories, a cup of fat-free milk only 85—and with almost double the calcium.

The fifth characteristic of a nutritious diet is **moderation.** Foods rich in fat and sugar provide enjoyment and energy but relatively few nutrients. In addition, they promote weight gain when eaten in excess. A person who practices moderation eats such foods only on occasion and regularly selects foods low in fat and sugar, a practice that automatically improves nutrient density. Returning to the example of ice cream and fat-free milk, the milk not only offers more calcium for less energy, but it contains far less fat than the ice cream.

Finally, the sixth characteristic of a nutritious diet is **variety:** the foods chosen differ from one day to the next. A diet may have all the virtues just described and still lack variety, if a person eats the same foods day after day. People should select foods from each of the food groups daily and vary their choices within each food group from day to day for a couple of reasons. First, different foods within the same group contain different arrays of nutrients. Among the fruits, for example, strawberries are especially rich in vitamin C whereas apricots are rich in vitamin A. Second, no food is guaranteed entirely free of substances that, in excess, could be harmful. The strawberries might contain trace amounts of one contaminant,

TABLE 1-3	Leading Causes of Death in the United States, 2003

The four diseases shaded red are nutrition related.[a] The fifth-ranked cause, motor vehicle and other accidents, shaded green, is alcohol related.

1. Heart disease
2. Cancers
3. Strokes
4. Chronic lower respiratory diseases
5. Motor vehicle and other accidents
6. Diabetes mellitus
7. Pneumonia and influenza
8. Alzheimer's disease
9. Kidney disease
10. Infections of the blood

[a]Hypertension (high blood pressure), a nutrition-related disease, is ranked number 13.
Source: National Center for Health Statistics, 2005.

adequacy: the characteristic of a diet that provides all the essential nutrients, fiber, and energy necessary to maintain health and body weight.

balance: the dietary characteristic of providing foods of a number of types in proportion to each other, such that foods rich in some nutrients do not crowd out foods that are rich in other nutrients.

kcalorie (energy) control: management of food energy intake.

nutrient density: a measure of the nutrients a food provides relative to the energy it provides. The more nutrients and the fewer kcalories, the higher the nutrient density.

moderation: providing enough, but not too much, of a substance.

variety (dietary): eating a wide selection of foods within and among the major food groups (the opposite of monotony).

the apricots another. By alternating fruit choices, a person will ingest very little of either contaminant.

DIETARY GUIDELINES FOR AMERICANS

The *Dietary Guidelines for Americans 2005* originated from convincing scientific evidence that nutritious food, kcalorie control, and physical activity promote health and reduce the risk of chronic disease.[12] In general, the *Guidelines* answer the question, What should an individual eat to stay healthy? Table 1-4 presents the nine *Guideline* topics with their key recommendations. The first three topics focus on choosing nutrient-rich foods within energy needs, maintaining a healthy body weight, and engaging in regular physical activity. The fourth topic, "Food Groups to Encourage," focuses on the selection of a variety of fruits and vegetables, whole grains, and milk. The next four topics advise people to choose sensibly in their use of fats, carbohydrates, salt, and alcoholic beverages for those who partake. Finally, consumers are reminded to keep food safe. Together, the *Dietary Guidelines* point the way toward better health. Table 1-5 (p. 16) presents Canada's *Guidelines for Healthy Eating*.

> ### REVIEW NOTES
>
> A well-planned diet delivers adequate nutrients, a balanced array of nutrients, and an appropriate amount of energy.
>
> A well-planned diet is based on nutrient-dense foods, moderate in substances that can be detrimental to health, and varied in its selections.
>
> The *Dietary Guidelines* apply these principles, offering practical advice on how to eat for good health.

FITNESS GUIDELINES

The *Dietary Guidelines for Americans 2005* (see Table 1-4) encourage people not only to make wiser food choices but also to balance kcalories from food with kcalories expended and, in general, to be more physically active to promote health and reduce the risk of chronic disease. People don't have to run marathons to reap the health rewards of physical activity.[13] The American College of Sports Medicine (ACSM), the *Dietary Guidelines for Americans 2005*, and the DRI committee specify that for health's sake, people need to spend an accumulated minimum of 30 minutes in some sort of physical activity on most days of each week.[14] All three authorities, however, advise that 30 minutes of physical activity each day for adults is not enough to maintain a healthy body weight and recommend at least 60 minutes of moderately intense activity such as walking or jogging each day. The hour or more of activity can be split into shorter sessions throughout the day—two 30-minute sessions or four 15-minute sessions, for example.[15]

For many people, the health benefits of regular, moderate physical activity are reward enough. Others, however, seek the kinds and amount of physical activity that will improve their physical fitness or their performance in sports, as well as benefit health. The ACSM has issued recommendations for the quantity and quality of physical activity needed to develop and maintain fitness in healthy adults (see Table 1-6 on p. 16). The kinds and amounts of physical activity that improve physical fitness also provide still greater health benefits (further reduction of cardiovascular disease risk and improved body composition, for example).[16]

People who regularly engage in just moderate physical activity live longer on average than those who are physically inactive.[17] Yet, despite an increasing awareness of the health benefits that physical activity confers, 25 percent of adults in the United States are completely inactive.[18] A sedentary lifestyle ranks with smoking and obesity as a powerful risk factor for developing the major killer diseases of

TABLE 1-4	Key Recommendations of the *Dietary Guidelines for Americans 2005*

Adequate Nutrients within Energy Needs

Consume a variety of nutrient-dense foods and beverages within and among the basic food groups; limit intakes of saturated and *trans* fats, cholesterol, added sugars, salt, and alcohol.

Meet recommended intakes within energy needs by adopting a balanced eating pattern, such as the USDA Food Guide (Figure 1-4, pp. 18-19).

Weight Management

To maintain body weight in a healthy range, balance kcalories from foods and beverages with kcalories expended (see Chapter 6).

To prevent gradual weight gain over time, make small decreases in food and beverage kcalories and increase physical activity.

Physical Activity

Engage in regular physical activity and reduce sedentary activities to promote health, psychological well-being, and a healthy body weight.

To reduce the risk of chronic disease in adulthood, engage in at least 30 minutes of moderate-intensity physical activity, above usual activity, at home or work on most days of the week.

For most people, greater health benefits can be obtained by engaging in physical activity of more vigorous intensity or longer duration.

Achieve physical fitness by including cardiovascular conditioning, stretching exercises for flexibility, and resistance exercises or calisthenics for muscle strength and endurance.

Food Groups to Encourage

Consume a sufficient amount of fruits, vegetables, milk and milk products, and whole grains while staying within energy needs.

Select a variety of fruits and vegetables each day, including selections from all five vegetable subgroups (dark green, orange, legumes, starchy vegetables, and other vegetables) several times a week. Make at least half of the grain selections whole grains. Select fat-free or low-fat milk products.

Fats

Consume less than 10 percent of kcalories from saturated fats and less than 300 milligrams of cholesterol per day, and keep *trans* fat consumption as low as possible (see Chapter 4).

Keep total fat intake between 20 and 35 percent of kcalories; choose from mostly polyunsaturated and monounsaturated fat sources such as fish, nuts, and vegetable oils.

Select and prepare foods that are lean, low fat, or fat-free and low in saturated and/or *trans* fat.

Carbohydrates

Choose fiber-rich fruits, vegetables, and whole grains often.

Choose and prepare foods and beverages with little added sugars (see Chapter 3).

Reduce the incidence of dental caries by practicing good oral hygiene and consuming sugar- and starch-containing foods and beverages less frequently.

Sodium and Potassium

Choose and prepare foods with little salt (less than 2300 milligrams sodium or approximately 1 teaspoon salt). At the same time, consume potassium-rich foods, such as fruits and vegetables (see Chapter 9).

Alcoholic Beverages

Those who choose to drink alcoholic beverages should do so sensibly and in moderation.

Some individuals should not consume alcoholic beverages (see Nutrition in Practice 19).

Food Safety

To avoid microbial foodborne illness, keep foods safe: clean hands, food contact surfaces, and fruits and vegetables; separate raw, cooked, and ready-to-eat foods; cook foods to a safe internal temperature; chill perishable food promptly; and defrost food properly.

Avoid unpasteurized milk and products made from it; raw or undercooked eggs, meat, poultry, fish, and shellfish; unpasteurized juices; and raw sprouts.

Note: These guidelines are intended for adults and healthy children ages 2 and older.

Source: The *Dietary Guidelines for Americans 2005,* available at **www.healthierus.gov/dietaryguidelines**.

TABLE 1-5	Canada's *Guidelines for Healthy Eating*

Enjoy a variety of foods.

Emphasize cereals, breads, other grain products, vegetables, and fruits.

Choose lower-fat dairy products, leaner meats, and foods prepared with little or no fat.

Achieve and maintain a healthy body weight by enjoying regular physical activity and healthy eating.

Limit salt, alcohol, and caffeine.

Source: These guidelines derive from *Action Towards Healthy Eating—Canada's Guidelines for Healthy Eating and Recommended Strategies for Implementation.*

our time—cardiovascular disease, some forms of cancer, stroke, diabetes, and hypertension.[19] Every year an estimated $24 billion is spent on health care costs attributed to physical inactivity in the United States. Therefore, one of the most important challenges health care professionals face is to motivate more people to become physically active. To motivate others, health care professionals must first become more physically active themselves, thereby enhancing their own health. Second, they can include regular physical activity as a component of therapy for their clients. Table 1-7 summarizes the benefits fitness confers.

REVIEW NOTES

Regular physical activity promotes health and reduces risk of chronic disease.

The ACSM, the *Dietary Guidelines for Americans 2005,* and the DRI committee recommend at least 30 minutes of physical activity each day for health benefits and 60 minutes or more for maintaining body weight.

The ACSM has issued recommendations for physical activity to develop and maintain physical fitness.

THE USDA FOOD GUIDE

To help people achieve the goals set forth by the *Dietary Guidelines for Americans 2005* (Table 1-4), the USDA provides a **food group plan**—the **USDA Food Guide**—

TABLE 1-6	Guidelines for Physical Fitness		
	**Cardiorespiratory**	**Strength**	**Flexibility**
Type of activity	Aerobic activity that uses large-muscle groups and can be maintained continuously	Resistance activity that is performed at a controlled speed and through a full range of motion	Stretching activity that uses the major muscle groups
Frequency	3 to 5 days per week	2 to 3 days per week	2 to 3 days per week
Intensity	55 to 90% of maximum heart rate	Enough to enhance muscle strength and improve body composition	Enough to develop and maintain a full range of motion
Duration	20 to 60 minutes	8 to 12 repetitions of 8 to 10 different exercises (minimum)	4 repetitions of 10 to 30 seconds per muscle group (minimum)
Examples	Running, cycling, swimming, in-line skating, rowing, power walking, cross-country skiing, kickboxing, jumping rope; sports activities such as basketball, soccer, racquetball, tennis, volleyball	Pull-ups, push-ups, weight lifting, pilates	Yoga

TABLE 1-7	Benefits of Fitness

More restful sleep. Rest and sleep occur naturally after periods of physical activity. During rest, the body repairs injuries, disposes of wastes generated during activity, and builds new physical structures.

Better nutritional health. Physical activity expends energy and thus allows people to eat more food. If they choose wisely, active people will consume more nutrients and be less likely to develop nutrient deficiencies.

Improved body composition. A balanced program of physical activity limits body fat and increases or maintains lean tissue. Thus physically active people have relatively less body fat than sedentary people at the same body weight.*[a]

Improved bone density. Weight-bearing physical activity builds bone strength and protects against osteoporosis.[b]

Enhanced resistance to colds and other infectious diseases. Fitness enhances immunity.[†c]

Lower risks of some types of cancers. Lifelong physical activity may help to protect against colon cancer, breast cancer, and some other cancers.[d]

Stronger circulation and lung function. Physical activity that challenges the heart and lungs strengthens both the circulatory and the respiratory system.

Lower risks of cardiovascular disease. Physical activity lowers blood pressure, slows resting pulse rate, lowers total blood cholesterol, and raises HDL cholesterol, thus reducing the risks of heart attacks and strokes.[e] Some research suggests that physical activity may reduce the risk of cardiovascular disease in another way as well—by reducing intra-abdominal fat stores.[f]

Lower risks of type 2 diabetes. Physical activity normalizes glucose tolerance.[g] Regular physical activity reduces the risk of developing type 2 diabetes and benefits those who already have the condition.

Reduced risk of gallbladder disease (women). Regular physical activity reduces women's risk of gallbladder disease—perhaps by facilitating weight control and lowering blood lipid levels.[h]

Lower incidence and severity of anxiety and depression. Physical activity may improve mood and enhance the quality of life by reducing depression and anxiety.[i]

Stronger self-image. The sense of achievement that comes from meeting physical challenges promotes self-confidence.

Longer life and higher quality of life in the later years. Active people have a lower mortality rate than sedentary people.[j] Even a two-mile walk daily can add years to a person's life. In addition to extending longevity, physical activity supports independence and mobility in later life by reducing the risk of falls and minimizing the risk of injury should a fall occur.

*The reference notes for this table are on p. 34.

[†]Moderate physical activity can stimulate immune function. Intense, vigorous, prolonged activity such as marathon running, however, may compromise immune function.

Physical activity helps you look good, feel good, and have fun, and it brings many long-term health benefits as well.

that builds a diet from categories of foods that are similar in vitamin and mineral content. Thus each group provides a set of nutrients that differs somewhat from the nutrients supplied by the other groups. Selecting foods from each of the groups eases the task of creating an adequate and balanced diet.

Figure 1-4 (on pp. 18–19) presents the USDA Food Guide. The USDA Food Guide assigns foods to five major food groups and recommends daily amounts of foods from each group to meet nutrient needs. The judicious use of oils, solid fats, and added sugars is also described in the USDA Food Guide. In addition to presenting the food groups, the figure indicates the most notable nutrients of each group and lists some foods within each group sorted by nutrient density.

The **DASH** Eating Plan, presented in Chapter 21, is another dietary pattern that meets the goals of the *Dietary Guidelines for Americans 2005*.

Five food groups:

- Fruits.
- Vegetables.
- Grains.
- Meat, poultry, fish, legumes, eggs, and nuts.
- Milk, yogurt, and cheese.

Chapter 11 provides a food guide for young children, and Appendix B presents Canada's food group plan, the *Food Guide to Healthy Eating*.

food group plan: a diet-planning tool that sorts foods into groups based on nutrient content and then specifies that people should eat certain amounts of food from each group.

USDA Food Guide: the USDA's food group plan for ensuring dietary adequacy that assigns foods to five major food groups.

FIGURE 1-4 USDA Food Guide, 2005

> **Key:**
> ● Foods generally high in nutrient density (choose most often)
> ▲ Foods lower in nutrient density (limit selections)

GRAINS

© Polara Studios, Inc.

Make at least half of the grain selections whole grains.

These foods contribute folate, niacin, riboflavin, thiamin, iron, magnesium, selenium, and fiber.

> 1 oz grains is equivalent to 1 slice bread; ½ c cooked rice, pasta, or cereal; 1 oz dry pasta or rice; 1 c ready-to-eat cereal; 3 c popped popcorn.

● Whole grains (barley, brown rice, bulgur, millet, oats, rye, wheat) and whole-grain, low-fat breads, cereals, crackers, and pastas; popcorn.

● Enriched bagels, breads, cereals, pastas (couscous, macaroni, spaghetti), pretzels, rice, rolls, tortillas.

▲ Biscuits, cakes, cookies, cornbread, crackers, croissants, doughnuts, French toast, fried rice, granola, muffins, pancakes, pastries, pies, presweetened cereals, taco shells, waffles.

VEGETABLES

© Polara Studios, Inc.

Choose a variety of vegetables from all five subgroups several times a week.

These foods contribute folate, vitamin A, vitamin C, vitamin E, magnesium, potassium, and fiber.

> ½ c vegetables is equivalent to ½ c cut-up raw or cooked vegetables; ½ c cooked legumes; ½ c vegetable juice; 1 c raw, leafy greens.

● Dark green vegetables: Broccoli and leafy greens such as arugula, beet greens, bok choy, collard greens, kale, mustard greens, romaine lettuce, spinach, turnip greens.

● Orange and deep yellow vegetables: Carrots, carrot juice, pumpkin, sweet potatoes, winter squash (acorn, butternut).

● Legumes: Black beans, black-eyed peas, garbanzo beans (chickpeas), kidney beans, lentils, navy beans, pinto beans, soybeans and soy products such as tofu, split peas.

● Starchy vegetables: Cassava, corn, green peas, hominy, lima beans, potatoes.

● Other vegetables: Artichokes, asparagus, bamboo shoots, bean sprouts, beets, brussels sprouts, cabbages, cactus, cauliflower, celery, cucumbers, eggplant, green beans, iceberg lettuce, mushrooms, okra, onions, peppers, seaweed, snow peas, tomatoes, vegetable juices, zucchini.

▲ Baked beans, candied sweet potatoes, coleslaw, french fries, potato salad, refried beans, scalloped potatoes, tempura vegetables.

FRUITS

© Polara Studios, Inc.

Consume a variety of fruits and no more than one-third of the recommended intake as fruit juice.

These foods contribute folate, vitamin A, vitamin C, potassium, and fiber.

> ½ c fruit is equivalent to ½ c fresh, frozen, or canned fruit; 1 small fruit; ¼ c dried fruit; ½ c fruit juice.

● Apples, apricots, avocados, bananas, blueberries, cantaloupe, cherries, grapefruit, grapes, guava, kiwi, mango, oranges, papaya, peaches, pears, pineapples, plums, raspberries, strawberries, watermelon; dried fruit; unsweetened juices.

▲ Canned or frozen fruit in syrup; juices, punches, ades, and fruit drinks with added sugars; fried plantains.

MILK, YOGURT, AND CHEESE

© Polara Studios, Inc.

Make fat-free or low-fat choices. Choose lactose-free products or other calcium-rich foods if you don't consume milk.

These foods contribute protein, riboflavin, vitamin B_{12}, calcium, magnesium, potassium, and, when fortified, vitamin A and vitamin D.

> **1 c milk is equivalent to 1 c fat-free milk or yogurt; 1½ oz fat-free natural cheese; 2 oz fat-free processed cheese.**

● Fat-free milk and fat-free milk products such as buttermilk, cheeses, cottage cheese, yogurt; fat-free fortified soy milk.

△ 1% low-fat milk, 2% reduced-fat milk, and whole milk; low-fat, reduced-fat, and whole-milk products such as cheeses, cottage cheese, and yogurt; milk products with added sugars such as chocolate milk, custard, ice cream, ice milk, milk shakes, pudding, sherbet; fortified soy milk.

MEAT, POULTRY, FISH, LEGUMES, EGGS, AND NUTS

© Polara Studios, Inc.

Make lean or low-fat choices. Prepare them with little, or no, added fat.

Meat, poultry, fish, and eggs contribute protein, niacin, thiamin, vitamin B_6, vitamin B_{12}, iron, magnesium, potassium, and zinc; legumes and nuts are notable for their protein, folate, thiamin, vitamin E, iron, magnesium, potassium, zinc, and fiber.

> **1 oz meat is equivalent to 1 oz cooked lean meat, poultry, or fish; 1 egg; ¼ c cooked legumes or tofu; 1 tbs peanut butter; ½ oz nuts or seeds.**

● Poultry (no skin), fish, shellfish, legumes, eggs, lean meat (fat-trimmed beef, game, ham, lamb, pork); low-fat tofu, tempeh, peanut butter, nuts or seeds.

△ Bacon; baked beans; fried meat, fish, poultry, eggs, or tofu; refried beans; ground beef; hot dogs; luncheon meats; marbled steaks; poultry with skin; sausages; spare ribs.

OILS

Matthew Farruggio

Select the recommended amounts of oils from among these sources.

These foods contribute vitamin E and essential fatty acids (see Chapter 4), along with abundant kcalories.

> **1 tsp oil is equivalent to 1 tbs low-fat mayonnaise; 2 tbs light salad dressing; 1 tsp vegetable oil; 1 tsp soft margarine.**

● Liquid vegetable oils such as canola, corn, flaxseed, nut, olive, peanut, safflower, sesame, soybean, and sunflower oils; mayonnaise, oil-based salad dressing, soft *trans*-free margarine.

● Unsaturated oils that occur naturally in foods such as avocados, fatty fish, nuts, olives, and shellfish.

SOLID FATS AND ADDED SUGARS

Matthew Farruggio

Limit intakes of food and beverages with solid fats and added sugars.

Solid fats deliver saturated fat and *trans* fat, and intake should be kept low. Solid fats and added sugars contribute abundant kcalories but few nutrients, and intakes should not exceed the discretionary kcalorie allowance—kcalories to meet energy needs after all nutrient needs have been met with nutrient-dense foods. Alcohol also contributes abundant kcalories but few nutrients, and its kcalories are counted among discretionary kcalories. See Table 1-8 for some discretionary kcalorie allowances.

△ Solid fats that occur in foods naturally such as milk fat and meat fat (see △ in previous lists).

△ Solid fats that are often added to foods such as butter, cream cheese, hard margarine, lard, sour cream, and shortening.

△ Added sugars such as brown sugar, candy, honey, jelly, molasses, soft drinks, sugar, and syrup.

△ Alcoholic beverages include beer, wine, and liquor.

A portion of grains is 1 ounce, yet most bagels today weigh 4 ounces or more—meaning that a single bagel can easily supply four or more portions of grains, not one as many people assume.

Chapter 6 explains how to determine energy needs.

Reminder: *Phytochemicals* are the nonnutrient compounds found in plant-derived foods that have biological activity in the body.

legumes (lay-GYOOMS, LEG-yooms): plants of the bean and pea family, with seeds that are rich in protein compared with other plant-derived foods.

Recommended amounts for fruits, vegetables, and milk are given in cups and those for grains and meats, in ounces. Figure 1-4 provides equivalent measures for foods that are not readily measured in cups and ounces. For example, users learn that 1 ounce of grains is equivalent to 1 slice of bread or $^1/_2$ cup of cooked rice.

Recommended Daily Food Amounts All food groups offer valuable nutrients, and people should make selections from each group daily. Table 1-8 specifies the amounts of food needed from each group daily to create a healthful diet for several energy (kcalorie) levels. Estimated daily kcalorie needs for sedentary and active men and women are shown in Table 1-9. For example, a person needing 2000 kcalories a day would select 2 cups of fruit; $2^1/_2$ cups of vegetables (dispersed among the vegetable subgroups); 6 ounces of grain foods (with at least half coming from whole grains); $5^1/_2$ ounces of meat, poultry, or fish, or the equivalents of **legumes**, eggs, seeds, or nuts; and 3 cups of milk or yogurt, or the equivalent of cheese or fortified soy products. Additionally, a small amount of unsaturated oil, such as vegetable oil, or the oils of nuts, olives, or fatty fish, is required to supply needed nutrients.

All vegetables provide an array of vitamins, fiber, and the mineral potassium, but some vegetables are especially good sources of certain nutrients and beneficial phytochemicals. For this reason, the USDA Food Guide sorts the vegetable group into subgroups. The dark green vegetables deliver the B vitamin folate; the orange vegetables provide abundant vitamin A; legumes supply iron and protein; the starchy vegetables contribute abundant carbohydrate energy; and the other vegetables fill in the gaps and add more of these same nutrients.

In a 2000-kcalorie diet, then, the recommended $2^1/_2$ cups of daily vegetables should be spread among the subgroups over a week's time, as shown in Table 1-10. In other words, eating $2^1/_2$ cups of potatoes or even spinach every day for seven days would *not* meet the recommended vegetable intakes. Potatoes and spinach make excellent choices when consumed in balance with other vegetables from other subgroups. Intakes of vegetables are appropriately averaged over a week's time—it isn't necessary to include every subgroup every day.

TABLE 1-8	Recommended Daily Amounts from Each Food Group							
FOOD GROUP	**1600 kcal**	**1800 kcal**	**2000 kcal**	**2200 kcal**	**2400 kcal**	**2600 kcal**	**2800 kcal**	**3000 kcal**
Fruits	$1^1/_2$ c	$1^1/_2$ c	2 c	2 c	2 c	2 c	$2^1/_2$ c	$2^1/_2$ c
Vegetables	2 c	$2^1/_2$ c	$2^1/_2$ c	3 c	3 c	$3^1/_2$ c	$3^1/_2$ c	4 c
Grains	5 oz	6 oz	6 oz	7 oz	8 oz	9 oz	10 oz	10 oz
Meat and legumes	5 oz	5 oz	$5^1/_2$ oz	6 oz	$6^1/_2$ oz	$6^1/_2$ oz	7 oz	7 oz
Milk	3 c	3 c	3 c	3 c	3 c	3 c	3 c	3 c
Oils	5 tsp	5 tsp	6 tsp	6 tsp	7 tsp	8 tsp	8 tsp	10 tsp
Discretionary kcalorie allowance	132 kcal	195 kcal	267 kcal	290 kcal	362 kcal	410 kcal	426 kcal	512 kcal

Notable Nutrients As Figure 1-4 notes, each food group contributes key nutrients. This feature provides flexibility in diet planning: a person can select any food from a food group and receive similar nutrients. For example, a person can choose milk, cheese, or yogurt and receive the same key nutrients the milk group offers. Legumes are included in the meat, fish, and poultry group because these foods are notable for their contributions of protein, iron, and zinc. Legumes are unique, however, because they are also grouped with the vegetables by virtue of their fiber and vitamins. Importantly, foods provide not only the nutrients for which each group is noted, but small amounts of other nutrients and phytochemicals as well.

The USDA Food Guide encourages greater consumption from certain food groups to provide the nutrients most often lacking in the diets of Americans. In general, most people need to eat:

- *More* dark green vegetables, orange vegetables, legumes, fruits, whole grains, and low-fat milk and milk products.
- *Less* refined grains, total fats (especially saturated fat, *trans* fat, and cholesterol), added sugars, and total kcalories.

Nutrient Density The USDA Food Guide provides a foundation for a healthy diet by emphasizing nutrient-dense options within each food group. By consistently selecting nutrient-dense foods, a person can both obtain all the nutrients needed and keep kcalories under control. In contrast, eating foods that are low in nutrient density makes it difficult to get enough nutrients without exceeding energy needs and gaining weight. For this reason, consumers should select low-fat foods from each group and foods without added fats or sugars—for example, fat-free milk instead of whole milk, baked chicken without the skin instead of hot dogs, green beans instead of french fries, orange juice instead of fruit punch, and whole-wheat bread instead of biscuits. Notice that Figure 1-4 provides a key indicating which foods *within each group* are high or low in nutrient density. Oil is a notable exception: even though oil is pure fat and therefore rich in kcalories, a small amount of oil from sources such as nuts, fish, or vegetable oils is necessary every day to provide nutrients lacking from other foods. Consequently these high-fat foods are listed among the nutrient-dense foods (see Nutrition in Practice 4 to learn why).

TABLE 1-9	Estimated Daily kCalorie Needs for Adults	
	Sedentary[a]	Active[b]
Women		
19–30 yr	2000	2400
31–50 yr	1800	2200
51+ yr	1600	2100
Men		
19–30 yr	2400	3000
31–50 yr	2200	2900
51+ yr	2000	2600

Note: In addition to gender, age, and activity level, energy needs vary with height and weight (see Chapter 6).

[a]*Sedentary* describes a lifestyle that includes only the activities typical of day-to-day life.

[b]*Active* describes a lifestyle that includes physical activity equivalent to walking more than 3 miles per day at a rate of 3 to 4 miles per hour, in addition to the activities typical of day-to-day life. kCalorie values for active people reflect the midpoint of the range appropriate for age and gender, but within each group, older adults may need fewer kcalories and younger adults may need more.

The USDA nutrients of concern are vitamins A, C, and E; the minerals calcium, magnesium, and potassium; and fiber.

TABLE 1-10	Recommended Weekly Amounts from the Vegetable Subgroups							
Vegetable Subgroups	1600 kcal	1800 kcal	2000 kcal	2200 kcal	2400 kcal	2600 kcal	2800 kcal	3000 kcal
Dark green	2 c	3 c	3 c	3 c	3 c	3 c	3 c	3 c
Orange and deep yellow	1½ c	2 c	2 c	2 c	2 c	2½ c	2½ c	2½ c
Legumes	2½ c	3 c	3 c	3 c	3 c	3½ c	3½ c	3½ c
Starchy	2½ c	3 c	3 c	6 c	6 c	7 c	7 c	9 c
Other	5½ c	6½ c	6½ c	7 c	7 c	8½ c	8½ c	10 c

Table 1-8 specifies the recommended amounts of total vegetables per *day*. This table shows those amounts dispersed among five vegetable subgroups per *week*.

This cola and bunch of grapes illustrate nutrient density. Each provides about 150 kcalories (mainly from carbohydrate), but the grapes offer a trace of protein, some vitamins, minerals, and fiber along with the energy; the cola beverage offers only "empty" kcalories. Grapes, or any fruit for that matter, are more nutrient dense than cola beverages.

Discretionary kCalorie Allowance At each kcalorie level, people who consistently choose nutrient-dense foods may be able to meet their nutrient needs without consuming their full allotment of kcalories. This difference between the kcalories needed to supply nutrients and those needed for energy—known as the **discretionary kcalorie allowance**—is illustrated in Figure 1-5.

Table 1-8 includes the discretionary kcalorie allowance for several kcalorie levels. A person with discretionary kcalories available might choose to:

- Eat more nutrient-dense foods, such as an extra serving of skinless chicken or a second ear of corn.
- Select a few foods with fats or added sugars, such as reduced-fat milk or sweetened cereal.
- Add a little fat or sugar to foods, such as butter or jelly on toast.
- Consume some alcohol. (Nutrition in Practice 19 explains why this may not be a good choice for some individuals.)

Alternatively, a person wanting to lose weight might choose to:

- *Not* use the kcalories available from the discretionary kcalorie allowance.

Compared to physically active adults, sedentary adults have lower kcalorie needs and so have a lower discretionary kcalorie allowance. Physically active adults expend more energy and therefore have a higher energy need and a greater discretionary allowance.

Added fats and sugars are always counted as discretionary kcalories. The kcalories from the fat in higher-fat milks and meats are also counted among discretionary kcalories. It helps to think of fat-free milk as "milk," and whole milk or reduced-fat milk as "milk with added fat." Similarly, "meats" should be the leanest; other cuts are "meats with added fat." Puddings and other desserts made from whole milk provide discretionary kcalories from both the sugar added to sweeten them and the naturally occurring fat in the whole milk they contain. Even fruits, vegetables, and grains can carry discretionary kcalories into the diet in the form of peaches canned in syrup, french fries, or high-fat crackers.

Discretionary kcalories must be counted separately from the kcalories of the nutrient-dense foods of which they may be a part. A fried chicken leg, for example, provides discretionary kcalories from two sources: the naturally occurring fat of the chicken skin and the added fat absorbed during frying. The kcalories of the skinless chicken underneath are not discretionary kcalories—they are necessary to provide the nutrients of chicken.

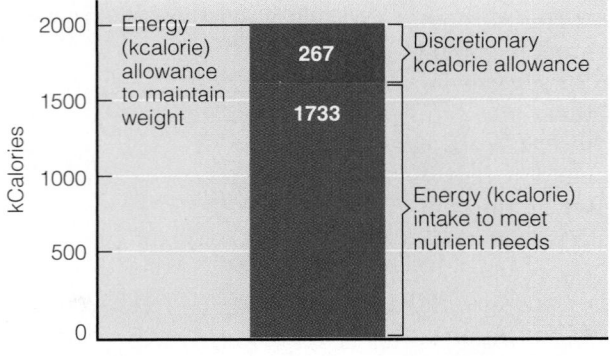

FIGURE 1-5 Discretionary kCalorie Allowance in a 2000-kCalorie Diet

For quick and easy estimates, visualize each portion as being about the size of a common object:

- 1/4 c dried fruit = a golf ball.
- 3 oz meat = a deck of cards.
- 1 1/2 oz cheese = 6 stacked dice or the size of a nine-volt battery.
- 1/2 c ice cream = a racquetball.
- 4 small cookies = 4 poker chips.

discretionary kcalorie allowance: the kcalories remaining in a person's energy allowance after consuming enough nutrient-dense foods to meet all nutrient needs for a day.

Portion Control To control kcalories, the diet planner must also learn to control food portions (USDA serving equivalents were listed in Figure 1-4). The trend in the United States has been toward consuming larger food portions, especially of foods rich in fat and sugar. At the same time, body weights have been steadily creeping upward, suggesting an increasing need to control portion sizes. In contrast to the random-size (and often large) helpings offered in restaurants, fast-food franchises, and elsewhere, the quantities recommended in the USDA Food Guide are specific, precise, and reliable for delivering certain amounts of key nutrients in food. The margin offers some tips for estimating portion sizes.

Mixtures of Foods Some foods—such as casseroles, soups, and sandwiches—fall into two or more food groups. With a little practice, users can learn to divide these foods into food groups. From the USDA Food Guide point of view, a taco represents four different food groups: the taco shell from the grains group; the onions, lettuce, and tomatoes from the "other vegetable" group; the ground beef from the meat group; and the cheese from the milk group.

Vegetarian Food Guide Vegetarian diets rely mainly on plant foods: grains, vegetables, legumes, fruits, seeds, and nuts. Some vegetarian diets include eggs, milk products, or both. People who do not eat meats or milk products can still use the USDA Food Guide to create an adequate diet.[20] The food groups are similar, and the recommended amounts of foods remain the same. Vegetarians select *meat alternates* from the meat group—foods such as legumes, seeds, nuts, tofu, and, for those who eat them, eggs. Legumes and at least one cup of dark green leafy vegetables help to supply the iron that meats usually provide. Vegetarians who do not drink cow's milk can use soy "milk"—a product made from soybeans that provides similar nutrients if it has been fortified with calcium, vitamin D, and vitamin B_{12}. Nutrition in Practice 5 presents a Food Guide for Vegetarians, defines vegetarian terms, and provides more information on vegetarian diet planning.

Ethnic Food Choices People can use the USDA Food Guide and still enjoy a diverse array of culinary styles by sorting ethnic foods into their appropriate food groups. For example, a person eating Mexican foods would find tortillas in the grains group, jicama in the vegetable group, and guava in the fruit group. Table 1-1 (p. 4) features ethnic food choices.

MYPYRAMID

The USDA created an educational tool called MyPyramid to illustrate the concepts presented in the *Dietary Guidelines* and the USDA Food Guide. Figure 1-6 presents

FIGURE 1-6 MyPyramid: Steps to a Healthier You
Source: USDA, 2005.

The multiple colors of the pyramid illustrate variety: each color represents one of the five food groups, plus one for oils. Different widths of colors suggest the proportional contribution of each food group to a healthy diet.

A person climbing steps reminds consumers to be physically active each day.

The narrow slivers of color at the top imply moderation in foods rich in solid fats and added sugars.

The broad bases at the bottom represent nutrient-dense foods that should make up the bulk of the diet.

Greater intakes of grains, vegetables, fruit, and milk are encouraged by the broad bases of orange, green, red, and blue.

GRAINS   VEGETABLES   FRUITS   OIL   MILK   MEAT & BEANS

a graphic image of MyPyramid, which was designed to encourage consumers to make healthy food and physical activity choices every day.

An abundance of material supporting MyPyramid is available to consumers who want to find the kinds and amounts of foods to eat each day (**www.MyPyramid .gov**). In addition to creating a personal plan, consumers can find tips to help them improve their diet and lifestyle by taking small steps each day.

REVIEW NOTES

Food group plans such as the USDA Food Guide serve as the basis for planning adequate, balanced, and varied diets.

Each food group contributes key nutrients, a feature that provides flexibility in diet planning.

The USDA Food Guide emphasizes nutrient-dense foods within each group.

The discretionary kcalorie allowance is the difference between the kcalories needed to meet nutrient needs and those needed for energy.

MyPyramid is an educational tool used to illustrate the concepts presented in the *Dietary Guidelines* and the USDA Food Guide.

Food Labels

Today, consumers know more about the links between diet and disease than they did in the past, and they are demanding still more information on disease prevention. Many people rely on food labels to help them select foods with less saturated fat, *trans* fat, cholesterol, and sodium and more vitamins, minerals, and dietary fiber. Figure 1-7 illustrates the requirements for label information. Most food labels must conform to all these requirements. Exceptions include plain coffee, tea, spices, and other foods contributing few nutrients; foods produced by small businesses; packages with fewer than 12 square inches of surface area; and those prepared and sold in the same establishment as long as the foods do not make nutrient or health claims.*

The Ingredient List All foods must list all ingredients on the label in descending order of predominance by weight. Knowing that the first ingredient predominates by weight, consumers can glean much information. Compare these products, for example:

- A beverage powder that contains "sugar, citric acid, natural flavors . . ." versus a juice that contains "water, tomato concentrate, concentrated juices of carrots, celery. . . ."
- A cereal that contains "puffed milled corn, sugar, corn syrup, molasses, salt . . ." versus one that contains "100 percent rolled oats."

In each comparison, consumers can tell that the second product is the more nutrient dense.

Serving Sizes Because labels present nutrient information per serving, they must identify the size of a serving. The Food and Drug Administration (FDA) has established specific serving sizes for various foods and requires that all labels for a given product use the same serving size. For example, the serving size for all ice creams is $1/2$ cup and for all beverages, 8 fluid ounces. This facilitates comparison shopping. Consumers can see at a glance which brand has more or fewer kcalories or grams of fat, for example. Standard serving sizes are expressed in both

---

* For example, restaurants need not provide complete nutrition information unless they are making "heart-healthy" claims for menu items.

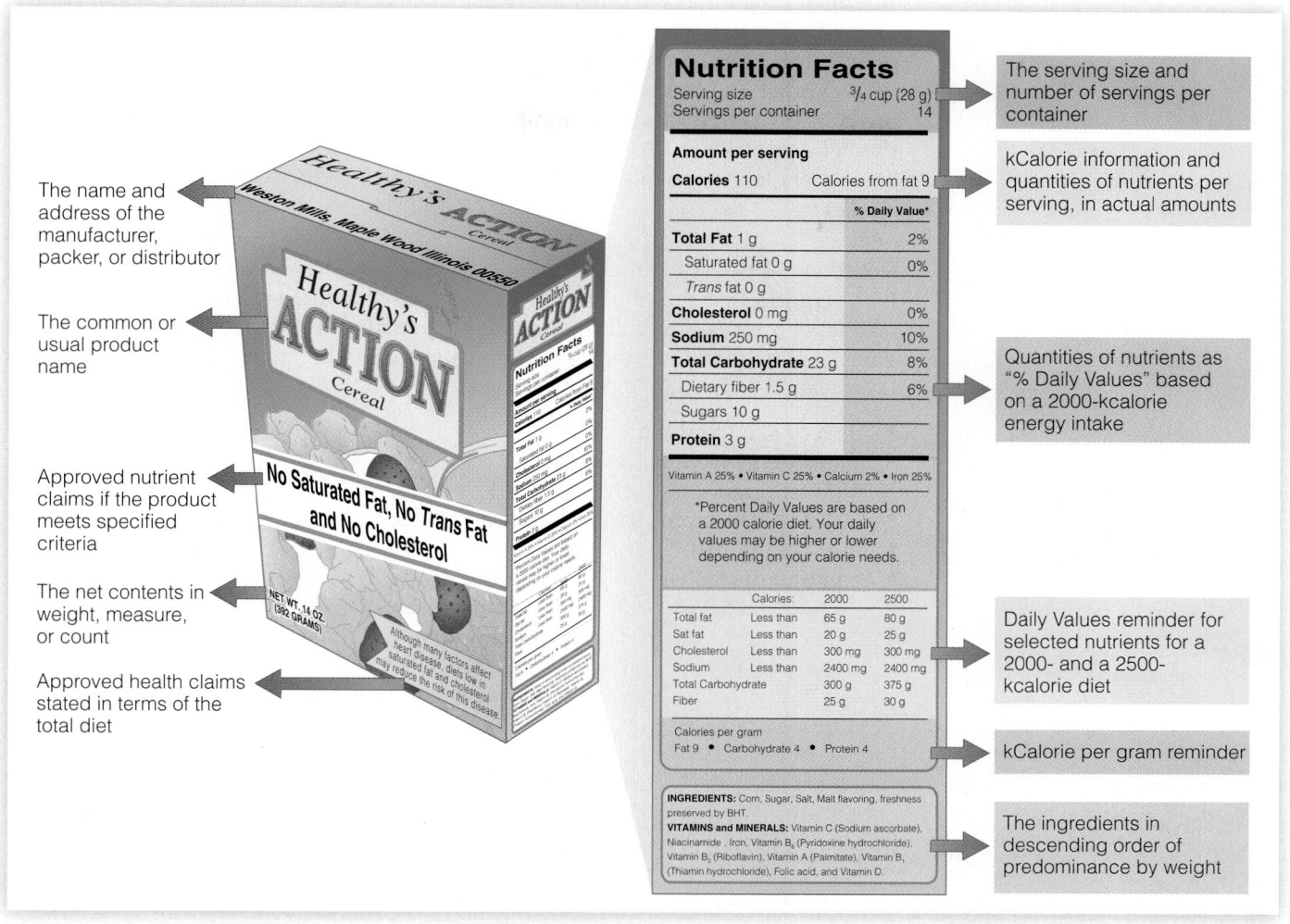

The name and address of the manufacturer, packer, or distributor

The common or usual product name

Approved nutrient claims if the product meets specified criteria

The net contents in weight, measure, or count

Approved health claims stated in terms of the total diet

The serving size and number of servings per container

kCalorie information and quantities of nutrients per serving, in actual amounts

Quantities of nutrients as "% Daily Values" based on a 2000-kcalorie energy intake

Daily Values reminder for selected nutrients for a 2000- and a 2500-kcalorie diet

kCalorie per gram reminder

The ingredients in descending order of predominance by weight

FIGURE 1-7 Example of a Food Label

common household measures, such as cups, and metric measures, such as milliliters, to accommodate users of both types of measures.

When examining the nutrition information on a food label, consumers need to consider how the serving size compares with the actual quantity eaten. If it is not the same, they will need to adjust the quantities accordingly. For example, if the serving size is four cookies and you only eat two, then you need to cut the nutrient and kcalorie values in half; similarly, if you eat eight cookies, then you need to double the values. Notice, too, that small bags or individually wrapped items, such as chips or candy bars, may contain more than a single serving. The number of servings per container is listed just below the serving size.

The Daily Values To help consumers evaluate the information found on labels, the FDA created a set of nutrient standards called the **Daily Values** specifically for use on food labels. The Daily Values do two things: they set adequacy standards for nutrients that are desirable in the diet such as protein, vitamins, minerals, and fiber, and they also set moderation standards for other nutrients that must be limited, such as fat, saturated fat, cholesterol, and sodium.

The "% Daily Value" column on a label provides a ballpark estimate of how individual foods contribute to the total diet. It compares key nutrients in a serving of food with the daily goals of a person consuming 2000 kcalories. Most labels list, at the bottom, Daily Values for both a 2000-kcalorie and a 2500-kcalorie diet, but the "% Daily Value" column on all labels applies only to a 2000-kcalorie diet.

Daily Values: reference values developed by the FDA specifically for use on food labels.

Although the Daily Values are based on a 2000-kcalorie diet, people's actual energy intakes vary widely; some people need fewer kcalories, and some people need many more. This makes the Daily Values most useful for comparing one food with another and less useful as nutrient intake targets for individuals. By examining a food's general nutrient profile, however, a person can determine whether the food contributes "a little" or "a lot" of a nutrient, whether it contributes "more" or "less" than another food, and how well it fits into the consumer's overall diet.

Nutrition Facts In addition to the serving size and the servings per container, the FDA requires that the "Nutrition Facts" panel on a label present nutrient information in two ways—in quantities (such as grams) and as percentages of the Daily Values. The Nutrition Facts panel must provide the nutrient amount, percent Daily Value, or both for the following:

- Total food energy (kcalories).
- Food energy from fat (kcalories).
- Total fat (grams and percent Daily Value).
- Saturated fat (grams and percent Daily Value).
- *Trans* fat (grams).
- Cholesterol (milligrams and percent Daily Value).
- Sodium (milligrams and percent Daily Value).
- Total carbohydrate, including starch, sugar, and fiber (grams and percent Daily Value).
- Dietary fiber (grams and percent Daily Value).
- Sugars (grams), including both those naturally present in and those added to the food.
- Protein (grams).

The labels must also present nutrient content information as a percentage of the Daily Values for the following vitamins and minerals:

- Vitamin A.
- Vitamin C.
- Iron.
- Calcium.

The FDA developed the Daily Values for use on food labels because comparing nutrient amounts against a standard helps make them meaningful to consumers. A person might wonder, for example, whether 1 milligram of iron or calcium is a little or a lot. As Table 1-11 shows, the Daily Value for iron is 18 milligrams, so 1 milligram of iron is enough to take notice of: it is over 5 percent. But the Daily Value for calcium on food labels is 1000 milligrams, so 1 milligram of calcium is essentially nothing.

Nutrient Claims The FDA defines the **nutrient claims** a label may use to describe the contents of a product (see Table 1-12). Definitions include the conditions under which each term can be used. For example, in addition to having less than 2 milligrams of cholesterol, a "cholesterol-free" product may not contain more than 2 grams of saturated fat and *trans* fat combined per serving.

Some descriptions imply that a food contains, or does not contain, a nutrient. Implied claims are prohibited unless they meet specified criteria. For example, a claim

nutrient claims: statements that characterize the quantity of a nutrient in a food.

TABLE 1-11	Daily Values for Food Labels	
Food labels must present the "% Daily Value" for these nutrients.		
Food Component	Daily Value	Calculation Factors
Fat	65 g	30% of kcalories
Saturated fat	20 g	10% of kcalories
Cholesterol	300 mg	—
Carbohydrate (total)	300 g	60% of kcalories
Fiber	25 g	11.5 g per 1000 kcalories
Protein	50 g	10% of kcalories
Sodium	2400 mg	—
Potassium	3500 mg	—
Vitamin C	60 mg	—
Vitamin A	1500 µg	—
Calcium	1000 mg	—
Iron	18 mg	—

Note: Daily Values were established for adults and children over 4 years old. The values for energy-yielding nutrients are based on 2000 kcalories a day. For fiber, the Daily Value was rounded up from 23.

TABLE 1-12	Terms Used on Food Labels

General Terms

free, without, no, zero none or a trivial amount. *Calorie free* means containing fewer than 5 kcalories per serving; *sugar free* or *fat free* means containing less than half a gram per serving.

fresh raw, unprocessed, or minimally processed with no added preservatives.

good source 10 to 19% of the Daily Value per serving.

healthy low in fat, saturated fat, *trans* fat, cholesterol, and sodium and containing at least 10% of the Daily Value for vitamin A, vitamin C, iron, calcium, protein, or fiber.

high in 20% or more of the Daily Value for a given nutrient per serving; synonyms include "rich in" or "excellent source."

less, fewer, reduced containing at least 25% less of a nutrient or kcalories than a reference food. This may occur naturally or as a result of altering the food. For example, pretzels, which are usually low in fat, can claim to provide less fat than potato chips, a comparable food.

light this descriptor has three meanings on labels:
1. A serving provides one-third fewer kcalories or half the fat of the regular product.
2. A serving of a low-kcalorie, low-fat food provides half the sodium normally present.
3. The product is light in color and texture, so long as the label makes this intent clear, as in "light brown sugar."

more, extra at least 10% more of the Daily Value than in a reference food. The nutrient may be added or may occur naturally.

Energy Terms

low calorie 40 kcalories or fewer per serving.

reduced calorie at least 25% lower in kcalories than a "regular," or reference, food.

calorie free fewer than 5 kcalories per serving.

Fat Terms (meat and poultry products)

extra lean
less than 5 g of fat *and*
less than 2 g of saturated fat and *trans* fat combined, *and*
less than 95 mg of cholesterol per serving.

lean[a]
less than 10 g of fat *and*
less than 4 g of saturated fat and *trans* fat combined, *and*
less than 95 mg of cholesterol per serving.

Fat and Cholesterol Terms (all products)

cholesterol free
less than 2 mg of cholesterol *and*
2 g or less saturated fat and *trans* fat combined per serving.

fat free less than 0.5 g of fat per serving.

less saturated fat 25% or less saturated fat and *trans* fat combined than the comparison food.

low cholesterol
20 mg or less of cholesterol *and*
2 g or less saturated fat per serving.

low fat 3 g or less fat per serving.[b]

low saturated fat 1 g or less saturated fat and less than 0.5 g of *trans* fat per serving.

percent fat free may be used only if the product meets the definition of *low fat* or *fat free.* Requires disclosure of grams of fat per 100 g food.

(continued)

TABLE 1-12   Terms Used on Food Labels *(continued)*

Fat and Cholesterol Terms (all products)

reduced or **less cholesterol**
at least 25% less cholesterol than a reference food *and*
2 g or less saturated fat per serving.

reduced saturated fat
at least 25% less saturated fat *and*
reduced by more than 1 g saturated fat per serving compared with a reference food.

saturated fat free
less than 0.5 g of saturated fat *and*
less than 0.5 g of *trans* fat.

***trans* fat free**
less than 0.5 g of *trans* fat *and*
less than 0.5 g of saturated fat per serving.

Fiber Terms

high fiber 5 g or more per serving. (Foods making high-fiber claims must fit the definition of low fat, or the level of total fat must appear next to the high-fiber claim.)

good source of fiber 2.5 g to 4.9 g per serving.

more or **added fiber** at least 2.5 g more per serving than a reference food.

Sodium Terms

low sodium 140 mg or less sodium per serving.

reduced sodium at least 25% lower in sodium than the regular product.

sodium free less than 5 mg per serving.

very low sodium 35 mg or less sodium per serving.

[a]The word *lean* as part of the brand name (as in "Lean Supreme") indicates that the product contains fewer than 10 grams of fat per serving.

[b]Exceptions may be granted for foods with low levels of saturated and *trans* fats.

that a product "contains no oil" implies that the food contains no fat. If the product is truly fat-free, then it may make the no-oil claim, but if it contains another source of fat, such as butter, it may not.

Health Claims Until recently, the FDA held manufacturers to the highest standards of scientific evidence before allowing them to place **health claims** on food labels. When a label stated, "diets low in sodium may reduce the risk of high blood pressure," for example, consumers could be sure that the FDA had examined much scientific evidence and found substantial support for the claim. Such reliable health claims make up the FDA's "A" list and still appear on some food labels (see Table 1-13).

The FDA recently created a new ranking system that includes three new categories of allowable health claims that are supported by evidence that is less conclusive.[21] The new system assigns each claim a letter grade reflecting the degree to which the claim is backed by science (see Table 1-14). The FDA can no longer demand that only health claims with the highest degree of scientific support appear on food labels.

Under the new system, the reliable health claims shown in Table 1-13 receive an "A" grade. Claims with a "B," "C," or "D" grade are called "qualified" health claims because they must bear a statement explaining the degree of scientific evidence backing them up. The FDA is relying on consumers to notice these printed disclaimers.

health claims: statements that characterize the relationship between a nutrient or other substance in food and a disease or health-related condition.

TABLE 1-13	Food Label Health Claims—The "A" List

Calcium and reduced risk of osteoporosis

Sodium and reduced risk of hypertension

Dietary saturated fat and cholesterol and reduced risk of coronary heart disease

Dietary fat and reduced risk of cancer

Fiber-containing grain products, fruits, and vegetables and reduced risk of cancer

Fruits, vegetables, and grain products that contain fiber, particularly soluble fiber, and reduced risk of coronary heart disease

Fruits and vegetables and reduced risk of cancer

Folate and reduced risk of neural tube defects

Sugar alcohols and reduced risk of tooth decay

Soluble fiber from whole oats and from psyllium seed husk and reduced risk of heart disease

Soy protein and reduced risk of heart disease

Whole grains and reduced risk of heart disease and certain cancers

Plant sterol and plant stanol esters and heart disease

Potassium and reduced risk of hypertension and stroke

TABLE 1-14	The FDA's Health Claims Report Card

Grade	Level of Confidence in Health Claim	Required Label Disclaimers
A	High: Significant scientific agreement	These health claims do not require disclaimers; see Table 1-13 for examples.
B	Moderate: Evidence is supportive, but not conclusive	"[Health claim.] Although there is scientific evidence supporting this claim, the evidence is not conclusive."
C	Low: Evidence is limited and not conclusive	"Some scientific evidence suggests [health claim]. However, FDA has determined that this evidence is limited and not conclusive."
D	Very low: Little scientific evidence supporting this claim	"Very limited and preliminary scientific research suggests [health claim]. FDA concludes that there is little scientific evidence supporting this claim."

Structure-Function Claims Consumers need to be aware that a different kind of claim, known as a "structure-function" claim, may also appear on food or dietary supplement labels. **Structure-function claims** are statements about a food substance's effect on a structure or function of the body—for example, "antioxidants support heart health." Structure-function claims are not required to have FDA approval, but the claims may not refer to the reduction of disease risk. The claims must be carefully worded to avoid any mention of a specific disease. Typical claims are "slows aging," "improves memory," and "builds strong bones." To make a more specific claim such as "prevents osteoporosis," the manufacturer would have to submit to the rigorous requirements for health claims or meet the even stricter safety and efficacy standards applied to drugs. These rules ensure that consumers can have confidence that when a claim names a specific disease, there is substantial scientific agreement that the food, in the context of a healthy diet, may help protect against that disease.

structure-function claims: statements that describe how a product may affect a structure or function of the body; for example, "calcium builds strong bones." Structure-function claims do not require FDA authorization.

Frequently unsafe:
- Raw milk and milk products.
- Raw or undercooked seafood, meat, poultry, or eggs.
- Raw sprouts or scallions.

Occasionally unsafe:
- Hamburgers.
- Salad bar items.
- Sandwiches.
- Soft cheeses (Mexican style, feta, brie, camembert, blue-veined).
- Unpasteurized fruit juices and ciders.
- Unwashed berries and grapes.

Rarely unsafe:
- Peeled fruit.
- Steaming-hot foods.
- High-sugar foods.

foodborne illness: illness transmitted to human beings through food and water, caused either by an infectious agent (foodborne infection) or a poisonous substance (food intoxication); commonly known as *food poisoning*.

REVIEW NOTES

Food labels provide consumers with information they need to select foods that will help them meet their nutrition and health goals.

Daily Values are a set of nutrient standards created by the FDA for use on food labels.

Health claims that are graded "A" are backed by the highest standards of scientific evidence. Health claims with a "B," "C," or "D" grade are supported by less conclusive scientific evidence than those graded with an "A."

Food Safety

Consumers not only want to know which foods to eat for health, they also have concerns about the safety of their food. Each year in the United States, an estimated 76 million people become ill from **foodborne illness,** and about 5000 of them die.[22] Even normally mild foodborne illnesses can be lethal for people who are ill or malnourished; have a compromised immune system; are in a health care facility; have liver or stomach illnesses; or are pregnant, very old, or very young.[23] The *Dietary Guidelines for Americans 2005* advise people to take preventive steps to minimize their chances of contracting foodborne illnesses. The margin identifies foods most commonly implicated in foodborne illnesses, and the accompanying "How to" offers tips to prevent foodborne illnesses.

HOW TO   Prevent Foodborne Illnesses

Most foodborne illnesses can be prevented by following four simple rules: keep a clean kitchen, avoid cross-contamination, keep hot foods hot, and keep cold foods cold.

Keep a Clean Kitchen

- Wash fruits and vegetables in a clean sink with a scrub brush and warm water; store washed and unwashed produce separately.
- Use hot, soapy water to wash hands, utensils, dishes, nonporous cutting boards, and countertops before handling food and between tasks when working with different foods. Use a bleach solution on cutting boards (one capful per gallon of water).
- Cover cuts with clean bandages before food preparation; dirty bandages carry harmful microorganisms.
- Mix foods with utensils, not hands; keep hands and utensils away from mouth, nose, and hair.
- Anyone may be a carrier of bacteria and should avoid coughing or sneezing over food. A person with a skin infection or infectious disease should not prepare food.
- Wash or replace sponges and towels regularly.
- Clean up food spills and crumb-filled crevices.

Avoid Cross-Contamination

- Wash all surfaces that have been in contact with raw meats, poultry, eggs, fish, and shellfish before reusing.
- Serve cooked foods on a clean plate. Separate raw foods from those that have been cooked.
- Don't use marinade that was in contact with raw meat for basting or sauces.

Keep Hot Foods Hot

- When cooking meats or poultry, use a thermometer to test the internal temperature. Insert the thermometer between the thigh and the body of a turkey or into the thickest part of other meats, making sure the tip of the thermometer is not in contact with bone or the pan. Cook to the temperature indicated for that particular meat; cook hamburgers to at least medium well-done. If you have safety questions, call the USDA Meat and Poultry Hotline: (800) 535-4555.
- Cook stuffing separately, or stuff poultry just prior to cooking.
- Do not cook large cuts of meat or turkey in a microwave oven; it leaves some parts undercooked while overcooking others.
- Cook eggs before eating them (soft-boiled for at least $3\frac{1}{2}$ minutes; scrambled until set, not runny; fried for at least 3 minutes on one side and 1 minute on the other).
- Cook seafood thoroughly. If you have safety questions about seafood, call the FDA hotline: (800) FDA-4010.
- When serving foods, maintain temperatures at 140°F or higher.
- Heat leftovers thoroughly to at least 165°F.

Keep Cold Foods Cold

- When running errands, stop at the grocery store last. When you get home, refrigerate the perishable groceries (such as meats and dairy products) immediately. Do not leave perishables in the car any longer than it takes for ice cream to melt.

- Put packages of raw meat, fish, or poultry on a plate before refrigerating to prevent juices from dripping on food stored below.
- Buy only foods that are solidly frozen in store freezers.
- Keep cold foods at 40°F or less; keep frozen foods at 0°F or less (keep a thermometer in the refrigerator).
- Marinate meats in the refrigerator, not on the counter.
- Refrigerate leftovers promptly; use shallow containers to cool foods faster; use leftovers within three to four days.
- Thaw meats or poultry in the refrigerator, not at room temperature. If you must hasten thawing, use cool water (changed every 30 minutes) or a microwave oven.
- Freeze meat, fish, or poultry immediately if not planning to use within a few days.

In General

- Do not reuse disposable containers; use nondisposable containers or recycle instead.
- Do not taste food that is suspect. "If in doubt, throw it out."
- Throw out foods with danger-signaling odors. Be aware, though, that most food-poisoning bacteria are odorless, colorless, and tasteless.
- Do not buy or use items that have broken seals or mangled packaging; such containers cannot protect against microbes, insects, spoilage, or even vandalism. Check safety seals, buttons, and expiration dates.
- Follow label instructions for storing and preparing packaged and frozen foods; throw out foods that have been thawed or refrozen.
- Discard foods that are discolored, moldy, or decayed or that have been contaminated by insects or rodents.

For Specific Food Items

- *Canned goods.* Carefully discard food from cans that leak or bulge so that other people and animals will not accidentally ingest it; before canning, seek professional advice from the USDA Extension Service (check your phone book under U.S. government listings, or ask directory assistance).
- *Milk and cheeses.* Use only pasteurized milk and milk products. Aged cheeses, such as cheddar and swiss, do well for an hour or two without refrigeration, but should be refrigerated or stored in an ice chest for longer periods.
- *Eggs.* Use clean eggs with intact shells. Do not eat eggs, even pasteurized eggs, raw; raw eggs are commonly found in Caesar salad dressing, eggnog, cookie dough, hollandaise sauce, and key lime pie. Cook eggs until whites are firmly set and yolks begin to thicken.
- *Honey.* Honey may contain dormant bacterial spores, which can awaken in the human body to produce botulism. In adults, this poses little hazard, but infants under one year of age should never be fed honey. Honey can accumulate enough toxin to kill an infant; it has been implicated in several cases of sudden infant death. (Honey can also be contaminated with environmental pollutants picked up by the bees.)
- *Mayonnaise.* Commercial mayonnaise may actually help a food to resist spoilage because of the acid content. Still, keep it cold after opening.
- *Mixed salads.* Mixed salads of chopped ingredients spoil easily because they have extensive surface area for bacteria to invade, and they have been in contact with cutting boards, hands, and kitchen utensils that easily transmit bacteria to food (regardless of their mayonnaise content). Chill them well before, during, and after serving.
- *Picnic foods.* Choose foods that last without refrigeration such as fresh fruits and vegetables, breads and crackers, and canned spreads and cheeses that can be opened and used immediately. Pack foods cold, layer ice between foods, and keep foods out of water.
- *Seafood.* Buy only fresh seafood that has been properly refrigerated or iced. Cooked seafood should be stored separately from raw seafood to avoid cross-contamination.

SELF CHECK

1. When people eat the foods typical of their families or geographic region, their choices are influenced by:
   a. occupation.
   b. nutrition.
   c. emotional state.
   d. ethnic heritage or tradition.

2. The energy-yielding nutrients are:
   a. fats, minerals, and water.
   b. minerals, proteins, and vitamins.
   c. carbohydrates, fats, and vitamins.
   d. carbohydrates, fats, and proteins.

3. The inorganic nutrients are:
   a. proteins and fats.
   b. vitamins and minerals.
   c. minerals and water.
   d. vitamins and proteins.

4. Alcohol is not a nutrient because
   a. the body derives no energy from it.
   b. it is organic.
   c. it is converted to body fat.
   d. it does not contribute to the body's growth or repair.

5. DRI stands for:
   a. Daily Recommended Intakes.
   b. Dietary Requirements for Individuals.
   c. Dietary Reference Intakes.
   d. Daily Recommendations for Individuals.

6. Which of the following is consistent with the *Dietary Guidelines for Americans 2005*?
   a. Limit intakes of fruits, vegetables, and whole grains.
   b. Engage in regular physical activity and reduce sedentary activities to promote health, psychological well-being, and a healthy body weight.
   c. Choose a diet with plenty of whole-milk products.
   d. Eat an abundance of foods to ensure nutrient adequacy.

7. In a food group plan such as the USDA Food Guide, foods within a given food group are similar in their contents of:
   a. energy.
   b. proteins and fibers.
   c. vitamins and minerals.
   d. carbohydrates and fats.

8. A slice of apple pie supplies 350 kcalories with 3 grams of fiber; an apple provides 80 kcalories and the same 3 grams of fiber. This is an example of:
   a. kcalorie control.
   b. nutrient density.
   c. variety
   d. essential nutrients

9. Which of the following statements does *not* apply to a person's discretionary kcalorie allowance?
   a. Compared to physically active adults, sedentary adults have a lower discretionary kcalorie allowance.
   b. It is the difference between the kcalories needed to supply nutrients and those needed for energy.
   c. Compared to physically active adults, sedentary adults have a higher discretionary kcalorie allowance.
   d. A person with discretionary kcalories available can, if he or she chooses to, select a few foods with added fat or sugar.

10. Food labels list ingredients in:
    a. alphabetical order.
    b. ascending order by predominance by weight.
    c. descending order of predominance by weight.
    d. manufacturer's order of preference.

Answers to these questions appear in Appendix H.

CLINICAL APPLICATIONS

1. Make a list of the foods and beverages you've consumed in the last two days. Look at each item on your list and consider why you chose the particular food or beverage you did. Did you eat cereal for breakfast because that's what you always eat (habit), or because it was the easiest, quickest food to prepare (convenience)? Did you put fat-free milk on the cereal because you want to control your energy intake (nutrition)? In going down your list, you may be surprised to discover exactly why you chose certain foods.

2. As a nurse, you can uncover clues about a client's food choices by paying close attention. You may be surprised to discover why a client chooses certain foods, but you can then use this knowledge to serve the best interests of the client. For example, an elderly, undernourished widower may eat the same sandwich for lunch every day. In talking with the client, you discover that is what he and his wife fixed together each day. Consider ways you might be able to help the client learn to eat other foods and vary his choices.

3. Using the list of foods and beverages from exercise #1, compare you day's intake with the USDA Food Guide. Did you vary your choices within each food group? Did your intake match the daily recommended amounts from each group? If not, list some changes you could have made to meet the recommendations.

NUTRITION ON THE NET

For further study of the topics in this chapter, access these websites.

Find updates and quick links to these and other nutrition-related sites at our website: **www.wadsworth.com/nutrition**

Search for "nutrition" at the U.S. Government health information site: **www.healthfinder.gov**

Review the Dietary Reference Intakes: **www.nap.edu/reading room**

Review nutrient recommendations from the Food and Agriculture Organization and the World Health Organization: **www.fao.org** and **www.who.org**

Learn more about the *Dietary Guidelines for Americans:* **www.health.gov/dietaryguidelines/dga2005/recommendations.htm**

View Canadian information on nutrition guidelines and food labels at: **www.hc-sc.gc.ca**

Visit MyPyramid: **www.MyPyramid.gov**

See food pyramids for various ethnic groups at Oldways Preservation and Exchange Trust: **www.oldwayspt.org**

View Healthy People 2010: **web.health.gov/healthypeople**

Review the Canadian National Plan of Action for Nutrition: **www.hc-sc.ca/datahpsb/npu**

Learn about NHANES: **www.cdc.gov/nchs/nhanes.htm**

Learn more about food labeling from the Food and Drug Administration: **www.cfsan.fda.gov** or **vm.cfsan.fda.gov/label.html**

NOTES

[1] A. Mennella, M. Y. Pepino, and D. R. Reed, Genetic and environmental determinants of bitter perception and sweet preferences, *Pediatrics* 115 (2005): e216; A. Drewnowski, S. A. Henderson, and A. Barratt-Fornell, Genetic taste markers and food preferences, *Drug Metabolism and Disposition* 29 (2001): 535–538.

[2] W. Lockeretz and K. A. Merrigan, Selling to the eco-conscious food shopper, *Nutrition Today* 40 (2005): 45–49.

[3] J. E. Tillotson, Fast-casual dining: Our next eating passion? *Nutrition Today* 38 (2003): 91–94; J. E. Tillotson, Our ready-prepared ready-to-eat nation, *Nutrition Today* 37 (2002): 36–38.

[4] Tillotson, 2002.

[5] K. Bertrand, Microwavable foods satisfy need for speed and palatability, *Food Technology* 59 (2005): 30–34.

[6] Position of the American Dietetic Association: Functional foods, *Journal of the American Dietetic Association* 104 (2004): 814–826.

[7] Standing Committee on the Scientific Evaluation of Dietary Reference Intakes, Food and Nutrition Board, Institute of Medicine, *Dietary Reference Intakes for Water, Potassium, Sodium, Chloride, and Sulfate* (Washington, D.C.: National Academies Press, 2005); Standing Committee on the Scientific Evaluation of Dietary Reference Intakes, Food and Nutrition Board, Institute of Medicine, *Dietary Reference Intakes for Energy, Carbohydrate, Fiber, Fat, Fatty Acids, Cholesterol, Protein, and Amino Acids* (Washington, D.C.: National Academies Press, 2005); Standing Committee on the Scientific Evaluation of Dietary Reference Intakes, Food and Nutrition Board, Institute of Medicine, *Dietary Reference Intakes for Vitamin A, Vitamin K, Arsenic, Boron, Chromium, Copper, Iodine, Iron, Manganese, Molybdenum, Nickel, Silicon, Vanadium, and Zinc* (Washington, D.C.: National Academy Press, 2001); Standing Committee on the Scientific Evaluation of Dietary Reference Intakes, Food and Nutrition Board, Institute of Medicine, *Dietary Reference Intakes for Vitamin C, Vitamin E, Selenium, and Carotenoids* (Washington, D.C.: National Academy Press, 2000); Standing Committee on the Scientific Evaluation of Dietary Reference Intakes, Food and Nutrition Board, Institute of Medicine, *Dietary Reference Intakes for Thiamin, Riboflavin, Niacin, Vitamin B₆, Folate, Vitamin B₁₂, Pantothenic Acid, Biotin, and Choline* (Washington, D.C.: National Academy Press, 1998); Standing Committee on the Scientific Evaluation of Dietary Reference Intakes, Food and Nutrition Board, Institute of Medicine, *Dietary Reference Intakes for Calcium, Phosphorus, Magnesium, Vitamin D, and Fluoride* (Washington, D.C.: National Academy Press, 1997).

[8] Subcommittee on Interpretation and Uses of Dietary Reference Intakes and the Standing Committee on the Scientific Evaluation of Dietary Reference Intakes, *Applications in Dietary Planning* (Washington, D.C.: National Academies Press, 2003).

[9] J. Dwyer and coauthors, Collection of food and dietary supplement intake data: What we eat in America—NHANES, *Journal of Nutrition* 133 (2003): 590S–600S.

[10] Dwyer and coauthors, 2003.

[11] Centers for Disease Control, Physical activity and good nutrition: Essential elements to prevent chronic diseases and obesity, 2004, available at **www.cdc.gov/nccdphp/dnpa**; D. Yach and coauthors, The global burden of chronic diseases: Overcoming impediments to prevention and control, *Journal of the American Medical Association* 291 (2004): 2616–2622; K. T. B. Knoops and coauthors, Mediterranean diet, lifestyle factors, and 10-year mortality in elderly European men and women, *Journal of the American Medical Association* 292 (2004): 1433–1439; D. Lee, L. M. Steffen, and D. R. Jacobs, Association between serum γ-glutamyltransferase and dietary factors: The Coronary Artery Risk Development in Young Adults (CARDIA) Study, *American Journal of Clinical Nutrition* 79 (2004): 600–605.

[12] U.S. Department of Agriculture and U.S. Department of Health and Human Services, *Nutrition and Your Health: Dietary Guidelines for Americans 2005*, 6th ed., Home and Garden Bulletin no. 232 (Washington, D.C.: 2005), available online at **www.usda.gov/cnpp** or call (888) 878-3256.

[13] J. Myers and coauthors, Exercise capacity and mortality among men referred for exercise testing, *New England Journal of Medicine* 346 (2002): 793–801; I. M. Lee and R. S. Paffenbarger, Associations of light, moderate, and vigorous intensity physical activity with longevity: The Harvard Alumni Study, *American Journal of Epidemiology* 151 (2000): 293–299.

[14] U.S. Department of Agriculture and U.S. Department of Health and Human Services, 2005; Standing Committee on the Scientific Evaluation of Dietary Reference Intakes, 2005, pp. 880-935; American College of Sports Medicine, *ACSM's Guidelines for Exercise Testing and Prescription*, 6th ed. (Philadelphia: Williams & Wilkins, 2000), pp. 137–164.

[15] M. Murphy and coauthors, Accumulating brisk walking for fitness, cardiovascular risk, and psychological health, *Medicine and Science in Sports and Exercise* 34 (2002): 1468–1474.

[16] J. E. Manson and coauthors, Walking compared with vigorous exercise for the prevention of cardiovascular events in women, *New England Journal of Medicine* 347 (2002): 716–725.

[17] T. R. Wessel and coauthors, Relationship of physical fitness vs body mass index with coronary artery disease and cardiovascular events in women, *Journal of the American Medical Association* 292 (2004): 1179–1187; Y. A. Kesaniemi and coauthors, Dose-response issues concerning physical activity and health: An evidence-based symposium, *Medicine and Science in Sports and Exercise* 33 (2001): 351S–358S; Lee and Paffenbarger, 2000.

[18] P. M. Barnes and C. A. Schoenborn, *Physical Activity among Adults: United States, 2000*, Advanced Data from Vital and Health Statistics document no. 333 (2003), available online from **www.cdc.gov/nchs/Default.htm**.

[19] K. R. Evenson and coauthors, The effect of cardiorespiratory fitness and obesity on cancer mortality in women and men, *Medicine and Science in Sports and Exercise* 35 (2003): 270–277; J. Dorn and coauthors, Lifetime physical activity and breast cancer risk in pre- and postmenopausal women, *Medicine and Science in Sports and Exercise* 35 (2003): 278–285; C. D. Lee and S. N. Blair, Cardiorespiratory fitness and stroke mortality in men, *Medicine and Science in Sports and Exercise* 34 (2002): 592–595.

[20] Position of the American Dietetic Association and Dietitians of Canada: Vegetarian diets, *Journal of the American Dietetic Association* 103 (2003): 748–765.

[21] J. E. Tillotson, Health claims 2005: A new tower of Babel? *Nutrition Today* 40 (2005): 88–91; U.S. Food and Drug Administration, FDA to encourage science-based labeling and competition for healthier dietary choices, *FDA News*, available at: **www.fda.gov/bbs/topics/NEWS/2003/NEW00923.html**.

[22] Centers for Disease Control and Prevention, Diagnosis and management of foodborne illnesses, *Morbidity and Mortality Weekly Report*, supplement, 53 (2004).

[23] Position of the American Dietetic Association: Food and water safety, *Journal of the American Dietetic Association* 103 (2003): 1203–1218.

Notes for Table 1-7

a. U. G. Kyle and coauthors, Physical activity and fat-free and fat mass by bioelectrical impedance in 3853 adults, *Medicine and Science in Sports and Exercise* 33 (2001): 576–584.

b. L. Metcalfe and coauthors, Postmenopausal women and exercise for prevention of osteoporosis: The Bone, Estrogen, Strength Training (BEST) Study, *ACSM's Health and Fitness Journal,* May/June, 2001, pp. 6–14; K. Delvaux and coauthors, Bone mass and lifetime physical activity in Flemish males: A 27-year follow-up study, *Medicine and Science in Sports and Exercise* 33 (2001): 1868–1875.

c. C. E. Matthews and coauthors, Moderate to vigorous physical activity and risk of upper-respiratory tract infection, *Medicine and Science in Sports and Exercise* 34 (2002): 1242–1248.

d. C. M. Friedenreich, Physical activity and cancer: Lessons learned from nutritional epidemiology, *Nutrition Reviews* 59 (2001): 349–357.

e. American College of Sports Medicine, Position stand, Exercise and hypertension, *Medicine and Science in Sports and Exercise* 36 (2004): 533–553; M. R. Carnethon and coauthors, Cardiorespiratory fitness in young adulthood and the development of cardiovascular disease risk factors, *Journal of the American Medical Association* 290 (2003): 3092–3100.

f. S. L. Wong and coauthors, Cardiorespiratory fitness is associated with lower abdominal fat independent of body mass index, *Medicine and Science in Sports and Exercise* 36 (2004): 286–291; A. Trichopoulou and coauthors, Physical activity and energy intake selectively predict the waist-to-hip ratio in men but not in women, *American Journal of Clinical Nutrition* 74 (2001): 574–578.

g. American Diabetes Association, Position statements: Physical activity/exercise and diabetes mellitus, *Diabetes Care* 26 (2003): S73–S77; R. M. van Dam and coauthors, Physical activity and glucose tolerance in elderly men: The Zutphen Elderly Study, *Medicine and Science in Sports and Exercise* 34 (2002): 1132–1136; K. J. Stewart, Exercise training and the cardiovascular consequences of type 2 diabetes and hypertension: Plausible mechanisms for improving cardiovascular health, *Journal of the American Medical Association* 288 (2002): 1622–1631.

h. G. Misciagna and coauthors, Diet, physical activity, and gallstones—a population-based, case-control study in southern Italy, *American Journal of Clinical Nutrition* 69 (1999): 120–126; M. F. Leitzmann and coauthors, Recreational physical activity and the risk of cholecystectomy in women, *New England Journal of Medicine* 341 (1999): 777–784.

i. W. J. Strawbridge and coauthors, Physical activity reduces the risk of subsequent depression for older adults, *American Journal of Epidemiology* 156 (2002): 328–334.

j. J. Myers and coauthors, Exercise capacity and mortality among men referred for exercise testing, *New England Journal of Medicine* 346 (2002): 793–801.

Finding the Truth about Nutrition

Nutrition and health receive so much attention on television, in the popular press, and on the Internet that it is easy to be overwhelmed with conflicting, confusing information. Determining whether nutrition information is accurate can be a challenging task. It is also an important task because nutrition affects a person both professionally and personally.

A person watches a nutrition report on television and then reads a conflicting report in the newspaper. Why do nutrition news reports and claims for nutrition products seem to contradict each other so often?

The problem of conflicting messages arises for several reasons:

- Popular media, often faced with tight deadlines and limited time or space to report new information, rush to present the latest "breakthrough" in a headline or a 60-second spot. They can hardly help omitting important facts about the study or studies that the "breakthrough" is based on.
- Despite tremendous advances in the last few decades, scientists still have much to learn about the human body and nutrition. Scientists themselves often disagree on their first tentative interpretations of new research findings, yet these are the very findings that the public hears most about.
- The popular media often broadcast preliminary findings in hopes of grabbing attention and boosting readership or television ratings.
- Commercial promoters turn preliminary findings into advertisements for products or supplements long before the findings have been validated—or disproved. The scientific process requires many experiments or trials to confirm a new finding. Seldom do promoters wait as long as they should to make their claims.
- Promoters are aware that consumers like to try new products or treatments even though they probably will not withstand the tests of time and scientific scrutiny.

So how can a person tell what claims to believe?

The Food and Nutrition Science Alliance (FANSA), whose partners include the American Dietetic Association (ADA), the American Society for Nutritional Sciences (ASNS), and the Institute of Food Technologists (IFT), attempts to help consumers distinguish valid from misleading nutrition information. FANSA has created a list of ten red flags for detecting "junk science" (see Table NP1-1).[1]

TABLE NP1-1 FANSA's Red Flags of Junk Science

1. Promises of a quick and easy fix.
2. Dire warnings of danger from a single product or regimen.
3. Too good to be true. The claim says what most people want to hear.
4. Enticingly simple conclusions drawn from a complex study.
5. Recommendations based on a single study.
6. Dramatic statements that are refuted by reputable scientific organizations.
7. Lists of "good" and "bad" foods.
8. Claims are made to help sell a product.
9. Claims or recommendations based on studies published without peer review.
10. Recommendations from studies that ignore differences among individuals or groups.

Source: Adapted from Position of the American Dietetic Association: Food and nutrition misinformation, *Journal of the American Dietetic Association* 102 (2002): 260–262.

Because nutrition misinformation harms the health and economic status of consumers, the ADA works with health care professionals and educators to present sound nutrition information to the public and to actively confront nutrition misinformation.[2] Table NP1-2 offers a list of credible sources of nutrition information.

TABLE NP1-2 Credible Sources of Nutrition Information

Professional health organizations, government health agencies, volunteer health agencies, and consumer groups provide consumers with reliable health and nutrition information. Credible sources of nutrition information include:

Professional health organizations, especially the American Dietetic Association's National Center for Nutrition and Dietetics (NCND) **www.eatright.org/ncnd.html**, also the Society for Nutrition Education **www.sne.org** and the American Medical Association **www.ama-assn.org**

Government health agencies such as the Federal Trade Commission (FTC) **www.ftc.gov**, the U.S. Department of Health and Human Services (HHS) **www.os.dhhs.gov**, the Food and Drug Administration (FDA) **www.fda.gov**, and the U.S. Department of Agriculture (USDA) **www.usda.gov**

Volunteer health agencies such as the American Cancer Society **www.cancer.org**, the American Diabetes Association **www.diabetes.org**, and the American Heart Association **www.americanheart.org**

Reputable consumer groups such as the Better Business Bureau **www.bbb.org**, the Consumers Union **www.consumersunion.org**, the American Council on Science and Health **www.acsh.org**, and the National Council Against Health Fraud **www.ncahf.org**

What about nutrition and health information found on the Internet? How does a person know whether the websites are reliable?

With hundreds of millions of websites on the Internet, searching for nutrition and health information can be daunting. The Internet offers no guarantee of the accuracy of the information found there, and much of it is pure fiction. Websites must be evaluated for their accuracy, just like every other source. Table NP1-3 provides clues to identifying reliable nutrition information sites and lists some credible sites.

The Federal Trade Commission (FTC), the Food and Drug Administration (FDA), and other law enforcement agencies have launched "Operation Cure.All" (see Table NP1-3) to take

TABLE NP1-3 Evaluating the Reliability of Websites

To judge whether an Internet site offers reliable nutrition information, answer the following questions.

Who is responsible for the site? Clues can be found in the three-letter "tag" that follows the dot in the site's name. For example, "gov" and "edu" indicate government and university sites, usually reliable sources of information.

Do the names and credentials of information providers appear? Is an editorial board identified? Many legitimate sources provide e-mail addresses or other ways to obtain more information about the site and the information providers behind it.

Are links with other reliable information sites provided? Reputable organizations almost always provide links with other similar sites because they want you to know of other experts in their area of knowledge. Caution is needed when you evaluate a site by its links, however. Anyone, even a quack, can link a webpage to a reputable site without the organization's permission. Doing so may give the quack's site the appearance of legitimacy, just the effect the quack is hoping for.

Is the site updated regularly? Nutrition information changes rapidly, and sites should be updated often.

Is the site selling a product or service? Commercial sites may provide accurate information, but they also may not, and their profit motive increases the risk of bias.

Does the site charge a fee to gain access to it? Many academic and government sites offer the best information, usually for free. Some legitimate sites do charge fees, but before paying up, check the free sites. Chances are good you'll find what you are looking for without paying.

Some credible websites include:

National Council Against Health Fraud
www.ncahf.org

Stephen Barrett's Quackwatch
www.quackwatch.com

Centers for Disease Control and Prevention's Current Health Related Hoaxes and Rumors
www.cdc.gov/hoax_rumors.htm

Federal Trade Commission's Operation Cure.All
www.ftc.gov/opa/2001/06/cureall.htm

action against fraudulent marketing of supplements and health products on the Internet.[3] The latest actions target unscrupulous companies that use the Internet to promote products to the most vulnerable consumers—those with diseases such as AIDS, Alzheimer's, and cancer. Of greatest concern are those products that not only make false promises but also are potentially dangerous. For example, herbal products touted as safe treatments for serious illnesses such as AIDS may interact with medications and impair the effectiveness of the medicines. The FTC advises consumers to be suspicious of:

- Claims that a product is "natural" or "nontoxic." "Natural" or "nontoxic" does not always mean safe.
- Claims that a product is a "scientific breakthrough," "miraculous cure," "secret ingredient," or "ancient remedy."
- Claims that a product cures a wide range of illnesses.
- Claims that use impressive-sounding medical terms.
- Claims of a "money-back" guarantee.

Everyone seems to be giving advice on nutrition. How can a person tell whom to listen to?

Registered dietitians (RDs) and nutrition professionals with advanced degrees (M.S., Ph.D.) are experts (see the glossary). These professionals are probably in the best position to answer a person's nutrition questions. On the other hand, "**nutritionists**" may be experts or quacks, depending on the state where they practice. Some states require people who use this title to meet strict standards. In other states, a "nutritionist" may be any individual who claims a career connection with the nutrition field.

Other purveyors of nutrition information may also lack credentials. A health food store owner may be in the nutrition business simply because it is a lucrative market. The owner may have a background in business or sales and no education in nutrition at all. Such a person is not qualified to provide nutrition information to customers. For accurate nutrition information, seek out a trained professional with a college education in nutrition—an expert in the field of **dietetics.**

What about nurses and other health care professionals?

All members of the health care team share responsibility for helping each client to achieve optimal health, but the registered dietitian is usually the primary nutrition expert. Each of the other team members has a related specialty. Some physicians are specialists in clinical nutrition and are also experts in the field. Other physicians, nurses, and **dietetic technicians** often assist dietitians in providing nutrition information and may help to administer direct nutrition care. Nurses play central roles in client care management and client relationships. Visiting nurses and home health care nurses may become intimately involved in clients' nutrition care at home, teaching them both theory and cooking techniques. Physical therapists can provide individualized exercise programs related to nutrition—for example, to help control obesity. Social workers may provide practical and emotional support.

What roles might these other health care professionals play in nutrition care?

Some of the responsibilities of the health care professional might be:

- Helping people understand why nutrition is important to them.
- Answering questions about food and diet.
- Explaining to clients how modified diets work.
- Collecting information about clients that may influence their nutritional health.
- Identifying clients at risk for poor nutrition status (see Chapter 13) and recommending or taking appropriate action.
- Recognizing when clients need extra help with nutrition problems (in such cases, the problems should be referred to a dietitian or physician).

Health care professionals may routinely perform these nutrition-related tasks:

- Obtaining diet histories.
- Taking weight and height measurements.
- Feeding clients who cannot feed themselves.
- Recording what clients eat or drink.
- Observing clients' responses and reactions to foods.
- Helping clients mark menus.
- Monitoring weight changes.
- Monitoring food and drug interactions.
- Encouraging clients to eat.
- Assisting clients at home in planning their diets and managing their kitchen chores.
- Alerting the physician or dietitian when nutrition problems are identified.
- Charting actions taken and communicating on these matters with other professionals as needed.

Thus, although the dietitian assumes the primary role as the nutrition expert on a health care team, other health care professionals play important roles in administering nutrition care.

Glossary of Terms Associated with Nutrition Experts

dietetic technicians: persons who have completed a two-year academic degree from an accredited college or university and an approved dietetic technician program. A **dietetic technician, registered (DTR)** has also passed a national examination and maintains registration through continuing professional education.

dietetics: the practical application of nutrition, including the assessment of nutrition status, recommendation of appropriate diets, nutrition education, and the planning and serving of meals.

nutritionists: persons who specialize in the study of nutrition. Some nutritionists are registered dietitians, but others are self-described experts whose training may be minimal or nonexistent. Some states make the term meaningful by allowing it to apply only to people who have master's (M.S.) or doctoral (Ph.D.) degrees from institutions accredited to offer such degrees in nutrition or related fields.

registered dietitians (RDs): dietitians who have graduated from a university or college after completing a program of dietetics that has been accredited by the American Dietetic Association (or Dietitians of Canada). The dietitian must serve in an approved internship or coordinated program to practice the necessary skills, pass the association registration examination, and maintain competency through continuing education. Many states require licensing for practicing dietitians. Licensed dietitians (LDs) have met all *state* requirements to offer nutrition advice.

Notes

[1]Position of the American Dietetic Association: Food and nutrition misinformation, *Journal of the American Dietetic Association* 106 (2006): 601–607.

[2]Position of the American Dietetic Association, 2002.

[3]Federal Trade Commission, "Operation Cure.All" wages new battle in ongoing war against Internet health fraud, **www.ftc.gov/opa/ 2001/06/cureall.htm**, site visited on April 30, 2005.

Digestion and Absorption

© Envision/Corbis

CHAPTER 2

The body's ability to transform the foods a person eats into the nutrients that fuel the body's work is quite remarkable. Yet most people probably give little, if any, thought to all the body does with food once it is eaten. This chapter offers the reader the opportunity to learn how the body digests, absorbs, and transports the nutrients and how it excretes the unwanted substances in foods.

One of the beauties of the digestive tract is that it is selective. Materials that are nutritive for the body are broken down into particles that can be absorbed into the bloodstream. Most of the nonnutritive materials are left undigested and pass out the other end of the digestive tract.

Anatomy of the Digestive Tract

The **gastrointestinal (GI) tract** is a flexible muscular tube extending from the mouth to the anus. Figure 2-1 on pages 40 and 41 traces the path followed by food from one end of the GI tract to the other. The accompanying glossary defines GI anatomical terms. In a sense, the human body surrounds the GI tract. Only when a nutrient or other substance passes through the cells of the digestive tract wall does it actually enter the body.

THE DIGESTIVE ORGANS

The process of **digestion** begins in the **mouth.** As you chew, your teeth crush and soften the food, while saliva mixes with the food mass and moistens it for comfortable swallowing. Saliva also helps dissolve the food so that you can taste it; only particles in solution can react with taste buds.

gastrointestinal (GI) tract: the digestive tract. The principal organs are the stomach and intestines. *gastro* = stomach

digestion: the process by which complex food particles are broken down to smaller absorbable particles.

GLOSSARY of GI Terms

These terms are listed in order from the beginning of the digestive tract to the end.

mouth: the oral cavity containing the tongue and teeth.
pharynx (FAIR-inks): the passageway leading from the nose and mouth to the larynx and esophagus, respectively.
epiglottis (epp-ee-GLOT-tiss): a cartilage structure in the throat that prevents fluid or food from entering the trachea when a person swallows.
epi = upon (over)
glottis = back of tongue
trachea (TRAKE-ee-uh): the windpipe; the passageway from the mouth and nose to the lungs.
esophagus (e-SOFF-uh-gus): the food pipe; the conduit from the mouth to the stomach.
lower esophageal sphincter (ee-SOF-ah-GEE-al SFINK-ter): the sphincter muscle at the junction between the esophagus and the stomach; also called the **cardiac sphincter.**
pyloric (pie-LORE-ic) sphincter: the sphincter muscle separating the stomach from the small intestine; also called the **pylorus** or **pyloric valve.**
pylorus = gatekeeper
gallbladder: the organ that stores and concentrates bile. When it receives the signal that fat is present in the duodenum, the gallbladder contracts and squirts bile through the bile duct into the duodenum.

pancreas: a gland that secretes enzymes and digestive juices into the duodenum. (This is its exocrine function; it also has the endocrine function of secreting insulin and other hormones into the blood.)
small intestine: a 10-foot length of small-diameter (1-inch) intestine that is the major site of digestion of food and absorption of nutrients.
duodenum (doo-oh-DEEN-um or doo-ODD-ah-num): the top portion of the small intestine (about "12 fingers' breadth" long, in ancient terminology).
duodecim = twelve
jejunum (je-JOON-um): the first two-fifths of the small intestine beyond the duodenum.
ileum (ILL-ee-um): the last segment of the small intestine.
ileocecal (ill-ee-oh-SEEK-ul) valve: the sphincter muscle separating the small and large intestines.
large intestine or **colon:** the last portion of the intestine, which absorbs water. Its main segments are the ascending colon, the transverse colon, the descending colon, and the sigmoid colon.
sigmoid = shaped like the letter S
rectum: the muscular terminal part of the GI tract extending from the sigmoid colon to the anus. The rectum stores waste prior to elimination.
anus (AY-nus): the terminal sphincter muscle of the GI tract.

FIGURE 2-1 The Gastrointestinal Tract

FIBER	CARBOHYDRATE

Mouth

The mechanical action of the mouth and teeth crushes and tears fiber in food and mixes it with saliva to moisten it for swallowing.

The salivary glands secrete a watery fluid into the mouth to moisten the food. The salivary enzyme amylase begins digestion:

$$\text{Starch} \xrightarrow{\text{amylase}} \text{small polysaccharides, maltose.}$$

Stomach

Fiber is unchanged.

Stomach acid and enzymes start to digest salivary enzymes, halting starch digestion. To a small extent, stomach acid hydrolyzes maltose and sucrose.

Small intestine

Fiber is unchanged.

The pancreas produces enzymes and releases them through the pancreatic duct into the small intestine:

$$\text{Polysaccharides} \xrightarrow{\text{pancreatic amylase}} \text{disaccharides.}$$

Then enzymes on the surfaces of the small intestinal cells break disaccharides into monosaccharides, and the cells absorb them:

$$\text{Maltose} \xrightarrow{\text{maltase}} \text{glucose} + \text{glucose.}$$

$$\text{Sucrose} \xrightarrow{\text{sucrase}} \text{fructose} + \text{glucose.}$$

$$\text{Lactose} \xrightarrow{\text{lactase}} \text{galactose} + \text{glucose.}$$

Large intestine (colon)

Most fiber passes intact through the digestive tract to the large intestine. Here, bacterial enzymes digest some fiber:

$$\text{Some fiber} \xrightarrow{\text{bacterial enzymes}} \text{fatty acids, gas.}$$

Fiber holds water; regulates bowel activity; and binds cholesterol and some minerals, carrying them out of the body as it is excreted with feces.

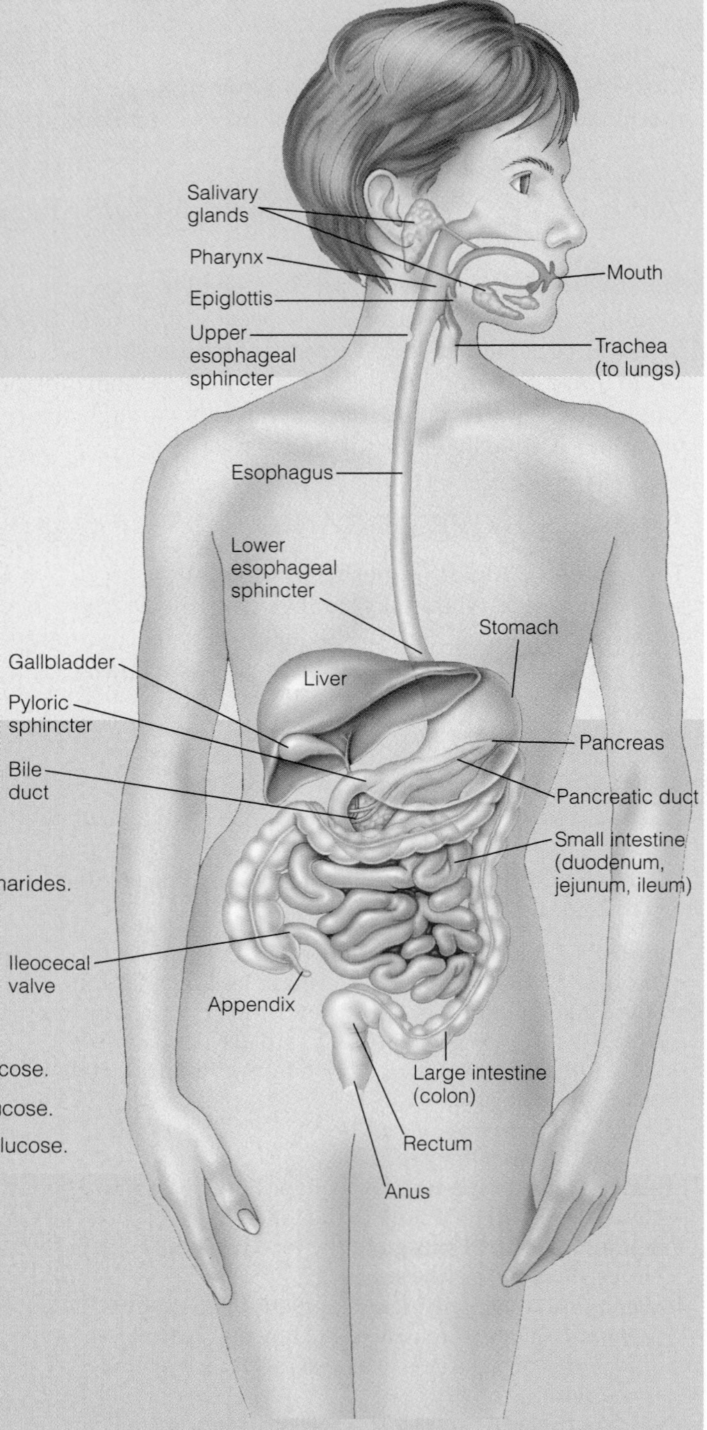

Salivary glands
Pharynx
Epiglottis
Upper esophageal sphincter
Mouth
Trachea (to lungs)
Esophagus
Lower esophageal sphincter
Stomach
Gallbladder
Liver
Pyloric sphincter
Bile duct
Pancreas
Pancreatic duct
Small intestine (duodenum, jejunum, ileum)
Ileocecal valve
Appendix
Large intestine (colon)
Rectum
Anus

FAT	PROTEIN	VITAMINS	MINERALS AND WATER
Mouth Glands in the base of the tongue secrete a fat-digesting enzyme known as lingual lipase. Some hard fats begin to melt as they reach body temperature.	Chewing and crushing moisten protein-rich foods and mix them with saliva to be swallowed.	No action.	The salivary glands add water to disperse and carry food.
Stomach The acid-stable lingual lipase splits one bond of triglycerides to produce diglycerides and fatty acids. The degree of hydrolysis is slight for most fats but may be appreciable for milk fats. The stomach's churning action mixes fat with water and acid. A gastric lipase accesses and hydrolyzes a very small amount of fat.	Stomach acid uncoils protein strands and activates stomach enzymes: $\text{Protein} \xrightarrow[\text{HCl}]{\text{pepsin}} \text{smaller polypeptides.}$	Intrinsic factor (see Chapter 8) attaches to vitamin B_{12}.	Stomach acid (HCl) acts on iron to reduce it, making it more absorbable. The stomach secretes enough watery fluid to turn a moist, chewed mass of solid food into liquid chyme.
Small intestine Bile flows in from the liver and gallbladder (via the common bile duct): $\text{Fat} \xrightarrow{\text{bile}} \text{emulsified fat.}$ Pancreatic lipase flows in from the pancreas (via the pancreatic duct): $\text{Emulsified fat} \xrightarrow[\text{lipase}]{\text{pancreatic}}$ monoglycerides, glycerol, fatty acids (absorbed).	$\xrightarrow[\text{and intestinal proteases}]{\text{pancreatic}}$ $\xrightarrow[\text{dipeptidases and tripeptidases}]{\text{intestinal}} \text{amino acids (absorbed)}$	Bile emulsifies fat-soluble vitamins and aids in their absorption with other fats. Water-soluble vitamins are absorbed.	The small intestine, pancreas, and liver add enough fluid so that approximately 2 gallons are secreted into the intestine in a day. Many minerals are absorbed. Vitamin D aids in the absorption of calcium.
Large intestine Some fat and cholesterol, trapped in fiber, exit in feces.		Bacteria produce vitamin K, which is absorbed.	More minerals and most of the water are absorbed.

The tongue allows you not only to taste food but also to move food around the mouth, facilitating chewing and swallowing. When you swallow a mouthful of food, it passes through the **pharynx,** a short tube that is shared by both the **digestive system** and the respiratory system.

Mouth to the Esophagus Once a mouthful of food has been swallowed, it is called a **bolus.** Each bolus first slides across your **epiglottis,** bypassing the entrance to your lungs. During each swallow, the epiglottis closes off your **trachea,** the air passageway to the lungs, so that you do not choke.

Esophagus to the Stomach Next, the bolus slides down the **esophagus,** which conducts it through the diaphragm to the stomach. At the junction with the stomach, the esophagus is surrounded by a band of muscle called a **sphincter.** This **lower esophageal sphincter** closes behind the bolus so that it cannot slip back. The stomach retains the bolus for a while, adds juices to it (gastric juices are discussed on p. 46), and transforms it into a semiliquid mass called **chyme.** Then, bit by bit, the stomach releases the chyme through another sphincter, the **pyloric sphincter,** which opens into the **small intestine** and then closes after the chyme passes through.

The Small Intestine At the beginning of the small intestine, the chyme passes by an opening from the common bile duct, which secretes digestive fluids into the small intestine from two organs outside the GI tract—the **gallbladder** and the **pancreas.** The chyme travels on down the small intestine through its three segments—the **duodenum,** the **jejunum,** and the **ileum.** Together, the segments amount to a total of about 10 feet of tubing coiled within the abdomen.* Digestion is completed within the small intestine.

The Large Intestine (Colon) Having traveled the length of the small intestine, what remains of the intestinal contents passes through another sphincter, the **ileocecal valve,** into the beginning of the **large intestine (colon)** in the lower right-hand side of the abdomen. The colon's contents travel up the right-hand side of the abdomen, across the front to the left-hand side, down to the lower left-hand side, and finally below the other folds of the intestines to the back side of the body above the **rectum.**

The Rectum As the intestinal contents pass through to the rectum, the colon withdraws water, leaving semisolid waste. The strong muscles of the rectum hold back this waste until it is time to defecate. Then the rectal muscles relax, and the last sphincter in the system, the **anus,** opens to allow the wastes to pass. Thus food follows the path shown in the margin.

THE INVOLUNTARY MUSCLES AND THE GLANDS

You are usually unaware of all the activity that goes on between the time you swallow and the time you defecate. As is the case with so much else that happens in the body, the muscles and **glands** of the digestive tract meet internal needs without your having to exert any conscious effort to get the work done.

People consciously chew and swallow, but even in the mouth there are some processes over which you have no control. The salivary glands secrete just enough saliva to moisten each mouthful of food so that it can pass easily down your esophagus.

Gastrointestinal Motility Once you have swallowed, materials are moved through the rest of the GI tract by involuntary muscular contractions. This motion, known

---

The path of food through the digestive tract:

- Mouth.
- Esophagus.
- Lower esophageal sphincter (or cardiac sphincter).
- Stomach.
- Pyloric sphincter.
- Duodenum (common bile duct enters here), jejunum, ileum.
- Ileocecal valve.
- Large intestine (colon).
- Rectum.
- Anus.

digestive system: all the organs and glands associated with the ingestion and digestion of food.

bolus (BOH-lus): the portion of food swallowed at one time.

sphincter (SFINK-ter): a circular muscle surrounding, and able to close, a body opening. *sphincter* = band (binder)

chyme (KIME): the semiliquid mass of partly digested food expelled by the stomach into the duodenum (the top portion of the small intestine).

glands: single cells or groups of cells that secrete materials for special uses in the body. Glands may be *exocrine glands,* secreting their materials "out" (into the digestive tract or onto the surface of the skin), or *endocrine glands,* secreting their materials "in" (into the blood).

    exo = outside
    endo = inside
    krine = to separate

---

*The small intestine is almost two and a half times shorter in living adults than it is at death, when muscles are relaxed and elongated.

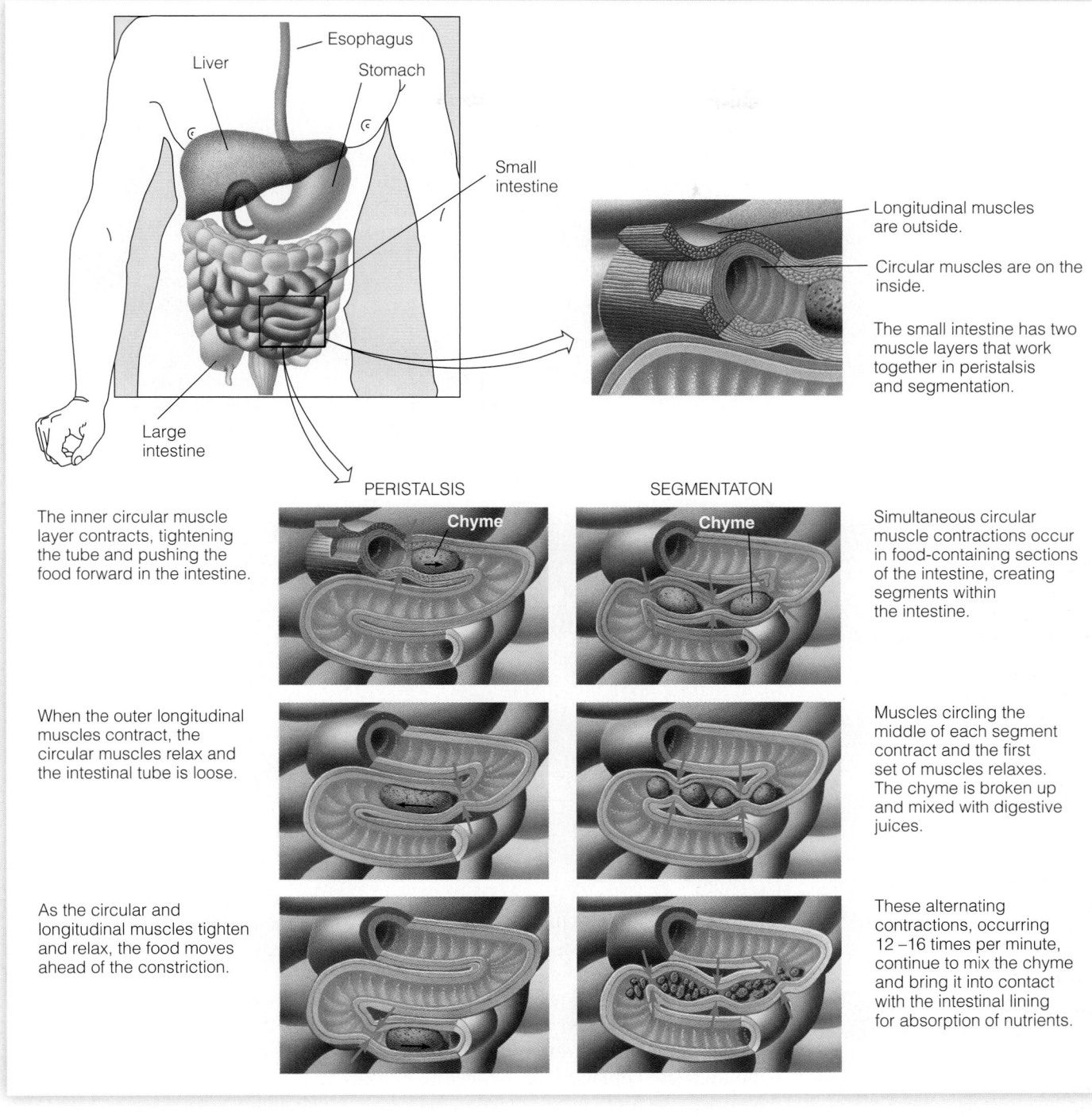

Liver

Esophagus

Stomach

Small intestine

Large intestine

Longitudinal muscles are outside.

Circular muscles are on the inside.

The small intestine has two muscle layers that work together in peristalsis and segmentation.

PERISTALSIS

SEGMENTATON

Chyme

Chyme

The inner circular muscle layer contracts, tightening the tube and pushing the food forward in the intestine.

Simultaneous circular muscle contractions occur in food-containing sections of the intestine, creating segments within the intestine.

When the outer longitudinal muscles contract, the circular muscles relax and the intestinal tube is loose.

Muscles circling the middle of each segment contract and the first set of muscles relaxes. The chyme is broken up and mixed with digestive juices.

As the circular and longitudinal muscles tighten and relax, the food moves ahead of the constriction.

These alternating contractions, occurring 12–16 times per minute, continue to mix the chyme and bring it into contact with the intestinal lining for absorption of nutrients.

as **gastrointestinal motility**, consists of two types of movement, peristalsis and segmentation (see Figure 2-2). Peristalsis propels, or pushes; segmentation mixes, with more gradual pushing.

Peristalsis Peristalsis begins when the bolus enters the esophagus. The entire GI tract is ringed with circular muscles. Surrounding these rings of muscle are longitudinal muscles. When the rings tighten and the long muscles relax, the tube is constricted. When the rings relax and the long muscles tighten, the tube bulges. These actions follow each other continuously and push the intestinal contents along. If you have ever watched a bolus of food pass along the body of a snake, you have a good picture of how these muscles work. The waves of contraction

FIGURE 2-2 Peristalsis and Segmentation

gastrointestinal motility: spontaneous motion in the digestive tract accomplished by involuntary muscular contractions.

peristalsis (peri-STALL-sis): successive waves of involuntary muscular contractions passing along the walls of the GI tract that push the contents along.
peri = around
stellein = wrap

FIGURE 2-3 Stomach Muscles
The stomach has three layers of muscles.

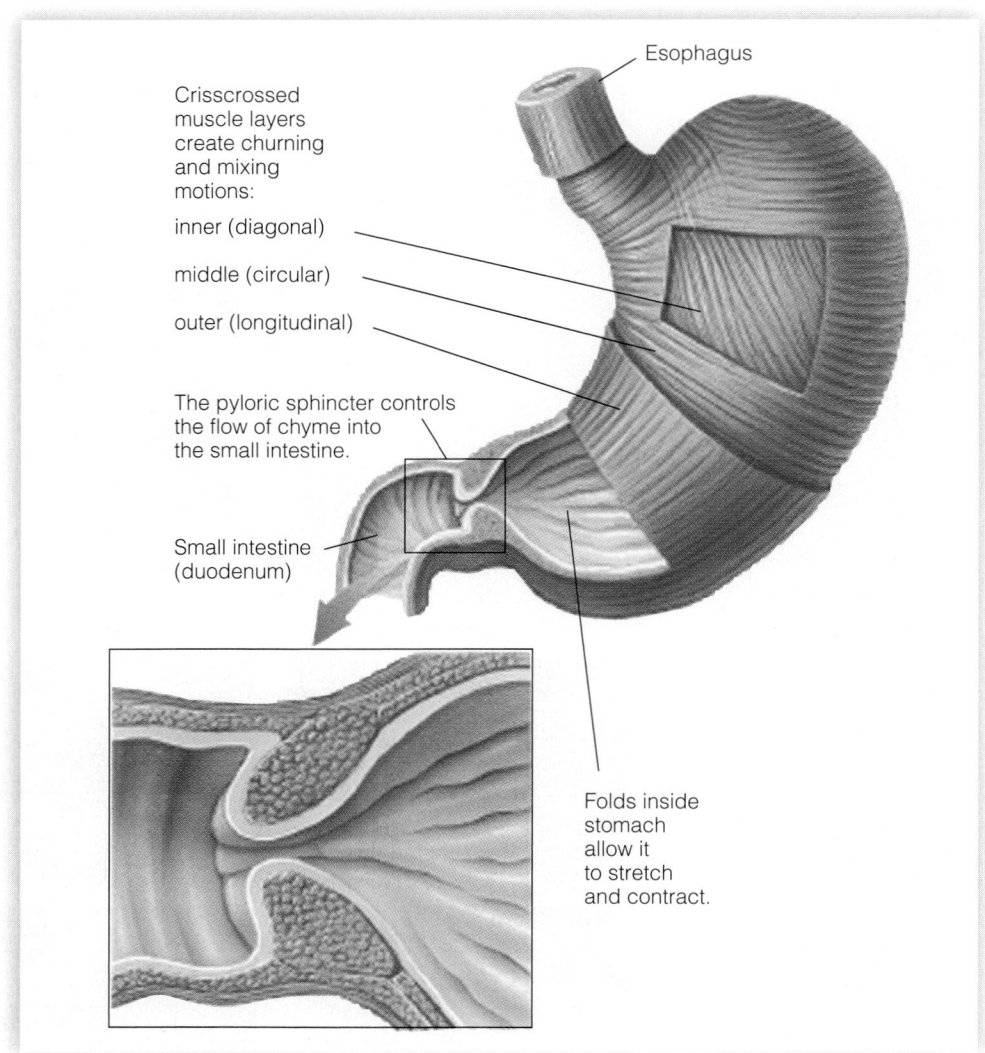

Crisscrossed muscle layers create churning and mixing motions:

inner (diagonal)

middle (circular)

outer (longitudinal)

The pyloric sphincter controls the flow of chyme into the small intestine.

Small intestine (duodenum)

Esophagus

Folds inside stomach allow it to stretch and contract.

ripple through the GI tract at varying rates and intensities depending on the part of the GI tract and on whether food is present. Peristalsis, aided by the sphincter muscles located at key places, keeps things moving along.

Segmentation The intestines not only push but also periodically squeeze their contents as if a string tied around the intestines were being pulled tight. This motion, called **segmentation,** forces the contents back a few inches, mixing them and promoting close contact with the digestive juices and the absorbing cells of the intestinal walls before letting the contents slowly move along again.

Liquefying Process Besides forcing the intestinal contents along, the muscles of the GI tract help to liquefy them to chyme so that the digestive juices will have access to all their nutrients. The mouth initiates this liquefying process by chewing, adding saliva, and stirring with the tongue to reduce the food to a coarse mash suitable for swallowing. The stomach then further mixes and kneads the food.

Stomach Action The stomach has the thickest walls and strongest muscles of all the GI tract organs. In addition to circular and longitudinal muscles, the stomach has a third layer of diagonal muscles that also alternately contract and relax (see Figure 2-3). These three sets of muscles work to force the chyme downward, but the pyloric sphincter usually remains tightly closed so that the stomach's contents

segmentation: a periodic squeezing or partitioning of the intestine by its circular muscles that both mixes and slowly pushes the contents along.

are thoroughly mixed and squeezed before being released. Meanwhile, the gastric glands are adding juices. When the chyme is thoroughly liquefied, the pyloric sphincter opens briefly, about three times a minute, to allow small portions through. At this point, the intestinal contents no longer resemble food in the least.

REVIEW NOTES

As Figure 2-1 shows, food enters the mouth and travels down the esophagus and through the lower esophageal sphincter to the stomach, then through the pyloric sphincter to the small intestine, on through the ileocecal valve to the large intestine, past the appendix to the rectum, ending at the anus.

The wavelike contractions of peristalsis and the periodic squeezing of segmentation keep things moving at a reasonable pace.

The Process of Digestion

One person eats nothing but vegetables, fruits, and nuts; another, nothing but meat, milk, and potatoes. How is it that both people wind up with essentially the same body composition? It all comes down to the body rendering food—whatever it is to start with—into the basic units that make up carbohydrate, fat, and protein. The body absorbs these units and builds its tissues from them.

To digest food, five different body organs secrete digestive juices: the salivary glands, the stomach, the small intestine, the liver (via the gallbladder), and the pancreas. These secretions enter the GI tract at various points along the way, bringing an abundance of water and a variety of enzymes. Each of the juices has a turn to mix with the food and promote its breakdown to small units that can be absorbed into the body. The accompanying glossary defines some of the digestive glands and their juices.

Enzymes are formally introduced in Chapter 5, but for now a simple definition will suffice. An *enzyme* is a protein that facilitates a chemical reaction—making a compound, breaking down a compound, changing the arrangement of a compound, or exchanging parts. Enzymes themselves are not changed by the reactions they facilitate.

GLOSSARY   *Digestive Glands and Their Secretions*

These terms are listed in order from the beginning of the digestive tract to the end.

salivary glands: exocrine glands that secrete saliva into the mouth.

saliva: the secretion of the salivary glands. The principal enzyme is salivary amylase.

amylase (AM-uh-lace): an enzyme that splits amylose (a form of starch). Amylase is a carbohydrase. The ending *-ase* indicates an enzyme; the root tells what it digests. Other examples: protease, lipase.

gastric glands: exocrine glands in the stomach wall that secrete gastric juice into the stomach.
gastro = stomach

gastric juice: the digestive secretion of the gastric glands containing a mixture of water, hydrochloric acid, and enzymes. The principal enzymes are pepsin (acts on proteins) and lipase (acts on emulsified fats).

hydrochloric acid (HCl): an acid composed of hydrogen and chloride atoms; normally produced by the gastric glands.

mucus (MYOO-cuss): a mucopolysaccharide (a relative of carbohydrate) secreted by cells of the stomach wall that protects the cells from exposure to digestive juices (and other destructive agents). The cellular lining of the stomach wall with its coat of mucus is known as the *mucous membrane*. (The noun is *mucus;* the adjective is *mucous*.)

pepsin: a protein-digesting enzyme (gastric protease) in the stomach. It circulates as a precursor, pepsinogen, and is converted to pepsin by the action of stomach acid.

intestinal juice: the secretion of the intestinal glands; contains enzymes for the digestion of carbohydrate and protein and a minor enzyme for fat digestion.

bile: an emulsifier that prepares fats and oils for digestion; made by the liver, stored in the gallbladder, and released into the small intestine when needed.

pancreatic (pank-ree-AT-ic) juice: the exocrine secretion of the pancreas, containing enzymes for the digestion of carbohydrate, fat, and protein. Juice flows from the pancreas into the small intestine through the pancreatic duct. The pancreas also has an endocrine function, the secretion of insulin and other hormones.

bicarbonate: an alkaline secretion of the pancreas; part of the pancreatic juice. (Bicarbonate also occurs widely in all cell fluids.)

DIGESTION IN THE MOUTH

Digestion of carbohydrate begins in the mouth, where the **salivary glands** secrete **saliva,** which contains water, salts, and enzymes (including salivary **amylase**) that break the bonds in the chains of starch. Saliva also protects the tooth surfaces and linings of the mouth, esophagus, and stomach from attack by molecules that might harm them. The enzymes in the mouth do not, for the most part, affect the fats, proteins, vitamins, minerals, and fiber that are present in the foods people eat (review Figure 2-1).

DIGESTION IN THE STOMACH

Gastric juice, secreted by the **gastric glands,** is composed of water, enzymes, and **hydrochloric acid.** The acid is so strong that it burns the throat if it happens to reflux into the upper esophagus and mouth. The strong acidity of the stomach prevents bacterial growth and kills most bacteria that enter the body with food. You might expect that the stomach's acid would attack the stomach itself, but the cells of the stomach wall secrete **mucus,** a thick, slimy, white polysaccharide that coats and protects the stomach's lining.

The major digestive event in the stomach is the initial breakdown of proteins. Other than being crushed and mixed with saliva in the mouth, nothing happens to protein until it comes in contact with the gastric juices in the stomach. There, the acid helps to uncoil (denature) the protein's tangled strands so that the stomach enzymes can attack the bonds. Both the enzyme **pepsin** and the stomach acid itself act as catalysts in the process. Minor events are the digestion of some fat by a gastric lipase, the digestion of sucrose (to a very small extent) by the stomach acid, and the attachment of a protein carrier to vitamin B_{12}.

The stomach enzymes work most efficiently in the stomach's strong acid, but salivary amylase, which is swallowed with food, does not work in acid this strong. Consequently, the digestion of starch gradually ceases as the acid penetrates the bolus. In fact, salivary amylase becomes just another protein to be digested. The amino acids in amylase end up being absorbed and recycled into other body proteins.

DIGESTION IN THE SMALL AND LARGE INTESTINES

By the time food leaves the stomach, digestion of all three energy-yielding nutrients has begun, but the process gains momentum in the small intestine. There, the pancreas and the liver contribute additional digestive juices through the duct leading into the duodenum, and the small intestine adds **intestinal juice.** These juices contain digestive enzymes, bicarbonate, and bile.

Digestive Enzymes Pancreatic juice contributes enzymes that digest fats, proteins, and carbohydrates. Glands in the intestinal wall also secrete digestive enzymes. (Review the glossary of digestive glands and their secretions on p. 45 for details.)

Bicarbonate The pancreatic juice also contains sodium **bicarbonate,** which neutralizes the acidic chyme as it enters the small intestine. From this point on, the contents of the digestive tract are neutral or slightly alkaline—the enzymes of both the intestine and the pancreas work best in this environment.

Bile Bile is secreted continuously by the liver and is concentrated and stored in the gallbladder. The gallbladder squirts bile into the duodenum whenever fat arrives there. Bile is not an enzyme but an **emulsifier** that brings fats into suspension in water (see Figure 2-4). After the fats are emulsified, enzymes can work on them, and they can be absorbed. Thanks to all these secretions, all three energy-yielding nutrients are digested in the small intestine.

emulsifier: a substance that mixes with both fat and water and that disperses the fat in the water, forming an emulsion.

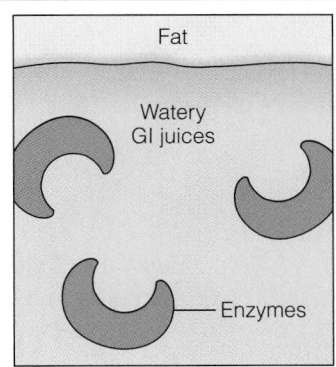 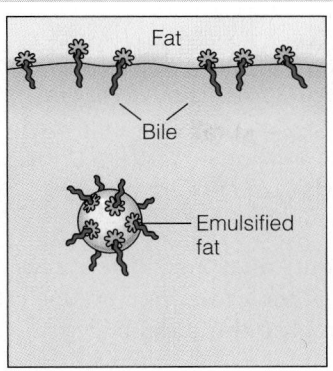 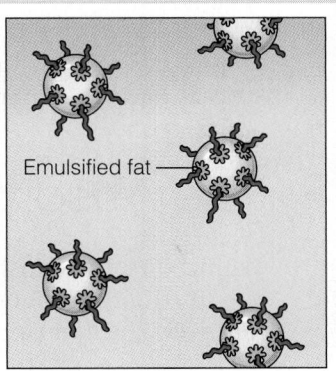 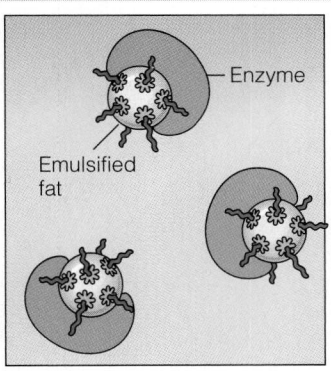

In the stomach, the fat and watery GI juices tend to separate. The enzymes are in the water and can't get at the fat.

When fat enters the small intestine, the gallbladder secretes bile. Bile has an affinity for both fat and water, so it can bring the fat into the water.

Bile's emulsifying action converts large fat globules into small droplets that repel each other.

After emulsification, the enzymes have easy access to the fat droplets.

FIGURE 2-4 Emulsification of Fat by Bile
Like bile, detergents are emulsifiers and work the same way, which is why they are effective at removing grease spots from clothes. Molecule by molecule, the grease is dissolved out of the spot and suspended in water, where it can be rinsed away.

The Rate of Digestion The rate of digestion of the energy nutrients depends on the contents of the meal. If the meal is high in simple sugars, digestion proceeds fairly rapidly. On the other hand, if the meal is rich in fat, digestion is slower.

Protective Factors The intestines contain bacteria that produce a variety of vitamins, including biotin and vitamin K (although bacteria alone cannot meet the need for these vitamins). The GI bacteria also protect people from infections. Provided that the normal **intestinal flora** are thriving, infectious bacteria have a hard time getting established and launching an attack on the system. In addition, the small intestine and the entire GI tract manufacture and maintain a strong arsenal of defenses against foreign invaders. Several different types of defending cells are present there and confer specific immunity against intestinal diseases.

The Final Stage The story of how food is broken down into nutrients that can be absorbed is now nearly complete. The three energy-yielding nutrients—carbohydrate, fat, and protein—are disassembled to basic building blocks before they are absorbed. Most of the other nutrients—vitamins, minerals, and water—are absorbed as they are. Undigested residues, such as some fibers, are not absorbed but continue through the digestive tract as a semisolid mass that stimulates the muscles of the GI tract, helping them remain strong and able to perform peristalsis efficiently. Fiber also retains water, keeping the stools soft, and carries some bile acids, sterols, and fat out of the body. Drinking plenty of water in conjunction with eating foods high in fiber supplies fluid for the fiber to take up. This is the basis for the recommendation to drink water and eat fiber-rich foods to relieve constipation.

The process of absorbing the nutrients into the body is discussed in the next section. For the moment, let us assume that the digested nutrients simply disappear from the GI tract as they are ready. Virtually all nutrients are gone by the time the contents of the GI tract reach the end of the small intestine. Little remains but water, a few salts and body secretions, and undigested materials such as fiber. These enter the large intestine (colon).

In the colon, intestinal bacteria degrade some of the fiber to simpler compounds. The colon itself retrieves from its contents the materials that the body is designed to recycle—water and dissolved salts. The waste that is finally excreted has little or nothing of value left in it. The body has extracted all that it can use from the food.

Mayonnaise, made from vinegar and oil, would separate as other vinegar-and-oil salad dressings do if food chemists did not blend the vinegar and oil with a third ingredient—an emulsifier. The emulsifier mixes well with the fatty oil and the watery vinegar. In the case of mayonnaise, the emulsifier is lecithin from egg yolks.

intestinal flora: the bacterial inhabitants of the GI tract.

flora = plant growth

REVIEW NOTES

To digest food, the salivary glands, stomach, pancreas, liver (via the gallbladder), and small intestine deliver fluids and digestive enzymes.

The Absorptive System

Within three or four hours after you have eaten a meal, your body must find a way to absorb millions of molecules one by one. The absorptive system is ingeniously designed to accomplish this task.

THE SMALL INTESTINE

Most absorption takes place in the small intestine. The small intestine is a tube about 10 feet long and about an inch across, yet it provides a surface comparable in area to a tennis court. When nutrient molecules make contact with this surface, they are absorbed and carried off to the liver and other parts of the body.

Villi and Microvilli How does the intestine manage to provide such a large absorptive surface area? Its inner surface looks smooth, but viewed through a microscope, it turns out to be wrinkled into hundreds of folds. Each fold is covered with thousands of fingerlike projections called **villi.** The villi are as numerous as the hairs on velvet fabric. A single villus, magnified still more, turns out to be composed of several hundred cells, each covered with microscopic hairs called **microvilli** (see Figure 2-5).

The villi are in constant motion. A thin sheet of muscle lines each villus so that it can wave, squirm, and wiggle like the tentacles of a sea anemone. Any nutrient molecule small enough to be absorbed is trapped among the microvilli and drawn into a cell beneath them. Some partially digested nutrients are caught in the microvilli, digested further by enzymes there, and then absorbed into the cells.

Specialization in the Intestinal Tract As you can see, the intestinal tract is beautifully designed to perform its functions. A further refinement of the system is that the cells of successive portions of the tract are specialized to absorb different nutrients. The nutrients that are ready for absorption early are absorbed near the top of the tract; those that take longer to be digested are absorbed further down. The rate at which the nutrients travel through the GI tract is finely adjusted to maximize their availability to the appropriate absorptive segment of the tract when they are ready. The lowly "gut" turns out to be one of the most elegantly designed organ systems in the body.

The Myth of "Food Combining" Some popular fad diets advocate that people should avoid eating certain food combinations (for example, fruit and meat) at the same meal, because the digestive system cannot handle more than one task at a time. This is a myth. The art of "food combining" (which actually emphasizes "food separating") is based on this idea, and it represents faulty logic and a gross underestimation of the body's capabilities. In fact, the contrary is often true; foods eaten together can enhance each other's use by the body. For example, vitamin C in a pineapple or other citrus fruit can enhance the absorption of iron from a meal of chicken and rice or other iron-containing foods. Many other instances of mutually beneficial interactions are presented in later chapters.

ABSORPTION OF NUTRIENTS

Once a molecule has entered a cell in a villus, the next step is to transmit it to a destination elsewhere in the body by way of the body's two transport systems—

villi (VILL-ee or VILL-eye): fingerlike projections from the folds of the small intestine. The singular form is **villus.**
   villus = shaggy hair

microvilli (MY-cro-VILL-ee or MY-cro-VILL-eye): tiny, hairlike projections on each cell of every villus that can trap nutrient particles and transport them into the cells. The singular form is **microvillus.**

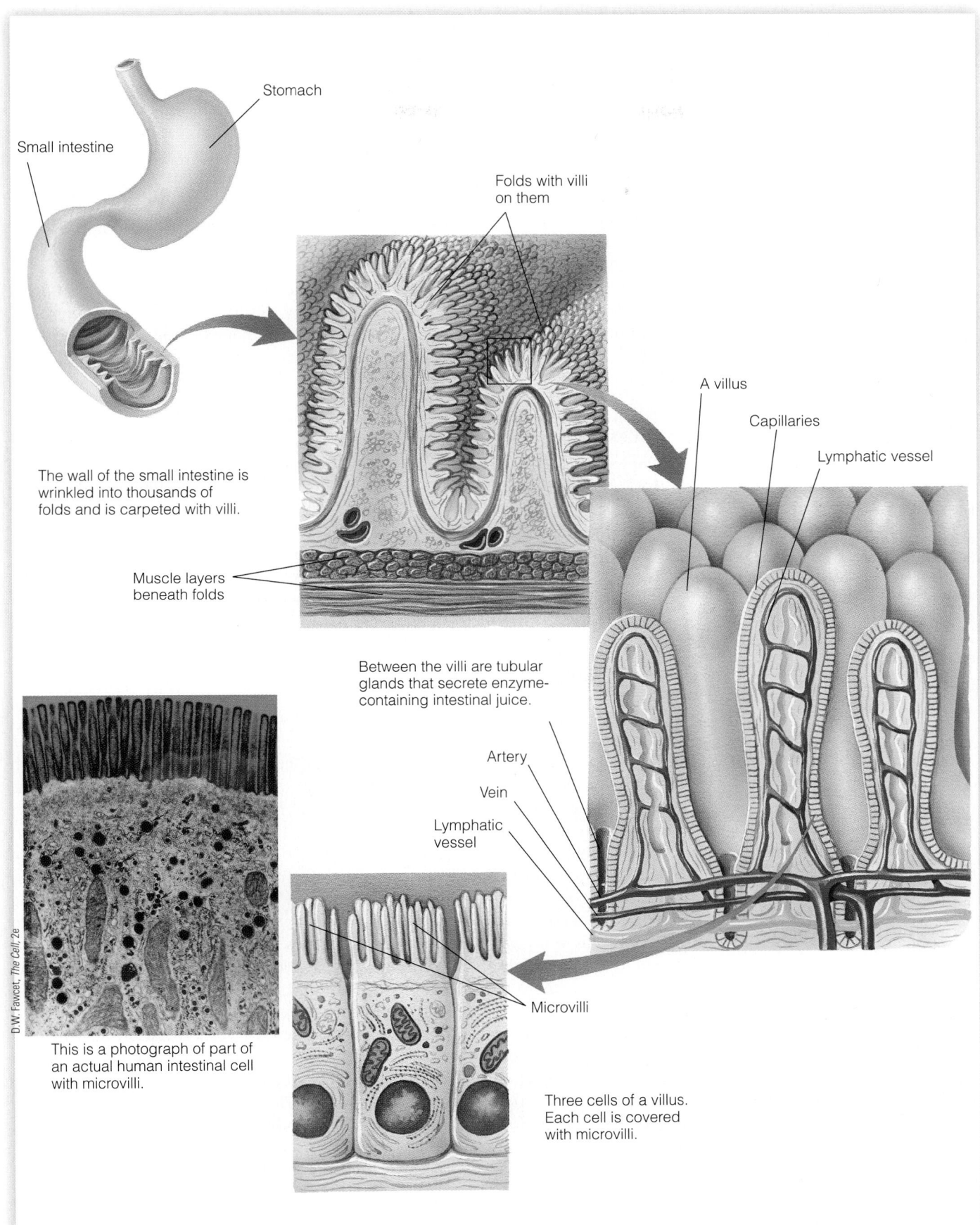

Stomach

Small intestine

Folds with villi on them

The wall of the small intestine is wrinkled into thousands of folds and is carpeted with villi.

Muscle layers beneath folds

A villus

Capillaries

Lymphatic vessel

Between the villi are tubular glands that secrete enzyme-containing intestinal juice.

Artery

Vein

Lymphatic vessel

D.W. Fawcet, *The Cell, 2e*

This is a photograph of part of an actual human intestinal cell with microvilli.

Microvilli

Three cells of a villus. Each cell is covered with microvilli.

FIGURE 2-5 The Small Intestinal Villi

lymphatic system: a loosely organized system of vessels and ducts that conveys the products of digestion toward the heart.

lymph (LIMF): the body fluid found in lymphatic vessels. Lymph consists of all the constituents of blood except red blood cells.

triglycerides (try-GLISS-er-rides): one of the three main classes of lipids: the chief form of fat in foods and the major storage form of fat in the body; composed of glycerol with three fatty acids attached.

   tri = three
   glyceride = a compound of glycerol

chylomicrons (kye-lo-MY-crons): the lipoproteins that transport lipids from the intestinal cells into the body. The cells of the body remove the lipids they need from the chylomicrons, leaving chylomicron remnants to be picked up by the liver cells.

lipoprotein: a cluster of lipids associated with proteins that serves as a transport vehicle for lipids in the lymph and blood.

artery: a vessel that carries blood away from the heart.

capillaries: small vessels that branch from an artery. Capillaries connect arteries to veins. Oxygen, nutrients, and waste materials are exchanged across capillary walls.

vein: a vessel that carries blood back to the heart.

The blood arriving at the intestines flows through the **mesentery** (MEZ-en-terry), a strong, flexible membrane that surrounds and supports the abdominal organs. *mes* = middle

The vein that collects blood from the mesentery and conducts it to capillaries in the liver is the **portal vein.** *portal* = gateway

The vein that collects blood from the liver capillaries and returns it to the heart is the **hepatic vein.** *hepat* = liver

The artery that delivers oxygen-rich blood from the heart and lungs to the liver is the **hepatic artery.**

the bloodstream and the **lymphatic system.** As Figure 2-5 shows, both systems supply vessels to each villus. Through these vessels, the nutrients leave the cell and enter either the **lymph** or the blood. In either case, the nutrients end up in the blood, at least for a while. The water-soluble nutrients (and the smaller products of fat digestion) are released directly into the bloodstream by way of the capillaries, but the larger fats and the fat-soluble vitamins find direct access into the capillaries impossible because these nutrients are insoluble in water (and blood is mostly water). They require some packaging before they are released.

The intestinal cells assemble the products of fat digestion into larger molecules called **triglycerides.** These triglycerides, fat-soluble vitamins (when present), and other large lipids (cholesterol and the phospholipids) are then packaged for transport. They cluster together with special proteins to form **chylomicrons,** one kind of **lipoprotein** (lipoproteins are described beginning on p. 51). Finally, the cells release the chylomicrons into the lymphatic system. They can then glide through the lymph spaces until they arrive at a point of entry into the bloodstream near the heart.

> **REVIEW NOTES**
>
> The many folds and villi of the small intestine dramatically increase its surface area, facilitating nutrient absorption.
>
> Nutrients pass through the cells of the villi and enter either the blood (if they are water soluble or small fat fragments) or the lymph (if they are fat soluble).

Transport of Nutrients

Once a nutrient has entered the bloodstream or the lymphatic system, it may be transported to any part of the body, from the tips of the toes to the roots of the hair, where it becomes available to any of the cells. The circulatory systems are arranged to deliver nutrients wherever they are needed.

THE VASCULAR SYSTEM

The vascular or blood circulatory system is a closed system of vessels through which blood flows continuously in a figure eight, with the heart serving as a pump at the crossover point. On each loop of the figure eight, blood travels a simple route: heart to arteries to capillaries to veins to heart.

The routing of the blood through the digestive system is different, however. The blood is carried to the digestive system (as it is to all organs) by way of an **artery,** which (as in all organs) branches into **capillaries** to reach every cell. Blood leaving the digestive system, however, goes by way of a **vein,** not back to the heart, but to the liver. This vein again branches into capillaries so that every cell of the liver has access to the newly absorbed nutrients that the blood is carrying. Blood leaving the liver then returns to the heart by way of another vein. The route is thus heart to arteries to capillaries (in intestines) to vein to capillaries (in liver) to vein to heart.

An anatomist studying this system knows there must be a reason for this special arrangement. The liver is located in the circulation system at the point where it will have the first chance at the materials absorbed from the GI tract. In fact, the liver is the body's major metabolic organ (see Figure 2-6) and must prepare the absorbed nutrients for use by the rest of the body. Furthermore, the liver stands as a gatekeeper to waylay intruders that might otherwise harm the heart or brain. Chapter 19 offers more information about this noble organ.

THE LYMPHATIC SYSTEM

The lymphatic system is a one-way route for fluids to travel from tissue spaces into the blood. The lymphatic system has no pump; instead, lymph is squeezed

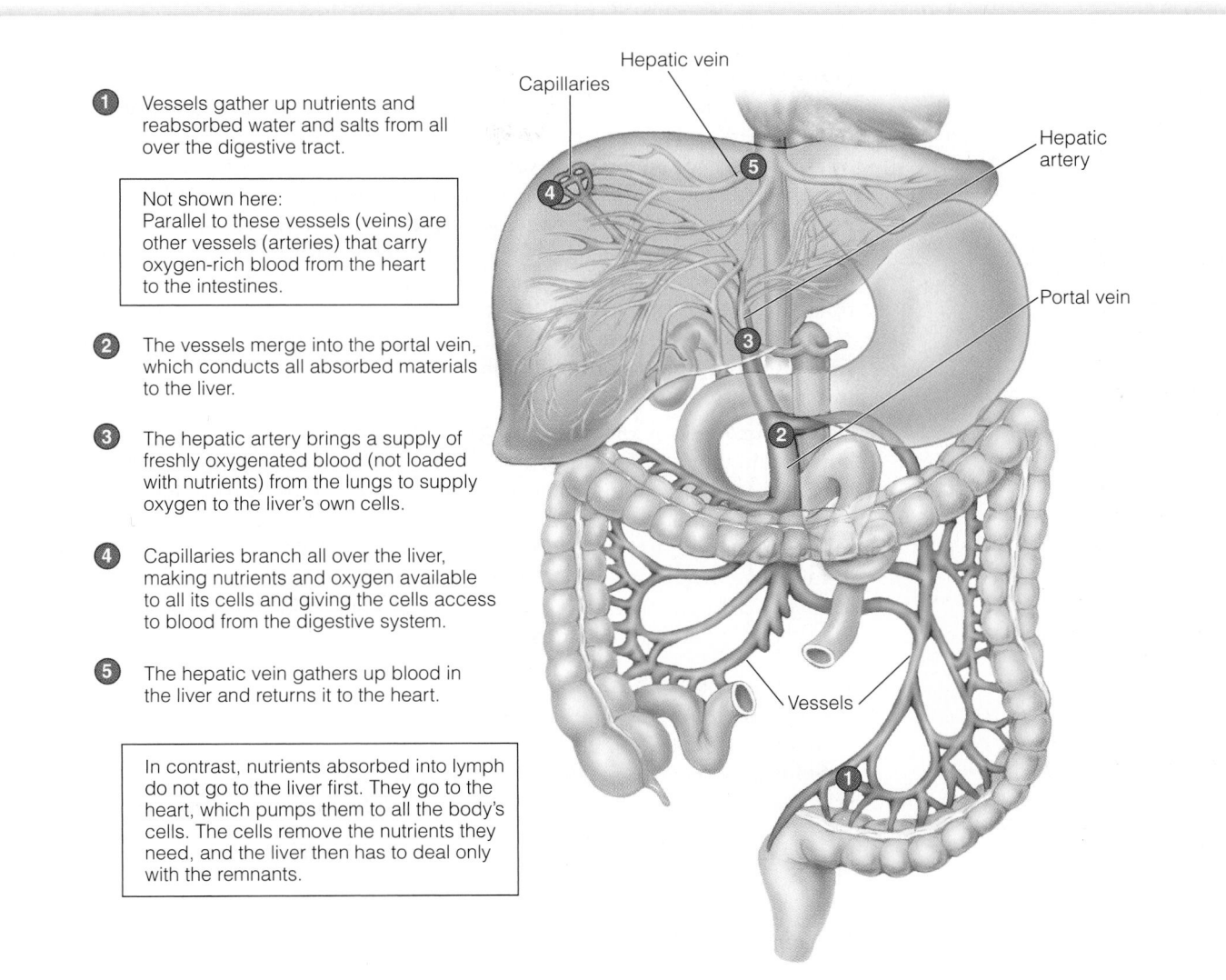

1 Vessels gather up nutrients and reabsorbed water and salts from all over the digestive tract.

Not shown here:
Parallel to these vessels (veins) are other vessels (arteries) that carry oxygen-rich blood from the heart to the intestines.

2 The vessels merge into the portal vein, which conducts all absorbed materials to the liver.

3 The hepatic artery brings a supply of freshly oxygenated blood (not loaded with nutrients) from the lungs to supply oxygen to the liver's own cells.

4 Capillaries branch all over the liver, making nutrients and oxygen available to all its cells and giving the cells access to blood from the digestive system.

5 The hepatic vein gathers up blood in the liver and returns it to the heart.

In contrast, nutrients absorbed into lymph do not go to the liver first. They go to the heart, which pumps them to all the body's cells. The cells remove the nutrients they need, and the liver then has to deal only with the remnants.

Hepatic vein

Capillaries

Hepatic artery

Portal vein

Vessels

FIGURE 2-6 The Liver

from one portion of the body to another like water in a sponge, as muscles contract and create pressure here and there. Ultimately, the lymph collects in a large duct behind the heart. This duct terminates in a vein that conducts the lymph into the heart. Thus some materials from the GI tract enter the lymphatic system before entering the bloodstream.

TRANSPORT OF LIPIDS: LIPOPROTEINS

Within the circulatory system, lipids always travel from place to place bundled with protein, that is, as lipoproteins. When physicians measure a person's blood lipid profile, they are interested in both the types of fat present (such as triglycerides and cholesterol) and the types of lipoproteins that carry them.

VLDL, LDL, and HDL As mentioned earlier, chylomicrons transport newly absorbed (*diet-derived*) lipids from the intestinal cells to the rest of the body. As chylomicrons circulate through the body, cells remove their lipid contents, so the chylomicrons get smaller and smaller. The liver picks up these chylomicron remnants. When necessary, the liver can assemble different lipoproteins, which are known as **very-low-density lipoproteins (VLDL).** As the body's cells remove

The duct that conveys lymph toward the heart is the **thoracic** (thor-ASS-ic) **duct.** The **subclavian vein** connects this duct with the right upper chamber of the heart, providing a passageway by which lymph can be returned to the vascular system.

very-low-density lipoproteins (VLDL): the type of lipoproteins made primarily by liver cells to transport lipids to various tissues in the body; composed primarily of triglycerides.

triglycerides from the VLDL, the proportions of their lipid and protein contents shift. As this occurs, VLDL become cholesterol-rich **low-density lipoproteins (LDL).** Cholesterol returning to the liver for metabolism or excretion from other parts of the body is packaged in lipoproteins known as **high-density lipoproteins (HDL).**

The density of lipoproteins varies according to the proportions of lipids and protein they contain. The more lipid in the lipoprotein molecule, the lower the density; the more protein, the higher the density. Both LDL and HDL carry lipids around in the blood, but LDL are larger, lighter, and filled with more lipid; HDL are smaller, denser, and packaged with more protein. LDL deliver cholesterol and triglycerides from the liver to the tissues; HDL scavenge excess cholesterol from the tissues and return it to the liver for metabolism or disposal. Figure 2-7 shows the relative sizes and composition of the lipoproteins.

Health Implications of LDL and HDL The distinction between LDL and HDL has implications for the health of the heart and blood vessels. Elevated LDL concentrations in the blood are associated with a high risk of heart disease, and elevated HDL concentrations are associated with a low risk.[1] These associations explain why some people refer to LDL as "bad" cholesterol and HDL as "good" cholesterol. Keep in mind, though, that there is only *one* kind of cholesterol molecule; the differences between LDL and HDL reflect *proportions* of lipids and proteins within them—not the type of cholesterol. Factors that improve the LDL-to-HDL ratio include:

- Weight control (see Chapter 7).
- Polyunsaturated or monounsaturated, instead of saturated, fatty acids in the diet (see Chapter 4).
- Soluble fibers (see Chapter 3).
- Physical activity.

Lipoproteins and heart disease are discussed in Chapter 21.

> **REVIEW NOTES**
>
> Nutrients leaving the digestive system via the blood are routed directly to the liver before being transported to the body's cells. Those leaving via the lymphatic system eventually enter the vascular system but bypass the liver at first.
>
> Within the circulatory system, lipids travel bundled with proteins as lipoproteins. Different types of lipoproteins include chylomicrons, very-low-density lipoproteins (VLDL), low-density lipoproteins (LDL), and high-density lipoproteins (HDL).
>
> Elevated blood concentrations of LDL are associated with a high risk of heart disease. Elevated HDL are associated with a low risk of heart disease.

The System at Its Best

The GI tract is the first organ in the body to deal with the nutrients that will ultimately maintain the health and nutrition status of the whole body. The intricate architecture of the GI tract makes it sensitive and responsive to conditions in its environment. One condition indispensable to its performance is its own good health. Such lifestyle factors as sleep, physical activity, state of mind, and nutrition affect GI tract health. Adequate sleep allows for repair and maintenance of tissue. Physical activity promotes healthy muscle tone and may protect against cancer of the colon.[2] Mental state profoundly affects digestion and absorption through the activity of nerves and hormones that help regulate these processes. A relaxed, peaceful attitude during a meal enhances digestion and absorption.

low-density lipoproteins (LDL): the type of lipoproteins derived from VLDL as cells remove triglycerides from them. LDL carry cholesterol and triglycerides from the liver to the cells of the body and are composed primarily of cholesterol.

high-density lipoproteins (HDL): the type of lipoproteins that transport cholesterol back to the liver from peripheral cells; composed primarily of protein.

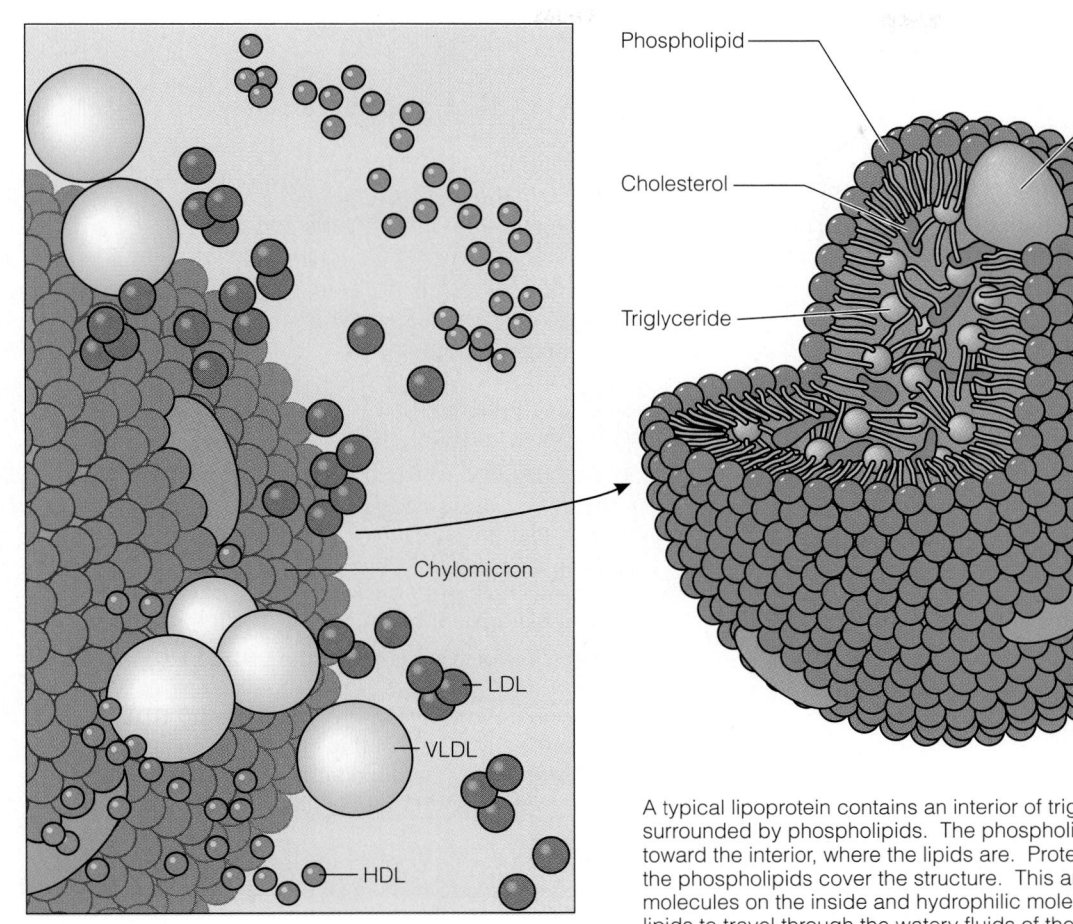

This solar system of lipoproteins shows their relative sizes. Notice how large the fat-filled chylomicron is compared with the others and how the others get progressively smaller as their proportion of fat declines and protein increases.

A typical lipoprotein contains an interior of triglycerides and cholesterol surrounded by phospholipids. The phospholipids' fatty acid "tails" point toward the interior, where the lipids are. Proteins near the outer ends of the phospholipids cover the structure. This arrangement of hydrophobic molecules on the inside and hydrophilic molecules on the outside allows lipids to travel through the watery fluids of the blood.

Chylomicrons contain so little protein and so much triglyceride that they are the lowest in density.

Very-low-density lipoproteins (VLDL) are half triglycerides, accounting for their low density.

Low-density lipoproteins (LDL) are half cholesterol, accounting for their implication in heart disease.

High-density lipoproteins (HDL) are half protein, accounting for their high density.

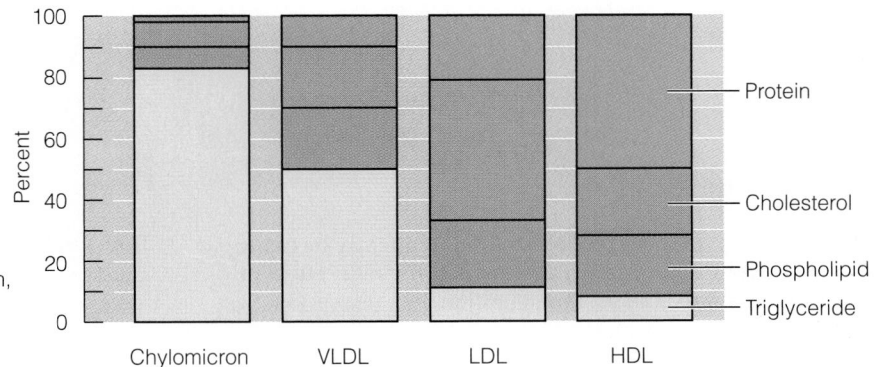

FIGURE 2-7 The Lipoproteins

SELF CHECK

1. Once food is swallowed, it travels through the digestive tract in this order:
   a. esophagus, stomach, larger intestine, liver.
   b. esophagus, stomach, small intestine, large intestine.
   c. small intestine, stomach, esophagus, large intestine.
   d. small intestine, large intestine, stomach, esophagus.

2. Once chyme travels the length of the small intestine, it passes through the ileocecal valve at the beginning of the:
   a. large intestine.
   b. stomach.
   c. esophagus.
   d. jejunum.

3. The periodic squeezing or partitioning of the intestine by its circular muscles that both mixes and slowly pushes the contents along is known as:
   a. secretion.
   b. absorption.
   c. peristalsis.
   d. segmentation.

4. An enzyme in saliva begins the digestion of:
   a. starch.
   b. vitamins.
   c. protein.
   d. minerals.

5. Bile is:
   a. an enzyme that splits starch.
   b. an alkaline secretion of the pancreas.
   c. an emulsifier made by the liver that prepares fats and oils for digestion.
   d. a stomach secretion containing water, hydrochloric acid, and the enzymes pepsin and lipase.

6. Which nutrient passes through the large intestine mostly unabsorbed?
   a. fiber
   b. vitamins

   c. minerals
   d. starch

7. The two major nutrient transport systems in the body are:
   a. LDL and HDL
   b. digestion and absorption.
   c. lipoproteins and chylomicrons.
   d. the vascular and lymphatic systems.

8. Within the circulatory system, lipids always travel from place to place bundled with proteins as:
   a. microvilli.
   b. chylomicrons.
   c. lipoproteins.
   d. phospholipids.

9. Elevated LDL concentrations in the blood are associated with:
   a. a high-protein diet.
   b. a low risk of diabetes.
   c. too much physical activity.
   d. a high risk of heart disease.

10. Three factors that improve the LDL-to-HDL ratio include:
    a. polyunsaturated fat, rest, and dietary HDL.
    b. antioxidants, insoluble fibers, and dietary HDL.
    c. saturated fat, antioxidants, and insoluble fibers.
    d. weight control, soluble fibers, and physical activity.

Answers to these questions can be found in Appendix H.

CLINICAL APPLICATIONS

1. People who experience malabsorption frequently have the most difficulty digesting fat. Considering the differences in fat, carbohydrate, and protein digestion and absorption, can you offer an explanation?

2. How might you explain the importance of dietary fiber to a client who frequently experiences constipation?

NUTRITION ON THE NET

For further study of the topics in this chapter, access these websites.

Find updates and quick links to these and other nutrition-related sites at our website: **www.wadsworth.com/nutrition**

Visit the Center for Digestive Health and Nutrition: **www.gihealth.com**

Visit the Digest This! section of the American College of Gastroenterology: **www.acg.gi.org**

NOTES

[1] Expert Panel on Detection, Evaluation, and Treatment of High Blood Cholesterol in Adults (Adult Treatment Panel III), *Third Report of the National Cholesterol Education Program (NCEP)*, NIH publication no. 02-5215 (Bethesda, Md.: National Heart, Lung, and Blood Institute, 2002) p. II-1–II-61.

[2] Y. Mao and coauthors, Physical inactivity, energy intake, obesity, and the risk of rectal cancer in Canada, *International Journal of Cancer* 105 (2003): 831–837; National Cancer Policy Board, Institute of Medicine, S. J. Curry, T. Byers, and M. Hewitt, eds., Fulfilling the Potential of Cancer Prevention and Early Detection (Washington, D.C.: National Academies Press, 2003), pp. 58–61.

Hunger and Community Nutrition

One person in every seven worldwide experiences persistent hunger—not the healthy appetite triggered by anticipation of a hearty meal, but the painful sensation caused by a lack of food. Tens of thousands die of starvation each day: one every two seconds.

In the United States, where most people enjoy a life of relative abundance, about one in every ten households has one or more members who experience pain from hunger caused by lack of food. In these households, 13 million children do not know where their next meal is coming from, or when it will come.[1] Given the enormous wealth and economic growth in this country, do these numbers surprise you? The limited or uncertain availability of nutritionally adequate and safe foods is known as **food insecurity** and is a major problem in our nation today.[2] The "How to" describes how national surveys identify food insecurity in the United States, and Figure NP2-1 presents the most recent findings; the glossary (p. 59) defines related terms. Surveys like these provide crude, but necessary, data to estimate the degree of hunger in this country.

Why is hunger a problem in developed countries such as the United States where food is abundant?

Hunger has many causes, but in developed countries, the primary cause is **food poverty.** People are hungry not because there is no food nearby to purchase, but because they lack sufficient money with which to buy nutritious food and pay for other necessities, such as housing, clothing, medicines, and utilities. More than 12 percent of the population of the United States lives in poverty.[3] Even those above the poverty line may not have **food security.** Physical and mental illnesses and disabilities, sudden job losses, and high living expenses threaten their financial stability. Further contributing to food poverty are other problems such as abuse of alcohol and other drugs; lack of awareness of available food assistance pro-

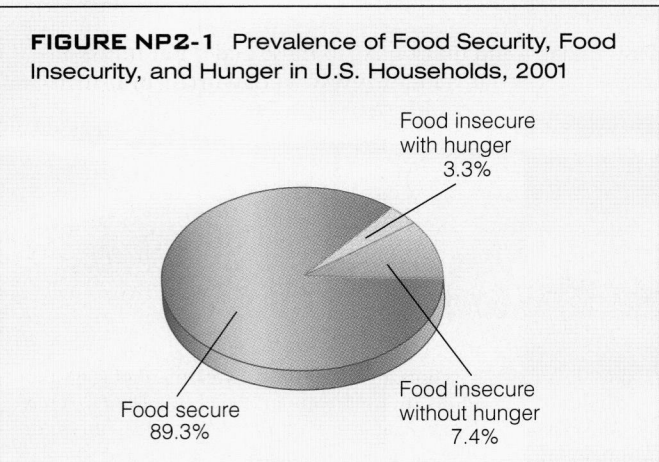

FIGURE NP2-1 Prevalence of Food Security, Food Insecurity, and Hunger in U.S. Households, 2001

Food insecure with hunger 3.3%

Food secure 89.3%

Food insecure without hunger 7.4%

Source: Economic Research Service, U.S. Department of Agriculture, www.ers.usda.gov/publications/fanrr29, posted October 2002 and visited on January 19, 2006.

grams; and the reluctance of people, particularly the elderly, to accept what they perceive as "welfare" or "charity." Lack of resources remains the major cause of food poverty, and solving this problem would do a lot to relieve hunger.

In the United States, food poverty and hunger reach into many segments of society, affecting not only the chronic poor (migrant workers, the unskilled and unemployed, the homeless, and some elderly) but also the so-called working poor. Some are displaced farm families. Some are former blue-collar and white-collar workers forced out of their trades and professions into minimum-wage jobs. These people outnumber the chronic poor, and they are not on welfare—they have jobs, but the pay is too low to meet their needs. Families with incomes below a certain level are simply unable to buy sufficient amounts of nourishing foods, even if they are skilled in food shopping. For many of the children in these families, school lunch is their only meal of the day. Other-

HOW TO · Identify Food Insecurity in a U.S. Household

Questions like these are asked on surveys to determine the extent of food insecurity in a household. The more questions answered "Yes," the more intense the hunger the household is experiencing.

- Do you often go hungry?
- Do you often have too little food to eat because you have no money, transportation, or kitchen appliances (stove, refrigerator)?
- Do you ever rely on nutritionally inferior foods to feed yourself or your children because you lack any of these resources?

- Do you ever eat less than you feel you should because you lack any of these resources?
- Do you ever skip meals or cut the size of meals because you lack any of these resources?
- Do you ever rely on neighbors, friends, relatives, or schools to feed any of your children because there is not enough food in the house?
- Do your children ever say they are hungry because there is not enough food in the house?
- Do you or any of your children ever go to bed hungry because there is not enough food in the house?

These people and many others like them in the United States face food insecurity daily.

food and nutrition security for all residents of the United States.[5] Many federal and local programs aim to prevent or relieve malnutrition and hunger in the United States.

An extensive network of federal assistance programs provides life-giving food daily to millions of U.S. citizens. One out of every six Americans receives food assistance of some kind, at a total cost of almost $40 billion per year. Even so, the programs are not fully successful in preventing hunger, but they do seem to improve the nutrient intakes of those who participate. Programs described in the life cycle chapters include the WIC program for low-income pregnant women, breastfeeding mothers, and their young children (Chapter 10); the school lunch and breakfast programs for children (Chapter 11); and the food assistance programs for older adults such as congregate meals and Meals on Wheels (Chapter 12).

The centerpiece of food programs for low-income people in the United States is the Food Stamp Program, administered by the U.S. Department of Agriculture (USDA). The USDA issues food stamp coupons or debit cards through state agencies to households—people who buy and prepare food together. The amount a household receives depends on its size and income. Recipients may use the coupons or cards like cash to purchase food and food-bearing plants and seeds, but not to buy tobacco, cleaning items, alcohol, or other nonfood items. The accompanying "How to" offers shopping tips for those on a limited budget.

wise they go hungry, waiting for an adult to find money for food. Not surprisingly, these children perform poorly in school and in social situations.[4]

What U.S. food programs are directed at relieving hunger in the United States?

The American Dietetic Association (ADA) calls for aggressive action to bring an end to domestic hunger and to achieve

HOW TO Plan Healthy, Thrifty Meals

Chapter 1 introduced the USDA Food Guide and principles for planning a healthy diet. Meeting that goal on a limited budget adds to the challenge. To save money and spend wisely, plan and shop for healthy meals with the following tips in mind:

Planning

- Make a grocery list before going to the store to avoid expensive "impulse" items. Do not shop when hungry.
- Use leftovers.
- Center meals on rice, noodles, and other grains.
- Use small quantities of meat, poultry, fish, or eggs.
- Use legumes instead of meat, poultry, fish, or eggs several times a week.
- Use cooked cereals such as oatmeal instead of ready-to-eat breakfast cereals.
- Cook large quantities when time and money allow.
- Check for sales and clip coupons for products you need; plan meals to take advantage of sale items.

Shopping

- Buy day-old bread and other products from the bakery outlet.
- Select whole foods instead of convenience foods (potatoes instead of instant mashed potatoes, for example).
- Try store brands.
- Buy fresh produce that is in season; buy canned or frozen items at other times.
- Buy only the amount of fresh foods that you will eat before it spoils. Buy large bags of frozen items or dry goods; when cooking, take out the amount needed and store the remainder.
- Buy fat-free dry milk; mix and refrigerate quantities needed for a day or two. Buy fresh milk by the gallon or half-gallon.
- Buy less expensive cuts of meat. Chuck and bottom round roast are usually inexpensive; cover during cooking and cook long enough to make meat tender. Buy whole chickens instead of pieces.
- Compare the unit price (cost per ounce, for example) of similar foods so that you can select the least expensive brand or size.
- Buy nonfood items such as toilet paper and laundry detergent at discount stores instead of grocery stores.

For daily menus and recipes for healthy, thrifty meals, visit the USDA Center for Nutrition Policy and Promotion: **www.usda.gov/cnpp**

The Food Stamp Program is the largest of the federal food assistance programs, both in amount of money spent and in number of people participating. Over 17 million people receive food stamps at a cost of over $20 billion per year; more than half of the recipients are children.[6]

Although food assistance programs improve nutrient intakes significantly, hunger continues to plague the United States. Of the estimated 2 million homeless people in the United States who are eligible for food assistance, only 15 percent of single adults and 50 percent of families receive food stamps.

Why do nurses need to know about food assistance programs?

Public health nurses are generally well acquainted with food assistance programs, and often many of their clients receive such assistance. Regardless of the setting in which you see clients, encourage those who may be having financial problems to talk with a social worker who can assess their eligibility for food assistance programs. Remember to approach the subject in a nonjudgmental and tactful manner—the client may feel uncomfortable about seeking assistance. Be proactive—don't assume that someone else is already addressing the problem.

Are there other programs aimed at reducing hunger in the United States?

Efforts to resolve the problem of hunger in the United States do not depend solely on federal assistance programs. National **food recovery** programs have made a dramatic difference; the largest program, America's Second Harvest, coordinates the efforts of more than 250 **food pantries, emergency kitchens,** and homeless shelters in providing more than 1 billion pounds of food to 45,000 local agencies that feed 26 million people a year. Table NP2-1 lists addresses, phone numbers, and websites for America's Second Harvest and other hunger relief organizations.

Each year, an estimated one-fifth of our food supply is wasted in fields, commercial kitchens, grocery stores, and restaurants—that's enough food to feed 49 million people. Food recovery programs collect and distribute good food that would otherwise go to waste. Volunteers might pick corn left in an already harvested field, a grocer might deliver ripe bananas to a local **food bank,** and a caterer might take left-

TABLE NP 2-1 Hunger Relief Organizations

Action without Borders 350 Fifth Ave., Suite 6614 New York, NY 10118 (212) 843-3973 **www.idealist.org**	Food Research and Action Center 1875 Connecticut Ave. Suite 540 Washington, D.C. 20009 **www.frac.org**	United Nations International Children's Emergency Fund (UNICEF) 3 United Nations Plaza New York, NY 10017-4414 (212) 326-7035 **www.unicef.org**
America's Second Harvest 35 E. Wacker Dr. #2000 Chicago, IL 60601 (800) 771-2303 **www.secondharvest.org**	OXFAM America 26 West St. Boston, MA 02111-1206 (800) 77-OXFAM or (800) 776-9326 **www.oxfam.america.org**	United Nations World Food Program Via Cesare Giulio Viola, 68 Parco dé Medici Rome, Italy 00148 **www.wfp.org**
Bread for the World 50 F St. NW, Suite 500 Washington, DC 20010 (800) 82-BREAD or (800) 822-7323 (202) 639-9400; fax (202) 639-9401 **www.bread.org**	Pan American Health Organization 525 23rd St. NW Washington, DC 20037 (202) 974-3000 **www.paho.org**	World Health Organization (WHO) 525 23rd St. NW Washington, DC 20037 (202) 861-3200 **www.who.org**
Children's Hunger Relief Fund 182 Farmer's Lane Suite 200 Santa Rosa, CA 95405 (888) 781-1585 **www.childrenshungerrelief.org**	Society of St. Andrew 3383 Sweet Hollow Rd. Big Island, VA 24526 (800) 333-4597 **www.endhunger.org**	World Hunger Year 505 Eighth Ave., 21st Floor New York, NY 10018-6582 (800) GleanIt **www.worldhungeryear.org**
Congressional Hunger Center 229 1/2 Pennsylvania Ave. Washington, DC 20003 (202) 547-7022 **www.hungercenter.org**	United Nations Food and Agriculture Organization (FAO) 1001 22nd St. NW, Suite 300 Washington, DC 20437 (202) 653-2400 **www.fao.org**	

Feeding the hungry in the United States.

Community-based efforts to feed citizens include food pantries that provide groceries.

over chicken salad to a community shelter, for example. All of these efforts help to feed the hungry in the United States.

What about local efforts and community nutrition programs?

Food recovery programs depend on volunteers. Concerned citizens work through local agencies and churches to feed the hungry. Community-based food pantries provide groceries, and soup kitchens serve prepared meals. Meals often deliver adequate nourishment, but most homeless people receive fewer than one and a half meals a day, so many are still inadequately nourished. Nurses and other health care professionals can serve as valuable members of community groups seeking to provide food assistance.

Glossary

emergency kitchens: programs that provide prepared meals to be eaten on site; often called *soup kitchens*.

food bank: a facility that collects and distributes food donations to authorized organizations feeding the hungry.

food insecurity: limited or uncertain access to foods of sufficient quality or quantity to sustain a healthy and active life.

food pantries: programs that provide groceries to be prepared and eaten at home.

food poverty: hunger resulting from inadequate access to available food for various reasons, including inadequate resources, political obstacles, social disruptions, poor weather conditions, and lack of transportation.

food recovery: collecting wholesome food for distribution to low-income people who are hungry. Four common methods of food recovery are:
  Field gleaning: collecting crops from fields that either have already been harvested or are not profitable to harvest.
  Perishable food rescue or salvage: collecting perishable produce from wholesalers and markets.
  Prepared food rescue: collecting prepared foods from commercial kitchens.
  Nonperishable food collection: collecting processed foods from wholesalers and markets.

food security: certain access to enough food for all people at all times to sustain a healthy and active life.

Notes

[1] Federal Interagency Forum on Child and Family Statistics, *America's Children: Key National Indicators of Well-Being*, 2003, available at **www.childstats.gov**.

[2] Position of the American Dietetic Association: Food insecurity and hunger in the United States, *Journal of the American Dietetic Association* 106 (2006): 446–458.

[3] U.S. Census Bureau, *America's Community Survey 2003*, available at **www.census.gov/acs**.

[4] K. Alaimo, C. M. Olson, and E. A. Frongillo, Jr., Food insufficiency and American school-aged children's cognitive, academic, and psychosocial development, *Pediatrics* 108 (2001): 44–53.

[5] Position of the American Dietetic Association, 2006.

[6] USDA Food and Nutrition Service website, **www.fns.usda.gov/fcs**.

Carbohydrates

CHAPTER 3

Most people would like to feel good all the time. Part of the secret of feeling well is replenishing the body's energy supply with food. That means choosing foods that contain the energy nutrients—carbohydrate and fat, primarily. But which to choose?

Carbohydrate is the preferred energy source for many of the body's functions. As long as carbohydrate is available, the human brain depends exclusively on it as an energy source. Athletes eat a "high-carb" diet to store as much muscle fuel as possible, and dietary recommendations urge people to eat carbohydrate-rich foods for better health. Many people, however, mistakenly think of carbohydrate-rich foods as "fattening" and avoid them. In truth, people who wish to lose fat and to maintain lean tissue and the health of the body can best do so by being physically active, paying close attention to portion sizes, and designing a diet based on foods that supply carbohydrate in balance with other energy nutrients.[1] Most unrefined plant foods—grains, vegetables, legumes, and fruits—provide ample carbohydrate and fiber with little or no fat. Milk is the only animal-derived food that contains significant amounts of carbohydrate.

Carbohydrate shares its fuel-providing responsibility with fat. Fat, however, normally is not used as fuel by the brain and central nervous system, and diets high in certain types of fat are associated with chronic diseases. The other energy sources available to the body—protein and alcohol—offer no advantage as fuels. Protein is best left to serve its own diverse functions, as discussed in Chapter 5. Alcohol, of course, has well-known undesirable side effects when used in excess. Alcohol and its relationships with health and disease are the subject of Nutrition in Practice 19.

The Chemist's View of Carbohydrates

The dietary **carbohydrates** include the **simple sugars** and starch and fiber. Chemists describe the simple sugars as:

- **Monosaccharides** (single sugars).
- **Disaccharides** (double sugars).

Starch and fiber are:*

- **Polysaccharides**—compounds composed of chains of monosaccharide units.

All of these carbohydrates are composed of the simple sugar **glucose** and other compounds that are much like glucose in composition and structure. Figure 3-1 shows the chemical structure of glucose.

MONOSACCHARIDES

Three monosaccharides are important in nutrition: glucose, fructose, and galactose. All three monosaccharides have the same number and kinds of atoms but in different arrangements.

Glucose Most cells depend on glucose for their fuel to some extent, and the cells of the brain and the rest of the nervous system depend almost exclusively on glucose for their energy. The body can obtain this glucose from carbohydrates. To function optimally, the body must maintain blood glucose within limits that allow the cells to nourish themselves. If blood glucose falls below normal, the person may become dizzy and weak; if it rises substantially above normal, the person may become fatigued. Left untreated, fluctuations to the extremes—either high or low—can be fatal. Blood glucose **homeostasis** is regulated primarily by two

*Monosaccharides and disaccharides (sugars) are sometimes called *simple carbohydrates*, and the polysaccharides (starch and fiber) are sometimes called *complex carbohydrates*.

Grains, vegetables, legumes, fruits, and milk offer ample carbohydrate.

Chapter 4 describes the roles of fats in health and disease.

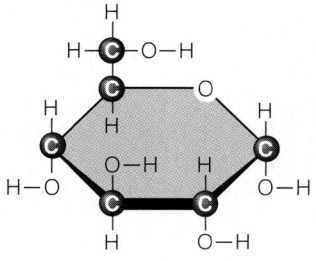

FIGURE 3-1 Chemical Structure of Glucose
On paper, the structure of glucose has to be drawn flat, but in nature the five carbons and oxygen are roughly in a plane, with the H, OH, and CH_2OH extending out above and below it.

carbohydrates: energy nutrients, composed of monosaccharides.
  carbo = carbon   *hydrate* = water

simple sugars: the monosaccharides (glucose, fructose, and galactose) and the disaccharides (sucrose, lactose, and maltose).

monosaccharides (mon-oh-SACK-uh-rides): single sugar units.
  mono = one   *saccharide* = sugar

disaccharides (dye-SACK-uh-rides): pairs of sugar units bonded together.
  di = two

polysaccharides: long chains of monosaccharide units arranged as starch, glycogen, or fiber.
  poly = many

glucose: a monosaccharide, the sugar common to all disaccharides and polysaccharides; also called *blood sugar* or *dextrose*.

homeostasis (HOME-ee-oh-STAY-sis): the maintenance of constant internal conditions (such as chemistry, temperature, and blood pressure) by the body's control system.
  homeo = the same   *stasis* = staying

Diabetes, a disorder characterized by elevated blood glucose, is the topic of Chapter 20.

hormones: **insulin**, which moves glucose from the blood into the cells, and **glucagon**, which brings glucose out of storage when blood glucose falls (as occurs between meals).

Fructose Fructose is the sweetest of the sugars. Fructose occurs naturally in fruits, honey, and saps. Other sources include soft drinks, ready-to-eat cereals, and other products sweetened with **high-fructose corn syrup**. Glucose and fructose are the most common monosaccharides in nature.

Galactose The third single sugar, **galactose**, occurs mostly as part of lactose, a disaccharide also known as milk sugar. During digestion galactose is freed as a single sugar.

DISACCHARIDES

In disaccharides, pairs of single sugars are linked together. Three disaccharides are important in nutrition: maltose, sucrose, and lactose. All three have glucose as one of their single sugars. As Table 3-1 shows, the other monosaccharide is either another glucose (in maltose), or fructose (in sucrose), or galactose (in lactose). The shapes of the sugars in Table 3-1 reflect their chemical structures as drawn on paper.

Sucrose Sucrose (table, or white, sugar) is the most familiar of the three disaccharides and is what people mean when they speak of "sugar." This sugar is usually obtained by refining the juice from sugar beets or sugarcane to provide the brown, white, and powdered sugars available in the supermarket, but it occurs naturally in many fruits and vegetables.

When a person eats a food containing sucrose, enzymes in the digestive tract split the sucrose into its glucose and fructose components. Because the body can convert fructose to glucose, one molecule of sucrose can ultimately yield two molecules of glucose.

Lactose Lactose is the principal carbohydrate of milk. Most human infants are born with the digestive enzymes necessary to split lactose into its two monosaccharide parts, glucose and galactose, so as to absorb it. Breast milk thus provides a simple, easily digested carbohydrate that meets an infant's energy needs; many formulas do, too, because they are made from milk.

Maltose The third disaccharide, **maltose**, is a plant sugar that consists of two glucose units. Maltose is produced whenever starch breaks down—as happens in plants when they break down their stored starch for energy and start to sprout and in human beings during carbohydrate digestion.

insulin: a hormone secreted by the pancreas in response to high blood glucose. It promotes cellular glucose uptake for use or storage.

glucagon (GLOO-ka-gon): a hormone that is secreted by special cells in the pancreas in response to low blood glucose concentration and elicits release of glucose from storage.

fructose: a monosaccharide; sometimes known as *fruit sugar*. It is abundant in fruits, honey, and saps.
  fruct = fruit

high-fructose corn syrup (HFCS): the predominant sweetener used in processed foods today. HFCS is mostly fructose; glucose makes up the balance.

galactose: a monosaccharide; part of the disaccharide lactose.

sucrose: a disaccharide composed of glucose and fructose; commonly known as *table sugar, beet sugar*, or *cane sugar*.
  sucro = sugar

lactose: a disaccharide composed of glucose and galactose; commonly known as *milk sugar*.
  lact = milk

maltose: a disaccharide composed of two glucose units; sometimes known as *malt sugar*.

TABLE 3-1	The Major Sugars	
Monosaccharides		**Disaccharides**
Glucose		Sucrose (glucose + fructose)
Fructose		Lactose (glucose + galactose)
Galactose		Maltose (glucose + glucose)
(found only as part of lactose)		

POLYSACCHARIDES

Unlike the sugars, which contain the three monosaccharides—glucose, fructose, and galactose—in different combinations, the polysaccharides are composed almost entirely of glucose (and, in some cases, other monosaccharides). Three types of polysaccharides are important in nutrition: starch, glycogen, and fibers.

Glycogen is a storage form of energy for human beings and animals; starch plays that role in plants; and fibers provide structure in stems, trunks, roots, leaves, and skins of plants. Both glycogen and starch are built entirely of glucose units; fibers are composed of a variety of monosaccharides and other carbohydrate derivatives.

Starch Starch is a long, straight or branched chain of hundreds of glucose units linked together. These giant molecules are packed side by side in a rice grain or potato root—as many as a million per cubic inch of food. When a person eats the plant, the body splits the starch into glucose units and uses the glucose for energy.

All starchy foods are plant foods. Grains are the richest food source of starch. In most human societies, people depend on a staple grain for much of their food energy: rice in Asia; wheat in Canada, the United States, and Europe; corn in much of Central and South America; and millet, rye, barley, and oats elsewhere. A second important source of starch is the legume (bean and pea) family. Legumes include peanuts and "dry" beans such as butter beans, kidney beans, "baked" beans, black-eyed peas (cowpeas), chickpeas (garbanzo beans), and soybeans. Root vegetables (tubers) such as potatoes and yams are a third major source of starch, and in many non-Western societies, they are the primary starch sources. Grains, legumes, and tubers not only are rich in starch but may also contain abundant dietary fiber, protein, and other nutrients.

Glycogen Glycogen molecules, which are also made of chains of glucose, are more highly branched than starch molecules. Glycogen is found in meats only to a limited extent, and not at all in plants.* For this reason, glycogen is not a significant food source of carbohydrate, but it does play an important role in the body. The human body stores much of its glucose as glycogen in the liver and muscles.

After a meal, as blood glucose rises, the pancreas is the first organ to respond. It releases the hormone insulin, which signals the body's tissues to take up surplus glucose. Muscle and liver cells use some of this excess glucose to build glycogen. The muscles hoard two-thirds of the body's total glycogen and use it just for themselves during physical activity. The brain stores a tiny fraction of the total, thought to provide an emergency glucose reserve sufficient to fuel the brain for an hour or two in severe glucose deprivation.[2] The liver stores the remainder and is generous with its glycogen, making it available as blood glucose for the brain or other tissues when the supply runs low.

Fibers Dietary fibers are the structural parts of plants and thus are found in all plant-derived foods—vegetables, fruits, whole grains, and legumes.† Most dietary fibers are polysaccharides—chains of sugars—just as starch is, but in fibers the sugar units are held together by bonds that human digestive enzymes cannot break. Consequently, most dietary fibers pass through the body providing little or no energy for its use. Figure 3-2 (p. 64) shows the difference between starch and the plant fiber cellulose. In addition to cellulose, fibers include the polysaccharides hemicellulose, pectins, gums, and mucilages, as well as the nonpolysaccharide lignins.

Cellulose is the main constituent of plant cell walls, so it is found in all vegetables, fruits, and legumes. Hemicellulose is the main constituent of cereal fibers.

---

*Glycogen in animal muscles rapidly breaks down after slaughter.
†The DRI committee has proposed these fiber definitions: the term *dietary fibers* refers to naturally occurring fibers in intact foods, and *functional fibers* refers to added fibers that have health benefits; *total fiber* refers to the sum of fibers from both sources.

The short chains of glucose units that result from the breakdown of starch are known as *dextrins*. The word sometimes appears on food labels because dextrins can be used as thickening agents in foods.

starch: a plant polysaccharide composed of glucose and digestible by human beings.

glycogen (GLY-co-gen): a polysaccharide composed of glucose, made and stored by liver and muscle tissues of human beings and animals as a storage form of glucose. Glycogen is not a significant food source of carbohydrate and is not counted as one of the polysaccharides in foods.

dietary fibers: a general term denoting in plant foods the polysaccharides cellulose, hemicellulose, pectins, gums, and mucilages, as well as the nonpolysaccharide lignins, that are not digested by human digestive enzymes, although some are digested by GI tract bacteria.

FIGURE 3-2 Starch and Cellulose Molecules Compared (Small Segments)

The bonds that link the glucose units together in cellulose are different from the bonds in starch (and glycogen). Human enzymes cannot digest cellulose.

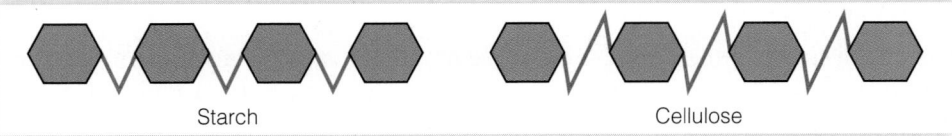

Starch   Cellulose

Starch- and fiber-rich foods are the foods to emphasize.

Soluble, viscous, fermentable fibers are often gummy or add thickness to foods.

Insoluble, nonviscous, less fermentable fibers are often tough, stringy, or gritty in foods.

resistant starches: starches that escape digestion and absorption in the small intestine of healthy people.

soluble fibers: indigestible food components that readily dissolve in water and often impart gummy or gel-like characteristics to foods. An example is pectin from fruit, which is used to thicken jellies.

viscous: a gel-like consistency.

insoluble fibers: the tough, fibrous structures of fruits, vegetables, and grains; indigestible food components that do not dissolve in water.

added sugars: sugars and syrups added to a food for any purpose, such as to add sweetness or bulk or to aid in browning (baked foods). Also called *carbohydrate sweeteners,* they include glucose, fructose, corn syrup, concentrated fruit juice, and other sweet carbohydrates.

Pectins are abundant in vegetables and fruits, especially citrus fruits and apples. The food industry uses pectins to thicken jelly and keep salad dressing from separating. Gums and mucilages have similar structures and are used as additives or stabilizers by the food industry. Lignins are the tough, woody parts of plants; few foods people eat contain much lignin.

A few starches are classified as fibers. Known as **resistant starches,** these starches escape digestion and absorption in the small intestine. Starch may resist digestion for several reasons, including the individual's efficiency in digesting starches and the food's physical properties. Resistant starch is common in whole legumes, raw potatoes, and unripe bananas.

Although cellulose and other dietary fibers are not broken down by human enzymes, some fibers can be digested by bacteria in the human digestive tract. Bacterial digestion of fibers can generate some absorbable products that can yield energy when metabolized. Food fibers, therefore, can contribute some energy (1.5 to 2.5 kcalories per gram), depending on the extent to which they break down in the body.

Fibers are divided into two general groups by their chemical and physical properties. In the first group are fibers that dissolve in water (**soluble fibers**). These form gels (are **viscous**) and are easily digested by bacteria in the human large intestine (are easily fermented). Commonly found in barley, legumes, fruits, oats, and vegetables, these fibers are often associated with lower risks of chronic diseases (as discussed in a later section). In foods, soluble fibers add pleasing consistency, such as the pectin that puts the gel in jelly and the gums that are added to salad dressings and other foods to thicken them.

Other fibers do not dissolve in water (**insoluble fibers**), do not form gels (are not viscous), and are less readily fermented. Insoluble fibers, such as cellulose and many hemicelluloses, are found in the outer layers of whole grains (bran), the strings of celery, the hulls of seeds, and the skins of corn kernels. These fibers retain their structure and rough texture even after hours of cooking. In the body, they aid the digestive system by easing elimination.[3]

REVIEW NOTES

Carbohydrate is the body's preferred energy source. Six simple sugars are important in nutrition: the three monosaccharides (glucose, fructose, and galactose) and the three disaccharides (sucrose, lactose, and maltose).

The three disaccharides are pairs of monosaccharides; each contains glucose paired with one of the three monosaccharides. The polysaccharides (chains of monosaccharides) are glycogen, starches, and dietary fibers.

Both glycogen and starch are storage forms of glucose—glycogen in the body and starch in plants—and both yield energy for human use.

The dietary fibers also contain glucose (and other monosaccharides), but their bonds cannot be broken by human digestive enzymes, so they yield little, if any, energy.

Health Effects of Sugars and Alternative Sweeteners

Fiber-rich carbohydrate foods such as vegetables, whole grains, legumes, and fruits should predominate in people's diets; concentrated sweets such as candy, cola beverages and other soft drinks, cookies, pies, cakes, and other foods with **added sugars**

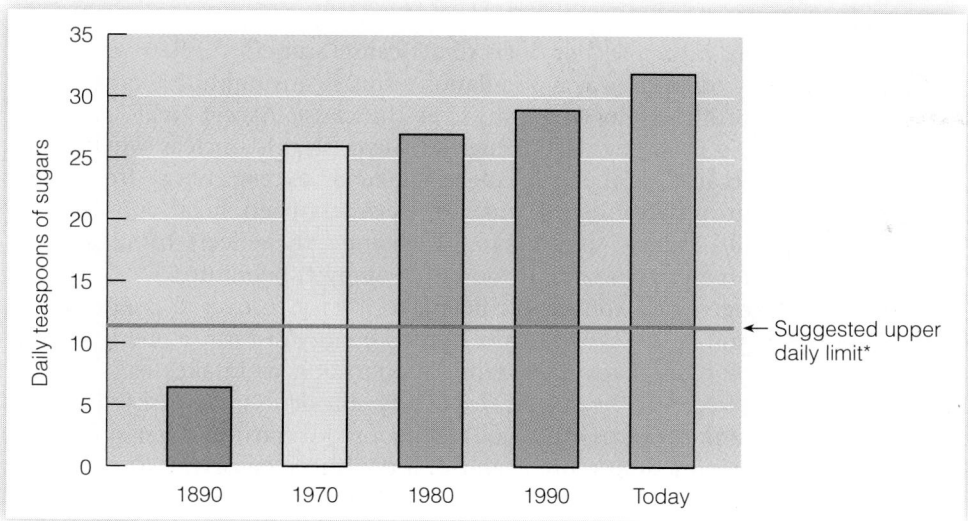

FIGURE 3-3 Added Sugars: Average Supply Per Person in the United States, 1890–Present

*The World Health Organization recommends an upper limit of about 12 teaspoons in an adequate 2200-kcalorie diet.

Source: Data from J. Putnam and S. Haley, Estimating consumption of caloric sweeteners, *Amber Waves*, April 2003, available at **www.ers.usda.gov/Amberwaves/April03/Indicators/Behind.Data.htm**.

add kcalories, but few, if any, other nutrients. These foods therefore contribute discretionary kcalories to the diet.

The **naturally occurring sugars** in vegetables, fruits, grains, and milk are acceptable because they are accompanied by many nutrients. People who want to limit their use of sugar may choose from two sets of alternative sweeteners: sugar alcohols and artificial sweeteners.

SUGARS

Estimates are that each man, woman, and child in the United States consumes over 100 pounds of sugar per year, or a little less than 2 pounds per week. The steady upward trend in consumption of added sugars shown in Figure 3-3 is largely the result of a dramatic increase in consumption of commercially prepared foods and beverages that contain added sugars. Food manufacturers are adding the sugars to foods during processing. By one account, consumption of regular soft drinks and sugar-sweetened fruit drinks is responsible for 80 percent of this increase.[4] Desserts and jams and jellies make up most of the remainder. In contrast, people are adding less sugar in the kitchen. The committee that drafted the *Dietary Guidelines for Americans 2005* offers clear advice on added sugars: treat them as discretionary kcalories.[5] In other words, fill the day's nutrient needs with foods contributing little or no added sugar. Most people can afford only a little added sugar in their diets if they are to meet nutrient needs within kcalorie limits.

The U.S. population is not alone in increasing sugar consumption. The trend is a worldwide phenomenon. In response, the World Health Organization has also taken a stand on sugar intake: consume no more than 10 percent of total kcalories from added sugars.*[6]

The increase in sugar consumption has raised many questions about sugar's effects on health. Many accusations have been made against sugar, but the Food and Drug Administration (FDA) and the National Academy of Sciences have concluded that in the amounts people currently consume, sugar poses no major health risk. Still, some controversy continues. A brief description of accusations pertaining to sugar's effects on health may help clarify the issues. Most commonly, reports accuse sugar of causing obesity, hyperactivity and aggressive behavior in children, and **dental caries.**

*The World Health Organization uses the term *free sugars* to mean all monosaccharides and disaccharides added to foods by the manufacturer, cook, or consumer plus the naturally occurring sugars in honey, syrups, and fruit juices.

The *USDA Food Guide* distinguishes between naturally occurring and added sugars, designating the kcalories from added sugars as discretionary kcalories.

Sugary soft drinks are the leading source of added sugars in the United States; cakes, cookies, pies, and other baked goods come next; and sweetened fruit drinks and punches follow closely behind.

Professionally, hyperactivity is called attention-deficit/hyperactivity disorder (see Chapter 11).

naturally occurring sugars: sugars that are not added to a food but are present as its original constituents, such as the sugars of fruit or milk.

dental caries: the gradual decay and disintegration of a tooth.

Sugar and Obesity On the first accusation—that sugar causes obesity—the evidence shows a supportive role, but not a direct cause-and-effect relationship. The incidence of obesity often rises as a population's sugar consumption increases, but this evidence is insufficient to name sugar as the cause. Sweet treats are often high in kcalories and in fat, as well as high in sugar, so it is unclear whether the obesity develops because of the high sugar intake or excess energy from other sources. In addition, physical activity usually declines when sugar consumption increases. Population studies simply cannot separate the effects of sugar from those of too great an energy intake or too little energy expenditure.

Concentrated sweets and soft drinks do make it easy to exceed energy needs quickly, however, and an excessive intake of food energy can cause obesity. In the United States, obesity has reached epidemic proportions, as intakes of both food energy and sugar-sweetened beverages, especially soft drinks, have skyrocketed.[7] Adolescents who drink upwards of 26 ounces (about two cans) of sugar-sweetened soft drinks daily consume 400 more kcalories a day than teens who abstain from such drinks. Soft drinks have been the focus of attention because the increase in intake of added sugars is accounted for largely by a dramatic rise in consumption of high-fructose corn syrup, mostly in soft drinks. Between 1977 and 2001, as people grew fatter, their intake of kcalories from soft drinks nearly tripled and that from fruit drinks and punches doubled.[8] Thus, to the extent that sugar contributes to an excessive energy intake, it can play a role in the development of obesity.

> Chapter 11 offers further discussion of children's nutrition.

Sugar and Behavior The accusation that sugar causes hyperactive or aggressive behavior in children has not been proved. Scientific research has failed to demonstrate any consistent effect of sugar on behavior in either normal or hyperactive children.[9] If sugar is related to behavior problems in children, it may be because the sugary foods replace nutrient-dense foods in children's diets, making nutrient deficiencies likely. Many different nutrient deficiencies adversely affect behavior. A lack of nutrients in children's diets, not sugar itself, can in some cases contribute to undesirable behavior.

> *Dietary Guidelines for Americans 2005*: Reduce the incidence of dental caries by practicing good oral hygiene and consuming sugar- and starch-containing foods and beverages less frequently.

Sugar and Dental Caries As to whether sugar contributes to dental caries, the evidence says yes. Any carbohydrate-containing food, including bread, bananas, or milk, as well as sugar, can support bacterial growth in the mouth.[10] These bacteria produce the acid that eats away tooth enamel. Of major importance is the length of time the food stays in the mouth. This, in turn, depends on the composition of the food, how sticky the food is, how often a person eats the food, and especially whether the teeth are brushed afterward. Total sugar intake still plays a major role in caries incidence; populations whose diets provide no more than 10 percent of kcalories from sugar have a low prevalence of dental caries.[11] Nutrition in Practice 3 discusses nutrition and dental health.

> The USDA Food Guide suggests:*
> - 3 tsp for 1600 kcal.
> - 5 tsp for 1800 kcal.
> - 8 tsp for 2000 kcal.
> - 9 tsp for 2200 kcal.
> - 12 tsp for 2400 kcal.

Recommended Sugar Intakes Moderate sugar intakes—enough for pleasure, but not enough to displace more nutritious foods—are not harmful. Sugar is a delicious, concentrated source of food energy, but it contains no protein, vitamins, or minerals. Eaten in the place of nutrient- and fiber-rich foods, it makes malnutrition likely.

The *Dietary Guidelines for Americans 2005* urge people to "choose and prepare foods and beverages with little added sugars." The USDA Food Guide suggests that, within kcalorie limits, small amounts of added sugars can be enjoyed as part of the discretionary kcalorie allowance in a nutrient-dense diet that provides no more than 30 percent of kcalories from fat. The USDA Food Guide recommendations represent about 5 to 10 percent of the day's total energy intake. As noted earlier, the

*The amounts of added sugars suggested here are *not* specific recommendations for amounts of added sugars to consume, but rather represent the amounts that can be included in the diet at each kcalorie level.

World Health Organization agrees that people should restrict their consumption of added sugars to 10 percent or less of total energy.

The DRI committee did not set a Tolerable Upper Intake Level for added sugars, but as mentioned, excessive intakes can interfere with sound nutrition and dental health. Few people can eat lots of sugary treats and still meet all of their nutrient needs without exceeding their kcalorie allowance. Instead, the DRI committee suggests, as a rather high maximum, that added sugars should account for no more than 25 percent of the day's total energy intake.[12] For a person consuming 2000 kcalories a day, 25 percent represents 500 kcalories from added sugars—quite a lot of sugar. Perhaps an athlete in training whose energy needs are high can afford the added sugars from sports drinks without compromising nutrient intake, but most people would do better following recommendations to limit added sugar consumption to less than 10 percent of the day's total energy intake.

Recognizing Sugars People often fail to recognize sugar in all its forms and so do not realize how much they consume. To help your clients estimate how much sugar they consume, tell them to treat all of the following concentrated sweets as equivalent to 1 teaspoon of white sugar (4 grams of carbohydrate):

- 1 teaspoon brown sugar, candy, jam, jelly, any corn sweetener, syrup, honey, molasses, or maple sugar.
- 1 tablespoon catsup.
- 1 1/2 ounces carbonated soft drink.

These portions of sugar all provide about the same number of kcalories. Some are closer to 10 kcalories (for example, 14 kcalories for molasses), while some are over 20 (22 kcalories for honey), so an average figure of 16 kcalories is an acceptable approximation. The accompanying glossary presents the multitude of names that denote sugar on food labels.

For perspective, each of these sources of concentrated sugars provides about 500 kcal:

- 40 oz cola.
- 1/2 c honey.
- 125 jelly beans.
- 23 marshmallows.
- 30 tsp sugar.

GLOSSARY of Sugars

brown sugar: white sugar with molasses added, 95% pure sucrose.

concentrated fruit juice sweetener: a concentrated sugar syrup made from dehydrated, deflavored fruit juice, commonly grape juice; used to sweeten products that can then claim to be "all fruit."

confectioner's sugar: finely powdered sucrose, 99.9% pure.

corn sweeteners: corn syrup and sugar solutions derived from corn.

corn syrup: a syrup, mostly glucose, partly maltose, produced by the action of enzymes on cornstarch.

dextrose: an older name for glucose.

evaporated cane juice: raw sugar from which impurities have been removed.

fructose, galactose, glucose: the monosaccharides.

granulated sugar: common table sugar, crystalline sucrose, 99.9% pure.

high-fructose corn syrup (HFCS): a mixture of primarily fructose and glucose; the predominant sweetener used in processed foods today.

honey: a concentrated solution primarily composed of glucose and fructose, produced by enzymatic digestion of the sucrose in nectar by bees.

invert sugar: a mixture of glucose and fructose formed by the splitting of sucrose in an industrial process. Sold only in liquid form and sweeter than sucrose, invert sugar forms during certain cooking procedures and works to prevent crystallization of sucrose in soft candies and sweets.

lactose, maltose, sucrose: the disaccharides.

levulose: an older name for fructose.

maple sugar: a concentrated solution of sucrose derived from the sap of the sugar maple tree, mostly sucrose. This sugar was once common but is now usually replaced by sucrose and artificial maple flavoring.

molasses: a syrup left over from the refining of sucrose from sugarcane; a thick, brown syrup. The major nutrient in molasses is iron, a contaminant from the machinery used in processing it.

raw sugar: the first crop of crystals harvested during sugar processing. Raw sugar cannot be sold in the United States because it contains too much filth (dirt, insect fragments, and the like). Sugar sold as "raw sugar" is actually evaporated cane juice.

turbinado: (ter-bih-NOD-oh) **sugar:** raw sugar from which the filth has been washed; legal to sell in the United States.

white sugar: pure sucrose, produced by dissolving, concentrating, and recrystallizing raw sugar.

You receive about the same amount and kinds of sugars from an orange as from a tablespoon of honey, but the packaging makes a big nutrition difference.

A 12 oz can of cola contains the equivalent of 8 tsp sugar.

People often ask: What is the difference between honey and white sugar? Is honey, by virtue of being natural, more nutritious? Honey, like white sugar, contains glucose and fructose. The difference is that in white sugar, the glucose and fructose are bonded together in pairs, whereas in honey some of them are paired and some are free single sugars. When you eat either white sugar or honey, though, your body breaks all of the sugars apart into single sugars. It ultimately makes no difference, then, whether you eat single sugars linked together, as in white sugar, or the same sugars unlinked, as in honey; they will end up as single sugars in your body. True, honey contains trace amounts of a few vitamins and minerals, but to say that honey is nutritious is misleading.

Some sugar sources are more nutritious than others, though. Consider a fruit such as an orange. The orange provides the same sugars and about the same energy as a tablespoon of sugar or honey, but the packaging makes a big difference in nutrient density. The sugars of the orange are diluted in a large volume of fluid that contains valuable vitamins and minerals, and the flesh and skin of the orange are supported by fibers that also offer health benefits. A tablespoon of honey offers no such bonuses.

Of course, a cola beverage, containing many teaspoons of sugar, offers no advantages either. Table 3-2 shows sample nutrients supplied by some sugar sources and other foods; note the "0s" and "traces" by honey, sugar, and the cola beverage and the substantial numbers by the others.

TABLE 3-2	Sample Nutrients in Sugars and Other Foods

The indicated portion of any of these foods provides approximately 100 kcalories. Notice that for a similar number of kcalories and grams of carbohydrate, milk, legumes, fruits, grains and vegetables offer more of the other nutrients than do the sugars.

	Size of 100 kcal Portion	Carbohydrate (g)	Protein (g)	Calcium (mg)	Iron (mg)	Vitamin A (μg)	Vitamin C (mg)
Foods							
Milk, 1% low-fat	1 c	12	8	300	0.1	144	2
Kidney beans	½ c	20	7	30	1.6	0	2
Apricots	6	24	2	30	1.1	554	22
Bread, whole wheat	1½ slices	20	4	30	1.9	0	0
Broccoli, cooked	2 c	20	12	188	2.2	696	148
Sugars							
Sugar, white	2 tbs	24	0	trace	trace	0	0
Molasses, blackstrap	2½ tbs	28	0	343	12.6	0	0.1
Cola beverage	1 c	26	0	6	trace	0	0
Honey	1½ tbs	26	trace	2	0.2	0	trace

TABLE 3-3	Sugar Alcohols		
Sugar Alcohols	Relative Sweetness*	Energy (kcal/g)	Approved Uses
Isomalt	0.5	2.0	Candies, chewing gum, ice cream, jams and jellies, frostings, beverages, baked goods
Lactitol	0.4	2.0	Candies, chewing gum, frozen dairy desserts, jams and jellies, frostings, baked goods
Maltitol	0.9	2.1	Particularly good for candy coating
Mannitol	0.7	1.6	Bulking agent, chewing gum
Sorbitol	0.5	2.6	Special dietary foods, candies, gums
Xylitol	1.0	2.4	Chewing gum, candies, pharmaceutical and oral health products

*The relative sweetness depends on the temperature, acidity, and other flavors of the foods in which the substance occurs. The sweetness of pure sucrose is the standard with which the approximate sweetness of sugar substitutes is compared.

REVIEW NOTES

Sugars pose no major health threat except for an increased risk of dental caries.

Excessive sugar intakes may displace needed nutrients and fiber and may contribute to obesity. A person deciding to limit daily sugar intake should recognize that not all sugars need to be restricted, just concentrated sweets with added sugars, which are high in kcalories and relatively lacking in other nutrients. Sugars that occur naturally in fruits, vegetables, and milk are acceptable.

ALTERNATIVE SWEETENERS: SUGAR ALCOHOLS

The **sugar alcohols** are carbohydrates, but they yield slightly less energy (2 to 3 kcalories per gram) than sucrose (4 kcalories per gram) because they are not absorbed completely.[13] The sugar alcohols are sometimes called **nutritive sweeteners** because they do yield some energy.

The sugar alcohols occur naturally in fruits; they are also used by manufacturers to provide sweetness and bulk to cookies, sugarless gum, hard candies, and jams and jellies. Their sweetness relative to sugar is shown in Table 3-3. Unlike sucrose, sugar alcohols are fermented in the large intestine by intestinal bacteria. Consequently, side effects such as gas, abdominal discomfort, and diarrhea make the sugar alcohols less attractive than the artificial sweeteners.

The advantage of using sugar alcohols is that they do not contribute to dental caries. Bacteria in the mouth metabolize sugar alcohols much more slowly than sucrose, thereby inhibiting the production of acids that promote caries formation. They are therefore valuable in chewing gums, breath mints, and other products that people keep in their mouths a while. The FDA allows food labels to carry a health claim (see p. 29 in Chapter 1) about

sugar alcohols: sugarlike compounds. Like sugars, they are sweet to taste but yield 2 to 3 kcal per gram, slightly less than sucrose. Examples are maltitol, mannitol, sorbitol, isomalt, lactitol, and xylitol.

nutritive sweeteners: sweeteners that yield energy, including both the sugars and the sugar alcohols.

These chewing gums contain sugar alcohols, which are better than sugar for the teeth, but are not kcalorie-free.

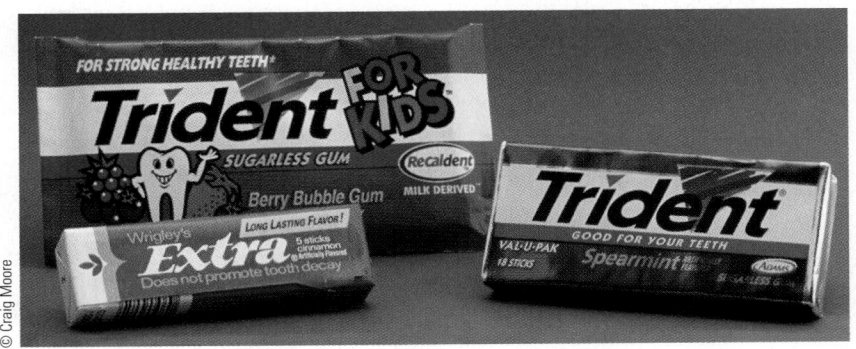

© Craig Moore

the relationship between sugar alcohols and the nonpromotion of dental caries as long as the FDA criteria for sugar-free status and other criteria are met.

ALTERNATIVE SWEETENERS: ARTIFICIAL SWEETENERS

The **artificial sweeteners** are not carbohydrates. They yield virtually no energy in the amounts typically used and are sometimes called nonnutritive sweeteners. Like the sugar alcohols, artificial sweeteners make foods taste sweet without promoting tooth decay. The accompanying glossary and Table 3-4 offer details on artificial sweeteners of interest.

Saccharin Saccharin has been used for more than 100 years in the United States and is currently used by millions of people, mainly in soft drinks and as the tabletop sweetener Sweet'N Low or SugarTwin. Questions about the safety of saccharin arose in 1977 when experiments suggested that it caused bladder tumors in second-generation rats fed high doses. The FDA proposed banning saccharin as a result. Public outcry in favor of saccharin was so loud that Congress declared a moratorium on the ban, a moratorium that has been repeatedly extended. In 1991, the FDA withdrew its proposal to ban saccharin, but labels still had to carry a consumer warning about saccharin as a cancer hazard. Saccharin remained on the government's roster of "anticipated carcinogens" until the year 2000. At that time, government officials reviewed the research and reversed their opinion, removing saccharin from the carcinogen list and freeing it from the labeling requirement.[14] Opponents to these changes, however, maintain that saccharin causes cancer in mice and rats and so should be avoided.

Does saccharin cause cancer? The largest population study to date, involving 9000 men and women, showed overall that saccharin use did not raise the risk of cancer. Among certain small groups of the population, however, such as those who both smoked heavily and used saccharin, the risk of bladder cancer was slightly greater. Other studies involving more than 5000 people with bladder cancer showed no association between bladder cancer and saccharin use. Common sense dictates that consuming large amounts of saccharin is probably not safe, but at moderate intake levels, saccharin is currently assumed to be safe for most people. The FDA has set **Acceptable Daily Intake (ADI)** levels for the artificial sweeteners used in the United States.

artificial sweeteners: noncarbohydrate, nonkcaloric synthetic sweetening agents; sometimes called **nonnutritive sweeteners**.

Acceptable Daily Intake (ADI): the amount of an artificial sweetener that individuals can safely consume each day over the course of a lifetime without adverse effect. It includes a 100-fold safety factor.

GLOSSARY of Artificial Sweeteners

acesulfame (AY-sul-fame) **potassium,** also called **acesulfame-K:** a zero-kcalorie sweetener approved by the FDA and Health Canada.

alitame: a nonkcaloric sweetener formed from the amino acids L-aspartic acid and L-alanine. In the United States, the FDA is considering its approval.

aspartame: a compound of phenylalanine and aspartic acid that tastes like the sugar sucrose but is much sweeter. It is used in both the United States and Canada.

cyclamate: a zero-kcalorie sweetener under consideration for use in the United States and used with restrictions in Canada.

neotame (NEE-oh-tame): an artificial sweetener composed of two amino acids (phenylalanine and aspartic acid)

linked in such a way as to make them indigestible by human enzymes.

saccharin: a zero-kcalorie sweetener used freely in the United States but restricted in Canada.

sucralose: a nonkcaloric sweetener derived from a chlorinated form of sugar that travels through the digestive tract unabsorbed. Approved for use in the United States and Canada.

tagatose: an incompletely absorbed monosaccharide sweetener derived from lactose with a kcaloric value of 1.5 kcalories per gram. About 80% of the ingested tagatose travels to the large intestine where bacterial colonies ferment it. Tagatose is not readily used by mouth bacteria, and so does not promote dental caries.

TABLE 3-4	U.S.-Approved Sweeteners			
Artificial Sweeteners	Energy (kcal/g)	Acceptable Daily Intake (ADI)	Average Amount to Replace 1 tsp Sucrose[a]	Approved Uses
Saccharin (SugarTwin, Sweet'N Low, others)	0	5 mg/kg body weight (341 mg for a 150 lb person)	12 mg	Tabletop sweeteners, wide range of foods, beverages, cosmetics, and pharmaceutical products
Aspartame (NutraSweet, Equal, others)	4	50 mg/kg body weight (3409 mg for a 150 lb person)	18 mg	General-purpose sweetener in all foods and beverages. Warning to population with PKU
Acesulfame-potassium (Sunette, Sweet One)	0	15 mg/kg body weight (1023 mg for a 150 lb person)	25 mg	Alcoholic beverages, baked goods, candies, chewing gum, desserts, gelatins, puddings, tabletop sweeteners
Sucralose (Splenda)	0	5 mg/kg body weight (341 mg for a 150 lb person)	6 mg	Baked goods, carbonated beverages, chewing gum, coffee and tea, dairy products, frozen desserts, fruit spreads, salad dressing, syrups, tabletop sweeteners
Neotame	0	18 mg/day	0.5 μg	Baked goods, beverages (nonalcoholic), candies, chewing gum, frostings, frozen desserts, gelatins, puddings, jams and jellies, syrups
Tagatose	1.5	7.5 g/day	1 tsp	Bakery products, beverages, cereals, chewing gum, confections, dairy products, dietary supplements, health bars, tabletop sweetener

[a]Rounded values

Aspartame **Aspartame** is the active ingredient in NutraSweet, which is used in many commercially prepared foods, and in Equal, a tabletop sweetener. Aspartame is 200 times sweeter than sucrose and was approved by the FDA in 1981. Aspartame is one of the most studied of all food additives: extensive animal and human studies document its safety. Long-term consumption of aspartame is safe and is not associated with any adverse health effects.[15] Aspartame is approved for use in more than 100 countries.

The FDA's approval of aspartame is based on the assumption that no one will consume more than the ADI of 50 milligrams per kilogram of body weight in a day. This daily intake is indeed a lot: for a 132-pound person, it adds up to 80 packets of Equal or 15 soft drinks sweetened only with aspartame. Most users of aspartame consume between 2 and 10 milligrams per kilogram of body weight per day. Still, a child who drinks a quart of Kool-Aid sweetened with aspartame on a hot day, and also has pudding, chewing gum, cereal, and other products sweetened with aspartame, takes in more than the ADI. Although this presents no proven hazard, it seems wise to offer children other foods so as not to exceed the limit. Infants or toddlers under two years old should not be fed artificially sweetened foods and drinks.

Aspartame and PKU Although aspartame is considered safe for most people, individuals with the metabolic disorder phenylketonuria (PKU) are an exception. The labels of products that contain aspartame must include information for individuals with PKU. Aspartame contains the amino acid phenylalanine, and people with PKU cannot dispose of it efficiently (see Nutrition in Practice 17).

Acesulfame Potassium The FDA approved **acesulfame potassium (acesulfame-K)** in 1988 after reviewing more than 90 safety studies, conducted over 15 years. Marketed under the trade names Sunette and Sweet One, acesulfame-K is about as sweet as aspartame. It is used in chewing gum, beverages, instant coffee and tea, gelatins, and puddings, as well as for table use. Unlike aspartame, acesulfame-K holds up well during cooking.

Sucralose **Sucralose** received FDA approval in 1998 for use as a sweetener in the United States. Sucralose is the only artificial sweetener made from sucrose. Many years of testing have deemed sucralose safe to use and, specifically, not a cause of cancer. Sucralose is not recognized by the body as sugar and therefore passes through unchanged. Sucralose is heat stable and so is useful for cooking and baking; it is used in commercially prepared products and as the tabletop sweetener Splenda. Its sugarlike taste and versatility are earning sucralose some popularity among consumers.

Neotame In 2002, the FDA approved **neotame** on the strength of findings from over 110 safety studies conducted on both animals and human beings. Neotame is so intensely sweet—about 7,000 to 13,000 times sweeter than sugar—that very little is needed to sweeten foods and beverages. Currently, neotame is available to food manufacturers for sweetening processed foods but not to consumers for home use.

A chemical cousin of aspartame, neotame also contains the amino acids phenylalanine and aspartic acid. Unlike aspartame, however, neotame is indigestible. Whereas digestive enzymes can separate the amino acids in aspartame, neotame's chemistry is sufficiently different to block the enzymes' action. This slight chemical difference makes neotame a better choice for people with phenylketonuria because it provides no phenylalanine to the body.

Tagatose The FDA has granted **tagatose**, a relative of fructose, the status of "generally recognized as safe," making it available as a lower-kcalorie sweetener for foods, beverages, confections, dietary supplements, and other uses. Tagatose is derived from lactose, but unlike fructose or lactose, 80 percent of tagatose remains unabsorbed until it reaches the large intestine. There the normal bacterial colonies of the intestine ferment tagatose, releasing gases and small products that are absorbed. At high doses, tagatose causes flatulence, rumbling, and loose stools. Otherwise, no adverse side effects have been noted by the maker.

Other Artificial Sweeteners FDA approval for two other sweeteners—**cyclamate** and **alitame**—is still pending. To date, no safety issues have been raised for alitame. Cyclamate, on the other hand, has been battling safety issues for 50 years. Approved by the FDA in 1949, cyclamate was banned in 1970 because of evidence indicating that it caused bladder cancer in rats. In 1985, the National Academy of Sciences concluded that evidence to date indicated that cyclamate did not cause cancer in human beings but that further studies were warranted. In Canada, cyclamate is restricted to use as a tabletop sweetener on the advice of a physician. In the United States, the FDA is currently reviewing a petition to reapprove the use of cyclamate.

Artificial Sweeteners and Weight Control Many people eat and drink products sweetened with artificial sweeteners to help them control weight. Does this work? Ironically, a few studies have reported that after consuming such products, people experience heightened feelings of hunger. Despite these reports, most studies find that artificial sweeteners do not heighten feelings of hunger, enhance food intake, or cause weight gain in people. When people reduce their energy intakes by replacing sugar in their diets with artificial sweeteners and then compensate for the reduced energy at later meals, energy intake may stay the same or increase. Using artificial sweeteners will not automatically lower energy intake; to successfully control energy intake, a person needs to make informed diet and activity decisions throughout the day.

Health Effects of Starch and Dietary Fibers

Despite dietary recommendations that people should eat generous servings of starch- and fiber-rich carbohydrate foods for their health, many people still believe that carbohydrate is the "fattening" component of foods. Gram for gram, carbohydrates contribute fewer kcalories to the body than do dietary fats, so a moderate diet based on starch and fiber-rich carbohydrate foods is likely to be lower in kcalories than a diet based on high-fat foods.

For health's sake, most people should increase their intakes of carbohydrate-rich foods such as whole grains, vegetables, legumes, and fruits—foods noted for their starch, fiber, and naturally occurring sugars. In addition, most people should also limit their intakes of foods with added sugars and the types of fats associated with heart disease (see Chapter 4). A diet that emphasizes whole grains, vegetables, legumes, and fruits is almost invariably moderate in food energy, low in fats that can harm health, and high in dietary fiber, vitamins, and minerals. All these factors working together can help reduce the risks of obesity, cancer, cardiovascular disease, diabetes, dental caries, gastrointestinal disorders, and malnutrition.

It is difficult to sort out which carbohydrates contribute to which health benefits. Starch and fibers almost always occur together in foods (except refined foods), so it is hard to distinguish their effects. Some health effects appear to be especially closely associated with fibers, however.

CARBOHYDRATES: DISEASE PREVENTION AND RECOMMENDATIONS

Fiber-rich carbohydrate foods benefit health in many ways. Foods such as whole grains, legumes, vegetables, and fruits supply valuable vitamins, minerals, and phytochemicals, along with abundant dietary fiber and little or no fat. The following paragraphs describe some of the health benefits of diets that emphasize a variety of these foods each day.

Heart Disease Diets rich in whole grains, legumes, and vegetables, especially those rich in whole grains, may protect against heart disease, although sorting out the exact reasons why has proved difficult.[16] Such diets are low in animal fat and cholesterol and high in dietary fibers, vegetable proteins, and phytochemicals—all factors associated with a lower risk of heart disease. Relatively certain is that viscous fibers (such as oat bran, barley, and legumes) lower blood cholesterol by binding cholesterol compounds and carrying them out of the body with the feces.[17] High-fiber foods may also lower blood cholesterol indirectly by displacing fatty, cholesterol-raising foods from the diet. Even when dietary fat intake is low, research shows that high intakes of soluble fiber (such as that in apples and oat bran) exert separate and significant blood cholesterol–lowering effects.

The role of animal fat and cholesterol in heart disease is discussed in Chapter 4. The role of vegetable proteins in heart disease is presented in Chapter 5. The benefits of phytochemicals in disease prevention are presented in Nutrition in Practice 8.

Diabetes is the topic of Chapter 20.

Diabetes Some fibers delay the passage of nutrients from the stomach to the small intestine. This delay slows glucose absorption, thus eliciting a moderate insulin response and a moderate rise in blood glucose. This slow, sustained rise in blood glucose is desirable. The term **glycemic effect** describes the effect of food on blood glucose; a high glycemic effect reflects fast glucose absorption and a surge in blood glucose, which is undesirable. A study of more than 90,000 women showed that women whose diets had the highest glycemic effect and the lowest cereal fiber content were most vulnerable to **type 2 diabetes** independent of other dietary constituents or known risk factors.[18]

GI Health Fibers that both enlarge and soften stools such as cellulose (in cereal brans, fruits, and vegetables) ease elimination for the rectal muscles and thereby alleviate or prevent constipation and hemorrhoids. Other fibers help to solidify watery stools.

Some fibers (again, such as cereal bran) help keep the contents of the intestinal tract moving easily. This action helps prevent compaction of the intestinal contents, which could obstruct the appendix and permit bacteria to invade and infect it.

In addition, dietary fibers stimulate the muscles of the gastrointestinal (GI) tract so that they retain their health and tone. This prevents the muscles from becoming weak and the lining of the digestive tract from bulging out in places, as occurs in diverticulosis. Insoluble fiber, particularly cellulose, seems to be most beneficial in lowering the risk of diverticulosis.[19] Chapter 18 describes diverticulosis.

Cancer Many, but not all, research studies suggest that increasing dietary fiber protects against colon cancer.[20] On completing a study of almost 520,000 people, researchers concluded that doubling the naturally occurring fiber in diets of populations with low fiber intakes could reduce the risks of colon cancer by 40 percent.[21] Importantly, the study focused on fiber-rich diets, not fiber supplements or additives. Fiber supplements and additives lack valuable nutrients and phytochemicals that also help protect against cancer.

The relationship between a fiber-rich diet and colon cancer is under investigation. One focus of research is fiber's ability to dilute and speed removal of potential cancer-causing agents from the large intestine (colon). In addition, some fibers stimulate bacterial fermentation of fiber in the colon, a process that produces small fatlike molecules that lower the pH.[22] These small fatlike molecules and the lower pH inhibit cancer growth in the colon.[23]

pH is a unit of measure expressing a substance's acidity or alkalinity (Chapter 5 provides a more detailed definition).

Recommended amounts of foods from each group for varying kcalorie levels are shown in Table 1-8 on p. 20.

Other processes may also be at work. As research progresses, the *Dietary Guidelines for Americans 2005* and the USDA Food Guide recommend a high-fiber diet that includes $4^{1}/_{2}$ cups of fruits and vegetables and at least 3 ounces of whole grains each day for those whose energy intakes are about 2000 kcalories per day.

Weight Management Fiber-rich foods tend to be low in fats and added sugars and therefore promote weight loss by delivering less energy per bite. High-fiber foods also promote a feeling of fullness as they absorb water. In addition, soluble fibers in a meal slow the movement of food through the upper digestive tract, so a person feels fuller longer.[24] In a study of more than 27,000 men, those who ate the most whole-grain foods, such as certain cooked and cold breakfast cereals, whole-wheat bread, brown rice, and popcorn, gained the least amount of weight over eight years.[25] The researchers observed a dose-response relationship: for every 40-gram increase in whole grains from all foods per day, weight gain was reduced by about a pound. This inverse relationship between consumption of whole-grain foods and weight gain also seems to hold true for women. Over a span of 12 years, women who ate more whole-grain foods gained significantly less weight than those who ate less whole-grain foods.[26] In contrast, intake of refined grain foods was positively associated with weight gain.

glycemic effect: a measure of the extent to which a food raises the blood glucose concentration and elicits an insulin response, as compared with pure glucose.

type 2 diabetes: the more common type of diabetes in which the fat cells resist insulin.

TABLE 3-5	Characteristics, Food Sources, and Health Effects of Fibers		
Fiber Characteristics	**Major Food Sources**	**Actions in the Body**	**Health Benefits**
Viscous, soluble, more fermentable			
• Gums • Pectins • Psyllium • Some hemicelluloses	Barley, oats, oat bran, rye, fruits (apples, citrus), legumes (especially young green peas and black-eyed peas), seaweeds, seeds and husks, vegetables; fibers used as food additives.	• Lower blood cholesterol by binding bile. • Slow glucose absorption. • Slow transit of food through upper GI tract, lending satiety. • Hold moisture in stools, softening them. • Yield small fatlike molecules after fermentation that the large intestine can use for energy.	• Lower risk of heart disease. • Lower risk of diabetes.
Nonviscous, insoluble, less fermentable			
• Cellulose • Lignin • Resistant starch • Many hemicelluloses	Brown rice, fruits, legumes, seeds, vegetables (cabbage, carrots, brussels sprouts), wheat bran, whole grains; extracted fibers used as food additives.	• Increase fecal weight and speed fecal passage through large intestine. • Provide bulk and feelings of fullness.	• Alleviate constipation. • Lower risks of diverticulosis, hemorrhoids, and appendicitis. • May help with weight management.

Many weight-loss products on the market today contain bulk-inducing fibers such as methylcellulose, but buying pure fiber compounds like this is neither necessary nor advisable. Besides adding bulk to the diet, high-fiber foods supply other nutrients as well. Most experts agree that the health benefits attributed to fiber may come from other constituents of fiber-containing foods, and not from the fiber alone.[27] Therefore, to use fiber in a weight-loss plan, select fresh fruits, vegetables, legumes, and whole-grain foods. As a bonus, fiber-rich foods are often economical as well as nutritious. Table 3-5 summarizes fibers and their health benefits.

Harmful Effects of Excessive Fiber Intake Despite fiber's benefits to health, when too much fiber is consumed, some minerals may bind to it and be excreted with it without becoming available for the body to use. When mineral intake is adequate, however, a reasonable intake of high-fiber foods does not seem to compromise mineral balance.

People with marginal intakes who eat mostly high-fiber foods may not be able to take in enough food to meet energy or nutrient needs. The malnourished, the elderly, and young children adhering to all-plant (vegan) diets are especially vulnerable to this problem. Fibers also carry water out of the body and can cause dehydration. Advise clients to add an extra glass or two of water to go along with the fiber added to their diets. Athletes may want to avoid bulky, fiber-rich foods just prior to competition. Adding purified fibers, such as oat bran or wheat bran, to foods can be taken to extremes. Too much fiber and too little fluid can obstruct the GI tract. Also, purified fiber may not affect the body the same way as the fiber in its original food product.

Recommendations The DRI committee advises that carbohydrates should contribute about half (45 to 65 percent) of the energy requirement. A person consuming 2000 kcalories a day should therefore have 900 to 1300 kcalories of carbohydrate, or between 225 and 325 grams. This amount is more than adequate to meet the RDA for carbohydrate, which is set at 130 grams per day based on the average minimum amount of glucose used by the brain.[28]

When it established the Daily Values that appear on food labels, the FDA used a 60 percent of kcalories guideline in setting the Daily Value for carbohydrate at

45% of 2000 kcal:
2000 × .45 = 900 kcal.
900 kcal ÷ 4 kcal/g = 225 g.

65% of 2000 kcal:
2000 × .65 = 1300 kcal.
1300 kcal ÷ 4 kcal/g = 325 g.

RDA for carbohydrate:
• 130 g/day.
• 45 to 65% of energy intake.

Daily Value:

- 25 g fiber (based on 11.5g/1000 kcal).

Fiber AI:

- 14 g/1000 kcal/day.
- Men:
  19–50 yr: 38 g/day.
  51+ yr: 30 g/day.
- Women:
  19–50 yr: 25 g/day.
  51+ yr: 21 g/day.

Reminder: An *AI (Adequate Intake)* is used as a guide for nutrient intake when an RDA cannot be established (see Chapter 1).

300 grams per day. For most people, this means increasing total carbohydrate intake. To this end, as mentioned earlier, the *Dietary Guidelines for Americans 2005* encourage people to choose fiber-rich fruits, vegetables, and whole grains often.

Recommendations for fiber encourage the same foods just mentioned: whole grains, vegetables, fruits, and legumes, which also provide vitamins, minerals, and phytochemicals. The FDA set the Daily Value for fiber at 25 grams or 11.5 grams per 1000-kclaorie intake. The DRI recommendation is slightly higher—14 grams per 1000-kcalorie intake. Similarly, the American Dietetic Association suggests 20 to 35 grams of fiber daily, which is about twice the average intake in the United States.[29]

As health care professionals, you can advise your clients that an effective way to add dietary fiber while lowering fat is to substitute plant sources of proteins (legumes) for some of the animal sources of protein (meats and cheeses) in the diet. Another way to add fiber is to encourage clients to consume the recommended amounts of fruits and vegetables each day. People choosing high-fiber foods are wise to seek out a variety of fiber sources and to drink extra fluids to help the fiber do its job. Many foods provide fiber in varying amounts as Figure 3-4 shows.

As mentioned earlier, too much fiber is no better than too little. The World Health Organization recommends an upper limit of 40 grams of dietary fiber a day.

REVIEW NOTES

A diet rich in starches and dietary fibers helps prevent heart disease, diabetes, GI disorders, and possibly some types of cancer. It also supports efforts to manage body weight.

For these reasons, recommendations urge people to eat plenty of whole grains, vegetables, legumes, and fruits—enough to provide 45 to 65 percent of the daily energy from carbohydrate.

CARBOHYDRATES: FOOD SOURCES

A day's meals based on the USDA Food Guide not only meet carbohydrate recommendations but provide abundant fiber, too. Grains, vegetables, fruits, and legumes deliver fiber to the diet and, like milk as well, are noted for their valuable energy-yielding starches and dilute sugars. Each class of foods makes its own typical carbohydrate contribution. The USDA Food Guide in Chapter 1 can help you and your clients obtain carbohydrate-rich foods.

Grains Most foods in this group—a slice of whole-wheat bread, half an English muffin or bagel, a 6-inch tortilla, or $^1/_2$ cup of rice, pasta, or cooked cereal—provide about 15 grams of carbohydrate, mostly as starch.* Most grain choices should be low in fat and sugar. When extra kcalories are needed to meet energy needs, some selections higher in fat, preferably unsaturated fat (see Chapter 4), and sugar can supply discretionary kcalories. These choices include biscuits, muffins, and snack crackers.

Vegetables Some vegetables are major contributors of starch in the diet. Just a small white or sweet potato or $^1/_2$ cup of cooked dry beans, corn, peas, plantain, or winter squash provides 15 grams of carbohydrate, as much as in a slice of bread, though as a mixture of sugars and starch. A $^1/_2$-cup portion of carrots, okra, onions, tomatoes, cooked greens, or most other nonstarchy vegetables or a cup of salad greens provides about 5 grams as a mixture of starch and sugars. Each of these foods also contributes a little protein, some fiber, and no fat.

Fruits The size of a typical serving of fruit varies depending on the form of the fruit: $^1/_2$ cup of juice; a small banana, apple, or orange; $^1/_2$ cup of most canned or fresh fruit; or $^1/_4$ cup of dried fruit. A typical fruit serving contains an average of

*Gram values in this section are adapted from the U.S. Food Exchange System.

FIGURE 3-4 Fiber in Selected Foods

Grains

Whole-grain products provide 1 to 2 g of fiber or more per serving:

- 1 slice whole-wheat or rye bread (1 g).

- 1 slice pumpernickel bread (2 g).

- $\frac{1}{2}$ c ready-to-eat 100% bran cereal (10 g).

- $\frac{1}{2}$ c cooked barley, bulgur, grits, oatmeal (2 to 3 g).

Vegetables

Most vegetables contain 2 to 3 g of fiber per serving:

- 1 c raw bean sprouts.

- $\frac{1}{2}$ c cooked broccoli, brussels sprouts, cabbage, carrots, cauliflower, collards, corn, eggplant, green beans, green peas, kale, mushrooms, okra, parsnips, potatoes, pumpkin, spinach, sweet potatoes, swiss chard, winter squash.

- $\frac{1}{2}$ c chopped raw carrots, peppers.

Fruits

- Fresh, frozen, and dried fruits have about 2 g of fiber per serving:

- 1 medium apple, banana, kiwi, nectarine, orange, pear.

- $\frac{1}{2}$ c applesauce, blackberries, blueberries, raspberries, strawberries.

- Fruit juices contain very little fiber.

Legumes and Nuts

Many legumes provide about 8 g of fiber per serving:

- $\frac{1}{2}$ c cooked baked beans, black beans, black-eyed peas, kidney beans, navy beans, pinto beans.

Some legumes provide about 5 g of fiber per serving:

- $\frac{1}{2}$ c cooked garbanzo beans, great northern beans, lentils, lima beans, split peas.

Most nuts and seeds provide 1 to 3 g of fiber per serving:

- 1 oz almonds, cashews, hazelnuts, peanuts, pecans, pumpkin seeds, sunflower seeds.

about 15 grams of carbohydrate, mostly as sugars, including the fruit sugar fructose. Fruits vary greatly in their water and fiber contents, and therefore their sugar concentrations vary also. No more than a third of the day's fruit should come from juice. With the exception of avocado, which is high in fat, the fruits contain insignificant amounts of fat and protein.

Milk, Cheese, and Yogurt One cup of milk or yogurt or the equivalent (1 cup of buttermilk, $^1/_3$ cup of dry milk powder, or $^1/_2$ cup of evaporated milk) provides a generous 12 grams of carbohydrate. Among cheeses, cottage cheese provides about 6 grams of carbohydrate per cup, while most other types contain little, if any, carbohydrate. These foods also contribute high-quality protein as well as several important vitamins and minerals. Calcium-fortified soy beverages are options for providing calcium and about the same amount of carbohydrate as milk. All milk products vary in fat content, an important consideration in choosing among them; Chapter 4 provides the details.

Cream and butter, although dairy products, are not equivalent to milk because they contain little or no carbohydrate and insignificant amounts of the other nutrients important in milk. They are appropriately placed with the solid fats and added sugars.

Meat, Poultry, Fish, Legumes, Eggs, and Nuts With two exceptions, foods of this group provide almost no carbohydrate to the diet. The exceptions are nuts, which provide a little starch and fiber along with their abundant fat, and dry beans, which are excellent sources of both starch and fiber. Just a $^1/_2$-cup serving of beans provides 15 grams of carbohydrate, an amount equal to the richest carbohydrate sources. Among sources of fiber, beans and other legumes are outstanding, providing as much as 8 grams in $^1/_2$ cup. The carbohydrate content of a diet can be determined by using a nutrient composition table such as that found in Appendix A, the exchange list system described in Chapter 20, or a computer diet analysis program.

CARBOHYDRATES: FOOD LABELS AND HEALTH CLAIMS

Food labels list the amount, in grams, of total carbohydrate—including starch, fibers, and sugars—per serving. Fiber grams are also listed separately, as are the grams of sugars. (With this information, consumers can calculate starch grams by subtracting the grams of fibers and sugars from the total carbohydrate.) Sugars on the Nutrition Facts panel of a food label reflect both added sugars and those that occur naturally in foods. Total carbohydrate and dietary fiber are also expressed in the "% Daily Values" column for a person consuming 2000 kcalories; there is no Daily Value for sugars.

The FDA authorizes four health claims on food labels concerning fiber-rich carbohydrate foods. One is for "fiber-containing grain products, fruits, and vegetables and reduced risk of cancer." Another is for "fruits, vegetables, and grain products that contain fiber, and reduced risk of coronary heart disease," a third is for "soluble fiber from whole oats and from psyllium seed husk and reduced risk of coronary heart disease," and a fourth is for "whole grains and reduced risk of heart disease and certain cancers." Chapter 1 describes the criteria foods must meet to bear these health claims.

REVIEW NOTES

Grains, vegetables, fruits, and legumes contribute dietary fiber to the diet and, like milk, also contribute energy-yielding starches and dilute sugars.

Food labels list grams of total carbohydrate and also provide separate listings of grams of fiber and sugar.

SELF CHECK

1. Carbohydrates are found in virtually all foods **except:**
   a. milks.
   b. meats.
   c. breads.
   d. vegetables.

2. Polysaccharides include:
   a. galactose, starch, and glycogen.
   b. starch, glycogen, and fiber.
   c. lactose, maltose, and glycogen.
   d. sucrose, fructose, and glucose.

3. The chief energy source of the body is:
   a. sucrose.
   b. starch.
   c. glucose.
   d. fructose.

4. The primary form of stored glucose in animals is:
   a. glycogen.
   b. cellulose.
   c. starch.
   d. lactose.

5. The polysaccharide that helps form the cell walls of plants is:
   a. cellulose.
   b. starch.
   c. glycogen.
   d. lactose.

6. Which of the following items may denote sugar on food labels?
   a. corn syrup
   b. aspartame
   c. xylitol
   d. cellulose

7. The two types of alternative sweeteners are:
   a. saccharin and cyclamate.
   b. sugar alcohols and artificial sweeteners.
   c. sorbitol and xylitol.
   d. sucrose and fructose.

8. A diet high in carbohydrate-rich foods such as whole grains, vegetables, fruits, and legumes is:
   a. most likely low in fat.
   b. most likely low in fiber.
   c. most likely poor in vitamins and minerals.
   d. most likely disease promoting.

9. A fiber-rich diet may help to prevent or control:
   a. diabetes.
   b. heart disease.
   c. constipation.
   d. all of the above.

10. The DRI fiber recommendation is:
    a. 10 grams per 1000 kcalories.
    b. 15 to 25 grams per day.
    c. 14 grams per 1000 kcalories.
    d. 40 to 55 grams per day.

Answers are in Appendix H.

CLINICAL APPLICATIONS

1. Considering the health benefits of carbohydrate-rich foods, especially those that provide starch and fiber, what suggestions would you offer to a client who reports the following:
   - Eats only three servings of refined, sugary breads or cereals each day.
   - Eats one serving of vegetables (usually french fries) each day.
   - Drinks fruit juice once a day, but never eats fruit.
   - Eats cheese at least twice a day, but does not drink milk.
   - Eats large servings of meat at least twice a day.
   - Eats hard candy two or three times a day.

NUTRITION ON THE NET

For further study of the topics of this chapter, access these websites.

Find updates and quick links to these and other nutrition-related sites at our website: **www.wadsworth.com/nutrition**

Search for "artificial sweeteners" at the U.S. Government health information site: **www.healthfinder.gov**

Search for "sugars" and "fiber" at the International Food Information Council site: **www.ific.org**

Learn more about dental caries from the American Dental Association and the National Institute of Dental Research: **www.ada.org** and **www.nidr.nih.gov**

Learn more about diabetes from the American Diabetes Association, the Canadian Diabetes Association, and the National Diabetes Information Clearinghouse: **www.diabetes.org**, **www.diabetes.ca**, and **www.niddk.nih.gov**

NOTES

[1] U.S. Food and Drug Administration, *Counting Calories: Report of the Working Group on Obesity*, March 12, 2004, available at **www.cfsan.fda.gov/~dms/owg-rpt.html**; Standing Committee on the Scientific Evaluation of Dietary Reference Intakes, Food and Nutrition Board, Institute of Medicine, *Dietary Reference Intakes for Energy, Carbohydrate, Fiber, Fat, Fatty Acids, Cholesterol, Protein, and Amino Acids* (Washington, D.C.: National Academies Press, 2005), pp. 265–358; A. Trichopoulou and coauthors, Lipid, protein, and carbohydrate intake in relation to body mass index, *European Journal of Clinical Nutrition* 56 (2002): 37–43.

[2] R. Gruetter, Glycogen: The forgotten cerebral energy store, *Journal of Neuroscience Research* 74 (2003): 179–183; I. Y. Choi, E. R. Seaquist, and R. Gruetter, Effect of hypoglycemia on brain glycogen metabolism in vivo, *Journal of Neuroscience Research* 72 (2003): 25–32.

[3] B. V. McCleary, Dietary fibre analysis, *Proceedings of the Nutrition Society* 62 (2003): 3–9.

[4] B. M. Popkin and S. J. Nielsen, The sweetening of the world's diet, *Obesity Research* 11 (2003): 1325–1332.

[5] U.S. Department of Agriculture and U.S. Department of Health and Human Services, *2005 Dietary Guidelines Committee Report*, 2005, available online at **www.healthierus.gov/dietaryguidelines**.

[6] M. Nestle, as quoted by O. Dyer, U.S. government rejects WHO's attempts to improve diet, *British Medical Journal* 328 (2004): 185.

[7] M. B. Schulze and coauthors, Sugar-sweetened beverages, weight gain, and incidence of type 2 diabetes in young and middle-aged women, *Journal of the American Medical Association* 292 (2004): 927–934; A. M. Coulston and R. K. Johnson, Sugar and sugars: Myths and realities, *Journal of the American Dietetic Association* 102 (2002): 351–353.

[8] G. A. Bray, S. J. Nielsen, and B. M. Popkin, Consumption of high-fructose corn syrup in beverages may play a role in the epidemic of obesity, *American Journal of Clinical Nutrition* 79 (2004): 537–543; S. J. Nielsen and B. M. Popkin, Changes in beverage intake between 1977 and 2001, *American Journal of Preventive Medicine* 27 (2004): 205–210.

[9] Standing Committee on the Scientific Evaluation of Dietary Reference Intakes, 2005, pp. 294–338.

[10] Position of the American Dietetic Association: Oral health and nutrition, *Journal of the American Dietetic Association* 103 (2003): 615–625.

[11] Joint WHO/FAO Expert Consultation, *Diet, Nutrition, and the Prevention of Chronic Diseases* (Geneva, Switzerland: World Health Organization, 2003), p. 119.

[12] Standing Committee on the Scientific Evaluation of Dietary Reference Intakes, 2005, p. 770.

[13] Position of the American Dietetic Association: Use of nutritive and nonnutritive sweeteners, *Journal of the American Dietetic Association* 104 (2004): 255–275.

[14] Fact Sheet: The Report on Carcinogens, 9th ed., National Institutes of Health News Release, available at **www.nih.gov/news/pr/may2000/niehs-15.htm.**

[15] Position of the American Dietetic Association, 2004.

[16] M. K. Jensen and coauthors, Intakes of whole grains, bran, and germ and the risk of coronary heart disease in men, *American Journal of Clinical Nutrition* 80 (2004): 1492–1499; F. B. Hu and W. C. Willett, Optimal diets for prevention of coronary heart disease, *Journal of the American Medical Association* 288 (2002): 2569–2578; N. M. McKeown and coauthors, Whole-grain intake is favorably associated with metabolic risk factors for type 2 diabetes and cardiovascular disease in the Framingham Offspring Study, *American Journal of Clinical Nutrition* 76 (2002): 390–398.

[17] Position of the American Dietetic Association, Health implications of dietary fiber, *Journal of the American Dietetic Association* 102 (2002): 993–1000; L. Van Horn and N. Ernst, A summary of the science supporting the new National Cholesterol Education program dietary recommendations: What dietitians should know, *Journal of the American Dietetic Association* 101 (2001): 1148–1154; L. Brown and coauthors, Cholesterol-lowering effects of dietary fiber: A meta-analysis, *American Journal of Clinical Nutrition* 69 (1999): 30–42.

[18] M. B. Schulze and coauthors, Glycemic index, glycemic load, and dietary fiber intake and incidence of type 2 diabetes in younger and middle-aged women, *American Journal of Clinical Nutrition* 80 (2004): 348–356.

[19] E. Cunningham and W. Marcason, What role does fiber play in diverticular disease? *Journal of the American Dietetic Association* 102 (2002): 225.

[20] J. G. Muir and coauthors, Combining wheat bran with resistant starch has more beneficial effects on fecal indexes than does wheat bran alone, *American Journal of Clinical Nutrition* 79 (2004):

1020–1028; Editorial board, The facts on fiber and colon cancer, *Nutrition & the M.D.* March 2004, pp. 6–7; U. Peters and coauthors, Dietary fibre and colorectal adenoma in a colorectal cancer early detection programme, *Lancet* 361 (2003): 1491–1495; S. A. Bingham and coauthors, Dietary fibre in food and protection against colorectal cancer in the European Prospective Investigation into Cancer and Nutrition (EPIC): An observational study, *Lancet* 361 (2003): 1496–1501; T. Asano and R. S. McLeod, Dietary fibre for the prevention of colorectal adenomas and carcinomas, *Cochrane Database of Systematic Reviews* 2 (2002): CD003430.

[21] Bingham and coauthors, 2003.

[22] Muir and coauthors, 2004.

[23] Muir and coauthors, 2004; A. Andoh, T. Tsujikasw, and Y. Fujiyama, Role of dietary fiber and short-chain fatty acids in the colon, *Current Pharmaceutical Design* 9 (2003): 347–358; I. McMillan and coauthors, Opposing effects of butyrate and bile acids on apoptosis of human colon adenoma cells: Differential activation of PKC and MAP kinases, *British Journal of Cancer* 88 (2003): 748–753.

[24] A. Sparti and coauthors, Effect of diet high or low in unavailable and slowly digestible carbohydrates on the pattern of 24-h substrate oxidation and feelings of hunger in humans, *American Journal of Clinical Nutrition* 72 (2000): 1461–1468.

[25] P. Koh-Banerjee and coauthors, Changes in whole-grain, bran, and cereal fiber consumption in relation to 8-y weight gain among men, *American Journal of Clinical Nutrition* 80 (2004): 1237–1245.

[26] S. Liu and coauthors, Relation between changes in intakes of dietary fiber and grain products and changes in weight and development of obesity among middle-aged women, *American Journal of Clinical Nutrition* 78 (2003): 920–927.

[27] Standing Committee on the Scientific Evaluation of Dietary Reference Intakes, 2005, pp. 391–399.

[28] Standing Committee on the Scientific Evaluation of Dietary Reference Intakes, 2005, p. 265.

[29] Position of the American Dietetic Association, 2002.

Nutrition and Dental Health

Chapter 3 emphasized the health benefits of eating carbohydrate-rich foods, especially those containing starch and fiber. Carbohydrates may support overall health, but they do not necessarily promote dental health. The carbohydrates people eat and the times they eat them play a major role in the development of dental caries—a pervasive health problem throughout the world.

What is dental caries?

Dental caries is an infectious oral disease that develops in the tooth **enamel** (see the accompanying glossary and Figure NP3-1). Caries develops when bacteria that reside in the **dental plaque** consume and metabolize carbohydrates, producing acids that attack the tooth enamel. Thus at least two main ingredients are required to make dental caries: bacteria and carbohydrates. In addition, factors such as heredity, nutrition status during early tooth development, dental hygiene practices, and fluoride intake influence a person's susceptibility to caries. Poor nutrition during pregnancy, infancy, or early childhood can impair the development of healthy teeth, making caries likely.[1] Table NP3-1 shows the effects of specific nutrient deficiencies on tooth development.

How do carbohydrate-rich foods promote caries development?

The bacteria that promote dental caries thrive on food particles that contain carbohydrate. Both sugar and starch can support bacterial growth. Equally important is the length of time the food stays in the mouth, and this depends on how soon the teeth are brushed after eating and how sticky the food is. The damage a food does relates to both its carbohydrate content and its stickiness. For example, raisins and granola, which adhere to the teeth, cause more caries than a food that is easily rinsed off such as a sugary beverage.

Sugar can be eaten without inviting tooth decay if it is removed from tooth surfaces promptly. Bacterial action is maximal in the first 20 minutes after the first contact. If immediate brushing is not possible, water or other beverages swished in the mouth after a meal can effectively rinse the teeth. Once-a-day flossing may also effectively control formation of caries, regardless of the carbohydrate content of the diet. Some people may never get caries because they have inherited resistance to them.

Doesn't saliva rinse the mouth and protect the teeth?

Yes, and some foods stimulate more saliva flow than others. Saliva protects against caries formation in several ways. In addition to rinsing the mouth, it also dilutes the caries-causing acid produced by bacteria, exerts antibacterial action, and provides protective minerals such as calcium and phosphorus that promote the repair (remineralization) of tooth enamel.[2]

TABLE NP3-1 Nutrient Deficiencies Affecting Tooth Development

Nutrient Deficiency	Effect on Tooth Development
Protein	Small, irregularly shaped teeth; delayed eruption; high caries susceptibility
Vitamin C	Disturbance of dentin formation
Vitamin A	Disturbance of enamel formation, delayed eruption
Vitamin D	Poor mineralization, pitting, striations
Calcium	Poor mineralization
Phosphorus	Poor mineralization
Magnesium	Enamel underdeveloped
Iron	High caries susceptibility
Zinc	High caries susceptibility
Fluoride	High caries susceptibility

FIGURE NP3-1 A Tooth

The inner layer of dentin is bonelike material. The outer layer is enamel, which is harder than bone. Caries begin when acid dissolves the enamel that covers the tooth. If it is not repaired, the decay may penetrate the dentin and spread into the pulp of the tooth, causing inflammation and an abscess.

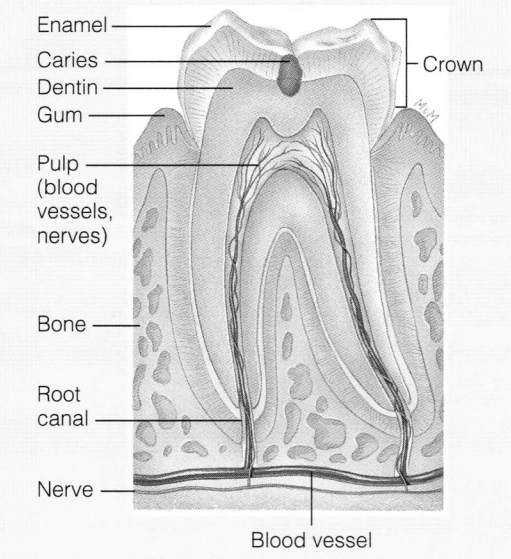

Foods that elicit saliva flow may therefore defend against caries formation, but not all of them are protective; foods that also contain sugar may promote acid formation. Apples are an example; they stimulate saliva flow, but they also release sugar, so they have both caries-preventing and caries-promoting effects. Clearly, many different factors influence caries development, making it difficult to predict exactly which foods are **cariogenic.**

Do any foods prevent caries?

Yes. Some foods stimulate saliva flow and do not contribute to acid formation in the mouth: cheese is an example. Such foods are good choices to eat at the end of a meal. Cheese is a powerful saliva stimulant and does not promote acid formation, so it reduces the cariogenicity of a meal. Furthermore, the high calcium and phosphorus content of cheese supports dental health.

High-fiber foods are, in general, anticariogenic, especially if their sugar content is low. For example, raw vegetables do not stick to the teeth, and because they require vigorous chewing, they stimulate saliva flow. Cocoa products (including chocolate), coffee, tea, and beer all contain tannin, an acid that prevents caries formation. Table NP3-2 lists some dietary recommendations for controlling dental caries.

Besides foods, what other factors protect against dental caries development?

Research shows that when fluoride is added to the water supply, the children in the community have fewer dental caries than children who drink nonfluoridated water. Water fluoridation is the most effective, least expensive way to provide dental care to everyone.[3] Fluoride increases the mineralization of teeth, helps prevent tooth decay, and promotes tooth enamel remineralization throughout life.[4] The following recommendations will maximize protection against dental caries:

- Use sugars sparingly; watch for hidden sugars in foods; and use low-sugar or sugar-free products whenever possible.
- Restrict sweets to mealtimes.
- After eating a meal or a between-meal snack, brush and floss or, at least, rinse with water.
- Limit the time that teeth are exposed to sticky foods.
- In any case, brush at least twice daily, and floss at least once daily.

- Visit a dentist regularly; repair damaged teeth.
- Drink fluoridated water; provide infants and children with fluoride supplements when such water is not available.
- Eat a balanced diet composed of a variety of foods that will maintain adequate nutrition status.
- Eat foods that are rich in calcium and phosphorus.
- Eat a variety of firm, fibrous foods that will stimulate saliva flow.

In summary, learning and practicing sound dental hygiene habits, as well as developing eating habits that are consistent with both dental health and nutritional health, will serve a person throughout life.

Glossary of Dental Caries Terms

cariogenic (KARE-ee-oh-JEN-ik): conducive to dental decay.
dental caries (KARE-eez): the gradual decay and disintegration of a tooth.
dental plaque (PLACK): a gummy mass of bacteria that grows on teeth and can lead to dental caries and gum disease.
enamel: the hard, white, dense substance that forms a covering for the crown of a tooth.

Notes

[1] Position of the American Dietetic Association: Oral health and nutrition, *Journal of the American Dietetic Association* 103 (2003): 615–625.
[2] D.P. DePaola and C. F. Schachtele, Diet and oral health, in M. H. Stipanuk, *Biochemical and Physiological Aspects of Human Nutrition* (Philadelphia: W.B. Saunders, 2000), pp. 866–881.
[3] Position of the American Dietetic Association: The impact of fluoride on health, *Journal of the American Dietetic Association* 105 (2005): 1620–1628.
[4] Position of the American Dietetic Association, 2005.

TABLE NP3-2 Dietary Recommendations for Controlling Dental Caries

Food Group	Low Cariogenicity (Use When Teeth Cannot Be Brushed Immediately)	High Cariogenicity (Do Not Use unless Followed by Prompt and Thorough Dental Hygiene)
Milk/yogurt/cheese	Milk, cheese, plain yogurt, cottage cheese	Ice cream, ice milk, milk shakes, fruited yogurts, eggnog
Meats/poultry/fish/ legumes/eggs/nuts	Meat, fish, poultry, eggs, legumes, nuts	Peanut butter with added sugar, luncheon meats with added sugar, meats with sugared glazes
Fruits[a]	Fresh, packed in water	Dried (raisins, figs, dates), packed in syrup or juice, jams, jellies, preserves, fruit juices and drinks
Vegetables	Most vegetables	Candied sweet potatoes, glazed carrots
Grains[b]	Popcorn, toast, hard rolls, pretzels, pizza, bagels	Cookies, sweet rolls, pies, cakes, dry sugared cereals as between-meal snacks, doughnuts, potato chips, granola bars
Other	Sugarless gum, coffee or tea without sugar	Sugared soft drinks, candy, fudge, caramels, honey, sugars, syrups

[a] Tiny particles of bananas can get lodged between teeth and decompose, increasing risk of caries.
[b] Tiny particles of breads, crackers, and chips can also become lodged in teeth, promoting caries formation.

CONTENTS

Lipids

CHAPTER 4

Most people know that too much fat, especially certain kinds of fat, in the diet imposes health risks, but they may be surprised to learn that too little does, too. People in the United States, however, are more likely to eat too much fat than too little.

Fat is a member of the class of compounds called **lipids**. The lipids in foods and in the human body include triglycerides (**fats** and **oils**), phospholipids, and sterols.

Roles of Body Fat

Lipids perform many tasks in the body, but most importantly, they provide energy. A constant flow of energy is so vital to life that, in a pinch, any other function is sacrificed to maintain it. Chapter 3 described one safeguard against such an emergency—the stores of glycogen in the liver that provide glucose to the blood whenever the supply runs short. The body's stores of glycogen are limited, however. In contrast, the body's capacity to store fat for energy is virtually unlimited due to the fat-storing cells of the **adipose tissue.** The fat cells of the adipose tissue readily take up and store fat, growing in size as they do so. When extra energy storage is needed, new fat cells are readily produced. Fat cells are more than just storage depots, however; fat cells produce enzymes and secrete hormones that help regulate the appetite and influence other body functions.[1] Figure 4-1 shows a fat cell.

The fat stored in fat cells supplies 60 percent of the body's ongoing energy needs during rest. During some types of physical activity or prolonged periods of food deprivation, fat stores may make an even greater energy contribution. The brain and nerves, however, need their energy as glucose, and the body cannot convert fat to glucose. After a long period of glucose deprivation (during fasting or starvation), brain and nerve cells develop the ability to derive about half of their energy from a special form of fat known as **ketones**, but they still require glucose as well. This means that people wanting to lose weight need to eat a certain minimum amount of carbohydrate to meet their energy needs, even when they are limiting their food intakes.

Body fat supplies much of the fuel that muscles need to do their work.

Chapter 6 discusses fat use during fasting.

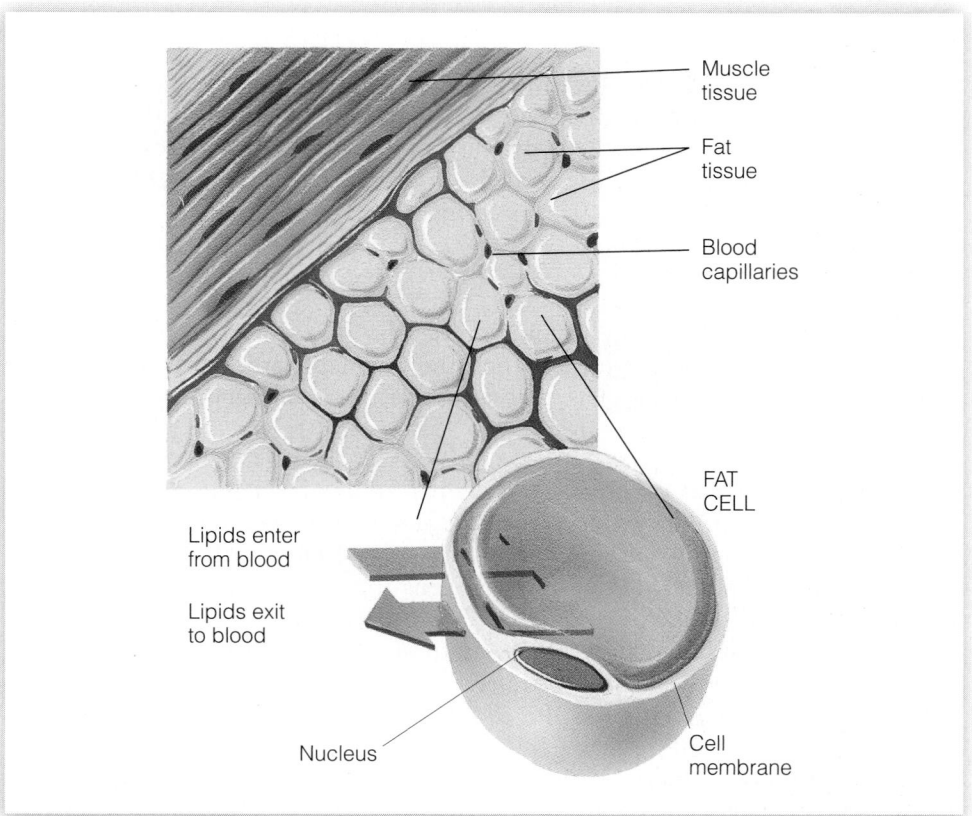

Muscle tissue

Fat tissue

Blood capillaries

FAT CELL

Lipids enter from blood

Lipids exit to blood

Nucleus

Cell membrane

FIGURE 4-1 A Fat Cell
Within the fat, or adipose, cell, lipid is stored in a droplet. This droplet can greatly enlarge, and the fat cell membrane will expand to accommodate its swollen contents.

lipids: a family of compounds that includes triglycerides (fats and oils), phospholipids, and sterols.

fats: lipids that are solid at room temperature (70°F or 25°C).

oils: lipids that are liquid at room temperature (70°F or 25°C).

adipose tissue: the body's fat, which consists of masses of fat-storing cells called adipose cells.

ketones (KEY-tones): acidic, fat-related compounds formed from the incomplete breakdown of fat when carbohydrate is not available; technically known as *ketone bodies*.

TABLE 4-1	The Functions of Fats in the Body

Fats in the body:

- Provide energy.
- Insulate the body against temperature extremes.
- Protect the body's vital organs from shock.
- Form the major material of cell membranes.

In addition to supplying energy, fat serves other roles in the body. Natural oils in the skin provide a radiant complexion; in the scalp, they help nourish the hair and make it glossy. The layer of fat beneath the skin insulates the body from extremes of temperature. A pad of hard fat beneath each kidney protects it from being jarred and damaged, even during a motorcycle ride on a bumpy road. The soft fat in a woman's breasts protects her mammary glands from heat and cold and cushions them against shock. The fat embedded in muscle tissue shares with muscle glycogen the task of providing energy when the muscles are active. The phospholipids and the sterol cholesterol are cell membrane constituents that help maintain the structure and health of all cells. Table 4-1 summarizes the major functions of fats in the body.

> **REVIEW NOTES**
>
> Lipids in the body not only serve as energy reserves but also protect the body from temperature extremes, cushion the vital organs, and provide the major material of cell membranes.

The Chemist's View of Lipids

The diverse and vital functions that lipids play in the body reveal why eating too little fat can be harmful. As mentioned earlier, though, too much fat in the diet seems to be the greater problem for most people. To understand both the beneficial and harmful effects that fats exert on the body, a closer look at the structure and function of members of the lipid family is in order.

TRIGLYCERIDES

When people talk about fat—for example, "I'm too fat" or "That meat is fatty"—they are usually referring to triglycerides. Among lipids, **triglycerides** predominate—both in the diet and in the body. The name *triglyceride* almost explains itself: three (*tri*) **fatty acids** attached to a **glycerol** "backbone." Figure 4-2 shows how three fatty acids combine with glycerol to make a triglyceride.

FATTY ACIDS

When energy from any energy-yielding nutrient is to be stored as fat, the nutrient is first broken into small fragments. Then the fragments are linked together into chains known as fatty acids. The fatty acids are then packaged, three at a time, with glycerol to make triglycerides.

Chain Length and Saturation Fatty acids may differ from one another in two ways—in chain length and in degree of saturation. The chain length refers to the number of carbons in a fatty acid. Saturation also refers to its chemical structure—specifically, to the number of hydrogens the carbons in the fatty acid are holding. If every available carbon is filled to capacity with hydrogen atoms, the chain is called a **saturated fatty acid.** A saturated fatty acid is fully loaded with hydrogens and has only single bonds between the carbons. The first zigzag structure in Figure 4-3 represents a saturated fatty acid.

Unsaturated Fatty Acids In some fatty acids, including most of those in plants and fish, hydrogens are missing in the fatty acid chains. The places where the

triglycerides (try-GLISS-er-rides): one of the main classes of lipids; the chief form of fat in foods and the major storage form of fat in the body; composed of glycerol with three fatty acids attached.
   tri = three
   glyceride = a compound of glycerol

fatty acids: organic compounds composed of a chain of carbon atoms with hydrogens attached and an acid group at one end.

glycerol (GLISS-er-ol): an organic compound, three carbons long, that can form the backbone of triglycerides and phospholipids.

saturated fatty acid: a fatty acid carrying the maximum possible number of hydrogen atoms (having no points of unsaturation).

hydrogens are missing are called points of unsaturation, and a chain containing such points is called an **unsaturated fatty acid.** An unsaturated fatty acid has at least one double bond between its carbons. If there is one point of unsaturation, the chain is a **monounsaturated fatty acid.** The second structure in Figure 4-3 is an example. If there are two or more points of unsaturation, then the fatty acid is a **polyunsaturated fatty acid** (see the third structure in Figure 4-3).

Hard and Soft Fat A triglyceride can contain any combination of fatty acids—long chain or short chain and saturated, monounsaturated, or polyunsaturated. The degree of saturation of the fatty acids in a fat influences the health of the body (discussed in a later section) and the characteristics of foods. Fats that contain the shorter-chain or the more unsaturated fatty acids are softer at room temperature and melt more readily. A comparison of three fats—lard (which comes from pork), chicken fat, and safflower oil—illustrates these differences: lard is the most saturated and the hardest; chicken fat is less saturated and somewhat soft; and safflower oil, which is the most unsaturated, is a liquid at room temperature.

Stability Saturation also influences stability. Fats can become **rancid** when exposed to oxygen. Polyunsaturated fatty acids spoil most readily because their double bonds are unstable. The **oxidation** of unsaturated fats produces a variety of compounds that smell and taste rancid; saturated fats are more resistant to oxidation

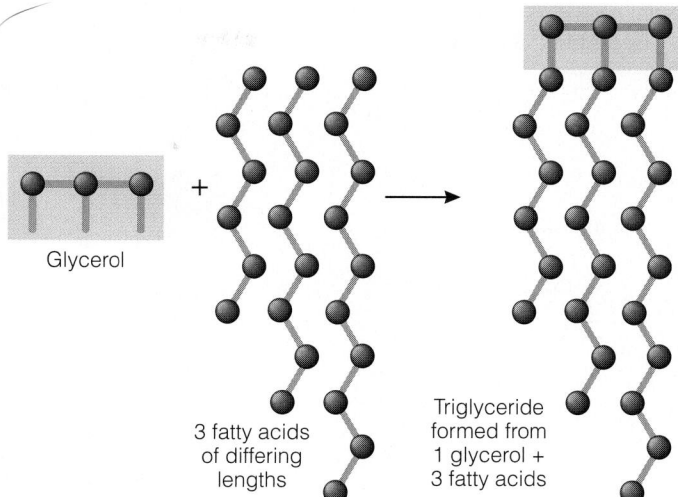

Glycerol

3 fatty acids of differing lengths

Triglyceride formed from 1 glycerol + 3 fatty acids

FIGURE 4-2 Triglyceride Formation
Glycerol, a small, water-soluble compound, plus three fatty acids, equals a triglyceride.

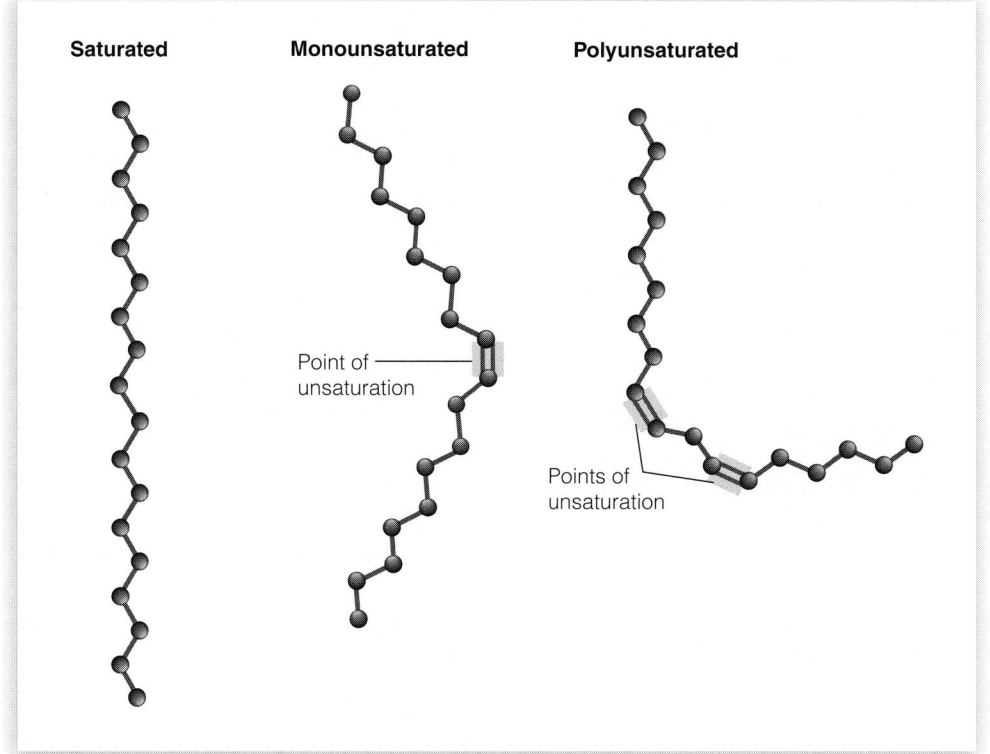

Saturated

Monounsaturated

Polyunsaturated

Point of unsaturation

Points of unsaturation

FIGURE 4-3 Three Types of Fatty Acids
The more carbon atoms in a fatty acid, the longer it is. The more hydrogen atoms attached to those carbons, the more saturated the fatty acid is.

unsaturated fatty acid: a fatty acid with one or more points of unsaturation where hydrogens are missing (includes monounsaturated and polyunsaturated fatty acids).

monounsaturated fatty acid (MUFA): a fatty acid that has one point of unsaturation; for example, the oleic acid found in olive oil.

polyunsaturated fatty acids (PUFA): fatty acids with two or more points of unsaturation. For example, linoleic acid has two such points, and linolenic acid has three. Thus polyunsaturated *fat* is composed of triglycerides containing a high percentage of PUFA.

rancid: the term used to describe fats when they have deteriorated, usually by oxidation. Rancid fats often have an "off" odor.

oxidation (OKS-ee-day-shun): the process of a substance combining with oxygen.

FIGURE 4-4 *Cis-* and *Trans-*
Fatty Acids Compared

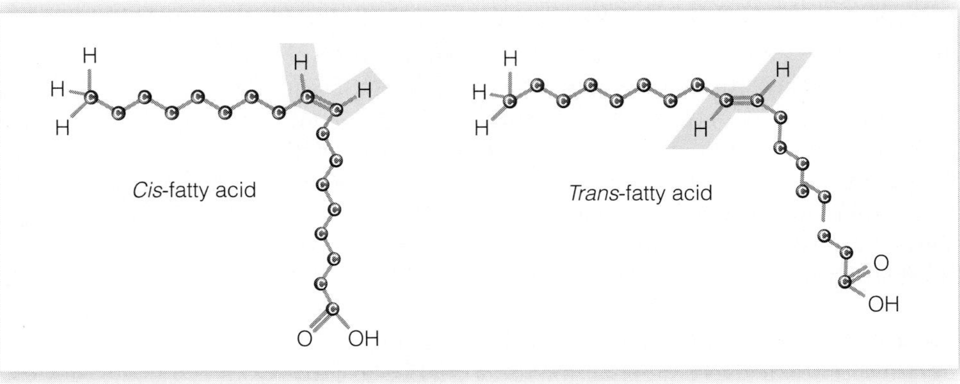

Cis-fatty acid   *Trans*-fatty acid

and thus less likely to become rancid. Other types of spoilage can occur due to microbial growth, however.

Manufacturers can protect fat-containing products against rancidity in three ways—none of them perfect. First, products may be sealed airtight and refrigerated—an expensive and inconvenient storage system. Second, manufacturers may add **antioxidants** to compete for the oxygen and thus protect the oil (examples are the additives **BHA** and **BHT** and vitamins C and E).* Third, manufacturers may saturate some or all of the points of unsaturation by adding hydrogen atoms—a process known as hydrogenation.

Hydrogenation Hydrogenation offers two advantages: it protects against oxidation (thereby prolonging shelf life) and also alters the texture of foods by increasing the solidity of fats. When partially hydrogenated, vegetable oils become spreadable margarine. Hydrogenated fats make piecrusts flaky and puddings creamy. A disadvantage is that hydrogenation makes polyunsaturated fats more saturated. Consequently, any health advantages of using polyunsaturated fats instead of saturated fats are lost in hydrogenation.

***Trans*-Fatty Acids** Another disadvantage of hydrogenation is that some of the molecules that remain unsaturated after processing change shape from *cis* to *trans*. In nature, most unsaturated fatty acids are *cis*-fatty acids—meaning that the hydrogens next to the double bonds are on the same side of the carbon chain. Only a few fatty acids in nature (notably in milk and butter) are ***trans*-fatty acids**—meaning that the hydrogens next to the double bonds are on opposite sides of the carbon chain (see Figure 4-4). These arrangements result in different configurations for the fatty acids, and this difference affects function: in the body, *trans*-fatty acids behave more like saturated fats than like unsaturated fats. The relationship between *trans*-fatty acids and heart disease has been the subject of much recent research, as a later section describes.

Essential Fatty Acids The human body can synthesize all the fatty acids it needs from carbohydrate, fat, or protein except for two—**linoleic acid** and **linolenic acid**. Both linoleic acid and linolenic acid are polyunsaturated fatty acids. Because they cannot be made from other substances in the body, they must be obtained from food and are therefore called **essential fatty acids**. Linoleic acid and linolenic acid are found in small amounts in plant oils, and the body readily stores them, making deficiencies unlikely. From both of these essential fatty acids, the body makes important substances that help regulate a wide range of body functions: blood pressure, clot formation, blood lipid concentration, the immune

antioxidants: compounds that protect others from oxidation by being oxidized themselves.

BHA, BHT: preservatives commonly used to slow the development of "off" flavors, odors, and color changes caused by oxidation.

hydrogenation (high-dro-gen-AY-shun): a chemical process by which hydrogens are added to monounsaturated or polyunsaturated fats to reduce the number of double bonds, making the fats more saturated (solid) and more resistant to oxidation (protecting against rancidity). Hydrogenation produces *trans*-fatty acids.

trans-fatty acids: fatty acids in which the hydrogens next to the double bond are on opposite sides of the carbon chain.

linoleic acid, linolenic acid: polyunsaturated fatty acids that are essential for human beings.

essential fatty acids: fatty acids that the body requires but cannot make in amounts sufficient to meet its physiological needs.

*BHA is butylated hydroxyanisole; BHT is butylated hydroxytoluene.

response, the inflammatory response to injury, and many others.[2] These two essential nutrients also serve as structural components of cell membranes.

Linoleic Acid: An Omega-6 Fatty Acid Linoleic acid is an **omega-6 fatty acid,** found in the seeds of plants and in the oils produced from the seeds. Any diet that contains vegetable oils, seeds, nuts, and whole-grain foods provides enough linoleic acid to meet the body's needs. Researchers have long known and appreciated the importance of the omega-6 fatty acid family.

Linolenic Acid and Other Omega-3 Fatty Acids Linolenic acid belongs to a family of polyunsaturated fatty acids known as **omega-3 fatty acids,** a family that also includes **EPA** and **DHA.** EPA and DHA are found primarily in fish oils. As mentioned, the human body cannot make linolenic acid, but given dietary linolenic acid, it can make EPA and DHA, although the process is slow.

The importance of omega-3 fatty acids was first recognized during the 1980s when research began to unveil impressive roles for EPA and DHA in metabolism and disease prevention. The brain has a high content of DHA, and both EPA and DHA are needed for normal brain development.[3] DHA is also especially active in the rods and cones of the retina of the eye.[4] Today, researchers know that these omega-3 fatty acids are essential for normal growth and development and that they may play an important role in the prevention and treatment of heart disease, diabetes, hypertension, arthritis, and cancer.[5]

PHOSPHOLIPIDS

Up to now, this discussion has focused on one class of lipids, the triglycerides (fats and oils) and their component parts, the fatty acids (see Table 4-2). Two other classes of lipids, the **phospholipids** and sterols, make up only 5 percent of the lipids in the diet, but they are nevertheless worthy of attention. Among the phospholipids, the lecithins are of particular interest.

Structure of Phospholipids Like the triglycerides, the **lecithins** and other phospholipids have a backbone of glycerol; they differ from the triglycerides in having only two fatty acids attached to the glycerol. In place of the third fatty acid, they have a phosphate group (a phosphorus-containing acid) and a molecule of **choline** or a similar compound. The fatty acids make phospholipids soluble in fat; the phosphate group enables them to dissolve in water. Such versatility benefits the food industry, which uses phospholipids as emulsifiers to mix fats with water in such products as mayonnaise and candy bars.

Roles of Phospholipids Lecithins and other phospholipids are important constituents of cell membranes. They also act as emulsifiers in the body, helping to keep other fats in solution in the watery blood and body fluids.

Phospholipids: Not Essential Lecithins periodically receive attention in the popular press. People may hear that lecithins are a major constituent of cell membranes (true), that the functioning of all cells depends on the integrity of the cell membranes (true), and that consumers must therefore take lecithin supplements (false). The body digests lecithins before it absorbs them, so the lecithins people eat do not reach the body tissues intact. Instead, the lecithins used for building cell membranes are made from scratch by the liver. In other words, the lecithins are not essential nutrients.

TABLE 4-2	The Lipid Family

Triglycerides (fats and oils)

- Glycerol (1 per triglyceride)

- Fatty acids (3 per triglyceride)
  Saturated
  Monounsaturated
  Polyunsaturated
  Omega-6
  Omega-3

Phospholipids (such as the lecithins)

Sterols (such as cholesterol)

Chemists use the term *omega,* the last letter of the Greek alphabet, to refer to the position of the last double bond in a fatty acid.

Reminder: *Emulsifiers* are substances that mix with both fat and water and are able to dispense the fat in the water, forming an emulsion.

omega-6 fatty acid: a polyunsaturated fatty acid with its endmost double bond six carbons back from the end of its carbon chain; long recognized as important in nutrition. Linoleic acid is an example.

omega-3 fatty acids: polyunsaturated fatty acids in which the endmost double bond is three carbons back from the end of the carbon chain; relatively newly recognized as important in nutrition. Linolenic acid is an example.

EPA, DHA: omega-3 fatty acids made from linolenic acid. The full name for EPA is *eicosapentaenoic* (EYE-cosa-PENTA-ee-NO-ick) *acid.* The full name for DHA is *docosahexaenoic* (DOE-cosa-HEXA-ee-NO-ick) *acid.*

phospholipids: one of the three main classes of lipids; compounds similar to triglycerides but with choline (or another compound) and a phosphorus-containing acid in place of one of the fatty acids.

lecithins: one type of phospholipid.

choline: a nonessential nutrient that can be made in the body from an amino acid.

STEROLS

Sterols are large, complex molecules consisting of interconnected rings of carbon. Cholesterol is the most familiar sterol, but others, such as vitamin D and the sex hormones (for example, testosterone), are important, too.

Sterols in Foods Both plant and animal-derived foods contain sterols, but only animal-derived foods contain significant amounts of cholesterol: meats, eggs, fish, poultry, and dairy products. Organ meats, such as liver and kidneys, and eggs are richest in cholesterol; cheeses and meats have less. Shellfish contain many sterols, but much less cholesterol than was previously thought.

Cholesterol Synthesis Like the lecithins, cholesterol can be made by the body, so it is not an essential nutrient. Your liver is manufacturing it now, as you read. The raw materials that the liver uses to make cholesterol can all be taken from glucose or fatty acids. In other words, cholesterol can be made from either carbohydrate or fat. Most of the body's cholesterol ends up in the membranes of cells, where it performs vital structural and metabolic functions.

Cholesterol's Two Routes in the Body After being made, cholesterol leaves the liver by two routes:

Reminder: *Bile* is a compound made by the liver from cholesterol and stored in the gallbladder. Bile prepares fat for digestion.

1. It may be made into bile, stored in the gallbladder, and delivered to the intestine.
2. It may travel, via the bloodstream, to all the body's cells.

The bile that is made from cholesterol in the liver is released into the intestine to aid in the digestion and absorption of fat. After bile does its job, most of it is reabsorbed into the body and recycled; the rest is excreted in the feces.

Cholesterol Excreted While bile is in the intestine, some of it may be trapped by soluble fibers or by some medications, which carry it out of the body in feces. The excretion of bile reduces the total amount of cholesterol remaining in the body.

Both the intestine and the liver make lipoproteins. Chapter 2 tells the story of lipid transport.

Cholesterol Transport Recall from Chapter 2 that some cholesterol, packaged with other lipids and protein, leaves the liver via the arteries and is transported to the body tissues by the blood. These packages of lipids and proteins are called lipoproteins. As the lipoproteins travel through the body, tissues can extract lipids from them. Cholesterol's harmful effects in the body occur when it forms deposits in the artery walls. These deposits contribute to **atherosclerosis**, a disease that can cause heart attacks and strokes.

sterols: one of the three main classes of lipids; include cholesterol, vitamin D, and the sex hormones (such as testosterone).

atherosclerosis (ath-er-oh-scler-OH-sis): a type of artery disease characterized by accumulations of lipid-containing material on the inner walls of the arteries (see Chapter 21).

REVIEW NOTES

Table 4-2 summarizes the members of the lipid family.

The predominant lipids both in foods and in the body are triglycerides, which have glycerol backbones with three fatty acids attached.

Fatty acids vary in the length of their carbon chains and their degree of saturation. Those that are fully loaded with hydrogens are saturated; those that are missing hydrogens and therefore have double bonds are unsaturated (monounsaturated or polyunsaturated).

Most triglycerides contain more than one type of fatty acid.

Fatty acid saturation affects the physical characteristics and storage properties of fats.

Hydrogenation, which makes polyunsaturated fats more saturated, gives rise to *trans*-fatty acids, altered fatty acids that may have health effects similar to those of saturated fatty acids.

Linoleic acid and linolenic acid are essential nutrients. In addition to serving as structural parts of cell membranes, they make powerful substances that help regulate blood pressure, blood clot formation, and the immune response.

Phospholipids, including the lecithins, have a unique chemical structure that allows them to be soluble in both water and fat.

In the body, phospholipids are major constituents of cell membranes; the food industry uses phospholipids as emulsifiers.

Sterols include cholesterol, bile, vitamin D, and the sex hormones.

Only animal-derived foods contain significant amounts of cholesterol.

Health Effects and Recommended Intakes of Fats

Of all the dietary factors related to chronic diseases prevalent in developed countries, high intakes of certain fats are by far the most significant. The person who chooses a diet too high in saturated fats or *trans* fats invites the risk of **cardiovascular disease (CVD)**, and heart disease is the number one killer of adults in the United States and Canada. As for cancer, evidence is less compelling than for heart disease, but it does suggest that a diet high in certain kinds of fat is associated with a greater-than-average risk of developing some types of cancer.[6] Conversely, some research suggests that omega-3 fatty acids from fish may protect against some cancers.[7] Obesity carries serious risks to health, and the high energy density of fatty foods makes it easy for people to exceed their energy needs and so gain unneeded weight. The links between diet and disease are the focus of much research. Diet and disease prevention is the topic of Nutrition in Practice 12. Some points about fats and heart health are presented here because they underlie dietary recommendations concerning fats.

FATS AND HEART HEALTH

As noted earlier and described in Chapter 2, cholesterol travels in the blood within lipoproteins. Two of the lipoproteins, LDL and HDL, play major roles with regard to heart health and are the focus of most recommendations made for reducing the risk of heart disease. A high blood LDL cholesterol concentration is a predictor of the likelihood of suffering a fatal heart attack or stroke, and the higher the LDL, the earlier the episode is expected to occur. Conversely, high HDL cholesterol signifies a *lower* disease risk.

Most people realize that elevated blood cholesterol is an important risk factor for heart disease. Most people may not realize, though, that cholesterol in *food* is not the main influential factor in raising *blood* cholesterol.

Saturated Fats and Blood Cholesterol The main dietary factors associated with elevated blood LDL cholesterol are high saturated fat and high *trans* fat intakes.* LDL cholesterol indicates a risk of heart disease because high LDL concentrations promote the uptake of cholesterol in the blood vessel walls.

Fats from animal sources are the main contributors of saturated fats in most people's diets. Some vegetable fats (coconut oil, palm kernel oil, and palm oil) and hydrogenated fats such as shortening or stick margarine provide smaller amounts of saturated fats. To minimize intake of saturated fat, most people need to eat less meat. When eating meat, choose the leanest cuts and trim away the visible fat.[8] Selecting fat-free milk and using nonhydrogenated margarine and unsaturated cooking oil such as olive oil, safflower oil, or canola oil are other simple changes that can dramatically lower saturated fat intake and heart disease risk.[9]

*It should be noted that not all saturated fatty acids have the same cholesterol-raising effect. Stearic acid, an 18-carbon fatty acid, does not seem to raise blood cholesterol.

Nutrition and cancer is a topic of Chapter 23.

The health risks of obesity are described in Chapter 6, and Chapter 7 focuses on weight management.

Nutrition and heart disease is the topic of Chapter 21.

Major sources of saturated fats:
- Whole milk (and even reduced-fat milk), cheese, butter, cream, cream cheese, sour cream, and ice cream.
- Fatty cuts of beef and pork, and processed meats such as bacon, sausage, and hot dogs.
- Tropical oils (coconut, palm, and palm kernel) and products that contain them such as cakes, cookies, doughnuts, and pastries.
- Shortening and lard.

Nutrition in Practice 4 examines various types of fats and their roles in supporting or harming heart health.

cardiovascular disease (CVD): a general term for all diseases of the heart and blood vessels (see Chapter 21).

The words *hydrogenated vegetable oil or shortening* in an ingredients list indicate *trans*-fatty acids in the product.

Trans-Fatty Acids and Blood Cholesterol Consuming fats with *trans*-fatty acids poses a risk to the health of the heart and arteries by raising LDL and lowering HDL cholesterol, and by producing inflammation.[10] When news of *trans*-fatty acids' effects on heart health was first emerging, some people hastily switched from using margarine back to butter, believing oversimplified reports that margarine provided no heart health advantage over butter. It is true that most margarines and virtually all shortenings are made from mostly hydrogenated fats and therefore contain substantial *trans*-fatty acids—up to 40 percent by weight. Some margarines, however, especially the soft or liquid varieties, are made from unhydrogenated oils. These have long proved to be less likely to elevate blood cholesterol than the saturated fats of butter. When oils (but not hydrogenated oils) are the first ingredient listed on a margarine label, that margarine is probably low in both *trans*-fatty acids and saturated fat. In general, margarine contributes less *trans* fat to the average diet than do other contributors such as processed foods.

In addition to soft and liquid margarine choices, some margarines are now specially formulated to contain few or no *trans*-fatty acids. Other types contain **sterol esters**, a functional food ingredient that reduces blood cholesterol when consumed as part of a diet low in saturated fat.* Sterol esters are not recognized by the intestine and therefore are not absorbed, and they also block the absorption of cholesterol. Simply adding the margarine to a high-fat diet is unlikely to bring benefits, however. Sterol esters work only when people cut their fat intakes as well. Drawbacks include the price (three or four times higher than regular margarine), a high fat content (the full-fat kind equals the fat in regular margarine), and an unproven record of safety for use by certain populations, such as growing children.

Many foods other than margarine also contribute *trans*-fatty acids. Fast foods, chips, baked goods, and other commercially prepared foods are high in fats containing up to 50 percent *trans*-fatty acids. Fast-food chains fry foods in hydrogenated vegetable oil that contains abundant *trans*-fatty acids, although healthier commercial fats are now being developed. Overall, consumers are now eating more fats containing *trans*-fatty acids than ever before, amounting to about 3 percent of daily kcalories, and they are eating these *trans*-fatty acids in the form of processed foods (see the margin list).[11]

As of 2006, all food labels must list grams of *trans* fat in foods to help consumers make informed choices.[12] Reducing total fat and replacing both saturated and *trans* fats with monounsaturated and polyunsaturated fats may be the wisest strategies for preventing heart disease. To this end, many manufacturers are reformulating foods to reduce their contents of harmful *trans* fats. At the same time, as a later section describes, food scientists are perfecting fat replacers intended to eliminate added fats altogether.

Major food sources of *trans* fats:
- Cakes, cookies, pies, doughnuts, and crackers.
- Meat and dairy products.
- Hard margarine.
- Fried potatoes and other commercial fried foods such as chicken.
- Potato chips and corn chips.
- Shortening.

Dietary Cholesterol and Blood Cholesterol Dietary cholesterol has also been implicated in raising blood cholesterol and increasing the risk of heart disease, although its effect is not as strong as that of saturated fat or *trans* fat. Still, health experts advise limiting cholesterol intake.

Recall that cholesterol is found primarily in foods derived from animals. Consequently, eating less fat from meats, eggs, and milk products helps lower dietary cholesterol intake (as well as total and saturated fat intakes).

Major sources of cholesterol:
- Eggs.
- Meat, poultry, and shellfish.
- Cheese.
- Milk.

Monounsaturated Fatty Acids and Blood Cholesterol Replacing saturated and *trans* fats with monounsaturated fat such as olive oil may be an effective dietary strategy to prevent heart disease. The lower rates of heart disease among people in the Mediterranean region of the world are often attributed to their liberal use of olive oil, a rich source of monounsaturated fatty acids.[13] Olive oil also delivers

sterol esters: compounds, derived from plants, that belong to the sterol family of lipids and have been shown experimentally to reduce blood cholesterol when consumed in place of other fats in a low–saturated fat diet.

---

*Two brand names of margarines with sterol esters currently on the market are *Benecol* and *Take Control*.

valuable phytochemicals that help to protect against heart disease. Nutrition in Practice 4 examines the role of olive oil and other fats in supporting or harming heart health.

Polyunsaturated Fatty Acids, Blood Cholesterol, and Heart Disease Risk Polyunsaturated fatty acids (PUFA) of the omega-6 and omega-3 families are potent protectors against heart disease. The primary omega-6 fatty acid, linoleic acid, which is found in vegetable oils such as corn and safflower oil, exerts most of its beneficial effect by lowering both total blood cholesterol and LDL cholesterol.[14] In fact, linoleic acid is the only dietary fatty acid that effectively lowers LDL cholesterol.[15] Possibly, linoleic acid works by slowing production of LDL and by enhancing LDL clearance.

The omega-3 fatty acids, EPA and DHA, which are found mainly in fatty fish, exert their beneficial effects by influencing the function of both the heart and the blood vessels. Specifically, EPA and DHA protect heart health by:[16]

- Lowering blood triglycerides.
- Preventing blood clots.
- Protecting against irregular heartbeats.
- Lowering blood pressure.
- Defending against inflammation.

The primary member of the omega-3 family, linolenic acid, may benefit heart health as well, but evidence for this effect is much less certain than for EPA and DHA.

Balance Omega-6 and Omega-3 Intakes The U.S. diet is high in omega-6 fatty acids (due to the increased production and use of vegetable cooking oils such as corn and cottonseed oils) and low in omega-3 fatty acids. Experts recommend a more balanced intake. The best way for people to increase their intakes of omega-3 fatty acids is to follow the advice of the American Heart Association: eat at least two servings of fatty fish (see Table 4-3) each week.[17] Even one fish meal per week has been associated with a reduced risk of heart disease.[18]

Greater heart health benefits can be expected when fish is grilled, baked, or broiled, partly because the varieties prepared this way often contain more EPA and DHA than species used for fried fish in fast-food restaurants and frozen products. Additionally, benefits are attained by avoiding commercial frying fats, which

Grilling or broiling fish, instead of frying them, preserves their beneficial omega-3 fatty acids while adding little or no saturated fat.

TABLE 4-3	Food Sources of Omega-6 and Omega-3 Fatty Acids
Omega-6	
Linoleic acid	Leafy vegetables, seeds, nuts, grains, vegetable oils (corn, cottonseed, safflower, sesame, soybean, sunflower), poultry fat
Omega-3	
Linolenic acid[a]	Oils (canola, flaxseed, soybean, walnut, wheat germ; liquid or soft margarine made from canola or soybean oil) Nuts and seeds (butternuts, flaxseeds, walnuts, soybean kernels) Vegetables (soybeans)
EPA and DHA	Human milk Fatty coldwater fish[b] (Mackerel, salmon, bluefish, mullet, sablefish, menhaden, anchovy, herring, lake trout, sardines, tuna)

[a]Alpha-linolenic acid. Also found in the seed oil of the herb *evening primrose*.

[b]All of these fish except tuna provide at least 1 gram of omega-3 fatty acids in 100 grams of fish (3.5 ounces); the fish oil content of each species varies with the season and site of harvest. Tuna provides fewer omega-3 fatty acids, but because it is commonly consumed, its contribution can be significant.

are often laden with *trans* fat and saturated fat.[19] Further benefits arise when fish replaces high-fat meats or other foods rich in saturated fats in several meals each week. Although not every study supports a benefit from eating fish and fish oil, results from many population studies and controlled clinical trials support a recommendation to eat fish.[20]

Some species of fish and shellfish however, may contain significant levels of mercury or other environmental contaminants.[21] Most healthy people can safely consume most species of ocean fish several times a week, but some people face greater risks. Women who may become pregnant, pregnant and lactating women, and children are more sensitive to contaminants than others, but even they can benefit from safer fish varieties within recommended limits (see Chapter 10 for details). For everyone, consuming a variety of different types of fish is a good idea to minimize exposure to any single toxin that may accumulate in a favored species. The margin lists the species most heavily contaminated with mercury and those that are low in mercury. As for freshwater fish, consumers in some areas should check local advisories about the safety of fish caught by family and friends.

Omega-3 Supplements Omega-3 fatty acid supplements are aggressively marketed as a cure-all for many different diseases without regard for consumer safety. The Food and Drug Administration (FDA) does not permit labels to claim that fish oil supplements can prevent or cure diseases, but it does allow the claim that research is suggestive but inconclusive regarding heart disease. Experts agree that adding *fish* to the diet two or three times per week may help to prevent disease, but the idea that fish oil *supplements* are beneficial and safe in any amount is erroneous.

Most importantly, high intakes of omega-3 fatty acids may increase bleeding time, raise LDL cholesterol, and suppress immune function.[22] Fish oil supplements themselves may carry hazards. They may contain high levels of the two most potentially toxic vitamins, A and D. Fish oil supplements are made from fish skins and livers that may have accumulated toxic concentrations of pesticides, heavy metals such as mercury, and other industrial contaminants. Unless the oils are refined to eliminate them, such contaminants can become further concentrated in the pills.[23] Furthermore, omega-3 and omega-6 fatty acids compete for the same slots in the body. Consequently, taking supplements of one can interfere with normal body functions that depend on a proper balance between the two. Finally, supplements may not contain the quantities of omega-3 fatty acids that the labels claim.

RECOMMENDATIONS

Some fat in the diet is essential for good health, but too much fat, especially saturated fat and *trans* fat, increases the risks of chronic diseases. Defining the exact level of total fat intake at which risk of inadequacy or prevention of disease occurs is not possible; for this reason, no RDA or upper limit has been set.[24] Instead, the recommendation for total fat is set at 20 to 35 percent of the daily energy intake. The top end of this range is slightly higher than previous recommendations. Diets with up to 35 percent of kcalories from fat can be compatible with good health if energy intake is reasonable and saturated fat intake is low. When total fat intake exceeds 35 percent of kcalories, saturated fat intakes can increase to unhealthy levels. Fat and oil intakes below 20 percent of kcalories increase the risk of inadequate essential fatty acid intakes. The FDA established Daily Values on food labels using 30 percent of energy intake as the guideline for fat.

Part of the allowance for total fat should provide for the essential fatty acids—linoleic acid and linolenic acid. Recommendations suggest that linoleic acids provide 5 to 10 percent of the daily energy intake and linolenic acid, 0.6 to 1.2 percent.

Saturated fats, *trans* fats, and cholesterol increase total blood cholesterol and LDL cholesterol and therefore the risk of heart disease. Even low intakes of each

---

Margin notes:

- Fish most heavily contaminated with mercury: shark, swordfish, king mackerel, and tilefish (also called golden bass or golden snapper).
- Fish or shellfish low in mercury: shrimp, canned light tuna, salmon, pollock, and catfish. Canned albacore ("white") tuna contains more mercury than light tuna.

DRI for fat:
- 20 to 35% of energy intake.

Dietary Guidelines for Americans 2005:
- Keep total fat intake between 20 and 35% of kcalories, with most fats coming from sources of polyunsaturated and monounsaturated fatty acids, such as fish, nuts, and vegetable oils.

Daily Value:
- 65 g fat (based on 30% of 2000 kcal diet).

Linoleic acid AI:
- 5 to 10% of energy intake.

Men:
- 19–50 yr: 17 g/day.
- 51+ yr: 14 g/day.

Women:
- 19–50 yr: 12 g/day.
- 51+ yr: 11 g/day.

Linolenic acid AI:
- 0.6 to 1.2% of energy intake.

Men: 1.6 g/day.

Women: 1.1 g/day.

may elevate heart disease risk.[25] Thus recommendations urge people to eat diets that are low in saturated fat, *trans* fat, and cholesterol. Specifically, consume less than 10 percent of kcalories from saturated fat, keep *trans* fat intakes as low as possible, and consume less than 300 milligrams of cholesterol each day.[26] To help consumers meet these goals, the FDA established Daily Values on food labels using 10 percent of energy intake for saturated fat; the Daily Value for cholesterol is 300 milligrams, regardless of energy intake. There is no Daily Value for *trans* fat.

Nutrition in Practice 4 suggests substituting monounsaturated or polyunsaturated fats for harmful saturated fats. No beneficial change in blood lipids occurs when monounsaturated or polyunsaturated fat is *added* to a diet rich in saturated or *trans* fat, however. The best diet for heart health is also rich in fruits, vegetables, nuts, and whole grains that offer many health advantages by supplying abundant nutrients, fiber, and phytochemicals.

REVIEW NOTES

High intakes of saturated or *trans* fats contribute to heart disease, obesity, and other health problems.

High blood cholesterol, specifically, poses a risk of heart disease, and high intakes of saturated fat contribute most to high blood cholesterol. High intakes of *trans*-fatty acids also appear to raise blood cholesterol. Cholesterol in foods presents less of a risk.

Polyunsaturated fatty acids of the omega-6 and omega-3 families protect against heart disease.

When monounsaturated fat such as olive oil replaces saturated and *trans* fats in the diet, the risk of heart disease may be lessened.

Though some fat in the diet is necessary, health authorities recommend a diet moderate in total fat, and low in saturated fat, *trans* fat, and cholesterol.

Dietary Guidelines for Americans 2005:

- Consume less than 10% of kcalories from saturated fatty acids and less than 300 mg per day of cholesterol, and keep *trans*-fatty acid consumption as low as possible.

Maximum saturated fat intakes set by the *Dietary Guidelines for Americans 2005:*

1600 kcal diet: 18 g.
2000 kcal diet: 20 g.
2200 kcal diet: 24 g.
2500 kcal diet: 25 g.
2800 kcal diet: 31 g.

Daily Values:

- 20 g saturated fat (based on 10% of 2000 kcal diet).
- 300 mg cholesterol.

Fats in Foods

Fats are important in foods as well as in the body. Many of the compounds that give foods their flavors and aromas are found in fats and oils. The delicious aromas associated with bacon, ham, and other meats, as well as with onions being sautéed, come from fats. Fats also influence the texture of many foods, enhancing smoothness, creaminess, moistness, or crispness. In addition, four vitamins—A, D, E, and K—are soluble in fat. When the fat is removed from a food, many fat-soluble compounds, including these vitamins, are also removed. Table 4-4 summarizes the roles of fats in foods.

Fats are also an important part of most people's ethnic or national cuisines. Each culture has its own favorite food sources of fats and oils. In Canada, canola oil (also known as rapeseed oil) is widely used. In the Mediterranean area, Greeks, Italians, and Spaniards rely heavily on olive oil. Both canola oil and olive oil are rich sources of monounsaturated fatty acids. Asians use the polyunsaturated oil of soybeans. Jewish people traditionally employ chicken fat. Everywhere in North America, butter and margarine are widely used.

FINDING THE FATS IN FOODS

The remainder of this chapter and the Nutrition in Practice show you how to choose fats wisely with the goals of providing optimal health and pleasure in eating. To achieve such goals, you need to know which foods offer

TABLE 4-4	The Functions of Fats in Foods
Fats in foods:	
• Contribute flavor and aroma.	
• Influence the texture, adding creaminess, smoothness, moistness, or crispness.	
• Help make foods tender.	
• Carry fat-soluble vitamins.	

unsaturated fats that provide the essential fatty acids and which offer harmful saturated and *trans* fats. Perhaps most important for many people is learning to control portion sizes, particularly portions of fatty foods that can pack hundreds of kcalories into just a few bites.

The *Dietary Guidelines for Americans 2005* urge that, beyond a healthy minimum, people should limit their fat intakes in order to limit intakes of saturated fat, *trans* fat, and kcalories. To do this requires consistently making nutrient-dense choices, such as fat-free milk, fat-free cheese, and the leanest meats, and refraining from adding solid fats, such as butter, hard margarine, or shortening, to foods during preparation or at the table. Higher-fat foods may be included in the diet, but the fat kcalories they provide must fit within a person's discretionary kcalorie allowance. Exceptions are the fats of fatty fish, nuts, and vegetable oils: they provide EPA and DHA, linoleic acid and linolenic acid, and vitamin E needed for a healthful diet. Within kcalorie limits, the kcalories these fats provide are necessary, not discretionary. Many people, especially sedentary people, have few or no discretionary kcalories to spend on high-fat foods.

Added Fats in Foods A dollop of dessert topping, a spread of butter on bread, oil or shortening in a recipe, dressing on a salad—all of these are examples of *added* fats. Indeed, all sorts of fats can be added to foods during commercial or home preparation or at the table. The following amounts of these fats contain about 5 grams of pure fat, providing 45 kcalories and negligible protein and carbohydrate:

- 1 teaspoon of oil or shortening.
- 1½ teaspoons of mayonnaise, butter, or margarine.
- 1 tablespoon of regular salad dressing, cream cheese, or heavy cream.
- 1½ tablespoons of sour cream.

The majority of added fats in the diet are invisible. They are the hidden fats of fried foods and baked goods, sauces and mixed dishes, and dips and spreads. Other invisible fats include the fats in the marbling of meat and the fat ground into lunch meats and hamburger.

Milk, Yogurt, and Cheese The fat in whole milk is about 63 percent saturated fat; the cholesterol content is 24 milligrams per cup for whole milk or 5 milligrams for fat-free milk. Thus choosing fat-free in place of whole milk reduces your intake of cholesterol as well as of saturated fat.

Note that cream and butter do not appear in the milk group. Milk and yogurt are rich in calcium and protein, but cream and butter are not. Cream and butter are fats, as are whipped cream, sour cream, and cream cheese, so they are grouped together with the solid fats. That is why the food group that includes milk is carefully labeled the "milk, yogurt, and cheese group," not the "dairy group." Cheeses are major contributors of saturated fat in people's diets.

Meat, Poultry, Fish, Legumes, Eggs, and Nuts Meats conceal a good deal of the fat—and much of the saturated fat—that people consume. To help "see" the fat in meats, it is useful to think of them in four categories according to their fat contents: very lean, lean, medium-fat, and high-fat meats, as the exchange lists do in Appendix C. Meats in all four categories contain about equal amounts of protein, but their fat contents differ and their saturated fat and kcalorie amounts vary significantly. Table 1-8 on page 20 in Chapter 1 provided some definitions concerning the fat contents of meats.

The 2005 USDA Food Guide suggests that most adults limit a day's intake of meats or equivalents to about 5 to 7 ounces. For comparison, the smallest fast-food hamburger weighs about 3 ounces. A steak served in a restaurant often runs 8, 12, or 16 ounces, more than a whole day's meat allowance. You may have to weigh a serving or two of meat to see how much you are eating.

Marginal notes:

Table 1-8 of Chapter 1 specifies how much oil is required to meet nutrient needs at several kcalorie levels.

The concept of discretionary kcalories was discussed in Chapter 1.

1 c whole milk:
- 8 g fat.
- 5 g saturated fat.
- 24 mg cholesterol.

1 c reduced-fat milk:
- 5 g fat.
- 2 g saturated fat.
- 20 mg cholesterol.

1 c low-fat milk:
- 2 g fat.
- 1.5 g saturated fat.
- 10 mg cholesterol.

1 c fat-free milk:
- 0 g fat.
- 0 g saturated fat.
- 5 mg cholesterol.

People think of meat as protein food, but calculation of its nutrient content reveals a surprising fact. A big (4-ounce), fast-food hamburger sandwich contains 23 grams of protein and 20 grams of fat. Because protein offers 4 kcalories per gram and fat offers 9, the sandwich provides 92 kcalories from protein and about twice that amount from fat. The kcalorie total, counting carbohydrates from the bun and condiments, is over 400 kcalories, with more than 50 percent of them from fat. Hot dogs, fried chicken sandwiches, and fried fish sandwiches are also high-fat choices. Because so much of the energy in a meat eater's diet is hidden from view, people can easily overeat on high-fat food, making weight control difficult.

When choosing beef or pork, look for lean cuts named *loin* or *round* from which the fat can be trimmed. Eat small portions, too. As for chicken and turkey, these meats are naturally lean, but commercial processing and frying add fats, especially in "patties," "nuggets," "fingers," or "wings." Chicken wings are mostly skin, and a chicken stores most of its fat just under its skin. The tastiest wing snacks have also been fried in cooking fat (often a hydrogenated, saturated type with *trans*-fatty acids), smothered with a buttery, spicy sauce, and then dipped in blue cheese dressing, making wings an extraordinarily high-fat snack. People who snack on wings may want to plan on eating low-fat foods at several other meals to balance them out.

Vegetables, Fruits, and Grains Choosing vegetables, fruits, whole grains, and legumes also helps lower the saturated fat, cholesterol, and total fat content of the diet. Most vegetables and fruits naturally contain little or no fat; avocados and olives are exceptions, but most of their fat is unsaturated, which is not harmful to heart health. Most grains contain only small amounts of fat. Some grain *products* such as fried taco shells, croissants, and biscuits are high in saturated fat, so consumers need to read food labels. Similarly, many people add butter, margarine, or cheese sauce to grains and vegetables, which raises their saturated and *trans* fat contents. Because fruits are often eaten without added fat, a diet that includes several servings of fruit daily can help a person meet the dietary recommendations for fat.

A diet rich in vegetables, fruits, whole grains, and legumes offers abundant vitamin C, folate, vitamin A, vitamin E, and dietary fiber—all important in supporting health. Consequently, such a diet protects against disease by both reducing saturated fat, cholesterol, and total fat and by increasing nutrients. It also provides valuable phytochemicals that help defend against heart disease.

CUTTING FAT INTAKE AND CHOOSING UNSATURATED FATS

Knowing which foods contain the most fat is the first step toward meeting the recommendation to limit dietary fat in general and saturated fat in particular. As a general rule, a person who eats meat and wishes to reduce both saturated fat and cholesterol intake can eat fewer high-fat meats and dairy foods, fewer eggs, and more poultry (without the skin), fish, and fat-free dairy products. A vegetarian who eats dairy products and eggs can shift to fat-free milk and fat-free cheeses and limit butter and egg intake. Vegetarians who omit animal-derived foods generally eat less saturated fat and little or no cholesterol because plant foods do not contain significant amounts of cholesterol. The accompanying "How to" (p. 98) offers strategies for making heart-healthy choices, food group by food group.

Fats and kCalories Removing fat from food also removes energy as Figure 4-5 (p. 99) shows. A pork chop with the fat trimmed to within a half-inch of the lean provides 340 kcalories; with the fat trimmed off completely, it supplies 230 kcalories. A baked potato with butter and sour cream (1 tablespoon each) has 350 kcalories; a plain baked potato has 220 kcalories. The single most effective step you can take to reduce the energy value of a food is to eat it with less fat.

At room temperature, unsaturated fats (such as those found in oil) are usually liquid, whereas saturated fats (such as those found in butter) are solid.

Choosing Unsaturated Fats When a person does eat fats, those to choose are the unsaturated ones. Remember, the softer a fat is, the more unsaturated it is. Generally speaking, vegetable and fish oils are rich in polyunsaturates, olive oil and canola oil are rich in monounsaturates, and the harder fats—animal fats—are more saturated (see Figure 4-6).

Don't Overdo Fat Restriction Some people actually manage to eat too *little* fat—to their detriment. Among them are young women and men with eating disorders, described in Nutrition in Practice 7. As a practical guideline, it is wise to include the equivalent of at least a teaspoon of fat in every meal.

 HOW TO *Make Heart-Healthy Choices—by Food Group*

Breads and Cereals

- Select breads, cereals, and crackers that are low in saturated and *trans* fat (for example, bagels instead of croissants).
- Prepare pasta with a tomato sauce instead of a cheese or cream sauce.
- Use fruit butters or jellies instead of butter or margarine on bread.

Vegetables and Fruits

- Enjoy the natural flavor of steamed vegetables for dinner and fruits for dessert.
- Eat at least two vegetables (in addition to a salad) with dinner.
- Snack on raw vegetables or fruits instead of high-fat items like potato chips.
- Buy frozen vegetables without sauce.

Milk and Milk Products

- Switch from whole milk to reduced-fat, from reduced-fat to low-fat, and from low-fat to fat-free (nonfat) milk.
- Use fat-free and low-fat cheeses (such as part-skim ricotta and low-fat mozzarella) instead of regular cheeses.
- Use fat-free or low-fat yogurt or sour cream instead of regular sour cream.
- Use evaporated fat-free milk instead of cream.
- Enjoy fat-free frozen yogurt, sherbet, or ice milk instead of ice cream.

Meat and Meat Alternates

- Fat adds up quickly, even with lean meat; limit intake to about 6 ounces (cooked weight) daily.
- Eat at least two servings of fish per week (particularly fish such as mackerel, lake trout, herring, sardines, and salmon).
- Choose fish, poultry, or lean cuts of pork or beef; look for unmarbled cuts named *round* or *loin* (eye of round, top round, bottom round, round tip, tenderloin, sirloin, center loin, and top loin).
- When choosing processed meats such as lunch meats and hot dogs, choose those that are low in saturated fat and cholesterol.
- Trim the fat from pork and beef; remove the skin from poultry.

- Grill, roast, broil, bake, stir-fry, stew, or braise meats; don't fry. When possible, place food on a rack so that fat can drain.
- Use lean ground turkey or lean ground beef in recipes; brown ground meats without added fat, then drain off fat.
- Select tuna, sardines, and other canned meats packed in water; rinse oil-packed items with hot water to remove much of the fat.
- Fill kabob skewers with lots of vegetables and slivers of meat; create main dishes and casseroles by combining a little meat, fish, or poultry with a lot of pasta, rice, or vegetables.
- Use legumes often.
- Eat a meatless meal or two daily.
- Use egg substitutes in recipes instead of whole eggs or use two egg whites in place of each whole egg.

Fats and Oils

- Use butter or margarine sparingly; select soft margarines instead of hard margarines.
- Limit use of lard and meat fat.
- Limit use of products made with coconut oil, palm kernel oil, and palm oil (read labels on bakery goods, processed foods, popcorn oils, and nondairy creamers).
- Reduce use of hydrogenated shortenings and margarines and products that contain them (read labels on crackers, cookies, and other commercially prepared baked goods).

Miscellaneous

- Use a nonstick pan or coat the pan lightly with vegetable oil.
- Refrigerate soups and stews; when the fat solidifies, remove it.
- Use wine; lemon, orange, or tomato juice; herbs; spices; fruits; or broth instead of butter or margarine when cooking.
- Stir-fry in a small amount of oil; add moisture and flavor with broth, tomato juice, or wine.
- Use variety to enhance enjoyment of the meal: vary colors, textures, and temperatures—hot cooked versus cool raw foods—and use garnishes to complement food.

Source: Adapted from Expert Panel on Detection, Evaluation, and Treatment of High Blood Cholesterol in Adults (Adult Treatment Panel III), *Third Report of the National Cholesterol Education Program (NCEP)*, NIH publication no. 02-5215 (Bethesda, Md.: National Heart, Lung, and Blood Institute, 2002), pp. V-25–V-27.

Pork chop with fat (340 kcal, 19 g fat, 7 g saturated fat).

Potato with 1 tbs butter and 1 tbs sour cream (350 kcal, 14 g fat, 10 g saturated fat).

Whole milk, 1 c (150 kcal, 8 g fat, 5 g saturated fat).

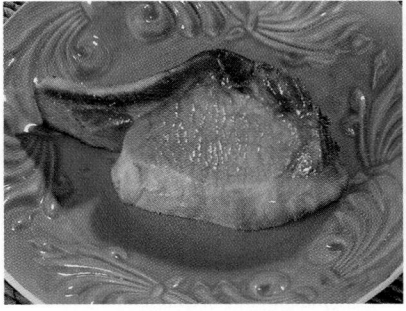

Pork chop with fat trimmed off (230 kcal, 9 g fat, 3 g saturated fat).

Plain potato (220 kcal, <1 g fat, 0 g saturated fat).

Fat-free milk, 1 c (90 kcal, <1 g fat, <1 g saturated fat).

© Polara Studios, Inc. (all)

FIGURE 4-5 Cutting Fat Cuts kCalories—and Saturated Fat

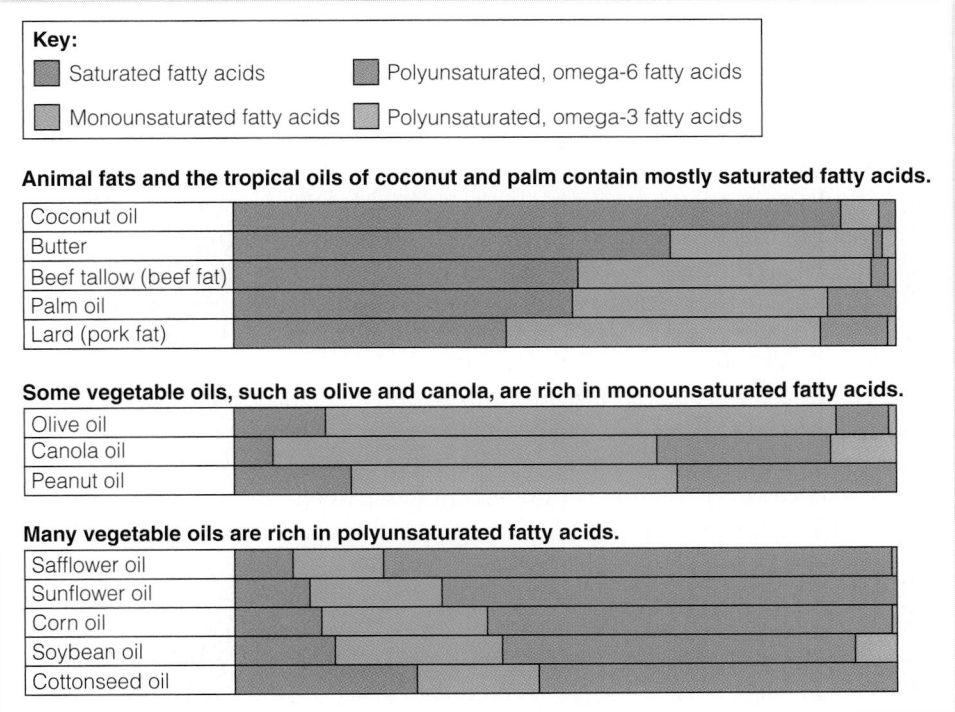

FIGURE 4-6 Comparison of Dietary Fats
Most fats are a mixture of saturated, monounsaturated, and polyunsaturated fatty acids.

Cautions People who wish to make choices consistent with current recommendations need to learn how to read food labels, limit fat in general, and seek out the polyunsaturated and monounsaturated fats in preference to the saturated ones. Remember, however: vegetable fat or vegetable oil doesn't always mean unsaturated fat. Both coconut oil and palm oil, for example, which are often used in nondairy creamers, are saturated fats, and both raise blood cholesterol.

Fat Replacers Today, consumers can choose from thousands of fat-reduced products. Many bakery goods, lunch meats, cheeses, spreads, frozen desserts, and other products made with **fat replacers** offer less than half a gram of fat, saturated fat, and *trans* fat in a serving. Some of these products contain **artificial fats**, and others use conventional ingredients in unconventional ways to reduce fats and kcalories. Among the latter, manufacturers can:

- water or whip air into foods.
- Add fat-free milk to creamy foods.
- Use lean meats and soy protein to replace high-fat meats.
- Bake foods instead of frying them.

Common food ingredients such as fibers, sugars, or proteins can also take the place of fats in some foods. In particular, fat replacers made from oats or barley not only cut down on fats in foods but introduce beneficial viscous fibers, while imparting desirable tastes and textures associated with real fats.[27] Products made from sugars or proteins still provide kcalories, but far fewer kcalories from fats. Manufactured fat replacers consist of chemical derivatives of carbohydrate, protein, or fat, or modified versions of foods rich in those constituents.

A familiar example of an artificial fat that has been approved for use in snack foods such as potato chips, crackers, and tortilla chips is **olestra**. Olestra's chemical structure is similar to that of a regular fat (a triglyceride) but with important differences. A triglyceride is composed of a glycerol molecule with three fatty acids attached, whereas olestra is made of a sucrose molecule with six to eight fatty acids attached. Enzymes in the digestive tract cannot break the bonds of olestra, so unlike sucrose or fatty acids, olestra passes through the system unabsorbed.

The FDA's evaluation of olestra's safety addressed two questions. First, is olestra toxic? Research on both animals and humans supports the safety of olestra as a partial replacement for dietary fats and oils, with no reports of cancer or birth defects. Second, does olestra affect either nutrient absorption or the health of the digestive tract? When olestra passes through the digestive tract unabsorbed, it binds with the fat-soluble vitamins A, D, E, and K and carries them out of the body, robbing the person of these valuable nutrients. To compensate for these losses, the FDA requires the manufacturer to fortify olestra with vitamins A, D, E, and K. Saturating olestra with these vitamins does not make the product a good source of vitamins, but it does block olestra's ability to bind with the vitamins from other foods. An asterisk in the ingredients list informs consumers that these added vitamins are "dietarily insignificant."

Some consumers experience digestive distress with olestra consumption: cramps, gas, bloating, and diarrhea. The FDA initially required a label warning stating that "olestra may cause abdominal cramping and loose stools" and that it "inhibits the absorption of some vitamins and other nutrients," but recently concluded that such a statement is no longer warranted.

Consumers need to keep in mind that low-fat and fat-free foods still deliver kcalories. Decades ago, consumers hailed the arrival of artificial sweeteners as a weight-loss wonder, but in reality, kcalories saved by using artificial sweeteners were readily replaced by kcalories from other foods. Alternatives to fat can help to lower energy intake and support weight loss only when they actually *replace* fat and energy in the diet.

fat replacers: ingredients that replace some or all of the functions of fat in foods and may or may not provide energy.

artificial fats: zero-energy fat replacers that are chemically synthesized to mimic the sensory and cooking qualities of naturally occurring fats but are totally or partially resistant to digestion.

olestra: a synthetic fat made from sucrose and fatty acids that provides zero kcalories per gram; also known as *sucrose polyester*.

Read Food Labels Labels list total fat, *trans* fat, saturated fat, and cholesterol contents of foods, as well as fat kcalories, per serving. Because each package provides information for a single serving and serving sizes are standardized, consumers can easily compare similar products.

REVIEW NOTES

Fats in foods contribute to sensory appeal—enhancing the flavor, aroma, and texture of foods.

Fats in foods deliver fat-soluble vitamins, energy, and essential fatty acids.

Though some fat in the diet is necessary, the *Dietary Guidelines for Americans 2005* recommend limiting fat intakes in order to limit intakes of saturated fat and *trans* fat.

Fats added to foods during preparation or at the table are a major source of fat in the diet.

The choice between whole and fat-free milk products can make a big difference to the fat, saturated fat, and cholesterol content of a diet.

Meats account for a large proportion of the hidden fat and saturated fat in many people's diets.

Most people consume more meat than is recommended.

Most vegetables and fruits naturally contain little or no fat.

Grain products such as croissants and biscuits can be high in saturated fat, so consumers need to read food labels to learn which foods in this group contain fats.

Consumers today can choose from an array of fat-reduced products, and many bakery goods and other foods made with fat replacers offer less than half a gram of fat, saturated fat, and *trans* fat in a serving.

Some products use artificial fats such as olestra, while others use conventional ingredients such as water or fat-free milk to reduce fat and kcalories.

Food labels list total fat, saturated fat, cholesterol, and *trans* fat, as well as fat kcalories, per serving.

Chapters 3 and 4 have looked briefly at the two major energy fuels in the body—carbohydrate and fat. When used for energy, each has desirable characteristics. The glucose derived from carbohydrate is needed by the brain and nerve tissues and is easily used for energy in other cells. Fat is a particularly useful fuel because the body stores it efficiently and in generous amounts. Chapter 5 looks at protein, a nutrient that can be used as fuel, but whose primary role is to provide machinery for getting things done.

SELF CHECK

1. Three classes of lipids in the body are:
   a. triglycerides, fatty acids, and cholesterol.
   b. triglycerides, phospholipids, and sterols.
   c. fatty acids, phospholipids, and cholesterol.
   d. glycerol, fatty acids, and triglycerides.

2. A triglyceride consists of:
   a. three glycerols attached to a lipid.
   b. three fatty acids attached to a glucose.
   c. three fatty acids attached to a glycerol.
   d. three phospholipids attached to a cholesterol.

3. A fatty acid that has the maximum possible number of hydrogen atoms is known as a(n):
   a. saturated fatty acid.
   b. monounsaturated fatty acid.
   c. PUFA.
   d. essential fatty acid.

4. The difference between *cis*- and *trans*-fatty acids is:
   a. the number of double bonds.
   b. the length of their carbon chains.
   c. the location of the first double bond.
   d. the configuration around the double bond.

5. Essential fatty acids:
   a. are used to make substances that regulate blood pressure, among other functions.
   b. can be made from carbohydrates.
   c. include lecithin and cholesterol.
   d. cannot be found in commonly eaten foods.

6. Lecithins and other phospholipids in the body function as:
   a. emulsifiers.
   b. enzymes.
   c. temperature regulators.
   d. shock absorbers.

7. To minimize saturated fat intake and lower the risk of heart disease, most people need to:
   a. eat less meat.
   b. select fat-free milk.
   c. use nonhydrogenated margarines and cooking oils such as olive oil or canola oil.
   d. all of the above.

8. To include omega-3 fatty acids in the diet, the American Heart Association recommends eating:
   a. cholesterol-free margarine.
   b. fish oil supplements.
   c. hydrogenated margarine.
   d. at least two fish meals per week.

9. Some examples of foods with hidden fats are:
   a. cheese, lettuce, and fruit juices.
   b. fried foods, sauces, dips, and lunch meats.
   c. fish, rice, and potatoes.
   d. baked potatoes, vegetables, and fruits.

10. Generally speaking, vegetable and fish oils are rich in:
    a. polyunsaturated fat.
    b. saturated fat.
    c. cholesterol.
    d. *trans*-fatty acids.

Answers to these questions are found in Appendix H.

CLINICAL APPLICATIONS

1. The connection between the overconsumption of fats, especially saturated and *trans* fat, and chronic diseases (obesity, diabetes, cancer, and cardiovascular disease) underscores the importance of being alert to a client's fat intake. What advice would you offer a client who reports the following?
   - Eats two or more 6-ounce servings of meat each day.
   - Drinks whole milk and eats regular cheddar cheese each day.
   - Eats four to five servings of breads and cereals each day, including a bagel with cream cheese for breakfast, a bologna sandwich on white bread for lunch, and biscuits or cornbread with butter to accompany dinner.
   - Eats one serving of fruit each day and eats vegetables only on occasion.

2. Make a list of foods, beverages, and seasonings that your client can substitute for foods high in saturated fat.

NUTRITION ON THE NET

For further study of the topics in this chapter, access these websites.

Find updates and quick links to these and other nutrition-related sites at our website: **www.wadsworth.com/nutrition**

Search for cholesterol and dietary fat at the U.S. Government health information site: **www.healthfinder.gov**

Search for "fat" at the International Food Information Council site: **www.ific.org**

NOTES

[1]G. Fruhbeck, The adipose tissue as a source of vasoactive factors, *Current Medicinal Chemistry Cardiovascular and Hematological Agents* 3 (2004): 197–208; R. L. Bradley, K. A. Cleveland, and B. Cheatham, The adipocyte as a secretory organ: Mechanisms of vesicle transport and secretory pathways, *Recent Progress in Hormone Research* 56 (2001): 329–358.

[2]K. C. Hayes, Relative cardiovascular benefits of n-6 and n-3 fatty acids, *Nutrition and the M.D.*, January 2005, pp. 1–4; H. Tapiero and coauthors, Polyunsaturated fatty acids (PUFA) and eicosanoids in human health and pathologies, *Biomedicine and Pharmacotherapy* 56 (2002): 215–222.

[3]J. M. Alessandri and coauthors, Polyunsaturated fatty acids in the central nervous system: Evolution of concepts and nutritional implications throughout life, *Reproduction, Nutrition, Development* 44 (2004): 509–538; R. Uauy and P. Mena, Lipids and neurodevelopment, *Nutrition Reviews* 59 (2001): S34–S46.

[4]D. R. Hoffman and coauthors, Maturation of visual acuity is accelerated in breast-fed term infants fed baby food containing DHA-enriched egg yolk, *Journal of Nutrition* 134 (2004): 2307–2313; Alessandri and coauthors, 2004; Uauy and Mena, 2001.

[5]Hayes, 2005; J. A. Nettleton and R. Katz, n-3 long-chain polyunsaturated fatty acids in type 2 diabetes: A review, *Journal of the American Dietetic Association* 105 (2005): 428–440; V. Wijendran and K. C. Hayes, Dietary n-6 and n-3 fatty acid balance and cardiovascular health, *Annual Review of Nutrition* 24 (2004): 597–615; M. F. Leitzmann and coauthors, Dietary intake of n-3 and n-6 fatty acids and the risk of prostate cancer, *American Journal of Clinical Nutrition* 80 (2004): 204–216; Tapiero and coauthors, 2002; P. M. Kris-Etherton and coauthors, Fish consumption, fish oil, omega-3 fatty acids, and cardiovascular disease, *Circulation* 106 (2002): 2747–2757.

[6]R. Stoeckli and U. Keller, Nutritional fats and the risk of type 2 diabetes and cancer, *Physiology and Behavior* 83 (2004): 611–615; Leitzmann and coauthors, 2004; S. A. Bingham and coauthors, Are imprecise methods obscuring a relation between fat and breast cancer? *Lancet* 362 (2003): 212–214; C. E. Spiegelman and coauthors, Premenopausal fat intake and risk of breast cancer, *Journal of the National Cancer Institute* 95 (2003): 1079–1085.

[7]W. E. Hardman, (n-3) fatty acids and cancer therapy, *Journal of Nutrition* 134 (2004): 3427S–3430S; National Cancer Policy Board, Institute of Medicine, S. J. Curry, T. Byers, and M. Hewitt, eds., *Fulfilling the Potential of Cancer Prevention and Early Detection* (Washington, D.C.: National Academies Press, 2003), p. 77; P. D. Terry, T. E. Rohan, and A. Wolk, Intakes of fish and marine fatty acids and the risks of cancers of the breast and prostate and of other hormone-related cancers: A review of the epidemiologic evidence, *American Journal of Clinical Nutrition* 77 (2003): 532–543.

[8]Standing Committee on the Scientific Evaluation of Dietary Reference Intakes, Food and Nutrition Board, Institute of Medicine, *Dietary Reference Intakes for Energy, Carbohydrate, Fiber, Fat, Fatty Acids, Cholesterol, Protein, and Amino Acids* (Washington, D.C.: National Academies Press, 2005): pp. 835–836.

[9]Expert Panel on Detection, Evaluation, and Treatment of High Blood Cholesterol in Adults (Adult Treatment Panel III), *Third Report of the National Cholesterol Education Program (NCEP)*, NIH publication no. 02-5215 (Bethesda, Md.: National Heart, Lung, and Blood Institute, 2002), pp. V-25 to V-27.

[10]Standing Committee on the Scientific Evaluation of Dietary Reference Intakes, 2005, pp. 494–505; D. J. Baer and coauthors, Dietary fatty acids affect plasma markers of inflammation in healthy men fed controlled diets: A randomized crossover study, *American Journal of Clinical Nutrition* 79 (2004): 969–973; D. Mozaffarian and coauthors, Dietary intake of *trans* fatty acids and systemic inflammation in women, *American Journal of Clinical Nutrition* 79 (2004): 606–612.

[11]S. L. Elias and S. M. Innis, Bakery foods are the major dietary source of *trans*-fatty acids among pregnant women with diets providing 30 percent energy from fats, *Journal of the American Dietetic Association* 102 (2002): 46–51.

[12]Revealing *trans* fats, *FDA Consumer*, September/October 2003, pp. 20–26.

[13]D. B Panagiotakos and coauthors, Can a Mediterranean diet moderate the development and clinical progression of coronary heart disease? *Medical Science Monitor* 10 (2004): RA193–198; A. H. Stark and Z. Madar, Olive oil as a functional food: Epidemiology and nutritional approaches, *Nutrition Reviews* 60 (2002): 170–176.

[14]V. Wijendran and K. C. Hayes, Dietary n-6 and n-3 fatty acid balance and cardiovascular health, *Annual Review of Nutrition* 24 (2004): 597–615.

[15]Wijendran and Hayes, 2004.

[16]Wijendran and Hayes, 2004; A. T. Erkkila and coauthors, Fish intake is associated with a reduced progression of coronary artery atherosclerosis in postmenopausal women with coronary artery disease, *American Journal of Clinical Nutrition* 80 (2004): 626–632; P. M. Kris-Etherton and coauthors, Fish consumption, fish oil, omega-3 fatty acids, and cardiovascular disease, *Circulation* 106 (2002): 2747–2757; F. B. Hu and coauthors, Fish and omega-3 fatty acid intake and risk of coronary heart disease in women, *Journal of the American Medical Association* 287 (2002): 1815–1821; C. M. Albert and coauthors, Blood levels of long-chain n-3 fatty acids and the risk of sudden death, *New England Journal of Medicine* 346 (2002): 1113–1118.

[17]R. M. Krauss and coauthors, American Heart Association Scientific Statement, AHA dietary guidelines, Revision 2000: A statement for healthcare professionals from the Nutrition Committee of the American Heart Association, available at **http://circ.ahajournals .org/cgi/content/full/4304635102**.

[18]Erkkila and coauthors, 2004.

[19]Kris-Etherton and coauthors, 2002.

[20]D. Mozaffarian, R. N. Lemaitre, and L. H. Kuller, Fish consumption and stroke risk in elderly individuals, The Cardiovascular Health Study, *Archives of Internal Medicine* 165 (2005): 200–206; Hayes, 2005; Wijendran and Hayes, 2004; Erkkila and coauthors, 2004.

[21]Backgrounder for the 2004 FDA/EPA Consumer Advisory: What you need to know about mercury in fish and shellfish, 2004, available at **www.fda.gov/oc/opacom/hottopics/mercury/ backgrounder.html**.

[22]Standing Committee on the Scientific Evaluation of Dietary Reference Intakes, 2005, pp. 486–494; H. E. Theobald and coauthors, LDL cholesterol-raising effect of low-dose docosahexaenoic acid in middle-aged men and women, *American Journal of Clinical Nutrition* 79 (2004): 558–563; S. Kew and coauthors, Effects of oils rich in eicosapentaenoic and docosahexaenoic acids on immune cell composition and function in healthy humans, *American Journal of Clinical Nutrition* 79 (2004): 674–681; S. Bechoua and coauthors, Influence of very low dietary intake of marine oil on some functional aspects of immune cells in healthy elderly people, *British Journal of Nutrition* 89 (2003): 523–532.

[23]Backgrounder for the 2004 FDA/EPA Consumer Advisory: What you need to know about mercury in fish and shellfish, 2004.

[24]Standing Committee on the Scientific Evaluation of Dietary Reference Intakes, 2005, pp. 769–770.

[25]Standing Committee on the Scientific Evaluation of Dietary Reference Intakes, 2005, pp. 835–836.

[26]U.S. Department of Agriculture and U.S. Department of Health and Human Services, *Dietary Guidelines for Americans 2005*, 6th ed., available online at **www.healthierus.gov**.

[27]Position of the American Dietetic Association: Fat replacers, *Journal of the American Dietetic Association* 105 (2005): 266–275.

Figuring Out Fats

To consumers, advice about dietary fat appears to change almost daily. "Eat less fat." "Eat more fatty fish." "Give up butter—use margarine instead." "Give up margarine—replace it with olive oil." "Steer clear of saturated." "Seek out omega-3." "Stay away from *trans*." "Stick with mono- and polyunsaturated." No wonder people feel confused about dietary fat. This Nutrition in Practice begins with a look at the latest dietary fat guidelines. It continues by identifying which foods provide which fats. It closes with strategies to help consumers choose the right amounts of the right kinds of fats for a healthy diet.

Why do today's fat messages seem to change constantly and become more confusing?

The confusion stems in part from the complexities of fat and in part from the nature of recommendations. As Chapter 4 explained, "dietary fat" refers to several kinds of fats; some fats support health whereas others damage it, and foods typically provide a mixture of fats in varying proportions. It has taken researchers decades to sort through the relationships among the various kinds of fat and their roles in supporting or harming health. Translating these research findings into dietary recommendations is a challenging process. Too little information can mislead consumers, but too much detail can overwhelm them. As scientific understanding has grown, recommendations have evolved to become less general and more specific. Recommendations may seem to "change constantly and become more confusing," but in fact they are becoming more meaningful.

How exactly have dietary recommendations for fat changed to become more meaningful for consumers?

Dietary recommendations for fat have changed by shifting the emphasis from lowering total fat, in general, to limiting saturated and *trans* fat, specifically. For decades, health experts urged consumers to limit total fat intake to 30 percent or less of energy intake. This advice was straightforward—cut the fat, improve your health. Health experts recognized that saturated fats and *trans* fats were the ones that raise blood cholesterol, but they reasoned that by limiting total fat, saturated and *trans* fat intake would decline as well. People were simply advised to cut back on all fat so that they would cut back on saturated and *trans* fat. Such advice may have oversimplified the message and unnecessarily restricted total fat.

But low-fat diets have been recommended for years to help people manage weight and reduce the risk of heart disease. Are you saying that low-fat diets are no longer recommended?

Low-fat diets have a place in treatment plans for people with elevated blood lipids or heart disease, but researchers question the wisdom of such diets for healthy people as a means of controlling weight and preventing diseases.[1] Several problems accompany low-fat diets. For one, many people find low-fat diets difficult to maintain over time. For another, low-fat diets are not necessarily low-kcalorie diets; if energy intake exceeds energy needs, weight gain follows, and obesity brings a host of health problems, including heart disease. For still another, diets extremely low in fat may exclude fatty fish, nuts, seeds, and vegetable oils—all valuable sources of many essential fatty acids, phytochemicals, vitamins, and minerals. Importantly, the fats from these sources protect against heart disease, as later sections explain.

How have today's recommendations for fat been revised?

Today, health experts have revised dietary recommendations to acknowledge that not all fats have damaging health consequences. In fact, higher intakes of some kinds of fats (for example, the omega-3 fatty acids) support good health. Instead of urging people to cut back on all fats, current recommendations suggest carefully replacing the "bad" saturated fats with the "good" unsaturated fats and enjoying them in moderation.[2] The goal is to create a diet moderate in kcalories that provides enough of the fats that support good health, but not too much of those that harm health. (Turn to pp. 91–95 for a review of the health consequences of each type of fat.)

With these findings and goals in mind, the DRI committee concluded that a diet containing 20 to 35 percent of energy intake from fat, but reduced in saturated fat and *trans* fat and moderate in energy, is compatible with low rates of heart disease, diabetes, obesity, and cancer.[3] The *Dietary Guidelines for Americans 2005* make clear that the human body has no need of dietary saturated fat or *trans* fat, so the less consumed the better, as long as the diet is adequate in nutrients, including the essential fatty acids.[4]

How can people distinguish between the fats in foods that support health and those that might harm it?

Asking consumers to limit their total fat intake was less than perfect advice, but it was straightforward—find the fat and cut back. Asking consumers to keep their intakes of saturated fats, *trans* fats, and cholesterol low and to use monounsaturated and polyunsaturated fats instead may be more on target with heart health, but it also makes diet planning more complicated. To make appropriate selections, consumers must first learn which foods contain which fats. For example, avocados, bacon, walnuts, potato chips, and mackerel are all high-fat foods, yet some of these foods have detrimental effects on heart health when consumed in excess, whereas others seem neutral or even beneficial.

Is there evidence to clarify why some high-fat foods are compatible with a heart-healthy diet and others are not?

Yes. The traditional diets of Greece and other countries in the Mediterranean region are exemplary in their use of "good" fats, especially olives and their oil. A classic study of the world's people, the Seven Countries Study, found that death rates from heart disease were strongly associated with diets high in saturated fats, but only weakly linked with total fat.[5] In fact, the two countries with the highest fat intakes, Finland and the Greek island of Crete, had the highest (Finland) and lowest (Crete) rates of heart disease deaths. In both countries, the people consumed 40 percent or more of their kcalories from fat. Clearly, a high-fat diet was not the primary problem, so researchers refocused their attention on the type of fat. They began to notice the benefits of olive oil.

When olive oil replaces saturated fats, such as those of butter, coconut oil or palm oil, hydrogenated stick margarine, lard, or shortening, it may offer numerous health benefits.[6] Olive oil helps to protect against heart disease by:

- Lowering total and LDL cholesterol and not lowering HDL cholesterol or raising triglycerides.[7]
- Reducing LDL cholesterol's susceptibility to oxidation.[8]
- Lowering blood-clotting factors.[9]
- Providing phytochemicals that act as antioxidants (see Nutrition in Practice 8).[10]
- Lowering blood pressure.[11]

The labels of olive oil and foods containing it are now allowed to state the following qualified health claim: "limited and not conclusive scientific evidence suggests that eating 2 tablespoons of olive oil daily may reduce the risk of coronary heart disease due to the monounsaturated fat in olive oil. To achieve this possible benefit, olive oil is to replace a similar amount of saturated fat and not increase the total number of kcalories you eat in a day. One serving of this product contains X grams of olive oil."

When compared with other fats, olive oil seems to be a wise choice, but controlled clinical trials are too scarce to support population-wide recommendations to switch to a high-fat diet rich in olive oil.[12] Importantly, olive oil is not a magic potion; drizzling it on foods does not make them healthier. Like other fats, olive oil delivers 9 kcalories per

Matthew Farruggio

Olives and their oil may benefit heart health.

gram, which can contribute to weight gain in people who fail to balance their energy intake with their energy output. Its role in a healthy diet is to *replace* the saturated fats. Other vegetable oils, such as canola or safflower oil, in their liquid unhydrogenated states, are also generally low in saturated fats and high in unsaturated fats, but they lack the phytochemicals that olive oil provides. When choosing olive oils, go for the darker "extra virgin" kind because it contains the highest levels of potentially beneficial phytochemicals.

Good, olive oil may help protect against heart disease; are there other food fats that may also be protective?

Posssibly so. Tree nuts and peanuts are traditionally excluded from low-fat diets, and for good reason. Nuts provide up to 80 percent of their kcalories from fat, and a quarter cup (about an ounce) of mixed nuts provides over 200 kcalories. In a recent review of the literature, however, researchers found that people who ate a 1-ounce serving of nuts on five or more days a week had a reduced risk of heart disease compared with people consuming no nuts.[13] A smaller positive association was noted for any amount greater than one serving of nuts a week. The nuts were those commonly eaten in the United States: almonds, Brazil nuts, cashews, hazelnuts, macadamia nuts, pecans, pistachios, walnuts, and even peanuts. On average, these nuts contain mostly monounsaturated fat (59 percent), some polyunsaturated fat (27 percent), and little saturated fat (14 percent).[14]

Research has shown a benefit from walnuts and almonds in particular. In study after study, walnuts, when substituted for other fats in the diet, produce favorable effects on blood

Stay mindful of kcalories when snacking on nuts.

lipids—even in people with elevated total and LDL choles-terol.[15] Results are similar for almonds.

Studies on peanuts, macadamia nuts, pecans, and pista-chios follow suit, indicating that including nuts may be a wise strategy against heart disease. Nuts may protect against heart disease because they provide:

- Monounsaturated and polyunsaturated fats in abundance, but few saturated fats.
- Fiber, vegetable protein, and other valuable nutrients, including the antioxidant vitamin E.
- Phytochemicals that act as antioxidants (see Nutrition in Practice 8).

In addition to their heart benefits, nuts may also benefit other body organs—people who frequently consume nuts and other healthy fats suffer fewer gallbladder problems.[16]

Before advising your clients to include nuts in their diets, a caution is in order. As mentioned, most of the energy nuts provide comes from fats. Consequently, they deliver many kcalories per bite. In studies examining the effects of nuts on heart disease, researchers carefully adjust diets to make room for the nuts without increasing the total kcalories—that is, they use nuts *instead of, not in addition to,* other foods (such as meats, potato chips, oils, margarine, and butter). People who do not make similar replacements could end up gaining weight if they simply add nuts on top of their regular diets. Weight gain, in turn, elevates blood lipids and raises the risks of heart disease.

What about fish? I have a friend whose doctor told her that eating fish is good for the heart. Is this true?

Yes. The preceding chapter made clear that fish oils hold the potential to improve health, and particularly the health of the heart. Research studies have provided strong evidence that increasing omega-3 fatty acids in the diet supports heart health and lowers the risk of death from heart disease.[17] For this reason, the American Heart Association and other author-

Fish is a good source of the omega-3 fatty acids.

ities recommend including two fatty fish servings a week in a heart-healthy diet. People who eat some fish each week can lower their risks of heart attack and stroke.[18] Table 4-3 on p. 93 lists fish that provide at least 1 gram of omega-3 fatty acids per serving.

Fish is the best source of EPA and DHA in the diet, but it is also a major source of mercury and other environmental contaminants. Most fish contain at least trace amounts of mercury, but tilefish, swordfish, king mackerel, and shark have especially high levels. Freshwater fish may contain PCBs and other pollutants, so local advisories warn sport fishers of species that can pose problems. The chapter listed safer species of fish. To minimize risks while obtaining fish benefits, vary your choices among fatty fish species often.

If olive oil, nuts, and fatty fish are protective against heart disease, which fats are harmful?

The number one dietary determinant of LDL cholesterol is saturated fat. Figure NP4-1 shows that each 1 percent increase in energy from saturated fatty acids in the diet may produce a 2 percent jump in heart disease risk by elevating blood LDL cholesterol. Conversely, reducing saturated fat intake by 1 percent can be expected to produce a 2 percent drop in heart disease risk by the same mechanism. Even a 2 percent drop in LDL represents a significant improvement for the health of the heart.[19] Like saturated fats, *trans* fats also

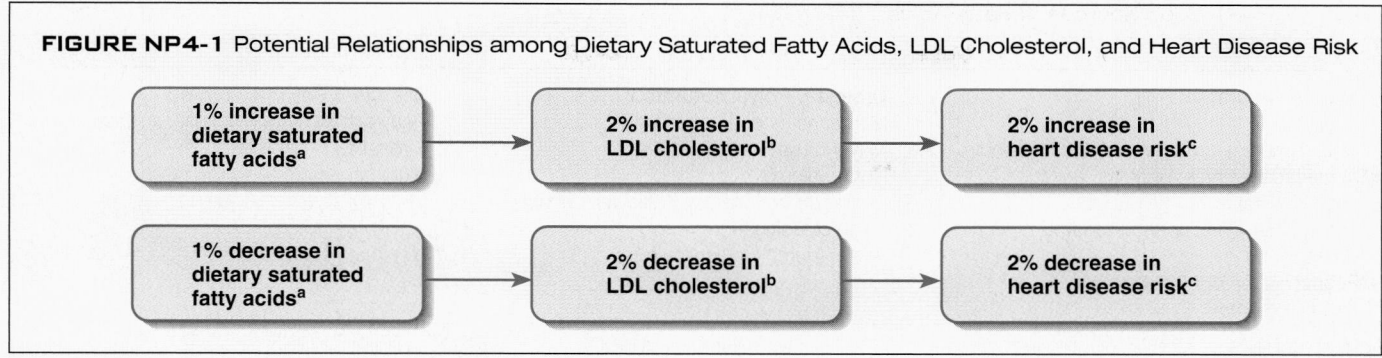

FIGURE NP4-1 Potential Relationships among Dietary Saturated Fatty Acids, LDL Cholesterol, and Heart Disease Risk

[a]Percentage of change in total dietary energy from saturated fatty acids.

[b]Percentage of change in blood LDL cholesterol.

[c]Percentage of change in an individual's risk of heart disease; the percentage of change in risk may increase when blood lipid changes are sustained over time.

Source: Expert Panel on Detection, Evaluation, and Treatment of High Blood Cholesterol in Adults (Adult Treatment Panel III), *Third Report of the National Cholesterol Education Program (NCEP)*, NIH publication no. 02-5215 (Bethesda, Md: National Heart, Lung, and Blood Institute, 2002), pp. V-8 and II-4.

raise heart disease risk by elevating LDL cholesterol. A heart-healthy diet limits foods rich in these two types of fat.

Which foods are highest in saturated and *trans* fats?

The major sources of saturated fats in the U.S. diet are fatty meats, whole-milk products, tropical oils, and products made from any of these foods. To limit saturated fat intake, consumers must choose carefully among these high-fat foods. Over a third of the fat in most meats is saturated. Similarly, over half of the fat is saturated in whole milk and other high-fat dairy products, such as cheese, butter, cream, Half-and-Half, cream cheese, sour cream, and ice cream. Consumers rarely use the tropical oils of palm, palm kernel, and coconut in the kitchen, but these oils are used heavily by food manufacturers and so are commonly found in many commercially prepared foods.

When choosing meats, milk products, and commercially prepared foods, look for those lowest in saturated fat. Labels provide a useful guide for comparing products in this regard, and Appendix A lists the saturated fat in several thousand foods.

Even with careful selections, a nutritionally adequate diet will provide some saturated fat. Zero saturated fat is not possible even when experts design menus with the mission to keep saturated fat as low as possible.[20] Diets based on fruits, vegetables, legumes, nuts, soy products, and whole grains can, and often do, deliver less saturated fat than diets that depend heavily on animal-derived foods, however.

As for *trans* fats, Chapter 4 explained that solid shortening and margarine are made from vegetable oil that has been hardened through hydrogenation. This process both saturates some of the unsaturated fatty acids and introduces *trans*-fatty acids. Many convenience foods contain *trans* fats. Table NP4-1 (p. 108) summarizes which foods provide which fats. Substituting unsaturated fats for saturated fats at each meal and snack can help protect against heart disease. Table NP4-2 (p. 108) provides several examples and shows how such sub-stitutions can lower saturated fat and raise unsaturated fat—even when total fat and kcalories remain unchanged.

So it seems that some fats are "good" and others "bad" from the body's point of view. Is that right?

The saturated and *trans* fats do indeed seem mostly bad for the health of the heart. Aside from providing energy, which unsaturated fats can do equally well, saturated and *trans* fats bring no indispensable benefits to the body. Furthermore, no harm can come from consuming diets low in them.

In contrast, the unsaturated fats are mostly good for the health of the heart when consumed in moderation. To date, their one proven fault seems to be that they, like all fats, provide abundant energy to the body and so may promote obesity if they drive kcalorie intakes higher than energy needs.[21] Obesity, in turn, often begets many body ills, as Chapter 7 makes clear.

When judging foods by their fatty acids, keep in mind that the fat in foods is a mixture of "good" and "bad," providing both saturated and unsaturated fatty acids. Even predominantly monounsaturated olive oil delivers some saturated fat. Consequently, even when a person chooses foods with mostly unsaturated fats, saturated fat can still add up if total fat is high. For this reason, fat must be kept below 35 percent of total kcalories if the diet is to be moderate in saturated fat.

Additionally, food manufacturers may come to the assistance of consumers wishing to avoid the health threats from saturated and *trans* fats. A margarine maker has announced that it will no longer offer products containing *trans* fats; a major snack manufacturer will soon reduce the saturated and *trans* fats in some of its products and offer snack foods in single-serving packages. A new oil with no *trans* fats, less saturated fat, and more unsaturated fat has been introduced for commercial production of french fries. Other companies are likely to follow if consumers respond favorably.

Basing one's diet on vegetables, fruits, and legumes as part of a balanced daily diet is a good idea, as is *replacing* saturated

TABLE NP4-1 Food Sources of Fatty Acids

Healthful Fatty Acids		
Monounsaturated	**Omega-6 Polyunsaturated**	**Omega-3 Polyunsaturated**
Avocado	Margarine (nonhydrogenated)	Fatty fish (herring, mackerel, salmon, tuna)
Nuts (almonds, cashews, filberts, hazelnuts, macadamia nuts, peanuts, pecans, pistachios)	Mayonnaise	Flaxseed
	Nuts (walnuts)	Nuts
Oils (canola, olive, peanut, sesame)	Oils (corn, cottonseed, safflower, soybean)	
Olives	Salad dressing	
Peanut butter (old-fashioned)	Seeds (pumpkin, sunflower)	
Seeds (sesame)		

Harmful Fatty Acids	
Saturated	**Trans**
Bacon	Commercial baked goods, including cookies, cakes, pies, or other products
Butter	made with margarine or vegetable shortening
Cheese	Fried foods, particularly restaurant and fast foods such as french fries and chicken
Chocolate	Many fried or processed snack foods, including microwave popcorn, chips, and crackers
Coconut	Margarine (hydrogenated or partially hydrogenated)
Cream, Half-and-Half	Nondairy creamers
Cream cheese	Shortening
Lard	
Meat	
Milk fat (whole-milk products)	
Oils (coconut, palm, palm kernel)	
Shortening	
Sour cream	

Note: Keep in mind that foods contain a mixture of fatty acids; see Figure 4-6, p. 99.

TABLE NP4-2 Replacing Saturated Fat with Unsaturated Fat

Examples of ways to replace saturated fats with unsaturated fats include sautéing foods in olive oil instead of butter, garnishing salads with sunflower seeds instead of bacon, snacking on mixed nuts instead of potato chips, using avocado instead of cheese on a sandwich, and eating salmon instead of steak. Portion sizes have been adjusted so that each of these foods provides approximately 100 kcalories. Notice that for a similar number of kcalories and grams of fat, the first choices offer less saturated fat and more unsaturated fat.

	Total Fat (g)	Saturated Fat (g)	Unsaturated Fat (g)
Olive oil vs. butter	11 vs. 11	2 vs. 7	9 vs. 4
Sunflower seeds vs. bacon	8 vs. 9	1 vs. 3	7 vs. 6
Mixed nuts vs. potato chips	9 vs. 7	1 vs. 2	8 vs. 5
Avocado vs. cheese	10 vs. 8	2 vs. 4	8 vs. 4
Salmon vs. steak	4 vs. 5	1 vs. 2	3 vs. 3
Totals	**42 vs. 40**	**7 vs. 18**	**35 vs. 22**

Note: Portion sizes that provide approximately 100 kcalories: 1 tbs olive oil, 1 tbs butter, 2 tbs dry roasted sunflower seeds, 2 slices cooked bacon, 2 tbs dry roasted mixed nuts, 10 potato chips, 6 slices avocado, 1 slice cheddar cheese, 2 oz salmon, and 1½ oz steak.

fats such as butter, shortening, and meat fat with unsaturated fats like olive oil and the oils from nuts and fish. These foods provide vitamins, minerals, and phytochemicals—all valuable in protecting the body's health. To further protect health, you may want to reduce fats from convenience foods and fast foods; choose small portions of meats, fish, and poultry; and include fresh foods from all the groups each day. Take care to select portion sizes that will best meet your energy needs. Also, be physically active each day.

Notes

[1] F. B. Hu, J. E. Manson, and W. C. Willett, Types of dietary fat and risk of coronary heart disease: A critical review, *Journal of the American College of Nutrition* 20 (2001): 5–19.

[2] Expert Panel on Detection, Evaluation, and Treatment of High Blood Cholesterol in Adults (Adult Treatment Panel III), *Third Report of the National Cholesterol Education Program (NCEP)*, NIH publication no. 02-5215 (Bethesda, Md.: National Heart, Lung, and Blood Institute, 2002); Standing Committee on the Scientific Evaluation of Dietary Reference Intakes, Food and Nutrition Board, Institute of Medicine, *Dietary Reference Intakes for Energy, Carbohydrate, Fiber, Fat, Fatty Acids, Cholesterol, Protein, and Amino Acids* (Washington, D.C.: National Academies Press, 2005).

[3] Standing Committee on the Scientific Evaluation of Dietary Reference Intakes, 2005, pp. 769–770.

[4] U.S. Department of Agriculture and U.S. Department of Health and Human Services, *Dietary Guidelines for Americans 2005,* 6th ed., available online at **www.healthierus.gov**.

[5] A. Keys, *Seven Countries: A Multivariate Analysis of Death and Coronary Heart Disease* (Cambridge: Harvard University Press, 1980).

[6] F. Visioli and coauthors, Virgin Olive Oil Study (VOLOS): Vasoprotective potential of extra virgin olive oil in mildly dislipidemic patients, *European Journal of Nutrition* 44 (2005): 121–127; A. H.

Stark and Z. Madar, Olive oil as a functional food: Epidemiology and nutritional approaches, *Nutrition Reviews* 60 (2002): 170–176.

[7] P. M. Kris-Etherton and coauthors, High-monounsaturated fatty acid diets lower both plasma cholesterol and triacyglycerol concentrations, *American Journal of Clinical Nutrition* 70 (1999): 1009–1015.

[8] R. L. Hargrove and coauthors, Low fat and high monounsaturated fat diets decrease human low density lipoprotein oxidative susceptibility in vitro, *Journal of Nutrition* 131 (2001): 1758–1763.

[9] C. M. Williams, Beneficial nutritional properties of olive oil: Implications for postprandial lipoproteins and factor VII, *Nutrition, Metabolism, and Cardiovascular Diseases* 11 (2001): 51–56; J. P. De La Cruz and coauthors, Antithrombotic potential of olive oil administration in rabbits with elevated cholesterol, *Thrombosis Research* 100 (2000): 305–315.

[10] Visioli and coauthors, 2005; F. Visioli and C. Galli, Biological properties of olive oil phytochemicals, *Critical Reviews in Food Science and Nutrition* 42 (2002): 209–221; M. N. Vissers and coauthors, Olive oil phenols are absorbed in humans, *Journal of Nutrition* 132 (2002): 409–417; M. Fito and coauthors, Protective effect of olive oil and its phenolic compounds against low density lipoprotein oxidation, *Lipids* 35 (2000): 633–638; R. W. Owen and coauthors, The antioxidant/anticancer potential of phenolic compounds isolated from olive oil, *European Journal of Cancer* 36 (2000): 1235–1247.

[11] L. A. Ferrara and coauthors, Olive oil and reduced need for antihypertensive medications, *Archives of Internal Medicine* 160 (2000): 837–842.

[12] L. Van Horn and N. Ernst, A summary of the science supporting the new National Cholesterol Education program dietary recommendations: What dietitians should know, *Journal of the American Dietetic Association* 101 (2001): 1148–1154.

[13] P. M. Kris-Etherton and coauthors, The effects of nuts on coronary heart disease risk, *Nutrition Reviews* 59 (2001): 103–111.

[14] F. B. Hu and M. J. Stampfer, Nut consumption and risk of coronary heart disease: A review of epidemiologic evidence, *Current Atherosclerosis Reports* 1 (1999): 204–209.

[15] E. B. Feldman, The scientific evidence for a beneficial health relationship between walnuts and coronary heart disease, *Journal of Nutrition* 132 (2002): 1062S–1101S; D. Zambón and coauthors, Substituting walnuts for monounsaturated fat improves the serum lipid profile of hypercholesterolemic men and women: A randomized crossover trial, *Annals of Internal Medicine* 132 (2000): 538–546.

[16] C.-J. Tsai and coauthors, Frequent nut consumption and decreased risk of cholecystectomy in women, *American Journal of Clinical Nutrition* 80 (2004): 76–81; C.-J. Tsai and coauthors, A prospective cohort study of nut consumption and the risk of gallstone disease in men, *American Journal of Epidemiology* 160 (2004): 961–968.

[17] K. C. Hayes, Relative cardiovascular benefits of n-6 and n-3 fatty acids, *Nutrition and the M.D.*, January 2005, pp. 1–4; V. Wijendran and K. C. Hayes, Dietary n-6 and n-3 fatty acid balance and cardiovascular health, *Annual Review of Nutrition* 24 (2004): 597–615; A. T. Erkkila and coauthors, Fish intake is associated with a reduced progression of coronary artery atherosclerosis in postmenopausal women with coronary artery disease, *American Journal of Clinical Nutrition* 80 (2004): 626–632; P. M. Kris-Etherton and coauthors, Fish consumption, fish oil, omega-3 fatty acids, and cardiovascular disease, *Circulation* 106 (2002): 2747–2757; F. B. Hu and coauthors, Fish and omega-3 fatty acid intake and risk of coronary heart disease in women, *Journal of the American Medical Association* 287 (2002): 1815–1821; C. M. Albert and coauthors, Blood levels of long-chain n-3 fatty acids and the risk of sudden death, *New England Journal of Medicine* 346 (2002): 1113–1118.

[18] D. Mozaffarian, R. N. Lemaitre, and L. H. Kuller, Fish consumption and stroke risk in elderly individuals, *Archives of Internal Medicine* 165 (2005): 200–206; Wijendran and Hayes, 2004; H. Iso and coauthors, Intake of fish and omega-3 acids and risk of stroke in women, *Journal of the American Medical Association* 285 (2001): 304–312.

[19] Expert Panel on Detection, Evaluation, and Treatment of High Blood Cholesterol in Adults (Adult Treatment Panel III), 2002, p. V–8.

[20] Standing Committee on the Scientific Evaluation of Dietary Reference Intakes, 2005, p. 835.

[21] Standing Committee on the Scientific Evaluation of Dietary Reference Intakes, 2005, pp. 796–797.

CONTENTS

Proteins and Amino Acids

© PhotoEdit

CHAPTER 5

People think of proteins as bodybuilding nutrients, the material of strong muscles, and rightly so. No new living tissue can be built without them. Some proteins form structures such as muscle, bone, skin, and other tissues. Other proteins do the cells' work. The energy to fuel that work comes primarily from carbohydrates and fats.

The Chemist's View of Proteins

Proteins are chemical compounds that contain the same atoms as carbohydrates and lipids—carbon (C), hydrogen (H), and oxygen (O)—but proteins are different in that they also contain nitrogen (N) atoms. These nitrogen atoms give the name *amino* (nitrogen containing) to the amino acids that form the links in the chains we call proteins.

THE STRUCTURE OF PROTEINS

About 20 different **amino acids** may appear in proteins.* All amino acids share a common chemical "backbone," and it is these backbones that are linked together to form proteins. Each amino acid also carries a side group, which varies from one amino acid to another (see Figure 5-1). The side group makes the amino acids differ in size, shape, and electrical charge. The side groups on amino acids are what make proteins so varied in comparison with either carbohydrates or lipids.

Protein Chains The 20 amino acids can be linked end-to-end in a virtually infinite variety of sequences to form proteins. When two amino acids bond together, the resulting structure is known as a **dipeptide.** Three amino acids bonded together form a **tripeptide.** As additional amino acids join the chain, the structure becomes a **polypeptide.** Most proteins are a few dozen to several hundred amino acids long.

Protein Shapes Polypeptide chains twist into complex shapes. Each amino acid has special characteristics that attract it to, or repel it from, the surrounding fluids and other amino acids. Because of these interactions, polypeptide chains fold and intertwine into intricate coils (see Figure 5-2, p. 112) and other shapes. The amino acid sequence of a protein determines the specific way the chain will fold.

Protein Functions The dramatically different shapes of proteins enable them to perform different tasks in the body. Some, such as hemoglobin in the blood (see Figure 5-3, p. 112), are globular in shape; some are hollow balls that can carry and store materials within them; and some, such as those that form tendons, are more than ten times as long as they are wide, forming stiff, sturdy, rodlike structures.

proteins: compounds made from strands of amino acids composed of carbon, hydrogen, oxygen, and nitrogen atoms. Some amino acids also contain sulfur atoms.

amino (a-MEEN-oh) **acids:** building blocks of protein. Each has a hydrogen atom, an amino group, and an acid group attached to a central carbon, which also carries a distinctive side chain.
   amino = containing nitrogen

dipeptide: two amino acids bonded together.
   di = two
   peptide = amino acid

tripeptide: three amino acids bonded together.
   tri = three

polypeptide: ten or more amino acids bonded together. An intermediate strand of between four and ten amino acids is an *oligopeptide.*
   poly = many
   oligo = few

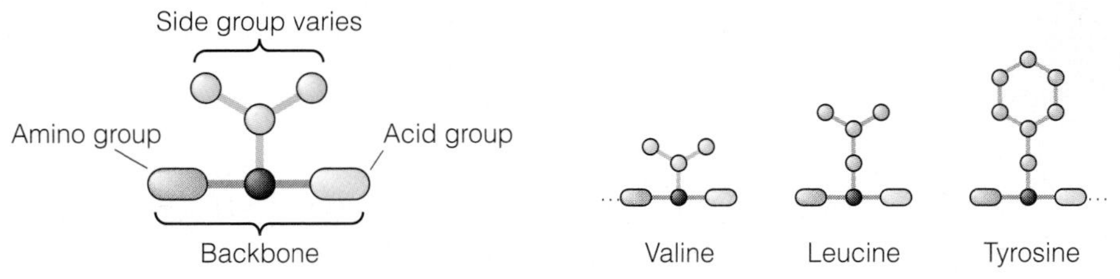

FIGURE 5-1 Amino Acid Structure and Examples of Amino Acids

Side group varies

Amino group   Acid group

Backbone

Valine   Leucine   Tyrosine

All amino acids have a "backbone" made of an amino acid group (which contains nitrogen) and an acid group. The side group varies from one amino acid to the next.

Note that the side group is a unique structure that differentiates one amino acid from another.

*Besides the 20 common amino acids, which can all be components of proteins, others occur individually (for example, ornithine).

FIGURE 5-2 The Coiling and Folding of a Protein Molecule

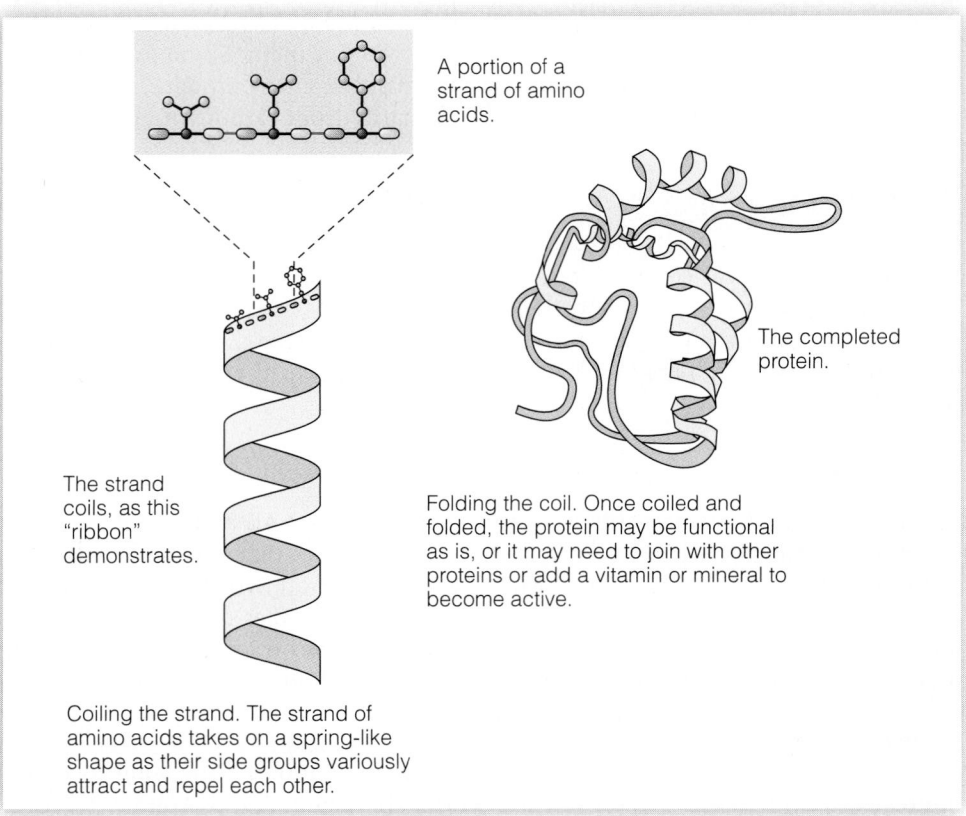

A portion of a strand of amino acids.

The strand coils, as this "ribbon" demonstrates.

The completed protein.

Folding the coil. Once coiled and folded, the protein may be functional as is, or it may need to join with other proteins or add a vitamin or mineral to become active.

Coiling the strand. The strand of amino acids takes on a spring-like shape as their side groups variously attract and repel each other.

nonessential amino acids: amino acids that the body can synthesize.

essential amino acids: amino acids that the body cannot synthesize in amounts sufficient to meet physiological need; also called *indispensable amino acids.* Nine amino acids are known to be essential for human adults:

- *histidine* (HISS-tuh-deen).
- *isoleucine* (eye-so-LOO-seen).
- *leucine* (LOO-seen).
- *lysine* (LYE-seen).
- *methionine* (meh-THIGH-oh-neen).
- *phenylalanine* (fen-il-AL-uh-neen).
- *threonine* (THREE-oh-neen).
- *tryptophan* (TRIP-toe-fane, TRIP-toe-fan).
- *valine* (VAY-leen).

conditionally essential amino acid: an amino acid that is normally nonessential but must be supplied by the diet in special circumstances when the need for it becomes greater than the body's ability to produce it.

Some researchers refer to essential amino acids as **indispensable** and to nonessential amino acids as **dispensable.**

FIGURE 5-3 The Structure of Hemoglobin

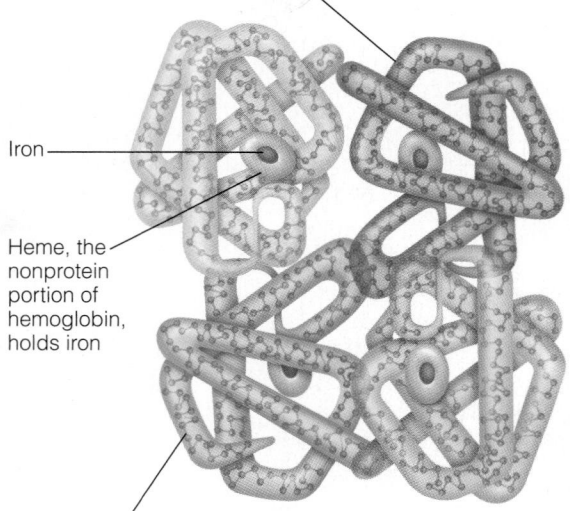

One of the four highly folded polypeptide chains that forms the globular hemoglobin protein

Iron

Heme, the nonprotein portion of hemoglobin, holds iron

The amino acid sequence determines the shape of the polypeptide chain

ESSENTIAL AMINO ACIDS

Proteins in foods do not provide body proteins directly, but rather supply the amino acids from which the body makes its own proteins. More than half of the amino acids are **nonessential,** meaning that the body can make them for itself. Proteins in foods usually deliver these amino acids, but it is not essential that they do so. There are other amino acids that the body cannot make at all, however, and some that it cannot make fast enough to meet its needs. The proteins in foods must supply these nine amino acids to the body; they are therefore called **essential amino acids.**

Sometimes a nonessential amino acid can become essential. During illness or conditions of trauma, or in other special circumstances such as premature birth, the need for an amino acid that is normally nonessential may become greater than the body's ability to produce it. In such circumstances, that amino acid becomes a **conditionally essential amino acid.** Research suggests that glutamine, normally a nonessential amino acid, may be a conditionally essential amino acid for some critically ill people.[1]

REVIEW NOTES

Chemically speaking, proteins are more complex than carbohydrates or lipids; proteins are made of some 20 different amino acids, 9 of which the body cannot make (they are essential).

Each amino acid contains a central carbon atom with an amino group, an acid group, a hydrogen atom, and a unique side group attached to it.

The distinctive sequence of amino acids in each protein determines its shape and function.

Proteins in the Body

What distinguishes you chemically from any other human being are minute differences in your particular body proteins (enzymes, antibodies, and others). These differences are determined by your proteins' amino acid sequences, which are written into the genes you inherited from your parents and ancestors. The genes direct the making of all the body's proteins.

The human body contains 30,000 or more different kinds of proteins. The roles of more than 3000 of these proteins are now known. Only a few of the many roles proteins play are described here, but these should serve to illustrate proteins' versatility, uniqueness, and importance.

Enzymes Enzymes are catalysts that are essential to all life processes. Enzymes in the cells of plants or animals put together the pairs of sugars that make disaccharides and the long strands of sugars that make starch, cellulose, and glycogen. Enzymes also dismantle these compounds to free their constituent parts and release energy. Enzymes also assemble and disassemble lipids, assemble all other compounds that the body makes, and disassemble all compounds that the body can use for building tissue and other metabolic work. As Figure 5-4 shows, enzymes themselves are not altered by the reactions they facilitate. All enzymes are proteins, and when amino acids have to be put together to make proteins, it is enzymes that put them together, too. In other words, these proteins can even make other proteins.

The protein story moves in a circle. To follow the circle in nutrition, start with a person eating food proteins. The food proteins are broken down into amino acids by digestive enzymes, which themselves are proteins. The amino acids enter the cells of the body, where other proteins (enzymes) put the amino acids together in long chains whose sequences are specified by the genes. The chains fold and twist back on themselves to form proteins, and some of these proteins become enzymes themselves. Some of these enzymes break apart compounds; others put compounds together. Day by day, in billions of reactions, these processes repeat themselves, and life goes on. Only living systems can achieve such self-renewal. A toaster cannot produce another toaster; a car cannot fix a broken-down car. Only living creatures and the parts they are composed of—the cells—can duplicate and repair themselves.

Fluid and Electrolyte Balance Proteins help maintain the body's **fluid and electrolyte balance.** As Figure 5-5 shows, the body's fluids are contained in three major body compartments: (1) the spaces inside the blood vessels; (2) the spaces within the cells; and (3) the spaces between the cells (the interstitial spaces outside the

enzymes: protein catalysts. A catalyst is a compound that facilitates chemical reactions without itself being changed in the process.

fluid and electrolyte balance: maintenance of the necessary amounts and types of fluid and minerals in each compartment of the body fluids.

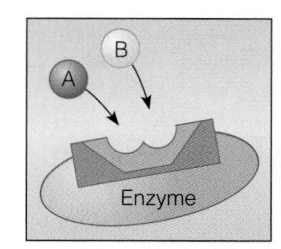 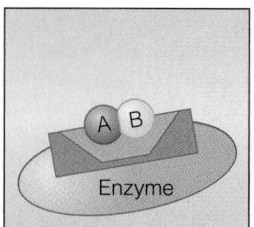 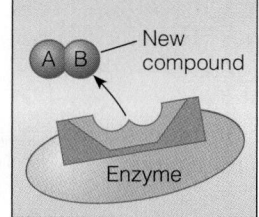

The separate compounds, A and B, are attracted to the enzyme's active site, making a reaction likely.

The enzyme forms a complex with A and B.

The enzyme is unchanged, but A and B have formed a new compound, AB.

FIGURE 5-4 Enzyme Action
Each enzyme facilitates a specific chemical reaction. In this diagram, an enzyme enables two compounds to make a more complex structure, but the enzyme itself remains unchanged.

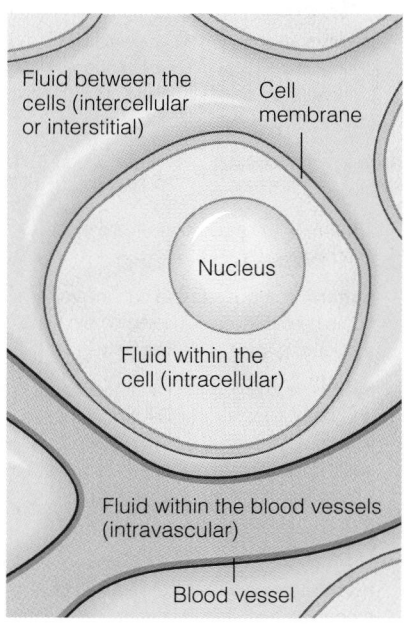

FIGURE 5-5 One Cell and its Associated Fluids

Minerals are helper nutrients. The attraction of protein and mineral particles to water is due to osmotic pressure (see Chapter 9).

blood vessels). Fluids flow back and forth between these compartments, and proteins in the fluids, together with minerals, help to maintain the needed distribution of these fluids.

Proteins are able to help determine the distribution of fluids in living systems for two reasons: first, proteins cannot pass freely across the membranes that separate the body compartments, and second, they are attracted to water. A cell that "wants" a certain amount of water in its interior space cannot move the water around directly, but it can manufacture proteins, and these proteins will hold water. Thus the cell can use proteins to help regulate the distribution of water indirectly. Similarly, the body makes proteins for the blood and the interstitial (intercellular) spaces. These proteins help maintain the fluid volume in those spaces. Excess fluid accumulation in the interstitial spaces is called **edema.**

Not only is the quantity of the body fluids vital to life, but so is their composition. Special transport proteins in the membranes of cells continuously transfer substances into and out of cells to maintain balance. For example, sodium is concentrated outside the cells, and potassium is concentrated inside. The balance of these two minerals is critical to nerve transmission and muscle contraction. Any disturbance in this balance triggers a major medical emergency. Such imbalances can cause irregular heartbeats, kidney failure, muscular weakness, and even death.

edema (eh-DEEM-uh): the swelling of body tissue caused by leakage of fluid from the blood vessels and accumulation of the fluid in the interstitial spaces.

acids: compounds that release hydrogen ions in a solution.

bases: compounds that accept hydrogen ions in a solution.

acid-base balance: the balance maintained between acid and base concentrations in the blood and body fluids.

pH: the concentration of hydrogen ions. The lower the pH, the stronger the acid. Thus pH 2 is a strong acid; pH 6 is a weak acid; pH 7 is neutral; and a pH above 7 is alkaline.

denaturation (dee-nay-cher-AY-shun): the change in a protein's shape brought about by heat, acid, or other agents. Past a certain point, denaturation is irreversible.

acidosis: too much acid in the blood and body fluids.

alkalosis: too much base in the blood and body fluids.

buffers: compounds that can reversibly combine with hydrogen ions to help keep a solution's acidity or alkalinity constant.

antibodies: large proteins of the blood and body fluids, produced in response to invasion of the body by unfamiliar molecules (mostly proteins) called *antigens.* Antibodies inactivate the invaders and so protect the body.
    anti = against

hormones: chemical messengers. Hormones are secreted by a variety of glands in the body in response to altered conditions. Each travels to one or more target tissues or organs and elicits specific responses to restore normal conditions.

Acid-Base Balance Proteins also help maintain the balance between **acids** and **bases** within the body's fluids. Normal body processes continually produce acids and bases, which must be carried by the blood to the kidneys and lungs for excretion. The blood must do this without upsetting its own **acid-base balance.** Blood **pH** is one of the most tightly controlled conditions in the body. If the blood becomes too acidic, vital proteins may undergo **denaturation,** losing their shape and ability to function. A similar situation arises when the balance tips too far toward base. These imbalances are known as **acidosis** and **alkalosis,** respectively, and both can be fatal. Figure 5-6 shows the normal and abnormal pH ranges of body fluids, as well as the pH's of some common substances.

Proteins such as albumin in the blood help to prevent acid-base imbalances. In a sense, the proteins protect one another by gathering up extra acid (hydrogen) ions when there are too many in the surrounding medium and by releasing them when there are too few. By accepting and releasing hydrogen ions, proteins act as **buffers,** maintaining the acid-base balance of the blood and body fluids.

Antibodies Other proteins in the blood—the **antibodies**—defend against viruses, bacteria, and other disease agents. The antibodies work so efficiently that if a million bacterial cells are injected into the skin of a healthy person, fewer than ten are likely to survive for five hours, which explains why most diseases never have a chance to get started. Without sufficient protein, the body cannot maintain its resistance to disease.

Hormones The blood also carries messenger molecules known as **hormones,** and *some* hormones are proteins. (Recall that some hormones are sterols, members of the lipid family.) Among the proteins that act as hormones are glucagon and insulin. Hormones have many profound effects, which will become evident in subsequent chapters.

Transport Proteins Some proteins move about in the body fluids, transporting nutrients and other molecules from one organ to another. The protein hemoglobin, which carries oxygen from the lungs to the body's cells, is a prime example. The lipoproteins transport lipids around the body, as described in Chapter 2. In addition, special proteins also carry vitamins and minerals.

Growth, Maintenance, and Repair The body uses amino acids to build the proteins of all its new tissues. The new tissues may be in an embryo, in a growing child, or in new hair and nails. Proteins also help replace worn-out cells in every-

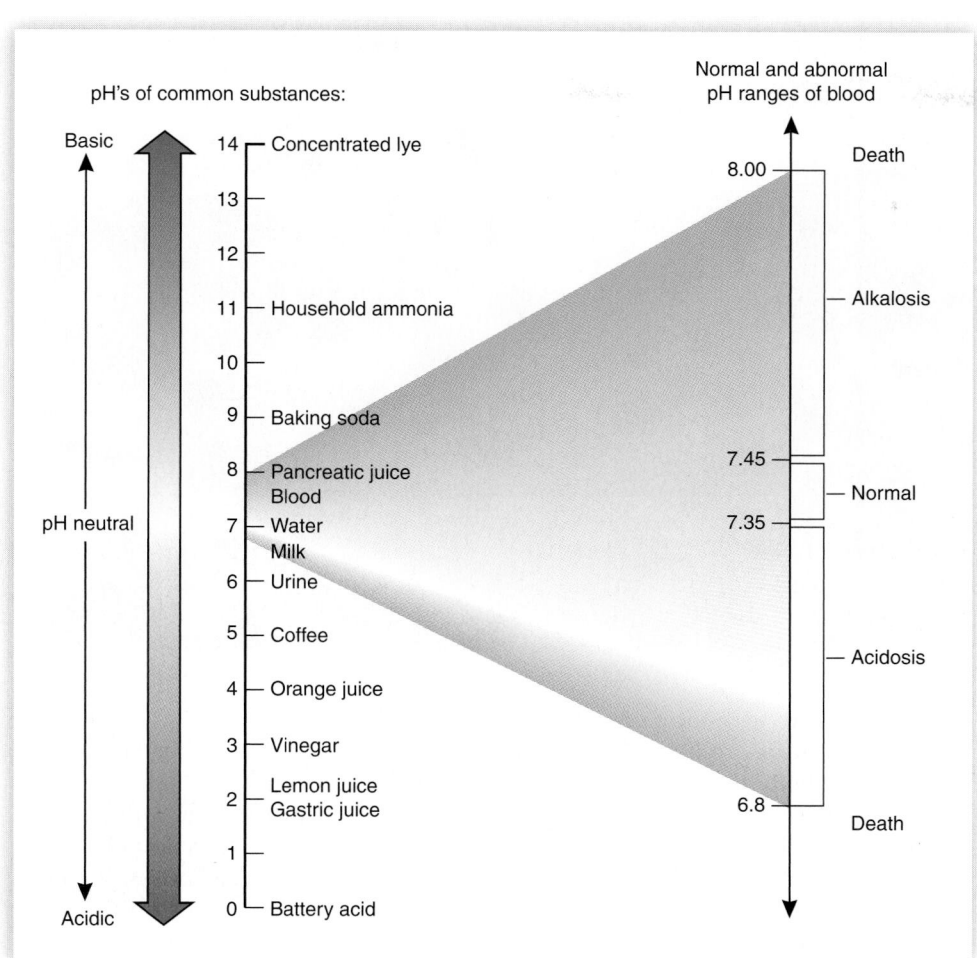

pH's of common substances:

Normal and abnormal
pH ranges of blood

Basic

14 — Concentrated lye

13 —

12 —

11 — Household ammonia

10 —

9 — Baking soda

8 — Pancreatic juice
Blood

pH neutral   7 — Water
Milk

6 — Urine

5 — Coffee

4 — Orange juice

3 — Vinegar

2 — Lemon juice
Gastric juice

1 —

Acidic   0 — Battery acid

8.00 — Death

— Alkalosis

7.45 —
7.35 — Normal

— Acidosis

6.8 — Death

FIGURE 5-6 The pH Scale
A substance's acidity or alkalinity is measured in pH units. Each step down the scale indicates a tenfold increase in the concentration of hydrogen ions. Notice how small the range of normal blood pH is.

one's body all the time. For example, the millions of cells that line the intestinal tract live for three to five days; they are constantly being shed and must be replaced. The cells of the skin die and rub off, and new ones grow from underneath. The body uses proteins to repair damaged tissues, too. The protein collagen serves as the mending material of torn tissue, forming scars to hold the separated parts together.

Both inside and outside the body, then, cells constantly make and break down their proteins. When proteins break down, their component amino acids are liberated within the cells or released into the general circulation. Some of these amino acids may be promptly recycled into other proteins; others may be stripped of their nitrogen and used for energy. By reusing amino acids to build proteins, however, the body conserves and recycles a valuable commodity. The entire process of breakdown, recovery, and synthesis is called **protein turnover.**

People need to eat protein-rich foods every day to replace the protein they continuously lose. If the body is growing, it needs more protein than is necessary just for maintenance. Children end each day with more blood cells, more muscle cells, and more skin cells than they had at the beginning of the day. So protein is needed both for routine maintenance (replacement) and for growth (addition).

Providing Energy and Glucose Even though amino acids are needed to do the work that only they can perform—build vital proteins—they will be sacrificed to provide energy and glucose if need be. For most people eating a normal, mixed diet, protein provides about 10 to 15 percent of the daily need for energy.[2] Under conditions of inadequate energy or carbohydrate, protein use speeds up.[3] Keeping energy and glucose available is one of the body's highest priorities: without

protein turnover: the continuous breakdown and synthesis of body proteins involving the recycling of amino acids.

Growing children end each day with more bone, blood, muscle, and skin cells than they had at the beginning of the day.

© Ariel Skelley/Corbis

energy, cells die; without glucose, the brain and nervous system falter. When glucose or fatty acids are limited, cells are forced to use amino acids for energy and glucose. The body does not make a specialized storage form of protein as it does for carbohydrate and fat. Glucose is stored as glycogen in the liver and muscles, fat as triglycerides in the adipose tissue, but body protein is available only as the working and structural components of the tissues. When the need arises, the body dismantles its tissue proteins and uses them for energy. Thus, over time, energy deprivation (starvation) always incurs wasting of lean body tissue as well as fat loss.

REVIEW NOTES

The list of protein functions discussed here and summarized in Table 5-1 is by no means exhaustive. Nevertheless, it does give some sense of the immense variety of proteins and their importance in the body.

TABLE 5-1	Summary of Functions of Proteins

Enzymes. Proteins facilitate chemical reactions.

Fluid and electrolyte balance. Proteins help to maintain the distribution and composition of various body fluids.

Acid-base balance. Proteins help maintain the acid-base balance of body fluids by acting as buffers.

Antibodies. Proteins act against disease agents to fight diseases.

Hormones. Proteins regulate body processes. Some, but not all, hormones are made of protein.

Transportation. Proteins transport substances such as lipids, minerals, and oxygen around the body.

Growth and maintenance. Proteins form integral parts of most body structures such as skin, tendons, ligaments, membranes, muscles, organs, and bones. As such, they support the growth and repair of body tissues.

Energy. Proteins provide some fuel for the body's energy needs.

Protein and Health

During the time that scientists have been studying nutrition, no nutrient has been more intensely scrutinized than protein. As you know by now, it is indispensable to life. And it should come as no surprise that protein deficiency can have devastating effects on people's health. But, as with the other nutrients, protein in excess can be harmful, too, so this section also discusses the consequences of protein excess.

PROTEIN-ENERGY MALNUTRITION

When people are deprived of food and suffer an energy deficit, they degrade their own body protein for energy and indirectly suffer a protein deficiency, as well as an energy deficiency. Because protein and energy deprivation go hand in hand, public health officials have adopted an abbreviation for the overlapping pair: **protein-energy malnutrition (PEM)**. PEM often strikes early in childhood, but it endangers many adults as well. PEM is the most widespread form of malnutrition in the world today. More than half of the 11 million children who die each day are malnourished and suffer from infectious diseases.[4] PEM is prevalent in Africa, Central America, South America, the Middle East, and South and East Asia; but developed countries including the United States are not immune to it. PEM is common among some population groups in the United States: impoverished people living on U.S. Indian reservations, in inner cities, and in rural areas; many elderly people; homeless children; and those suffering from the eating disorder anorexia nervosa.

Of all population groups, children are most seriously affected by malnutrition. Children who are thin for their heights may have recently developed PEM, whereas children who are short for their ages may have experienced PEM for extended periods of time. Stunted growth due to PEM is easy to overlook because a small child may look quite normal, but it may be the most common sign of malnutrition in the developing countries. Experts estimate that 182 million children are stunted, shorter than they should be for their age.[5]

Marasmus and Kwashiorkor PEM takes two different forms, with some cases exhibiting a combination of the two. In one form, the person is shriveled and emaciated—this disease is called **marasmus**. In the second, a swollen belly and skin rash are present, and the disease is named **kwashiorkor**. In the combination, some features of each type are present. Marasmus reflects a chronic inadequate food intake and therefore inadequate energy, vitamins, and minerals, as well as too little protein. Kwashiorkor may result from severe acute malnutrition, with too little protein to support body functions.

Marasmus Marasmus commonly occurs in children from 6 to 18 months of age in all the overpopulated and impoverished areas of the world. Children in impoverished nations subsist on a weak cereal drink that supplies scant energy and protein of low quality: such food can barely sustain life, much less support growth. Consequently, marasmic children look like little old people—just skin and bones.

Without adequate nutrition, muscles, including the heart muscle, waste and weaken. Because the brain normally grows to almost its full adult size within the first two years of life, marasmus impairs brain development and learning ability. Reduced synthesis of key hormones leads to a metabolism so slow that body temperature drops below normal. There is little or no fat under the skin to insulate against cold. Some hospital workers find that the primary need of marasmic children is to be clothed, covered, and kept warm.

The starving child faces this threat to life by engaging in as little activity as possible—not even crying for food. The body gathers all its forces to meet the crisis,

PEM can be a consequence of many different conditions. PEM has been recognized in people with many chronic diseases such as cancer and AIDS and in those who have severe stresses such as burns or extensive infections (see Chapter 16). The consequences of PEM as a world malnutrition problem are considered here; the problems associated with PEM and illness are described throughout later chapters.

protein-energy malnutrition (PEM): a deficiency of protein and food energy; the world's most widespread malnutrition problem, including both marasmus and kwashiorkor.

mal = bad, poor

marasmus (ma-RAZZ-mus): the most common form of severe PEM before one year of age. Marasmus is characterized by generalized muscle wasting associated with extreme deprivation, or impaired absorption, of energy, protein, vitamins, and minerals.

kwashiorkor (kwash-ee-OR-core or kwash-ee-or-CORE): a severe form of PEM that occurs more frequently after 18 months of age. Kwashiorkor is characterized by failure to grow and develop, changes in the pigmentation of the hair and skin, edema, and fatty liver. Kwashiorkor is associated with inadequate protein intake and infections.

so it cuts down on any expenditure of energy not needed for the heart, lungs, and brain to function. Growth ceases; the child is no larger at age four than at age two. Digestive enzymes are in short supply, the digestive tract lining deteriorates, and absorption fails. The child cannot assimilate what little food is eaten.

When two variables interact so that each increases the other, **synergism** (SIN-er-jiz-um) is said to be occurring. Malnutrition and infection are a deadly combination because they work in this way.

syn = with, together

ergism = work

Blood proteins, including hemoglobin, are no longer synthesized. Antibodies to fight off invading bacteria are degraded to provide amino acids for other uses, rendering the child vulnerable to infection. Then **dysentery,** an infection of the digestive tract, causes diarrhea, further depleting the body of nutrients. In the marasmic child, once infection has set in, kwashiorkor often follows and the immune response weakens further.[6] The infection that occurs with malnutrition is responsible for two-thirds of the deaths of young children in developing countries.

If caught in time, a child's starvation may be reversed by careful nutrition therapy. In severe cases, fluid balances are most critical. Diarrhea will have depleted the body's potassium and disturbed other electrolyte balances. Careful correction of fluid and electrolyte imbalances usually raises the blood pressure and strengthens the heart. After the first 24 to 48 hours, protein and energy may be given in small quantities, with intakes gradually increased as tolerated. Years after PEM is corrected, however, a child may still experience deficits in thinking and achievement in school compared with well-nourished peers.

dysentery (DIS-en-terry): an infection of the gastrointestinal tract caused by an amoeba or bacterium that gives rise to severe diarrhea.

dys = bad

entery = intestine

fatty liver: an accumulation of fat in the liver. In PEM, fat accumulates in the liver because no protein is available to form the lipoproteins that normally escort fat molecules in the blood (see Chapter 19).

Kwashiorkor Kwashiorkor was originally a Ghanaian word meaning a "sickness that infects the first child when the second child is born." If you consider how kwashiorkor often develops, you can easily see how the Ghanaians arrived at this name for the disease. When a mother who has been nursing her first child bears a second child, she weans the first child and puts the second one on the breast. The first child, suddenly switched from nutrient-dense, protein-rich breast milk to a starchy, protein-poor gruel, soon begins to sicken and die. Kwashiorkor typically sets in at about the age of two. Though rare in the United States, kwashiorkor is not entirely unknown, usually occurring when children are fed ill-conceived vegetarian or "anti-allergy" diets or given a protein-poor "health-food" rice drink instead of cow's milk.[7]

Some symptoms of kwashiorkor resemble those of marasmus (see Table 5-2). Proteins and hormones that previously maintained fluid balance diminish, and fluid leaks into the interstitial spaces. The child's limbs and belly become swollen with edema, a distinguishing feature of kwashiorkor; **fatty liver** develops due to a

In the photo on the left, the extreme loss of muscle and fat characteristic of marasmus is apparent in the child's matchstick arms and legs. In contrast, the edema and enlarged liver characteristic of kwashiorkor are apparent in the swollen bellies of the children in the photo on the right.

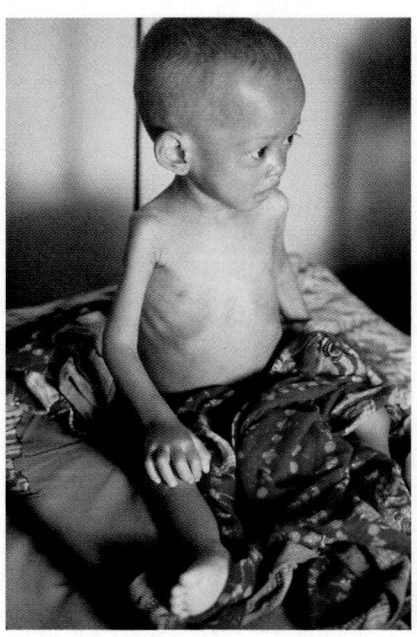
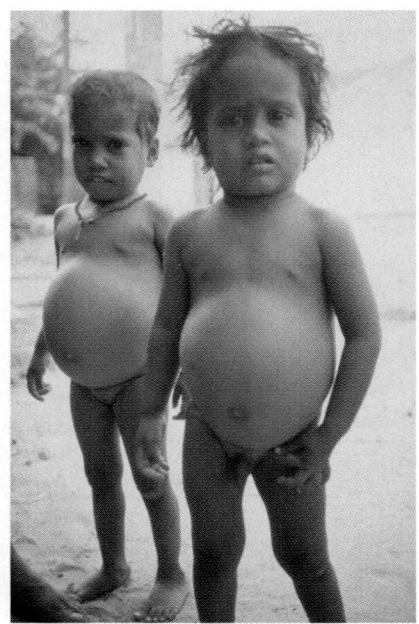

FAO/7752/F. Botts

FAO/13670/G. Kent

TABLE 5-2	Features of Marasmus and Kwashiorkor in Children

Separating PEM into two classifications oversimplifies the condition, but at the extremes, marasmus and kwashiorkor exhibit marked differences. Marasmus-kwashiorkor mix presents symptoms common to both marasmus and kwashiorkor. In all cases, children are likely to develop diarrhea, infections, and multiple nutrient deficiencies.

Marasmus	Kwashiorkor
Infancy (less than 2 yr)	Older infants and young children (1 to 3 yr)
Severe deprivation or impaired absorption of protein, energy, vitamins, and minerals	Inadequate protein intake or, more commonly, infections
Develops slowly; chronic PEM	Rapid onset; acute PEM
Severe weight loss	Some weight loss
Severe muscle wasting with fat loss	Some muscle wasting, with retention of some body fat
Growth: <60% weight-for-age	Growth: 60 to 80% weight-for-age
No detectable edema	Edema
No fatty liver	Enlarged, fatty liver
Anxiety, apathy	Apathy, misery, irritability, sadness
Appetite may be normal or impaired	Loss of appetite
Hair is sparse, thin, and dry; easily pulled out	Hair is dry and brittle; easily pulled out; changes color; becomes straight
Skin is dry, thin, and wrinkled	Skin develops lesions

lack of the protein carriers that transport fat out of the liver. The child's hair loses its color; the skin becomes patchy and scaly, sometimes with ulcers and sores that fail to heal.

Clearly, preventing marasmus and kwashiorkor from developing in the first place is more desirable than attempting to treat the conditions after they have set in. Experts assure us that we possess the knowledge, technology, and resources to end the hunger that leads to PEM. Programs that have involved the local people in the process of identifying the problem and devising its solution have met with some success. But until those who have the food, technology, and resources make fighting hunger a priority, the war on hunger will not be won.

PROTEIN EXCESS

While many of the world's people struggle to obtain enough food and enough protein to survive, in the developed nations protein is so abundant that problems of protein excess are seen. Overconsumption of protein offers no benefits and may pose health risks. For example, diets high in protein-rich foods are often associated with obesity and its accompanying health risks.[8] Animal-protein sources in particular can be high in saturated fat, a known contributor to atherosclerosis and heart disease. In a study of more than 40,000 women, the higher the intake of protein from red meat and dairy products, the greater the risk of heart disease.[9] Despite a correlation between animal-protein intake and heart disease, however, no independent effect of animal protein itself on heart health has been demonstrated. Conversely, substituting plant protein for animal protein improves indicators of heart disease, as this chapter's Nutrition in Practice explains.[10] The higher a person's intake of protein-rich foods such as meat and milk, the more likely that fruits, vegetables, and grains will be crowded out, making the diet inadequate in other nutrients.

Excretion of the end products of protein metabolism depends, in part, on an adequate fluid intake and healthy kidneys. A high protein intake increases the work of the kidneys but does not appear to cause kidney disease. A high-protein diet does worsen existing kidney disease and may accelerate a decline in only mildly impaired kidneys.[11] One of the most effective ways to slow the progression of kidney disease is to restrict dietary protein.

Research clearly shows that high protein intakes increase urinary calcium excretion.[12] It is less clear, however, whether high protein intakes deplete the bones of calcium and thereby compromise bone health.[13] It has been speculated that the effects of protein intake on bone health may depend on dietary calcium intake.[14] Increasing calcium intake may compensate for the effects of protein on urinary calcium excretion. In a three-year study of healthy elderly men and women, protein intake was positively associated with a gain in bone density in those who received calcium supplements (500 milligrams per day).[15] In contrast, the unsupplemented group lost bone over the three-year period. The researchers concluded that in many elderly men and women, bone health may be improved by increasing protein intake as long as calcium (and vitamin D) intakes meet recommendations. In establishing protein recommendations, the DRI committee considered protein's effect on bone health but did not find sufficient evidence to set a Tolerable Upper Intake Level.[16]

> **REVIEW NOTES**
>
> Protein deficiencies arise from both energy-poor and protein-poor diets and lead to the devastating diseases of marasmus and kwashiorkor.
>
> Together, these diseases are known as PEM (protein-energy malnutrition), a major form of malnutrition causing death in children worldwide.
>
> Excesses of protein offer no advantage; in fact, overconsumption of protein-rich foods may incur health problems as well.

AMINO ACID SUPPLEMENTS

In view of the high protein intakes of people in the United States and other developed nations, it is surprising that many people feel compelled to take protein and amino acid supplements. Why do people take protein or amino acid supplements? Athletes take them hoping to build muscle. Dieters take them to spare their bodies' protein while losing weight. About 20 years ago, popular reports that the amino acid tryptophan could relieve pain and cure depression and insomnia led to widespread public use. More than 1500 people who elected to take tryptophan developed an illness called EMS (short for *eosinophilia-myalgia syndrome*). EMS is characterized by severe muscle and joint pain, limb swelling, an elevated white blood cell count, extremely high fever, and, in some cases, death. Contaminants in the supplement were thought to be the cause of the disease, and the Food and Drug Administration (FDA) issued a recall of all products containing tryptophan except for specific medical formulas.

The DRI committee reviewed the available research on amino acids but with next to no safety research in existence, the committee was unable to set Tolerable Upper Intake Levels for supplemental doses. Until research becomes available, no level of amino acid supplementation can be assumed to be safe for all people. Growth or altered metabolism makes the following groups of people especially likely to suffer harm from amino acid supplements:

- All women of childbearing age.
- Pregnant or lactating women.
- Infants, children, and adolescents.

- Elderly people.
- People with inborn errors of metabolism that affect their bodies' handling of amino acids.
- Smokers.
- People on low-protein diets for therapeutic reasons.
- People with chronic or acute mental or physical illnesses who take amino acids without medical supervision.

Also, because weight lifters and bodybuilders may take frequent, massive doses of amino acids, they may suffer harm from the supplements while believing false promises of benefits. A review of the literature has concluded that not enough research exists to support recommending long-term consumption of amino acid supplements by healthy people.[17] Anyone considering taking amino acid supplements should check with a physician first.

PROTEIN RECOMMENDATIONS AND NITROGEN BALANCE

The committee that established the RDA states that a generous daily protein allowance for a healthy adult is 0.8 gram per kilogram (2.2 pounds) of healthy body weight. The protein RDA is adjusted to cover additional needs for building new tissue and so is slightly higher for infants, children, and pregnant and lactating women.

In setting the RDA, the committee assumes that the protein eaten will be of good quality, that it will be consumed together with adequate energy from carbohydrate and fat, and that other nutrients in the diet will be adequate. The committee also assumes that the RDA will be applied only to healthy individuals with no unusual alteration of protein metabolism. Most people in the United States receive much more protein than they need.

Protein on Food Labels All food labels must state the *quantity* of protein in grams. The "percent Daily Value" for protein is not mandatory on all labels, but it is required whenever a food makes a protein claim or is intended for consumption by children under four years old.*

Nitrogen Balance Underlying the protein recommendation are **nitrogen balance** studies, which compare nitrogen lost by excretion with nitrogen eaten in food. If the body maintains the same amount of protein in its tissues from day to day, it is in nitrogen equilibrium (zero nitrogen balance). If the body adds protein, it is in positive nitrogen balance; if it loses protein, it is in negative nitrogen balance.

Normally, healthy adults are in nitrogen equilibrium; that is, their nitrogen intakes equal their nitrogen outputs. In other words, protein intake from food balances with nitrogen excretion in the urine, feces, and sweat. Growing children, adolescents, and pregnant women are in positive nitrogen balance because they are adding new blood, bone, and muscle cells to their bodies. People who are fasting or starving and people who suffer from traumas, such as burns (see Chapter 16), are in negative nitrogen balance because their bodies are forced to use protein for energy.

> **REVIEW NOTES**
>
> Normal, healthy people do not need amino acid or protein supplements.
>
> Optimally, an adult's diet will be adequate in energy from carbohydrate and fat and will deliver at least 0.8 grams of protein per kilogram of healthy body weight each day.

*For labeling purposes, the Daily Values for protein are as follows: for infants, 14 grams; for children under age four, 16 grams; for older children and adults, 50 grams; for pregnant women, 60 grams; and for lactating women, 65 grams.

RDA for protein (adult) = 0.8 g/kg.
- 10 to 35% of energy intake.

To calculate your protein need:

1. Find your body weight in pounds (or kilograms).
2. Convert pounds to kilograms, if necessary (pounds divided by 2.2).
3. Multiply kilograms by 0.8 to find total grams of protein recommended.

For example:

1. Weight = 150 lb.
2. 150 lb ÷ 2.2 lb/kg = 68 kg.
3. 68 kg × 0.8 g/kg = 54 g.

Daily Values:
- 50 g protein (based on 10% of 2000 kcal diet).

Nitrogen equilibrium (zero nitrogen balance): N in = N out.

Positive nitrogen balance: N in > N out.

Negative nitrogen balance: N in < N out.

nitrogen balance: the amount of nitrogen consumed (N in) as compared with the amount of nitrogen excreted (N out) in a given period of time. The laboratory scientist can estimate the protein in a sample of food, body tissue, or excreta by measuring the nitrogen in it.

Protein in Foods

In the United States and Canada, where nutritious foods are abundant, most people eat protein in such large quantities that they receive all the amino acids they need. In countries where food is scarce and the people eat only marginal amounts of protein-rich foods, however, the *quality* of the protein becomes crucial.

PROTEIN QUALITY

The protein quality of the diet determines, in large part, how well children grow and how well adults maintain their health. Put simply, **high-quality proteins** provide enough of all the essential amino acids needed to support the body's work, and low-quality proteins don't. Two factors influence protein quality—the protein's digestibility and its amino acid composition.

Digestibility As explained earlier, proteins must be digested before they can provide amino acids. **Protein digestibility** depends on such factors as the protein's source and the other foods eaten with it. The digestibility of most animal proteins is high (90 to 99 percent); plant proteins are less digestible (70 to 90 percent for most, but over 90 percent for soy).

Amino Acid Composition To make proteins, cells must have all the needed amino acids available simultaneously. The liver can produce any nonessential amino acid that may be in short supply so that the cells can continue linking amino acids into protein strands. If an essential amino acid is missing, however, a cell must dismantle its own proteins to obtain it. Therefore, to prevent protein breakdown, dietary protein must supply at least the nine essential amino acids plus enough nitrogen-containing amino groups and energy for the synthesis of the others. If the diet supplies too little of any essential amino acid, protein synthesis will be limited. The body makes whole proteins only; if one amino acid is missing, the others cannot form a "partial" protein. An essential amino acid that is available in the shortest supply relative to the amount needed to support protein synthesis is called a **limiting amino acid.**

High-Quality Proteins A high-quality protein contains all the essential amino acids in amounts adequate for human use; it may or may not contain all the others. Generally, proteins derived from animal foods (meats, fish, poultry, cheese, eggs, yogurt, and milk) are high quality, although gelatin is an exception. Proteins derived from plant foods (legumes, grains, and vegetables) tend to be limiting in one or more essential amino acids. Some plant proteins are notoriously low quality—for example, corn protein. Others are high quality—for example, soy protein. As discussed in Nutrition in Practice 5, the educated vegetarian can design a diet that is adequate in protein by choosing a variety of legumes, whole grains, nuts, and vegetables. Table 5-3 lists the protein contents of foods based on the food groups of the USDA Food Guide in Chapter 1. Fruits are not included in Table 5-3 because they contribute only small amounts of protein.

Complementary Proteins If the body does not receive all the essential amino acids it needs, the supply of essential amino acids will dwindle until body organs are compromised. Obtaining enough essential amino acids presents no problem to people who regularly eat high-quality proteins, such as those of meat, fish, poultry, cheese, eggs, milk, and many soybean products. The proteins of these foods contain ample amounts of all the essential amino acids. An equally sound choice is to eat two different protein foods from plants so that each supplies the amino acids missing in the other. In this strategy, the

high-quality proteins: dietary proteins containing all the essential amino acids in relatively the same amounts that human beings require. They may also contain nonessential amino acids.

protein digestibility: a measure of the amount of amino acids absorbed from a given protein intake.

limiting amino acid: an essential amino acid that is present in dietary protein in the shortest supply relative to the amount needed for protein synthesis in the body.

Vegetarians obtain their protein from whole grains, legumes, nuts, vegetables, and, in some cases, eggs and milk products.

© Polara Studios Inc.

	Ile	Lys	Met	Trp
Legumes	�	▓		
Grains			▓	▓
Together	▓	▓	▓	▓

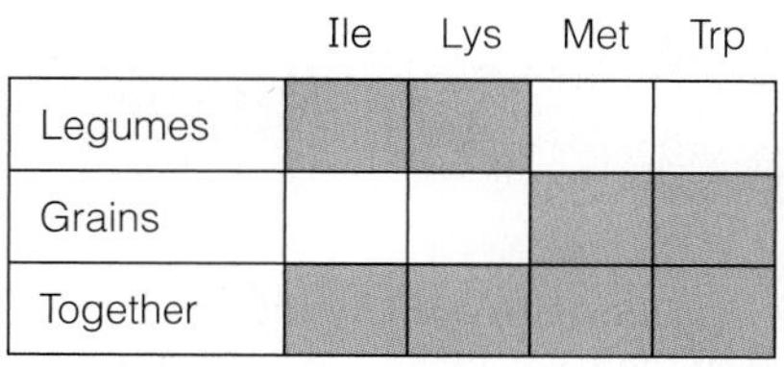

FIGURE 5-7 Complementary Proteins

In general, legumes provide plenty of isoleucine (Ile) and lysine (Lys) but fall short in methionine (Met) and tryptophan (Trp). Grains have the opposite strengths and weaknesses, making them a perfect match for legumes.

TABLE 5-3	Protein-Containing Foods

Milk, Yogurt, and Cheese

Each of the following provides about 8 grams of protein:

1 c milk, buttermilk, or yogurt (choose low-fat or fat-free).

1 oz regular cheese (for example, cheddar or swiss; choose low-fat)

$^1/_4$ c cottage cheese (choose low-fat or fat-free).

Meat, Poultry, Fish, Legumes, Eggs, and Nuts

Each of the following provides about 7 grams of protein:

1 oz meat, poultry, or fish (choose lean meats to limit saturated fat intake).

$^1/_2$ c legumes (navy beans, pinto beans, black beans, lentils, soybeans, and other dried beans and peas).

1 egg.

$^1/_2$ c tofu (soybean curd).

2 tbs peanut butter.

1 to 2 oz nuts or seeds.

Grains

Each of the following provides about 3 grams of protein:

1 slice of bread.

$^1/_2$ c cooked rice, pasta, cereals, or other grain foods.

Vegetables

Each of the following provides about 2 grams of protein:

$^1/_2$ c cooked vegetables.

1 c raw vegetables.

two protein-rich foods are combined to yield **complementary proteins** (see Figure 5-7)—proteins containing all the essential amino acids in amounts sufficient to support health. This concept is illustrated in Figure NP5-1 (p. 127) in this chapter's Nutrition in Practice. The two proteins need not even be eaten together, as long as the day's meals supply them both, and the diet provides enough energy and total protein from a variety of sources.

PROTEIN SPARING

Dietary protein—no matter how high the quality—will not be used efficiently and will not support growth when energy from carbohydrate and fat is lacking. The body assigns top priority to meeting its energy need and, if necessary, will break down protein to meet this need. After stripping off and excreting the nitrogen from the amino acids, the body will use the remaining carbon skeletons in much the same way it uses those from glucose or fat. A major reason why people must have ample carbohydrate and fat in the diet is to prevent this wasting of protein.

Reminder: Carbohydrate and fat allow amino acids to be used to build body proteins. This is known as the *protein-sparing effect* of carbohydrate and fat.

REVIEW NOTES

A diet inadequate in any of the essential amino acids limits protein synthesis.

The best guarantee of amino acid adequacy is to eat foods containing high-quality proteins or combinations of foods containing complementary proteins so that each can supply the amino acids missing in the other.

Vegetarians who consume no foods of animal origin can meet their protein needs by eating a variety of whole grains, legumes, seeds, nuts, and vegetables.

complementary proteins: two or more proteins whose amino acid assortments complement each other in such a way that the essential amino acids missing from one are supplied by the other.

SELF CHECK

1. Proteins are chemically different from carbohydrates and fats because they also contain:
   a. iron.
   b. sodium.
   c. nitrogen.
   d. phosphorus.

2. The basic building blocks for protein are:
   a. side groups.
   b. amino acids.
   c. glucose units.
   d. saturated bonds.

3. Enzymes are proteins that, among other things:
   a. defend the body against disease.
   b. regulate fluid and electrolyte balance.
   c. facilitate chemical reactions by changing themselves.
   d. help assemble disaccharides into starch, cellulose, or glycogen.

4. Functions of proteins in the body include:
   a. supplying omega-3 fatty acids for growth, lowering serum cholesterol, and helping with weight control.
   b. supplying fiber to aid digestion, digesting cellulose, and providing the main fuel source for muscles.
   c. protecting organs against shock, helping the body use carbohydrate efficiently, and providing triglycerides.
   d. supporting growth and maintenance, supplying hormones to regulate body processes, and maintaining fluid and electrolyte balance.

5. The swelling of body tissue caused by the leakage of fluid from the blood vessels into the interstitial spaces is called:
   a. edema.
   b. anemia.
   c. acidosis.
   d. sickle-cell anemia.

6. Major proteins in the blood that protect against bacteria and other disease agents are called:
   a. acids.
   b. buffers.
   c. antigens.
   d. antibodies.

7. Marasmus can be distinguished from kwashiorkor because in marasmus:
   a. only adults are victims.
   b. the cause is usually an infection.
   c. severe wasting of body fat and muscle occurs.
   d. the limbs and face swell with edema, and the belly bulges with a fatty liver.

8. The RDA for protein for a healthy adult is ___ gram(s) per kilogram of appropriate or average body weight for height.
   a. 0.5
   b. 0.8
   c. 1.1
   d. 1.4

9. Generally speaking, from which of the following foods are complete proteins derived?
   a. milk, gelatin, and soy
   b. rice, potatoes, and eggs
   c. meats, fish, and poultry
   d. vegetables, grains, and fruits

10. An incomplete protein lacks one or more:
    a. hydrogen bonds.
    b. essential fatty acids.
    c. saturated fatty acids.
    d. essential amino acids.

Answers to these questions can be found in Appendix H.

CLINICAL APPLICATIONS

1. Considering the health effects of too little dietary protein, what suggestions would you have for a teenage girl who reports the following information about her food intake:
   • She never eats any meat or other animal-derived foods because she is a vegan. On a typical day, she eats toast and juice for breakfast; chips, a soft drink, and a piece of fruit for lunch; and a small serving of plain pasta with tomato sauce or steamed vegetables for dinner, along with a glass of water or tea.
   • She takes amino acid supplements because a friend told her that the only way to get amino acids if she doesn't eat meat is to take them as supplements.

2. Considering the health effects of excess dietary protein, what advice would you have for a college athlete who tells you he wants to bulk up his muscles and reports the following information about his food intake:
   • He eats large servings of meat (usually red meat) at least twice a day. He drinks whole milk two or three times a day and eats eggs and bacon for breakfast almost every day.
   • He avoids breads, cereals, and pasta in order to save room for protein-rich foods such as meat, milk, and eggs.
   • He eats a piece of fruit once in a while but seldom eats vegetables because they are too time-consuming to prepare.

NUTRITION ON THE NET

For further study of the topics in this chapter, access these websites.

Find updates and quick links to these and other nutrition-related sites at our website: **www.wadsworth.com/nutrition**

To learn more about protein in foods, visit the American Dietetic Association (ADA) site: **www.eatright.org**

For more on vegetarian diets, search the Food and Drug Administration (FDA) site for foods: **www.fda.gov**

Search among thousands of current scientific and medical abstracts for any topic related to protein at: **www.ncbi.nlm .nih.gov/PubMed/**

Learn more about protein-energy malnutrition and world hunger from the World Health Organization's Nutrition Programme: **www.who.int**

NOTES

[1] T. R. Ziegler and coauthors, Trophic and cytoprotective nutrition for intestinal adaptation, mucosal repair, and barrier function, *Annual Review of Nutrition* 23 (2003): 229–261.

[2] Standing Committee on the Scientific Evaluation of Dietary Reference Intakes, Food and Nutrition Board, Institute of Medicine, *Dietary Reference Intakes for Energy, Carbohydrate, Fiber, Fat, Fatty Acids, Cholesterol, Protein, and Amino Acids* (Washington, D.C.: National Academies Press, 2005), p. 605.

[3] Standing Committee on the Scientific Evaluation of Dietary Reference Intakes, 2005, p. 605.

[4] World Health Organization, Protein-energy malnutrition, 2006, **www.wpro.who.int/health_topics/protein_energy**.

[5] World Health Organization, 2006.

[6] M. Reid and coauthors, The acute-phase protein response to infection in edematous and nonedematous protein-energy malnutrition, *American Journal of Clinical Nutrition* 76 (2002): 1409–1415.

[7] T. Liu and coauthors, Kwashiorkor in the United States: Fad diets, perceived and true milk allergy, and nutritional ignorance, *Archives of Dermatology* 137 (2001): 630–636; N. F. Carvalho and coauthors, Severe nutritional deficiencies in toddlers resulting from health food milk alternatives (electronic article), *Pediatrics* 107 (2001): e46.

[8] A. Trichopoulou and coauthors, Lipid, protein, and carbohydrate intake in relation to body mass index, *European Journal of Clinical Nutrition* 56 (2002): 37–43.

[9] L. E. Kelemen and coauthors, Associations of dietary protein with disease and mortality in a prospective study of postmenopausal women, *American Journal of Epidemiology* 161 (2005): 239–249.

[10] S. Tonstad, K. Smerud, and L. Hoie, A comparison of the effects of 2 doses of soy protein or casein on serum lipids, serum lipoproteins, and plasma total homocysteine in hypercholesterolemic subjects, *American Journal of Clinical Nutrition* 76 (2002): 78–84.

[11] E. L. Knight and coauthors, The impact of protein intake on renal function decline in women with normal renal function or mild renal insufficiency, *Annals of Internal Medicine* 138 (2003): 460–467.

[12] B. Dawson-Hughes, Calcium and protein in bone health, *The Proceedings of the Nutrition Society* 62 (2003): 505–509; F. Ginty, Dietary protein and bone health, *The Proceedings of the Nutrition Society* 62 (2003): 867–876; J. E. Kerstetter, K. O. O'Brien, and K. L. Insogna, Low protein intake: The impact on calcium and bone homeostasis in humans, *Journal of Nutrition* 133 (2003): 855S–861S.

[13] Dawson-Hughes, 2003; Ginty, 2003; Kerstetter, O'Brien, and Insogna, 2003.

[14] B. Dawson-Hughes, Interaction of dietary calcium and protein in bone health in humans, *Journal of Nutrition* 133 (2003): 852S–854S.

[15] B. Dawson-Hughes and S. S. Harris, Calcium intake influences the association of protein intake with rates of bone loss in elderly men and women, *American Journal of Clinical Nutrition* 75 (2002): 773–779.

[16] Standing Committee on the Scientific Evaluation of Dietary Reference Intakes, 2005, p. 694.

[17] P. Garlick, Assessment of the safety of glutamine and other amino acids, *Journal of Nutrition* 131 (2001): 2556S–2561S.

Vegetarian Diets

Eating patterns all along the continuum of dietary choices—from one end, where people eat no foods of animal origin, to the other end, where they eat generous quantities of meat every day—can support or compromise nutritional health. The quality of the diet depends not on whether it consists of all plant foods or centers on meat, but on whether the eater's food choices are based on sound nutrition principles: adequacy of nutrient intakes, balance and variety of foods chosen, appropriate energy intake, and moderation in intakes of substances such as fat, sodium, alcohol, and caffeine that are harmful when consumed in excess. As mentioned in Chapter 4, however, because vegetarian diets exclude at least some animal-derived foods, they are usually lower in saturated fat and cholesterol than many meat-based diets.

People choose to exclude meat and other animal-derived foods from their diets for various reasons—concern for animal welfare, beliefs about health, religious convictions, or convenience. Some believe that vegetarianism is better for the environment; some, that it is healthier; and some, that it is less costly than the meat-eating alternative. Some just like it better. Whatever the reasons, vegetarians and health professionals who work with them should be aware of the nutrition and health implications of vegetarian diets.

Because vegetarian diets vary in both the types and the amounts of animal-derived foods they include, these differences must be considered when evaluating the health status of vegetarians. The glossary on p. 129 defines the various kinds of vegetarian diets.

Are vegetarian diets nutritionally sound?

The American Dietetic Association takes the position that well-planned vegetarian diets offer nutrition and health benefits to adults in general.[1] Research suggests that meat-eating adults who switch to vegetarian diets reduce their risks of heart disease, hypertension, diabetes, some types of cancer, and obesity.[2]

What should be my main concerns when planning a nutritionally sound vegetarian diet?

A vegetarian diet planner faces the same task as other diet planners—obtaining a variety of foods that provide all the needed nutrients within an energy allowance that maintains a healthy body weight. The challenge is to do so using at least one less food group. Since all vegetarians omit meat and some omit other animal-derived foods, protein, the nutrient that meat is famous for, merits some discussion here.

Isn't protein a problem in vegetarian diets?

No, protein is not the problem it was once thought to be in vegetarian diets. People who include animal-derived foods such as milk and eggs in their diets need not worry at all about protein deficiency. Even for those who eat only plant-derived foods, protein intakes are usually satisfactory as long as energy intakes are adequate and protein sources are varied.[3] A mixture of proteins from whole grains, legumes, seeds, nuts, and vegetables can provide adequate amounts of high-quality protein.

The idea persists that **vegans** must carefully combine their plant-protein foods in order to obtain the protein they need, but this is not necessary. Plant foods can provide more than enough high-quality protein and can sustain people in good health, as long as the diet supplies sufficient energy and does not include too many empty-kcalorie foods. As Chapter 5 explained, however, a meal delivers higher-quality protein when it combines two or more different individual plant-protein sources. Figure NP5-1 shows combinations of plant proteins that provide higher-quality protein than the individual foods alone could supply.

What sorts of food energy intakes do vegetarian diets provide?

Researchers find that vegetarians as a group are closer to a healthy body weight than nonvegetarians.[4] Because obesity impairs health in a number of ways, vegetarians therefore have a health advantage. Vegetarian diets tend to be high in starch- and fiber-rich carbohydrates and low in saturated fat, characteristics that are consistent with current dietary recommendations aimed at reducing the incidence of obesity and other degenerative diseases in this country.

Not all vegetarians fit the average pattern, though. Obesity does threaten vegetarians who include milk, eggs, and cheese in their diets. They can easily consume excess saturated fat and food energy and so must be careful to select fat-free and low-fat dairy foods and to avoid relying too heavily on these foods in general.

In contrast, people who exclude all animal-derived foods (vegans) may have trouble obtaining *enough* food energy. This is especially true for children. Vegan diets can fail to provide food energy sufficient to support the growth of a child within a bulk of food small enough for the child to eat. Frequent meals of fortified breads, cereals, or pastas with legumes, nuts, nut butters, and sources of unsaturated fats can help to meet protein and energy needs in a smaller volume at each sitting.[5] Figure NP5-2 offers a suggested vegetarian food guide pyramid.

Tell me about vitamins and minerals. Does a person eating a vegetarian diet need to take vitamin supplements?

That depends on the kind of vegetarian diet. The diet of **lacto-ovo vegetarians** can be complete in all vitamins, but for vegans, several vitamins may be a problem. One such vitamin

FIGURE NP5-1 Nonmeat Mixtures that Provide High-Quality Protein

Vegetarians who eat no foods from animal sources select foods from two or more of these columns to create high-quality protein combinations:

Grains	Legumes	Seeds and Nuts	Vegetables
Barley	Dried beans (pinto, kidney, navy, etc.)	Almonds	Broccoli
Bulgur	Dried lentils	Cashews	Cabbage
Oats	Dried peas	Nut butters	Peppers
Pasta	Peanuts	Sesame seeds	Spinach
Rice	Soy products	Sunflower seeds	Squash
Whole-grain breads		Walnuts	

Black beans and rice, a favorite Hispanic combination.

Tofu and stir-fried vegetables with rice, an Asian dish.

© Polara Studios, Inc. (both)

FIGURE NP5-2 Vegetarian Food Guide Pyramid

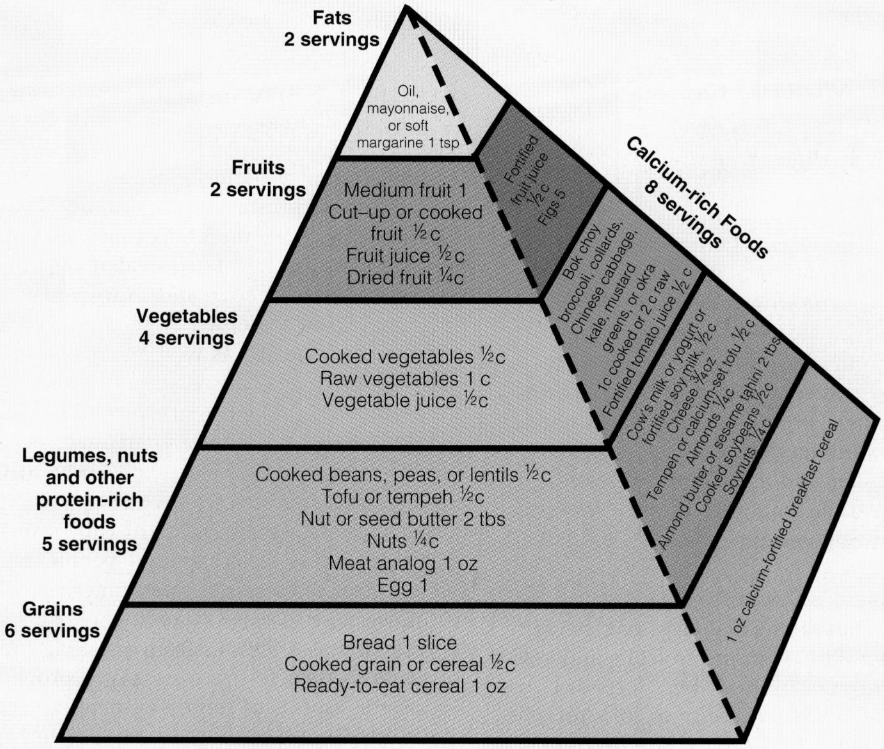

Fats
2 servings

Oil, mayonnaise, or soft margarine 1 tsp

Fruits
2 servings

Medium fruit 1
Cut-up or cooked fruit ½c
Fruit juice ½c
Dried fruit ¼c

Fortified fruit juice ½c
Figs 5

Calcium-rich Foods
8 servings

Bok choy, broccoli, collards, Chinese cabbage, kale, mustard greens, or okra 1 c cooked or 2 c raw
Fortified tomato juice ½ c
Cow's milk or yogurt or fortified soy milk ½ c
Cheese ¾ oz
Tempeh or calcium-set tofu ½ c
Almonds ¼ c
Almond butter or sesame tahini 2 tbs
Cooked soybeans ½ c
Soynuts ¼ c
1 oz calcium-fortified breakfast cereal

Vegetables
4 servings

Cooked vegetables ½c
Raw vegetables 1 c
Vegetable juice ½c

Legumes, nuts and other protein-rich foods
5 servings

Cooked beans, peas, or lentils ½c
Tofu or tempeh ½c
Nut or seed butter 2 tbs
Nuts ¼c
Meat analog 1 oz
Egg 1

Grains
6 servings

Bread 1 slice
Cooked grain or cereal ½c
Ready-to-eat cereal 1 oz

Source: Reprinted from V. Messina, V. Melina, and A. R. Mangels, A new food guide for North American Vegetarians, *Journal of the American Dietetic Association* 103 (2003): 771–775 with permission from the American Dietetic Association.

is B_{12}. Because vitamin B_{12} occurs only in animal-derived foods, supplements are necessary to prevent deficiency. Women who have adhered to all-plant diets for many years are especially likely to have low vitamin B_{12} stores. Pregnant vegan women, whose needs for vitamin B_{12} are especially high, find it virtually impossible to maintain adequate vitamin B_{12} status without taking supplements or including a reliable food source of the nutrient.

A vitamin B_{12} deficiency can take a long time to develop in adults because up to four years' worth of the vitamin can be stored in the body. But when the deficiency sets in, it does severe damage to the nervous system (see Chapter 8). In infants, deficiencies set in more rapidly and so threaten their nervous systems earlier. All vegan mothers must be sure to take the appropriate supplements or to use vitamin B_{12}–fortified products such as soy milk or breakfast cereals.

What other vitamins do vegans need?

Another vitamin of concern is vitamin D. The milk drinker is protected, provided the milk is fortified with vitamin D, but there is no practical source of vitamin D in plant foods. Regular exposure to the sun will prevent a deficiency, but vegans who are homebound or live in a northern climate or smoggy city probably should take vitamin D supplements to defend against bone loss. This is particularly important for infants, children and older adults.

Riboflavin, another vitamin often obtained from milk, is not a problem for the vegan who eats dark greens frequently in ample servings. The vegan who doesn't consume a lot of greens, however, may not meet riboflavin needs. Nutritional yeast is a rich source of riboflavin for the vegetarian.

So, on a vegan diet, vitamin B_{12}, vitamin D, and riboflavin can be problems if a person is not careful. What about minerals?

For *all* vegetarians, not just the vegan, two minerals may be of concern—iron and zinc. Legumes are an important source of iron in the vegetarian diet. The iron in legumes, however, is not as absorbable as that in meat. In fact, people absorb three times as much iron from a meal that includes meat as from one that does not. For this reason, the iron recommendation for adult vegetarian men, premenopausal women, and adolescent girls is almost double the recommendation for meat eaters of the same gender and age.[6] Because vitamin C in fruits and vegetables can triple iron absorption from other foods eaten at the same meal, vegetarian meals should be rich in foods offering vitamin C.

Zinc may also be a problem nutrient for vegetarians. It is widespread in plant foods, but its availability may be hindered by the fibers and other binders found in fruits and vegetables. The zinc needs of vegetarians and the effects of mineral binders are subjects of intensive study at the present time. While research continues, vegetarians are advised to eat varied diets that include legumes such as navy beans and kidney beans, zinc-enriched cereals, and whole-grain breads well leavened with yeast, which improves the availability of their minerals. For those who include seafood in their diets, oysters, crabmeat, and shrimp are rich in zinc.

What about calcium for the vegan?

Good thinking. Yes, calcium is of concern. The milk-drinking vegetarian is protected from deficiency, but the vegan must find other sources of calcium. Some good calcium sources are regular and ample servings of dark green leafy vegetables such as kale and collards; legumes; calcium-fortified foods such as breakfast cereals, soy milk, and orange juice, some nuts such as almonds; and certain seeds such as sesame seeds. The choices should be varied because binders in some of these foods may hinder calcium absorption. The vegan is urged to use calcium-fortified soy milk in ample quantities regularly. This is especially important for children. Infant formula based on soy is fortified with calcium and can easily be used in cooking foods, even for adults.

Do vegetarian diets provide adequate amounts of the essential fatty acids?

Vegetarian diets typically provide enough omega-6 fatty acids but lack omega-3 fatty acids. This imbalance slows production of EPA and DHA in the body, and without fish or eggs in the diet, intake of EPA and DHA falls short as well. To compensate for this inadequacy, vegetarians need to include good sources of linolenic acid, such as flaxseed, walnuts, soybeans, and their oils, in their diets daily.

Are there any other health advantages to the vegetarian diet?

Yes. Vegetarian protein foods are often higher in fiber, richer in certain vitamins and minerals, and lower in fat, especially saturated fat, than meats. Vegetarians can enjoy a nutritious diet low in saturated fat provided that they limit foods such as butter, cream cheese, and sour cream. If vegetarians follow the guidelines presented here and plan carefully, they can support their health as well as, or perhaps better than, non-vegetarians.

Abundant evidence supports the idea that vegetarians may actually be healthier than meat eaters. Informed vegetarians are not only more likely to be at the desired weights for their heights, but to have lower blood cholesterol levels, lower rates of certain kinds of cancer, better digestive function, and more. Even among people who are health conscious, generally vegetarians experience fewer deaths from cardiovascular disease than meat eaters do. Because many vegetarians also abstain from smoking and the consumption of alcohol, dietary practices alone probably do not account for all the aspects of improved health. Clearly, however, they contribute significantly to it.

Glossary of Vegetarian Terms

lacto-ovo vegetarians: people who include milk or milk products and eggs, but omit meat, fish, shellfish, and poultry from their diets.

lacto-vegetarians: people who include milk or milk products, but exclude meat, poultry, fish, shellfish, and eggs from their diets.

semivegetarians: people who include some, but not all, groups of animal-derived foods in their diets; they usually exclude meat and may occasionally include poultry, fish, and shellfish; also called *partial vegetarians*.

vegans: people who exclude all animal-derived foods (including meat, poultry, fish, shellfish, eggs, cheese, and milk) from their diets; also called *strict vegetarians* or *total vegetarians*.

Notes

[1] Position of the American Dietetic Association and Dietitians of Canada: Vegetarian diets, *Journal of the American Dietetic Association* 103 (2003): 748–765.

[2] S. E. Berkow and N. D. Barnard, Blood pressure regulation and vegetarian diets, *Nutrition Reviews* 63 (2005): 1–8; J. Sabate, The contribution of vegetarian diets to health and disease: A paradigm shift? *American Journal of Clinical Nutrition* 78 (2003): 502S–507S; P. N. Singh, J. Sabate, and G. E. Fraser, Does low meat consumption increase life expectancy in humans? *American Journal of Clinical Nutrition* 78 (2003): 526S–532S; F. B. Hu, Plant-based foods and prevention of cardiovascular disease: An overview, *American Journal of Clinical Nutrition* 78 (2003): 544S–551S.

[3] Position of the American Dietetic Association and Dietitians of Canada, 2003.

[4] E. H. Haddad and J. S. Tanzman, What do vegetarians in the United States eat? *American Journal of Clinical Nutrition* 78 (2003): 626S–632S; N. Brathwaite and coauthors, Obesity, diabetes, hypertension, and vegetarian status among Seventh-Day Adventists in Barbados: Preliminary results, *Ethnicity and Disease* 13 (2003): 34–39.

[5] Position of the American Dietetic Association and Dietitians of Canada, 2003.

[6] Standing Committee on the Scientific Evaluation of Dietary Reference Intakes, Food and Nutrition Board, Institute of Medicine, *Dietary Reference Intakes for Vitamin A, Vitamin K, Arsenic, Boron, Chromium, Copper, Iodine, Iron, Manganese, Molybdenum, Nickel, Silicon, Vanadium, and Zinc* (Washington, D.C.: National Academy Press, 2001), pp. 9–45.

CONTENTS

Energy Balance and Body Composition

© Envision/Corbis

CHAPTER 6

The body manages its energy supply with amazing precision. Consider that most people maintain their weight within about a 10- to 20-pound range throughout their lives. How do they do this? How does the body manage excess energy? And how does it manage to do without food for prolonged periods—as when someone is starving or fasting? The answers to these questions lie in understanding how the body adapts to energy imbalances, the focus of the first part of this chapter. The chapter then describes energy balance, body composition, and the health risks associated with too much body fat. The next chapter offers strategies toward solving the problems of too much or too little body fat.

The Body's Energy Budget

Each year the average person takes in and expends close to a million kcalories, maintaining a stable weight for years on end. In other words, the body's energy budget is balanced. Many people, however, eat too much or exercise too little and get fat; others eat too little or exercise too much and get thin. This section examines metabolism from the perspective of the energy budget, looking first at the two forms of unbalanced budget—feasting and fasting—and then at a balanced budget.

THE ECONOMICS OF FEASTING

Everyone knows that when people consume more energy than they expend, much of the excess is stored as body fat. Fat can be made from an excess of any energy-yielding nutrient. In addition, excess energy from alcohol is also stored as fat.[1] Alcohol has also been shown to slow down the body's use of fat for fuel, causing more fat to be stored.[2] The fat cells of the adipose tissue enlarge as they fill with fat, as Figure 6-1 shows.

Excess Carbohydrate Surplus carbohydrate (glucose) is first stored as glycogen in the liver and muscles, but the glycogen-storing cells have limited capacity. Once glycogen stores are filled, most of the additional carbohydrate is burned for energy, displacing the body's use of fat for energy and allowing body fat to accumulate. Thus excess carbohydrate can contribute to obesity.

Excess Fat Surplus dietary fat contributes easily to the body's fat stores. After a meal, fat is routed to the body's adipose tissue where it is stored until needed for energy. Thus excess fat from food easily adds to body fat.

Excess Protein Surplus protein may also contribute to body fat. If not needed to build body protein (as in response to physical activity) or to meet energy needs, amino acids will lose their nitrogens and be converted, through intermediates, to

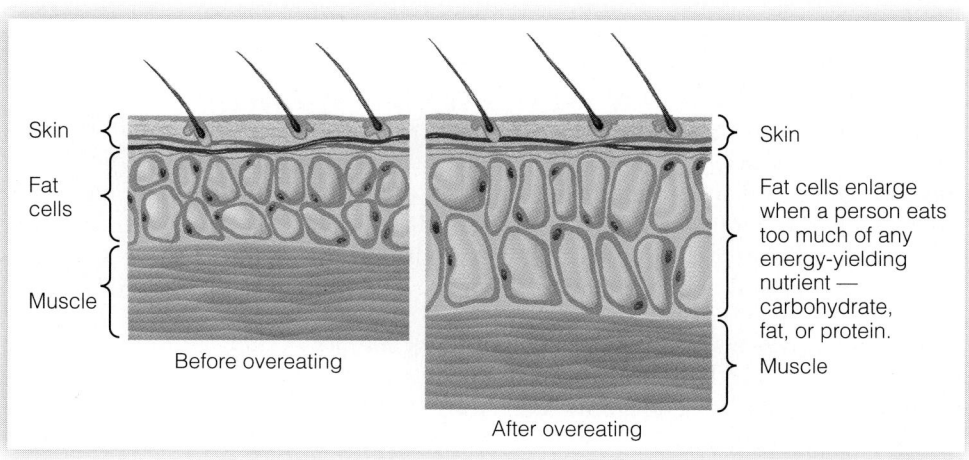

Skin

Fat cells

Muscle

Before overeating

After overeating

Skin

Fat cells enlarge when a person eats too much of any energy-yielding nutrient — carbohydrate, fat, or protein.

Muscle

FIGURE 6-1 Fat Cell Enlargement

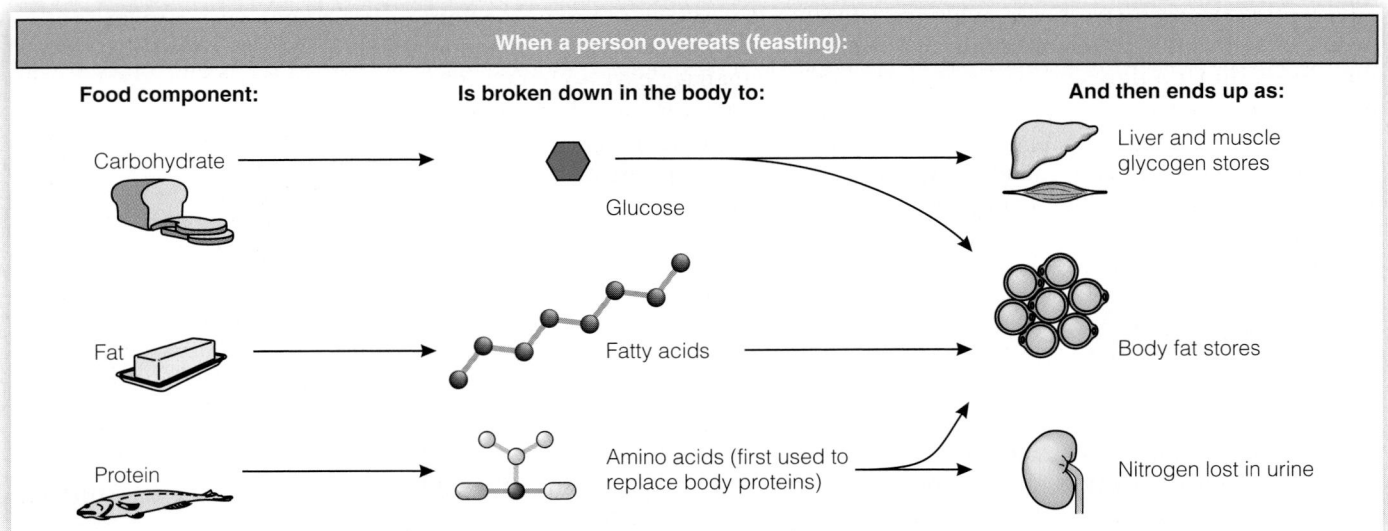

When a person overeats (feasting):

| Food component: | Is broken down in the body to: | And then ends up as: |

Carbohydrate → Glucose → Liver and muscle glycogen stores

Fat → Fatty acids → Body fat stores

Protein → Amino acids (first used to replace body proteins) → Nitrogen lost in urine

FIGURE 6-2 Feasting

When people overeat, they store energy.

triglycerides. These, too, swell the fat cells and add to body weight. Figure 6-2 shows the metabolic events of feasting.

> **REVIEW NOTES**
>
> Too little physical activity encourages body fat accumulation.
>
> Any food can make you fat if you eat too much of it. A net excess of energy is almost all stored in the body as fat in adipose tissue.
>
> Alcohol both delivers kcalories and encourages storage of body fat.
>
> Fat from food is particularly easy for the body to store as adipose tissue.
>
> Protein is not stored in the body except in response to physical activity; otherwise it is present only as working tissue. If not needed to meet energy needs, excess protein can be converted to fat.
>
> In short, excess energy from carbohydrate, fat, protein, and alcohol leads to storage of body fat.

THE ECONOMICS OF FASTING

The body spends energy all the time. Even when a person is asleep and totally relaxed, the cells of many organs are hard at work. In fact, this cellular work, which maintains all life processes, represents about two-thirds of the total energy a sedentary person spends in a day. (The other one-third is the work that a person's muscles do voluntarily during waking hours.)

In *fasting*, a person voluntarily stops eating food, whereas in *starvation* the failure to eat is involuntary. The body, however, makes no distinction between the two—metabolically, fasting and starvation are identical.

Energy Deficit The body's top priority is to meet the energy needs for this ongoing cellular activity. Its normal way of doing so is by periodic refueling, that is, by eating several times a day. When food is not available, the body uses fuel reserves from its own tissues. If people choose not to eat, we say they are fasting; if they have no choice (as in a famine), we say they are starving. In the body, however, no metabolic difference exists between fasting and starving. In either case, the body is forced to switch to a wasting metabolism, drawing on its stores of carbohydrate and fat and, within a day or so, on its vital protein tissues as well.

Glycogen Used First As a fast or period of starvation begins, glucose from the liver's stored glycogen and fatty acids from the body's stored fat both flow into the cells to fuel their work. Several hours later, most of the glucose has been used up—liver glycogen is exhausted. Low blood glucose concentrations serve as a signal to promote further fat breakdown.

Glucose Needed for the Brain At this point, a few hours into a fast, most of the cells are depending on fatty acids to continue providing fuel. But the nervous system (brain and nerves) and red blood cells cannot use fatty acids; they still need glucose. Even if other energy sources are available, glucose has to be present to permit the brain's energy-metabolizing machinery to work. Normally, the nervous system consumes a little more than half of the total glucose used each day—about 400 to 600 kcalories' worth.

Protein Breakdown and Ketosis Because fat stores cannot provide the glucose needed by the brain and nerves, body protein tissues (such as liver and muscle) always break down to some extent during fasting. In the first few days of a fast, body protein provides about 90 percent of the needed glucose, and glycerol provides about 10 percent.* If body protein losses were to continue at this rate, death would ensue within about three weeks. As the fast continues, however, the body finds a way to use its fat to fuel the brain. It adapts by condensing together fragments derived from fatty acids to produce ketone bodies, which can serve as fuel for some brain cells. Ketone body production rises until, after several weeks of fasting, it is meeting much of the nervous system's energy needs. Still, many areas of the brain rely exclusively on glucose, and body protein continues to be sacrificed to produce it. Figure 6-3 shows the metabolic events that occur during fasting.

Reminder: The liver releases glucose, and the fat cells release fat to fuel the body's cells, but the brain can use only glucose.

Reminder: *Ketone bodies* are acidic, fat-related compounds formed from the incomplete breakdown of fat when carbohydrate is not available. Small amounts of ketone bodies are normally produced during energy metabolism, but when their blood concentration rises, they spill into the urine. The combination of a high blood concentration of ketone bodies (*ketonemia*) and ketone bodies in the urine (*ketonuria*) is called *ketosis.*

Fasting = living on the body's fat and protein.

In fasting, muscle and lean tissues give up protein to supply amino acids for conversion to glucose. This glucose, with ketone bodies produced from fat, fuels the brain's activities.

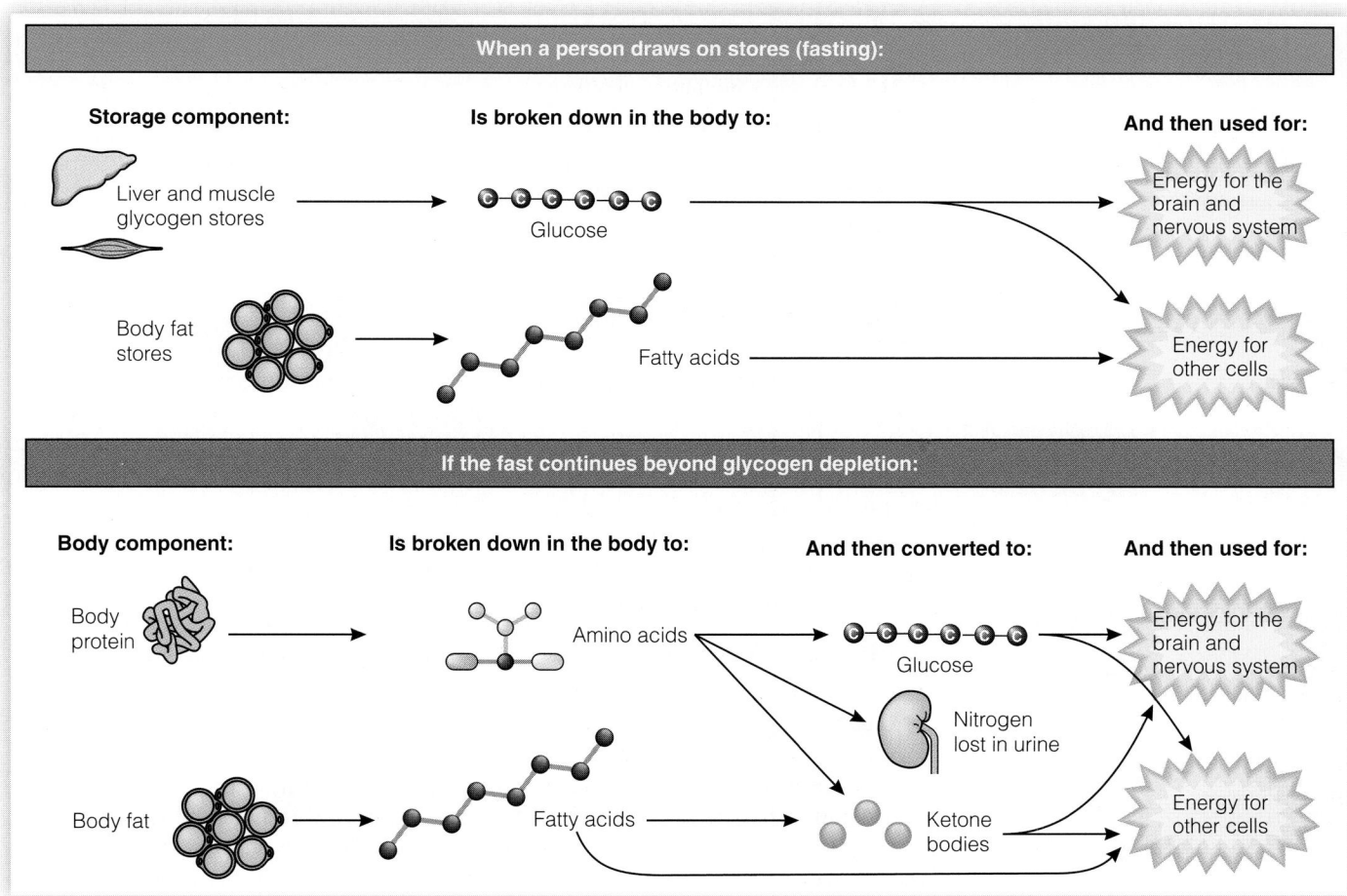

FIGURE 6-3 Fasting
When people are fasting, they draw on stored energy.

*The small glycerol portion of a triglyceride can yield glucose, but glycerol represents only about 5 percent of the weight of a triglyceride molecule. Thus fat is an inefficient source of glucose. About 95 percent of a triglyceride (the fatty acids attached to the glycerol) cannot be converted to glucose.

Slowed Metabolism As fasting continues and the body is shifting to partial dependence on ketone bodies for energy, the body simultaneously reduces its energy output (metabolic rate) and conserves both fat and lean tissue. Because of the slowed metabolism, energy use falls to a bare minimum.

Hazards of Fasting The body's adaptations to fasting are sufficient to maintain life for a long period. Mental alertness need not be diminished. Even physical energy may remain unimpaired for a surprisingly long time. Still, fasting is not without its hazards. Among the many changes that take place in the body are:

- Wasting of lean tissues.
- Impairment of disease resistance.
- Lowering of body temperature.
- Disturbances of the body's fluid and electrolyte balances.

For the person who wants to lose weight, fasting is not the best way to go. The body's lean tissue continues to be degraded, sometimes amounting to as much as 50 percent of the weight lost during the first week. Over the long term, a diet only moderately restricted in energy promotes primarily *fat* loss and the retention of more lean tissue than a severely restricted fast.

> **REVIEW NOTES**
>
> When fasting, the body makes a number of adaptations: increasing the breakdown of fat to provide energy for most of the cells, using glycerol and amino acids to make glucose for the red blood cells and central nervous system, producing ketones to fuel the brain, and slowing metabolism.
>
> All of these measures conserve energy and minimize losses.
>
> Over the long term, a diet moderately restricted in energy promotes primarily fat loss and the retention of lean tissue.

Energy Balance

If a person maintains a healthy weight over time, the person has a balanced energy budget. Food energy intake has equaled energy expenditure; that is, deposits of fat made at one time have been compensated for by withdrawals made at another. In other words, the body uses fat as a savings account for energy. But, unlike money, having more fat is not better; there is an optimum.

A day's energy balance can be stated like this: Change in energy stores equals the food energy taken in (kcalories) minus the energy spent on metabolism and physical activities (kcalories). More simply:

Change in energy stores = energy in (kcalories) − energy out (kcalories).

ENERGY IN

The energy in food and beverages is the only contributor to the "energy in" side of the energy balance equation. Before you can decide how much food will supply the energy you or one of your clients needs in a day, you must first become familiar with the amounts of energy in foods and beverages. One way to do so is to look up the kcalories provided by various foods and beverages in the Table of Food Composition (Appendix A). Alternatively, computer programs can readily provide this information for those with computer access.

For example, an apple provides 70 kcalories from carbohydrate; a candy bar supplies about 250 kcalories mostly from fat and carbohydrate. You may already

1 lb body fat = 3500 kcal.

know that for each 3,500 kcalories you eat in excess of expenditures, you store approximately 1 pound of body fat.

ENERGY OUT

The body spends energy in two major ways: to fuel its **basal metabolism** and to fuel its **voluntary activities.** People can change their voluntary activities to spend more or less energy in a day, and over time they can also change their basal metabolism by building up the body's metabolically active lean tissue, as explained in Chapter 7.

Basal Metabolism Basal metabolism supports the body's work that goes on all the time without the person's conscious awareness. The beating of the heart, the inhaling and exhaling of air, the maintenance of body temperature, and the transmission of nerve and hormonal messages to direct these activities are the basal processes that maintain life.

The **basal metabolic rate (BMR)** is the rate at which the body spends energy for these maintenance activities. This rate may vary from person to person and may vary for a single individual with a change in circumstance or physical condition. The rate is slowest when a person is sleeping undisturbed, but it is usually measured when the person is awake, lying still, in a room with a comfortable temperature after a restful sleep and an overnight (12- to 14-hour) fast. A similar measure of energy output—called the **resting metabolic rate (RMR)**—is slightly higher than the BMR because its criteria for recent food intake and physical activity are not as strict.

Table 6-1 (p. 136) summarizes the factors that raise and lower the BMR. For the most part, the BMR is highest in people who are growing (children and pregnant women) and in those with considerable lean body mass (physically fit people and males). One way to increase the BMR, then, is to maximize lean body tissue by participating regularly in endurance and strength-building activities. The BMR is also fast in people who are tall and so have a large surface area for their weight, in people with fever or under stress, in people taking certain medications, and in people with highly active thyroid glands. The BMR is slowed by loss of lean tissue and by depression of thyroid hormone activity due to disease, inactivity, fasting, or malnutrition.

Basal Metabolic Needs As Figure 6-4 shows, basal metabolic needs are surprisingly large. Depending on activity level, a person whose total energy needs are 2000 kcalories a day may spend 1000 to 1300 of them to support basal metabolism.

Energy for Activities The number of kcalories spent on voluntary activities depends on three factors: muscle mass, body weight, and activity. The larger the muscle mass required for the activity and the heavier the weight of the body part being moved, the more kcalories are spent. The activity's duration, frequency, and intensity also influence energy costs: the longer, the more frequent, and the more intense the activity, the more kcalories expended. Table 6-2 (p. 137) gives average energy expenditures for people of different body weights engaged in various activities and shows that a heavy person usually uses more energy per minute than a light person does in the same activity.

Energy to Manage Food When food is taken into the body, many cells that have been dormant become active. The muscles that move the food through the intestinal tract speed up their rhythmic contractions, and the cells that manufacture and secrete digestive juices begin their tasks. All these and other cells need extra energy as they participate in the digestion, absorption, and

basal metabolism: the energy needed to maintain life when a person is at complete digestive, physical, and emotional rest. Basal metabolism is normally the largest part of a person's daily energy expenditure.

voluntary activities: the component of a person's daily energy expenditure that involves conscious and deliberate muscular work—walking, lifting, climbing, and other physical activities. Voluntary activities normally require less energy in a day than basal metabolism does.

basal metabolic rate (BMR): the rate of energy use for metabolism under specified conditions: after a 12-hour fast and restful sleep, without any physical activity or emotional excitement, and in a comfortable setting. It is usually expressed as kcalories per kilogram of body weight per hour.

resting metabolic rate (RMR): a measure of the energy use of a person at rest in a comfortable setting; similar to the BMR but with less stringent criteria for recent food intake and physical activity. Consequently, the RMR is slightly higher than the BMR.

FIGURE 6-4 Components of Energy Expenditure
In most people, basal metabolism represents the person's largest expenditure of energy, followed by physical activity and the thermic effect of food.

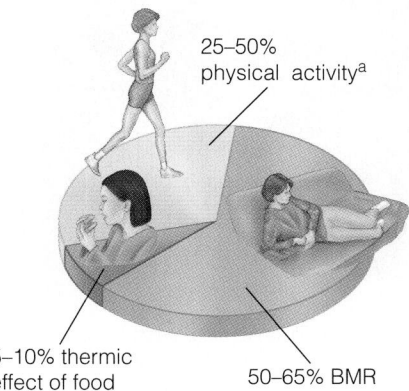

25–50% physical activity[a]

5–10% thermic effect of food

50–65% BMR

[a] For a sedentary person, physical activities may account for less than half as much energy as basal metabolism, whereas a very active person's activities may equal the energy cost of basal metabolism.

Source: Data from the Standing Committee on the Scientific Evaluation of Dietary Reference Intakes, Food and Nutrition Board, Institute of Medicine, *Dietary Reference Intakes for Energy, Carbohydrate, Fiber, Fat, Fatty Acids, Cholesterol, Protein, and Amino Acids* (Washington, D.C.: National Academies Press, 2005), p. 115.

TABLE 6-1	Factors That Affect the BMR
Factor	**Effect on BMR**
Age	Lean body mass diminishes with age, slowing the BMR.[a]
Height	In tall, thin people, the BMR is higher.[b]
Growth	In children and pregnant women, the BMR is higher.
Body composition (gender)	The more lean tissue, the higher the BMR (which is why males usually have a higher BMR than females). The more fat tissue, the lower the BMR.
Fever	Fever raises the BMR.[c]
Stresses	Stresses (including many diseases and certain drugs) raise the BMR.
Environmental temperature	Both heat and cold raise the BMR.
Fasting/starvation	Fasting/starvation lowers the BMR.[d]
Malnutrition	Malnutrition lowers the BMR.
Hormones (gender)	The thyroid hormone thyroxin, for example, can speed up or slow down the BMR.[e] Premenstrual hormones slightly raise the BMR.
Smoking	Nicotine increases energy expenditure.
Caffeine	Caffeine increases energy expenditure.
Sleep	BMR is lowest when sleeping.

[a] The BMR begins to decrease in early adulthood (after growth and development cease) at a rate of about 2 percent/decade. A reduction in voluntary activity as well brings the total decline in energy expenditure to 5 percent/decade.

[b] If two people weigh the same, the taller, thinner person will have the faster metabolic rate, reflecting the greater skin surface, through which heat is lost by radiation, in proportion to the body's volume.

[c] Fever raises the BMR by 7 percent for each degree Fahrenheit.

[d] Prolonged starvation reduces the total amount of metabolically active lean tissue in the body, although the decline occurs sooner and to a greater extent than body losses alone can explain. More likely, the neural and hormonal changes that accompany fasting are responsible for changes in the BMR.

[e] The thyroid gland releases hormones that travel to the cells and influence cellular metabolism. Thyroid hormone activity can speed up or slow down the rate of metabolism by as much as 50 percent.

Image Source/Getty Images

Physical activity spends energy and benefits health in many ways.

Note that Table 6-1 listed these factors among those that influence BMR and consequently energy expenditure.

thermic effect of food: an estimation of the energy required to process food (digest, absorb, transport, metabolize, and store ingested nutrients).

metabolism of food. This cellular activity produces heat and is known as the **thermic effect of food.** The thermic effect of food is generally thought to represent about 10 percent of the total food energy taken in. For purposes of rough estimates, though, the thermic effect of food is not always included.

ESTIMATING ENERGY REQUIREMENTS

In estimating energy requirements, the DRI committee developed equations that consider how the following factors influence energy expenditure:

- *Gender.* In general, women have a lower BMR than men, in large part because men typically have more lean body mass. In addition, menstrual hormones influence the BMR in women, raising it just prior to menstruation. Two sets of energy equations—one for men and one for women—were developed to accommodate the influence of gender on energy expenditure.

TABLE 6-2	Energy Spent on Various Activities

The values listed in this table reflect both the energy spent in physical activity *and* the amount used for BMR.

Activity	kCal/lb/min[a]	kCalories per Minute at Different Body Weights				
		110 lb	125 lb	150 lb	175 lb	200 lb
Aerobic dance (vigorous)	.062	6.8	7.8	9.3	10.9	12.4
Basketball (vigorous, full court)	.097	10.7	12.1	14.6	17.0	19.4
Bicycling						
13 mph	.045	5.0	5.6	6.8	7.9	9.0
15 mph	.049	5.4	6.1	7.4	8.6	9.8
17 mph	.057	6.3	7.1	8.6	10.0	11.4
19 mph	.076	8.4	9.5	11.4	13.3	15.2
21 mph	.090	9.9	11.3	13.5	15.8	18.0
23 mph	.109	12.0	13.6	16.4	19.0	21.8
25 mph	.139	15.3	17.4	20.9	24.3	27.8
Canoeing, flat water, moderate pace	.045	5.0	5.6	6.8	7.9	9.0
Cross-country skiing						
8 mph	.104	11.4	13.0	15.6	18.2	20.8
Golf (carrying clubs)	.045	5.0	5.6	6.8	7.9	9.0
Handball	.078	8.6	9.8	11.7	13.7	15.6
Horseback riding (trot)	.052	5.7	6.5	7.8	9.1	10.4
Rowing (vigorous)	.097	10.7	12.1	14.6	17.0	19.4
Running						
5 mph	.061	6.7	7.6	9.2	10.7	12.2
6 mph	.074	8.1	9.2	11.1	13.0	14.8
7.5 mph	.094	10.3	11.8	14.1	16.4	18.8
9 mph	.103	11.3	12.9	15.5	18.0	20.6
10 mph	.114	12.5	14.3	17.1	20.0	22.9
11 mph	.131	14.4	16.4	19.7	22.9	26.2
Soccer (vigorous)	.097	10.7	12.1	14.6	17.0	19.4
Studying	.011	1.2	1.4	1.7	1.9	2.2
Swimming						
20 yd/min	.032	3.5	4.0	4.8	5.6	6.4
45 yd/min	.058	6.4	7.3	8.7	10.2	11.6
50 yd/min	.070	7.7	8.8	10.5	12.3	14.0
Table tennis (skilled)	0.45	5.0	5.6	6.8	7.9	9.0
Tennis (beginner)	.032	3.5	4.0	4.8	5.6	6.4

(continued)

TABLE 6-2   Energy Spent on Various Activities *(continued)*

Activity	kCal/lb/min[a]	kCalories per Minute at Different Body Weights				
		110 lb	125 lb	150 lb	175 lb	200 lb
Walking (brisk pace)						
3.5 mph	.035	3.9	4.4	5.2	6.1	7.0
4.5 mph	.048	5.3	6.0	7.2	8.4	9.6
Weight lifting						
light-to-moderate effort	.024	2.6	3.0	3.6	4.2	4.8
vigorous effort	.048	5.2	6.0	7.2	8.4	9.6
Wheelchair basketball	.084	9.2	10.5	12.6	14.7	16.8
Wheeling self in wheelchair	.030	3.3	3.8	4.5	5.3	6.0

[a] To calculate kcalories spent per minute of activity for your own body weight, multiply kcal/lb/min by your exact weight and then multiply that number by the number of minutes spent in the activity. For example, if you weigh 142 pounds, and you want to know how many kcalories you spent doing 30 minutes of vigorous aerobic dance: $0.062 \times 142 = 8.8$ kcalories per minute; 8.8×30 (minutes) $= 264$ total kcalories spent.

- *Growth.* The BMR is high in people who are growing. For this reason, pregnant women and children have their own sets of energy equations.
- *Age.* The BMR declines during adulthood as lean body mass diminishes. Physical activities tend to decline as well, bringing the average reduction in energy expenditure to about 5 percent per decade. The decline in the BMR that occurs when a person becomes less active reflects the loss of lean body mass and may be prevented with ongoing physical activity. Because age influences energy expenditure, it is also factored into the energy equations.
- *Physical activity.* Using individual values for various physical activities (as in Table 6-2) is time-consuming and impractical for estimating the energy needs of a population. Instead, various activities are clustered according to the typical intensity of a day's efforts. Energy equations include a physical activity factor for various levels of intensity for each gender.
- *Body composition and body size.* The BMR is high in people who are tall and so have a large surface area. Similarly, the more a person weighs, the more energy is expended on basal metabolism. For these reasons, the energy equations include a factor for both height and weight.

Appendix D presents tables that provide a shortcut to estimating total energy expenditure and instructions to help you determine the appropriate physical activity factor to use in the equations.

As just explained, energy needs vary between individuals depending on such factors as gender, growth, age, physical activity, and body composition. Even when two people are similarly matched, however, their energy needs will still differ because of genetic differences. Perhaps one day genetic research will reveal how to estimate requirements for each individual. For now, the "How to" provides instructions on calculating estimated energy requirements using the DRI equations and physical activity factors.

REVIEW NOTES

A person takes in energy from food and, on average, spends most of it on basal metabolic activities, some of it on physical activities, and about 10 percent on the thermic effect of food.

Because energy requirements vary from person to person, such factors as age, gender, and weight must be considered when calculating energy spent on basal metabolism, and the intensity and duration of the activity must be taken into account when calculating expenditures on physical activities.

Body Weight and Body Composition

The body's weight reflects its composition—the proportions of its bone, muscle, fat, fluid, and other tissue. All of these body components can vary in quantity and quality: the bones can be dense or porous, the muscles can be well developed or underdeveloped, fat can be abundant or scarce, and so on. By far the most variable tissue, though, is body fat. More than any other component, fat responds to changes in food intake and physical activity, so it is fat that is usually the target of efforts at weight management.

DEFINING HEALTHY BODY WEIGHT

How much should a person weigh? How can a person know if her weight is appropriate for her height and age? How can a person know if his weight is jeopardizing his health? Questions such as these seem so simple, yet the answers can be complex—and different depending on whom you ask.

For *most* people, the actual energy requirement falls within these ranges:
- For men, EER ± 200 kcal.
- For women, EER ± 160 kcal.

 **HOW TO** Estimate Energy Requirements

To determine your estimated energy requirements (EER), use the appropriate equation, inserting your age in years, weight (wt) in kilograms, height (ht) in meters, and physical activity (PA) factor from the accompanying table. (To convert pounds to kilograms, divide by 2.2; to convert inches to meters, divide by 39.37.)

- For men 19 years and older:

  EER = 662 − 9.53 × age + PA × [(15.91 × wt) + (539.6 × ht)].

- For women 19 years and older:

  EER = 354 − 6.91 × age + PA × [(9.36 × wt) + (726 × ht)].

For example, consider an active 30-year-old male who is 5 feet 11 inches tall and weighs 178 pounds. First, he converts his weight from pounds to kilograms and his height from inches to meters, if necessary:

$$178 \text{ lb} \div 2.2 = 80.9 \text{ kg.}$$
$$71 \text{ in} \div 39.37 = 1.8 \text{ m.}$$

Next, he considers his level of daily physical activity and selects the appropriate PA factor from the accompanying table (in this example, 1.25 for an active male). Then, he inserts his age, PA factor, weight, and height into the appropriate equation:

$$\text{EER} = 662 − 9.53 \times 30 + 1.25 \times [(15.91 \times 80.9) + (539.6 \times 1.8)].$$

(A reminder: do calculations within the parentheses first, and multiplication before addition and subtraction.) He calculates:

$$\text{EER} = 662 − 9.53 \times 30 + 1.25 \times (1287 + 971).$$
$$\text{EER} = 662 − 9.53 \times 30 + 1.25 \times 2258.$$
$$\text{EER} = 662 − 286 + 2823.$$
$$\text{EER} = 3199.$$

The estimated energy requirement for an active 30-year-old male who is 5 feet 11 inches tall and weighs 178 pounds is about 3200 kcalories/day. His actual requirement probably falls within a range of 200 kcalories above and below this estimate.

Physical Activity (PA) Factors for EER Equations

	Men	Women	Physical Activity
Sedentary	1.0	1.0	Only those physical activities required for normal independent living
			For an average-weight person, activities equivalent to walking at a pace of 2–4 mph for the following distances:
Low active	1.11	1.12	1.5 to 3.0 miles/day
Active	1.25	1.27	3 to 10 miles/day
Very active	1.48	1.45	10 or more miles/day

TABLE 6-3	Tips for Accepting a Healthy Body Weight

Value yourself and others for human attributes other than body weight. Realize that prejudging people by weight is as harmful as prejudging them by race, religion, or gender.

Use positive, nonjudgmental descriptions of your body.

Accept positive comments from others.

Focus on your whole self including your intelligence, social grace, and professional and scholastic achievements.

Accept that no magic diet exists.

Stop dieting to lose weight. Adopt a lifestyle of healthy eating and physical activity permanently.

Follow the USDA Food Guide. Never restrict food intake below the minimum levels that meet nutrient needs.

Become physically active, not because it will help you get thin but because it will make you feel good and enhance your health.

Seek support from loved ones. Tell them of your plan for a healthy life in the body you have been given.

Seek professional counseling, *not* from a weight-loss counselor, but from someone who can help you make gains in self-esteem without weight as a factor.

Join with others to fight weight discrimination and fashion stereotypes. (Search your local paper, or see p. 147 for names of support groups.)

The Criterion of Fashion In asking what is an ideal body weight, people often mistakenly turn to fashion for the answer. Without a doubt, our society sets unrealistic ideals for body weight, especially for women. Magazines, movies, and television all convey the message that to be thin is to be beautiful and healthy. As a result, the media have a great influence on the weight concerns and dieting patterns of people of all ages, but most tragically on young, impressionable children and adolescents.[3] Even five-year-olds are concerned about their body weight.[4] One-half of preteen girls and one-third of preteen boys are dissatisfied with their body weight and shape.[5]

Importantly, perceived body image has little to do with actual body weight or size. People of all shapes, sizes, and ages—including extremely thin fashion models with anorexia nervosa and fitness instructors with ideal body composition—have learned to be unhappy with their "overweight" bodies. Such dissatisfaction can lead to damaging behaviors, such as starvation diets, diet pill abuse, and failure to seek health care.[6] The first step toward making healthy changes may be self-acceptance. Keep in mind that fashion is fickle; body shapes that our society values change with time. Furthermore, body shapes that our society values differ from those of other societies. The standards defining "ideal" are subjective and frequently have little in common with health. Table 6-3 offers some tips for adopting health as an ideal, rather than society's misconceived image of beauty.

The Criterion of Health Even if our society were to accept fat as beautiful, obesity would still be a major risk factor for several life-threatening diseases, as discussed later in the chapter. For this reason, the most important criterion for determining how much a person should weigh and how much body fat a person needs is not appearance but good health and longevity. A range of healthy body weights has been identified using a common measure of weight and height—the body mass index.

Body Mass Index The body mass index (BMI) describes relative weight for height in people older than 20 years and often correlates with degree of body fatness and disease risks. The margin provides the equation that is used to derive BMI values. A person who takes measurements in pounds and inches can convert them to metric units or can use this modified equation:

$$BMI = \frac{weight\ (lb)}{height\ (in)^2} \times 703.$$

The inside back cover shows weights for various heights using the BMI to define underweight, healthy weight, overweight, and obesity.

As the BMI table shows, healthy weight falls between a BMI of 18.5 and 24.9. Most people with a BMI within this range have few of the health risks typically associated with too-low or too-high body weight. Risks increase as BMI falls below 18.5 or rises above 24.9 (see Figure 6-5), reflecting the reality that both underweight and overweight impair health status. Figure 6-6 presents visual images associated with various BMI values.

The BMI values are most accurate in assessing degrees of obesity and are less useful for evaluating nonobese people's body fatness. BMI values fail to reveal two valuable pieces of information in assessing disease risk. They don't reveal how much of the weight is fat, and they don't indicate where the fat is located. For this knowledge, measures of body composition are needed.

BODY COMPOSITION

For many people, being overweight compared with the standard means that they are over*fat*. This is not the case, though, for athletes with dense bones and well-

$$BMI = \frac{weight\ (kg)}{height\ (m)^2}.$$

To convert pounds to kilograms, divide by 2.2.

To convert inches to meters, divide by 39.37.

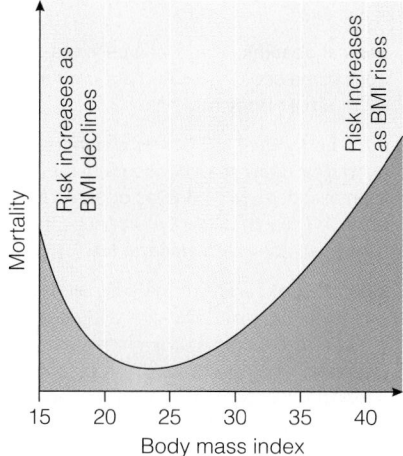

FIGURE 6-5 Body Mass Index and Mortality

This J-shaped curve describes the relationship between body mass index (BMI) and mortality and shows that both underweight and overweight present risks of a premature death.

FIGURE 6-6 Silhouettes and BMI (Actual BMI Shown)

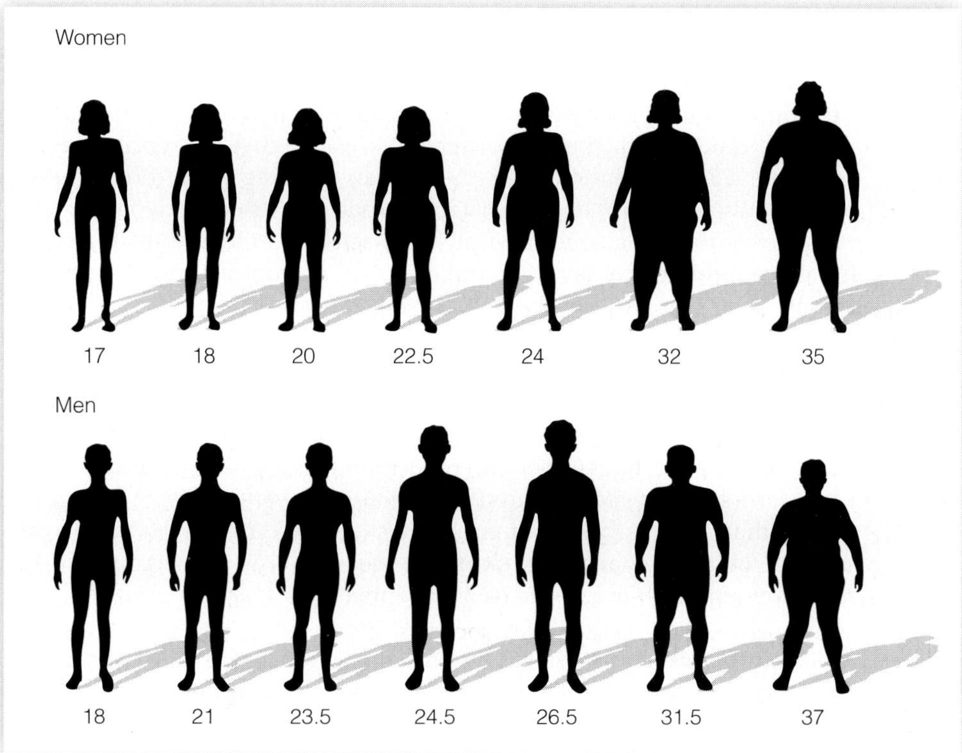

Women

17   18   20   22.5   24   32   35

Men

18   21   23.5   24.5   26.5   31.5   37

Source: Reprinted from "The Body Test" (1988). © Dietitians of Canada.

body mass index (BMI): an index of a person's weight in relation to height; determined by dividing the weight (in kilograms) by the square of the height (in meters).

FIGURE 6-7 Intra-abdominal Fat and Subcutaneous Fat

The fat deep within the body's abdominal cavity may pose an especially high risk to health.

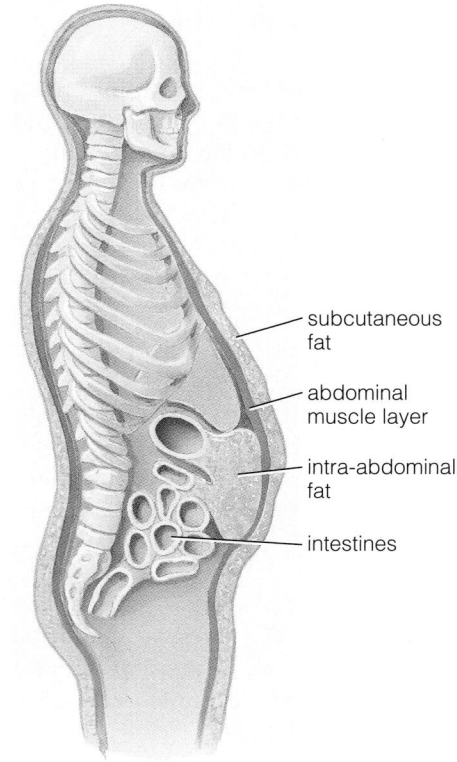

subcutaneous fat

abdominal muscle layer

intra-abdominal fat

intestines

central obesity: excess fat around the trunk of the body; also called **abdominal fat** or **upper-body fat.**

intra-abdominal fat: fat stored within the abdominal cavity in association with the internal abdominal organs, as opposed to fat stored directly under the skin (subcutaneous fat); also called **visceral fat.**

skinfold measure: a clinical estimate of total body fatness in which the thickness of a fold of skin on the back of the arm (over the triceps muscle), below the shoulder blade (subscapular), or in other places is measured with a caliper.

waist circumference: a measurement used to assess a person's abdominal fat.

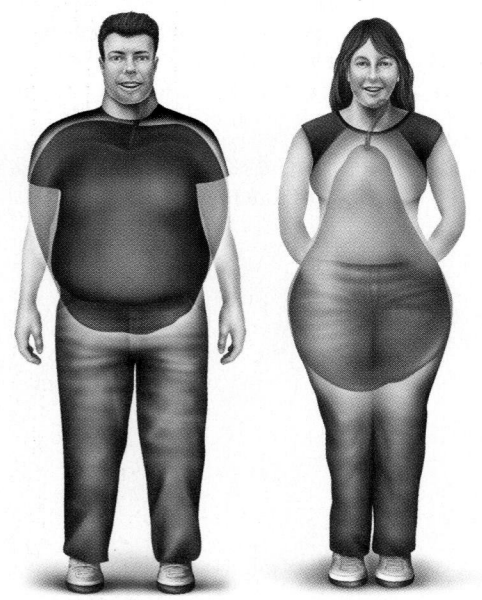

FIGURE 6-8 "Apple" and "Pear" Body Shapes Compared

Upper-body fat is more common in men than in women and is closely associated with heart disease, stroke, diabetes, hypertension, and some types of cancer. In contrast, lower-body fat is more common in women than in men and is not usually associated with chronic diseases. Popular articles sometimes call bodies with upper-body fat "apples" and those with lower-body fat, "pears." Researchers sometimes refer to upper-body fat as "android" (manlike) obesity and to lower-body fat as "gynoid" (womanlike) obesity.

developed muscles; they may be over-*weight* but carry little body fat. Conversely, inactive people may seem to have acceptable weights but still carry too much body fat. In addition, the distribution of fat on the body may be even more critical than overfatness alone.

Central Obesity Even more than total body fat, fat that collects deep within the central abdominal area of the body may be especially likely to lead to diabetes, stroke, hypertension, and coronary artery disease (see Figure 6-7).[7] The risk of death from all causes may be higher for those with **central obesity** than for those whose fat accumulates elsewhere in the body. Unlike the subcutaneous fat layers lying just beneath the skin of the abdomen and elsewhere, **intra-abdominal fat,** when mobilized, goes directly to the liver, where it is made into triglyceride-carrying very-low-density lipoprotein (VLDL).[8] Fat from elsewhere may arrive in the liver eventually, but it takes a circuitous route that first allows other tissues the chance to pull it from the circulation and metabolize it.

Abdominal fat creates the "apple" profile of central obesity. Fat around the hips and thighs creates more of a "pear" profile (see Figure 6-8). Abdominal fat is common in women past menopause and even more common in men. Even when total body fat is similar, men have more abdominal fat than either premenopausal or postmenopausal women. For those women with abdominal fat, the risks of cardiovascular disease and mortality are increased, just as they are for men. Smokers, too, may carry more of their body fat centrally. A smoker may weigh less than the average nonsmoker, but the smoker's waist circumference may be greater, leading researchers to think that smoking may directly affect body fat distribution.[9] Two other factors that may affect body fat distribution are intakes of alcohol and physical activity. Moderate-to-high alcohol consumption may favor central obesity.[10] In contrast, regular physical activity seems to prevent abdominal fat accumulation.[11]

Skinfold Measures Skinfold measurements provide an accurate estimate of total body fat and a fair assessment of the fat's location. About half of the fat in the body lies directly beneath the skin, so the thickness of this subcutaneous fat is assumed to reflect total body fat. Measures taken from central-body sites (around the abdomen) better reflect changes in fatness than those taken from upper sites (arm and back). A skilled assessor can obtain an accurate **skinfold measure** and then compare the measurement with standards (see Appendix E).

Waist Circumference Waist circumference serves as an indicator of abdominal fatness (see Figure 6-9). Previously, a ratio of waist-to-hip measurements served this purpose, but waist circumference alone has been deemed a valid indicator for both men and women. In general, women with a waist circumference greater than 35 inches and men with a waist circumference greater than 40 inches have a high risk of central obesity–related health problems.[12]

HOW MUCH BODY FAT IS TOO MUCH?

People often ask exactly how much fat is too fat for health. Ideally, a person has enough fat to meet basic needs but not so much as to incur health risks. The ideal amount of body fat depends partly on the person. A man with a BMI within the recommended range may have between 13 and 21 percent body fat; a woman, because of her greater quantity of indispensable fat, 23 to 31 percent.

Many athletes have a lower percentage of body fat—just enough fat to provide fuel, insulate and protect the body, assist in nerve impulse transmissions, and support normal hormone activity, but not so much as to burden the muscles. For athletes, then, body fat might be 5 to 10 percent for men and 15 to 20 percent for women.

For an Alaskan fisherman, a higher-than-average percentage of body fat is probably beneficial because fat helps prevent heat loss in cold weather. A woman starting a pregnancy needs sufficient body fat to support conception and fetal growth. Below a certain threshold for body fat, individuals may become infertile, develop depression, experience abnormal hunger regulation, or be unable to keep warm. These thresholds differ for each function and for each individual; much remains to be learned about them.

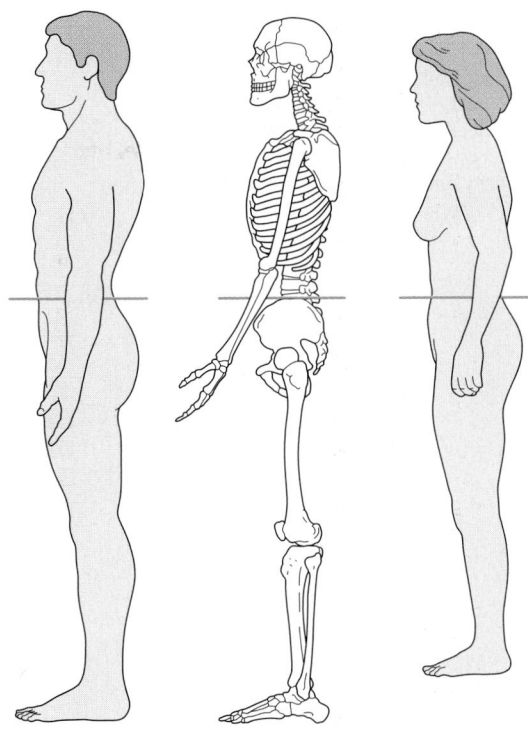

FIGURE 6-9 Measuring Waist Circumference
Using a nonstretching tape measure, measure the body around the point just above the iliac crest. Take the measure at the end of a normal expiration. A healthy waist circumference for men is no larger than 102 centimeters (40 inches); for women, no larger than 88 centimeters (35 inches).

Source: National Heart, Lung, and Blood Institute, National Institutes of Health, *The Practical Guide: Identification, Evaluation, and Treatment of Overweight and Obesity in Adults,* NIH publication no. 00-4084 (Washington, D.C.: Government Printing Office, 2000), p. 9.

> **REVIEW NOTES**
>
> Clearly, the most important criterion of appropriate fatness is health.
>
> Current standards for body weight are based on the body mass index (BMI), which describes a person's weight in relation to height.
>
> Health risks increase with a BMI below 18.5 or above 24.9.
>
> Central obesity, in which excess fat is distributed around the trunk of the body, presents greater health risks than excess fat distributed on the lower body.
>
> Researchers use a number of techniques to assess body composition, including waist circumference and skinfold measures.

overweight: overfatness of a moderate degree; defined as a body mass index (BMI) of 25.0 through 29.9.

Risks of Overweight and Obesity

Despite our nation's preoccupation with body image and weight loss, the prevalence of overweight and obesity continues to rise dramatically (see Figure 6-10, p. 144).[13] During the previous decade, obesity increased in every state, in both genders, and across all ages, races, and education levels. Almost 65 percent of U.S. adults are **overweight** (BMI of 25 or greater) or dangerously obese (BMI of 30 or greater).[14] The prevalence of overweight is especially high among women, the poor, and some ethnic groups. In short, obesity is a major public health problem that is becoming more and more prevalent.

Health Risks of Obesity The growing prevalence of overweight and obesity is a matter of concern because both conditions present risks to health. Indeed, the health risks of **obesity** are so many that it has been declared a disease. For example, excess weight contributes to up to half of all cases of hypertension, thereby increasing the risk of heart attack and stroke.[15] Often weight loss alone can normalize the blood pressure of an overfat person; some people with hypertension

© Rick Schaff

At 6 feet 3 inches tall and 245 pounds, Mike O'Hearn has a BMI greater than 30 and would be considered overweight by most weight-for-height standards. Yet he is clearly not overfat. In fact, his body fat is only 8 percent.

FIGURE 6-10 The Increasing Prevalence of Obesity among U.S. Adults by State

In 1991, only four states had an obesity rate (BMI ≥ 30) equal to or greater than 15 percent. By 2003, all states reported an obesity rate of at least 15 percent, and the trend is steadily upward.

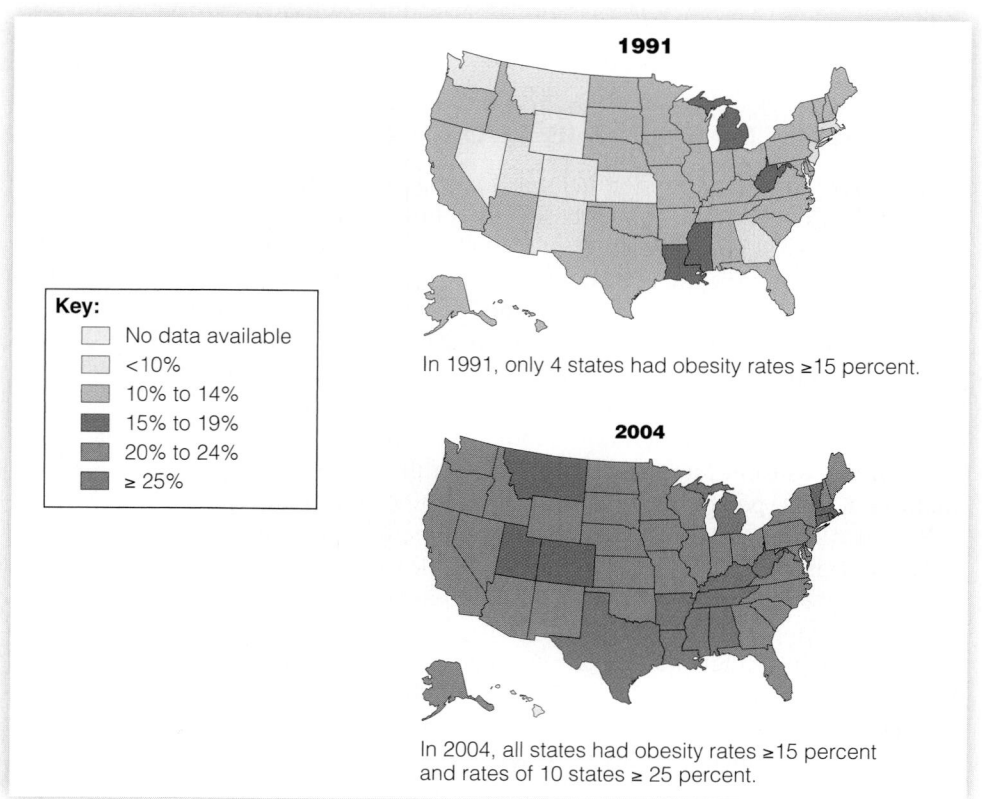

1991

Key:
- No data available
- <10%
- 10% to 14%
- 15% to 19%
- 20% to 24%
- ≥ 25%

In 1991, only 4 states had obesity rates ≥15 percent.

2004

In 2004, all states had obesity rates ≥15 percent and rates of 10 states ≥ 25 percent.

Source: U.S. Obesity Trends 1985 to 2003, Centers for Disease Control and Prevention, available at www.cdc.gov/nccdphp/dnpa/obesity/trend/maps/slide/001.htm.

can tell you the exact weight at which their blood pressure begins to rise. Weight gain can also precipitate diabetes in genetically susceptible people.[16]

Besides diabetes and hypertension, other risks threaten obese adults. Among them are high blood lipids, cardiovascular disease, sleep apnea (abnormal ceasing of breathing during sleep), osteoarthritis, abdominal hernias, some cancers, varicose veins, gout, gallbladder disease, kidney stones, respiratory problems (including Pickwickian syndrome, a breathing blockage linked with sudden death), liver malfunction, complications in pregnancy and surgery, flat feet, and even a high accident rate.[17] Each year these obesity-related illnesses cost our nation billions of dollars. The cost in terms of lives is also great. People with lifelong obesity are twice as likely to die prematurely as others. In the United States, obesity is second only to tobacco use as the most significant cause of preventable death.

The precise level at which excess weight begins to present risks has not been established. Some evidence indicates that being even moderately overweight aggravates the risk of heart disease, assuming that the excess weight is fat, not muscle.[18] Some obese people, however, seem to remain healthy and live long despite their body fatness. Physical fitness, despite body fatness, may protect against early death from disease.[19] It may be that genetics and other risk factors such as smoking help determine who among the overweight will be susceptible to diseases and who will stay healthy. Still, the majority of obese people do develop associated health problems. To help in identifying those most at risk, obesity experts have developed guidelines, described next.

National Guidelines for Identifying Those at Risk from Obesity The U.S. guidelines for identifying and evaluating the risks to health from overweight and obesity rely on three indicators. The first indicator is a person's BMI (see Table 6-4). As a general guideline, overweight for adults is defined as BMI of 25.0 through 29.9 and obesity as BMI equal to or greater than 30.

obesity: overfatness with adverse health effects, as determined by reliable measures and interpreted with good medical judgment. Obesity is officially defined as a body mass index (BMI) of 30 or higher.

TABLE 6-4	Disease Risks Based on BMI and Waist Circumference[a]	

The degree of risk is heightened by the presence of specific disease or other risk factors, such as elevated blood LDL cholesterol or smoking (see the margin on this page).

BMI	Waist ≤ 40 Inches (Men) or ≤ 35 Inches (Women)	Waist > 40 Inches (Men) or > 35 Inches (Women)
18.5 or less (Underweight)	Low	—
18.5–24.9 (Normal)	Low	—
25.0–29.9 (Overweight)	Increased	High
30.0–34.9 (Obese, class I)	High	Very high
35.0–39.9 (Obese, class II)	Very high	Very high
40 or greater (Extremely obese, class III)	Extremely high	Extremely high

[a]Risk for type 2 diabetes, hypertension, and cardiovascular disease.

Source: National Heart, Lung, and Blood Institute, National Institutes of Health, *The Practical Guide: Identification, Evaluation, and Treatment of Overweight and Obesity in Adults*, NIH publication no. 00-4084 (Washington, D.C.: Government Printing Office, 2000).

The second indicator is waist circumference, which, as discussed earlier, reflects the degree of intra-abdominal fatness in proportion to body fatness. As Table 6-4 shows, women with a waist circumference greater than 35 inches, and men with a waist circumference greater than 40 inches, are at greater risk of type 2 diabetes, hypertension, and cardiovascular disease than women or men with waist circumferences equal to, or below, these measures. In other words, waist circumference is an independent predictor of disease risk.

The third indicator is the person's disease risk profile. The categories of disease risk listed in Table 6-4 reflect disease risk *relative* to risk at normal weight. Relative risk is not the same as *absolute* risk, which is determined by the presence of certain obesity-related diseases or risk factors for disease.[20] People who have one or more of the diseases listed in the margin, or three or more of the cardiovascular disease risk factors listed, have a very high absolute risk for disease complications and mortality that requires aggressive treatment to manage the disease or modify the risk factors.

Other Risks of Obesity Although some obese people seem to escape health problems, few in our society can avoid the social and economic handicaps. Our society places enormous value on thinness. Obese people are less sought after for romance, less often hired, and less often admitted to college. They pay higher insurance premiums, and they pay more for clothing. This is especially true for women. In contrast, people with other chronic conditions such as asthma, diabetes, and epilepsy do not differ socially or economically from nonoverweight people. These social and economic disadvantages have psychological consequences, too—fat people often feel rejected and embarrassed, and this hurts self-esteem.

REVIEW NOTES

The health risks of obesity are many and serious.

Guidelines for identifying the health risks of overweight and obesity are based on a person's BMI, waist circumference, and disease risk profile.

Obesity also incurs social, economic, and psychological risks.

The National Heart, Lung, and Blood Institute states that aggressive treatment is urgently needed for a clinically obese person (BMI ≥ 30) who also has any of the following:

- Established cardiovascular disease (CVD).
- Established type 2 diabetes or impaired glucose tolerance (prediabetes).
- Sleep apnea, including temporary stopping of breathing.

The same urgency for treatment exists for an obese person with any *three* of the following CVD risk factors:

- Hypertension.
- Smoking.
- High LDL cholesterol.
- Low HDL cholesterol.
- Sedentary lifestyle.
- Age older than 45 years (men) or 55 years (women).
- Heart disease of an immediate family member before age 55 (male) or 65 (female).

SOURCE: National Heart, Lung, and Blood Institute, National Institutes of Health, *The Practical Guide: Identification, Evaluation, and Treatment of Overweight and Obesity in Adults*, NIH publication no. 00-4084 (Washington, D.C.: Government Printing Office, 2000).

SELF CHECK

1. As carbohydrate and fat stores are depleted during fasting or starvation, the body then uses ___ as its fuel source.
   a. alcohol
   b. protein
   c. glucose
   d. triglycerides

2. When carbohydrate is not available to provide energy for the brain, as in starvation, the body produces ketone bodies from:
   a. glucose.
   b. glycerol.
   c. fatty acid fragments.
   d. amino acids.

3. Three hazards of fasting are:
   a. water weight loss, decrease in mental alertness, and wasting of lean tissue.
   b. water weight gain, impairment of disease resistance, and lowering of body temperature.
   c. water weight gain, decrease in mental alertness, and impairment of disease resistance.
   d. wasting of lean tissue, impairment of disease resistance, and disturbances of the body's salt and water balance.

4. Two activities that contribute to the basal metabolic rate are:
   a. walking and running.
   b. maintenance of heartbeat and running.
   c. maintenance of body temperature and walking.
   d. maintenance of heartbeat and body temperature.

5. Three factors that affect the body's basal metabolic rate are:
   a. height, weight, and energy intake.
   b. age, body composition, and height.
   c. fever, body composition, and altitude.
   d. weight, fever, and environmental temperature.

6. The largest component of energy expenditure is:
   a. basal metabolism.
   b. physical activity.
   c. indirect calorimetry.
   d. thermic effect of food.

7. Which of the following reflects height and weight?
   a. body mass index
   b. central obesity
   c. waist circumference
   d. body composition

8. The BMI range that correlates with the fewest health risks is:
   a. 16.5 to 20.9.
   b. 18.5 to 24.9.
   c. 25.5 to 30.9.
   d. 30.5 to 34.9.

9. The profile of central obesity is sometimes referred to as a(an):
   a. beer.
   b. pear.
   c. apple.
   d. potato.

10. Which of the following health risks is **not** associated with being overweight?
    a. hypertension
    b. heart disease
    c. type 1 diabetes
    d. gallbladder disease

Answers to these questions can be found in Appendix H.

CLINICAL APPLICATIONS

1. Compare the energy a person might spend on various physical activities. Refer to Table 6-2 on pp. 137–38, and compute how much energy a person who weighs 142 pounds would spend doing each of the following activities. An example using aerobic dance has been provided for you. You may want to compare various activities based on your own weight.

   30 minutes of vigorous aerobic dance:

   0.062 kcal/lb/min × 142 lb = 8.8 kcal/min.

   8.8 kcal/min × 30 min = 264 kcal.

   a. 2 hours of golf, carrying clubs.
   b. 20 minutes running at 9 mph.
   c. 45 minutes of swimming at 20 yd/min
   d. 1 hour of walking at 3.5 mph.

2. Using the "How to" on p. 139 as a guide, determine your estimated energy requirements (EER).

Answers

1. (a.) 0.045 kcal/lb/min × 142 lb = 6.4 kcal/min, 6.4 kcal/min × 120 min = 768 kcal; (b) 0.103 kcal/lb/min × 142 lb = 14.6 kcal/min, 14.6 kcal/min × 20 min = 292 kcal; (c) 0.032 kcal/lb/min × 142 lb = 4.5 kcal/min, 4.5 kcal/min × 45 = 202.5 kcal; (d) 0.035 kcal/lb/min × 142 lb = 5 kcal/min, 5 kcal/min × 60 min = 300 kcal.

NUTRITION ON THE NET

For further study of the topics in this chapter, access these websites.

Find updates and quick links to these and other nutrition-related sites at our website: **www.wadsworth.com/nutrition**

Obtain food composition data from the USDA Nutrient Data Laboratory: **www.nal.usda.gov/fnic/foodcomp**

Search for "obesity" and "weight control" at the U.S. Government health information site: **www.healthfinder.gov**

Review the Clinical Guidelines on the Identification, Evaluation, and Treatment of Overweight and Obesity in Adults: **www.nhlbi.nih.gov/guidelines/obesity/ob_home.htm**

Learn about the 10,000 Steps Program at Shape Up America: **www.shapeup.org**

Visit the special web pages and interactive applications for Healthy Weight: **www.nhlbi.nih.gov/subsites/index.htm**

Find helpful information on achieving and maintaining a healthy weight from the Calorie Control Council: **www.caloriecontrol.org**

Learn how to end size discrimination and improve the quality of life for fat people from the National Association to Advance Fat Acceptance: **www.naafa.org**

Consider ways to live a healthy life at any weight: **www.bodypositive.com**.

NOTES

[1] S. G. Wannamethee, A. G. Shaper, and P. H. Whincup, Alcohol and adiposity: Effects of quantity and type of drink and time relation with meals, *International Journal of Obesity and Related Metabolic Disorders,* online print, August 2, 2005; R. A. Breslow and B. A. Smothers, Drinking patterns and body mass index in never smokers, *American Journal of Epidemiology* 161 (2005): 368–376; A. Raben and coauthors, Meals with similar energy densities but rich in protein, fat, carbohydrate, or alcohol have different effects on energy expenditure and substrate metabolism but not on appetite and energy intake, *American Journal of Clinical Nutrition* 77 (2003): 91–100.

[2] Raben and coauthors, 2003.

[3] L. C. Andrist, Media images, body dissatisfaction, and disordered eating in adolescent women, *American Journal of Maternal Child Nursing* 28 (2003): 119–123; K. K. Davison, C. N. Markey, and L. L. Birch, A longitudinal examination of patterns in girls' weight concerns and body dissatisfaction from ages 5–9 years, *International Journal of Eating Disorders* 33 (2003): 320–332; J. Wardle, J. Waller, and E. Fox, Age of onset and body dissatisfaction in obesity, *Addictive Behaviors,* 27 (2002): 561–573; A. E. Field and coauthors, Peer, parent, and media influences on the development of weight concerns and frequent dieting among preadolescent and adolescent girls and boys, *Pediatrics* 107 (2001): 54–60.

[4] J. A. Shunk and L. L. Birch, Girls at risk for overweight at age 5 are at risk for dietary restraint, disinhibited overeating, weight concerns, and greater weight gain from 5 to 9 years, *Journal of the American Dietetic Association* 104 (2004): 1120–1126; Davison, Markey, and Birch, 2003.

[5] H. Truby and S. J. Paxton, Development of the Children's Body Image Scale, *British Journal of Clinical Psychology* 41 (2002): 185–203; H. A. Hausenblas and coauthors, Body image in middle school children, *Eating and Weight Disorders* 7 (2002): 244–248.

[6] C. A. Drury and M. Louis, Exploring the association between body weight, stigma of obesity, and health care avoidance, *Journal of the American Academy of Nurse Practitioners* 14 (2002): 554–561.

[7] Y. Wang and coauthors, Comparison of abdominal adiposity and overall obesity in predicting risk of type 2 diabetes among men, *American Journal of Clinical Nutrition* 81 (2005): 555–563; G. De Simone and coauthors, Body composition and fat distribution influence systemic hemodynamics in the absence of obesity: The HyperGEN Study, *American Journal of Clinical Nutrition* 81 (2005): 757–761; I. Janssen, P. T. Katzmarzyk, and R. Ross, Waist circumference and not body mass index explains obesity-related health risk, *American Journal of Clinical Nutrition* 79 (2004): 379–384; S. Zhu and coauthors, Waist circumference and obesity-associated risk factors among whites in the Third National Health and Nutrition Survey: Clinical action thresholds, *American Journal of Clinical Nutrition* 76 (2002): 743–749.

[8] F. X. Pi-Sunyer, The epidemiology of central fat distribution in relation to disease, *Nutrition Reviews* 62 (2004): S120–S126.

[9] D. Canoy and coauthors, Cigarette smoking and fat distribution in 21,828 British men and women: A population-based study, *Obesity Research* 13 (2005): 1466–1475.

[10] Wannamethee, Shaper, and Whincup, 2005; J. M. Dorn and coauthors, Alcohol drinking patterns differentially affect central adiposity as measured by abdominal height in women and men, *Journal of Nutrition* 133 (2003): 2655–2662.

[11] C. A. Holcomb, D. L. Heim, and T. M. Loughin, Physical activity minimizes the association of body fatness with abdominal obesity in white, premenopausal women: Results from the Third National Health and Nutrition Examination Survey, *Journal of the American Dietetic Association* 104 (2004): 1859–1862.

[12] Janssen, Katzmarzyk, and Ross, 2004; Zhu and coauthors, 2002.

[13] G. L. Blackburn and W. A. Walker, Science-based solutions to obesity: What are the roles of academia, government, industry, and health care? *American Journal of Clinical Nutrition* 82 (2005): 207S–210S.

[14] Food and Drug Administration, *Counting Calories: Report of the Working Group on Obesity,* March 12, 2004, available at **www.cfsan.fda.gov/~dms/owg-rpt.html**; A. A. Hedley and coauthors, Prevalence of overweight and obesity among US children, adolescents, and adults, 1999–2002, *Journal of the American Medical Association* 291 (2004): 2847–2850.

[15] U.S. Preventive Services Task Force, 2003.

[16] A. R. Weinstein and coauthors, Relationship of physical activity vs body mass index with type 2 diabetes in women, *Journal of the*

American Medical Association 292 (2004): 1188–1194; American Diabetes Association Position Statement: Evidence-based nutrition principles and recommendations for the treatment and prevention of diabetes and related complications, *Journal of the American Dietetic Association* 102 (2002): 109–118; J. Tuomilehto and coauthors, Prevention of type 2 diabetes mellitus by changes in lifestyle among subjects with impaired glucose tolerance, *New England Journal of Medicine* 344 (2001): 1343–1350.

[17]G. R. Dagenais and coauthors, Prognostic impact of body weight and abdominal obesity in women and men with cardiovascular disease, *American Heart Journal* 149 (2005): 54–60; E. N. Taylor, M. J. Stampfer, and G. C. Curhan, Obesity, weight gain, and the risk of kidney stones, *Journal of the American Medical Association* 293 (2005): 455–462; C. Tsai and coauthors, Prospective study of abdominal adiposity and gallstone disease in US men, *American Journal of Clinical Nutrition* 80 (2004): 38–44; E. E. Calle and coauthors, Overweight, obesity, and mortality from cancer in a prospectively studied cohort of U.S. adults, *New England Journal of Medicine* 348 (2003): 1625–1638.

[18]J. K. Alexander, Obesity and coronary heart disease, *American Journal of Medical Sciences* 321 (2001): 215–224.

[19]J. M. Jakicic and A. D. Otto, Physical activity considerations for the treatment and prevention of obesity, *American Journal of Clinical Nutrition* 82 (2005): 226S–229S; P. T. Katzmarzyk and coauthors, Fitness, fatness, and estimated coronary heart disease risk: The HERITAGE Family Study, *Medicine and Science in Sports and Exercise* 33 (2001): 585–590.

[20]National Heart, Lung, and Blood Institute, National Institutes of Health, *The Practical Guide: Identification, Evaluation, and Treatment of Overweight and Obesity in Adults,* NIH publication no. 00-4084 (Washington, D.C.: Government Printing Office, 2000).

Fad Diets

To paraphrase William Shakespeare, "a fad diet by any other name would still be a fad diet." And the names are legion: the Atkins New Diet Revolution, the South Beach Diet, the Eat Right 4 Your Type Diet, the Ultimate Weight Solution Diet, the Zone Diet.* Year after year, "new and improved" diets appear on bookstore shelves and circulate among friends. Sometimes fad diets seem to work for a while, but more often than not, their success is short-lived. Then another fad diet takes the spotlight. Here's how Dr. K. Brownell, an obesity researcher at Yale University, describes this phenomenon: "When I get calls about the latest diet fad, I imagine a trick birthday cake candle that keeps lighting up and we have to keep blowing it out."

Why don't health professionals speak out against the relentless promotion of fad diets?

Realizing that many fad diets do not offer a safe and effective plan for weight loss, health professionals speak out, but they never get the candle blown out permanently. New fad diets can keep making outrageous claims because no one requires their advocates to prove what they say. Fad diet gurus do not have to conduct credible research on the benefits or dangers of their diets. They can simply make recommendations and then later, if questioned, search for bits and pieces of research that support the conclusions they have already reached. That's

© Geri Engberg

*The following sources offer comparisons and evaluations of various fad diets for your review: B. Liebman, Weighing the diet books, *Nutrition Action Healthletter*, January/February 2004, pp. 1–8; S. T. St. Jeor and coauthors, Dietary protein and weight reduction: A statement for healthcare professionals from the nutrition committee of the Council on Nutrition, Physical Activity, and Metabolism of the American Heart Association, *Circulation* 104 (2001): 1869–1874.

backwards. Diet and health recommendations should *follow* years of sound research that has been reviewed by panels of scientists *before* being offered to the public.

How can the promoters of fad diets get away with exaggerated claims?

Because anyone can publish anything—in books or on the Internet—peddlers of fad diets can make unsubstantiated statements that fall far short of the truth, but sound impressive to the uninformed. They often offer distorted bits of legitimate research. They may start with one or more actual facts, but then leap from one erroneous conclusion to the next. Anyone who wants to believe these claims has to wonder how the thousands of scientists working on obesity research over the past century could possibly have missed such obvious connections. Table NP6-1 (p. 150) presents some of the claims and truths of fad diets.

Most of the popular diets I hear about seem to target carbohydrates—eat less carbohydrate, more protein, or eat "good carbs" instead of "bad carbs." Is this true?

Yes, to a point, although once you start sifting through the stacks of diet books available today, you begin to realize that fad diets come in almost as many shapes and sizes as the people who search them out. Some restrict fats or carbohydrates, some limit portion sizes, some focus on which foods can or cannot be eaten with other foods, and some claim that your genetic type or blood type determines which foods you should or should not eat to manage your weight and prevent disease. Table NP6-2 (p. 151) compares some of the more popular diets, and shows that many of them focus on restricting carbohydrate intake. Exceptions include diets such as the Ornish Diet and the Pritikin Program, which advocate a *high*-carbohydrate, very low-fat diet. These diets are not as popular today as they once were because information about eating diets high in carbohydrate-rich foods that support health and restricting certain kinds of fat-rich foods is available at no cost by way of national recommendations and guidelines such as the USDA Food Guide and the *Dietary Guidelines for Americans 2005.*

Thus the most popular diets today, regardless of what their names are, espouse a low-carbohydrate diet. Most low-carbohydrate diets by design are relatively high in protein. Some of the diets severely restrict carbohydrate while emphasizing protein (low-carbohydrate, high-protein), whereas others focus more on replacing certain "bad" carbohydrates with "good" carbohydrates (carbohydrate-modified), while still restricting overall carbohydrate intake to some degree.

TABLE NP6-1 The Claims and Truths of Fad Diets

The Claim:	You can lose weight with "exceptionally easy rules."
The Truth:	Most fad diet plans have complicated rules that require you to calculate protein requirements, count carbohydrate grams, combine certain foods, time meal intervals, purchase special products, plan daily menus, and measure serving sizes.
The Claim:	You can lose weight by eating a specific ratio of carbohydrates, protein, and fat.
The Truth:	Weight loss depends on spending more energy than you take in.
The Claim:	This "revolutionary diet" can "reset your genetic code."
The Truth:	You inherited your genes and cannot alter your genetic code.
The Claim:	High-protein diets are popular, selling more than 20 million books, because they work.
The Truth:	Weight-loss books are popular because people grasp for quick fixes and simple solutions to their weight problems. If book sales were an indication of weight-loss success, we would be a lean nation—but they're not, and neither are we.
The Claim:	People gain weight on low-fat diets.
The Truth:	People can gain weight on low-fat diets if they overindulge in carbohydrates and proteins while cutting fat; low-fat diets are not necessarily low-kcalorie diets. But people can also lose weight on low-fat diets if they cut kcalories as well as fat.
The Claim:	High-protein diets energize the brain.
The Truth:	The brain depends on glucose for its energy; the primary dietary source of glucose is carbohydrate, not protein.
The Claim:	Thousands of people have been successful with this plan.
The Truth:	Authors of fad diets have not published their research findings in scientific journals. Success stories are anecdotal and failures are not reported.
The Claim:	Carbohydrates raise blood glucose levels, triggering insulin production and fat storage.
The Truth:	Insulin promotes fat storage when energy intake exceeds energy needs. Furthermore, insulin is only one hormone involved in the complex processes of maintaining the body's energy balance and health.
The Claim:	You'll lose weight fast without counting kcalories or exercising because the diet alters metabolism.
The Truth:	No known trick of metabolism produces weight loss without diet or exercise.

Matthew Faruggio

Low-carbohydrate, high-protein meals overemphasize meat, fish, poultry, eggs, and cheeses, and shun breads, pastas, fruits, and vegetables.

What do you mean by "bad" carbohydrates and "good" carbohydrates?

This is a current trend in fad diet books and, in short, here are the claims:

- "Bad" carbohydrates, such as sugar, white flour, and potatoes, cause a rapid rise in blood sugar, which raises blood insulin concentrations.
- High blood insulin concentrations make people gain weight because insulin promotes fat storage.
- Or, people gain weight because high blood insulin concentrations lower blood sugar so much that people become hungry and so eat more food.
- "Good" carbohydrates such as whole grains, vegetables, and beans cause a slow rise in blood sugar and a moderate insulin response and do not promote weight gain or hunger.

Now, here is what research reveals so far:

- The "good" carb versus "bad" carb diet premise is based on the glycemic effect of foods discussed in Chapters 3 and 20. The glycemic effect of a food depends on how the food is ripened, processed, and cooked; the time of day the food is eaten; the other foods eaten with it; and the presence or absence of certain diseases such as type 2 diabetes in the person eating the food.[1] Thus the glycemic effect of a particular food varies—fad diet books mislead people by claiming that each food has a set glycemic effect.
- All whole-grain foods do not have a low glycemic effect, and all refined grains do not have a high glycemic effect. For example, pasta has a low glycemic effect whether it is whole wheat or white, but thin pasta has a higher glycemic effect than thick pasta. Thus fad diet books mislead consumers in this respect as well.

TABLE NP6-2 Popular Diets Compared

Diet	Major Premise Promoted	Strong Point(s)	Weak Point(s)
High-Carbohydrate, Low-Fat			
Ornish Diet	• By strictly limiting fat (both animal and vegetable), you eat fewer kcalories without eating less food.	• High-fiber, low-fat foods in this plan can lower cholesterol and blood pressure.	• So little fat that essential fatty acids may be lacking. • Excludes fish and olive oil, which may protect against heart disease.
Pritikin Program	• By eating low-fat, mainly plant-based foods, you can eat more food and still feel satisfied.	• No food group is completely eliminated in this high-fiber, low-fat diet program. • Some use of foods rich in omega-3 fatty acids is encouraged.	• For some people, very low-fat diets may be unsatisfying and therefore difficult to adhere to.
Low-Carbohydrate, High-Protein			
Atkins Diet	• People are overweight or obese because they have metabolic imbalances caused by eating too many carbohydrates. • By restricting carbohydrates, these metabolic imbalances can be corrected. • You can lose weight without lowering kcalorie intake.	• Quick, short-term weight loss is achieved.	• Restricts carbohydrates to a level that induces ketosis. • Ketosis can cause nausea, light-headedness, and fatigue. • Ketosis can worsen existing medical problems such as kidney disease. • A diet high in fat such as Atkins can increase the risk of heart disease and some cancers.
Low-Carbohydrate			
Zone Diet	• Eating the correct proportions of carbohydrate, fat, and protein leads to hormonal balance, weight loss, disease prevention, and increased vitality.	• Promotes weight loss because it is a low-kcalorie diet.	• The diet is rigid, restrictive, and complicated, making it difficult for most people to follow accurately. • The overblown health claims of the diet's proponents are based on mis-interpreted science and remain unsubstantiated.
Carbohydrate-Modified			
South Beach Diet	• Eating "good carbohydrates" such as vegetables, whole-wheat pastas, and brown rice will main-tain satiety and resist cravings for "bad carbohydrates" such as white rice and potatoes.	• The diet encourages consump-tion of vegetables, lean meats, and fish, and the use of mono-unsaturated oils when cooking. • Restricts fatty meats and cheeses as well as sweets. • The diet emphasizes mostly healthy foods.	• Starchy carbohydrates and all fruits are completely excluded during the first two weeks. • The benefits of exercise for health are not emphasized.
The Ultimate Weight Solution Diet	• Foods that require a great deal of effort to prepare and to eat are nutrient dense. • Eating these kinds of foods (raw vegetables, vegetable soup, whole grains, beans, meat, poultry, and fish) will lead to weight loss. • Foods that take little effort to prepare and to eat provide excess kcalories relative to nutrients. • Eating these kinds of foods (fast foods, puddings, high-kcalorie convenience foods, easy-to-prepare processed foods) leads to uncon-trolled eating and weight gain.	• The diet encourages consumption of lean meats and fish, whole grains, vegetables, and fruit, and low-fat milk, yogurt, and cheese. • Restricts fatty meats and cheeses as well as sweets. • The diet emphasizes mostly healthy foods. Exercise is encouraged.	• Confusing as to exactly what to eat or how much.
Metabolic Type			
Eat Right 4 Your Type	• Your blood type determines which foods you should eat or not eat.	• None	• Food groups or individual foods are excluded, depending on blood type. • No scientific data on the relationship between blood type and food choices.

- There is no clear evidence that higher blood insulin concentrations enhance food intake or weight gain in healthy people.[2]
- Finally, evidence that foods with a low glycemic effect promote weight loss is lacking. A review of the evidence thus far concludes that the ideal long-term study has not yet been conducted.[3]

Are the diets that promote "good" carbs versus "bad" carbs dangerous to health?

In general, no. In fact, most of the foods that such diets promote are healthy foods—lean meats, fat-free or low-fat milk and yogurt, vegetables, whole grains, beans, and fruit. The *Dietary Guidelines for Americans, 2005* encourage consumers to eat the same foods. The *Dietary Guidelines* also advise consumers to eat less saturated fat, however, and some low-carbohydrate diets can be high in saturated fat. Like some of the carbohydrate-modified diets, the *Dietary Guidelines* also encourage people to eat a diet high in fiber-rich carbohydrate foods, rather than one that is restricted in such foods. The diet books claim that people lose weight because they switch from eating some types of carbohydrate foods to eating others, when, in truth, people lose weight on the diets because they are eating fewer kcalories.

Some of the diets exclude or restrict healthful foods such as carrots and bananas because the importance of the glycemic effect of a single food is inflated. Worse, because some foods are restricted or excluded, fad diet books often encourage readers to purchase special supplements, "energy bars," or other products to go along with the diets. Nevertheless, carbohydrate-modified diets such as the South Beach Diet or Dr. Phil's Ultimate Weight Solution Diet do include mostly healthy foods.

What is the appeal of very low-carbohydrate, high-protein diets?

Probably the greatest appeal of a very low-carbohydrate, high-protein diet such as the Atkins Diet is that it turns current diet recommendations upside down. Foods such as meats and milk products that need to be selected carefully to limit saturated fat can now be eaten with abandon. Grains, legumes, vegetables, and fruits that people are told to eat in abundance can now be ignored. For some people, this is a dream come true: steaks without the potatoes, ribs without the coleslaw, and meatballs without the pasta. Who can resist the promise of weight loss while eating freely from a list of favorite foods?

I've heard several people say that they successfully lose weight on these diets. Is this true?

If low-carbohydrate, high-protein diets were as successful as some people claim, then consumers who tried them would lose lots of weight, and our obesity problems would be solved. Obviously, this is not the case. Similarly, if high-protein diets were as worthless as others claim, then consumers would eventually stop pursuing them. Clearly, this is not happening either. These diets have enough going for them that they work for some people at least for a short time, but they fail to produce long-lasting results for most people. Studies report that people following low-carbohydrate, high-protein diets do lose weight.[4] In fact, they lose more than people following conventional high-carbohydrate, low-fat diets—but only for the first six months. Their later gains make up the difference, so total weight loss is no different after one year.[5]

People who have followed very low-carbohydrate, high-protein diet plans for several months have lost weight—at least temporarily. But can these diets also be harmful?

Diets that overemphasize protein and fall short on carbohydrate may not harm healthy people if used for only a little while, but they cannot support optimal health for long. Some fad diets focus so intently on promoting protein and curbing carbohydrate that they fail to account for the fat that accompanies many high-protein foods. A breakfast of bacon and eggs, lunch of ham and cheese, and dinner of barbecued short ribs would provide 100 grams of protein—and 121 grams of fat! Yet this day's meals, even with a snack of peanuts, provide only 1600 kcalories. Without careful selection, protein-rich diets can be extraordinarily high in saturated fat and cholesterol—dietary factors that raise LDL cholesterol and the risks for heart disease.

Overall, studies report that people following low-carbohydrate, high-protein diets have little or no change in blood pressure or blood lipids—risk factors for heart disease.[6] Some researchers speculate that the weight loss that occurs on these diets offsets the adverse effects of a diet high in saturated fat and low in fruits and vegetables.[7] Others point out that different sources of protein have different effects on risk factors for heart disease.[8] For example, the effects of white meat from chicken or fish differ from those of red meat. Diets containing large amounts of red meat appear to increase the risk of heart disease. In contrast, replacing animal sources of protein with plant sources of protein may benefit health.

What are the other drawbacks to low-carbohydrate, high-protein diets?

The quality of the diet suffers when carbohydrates are restricted.[9] Without fruits, vegetables, and whole grains, high-protein diets lack not only carbohydrate but fiber, vitamins, minerals, and phytochemicals as well—all dietary factors protective against disease. To help shore up some of these inadequacies, fad diets often recommend a daily supplement. Conveniently, many of the companies selling fad diets also peddle these supplements. But, as Nutrition in Practice 9 explains, foods offer many more health benefits than any supplement can provide. Quite simply, if the diet is inadequate, it needs to be improved, not supplemented.

Is it true that people don't get hungry on low-carbohydrate, high-protein diets, and if so, is there any scientific basis for this?

Of the three energy-yielding nutrients, protein is the most satiating. Consequently, high-protein meals may suppress hunger and delay the start of the next meal. Furthermore, studies have reported that people tend to eat less after a high-protein meal than after a low-protein one.[10] In one study, when protein intake increased from 15 percent of total energy to 30 percent, but carbohydrate was held constant at 50 percent of total energy, people decreased their energy intakes and lost body weight and body fat.[11] The researchers suggest that less emphasis should be placed on carbohydrate restriction, without regard for the accompanying increase in fat intake that often occurs with low-carbohydrate, high-protein diets. Other research shows that when *fat* is held to 30 percent of total energy, and protein is either 12 percent or 25 percent of total energy, greater weight loss and greater loss of intra-abdominal fat occur on the higher-protein diet.[12] In real-life situations, though, there is a strong association between a person's protein intake and BMI—the higher the intake, the higher the BMI.[13] This association remains apparent even after adjusting for energy intake and physical activity.

Thus, even though guidelines from the DRI committee include higher protein intakes (10 to 35 percent of total energy) than recommended previously, long-term studies of high-protein intakes are needed to ascertain the health consequences of such diets. One such study is under way as this text goes to print: The DiOGenes (Diet, Obesity, and Genes) project is examining the interactions among a high dietary protein intake, the glycemic effect of foods, and genetic and behavioral factors in preventing weight gain and regain.[14] The study focuses on about 700 overweight or obese adults and their children in eight different countries across Europe and may involve the United States as well.

Is a person's metabolism altered while following a low-carbohydrate, high-protein diet?

When a person consumes a low-carbohydrate diet, a metabolism similar to that of fasting prevails (see Chapter 6 for a review of fasting). With little dietary carbohydrate coming in, the body uses its glycogen stores to provide glucose for the cells of the brain, nerves, and blood. Once the body depletes its glycogen reserves, it begins making glucose from the amino acids of protein. A low-carbohydrate diet may provide abundant protein from food, but the body still uses some protein from body tissues.

Low-carbohydrate diets also induce ketosis, and ketones can be detected in the urine. Ketones form whenever glucose is lacking and fat breakdown is incomplete. Many fad diets regard ketosis as the key to losing weight, but a study comparing weight-loss diets found no relation between ketosis and weight loss.[15] People in ketosis may experience a loss of appetite and a dramatic weight loss within the first few days. They would be disillusioned if they were aware that much of this weight loss reflects the loss of glycogen and protein together with large quantities of body fluids and important minerals.[16] They need to learn to appreciate the difference between loss of *fat* and loss of *weight*. Fat losses on ketogenic diets are no greater than on other diets providing the same number of kcalories. Once the dieter returns to well-balanced meals that provide adequate energy, carbohydrate, fat, protein, vitamins, and minerals, the body avidly retains these needed nutrients. The weight will return, quite often to a level higher than the starting point. Table NP6-3 lists other consequences of a ketogenic diet.

Why are consumers so obsessed with diet books and products?

With over half of our nation's adults overweight and many more concerned about their weight, the market for a

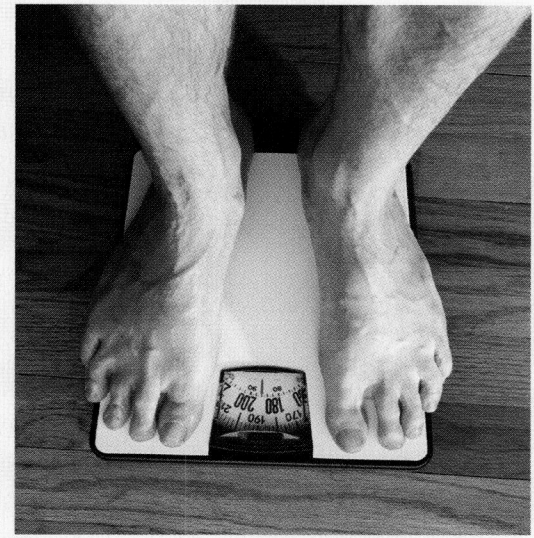

1998 Photo Disc Inc.

The cautious consumer distinguishes between loss of fat and loss of weight.

TABLE NP6-3 Adverse Side Effects of Low-Carbohydrate, Ketogenic Diets

Nausea
Fatigue (especially if physically active)
Constipation
Low blood pressure
Elevated uric acid (which may exacerbate kidney disease and cause inflammation of the joints in those predisposed to gout)
Stale, foul taste in the mouth (bad breath)
In pregnant women, fetal harm and stillbirth

weight-loss book, product, or program is huge (no pun intended). Americans spend an estimated $33 billion a year on weight-loss books and products. Even a plan that offers only minimal weight-loss success easily attracts a following. Carbohydrate-modified and low-carbohydrate, high-protein diet plans offer a little success to some people for a short time. Here's why.

Who wants to count kcalories? Even experienced dieters find counting kcalories burdensome, not to mention time-worn. They want a new, easy way to lose weight, and carbohydrate-modified and low-carbohydrate, high-protein diet plans seem to offer this boon. But, though these diets often claim to disregard kcalories, their design typically ensures a low energy intake. Most of the sample menu plans provided by these diets, especially in the early stages, are designed to deliver an average of 1200 kcalories a day.

Weight loss occurs because of the low energy intake—not the proportion of energy nutrients.[17] Success, then, depends on the restricted intake, not on protein's magical powers or carbohydrate's evil forces. This is an important point. Any diet can produce weight loss, at least temporarily, if intake is restricted. The real value of a diet is determined by its ability to maintain weight loss and support good health over the long term. The goal is not simply weight loss, but health gains—and whether carbohydrate-modified or low-carbohydrate, high-protein diets can support optimal health over time remains unknown. Table NP6-4 offers guidelines for identifying fad diets and other weight-loss scams; it includes the hallmarks of a reasonable weight-loss program as well.

Many people seem confused about what kinds of dietary changes they need to make to lose weight. Perhaps diet books that tell them what to do are simply what they are looking for. Do you agree?

Yes. Most people need specific instructions and examples to make dietary changes. Popular diets offer dieters a plan. The user doesn't have to decide what foods to eat, how to prepare them, or how much to eat. Unfortunately, these instructions serve short-term weight-loss needs only. They do not provide for long-term changes in lifestyle that will support weight maintenance or health goals.

Chapter 7 includes reasonable approaches to weight management and concludes that the ideal diet is one you can live with for the rest of your life. Keep that criterion in mind when you evaluate the next "latest and greatest weight-loss diet" that comes along.

TABLE NP6-4 Guidelines for Identifying Fad Diets and Other Weight-Loss Scams

1. They promise dramatic, rapid weight loss. Weight loss should be gradual and not exceed 2 pounds per week.
2. They promote diets that are nutritionally unbalanced or extremely low in kcalories. Diets should provide:
• A reasonable number of kcalories (not fewer than 1000 kcalories per day for women and 1200 kcalories per day for men).
• Enough, but not too much, protein (between the RDA and twice the RDA)
• Enough, but not too much, fat (between 20 and 35 percent of daily energy intake from fat).
• Enough carbohydrate to spare protein and prevent ketosis (at least 100 grams per day) and 20 to 30 grams of fiber from food sources.
• A balanced assortment of vitamins and minerals from a variety of foods from each of the food groups.
• At least 1 liter (about 1 quart) of water daily or 1 milliliter per kcalorie daily—whichever is more.
3. They use liquid formulas rather than foods. Foods should accommodate a person's ethnic background, taste preferences, and financial means.
4. They attempt to make clients dependent upon special foods or devices. Programs should teach clients how to make good choices from the conventional food supply.
5. They fail to encourage permanent, realistic lifestyle changes. Programs should provide physical activity plans that involve spending at least 300 kcalories a day and behavior-modification strategies that help to correct poor eating habits.
6. They misrepresent salespeople as "counselors" supposedly qualified to give guidance in nutrition and/or general health. Even if adequately trained, such "counselors" would still be objectionable because of the obvious conflict of interest that exists when providers profit directly from products they recommend and sell.
7. They collect large sums of money at the start or require that clients sign contracts for expensive, long-term programs. Programs should be reasonably priced and run on a pay-as-you-go basis.
8. They fail to inform clients of the risks associated with weight loss in general or the specific program being promoted. They should provide information about dropout rates, the long-term success of their clients, and possible side effects.
9. They promote unproven or spurious weight-loss aids such as human chorionic gonadotropin hormone (HCG), starch blockers, diuretics, sauna belts, body wraps, passive exercise, ear stapling, acupuncture, electric muscle-stimulating (EMS) devices, spirulina, amino acid supplements (e.g., arginine, ornithine), glucomannan, methylcellulose (a "bulking agent"), "unique" ingredients, and so forth.
10. They fail to provide for weight maintenance after the program ends.

Notes

[1] F. X. Pi-Sunyer, Glycemic index and disease, *American Journal of Clinical Nutrition* 76 (2002): 290S–298S.

[2] Pi-Sunyer, 2002.

[3] A. Raben, Should obese patients be counselled to follow a low-glycemic index diet? No. *Obesity Reviews* 3 (2002): 245–256.

[4] A. Astrup, Atkins and other low-carbohydrate diets: Hoax or an effective tool for weight loss? *Lancet* 364 (2004): 897–899; E. C. Westman and coauthors, Effect of 6-month adherence to a very low carbohydrate diet program, *American Journal of Medicine* 113 (2002): 30–36.

[5] L Stern and coauthors, The effects of low-carbohydrate versus conventional weight loss diets in severely obese adults: One-year follow-up of a randomized trial, *Annals of Internal Medicine* 140 (2004): 778–785; F. F. Samaha and coauthors, A low-carbohydrate diet as compared with a low-fat diet in severe obesity, *New England Journal of Medicine* 348 (2003): 2074–2081; B. J. Brehm and coauthors, A randomized trial comparing a very low-carbohydrate diet and a calorie-restricted low fat diet on body weight and cardiovascular risk factors in healthy women, *Clinics in Endocrinology and Metabolism* 88 (2003): 1617–1623; G. D. Foster and coauthors, A randomized trial of a low-carbohydrate diet for obesity, *New England Journal of Medicine* 348 (2003): 2082–2090.

[6] D. M. Bravata and coauthors, Efficacy and safety of low-carbohydrate diets, *Journal of the American Medical Association* 289 (2003): 1837–1850.

[7] Foster and coauthors, 2003.

[8] F. B. Hu, Protein, body weight, and cardiovascular health, *American Journal of Clinical Nutrition* 82 (2005): 242S–247S.

[9] E. T. Kennedy and coauthors, Popular diets: Correlation to health, nutrition, and obesity, *Journal of the American Dietetic Association* 101 (2001): 411–420.

[10] A. Astrup, The satiating power of protein—A key to obesity prevention? *American Journal of Clinical Nutrition* 82 (2005): 1–2; D. S. Weigle and coauthors, A high-protein diet induces sustained reductions in appetite, ad libitum caloric intake, and body weight despite compensatory changes in diurnal plasma leptin and ghrelin concentrations, *American Journal of Clinical Nutrition* 82 (2005): 41–48.

[11] Weigle and coauthors, 2005.

[12] A. Due and coauthors, Effect of normal-fat diets, either medium or high in protein, on body weight in overweight subjects: A randomised 1-year trial, *International Journal of Obesity and Related Metabolic Disorders* 28 (2004): 1283–1290.

[13] A. Trichopoulou and coauthors, Lipid, protein and carbohydrate intake in relation to body mass index, *European Journal of Clinical Nutrition* 56 (2002): 37–43.

[14] W. H. M. Saris and A. Harper, DiOGenes: A multidisciplinary offensive focused on the obesity epidemic, *Obesity Reviews* 6 (2005): 175–176.

[15] Foster and coauthors, 2003.

[16] S. T. St. Jeor and coauthors, Dietary protein and weight reduction: A statement for healthcare professionals from the nutrition committee of the Council on Nutrition, Physical Activity, and Metabolism of the American Heart Association, *Circulation* 104 (2001): 1869–1874.

[17] D. K. Layman and coauthors, A reduced ratio of dietary carbohydrate to protein improves body composition and blood lipid profiles during weight loss in adult women, *Journal of Nutrition* 133 (2003): 411–417; Bravata and coauthors, 2003; M. R. Freedman, J. King, and E. Kennedy, Popular diets: A scientific review, *Obesity Research* 9 (2001): 1S–5S.

CONTENTS

Weight Management: Overweight and Underweight

CHAPTER 7

Are you pleased with your body weight? If you answered yes, you are a rare individual. Nearly all people in our society think they should weigh more or less (mostly less) than they do. Usually, their primary reason is appearance, but they often perceive, correctly, that their weight is also related to physical health. Chapter 6 addressed the health risks of being overweight. As discussed in a later section of this chapter, health risks also accompany underweight.

Overweight and underweight both result from unbalanced energy budgets. The simple picture is as follows. Overweight people have consumed more food energy than they have spent and have banked the surplus in their body fat. To reduce body fat, overweight people need to spend more energy than they take in from food. In contrast, underweight people have consumed too little food energy to support their activities and so have depleted their bodies' fat stores and possibly some of their lean tissues as well. To gain weight, they need to take in more food energy than they expend.

This chapter's missions are to present strategies toward solving the problems of excessive and deficient body fatness and to point out how appropriate body composition, once achieved, can be maintained. The chapter emphasizes overweight because it has been more intensively studied and is a more widespread health problem in the developed countries.

Causes of Obesity

Henceforth, this chapter will use the term *obesity* to refer to excess body fat. Excess body fat accumulates when people take in more food energy than they spend. Why do they do this? Is it genetic? Metabolic? Psychological? Behavioral? All of these? Most likely, obesity has many interrelated causes: many experts in the field speak of several different *obesities*.

Genetics and Weight A person's genetic makeup influences the body's tendency to consume or store too much energy or to burn too little.[1] When both parents are obese, the chances that their children will be obese are quite high (up to 80 percent), whereas when neither parent is obese, the chances are relatively small (less than 10 percent). Adoption studies find a similarity in obesity between biological parents and their natural children, but not between adoptive parents and their adopted children.

To determine the relative contributions of genetic and environmental factors to body weight, one group of researchers studied identical and fraternal twins, some of whom were reared together and some apart. Like previous studies, this study found that identical twins were twice as likely to have similar weights as fraternal twins were—even when reared apart. These findings suggest an important role for genetics in determining a person's *susceptibility* to obesity.[2]

Genetics also influences the way energy is *stored*. When identical twins were given an extra 1000 kcalories a day for 100 days, some pairs gained less than 10 pounds whereas others gained up to 30 pounds. Within each pair, the amount of weight gained, percentage of body fat, and distribution of fat were similar.

Genetic factors also influence how much energy the body *spends*. For example, the differences in basal metabolic rate (BMR) between individuals are greater than can be explained by age, gender, and body composition alone. Similarities within families suggest a genetic influence on BMR. A low metabolic rate is a major risk factor for weight gain.

Lipoprotein Lipase Some of the research investigating genetic influence on obesity focuses on the enzyme **lipoprotein lipase (LPL),** which promotes fat storage in fat cells and muscle cells. People with high LPL activity are especially efficient at storing fat. Obese people generally have much more LPL activity in their fat cells than lean people do.

lipoprotein lipase (LPL): an enzyme mounted on the surface of fat cells (and other cells). It hydrolyzes triglycerides in the blood into fatty acids and glycerol for absorption into the cells. There they are metabolized or reassembled for storage.

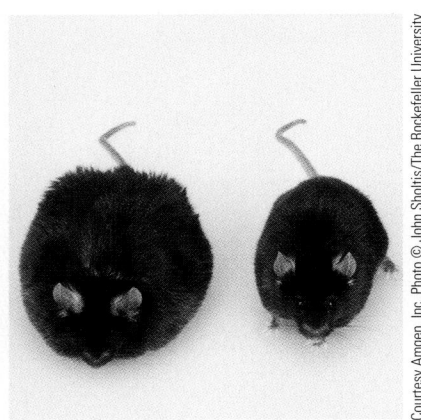

Courtesy Amgen, Inc. Photo © John Shotis/The Rockefeller University

The mouse on the left is genetically obese—it lacks the gene for producing leptin. The mouse on the right is *also* genetically obese, but because it receives leptin, it eats less, expends more energy, and is less obese than it would be had it not received the leptin.

Genes instruct cells to make proteins, and each protein performs a unique function.

Anorexia nervosa is an eating disorder discussed in this chapter's Nutrition in Practice.

leptin: a hormone produced by fat cells under the direction of the *ob* gene. It decreases appetite and increases energy expenditure.

   leptos = thin

ghrelin (GRELL-in): a hormone produced primarily by the stomach cells. It signals the hypothalamus of the brain to stimulate appetite and food intake.

Leptin Researchers have discovered an obesity gene in humans called *ob*. The obesity gene codes for the protein **leptin.** Leptin is a hormone primarily produced and secreted by the fat cells in proportion to the amount of fat stored.[3] A gain in body fatness stimulates the production of leptin, which, by way of the hypothalamus, suppresses the appetite, increases energy expenditure, and produces fat loss. Fat loss produces the opposite effect—suppression of leptin production, increased appetite, and decreased energy expenditure. As the accompanying photo shows, mice with a defective *ob* gene do not produce leptin and can weigh up to three times as much as normal mice. When injected with leptin, the mice lose weight. (Because leptin is a protein, it would be destroyed during digestion if given orally; consequently, it must be given by injection.)

Although it is extremely rare, researchers have identified a genetic deficiency of leptin in human beings as well. An error in the gene that codes for leptin was discovered in two extremely obese children whose blood levels of leptin were barely detectable. Without leptin, the children had little appetite control; they were constantly hungry and ate considerably more than their siblings or peers. Given daily injections of leptin, these children lost a substantial amount of weight, confirming leptin's role in regulating appetite and body weight.

Most obese people do not have leptin deficiency, however. In fact, in obese people, the more body fat, the more leptin.[4] Researchers speculate that leptin rises in an effort to suppress appetite and inhibit fat storage when fat cells are ample. Obese people with elevated leptin concentrations may be resistant to its appetite-suppressing effect.[5] The absence of or resistance to leptin in obesity parallels the scenario of insulin in diabetes: some people have an insulin deficiency (type 1), whereas many others have elevated insulin but are resistant to its glucose-storing effect (type 2).

Research on leptin is ongoing, and scientists are exploring the possibility that leptin may one day help treat human obesity.[6] So far, though, efforts to treat obesity with leptin have been disappointing. Even if leptin never proves useful as an antiobesity drug, however, its discovery has contributed much to our understanding of the complexities of the human body. For example, scientists no longer view adipose tissue as a metabolically sluggish storage depot for fat, but rather as a hormonally active regulatory tissue with widespread effects on the body.[7] In addition to its appetite function, leptin may have roles in immunity, reproduction, bone formation, and sexual maturation.[8]

Ghrelin Researchers recently discovered another protein that also acts as a hormone primarily in the hypothalamus, but it works in the opposite direction of leptin. Known as **ghrelin,** this protein is synthesized and secreted primarily by the stomach cells and promotes a positive energy balance by stimulating appetite and promoting efficient energy storage.[9] The role ghrelin plays in regulating food intake and body weight is the subject of much intense research.[10] Pharmaceutical companies are eager to develop products that mimic ghrelin to treat wasting conditions, as well as products that oppose ghrelin's actions to treat obesity.

Ghrelin powerfully triggers the desire to eat. Blood levels of ghrelin typically rise before a meal and fall rapidly after it—reflecting the hunger and satiety that precede and follow eating.[11] Furthermore, the rise and fall of ghrelin concentrations are dose-dependent on the number of kcalories ingested.[12] In other words, large meals suppress ghrelin (and hunger) to a greater extent than small meals do.[13]

In general, fasting blood levels of ghrelin correlate inversely with body weight: lean people have high ghrelin levels and obese people have low levels.[14] Interestingly, while ghrelin levels are high in underweight people, they are exceptionally high in anorexia nervosa and return to normal with nutrition intervention—indicating that both body weight and nutrition status influence ghrelin levels.[15] Also noteworthy, ghrelin levels are markedly high in Prader-Willi syndrome, a rare inherited disorder often characterized by severe obesity, and they remain elevated

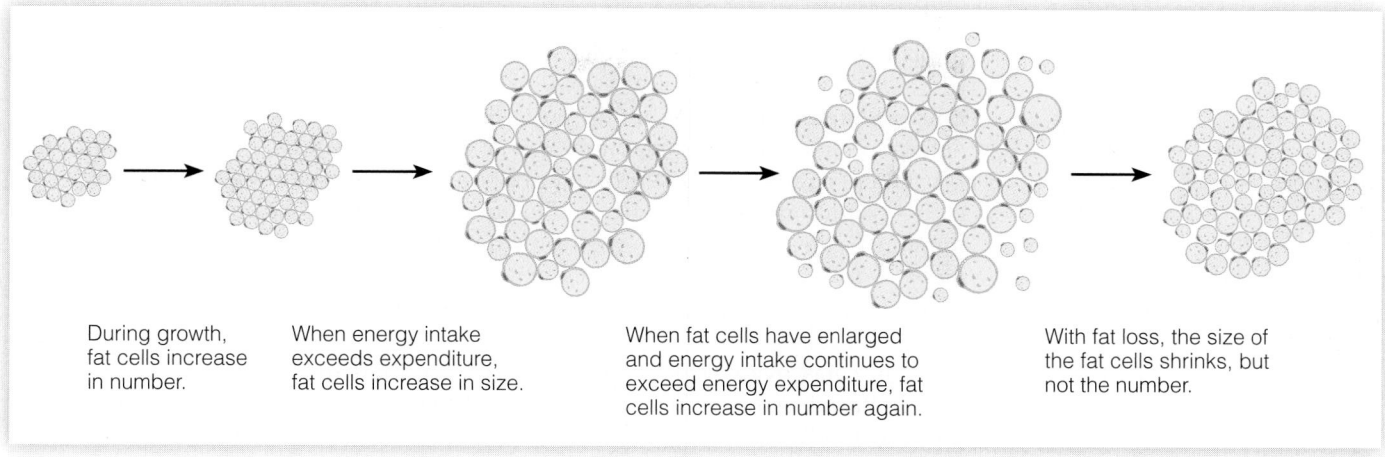

During growth, fat cells increase in number.

When energy intake exceeds expenditure, fat cells increase in size.

When fat cells have enlarged and energy intake continues to exceed energy expenditure, fat cells increase in number again.

With fat loss, the size of the fat cells shrinks, but not the number.

FIGURE 7-1 Fat Cell Development

Fat cells are capable of increasing their size by 20-fold and their number by several thousandfold.

even after a meal, which helps to explain the excessive appetite and food consumption commonly seen in this disorder.[16] Similarly, ghrelin levels in obese people do not seem to decline as much after a meal, as they do for lean people.[17]

Ghrelin fights to maintain a stable body weight.[18] On average, ghrelin levels are high whenever the body is in negative energy balance, as occurs during low-kcalorie diets, for example. This response may help explain why weight loss is so difficult to maintain. Ghrelin levels decline again whenever the body is in positive energy balance, as occurs with weight gains.

Like leptin, ghrelin plays roles in the body beyond energy regulation. In fact, it was first recognized for its participation in growth hormone activity.[19] The focus of intense, ongoing ghrelin research, however, is on its actions in stimulating appetite and short-term food intake and in regulating long-term body weight.

Fat Cell Development Another cause of obesity may be the development of excess fat cells during childhood. The amount of fat on a person's body reflects both fat cell *number* and fat cell *size*. The number of fat cells increases most rapidly during the growing years of late childhood and early puberty. Fat cell number increases more rapidly in obese children than in lean children, and obese children entering their teen years may already have as many fat cells as do adults of normal weight.

Fat cells can also expand in size. After they reach their maximum size, more cells can develop to store more fat. Thus obesity develops when a person's fat cells increase in number, in size, or quite often both. Figure 7-1 illustrates fat cell development.

With fat loss, the size of the fat cells shrinks, but not their number. For this reason, people with extra fat cells may tend to regain lost weight rapidly. Prevention of obesity, then, is most critical during the growing years when fat cell number is increasing.

Set-Point Theory One popular theory of why a person may store too much fat is the **set-point theory.** The set-point theory proposes that body weight, like body temperature, is physiologically regulated. Researchers have noted that many people who lose weight on reducing diets quickly regain all their lost weight. This suggests that somehow the body chooses a weight that it wants to be and defends that weight by regulating eating behaviors and hormonal actions. Research confirms that the body adjusts its metabolism whenever it gains or loses weight—in the direction that returns to the initial body weight: energy expenditure increases with weight gain and decreases with weight loss. These changes in energy expenditure are greater than those predicted based on body composition and help to explain why it is so difficult for an obese person to maintain weight losses. An individual's

set-point theory: the theory that proposes that the body tends to maintain a certain weight by means of its own internal controls.

set point for body weight may be adjustable, shifting over the life span in response to physiological changes and to genetic, dietary, and other factors.

Environmental Stimuli As discussed earlier, genetic factors play a partial role in determining a person's susceptibility to obesity, but they do not fully explain obesity. Obesity rates have risen dramatically during recent decades wherever people enjoy conditions of prosperity and abundance, but the human gene pool has remained unchanged. The environment must therefore play a role as well. In other words, although a person's genetic inheritance may make obesity likely, lifestyle and environmental conditions will determine the extent to which the disease is expressed.[20]

People may overeat in response to stimuli in their surroundings—primarily, the availability of many delectable foods. Most people in the United States find high-kcalorie foods readily available, relatively inexpensive, heavily advertised, and reasonably tasty. Food is available everywhere, all the time—thanks largely to fast food. Fast-food restaurants line the highways and crowd out mom and pop restaurants and businesses in small towns and big cities. Convenience stores, service stations, malls, airports, and even schools offer fast food as well. Most alarming are the extraordinarily large serving sizes and ready-to-go meals offered in supersize combinations.[21] People buy the large sizes and combinations, perceiving them to be a good value, but then they eat more than they need. Research shows that people eat more if they're served more.[22] And portion sizes of virtually all foods and beverages have increased markedly in the past several decades, most notably at fast-food restaurants.[23] The increase in portion sizes parallels the growing prevalence of overweight and obesity in the United States, beginning in the 1970s, rising sharply in the 1980s and 1990s, and continuing today.[24]

Fast food is often high in fat, and fat is perceived as especially palatable. People can easily overeat fat-rich foods because their delicious tastes stimulate eating and each bite of food is kcalorie dense. Not only does fat deliver more than twice as many kcalories, gram for gram, as protein and carbohydrate, but it also seems to be stored preferentially by the body, and with great efficiency in many people. A steady diet of high-kcalorie, high-fat fast food, then, encourages obesity.[25]

The combination of large portions and energy-dense foods strikes a double blow. Reducing portion sizes is helpful, but the real kcalorie savings come from lowering the energy density.[26] Satisfying portions of foods with low energy density such as fruits and vegetables can help with weight loss.[27]

Learned Behavior Psychological stimuli also trigger inappropriate eating behaviors in some people. Appropriate eating behavior is a response to **hunger.** Hunger is a drive programmed into people by their heredity. **Appetite,** in contrast, is learned and can lead people to ignore hunger or to overrespond to it. Hunger is physiological, whereas appetite is psychological, and the two do not always coincide.

Food behavior is also intimately connected to deep emotional needs such as the primitive fear of starvation. Yearnings, cravings, and addictions with profound psychological significance can express themselves in people's eating behavior. An emotionally insecure person might eat rather than call a friend and risk rejection. Another person might eat to relieve boredom or to ward off depression.

Physical Inactivity The possible causes of obesity mentioned so far all relate to the input side of the energy equation. What about output? People may be obese not because they eat too much, but because they spend too little energy. More than one-third of the overweight population report no physical activity during their leisure time. Obese people observed closely are

hunger: the physiological need to eat, experienced as a drive to obtain food; an unpleasant sensation that demands relief.

appetite: the psychological desire to eat; a learned motivation that is experienced as a pleasant sensation that accompanies the sight, smell, or thought of appealing foods.

Lack of physical activity fosters obesity.

often seen to eat less than lean people, but they are sometimes so extraordinarily inactive that they still manage to accumulate an energy surplus. Reducing their food intake further would jeopardize health and incur nutrient deficiencies. Physical activity, then, is a necessary component of nutritional health. People must be physically active if they are to eat enough food to deliver all the nutrients needed without unhealthy weight gain.

Our environment, however, often encourages inactivity. One hundred years ago, 30 percent of the energy used in farm and factory work came from muscle power; today, only 1 percent does. Modern technology has replaced physical activity at home, at work, and in transportation. Inactivity contributes to obesity and poor health.[28] In turn, television watching may contribute most to physical inactivity.

Watching television contributes to obesity in several ways.[29] First, television viewing requires little energy beyond the resting metabolic rate. Second, it replaces time spent in more vigorous activities. Third, television influences family food purchases; viewers are more likely to engage in between-meal snacking and eat the high-kcalorie foods most heavily advertised on programs.

REVIEW NOTES

Genetics, fat cell development, set point, and overeating all offer possible, but still incomplete, explanations of obesity.

Most likely, obesity has not one cause, but different causes and combinations of causes in different people.

Some causes may be within a person's control, and some may be beyond it.

Like all the other "causes" of obesity, inactivity alone fails to explain it fully.

Obesity Treatment: Who Should Lose?

An estimated 35 to 45 percent of all U.S. women (and 20 to 30 percent of all U.S. men) are trying to lose weight at any given time, spending up to $40 billion each year to do so. Some of these people do not even need to lose weight. Others need to lose weight but are not successful. Still others are successful. Despite the general notion that practically no one maintains weight loss over the long term, research shows otherwise. Approximately 20 percent of overweight people do achieve successful long-term weight loss, meaning that they "intentionally lost at least 10 percent of their body weight and kept it off at least one year."[30]

Many people assume that every overweight person can achieve slenderness and should pursue that goal. Consider, however, that most overweight people cannot become slender. People vary in their weight tendencies just as they vary in their potentials for height and degrees of health. The question of whether a person should lose weight depends on many factors: the extent of overweight, age, health, and genetics, to name a few. Weight-loss advice, then, does not apply equally to all overweight people. Some people may risk more in the process of losing weight than in remaining overweight. Others may reap significant health benefits with just modest weight loss.

REVIEW NOTES

Weight-loss advice does not apply equally to all overweight people.

Some people may risk more through misguided efforts to lose weight than by remaining overweight, whereas others may benefit from just a modest weight loss.

Inappropriate Obesity Treatments

The risks people incur in attempting to lose weight often depend on how they go about it. Weight-loss plans and obesity treatments abound—some are adequate, but many are ineffective and possibly dangerous. Fad diets are the topic of Nutrition in Practice 6. This section addresses other inappropriate obesity interventions. Aggressive approaches to obesity for those obese people who face high risks of medical problems and must lose weight rapidly are discussed in the next section. Reasonable approaches to overweight for those seeking safe, gradual weight loss are saved for the last part of the obesity treatment discussion.

OVER-THE-COUNTER WEIGHT-LOSS SUPPLEMENTS AND DRUGS

Millions of people in the United States use nonprescription weight-loss products, spending close to $2 billion a year for such products.[31] Most of the people who use over-the-counter weight-loss supplements are women, especially young obese women, but almost 10 percent are of normal weight.[32] Promoters and marketers of weight-loss products make all kinds of claims for their products with only one intention—profit. Such claims as "eat all you want and lose weight," "take 3 pills before bedtime and watch the fat disappear," "blocks carbs," "blocks fat," and many more, lure people into believing that maybe this time a product will really work.

In 2000, the FDA recommended that manufacturers voluntarily discontinue marketing over-the-counter products containing phenylpropanolamine, an ingredient commonly used in products to suppress appetite.* Reported side effects include dry mouth, rapid pulse, nervousness, sleeplessness, hypertension, irregular heartbeats, kidney failure, seizures, and strokes.

As this text goes to print, the FDA is considering whether to approve what could become the first nonprescription weight-loss drug for use in the United States. The drug is a low-dose version of the prescription drug orlistat, discussed later in the chapter.

Read labels of over-the-counter products to determine if they contain **phenylpropanolamine** (fen-ill-pro-pa-NOLE-a-mean).

HERBAL PRODUCTS AND DIETARY SUPPLEMENTS

In their search for weight-loss magic, some consumers turn to "natural" herbal products and dietary supplements, even though few have proved to be effective.[33] People falsely believe that "natural" herbs are not harmful to the body, but many herbs contain toxins. Belladonna and hemlock are infamous examples, but many lesser-known herbs, such as sassafras, contain toxins as well. Furthermore, because herbs are marketed as "dietary supplements," manufacturers need not present scientific evidence of their safety or effectiveness to the FDA before marketing them. Evidence about their safety is gathered only through reports of consumers who sicken or die after using the remedies.

A now familiar example is ephedra (also called ma huang), an herb that showed promise as a weight-loss drug in preliminary studies. Immediately, ephedra-containing products for dieters and athletes flooded the market. Many consumers of these products reported ill effects including cardiac arrest, abnormal heartbeats, hypertension, strokes, and seizures; the supplements have been linked to some deaths as well.[34] For this reason, the FDA has banned the sale of dietary supplements containing ephedra and its active constituent, ephedrine.† Table 7-1 presents

Ephedrine is an amphetamine-like substance extracted from the Chinese ephedra herb ma huang.

*Phenylpropanolamine is not commercially available in Canada.
†Ma huang (ephedrine) is illegal in Canada.

| TABLE 7-1 | Selected Herbal and Other Dietary Supplements Marketed for Weight Loss |

Product	Manufacturers' Claims	Research Findings	Adverse Effects
Chitosan[a] (pronounced KITE-oh-san; derived from chitin, the substance that forms the hard shells of lobsters, crabs, and other crustaceans)	Binds to dietary fat, preventing digestion and absorption	Ineffective	Impaired absorption of fat-soluble vitamins
Chromium (trace mineral)	Eliminates body fat	Ineffective; weight gain reported when not accompanied by exercise.	Headaches, sleep disturbances, and mood swings; hexavalent form is toxic and carcinogenic.
Conjugated linoleic acid (CLA; a group of fatty acids related to linoleic acid, but with different cis- and trans-configurations)	Reduces body fat and suppresses appetite	Some evidence in animal studies, but ineffective in human studies	None known
Ephedrine[b] (amphetamine-like substance derived from the Chinese ephedra herb ma huang)	Speeds body's metabolism	Weight loss and dangerous side effects	Insomnia, tremors, heart attacks, strokes, and death; the FDA has banned the sale of these products.
Hydroxycitric acid[c] (active ingredient derived from the rind of the tropical fruit garcinia cambogia)	Inhibits the enzyme that converts citric acid to fat; suppresses appetite	Ineffective	Toxicity symptoms reported in animal studies; headache, respiratory and gastrointestinal symptoms in humans
Pyruvate[d] (3-carbon compound produced during glycolysis)	Speeds body's metabolism	Modest weight loss with high doses	GI distress
Triiodothyroacetic acid[e] (TRIAC, a potent thyroid hormone)	Speeds up body's metabolism	Weight loss and dangerous side effects	Diarrhea, fatigue, drowsiness, insomnia, nervousness, sweating, heart attacks, and strokes; FDA warning issued
Yohimbine (derived from the bark of a West African tree)	Promotes weight loss	Ineffective	Nervousness, insomnia, anxiety, dizziness, tremors, headaches, nausea, vomiting, hypertension

Note: The FDA has not approved the use of any of these products; most products are used in conjunction with a 1000- to 1800-kcalorie diet.

[a] Marketed under the trade names Chitorich, Exofat, Fat Breaker, Fat Blocker, Fat Magnet, Fat Trapper, and Fatsorb.

[b] Marketed under the trade names Diet Fuel, Metabolife, and Nature's Nutrition Formula One.

[c] Marketed under the trade names Ultra Burn, Citralean, CitraMax, Citrin, Slim Life, Brindleslim, Medislim, and Beer Belly Busters.

[d] Marketed under the trade names Exercise in a Bottle, Pyruvate Punch, Pyruvate-c, and Provate.

[e] Marketed under the trade name Triax Metabolic Accelerator.

the claims and the dangers behind ephedrine and several other weight-loss supplements.[35]

Herbal laxatives containing senna, aloe, rhubarb root, cascara, castor oil, and buckthorn (or various combinations) are commonly sold as "dieter's tea." Such concoctions cause nausea, vomiting, diarrhea, cramping, and fainting and may have contributed to the deaths of four women who had drastically reduced their food intakes. Consumers mistakenly believe that laxatives will diminish nutrient absorption and reduce kcalorie intake, but remember that absorption occurs primarily in the upper small intestine and these laxatives act on the lower large intestine. Nutrition in Practice 23 explores the possible benefits and potential dangers of herbal products and other alternative therapies. Anyone using dietary supplements for weight loss should first consult a physician.

OTHER GIMMICKS

Other gimmicks don't help with weight loss either. Hot baths do not speed up metabolism so that pounds can be lost in hours. Steam and sauna baths do not melt the fat off the body, although they may dehydrate people so that they lose water weight. Brushes, sponges, wraps, creams, and massages intended to move, burn, or break up **"cellulite"** do nothing of the kind, because there is no such thing as cellulite.

> **REVIEW NOTES**
>
> When pursued via unwise weight-loss techniques, such as over-the-counter drugs and herbal supplements, weight-loss efforts can be physically and psychologically damaging.

Aggressive Treatments of Obesity

For some obese people, the medical problems caused by their obesity demand treatment approaches that may, themselves, incur some risks. The health benefits to be gained by weight loss, however, may make these risks worth taking.

OBESITY DRUGS

Several prescription medications for weight loss have been tried over the years. When used as part of a long-term comprehensive weight-loss program, medications can help obese people to lose approximately 10 percent of their weight and maintain that loss for at least a year. Because weight regain commonly occurs with the discontinuation of drug therapy, treatment is long term. And the long-term use of medications poses risks. Medical experts do not yet know whether a person would benefit more from maintaining a 20-pound excess or from taking a medication for a decade to keep the 20 pounds off.

The challenge, then, is to develop an effective medication that can be used over time without adverse side effects or the potential for abuse. No such medication currently exists.[36] Two prescription medications, however, are currently in use; others, including leptin, are being studied.

Sibutramine Sibutramine is an appetite suppressant that works on the brain's neurotransmitters.* The drug is most effective when used in combination with a reduced-kcalorie diet and increased physical activity. Sibutramine enhances satiety and elevates energy expenditure.[37] Side effects include dry mouth, rapid heart rate, insomnia, headache, and high blood pressure.[38] The FDA cautions those with hypertension against using it. Anyone taking sibutramine should monitor blood pressure carefully. As more information becomes known about the molecular chemistry of appetite control, safer appetite suppressants may be developed.

Orlistat Orlistat takes a different approach to weight loss.† Not an appetite suppressant, orlistat inhibits the action of fat-digesting enzymes in the gastrointestinal (GI) tract and so reduces fat digestion and absorption by about 30 percent.[39] As a result of absorbing less fat, people often lose weight. The problem with undigested fat is that it leaves the digestive tract intact, carrying with it fat-soluble vitamins and phytochemicals that would otherwise have been absorbed by the body. Possible side effects of orlistat resemble those of the artificial fat olestra—diarrhea and digestive distress.[40]

*Sibutramine's trade name is Meridia.
†Orlistat's trade name is Xenical.

cellulite (SELL-you-light or SELL-you-leet): supposedly, a lumpy form of fat; actually, a fraud. Fatty areas of the body may appear lumpy when the strands of connective tissue that attach the skin to underlying muscles pull tight where the fat is thick. The fat itself is the same as fat anywhere else in the body. If the fat in these areas is lost, the lumpy appearance disappears.

SURGERY

Surgery as an approach to weight loss is justified in some specific cases of **clinically severe obesity.** Two procedures, **gastric bypass** and **gastric banding,** have gained wide acceptance. Both procedures limit food intake by effectively reducing the capacity of the stomach and suppressing hunger by reducing the production of ghrelin.[41] The results are dramatic: most people achieve a lasting weight loss of more than 50 percent of their excess body weight.[42] Research shows that weight-loss surgery also helps to improve blood lipids, diabetes, sleep apnea, and hypertension.[43]

Laparoscopic weight-loss surgery techniques are now used to perform gastric bypass and other weight-loss surgeries. Laparoscopic weight-loss surgery produces significant, long-term weight loss, shortens recovery, and is less invasive than open surgery.[44]

The long-term safety and effectiveness of gastric surgery depend, in large part, on compliance with dietary instructions. Common immediate postsurgical complications include infections, nausea, vomiting, and dehydration; in the long term, vitamin and mineral deficiencies and psychological problems are common. Lifelong medical supervision is necessary for those who choose the surgical route, but in suitable candidates, the health benefits of weight loss may prove worth the risks.[45]

> ### REVIEW NOTES
> Obese people with high risks of medical problems may need aggressive treatment, including drugs or surgery.

Reasonable Strategies for Weight Loss

The *Dietary Guidelines for Americans 2005* suggest that those who need to lose weight should "aim for a slow, steady weight loss by decreasing kcalorie intake while maintaining an adequate nutrient intake and increasing physical activity." Modest weight loss, even when a person is still overweight, can improve control of diabetes and reduce the risks of heart disease by lowering blood pressure and blood cholesterol, especially for those with abdominal fat.

Of course, the same eating and activity habits that improve health often lead to a healthier body weight and composition as well. Successful weight loss, then, is defined not by pounds lost, but by health gained. People less concerned with disease risks may prefer to set goals for personal fitness, such as being able to play with children or climb stairs without becoming short of breath.

Whether the goal is health or fitness, weight-loss expectations need to be reasonable. Unreachable targets ensure frustration and failure. Setting reasonable goals helps to achieve the desired result in managing weight. For example, obese people who must reduce their weight to lower their disease risks might set three broad goals:

1. Reduce body weight by about 10 percent over half a year's time.
2. Maintain a lower body weight over the long term.
3. At a minimum, prevent further weight gain.

clinically severe obesity: a BMI of 40 or greater or a BMI of 35 or greater and one or more serious conditions such as hypertension. Another term used to describe the same condition is *morbid obesity.*

gastric bypass: surgery that restricts stomach size and reroutes food from the stomach to the lower part of the small intestine; creates a chronic, lifelong state of malabsorption by preventing normal digestion and absorption of nutrients.

gastric banding: a surgical means of producing weight loss by restricting stomach size with a constricting band; used in people whose severe obesity brings extreme health risks.

laparoscopic weight-loss surgery: a procedure in which surgeons gain access to the abdomen via several small incisions. A tiny video camera is inserted through one of the incisions and surgical instruments through the others. The surgeons watch their work on a large-screen monitor.

© Lori Adamski Peek/Stone/Getty Images

A healthy body contains enough lean tissue to support health and the right amount of fat to meet body needs.

TABLE 7-2	Recommendations for a Weight-Loss Diet
Nutrient	**Recommended Intake**
kCalories	Approximately 500 to 1000 kcalories per day reduction from usual intake
Total fat	30% or less of total kcalories
Saturated fatty acids[a]	8 to 10% of total kcalories
Monounsaturated fatty acids	Up to 15% of total kcalories
Polyunsaturated fatty acids	Up to 10% of total kcalories
Cholesterol[a]	< 300 mg per day
Protein[b]	Approximately 15% of total kcalories
Carbohydrate[c]	55% or more of total kcalories
Sodium chloride[d]	No more than 2300 mg of sodium or approximately 6 g of sodium chloride (salt) per day
Calcium	1000 to 1500 mg per day
Fiber[c]	20 to 30 g per day

[a]People with high blood cholesterol should aim for less than 7 percent kcalories from saturated fat and 200 milligrams cholesterol per day.

[b] Protein should be derived from plant sources and lean sources of animal protein.

[c] Carbohydrates and fiber should be derived from vegetables, fruits, and whole grains.

[d] Value from the *Dietary Guidelines for Americans 2005;* the Tolerable Upper Intake Level for sodium is 2300 milligrams.

Source: National Heart, Lung, and Blood Institute, National Institutes of Health, *The Practical Guide: Identification, Evaluation, and Treatment of Overweight and Obesity in Adults,* NIH publication no. 00-4084 (Washington, D.C.: Government Printing Office, 2000), p. 27.

Such goals may be achieved or even exceeded, providing a sense of accomplishment instead of disappointment.

A HEALTHFUL EATING PLAN

Contrary to the claims of many fad diets, no particular eating plan is magical, and no specific food must be either included or avoided for weight management. You are the one who will have to live with the plan, so you had better be involved in its planning. The diet is successful only if you can maintain a healthy weight. Think of it as an eating plan that you will adopt for life. It must consist of foods that you like, that are available to you, and that are within your means.

A Realistic Energy Intake The main characteristic of a weight-loss diet is that it provides less energy than the person needs to maintain present body weight. If food energy is restricted too severely, dieters may not receive sufficient nutrients and may lose lean tissue. Rapid weight loss usually means excessive loss of lean tissue, a lower BMR, and a rapid weight gain to follow. Restrictive eating may also set in motion the unhealthy behaviors of eating disorders (described in Nutrition in Practice 7).

Table 7-2 outlines the recommendations of a weight-loss diet. Energy intake should provide nutritional adequacy without excess—that is, somewhere between deprivation and complete freedom to eat whatever, whenever. A reasonable sug-

TABLE 7-3	Daily Amounts from Each Food Group for 1000- to 1600-kCalorie Diets			
FOOD GROUP	**1000 kcal**	**1200 kcal**	**1400 kcal**	**1600 kcal**
Fruit	1 c	1 c	1¹/₂ c	1¹/₂ c
Vegetables	1 c	1¹/₂ c	1¹/₂ c	2 c
Grains	3 oz	4 oz	5 oz	5 oz
Meat and legumes	2 oz	3 oz	4 oz	5 oz
Milk	3 c	3 c	3 c	3 c
Oils	3 tsp	3 tsp	3 tsp	4 tsp

Note: The USDA Food Guide patterns for 1000-, 1200-, and 1400-kcalorie diets were designed for children and provide 2 cups of milk. They are modified here to include an additional cup of milk, as 3 cups per day is recommended for all adults. The discretionary kcalorie allowance for these patterns is about 100 kcalories.

gestion is that an adult needs to increase activity and reduce food intake to create a deficit of 500 to 1000 kcalories per day. Such a deficit produces a weight loss of 1 to 2 pounds per week.[46] This amounts to an intake of 1000 to 1200 kcalories per day for most women and 1200 to 1600 kcalories per day for most men.[47] Table 7-3 suggests daily food amounts from which to build balanced 1000- to 1600-kcalorie diets. Diets providing energy intakes lower than 800 kcalories, called very low-calorie diets (VLCD), are notoriously unsuccessful at achieving lasting weight loss and can be dangerous, and so are not recommended.

Nutritional Adequacy Nutritional adequacy is difficult to achieve on fewer than 1200 kcalories a day, and most healthy adults need never consume any less than that. A plan that provides an adequate intake supports a healthier and more successful weight loss than a restrictive plan that creates feelings of starvation and deprivation, which can lead to an irresistible urge to binge.

Take a look at the 1200-kcalorie diet in Table 7-3. Such an intake would allow most people to lose weight and still meet their nutrient needs with careful, nutrient-dense food selections. (Women might need calcium or iron supplements.) Keep in mind that well-balanced diets that emphasize fruits, vegetables, whole grains, lean meats or meat alternates, and low-fat or fat-free milk products offer many health rewards even when they don't result in weight loss.

Small Portions Overweight people usually need to learn to eat less food at each meal—one piece of chicken for dinner instead of two, a teaspoon of butter on the vegetables instead of a tablespoon, and one cookie for dessert instead of six. Chew foods slowly and thoroughly. The goal is to eat enough food for energy, nutrients, and pleasure, but not more. This amount should leave a person feeling satisfied— not necessarily full. Keep in mind that even low-fat foods can deliver a lot of kcalories when a person eats large quantities.

Balancing Carbohydrates, Fats, and Protein Healthy diets based on abundant fresh fruits and vegetables, low-fat milk products, legumes, lean meats, fish, poultry, or meat alternates, and whole grains are high in carbohydrates, adequate in fiber and protein, and low in the kinds of fats associated with diseases. They are also best for managing weight. Earlier chapters described the importance of each of the energy-yielding nutrients to health. Therefore, diets for weight management should provide all three within the ranges recommended by the DRI committee (see the margin note).

1 lb body fat = 3500 kcal.

To lose a pound a week, cut 500 kcal/day.

The DRI committee recommends:
- 45 to 65% kcalories from carbohydrate.
- 20 to 35% kcalories from fat.
- 10 to 35% kcalories from protein.

Selecting grapes with their high water content instead of raisins increases the volume and cuts the energy intake in half.

Even at the same weight and similar serving sizes, the fiber-rich broccoli delivers twice the fiber of the potatoes for about one-fourth the energy.

By selecting the water-packed tuna (on the right) instead of the oil-packed tuna, a person can enjoy the same amount for fewer kcalories.

Matthew Farruggio (all)

FIGURE 7-2 Energy Density
Decreasing the energy density (kcal/g) of foods allows a person to eat satisfying portions while still reducing energy intake. To lower energy density, select foods high in water or fiber and low in fat.

Wholesome, high-fiber, unprocessed or lightly processed foods offer bulk and satiety for fewer kcalories than smooth, quickly consumed refined foods. Thus choosing whole grains and fiber-rich vegetables in place of most refined grains and added fats and sugars benefits both weight and nutrition. Choose fats sensibly by avoiding sources of saturated and *trans* fats and including enough of the health-supporting fats (details in Nutrition in Practice 4) to provide satiety but not so much as to oversupply kcalories. Lean meats or other low-fat protein sources also provide satiety. Limit these foods but don't eliminate them.

Lower Energy Density As discussed earlier, to lower energy intake, people can choose smaller portion sizes, or they can reduce the energy density of the foods they eat. Research shows that eating satisfying portions of food that are low in energy density (such as fruits, vegetables, and broth-based soups) maintains satiety while reducing energy intake.[48] Foods containing substantial water or fiber and those low in fat help to lower a meal's energy density (see Figure 7-2). The Clinical Applications feature at the end of this chapter describes how to calculate the energy density of foods.

Milk and Milk Products Some research supports a role for calcium, especially calcium from milk and milk products, in improving body composition during weight loss.[49] kCalorie-restricted diets that include at least three servings of milk products a day have been shown to enhance the fat loss that accompanies weight loss.[50] In one study, researchers compared two groups of obese women. Both groups consumed a kcalorie-restricted diet, but one group ate three servings of yogurt each day, while the control group did not. Carbohydrate, fat, protein, and fiber were held constant. The yogurt group consumed 1100 milligrams of calcium each day, while the control group consumed about 500 milligrams. Both groups lost weight, but the yogurt group lost significantly more fat and less lean tissue than the control group.[51] In another study, increasing calcium intake from dairy products—without restricting kcalories—had no effect on body weight or body fat in healthy normal-weight women.[52] The authors of this study point out that consumption of fat-free and low-fat milk products did not decrease *or* increase body weight or body fat in young women. Therefore, weight-conscious women who avoid milk and milk products can be encouraged to include them each day to optimize bone mass and overall health.

Sugar and Alcohol A person trying to achieve or maintain a healthy weight needs to pay attention to sugar and alcohol, as well as fat. Using them for pleasure on occasion is compatible with health as long as most daily choices are of nutrient-dense foods.

Meal Spacing Three meals a day is standard in our society, but no law says you can't have four or five—be sure they are smaller, of course. People who eat small, frequent meals are reported to be successful at weight loss and maintenance. Make sure that mild hunger, not appetite, is prompting you to eat. Eat regularly, and eat before you become extremely hungry.

Adequate Water Learn to satisfy thirst with water. Water fills the stomach between meals and meets the fluid needs that were formerly met by eating extra food (remember that food provides water). Water also helps the GI tract adapt to a high-fiber diet.

PHYSICAL ACTIVITY

Either dieting or physical activity alone can produce some weight loss. Clearly, however, the combination is most effective.[53] People who combine diet and physical activity are more likely to lose more fat, retain more muscle, and regain less weight than those who only diet.[54] To prevent weight gain and support weight loss, the *Dietary Guidelines for Americans 2005,* the DRI committee, and other fitness experts advise 60 minutes of moderately intense physical activity (walking/jogging at 4 to 5 miles per hour) a day in addition to the activities of daily life.[55] Physical activity may also help counteract the negative effects of excess body weight on health.[56] For example, physical activity reduces abdominal obesity, and this change improves blood pressure, insulin resistance, and fitness of the heart and lungs, even without weight loss.[57]

Energy Expenditure Physical activity makes many contributions to weight loss and maintenance. For one thing, it directly increases energy output by the muscles and cardiovascular system. A 150-pound person walking a brisk 4 miles per hour for 30 minutes spends an extra 185 kcalories on that activity. A football player may spend several thousand extra kcalories on a day of heavy training.

BMR Activity also contributes to energy output in an indirect way—by speeding up basal metabolism.[58] It does this both immediately and over the long term. On any given day, after intense and prolonged exercise, basal metabolism remains elevated for several hours. Over the long term, daily vigorous activity for many weeks gradually shifts body composition toward more lean tissue, which is more active metabolically than fat tissue. The ongoing metabolic rate rises accordingly, and this makes a contribution toward continued weight loss or maintenance.

The raised metabolic rate continues for as long as the person is physically active on a regular basis. The more energy expended in metabolic activities, the greater the energy requirement. This means that a person can eat more without gaining weight.

Appetite Control Physical activity also helps to control appetite. People think that exercising will make them want to eat, but this is not entirely true. Yes, active people do have healthy appetites, but immediately after a good workout, most people do not feel like eating. They want to shower and may be thirsty, but they do not want to eat. The reason is that the body has responded to the stress of activity by mobilizing fuels from storage: glucose and fatty acids are abundant in the blood. At the same time, the body has suppressed its digestive functions. Hard physical work and eating are not compatible.

Psychological Benefits Physical activity helps especially to curb the inappropriate appetite that prompts a person to eat when bored, anxious, or depressed. Weight-management programs encourage people to go out and be active when they're tempted to eat but are not really hungry.

Being active—even if overweight—is healthier than being sedentary. With a BMI of 36, aerobics instructor Jennifer Portnick is considered obese, but her daily workout routine helps to keep her in good health.

Dietary Guidelines for Americans 2005:

- To help manage body weight and prevent gradual unhealthy body weight gain in adulthood: Engage in approximately 60 minutes of activity of moderate-to-vigorous intensity on most days of the week while not exceeding kcaloric intake requirements.

DRI for physical activity: 60 min/day (moderate intensity).

Benefits of physical activity in a weight-management program:

- Improved body composition.
- Favorable effects on disease risks.
- Short-term increase in energy expenditure (from exercise and from a slight rise in BMR).
- Long-term increase (slight) in BMR.
- Appetite control.
- Stress reduction and control of stress eating.
- Physical, and therefore psychological, well-being.
- High self-esteem.

Physical activity also helps to reduce stress. Since stress itself is a cue to inappropriate eating behavior for many people, activity can help here, too.

Activity offers still more psychological advantages. The fit person looks and feels healthy, and high self-esteem accompanies these benefits. High self-esteem tends to support a person's resolve to persist in a weight-control effort, rounding out a beneficial cycle.

Choosing Activities What kind of physical activity is best? People seeking to lose weight should choose activities that they enjoy and are willing to do regularly. Health care professionals frequently advise people who want to manage their body weight and lose fat to engage in activities of low-to-moderate intensity for a long duration, such as an hour-long fast-paced walk. The reasoning behind such advice is that people exercising at low-to-moderate intensity are likely to stick with their activity for longer times and are less likely to injure themselves. People who regularly engage in more *vigorous* physical activities (fast bicycling or endurance running, for example), however, have less body fat than those who engage in moderately intense activities. The conditioned body that is adapted to strenuous and prolonged aerobic activity uses more fat all day long, not just during activity. The bottom line on physical activity and weight and/or fat loss seems to be that total energy expenditure is the main factor, regardless of how a person does it.

In addition to activities such as walking or cycling, there are hundreds of ways to incorporate energy-spending activities into daily routines: take the stairs instead of the elevator, walk to the neighbor's apartment instead of making a phone call, and rake the leaves instead of using a blower. These activities burn only a few kcalories each, but over a year's time they become significant.

Spot Reducing People sometimes ask about "spot reducing." Unfortunately, no one part of the body gives up fat in preference to another. Fat cells all over the body release fat in response to demand, and the fat is then used by whatever muscles are active. No exercise can remove the fat from any one particular area—and, incidentally, neither can a massage machine that claims to break up fat on trouble spots.

Physical activity can help with trouble spots in another way, though. Strengthening muscles in a trouble area can help to improve their tone; stretching to gain flexibility can help with posture problems. Thus cardiorespiratory endurance, strength, and flexibility workouts all have a place in fitness programs.

BEHAVIOR AND ATTITUDE

Behavior-modification therapy provides ways to overcome barriers to making dietary changes and increasing physical activity. Behavior-modification therapy does more than help people decide which behaviors to change, it also teaches them how to change.[59] Behavior and attitude are important supporting factors in achieving and maintaining appropriate body weight and composition. Changing the behaviors of overeating and underexercising that lead to, and perpetuate, obesity requires time and effort. A person must commit to take action.

Becoming Aware of Behaviors A person who is aware of all the behaviors that create a problem has a head start on developing a solution. First, the person needs to establish a baseline (a record of present eating and physical activity behaviors) against which to measure future progress. It is best to keep a diary (see Figure 7-3) that includes the time and place of meals and snacks, the type and amount of foods eaten, the persons present when food is eaten, and a description of the individual's feelings when eating. The diary should also record physical activities: the kind, the intensity level, the duration, and the person's feelings about them. These entries will help the individual identify possible behaviors to change.

behavior modification: the changing of behavior by the manipulation of *antecedents* (cues or environmental factors that trigger behavior), the behavior itself, and *consequences* (the penalties or rewards attached to behavior).

FIGURE 7-3 Food Record
The entries in a food record should include the times and places of meals and snacks, the types and amounts of foods eaten, and a description of the individual's feelings when eating. The diary should also record physical activities: the kind, the intensity level, the duration, and the person's feelings about them.

Time	Place	Activity or food eaten	People present	Mood
10:30–10:40	School vending machine	6 peanut butter crackers and 12 oz. cola	by myself	Starved
12:15–12:30	Restaurant	Sub sandwich and 12 oz. cola	friends	relaxed & friendly
3:00–3:45	Gym	Weight training	work out partner	tired
4:00–4:10	Snack bar	Small frozen yogurt	by myself	OK

Making Small Changes The accompanying "How to" describes behavioral strategies to support weight management. A particularly attractive feature of these strategies is that they do not involve blaming oneself or putting oneself down—an important element in fostering self-esteem.

HOW TO   Apply Behavior Modification to Manage Body Fatness

1. Eliminate inappropriate cues:
   - Don't buy problem foods.
   - Eat only in one room at the designated time.
   - Shop when not hungry.
   - Avoid vending machines, fast-food restaurants, and convenience stores.
   - Turn off television, video games, and computers.
2. Suppress the cues you cannot eliminate:
   - Serve individual plates; don't serve "family style."
   - Make small portions look large by spreading them over the plate.
   - Create obstacles to consuming problem foods—wrap them and freeze them, making them less quickly accessible.
   - Control deprivation; plan and eat regular meals.
   - Plan the time spent in sedentary activities, such as watching television or using a computer—don't use these activities just to fill time.
3. Strengthen cues to appropriate behaviors:
   - Share appropriate foods with others.
   - Store appropriate foods in convenient spots in the refrigerator.
   - Learn appropriate portion sizes.
   - Plan appropriate snacks.
   - Keep sports and play equipment by the door.
4. Repeat desired behaviors:
   - Slow down eating—put down utensils between bites.
   - Always use utensils.
   - Leave some food on your plate.
   - Move more—shake a leg, pace, stretch often.
   - Join groups of active people and participate.
5. Arrange negative consequences for negative behavior:
   - Ask that others respond neutrally to your deviations (make no comments—even negative attention is a reward).
   - If you slip, don't punish yourself.
6. Reward yourself personally and immediately for positive behaviors:
   - Buy tickets to sports events, movies, concerts, or other nonfood amusement.
   - Indulge in a new small purchase.
   - Get a massage; buy some flowers.
   - Take a hot bath; read a good book.
   - Treat yourself to a lesson in a new active pursuit such as horseback riding, handball, or tennis.
   - Praise yourself; visit friends.
   - Nap; relax.

Maintaining Weight Finally, be aware that it can be hard to maintain weight loss. On arriving at the goal weight after months of self-discipline and new-habit formation, the victorious weight loser must not "celebrate" by resuming old eating habits. Membership in an ongoing weight-control organization and regular, continued physical activity can provide indispensable support for the formerly overweight person who wants to remain trim.

Personal Attitude For many people, overeating and being overweight may have become an integral part of their identity. Changing diet and activity behaviors without attention to a person's self-concept invites failure.

Many people overeat to cope with the stresses of life. To break out of that pattern, they must first identify the particular stressors that trigger their urges to overeat. Then, when faced with these situations, they must learn to practice problem-solving skills. When the problems that trigger the urge to overeat are dealt with in alternative ways, people may find that they eat less. The message is that sound emotional health supports the ability to take care of health in all ways—including nutrition, weight management, and fitness.

> **REVIEW NOTES**
>
> A person who adopts a lifelong "eating plan for good health" rather than a "diet for weight loss" will be more likely to keep the lost weight off. Table 7-4 offers several tips for successful weight management.
>
> Physical activity should be an integral part of a weight-management program.
>
> Physical activity can increase energy expenditure, improve body composition, help control appetite, reduce stress and stress eating, and enhance physical and psychological well-being.
>
> Behavior modification provides ways to overcome barriers to successful weight management.

Underweight

Underweight is far less prevalent than overweight, affecting no more than 5 percent of U.S. adults. The health risks associated with underweight are fewer than those that accompany overweight. Both underweight women and those who have lost a significant amount of weight, however, are more susceptible to osteoporosis. Underweight women may become infertile or may give birth to unhealthy infants. An underweight woman can improve her chances of bearing a healthy infant by gaining weight prior to conception, during pregnancy, or both.

Underweight becomes more hazardous when accompanied by undernutrition. An inadequate supply of nutrients and energy leaves the body underprepared to handle its many metabolic and physical tasks. A person without reserves has a particularly difficult battle when faced with medical stresses such as surgery or the wasting diseases of cancer and AIDS. Thus underweight people are urged to gain lean tissue and body fat (as an energy reserve) and to acquire protective amounts of all the nutrients that can be stored.

An extreme underweight condition known as anorexia nervosa is sometimes seen in young people who exercise unreasonable self-denial in order to control their weight. They go to such extremes that they become severely undernourished and underweight. The distinguishing feature of a person with anorexia nervosa, as opposed to other thin people, is that the starvation is intentional. Anorexia nervosa is a major eating disorder seen in our society today. Another is bulimia nervosa—compulsive overeating, usually with purging. Eating disorders are the subject of the Nutrition in Practice that follows this chapter.

TABLE 7-4	Weight Management Strategies

In General

Focus on healthy eating and activity habits, not on weight losses or gains.

Adopt reasonable expectations about health and fitness goals and about how long it will take to achieve them.

Make nutritional adequacy a high priority.

Learn, practice, and follow a healthful eating plan for the rest of your life.

Participate in some form of physical activity regularly.

Adopt permanent lifestyle changes to achieve and maintain a healthy weight.

For Weight Loss

Energy out should exceed energy in by about 500 kcalories/day. Increase your physical activity enough to spend more energy than you consume from foods.

Emphasize foods with a low energy density and a high nutrient density.

Eat small portions. Share a restaurant meal with a friend or take home half for lunch tomorrow.

Eat slowly.

Limit high-fat foods. Make legumes, whole grains, vegetables, and fruits central to your diet plan.

Limit low-fat treats to the serving size on the label.

Limit concentrated sweets and alcoholic beverages.

Drink a glass of water before you begin to eat and another while you eat. Drink plenty of water throughout the day (8 glasses or more a day).

Keep a record of diet and exercise habits; it reveals problem areas, the first step toward improving behaviors.

Learn alternative ways to deal with emotions and stresses.

Attend support groups regularly or develop supportive relationships with others.

For Weight Gain

Energy in should exceed energy out by at least 500 kcalories/day. Increase your food intake enough to store more energy than you spend in exercise. Exercise and eat to build muscles.

Expect weight gain to take time (1 pound per month would be reasonable).

Emphasize energy-dense foods.

Eat at least three meals a day.

Eat large portions of foods and expect to feel full.

Eat snacks between meals.

Drink plenty of juice and milk.

REVIEW NOTES

Both the incidence of underweight and the health problems associated with it are less prevalent than overweight and its associated problems.

Strategies for Weight Gain

Weight gain, like weight loss, is an individual matter. People who are healthy at their present weights may stay there; those who are unhealthy might try to gain.

Some people are unalterably thin by reasons of genetics or early physical influences. Those who wish to gain weight for appearance's sake or to improve athletic performance should be aware that a healthful weight can be achieved only through physical activity, particularly strength training, combined with a high energy intake. Eating many high-kcalorie foods can bring about weight gain, but it will be mostly fat, and this can be as detrimental to health as being slightly underweight. In an athlete, such a weight gain can impair performance. Therefore, in weight gain, as in weight loss, physical activity is an essential component of a sound plan.

Physical Activity to Build Muscles The person who wants to gain weight should use **weight training** primarily. As activity is increased, energy intake must be increased to support that activity. Eating extra food will then support a gain of both muscle and fat.

Energy-Dense Foods Energy-dense foods (the very ones eliminated from a successful weight-loss diet) hold the key to weight gain. Pick the highest-kcalorie items from each food group—that is, milk shakes instead of fat-free milk, peanut butter instead of lean meat, avocados instead of cucumbers, and whole-wheat muffins instead of whole-wheat bread. Because fat contains more than twice as many kcalories per teaspoon as sugar does, fat adds kcalories without adding much bulk.

Be aware that health experts recommend a moderate-fat diet for the general U.S. population because the general population is overweight and at risk for heart disease. Consumption of excessive fat is not healthy for most people, of course, but may be essential for an underweight individual who needs to gain weight. An underweight person who is physically active and eating a nutritionally adequate diet can afford a few extra kcalories from fat.

Three Meals Daily People wanting to gain weight should eat at least three hearty meals a day. Many people who are underweight have simply been too busy (sometimes for months) to eat enough to gain or maintain weight. Therefore, they need to make meals a priority and plan them in advance. Taking time to prepare and eat each meal can help, as can learning to eat more food within the first 20 minutes of a meal before you begin to feel full. Another suggestion is to eat meaty appetizers or the main course first and leave the soup or salad until later.

Large Portions It is also important for the underweight person to learn to eat more food at each meal: have two sandwiches for lunch instead of one, drink milk from a larger glass, and eat cereal from a larger bowl. Expect to feel full. Most underweight individuals are accustomed to small quantities of food. When they begin eating significantly more, they feel uncomfortable. This is normal and passes over time.

Extra Snacks Because a substantially higher energy intake is needed each day, in addition to eating more food at each meal, it is necessary to eat more frequently. Between-meal snacking offers a solution. For example, a student might make three sandwiches in the morning and eat them between classes in addition to the day's three regular meals.

Juice and Milk Beverages provide an easy way to increase energy intake. Consider that 6 cups of cranberry juice add almost 1000 kcalories to the day's intake.

weight training: the use of free weights or weight machines to provide resistance for developing muscle strength and endurance; also called *resistance training*. A person's own body weight may also be used to provide resistance as when a person does push-ups, pull-ups, or abdominal crunches.

Kcalories can be added to milk by mixing in powdered milk or packets of instant breakfast.

For people who are underweight due to illness, concentrated liquid formulas are often recommended because a weak person can swallow them easily. A registered dietitian can recommend high-protein, high-kcalorie formulas to help the underweight person maintain or gain weight. Used in addition to regular meals, these formulas can help considerably.

REVIEW NOTES

To gain weight, a person must train physically and increase energy intake by selecting energy-dense foods, eating regular meals, taking larger portions, and consuming extra snacks and beverages. Review the tips for weight gain in Table 7-4.

SELF CHECK

1. Two causes of obesity in humans are:
   a. set-point theory and BMI.
   b. genetics and physical inactivity.
   c. genetics and low-carbohydrate diets.
   d. mineral imbalances and fat cell imbalance.

2. The protein produced by the fat cells under the direction of the *ob* gene is called:
   a. leptin.
   b. orlistat.
   c. sibutramine.
   d. lipoprotein lipase.

3. All of the following describe the behavior of fat cells *except:*
   a. the number decreases when fat is lost from the body.
   b. the storage capacity for fat depends on both cell number and cell size.
   c. the size is larger in obese people than in normal-weight people.
   d. the number increases several-fold during the growth years and tapers off when adult status is reached.

4. The obesity theory that suggests the body chooses to be at a specific weight is the:
   a. fat cell theory.
   b. enzyme theory.
   c. set-point theory.
   d. external cue theory.

5. The biggest problem associated with the use of prescription drugs in the treatment of obesity is:
   a. cost
   b. the necessity for long-term dosage.
   c. ineffectiveness.
   d. adverse side effects.

6. Television watching contributes to obesity for all of the following reasons, *except:*
   a. it promotes inactivity.
   b. it promotes between-meal snacking.
   c. it replaces time that could be spent eating.
   d. it gives high exposure to energy-dense foods featured in the advertisements.

7. What is the best approach to weight loss?
   a. Avoid foods containing carbohydrates.
   b. Eliminate all fats from the diet and decrease water intake.
   c. Greatly increase protein intake to prevent body protein loss.
   d. Reduce daily energy intake and increase energy expenditure.

8. To help prevent body fat gain, the DRI committee suggests daily, moderately intense, physical activities totaling:
   a. 20 minutes.
   b. 60 minutes.
   c. 1½ hours.
   d. 3 hours.

9. Suggestions to change behaviors for successful weight control include:
   a. shop only when hungry.
   b. eat in front of the television for distraction.
   c. learn appropriate portion sizes.
   d. eat quickly.

10. Which strategy would *not* help an underweight person to gain weight?
    a. Exercise.
    b. Drink plenty of water.
    c. Eat snacks between meals.
    d. Eat large portions of foods.

Answers to these questions can be found in Appendix H.

CLINICAL APPLICATIONS

1. Chapter 1 discussed the nutrient density of foods—their nutrient contribution per kcalorie. Another way to evaluate foods is to consider their energy density—their energy contribution per gram:

   • A carrot weighing 72 grams delivers 31 kcalories.

   • To calculate the energy density, divide kcalories by grams: 31 kcalories divided by 72 grams = 0.43 kcalories per gram.

2. Do the same for french fries weighing 50 grams and contributing 167 kcalories: 167 kcalories divided by 50 grams = _____ kcalories per gram.

   • The more kcalories per gram, the greater the energy density.

   • Which food is more energy dense? The conclusion is no surprise, but understanding the mathematics may offer valuable insight into the concept of energy density.

   • French fries are more energy dense, providing 3.34 kcalories per gram. They provide more energy per gram—and per bite.

Matthew Farruggio

3. Considering a food's energy density is especially useful in planning diets for weight management. Foods with a high energy density can help with weight gain, whereas those with a low energy density can help with weight loss. Give some examples of foods that you might suggest for a client who wants to gain weight and some that might be appropriate for a client who is trying to lose weight.

NUTRITION ON THE NET

For further study of the topics in this chapter, access these websites.

Find updates and quick links to these and other nutrition-related sites at our website: **www.wadsworth.com/ nutrition**

Search for obesity at: **www.ilsi.org**

Information on a variety of obesity topics is available at the American Obesity Association website: **www.obesity.org/**

Search for obesity at: **www.eatright.org/**

Peruse the offerings of the Division of Nutrition and Physical Activity, National Center for Chronic Disease Prevention and Health Promotion, at this website: **www.cdc.gov/nccdphp/ dnpa**

Many materials to help teach others about obesity are available at the Weight Control Information Network: **www .niddk.nih.gov/health/nutrit/win.htm**

Learn about the drugs used for weight loss from the Center for Drug Evaluation and Research: **www.fda.gov/cder**

Learn about weight control and the WIN program from the Weight-Control Information Network: **www.niddk.nih.gov/ health/nutrit/win.htm**

Visit weight-loss support groups, such as Take Off Pounds Sensibly (TOPS), Overeaters Anonymous (OA), and Weight Watchers: **www.tops.org**, **www.oa.org**, and **www.weight watchers.com**

Read the latest materials on obesity diagnosis and treatment here: **www.nhlbi.nih.gov/guidelines/obesity/ practgde.htm**

The Partnership for Healthy Weight Management offers practical advice on starting a weight-loss program at this site: **www.consumer.gov/weightloss/**

NOTES

[1]H. N. Lyon and J. N. Hirschhorn, Genetics of common forms of obesity: A brief overview, *American Journal of Clinical Nutrition* 82 (2005): 215S–217S; R. J. F. Loos and T. Rankinen, Gene-diet interactions on body weight changes, *Journal of the American Dietetic Association* 105 (2005): 29S–34S; S. Tholin and coauthors, Genetic and environmental influences on eating behavior: The Swedish Young Male Twins Study, *American Journal of Clinical Nutrition* 81 (2005): 564–569; R. J. F. Loos and C. Bouchard, Obesity—Is it a genetic disorder? *Journal of Internal Medicine* 254 (2003): 401–425.

[2]Lyon and Hirschhorn, 2005; Loos and Rankinen, 2005; Tholin and coauthors, 2005; Loos and Bouchard, 2003.

[3]D. E. Cummings and M. W. Schwartz, Genetics and pathophysiology of human obesity, *Annual Review of Medicine* 54 (2003): 453–471.

[4]V. Popovic and L. H. Duntas, Brain somatic cross-talk: Ghrelin, leptin, and ultimate challenges of obesity, *Nutritional Neuroscience* 8 (2005): 1–5.

[5]Popovic and Duntas, 2005.

[6]M. S. Westerterp-Plantenga and coauthors, Effects of weekly administration of pegylated recombinant human OB protein on appetite profile and energy metabolism in obese men, *American Journal of Clinical Nutrition* 74 (2001): 426–434.

[7]G. Fruhbeck, The adipose tissue as a source of vasoactive factors, *Current Medicinal Chemistry Cardiovascular and Hematological Agents* 3 (2004): 197–208; R. L. Bradley, K. A. Cleveland, and B. Cheatham, The adipocyte as a secretory organ: Mechanism of vesicle transport and secretory pathways, *Recent Progress in Hormone Research* 56 (2001): 329–358; R. B. S. Harris, Leptin—Much more than a satiety signal, *Annual Review of Nutrition* 20 (2000): 45–75.

[8]P. Fietta, Focus on leptin, a pleiotropic hormone, *Minerva Medica* 96 (2005): 65–75; J. Harvey and M. L. Ashford, Leptin in the CNS: Much more than a satiety signal, *Neuropharmacology* 44 (2003): 845–854; S. Takeda, F. Elefteriou, and G. Karsenty, Common endocrine control of body weight, reproduction, and bone mass, *Annual Review of Nutrition* 23 (2003): 403–411.

[9]D. E. Cummings, K. E. Foster-Schubert, and J. Overduin, Ghrelin and energy balance: Focus on current controversies, *Current Drug Targets* 6 (2005): 153–169.

[10]Cummings, Foster-Schubert, and Overduin, 2005; Popovic and Duntas, 2005; J. Eisenstein and A. Greenberg, Ghrelin: Update 2003, *Nutrition Reviews* 61 (2003): 101–104.

[11]Cummings, Foster-Schubert, and Overduin, 2005; J. Orr and B. Davy, Dietary influences on peripheral hormones regulating energy intake: Potential applications for weight management, *Journal of the American Dietetic Association* 105 (2005): 1115–1124.

[12]H. S. Callahan and coauthors, Postprandial suppression of plasma ghrelin level is proportional to ingested caloric load but does not predict intermeal interval in humans, *Journal of Clinical Endocrinology and Metabolism* 89 (2004): 1319–1324.

[13]Cummings, Foster-Schubert, and Overduin, 2005.

[14]Popovic and Duntas, 2005; M. Tanaka and coauthors, Habitual binge/purge behavior influences circulating ghrelin levels in eating disorders, *Journal of Psychiatric Research* 37 (2003): 17–22.

[15]M. Tanaka and coauthors, Effect of nutritional rehabilitation on circulating ghrelin and growth hormone levels in patients with anorexia nervosa, *Regulatory Peptides* 122 (2004): 163–168; V. Tolle and coauthors, Balance in ghrelin and leptin plasma levels in anorexia nervosa patients and constitutionally thin women, *Journal of Clinical Endocrinology and Metabolism* 88 (2003): 109–116.

[16]A. M. Haqq and coauthors, Serum ghrelin levels are inversely correlated with body mass index, age, and insulin concentrations in normal children and are markedly increased in Prader-Willi syndrome, *Journal of Clinical Endocrinology and Metabolism* 88 (2003): 174–178; A. DelParigi and coauthors, High circulating ghrelin: A potential cause for hyperphagia and obesity in Prader-Willi syndrome, *Journal of Clinical Endocrinology and Metabolism* 87 (2002): 5461–5464.

[17]N. Tentolouris and coauthors, Differential effects of high-fat and high-carbohydrate content isoenergetic meals on plasma active ghrelin concentrations in lean and obese women, *Hormone and Metabolic Research* 36 (2004): 559–563; P. J. English and coauthors, Food fails to suppress ghrelin levels in obese humans, *Journal of Clinical Endocrinology and Metabolism* 87 (2002): 2984.

[18]D. E. Cummings and coauthors, Plasma ghrelin levels after diet-induced weight loss or gastric bypass surgery, *New England Journal of Medicine* 346 (2002): 1623–1630.

[19]Cummings, Foster-Schubert, and Overduin, 2005.

[20]Loos and Rankinen, 2005.

[21]L. R. Young and M. Nestle, The contribution of expanding portion sizes to the US obesity epidemic, *American Journal of Public Health* 92 (2002): 246–249.

[22]B. J. Rolls and coauthors, Increasing the portion size of a packaged snack increases energy intake in men and women, *Appetite* 42 (2004): 63–69.

[23]S. J. Nielsen and B. M. Popkin, Patterns and trends in food portion sizes, 1977–1998, *Journal of the American Medical Association* 289 (2003): 450–453.

[24]Young and Nestle, 2002.

[25]S. A. Bowman and B. T. Vinyard, Fast food consumption of U.S. adults: Impact on energy and nutrient intakes and overweight status, *Journal of the American College of Nutrition* 23 (2004): 163–168.

[26]B. J. Rolls, A. Drewnowski, and J. H. Ledikwe, Changing the energy density of diet as a strategy for weight management, *Journal of the American Dietetic Association* 105 (2005): 98S–103S.

[27]J. A. Ello-Martin, J. H. Ledikwe, and B. J. Rolls, The influence of food portion size and energy density on energy intake: Implications for weight management, *American Journal of Clinical Nutrition* 82 (2005): 236S–241S.

[28]J. M. Jakicic and A. D. Otto, Physical activity considerations for the treatment and prevention of obesity, *American Journal of Clinical Nutrition* 82 (2005): 226S–229S; L. L. Frank and coauthors, Effects of exercise on metabolic risk variables in overweight postmenopausal women, *Obesity Research* 13 (2005): 615–625; D. L. Thompson, J. Rakow, and S. M. Perdue, Relationship between accumulated walking and body composition in middle-aged women, *Medicine and Science in Sports and Exercise* 36 (2004): 911–914; R. Jurca and coauthors, Associations of muscle strength and aerobic fitness with metabolic syndrome in men, *Medicine and Science in Sports and Exercise* 36 (2004): 1301–1307; R. G. Ketelhut, I. W. Franz, and J. Scholze, Regular exercise as an effective approach in antihypertensive therapy, *Medicine and Science in Sports and Exercise* 36 (2004): 4–8; F. B. Hu and coauthors, Adiposity as compared with physical activity in predicting mortality among women, *New England Journal of Medicine* 351 (2004): 2694–2703; S. L. Wong and coauthors, Cardiorespiratory fitness is associated with lower abdominal fat independent of body mass index, *Medicine and Science in Sports and Exercise* 36 (2004): 286–291.

[29]F. B. Hu and coauthors, Television watching and other sedentary behaviors in relation to risk of obesity and type 2 diabetes mellitus in women, *Journal of the American Medical Association* 289 (2003): 1785–1791.

[30]R. R. Wing and S. Phelan, Long-term weight loss maintenance, *American Journal of Clinical Nutrition* 82 (2005): 222S–225S; R. R. Wing and J. O. Hill, Successful weight loss maintenance, *Annual Review of Nutrition* 21 (2001): 323–341.

[31]NBJ's annual overview of the nutrition industry VII, *Nutrition Business Journal* 7 (2002): 1–10, as cited in P. R. Thomas, Dietary supplements for weight loss? *Nutrition Today* 40 (2005): 6–12.

[32]H. M. Blanck, L. K. Khan, and M. K. Serdula, Use of nonprescription weight loss products: Results from a multistate survey, *Journal of the American Medical Association* 286 (2001): 930–935.

[33]P. R. Thomas, Dietary supplements for weight loss, *Nutrition Today* 40 (2005): 6–12; M. H. Pittler and E. Ernst, Dietary supplements for body-weight reduction: A systematic review, *American Journal of Clinical Nutrition* 79 (2004): 529–536.

[34]P. G. Shekelle and coauthors, Efficacy and safety of ephedra and ephedrine for weight loss and athletic performance: A meta-analysis, *Journal of the American Medical Association* 289 (2003): 1537–1545.

[35]G. Egger, D. Cameron-Smith, and R. Stanton, The effectiveness of popular, non-prescription weight loss supplements, *Medical Journal of Australia* 171 (1999): 604–608; A. Sarubin, *The Health Professional's Guide to Popular Dietary Supplements* (Chicago: The American Dietetic Association, 1999); S. Foster and V. E. Tyler, *Tyler's Honest Herbal—A Sensible Guide to the Use of Herbs and Related Remedies* (New York: Haworth Herbal Press, 1999).

[36]F. Greenway, Another type of intervention: Treating obesity with medication, *Journal of the American Dietetic Association* 105 (2005): 895–898; Position of the American Dietetic Association: Weight management, *Journal of the American Dietetic Association* 102 (2002): 1145–1155.

[37]S. B. Moyers, Medications as adjunct therapy for weight loss: Approved and off-label agents in use, *Journal of the American Dietetic Association* 105 (2005): 948–959.

[38]Greenway, 2005.

[39]Greenway, 2005; Position of the American Dietetic Association, 2002.

[40]Moyers, 2005.

[41]Cummings, Foster-Schubert, and Overduin, 2005.

[42]G. L. Blackburn, Solutions in weight control: Lessons from gastric surgery, *American Journal of Clinical Nutrition* 82 (2005): 248S–252S; H. Buchwald and coauthors, Bariatric surgery: A systematic review and meta-analysis, *Journal of the American Medical Association* 292 (2004): 1724–1737.

[43]Buchwald and coauthors, 2004.

[44]Blackburn, 2005.

[45]R. E. Brolin, Bariatric surgery and long-term control of morbid obesity, *Journal of the American Medical Association* 288 (2002): 2793–2796.

[46]National Heart, Lung, and Blood Institute, National Institutes of Health, *The Practical Guide: Identification, Evaluation, and Treatment of Overweight and Obesity in Adults,* NIH publication no. 00-4084 (Washington, D.C.: Government Printing Office, 2000), p. 23.

[47]National Heart, Lung, and Blood Institute, National Institutes of Health, 2000, p. 3.

[48]Ello-Martin, Ledikwe, and Rolls, 2005.

[49]M. B. Zemel, Role of calcium and dairy products in energy partitioning and weight management, *American Journal of Clinical Nutrition* 79 (2004): 907S–912S; R. J. Loos and coauthors, Calcium intake is associated with adiposity in black and white men and white women of the *Heritage* Family Study, *Journal of Nutrition* 134 (2004): 1772–1778.

[50]B. Zemel and coauthors, Increasing dairy calcium intake reduces adiposity in obese African-American adults, *Circulation* (supplement) 106 (2002): II-610 (abstract).

[51]M. B. Zemel and coauthors, Dairy augmentation of total and central fat loss in obese subjects, *International Journal of Obesity* 88 (2003): 388–390.

[52]C. W. Gunther and coauthors, Dairy products do not lead to alterations in body weight or fat mass in young women in a 1-year intervention, *American Journal of Clinical Nutrition* 81 (2005): 751–756.

[53]Jakicic and Otto, 2005; Wing and Phelan, 2005; Position of the American Dietetic Association, 2002.

[54]Wing and Phelan, 2005; American College of Sports Medicine Position Stand, Appropriate intervention strategies for weight loss and prevention of weight regain for adults, *Medicine and Science in Sports and Exercise* 33 (2001): 2145–2156.

[55]Standing Committee on the Scientific Evaluation of Dietary Reference Intakes, Food and Nutrition Board, Institute of Medicine, *Dietary Reference Intakes for Energy, Carbohydrate, Fiber, Fat, Fatty Acids, Cholesterol, Protein, and Amino Acids* (Washington, D.C.: National Academies Press, 2005), p. 880; U.S. Department of Agriculture and U.S. Department of Health and Human Services, *2005 Dietary Guidelines Committee Report,* 2005, available online at **www.healthierus.gov/dietaryguidelines/**; S. N. Blair, M. J. LaMonte, and M. Z. Nichaman, The evolution of physical activity recommendations: How much is enough? *American Journal of Clinical Nutrition* 79 (2004): 913S–920S; American College of Sports Medicine, Position Stand, 2001.

[56]Jakicic and Otto, 2005.

[57]Wong and coauthors, 2004; Ketelhut, Franz, and Scholze, 2004.

[58]Standing Committee on the Scientific Evaluation of Dietary Reference Intakes, 2005, p. 115.

[59]G. D. Foster, A. P. Makris, and B. A. Bailer, Behavioral treatment of obesity, *American Journal of Clinical Nutrition* 82 (2005): 230S–235S.

Eating Disorders

More than 5 million people in the United States, primarily girls and young women, suffer from some form of an **eating disorder** (see the glossary on p. 185 for the relevant terms).[1] About 5 percent of females and 1 percent of males have **anorexia nervosa, bulimia nervosa,** or **binge eating disorder.** Many more suffer from other related conditions that do not meet the strict criteria for anorexia nervosa, bulimia nervosa, or binge eating disorder but still imperil a person's well-being. Characteristics of disordered eating such as restrained eating, binge eating, purging, fear of fatness, and distortion of body image are common, especially among young middle-class girls. In most other societies, these behaviors and attitudes are much less prevalent.

Why do so many young people in our society suffer from eating disorders?

Excessive pressure to be thin is at least partly to blame. In one national survey of more than 6700 adolescents in grades 5 through 12, almost half of the girls and a fifth of the boys reported having dieted to lose weight.[2] By making thinness the ideal, society pushes people to view a healthy body of normal weight as too fat. Healthy people then take unhealthy actions to lose weight. Severe restriction of food intake may create intense hunger that leads to binges. Research confirms this theory, showing that unhealthy or dangerous diets often precede binge eating in adolescent girls.[3] Energy restriction followed by bingeing can set in motion a pattern of **weight cycling,** which may make weight loss and maintenance more difficult over time.

People who attempt extreme weight loss are dissatisfied with their bodies to begin with; they may also be depressed or suffer social anxiety. As weight loss becomes more and more difficult, psychological problems worsen, and the likelihood of developing full-blown eating disorders intensifies.

People with anorexia nervosa suffer from an extreme preoccupation with weight loss that seriously endangers their health and even their lives. People with bulimia nervosa engage in episodes of binge eating alternating with periods of severe dieting or self-starvation. Some bulimics also follow binge eating with self-induced vomiting, laxative abuse, or diuretic abuse to "undo the damage."

Are there other groups, besides girls and young women, who are vulnerable to anorexia nervosa and bulimia nervosa?

Yes. Athletes who participate in sports that emphasize leanness are at special risk for developing eating disorders.[4] Athletes must often meet stringent weight requirements to compete in their sport. Many athletes report that they engage in behaviors that are typical of people with eating disorders. Female competitors often report being terrified of becoming fat, being obsessed with food, and using laxatives in attempting to control weight. Dancers, jockeys, wrestlers, distance runners, bodybuilders, divers, figure skaters, gymnasts, and others whose body weight and appearance are frequently judged in comparison with an "ideal" are especially prone to develop problems.[5] These athletes may engage in extreme weight-loss practices such as overtraining, prolonged fasting, vomiting, taking diet pills, and using steam baths and saunas to induce sweating.

Men account for about 1 in 20 cases in the general population, but among male athletes and dancers, eating disorders are much more common. Male teenagers normally average about 15 percent of body weight as fat, but some high school athletes strive to carry only 5 percent or so of their body weight as fat.

Wrestlers, for example, are required to "make weight" to compete in the lowest weight class to face the smallest possible opponents. To that end, wrestlers starve themselves, don rubber suits, sweat in steam rooms, and take diuretics to shed water weight before weighing in for competition. These practices were responsible for the deaths of three college athletes in recent years and have caused untold misery and harm to many others. Athletes engaging in these practices actually compromise their athletic abilities. The diminished anaerobic strength, reduced endurance, decreased oxygen capacity, and general weakness caused by food deprivation and dehydration can impair performance, an effect lasting days after food and water are replenished.

Even among athletes, however, women are most vulnerable to developing eating disorders. Many female athletes appear healthy, but in fact may easily develop the three interrelated components of the **female athlete triad:** disordered eating, **amenorrhea** (the absence of three or more consecutive menstrual cycles), and osteoporosis.

How does the female athlete triad develop?

Many athletic women engage in self-destructive eating behaviors (disordered eating) because they and their coaches have adopted unsuitable weight standards. An athlete's body must be heavier for height than a nonathlete's body because the athlete's bones and muscles are denser. Weight standards that may be appropriate for others are inappropriate for athletes. Measures such as skinfold measures yield more useful information about body composition.

Many young female athletes severely restrict energy intakes to improve performance, enhance the aesthetic appeal of their performance, or meet the weight guidelines of their specific sports. They fail to realize that the loss of lean tissue that accompanies energy restriction actually impairs their physical performance. Risk factors for the female athlete triad include the following:

• Young age (adolescence).
• Pressure to excel at a chosen physical activity.

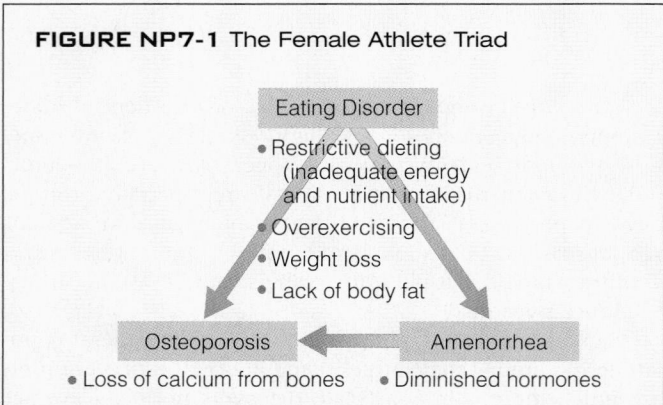

FIGURE NP7-1 The Female Athlete Triad

Eating Disorder
- Restrictive dieting (inadequate energy and nutrient intake)
- Overexercising
- Weight loss
- Lack of body fat

Osteoporosis
- Loss of calcium from bones

Amenorrhea
- Diminished hormones

- Focus on achieving or maintaining an "ideal" body weight or body fat percentage.
- Participation in endurance sports or competitions that judge performance on aesthetic appeal such as gymnastics, figure skating, or dance.
- Dieting at an early age.
- Unsupervised dieting.

As for amenorrhea, its prevalence among premenopausal women in the United States is about 2 to 5 percent overall, but among female athletes it may be as high as 66 percent. Contrary to previous notions, amenorrhea is not a normal adaptation to strenuous physical training: it is a symptom of something going wrong. Amenorrhea is characterized by low blood estrogen, infertility, and often bone mineral losses.

In general, weight-bearing physical activity, dietary calcium, and the hormone estrogen protect against the bone loss of osteoporosis, but in women with disordered eating and amenorrhea, strenuous activity may impair bone health.[6] Vigorous training combined with low food energy intakes and other life stresses may promote bone loss even without obvious menstrual irregularities.[7] Such bone losses may increase the risks of stress fractures today and osteoporosis in later life (see Figure NP7-1). Many underweight young athletes have decreased bone density, similar to that of 50- to 60-year-old women, when they should have dense, strong bones. Young athletes should be encouraged to consume at least 1300 milligrams of calcium each day, to eat nutrient-dense foods, and to obtain enough food energy to cover the energy expended in physical activity.

What can be done to prevent eating disorders in athletes and dancers?

To prevent eating disorders in athletes and dancers, both the performers and their coaches must be educated about links between inappropriate body weight ideals, improper weight-loss techniques, eating disorder development, adequate nutrition, and safe weight-control methods. Coaches and dance instructors should never encourage unhealthy weight loss to qualify for competition or to conform with distorted artistic ideals. Frequent weighings can push young people who are striving to lose weight into a cycle of starving to confront the scale, then bingeing uncontrollably afterwards. The erosion of self-esteem that accompanies these events can interfere with the normal psychological development of the teen years and set the stage for serious problems later on.

The accompanying "How to" provides some suggestions to help athletes and dancers protect themselves against developing eating disorders. The next sections describe eating disorders that anyone, athlete or nonathlete, may experience.

HOW TO   Combat Eating Disorders

The following guidelines may be useful in combating eating disorders:

- Never restrict food intakes to below the amounts suggested for adequacy by the USDA Food Guide.
- Eat frequently. People often do not eat frequent meals because of time constraints, but eating can be incorporated into other activities, such as snacking while studying or commuting. The person who eats frequently never gets so hungry as to allow hunger to dictate food choices.
- If not at a healthy weight, establish a reasonable weight goal based on a healthy body composition.
- Allow a reasonable time to achieve the goal. A reasonable loss of excess fat can be achieved at the rate of about 1 percent of body weight per week.

- Establish a weight-maintenance support group with people who share interests.

Specific guidelines for athletes and dancers include:
- Replace weight-based goals with performance-based goals.
- Remember that eating disorders impair physical performance. Seek confidential help in obtaining treatment if needed.
- Restrict weight-loss activities to the off-season.
- Focus on proper nutrition as an important facet of your training—as important as proper technique.

What are the characteristics of anorexia nervosa?

Most anorexia nervosa victims are females who come from middle- or upper-class families. Family patterns often include parents who oppose one another's authority and who vacillate between defending and condemning the anorexic child's behavior, confusing the child and disrupting normal parental control. The family values achievement and outward appearances more than an inner sense of self-worth and self-actualization.

The person with anorexia nervosa is often a perfectionist who works hard to please her parents. She may identify so strongly with her parents' ideals and goals for her that she sometimes feels she has no identity of her own. She is respectful of authority but sometimes feels like a robot, and she may act that way, too: polite but controlled, rigid, and unspontaneous. She earnestly desires to control her own destiny, but she feels controlled by others. When she does not eat, she gains control.

Although families of children with anorexia nervosa often have problems, blame is a useless concept, and parents may suffer deeply from being blamed for their child's illness. In truth, no one knows what causes anorexia nervosa, and it may turn out that the cause is a physical one, and not related to parenting. Rather than judging parents, a more useful tactic is to identify the family's strong points and resources and to prepare them for the job of helping their ill child benefit from treatment.

How does a person know when dieting is going too far?

When a person loses weight to well below the average for her height and is no longer slim, but too slim, and still doesn't stop, she has gone too far. Regardless of how thin she is, she looks in the mirror and sees herself as fat. Central to the

© Tony Freeman/PhotoEdit, Inc.

Women with anorexia nervosa see themselves as fat, even when they are dangerously underweight.

diagnosis of anorexia nervosa is a distorted body image that overestimates body fatness. Table NP7-1 shows the criteria that professionals use to diagnose anorexia nervosa. Anorexia nervosa resembles an addiction. The characteristic behavior is obsessive and compulsive. Before drawing conclusions about someone who is extremely thin or who eats very little, remember that diagnosis of anorexia nervosa requires professional assessment.

TABLE NP7-1 Criteria for Diagnosis of Anorexia Nervosa

A person with anorexia nervosa demonstrates the following:
A. Refusal to maintain body weight at or above a minimal normal weight for age and height, e.g., weight loss leading to maintenance of body weight less than 85 percent of that expected; or failure to make expected weight gain during period of growth, leading to body weight less than 85 percent of that expected.
B. Intense fear of gaining weight or becoming fat, even though underweight.
C. Disturbance in the way in which one's body weight or shape is experienced; undue influence of body weight or shape on self-evaluation, or denial of the seriousness of the current low body weight.
D. In females past puberty, amenorrhea, i.e., the absence of at least three consecutive menstrual cycles. (A woman is considered to have amenorrhea if her periods occur only following hormone, e.g., estrogen, administration.)
Two types:
• **Restricting type:** During the episode of anorexia nervosa, the person does not regularly engage in binge eating or purging behavior (i.e., self-induced vomiting or the misuse of laxatives, diuretics, or enemas).
• **Binge eating/purging type:** During the episode of anorexia nervosa, the person regularly engages in binge eating or purging behavior (i.e., self-induced vomiting or the misuse of laxatives, diuretics, or enemas).

Source: Reprinted with permission from the *Diagnostic and Statistical Manual of Mental Disorders*, 4th ed., Text Revision (Copyright 2000). American Psychiatric Association.

What is the harm in being very thin?

Anorexia nervosa damages the body much as starvation does. In young people, growth ceases and normal development falters. They lose so much lean tissue that basal metabolic rate slows.[8] Additionally, the heart pumps inefficiently and irregularly, the heart muscle becomes weak and thin, the heart chambers diminish in size, and the blood pressure falls.[9] Electrolytes that help to regulate heartbeat become unbalanced. Many deaths from heart failure occur in people with anorexia.

Starvation brings other physical consequences as well: impaired immune response, anemia, and a loss of digestive function that worsens malnutrition. Digestive functioning becomes sluggish, the stomach empties slowly, and the lining of the intestinal tract shrinks. The ailing digestive tract fails to sufficiently digest any food the victim may eat. The pancreas slows its production of digestive enzymes. The person may suffer from diarrhea, further worsening malnutrition.

What kind of treatment helps people with anorexia nervosa?

Treatment of anorexia nervosa requires a multidisciplinary approach that addresses two sets of issues and behaviors: those relating to food and weight, and those involving relationships with oneself and others.[10] Teams of physicians, nurses, psychiatrists, family therapists, and dietitians work together to treat people with anorexia nervosa. Appropriate diet is crucial for normalizing body weight and must be tailored individually to each client's needs. Seldom are clients willing to eat for themselves, but if they are, chances are they can recover without other interventions.

Professionals classify clients based on the risks posed by the degree of malnutrition present. Clients with low risks may benefit from family counseling, **cognitive therapy,** behavior

modification, and nutrition guidance; those with greater risks may also need other forms of psychotherapy and supplemental formulas to provide extra nutrients and energy.

High-risk clients may require hospitalization and may need to be force-fed by tube at first to forestall death. This step causes psychological trauma. Medications are commonly prescribed, but to date, they play a limited role in treatment.

Denial runs high among those with anorexia nervosa. Few seek treatment on their own. Almost half of the women who are treated can maintain their body weight within 15 percent of healthy weight; at that weight, many of them begin menstruating again. The other half have poor or fair treatment outcomes, and two-thirds of those treated fight an ongoing mental battle with recurring morbid thoughts about food and body weight. Many relapse into abnormal eating behaviors to some extent. About 5 percent die during treatment, 1 percent by suicide.[11]

How does bulimia nervosa differ from anorexia nervosa?

Bulimia nervosa is distinct from anorexia nervosa and is more prevalent. More men suffer from bulimia nervosa than from anorexia, but bulimia is still more common in women. The secretive nature of bulimic behaviors makes recognition of the problem difficult, but once it is recognized, diagnosis is based on the criteria listed in Table NP7-2.

The typical person with bulimia is well educated, in her early twenties, and close to ideal body weight. She is a high achiever, with a strong feeling of dependence on her parents. She experiences considerable social anxiety and has difficulty establishing personal relationships. She is sometimes depressed and often exhibits impulsive behavior.

Like the person with anorexia nervosa, the person with bulimia spends much time thinking about her body weight

TABLE NP7-2 Criteria for Diagnosis of Bulimia Nervosa

A person with bulimia nervosa demonstrates the following:
A. Recurrent episodes of binge eating. An episode of binge eating is characterized by both of the following:
1. eating, in a discrete period of time (e.g., within any two-hour period), an amount of food that is definitely larger than most people would eat during a similar period of time and under similar circumstances, and
2. a sense of lack of control over eating during the episode (e.g., a feeling that one cannot stop eating or control what or how much one is eating).
B. Recurrent inappropriate compensatory behavior in order to prevent weight gain, such as self-induced vomiting; misuse of laxatives, diuretics, enemas, or other medications; fasting; or excessive exercise.
C. Binge eating and inappropriate compensatory behaviors that both occur, on average, at least twice a week for three months.
D. Self-evaluation unduly influenced by body shape and weight.
E. The disturbance does not occur exclusively during episodes of anorexia nervosa.
Two types:
• **Purging type:** The person regularly engages in self-induced vomiting or the misuse of laxatives, diuretics, or enemas.
• **Nonpurging type:** The person uses other inappropriate compensatory behaviors, such as fasting or excessive exercise, but does not regularly engage in self-induced vomiting or the misuse of laxatives, diuretics, or enemas.

Source: Reprinted with permission from the *Diagnostic and Statistical Manual of Mental Disorders,* 4th ed., Text Revision (Copyright 2000). American Psychiatric Association, 2000.

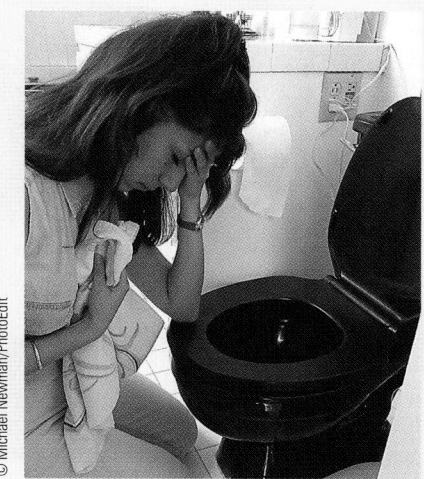

For many people with bulimia, guilt, depression, and self-condemnation follow a binge eating episode.

and food. Her preoccupation with food manifests itself in secretive binge eating episodes followed by self-induced vomiting, fasting, or the use of laxatives or diuretics. Such behaviors typically begin in late adolescence after a long series of various unsuccessful weight-reduction diets. People with bulimia commonly follow a pattern of restrictive dieting interspersed with bulimic behaviors and experience weight fluctuations of more than 10 pounds up and down over short periods of time.

Unlike the person with anorexia nervosa, the person with bulimia is aware of the consequences of her behavior, feels that it is abnormal, and is deeply ashamed of it. She feels inadequate and unable to control her eating, so she tends to be passive and to look to men for confirmation of her sense of self-worth. When she is rejected, either in reality or in her imagination, her bulimia becomes worse. If her depression deepens, she may seek solace in drug or alcohol abuse or other addictive behaviors. Many studies show a link between bulimia nervosa and drug and alcohol dependency.[12]

What exactly is binge eating?

Binge eating is unlike normal eating, and the food is not consumed for its nutritional value. The binge eater has a compulsion to eat. A typical binge occurs periodically, is done in secret, usually at night, and lasts an hour or more. A binge frequently follows a period of rigid dieting, so the binge eating is accelerated by hunger. During a binge, the person with bulimia may consume from a thousand to many thousands of kcalories of food. The food typically contains little fiber or water, has a smooth texture, and is high in sugar and fat, so it is easy to consume vast amounts rapidly with little chewing.

What are the consequences of binge eating?

After a binge, the person may use a **cathartic**—a strong laxative that can injure the lower intestinal tract. Or the person may induce vomiting, using an **emetic**—a drug intended as first aid for poisoning.

On first glance, purging seems to offer a quick and easy solution to the problems of unwanted kcalories and body weight. Many people perceive such behavior as neutral or even positive, when, in fact, bingeing and purging have serious physical consequences. Fluid and electrolyte imbalances caused by vomiting or diarrhea can lead to metabolic alkalosis, a condition characterized by apathy, confusion, and muscle spasms. Vomiting causes irritation and infection of the pharynx, esophagus, and salivary glands; erosion of the teeth; and dental caries. The esophagus may rupture or tear, as may the stomach. Overuse of emetics depletes potassium concentrations and can lead to death by heart failure.

What is the treatment for bulimia nervosa?

As for people with anorexia nervosa, a team approach provides the most effective treatment for people with bulimia nervosa. Bulimia nervosa is easier to treat than anorexia nervosa in many respects because it seems to be more of a chosen behavior. People with bulimia know that their behavior is abnormal, and many are willing to try to cooperate.

The goal of the dietary plan to treat bulimia is to help the client gain control, establish regular eating patterns, and restore nutritional health. Energy intake should not be severely restricted. The person needs to learn to eat a quantity of nutritious food sufficient to nourish her body and to satisfy hunger (at least 1600 kcalories a day). The person may also benefit from a regular exercise program.[13] The "How to" (p. 184) offers some ways to begin correcting bulimia nervosa.

Anorexia nervosa and bulimia nervosa are distinct eating disorders, yet they sometimes overlap. Anorexia victims may purge, and victims of both conditions share an overconcern with body weight and the tendency to drastically undereat. The two disorders can also appear in the same person, or one can lead to the other. Other people have eating disorders that fall short of anorexia nervosa or bulimia nervosa but share some of their features, such as fear of body fatness. One such condition is binge eating disorder (defined earlier in the glossary).

How does binge eating disorder differ from bulimia nervosa?

Up to half of all people who restrict eating to lose weight periodically binge without purging, including about one-third of obese people who regularly engage in binge eating. Obesity itself, however, does not constitute an eating disorder. Table NP7-3 (p. 184) lists the official diagnostic criteria for binge eating disorder.

Clinicians note differences between people with bulimia nervosa and those with binge eating disorder.[14] People with binge eating disorder consume less during a binge, rarely purge, and exert less restraint during times of dieting. Similarities also exist, including feeling out of control, disgusted, depressed, embarrassed, guilty, or distressed because of their

 **HOW TO** Establish Healthy Eating Patterns

The following advice has proved useful for people fighting bulimia nervosa.

Planning principles:

- Plan meals and snacks; record plans in a food diary prior to eating.
- Plan meals and snacks that require eating at the table and using utensils.
- Refrain from finger foods.
- Refrain from "dieting" or skipping meals.

Nutrition principles:

- Eat a well-balanced diet and regularly timed meals consisting of a variety of foods.
- Include raw vegetables, salad, or raw fruit at meals to prolong eating times.
- Choose whole-grain, high-fiber breads, pasta, rice, and cereals to increase bulk.

- Consume adequate fluid, particularly water.

Other tips:

- Choose meals that provide protein and fat for satiety, and bulky, fiber-rich carbohydrates for immediate feelings of fullness.
- Consume the amounts of food specified in the USDA Food Guide (pp. 18–19).
- For convenience (and to reduce temptation) select foods that naturally divide into portions. Select one potato, rather than rice or pasta that can be overloaded onto the plate; purchase yogurt and cottage cheese in individual containers; look for small packages of precut steak or chicken; choose frozen dinners with metered portions.
- Include 30 minutes of physical activity every day—exercise may be an important tool in controlling bulimia.

self-perceived gluttony.[15] Binge eating behavior responds more readily to treatment than other eating disorders, and resolving such behaviors can be a first step to authentic weight control. Successful treatment also improves physical health, mental health, and the chances of breaking the cycle of rapid weight losses and gains.

At so tender an age as 12 years, beautifully growing, normal-weight female youngsters are already worried that they are too fat. Most are "on diets." Magazines, newspapers, and television all present the message that to be thin is to be beautiful and happy. Anorexia nervosa and bulimia nervosa are not a form of rebellion against these unreasonable expectations, but rather the exaggerated acceptance of them. Body dissatisfaction is a primary factor in the development of eating disorders.[16] Perhaps a person's best defense against these disorders is to learn to appreciate his or her own uniqueness.

TABLE NP7-3 Criteria for Diagnosis of Binge Eating Disorder

A person with a binge eating disorder demonstrates the following:
A. Recurrent episodes of binge eating. An episode of binge eating is characterized by both of the following:
1. Eating, in a discrete period of time (e.g., within any two-hour period) an amount of food that is definitely larger than most people would eat in a similar period of time under similar circumstances.
2. A sense of lack of control over eating during the episode (e.g., a feeling that one cannot stop eating or control what or how much one is eating).
B. Binge eating episodes are associated with at least three of the following:
1. Eating much more rapidly than normal.
2. Eating until feeling uncomfortably full.
3. Eating large amounts of food when not feeling physically hungry.
4. Eating alone because of being embarrassed by how much one is eating.
5. Feeling disgusted with oneself, depressed, or very guilty after overeating.
C. The binge eating causes marked distress.
D. The binge eating occurs, on average, at least twice a week for six months.
E. The binge eating is not associated with the regular use of inappropriate compensatory behaviors (e.g., purging, fasting, excessive exercise) and does not occur exclusively during the course of anorexia nervosa or bulimia nervosa.

Source: Reprinted with permission from the *Diagnostic and Statistical Manual of Mental Disorders*, 4th ed., Text Revision (Copyright 2000). American Psychiatric Association.

Glossary

amenorrhea (ay-MEN-oh-REE-ah): the absence of or cessation of menstruation. **Primary amenorrhea** is menarche delayed beyond 16 years of age. **Secondary amenorrhea** is the absence of three to six consecutive menstrual cycles.

anorexia nervosa: an eating disorder characterized by a refusal to maintain a minimally normal body weight, self-starvation to the extreme, and a disturbed perception of body weight and shape; seen (usually) in adolescent girls and young women.

　anorexia = without appetite
　nervosa = of nervous origin

binge eating disorder: an eating disorder whose criteria are similar to those of bulimia nervosa, excluding purging or other compensatory behaviors.

bulimia (byoo-LEEM-ee-uh) **nervosa:** recurring episodes of binge eating combined with a morbid fear of becoming fat, usually followed by self-induced vomiting or purging.

cathartic: a strong laxative.

cognitive therapy: psychological therapy aimed at changing undesirable behaviors by changing underlying thought processes contributing to these behaviors. In anorexia nervosa, a goal is to replace false beliefs about body weight, eating, and self-worth with health-promoting beliefs.

eating disorder: a disturbance in eating behavior that jeopardizes a person's physical and psychological health.

emetic (em-ETT-ic): an agent that causes vomiting.

female athlete triad: a potentially fatal triad of medical problems: disordered eating, amenorrhea, and osteoporosis.

weight cycling: repeated rounds of weight loss and subsequent regain, with reduced ability to lose weight with each attempt; also called *yo-yo dieting*.

Notes

[1] Position of the American Dietetic Association: Nutrition intervention in the treatment of anorexia nervosa, bulimia nervosa, and eating disorders not otherwise specified (EDNOS), *Journal of the American Dietetic Association* 101 (2001): 810–819.

[2] D. Neumark-Sztainer and P. J. Hannan, Weight-related behaviors among adolescent girls and boys: Results from a national survey, *Archives of Pediatrics and Adolescent Medicine* 154 (2000): 569–577.

[3] A. E. Field and coauthors, Relation between dieting and weight change among preadolescents and adolescents, *Pediatrics* 112 (2003): 900–906; G. C. Patton and coauthors, Onset of adolescent eating disorders: Population-based cohort study over 3 years, *British Journal of Medicine* 318 (1999): 765–768.

[4] M. K. Torstveit and J. Sundgot-Borgen, The female athlete triad: Are elite athletes at increase risk? *Medicine and Science in Sports and Exercise* 37 (2005): 184–193; American Academy of Pediatrics, Committee on Sports Medicine and Fitness, Medical concerns in the female athlete, *Pediatrics* 106 (2000): 610–613.

[5] Torstveit and Sundgot-Borgen, 2005; American Academy of Pediatrics, 2000.

[6] C. L. Zanker and C. B. Cooke, Energy balance, bone turnover, and skeletal health in physically active individuals, *Medicine and Science in Sports and Exercise* 36 (2004): 1372–1381.

[7] K. L. Cobb and coauthors, Disordered eating, menstrual irregularity, and bone mineral density in female runners, *Medicine and Science in Sports and Exercise* 35 (2003): 711–719.

[8] A. Polito and coauthors, Basal metabolic rate in anorexia nervosa: Relation to body composition and leptin concentrations, *American Journal of Clinical Nutrition* 71 (2000): 1495–1502.

[9] C. Romano and coauthors, Reduced hemodynamic load and cardiac hypotrophy in patients with anorexia nervosa, *American Journal of Clinical Nutrition* 77 (2003): 308–312.

[10] Committee on adolescence, Identifying and treating eating disorders, *Pediatrics* 111 (2003): 204–211.

[11] B. Herzog and coauthors, Mortality in eating disorders: A descriptive study, *International Journal of Eating Disorders* 28 (2000): 20–26.

[12] B. Vastag, What's the connection? No easy answers for people with eating disorders and drug abuse, *Journal of the American Medical Association* 285 (2001): 1006–1007.

[13] J. Sundgot-Borgen and coauthors, The effect of exercise, cognitive therapy, and nutritional counseling in treating bulimia nervosa, *Medicine and Science in Sports and Exercise* 34 (2002): 190–195.

[14] A. E. Dingemans, M. J. Bruna, and E. F. Van Furth, Binge eating disorder: A review, *International Journal of Obesity and Related Metabolic Disorders* 26 (2002): 299–307.

[15] D. M. Ackard and coauthors, Overeating among adolescents: Prevalence and associations with weight-related characteristics and psychological health, *Pediatrics* 111 (2003): 67–74.

[16] J. Polivy and C. P. Herman, Causes of eating disorders, *Annual Review of Psychology* 53 (2002): 187–213.

The Vitamins

CHAPTER 8

Earlier chapters focused primarily on the energy-yielding nutrients—carbohydrate, fat, and protein. This chapter and the next one discuss the nutrients everyone thinks of when nutrition is mentioned—the vitamins and minerals.

The Vitamins—An Overview

The **vitamins** occur in foods in much smaller quantities than do the energy-yielding nutrients, and they themselves contribute no energy to the body. Instead, they serve mostly as facilitators of body processes. They are a powerful group of substances, as their absence attests. Vitamin A deficiency can cause blindness, a lack of niacin can cause dementia, and a lack of vitamin D can retard bone growth. The consequences of deficiencies are so dire and the effects of restoring the needed nutrients so dramatic that people spend billions of dollars each year on vitamin supplements to cure many different ailments. Vitamins certainly contribute to sound nutritional health, but supplements do not cure all ills. Actually, a vitamin can cure only the disease caused by a deficiency of that vitamin. The vitamins' roles in supporting optimal health extend far beyond preventing deficiency diseases, however. Emerging evidence points to relationships between low intakes of vitamins and chronic diseases such as cancer and heart disease.

A child once defined vitamins as "what, if you don't eat, you get sick." The description is both insightful and accurate. A more prosaic definition is that vitamins are potent, essential, nonkaloric, organic nutrients needed from foods in trace amounts to perform specific functions that promote growth, reproduction, and the maintenance of health and life. Two characteristics distinguish vitamins from energy nutrients:

1. Vitamins do not yield energy when broken down, but assist the enzymes that release energy from carbohydrate, fat, and protein.
2. Vitamins are needed in much smaller amounts than the energy nutrients.

As the individual vitamins were discovered, they were named or given letters, numbers, or both. This led to the confusion that still exists today. This chapter uses the names shown in Table 8-1; alternative names are given in Tables 8-2 and 8-3, which appear later in the chapter.

Bioavailability The availability of vitamins from foods depends on two factors: the quantity provided by a food and the amount absorbed and used by the body (the vitamin's **bioavailability**). Researchers analyze foods to determine their vitamin contents and publish the results in tables of food composition such as Appendix A. Determining the bioavailability of a vitamin is more difficult because it depends on many factors, including:

- The efficiency of digestion.
- A person's previous nutrient intake and nutrition status.
- Other foods eaten at the same time.
- The method of food preparation (raw or cooked, for example).
- The source of the nutrient (naturally occurring, synthetic, or fortified).

Experts consider these factors when estimating recommended intakes.

Precursors Some of the vitamins are available from foods in inactive forms known as **precursors,** or provitamins. Once inside the body, the precursor is converted to the active form of the vitamin. Thus, in measuring a person's vitamin intake, it is important to count both the amount of the actual vitamin and the potential amount available from its precursors.

Solubility Vitamins fall naturally into two classes—fat soluble and water soluble. The solubility of a vitamin confers on it many characteristics and

Carbohydrate, fat, and protein are sometimes referred to as *macronutrients* because they are measured in foods in gram amounts. Vitamins and minerals are sometimes referred to as *micronutrients* because they are measured in foods in milligram (mg) and microgram (μg) amounts.

vitamins: essential, nonkaloric, organic nutrients needed in tiny amounts in the diet.

bioavailability: the rate and extent to which a nutrient is absorbed and used.

precursors: compounds that can be converted into other compounds; with regard to vitamins, compounds that can be converted into active vitamins; also known as **provitamins.**

TABLE 8-1	Vitamin Names
Fat-Soluble Vitamins	
Vitamin A	
Vitamin D	
Vitamin E	
Vitamin K	
Water-Soluble Vitamins	

B vitamins
  Thiamin
  Riboflavin
  Niacin
  Pantothenic acid
  Biotin
  Vitamin B$_6$
  Folate
  Vitamin B$_{12}$	

Vitamin C	

determines how it is absorbed, transported, stored, and excreted. This discussion of vitamins begins with the fat-soluble vitamins.

REVIEW NOTES

Vitamins are essential, nonkcaloric nutrients that are needed in trace amounts in the diet to help facilitate body processes.

The Fat-Soluble Vitamins

The fat-soluble vitamins—A, D, E, and K—usually occur together in the fats and oils of foods, and the body absorbs them in the same way it absorbs lipids. Therefore, any condition that interferes with fat absorption can precipitate a deficiency of the fat-soluble vitamins. Once absorbed, fat-soluble vitamins are stored in the liver and fatty tissues until the body needs them. They are not readily excreted, and unlike most of the water-soluble vitamins, they can build up to toxic concentrations. Excesses of vitamins A and D from supplements can reach toxic levels especially easily.

The capacity to store fat-soluble vitamins affords a person some flexibility in dietary intake. When blood concentrations begin to decline, the body can retrieve the vitamins from storage. Thus a person need not eat a day's allowance of each fat-soluble vitamin every day but need only make sure that, over time, average daily intakes approximate recommended intakes. In contrast, most water-soluble vitamins must be consumed more regularly because the body does not store them to any great extent.

VITAMIN A AND BETA-CAROTENE

Vitamin A has the distinction of being the first fat-soluble vitamin to be recognized. Today, after almost a century of revelations, vitamin A and its plant-derived precursor, **beta-carotene,** continue to intrigue researchers with their diverse roles and profound effects on health.

Vitamin A is a versatile vitamin, with roles in gene expression, vision, cell differentiation (thereby maintaining the health of body linings and skin), immunity, and reproduction and growth.[1] Three different forms of vitamin A are active in the body: retinol, retinal, and retinoic acid. Each form of vitamin A performs specific tasks. Retinol supports reproduction and is the major transport and storage form of the vitamin; the cells convert retinol to retinal or retinoic acid as needed. Retinal is active in vision, and retinoic acid acts as a hormone, regulating cell differentiation, growth, and embryonic development. A special transport protein, **retinol-binding protein (RBP),** picks up retinol from the liver where it is stored and carries it in the blood.

Vitamin A's Role in Gene Expression Vitamin A exerts considerable influence on body functions through its regulation of the activities of the genes.[2] Genes direct the synthesis of proteins, including enzymes, and enzymes perform the metabolic work of the tissues (see Chapter 5). Hence, factors that influence gene expression also affect the metabolic activities of the tissues and the health of the body. Hundreds of genes have been suggested as regulatory targets of the retinoic acid form of vitamin A.[3]

Researchers have long known that simply possessing the genetic equipment needed to make a particular protein does not guarantee that the protein will be produced, any more than owning a car guarantees you a trip across town. To get the car rolling, you must also use the right key to trigger the events that start up its engine or turn it off at the appropriate time. Some dietary components, includ-

Measurement of the blood concentration of RBP is a sensitive test of vitamin A status.

vitamin A: a fat-soluble vitamin. Its three chemical forms are *retinol* (the alcohol form), *retinal* (the aldehyde form), and *retinoic acid* (the acid form).

beta-carotene: a vitamin A precursor made by plants and stored in human fat tissue; an orange pigment.

retinol-binding protein (RBP): the specific protein responsible for transporting retinol.

ing the retinoic acid form of vitamin A, are now known to be such keys—they help to activate or deactivate certain genes and thus affect the production of specific proteins.[4]

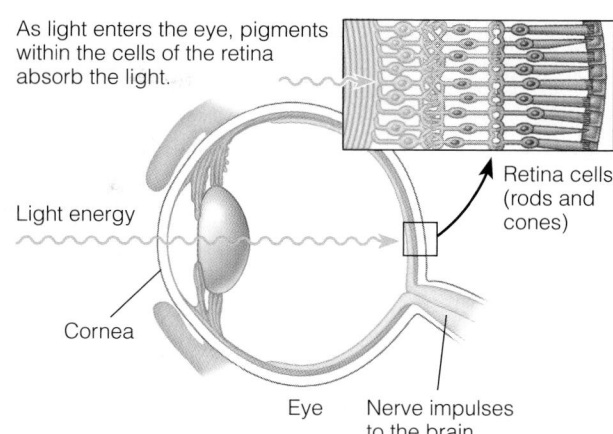

As light enters the eye, pigments within the cells of the retina absorb the light.

Light energy

Cornea

Retina cells (rods and cones)

Eye   Nerve impulses to the brain

FIGURE 8-1 Vitamin A's Role in Vision

Vitamin A's Role in Vision Vitamin A plays two indispensable roles in the eye. It helps maintain a healthy, crystal-clear outer window, the **cornea;** and it participates in the events of light detection at the **retina.** Figure 8-1 shows vitamin A's site of action inside the eye.

When vitamin A is lacking, the eye has difficulty adapting to changing light levels. At night, after the eye has adapted to darkness, a lag occurs before the eye can see again after a flash of bright light. This lag in the recovery of night vision is known as **night blindness.** Because night blindness is easy to test, it aids in the diagnosis of vitamin A deficiency. Night blindness is only a symptom, however, and may indicate a condition other than vitamin A deficiency.

Vitamin A's Role in Protein Synthesis and Cell Differentiation The role that vitamin A plays in vision is undeniably important, but only one-thousandth of the body's vitamin A is in the retina. Much more is in the skin and the linings of organs, where it works behind the scenes at the genetic level to promote protein synthesis and cell **differentiation.** The process of cell differentiation allows each type of cell to mature so that it is capable of performing a specific function.

All body surfaces, both inside and out, are covered by layers of cells known as **epithelial cells.** The **epithelial tissue** on the outside of the body is, of course, the skin. The epithelial tissues inside the body include the linings of the mouth, stomach, and intestines; the linings of the lungs and the passages leading to them; the lining of the bladder; the linings of the uterus and vagina; and the linings of the eyelids and sinus passageways. The epithelial tissues on the inside of the body must be kept smooth. To ensure that they are, the epithelial cells on their surfaces secrete a smooth, slippery substance (mucus) that coats the tissues and protects them from invasive microorganisms and other harmful particles. The **mucous membrane** that lines the stomach also shields its cells from digestion by gastric juices. Vitamin A, by way of its role in cell differentiation, helps to maintain the integrity of the epithelial cells.

Vitamin A's Role in Immunity Vitamin A's role in protecting the body's immune response was recognized more than 70 years ago when researchers noticed that vitamin A deficiency reduces resistance to infection.[5] Immune function depends on the growth, differentiation, and activation of the cells that defend the body against infectious agents.[6]

Vitamin A's Role in Reproduction, Growth, and Development Vitamin A is crucial to normal reproduction and growth. In men, vitamin A participates in sperm development, and in women, vitamin A promotes normal fetal growth and development.[7] During pregnancy, vitamin A is transferred to the fetus and is essential to the development of the nervous system, lungs, heart, kidneys, skeleton, eyes, and ears.[8]

Beta-Carotene's Role as an Antioxidant For many years scientists believed beta-carotene to be of interest solely as a vitamin A precursor. Eventually, though, researchers began to recognize that beta-carotene is an extremely effective **antioxidant** in the body. Antioxidants are compounds that protect other compounds (such as lipids in cell membranes) from attack by oxygen. Oxygen triggers the formation of compounds known as **free radicals** that can start chain reactions in cell membranes. If left uncontrolled, these chain reactions can damage cell

cornea (KOR-nee-uh): the hard, transparent membrane covering the outside of the eye.

retina (RET-in-uh): the layer of light-sensitive nerve cells lining the back of the inside of the eye; consists of rods and cones.

night blindness: the slow recovery of vision after exposure to flashes of bright light at night; an early symptom of vitamin A deficiency.

differentiation: the development of specific functions different from those of the original.

epithelial (ep-i-THEE-lee-ul) **cells:** cells on the surface of the skin and mucous membranes.

epithelial tissue: tissue composing the layers of the body that serve as selective barriers between the body's interior and the environment (examples are the cornea, the skin, the respiratory lining, and the lining of the digestive tract).

mucous membrane: membrane composed of mucus-secreting cells that lines the surfaces of body tissues. (Reminder: *Mucus* is the smooth, slippery substance secreted by these cells.)

antioxidant (anti-OX-ih-dant): a compound that protects other compounds from oxygen by itself reacting with oxygen. *Oxidation* is a potentially damaging effect of normal cell chemistry involving oxygen.
   anti = against
   oxy = oxygen

free radicals: highly reactive chemical forms that can cause destructive changes in nearby compounds, sometimes setting up a chain reaction.

structures and impair cell functions. Oxidative and free-radical damage to cells is suspected of instigating some early stages of cancer and heart disease. Research has identified links between oxidative damage and the development of many other diseases, including age-related blindness, Alzheimer's disease, arthritis, cataracts, diabetes, and kidney disease.[9]

Studies of populations suggest that people whose diets are low in foods rich in beta-carotene have higher incidences of certain types of cancer than those whose diets contain generous amounts of such foods. Based on findings that beta-carotene in foods may protect against cancer, researchers designed a study to determine the effects of beta-carotene *supplements* on the incidence of lung cancer among smokers. The researchers expected to see a beneficial effect, but instead they found that smokers taking the beta-carotene supplements suffered a *greater* incidence of lung cancer than those taking placebos. In a different study, the Physicians' Health Study, researchers tested the effects of beta-carotene supplements versus a placebo on cancer and heart disease risk in more than 20,000 male physicians for 12 years. Beta-carotene neither decreased nor increased cancer or heart disease risk. The lead researcher on this study (Dr. C. H. Hennekens), as well as on three other large-scale studies of antioxidant vitamins and disease risk, states that "the totality of evidence on beta-carotene supplements shows that well-nourished populations accrue no benefits on cancer or heart disease." He emphasizes the need for more research on antioxidant vitamins and disease risk and reminds consumers that so far, there is no "magic pill" for reducing the risk of cancer and heart disease. Positive changes in diet and lifestyle, however, can reduce risk. Beta-carotene is just one of many nutrients and compounds present in foods. As discussed in Nutrition in Practice 8, many other phytochemicals are also present in foods and may be responsible for some of the protective effects attributed to beta-carotene. Until more is known, eating beta-carotene–rich foods, not supplements, is in the best interests of health. Based on research so far, the DRI committee has not established a recommended intake value for beta-carotene.

Vitamin A Deficiency Up to a year's supply of vitamin A can be stored in the body, 90 percent of it in the liver. If a healthy adult were to stop eating vitamin A–rich foods, deficiency symptoms would not begin to appear until after stores were depleted, which would take one to two years. Then, however, the consequences would be profound and severe. Table 8-2, later in this chapter, lists some of them.

In vitamin A deficiency, cell differentiation and maturation are impaired. The epithelial cells flatten and begin to produce keratin—the hard, inflexible protein of hair and nails. In the eye, this process leads to drying and hardening of the cornea, which may progress to permanent blindness. Vitamin A deficiency is the major cause of childhood blindness in the world, causing more than half a million children to lose their sight every year.[10] More than 200 million children worldwide endure less severe forms of vitamin A deficiency, making them vulnerable to infectious diseases.

All body surfaces, both inside and out, maintain their integrity with the help of vitamin A. When vitamin A is lacking, cells of the skin harden and flatten, making it dry, rough, scaly, and hard. An accumulation of keratin makes a lump around each hair follicle (keratinization).

In the mouth, a vitamin A deficiency results in drying and hardening of the salivary glands, making them susceptible to infection. Secretions of mucus in the stomach and intestines are reduced, hindering normal digestion and absorption of nutrients. Infections of other mucous membranes also become likely.

Vitamin A's role in maintaining the body's defensive barriers may partially explain the relationship between vitamin A deficiency and susceptibility to infection.[11] In several studies, when children with measles complicated by diarrhea, infections such as pneumonia, or both were given vitamin A supplements, their

A *placebo* is an inactive substance used in research studies.

Reminder: *Phytochemicals* are nonnutrient compounds found in plant-derived foods that have biological activity in the body.

The progressive blindness caused by vitamin A deficiency is called **xerophthalmia** (zer-off-THAL-mee-uh).
  xero = dry
  ophthalm = eye

An early sign of xerophthalmia is **xerosis** (drying of the cornea); the last and most severe stage is **keratomalacia** (kerr-uh-to-mal-AY-shuh), or total blindness.
  malacia = softening, weakening

The accumulation of the hard material keratin around each hair follicle is **follicular hyperkeratosis.**

keratin (KERR-uh-tin): a water-insoluble protein; the normal protein of hair and nails. Keratin-producing cells may replace mucus-producing cells in vitamin A deficiency.

follicle (FOLL-i-cul): a group of cells in the skin from which a hair grows.

overall survival rates were significantly higher than those of similar children who did not receive vitamin A.

The evidence that vitamin A reduces the severity of measles and measles-related infections and diarrhea has prompted the World Health Organization (WHO) and UNICEF (the United Nations International Children's Emergency Fund) to make control of vitamin A deficiency a major goal in their quest to improve child survival throughout the developing world. The American Academy of Pediatrics recommends vitamin A supplementation for certain groups of measles-infected infants and children in the United States.

Vitamin A Toxicity When the body stores excess vitamin A, toxicity is possible. Normally, toxicity symptoms are likely only when animal-derived foods or supplements are the source of the excess vitamin, for in these sources the vitamin is already in its active form, called **preformed vitamin A.** Plant foods contain the vitamin only as beta-carotene, its inactive precursor form. The precursor does not convert to active vitamin A rapidly enough to cause toxicity.

Overdoses of vitamin A damage the same body systems that exhibit symptoms in vitamin A deficiency (see Table 8-2 later in the chapter). Children are most vulnerable to vitamin A toxicity because, being smaller, they need less than adults, and it is easy to give them too much in pill form. The availability of breakfast cereals, instant meals, fortified milk, and chewable candylike vitamins, each containing 100 percent or more of the recommended daily intake of vitamin A, makes it possible for a well-meaning parent to provide several times the daily allowance of the vitamin to a child within a few hours. Serious toxicity is seen in infants and young children when they are given more than ten times the recommended amount every day for weeks at a time.

Excessive vitamin A also poses a **teratogenic** risk.[12] In pregnant women, chronic use of vitamin A supplements providing three to four times the amount recommended for pregnancy can cause malformations of the fetus. The Tolerable Upper Intake Level of 3000 micrograms for women of childbearing age is based on the teratogenic effect of vitamin A.[13]

Excessive amounts of vitamin A over the years may weaken the bones and contribute to osteoporosis (see the later discussion of vitamin D for more on osteoporosis).[14] In some studies, people consuming large amounts of vitamin A either from supplements or from foods containing retinol have a significantly greater risk of hip fractures.[15] More research is needed to clarify the relationship between vitamin A intake and bone health, but such findings suggest that most people should not take vitamin A supplements. Even multivitamin supplements provide more vitamin A than most people need.

Certain vitamin A relatives are available by prescription as acne treatments. When applied directly to the skin surface, these preparations help relieve the symptoms of acne. Taking massive doses of vitamin A internally will *not* cure acne, however, and may cause the miseries itemized in Table 8-2. In most cases, foods are a better choice than supplements for needed nutrients. The best way to ensure a safe vitamin A intake is to eat generous servings of vitamin A–rich foods.

Beta-Carotene Conversion and Toxicity Nutrition scientists do not use micrograms to specify the quantity of beta-carotene in foods. Instead, they use a value known as **retinol activity equivalents (RAE),** which express the amount of retinol the body actually derives from a plant food after conversion. The body can make one unit of retinol from about 12 units of beta-carotene.

As mentioned earlier, beta-carotene from plant foods is not converted to the active form of vitamin A rapidly enough to be hazardous. It has, however, been known to turn people bright orange-yellow if they eat too much. Beta-carotene builds up in the fat just beneath the skin and imparts a yellowish cast (see Figure 8-2, p. 192).

preformed vitamin A: vitamin A in its active form.

teratogenic (ter-AT-oh-jen-ik): causing abnormal fetal development and birth defects.
 terato = monster
 genic = to produce

retinol activity equivalents (RAE): a measure of vitamin A activity; the amount of retinol that the body will derive from a food containing preformed retinol or its precursor beta-carotene.

Multivitamin supplements typically provide:
- 750 µg (2500 IU).
- 1500 µg (5000 IU).

For perspective, the RDA for vitamin A is 700 µg for women and 900 µg for men. The **IU (international unit)** was used earlier to express the amount of vitamins A, D, and E in foods and is still used on supplement labels.

1 µg RAE = 1 µg retinol.
 = 12 µg beta-carotene from food.

Yellowing of the skin caused by excess carotene in the blood is known as **carotenemia** (KAR-oh-teh-NEE-me-ah). Carotenemia can be distinguished from jaundice because the mucous membranes lining the eyelids do not turn yellow as they do in jaundice.

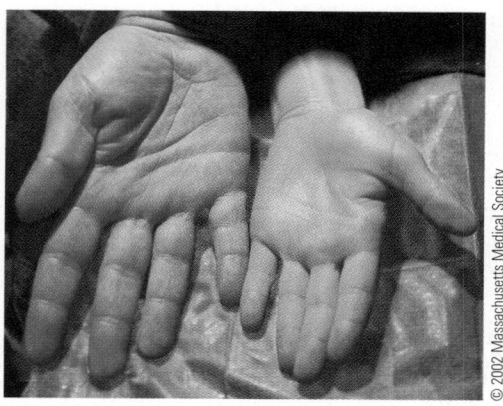

FIGURE 8-2 Symptom of Beta-Carotene Excess—Discoloration of the Skin
The hand on the right shows the skin discoloration that occurs when blood levels of beta-carotene rise in response to a low-kcalorie diet that features carrots, pumpkins, and orange juice. (The hand on the left belongs to someone else and is shown for comparison.)

Vitamin A in Foods Preformed vitamin A is found only in foods of animal origin. The richest sources of vitamin A are liver and fish oil, but milk, cheese, and fortified cereals are also good sources. Healthy people can eat vitamin A–rich foods in large amounts without risking toxicity with the possible exception of liver. Eating liver once every week or so is enough. Butter and eggs also provide some vitamin A to the diet.

Because vitamin A is fat soluble, it is lost when milk is skimmed. Fat-free milk is thus often fortified with vitamin A to compensate. Margarine is also usually fortified so as to provide the same amount of vitamin A as butter. Snapshot 8-1 shows a sampling of the richest food sources of both preformed vitamin A and beta-carotene.

Fast-food meals often lack vitamin A. When fast-food restaurants offer salads with cheese, carrots, and other vitamin A–rich foods, the nutritional quality of their meals greatly improves.

Beta-Carotene in Foods Many foods from plants contain beta-carotene, the orange pigment responsible for the bold colors of many fruits and vegetables. Carrots, sweet potatoes, pumpkins, cantaloupe, and apricots are all rich sources, and their bright orange color enhances the eye appeal of the plate. Another colorful group, *dark* green vegetables, such as spinach, other greens, and broccoli, owe their color to both chlorophyll and beta-carotene. The orange and green pigments together impart a deep, murky green color to the vegetables. Other colorful vegetables, such as iceberg lettuce, beets, and sweet corn, can fool you into thinking they contain beta-carotene, but these foods derive their color from other pigments and are poor sources of beta-carotene. As for "white" plant foods such as rice and potatoes, they have little or none. Recommendations to eat *dark* green or *deep* orange vegetables and fruits at least every other day help people to meet their vitamin A needs.

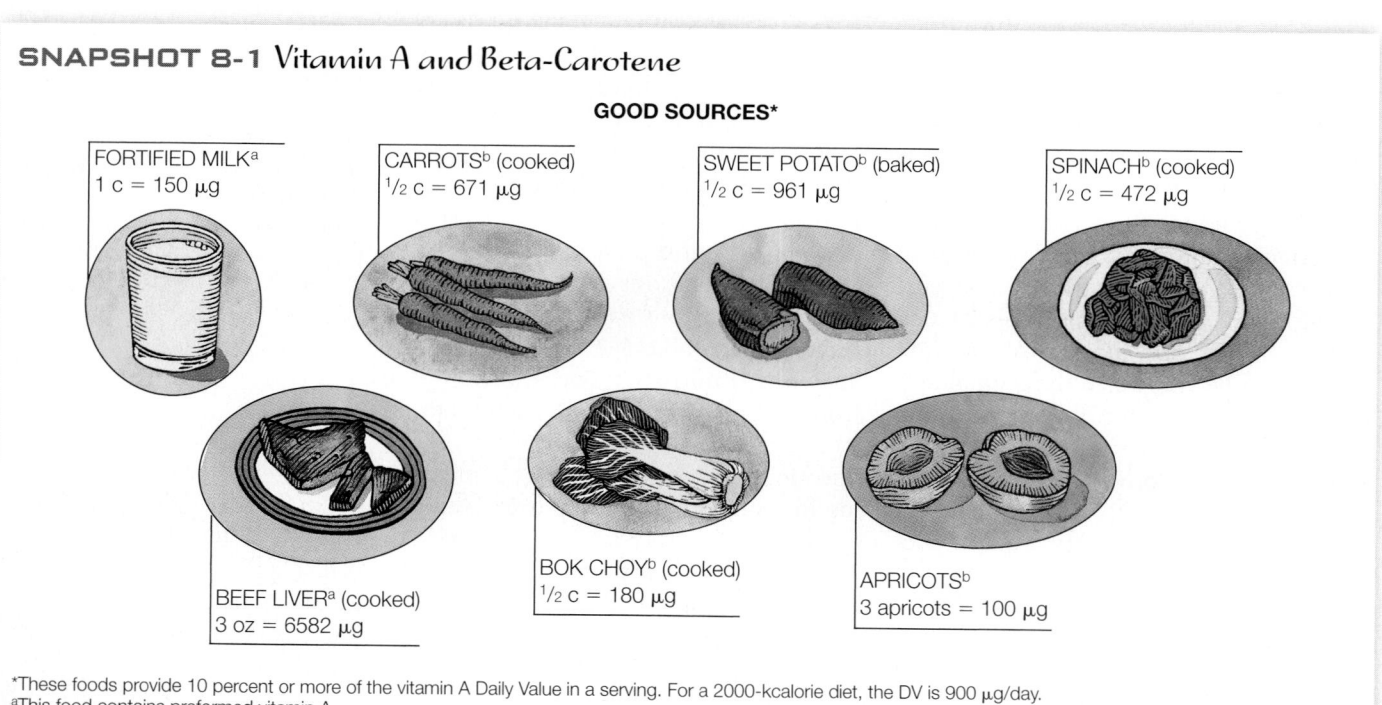

SNAPSHOT 8-1 *Vitamin A and Beta-Carotene*

GOOD SOURCES*

FORTIFIED MILK[a]
1 c = 150 μg

CARROTS[b] (cooked)
1/2 c = 671 μg

SWEET POTATO[b] (baked)
1/2 c = 961 μg

SPINACH[b] (cooked)
1/2 c = 472 μg

BEEF LIVER[a] (cooked)
3 oz = 6582 μg

BOK CHOY[b] (cooked)
1/2 c = 180 μg

APRICOTS[b]
3 apricots = 100 μg

*These foods provide 10 percent or more of the vitamin A Daily Value in a serving. For a 2000-kcalorie diet, the DV is 900 μg/day.
[a]This food contains preformed vitamin A.
[b]This food contains beta-carotene.

VITAMIN D

Vitamin D is different from all the other nutrients in that the body can synthesize it in significant quantities with the help of sunlight. Therefore, in a sense, vitamin D is not an essential nutrient. Given enough sun, people need no vitamin D from foods.

Vitamin D's Metabolic Conversions The liver manufactures a vitamin D precursor, which migrates to the skin where it is converted to a second precursor with the help of the sun's ultraviolet rays. Next, the liver and then the kidneys alter the second precursor to produce the active vitamin. Vitamin D precursors from plants require the same two conversions by the liver and kidneys to become active. The biological activity of the active vitamin is 500- to 1000-fold greater than that of its precursor. Diseases that affect either the liver or the kidneys may impair the transformations of precursor vitamin D to active vitamin D and therefore produce symptoms of vitamin D deficiency.

Vitamin D's Actions Although known as a vitamin, vitamin D is actually a hormone—a compound manufactured by one organ of the body that has effects on another. The best-known vitamin D target organs are the small intestine, the kidneys, and the bones, but scientists have discovered many other vitamin D target tissues, including the brain, the pancreas, the skin, and the reproductive organs.[16] These discoveries suggest that numerous additional functions for vitamin D may surface, including regulation of the immune system.[17] Research is hinting that to incur a deficit of vitamin D is to invite problems of many kinds, including high blood pressure, rheumatoid arthritis, inflammatory bowel disease, and even multiple sclerosis.[18]

Vitamin D's Roles in Bone Vitamin D is a member of a large, cooperative bone-making and maintenance team composed of nutrients and other compounds, including vitamins A, C, and K; the hormones parathormone (parathyroid hormone) and calcitonin; the protein collagen; and the minerals calcium, phosphorus, magnesium, and fluoride. Many of these interactions take place at the genetic level in ways that are under investigation.[19] Vitamin D's special role in bone growth is to make calcium and phosphorus available in the blood that bathes the bones. The bones grow denser and stronger as the minerals are deposited from the blood. Vitamin D acts in three ways to maintain blood concentrations of calcium and phosphorus: it stimulates their absorption from the gastrointestinal (GI) tract; it mobilizes calcium and phosphorus from bones into the blood; and it stimulates their retention by the kidneys.

Vitamin D Deficiency The symptoms of vitamin D deficiency are those of calcium deficiency, as shown in Table 8-2. The bones fail to calcify normally and may grow so weak that they become bent when they have to support the body's weight. A child with the vitamin D–deficiency disease **rickets** who is old enough to walk characteristically develops bowed legs, often the most obvious sign of the disease (see Figure 8-3, p. 194). Rickets was a major pediatric health problem in the United States before vitamin D fortification of commercially prepared milk was introduced decades ago. Unfortunately, rickets seems to be making a comeback among African American breastfed infants who are exposed to little sunlight and receive no vitamin D supplements and among toddlers fed unfortified soy and rice beverages instead of milk.[20] Worldwide, rickets afflicts a large number of children who receive inadequate food and little exposure to sunlight. The adult form of rickets is called **osteomalacia.**

Adolescents, who often abandon vitamin D–fortified milk in favor of soft drinks, may also prefer indoor pastimes such as computer games to outdoor activities during

Sunlight promotes vitamin D synthesis in the skin. Exposure to the sun should be moderate, however; excessive exposure may cause skin cancer.

The precursor of vitamin D made in the liver is 7-dehydrocholesterol, which is made from cholesterol. This is one of the body's many "good" uses for cholesterol.

The final, active vitamin is 1-25 dihydroxycholecalciferol or, more simply, dihydroxy vitamin D.

Chapter 9 offers a detailed discussion of the minerals.

rickets: the vitamin D–deficiency disease in children.

osteomalacia (os-tee-oh-mal-AY-shuh): a bone disease characterized by softening of the bones. Symptoms include bending of the spine and bowing of the legs. The disease occurs most often in adults with renal failure or malabsorption disorders.
   osteo = bone
   mal = bad (soft)

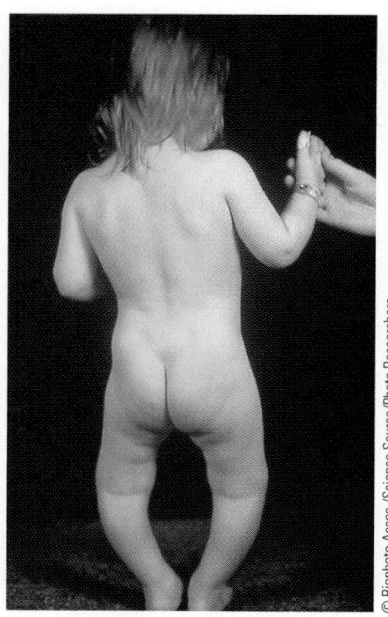

FIGURE 8-3 Rickets
This child has the bowed legs of the vitamin D–deficiency disease rickets.

Vitamin D activity was previously expressed in international units (IU), but as of 1980, it is expressed in micrograms of cholecalciferol, the active form of vitamin D. To convert, use the following factor:
• 100 IU = 2.5 μg.
• 400 IU = 10 μg.

Vitamin D AI:
• Adults (19–50 yr): 5 μg/day.
• Adults (51–70 yr): 10 μg/day.
• Adults (>70 yr): 15 μg/day.

osteoporosis (os-tee-oh-por-OH-sis): literally, porous bones; reduced density of the bones, also known as *adult bone loss*.

daylight hours. Such teens often lack vitamin D and so fail to develop the bone density needed to prevent bone loss in later life.[21]

Inadequate vitamin D is recognized as a risk factor in **osteoporosis** (reduced bone density). Without sufficient vitamin D, absorption of calcium is limited, and bone remodeling is impaired. This combination leads to a loss of bone mass.

Vitamin D Toxicity Whereas vitamin D deficiency depresses calcium absorption, blood calcium, and bone mineralization, an excess of vitamin D does the opposite, as shown in Table 8-2. It enhances calcium absorption, produces high blood calcium, and promotes return of bone calcium into the blood. The excess calcium then tends to precipitate in the soft tissues, forming stones, including kidney stones. Calcification may also harden the blood vessels and is especially dangerous in the major arteries of the heart and lungs, where it can cause death.

Vitamin D in excess is the most toxic of all the vitamins. The amounts of vitamin D in foods available in the United States and Canada are well within safe limits, but supplements containing the vitamin in concentrated form are not. Adults should use caution when taking vitamin D supplements and keep them out of the reach of children. The DRI committee has set a Tolerable Upper Intake Level for vitamin D at 50 micrograms per day (2000 IU on supplement labels).

Vitamin D from the Sun Most of the world's population relies on natural exposure to sunlight to maintain adequate vitamin D nutrition. The sun imposes no risk of vitamin D toxicity. Prolonged exposure to sunlight degrades the vitamin D precursor in the skin, preventing its conversion to the active vitamin. Even lifeguards on southern beaches are safe from vitamin D toxicity from the sun.

Prolonged exposure to sunlight has other undesirable consequences, however, such as premature wrinkling of the skin and the risk of skin cancer. These risks may be reduced by using sunscreens. Unfortunately, sunscreens with sun protection factors (SPF) of 8 and above also retard vitamin D synthesis.[22] A strategy to avoid this dilemma is to apply sunscreen after enough time has elapsed to provide sufficient vitamin D. For most people, exposing hands, face, and arms on a clear summer day for 10 minutes, a few times a week, should be sufficient to maintain vitamin D nutrition. Dark-skinned people require longer exposure than light-skinned people, but by three hours, vitamin D synthesis in heavily pigmented skin arrives at the same plateau as in fair skin after 30 minutes.

The ultraviolet rays from tanning lamps and tanning booths may also stimulate vitamin D synthesis, but the hazards outweigh any possible benefits. The Food and Drug Administration (FDA) warns that if the lamps are not properly filtered, people using tanning booths risk burns, damage to the eyes and blood vessels, and skin cancer.

The ultraviolet rays of the sun that promote vitamin D synthesis cannot penetrate clouds, smoke, smog, heavy clothing, window glass, or even window screens. In the United States and Canada, dark-skinned people who live in smoggy northern cities or who lack exposure to sunlight are at greatest risk of vitamin D deficiency.[23] A surprisingly high number of otherwise healthy northern adults have been found to have low blood levels of vitamin D, most often at the end of the winter season, even though they were drinking milk fortified with vitamin D.[24] It may be that milk does not deliver enough vitamin D to prevent a drop in blood levels through the winter months. In addition, people who are housebound or institutionalized and those who work at night may incur (over years) a vitamin D deficiency, as do many elderly people, who have limited exposure to sunlight and become less efficient at activating vitamin D as they age. For these people, dietary vitamin D is essential.[25] Because of the increased risk with age, the DRI committee set Adequate Intakes (AI) for vitamin D that increase over the years (see the margin).

Vitamin D in Foods Only a few animal foods, notably, eggs, liver, butter, some fatty fish, and fortified milk, supply significant amounts of vitamin D. For those who use margarine in place of butter, fortified margarine is a significant source. Infant formulas are fortified with vitamin D in amounts adequate for daily intake as long as infants consume at least 500 milliliters (15 ounces) of formula. Breast milk is low in vitamin D, so vitamin D supplements (5 micrograms daily) are recommended for infants who are breastfed exclusively and for those who do not receive at least 500 milliliters of vitamin D–fortified formula per day.[26] These sources, plus any exposure to the sun, provide infants with more than enough of this vitamin.

The fortification of milk with vitamin D is the best guarantee that children will meet their vitamin D needs and underscores the importance of milk in children's diets. Vitamin D supplements (5 micrograms daily) are recommended for children and adolescents who do not drink at least 2 cups of milk per day, are not regularly exposed to sunlight, or do not take a multivitamin supplement containing at least 5 micrograms of vitamin D.[27] Unlike milk, cheese and yogurt are not fortified with vitamin D. Vegans, and especially their children, may have low vitamin D intakes because few fortified plant sources exist. Exceptions include margarine and some soy milks. In the United States, breakfast cereals may be fortified with vitamin D, as their labels indicate.

VITAMIN E

More than 80 years ago, researchers discovered a compound in vegetable oils necessary for reproduction in rats. The compound was named **tocopherol,** which means "offspring." Eventually, the compound was named vitamin E. When chemists isolated four tocopherol compounds, they designated them by the first four letters of the Greek alphabet: alpha, beta, gamma, and delta. Of these, alpha-tocopherol is the gold standard for vitamin E activity; recommended intakes are based on it. Table 8-2 later in the chapter summarizes important information about vitamin E.

Vitamin E as an Antioxidant Like beta-carotene, vitamin E is a fat-soluble antioxidant. It protects other substances from oxidation by being oxidized itself. If there is plenty of vitamin E in the membranes of cells exposed to an oxidant, chances are this vitamin will take the brunt of any oxidative attack, protecting the lipids and other vulnerable components of the membranes. Vitamin E is especially effective in preventing the oxidation of the polyunsaturated fatty acids (PUFA), but it protects all other lipids (for example, vitamin A) as well.

Vitamin E exerts an especially important antioxidant effect in the lungs, where the cells are exposed to high concentrations of oxygen. Vitamin E also protects the lungs from air pollutants that are strong oxidants.

Some evidence suggests that vitamin E may also offer protection against heart disease by protecting LDL (low-density lipoproteins) from oxidation.[28] The oxidation of LDL encourages the development of atherosclerosis. Vitamin E may also benefit heart health by inhibiting several other processes involved in the development of heart disease.[29] Despite this theoretical understanding of how vitamin E may exert protection against heart disease, the results of controlled clinical studies in which human beings were given vitamin E supplements have been disappointing. Two extensive long-term studies, one with healthy people and one with people with vascular disease or diabetes, showed no benefit of vitamin E supplementation to heart health.[30] One review of vitamin E supplementation and heart disease concluded that in people with chronic diseases, vitamin E supplements may be harmful.[31] Another review of antioxidant vitamins and heart disease concluded that the data thus far do not support routine use of vitamin E supplements for the prevention of heart disease.[32] Several leading vitamin E researchers point out that clinical trials of vitamin E supplementation differ in many important ways, such as the selection of the subjects, the source of the vitamin, the dose of the vitamin, and the outcomes studied. Such differences may partly explain the inconsistent findings.

Reminder: *Oxidation* is a type of chemical reaction, so named because oxygen is one of the agents that often brings it about.

tocopherol (tuh-KOFF-er-ol): a general term for several chemically related compounds, one of which has vitamin E activity.

Future research that addresses such issues may clarify the relationship between vitamin E and heart disease. In the meantime, the American Heart Association and a U.S. Preventive Services Task Force of scientists concluded that insufficient support exists to recommend taking vitamin E or other antioxidant supplements to prevent cardiovascular disease.[33] The American Heart Association does support the consumption of antioxidant-rich fruits and vegetables, as well as whole grains and nuts, to reduce the risk of heart disease.[34]

Vitamin E Myths Although research continues to reveal possible roles for vitamin E, it has also clearly discredited claims that vitamin E improves athletic skill, enhances sexual performance, or cures sexual dysfunction in males. Vitamin E also does not prevent or cure hereditary **muscular dystrophy,** nor does it slow or prevent processes of aging, such as graying of the hair, wrinkling of the skin, or reduced activity of body organs.

Vitamin E Deficiency When blood concentrations of vitamin E fall below a certain critical level, the red blood cells tend to break open and spill their contents, probably because the PUFA in their membranes oxidize. This classic vitamin E–deficiency symptom, known as **erythrocyte hemolysis,** is seen in premature infants born before the transfer of vitamin E from the mother to the fetus that takes place in the last weeks of pregnancy. Vitamin E treatment corrects erythrocyte hemolysis.

The few symptoms of vitamin E deficiency that have been observed in adults include loss of muscle coordination and reflexes with impaired movement, vision, and speech. All of these symptoms may be caused by oxidative damage; vitamin E treatment corrects them.

In adults, vitamin E deficiency is usually associated with diseases, notably those that cause malabsorption of fat. These include diseases of the liver, gallbladder, and pancreas, as well as various hereditary diseases involving digestion and use of nutrients.

On rare occasions, vitamin E deficiencies develop in people without diseases. Most likely, such deficiencies occur after years of eating diets extremely low in fat; using fat substitutes, such as diet margarines and salad dressings, as the only sources of fat; or consuming diets composed of highly processed or "convenience" foods. Extensive heating in the processing of foods destroys vitamin E.

Vitamin E Toxicity Vitamin E supplement use has increased in recent years as its antioxidant action against disease has been recognized. As a result, signs of toxicity are now known or suspected, although vitamin E toxicity is not nearly as common, and its effects are not as serious, as vitamin A or vitamin D toxicity. Extremely high doses of vitamin E interfere with the blood-clotting action of vitamin K and enhance the action of anticoagulant medications, leading to hemorrhage.

Pooled results from studies involving almost 136,000 people revealed that those taking vitamin E in doses greater than 400 IU (268 milligrams) per day were at an increased risk of death from all causes compared with people taking smaller doses.[35] In contrast, three other reports of pooled results from vitamin E supplement trials found no evidence that vitamin E supplementation up to 800 IU (536 milligrams) increased or decreased mortality.[36] Other researchers point out that vitamin E supplements in amounts greater than the RDA are widely used in the United States and other industrialized countries and that reports of adverse effects are rare, thereby giving credence to the safety of supplemental vitamin E in amounts below the Tolerable Upper Intake Level (UL).[37] To err on the safe side, until more is known about the safety of vitamin E supplements, intakes should probably be kept low, and they certainly should not exceed the UL of 1000 milligrams of alpha-tocopherol per day. The UL for vitamin E is more than 65 times greater than the recommended intake for adults (15 milligrams).

Nutrition and upper GI disorders are discussed in Chapter 17. Nutrition and lower GI disorders are discussed in Chapter 18. Nutrition and liver disorders are discussed in Chapter 19.

muscular dystrophy (DIS-tro-fee): a hereditary disease in which the muscles gradually weaken, with the most debilitating effects occurring in the lungs. This disease should not be confused with *nutritional* muscular dystrophy, a vitamin E–deficiency disease of animals characterized by gradual paralysis of the muscles.

erythrocyte (er-REETH-ro-cite) **hemolysis** (he-MOLL-uh-sis): rupture of the red blood cells, caused by vitamin E deficiency.
  erythro = red
  cyte = cell
  hemo = blood
  lysis = breaking

Vitamin E in Foods Vitamin E is widespread in foods. About 20 percent of the vitamin E in the diet comes from vegetable oils and the products made from them, such as margarine, salad dressings, and shortenings. (Soybean oils and wheat germ oils have especially high concentrations of vitamin E.) Another 20 percent comes from fruits and vegetables although none of these is a good source by itself. Fortified cereals and other grain products contribute about 15 percent of vitamin E in the diet, and meats, poultry, fish, eggs, milk products, nuts, and seeds contribute smaller percentages. Because vitamin E is readily destroyed by heat processing and oxidation, fresh or lightly processed foods are the best sources of this vitamin.

Prior to 2000, values of the vitamin E in food reflected all of the different tocopherols and were expressed in "milligrams of tocopherol equivalents." These measures overestimated the amount of alpha-tocopherol. To estimate the alpha-tocopherol content of foods stated in tocopherol equivalents, multiply by 0.8.

Vegetable oils, some nuts and seeds such as almonds and sunflower seeds, and wheat germ are rich in vitamin E.

VITAMIN K

Vitamin K has long been known for its role in blood clotting, where its presence can make the difference between life and death. The vitamin also participates in the synthesis of several bone proteins.[38] Without vitamin K, the bones produce an abnormal protein that cannot bind to the minerals that normally form bones, so bone density is low.[39] Short-term vitamin K depletion increases the rate of bone turnover, and the rate of bone turnover is a major determinant of bone mineral density.[40] In young girls, better vitamin K status is associated with decreased bone turnover.[41] Thus vitamin K may influence the risk of fracture: people who consume abundant vitamin K, often in the form of green leafy vegetables, suffer fewer hip fractures than those with lower intakes.[42]

K stands for the Danish word *koagulation* (coagulation or "clotting").

Blood Clotting At least 13 different proteins and the mineral calcium are involved in making blood clot. Vitamin K is essential for the activation of seven of these proteins, among them prothrombin, the precursor of the enzyme thrombin (see Figure 8-4). When any of the blood-clotting factors is lacking, **hemorrhagic disease** results. If an artery or vein is cut or broken, bleeding goes unchecked. Of course, this is not to say that hemorrhaging is always caused by a vitamin K deficiency.

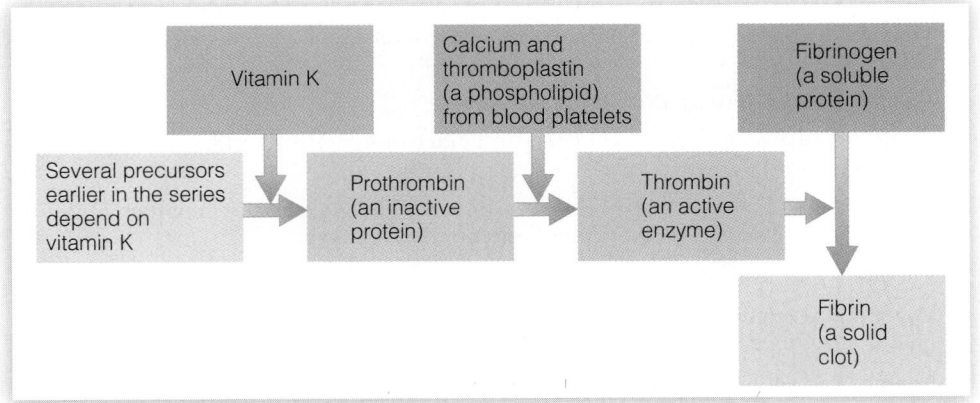

FIGURE 8-4 Blood-Clotting Process
When blood is exposed to air, foreign substances, or secretions from injured tissues, platelets (small, cell-like structures in the blood) release a phospholipid known as thromboplastin. Thromboplastin catalyzes the conversion of the inactive protein prothrombin to the active enzyme thrombin. Thrombin then catalyzes the conversion of the precursor protein fibrinogen to the active protein fibrin that forms the clot.

hemorrhagic (hem-oh-RAJ-ik) **disease:** the vitamin K–deficiency disease in which blood fails to clot.

© Polara Studios, Inc.

Notable food sources of vitamin K include milk, eggs, brussels sprouts, cabbage, and spinach.

Reminder: The bacterial inhabitants of the digestive tract are known as the *intestinal flora.*
flora = plant inhabitants

Vitamin K AI:
- Men: 120 μg/day.
- Women: 90 μg/day.

Intestinal Synthesis Like vitamin D, vitamin K can be obtained from a nonfood source. Bacteria in the intestinal tract synthesize vitamin K that the body can absorb, but people cannot depend on this source alone for their vitamin K.

Vitamin K Deficiency Vitamin K deficiency is rare, but it may occur in two circumstances. First, it may arise in conditions of fat malabsorption. Second, some medications interfere with vitamin K's synthesis and action in the body: antibiotics kill the vitamin K–producing bacteria in the intestine, and anticoagulant medications interfere with vitamin K metabolism and activity. When vitamin K deficiency does occur, it can be fatal.

Vitamin K for Newborns Newborn infants present a unique case of vitamin K nutrition. An infant is born with a **sterile** digestive tract, and some weeks pass before the vitamin K–producing bacteria become fully established in the infant's intestines. At the same time, plasma prothrombin concentrations are low (this helps prevent blood clotting during the stress of birth, which might otherwise be fatal). A single dose of vitamin K, usually in a water-soluble form, is recommended at birth to prevent hemorrhagic disease in the newborn.

Vitamin K Toxicity Vitamin K toxicity is rare, and no adverse effects have been reported with high intakes. Therefore, a Tolerable Upper Intake Level has not been established. High doses of vitamin K can reduce the effectiveness of anticoagulant medications used to prevent blood clotting. People taking these medications should eat vitamin K–rich foods in moderation and keep their intakes consistent from day to day.

Vitamin K in Foods Many foods contain ample amounts of vitamin K, notably, green leafy vegetables, members of the cabbage family, and liver. Other vegetables, milk, meats, eggs, cereal, and fruits provide smaller, but still significant, amounts.

REVIEW NOTES

The fat-soluble vitamins are vitamins A, D, E, and K.

Vitamin A is essential to gene expression, vision, cell differentiation and integrity of epithelial tissues, immunity, and reproduction and growth.

Vitamin A deficiency can cause blindness, sickness, and death and is a major problem worldwide.

Overdoses of vitamin A are possible and dangerous.

Vitamin D raises calcium and phosphorus levels in the blood. A deficiency can cause rickets in children or osteomalacia in adults.

Vitamin D is the most toxic of all the vitamins.

People exposed to the sun make vitamin D in their skin; fortified milk is an important food source.

Vitamin E acts as an antioxidant in cell membranes and is especially important in the lungs where cells are exposed to high concentrations of oxygen.

Vitamin E may protect against heart disease, but evidence is not conclusive yet.

Vitamin E deficiency is rare in healthy human beings. The vitamin is widely distributed in plant foods.

Vitamin K is necessary for blood to clot and for bone health.

The bacterial inhabitants of the digestive tract produce vitamin K, but people need vitamin K from foods as well.

Dark green, leafy vegetables are good sources of vitamin K.

Table 8-2 offers a complete summary of the fat-soluble vitamins.

sterile: free of microorganisms, such as bacteria.

TABLE 8-2	The Fat-Soluble Vitamins—A Summary			
Vitamin Name	**Chief Functions**	**Deficiency Symptoms**	**Toxicity Symptoms**	**Significant Sources**
Vitamin A (Retinol, retinal, retinoic acid; main precursor is beta-carotene)	Vision, maintenance of cornea, epithelial cells, mucous membranes, skin; bone and tooth growth; reproduction; regulation of gene expression; immunity	Infectious diseases, night blindness, blindness (xerophthalmia), keratinization	Reduced bone mineral density, liver abnormalities, birth defects	Retinol: milk and milk products; eggs; liver Beta-carotene: spinach and other dark, leafy greens; broccoli; deep orange fruits (apricots, cantaloupe) and vegetables (carrots, winter squashes, sweet potatoes, pumpkin)
Vitamin D (Calciferol, cholecalciferol, dihydroxy vitamin D; precursor is cholesterol)	Mineralization of bones (raises blood calcium and phosphorus by increasing absorption from digestive tract, withdrawing calcium from bones, stimulating retention by kidneys)	Rickets, osteomalacia	Calcium imbalance (calcification of soft tissues and formation of stones)	Synthesized in the body with the help of sunshine; fortified milk, margarine, butter, and cereals; eggs; liver; fatty fish (salmon, sardines)
Vitamin E (Alpha-tocopherol, tocopherol)	Antioxidant (stabilization of cell membranes, regulation of oxidation reactions, protection of polyunsaturated fatty acids [PUFA] and vitamin A)	Erythrocyte hemolysis, nerve damage	Hemorrhagic effects	Polyunsaturated plant oils (margarine, salad dressings, shortenings), green and leafy vegetables, wheat germ, whole-grain products, nuts, seeds
Vitamin K (Phylloquinone, menaquinone, naphthoquinone)	Synthesis of blood-clotting proteins and bone proteins	Hemorrhage	None known	Synthesized in the body by GI bacteria; green leafy vegetables; cabbage-type vegetables; milk; liver

The Water-Soluble Vitamins

The B vitamins and vitamin C are the water-soluble vitamins. These vitamins, found in the watery compartments of foods, are distributed into water-filled compartments of the body. They are easily absorbed into the bloodstream and are just as easily excreted if their blood concentrations rise too high. Thus the water-soluble vitamins are less likely to reach toxic concentrations in the body than are the fat-soluble vitamins. Foods never deliver excessive amounts of the water-soluble vitamins, but the large doses concentrated in vitamin supplements can reach toxic levels.

THE B VITAMINS

Despite advertisements that claim otherwise, the B vitamins do not give people energy. Carbohydrate, fat, and protein—the *energy-yielding* nutrients—supply the fuel for energy. The B vitamins help to burn that fuel but do not serve as fuel themselves.

Coenzymes The eight B vitamins were listed in Table 8-1. Each is part of an enzyme helper known as a **coenzyme.** Some B vitamins have other important functions in the body as well, but the roles these vitamins play as parts of coenzymes are the

coenzyme (co-EN-zime): a small molecule that works with an enzyme to promote the enzyme's activity. Many coenzymes have B vitamins as part of their structure.

co = with

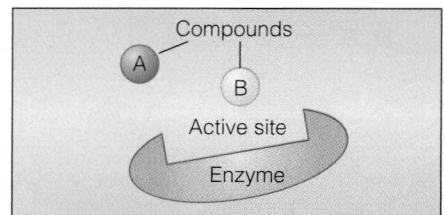

Without the coenzyme, compounds A and B don't respond to the enzyme.

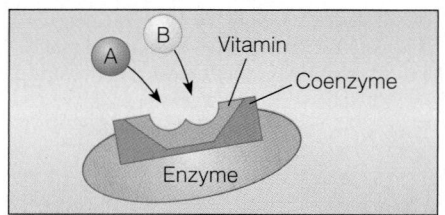

With the coenzyme in place, compounds A and B are attracted to the active site on the enzyme, and they react.

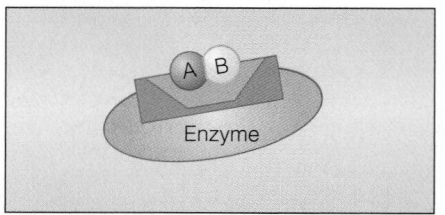

The reaction is completed with the formation of a new product. In this case the product is AB.

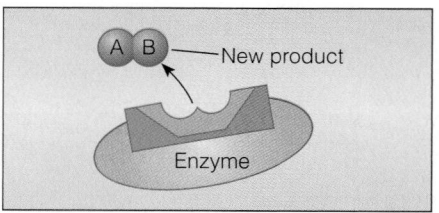

The product AB is released.

FIGURE 8-5 Coenzyme Action

beriberi: the thiamin-deficiency disease; characterized by loss of sensation in the hands and feet, muscular weakness, advancing paralysis, and abnormal heart action.

pellagra (pell-AY-gra): the niacin-deficiency disease. Symptoms include the "4 Ds": diarrhea, dermatitis, dementia, and, ultimately, death.
pellis = skin
agra = seizure

refined grain: a product from which the bran, germ, and husk have been removed, leaving only the endosperm.

best understood. A coenzyme is a small molecule that combines with an enzyme to make it active. With the coenzyme in place, a substance is attracted to the enzyme, and the reaction proceeds instantaneously. Figure 8-5 illustrates coenzyme action.

Active forms of five of the B vitamins—thiamin, riboflavin, niacin, pantothenic acid, and biotin—participate in the release of energy from carbohydrate, fat, and protein. A coenzyme containing vitamin B_6 assists enzymes that metabolize amino acids. The making of new cells depends on a folate coenzyme, and the making of this coenzyme depends on vitamin B_{12}.

The eight B vitamins play many specific roles in helping the enzymes to perform thousands of different molecular conversions in the body. They must be present in every cell continuously for the cells to function as they should. As for vitamin C, its primary role, discussed later, is as an antioxidant.

B Vitamin Deficiencies In academic and clinical discussions of the vitamins, different sets of deficiency symptoms are ascribed to each individual vitamin. Such clear-cut symptoms, however, are found only in laboratory animals that have been fed contrived diets that lack just one nutrient. In reality, a deficiency of any single B vitamin seldom shows up in isolation because people do not eat nutrients one by one; they eat foods containing mixtures of many nutrients. If a major class of foods is missing from the diet, all of the nutrients delivered by those foods will be lacking to various extents.

In only two cases have dietary deficiencies associated with single B vitamins been observed on a large scale in human populations. Diseases have been named for these deficiency states. One of them, **beriberi,** was first observed in Southeast Asia when the custom of polishing rice became widespread. Rice contributed 80 percent of the energy intake of the people in these areas, and rice bran was their principal source of thiamin. When the bran was removed to make the rice whiter, beriberi spread like wildfire.

The niacin-deficiency disease, **pellagra,** became widespread in the southern United States in the early part of the twentieth century among people who subsisted on a low-protein diet with a staple grain of corn. This diet was unusual in that it supplied neither enough niacin nor enough tryptophan, its amino acid precursor, to make the niacin intake adequate.

Even in the cases of beriberi and pellagra, the deficiencies were probably not pure. When foods were provided containing the one vitamin known to be needed, other vitamins that may have been in short supply came as part of the package.

Major deficiency diseases such as pellagra and beriberi no longer occur in the United States and Canada, but more subtle deficiencies of nutrients, including the B vitamins, are sometimes observed. When they do occur, it is usually in people whose food choices are poor because of poverty, ignorance, illness, or poor health habits such as alcohol abuse.

Interdependent Systems Table 8-3, at the end of this chapter, sums up a few of the better-established facts about B vitamin deficiencies. A look at the table will make another generalization possible. Different body systems depend to different extents on these vitamins. Processes in nerves and in their responding tissues, the muscles, depend heavily on glucose metabolism and hence on thiamin, so paralysis sets in when this vitamin is lacking; but thiamin is important in all cells, not just in nerves and muscles. Similarly, because the red blood cells and GI tract cells divide the most rapidly, two of the first symptoms of a deficiency of folate are a type of anemia and GI tract deterioration—but again, all systems depend on folate, not just these. The list of symptoms in Table 8-3 is far from complete.

B Vitamin Enrichment of Foods If the staple food of a region is made from **refined grain,** vitamin B deficiencies are especially likely. One way to protect people from deficiencies is to add nutrients to their staple food, a process known as **fortification** or **enrichment.** The enrichment of refined breads and cereals has drastically reduced the incidence of iron and B vitamin deficiencies.

The preceding discussion has shown both the great importance of the B vitamins in promoting normal, healthy functioning of all body systems and the severe consequences of deficiency. Now you may want to know how to be sure you and your clients are getting enough of these vital nutrients. The next sections present information on each B vitamin. While reading further, keep in mind that *foods* can provide all the needed nutrients and that supplements are a poor second choice. Some supplements are absurdly costly, but even if they are inexpensive, most people don't need them. Nutrition in Practice 9 discusses uses and choices of supplements in more detail.

THIAMIN

All cells use thiamin, which plays a critical role in their energy metabolism. Thiamin also occupies a special site on nerve cell membranes. Consequently, as mentioned earlier, thiamin is critical to the normal functioning of the nerves and muscles.

Thiamin Need As long as people consume enough food to meet their energy needs—and obtain that energy from nutritious foods—thiamin needs will be met. People who derive a large proportion of their energy from empty-kcalorie items, like sugar or alcohol, however, risk thiamin deficiency, a condition that seems to be reappearing as the population of malnourished and homeless people rises. A person who is fasting or who has adopted a very low-kcalorie diet needs as much thiamin as when eating enough to meet energy needs.

In developed countries today, abuse of alcohol may lead to a severe form of thiamin deficiency, Wernicke-Korsakoff syndrome.[43] Alcohol contributes energy but carries almost no nutrients with it and often displaces food. In addition, alcohol impairs absorption of thiamin from the digestive tract and hastens its excretion in urine, tripling the risk of deficiency. Wernicke-Korsakoff syndrome is characterized by symptoms that are almost indistinguishable from alcohol abuse itself: mental confusion, disorientation, loss of memory, jerky eye movements, and staggering gait. Unlike alcohol toxicity, the syndrome responds quickly to an injection of thiamin, and some experts recommend a precautionary dose for any patients suspected of having the syndrome.

Thiamin in Foods Thiamin occurs in small quantities in virtually all nutritious foods, but it is concentrated in only a few foods, of which pork is the most commonly eaten. A useful guideline for meeting thiamin needs is to keep empty-kcalorie foods to a minimum and to include ten or more different servings of nutritious foods each day, assuming that each serving will contribute, on the average, about 10 percent of needs. Foods chosen from the bread and cereal group should be either whole grain or enriched. Thiamin is not stored in the body to any great extent, so daily intake is important.

RIBOFLAVIN

Like thiamin, riboflavin facilitates energy production in the body. The needs of infants, children, and pregnant women rise rapidly during periods of active growth.

Riboflavin in Foods Unlike thiamin, riboflavin is not evenly distributed among the food groups. The major contributors of riboflavin to people's diets are milk, milk products, meats, and green vegetables (broccoli, turnip greens, asparagus,

Nutritious foods such as pork, legumes, sunflower seeds, and enriched and whole-grain breads are valuable sources of thiamin.

Note: The terms *fortified* and *enriched* may be used interchangeably.

Thiamin RDA:
- Men: 1.2 mg/day.
- Women: 1.1 mg/day.

Riboflavin RDA:
- Men: 1.3 mg/day.
- Women: 1.1 mg/day.

fortification: the addition to a food of nutrients that were either not originally present or present in insignificant amounts. Fortification can be used to correct or prevent a widespread nutrient deficiency, to balance the total nutrient profile of a food, or to restore nutrients lost in processing.

enrichment: the addition to a food of nutrients to meet a specified standard. In the case of refined bread or cereal, five nutrients have been added: thiamin, riboflavin, niacin, and folate in amounts approximately equivalent to, or higher than, those originally present and iron in amounts to alleviate the prevalence of iron-deficiency anemia.

Milk and milk products supply much (about 50 percent) of the riboflavin in people's diets, but meats, eggs, green vegetables, and enriched and whole-grain breads and cereals are good sources, too.

Niacin-rich foods include meat, fish, poultry, and peanut butter, as well as enriched breads and cereals and a few vegetables.

A food containing 1 mg of niacin and 60 mg of tryptophan contains the niacin equivalent of 2 mg, or 2 mg NE.

Niacin RDA:
• Men: 16 mg NE/day.
• Women: 14 mg NE/day.

When a normal dose of a nutrient clears up a deficiency condition, the effect is a **physiological** one. When a large dose of a nutrient overwhelms a body system and acts like a drug, the effect is a **pharmacological** one.

niacin equivalents (NE): the amount of niacin present in food, including the niacin that can theoretically be made from trypto-phan, its precursor, present in the food.

and spinach). The riboflavin richness of milk and milk products is a good reason to include these foods in every day's meals. No other commonly eaten food can make such a substantial contribution. People who omit milk and milk products from their diets can substitute generous servings of dark green, leafy vegetables. Among the meats, liver and heart are the richest sources, but all lean meats, as well as eggs, offer some riboflavin.

Effects of Light Riboflavin is light sensitive; the ultraviolet rays of the sun or of fluorescent lamps can destroy it. For this reason, milk is sold in cardboard or opaque plastic containers to protect the riboflavin in the milk from light. In con-trast, riboflavin is heat stable, so ordinary cooking does not destroy it.

NIACIN

Like thiamin and riboflavin, niacin participates in the energy metabolism of every body cell. Niacin is unique among the B vitamins in that the body can make it from protein. The amino acid tryptophan can be converted to niacin in the body: 60 milligrams of tryptophan yield 1 milligram of niacin. Recommended intakes are therefore stated in **niacin equivalents (NE)**, reflecting the body's ability to convert tryptophan to niacin.

Certain forms of niacin supplements in amounts ten times or more than the dietary recommendation cause "niacin flush," a dilation of the capillaries of the skin with perceptible tingling that, if intense, can be painful. The Tolerable Upper Intake Level (35 milligrams NE) is based on flushing as the critical adverse effect.

Niacin Used as Medication Physicians sometimes use diet and large doses of a form of niacin (nicotinic acid) to lower blood cholesterol in the treatment of ath-erosclerosis. When used this way, niacin leaves the realm of nutrition to become a pharmacological agent, a drug. As with any medication, self-dosing with niacin is ill advised; large doses may injure the liver and produce some symptoms of diabetes.[44]

Niacin in Foods Meat, poultry, and fish contribute about half the niacin equiva-lents most people consume; enriched breads and cereals contribute about a fourth. Among the vegetables, mushrooms, asparagus, and green leafy vegetables are the richest niacin sources. Niacin is less vulnerable to losses during food preparation and storage than other water-soluble vitamins. Being fairly heat-resistant, niacin can withstand reasonable cooking times, but like other water-soluble vitamins, it will leach into cooking water.

PANTOTHENIC ACID AND BIOTIN

Two other B vitamins—pantothenic acid and biotin—are also important in energy metabolism. Pantothenic acid was first recognized as a substance that stimulates growth. It is a component of a key enzyme that makes possible the release of energy from the energy nutrients. Pantothenic acid is involved in more than 100 different steps in the synthesis of lipids, neurotransmitters, steroid hormones, and hemoglobin. Biotin plays an important role in metabolism as a coenzyme that car-ries carbon dioxide. Emerging evidence indicates that biotin participates in other processes such as gene expression and cell signaling and in the structure of DNA-binding proteins in the cell nucleus.[45]

Pantothenic Acid and Biotin in Foods Both pantothenic acid and biotin are more widespread in foods than the other vitamins discussed so far. There seems to be no danger that people who consume a variety of foods will suffer deficiencies. Claims that pantothenic acid and biotin are needed in pill form to prevent or cure disease conditions are at best unfounded and at worst intentionally misleading.

Biotin Deficiency Biotin deficiencies are rare but have been reported in adults fed artificially by vein without biotin supplementation. Researchers can induce biotin deficiency in animals or human beings by feeding them raw egg whites, which contain a protein that binds biotin and prevents its absorption. Long-term use of anticonvulsant medication may also lead to biotin deficiency, as may alcohol abuse.[46]

VITAMIN B$_6$

Vitamin B$_6$ has been called the "sleeping giant" of vitamins. A surge of research interest in the last two decades has not only revealed new knowledge but has also raised new questions. For example, unlike the other water-soluble vitamins, vitamin B$_6$ is stored extensively in muscle tissue. Most recently, research interest has centered on a possible role for vitamin B$_6$ in the treatment of disease.

Metabolic Roles of Vitamin B$_6$ Vitamin B$_6$ has long been known to play roles in protein and amino acid metabolism. In the cells, vitamin B$_6$ helps to convert one kind of amino acid, which the cells have in abundance, to another, which they need in larger amounts. It also aids in the conversion of the amino acid tryptophan to niacin and plays important roles in the synthesis of hemoglobin and neurotransmitters (neurotransmitters are the communication molecules of the brain). Vitamin B$_6$ also assists in releasing stored glucose from glycogen and thus contributes to the regulation of blood glucose. Research suggests roles for vitamin B$_6$ in immune function, cognitive performance, and hormone response.[47] The association between vitamin B$_6$ and immune function is related to the critical role the vitamin plays in protein metabolism. Vitamin B$_6$ deficiency can significantly impair the immune response, perhaps by way of impaired antibody production.

Vitamin B$_6$ status may be related to cardiovascular disease risk. Elevated blood levels of the amino acid homocysteine correlate with a high incidence of heart disease.[48] Evidence suggests that low blood concentrations of vitamin B$_6$, Vitamin B$_{12}$, and folate are associated with elevated homocysteine concentrations.[49] In a study of more than 7000 people, the use of B vitamin supplements over a six-year period lowered homocysteine concentrations.[50] Whether raising blood concentrations of vitamins and lowering blood homocysteine levels reduces a person's risk of developing heart disease is unknown, however.[51] Higher dietary intakes of folate, vitamin B$_{12}$, and vitamin B$_6$ also correlate with lower blood values for other substances associated with heart disease and with a lower incidence of heart disease itself. The missing link is how, or even whether, B vitamin deficiencies or elevated homocysteine directly affects processes leading to heart disease.[52]

Vitamin B$_6$ Deficiency Besides a weakening immune response, vitamin B$_6$ deficiency is expressed in general symptoms, such as weakness, irritability, and insomnia. Other symptoms include a greasy, flaky dermatitis; anemia; and, in advanced cases, convulsions.

Vitamin B$_6$ Toxicity For years it was believed that vitamin B$_6$, like other water-soluble vitamins, could not reach toxic concentrations in the body. Toxic effects of vitamin B$_6$ became known when a physician reported them in women who had been taking more than 2 grams of vitamin B$_6$ daily (20 times the current Tolerable Upper Intake Level) for two months or more. Most of these women had been attempting to relieve premenstrual syndrome (PMS), the cluster of physical, emotional, and psychological symptoms that some women experience prior to menstruation. The first symptom of toxicity was numb feet; then the women lost sensation in their hands; then they became unable to walk. The women recovered after they discontinued the supplements.

Pantothenic acid AI:
• Adults: 5 mg/day.

Biotin AI:
• Adults: 30 μg/day.

The protein **avidin** in egg whites binds biotin.

Most protein-rich foods such as meat, fish, and poultry provide ample vitamin B₆; some vegetables and fruits are good sources, too.

Vitamin B₆ RDA:
- Adults (19–50): 1.3 mg/day.
- Adults (51– >70): 1.7 mg/day.

The two main types of neural tube defects are **spina bifida** (literally, "split spine") and **anencephaly** ("no brain").

Folate RDA:
- Adults: 400 μg/day.

Bread products, flour, corn grits, and pasta must be fortified with 140 μg per 100 g of food (about ¹/₂ c cooked food or 1 slice of bread).

neural tube defects (NTD): malformations of the brain, spinal cord, or both during embryonic development.

Vitamin B₆ Recommendations Because vitamin B₆ coenzymes play many roles in amino acid metabolism, previous RDA were expressed in terms of protein intakes; the current RDA for vitamin B₆, however, is not. Research does not support claims that large doses of vitamin B₆ enhance muscle strength or physical endurance.

Vitamin B₆ in Foods The richest food sources of vitamin B₆ are protein-rich meat, fish, and poultry. Potatoes, a few other vegetables, and some fruits are good sources, too. Foods lose vitamin B₆ when heated.

FOLATE

The B vitamin folate is active in cell division. During periods of rapid growth and cell division, such as pregnancy and adolescence, folate needs increase, and deficiency is especially likely. When a deficiency occurs, the replacement of the rapidly dividing cells of the blood and the GI tract falters. Not surprisingly, then, two of the first symptoms of a folate deficiency are a type of anemia and GI tract deterioration (see Table 8-3, p. 211).

Folate, Alcohol, and Drugs Of all the vitamins, folate appears to be the most vulnerable to interactions with alcohol and other drugs.[53] As Nutrition in Practice 19 describes, alcohol-addicted people risk folate deficiency because alcohol impairs folate's absorption and increases its excretion. Furthermore, as people's alcohol intakes rise, their folate intakes decline. Many medications, including aspirin, oral contraceptives, and anticonvulsants, also impair folate status. Smoking exerts a negative effect on folate status as well.

Folate and Neural Tube Defects Research studies confirm the importance of folate in preventing **neural tube defects (NTD).** The brain and spinal cord develop from the neural tube, and defects in its orderly formation during the early weeks of pregnancy may result in various central nervous system disorders and death. Folate supplements taken before conception and continued throughout the first trimester of pregnancy can prevent NTD. For this reason, the American Academy of Pediatrics and the Public Health Services recommend that all women of childbearing age who are capable of becoming pregnant take 0.4 milligram (400 micrograms) of folate daily.

Folate status improves more with supplementation or fortification than with a dietary intake that meets recommendations. Neural tube defects arise early in pregnancy before most women realize they are pregnant, and most women eat too few fruits and vegetables to supply even half the folate needed to prevent NTD. For these reasons, in the late 1990s, the FDA mandated that enriched grain products (flour, cornmeal, pasta, and rice) be fortified with an especially absorbable synthetic form of folate, folic acid. Since this fortification began, typical folate intakes from fortified foods have increased dramatically—by more than double the predicted levels—and observers report an almost 25 percent drop in the national incidence of NTD, even among women receiving late or no prenatal care.[54] Researchers expect to see declines in some other birth defects and miscarriages as well.[55] Folate fortification also raises safety concerns, however. High doses of folate can complicate the diagnosis of vitamin B₁₂ deficiency, as discussed later. The DRI committee set a Tolerable Upper Intake Level of 1000 micrograms per day from fortified foods or supplements.

Folate status, like vitamin B₆ status, may be related to cardiovascular disease risk.[56] As discussed earlier, elevated levels of homocysteine are associated with a greater risk of cardiovascular disease. One of folate's key roles in the body is to metabolize homocysteine. Without folate, homocysteine accumulates. Fortified foods and folate supplements raise blood folate and reduce homocysteine levels.[57]

SNAPSHOT 8-2 Folate

GOOD SOURCES*

BEEF LIVER (cooked)
3 oz = 221 μg

PINTO BEANS (cooked)
1/2 c = 146 μg

ASPARAGUS
1/2 c = 131 μg

AVOCADO
1/2 c = 45 μg

LENTILS (cooked)
1/2 c = 179 μg

SPINACH (raw)
1 c = 58 μg

ENRICHED CEREAL
(ready-to-eat)a
3/4 c = 82 μg

BEETS
1/2 c = 68 μg

*These foods provide 10 percent or more of the folate Daily Value in a serving. For a 2000-kcalorie diet, the DV is 400 μg/day.
aSome highly enriched cereals may provide 400 or more micrograms in a serving.

Folate in Foods As Snapshot 8-2 shows, the best food sources of folate are liver, legumes, beets, and leafy green vegetables (the vitamin's name suggests the word *foliage*). Among the fruits, oranges, orange juice, and cantaloupe are the best sources. With fortification, grain products are good sources of folate, too. Heat and oxidation during cooking and storage can destroy up to half of the folate in foods.

The difference in absorption between naturally occurring food folate and synthetic folate that enriches foods and is added to supplements necessitated a new unit of measurement for folate: the **dietary folate equivalents, or DFE.**[58] The DFE convert all forms of folate into units that are equivalent to the folate in foods. Most food labels and tables of food composition express folate values in micrograms, however, so the accompanying "How to" describes how to estimate dietary folate equivalents.

dietary folate equivalents (DFE): the amount of folate available to the body from naturally occurring sources, fortified foods, and supplements, accounting for differences in bioavailability from each source.

HOW TO *Estimate Dietary Folate Equivalents*

Folate is expressed in terms of DFE (dietary folate equivalents) because synthetic folate from supplements and fortified foods is absorbed at almost twice (1.7 times) the rate of naturally occurring folate from other foods. Use the following equation to calculate:

DFE = μg food folate + (1.7 × μg synthetic folate).

Consider, for example, a pregnant woman who takes a supplement and eats a bowl of fortified corn flakes, 2 slices of fortified bread, and a cup of fortified pasta. From the supplement and fortified foods, she obtains synthetic folate:

Supplement	100 μg folate
Fortified corn flakes	100 μg folate
Fortified bread	40 μg folate
Fortified pasta	60 μg folate
	300 μg folate

To calculate the DFE, multiply the amount of synthetic folate by 1.7:

$$300 \text{ μg} \times 1.7 = 510 \text{ mg DFE.}$$

Now add the naturally occurring folate from the other foods in her diet—in this example, another 90 micrograms of folate.

$$510 \text{ μg DFE} + 90 \text{ μg} = 600 \text{ μg DFE.}$$

Notice that if we had not converted synthetic folate from supplements and fortified foods to DFE, this woman's intake would appear to fall short of the 600 micrograms recommended for pregnancy (300 μg + 90 μg = 390 μg). But as our example shows, her intake does meet the recommendation. At this time, supplement and fortified food labels list folate in micrograms only, not micrograms DFE, making such calculations necessary.

VITAMIN B₁₂

Vitamin B_{12} and folate share a special relationship: vitamin B_{12} assists folate in cell division. Their roles intertwine, but each performs a specific task that the other cannot accomplish.

Vitamin B₁₂, Folate, and Cell Division   Vitamin B_{12} (in coenzyme form) stands by to accept carbon groups from folate as folate removes them from other compounds. The passing of these carbon groups from folate to vitamin B_{12} regenerates the active form of folate so that it can continue its dismantling tasks. In the absence of vitamin B_{12}, folate is trapped in its inactive, metabolically useless form, unable to do its job. When folate is either trapped due to a vitamin B_{12} deficiency or unavailable due to a deficiency of folate itself, cells that are growing most rapidly, notably, the blood cells, are the first to be affected. Thus a deficiency of either nutrient—vitamin B_{12} or folate—impairs maturation of the blood cells and produces anemia. The anemia is identifiable by microscopic examination of the blood, which reveals many large, immature red blood cells. Either vitamin B_{12} or folate will clear up the anemia.

Large-cell anemia is known as macrocytic or megaloblastic anemia.
macro = large
cyte = cell
mega = large
blast = immature cell

Vitamin B₁₂ and the Nervous System   Although either vitamin will clear up the anemia caused by vitamin B_{12} deficiency, if folate is given when vitamin B_{12} is needed, the result is disastrous, not to the blood but to the nervous system. The reason: vitamin B_{12} also helps maintain nerve fibers. A vitamin B_{12} deficiency can ultimately result in devastating neurological symptoms, undetectable by a blood test. A deceptive folate "cure" of the anemia in vitamin B_{12} deficiency allows the nerve deterioration to progress, leading to paralysis and permanent nerve damage. This interaction between folate and vitamin B_{12} raises safety concerns about the use of folate supplements and fortification of foods.

The way folate masks vitamin B_{12} deficiency underlines a point already made several times: it takes a skilled diagnostician to make a correct diagnosis. A person who self-diagnoses on the basis of a single observed symptom takes a serious risk.

Vitamin B₁₂ Absorption   Vitamin B_{12} requires an **"intrinsic factor"**—a compound made inside the body—for absorption from the intestinal tract into the bloodstream. This intrinsic factor is made in the stomach, where it attaches to the vitamin; the complex then passes to the small intestine and is gradually absorbed.

Loss of Intrinsic Factor   In some cases, intrinsic factor production becomes inadequate or ceases altogether—for example, after surgical removal of the stomach. Some people inherit a defective gene for intrinsic factor. Because vitamin B_{12} deficiency in the body may be caused either by a lack of the vitamin in the diet or by the body's inability to absorb the vitamin, a change in diet alone may not correct the deficiency. When absorption failure is the problem, vitamin B_{12} must be supplied by injection.

Vitamin B₁₂ RDA:
• Adults: 2.4 μg/day.

intrinsic factor: inside the system. Anemia that reflects a vitamin B_{12} deficiency caused by lack of intrinsic factor is known as **pernicious anemia.**

Vitamin B₁₂ in Foods   A unique characteristic of vitamin B_{12} is that it is found almost exclusively in foods derived from animals. People who eat meat are guaranteed an adequate intake, and lacto-ovo vegetarians (who use milk, cheese, and eggs) are also protected from deficiency. It is a myth, however, that fermented soy products, such as miso (a soybean paste), or sea algae, such as spirulina, provide vitamin B_{12} in its active form. Extensive research shows that the amounts of vitamin B_{12} listed on the labels of these plant products are inaccurate and misleading because the vitamin B_{12} in these products occurs in an inactive, unavailable form. Vegans must take vitamin B_{12} supplements or find other sources of active vitamin B_{12}. Some loss of vitamin B_{12} occurs when foods are heated in microwave ovens.[59]

Vitamin B₁₂ Deficiency in Vegans Vegans are at special risk for undetected vitamin B_{12} deficiency for two reasons: first, they receive none in their diets, and second, they consume large amounts of folate in the vegetables they eat. Because the body can store many times the amount of vitamin B_{12} used each day, a deficiency may take years to develop in a new vegetarian. When a deficiency does develop, though, it may progress to a dangerous extreme because the deficiency of vitamin B_{12} may be masked by the high folate intake.

Worldwide, vitamin B_{12} deficiency among vegetarians is a growing problem.[60] A pregnant or lactating vegetarian woman who eats no foods of animal origin should be aware that her infant can develop a vitamin B_{12} deficiency, even if the mother appears healthy. Breastfed infants born to vegan mothers with low concentrations of vitamin B_{12} in their breast milk can develop severe neurological symptoms such as seizures and cognitive problems.[61]

NON-B VITAMINS

Other compounds are sometimes inappropriately called B vitamins because, like the true B vitamins, they serve as coenzymes in metabolism. Even if they were essential, however, supplements would be unnecessary because these compounds are abundant in foods.

Inositol, Choline, and Carnitine Among the non-B vitamins are a trio of substances known as inositol, choline, and carnitine. Researchers are exploring the possibility that these substances may be essential. Thus far, only choline has been assigned an Adequate Intake value.

Choline AI:
- Men: 550 mg/day.
- Women: 425 mg/day.

Other Non-B Vitamins Other substances have also been mistaken for essential nutrients. They include para-aminobenzoic acid (PABA), bioflavonoids (vitamin P or hesperidin), and ubiquinone. Other names you may hear are "vitamin B_{15}" (a hoax) and "vitamin B_{17}" (laetrile, a fake cancer-curing drug and not a vitamin by any stretch of the imagination). There is, however, one other water-soluble vitamin of great interest and importance—vitamin C.

VITAMIN C

Three hundred years ago, any man who joined the crew of a seagoing ship knew he had only half a chance of returning alive—not because he might be slain by pirates or die in a storm, but because he might contract the dread disease **scurvy.** Then a physician with the British navy found that citrus fruits could cure the disease, and thereafter, all ships were required to carry lime juice for every sailor. (This is why British sailors are still called "limeys" today.) In the 1930s, the antiscurvy factor in citrus fruits was isolated from lemon juice and named **ascorbic acid.** Today, hundreds of millions of vitamin C pills are produced in pharmaceutical laboratories.

scurvy: the vitamin C–deficiency disease.

ascorbic acid: one of the two active forms of vitamin C. Many people refer to vitamin C by this name.
  a = without
  scorbic = having scurvy

Metabolic Roles of Vitamin C Vitamin C's action defies a simple, tidy description. It plays many important roles in the body, and its modes of action differ in different situations.

Vitamin C's Role in Collagen Formation The best-understood action of vitamin C is its role in helping to form **collagen,** the single most important protein of connective tissue. Collagen serves as the matrix on which bone is formed, the material of scars, and an important part of the "glue" that attaches one cell to another.

collagen: the characteristic protein of connective tissue.
  kolla = glue
  gennan = produce

This latter function is especially important in the artery walls, which must expand and contract with each beat of the heart, and in the walls of the capillaries, which are thin and fragile. Vitamin C also plays a role in the production of carnitine, important for transporting fatty acids within cells.

Vitamin C as an Antioxidant Vitamin C is also an important antioxidant. Recall that the antioxidants beta-carotene and vitamin E protect fat-soluble substances from oxidizing agents; vitamin C protects water-soluble substances the same way. By being oxidized itself, vitamin C regenerates already-oxidized substances such as iron and copper to their original, active form. In the intestines, it protects iron from oxidation and so enhances iron absorption. In the cells and body fluids, it helps to protect other molecules, including the fat-soluble compounds vitamin A, vitamin E, and the polyunsaturated fatty acids.

Vitamin C in Amino Acid Metabolism Vitamin C is also involved in the metabolism of several amino acids. Some of these amino acids end up being used to make hormones of great importance in body functioning, among them norepinephrine and thyroxine.

Role of Stress During stress, the adrenal glands release large quantities of vitamin C together with the stress hormones epinephrine and norepinephrine. What the vitamin has to do with the stress reaction is unclear, but it is known that stress increases vitamin C needs somewhat.

Vitamin C as a Possible Antihistamine Newspaper headlines touting vitamin C as a cure for colds and cancer have appeared frequently over the years. Some research suggests that vitamin C (2 grams per day for two weeks) may reduce the severity and duration of cold and allergy symptoms by reducing blood histamine concentrations. In other words, vitamin C acts as an antihistamine. If further research confirms vitamin C's antihistamine effect, its use may permit people to rely less heavily on antihistamine drugs when suffering from cold and allergy symptoms.

Vitamin C's Role in Cancer Prevention and Treatment The role of vitamin C in the prevention and treatment of cancer is still being studied. In a dozen or so different well-controlled studies, researchers identified individuals with and without cancer and assessed their dietary intakes of vitamin C. They found that people with high vitamin C intakes had lower risks of cancer than did people with low intakes. The correlation may reflect not just an association with vitamin C, but the broader benefits of a diet rich in fruits and vegetables and low in fat. It does not support the taking of vitamin C supplements to prevent or treat cancer.

Vitamin C Deficiency When intake of vitamin C is inadequate, the body's vitamin C pool dwindles, and **latent** scurvy appears. The blood vessels show the first deficiency signs. The gums around the teeth begin to bleed easily, and capillaries under the skin break spontaneously, producing pinpoint hemorrhages. Then the symptoms of **overt** scurvy appear. Muscles, including the heart muscle, may degenerate. The skin becomes rough, brown, scaly, and dry. Wounds fail to heal because scar tissue will not form without collagen. Bone rebuilding falters; the ends of the long bones become softened, malformed, and painful; and fractures occur. The teeth may become loose in the jawbone and fall out. Anemia and infections are common. Sudden death is likely, perhaps because of massive bleeding into the joints and body cavities.

latent: the period in the course of a disease when the condition is present but the symptoms have not begun to appear.
latens = lying hidden

overt: out in the open, full-blown.
ouvrire = to open

It takes only 10 or so milligrams of vitamin C a day to prevent overt scurvy, and not much more than that to cure it. Once diagnosed, scurvy is readily reversible with moderate doses, in the neighborhood of 100 milligrams per day. Such an intake is easily achieved by including vitamin C–rich foods in the diet.

Vitamin C Toxicity The easy availability of vitamin C in pill form and the publication of books recommending vitamin C to prevent everything from the common cold to life-threatening cancer have led thousands of people to take megadoses of vitamin C. Not surprisingly, instances of vitamin C causing harm have surfaced.

Some of the suspected toxic effects of vitamin C megadoses have not been confirmed, but others have been seen often enough to warrant concern. Nausea, abdominal cramps, and diarrhea are often reported. Several instances of interference with medical regimens are known. Large amounts of vitamin C excreted in the urine obscure the results of tests used to detect diabetes. People taking anticoagulants may unwittingly counteract the effect of these medications if they also take massive doses of vitamin C. Vitamin C megadoses can also enhance iron absorption too much, resulting in iron overload (see Chapter 9).

People with sickle-cell anemia may be especially vulnerable to megadoses of vitamin C. Those who have a tendency toward **gout,** as well as those who have a genetic abnormality that alters the way they metabolize vitamin C, are more prone to forming kidney stones if they take megadoses of vitamin C.

Recommended Intakes of Vitamin C The vitamin C RDA is 90 milligrams for men and 75 milligrams for women. These amounts are far higher than the 10 milligrams per day needed to prevent the symptoms of scurvy. In fact, they are close to the amount at which the body's pool of vitamin C is full to overflowing: about 100 milligrams per day.

Special Needs for Vitamin C As is true of all nutrients, unusual circumstances may raise vitamin C needs. Among the stresses known to do so are infections; burns; surgery; extremely high or low temperatures; toxic doses of heavy metals, such as lead, mercury, and cadmium; and the chronic use of certain medications, including aspirin, barbiturates, and oral contraceptives. Smoking, too, has adverse effects on vitamin C status. Cigarette smoke contains oxidants, which deplete this potent antioxidant. Accordingly, the vitamin C recommendation for smokers is set high, at 125 milligrams for men and 105 milligrams for women.

Safe Limits Few instances warrant the taking of more than 100 to 300 milligrams of vitamin C a day. The risks may not be great for adults who dose themselves with 1 to 2 grams a day, but those taking more than 2 grams, and especially those taking above 3 grams per day, should be aware of the distinct possibility of harm.[62]

Vitamin C in Foods The inclusion of intelligently selected fruits and vegetables in the daily diet guarantees a generous intake of vitamin C. Even those who wish to ingest amounts well above the RDA can easily meet their goals by eating certain foods (see Snapshot 8-3, p. 210). Citrus fruits are rightly famous for their vitamin C contents. Certain other fruits and vegetables are also rich sources: cantaloupe, strawberries, broccoli, and brussels sprouts. No animal foods other than organ meats, such as chicken liver and kidneys, contain vitamin C. The humble potato is an important source of vitamin C in Western countries, where potatoes are eaten so frequently that they make substantial vitamin C contributions overall. They provide

Doses of 10 to 30 or more times the recommended intake of a nutrient are termed **megadoses.** In the case of vitamin C, current recommendations are 75 mg/day for women and 90 mg/day for men. The Tolerable Upper Intake Level for vitamin C is 2000 mg/day.

The anticoagulants with which vitamin C interferes are warfarin and dicumarol.

gout (GOWT): a metabolic disease in which crystals of uric acid precipitate in the joints.

SNAPSHOT 8-3 Vitamin C

GOOD SOURCES*

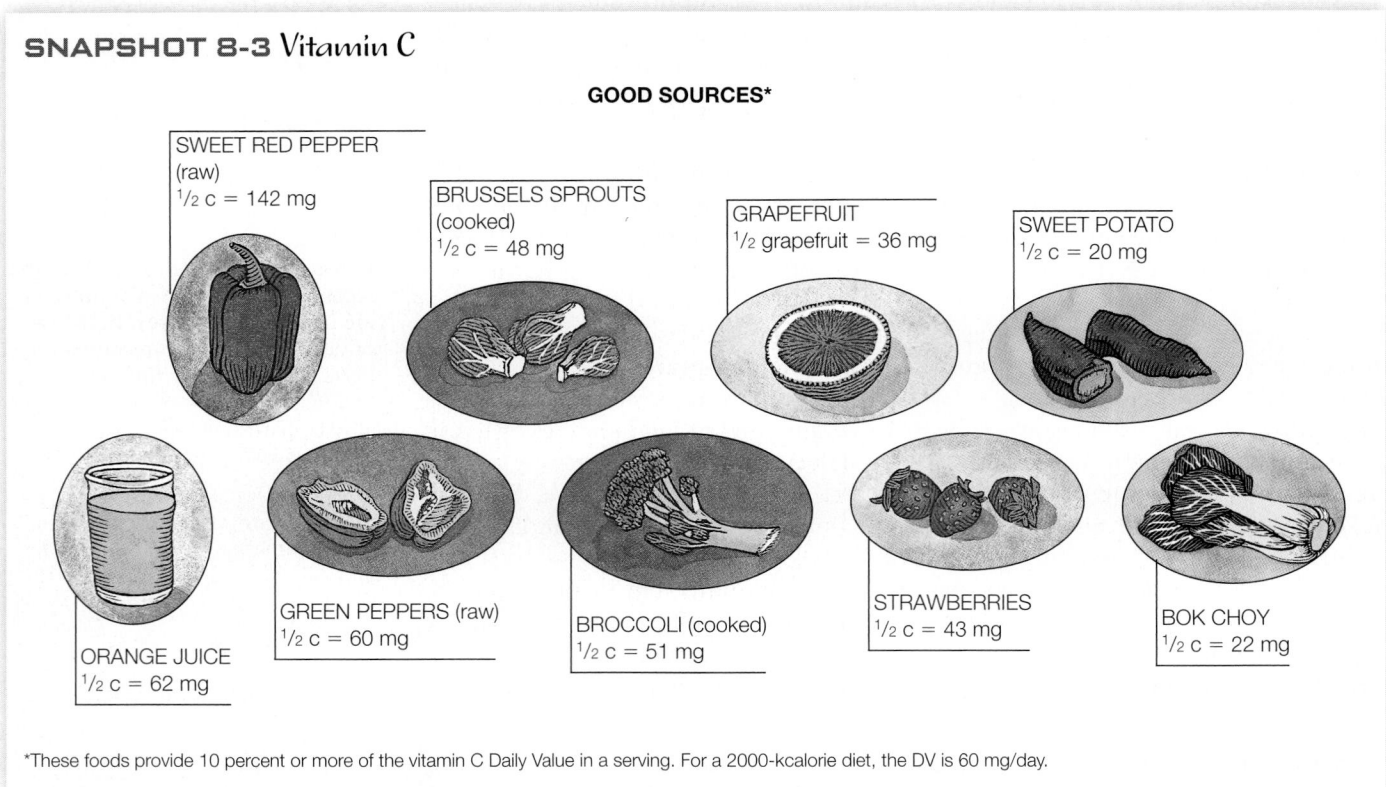

SWEET RED PEPPER (raw)
$1/2$ c = 142 mg

BRUSSELS SPROUTS (cooked)
$1/2$ c = 48 mg

GRAPEFRUIT
$1/2$ grapefruit = 36 mg

SWEET POTATO
$1/2$ c = 20 mg

ORANGE JUICE
$1/2$ c = 62 mg

GREEN PEPPERS (raw)
$1/2$ c = 60 mg

BROCCOLI (cooked)
$1/2$ c = 51 mg

STRAWBERRIES
$1/2$ c = 43 mg

BOK CHOY
$1/2$ c = 22 mg

*These foods provide 10 percent or more of the vitamin C Daily Value in a serving. For a 2000-kcalorie diet, the DV is 60 mg/day.

about 20 percent of all the vitamin C in the average diet. Vitamin C in foods is easily oxidized, so store cut produce and juices in airtight containers.

Iron is discussed in Chapter 9.

Vitamin C and Iron Absorption Eating foods containing vitamin C at the same meal with foods containing iron can double or triple the absorption of iron from those foods. This strategy is highly recommended for women and children, whose energy intakes are not large enough to guarantee that they will get enough iron from the foods they eat.

REVIEW NOTES

The B vitamins and vitamin C are the water-soluble vitamins.

Each B vitamin is part of an enzyme helper known as a coenzyme.

As parts of coenzymes, the B vitamins assist in the release of energy from glucose, amino acids, and fats and help in many other body processes.

Folate and vitamin B_{12} are important in cell division.

Vitamin C's primary role is as an antioxidant.

Historically, famous B vitamin–deficiency diseases are beriberi (thiamin) and pellagra (niacin). The vitamin C–deficiency disease is known as scurvy.

Table 8-3 summarizes functions, deficiency and toxicity symptoms, and food sources of the other water-soluble vitamins.

TABLE 8-3 The Water-Soluble Vitamins—A Summary

Vitamin Name	Chief Functions	Deficiency Symptoms	Toxicity Symptoms	Significant Sources
Thiamin (Vitamin B$_1$)	Part of a coenzyme used in energy metabolism	Beriberi (edema or muscle wasting), anorexia and weight loss, neurological disturbances, muscular weakness, heart enlargement and failure	None reported	Enriched, fortified, or whole-grain products; pork
Riboflavin (Vitamin B$_2$)	Part of coenzymes used in energy metabolism	Inflammation of the mouth, skin, and eyelids; sensitivity to light; sore throat	None reported	Milk products; enriched, fortified, or whole-grain products; liver
Niacin (Nicotinic acid, nicotinamide, niacinamide, vitamin B$_3$; precursor is dietary tryptophan, an amino acid)	Part of coenzymes used in energy metabolism	Pellagra (diarrhea, dermatitis, and dementia)	Niacin flush, liver damage, impaired glucose tolerance	Milk, eggs, meat, poultry, fish, whole-grain and enriched breads and cereals, nuts, and all protein-containing foods
Biotin	Part of a coenzyme used in energy metabolism	Skin rash, hair loss, neurological disturbances	None reported	Widespread in foods; GI bacteria synthesis
Pantothenic acid	Part of a coenzyme used in energy metabolism	Digestive and neurological disturbances	None reported	Widespread in foods
Vitamin B$_6$ (Pyridoxine, pyridoxal, pyridoxamine)	Part of coenzymes used in amino acid and fatty acid metabolism	Scaly dermatitis, depression, confusion, convulsions, anemia	Nerve degeneration, skin lesions	Meats, fish, poultry, potatoes, legumes, noncitrus fruits, fortified cereals, liver, soy products
Folate (Folic acid, folacin, pteroylglutamic acid)	Activates vitamin B$_{12}$; helps synthesize DNA for new cell growth	Anemia; smooth, red tongue; mental confusion; elevated homocysteine	Masks vitamin B$_{12}$ deficiency	Fortified grains, leafy green vegetables, legumes, seeds, liver
Vitamin B$_{12}$ (Cobalamin)	Activates folate; helps synthesize DNA for new cell growth; protects nerve cells	Anemia; nerve damage and paralysis	None reported	Foods derived from animals (meat, fish, poultry, shellfish, milk, cheese, eggs), fortified cereals
Vitamin C (Ascorbic acid)	Synthesis of collagen, carnitine, hormones, neurotransmitters; antioxidant	Scurvy (bleeding gums, pinpoint hemorrhages, abnormal bone growth, and joint pain)	Diarrhea, GI distress	Citrus fruits, cabbage-type vegetables, dark green vegetables (such as bell peppers and broccoli), cantaloupe, strawberries, lettuce, tomatoes, potatoes, papayas, mangoes

SELF CHECK

1. Which of the following vitamins are fat soluble?
   a. vitamins B, C, and E
   b. vitamins B, C, D, and E
   c. vitamins A, C, E, and K
   d. vitamins A, D, E, and K

2. Which of the following describes fat-soluble vitamins?
   a. They include thiamin, vitamin A, and vitamin K.
   b. They cannot be stored to any great extent and so must be consumed daily.
   c. Toxic levels can be reached by consuming citrus fruits and vegetables.
   d. They can be stored in the liver and fatty tissues and can build up toxic concentrations.

3. Night blindness and susceptibility to infection are the result of a deficiency of which vitamin?
   a. niacin
   b. vitamin C
   c. vitamin A
   d. vitamin K

4. Good sources of vitamin D include:
   a. eggs, fortified milk, and sunlight.
   b. citrus fruits, sweet potatoes, and spinach.
   c. leafy green vegetables, cabbage, and liver.
   d. breast milk, polyunsaturated plant oils, and citrus fruits.

5. Which of the following describes water-soluble vitamins?
   a. They include vitamins D and E.
   b. They are frequently toxic.
   c. They are stored extensively in tissues.
   d. They are easily absorbed and excreted.

6. A coenzyme is:
   a. a fat-soluble vitamin.
   b. an energy-yielding nutrient.
   c. a source of vitamin K.
   d. a molecule that combines with an enzyme to make it active.

7. Good food sources of folate include:
   a. citrus fruits, dairy products, and eggs.
   b. liver, legumes, and leafy green vegetables.
   c. dark green vegetables, corn, and cabbage.
   d. potatoes, broccoli, and whole-wheat bread.

8. Which vitamin is present only in foods of animal origin?
   a. riboflavin
   b. pantothenic acid
   c. vitamin B_{12}
   d. the inactive form of vitamin A

9. Which of the following nutrients is an antioxidant that protects water-soluble substances from oxidizing agents?
   a. beta-carotene
   b. thiamin
   c. vitamin C
   d. vitamin D

10. Eating foods containing vitamin C at the same meal can increase the absorption of which mineral?
    a. iron
    b. calcium
    c. magnesium
    d. folate

Answers to these questions appear in Appendix H.

CLINICAL APPLICATIONS

1. How might a vitamin deficiency weaken a client's resistance to disease?

2. Pull together information from Chapter 1 about the different food groups and the significant sources of vitamins shown in the photos and Snapshots throughout this chapter. Consider which vitamins might be lacking in the diet of a client who reports the following:

- Dislikes leafy green vegetables.
- Never uses milk, milk products, or cheese.
- Follows a very low-fat diet.
- Eats a fruit or vegetable once a day.

What additional information would help you pinpoint problems with vitamin intake?

NUTRITION ON THE NET

For further study of the topics in this chapter, access these websites. Be aware that many websites on the Internet are peddling vitamin supplements, not accurate information.

Find updates and quick links to these and other nutrition-related sites at our website: **www.wadsworth.com/ nutrition**

Search for "vitamins" at the American Dietetic Association: **www.eatright.org**

Review the Dietary Reference Intakes for the water-soluble vitamins: **www.nap.edu/readingroom**

Visit the World Health Organization to learn about "vitamin deficiencies" around the world: **www.who.int**

Search for "vitamins" at the U.S. Government health information site: **www.healthfinder.gov**

Learn more about neural tube defects from the Spina Bifida Association of America: **www.sbaa.org**

Read about Dr. Joseph Goldberger and his groundbreaking discovery linking pellagra to diet by searching for his name at: **www.nih.gov** or **www.pbs.org**

Learn how fruits and vegetables support a healthy diet rich in vitamins from the 5 A Day for Better Health program: **www.5aday.com**

NOTES

[1] A. C. Ross, Vitamin A and carotenoids, in M. E. Shils and coeditors, *Modern Nutrition in Health and Disease,* 10th ed. (Philadelphia: Lippincott, Williams, & Wilkins, 2006), pp. 351–375.

[2] Ross, 2006; J. Bastien and C. Rochette-Egly, Nuclear retinoid receptors and the transcription of retinoid-target genes, *Gene* 328 (2004): 1–16; J. E. Balmer and R. Blomhoff, Gene expression regulation by retinoic acid, *Journal of Lipid Research* 43 (2002): 1773–1808.

[3] Balmer and Blomhoff, 2002.

[4] G. L. Johanning and C. J. Piyathilake, Retinoids and epigenetic silencing in cancer, *Nutrition Reviews* 61 (2003): 284–289.

[5] Ross, 2006.

[6] Standing Committee on the Scientific Evaluation of Dietary Reference Intakes, Food and Nutrition Board, Institute of Medicine, *Dietary Reference Intakes for Vitamin A, Vitamin K, Arsenic, Boron, Chromium, Copper, Iodine, Iron, Manganese, Molybdenum, Nickel, Silicon, Vanadium, and Zinc* (Washington, D.C.: National Academy Press, 2001), pp. 82–161.

[7] C. Debier and Y. Larondelle, Vitamins A and E: Metabolism, roles, and transfer to offspring, *British Journal of Nutrition* 93 (2005): 153–174.

[8] Debier and Larondelle, 2005; M. Clagett-Dame and H. F. DeLuca, The role of vitamin A in mammalian reproduction and embryonic development, *Annual Review of Nutrition* 22 (2002): 347–381; Standing Committee on the Scientific Evaluation of Dietary Reference Intakes, 2001, pp. 84–86.

[9] P. I. Moreira and coauthors, Oxidative stress: The old enemy in Alzheimer's disease pathophysiology, *Current Alzheimer Research* 4 (2005): 403–408; F. Grodstein, J. Chen, and W. C. Willett, High-dose antioxidant supplements and cognitive function in community-dwelling elderly women, *American Journal of Clinical Nutrition* 77 (2003): 975–984; S. F. Clark, The biochemistry of antioxidants revisited, *Nutrition in Clinical Practice* 1 (2002): 5–17; M. J. Engelhart and coauthors, Dietary intake of antioxidants and risk of Alzheimer disease, *Journal of the American Medical Association* 287 (2002): 3223–3229.

[10] Standing Committee on the Scientific Evaluation of Dietary Reference Intakes, 2001.

[11] Ross, 2006; Debier and Larondelle, 2005.

[12] Standing Committee on the Scientific Evaluation of Dietary Reference Intakes, 2001, pp. 128–129.

[13] Standing Committee on the Scientific Evaluation of Dietary Reference Intakes, 2001, pp. 126–133.

[14] H. A. Jackson and A. H. Sheehan, Effect of vitamin A on fracture risk, *Annals of Pharmacotherapy,* October 25, 2005, e-pub ahead of print; P. Genaro and L. A. Martini, Vitamin A supplementation and risk of skeletal fracture, *Nutrition Reviews* 62 (2004): 65–72.

[15] K. Michaelsson and coauthors, Serum retinol levels and the risk of fractures, *New England Journal of Medicine* 348 (2003): 287–294; D. Feskanich and coauthors, Vitamin A intake and hip fractures among postmenopausal women, *Journal of the American Medical Association* 287 (2002): 47–54.

[16] A. W. Norman, Vitamin D, in B. A. Bowman and R. M. Russell, eds., *Present Knowledge in Nutrition,* 8th ed. (Washington, D.C.: International Life Sciences Institute Press, 2001), pp. 146–155.

[17] Norman, 2001; M. F. Holick, Vitamin D, in M. E. Shils and coeditors, *Modern Nutrition in Health and Disease,* 10th ed. (Philadelphia, Pa.: Lippincott, Williams, & Wilkins, 2006), pp. 376–395.

[18] Holick, 2006; M. T. Cantorna and coauthors, Vitamin D status, 1, 25–dihydroxyvitamin D_3, and the immune system, *American Journal of Clinical Nutrition* 80 (2004): 1717S–1720S; M. F. Holick, Vitamin D: Importance in the prevention of cancers, type 1 diabetes, heart disease, and osteoporosis, *American Journal of Clinical Nutrition* 79 (2004): 362–371; J. B. Zella and H. F. DeLuca, Vitamin D and autoimmune diabetes, *Journal of Cellular Biochemistry* 88 (2003): 216–222; A. Zitterman, Vitamin D in preventive medicine: Are we ignoring the evidence? *British Journal of Nutrition* 89 (2003): 552–572; I. A. van der Mei and coauthors, Past exposure to sun, skin phenotype, and risk of multiple sclerosis: Case-control study, *British Medical Journal* 327 (2003): 316–332; G. Wolf, Intestinal bile acids can bind to and activate the vitamin D receptor, *Nutrition Reviews* 60 (2002): 281–288.

[19] Holick, 2006.

[20] P. Weisberg and coauthors, Nutritional rickets among children in the United States: Review of cases reported between 1986 and 2003, *American Journal of Clinical Nutrition* 80 (2004): 1697S–1705S; N. F. Carvalho and coauthors, Severe nutritional deficiencies in toddlers resulting from health food milk alternatives, *Pediatrics* 107 (2001): E46; S. R. Kreiter and coauthors, Nutritional rickets in African American breast-fed infants, *Journal of Pediatrics* 137 (2000): 153–157.

[21] S. S. Sullivan and coauthors, Adolescent girls in Maine are at risk for vitamin D insufficiency, *Journal of the American Dietetic Association* 105 (2005): 971–974; M. K. M. Lehtonen-Veromaa and coauthors, Vitamin D and attainment of peak bone mass among peripubertal Finnish girls: a 3-y prospective study, *American Journal of Clinical Nutrition* 76 (2002): 1446–1453.

[22] Holick, 2006.

[23] M. S. Calvo and S. J. Whiting, Prevalence of vitamin D insufficiency in Canada and the United States: Importance to health status and efficacy of current food fortification and dietary supplement use, *Nutrition Reviews* 61 (2003): 107–113.

[24] Calvo and Whiting, 2003; V. Tangpricha and coauthors, Vitamin D insufficiency among free-living adults, *American Journal of Medicine* 112 (2002): 659–662.

[25] M. F. Holick, Sunlight and vitamin D for bone health and prevention of autoimmune diseases, cancers, and cardiovascular disease, *American Journal of Clinical Nutrition* 80 (2004): 1678–1688; R. P. Heaney and coauthors, Human serum 25-hydroxycholecalciferol response to extended oral dosing with cholecalciferol, *American Journal of Clinical Nutrition* 77 (2003): 204–210.

[26] L. M. Gartner, F. R. Greer, and the Section on Breastfeeding and Committee on Nutrition, Prevention of rickets and vitamin D deficiency: New guidelines for vitamin D intake, *Pediatrics* 111 (2003): 908–910.

[27] Gartner, Greer, and the Section on Breastfeeding and Committee on Nutrition, 2003.

[28] U. Singh, S. Devaraj, and I. Jialal, Vitamin E, oxidative stress, and inflammation, *Annual Review of Nutrition* 25 (2005): 151–174; Standing Committee on the Scientific Evaluation of Dietary Reference Intakes, Food and Nutrition Board, Institute of Medicine, *Dietary Reference Intakes for Vitamin C, Vitamin E, Selenium, and Carotenoids* (Washington, D.C.: National Academy Press, 2000), pp. 211–216.

[29] Singh, Devaraj, and Jialal, 2005.

[30] I. Lee and coauthors, Vitamin E in the primary prevention of cardiovascular disease and cancer, The Women's Health Study: A randomized controlled trial, *Journal of the American Medical Association* 294 (2005): 56–65; E. Lonn and coauthors, Effects of long-term vitamin E supplementation on cardiovascular events and cancer, *Journal of the American Medical Association* 293 (2005): 1338–1347.

[31]E. R. Miller and coauthors, Meta-analysis: High-dosage vitamin supplementation may increase all-cause mortality, *Annals of Internal Medicine* 142 (2005), e-pub available from **www.annals.org**.

[32]D. P. Vivekananthan and coauthors, Use of antioxidant vitamins for the prevention of cardiovascular disease: Meta-analysis of randomised trials, *Lancet* 361 (2003): 2017–2023.

[33]P. M. Kris-Etherton and coauthors, Antioxidant vitamin supplements and cardiovascular disease, *Circulation* 110 (2004): 637–641; U.S. Preventive Task Force, Routine vitamin supplementation to prevent cancer and cardiovascular disease: Recommendations and rationale, *Annals of Internal Medicine* 139 (2003): 51–55.

[34]Kris-Etherton and coauthors, 2004.

[35]Miller and coauthors, 2005.

[36]R. S. Eidelman and coauthors, Randomized trials of vitamin E in the treatment and prevention of cardiovascular disease, *Archives of Internal Medicine* 164 (2004): 1552–1556; P. G. Shekelle and coauthors, Effect of supplemental vitamin E for the prevention and treatment of cardiovascular disease, *Journal of General Internal Medicine* 19 (2004): 380–389; Vivekananthan and coauthors, 2003.

[37]J. N. Hathcock and coauthors, Vitamins E and C are safe across a broad range of intakes, *American Journal of Clinical Nutrition* 81 (2005): 736–745.

[38]J. W. Suttie, Vitamin K, in M. E. Shils and coeditors, *Modern Nutrition in Health and Disease,* 10th ed. (Philadelphia: Lippincott, Williams, & Wilkins, 2006), pp. 412–425.

[39]S. L. Booth and coauthors, Vitamin K intake and bone mineral density in women and men, *American Journal of Clinical Nutrition* 77 (2003): 512–516.

[40]K. D. Cashman, Vitamin K status may be an important determinanat of childhood bone health, *Nutrition Reviews* 63 (2005): 284–293; S. L. Booth and coauthors, Effects of a hydrogenated form of vitamin K on bone formation and resorption, *American Journal of Clinical Nutrition* 74 (2001): 783–790.

[41]H. J. Kalkwarf and coauthors, Vitamin K, bone turnover, and bone mass in girls, *American Journal of Clinical Nutrition* 80 (2004): 1075–1080.

[42]N. C. Binkley and coauthors, A high phylloquinone intake is required to achieve maximal osteocalcin γ-carboxylation, *American Journal of Clinical Nutrition* 76 (2002): 1055–1060; S. L. Booth and coauthors, Dietary vitamin K intakes are associated with hip fracture but not with bone mineral density in elderly men and women, *American Journal of Clinical Nutrition* 71 (2000): 1201–1208.

[43]R. F. Butterworth, Thiamin, in M. E. Shils and coeditors, *Modern Nutrition in Health and Disease,* 10th ed. (Philadelphia: Lippincott, Williams, & Wilkins, 2006), pp. 426–433.

[44]C. Bourgeois, D. Cercantes-Laurean, and J. Moss, Niacin, in M. E. Shils and coeditors, *Modern Nutrition in Health and Disease,* 10th ed. (Philadelphia: Lippincott, Williams, & Wilkins, 2006), pp. 426–433.

[45]J. Zempleni, Uptake, localization, and noncarboxylase roles of biotin, *Annual Review of Nutrition* 25 (2005): 175–196.

[46]D. M. Mock, Biotin, in M.E. Shils and coeditors, *Modern Nutrition in Health and Disease,* 10th ed. (Philadelphia: Lippincott, Williams, & Wilkins, 2006), pp. 498–506.

[47]A. D. Mackey, S. R. Davis, and J. F. Gregory, Vitamin B$_6$, in *Modern Nutrition in Health and Disease,* 10th ed. (Philadelphia: Lippincott, Williams, & Wilkins, 2006), pp. 452–461; K. L. Tucker and coauthors, High homocysteine and low B vitamins predict cognitive decline in aging men: The Veterans Affairs Normative Aging Study, *American Journal of Clinical Nutrition* 82 (2005): 627–635; J. Bryan, E. Calvaresi, and D. Hughes, Short-term vitamin B-12 or vitamin B-6 supplementation slightly affects memory performance but not mood in women of various ages, *Journal of Nutrition* 132 (2002): 1345–1356.

[48]The Homocysteine Studies Collaboration, Homocysteine and risk of ischemic heart disease and stroke, *Journal of the American Medical Association* 288 (2002): 2015–2022; S. E. Vollset, Plasma total homocysteine and cardiovascular and noncardiovascular mortality: The Hordaland Homocysteine Study, *American Journal of Clinical Nutrition* 74 (2001): 130–136.

[49]K. L. Tucker and coauthors, Breakfast cereal fortified with folic acid, vitamin B-6, and vitamin B-12 increases vitamin concentrations and reduces homocysteine concentrations: A randomized trial, *American Journal of Clinical Nutrition* 79 (2004): 805–811; G. Schnyder and coauthors, Effect of homocysteine-lowering therapy with folic acid, vitamin B12, and vitamin B6 on clinical outcome after percutaneous coronary intervention: The Swiss Heart Study: A randomized controlled trial, *Journal of the American Medical Association* 288 (2002): 973–979; K. M. Fairfield and R. H. Fletcher, Vitamins for chronic disease prevention in adults, *Journal of the American Medical Association* 287 (2002): 3116–3126; S. M. Saw and coauthors, Genetic, dietary, and other lifestyle determinants of plasma homocysteine concentrations in middle-aged and older Chinese men and women in Singapore, *American Journal of Clinical Nutrition* 73 (2001): 232–239; P. F. Jaques and coauthors, Determinants of plasma total homocysteine concentration in the Framingham offspring cohort, *American Journal of Clinical Nutrition* 73 (2001): 613–621.

[50]E. Nurk and coauthors, Changes in lifestyle and plasma total homocysteine: The Hordaland Homocysteine Study, *American Journal of Clinical Nutrition* 79 (2004): 812–819.

[51]F. V. Van Oort and coauthors, Folic acid and reduction of plasma homocysteine concentrations in older adults: A dose-response study, *American Journal of Clinical Nutrition* 77 (2003): 1318–1323; D. S. Wald, M. Law, and J. K. Morris, Homocysteine and cardiovascular disease: Evidence on causality from a meta-analysis, *British Medical Journal* 325 (2002): 1202–1208.

[52]S. Friso and coauthors, Low plasma vitamin B-6 concentrations and modulation of coronary artery disease risk, *American Journal of Clinical Nutrition* 79 (2004): 992–998.

[53]R. Carmel, Folic acid, in M.E. Shils and coeditors, *Modern Nutrition in Health and Disease,* 10th ed. (Philadelphia: Lippincott, Williams, & Wilkins, 2006), pp. 470–481.

[54]Centers for Disease Control and Prevention, Spina bifida and anencephaly before and after folic acid mandate—United States, 1995–1996 and 1999–2000, *Morbidity and Mortality Weekly Report* 53 (2004): 362–365; E. A. Yetley and J. I. Rader, Modeling the level of fortification and post-fortification assessments: U.S. experience, *Nutrition Reviews* 62 (2004): S50–S59.

[55]L. B. Bailey and R. J. Berry, Folic acid supplementation and the occurrence of congenital heart defects, orofacial clefts, multiple births, and miscarriage, *American Journal of Clinical Nutrition* 81 (2005): 1213S–1217S.

[56]S. Voutilainen and coauthors, Serum folate and homocysteine and the incidence of acute coronary events: The Kuopio Ischaemic Heart Disease Risk Factor Study, *American Journal of Clinical Nutrition* 80 (2004): 317–323; G. Schnyder and coauthors, Effect of homocysteine-lowering therapy with folic acid, vitamin B$_{12}$, and vitamin B$_6$ on clinical outcome after percutaneous coronary intervention, The Swiss Heart Study: A randomized controlled trial, *Journal of the American Medical Association* 288 (2002): 973–979; D. S. Wald, M. Law, and J. K. Morris, Homocysteine and cardiovascular disease: Evidence on causality from a meta-analysis, *British Medical Journal* 325 (2002): 2002–2008.

[57]Van Oort and coauthors, 2003; A. de Bree and coauthors, Association between B vitamin intake and plasma homocysteine concentration in the general Dutch population aged 20–65 y, *American Journal of Clinical Nutrition* 73 (2001): 1027–1033; J. A. Tice and coauthors, Cost-effectiveness of vitamin therapy to lower plasma homocysteine levels for the prevention of coronary heart disease—Effect of grain fortification and beyond, *Journal of the American Medical Association* 286 (2001): 936–943; D. L. McKay and coauthors, Multivitamin/mineral supplementation improves

plasma B-vitamin status and homocysteine concentration in healthy older adults consuming folate-fortified diet, *Journal of Nutrition* 130 (2000): 3090–3096.

[58]C. W. Suitor and L. B. Bailey, Dietary folate equivalents: Interpretation and application, *Journal of the American Dietetic Association* 100 (2000): 88–94.

[59]F. Watanabe and coauthors, Effects of microwave heating on the loss of vitamin B_{12} in foods, *Journal of Agricultural and Food Chemistry* 46 (1998): 206–210.

[60]S. P. Stabler and R. H. Allen, Vitamin B12 deficiency as a worldwide problem, *Annual Review of Nutrition* 24 (2004): 299–326.

[61]Carmel, 2006.

[62]Standing Committee on the Scientific Evaluation of Dietary Reference Intakes, 2000, p. 155.

Phytochemicals and Functional Foods

The wisdom of the familiar advice, "Eat your vegetables, they're good for you," stands on firmer scientific ground today than ever before as population studies around the world suggest that diets rich in vegetables and fruits protect against heart disease, cancer, and other chronic diseases.[1] We now know that the "goodness" of vegetables, fruits, and other whole foods such as legumes and grains comes not only from the nutrients they contain, but from the **nonnutrients** known as **phytochemicals** that they offer.

Vegetables, fruits, and other whole foods are the simplest examples of foods now known as **functional foods.** Functional foods provide health benefits beyond basic nutrition by altering one or more physiological processes. Modified foods, such as those that have been fortified with nutrients, phytochemicals, herbs, or other food components, also are functional foods.[2] Functional foods that fit this description include orange juice fortified with calcium, folate-enriched cereal, beverages with herbal additives, and margarine enhanced with **sterol esters.** This Nutrition in Practice begins with a look at the evidence concerning the effectiveness and safety of a few selected phytochemicals in the simplest of functional foods—vegetables, fruits, and other whole foods. Then the discussion turns to examine the most controversial of functional foods—novel foods to which phytochemicals have been added to promote health. How these foods fit into a healthy diet is still unclear.[3] The glossary (p. 221) defines the relevant terms.

What are phytochemicals, and what do they do?

Phytochemicals are nonnutrient compounds found in plants. In foods, phytochemicals impart tastes, aromas, colors, and other characteristics. They give hot peppers their burning sensation, garlic and onions their pungent flavor, chocolate its bitter tang, and tomatoes their dark red color. In the body, phytochemicals can have profound physiological effects, acting as antioxidants, mimicking hormones, and suppressing the development of diseases.[4] Notably, cancer and heart disease are linked to processes involving oxygen compounds in the body, and antioxidants are thought to oppose these actions. Table NP8-1 introduces the names, possible physiological effects, and food sources of phytochemicals.

Why are phytochemicals receiving so much attention these days, and what are some examples of those in the spotlight?

Diets rich in whole grains, legumes, vegetables, and fruits seem to be protective against heart disease and cancer, but identifying *the* specific foods or components of foods that are responsible is difficult.[5] Whenever bits of research news surface, however, new supplements appear—and terms like *antioxidants* and *phytochemicals* become buzzwords again. Meanwhile, scientists are conducting extensive research studies to discover phytochemical connections to disease prevention, but so far, solid evidence is generally lacking. Some of the likeliest candidates include **flavonoids** and **carotenoids** (including **lycopene**).

What are flavonoids, and which foods are they found in?

Flavonoids, a large group of phytochemicals known for their health-promoting qualities, are found in whole grains, vegetables, fruits, herbs, spices, teas, and red wine. A large body of population evidence spanning many countries reveals that deaths from cancer, heart disease, and heart attacks are less common where these foods are plentiful in the diet, where tea is a beverage, or where red wine is consumed in moderation.[6] Flavonoids are powerful antioxidants that may help to protect LDL against oxidation and reduce blood platelet stickiness, making blood clots less likely.[7] Nevertheless, more evidence is needed before any claims can be made for flavonoids themselves as the protective factors in foods, particularly when they are extracted from foods or herbs and sold as supplements.[8] Furthermore, studies evaluating potentially adverse health effects of flavonoids must be conducted as well.[9]

Flavonoids impart a bitter taste to foods, so manufacturers often refine away the natural flavonoids to please consumers who usually prefer milder flavors.[10] For example, the hearty taste of whole-wheat foods vanishes when whole wheat is refined into white flour by removing the tough brown parts that contain flavonoids. For white grape juice or white wine, manufacturers remove the red, flavonoid-rich grape skins to lighten the flavor and color of the product, while greatly reducing its beneficial flavonoid content. For example, one flavonoid of grapes and wine may have anticancer activity.*[11]

What about carotenoids?

In addition to flavonoids, fruits and vegetables are rich in carotenoids—the red and yellow pigments of plants. Some carotenoids, such as beta-carotene, are vitamin A precursors. Studies suggest that a diet rich in carotenoids is associated with a lower risk of heart disease.[12] Among the carotenoids that may defend against heart disease is lycopene.[13] Researchers are investigating a tentative link between low levels of lycopene in the blood and elevated incidence of heart disease, heart attack, and stroke.[14] Lycopene may also protect against certain types of cancer.

* The flavonoid is resveratrol.

TABLE NP8-1 A Sampling of Phytochemicals—Possible Effects and Food Sources

Name	Possible Effects	Food Sources
Capsaicin	May modulate blood clotting, may reduce the risk of fatal clots in heart and artery disease.	Hot peppers
Carotenoids (including beta-carotene, lutein, lycopene, and hundreds of related compounds)[a]	Act as antioxidants; possibly reduce risks of heart disease, age-related eye disease,[b] cancer, and other diseases.	Deeply pigmented fruits and vegetables (apricots, broccoli, cantaloupe, carrots, pumpkin, spinach, sweet potatoes, tomatoes)
Curcumin	May inhibit enzymes that activate carninogens.	Turmeric, a yellow-colored spice
Flavonoids (including flavones, flavonols, and isoflavones, catechins, others)[c,d]	Act as antioxidants; may scavenge carcinogens; bind to nitrates in the stomach, preventing conversion to nitrosamines; inhibit cell proliferation; flavonoids of blueberries may improve memory.	Berries, black tea, celery, chocolate, citrus fruits, green tea, olives, onions, oregano, purple grapes, purple grape juice, soybeans and soy products, vegetables, whole wheat, wine
Indoles	May trigger production of enzymes that block DNA damage from carcinogens; may inhibit estrogen action.	Broccoli and other cruciferous vegetables (brussels sprouts, cabbage, cauliflower), horseradish, mustard greens
Isothiocyanates (including sulforaphane)	May inhibit enzymes that activate carcinogens; trigger production of enzymes that detoxify carcinogens.	Broccoli and other cruciferous vegetables (brussels sprouts, cabbage, cauliflower), horseradish, mustard greens
Monoterpenes (including limonene)	May trigger enzyme production to detoxify carcinogens; may inhibit cancer promotion and cell proliferation.	Citrus fruit peels and oils
Organosulfur compounds (including allicin)	May speed production of carcinogen-destroying enzymes or slow production of carcinogen-activating enzymes.	Chives, garlic, leeks, onions
Phenolic acids[d] (including ellagic acid)	May trigger enzyme production to make carcinogens water soluble, facilitating excretion.	Coffee beans, fruits (apples, blueberries, cherries, grapes, oranges, pears, prunes, strawberries), oats, potatoes, soybeans
Phytic acid	Binds to minerals, preventing free-radical formation, possibly reducing cancer risk.	Whole grains
Phytoestrogens (members of the flavonoid family, genistein and diadzein)	May inhibit estrogen and produce these actions: inhibit cell replication in GI tract; reduce risk of breast, colon, ovarian, prostate, and other estrogen-sensitive cancers; reduce cancer cell survival; may reduce risk of osteoporosis. May also alter blood lipids favorably and reduce heart disease risk when consumed in soy foods.	Soybeans, soy flour, soy milk, tofu, textured vegetable protein, other legume products
Phytoestrogens (lignans)	Block estrogen activity in cells, possibly reducing the risk of cancer of the breast, colon, ovaries, and prostate.	Flaxseed, whole grains
Protease inhibitors	May suppress enzyme production in cancer cells, slowing tumor growth; inhibit hormone binding; inhibit malignant changes in cells.	Broccoli sprouts, potatoes, soybeans and other legumes, soy products
Resveratrol[e]	May offset artery-damaging effects of high-fat diets.	Red wine, peanuts
Saponins	May interfere with DNA replication, preventing cancer cells from multiplying; stimulate immune response.	Alfalfa sprouts, other sprouts, green vegetables, potatoes, tomatoes
Tannins[d]	May inhibit carcinogen activation and cancer promotion; act as antioxidants.	Black-eyed peas, grapes, lentils, red and white wine, tea

[a] Other carotenoids include alpha-carotene, beta-cryptoxanthin, and zeaxanthin.

[b] The age-related eye disease is macular degeneration.

[c] Other flavonoids of interest include ellagic acid and ferulic acid.

[d] A subset of the larger group *polyphenolic phytochemicals*.

[e] A member of the chemical group stilbene, which is a subset of the larger group *polyphenolic phytochemicals*.

What is lycopene, and what foods contain it?

Lycopene is a red pigment with powerful antioxidant activity found in guava, papaya, pink grapefruit, tomatoes (especially cooked tomatoes and tomato products), and watermelon. More than 80 percent of the lycopene consumed in the United States comes from tomato products such as tomato sauce, tomato juice, and catsup. Around the world, people who eat five or more tomato-containing meals per week are less likely to suffer from cancers of the esophagus, prostate, or stomach than those who avoid tomatoes.[15] Lycopene is a leading candidate for this protective effect.

Lycopene may inhibit the reproduction of cancer cells.[16] In one study, women who consumed a diet rich in fruits and vegetables had high blood lycopene concentrations and a greatly reduced concentration of an indicator of cervical cancer risk.[17] Studies also find an association between increased breast cancer risk and low intakes of lycopene and related compounds.[18]

Are lycopene supplements a good idea?

Stick with the tried and true. Eat more fruits and vegetables. Cancer research favors eating foods high in lycopene, but research does not support the consumption of purified supplements of lycopene. Recall from Chapter 8 that diets high in the carotenoid beta-carotene often correlate with low rates of lung cancer. When given to smokers in studies, however, purified beta-carotene in supplement form *increases* lung cancer rates. The only safe option is to eat lycopene-rich foods such as tomatoes and tomato products and avoid concentrated supplements of lycopene until safety studies are completed.

Do foods contain other phytochemicals that may help to protect people from cancer or other diseases?

Foods contain thousands of different phytochemicals, and so far only a few have been researched at all. There are still many questions about the phytochemicals that have been studied and only tentative answers about their roles in human health. For example, compared with people in the West, Asians living in Asia suffer less frequently from osteoporosis (adult bone loss); cancers, especially of the breast, colon, and prostate; and heart disease. Among many differences between the diets of the two regions, Asians consume far more soybeans and soy products such as **tofu** than do Westerners.

Soybeans contain phytochemicals known as **phytoestrogens.** Researchers suspect that the phytoestrogens of soy foods, their protein content, or a combination of these factors may be responsible for health effects in soy-eating peoples. Nevertheless, research, though ongoing, is limited. So far, we know with certainty that phytoestrogens are plant-derived chemical relatives of the human hormone estrogen, that they weakly mimic or modulate the hormone's effects on some body tissues, and that they also act as antioxidants. We also know that breast cancer, colon cancer, and prostate cancer are estrogen-sensitive—that is, they grow when exposed to estrogen. Whether any of these or other actions of phytoestrogens may alter the course of estrogen-sensitive cancers remains unknown, but recent results from breast cancer studies do not support the idea.[19] Until more is known, the safest way to obtain soy phytoestrogens is to include moderate amounts of soy-based foods in the diet as generations of Asian people have done through the ages.[20]

Other foods under study for potential health benefits include **flaxseed** and its oil. Flaxseed is found as a whole seed or ground meal, or as flaxseed oil. Flaxseed is of interest for its possible benefits to heart health because it is the richest known source of both the omega-3 fatty acid linolenic acid and the phytoestrogen **lignans,** as well as being a good source of soluble fiber.[21] Flaxseed oil, though rich in linolenic acid, does not contain fiber or lignans. Studies of the direct effect of giving flaxseed to people are lacking, and some risks are possible with its use. Large quantities of flaxseed can cause digestive distress, and severe allergic reactions to flaxseed have been reported.

What about other phytochemical supplements?

Even when people don't eat in the best interest of health, taking supplements of purified phytochemicals is not the way to go. Phytochemicals can alter body functions, sometimes powerfully. Researchers are just beginning to understand how a handful of phytochemicals work, and what is current today may change tomorrow. The body is equipped to handle phytochemicals in diluted form, mixed with all of the other constituents of foods, but it is not adapted to phytochemicals in concentrated form. Fruits, vegetables, legumes, whole grains, nuts, and seeds contain a wide array of beneficial nutrients and thousands of phytochemicals. The best way to reap the benefits of phytochemicals is by eating foods, not supplements (see Figure NP8-1).

How do whole foods compare with processed foods that have been enriched with phytochemicals?

Good question. The American food supply is being transformed by a proliferation of functional foods—foods claimed to provide health benefits beyond those of the traditional nutrients. In truth, all the foods mentioned so far in this Nutrition in Practice, from tomatoes to tofu, are functional—they stand out as having a potentially greater impact on the health of the body (see Table NP8-2, p. 220). Cranberries may protect against urinary tract infections because cranberries contain a phytochemical that dislodges bacteria from the tract.[22] Cooked tomatoes, as mentioned, provide lycopene, along with **lutein** (an antioxidant associated with healthy eye function), vitamin C (an antioxidant vitamin), and many other healthful attributes.[23] But this has not stopped food manufacturers from trying to create functional foods as well.

FIGURE NP8-1 An Array of Phytochemicals in a Variety of Fruits and Vegetables

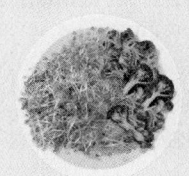

Broccoli and broccoli sprouts contain an abundance of sulforaphane.

An apple a day—rich in flavonoids.

The phytoestrogens genistein and diadzein are found in soybeans and soy products.

Garlic, leeks, and onions, with their abundant organosulfur compounds, may benefit health.

The phytochemicals of grapes, red wine, and peanuts include resveratrol.

Strawberries are a source of flavonoids .

Citrus fruits provide limonene.

The flavonoids in black tea differ from those in green tea, with different potential effects on health.

Cooked tomatoes are the best source of lycopene, a carotenoid that may reduce cancer risk.

The flavonoids in cocoa, chocolate, fruits and vegetables, legumes, and tea, may benefit health.

Spinach and other green, orange, red, and yellow fruits and vegetables contain carotenoids.

Flaxseed is the richest source of lignans.

Blueberries are among the richest sources of flavonoids.

Garlic,Citrus fruits, © EyeWire, Inc.; Flaxseed, Courtesy of Flax Council of Canada; Broccoli, Courtesy of Brassica Protection Products. Apples, Strawberries, Tomatoes, Blueberries, Black Tea, Soybeans, Grapes, Digital Imagery © 2001 PhotoDisc, Inc., Orange, Spinach, Cocoa by Matthew Farruggio

Source: Data from P. M. Kris-Etherton and coauthors, Bioactive compounds in nutrition and health—research methodologies for establishing biological function: The antioxidant and anti-inflammatory effects of flavonoids on atherosclerosis, *Annual Review of Nutrition* 24 (2004): 511–538.

As consumer demand for healthful foods continues to grow, so will the development of functional foods.[24]

What are some examples of manufactured functional foods?

Many processed foods become functional foods when they are fortified with nutrients or enhanced with phytochemicals or herbs (calcium-fortified orange juice, for example). Less frequently, an entirely new food is created, as in the case of a meat substitute made of mycoprotein—a protein derived from a fungus.[†25] This functional food not only provides dietary fiber, polyunsaturated fats, and high-quality protein, but it

† This mycoprotein product is marketed under the trade name Quorn (pronounced KWORN).

TABLE NP8-2 Categories of Functional Foods

These categories of functional foods have been identified by the American Dietetic Association:[a]
1. Foods with naturally occurring beneficial compounds, which qualify to bear on their labels an FDA "A" level health claim. These claims are listed in Table 1-13 on p. 29.
2. Foods with naturally occurring beneficial compounds, such as fish that provide fish oils, which have significant evidence supporting their benefits but lack the degree of certainty necessary for an "A" health claim.
3. Foods with biological plausibility for an effect, such as chocolate, garlic, and other foods discussed earlier in this Nutrition in Practice.
4. Foods fortified with nutrients or other constituents associated with the prevention or treatment of a disease, such as calcium-fortified orange juice marketed to reduce the risk of osteoporosis. Other examples include folate-enriched cereal to reduce birth defects and margarine enhanced with sterol esters to lower blood cholesterol.
5. Fortified foods marketed as "dietary supplements." Examples include beverages, candies, and bars enhanced with antioxidants, minerals, or herbs.

[a]Position of the American Dietetic Association: Functional foods, *Journal of the American Dietetic Association* 104 (2004): 814–826.

lowers LDL cholesterol, raises HDL cholesterol, improves glucose response, and prolongs satiety after a meal. Such a novel functional food raises the question—is it a food or a drug?

Isn't the distinction between a food and a drug pretty clear?

Not too long ago, most of us could agree on what was a food and what was a drug. Today, functional foods blur the distinctions. They have characteristics similar to both foods and drugs but do not fit neatly into either category. For example, food companies have already developed products with added phytochemicals. Consider margarine, for example.

Eating nonhydrogenated margarine sparingly instead of butter generously may lower blood cholesterol slightly over several months and clearly falls into the food category. Taking the drug Lipitor, on the other hand, lowers blood cholesterol significantly within weeks and clearly falls into the drug category. But margarine enhanced with sterol esters or stanol esters that lower blood cholesterol is in a gray area between the two. The margarine looks and tastes like a food, but it acts like a drug.

The use of functional foods as drugs creates a whole new set of diet-planning problems. Not only must foods provide an adequate intake of all the nutrients to support good health, but they must also deliver druglike ingredients to protect against disease. Like drugs used to treat chronic diseases, functional foods may need to be eaten several times a day for several months or years to have a beneficial effect. Sporadic users may be disappointed in the results. When used four times a day for four weeks, margarine enriched with sterol esters reduces cholesterol by 8 percent, much more than regular margarine does, but not nearly as much as the 32 percent reduction seen with cholesterol-lowering drugs.[26] For this reason, functional foods may be more useful for prevention and mild cases of disease than for intervention and more severe cases.

How do the costs of functional foods compare with those of conventional foods?

Foods and drugs differ dramatically in cost. Functional foods such as fruits and vegetables incur no added costs, of course, but foods that have been manufactured with added phytochemicals can be expensive, costing up to six times as much as their conventional counterparts. The price of functional foods typically falls between that of traditional foods and medicines.

What about using sterol-enhanced margarines to lower cholesterol or other manufactured functional foods to help prevent cancer?

To achieve a desired health effect, which is the better choice: to eat a food designed to affect some body function or simply to adjust the diet? Does it make more sense to use a margarine enhanced with a sterol ester that lowers blood cholesterol or simply to limit the amount of butter eaten?‡[27] Is it smarter to eat eggs enriched with omega-3 fatty acids or to restrict egg consumption? Might functional foods offer a sensible solution for improving our nation's health—if done correctly? Perhaps so—but there is a problem with functional foods: the food industry is moving too fast for either scientists or the Food and Drug Administration (FDA) to keep up. Consumers were able to buy soup with St. John's wort that claimed to enhance mood and fruit juice with echinacea that was supposed to fight colds while scientists were still conducting their studies on these ingredients. Research to determine the safety and effectiveness of these substances is still in progress. Until this work is complete, consumers are on their own in finding the answers to the following questions:[28]

- *Does it work?* Research is generally lacking and findings are often inconclusive.
- *How much does it contain?* Food labels are not required to list the quantities of added phytochemicals. Even if they were, consumers have no standard for comparison and cannot deduce whether the amounts listed are a little or a lot. Most importantly, until research is complete, food manufacturers do not know what amounts (if any) are most effective—or most toxic.
- *Is it safe?* Functional foods can act like drugs. They contain ingredients that can alter body functions and cause aller-

‡ Margarine products that lower blood cholesterol contain either sterol esters from vegetable oils, soybeans, and corn or stanol esters from wood pulp.

Functional foods currently on the market promise to "enhance mood," "promote relaxation and good karma," "increase alertness," and "improve memory," among other claims.

gies, drug interactions, drowsiness, and other side effects. Yet, unlike drug labels, food labels do not provide instructions for the dosage, frequency, or duration of treatment.

- *Has the FDA issued warnings about any of the ingredients?* Check the FDA's MedWatch website (**www.fda.gov/medwatch**) or call the FDA (1-888-INFO-FDA) to find out.
- *Is it healthy?* Adding phytochemicals to a food does not magically make it a healthy choice. A candy bar may be fortified with phytochemicals, but it is still made mostly of sugar and fat.

Critics suggest that the designation "functional foods" may be nothing more than a marketing tool. After all, even the most experienced researchers cannot yet identify the perfect combination of nutrients and phytochemicals to support optimal health. Yet manufacturers are freely experimenting with various concoctions as if they possessed that knowledge. Is it okay for them to sprinkle phytochemicals on fried snack foods and label them "functional," thus implying health benefits? Do we want our children receiving their nourishment from fortified caramel candies and chocolate cakes?

What is the final word regarding phytochemicals and functional foods?

Nature has elegantly designed foods to provide us with a complex array of dozens of nutrients and thousands of additional compounds that may benefit health—most of which we have yet to identify or understand. Over the years, we have taken those foods and first deconstructed them and then reconstructed them in an effort to "improve" them. With new scientific understandings of how nutrients—and the myriad of other compounds in foods—interact with genes, we may someday be able to design foods to meet the *exact* health needs of *each* individual.[29] If the present trend continues, then someday physicians may be able to prescribe the perfect foods to enhance a person's health, and farmers will be able to grow them. In the meantime, however, it seems clear that a moderate approach to phytochemicals and functional foods is warranted. People who eat the recommended amounts of a variety of fruits and vegetables may cut their risk of many diseases by as much as half. Replacing some meat with soy foods or other legumes may also lower heart disease and cancer risks. Beneficial constituents are widespread among foods. Take a no-nonsense approach where your health is concerned: Choose a wide variety of whole grains, legumes, fruits, and vegetables in the context of an adequate, balanced, and varied diet and receive all of the health benefits that these foods offer.

Glossary of Phytochemical and Functional Food Terms

carotenoids (kah-ROT-eh-noyds): pigments commonly found in plants and animals, some of which have vitamin A activity. The carotenoid with the greatest vitamin A activity is beta-carotene.
flavonoids (FLAY-von-oyds): yellow pigments in foods; phytochemicals that may exert physiological effects on the body.
flaxseed: the small brown seed of the flax plant; used in baking, cereals, or other foods and valued by industry as a source of linseed oil and fiber.
functional foods: foods that contain physiologically active compounds that provide health benefits beyond basic nutrition; sometimes called *designer foods* or *nutraceuticals.*
lignans: phytochemicals present in flaxseed, but not in flax oil, that are converted to phytosterols by intestinal bacteria and are under study for potential health benefits.
lutein (LOO-teen): a plant pigment of yellow hue; a phytochemical believed to play roles in eye functioning and health.
lycopene (LYE-koh-peen): a pigment responsible for the red color of tomatoes and other red-hued vegetables; a phytochemical that may act as an antioxidant in the body.
nonnutrients: compounds in foods that do not fit within the six classes of nutrients.
phytochemicals (FIGH-toe-CHEM-ih-cals): biologically active compounds of plants believed to confer resistance to diseases.
phytoestrogens: plant-derived compounds that have structural and functional similarities to human estrogen. Phytoestrogens include genistein, daidzein, and glycitein.
sterol esters: plant-derived compounds that have structural similarities to cholesterol and lower blood cholesterol by competing with cholesterol for absorption.
tofu: a white curd made of soybeans, popular in Asian cuisines, and considered to be a functional food.

Notes

[1] J. M. Genkinger and coauthors, Fruit, vegetables, and antioxidant intake and all-cause cancer, and cardiovascular disease mortality in a community-dwelling population in Washington County, Maryland, *American Journal of Epidemiology* 160 (2004): 1223–1333; L. A. Bazzano and coauthors, Fruit and vegetable intake and risk of cardiovascular disease in US adults: The First National Health and Nutrition Examination Survey Epidemiologic Follow-up Study, *American Journal of Clinical Nutrition* 76 (2002): 93–99; L. Arab and S. Steck, Lycopene and cardiovascular disease, *American Journal of Clinical Nutrition* 71 (2000): 1691S–1695S; L. LeMarchand and coauthors, Intake of flavonoids and lung cancer, *Journal of the National Cancer Institute* 92 (2000): 154–160; P. Knekt and coauthors, Quercetin intake and the incidence of cerebrovascular disease, *European Journal of Clinical Nutrition* 54 (2000): 415–417.

[2] Position of the American Dietetic Association: Functional foods, *Journal of the American Dietetic Association* 104 (2004): 814–826.

[3] C. H. Halsted, Dietary supplements and functional foods: 2 sides of a coin? *American Journal of Clinical Nutrition* 77 (2003): 1001S–1007S.

[4] A. Smeltzer and Y. S. Kim, The effects of bioactive food components on *p53* pathway in cancer prevention, *Nutrition Today* 40 (2005): 50–53; P. M. Kris-Etherton and coauthors, Bioactive compounds in foods: Their role in the prevention of cardiovascular disease and cancer, *American Journal of Medicine* 113 (2002): 71S–88S.

[5] Kris-Etherton and coauthors, 2002; A. L. Normen and coauthors, Plant sterol intakes and colorectal cancer risk in the Netherlands Cohort Study on Diet and Cancer, *American Journal of Clinical Nutrition* 74 (2001): 141–148; C. M. Steinmaus, S. Nunez, and A. H. Smith, Diet and bladder cancer: A meta-analysis of six dietary variables, *American Journal of Epidemiology* 151 (2000): 693–702.

[6] J. A. Vita, Polyphenols and cardiovascular disease: Effects on endothelial and platelet function, *American Journal of Clinical Nutrition* 81 (2005): 292S–297S; A. H. Wu and coauthors, Green tea and risk of breast cancer in Asian Americans, *International Journal of Cancer* 106 (2003): 574–579; J. M. Geleijnse and coauthors, Inverse association of tea and flavonoid intakes with incident myocardial infarction: The Rotterdam Study, *American Journal of Clinical Nutrition* 75 (2002): 880–886; P. Knekt and coauthors, Flavonoid intake and risk of chronic disease, *American Journal of Clinical Nutrition* 76 (2002): 560–568; LeMarchand and coauthors, 2000.

[7] M. J. Davies and coauthors, Black tea consumption reduces total and LDL cholesterol in mildly hypercholesterolemic adults, *Journal of Nutrition* 133 (2003): 3298S–3302S; B. Fuhrman and M. Aviram, Flavonoids protect LDL from oxidation and attenuate atherosclerosis, *Current Opinion in Lipidology* 12 (2001): 41–48.

[8] A. Scalbert, I. T. Johnson, and M. Saltmarsh, Polyphenols: Antioxidants and beyond, *American Journal of Clinical Nutrition* 81 (2005): 215S–217S; J. A. Ross and C. M. Kasum, Dietary flavonoids: Bioavailability, metabolic effects, and safety, *Annual Review of Nutrition* 22 (2002): 19–34.

[9] L. I. Mennen and coauthors, Risks and safety of polyphenol consumption, *American Journal of Clinical Nutrition* 81 (2005): 326S–329S.

[10] I. Lesschaeve and A. C. Noble, Polyphenols: Factors influencing their sensory properties and their effects on food and beverage preferences, *American Journal of Clinical Nutrition* 81 (2005): 330S–335S; A. Drewnowski and C. Gomez-Caneros, Bitter taste, phytonutrients, and the consumer: A review, *American Journal of Clinical Nutrition* 72 (2000): 1424–1435.

[11] P. Signorelli and R. Ghidoni, Resveratrol as an anticancer nutrient: Molecular basis, open questions and promises, *Journal of Nutritional Biochemistry* 16 (2005): 449–466; Y. Scheider and coauthors, Anti-proliferative effect of resveratrol, a natural component of grapes and wine, on human colonic cancer cells, *Cancer Letter* 158 (2000): 85–91.

[12] S. Liu and coauthors, Intake of vegetables rich in carotenoids and risk of coronary heart disease in men: The Physicians' Heart Study, *International Journal of Epidemiology* 30 (2001): 130–135.

[13] Arab and Steck, 2000.

[14] T. H. Rissanen and coauthors, Serum lycopene concentrations and carotid atherosclerosis: The Kuopio Ischaemic Heart Disease Risk Factor Study, *American Journal of Clinical Nutrition* 77 (2003): 133–138.

[15] E. Giovannucci and coauthors, A prospective study of tomato products, lycopene, and prostate cancer risk, *Journal of the National Cancer Institute* 94 (2002): 391–398: D. Herber and Q. Y. Lu, Overview of mechanisms of action of lycopene, *Experimental Biology and Medicine* 227 (2002): 920–923; T. M. Vogt and coauthors, Serum lycopene, other serum carotenoids, and risk of prostate cancer in US blacks and whites, *American Journal of Epidemiology* 155 (2002): 1023–1032; Q. Y. Lu and coauthors, Inverse associations between plasma lycopene and other carotenoids and prostate cancer, *Cancer Epidemiology, Biomarkers, and Prevention* 10 (2001): 749–756.

[16] P. Prakash, R. M. Russell, and N. I. Krinsky, In vitro inhibition of proliferation of estrogen-dependent and estrogen-independent human breast cancer cells treated with carotenoids or retinoids, *Journal of Nutrition* 131 (2001): 1574–1580.

[17] R. L. Sedjo and coauthors, Vitamin A, carotenoids, and risk of persistent oncogenic human papillomavirus infection, *Cancer Epidemiology, Biomarkers and Prevention* 11 (2002): 876–884.

[18] P. Toniolo and coauthors, Serum carotenoids and breast cancer, *American Journal of Epidemiology* 153 (2001): 1142–1147.

[19] L. Keinan-Boker and coauthors, Dietary phytoestrogens and breast cancer risk, *American Journal of Clinical Nutrition* 79 (2004): 282–288; A. Cassidy, Potential risks and benefits of phytoestrogen-rich diets, *International Journal of Vitamin Nutrition Research* 73 (2003): 120–126.

[20] C. Munro and coauthors, Soy isoflavones: A safety review, *Nutrition Reviews* 61 (2003): 1–33.

[21] L. T. Bloedon and P. O. Szapary, Flaxseed and cardiovascular risk, *Nutrition Reviews* 62 (2004): 18–27.

[22] A. B. Howell and B. Foxman, Cranberry juice and adhesion of antibiotic resistant uropathogens, *Journal of the American Medical Association* 287 (2002): 3082–3083.

[23] P. F. Jacques and coauthors, Long-term nutrient intake and early age-related nuclear lens opacities, *Archives of Ophthalmology* 119 (2001): 1009–1019; J. T. Landrum and R. A. Bone, Lutein, zeaxanthin, and the macular pigment, *Archives of Biochemistry and Biophysics* 385 (2001): 28–40.

[24] Position of the American Dietetic Association, 2004.

[25] T. Peregrin, Mycoprotein: Is America ready for a meat substitute derived from a fungus? *Journal of the American Dietetic Association* 102 (2002): 628.

[26] L. A. Simons, Additive effect of plant sterol-ester margarine and cerivastatin in lowering low-density lipoprotein cholesterol in primary hypercholesterolemia, *American Journal of Cardiology* 90 (2002): 737–740.

[27] P. Nestel and coauthors, Cholesterol-lowering effects of plant sterol esters and non-esterified stanols in margarine, butter and low-fat foods, *European Journal of Clinical Nutrition* 55 (2001): 1084–1090.

[28] C. Hasler and coauthors, How to evaluate the safety, efficacy, and quality of functional foods and their ingredients, *Journal of the American Dietetic Association* 101 (2001): 733–736.

[29] J. A. Milner, Functional foods and health: A US perspective, *British Journal of Nutrition* 88 (2002): S151–S158.

Water and the Minerals

© Royalty-Free/Corbis

CHAPTER 9

Water is the most indispensable nutrient of all.

The enzyme **renin** (REN-in), released by the kidneys in response to low blood pressure, aids the kidneys in retaining water through the **renin-angiotensin mechanism.**

water balance: the balance between water intake and water excretion that keeps the body's water content constant.

dehydration: the loss of water from the body that occurs when water output exceeds water input. The symptoms progress rapidly from thirst, to weakness, to exhaustion and delirium and end in death if not corrected.

water intoxication: the rare condition in which body water contents are too high. The symptoms may include confusion, convulsion, coma, and even death in extreme cases.

hypothalamus (high-poh-THALL-uh-mus): a part of the brain that helps regulate many body balances, including fluid balance.

pituitary (pit-TOO-ih-tary) **gland:** in the brain, the "king gland" that regulates the operation of many other glands.

antidiuretic hormone (ADH): a hormone released by the pituitary gland in response to high salt concentrations in the blood. The kidneys respond by reabsorbing water.

aldosterone (al-DOS-ter-own): a hormone secreted by the adrenal glands that stimulates the reabsorption of sodium by the kidneys; also regulates chloride and potassium concentrations.

The body's water cannot be considered separately from the minerals dissolved in it. A person can drink pure water, but in the body, water mingles with minerals to become fluids in which all life processes take place. This chapter begins by discussing the body's fluids and their chief minerals. The focus then shifts to other functions of the minerals.

Water and Body Fluids

Water constitutes about 60 percent of an adult's body weight and a higher percentage of a child's. Every cell in the body is bathed in a fluid of the exact composition that is best for that cell. The body fluids bring to each cell the ingredients it requires and carry away the end products of the life-sustaining reactions that take place within the cell's boundaries. The water in the body fluids:

- Carries nutrients and waste products throughout the body.
- Maintains the structure of large molecules such as proteins and glycogen.
- Participates in metabolic reactions.
- Serves as the solvent for minerals, vitamins, amino acids, glucose, and many other small molecules.
- Aids in maintaining the body's blood pressure and temperature.
- Maintains blood volume.
- Acts as a lubricant and cushion around joints and inside the eyes, spinal cord, and amniotic sac surrounding a fetus in the womb.

To support these and other vital functions, the body regulates its **water balance.**

WATER BALANCE

The cells themselves regulate the composition and amounts of fluids within and surrounding them. The entire system of cells and fluids remains in a delicate but firmly maintained state of dynamic equilibrium. Imbalances such as **dehydration** (see Table 9-1) and **water intoxication** can occur, but the body quickly restores the balance to normal if it can. The body controls both water intake and water excretion.

Water Intake Regulation The body can survive for only a few days without water. In healthy people, thirst governs water intake. Thirst is finely adjusted to ensure a water intake that meets the body's needs. When the blood becomes too concentrated (having lost water but not salt and other dissolved substances), the mouth becomes dry, and the brain center known as the **hypothalamus** initiates drinking behavior.

Thirst lags behind the lack of water. A water deficiency that develops slowly can switch on drinking behavior in time to prevent serious dehydration, but a deficiency that develops quickly may not. Also, thirst itself does not remedy a water deficiency; a person must pay attention to the thirst signal and take the time to get a drink. With aging, thirst sensations may diminish. Dehydration can threaten elderly people who do not develop the habit of drinking water regularly.

Water Excretion Regulation Water excretion is regulated by the brain and the kidneys. The cells of the brain's hypothalamus, which monitor blood salts, stimulate the **pituitary gland** to release **antidiuretic hormone (ADH)** whenever the salts are too concentrated, or the blood volume or blood pressure is too low. ADH stimulates the kidneys to reabsorb water rather than excrete it. Thus, the more water you need, the less you excrete.

If too much water is lost from the body, blood volume and blood pressure fall. Cells in the kidneys respond to the low blood pressure by releasing an enzyme. Through a complex series of events, involving the hormone **aldosterone,** this

TABLE 9-1   Threats from Mild and Severe Dehydration

Mild Dehydration (Loss of <5% Body Weight)	Severe Dehydration (Loss of >5% Body Weight)
Thirst	Pale skin
Sudden weight loss	Bluish lips and fingertips
Rough dry skin	Confusion; disorientation
Dry mouth, throat, body linings	Rapid, shallow breathing
Rapid pulse	Weak, rapid, irregular pulse
Low blood pressure	Thickening of blood
Lack of energy; weakness	Shock; seizures
Impaired kidney function	Coma; death
Reduced quantity of urine; concentrated urine	
Decreased mental functioning	
Decreased muscular work and athletic performance	
Fever or increased internal temperature	
Fainting	

Source: Standing Committee on the Scientific Evaluation of Dietary Reference Intakes, Food and Nutrition Board, Institute of Medicine, *Dietary Reference Intakes: Water, Potassium, Sodium, Chloride, and Sulfate* (Washington, D.C.: National Academies Press, 2005), pp. 90–122.

enzyme also causes the kidneys to retain more water. Again, the effect is that when more water is needed, less is excreted.

Minimum Water Needed These mechanisms can maintain water balance only if a person drinks enough water. The body must excrete a minimum of about 500 milliliters each day as urine—enough to carry away the waste products generated by a day's metabolic activities. Above this amount, excretion adjusts to balance intake, so the more a person drinks, the more dilute the urine becomes. In addition to urine, some water is lost from the lungs as vapor, some is excreted in feces, and some evaporates from the skin. A person's water losses from all of these routes total about $2\frac{1}{2}$ liters (about $2\frac{1}{2}$ quarts) a day on the average. Table 9-2 shows how fluid intake and output naturally balance out.

500 mL = about $\frac{1}{2}$ qt.

TABLE 9-2   Water Balance

Water Sources	Amount (mL)	Water Losses	Amount (mL)
Liquids	550 to 1500	Kidneys (urine)	500 to 1400
Foods	700 to 1000	Skin (sweat)	450 to 900
Metabolic water	200 to 300	Lungs (breath)	350
		GI tract (feces)	150
Total	1450 to 2800	Total	1450 to 2800

AI for *total* water:
- Men: 3.7 L/day.
- Women: 2.7 L/day.

Water Recommendations and Sources Water needs vary greatly depending on the foods a person eats, the environmental temperature and humidity, the person's activity level, and other factors. Accordingly, a general water requirement is difficult to establish. In the past, recommendations for adults were expressed in proportion to the amount of energy expended under normal environmental conditions. For the person who expends about 2000 kcalories a day, this works out to 2 to 3 liters, or about 8 to 12 cups. This recommendation is in line with the Adequate Intake (AI) for *total* water set by the DRI committee. Total water includes not only drinking water, but water in other beverages and in foods as well.[1]

Because a wide range of water intakes will prevent dehydration and its harmful consequences, the AI is based on average intakes. Strenuous physical activity and heat stress can increase water needs considerably, however.[2] In general, you can tell from the color of the urine whether a person needs more water. Pale yellow urine reflects appropriate dilution.[3]

The obvious dietary sources of water are water itself and other beverages, but nearly all foods also contain water. Most fruits and vegetables contain up to 90 percent water; many meats and cheeses contain at least 50 percent. The energy nutrients in foods also give up water during metabolism.

People often ask whether caffeine-containing beverages such as coffee, tea, or soda can help to meet water needs. When people who normally abstain from caffeine drink a caffeine-containing beverage, their urine output increases somewhat more than it would for a similar amount of water. This occurs because caffeine acts as a mild diuretic. Research is mixed on whether any but the highest caffeine intakes (4 or 5 cups of coffee) cause a net water deficit in the body; most people make up for small water losses by drinking additional fluid later. The DRI committee considered such findings in making its recommendations for water intake and concluded that "caffeinated beverages contribute to the daily total water intake similar to that contributed by non-caffeinated beverages."[4] In other words, it doesn't seem to matter whether people rely on caffeine-containing beverages or other beverages to meet their fluid needs.

In contrast, alcohol should not be used to meet fluid needs. As Nutrition in Practice 19 explains, alcohol has many adverse effects on health and nutrition status.

FLUID AND ELECTROLYTE BALANCE

Exceptions: A compound in which the positive ions are hydrogen ions (H$^+$) is an acid (example: hydrochloric acid, or H$^+$Cl$^-$); a compound in which the negative ions are hydroxyl ions (OH$^-$) is a base (example: potassium hydroxide, or K$^+$OH$^-$).

The simple statement that water follows salt describes the force that chemists call *osmosis*.

salts: compounds composed of charged particles (ions). An example of a salt is potassium chloride (K$^+$Cl$^-$).

electrolyte: a salt that dissolves in water and dissociates into charged particles called ions.

electrolyte solutions: solutions that can conduct electricity.

When mineral **salts** dissolve in water, they separate (dissociate) into charged particles known as ions, which can conduct electricity. For this reason, a salt that dissociates in water is known as an **electrolyte.** The body fluids, which contain water and partly dissociated salts, are **electrolyte solutions.**

The body's electrolytes are vital to the life of the cells and therefore must be closely regulated to help maintain the appropriate distribution of body fluids. The major minerals form salts that dissolve in the body fluids; the cells direct where these salts go; and the movement of the salts determines where the fluids flow because water follows salt. Cells use this force to move fluids back and forth across their membranes. Thanks to the electrolytes, water can be held in compartments where it is needed.

Proteins in the cell membranes move ions in or out of the cells. These protein pumps tend to concentrate sodium and chloride outside cells and potassium and other ions inside. By maintaining specific amounts of sodium outside and potassium inside, cells can regulate the exact amounts of water inside and outside their boundaries.

Healthy kidneys regulate the body's sodium, as well as its water, with remarkable precision. The intestinal tract absorbs sodium readily, and it travels freely in the blood, but the kidneys excrete unneeded amounts. The kidneys actually filter all of the sodium out of the blood; then, they return to the bloodstream the exact amount the body needs to retain. Thus the body's total electrolytes remain constant, while the urinary electrolytes fluctuate according to what is eaten.

In some cases, the body's mechanisms for maintaining fluid and electrolyte balances cannot compensate for a sudden loss of large amounts of fluid and electrolytes. Vomiting, diarrhea, heavy sweating, fever, burns, wounds, and the like may incur great fluid and electrolyte losses, precipitating an emergency that demands medical intervention.

Water follows salt. Notice the beads of "sweat," formed on the right-hand slices of eggplant, which were sprinkled with salt. Cellular water moves across each cell's membrane (water-permeable divider) toward the higher concentration of salt (dissolved particles) on the surface.

ACID-BASE BALANCE

The body uses ions not only to help maintain water balance but also to regulate the acidity (pH) of its fluids. Like proteins, electrolyte mixtures in the body fluids protect the body against changes in acidity by acting as buffers—substances that can accommodate excess acids or bases.

The body's buffer systems serve as a first line of defense against changes in the fluids' acid-base balance. The lungs, skin, gastrointestinal (GI) tract, and kidneys provide other defenses. Of these organ systems, the kidneys play the primary role in maintaining acid-base balance. Thus, disorders of the kidneys impair the body's ability to regulate its acid-base balance, as well as its fluid and electrolyte balances.

The body's responses to severe stress and trauma are discussed in Chapter 16.

Reminder: *Buffers* are compounds that help keep a solution's acidity or alkalinity constant; buffers are capable of neutralizing both acids and bases and thereby maintaining the original concentration of hydrogen ions (pH) in the solution.

REVIEW NOTES

Water makes up about 60 percent of the body's weight.

Water helps transport nutrients and waste products throughout the body, participates in metabolic reactions, acts as a solvent, assists in maintaining blood pressure and body temperature, acts as a lubricant and cushion around joints, and serves as a shock absorber.

To maintain water balance, intake from liquids, foods, and metabolism must equal losses from kidneys, skin, lungs, and feces.

Electrolytes help maintain the appropriate distribution of body fluids and help to maintain acid-base balance as well.

The Major Minerals

Table 9-3 (p. 228) lists the major and trace minerals in the body, and Figure 9-1 (p. 228) shows the amounts found in the body. As you can see, the most prevalent minerals are calcium and phosphorus, the chief minerals of bone. The distinction between the major and the trace minerals does not mean that one group is more important than the other. A deficiency of the few micrograms of iodine needed daily is just as serious as a deficiency of the several hundred milligrams of calcium. The major minerals are so named because they are present, and needed, in larger amounts in the body than the trace minerals.

Although all the major minerals influence the body's fluid balance, sodium, chloride, and potassium are most noted for that role. For this reason, these three minerals are discussed first. Each major mineral also plays other specific roles in the body. Sodium, potassium, calcium, and magnesium are critical to nerve transmission and muscle contractions. Phosphorus and magnesium are involved in

TABLE 9-3	The Major and Trace Minerals
Major Minerals	**Trace Minerals**
Calcium	Arsenic
Chloride	Boron
Magnesium	Chromium
Phosphorus	Cobalt
Potassium	Copper
Sodium	Fluoride
Sulfur	Iodine
	Iron
	Manganese
	Molybdenum
	Nickel
	Selenium
	Silicon
	Zinc

Sodium AI:
- 1500 mg/day (19–50 yr).
- 1300 mg/day (51–70 yr).
- 1200 mg/day (>70 yr).

Salt (sodium chloride) is about 40% sodium.

1 g salt contributes 400 mg sodium.

5 g salt = 1 tsp.

1 tsp salt contributes 2000 mg sodium.

energy metabolism. Calcium, phosphorus, and magnesium contribute to the structure of the bones. Sulfur helps determine the shape of proteins. Table 9-6 later in the chapter provides a summary of information about the major minerals.

SODIUM

Sodium is the principal electrolyte in the **extracellular fluid** (the fluid outside the cells) and the primary regulator of the extracellular fluid volume. When the blood concentration of sodium rises, as when a person eats salted foods, thirst prompts the person to drink water until the appropriate sodium-to-water ratio is restored. Sodium also helps maintain acid-base balance and is essential to muscle contraction and nerve transmission. Too much sodium, however, can contribute to high blood pressure.

Sodium Recommendations and Food Sources Diets rarely lack sodium, and even when intakes are low, the body adapts by reducing sodium losses in urine and sweat, thus making deficiencies unlikely. Sodium recommendations are set low enough to protect against high blood pressure, but high enough to allow an adequate intake of other nutrients. Because high sodium intakes correlate with high blood pressure, the Tolerable Upper Intake Level (UL) for adults is set at 2300 milligrams per day, slightly lower than the Daily Value used on food labels (2400 milligrams). The UL corresponds to slightly more than 1 teaspoon of salt (sodium chloride).

Cultures vary in their use of salt. In the United States, men consume an average of 3300 milligrams of sodium (equivalent to about 8 grams of salt) a day. Asian people, whose staple sauces and flavorings are based on soy sauce and monosodium glutamate (MSG), consume the equivalent of about 20 to 30 grams of salt per day. In China, Japan, and Korea, high blood pressure is as prevalent as in the United States or more so.

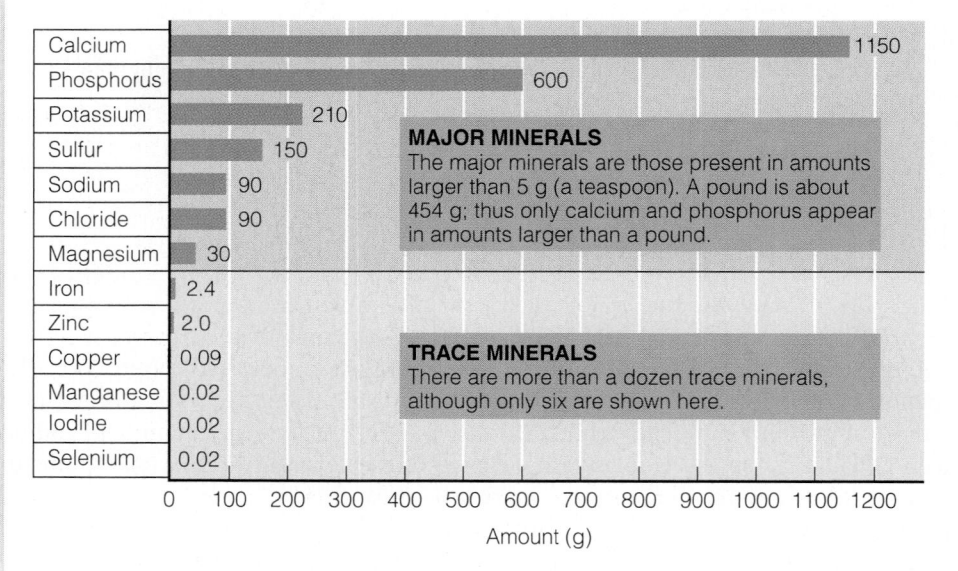

FIGURE 9-1 Amounts of Minerals in a 60-kilogram (132-pound) Human Body
Not only are the major minerals present in the body in larger amounts than the trace minerals, but they are also needed by the body in larger amounts. Recommended intakes for the major minerals are stated in *hundreds of milligrams* or *grams*, whereas those for the trace minerals are listed in *tens of milligrams* or even *micrograms*.

extracellular fluid: fluid residing outside the cells; includes the fluid between the cells (*interstitial fluid*), plasma, and the water of structures such as the skin and bones. Extracellular fluid accounts for about one-third of the body's water.

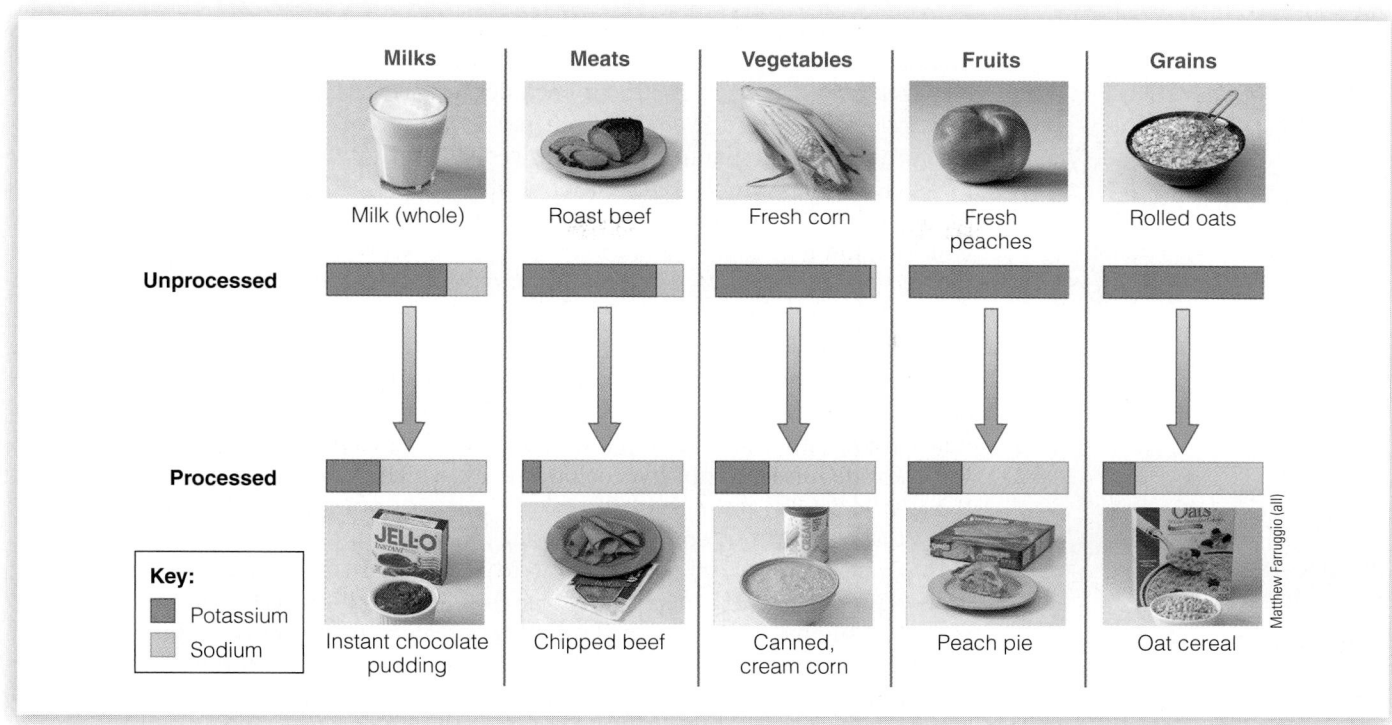

FIGURE 9-2 What Processing Does to the Sodium and Potassium Contents of Foods

People who eat foods high in salt often happen to be eating fewer potassium-containing foods at the same time. Note how potassium is lost and sodium is gained as foods become more processed, causing the potassium-to-sodium ratio to fall dramatically. Even when potassium isn't lost, the addition of sodium still lowers the potassium-to-sodium ratio. Limiting sodium intake may help in two ways, then—by lowering blood pressure in salt-sensitive individuals and by indirectly raising potassium intakes in all individuals.

Sodium intakes also vary widely. People who eat mostly processed foods have the highest sodium intakes, whereas those who eat mostly whole, unprocessed foods, such as fresh fruits and vegetables, have the lowest intakes. In fact, about three-fourths of the sodium in people's diets comes from salt added to foods by manufacturers. Figure 9-2 shows that processed foods contain not only more sodium but also less potassium than their less processed counterparts.

Sodium and Blood Pressure More than 60 million people in the United States have high blood pressure, or **hypertension,** or are taking antihypertensive medications.[5] Hypertension is a major risk factor for cardiovascular disease. As blood pressure rises, the risk of death from cardiovascular disease climbs steadily.[6]

For years, a high *sodium* intake was considered the primary factor responsible for high blood pressure. Then research pointed to *salt* (sodium chloride) as the dietary culprit. Salt has a greater effect on blood pressure than either sodium or chloride alone or in combination with other ions.[7] The relationship between salt intake and blood pressure is direct—the more salt a person eats, the higher the blood pressure goes.[8] This effect occurs more strongly among people who are **salt-sensitive.** People who tend to be salt-sensitive include those with hypertension, chronic kidney disease, or diabetes, African Americans, and people over 50 years of age because the blood pressure responds to salt more dramatically in older age.[9]

In fact, a salt-restricted diet lowers blood pressure in people without hypertension as well.[10] Because reducing salt intake causes no harm and diminishes the risk of hypertension and heart disease, the *Dietary Guidelines for Americans 2005* advise limiting daily *sodium* intakes to less than 2300 milligrams (approximately 1 teaspoon of salt). Higher intakes seem to be well tolerated in most healthy people, however. The "How to" (p. 230) offers strategies for cutting salt (and, therefore, sodium) intake.

hypertension: high blood pressure.

salt-sensitive: an individual who responds to a high salt intake with an increase in blood pressure or to a low salt intake with a decrease in blood pressure.

One diet plan, known as the DASH (Dietary Approaches to Stop Hypertension) diet, also lowers blood pressure.[11] The DASH approach emphasizes fruits, vegetables, and low-fat dairy products; includes whole grains, nuts, poultry, and fish; and calls for reduced intakes of red meat, butter, and other high-fat foods. The DASH diet in combination with a reduced sodium intake is even more effective at lowering blood pressure than either strategy alone. Chapter 21 offers a complete discussion of hypertension and the dietary recommendations for its prevention and treatment.

CHLORIDE

The chloride ion is the major negative ion of the extracellular fluids, where it occurs primarily in association with sodium. Like sodium, chloride is critical to maintaining fluid, electrolyte, and acid-base balances in the body. In the stomach, the chloride ion is part of hydrochloric acid, which maintains the strong acidity of the gastric fluids.

Salt is a major food source of chloride, and as with sodium, processed foods are a major contributor of this mineral to people's diets. Because salt contains a higher proportion of chloride (by weight) than sodium, chloride recommendations are slightly higher than, but still equivalent to, those of sodium. In other words, $3/4$ teaspoon of salt will deliver some sodium, more chloride, and still meet the AI for both.

Chloride AI:
- 2300 mg/day (19–50 yr).
- 2000 mg/day (51–70 yr).
- 1800 mg/day (>70 yr).

POTASSIUM

Potassium is the principal positively charged ion inside the body cells. It plays a major role in maintaining fluid and electrolyte balance and cell integrity. Potassium is also critical to keeping the heartbeat steady. The sudden deaths that occur in severe diarrhea and in children with kwashiorkor or people with eating disorders are likely due to heart failure caused by potassium loss.

Potassium Deficiency and Toxicity Potassium deficiency results more often from excessive losses than from deficient intakes. Deficiency arises in abnormal condi-

H O W   T O   Cut Salt Intake

Most people eat more salt (and therefore sodium) than they need, and some people can lower their blood pressure by avoiding highly salted foods and removing the saltshaker from the table. Foods eaten without salt may seem less tasty at first, but with repetition, people can learn to enjoy the natural flavors of many unsalted foods. Strategies to cut salt intake include:

- Cook with little or no added salt.
- Prepare foods with sodium-free spices such as basil, bay leaves, curry, garlic, ginger, mint, oregano, pepper, rosemary, and thyme; lemon juice; vinegar; or wine.
- Add little or no salt at the table; taste foods before adding salt.
- Read labels with an eye open for salt. (See the glossary on p. 27–28 for terms used to describe the sodium contents of foods on labels.)
- Select low-salt or salt-free products when available.

Use these foods sparingly:

- Foods prepared in brine, such as pickles, olives, and sauerkraut.
- Salty or smoked meats, such as bologna, corned or chipped beef, bacon, frankfurters, ham, lunch meats, salt pork, sausage, and smoked tongue.
- Salty or smoked fish, such as anchovies, caviar, salted and dried cod, herring, sardines, and smoked salmon.
- Snack items such as potato chips, pretzels, salted popcorn, salted nuts, and crackers.
- Condiments such as bouillon cubes; seasoned salts; MSG; soy, teriyaki, Worcestershire, and barbeque sauces; prepared horseradish, catsup, and mustard.
- Cheeses, especially processed types.
- Canned and instant soups.

tions such as diabetic acidosis, dehydration, or prolonged vomiting or diarrhea; potassium deficiency can also result from the regular use of certain medications, including **diuretics, steroids,** and **cathartics.** One of the earliest symptoms of deficiency is muscle weakness. Inadequate potassium intakes are possible with diets low in fresh fruits and vegetables, but out-and-out deficiencies of potassium are unlikely in healthy people.

Potassium toxicity does not result from overeating foods high in potassium; therefore, a Tolerable Upper Intake Level was not set. Toxicity can result from overconsumption of potassium salts or supplements and from certain diseases or medications.[12] Given more potassium than the body needs, the kidneys accelerate their excretion. If the GI tract is bypassed, however, and potassium is injected directly into a vein, it can stop the heart.

Potassium Recommendations and Food Sources In healthy people, almost any reasonable diet provides enough potassium to prevent the dangerously low blood potassium that indicates a severe deficiency. Potassium is abundant inside all living cells, both plant and animal, and because cells remain intact until foods are processed, the richest sources of potassium are *fresh* foods of all kinds—especially fruits and vegetables. A typical U.S. diet, however, with its low intakes of fruits and vegetables, provides only about half the recommended intake.[13] Although blood potassium may remain normal on such a diet, chronic diseases are more likely to occur. With low potassium intake, blood pressure increases, glucose tolerance is impaired, metabolic acidity increases, calcium losses from bones accelerate, and kidney stone formation becomes more likely.[14]

CALCIUM

Calcium occupies more space in this chapter than any other major mineral. Other minerals are revisited later in this book, where their key roles in heart disease and kidney disease are discussed. Calcium, though, deserves emphasis here in the normal nutrition part of the book because an adequate intake of calcium early in life helps grow a healthy skeleton and prevent bone disease in later life.

Calcium Roles in the Body Calcium owns the distinction of being the most abundant mineral in the body. Ninety-nine percent of the body's calcium is stored in the bones, where it plays two important roles. First, it is an integral part of bone structure. Second, it serves as a calcium bank available to the body fluids should a drop in blood calcium occur.

Calcium in Bone As bones begin to form, calcium salts form crystals on a matrix of the protein collagen. As the crystals become denser, they give strength and rigidity to the maturing bones. As a result, the long leg bones of children can support their weight by the time they have learned to walk. Figure 9-3 shows the lacy network of calcium-containing crystals in the bone.

Many people have the idea that bones are inert, like rocks. Not so. Bones continuously gain and lose minerals in an ongoing process of remodeling. Growing children gain more bone than they lose, and healthy adults maintain a reasonable balance. When withdrawals substantially exceed deposits, however, problems such as osteoporosis develop.

From birth to approximately age 20, the bones are actively growing by modifying their length, width, and shape (Figure 9-4, p. 232). This rapid growth phase overlaps with the next period of peak bone mass development, which occurs between the ages of 12 and 30. During this period, skeletal mass increases. Bones grow thicker and denser by remodeling, a maintenance and repair process involving the loss of existing bone and the deposition of new bone. In the final phase,

Fresh fruits and vegetables provide potassium in abundance.

Potassium AI:
• Adults: 4700 mg/day.

diuretics (dye-yoo-RET-ics): medications that promote the excretion of water through the kidneys. Not all diuretics increase the urinary loss of potassium. Some, called potassium-sparing diuretics, are less likely to result in a potassium deficiency (see Chapter 21).

steroids (STARE-oids): medications used to reduce tissue inflammation, to suppress the immune response, or to replace certain steroid hormones in people who cannot synthesize them.

cathartics (ca-THART-ics): strong laxatives.

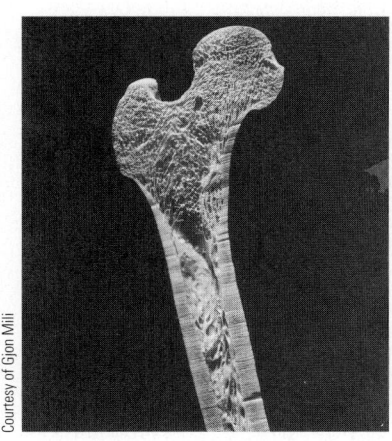

FIGURE 9-3 Cross Section of Bone
The lacy structural elements are *trabeculae* (tra-BECK-you-lee), which can be drawn on to replenish blood calcium.

FIGURE 9-4 Phases of Bone Development throughout Life
The active growth phase occurs from birth to approximately age 20. The next phase of peak bone mass development occurs between the ages of 12 and 30. The final phase, when bone resorption exceeds formation, begins between the ages of 30 and 40 and continues through the remainder of life.

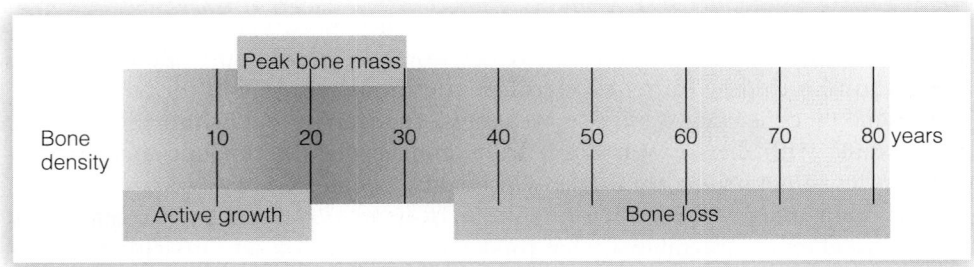

which begins between 30 and 40 years of age and continues throughout the remainder of life, bone loss exceeds new bone formation.

Calcium in Body Fluids The 1 percent of the body's calcium that circulates in the fluids as ionized calcium is vital to life. It helps regulate muscle contractions, transmit nerve impulses, clot blood, and secrete hormones, digestive enzymes, and neurotransmitters. In most of these processes, the calcium ion operates in two ways: it conveys signals received at the cell surface to the inside of the cell; and it activates the proteins involved in the function.[15] Calcium is a **cofactor** for several enzymes as well.

The regulators are hormones from the thyroid and parathyroid glands, as well as vitamin D. One hormone, **parathormone,** raises blood calcium. Another hormone, **calcitonin,** lowers blood calcium by inhibiting release of calcium from bone. The hormonelike **vitamin D** raises blood calcium by acting at the three sites listed.

Calcium Balance Blood calcium concentration is tightly controlled. Whenever blood calcium rises too high, a system of hormones and vitamin D promotes its deposit into bone. Whenever blood calcium falls too low, the regulatory system acts in three locations to raise it:

1. The small intestine absorbs more calcium.
2. The bones release more calcium.
3. The kidneys excrete less calcium.

Thus blood calcium rises to normal.

The calcium stored in bone provides a nearly inexhaustible source of calcium for the blood. Even in a calcium deficiency, blood calcium remains normal. Blood calcium changes only in response to abnormal regulatory control, not to diet. Blood calcium above normal causes **calcium rigor:** the muscles contract and cannot relax. Blood calcium below normal causes **calcium tetany**—also characterized by uncontrolled muscle contraction. These conditions are caused by a lack of vitamin D or by abnormal concentrations of the hormones that regulate calcium homeostasis.

Although a chronic *dietary* deficiency of calcium or a chronic deficiency due to poor absorption does not change blood calcium, it does deplete the savings account in the bones. Because this is an important concept, we repeat: it is the bones, not the blood, that are robbed by calcium deficiency.

Reminder: *Osteoporosis* is a condition characterized by reduced density of the bones. The bones become porous and fragile and fracture easily.

cofactor: a mineral element that, like a coenzyme, works with an enzyme to facilitate a chemical reaction.

calcium rigor: hardness or stiffness of the muscles caused by high blood calcium.

calcium tetany: intermittent spasms of the extremities due to nervous and muscular excitability caused by low blood calcium.

Calcium and Osteoporosis As mentioned earlier, bone mass peaks at the time of skeletal maturity (about age 30), and a high peak bone mass is the best protection against later age-related bone loss and fracture. Adequate calcium nutrition during the growing years is essential to achieving optimal peak bone mass.[16] Following menopause, women lose about 15 percent of their bone mass, as do middle-aged and older men. When bone loss has reached such an extreme that bones fracture under even common, everyday stresses, the condition is known as osteoporosis. Osteoporosis afflicts more than 40 million people in the United States, mostly women 50 years of age or older.[17] Men, however, are not immune to osteoporosis. Each year, a million and a half people—30 percent of them men—suffer broken hips, pelvises, legs, arms, hands, and ankles attributable to osteoporosis.

Both genetic and environmental factors contribute to osteoporosis. Table 9-4 summarizes risk and protective factors for osteoporosis. Osteoporosis is more prevalent in women than men for several reasons. First, women consume less dietary

TABLE 9-4   Risk Factors and Protective Factors for Osteoporosis

Risk Factors	Protective Factors
Older age	Younger age
Low body mass index (BMI)	High BMI
Caucasian, Asian, or Hispanic heritage	African American heritage
Cigarette smoking	No smoking
Alcohol consumption in excess	Alcohol consumption in moderation
Sedentary lifestyle	Regular weight-bearing exercise
Use of glucocorticoids or anticonvulsants	Use of diuretics
Female gender	Male gender
Maternal history of osteoporosis fracture or personal history of fracture	Bone density assessment and treatment (if necessary)
Estrogen deficiency in women (amenorrhea or menopause, especially early or surgically induced); testosterone deficiency in men	Use of estrogen therapy
Lifetime diet inadequate in calcium and vitamin D	Lifetime diet rich in calcium and vitamin D

calcium than men do. Second, at all ages, women's bone mass is lower than men's because women generally have smaller bodies. Finally, bone loss accelerates after menopause and exceeds men's losses for up to ten years.

In addition to calcium, many other minerals and vitamins, including phosphorus, magnesium, fluoride, and vitamin D, help to form and stabilize the structure of bones. Any or all of these elements are needed to prevent bone loss. The first, most obvious lines of defense, however, are to maintain a lifelong adequate intake of calcium and to "exercise it into place." Active bones are denser than sedentary bones.[18] Weight-bearing physical activity, such as walking, running, dancing, and weight training, prompts the bones to deposit minerals. It has long been known that when people are confined to bed, both their muscles and their bones lose strength. Muscle strength and bone strength go together: when muscles work, they pull on the bones, and both are stimulated to grow stronger.

Calcium and Hypertension Some evidence suggests that calcium may protect against hypertension.[19] For this reason, restricting salt intake to treat hypertension is narrow advice, especially considering the success of the DASH diet in lowering blood pressure. The DASH diet is not particularly low in sodium, but it is rich in calcium, as well as in potassium and magnesium. As mentioned earlier, the DASH diet, together with a reduced sodium intake, is more effective at lowering blood pressure than either strategy alone.

Calcium Recommendations As mentioned earlier, blood calcium concentration does not reflect calcium status. Calcium recommendations are therefore based on balance studies, which measure daily intake and excretion. An optimal calcium intake reflects the amount needed to retain the most calcium. The more calcium retained, the greater the bone density (within genetic limits) and, potentially, the lower the risk of osteoporosis. Calcium recommendations during adolescence are

Calcium AI:
- Adults (19–50 yr): 1000 mg/day.
- Adults (51 and older): 1200 mg/day.

TABLE 9-5

Suggested Minimum Daily Fluid Milk Intakes

Young children	2 cups
Adolescents	3 cups
Adults	3 cups
Pregnant or lactating women	3 cups
Women past menopause	3 cups

Apparently, all fibers in plant foods—cellulose, hemicellulose, pectin, and others—bind calcium to some extent, as do **phytate** and **oxalate.** Some plant foods contain the compounds phytate and oxalate, which are *binders* that combine with minerals to form complexes that the body cannot absorb.

set high (1300 milligrams) to help ensure that the skeleton will be strong and dense. Between the ages of 19 and 50, recommendations are lowered to 1000 milligrams a day. For those over 50, recommendations are raised again, to 1200 milligrams, to minimize bone loss. Some authorities advocate calcium recommendations as high as 1500 milligrams per day for women over 50. Many women have intakes well below recommendations, however.

Calcium in Foods Calcium is found most abundantly in a single food group— milk and milk products. For this reason, dietary recommendations advise daily consumption of low-fat or fat-free milk products. A cup of milk offers about 300 milligrams of calcium, so an adult who drinks 3 cups of milk a day (or eats the equivalent in yogurt) is well on the way to meeting daily calcium needs (see Table 9-5). The other dairy food that contains comparable amounts of calcium is cheese. One slice of cheese (1 ounce) contains about two-thirds as much calcium as a cup of milk. Cottage cheese, however, contains much less. Snapshot 9-1 shows foods that are rich in calcium, and the accompanying "How to" suggests ways of adding calcium to meals.

Some foods offer large amounts of calcium because of fortification. Calcium-fortified juice, high-calcium milk (milk with extra calcium added), and calcium-fortified cereals are examples. Some calcium-rich mineral waters provide as much as 500 milligrams of calcium per liter, and others provide slightly less.[20] Drinking calcium-rich water throughout the day offers a convenient way to meet both calcium and water needs.[21]

Among the vegetables, mustard greens, kale, parsley, watercress, and broccoli are good sources of available calcium. Some dark green, leafy vegetables— notably, spinach and Swiss chard—appear to be calcium-rich but actually provide very little, if any, calcium to the body. These foods contain binders that prevent calcium absorption.

Aided by vitamin D, the body is able to regulate its absorption of calcium by altering its production of the calcium-binding protein. More of this protein is

SNAPSHOT 9-1 *Calcium*

GOOD SOURCES*

SARDINES (with bones)
3 oz = 324 mg

MILK
1 c = 300 mg

TOFU (calcium set)
1/2 c = 275 mg

BLACK-EYED PEAS (cooked)
1/2 c = 105 mg

CHEDDAR CHEESE
1 1/2 oz = 306 mg

TURNIP GREENS (cooked)
1 c = 197 mg

WAFFLE (whole grain)
1 waffle = 196 mg

BROCCOLI[a] (cooked)
1 1/2 c = 93 mg

* These foods provide 10 percent or more of the calcium Daily Value in a serving. For a 2000-kcalorie diet, the DV is 1000 mg/day.
[a] Although broccoli, kale, and some other cooked green leafy vegetables fall short of supplying 10 percent of calcium in a serving, these foods are important sources of bioavailable calcium. Almonds also supply calcium. Other greens, such as spinach and chard, contain calcium in an unabsorbable form. Note that the amounts for green vegetables exceed the 1/2 c servings of the USDA Food Guide. Some calcium-rich mineral waters may also be good sources.

made if more calcium is needed. Infants and children absorb up to 60 percent of the calcium they ingest and pregnant women, about 50 percent. Other adults, who are not growing, absorb about 25 percent.

People may think that taking a calcium supplement is preferable to getting calcium from food, but foods offer important fringe benefits. For example, drinking 3 cups of milk fortified with vitamins A and D will supply substantial amounts of other nutrients. Furthermore, the vitamin D, and possibly other nutrients in the milk enhance calcium absorption. Some people absorb calcium better from milk and milk products than from even the most absorbable supplements. The National Institutes of Health concludes that foods are the best sources of calcium and recommends supplements only when intake from food is insufficient.

PHOSPHORUS

Phosphorus is the second most abundant mineral in the body. About 85 percent of it is found combined with calcium in the crystals of the bones and teeth. As part of one of the body's buffer systems (phosphoric acid), phosphorus is also found in all body tissues. Phosphorus is a part of DNA and RNA, the genetic material present in every cell. Thus phosphorus is necessary for all growth. Phosphorus also plays many key roles in the transfer of energy that occurs during cellular metabolism. Phosphorus-containing lipids (phospholipids) help transport other lipids in the blood. Phospholipids are also principal components of cell membranes.

Animal protein is the best source of phosphorus because the mineral is so abundant in the cells of animals. Diets that provide adequate energy and protein

 **HOW TO** *Add Calcium to Daily Meals*

For those who tolerate milk, many cooks slip extra calcium into meals by sprinkling a tablespoon or two of fat-free dry milk into almost everything. The added kcalorie value is small, and changes to the taste and texture of the dish are practically nil; yet each 2 tablespoons adds about 100 extra milligrams of calcium and moves people closer to meeting the recommendation to obtain 3 cups of milk each day.

Here are some more tips for including calcium-rich foods in your meals.

At Breakfast

- Choose calcium-fortified orange or vegetable juice.
- Serve tea or coffee, hot or iced with milk.
- Choose cereals, hot or cold, with milk.
- Cook hot cereals with milk instead of water; then mix in 2 tablespoons of fat-free dry milk.
- Make muffins or quick breads with milk and extra fat-free powdered milk.
- Add milk to scrambled eggs.
- Moisten cereals with flavored yogurt.

At Lunch

- Add low-fat cheeses to sandwiches, burgers, or salads.
- Use a variety of green vegetables, such as watercress or kale, in salads and on sandwiches.
- Drink fat-free milk or calcium-fortified soy milk as a beverage or in a smoothie.

- Drink calcium-rich mineral water as a beverage (studies suggest significant calcium absorption).
- Marinate cabbage shreds or broccoli spears in low-fat Italian dressing for an interesting salad that provides calcium.
- Choose coleslaw over potato and macaroni salads.
- Mix the mashed bones of canned salmon into salmon salad or patties.
- Eat sardines with their bones.
- Stuff potatoes with broccoli and low-fat cheese.
- Try pasta such as ravioli stuffed with low-fat ricotta cheese instead of meat.
- Sprinkle parmesan cheese on pasta salads.

At Supper

- Toss a handful of thinly sliced green vegetables, such as kale or young turnip greens, with hot pasta; the greens wilt pleasingly in the steam of the freshly cooked pasta.
- Serve a green vegetable every night and try new ones—how about kohlrabi? It tastes delicious when cooked like broccoli.
- Learn to stir-fry Chinese cabbage and other Asian foods.
- Try tofu (the calcium-set kind); this versatile food has inspired whole cookbooks devoted to creative uses.
- Add fat-free powdered milk to almost anything—meat loaf, sauces, gravies, soups, stuffings, casseroles, blended beverages, puddings, quick breads, cookies, brownies. Be creative.
- Choose frozen yogurt, ice milk, or custards for dessert.

Phosphorus RDA:
• Adults: 700 mg/day.

also supply adequate phosphorus. Dietary deficiencies are rare. A summary of facts about phosphorus appears in Table 9-6 later in the chapter.

MAGNESIUM

Magnesium barely qualifies as a major mineral. Only about 1 ounce of magnesium is present in the body of a 130-pound person, over half of it in the bones.[22] Most of the rest is in the muscles, heart, liver, and other soft tissues, with only 1 percent in the body fluids. Bone magnesium seems to be a reservoir to ensure that some will be on hand for vital reactions regardless of recent dietary intake.

Magnesium is critical to the operation of hundreds of enzymes. It acts in all the cells of the soft tissues, where it forms part of the protein-making machinery and is necessary for the release of energy. Magnesium also helps muscles to relax after contraction.

Magnesium Deficiency Magnesium deficiency can result from vomiting, diarrhea, alcohol abuse, or protein malnutrition; in people who have been fed incomplete fluids intravenously for too long after surgery; or in people using diuretics. A severe magnesium deficiency causes tetany, an extreme and prolonged contraction of the muscles similar to the calcium tetany described earlier. Magnesium deficiency may also be related to cardiovascular disease and hypertension.[23] Additionally, magnesium intakes may be related to the risk of colon cancer. In a study of more than 60,000 women over a 14-year period, those with the lowest magnesium intakes had the greatest risks of colon cancer compared with women who had the highest magnesium intakes.[24] Magnesium deficiency is thought to cause the hallucinations commonly experienced during withdrawal from alcohol intoxication.

Magnesium Toxicity Magnesium toxicity is most often reported in older adults who abuse magnesium-containing laxatives, antacids, and other medications. The consequences can be severe: lack of coordination, confusion, coma, and, in extreme cases, death.

Magnesium RDA:
• Men (19–30 yr): 400 mg/day.
  (31 and older): 420 mg/day.
• Women (19–30 yr): 310 mg/day.
  (31 and older): 320 mg/day.

Magnesium Intakes and Food Sources Dietary intakes of magnesium for both men and women in the United States average about three-quarters of the recommended intakes.[25] Dietary intake data, however, do not assess the magnesium contribution of water. In various parts of the country, the water contains both calcium and magnesium. Known as "hard" water, this water can contribute significantly to magnesium intakes.

Magnesium-rich food sources include dark green, leafy vegetables, nuts, legumes, whole-grain breads and cereals, seafood, chocolate, and cocoa (see Snapshot 9-2). Magnesium is easily lost from foods during processing, so unprocessed foods are the best choices.

SULFATE

Sulfate is the oxidized form of sulfur as it exists in food and water. The body requires sulfate for the synthesis of many important sulfur-containing compounds. Sulfur-containing amino acids play an important role in helping to shape strands of protein. The particular shape of a protein enables it to do its specific job, such as enzyme work. Skin, hair, and nails contain some of the body's more rigid proteins, which have high sulfur contents.

The sulfur-containing amino acids are methionine and cysteine. Cysteine in one part of a protein chain can bind to cysteine in another part of the chain by way of a sulfur-sulfur bridge, thus helping to stabilize the protein structure.

There is no recommended intake for sulfur, and no deficiencies are known. Only a person who lacks protein to the point of severe deficiency will lack the sulfur-containing amino acids.

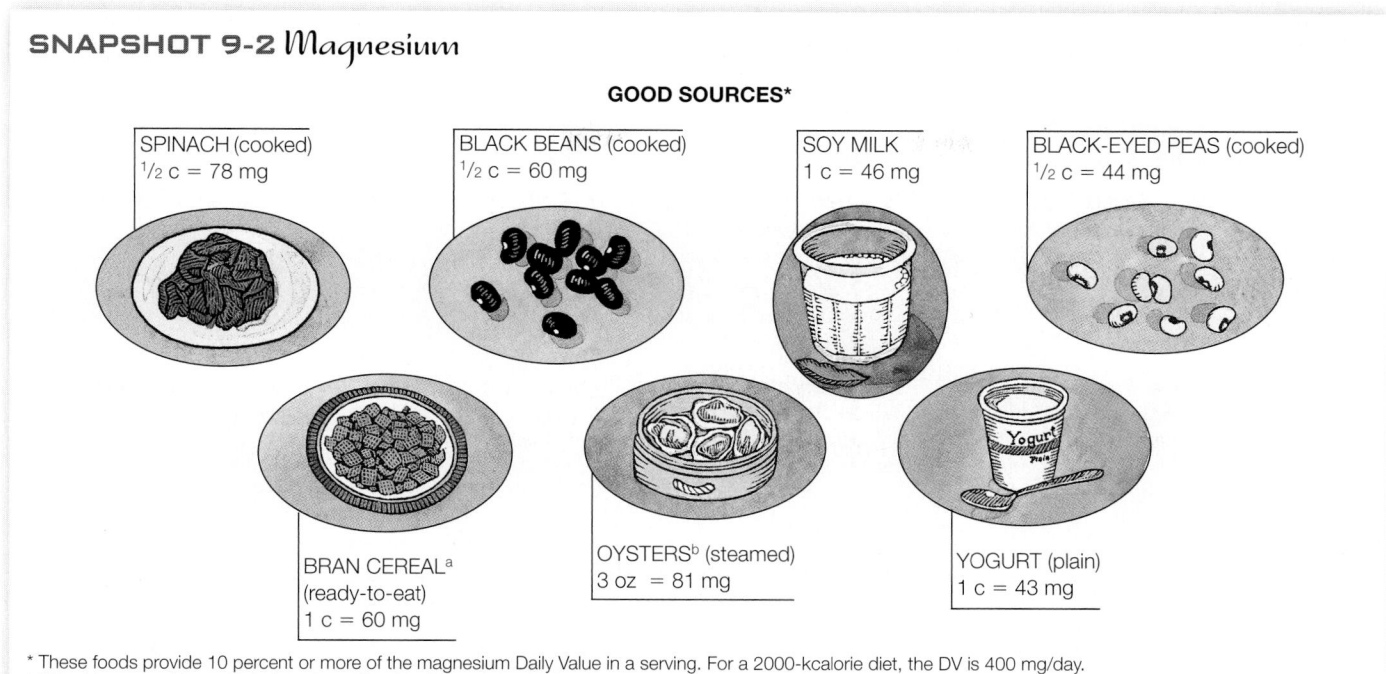

SNAPSHOT 9-2 *Magnesium*

GOOD SOURCES*

SPINACH (cooked)
½ c = 78 mg

BLACK BEANS (cooked)
½ c = 60 mg

SOY MILK
1 c = 46 mg

BLACK-EYED PEAS (cooked)
½ c = 44 mg

BRAN CEREAL[a]
(ready-to-eat)
1 c = 60 mg

OYSTERS[b] (steamed)
3 oz = 81 mg

YOGURT (plain)
1 c = 43 mg

* These foods provide 10 percent or more of the magnesium Daily Value in a serving. For a 2000-kcalorie diet, the DV is 400 mg/day.
[a] Wheat bran provides magnesium, but refined grain products are low in magnesium.
[b] Magnesium in oysters varies.

REVIEW NOTES

All of the major minerals influence the body's fluid balance, but sodium, chloride, and potassium are most noted for this role.

Excess sodium in the diet contributes to high blood pressure.

Most of the body's calcium is in the bones, where it provides a rigid structure and a reservoir of calcium for the blood.

Table 9-6 (p. 238) offers a summary of the major minerals and their functions.

The Trace Minerals

Figure 9-1, earlier in this chapter, shows how tiny the quantities of trace minerals in the human body are. If you could remove all of them from your body, you would have only a bit of dust, hardly enough to fill a teaspoon. Yet each of the trace minerals performs some vital role for which no substitute will do. A deficiency of any of them can be fatal, and an excess of many can be equally deadly.

Recommendations have been established for nine of the trace minerals. Others are recognized as essential nutrients for some animals but have not been proved to be required for human beings (see Table 9-7, p. 239). Still others are under study to determine whether they, too, perform indispensable roles in the body. Table 9-8 at the end of the chapter provides a summary of the trace minerals.

IRON

Every living cell—both plant and animal—contains iron. Most of the iron in the body is a component of the proteins **hemoglobin** in red blood cells and **myoglobin** in muscle cells. The iron in both hemoglobin and myoglobin helps them carry

hemoglobin: the oxygen-carrying protein of the red blood cells.
hemo = blood
globin = globular protein

myoglobin: the oxygen-carrying protein of the muscle cells.
myo = muscle

TABLE 9-6	The Major Minerals—A Summary			
Mineral Name	Chief Functions in the Body	Deficiency Symptoms	Toxicity Symptoms	Significant Sources
Sodium	With chloride and potassium (electrolytes), maintains cells' normal fluid balance and acid-base balance in the body. Also critical to nerve impulse transmission.	Muscle cramps, mental apathy, loss of appetite	Hypertension	Salt, soy sauce, processed foods
Chloride	Part of the hydrochloric acid found in the stomach and necessary for proper digestion.	Growth failure in children; muscle cramps, mental apathy, loss of appetite; can cause death (uncommon).	Normally harmless (the gas chlorine is a poison but evaporates from water); can cause vomiting.	Salt, soy sauce; moderate quantities in whole, unprocessed foods, large amounts in processed foods
Potassium	Facilitates reactions, including the making of protein; the maintenance of fluid and electrolyte balance; the support of cell integrity; the transmission of nerve impulses; and the contraction of muscles, including the heart.	Deficiency accompanies dehydration; causes muscular weakness, paralysis, and confusion; can cause death.	Causes muscular weakness; triggers vomiting; if given into a vein, can stop the heart.	All whole foods: meats, milk, fruits, vegetables, grains, legumes
Calcium	The principal mineral of bones and teeth. Also acts in normal muscle contraction and relaxation, nerve functioning, blood clotting, blood pressure, and immune defenses.	Stunted growth in children; adult bone loss (osteoporosis)	Constipation; increased risk of urinary stone formation and kidney dysfunction; interference with absorption of other minerals	Milk and milk products, oysters, small fish (with bones), tofu (bean curd), greens, legumes
Phosphorus	Important in cells' genetic material, in cell membranes as phospholipids, in energy transfer, and in buffering systems.	Phosphorus deficiency unknown	Excess phosphorus may cause calcium excretion.	All animal tissues
Magnesium	Another factor involved in bone mineralization, the building of protein, enzyme action, normal muscular contraction, transmission of nerve impulses, and maintenance of teeth.	Weakness; confusion; if extreme, convulsions, bizarre movements (especially of eyes and face), hallucinations, and difficulty in swallowing. In children, growth failure.[a]	Large doses taken in the form of the laxative Epsom salts cause diarrhea.	Nuts, legumes, whole grains, dark green vegetables, seafoods, chocolate, cocoa
Sulfur	A component of certain amino acids; part of the vitamins biotin and thiamin and the hormone insulin; combines with toxic substances to form harmless compounds; stabilizes protein shape by forming sulfur-sulfur bridges.	None known; protein deficiency would occur first.	Would occur only if sulfur amino acids were eaten in excess; this (in animals) depresses growth.	All protein-containing foods

[a] A still more severe deficiency causes tetany, an extreme, prolonged contraction of the muscles similar to that caused by low blood calcium.

and hold oxygen and then release it. Hemoglobin in the blood carries oxygen from the lungs to tissues throughout the body. Myoglobin holds oxygen for the muscles to use when they contract. As part of many enzymes, iron is vital to the processes by which cells generate energy. Iron is also needed to make new cells, amino acids, hormones, and neurotransmitters.

The special provisions the body makes for iron's handling show that it is a precious mineral to be tightly hoarded. For example, when a red blood cell dies, the liver saves the iron and returns it to the bone marrow, which uses it to build new red blood cells. Thus only tiny amounts of iron are lost, principally in urine, sweat, shed skin, and blood (if bleeding occurs).

Normally, only about 10 to 15 percent of dietary iron is absorbed; but if the body's supply is diminished or if the need increases for any reason (such as pregnancy), absorption increases. The body makes several provisions for absorbing iron. A special protein in the intestinal cells captures iron and holds it in reserve for release into the body as needed; another protein transfers the iron to a special iron carrier in the blood. The blood protein (**transferrin**) carries the iron to tissues throughout the body. When more iron is needed, more of these special proteins are produced so that more than the usual amount of iron can be absorbed and carried. If there is a surplus of iron, special storage proteins in the liver, bone marrow, and other organs store it.

Researchers are discovering genes that participate in the regulation of the body's iron at an extraordinary pace.[26] Along the way, new information about how the body controls the uptake, storage, and distribution of iron is emerging. For example, it has long been known that the liver is the main site of iron storage in the body, but only recently has research revealed that the liver may play a major role in regulating how much iron is absorbed from the GI tract and how much is released from storage.[27] Circulating levels of hepcidin, a hormone produced by the liver are inversely related to the efficiency of iron absorption—high blood concentrations of the hormone are associated with low iron absorption.[28]

TABLE 9-7	Trace Minerals
RDA	
Copper	
Iodine	
Iron	
Molybdenum	
Selenium	
Zinc	
Adequate Intake (AI)	
Chromium	
Fluoride	
Manganese	
Known Essential for Animals; Human Requirements under Study	
Arsenic	
Boron	
Nickel	
Silicon	
Vanadium	
Known Essential for Some Animals; No Evidence That Intake by Humans Is Ever Limiting; No Recommendation Necessary	
Cobalt	

Note: The evidence for requirements and essentiality is weak for the trace minerals cadmium, lead, lithium, and tin.

The storage proteins are **ferritin** (FERR-i-tin) and **hemosiderin** (heem-oh-SID-er-in).

Iron Deficiency Worldwide, **iron deficiency** is the most common nutrient deficiency, affecting an estimated 2 to 5 billion people.[29] In developing countries, many children and women of childbearing age suffer from **iron-deficiency anemia.** In the United States, iron deficiency is less prevalent, but it still affects about 10 percent of toddlers, adolescent girls, and women of childbearing age.[30]

Women are especially prone to iron deficiency during their reproductive years because of blood losses during menstruation. Pregnancy places further iron demands on women. Iron is needed to support the added blood volume, the growth of the fetus, and blood loss during childbirth. Infants (six months or older) and young children receive little iron from their high-milk diets, yet they need extra iron to support growth. The rapid growth of adolescence, especially for males, and the blood losses of menstruation for females also demand extra iron that a typical teen diet may not provide.

transferrin (trans-FERR-in): the body's iron-carrying protein.

iron deficiency: having depleted iron stores.

iron-deficiency anemia: a blood iron deficiency characterized by small, pale red blood cells; also called **microcytic hypochromic anemia.**
micro = small
cytic = cells
hypo = too little
chrom = color

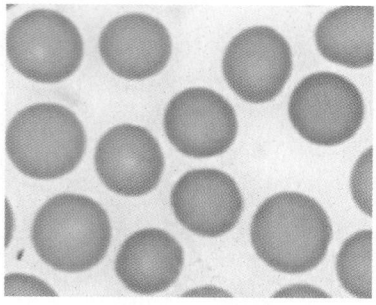

Normal red blood cells. Both size and color are normal.

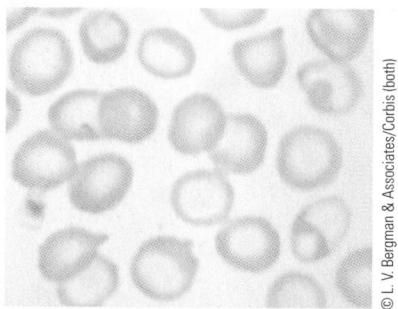

Blood cells in iron-deficiency anemia. These cells are small and pale because they contain less hemoglobin.

FIGURE 9-5 Normal and Anemic Blood Cells

Stages of iron deficiency:
- Iron stores diminish.
- Transport iron decreases.
- Hemoglobin production declines.

The effects of iron deficiency on children's behavior are revisited in Chapter 11.

erythrocyte protoporphyrin (PRO-toe-PORE-fe-rin): a precursor to hemoglobin

hematocrit (hee-MAT-oh-crit): measurement of the volume of the red blood cells packed by centrifuge in a given volume of blood.

© L. V. Bergman & Associates/Corbis (both)

Causes of Iron Deficiency The cause of iron deficiency is usually inadequate intake from ignorance of which foods to choose, from sheer lack of food altogether, or from high consumption of iron-poor foods. In the Western world, high sugar and fat intakes are often associated with low iron intakes. Blood loss is the primary nonnutritional cause, especially in poor regions of the world where parasitic infections of the GI tract may lead to blood loss.

Assessment of Iron Deficiency Iron deficiency develops in stages. This section provides a brief overview of how to detect these stages, and Appendix E provides more details. In the first stage of iron deficiency, iron stores diminish, as do levels of ferritin, an iron-storing protein. Measures of serum ferritin (in the blood) reflect iron stores and are most valuable in assessing iron status at this earliest stage.

The second stage of iron deficiency is characterized by a decrease in transport iron: levels of serum iron fall, and levels of the iron-carrying protein transferrin *increase* (an adaptation that enhances iron absorption). Together, these two measures can determine the severity of the deficiency—the more transferrin and the less iron in the blood, the more advanced the deficiency is. Transferrin saturation—the percentage of transferrin that is saturated with iron—decreases as iron stores decline.

The third stage of iron deficiency occurs when the lack of iron limits hemoglobin production. Now the hemoglobin precursor, **erythrocyte protoporphyrin**, begins to accumulate as hemoglobin and **hematocrit** values decline. Hemoglobin and hematocrit tests are easy, quick, and inexpensive, so they are the tests most commonly used in evaluating iron status; their usefulness is limited, however, because they are late indicators of iron deficiency. Furthermore, other nutrient deficiencies and medical conditions can influence their values.

Iron Deficiency and Anemia Iron deficiency and iron-deficiency anemia are not the same: people may be iron deficient without being anemic. The term *iron deficiency* refers to depleted body iron stores without regard to the degree of depletion or to the presence of anemia. The term *iron-deficiency anemia* refers to the severe depletion of iron stores that results in a low hemoglobin concentration. In iron-deficiency anemia, red blood cells are pale and small (see Figure 9-5). They can't carry enough oxygen from the lungs to the tissues. Without adequate iron, energy metabolism in the cells falters. The result is fatigue, weakness, headaches, apathy, pallor, and poor resistance to cold temperatures. Since hemoglobin is the bright red pigment of the blood, the skin of a fair person who is anemic may become noticeably pale. In a dark-skinned person, the tongue and eye lining, normally pink, will be very pale.

The fatigue that accompanies iron-deficiency anemia differs from the tiredness a person experiences from a simple lack of sleep. People with anemia feel fatigue only when they exert themselves. Iron supplementation can, over time, relieve the fatigue and improve the body's response to physical activity.[31]

Less severe iron deficiency produces symptoms, too. Long before the red blood cells are affected and anemia is diagnosed, a developing iron deficiency affects behavior. Even at slightly lowered iron levels, energy metabolism is impaired, reducing physical work capacity and productivity. Children deprived of iron become irritable, restless, and unable to pay attention. These symptoms are among the first to appear when the body's iron begins to fall and among the first to disappear when iron status is restored.

Iron Deficiency and Pica A curious symptom sometimes seen in iron-deficient individuals is an appetite for ice, clay, paste, or other nonnutritious substances. Some people have been known to eat as many as eight trays of ice in a day, for example. This behavior, known as **pica,** has been observed for years, especially in women and children of low-income groups who are deficient in either iron or zinc. After iron is given, pica clears up dramatically within days, long before the red blood cells respond.

Caution on Self-Diagnosis Low hemoglobin may reflect an inadequate iron intake, and if it does, the physician may prescribe iron supplements. Any nutrient deficiency or disease or agent that interferes with hemoglobin synthesis, disrupts hemoglobin function, or causes a loss of red blood cells can precipitate anemia, however.

Feeling fatigued, weak, and apathetic is thus a sign that something is wrong, but it does not indicate that a person should take iron supplements; it means that the person should consult a physician. In fact, taking iron supplements may be the worst possible thing a person can do, because such supplements can mask a serious medical condition, such as hidden bleeding from cancer or an ulcer. Furthermore, a person can waste precious time in not seeking treatment. Remember, don't self-diagnose.

Iron Overload Normally, the body protects itself against absorbing too much iron by setting up a block in the intestinal cells. The system can be overwhelmed, however, resulting in **iron overload.** Once considered rare, iron overload has increased in frequency over the last few decades.

Iron overload, known as **hemochromatosis,** is usually caused by a genetic disorder that enhances iron absorption. Other causes of iron overload include repeated blood transfusions, massive doses of supplementary iron, and other rare metabolic disorders.

Some of the signs and symptoms of iron overload are similar to those of iron deficiency: apathy, lethargy, and fatigue. Therefore, taking iron supplements before assessing iron status is clearly unwise; hemoglobin tests alone would fail to make the distinction.

Iron overload is characterized by tissue damage, especially in iron-storing organs such as the liver. Infections are likely because bacteria thrive on iron-rich blood. Symptoms are most severe in alcohol abusers because alcohol damages the intestine, further impairing its defenses against absorbing excess iron. Untreated hemochromatosis aggravates the risk of diabetes, liver cancer, heart disease, and arthritis.

Symptoms of iron overload appear earlier in men than in women and iron overload is twice as prevalent among men as iron deficiency. The fortification of many foods with iron makes it difficult to follow an iron-restricted diet.

Iron Poisoning The rapid ingestion of massive amounts of iron can cause sudden death. A leading cause of accidental poisoning in small children is ingestion of iron supplements or vitamins with iron. The American Academy of Pediatrics has urged the Food and Drug Administration (FDA) to improve the labeling and packaging of iron-containing drugs and supplements. As few as 6 to 12 tablets have caused death in a child. A child suspected of iron poisoning should be rushed to the hospital to have the stomach pumped. Thirty minutes can make a crucial difference.

Iron Recommendations The average diet in the United States provides only about 6 to 7 milligrams of iron in every 1000 kcalories. Men need 8 milligrams of iron each day; most men easily eat more than 2000 kcalories, so a man can meet his iron needs without special effort. The recommendation for women during childbearing years, however, is 18 milligrams. Because women have higher iron needs and typically consume fewer than 2000 kcalories per day, they have trouble achieving appropriate iron intakes. On the average, women receive only 12 to 13 milligrams of iron per day, not enough until after menopause. To meet their iron needs from foods, premenopausal women need to select iron-rich foods at every meal.

Binding proteins in the intestinal cells (*mucosal ferritin* and *mucosal transferrin*) capture and hold unneeded iron to be shed with the cells, thereby forming a **mucosal block** to iron absorption.

Iron RDA:
- Men (19 and older): 8 mg/day.
- Women (19–50 yr): 18 mg/day.
  (>50 yr): 8 mg/day.

pica (PIE-ka): a craving for nonfood substances; also known as *geophagia* (jee-oh-FAY-jee-uh) when referring to clay-eating behavior.
> *picus* = woodpecker or magpie
> *geo* = earth
> *phagein* = to eat

iron overload: toxicity from excess iron.

hemochromatosis (heem-oh-crome-a-TOE-sis): iron overload characterized by deposits of iron-containing pigment in many tissues, with tissue damage. Hemochromatosis is usually caused by a hereditary defect in iron metabolism.

SNAPSHOT 9-3 Iron

GOOD SOURCES*

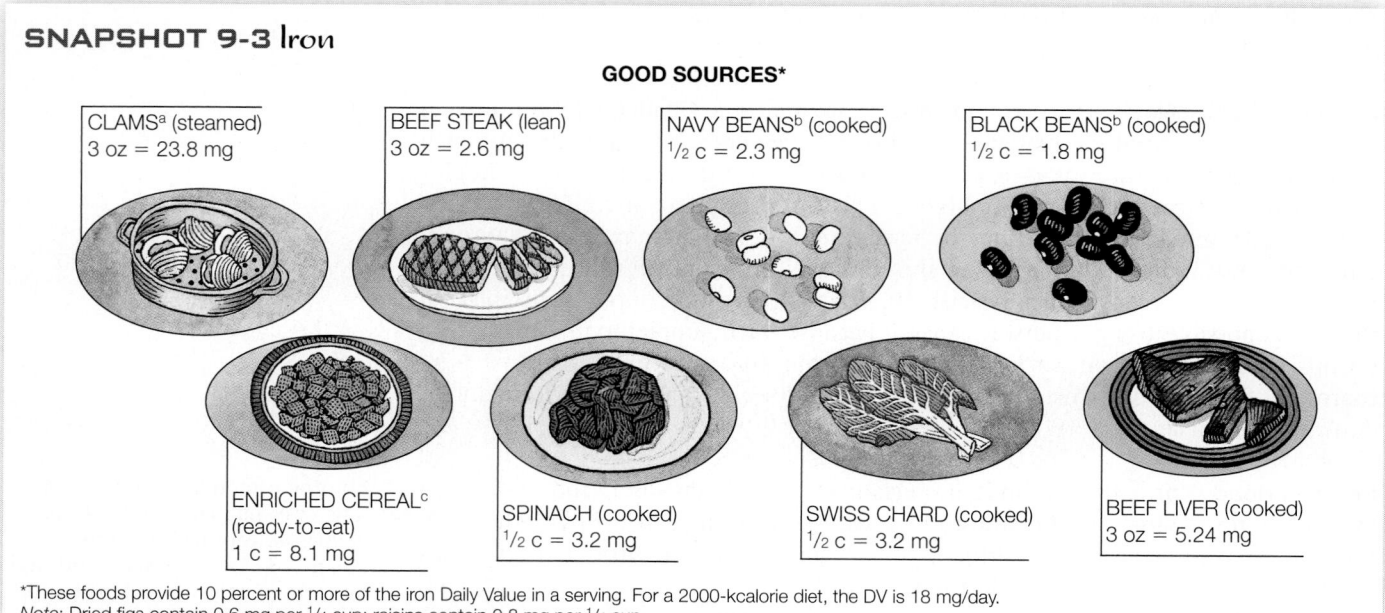

CLAMSᵃ (steamed)
3 oz = 23.8 mg

BEEF STEAK (lean)
3 oz = 2.6 mg

NAVY BEANSᵇ (cooked)
½ c = 2.3 mg

BLACK BEANSᵇ (cooked)
½ c = 1.8 mg

ENRICHED CEREALᶜ
(ready-to-eat)
1 c = 8.1 mg

SPINACH (cooked)
½ c = 3.2 mg

SWISS CHARD (cooked)
½ c = 3.2 mg

BEEF LIVER (cooked)
3 oz = 5.24 mg

*These foods provide 10 percent or more of the iron Daily Value in a serving. For a 2000-kcalorie diet, the DV is 18 mg/day.
Note: Dried figs contain 0.6 mg per ¼ cup; raisins contain 0.8 mg per ¼ cup.
ᵃ Some clams may contain less, but most types are iron-rich foods.
ᵇ Legumes contain phytates that reduce iron absorption. Soaking in water before cooking reduces phytates, and consuming legumes with vitamin C or meats increases iron absorption.
ᶜ Enriched cereals vary widely in iron content.

About 40% of the iron in meat, fish, and poultry is bound into molecules of **heme** (HEEM), the iron-holding part of the hemoglobin and myoglobin proteins. This **heme iron** is much more absorbable than **nonheme iron**.

Iron in Foods Iron occurs in two forms in foods, one of which is up to ten times more absorbable than the other. The most absorbable form is heme iron, which is bound into the iron-carrying proteins hemoglobin and myoglobin in meat, poultry, and fish. Heme iron contributes a small portion of the iron consumed by most people, but it is absorbed at a fairly constant rate of about 23 percent. The less absorbable form is nonheme iron, found in meats and also in plant foods. People absorb nonheme iron at a lower rate (2 to 20 percent); its absorption depends on several dietary factors and iron stores. Most of the iron people consume is nonheme iron from vegetables, grains, eggs, meat, fish, and poultry. Snapshot 9-3 shows the iron found in usual serving sizes of different foods.

Iron absorption from foods can be maximized by two substances that enhance iron absorption: MFP factor and vitamin C. Meat, fish, and poultry contain a factor (MFP factor) other than heme that promotes the absorption of iron. MFP factor even enhances the absorption of nonheme iron from other foods eaten at the same time. Vitamin C eaten in the same meal also doubles or triples nonheme iron absorption. Additionally, cooking with iron skillets can contribute iron to the diet. Some substances impair iron absorption; they include the **tannins** of tea and coffee, the calcium in milk, and the **phytates** that accompany fiber in legumes and whole-grain cereals.[32] The accompanying "How to" offers suggestions on obtaining adequate iron.

tannins: compounds in tea (especially black tea) and coffee that bind iron.

phytates: nonnutrient components of grains, legumes, and seeds. Phytates can bind minerals such as iron, zinc, calcium, and magnesium in insoluble complexes in the intestine, and the body excretes them unused.

ZINC

Zinc is a versatile trace mineral required as a cofactor by more than 200 enzymes. These zinc-requiring enzymes perform tasks in the eyes, liver, kidneys, muscles, skin, bones, and male reproductive organs. Zinc works with the enzymes that make genetic material; manufacture heme; digest food; metabolize carbohydrate, protein, and fat; liberate vitamin A from storage in the liver; and dispose of damaging free radicals. Zinc also interacts with platelets in blood clotting, affects thyroid hormone function, assists in immune function, and affects behavior and learning performance. Zinc is needed to produce the active form of vitamin A in visual pigments and is essential to wound healing, taste perception, the making of sperm, and fetal development. When zinc deficiency occurs, it impairs all these and other functions.

The body's handling of zinc differs from that of iron, but with some interesting similarities. For example, like iron, extra zinc that enters the body is held within the intestinal cells, and only the amount needed is released into the bloodstream. As with iron, zinc status influences the percentage of zinc absorbed from the diet; if more is needed, more is absorbed.

Zinc's main transport vehicle in the blood is the protein albumin. This may account for observations that serum zinc concentrations decline in conditions that lower plasma albumin concentrations—for example, pregnancy and malnutrition.

This chili dinner provides iron and MFP factor from meat, iron from legumes, and vitamin C from tomatoes. The combination of heme iron, nonheme iron, MFP factor, and vitamin C helps to achieve maximum iron absorption.

Zinc Deficiency Zinc deficiency in human beings was first reported in the 1960s from studies of growing children and male adolescents in Egypt, Iran, and Turkey. Their diets were typically low in zinc and high in fiber and phytates (which impair zinc absorption). The zinc deficiency was marked by dwarfism, or severe growth retardation, and arrested sexual maturation—symptoms that were responsive to zinc supplementation.

Since that time, zinc deficiency has been recognized elsewhere and is known to affect more than growth. It drastically impairs immune function, causes loss of appetite, and, during pregnancy, may lead to growth and developmental disorders.[33] Conditions other than poor diet that contribute to the development of zinc deficiency include loss of blood due to parasitic infections, climates that increase sweat losses, and the practice of clay eating.

Clay eating is a form of pica (see p. 240).

Pronounced zinc deficiency is not widespread in developed countries, but deficiencies do occur in the most vulnerable groups of the U.S. population—pregnant women, young children, the elderly, and the poor. Even mild zinc deficiency can result in metabolic changes such as impaired immune response, abnormal taste, and abnormal dark adaptation (zinc is required to produce the active form of vitamin A, retinal, in visual pigments).[34]

Some people are at greater risk of zinc deficiency than others. Pregnant teenagers need zinc for their own growth as well as for the developing fetus. Vegetarians

HOW TO   Add Iron to Daily Meals

The following set of guidelines can be used for planning an iron-rich diet:

- *Breads and cereals.* Use only whole-grain, enriched, and fortified products (iron is one of the enrichment nutrients).
- *Vegetables.* The dark green, leafy vegetables are good sources of vitamin C and iron. Eat vitamin C–rich vegetables often to enhance absorption of the iron from foods eaten with them.
- *Fruits.* Dried fruits, such as raisins, apricots, peaches, and prunes, are high in iron. Eat vitamin C–rich fruits often with iron-containing foods.

- *Milk and cheese.* Don't overdo foods from the milk group; they are poor sources of iron. But don't omit them either, because they are rich in calcium. Drink fat-free milk to free kcalories to be invested in iron-rich foods.
- *Meats.* Meat, fish, and poultry are excellent iron sources.
- *Meat alternates.* Include legumes frequently. A cup of peas or beans can supply up to 7 milligrams of iron.

SNAPSHOT 9-4 Zinc

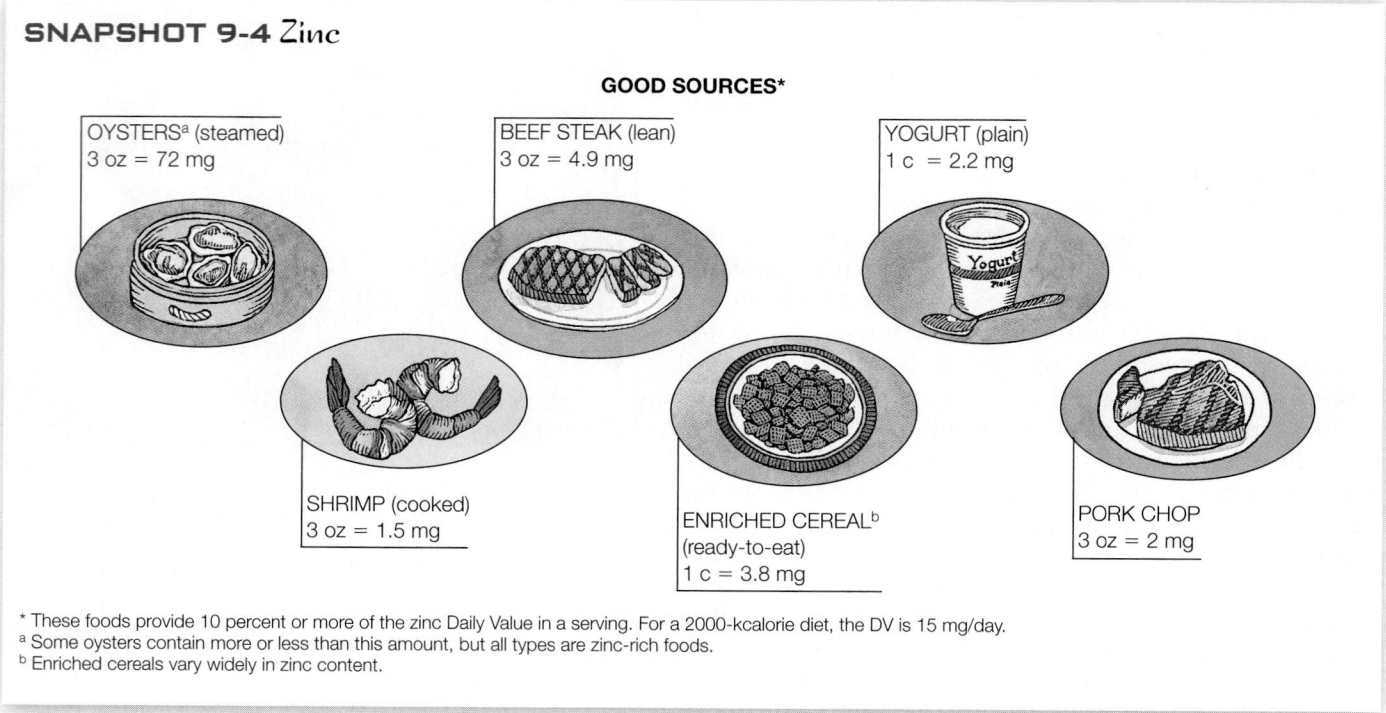

GOOD SOURCES*

OYSTERS^a (steamed)
3 oz = 72 mg

BEEF STEAK (lean)
3 oz = 4.9 mg

YOGURT (plain)
1 c = 2.2 mg

SHRIMP (cooked)
3 oz = 1.5 mg

ENRICHED CEREAL^b
(ready-to-eat)
1 c = 3.8 mg

PORK CHOP
3 oz = 2 mg

* These foods provide 10 percent or more of the zinc Daily Value in a serving. For a 2000-kcalorie diet, the DV is 15 mg/day.
^a Some oysters contain more or less than this amount, but all types are zinc-rich foods.
^b Enriched cereals vary widely in zinc content.

whose diets emphasize whole grains, legumes, and other plant foods may also be at risk. These foods, though rich in zinc, also contain phytate, a potent inhibitor of zinc absorption. Protein enhances zinc absorption, but most plant-protein foods also contain phytate. The DRI committee suggests that the dietary zinc requirement for vegetarians who exclude all animal-derived foods may be as much as 50 percent greater than the RDA, but so far evidence is insufficient to establish zinc recommendations based on the presence of other food components or nutrients.[35] Vegetarians who include cheese, eggs, or other animal protein in their diet absorb more zinc than those who exclude these foods.[36]

Zinc Toxicity Zinc can be toxic if consumed in large enough quantities. A high zinc intake is known to produce copper-deficiency anemia by inducing the intestinal cells to synthesize large amounts of a protein (metallothionein) that captures copper in a nonabsorbable form. Accidental consumption of high levels of zinc can cause vomiting, diarrhea, fever, exhaustion, and a host of other symptoms (see Table 9-8, later in the chapter). Large doses can even be fatal. The Tolerable Upper Intake Level for zinc for adults is 40 milligrams per day.

Zinc Recommendations and Food Sources The zinc recommendation for men is 11 milligrams per day; for women, 8 milligrams. Most people in the United States have zinc intakes that approximate recommendations.[37]

Zinc is most abundant in foods high in protein, such as shellfish (especially oysters), meats, and liver. In general, two ordinary servings a day of animal protein provide most of the zinc a healthy person needs. Milk, eggs, and whole-grain products are good sources of zinc if eaten in large quantities. For infants, breast milk is a good source of zinc, which is more efficiently absorbed from human milk than from cow's milk. Commercial infant formulas are fortified with zinc, of course. Snapshot 9-4 shows zinc-rich foods.

Zinc supplements are not recommended except for an accurately diagnosed zinc deficiency or when needed for use as a medication to displace other ions in

Zinc RDA:
• Men: 11 mg/day.
• Women: 8 mg/day.

unusual medical circumstances. Normally, it should be possible to obtain enough zinc from the diet.

SELENIUM

Selenium is an essential trace mineral that functions as part of a group of antioxidant enzymes called glutathione peroxidases. These enzymes prevent free-radical formation, thus blocking the damaging chain reaction before it begins. Glutathione peroxidases and vitamin E work in concert. If free radicals do form, and a chain reaction starts, vitamin E halts it. Selenium also plays roles in converting thyroid hormone to its active form.

Selenium and Cancer The question of whether selenium protects against the development of some cancers is under investigation.[38] Some research suggests that selenium supplements may reduce the incidence of some types of cancers, but given the potential for harm and the lack of additional evidence, recommendations to take selenium supplements would be premature.

Selenium Deficiency Selenium deficiency is associated with heart disease in children and young women living in regions of China where the soil and foods lack selenium. The heart disease is named *Keshan disease* for one of the provinces of China where it was studied.

Selenium Toxicity High doses of selenium are toxic. Selenium toxicity causes vomiting, diarrhea, loss of hair and nails, and lesions of the skin and nervous system.[39]

Selenium Recommendations and Intakes Anyone who eats a normal diet composed mostly of unprocessed foods need not worry about meeting selenium recommendations. Selenium is widely distributed in foods such as meats and shellfish and in vegetables and grains grown on selenium-rich soil. Some regions in the United States and Canada produce crops on selenium-poor soil, but people are protected from deficiency because they eat selenium-rich meat and supermarket foods transported from other regions.

IODINE

Iodine occurs in the body in minuscule amounts, but its principal role in human nutrition is well known, and the amount needed is well established. Iodine is an integral part of the thyroid hormones, which regulate body temperature, metabolic rate, reproduction, growth, the making of blood cells, nerve and muscle function, and more.

Iodine Deficiency When the iodine concentration in the blood is low, the cells of the thyroid gland enlarge in an attempt to trap as many particles of iodine as possible. If the gland enlarges until it is visible, the swelling is called a simple **goiter.** Globally, more than 2 billion people live in areas with iodine deficiency and risk its complications.[40] In most cases of goiter, the cause is iodine deficiency, but some people have goiter because they overconsume plants of the cabbage family and others that contain an antithyroid substance whose effect is not counteracted by dietary iodine.

In addition to causing sluggishness and weight gain, an iodine deficiency may have serious effects on fetal development. Severe thyroid hormone undersecretion during pregnancy causes the extreme and irreversible mental and physical retardation known as **cretinism.** A child

Selenium RDA:
- Adults: 55 μg/day.

goiter (GOY-ter): an enlargement of the thyroid gland due to an iodine deficiency, malfunction of the gland, or overconsumption of a thyroid antagonist. Goiter caused by iodine deficiency is *simple goiter.*

cretinism (CREE-tin-ism): an iodine-deficiency disease characterized by mental and physical retardation.

A thyroid antagonist that is found in food and causes *toxic goiter* is called a **goitrogen.**

© Bob Daemmrich/The Image Works

In iodine deficiency, the thyroid gland enlarges—a condition known as simple goiter.

with cretinism may have an IQ as low as 20 (100 is normal) and a face and body with many abnormalities. Iodine deficiency is one of the world's most common preventable causes of mental retardation and can be averted if the pregnant woman's deficiency is detected and treated in time.

Iodine Toxicity Excessive intakes of iodine can enlarge the thyroid gland, just as deficiencies can. Intakes in the United States are slightly above the recommended intake of 150 micrograms, but still below the Tolerable Upper Intake Level of 1100 micrograms per day for an adult.[41]

Iodine Sources The ocean is the world's major source of iodine. In coastal areas, seafood, water, and even iodine-containing sea mist are important iodine sources. Further inland, the amount of iodine in the diet is variable and generally reflects the amount present in the soil in which plants are grown or on which animals graze. In the United States and Canada, the use of iodized salt has largely wiped out the iodine deficiency that once was widespread.

The need for iodine is easily met by consuming seafood, vegetables grown in iodine-rich soil, and iodized salt. In the United States, you have to read the label to find out whether salt is iodized; in Canada, all table salt is iodized.

Iodine Used as a Medication An iodine-containing medication, **potassium iodide,** effectively blocks damage to the thyroid gland that could be caused during radiation emergencies, such as hostile attacks or malfunction of nuclear power plants. When given with the correct dosage and timing, potassium iodide can greatly reduce the likelihood of subsequent thyroid cancer development.[42] Given in the wrong dosage or with the wrong timing, however, potassium iodide is useless or toxic with overdoses. For this reason, concerned people who live near power plants are urged to rely on health professionals for guidance.

COPPER

The body contains about 100 milligrams of copper. About one-fourth is in the muscles; one-fourth is in the liver, brain, and blood; and the rest is in the bones, kidneys, and other tissues. The primary function of copper in the body is to serve as a constituent of enzymes.[43] The copper-containing enzymes have diverse metabolic roles: they catalyze the formation of hemoglobin, help manufacture the protein collagen, assist in the healing of wounds, and help maintain the sheaths around nerve fibers. One of copper's most vital roles is to help cells use iron. Like iron, copper is needed in many reactions related to respiration and energy metabolism.

Copper Deficiency Copper deficiency is rare but not unknown. It has been seen in premature infants and malnourished infants. High intakes of zinc interfere with copper absorption and can lead to deficiency.

Copper Toxicity Some genetic disorders create a copper toxicity. Copper toxicity from foods, however, is unlikely. The Tolerable Upper Intake Level for copper is set at 10,000 micrograms per day.

Copper Recommendations and Food Sources The RDA for copper is 900 micrograms per day, which is slightly below the average intake for adults in the United States.[44] The best food sources of copper are legumes, whole grains, seafood, nuts, and seeds.

MANGANESE

The human body contains a tiny 20 milligrams of manganese, mostly in the bones and glands. Manganese is a cofactor for many enzymes, helping to facilitate dozens

Iodine RDA:
• Adults: 150 μg/day.

potassium iodide: a medication approved by the FDA as safe and effective for the prevention of thyroid cancer caused by radioactive iodine known to be released during radiation emergencies.

Copper RDA:
• Adults: 900 μg/day.

Manganese AI:
• Men: 2.3 mg/day.
• Women: 1.6 mg/day.

of different metabolic processes. Deficiencies of manganese have not been noted in people, but toxicity may be severe. Miners who inhale large quantities of manganese dust on the job over prolonged periods show many symptoms of a brain disease, along with abnormalities in appearance and behavior. The Tolerable Upper Intake Level for manganese is 11 milligrams per day.

Manganese requirements are low, and plant foods such as nuts, whole grains, and leafy green, vegetables contain significant amounts of this trace mineral. Deficiencies are therefore unlikely.

FLUORIDE

Only a trace of fluoride occurs in the human body, but research demonstrates that where diets are high in fluoride during the growing years, crystalline deposits in bones and teeth are larger and harder. When bones and teeth become mineralized, first a crystal called hydroxyapatite forms from calcium and phosphorus. Then fluoride replaces the hydroxy portion of hydroxyapatite, forming **fluorapatite,** which makes the bones stronger and the teeth more resistant to decay. Once the teeth have erupted, the topical application of fluoride by way of toothpaste or mouth rinse continues to exert a caries-reducing effect.

Fluoride Deficiency Where fluoride is lacking in the water supply, the incidence of dental decay is high. Fluoridation of water to raise its fluoride concentration to 1 part per million is recommended as an important public health measure. Those fortunate enough to have had sufficient fluoride during the tooth-forming years of infancy and childhood are protected throughout life from dental decay. Dental problems are of great concern because they can lead to a multitude of other health problems affecting the whole body. Based on the accumulated evidence of its beneficial effects, water fluoridation has been endorsed by nearly 100 national and international organizations including the National Institute of Dental Health, the American Dietetic Association, the American Medical Association, the National Cancer Institute, and the Centers for Disease Control and Prevention.[45] In fact, the Centers for Disease Control and Prevention name water fluoridation as one of the ten most important public health measures of the twentieth century.[46] Despite fluoride's value, the introduction of fluoride to a community may encounter opposition. As of 2002, about one-third of the U.S. population (over 100 million people) was still without public water fluoridation.[47] Figure 9-6 shows the extent of fluoridation nationwide.

Fluoride Sources All normal diets include some fluoride, but drinking water, processed soft drinks and fruit juice made with fluoridated water, and fluoride toothpastes, gels, and oral rinses are the most common fluoride sources in the United States. Fish and tea may supply substantial amounts as well.

In some areas, the natural fluoride concentration in water is high, and too much fluoride can damage teeth, causing **fluorosis.** For this reason, a Tolerable Upper Intake Level has been established. In mild cases, the teeth develop small white specks; in severe cases, the enamel becomes pitted and permanently stained (see Figure 9-7). Fluorosis occurs only during tooth development and cannot be reversed, making its prevention a high priority.

CHROMIUM

Chromium is an essential mineral that participates in carbohydrate and lipid metabolism. Chromium enhances the activity of the hormone insulin.[48] Consequently, less insulin is needed to control blood glucose. When chromium is lacking, a diabetes-like condition may develop with elevated blood glucose and impaired glucose tolerance, insulin response, and glucagon response. In spite of

FIGURE 9-6 Percentage of State Populations with Access to Fluoridated Water through Public Water Systems

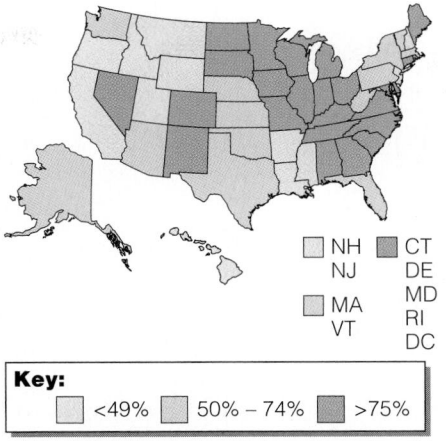

NH   CT
NJ   DE
MA   MD
VT   RI
     DC

Key:

<49%   50% – 74%   >75%

Source: Centers for Disease Control and Prevention, Recommendations for using fluoride to prevent and control dental caries in the United States, *Morbidity and Mortality Weekly Report* (supplement), August 17, 2001, p. 10.

fluorapatite (floor-APP-uh-tite): the stabilized form of bone and tooth crystal, in which fluoride has replaced the hydroxy portion of hydroxyapatite.

fluorosis (floor-OH-sis): mottling of the tooth enamel from ingestion of too much fluoride during tooth development.

Fluoride AI:
- Men: 4 mg/day.
- Women: 3 mg/day.

Fluoride Tolerable Upper Intake Level:
- Adults: 10 mg/day.

To prevent fluorosis:
- Monitor the fluoride content of the local water supply.
- Supervise children younger than six when they brush their teeth (to ensure that they don't swallow the toothpaste), and use only a pea-size amount of toothpaste.
- Use fluoride supplements only as prescribed by a physician.

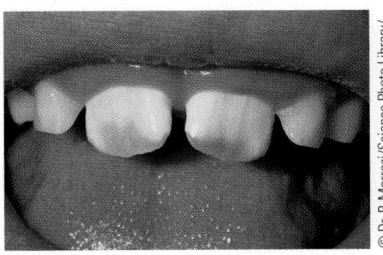

© Dr. P. Marrazi/Science Photo Library/ Photo Researchers Inc.

FIGURE 9-7 Fluoride Toxicity Symptom—The Mottled Teeth of Fluorosis

TABLE 9-8	The Trace Minerals—A Summary			
Mineral Name	Chief Functions in the Body	Deficiency Symptoms	Toxicity Symptoms	Significant Sources
Iron	Part of the protein hemoglobin, which carries oxygen in the blood; part of the protein myoglobin in muscles, which makes oxygen available for muscle contraction; necessary for the utilization of energy.	Anemia: weakness, pallor, headaches, reduced work productivity, inability to concentrate, impaired cognitive function (children), lowered cold tolerance	Iron overload: infections, liver injury, possible increased risk of heart attack, acidosis, bloody stools, shock	Red meats, fish, poultry, shellfish, eggs, legumes, dried fruits
Zinc	Part of the hormone insulin and many enzymes; involved in making genetic material and proteins, immune reactions, transport of vitamin A, taste perception, wound healing, the making of sperm, and normal fetal development.	Growth failure in children, sexual retardation, loss of taste, poor wound healing, eye lesions leading to impaired dark adaptation	Fever, nausea, vomiting, diarrhea, muscle incoordination, dizziness, anemia, accelerated atherosclerosis, kidney failure	Protein-containing foods: meats, fish, shellfish, poultry, grains, vegetables
Selenium	Assists a group of enzymes that break down reactive chemicals that harm cells.	Predisposition to heart disease characterized by cardiac tissue becoming fibrous (uncommon)	Nausea, abdominal pain, nail and hair changes, nerve damage	Seafoods, organ meats, other meats, whole grains and vegetables depending on soil content
Iodine	A component of two thyroid hormones, which help to regulate growth, development, and metabolic rate.	Goiter, cretinism	Depressed thyroid activity; goiterlike thyroid enlargement	Iodized salt; seafood; bread; plants grown in most parts of the country and animals fed those plants.
Copper	Necessary for the absorption and use of iron in the formation of hemoglobin; part of several enzymes.	Anemia, bone abnormalities (rare in human beings)	Vomiting, diarrhea, liver damage	Organ meats, seafood, nuts, seeds, whole grains, drinking water
Manganese	Facilitator, with enzymes, of many cell processes.	(In experimental animals): poor growth, nervous system disorders, reproductive abnormalities	Nervous system disorders	Widely distributed in foods
Fluoride	An element involved in the formation of bones and teeth; helps to make teeth resistant to decay.	Susceptibility to tooth decay	Fluorosis (discoloration of teeth), nausea, diarrhea, chest pain, itching, vomiting	Drinking water (if fluoride containing or fluoridated), tea, seafood
Chromium	Associated with insulin and required for the release of energy from glucose.	Diabetes-like condition marked by an inability to use glucose normally	None reported	Meat, unrefined foods

Small organic compounds that enhance insulin's activity are called **glucose tolerance factors (GTF).** Some glucose tolerance factors contain chromium.

these relationships, research findings conflict as to whether chromium supplements improve glucose or insulin response in diabetes.[49]

Chromium deficiency is unlikely, given the small amount of chromium required and its presence in a variety of foods. The more refined foods people eat, however, the less chromium they obtain from their diets. Unrefined foods such as liver, brewer's yeast, whole grains, nuts, and cheeses are the best sources.

OTHER TRACE MINERALS

molybdenum (mo-LIB-duh-num): a trace element.

An RDA has been established for one other trace mineral, molybdenum. **Molybdenum** functions as a working part of several metal-containing enzymes, some of which are giant proteins. Deficiencies or toxicities of molybdenum are unknown.

Several other trace minerals are known or suspected to contribute to the health of the body. Nickel is recognized as important for the health of many body tissues. Nickel deficiencies harm the liver and other organs. Silicon participates in bone calcification, at least in animals. Tin is necessary for growth in animals and probably in people as well. Cobalt is found in the large vitamin B_{12} molecule. The future may reveal that other trace minerals also play key roles. Even arsenic— famous as the death potion in many murder mysteries and known to be a carcinogen—may turn out to be an essential nutrient in tiny quantities.

Chromium AI:
- Men (19–50 yr.): 35 µg/day. (51 and older): 30 µg/day.
- Women (19–50 yr.): 25 µg/day. (51 and older): 20 µg/day.

Molybdenum RDA:
- Adults: 45 µg/day.

REVIEW NOTES

The body requires trace minerals in tiny amounts, and they function in similar ways—assisting enzymes all over the body.

Eating a diet that consists of a variety of foods is the best way to ensure an adequate intake of these important nutrients.

Many dietary factors, including the trace minerals themselves, affect the absorption and availability of these nutrients.

Table 9-8 offers a summary of facts about trace minerals in the body.

SELF CHECK

1. Which of the following body structures helps to regulate thirst?
   a. brain stem
   b. cerebellum
   c. optic nerve
   d. hypothalamus

2. Which of the following is *not* a function of water in the body?
   a. lubricant
   b. source of energy
   c. maintains protein structure
   d. participant in chemical reactions

3. Two situations in which a person may experience fluid and electrolyte imbalances are:
   a. vomiting and burns.
   b. diarrhea and cuts.
   c. broken bones and fever.
   d. heavy sweating and excessive carbohydrate intake.

4. Three-fourths of the sodium in people's diets comes from:
   a. fresh meats.
   b. home-cooked foods.
   c. frozen vegetables and meats.
   d. salt added to food by manufacturers.

5. Which mineral is critical to keeping the heartbeat steady and plays a major role in maintaining fluid and electrolyte balance?
   a. sodium
   b. calcium
   c. potassium
   d. magnesium

6. The two best ways to prevent age-related bone loss and fracture are to:
   a. take calcium supplements and estrogen.
   b. participate in aerobic activity and drink 8 glasses of milk daily.
   c. eat a diet low in fat and salt and refrain from smoking.
   d. maintain a lifelong adequate calcium intake and engage in weight-bearing physical activity.

7. Three good food sources of calcium are:
   a. milk, sardines, and broccoli.
   b. spinach, yogurt, and sardines.
   c. cottage cheese, spinach, and tofu.
   d. Swiss chard, mustard greens, and broccoli.

8. Foods high in iron that help prevent or treat anemia include:
   a. green peas and cheese.
   b. dairy foods and fresh fruits.
   c. homemade breads and most fresh vegetables.
   d. meat and dark green, leafy vegetables.

9. Two groups of people who are especially at risk for zinc deficiency are:
   a. Asians and children.
   b. infants and teenagers.
   c. smokers and athletes.
   d. pregnant adolescents and vegetarians.

10. A deficiency of ___ is one of the world's most common preventable causes of mental retardation.
    a. zinc.
    b. iodine.
    c. selenium.
    d. magnesium

Answers to these questions appear in Appendix H.

CLINICAL APPLICATIONS

1. Pull together information from Chapter 1 about the different food groups and the significant sources of minerals shown or discussed in this chapter. Consider which minerals might be lacking (or excessive) in the diet of a client who reports the following:

- Relies on highly processed foods, snack foods, and fast foods as mainstays of the diet.

- Never uses milk, milk products, or cheese.
- Dislikes leafy green vegetables.
- Never eats meat, fish, poultry, or even meat alternates such as legumes.

What additional information would help you pinpoint problems with mineral intake?

NUTRITION ON THE NET

For further study of the topics in this chapter, access these websites.

Find updates and quick links to these and other nutrition-related sites at our website: **www.wadsworth.com/nutrition**

Find information about mineral supplements: **http://dietary-supplements.info.nih.gov**

Search for "minerals" at the American Dietetic Association site: **www.eatright.org**

To learn about the quality of drinking water in your area, visit: **www.epa.gov/ebtpages**

Learn about sodium in foods and on food labels from the Food and Drug Administration: **www.fda.gov/fdac/foodlabel/sodium.html**

Find tips and recipes for including more milk in the diet: **www.whymilk.com**

Learn about the benefits of calcium from the National Dairy Council: **www.nationaldairycouncil.org**

Search for the individual minerals by name at the U.S. Government health information site: **www.healthfinder.org**

Learn more about iron overload from the Iron Overload Diseases Association: **www.ironoverload.org**

Learn more about iodine deficiency and thyroid disease from the American Thyroid Association: **www.thyroid.org**

NOTES

[1] Standing Committee on the Scientific Evaluation of Dietary Reference Intakes, Food and Nutrition Board, Institute of Medicine, *Dietary Reference Intakes for Water, Potassium, Sodium, Chloride, and Sulfate* (Washington, D.C.: National Academies Press, 2005), p. 73.

[2] M. N. Sawka, S. N. Cheuvront, and R. Carter, Human water needs, *Nutrition Reviews* 63 (2005): S30–S39.

[3] L. E. Armstrong, Hydration assessment techniques, *Nutrition Reviews* 63 (2005): S40–S54.

[4] Standing Committee on the Scientific Evaluation of Dietary Reference Intakes, 2005, pp. 133–134.

[5] L. E. Fields and coauthors, The burden of adult hypertension in the United States 1999 to 2000: A rising tide, *Hypertension* 44 (2004): 398–404.

[6] A. V. Chobanian and coauthors, National High Blood Pressure Education Program Coordinating Committee, Seventh Report of the Joint National Committee on Prevention, Detection, Evaluation,

and Treatment of High Blood Pressure, *Hypertension* 42 (2003): 1206–1252; L. E. Fields, Mortality from stroke and ischemic heart disease increases exponentially with blood pressure, *Hypertension* 43 (2004): e28; S. Lewington and coauthors; Prospective Studies Collaboration, Age-specific relevance of usual blood pressure to vascular mortality: A meta-analysis of individual data for one million adults in 61 prospective studies, *Lancet* 360 (2002): 1903–1913.

[7] T. A. Kotchen and J. M. Kotchen, Nutrition, diet, and hypertension, in M. E. Shils and coeditors, *Modern Nutrition in Health and Disease*, 10th ed. (Philadelphia: Lippincott, Williams, & Wilkins, 2006), pp. 1095–1107.

[8] U.S. Department of Agriculture and U.S. Department of Health and Human Services, *2005 Dietary Guidelines Committee Report*, 2005, available at **www.healthierus.gov/dietary guidelines**.

[9] Standing Committee on the Scientific Evaluation of Dietary Reference Intakes, 2005, pp. 269–272.

[10]F. M. Sacks and coauthors, Effects on blood pressure of reduced dietary sodium and the Dietary Approaches to Stop Hypertension (DASH) diet, *New England Journal of Medicine* 344 (2001): 3–10.

[11]Sacks and coauthors, 2001.

[12]M. S. Oh and J. Uribarri, Electrolytes, water, and acid-base balance, in M. E. Shils and coeditors, in M. E. Shils and coeditors, *Modern Nutrition in Health and Disease,* 10th ed. (Philadelphia: Lippincott, Williams, & Wilkins, 2006), pp. 149–193.

[13]Standing Committee on the Scientific Evaluation of Dietary Reference Intakes, 2005, pp. 186–268.

[14]Standing Committee on the Scientific Evaluation of Dietary Reference Intakes, 2005, pp. 186–268.

[15]C. M. Weaver and R. P. Heaney, Calcium, in M. E. Shils and coeditors, *Modern Nutrition in Health and Disease,* 10th ed., (Philadelphia: Lippincott, Williams, and Wilkins, 2006), pp. 194–210.

[16]R. P. Heaney, Bone biology in health and disease, in M. E. Shils and coeditors, *Modern Nutrition in Health and Disease,* 10th ed. (Philadelphia: Lippincott, Williams, and Wilkins, 2006), pp. 1314–1325; V. Matkovic and coauthors, Calcium supplementation and bone mineral density in females from childhood to young adulthood: A randomized controlled trial, *American Journal of Clinical Nutrition* 81 (2005): 175–188.

[17]U.S. Department of Health and Human Services, *Bone Health and Osteoporosis: A Report of the Surgeon General* (Rockville, Md.: Government Printing Office, 2004), available at **www.surgeongeneral.gov/library.**

[18]American College of Sports Medicine, Position Stand: Physical activity and bone health, *Medicine and Science in Sports and Exercise* 36 (2004): 1985–1996; T. Lloyd and coauthors, Lifestyle factors and the development of bone mass and bone strength in young women, *Journal of Pediatrics* 144 (2004): 776–782; S. M. Runyan and coauthors, Familial resemblance of bone mineralization, calcium intake, and physical activity in early-adolescent daughters, their mothers, and maternal grandmothers, *Journal of the American Dietetic Association* 103 (2003): 1320–1325.

[19]Kotchen and Kotchen, 2006.

[20]A. Azoulay, P. Garzon, and M. J. Eisenberg, Comparison of the mineral content of tap water and bottled waters, *Journal of General Internal Medicine* 16 (2001): 168–175.

[21]J. Guillemant and coauthors, Mineral water as a source of dietary calcium: Acute effects on parathyroid function and bone resorption in young men, *American Journal of Clinical Nutrition* 71 (2000): 999–1002.

[22]R. K. Rude and M. E. Shils, Magnesium, in M. E. Shils and coeditors, *Modern Nutrition in Health and Disease,* 10th ed. (Philadelphia: Lippincott, Williams, & Wilkins, 2006).

[23]Rude and Shils, 2006; S. H. Jee and coauthors, The effect of magnesium supplementation on blood pressure: A meta-analysis of randomized clinical trials, *American Journal of Hypertension* 15 (2002): 691–696.

[24]S. C. Larsson, L. Berfkvist, and A. Wolk, Magnesium intake in relation to risk of colorectal cancer in women, *Journal of the American Medical Association* 293 (2005): 86–89.

[25]E. S. Ford and A. H. Mokdad, Dietary magnesium intake in a national sample of US adults, *Journal of Nutrition* 133 (2003): 2879–2882.

[26]M. W. Hentze, M. U. Muckenthaler, and N. C. Andrews, Balancing acts: Molecular control of mammalian iron metabolism, *Cell* 117 (2004): 285–297.

[27]R. J. Wood and A. G. Ronnenberg, Iron, in M. E. Shils and coeditors, *Modern Nutrition in Health and Disease,* 10th ed. (Philadelphia: Lippincott, Williams, & Wilkins, 2006), pp. 248–270; R. E. Fleming and B. R. Bacon, Orchestration of iron homeostasis, *New England Journal of Medicine* 352 (2005): 1741–1744.

[28]Wood and Ronnenberg, 2006; Fleming and Bacon, 2005.

[29]Wood and Ronnenberg, 2006.

[30]K. C. White, Anemia is a poor predictor of iron deficiency among toddlers in the United States: For heme the bell tolls, *Pediatrics* 115 (2005): 315–320.

[31]T. Brownlie and coauthors, Marginal iron deficiency without anemia impairs aerobic adaptation among previously untrained women, *American Journal of Clinical Nutrition* 75 (2002): 734–742.

[32]Standing Committee on the Scientific Evaluation of Dietary Reference Intakes, Food and Nutrition Board, National Institute of Health, *Dietary Reference Intakes for Vitamin A, Vitamin K, Arsenic, Boron, Chromium, Copper, Iodine, Iron, Manganese, Molybdenum, Nickel, Silicon, Vanadium, and Zinc* (Washington, D.C.: National Academy Press, 2001), pp. 311–316.

[33]L. E. Caulfield, Maternal zinc deficiency and maternal and child health in Peru, *Nutrition Today* 39 (2004): 78–87.

[34]J. C. King and R. J. Cousins, Zinc, in M. E. Shils and coeditors, *Modern Nutrition in Health and Disease,* 10th ed. (Philadelphia: Lippincott, Williams, & Wilkins, 2006), pp. 271–285.

[35]Standing Committee on the Scientific Evaluation of Dietary Reference Intakes, 2001, pp. 479–480.

[36]J. R. Hunt, Bioavailability of iron, zinc, and other trace minerals from vegetarian diets, *American Journal of Clinical Nutrition* 78 (2003): 633S–639S.

[37]Standing Committee on the Scientific Evaluation of Dietary Reference Intakes, 2001, p. 442.

[38]A. J. Duffield-Lillico, I. Shureiqi, and S. M. Lippman, Can selenium prevent colorectal cancer? A signpost from epidemiology, *Journal of the National Cancer Institute* 96 (2004): 1645–1647.

[39]R. F. Burk and O. A. Levander, Selenium, in M. E. Shils and coeditors, *Modern Nutrition in Health and Disease,* 10th ed. (Philadelphia: Lippincott, Williams, & Wilkins, 2006), pp. 312–325.

[40]International Council for the Control of Iodine Deficiency, Iodine deficiency disorder, available online at **www.iccidd.org**. Site updated August 18, 2005; visited on November 28, 2005.

[41]Standing Committee on the Scientific Evaluation of Dietary Reference Intakes, 2001, pp. 278–282.

[42]Food and Drug Administration, Center for Drug Evaluation and Research, Guidance: Potassium iodide as a thyroid blocking agent in radiation emergencies, November 2001, available at **www.fda.gov/cder/guidance/index.htm**.

[43]J. R. Turnlund, Copper, in M. E. Shils and coeditors, *Modern Nutrition in Health and Disease,* 10th ed. (Philadelphia: Lippincott, Williams, & Wilkins, 2006), pp. 286–299.

[44]Standing Committee on the Scientific Evaluation of Dietary Reference Intakes, 2001, pp. 224–257.

[45]Position of the American Dietetic Association: The impact of fluoride on health, *Journal of the American Dietetic Association* 105 (2005): 1620–1628.

[46]D. P. DePaola and coauthors, Nutrition and dental medicine, in M. E. Shils and coeditors, *Modern Nutrition in Health and Disease,* 10th ed. (Philadelphia, PA.: Lippincott, Williams, & Wilkins, 2006), pp. 1152–1178; Position of the American Dietetic Association, 2005.

[47]Position of the American Dietetic Association, 2005.

[48]Standing Committee on the Scientific Evaluation of Dietary Reference Intakes, 2001, pp. 197–223.

[49]M. D. Althuis and coauthors, Glucose and insulin responses to dietary chromium supplements: A meta-analysis, *American Journal of Clinical Nutrition* 76 (2002): 148–155; D. Ghosh and coauthors, Role of chromium supplementation in Indians with type 2 diabetes mellitus, *Journal of Nutritional Biochemistry* 13 (2002): 690–697; S. M. Bahijri and A. M. Mufti, Beneficial effects of chromium in people with type 2 diabetes, and urinary chromium response to glucose load as a possible indicator of status, *Biological Trace Element Research* 85 (2002): 97–109.

Vitamin and Mineral Supplements

© Tom Carter/PhotoEdit

About 50 percent of U.S. adults collectively spend billions of dollars a year on vitamin and mineral supplements.[1] This trend is accelerating as scientists discover more and more links between nutrition and disease prevention. One out of three people takes some kind of vitamin or mineral supplement every day.[2] Some people take a single pill containing a multitude of nutrients; others take huge doses of single nutrients in an attempt to ward off diseases. In many cases, taking supplements is a costly but harmless practice; sometimes, it is both costly and harmful to health. The main message of this Nutrition in Practice is that most healthy people can get the nutrients they need from foods. Supplements cannot substitute for a healthy diet. For some people, however, certain nutrient supplements may be desirable. In some cases, they can correct deficiencies; in others, they can reduce the risks of disease.

Do foods really contain enough vitamins and minerals to supply all that most people need?

Emphatically, yes, for both healthy adults and children who choose a variety of foods. The USDA Food Guide described in Chapter 1 is the guide to follow to achieve adequate intakes. People who meet their nutrient needs from foods, rather than supplements, have little risk of deficiency or toxicity.

Do some people need supplements?

Yes, some people may suffer marginal nutrient deficiencies due to illness, alcohol or drug addiction, or other conditions that limit food intake.[3] People who may benefit from nutrient supplements in amounts consistent with the RDA include:

- People with nutrient deficiencies.
- People with low food energy intakes (less than 1200 kcalories per day), such as habitual dieters.
- People with illnesses that take away the appetite.
- People with illnesses that impair nutrient absorption, such as diseases of the gallbladder, pancreas, and digestive system.
- People taking medications that interfere with nutrient metabolism.
- People who are lactose intolerant, have milk allergies, or otherwise do not consume enough dairy products to forestall extensive bone loss need calcium.
- People with limited milk intake and sun exposure need vitamin D.
- Elderly people who may have difficulty chewing or swallowing and so do not eat enough food to meet nutrient needs.
- Women who bleed excessively during menstruation need iron supplements.
- People in certain stages of the life cycle who have increased nutrient needs (for example, infants need iron and fluoride, and some may need vitamin D; women of childbearing age need folate; pregnant women need iron; and the elderly need vitamin D).
- People who eat all-plant diets (vegans) and those with atrophic gastritis need vitamin B_{12}.
- Newborn infants need a single dose of vitamin K at birth under the direction of a physician.
- People who have infections or injuries or who have undergone major surgery. (The increased metabolic needs associated with these severe stresses are discussed in Chapter 16.)

Most adults can get all the nutrients they need by eating a varied diet of nutrient-dense foods. Nutrients are potentially toxic when taken in large doses, and individual tolerances vary depending on health and age. Whenever a health care professional finds a person's diet inadequate, the right corrective step is to improve the person's food choices and eating patterns, not to begin supplementation.

Why do so many people take supplements?

People frequently take supplements for mistaken reasons, such as "They give me energy" or "They make me strong." Other invalid reasons why people may take supplements include:

- Their feeling of insecurity about the nutrient content of the food supply.
- Their belief that extra vitamins and minerals will help them cope with stress.

- Their belief that supplements can enhance athletic performance or build lean body tissue without physical work.
- Their desire to prevent, treat, or cure symptoms or diseases ranging from the common cold to cancer.[4]

Ironically, many supplement users eat more nutrient-dense diets than nonusers and therefore need supplements less. In addition, little relationship exists between the nutrients people need and the ones they take in supplements. In fact, an argument against supplements is that they may lull people into a false sense of security. A person might eat irresponsibly, thinking, "My supplement will cover my needs."

Do antioxidant supplements prevent cancer and heart disease?

Again, it is better advice to eat a very nutritious diet. Evidence from population studies shows a correlation between low intakes of antioxidant nutrients such as vitamin C, vitamin E, beta-carotene and other carotenoids, and the mineral selenium and a high incidence of disease. Other studies show that low intakes of vegetables and fruits, the richest sources of antioxidant nutrients, are linked with an increased incidence of certain types of cancer.[5] More than 200 population studies have examined the effects of fruits and vegetables on cancer risk, and many show that people who eat more of these foods are less likely to develop certain cancers.[6] Findings from other types of studies, such as intervention studies and clinical trials, however, show weaker associations between fruit and vegetable intake and reduced risk of cancer.[7] Some researchers speculate that fruit and vegetable intake may play a smaller role in total cancer protection than previously thought, but fruits and vegetables do contain protective factors for specific cancers. For example, research suggests lycopene-rich tomatoes protect against prostate cancer.[8] When research combines many different fruits and vegetables, specific protective factors may not stand out. Many experts agree that the antioxidant vitamins in these foods are probably important protective factors, but they also note that other constituents of fruits and vegetables (see Table NP8-2 on p. 220) certainly have not been ruled out as contributing factors.

The way to apply this information is to eat nutritious foods. Before supplementation is recommended as a strategy to prevent cancer or other diseases, researchers must determine the optimal doses to reduce risk and the potential adverse effects of long-term supplementation.

When a person needs a vitamin-mineral supplement, what kind should be used?

Take your health care professional's advice, if it is offered. If you are selecting a supplement yourself, a single, balanced vitamin-mineral supplement should suffice. Choose the kind that provides all the nutrients in amounts less than, equal to, or very close to, the RDA (remember, you get some nutrients from foods). For those who require a higher dose, such as young women who need supplemental folate in the childbearing years, choose a supplement with just the needed

This symbol means that a supplement contains the nutrients stated and that it will dissolve in the system—the symbol does not guarantee safety or health advantages.

nutrient or in combination with a reasonable dose of others. Avoid any preparations that, in daily dose, provide more than the recommended intake of vitamin A, vitamin D, or any mineral or more than the Tolerable Upper Intake Level for any nutrient. In addition, avoid these:

- High doses of iron (more than 10 milligrams per day) except for menstruating women. People who menstruate need more iron, but people who don't, don't.
- "Organic" or "natural" preparations with added substances. They are no better than standard types, but they cost more.
- "High-potency" or "therapeutic dose" supplements. More is not better.
- Items not needed in human nutrition such as inositol and carnitine. These particular items won't harm you, but they reveal a marketing strategy that makes the whole mix suspect.

As for price, be aware that local or store brands may be just as good as or better than nationally advertised brands.[9] If they are less expensive, it may be because the price does not have to cover the cost of national advertising. Finally, be aware that if you see a USP symbol on the label, it means that the manufacturer has voluntarily paid an independent laboratory to test the product and affirm that it contains the ingredients listed and that it will dissolve or disintegrate in the digestive tract to make the ingredients available for absorption. The symbol does not imply that the supplement has been tested for safety or effectiveness with regard to health, however.

Can supplement labels help consumers make informed choices?

Yes, to some extent. To enable consumers to make more informed choices about nutrient supplements, the Food and Drug Administration (FDA), with the encouragement of the

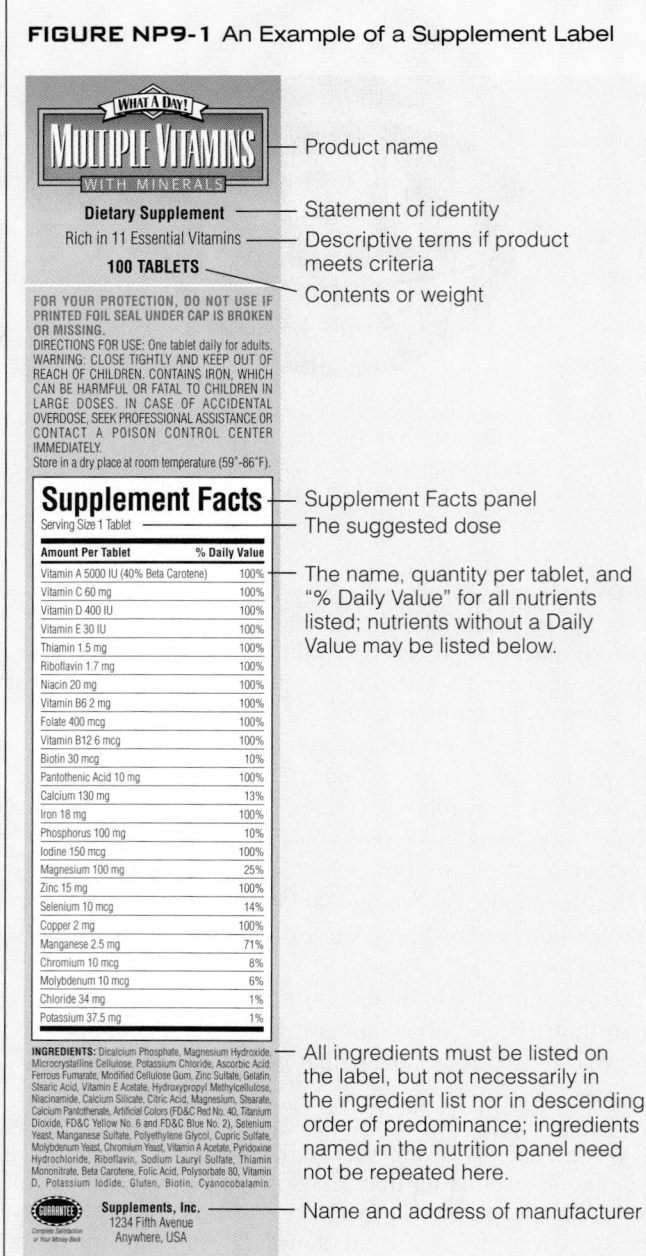

FIGURE NP9-1 An Example of a Supplement Label

Product name

Statement of identity

Descriptive terms if product meets criteria

Contents or weight

Supplement Facts panel

The suggested dose

The name, quantity per tablet, and "% Daily Value" for all nutrients listed; nutrients without a Daily Value may be listed below.

All ingredients must be listed on the label, but not necessarily in the ingredient list nor in descending order of predominance; ingredients named in the nutrition panel need not be repeated here.

Name and address of manufacturer

- Labels may make nutrient claims (as "high" or "low") according to specific criteria (for example, "an excellent source of vitamin C").
- The FDA authorizes health claims on supplement labels about the relationship between folate and the risk of neural tube defects, calcium and osteoporosis, soluble fiber from whole oats and psyllium husks and heart disease, and omega-3 fatty acids and heart disease.
- Supplement labels are not allowed to include health claims on a number of other nutrient-disease relationships that have been approved for foods.
- Products may not bear claims to diagnose, treat, cure, or relieve a specific disease.
- Labels may make structure-function claims (see Chapter 1) about the role a nutrient plays in the body, explain how the nutrient performs its function, and indicate that consuming the nutrient is associated with general well-being.
- Labels may claim a substance benefits common complaints such as memory loss or menstrual cramps without proof of effectiveness.

Note, however, that despite these requirements, in effect, the Dietary Supplement Health and Education Act resulted in the deregulation of the supplement industry. Unlike food additives or drugs, supplements do not need the FDA's approval before being marketed. Manufacturers alone decide whether their products are safe and effective. Should a problem arise, the burden falls to the FDA to prove that the supplement poses an unreasonable risk and should be removed from the market. To do this, the FDA needs scientific evidence of supplement safety from manufacturers and the records of adverse health effects reported by consumers, but just a few ethical companies provide this information voluntarily. Consumers can report adverse reactions from supplements directly to the FDA via its hotline or website, but most people are unaware of these options (see the note on this page).*

Notes

[1] A. E. Millen, K. W. Dodd, and A. F. Subar, Use of vitamin, mineral, nonvitamin, and nonmineral supplements in the United States: The 1987, 1992, and 2000 National Health Interview Survey Results, *Journal of the American Dietetic Association* 104 (2004): 942–950.

[2] Millen, Dodd, and Subar, 2004.

[3] Position of the American Dietetic Association: Fortification and nutritional supplements, *Journal of the American Dietetic Association* 105 (2005): 1300–1311.

[4] Practice Paper of the American Dietetic Association: Dietary supplements, *Journal of the American Dietetic Association* 105 (2005): 460–470.

*Consumers should report suspected harm from dietary supplements to their health providers or to the FDA's MedWatch program at (800) FDA-1088 or on the Internet at **www.fda.gov/medwatch/**.

American Dietetic Association (ADA), published labeling regulations for supplements. The Dietary Supplement Health and Education Act subjects supplements to the same general labeling requirements that apply to foods. Specifically:

- Nutrition labeling for dietary supplements is required. The nutrition panel on supplements is called "Supplement Facts" (see Figure NP9-1). The Supplement Facts panel lists the quantity and the percentage of the Daily Value for each nutrient in the supplement. Ingredients that have no Daily Value—for example, sugars and gelatin—appear in a list below the Supplement Facts panel.

[5]W. C. Willett and E. Giovannucci, Epidemiology of diet and cancer, in M.E. Shils and coeditors, *Modern Nutrition in Health and Disease,* 10th ed. (Philadelphia, PA.: Lippincott, Williams, and Wilkins, 2006), pp. 1267–1279; M. R. Forman and coauthors, Nutrition and cancer prevention: A multidisciplinary perspective on human trials, *Annual Review of Nutrition* 24 (2004): 223–254; H. Chen and coauthors, Dietary patterns and adenocarcinoma of the esophagus and distal stomach, *American Journal of Clinical Nutrition* 75 (2002): 137–144.

[6]Willett and Giovannucci, 2006; T. P. Giovannucci and coauthors, Fruit, vegetables, dietary fiber, and risk of colorectal cancer, *Journal of the National Cancer Institute* 93 (2001): 525–533.

[7]Willett and Giovannucci, 2006; P. Terry, J. B. Terry, and A. Wolk, Fruit and vegetable consumption in the prevention of cancer: An update, *Journal of Internal Medicine* 250 (2001): 280–290.

[8]E. Giovannucci and coauthors, A prospective study of tomato products, lycopene, and prostate cancer risk, *Journal of the National Cancer Institute* 94 (2002): 391–398: D. Herber and Q. Y. Lu, Overview of mechanisms of action of lycopene, *Experimental Biology and Medicine* 227 (2002): 920–923; T. M. Vogt and coauthors, Serum lycopene, other serum carotenoids, and risk of prostate cancer in US blacks and whites, *American Journal of Epidemiology* 155 (2002): 1023–1032; Q. Y. Lu and coauthors, Inverse associations between plasma lycopene and other carotenoids and prostate cancer, *Cancer Epidemiology, Biomarkers, and Prevention* 10 (2001): 749–756.

[9]B. Liebman, Spin the bottle: How to pick a multivitamin, *Nutrition Action Health Letter,* January/February 2003, pp. 3–9.

Nutrition through the Life Span: Pregnancy and Infancy

CHAPTER 10

All people need the same nutrients, but the amounts they need vary depending on their stage of life. This chapter focuses on nutrition in preparation for, and support of, pregnancy, lactation, and infancy. The next two chapters address the needs of children, adolescents, and older adults.

Pregnancy: The Impact of Nutrition on the Future

Both parents can prepare in advance for a healthy pregnancy.

The woman who enters pregnancy with full nutrient stores, sound eating habits, and a healthy body weight has done much to ensure an optimal pregnancy. Then, if she eats a variety of nutrient-dense foods during the pregnancy itself, her own and her infant's health will benefit further.

NUTRITION PRIOR TO PREGNANCY

A section on nutrition prior to pregnancy must, by its nature, focus mainly on women. A man's nutrition may affect his **fertility** and possibly the genetic contributions he makes to his children, but nutrition exerts its primary influence through the woman. Her body provides the environment for the growth and development of a new human being. Full nutrient stores *before* pregnancy are essential both to conception and to healthy infant development during pregnancy. In the early weeks of pregnancy, before many women are even aware that they are pregnant, significant developmental changes occur that depend on a woman's nutrient stores. In preparation for a healthy pregnancy, a woman can establish the following habits:

- *Achieve and maintain a healthy body weight.* Both underweight and overweight women, and their newborns, face increased risks of complications.
- *Choose an adequate and balanced diet.* Malnutrition reduces fertility and impairs the early development of an infant should a woman become pregnant.
- *Be physically active.* A woman who wants to be physically active *when* she is pregnant needs to become physically active *beforehand*.
- *Avoid harmful influences.* Both maternal and paternal ingestion of harmful substances (such as cigarettes, alcohol, drugs, or environmental contaminants) can alter genes or their expression, interfering with fertility and causing abnormalities.

Young adults who nourish and protect their bodies do so not only for their own sakes, but also for future generations.

PREPREGNANCY WEIGHT

Appropriate weight prior to pregnancy benefits pregnancy outcome. Being either underweight or overweight (see p. 267) presents medical risks during pregnancy and childbirth. Underweight women are therefore advised to gain weight before becoming pregnant, and overweight women to lose excess weight.

Underweight Infant birthweight correlates with prepregnancy weight and weight gain during pregnancy and is the most potent single predictor of the infant's future health and survival. An underweight woman has a high risk of having a **low-birthweight** infant, especially if she is unable to gain sufficient weight during pregnancy.[1] Compared with normal-weight infants, low-birthweight infants are more likely to contract diseases and nearly 40 times more likely to die in the first month of life. Impaired growth and development during pregnancy may have long-term health effects as well. Research suggests that when nutrient supplies fail to meet demands, permanent adaptations take place that may make obesity or chronic diseases such as hypertension more likely in later life (see Nutrition in

fertility: the capacity of a woman to produce a normal ovum periodically and of a man to produce normal sperm; the ability to reproduce.

low birthweight (LBW): a birthweight less than 5½ lb (2500 g); indicates probable poor health in the newborn and poor nutrition status of the mother during pregnancy. Normal birthweight for a full-term infant is 6½ to 9 lb (about 3000 to 4000 g).

Low-birthweight infants are of two different types. Some are **premature;** they are born early and are of a weight **appropriate for gestational age (AGA).** Others have suffered growth failure in the uterus; they may or may not be born early, but they are **small for gestational age (SGA).**

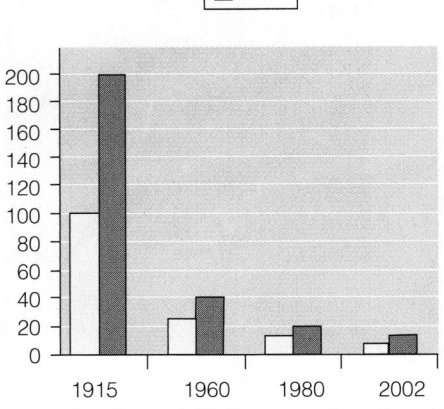

Source: Data from J. A. Martin and coauthors, Annual summary of vital statistics—2003, *Pediatrics* 115 (2005): 619–634.

FIGURE 10-1 Infant Mortality Decline over Time
The graph shows infant deaths per 1000 live births.

Neural tube defects and gestational diabetes are discussed in later sections.

uterus (YOO-ter-us): the womb, the muscular organ within which the infant develops before birth.

placenta (pla-SEN-tuh): an organ that develops inside the uterus early in pregnancy, in which maternal and fetal blood circulate in close proximity and exchange materials. The fetus receives nutrients and oxygen across the placenta; the mother's blood picks up carbon dioxide and other waste materials to be excreted via her lungs and kidneys.

gestation: the period of about 40 weeks (three trimesters) from conception to birth; the term of a pregnancy.

umbilical (um-BIL-ih-cul) **cord:** the rope-like structure through which the fetus's veins and arteries reach the placenta; the route of nourishment and oxygen into the fetus and the route of waste disposal from the fetus.

amniotic (am-nee-OTT-ic) **sac:** the "bag of waters" in the uterus in which the fetus floats.

lactation: production and secretion of breast milk for the purpose of nourishing an infant.

Practice 11).[2] Other hazards of low birthweight may include lower adult IQ and other brain impairments, short stature, and educational disadvantages.[3] Underweight women are therefore advised to gain weight before becoming pregnant and to strive to gain adequately during pregnancy.

Nutritional deficiency, coupled with low birthweight, is the underlying cause of more than half of all the deaths worldwide of children under five years of age. In the United States, the infant mortality rate in 2002 was 7.0 deaths per 1000 live births.[4] This rate, though higher than that of some other developed countries, has seen a significant steady decline over the past two decades and stands as a tribute to public health efforts aimed at reducing infant deaths (see Figure 10-1).

Not all cases of low birthweight reflect poor nutrition. Heredity, disease conditions, smoking, and drug (including alcohol) use during pregnancy all contribute.[5] Even with optimal nutrition and health during pregnancy, some women give birth to small infants for unknown reasons. But poor nutrition is the major factor in low birthweight—and an avoidable one, as later sections make clear.[6]

Overweight and Obesity Obese women are also urged to strive for healthy weights before pregnancy. The infant of an obese mother may be larger than normal and may be large even if born prematurely. The large early infant may not be recognized as premature and thus may not receive the special medical care required. The infant of an obese mother may be twice as likely to be born with a neural tube defect as others, but the reasons why are not known.[7] Obese women are more likely to require drugs to induce labor or require surgical intervention for the birth, and they suffer gestational diabetes, hypertension, and infections after the birth more often than do women of healthy weight.[8] In addition, both overweight and obese women have a greater risk of giving birth to infants with heart defects and other abnormalities.[9] An appropriate goal for the obese woman who wishes to become pregnant is to strive to attain a healthy prepregnancy body weight so as to minimize her medical risks and those of her future child.

HEALTHY SUPPORT TISSUES

A major reason that the mother's prepregnancy nutrition is so crucial is that it determines whether her **uterus** will be able to support the growth of a healthy **placenta** during the first month of **gestation**. The placenta is both a supply depot and a waste-removal system for the fetus. If the placenta works perfectly, the fetus wants for nothing; if it doesn't, no alternative source of sustenance is available, and the fetus will fail to thrive. Figure 10-2 shows the placenta, a mass of tissue in which maternal and fetal blood vessels intertwine and exchange materials. The two bloods never mix, but the barrier between them is notably thin. To grasp how thin, picture your hands as fetal blood vessels, skintight surgical gloves as the tissue-thin placenta, and finally, your gloved hands immersed in water as the pool of maternal blood. Across this thin barrier, nutrients and oxygen move from the mother's blood into the fetus's blood, and wastes move out of the fetal blood, to be excreted by the mother. The **umbilical cord** is the pipeline from the placenta to the fetus. The **amniotic sac** surrounds and cradles the fetus, cushioning it with fluids.

The placenta is an active metabolic organ with many responsibilities of its own. It actively gathers up hormones, nutrients, and protein molecules such as antibodies and transfers them into the fetal bloodstream. The placenta also produces a broad range of hormones that act in many ways to maintain pregnancy and prepare the mother's breasts for **lactation**.[10] A healthy placenta is essential for the developing fetus to attain its full potential.

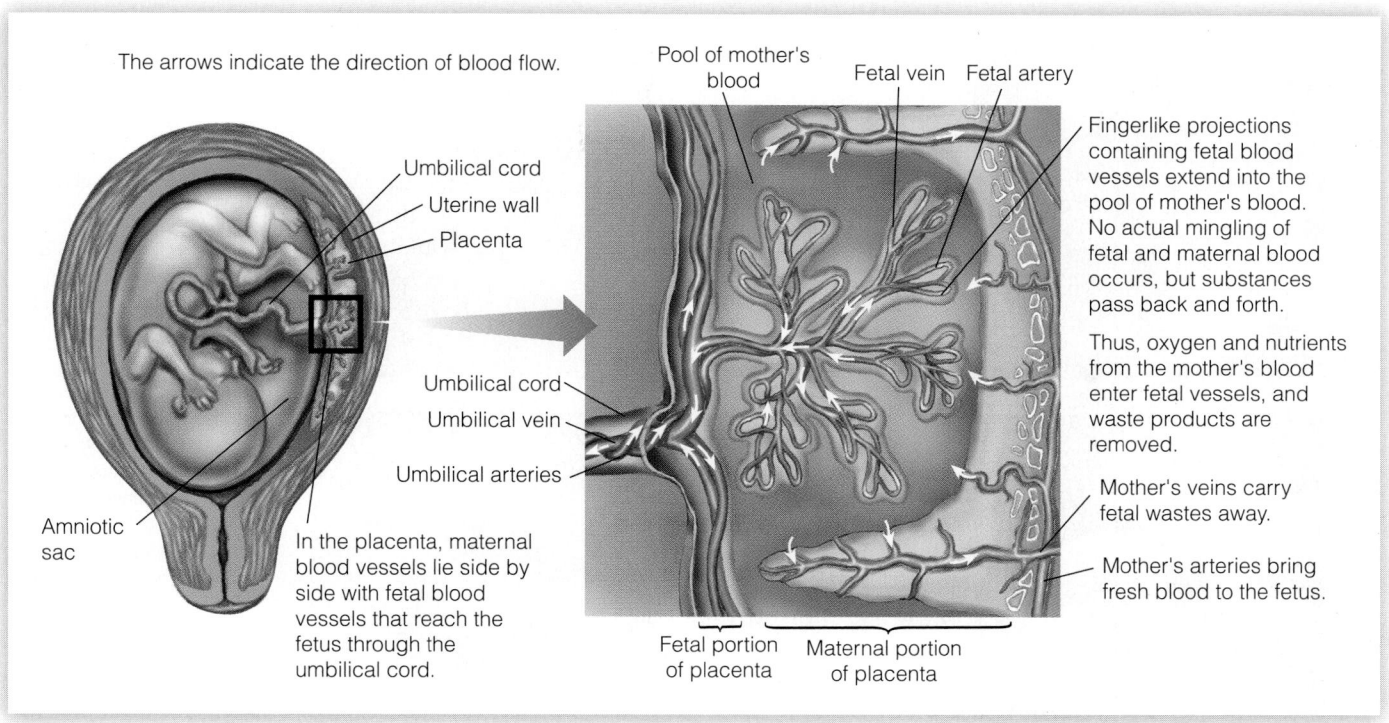

The arrows indicate the direction of blood flow.

Umbilical cord
Uterine wall
Placenta

Amniotic sac

In the placenta, maternal blood vessels lie side by side with fetal blood vessels that reach the fetus through the umbilical cord.

Umbilical cord
Umbilical vein
Umbilical arteries

Pool of mother's blood   Fetal vein   Fetal artery

Fingerlike projections containing fetal blood vessels extend into the pool of mother's blood. No actual mingling of fetal and maternal blood occurs, but substances pass back and forth.

Thus, oxygen and nutrients from the mother's blood enter fetal vessels, and waste products are removed.

Mother's veins carry fetal wastes away.

Mother's arteries bring fresh blood to the fetus.

Fetal portion of placenta   Maternal portion of placenta

FIGURE 10-2 The Placenta

REVIEW NOTES

Adequate nutrition before pregnancy establishes physical readiness and nutrient stores to support fetal growth.

Both underweight and overweight women should strive for appropriate body weights before pregnancy.

Newborns who weigh less than 5$^{1}/_{2}$ pounds face greater health risks than normal-weight infants.

The healthy development of the placenta depends on adequate nutrition before pregnancy.

THE EVENTS OF PREGNANCY

The newly fertilized **ovum,** called a **zygote,** begins as a single cell and divides into many cells during the days after fertilization. Within two weeks, the zygote embeds itself in the uterine wall in a process known as **implantation,** and the placenta begins to grow inside the uterus. Minimal growth in size takes place at this time, but it is a crucial period in development. Adverse influences such as smoking, drug abuse, and malnutrition at this time lead to failure to implant or to abnormalities such as neural tube defects that can cause the loss of the zygote, possibly before the woman knows she is pregnant.

The Embryo and Fetus During the next six weeks of development, the **embryo** registers astonishing physical changes (see Figure 10-3, p. 260). At eight weeks, the **fetus** has a complete central nervous system, a beating heart, a fully formed digestive system, well-defined fingers and toes, and the beginnings of facial features.

In the last seven months of pregnancy, the fetal period, the fetus grows 50 times heavier and 20 times longer. Critical periods of cell division and development occur

ovum (OH-vum): the female reproductive cell, capable of developing into a new organism upon fertilization; commonly referred to as an egg.

zygote (ZY-goat): the product of the union of ovum and sperm; so-called for the first two weeks after fertilization.

implantation: the stage of development in which the zygote embeds itself in the wall of the uterus and begins to develop; occurs during the first two weeks after conception.

embryo (EM-bree-oh): the developing infant from two to eight weeks after conception.

fetus (FEET-us): the developing infant from eight weeks after conception until its birth.

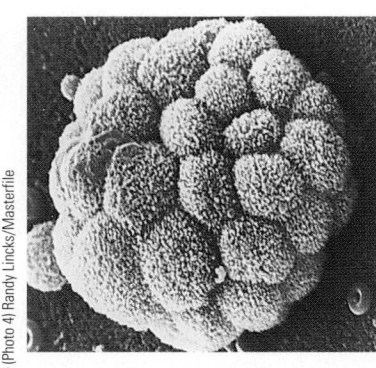

(1) A newly fertilized ovum is about the size of the period at the end of this sentence. This zygote at less than one week after fertilization is not much bigger and is ready for implantation.

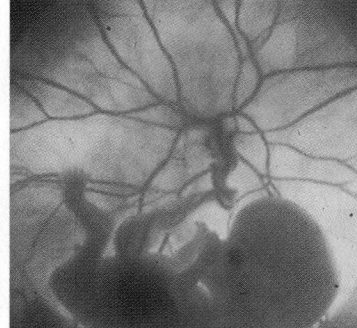

(3) A fetus after 11 weeks of development is just over an inch long. Notice the umbilical cord and blood vessels connecting the fetus with the placenta.

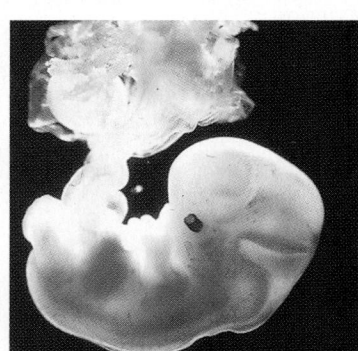

(2) After implantation, the placenta develops and begins to provide nourishment to the developing embryo. An embryo five weeks after fertilization is about 1/2 inch long.

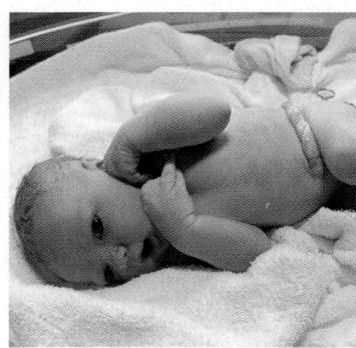

(4) A newborn infant after nine months of development measures close to 20 inches in length. The average birthweight is about 7 1/2 pounds. From eight weeks to term, this infant grew 20 times longer and 50 times heavier.

FIGURE 10-3 Stages of Embryonic and Fetal Development

in organ after organ. Most successful pregnancies last 39 to 41 weeks and produce a healthy infant weighing between 6 1/2 and 9 pounds. The 40 or so weeks of pregnancy are divided into thirds, each of which is called a **trimester.**

A Note about Critical Periods Each organ and tissue type grows with its own characteristic pattern and timing. The development of each takes place only at a certain time—the **critical period.** Whatever nutrients and other environmental conditions are necessary during this period must be supplied on time if the organ is to reach its full potential. If the development of an organ is limited during a critical period, recovery is impossible. For example, the fetus's heart and brain are well developed at 14 weeks; the lungs, 10 weeks later. Therefore, early malnutrition impairs the heart and brain; later malnutrition impairs the lungs.

The effects of malnutrition during critical periods of pregnancy are seen in defects of the nervous system of the embryo (explained later), in the child's poor dental health, and in the adolescent's and adult's vulnerability to infections and possibly higher risks of diabetes, hypertension, stroke, or heart disease.[11] The effects of malnutrition during critical periods are irreversible: Abundant and nourishing food, consumed after the critical time, cannot remedy harm already done.

Table 10-1 provides a list of factors that make nutrient deficiencies and complications likely during pregnancy. Notice that young age heads the list; a later section explains why pregnant adolescents are especially prone to malnutrition.

trimester: a period representing one-third of the term of gestation. A trimester is about 13 to 14 weeks.

critical period: a finite period during development in which certain events occur that will have irreversible effects on later developmental stages; usually a period of rapid cell division.

REVIEW NOTES

Placental development, implantation, and early critical periods depend on maternal nutrition before and during pregnancy.

TABLE 10-1	Factors Placing Pregnant Women at Nutritional Risk

Women likely to develop nutrient deficiencies include those who:

Are young (adolescents).

Have had many previous pregnancies (3 or more to mothers under age 20; 4 or more to mothers age 20 or older).

Have short intervals between pregnancies (<18 months).

Lack nutrition knowledge, have too little money to purchase adequate food, or have too little family support.

Ordinarily consume an inadequate diet due to food faddism, preferences, weight-loss "dieting," uninformed vegetarianism, or eating disorders.

Smoke cigarettes or use alcohol or drugs.

Are lactose intolerant or suffer chronic health conditions requiring special diets.

Are underweight or overweight at conception.

Are carrying twins or triplets.

Gain insufficient or excessive weight during pregnancy.

Have a low level of education.

NUTRIENT NEEDS DURING PREGNANCY

Nutrient needs during pregnancy increase more for certain nutrients than for others. Figure 10-4 (p. 262) shows the percentage increase in nutrient intakes recommended for pregnant women compared to nonpregnant women. To meet the high nutrient demands of pregnancy, a woman must make careful food choices, but her body will also do its part by maximizing nutrient absorption and minimizing losses.[12]

Energy, Carbohydrate, Protein, and Fat Energy needs vary with the progression of pregnancy. In the first trimester, the pregnant woman needs no additional energy, but as pregnancy progresses, her energy needs rise. She requires an additional 340 kcalories daily during the second trimester and an extra 450 kcalories each day during the third trimester.[13] Well-nourished pregnant women meet these demands for more energy in several ways: some eat more food, some reduce their activity, and some store less of their food energy as fat. A woman can easily meet the need for extra kcalories by selecting more nutrient-dense foods from the five food groups. Table 1-8 (p. 20) provides suggested eating patterns for several kcalorie levels, and Table 10-2 (p. 263) offers a sample menu for pregnant and lactating women.

If a woman chooses less nutritious options such as sugary soft drinks or fatty snack foods to meet energy needs, she will undoubtedly come up short on nutrients. The increase in the need for nutrients is even greater than that for energy, so the mother-to-be should choose nutrient-dense foods such as whole-grain breads and cereals, legumes, dark green vegetables, citrus fruits, low-fat milk and milk products, and lean meats, fish, poultry, and eggs.

Ample carbohydrate (ideally, 175 grams or more per day and certainly no less than 135 grams) is necessary to fuel the fetal brain and spare the protein needed for fetal growth. Fiber in carbohydrate-rich foods such as whole grains, vegetables, and fruit can help alleviate the constipation that many pregnant women experience.

The protein RDA for pregnancy is 25 grams per day higher than for nonpregnant women. Most women in the United States, however, do not need to add

Nutrient and energy intake recommendations for pregnant women are listed on the inside front cover.

FIGURE 10-4 Comparison of Nutrient Recommendations for Nonpregnant, Pregnant, and Lactating Women

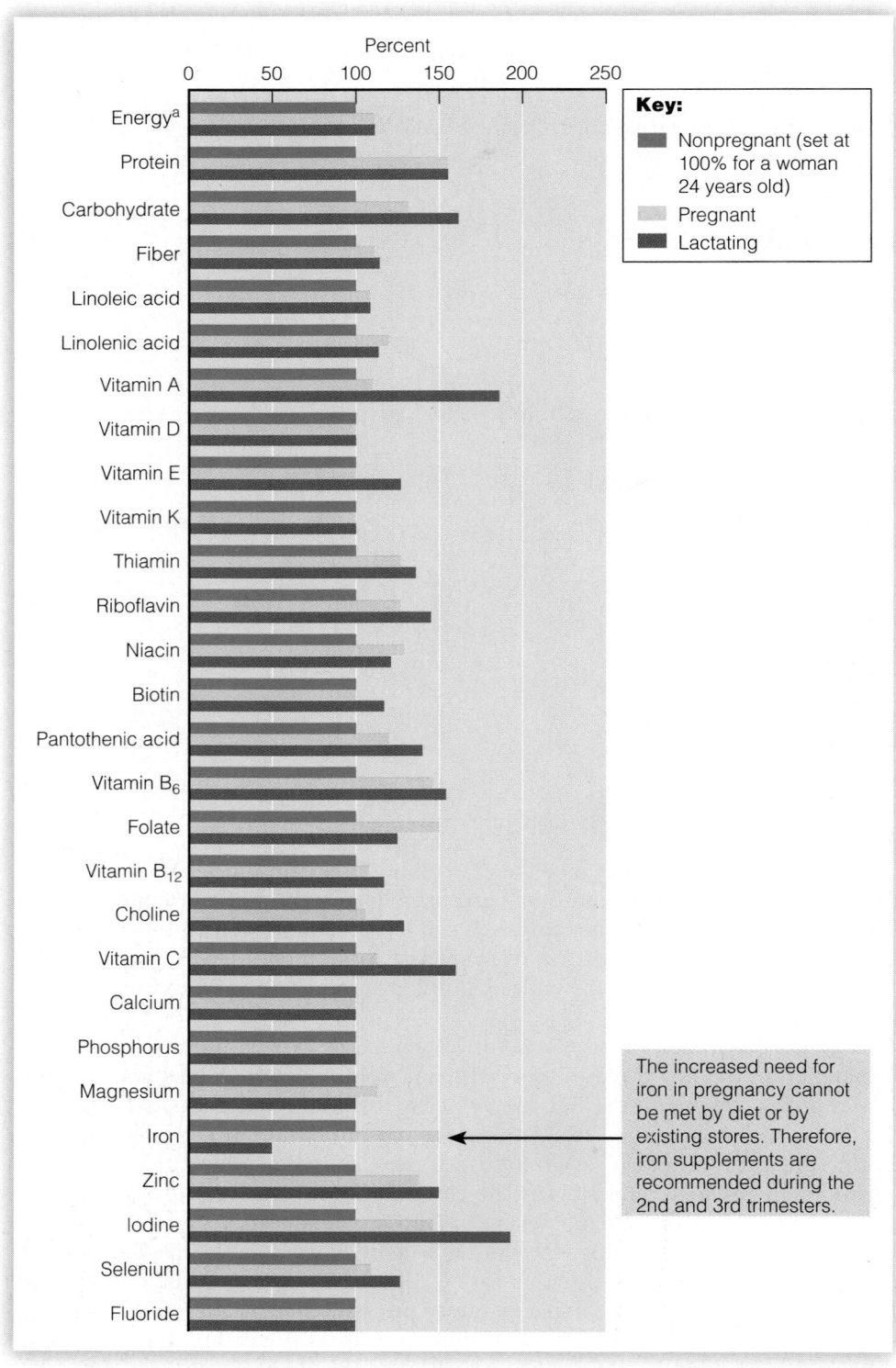

^aEnergy allowance during pregnancy is for 2nd trimester; energy allowance during the 3rd trimester is slightly higher; no additional allowance is provided during the 1st trimester. Energy allowance during lactation is for the first 6 months; energy allowance during the second 6 months is slightly higher.

protein-rich foods to their diets because they already exceed the recommended protein intake for pregnancy. Excess protein may also have adverse effects, as Chapter 5 explained.

Some vegetarian women limit or omit protein-rich meats, eggs, and dairy products from their diets. For them, meeting the recommendation for food energy each day and including several generous servings of plant-protein foods such as

TABLE 10-2	Sample Menu for Pregnant and Lactating Women

Breakfast

1 whole-wheat English muffin

2 tbs peanut butter

1 c low-fat vanilla yogurt

$1/2$ c fresh strawberries

1 c orange juice

Midmorning snack

$1/2$ c cranberry juice

1 oz pretzels

Lunch

Sandwich (tuna salad on whole-wheat bread)

$1/2$ carrot (sticks)

1 c low-fat milk

Dinner

Chicken cacciatore

   3 oz chicken

   $1/2$ c stewed tomatoes

1 c rice

$1/2$ c summer squash

$1^1/2$ c salad (spinach, mushrooms, carrots)

1 tbs salad dressing

1 slice Italian bread

2 tsp soft margarine

1 c low-fat milk

Note: This sample meal plan provides about 2500 kcalories (55 percent from carbohydrate, 20 percent from protein, and 25 percent from fat) and meets most of the vitamin and mineral needs of pregnant and lactating women.

legumes, tofu, whole grains, nuts, and seeds are imperative. Protein supplements during pregnancy can be harmful, and their use is discouraged.

The high nutrient requirements of pregnancy leave little room in the diet for excess energy from added purified fats such as oil, margarine, and butter. The essential fatty acids, however, are particularly important to the growth and development of the fetus.[14] The brain contains a substantial amount of lipid material and depends heavily on long-chain omega-3 and omega-6 fatty acids for its growth, function, and structure. (See Table 4-3 on p. 93 for a list of good food sources of the essential fatty acids.)

REVIEW NOTES

Pregnancy brings physiological adjustments that demand increased intakes of energy and nutrients.

A balanced diet that includes more nutrient-dense foods from each of the five food groups can help to meet these needs.

Of Special Interest: Folate and Vitamin B₁₂ The vitamins famous for their roles in cell reproduction—folate and vitamin B_{12}—are needed in large amounts during pregnancy. New cells are laid down at a tremendous pace as the fetus grows and develops. At the same time, because the mother's blood volume increases, the number of her red blood cells must rise, requiring more cell division and therefore more vitamins. To accommodate these needs, the recommendation for folate during pregnancy increases from 400 to 600 micrograms a day.

As described in Chapter 8, folate plays an important role in preventing neural tube defects. To review, the early weeks of pregnancy are a critical period for the formation and closure of the **neural tube** that will later develop to form the brain and spinal cord. By the time a woman suspects she is pregnant, usually around the sixth week of pregnancy, the embryo's neural tube normally has closed. A **neural tube defect (NTD)** occurs when the tube fails to close properly. In the United States, about 30 of every 100,000 newborns are born with a neural tube defect.[15] When the neural tube fails to close properly and brain development fails,

Folate RDA during pregnancy:
• 600 μg/day.

neural tube: the embryonic tissue that later forms the brain and spinal cord.

neural tube defect (NTD): a serious central nervous system birth defect that often results in lifelong disability or death.

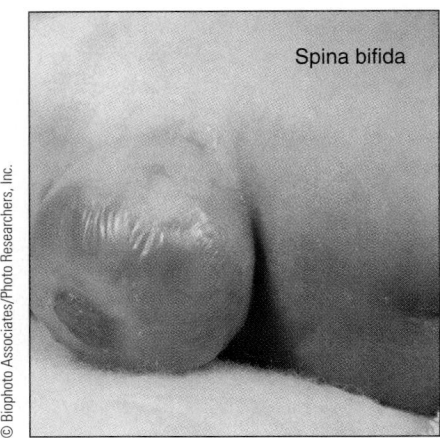

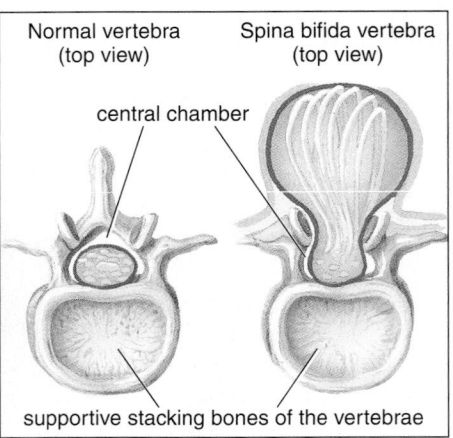

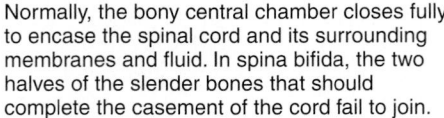

Normally, the bony central chamber closes fully to encase the spinal cord and its surrounding membranes and fluid. In spina bifida, the two halves of the slender bones that should complete the casement of the cord fail to join.

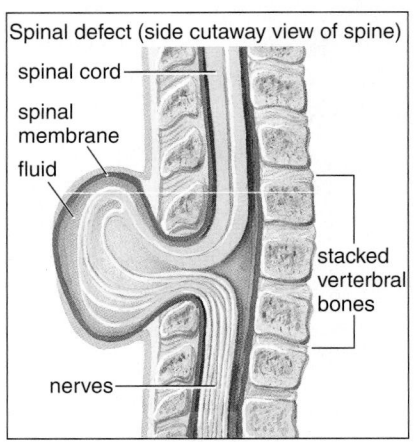

In the serious form shown here, membranes and fluid have bulged through the gap and nerves are exposed, invariably leading to some degree of paralysis and often to mental retardation.

FIGURE 10-5 Spina Bifida—A Neural Tube Defect

A pregnancy affected by a neural tube defect can occur in any woman, but these factors make it more likely:

- Inadequate folate intake.
- A previous pregnancy affected by a neural tube defect.
- Maternal diabetes (type 1).
- Maternal use of antiseizure medications.
- Maternal obesity.
- Exposure to high temperatures early in pregnancy (prolonged fever or hot tub use).
- Race/ethnicity (neural tube defects are more common among whites and Hispanics than among others).
- Low socioeconomic status.

Chapter 8 describes how excessive folate intakes can mask the symptoms of a vitamin B_{12} deficiency.

Vitamin B_{12} RDA during pregnancy:

- 2.6 µg/day.

anencephaly (AN-en-SEF-a-lee): an uncommon and always fatal type of neural tube defect; characterized by the absence of a brain.

   an = not (without)

   encephalus = brain

spina (SPY-nah) **bifida** (BIFF-ih-dah): one of the most common types of neural tube defects; characterized by the incomplete closure of the spinal cord and its bony encasement.

   spina = spine

   bifida = split

a rare but lethal defect known as **anencephaly** occurs. All infants with anencephaly die shortly after birth.

In a more common NTD, the spinal cord and backbone do not develop normally—and the result is **spina bifida** (see Figure 10-5). The membranes covering the spinal cord often protrude from the spine as a sac, and sometimes a portion of the spinal cord is contained in the sac. Spina bifida is often accompanied by varying degrees of paralysis, depending on the extent of spinal cord damage. Mild cases may not be noticed. Moderate cases may involve curvature of the spine, muscle weakness, mental handicaps, and other ills, while severe cases can lead to death.

To reduce the risk of neural tube defects, women who are capable of becoming pregnant should obtain 400 micrograms of folic acid daily from supplements, fortified foods, or both, *in addition* to eating folate-rich foods (see Table 10-3, p. 265). The DRI committee recommends intake of synthetic folate, called folic acid, in supplements and fortified food because it is absorbed better than the folate naturally present in foods. Foods that naturally contain folate are still important, however, because they contribute to folate intakes while providing other needed vitamins, minerals, fiber, and phytochemicals.

All enriched grain products (cereal, grits, pasta, rice, bread, and the like) sold commercially in the United States are fortified with folic acid. This measure has improved folate status in women of childbearing age and lowered the number of neural tube defects that occur each year.[16] Researchers expect to see declines in some other birth defects and miscarriages as well.[17] Folate fortification does raise one safety concern, however. The pregnant woman needs a greater amount of vitamin B_{12} to assist folate in the manufacture of new cells. Because high doses of folate complicate the diagnosis of a vitamin B_{12} deficiency, quantities of 1 milligram or more require a prescription.

People who eat meat, eggs, or dairy products receive all the vitamin B_{12} they need, even for pregnancy. Those who exclude all animal products from the diet need vitamin B_{12}-fortified foods or supplements.

REVIEW NOTES

Due to their key roles in cell reproduction, folate and vitamin B_{12} are needed in large amounts during pregnancy.

Folate plays an important role in preventing neural tube defects.

TABLE 10-3	Rich Folate Sources[a]
Natural Folate Sources	**Fortified Folate Sources**
Liver (3 oz) 221 μg	Multi-Grain Cheerios Plus cereal (1 c) 400 μg[b]
Lentils (1/2 c) 179 μg	Product 19 cereal (1 c) 400 μg[b]
Chickpeas or pinto beans (1/2 c) 145 μg	Total cereal (1 c) 400 μg[b]
Asparagus (1/2 c) 131 μg	Pasta, cooked (1 c) 110 μg
Spinach (1 c raw) 131 μg	Rice, cooked (1 c) 134 μg
Avocado (1/2 c) 45 μg	Bagel (1 small whole) 75 μg
Orange juice (1 c) 74 μg	Waffles, frozen (2) 36 μg
Beets (1/2 c) 68 μg	Bread, white (1 slice) 28 μg

[a] Folate amounts for these and 2000 other foods are listed in the Table of Food Composition in Appendix A.

[b] Folate in cereals varies; read the Nutrition Facts panel of the label.

Vitamin D and Calcium for Bones Vitamin D and the minerals involved in building the skeleton—calcium, phosphorus, and magnesium—are in great demand during pregnancy. Insufficient intakes may produce abnormal fetal bone development.

Intestinal absorption of calcium doubles early in pregnancy, when the mother's bones store the mineral. Later, as the fetal bones begin to calcify, there is a dramatic shift of calcium across the placenta.[18] Whether the calcium added to the mother's bones early in pregnancy is withdrawn to build the fetus's bones later is unclear. In the final weeks of pregnancy, more than 300 milligrams a day are transferred to the fetus. Recommendations to ensure an adequate calcium intake during pregnancy are aimed at conserving the mother's bone mass while supplying fetal needs.

For women whose prepregnancy calcium intakes are below recommendations, as most are, increased calcium intakes may be especially important. Milk products offer many advantages over supplements, as emphasized in earlier chapters. Because bones are still actively depositing minerals until about age 25, adequate calcium is especially important for young women. Pregnant women under age 25 who consume less than 600 milligrams of calcium a day need to increase their intakes of milk, cheese, yogurt, and other calcium-rich foods. Alternatively, and less preferably, they may need a daily supplement of 600 milligrams of calcium.

Women who exclude milk products need calcium-fortified foods such as soy milk. It is worth noting that not all soy milk is fortified with calcium; products fortified with calcium and vitamin D are recommended. Calcium-fortified orange juice offers folate and vitamin C as well as calcium.

Fluoride Mineralization of the fetus's teeth begins in the fifth month after conception. For this and for bone development, fluoride may be needed. Fluoride crosses the placenta, and whether the placenta can defend against excess intakes is questionable. Therefore, fluoride supplements are not recommended for pregnant women who drink fluoridated water. For women who live in communities without fluoridated water, a fluoride supplement may protect fetal teeth.

Calcium AI during pregnancy:
• 1300 mg/day (14 to 18 yr).
• 1000 mg/day (19 to 50 yr).

Phosphorus RDA during pregnancy:
• 1250 mg/day (14 to 18 yr).
• 700 mg/day (19 to 50 yr).

Magnesium RDA during pregnancy:
• 400 mg/day (14 to 18 yr).
• 350 mg/day (19 to 30 yr).
• 360 mg/day (31 to 50 yr).

Three cups of milk a day will supply 900 mg of calcium. For other food sources of calcium, see Chapter 9.

Fluoride AI during pregnancy:
• 3.0 mg/day.

REVIEW NOTES

All pregnant women, but especially those who are less than 25 years of age, need to pay special attention to calcium to ensure adequate intakes.

Fluoride supplements are not recommended for pregnant women who drink fluoridated water, but for those who live in communities where the water is not fluoridated, a fluoride supplement may protect fetal teeth.

Iron RDA during pregnancy:
- 27 mg/day.

In pregnancy, hemoglobin values of 12 g are not unusual, and 11 g is where the line defining "too low" is often drawn. Appendix E discusses more sensitive measures of iron status.

Food sources of iron:
- Liver, oysters.
- Red meat, fish, other meat.
- Dried fruits (raisins, prunes).
- Legumes (dried beans, peas, lima beans).
- Dark green vegetables.

Reminder: Vitamin C–rich foods enhance iron absorption from foods.

Zinc RDA during pregnancy:
- 12 mg/day (≤18 yr).
- 11 mg/day (19 to 50 yr).

Iron The body conserves iron especially well during pregnancy: menstruation ceases, and absorption of iron increases up to threefold due to a rise in the blood's iron-absorbing and iron-carrying protein transferrin. Still, iron needs are so high that stores dwindle during pregnancy.

The developing fetus draws heavily on the mother's iron stores to create stores of its own to last through the first four to six months of life. Even women with inadequate iron stores transfer significant amounts of iron to the fetus, suggesting that the iron needs of the fetus have priority over those of the mother.[19] Iron losses also occur with the bleeding that is inevitable at birth.

Few women enter pregnancy with adequate iron stores. Women who enter pregnancy with iron-deficiency anemia have a greater-than-normal risk of delivering low-birthweight or preterm infants.[20] For all women not taking supplements containing iron, a daily iron supplement containing 30 milligrams is recommended during the second and third trimesters of pregnancy. When a low hemoglobin or hematocrit is confirmed by a repeat test, more than 30 milligrams of iron may be prescribed. To enhance iron absorption, the supplement should be taken between meals and with liquids other than milk, coffee, or tea, which inhibit iron absorption.

Zinc Zinc is required for DNA and RNA synthesis and thus for protein synthesis. Severe zinc deficiency predicts a low infant birthweight.[21] Zinc is most abundant in foods of high-protein content, such as shellfish, meat, and nuts, but the presence of other trace elements and fiber in foods may adversely affect zinc absorption. For example, iron interferes with the body's absorption and use of zinc, so women taking iron supplements (more than 30 milligrams per day) may also need zinc supplements.

REVIEW NOTES

A daily iron supplement is recommended for all pregnant women during the second and third trimesters.

Iron interferes with zinc absorption, so women taking iron supplements (more than 30 milligrams per day) may need zinc supplements as well.

Nutrient Supplements Physicians often recommend daily multivitamin-mineral supplements for pregnant women. These prenatal supplements typically provide more folate, iron, and calcium than regular supplements. Prenatal supplements are especially beneficial for women who do not eat adequately and for those in high-risk groups: women carrying twins or triplets, cigarette smokers, and alcohol and drug abusers.[22] For these women, prenatal supplements may be of some help in reducing the risks of preterm delivery, low birthweights, and birth defects (see Figure 10-6). Be aware, however, that supplements cannot prevent the vast majority of destruction from tobacco, alcohol, and drugs, which continues unopposed, as later sections explain.

REVIEW NOTES

Women most likely to benefit from multivitamin-mineral supplements during pregnancy include those who do not eat adequately, those carrying twins or triplets, and those who smoke cigarettes or are alcohol or drug abusers.

FOOD ASSISTANCE PROGRAMS

Women of limited financial means may eat diets too low in calcium, iron, vitamins A and C, and protein. Often, they and their children need help in obtaining food and benefit from nutrition counseling. At the federal level, the **Special Supplemental Food Program for Women, Infants, and Children (WIC)** provides vouchers redeemable for nutritious foods, along with nutrition education to low-income pregnant and lactating women and their children.[23] The foods offered are milk and cheese, iron-fortified cereals, fruit or vegetable juices, carrots, eggs, dried beans, tuna fish, and peanut butter. These foods provide nutrients often lacking in diets of low-income women and children. For infants given infant formula, WIC also provides iron-fortified formula. WIC encourages mothers to breastfeed their infants, however, and offers incentives to those who do.[24]

More than 7 million people—most of them infants and young children—receive WIC benefits each month. Participation in the WIC program benefits both the nutrient status and the growth and development of infants and children.[25] WIC participation during pregnancy can effectively reduce infant mortality, low birthweight, and maternal and newborn medical costs.

The Food Stamp Program provides a debit card that can also help to stretch the low-income pregnant woman's grocery dollars. Many communities provide educational services and materials, including nutrition, food budgeting, and shopping information through the local agricultural extension service. Organizations such as the American Dietetic Association, the American Diabetes Association, and local hospitals also provide nutrition information.

> ### REVIEW NOTES
> Food assistance programs such as WIC can provide nutritious food for pregnant women of limited financial means.

WEIGHT GAIN

Women must gain weight during pregnancy—fetal and maternal well-being depend on it. Ideally, a woman will have begun her pregnancy at a healthy weight, but even more importantly, she will gain within the recommended weight range based on her prepregnancy body mass index (BMI). Table 10-4 presents recommended weight gains for pregnancy. For the normal-weight woman, the ideal pattern is about $3^{1}/_{2}$ pounds total during the first trimester and a pound per week thereafter. Pregnancy weight gains within the recommended ranges are associated with fewer

Prenatal Vitamins

Supplement Facts
Serving Size 1 Tablet

Amount Per Tablet	% Daily Value for Pregnant/ Lactating Women
Vitamin A 4000 IU	50%
Vitamin C 100 mg	167%
Vitamin D 400 IU	100%
Vitamin E 11 IU	37%
Thiamin 1.84 mg	108%
Riboflavin 1.7 mg	85%
Niacin 18 mg	90%
Vitamin B6 2.6 mg	104%
Folate 800 mcg	100%
Vitamin B12 4 mcg	50%
Calcium 200 mg	15%
Iron 27 mg	150%
Zinc 25 mg	167%

INGREDIENTS: calcium carbonate, microcrystalline cellulose, dicalcium phosphate, ascorbic acid, ferrous fumarate, zinc oxide, acacia, sucrose ester, niacinamide, modified cellulose gum, di-alpha tocopheryl acetate, hydroxypropyl methylcellulose, hydroxypropyl cellulose, artificial colors (FD&C blue no. 1 lake, FD&C red no. 40 lake, FD&C yellow no. 6 lake, titanium dioxide), polyethylene glycol, starch, pyridoxine hydrochloride, vitamin A acetate, riboflavin, thiamin mononitrate, folic acid, beta carotene, cholecalciferol, maltodextrin, gluten, cyanocobalamin, sodium bisulfite.

FIGURE 10-6 Example of a Prenatal Supplement Label Notice that vitamin A is reduced to guard against birth defects, while extra amounts of folate, iron, and other nutrients are provided to meet the specific needs of pregnant women.

A prenatal weight-gain grid (see Appendix E) plots the rate of weight gain during pregnancy.

TABLE 10-4	Recommended Weight Gains Based on Prepregnancy Weight
Prepregnancy Weight	**Recommended Weight Gain**
Underweight (BMI <18.5)	28 to 40 lb (12.5 to 18.0 kg)
Healthy weight (BMI 18.5 to 24.9)	25 to 35 lb (11.5 to 16.0 kg)
Overweight (BMI 25.0 to 29.9)	15 to 25 lb (7.0 to 11.5 kg)
Obese (BMI ≥30)	15 lb minimum (6.8 kg minimum)

Note: These classifications for BMI are slightly different from those developed in 1990 by the Committee on Nutritional Status during Pregnancy and Lactation for the publication *Nutrition during Pregnancy* (Washington, D.C.: National Academy Press). That committee acknowledged that because such classifications had not been validated by research on pregnancy outcome, "any cut off points will be arbitrary for women of reproductive age." For these reasons, it seems appropriate to use the values developed for adults in 1998 by the National Institutes of Health (see Chapter 6).

Special Supplemental Food Program for Women, Infants, and Children (WIC): a high-quality, cost-effective health care and nutrition services program administered by the U.S. Department of Agriculture for low-income women, infants, and children who are nutritionally at risk. WIC provides supplemental foods, nutrition education, and referrals to health care and other social services.

FIGURE 10-7 Components of Weight Gain during Pregnancy

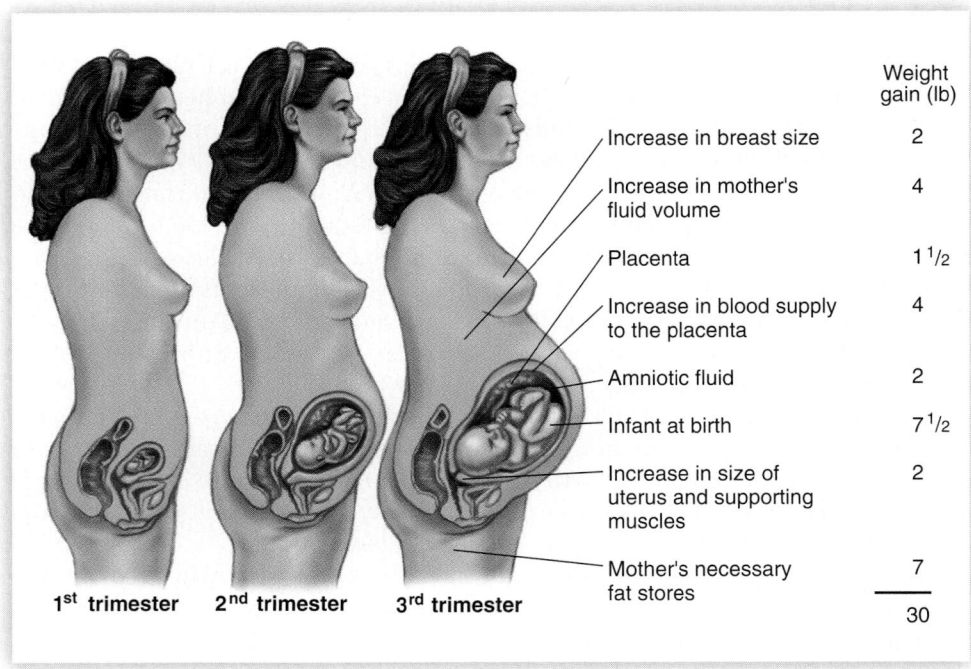

	Weight gain (lb)
Increase in breast size	2
Increase in mother's fluid volume	4
Placenta	1 1/2
Increase in blood supply to the placenta	4
Amniotic fluid	2
Infant at birth	7 1/2
Increase in size of uterus and supporting muscles	2
Mother's necessary fat stores	7
	30

1st trimester   2nd trimester   3rd trimester

surgical births, a greater number of healthy birthweights, and other positive outcomes for both mothers and infants, but many women do not gain within these ranges.

Dieting during pregnancy is not recommended. Even an obese woman should gain at least 15 pounds for the best chance of delivering a healthy infant.[26] Weight gain for a pregnant adolescent must be adequate enough to accommodate her own growth and that of her fetus. Women who are carrying twins must gain more still. A sudden, large weight gain is a danger signal, however, because it may indicate the onset of preeclampsia (discussed later in the chapter).

The weight the pregnant woman gains is nearly all lean tissue: placenta, uterus, blood, milk-producing glands, and, the fetus itself (see Figure 10-7). The fat she gains is needed later for lactation. Physical activity can help a pregnant woman cope with the extra weight, as the next section explains. Some weight is lost at delivery, but many women retain a few pounds with each pregnancy.

REVIEW NOTES

Weight gain is essential for a healthy pregnancy.

A woman's prepregnancy BMI, her own nutrient needs, and the number of fetuses she is carrying help to determine appropriate weight gain.

PHYSICAL ACTIVITY

Physical activity is important to the pregnant woman, not only to help her carry the extra weight of pregnancy without strain, but also to help ease her upcoming childbirth. Staying active during the course of a normal, healthy pregnancy improves the fitness of the mother-to-be, facilitates labor, helps to prevent or manage gestational diabetes, and reduces psychological stress.[27] Women who remain active during pregnancy report fewer discomforts throughout their pregnancies and retain habits that help in losing excess weight and regaining fitness after the birth.

Pregnant women should take care in choosing their physical activities, however. They should participate in "low-impact" activities and avoid sports in which

DO		DON'T
Do exercise regularly (at least three times a week).		Don't exercise vigorously after long periods of inactivity.
Do warm up with 5 to 10 minutes of light activity.		Don't exercise in hot, humid weather.
Do exercise for 20 to 30 minutes at your target heart rate.		Don't exercise when sick with fever.
Do cool down with 5 to 10 minutes of slow activity and gentle stretching.		Don't exercise while lying on your back after the first trimester of pregnancy or stand motionless for prolonged periods.
Do drink water before, after, and during exercise.		Don't exercise if you experience any pain or discomfort.
Do eat enough to support the additional needs of pregnancy plus exercise.		Don't participate in activities that may harm the abdomen or involve jerky, bouncy movements.
	Pregnant women can enjoy the benefits of physical activity.	Don't scuba dive.

© 2002 Tracy Frankel/Image Bank/Getty Images

FIGURE 10-8 Guidelines for Physical Activity during Pregnancy

they might fall or be hit by other people or objects. As is true for everyone, the frequency, duration, and intensity of the activity affect the likelihood of the benefits or risks. A pregnant woman should consult her health care provider before taking up additional activity. A few guidelines are offered in Figure 10-8. Several of the guidelines are aimed at preventing excessively high internal body temperature and dehydration, both of which can harm fetal development. To this end, a pregnant woman should also stay out of saunas, steam rooms, and hot whirlpools.

REVIEW NOTES

By remaining active throughout pregnancy, a woman can develop the strength she needs to carry the extra weight and maintain habits that will help her lose it after the birth.

COMMON NUTRITION-RELATED CONCERNS OF PREGNANCY

Food sensitivities, nausea, heartburn, and constipation are common during pregnancy. A few simple strategies can help alleviate maternal discomforts (see Table 10-5, p. 270).

Food Cravings and Aversions Some women develop cravings for, or aversions to, certain foods and beverages during pregnancy. Individual **food cravings** during pregnancy do not seem to reflect real physiological needs. In other words, a woman who craves pickles does not necessarily need salt. Similarly, cravings for ice cream are common during pregnancy but do not signify a calcium deficiency. **Food aversions** and cravings that arise during pregnancy are probably due to hormone-induced changes in taste and sensitivities to smells.

Nonfood Cravings Some pregnant women develop cravings for nonfood items such as laundry starch, clay, soil, or ice—a practice known as pica.[28] Pica may be practiced for cultural reasons that reflect a society's folklore; it is especially common among African American women. Pica is often associated with iron deficiency, but whether iron deficiency leads to pica or pica leads to iron deficiency is unclear. Eating clay or soil may interfere with iron absorption and displace iron-rich foods

food cravings: deep longings for particular foods.

food aversions: strong desires to avoid particular foods.

TABLE 10-5	Strategies to Alleviate Maternal Discomforts

To alleviate the nausea of pregnancy:

On waking, get up slowly.

Eat dry toast or crackers.

Chew gum or suck hard candies.

Eat small, frequent meals whenever hunger strikes.

Avoid foods with offensive odors.

When nauseated, do not drink citrus juice, water, milk, coffee, or tea.

To prevent or alleviate constipation:

Eat foods high in fiber.

Exercise daily.

Drink at least 8 glasses of liquids a day.

Respond promptly to the urge to defecate.

Use laxatives only as prescribed by a physician; avoid mineral oil—it carries needed fat-soluble vitamins out of the body.

To prevent or relieve heartburn:

Relax and eat slowly.

Eat small, frequent meals.

Drink liquids between meals.

Avoid spicy or greasy foods.

Sit up while eating.

Wait an hour after eating before lying down.

Wait 2 hours after eating before exercising.

from the diet. Furthermore, if the soil or clay contains environmental contaminants such as lead or parasites, health and nutrition suffer.

Morning Sickness The nausea of "morning" (actually, anytime) sickness is usually benign, although it is distressing to some women. It arises from the hormonal changes taking place early in pregnancy, ranges from mild queasiness to debilitating nausea, and afflicts more than half of all pregnant women. Many women complain that smells, especially cooking smells, make them sick. Thus minimizing odors is a key to alleviating morning sickness. Traditional strategies for quelling nausea are listed in Table 10-5, but many women benefit most from simply eating the foods they want when they feel like eating.

Heartburn Heartburn, a burning sensation in the lower esophagus near the heart, is common during pregnancy and is also benign. As the growing fetus puts increasing pressure on the woman's stomach, acid may back up and create a burning sensation in her throat. Tips to relieve heartburn are also listed in Table 10-5.

Constipation As the hormones of pregnancy alter muscle tone and the thriving infant crowds intestinal organs, an expectant mother may complain of constipation, another harmless but annoying condition. A high-fiber diet, physical activity, and plentiful fluids will help relieve this condition. Also, responding promptly to the urge to defecate can help. Laxatives should be used only as prescribed by the physician. Mineral oil should not be used because it interferes with the absorption of fat-soluble vitamins.

PROBLEMS IN PREGNANCY

Just as adequate nutrition and normal weight gain support the health of the mother and growth of the fetus, maternal diseases can have an adverse effect. If discovered early, many diseases can be controlled—another reason why early prenatal care is recommended. Some nutrition measures can help alleviate the most common problems encountered during pregnancy.

Gestational Diabetes Some women are prone to developing a pregnancy-related form of diabetes, **gestational diabetes.** Gestational diabetes usually resolves after the infant is born, but some women go on to develop diabetes (type 2) later in life, especially if they are overweight.[29] Gestational diabetes can lead to fetal or infant sickness or death. When it is identified early and managed properly, however, the most serious risks fall dramatically.[30] More commonly, gestational diabetes leads to surgical birth and high infant birthweight.[31] The American Diabetes Association recommends that all women be assessed for risk of gestational diabetes at their first prenatal examination. Those with elevated risks should undergo further testing and treatment.[32] Chapter 20 provides information about medical nutrition therapy for gestational diabetes.

Hypertension Hypertension complicates pregnancy and affects its outcome in different ways, depending on when the hypertension first develops and on how severe it becomes. Hypertension can be a preexisting chronic condition that develops before a woman becomes pregnant or a transient condition that develops during the pregnancy and subsides after childbirth. In some cases, hypertension that develops during pregnancy warns of the ominous disorder preeclampsia.

Preexisting Chronic Hypertension In addition to the health risks normally imposed by hypertension (heart attack and stroke), high blood pressure increases the risks of having a low-birthweight infant or of having the placenta separate from the wall of the uterus before the birth, resulting in stillbirth. Ideally, before a woman with hypertension becomes pregnant, her blood pressure will be under control.

Transient Hypertension of Pregnancy Some women first develop hypertension during the second half of pregnancy. Most often, the rise in blood pressure is mild and does not affect the pregnancy adversely. Blood pressure usually returns to normal during the first few weeks after childbirth. This transient hypertension of pregnancy differs from the life-threatening hypertension diseases of pregnancy—preeclampsia and eclampsia.

Preeclampsia Hypertension may signal the onset of **preeclampsia,** a condition characterized not only by high blood pressure but also by protein in the urine and fluid retention (edema). Preeclampsia, which affects less than 10 percent of pregnant women, usually occurs with first pregnancies and almost always appears after 20 weeks' gestation.[33] Symptoms typically regress within 48 hours of delivery. The edema of preeclampsia is a whole-body edema, distinct from the localized fluid retention women normally experience late in pregnancy.

Preeclampsia affects almost all of the woman's organs—the circulatory system, liver, kidneys, and brain. If it progresses, she may experience convulsions; when this occurs, the condition is called **eclampsia.** Maternal mortality during pregnancy is

Risk factors for gestational diabetes:
- Obesity.
- Personal history of gestational diabetes.
- Strong family history of diabetes.
- Glucose in the urine.

These racial and ethnic groups are more prone to gestational diabetes:
- Hispanic American.
- Native American.
- Asian American.
- African American.
- Pacific Islander.

The normal edema of pregnancy responds to gravity: blood pools in the ankles. The edema of preeclampsia is a generalized edema. The distinction helps with diagnosis.

Warning signs of preeclampsia:
- Hypertension.
- Protein in the urine.
- Upper abdominal pain.
- Severe and constant headaches.
- Swelling, especially of the face.
- Dizziness.
- Blurred vision.
- Sudden weight gain (1 lb/day).

gestational diabetes: the presence of abnormal glucose tolerance during pregnancy.

preeclampsia: a condition characterized by hypertension, fluid retention, and protein in the urine.

eclampsia: a severe complication during pregnancy in which convulsions occur.

rare in developed countries, but eclampsia is the most common cause. Preeclampsia demands prompt medical attention. Treatment focuses on regulating blood pressure and preventing convulsions.

> **REVIEW NOTES**
>
> Conditions such as gestational diabetes, hypertension, and preeclampsia can threaten the health and life of both mother and infant.
>
> Such conditions require medical and nutrition treatment.

PRACTICES TO AVOID

A general guideline for the pregnant woman is to eat a normal, healthy diet and practice moderation. A woman's daily choices during pregnancy take on enormous importance. Forewarned, pregnant women can choose to abstain from or avoid potentially harmful practices.

Cigarette Smoking One practice to be avoided during pregnancy is cigarette smoking. A surgeon general's warning states that parental smoking can kill an otherwise healthy fetus or newborn. Constituents of cigarette smoke, such as nicotine and cyanide, are toxic to a fetus. Research shows that smoking during pregnancy can cause damage to fetal chromosomes, which could lead to developmental defects or genetic disorders, including cancer.[34] Smoking also restricts the blood supply to the growing fetus and so limits the delivery of oxygen and nutrients and the removal of wastes. It slows growth, thus retarding physical development of the fetus, and it may cause behavioral or intellectual problems later.

A mother who smokes is more likely to have a complicated birth, and her infant is more likely to be of low birthweight.[35] The more a mother smokes, the smaller her infant will be. Of all preventable causes of low birthweight in the United States, smoking has the greatest impact. Sudden infant death syndrome (SIDS), the unexplained death that sometimes occurs in an otherwise healthy infant, has been linked to the mother's cigarette smoking during pregnancy.[36] Research suggests that even in women who do not smoke, exposure to **environmental tobacco smoke** (**ETS,** or secondhand smoke) during pregnancy increases the risk of low birthweight and the likelihood of SIDS.[37]

Unfortunately, an estimated one out of nine pregnant women smokes, and rates are even higher for unmarried women and those who have not graduated from high school.[38] In addition to the harms already described, cigarette (and cigar) smoking adversely affects the pregnant woman's nutrition status and thus impairs fetal nutrition and development. Smokers tend to have lower intakes of dietary fiber, vitamin A, beta-carotene, folate, and vitamin C.

Medicinal Drugs and Herbal Supplements Medicinal drugs taken during pregnancy can cause serious birth defects. Pregnant women should not take over-the-counter drugs or any medications not prescribed by a physician. Drug labels warn: "As with any drug, if you are pregnant or nursing a baby, seek the advice of a health professional before using this product." For aspirin and ibuprofen, there is an additional warning: "It is especially important not to use aspirin (or ibuprofen) during the last three months of pregnancy unless specifically directed to do so by a doctor because it may cause problems in the unborn child or excessive bleeding during delivery." Such warnings should be taken seriously.

Some pregnant women mistakenly consider herbal supplements to be safe alternatives to medicinal drugs and take them to relieve nausea, promote water loss, alleviate depression, help them sleep, or for other reasons. Some herbal products may be safe, but almost none have been tested for safety or effectiveness during pregnancy. Pregnant women should stay away from herbal supplements,

environmental tobacco smoke (ETS): the combination of exhaled smoke (mainstream smoke) and smoke from lighted cigarettes, pipes, or cigars (sidestream smoke) that enters the air and may be inhaled by other people.

teas, or other products unless their safety during pregnancy has been ascertained.[39] The American Dietetic Association's website lists more than 100 herbal supplements that may not be safe to use during pregnancy.* Nutrition in Practice 23 offers more information about herbal supplements and other alternative therapies.

Drugs of Abuse Research shows that women who abuse drugs such as marijuana and cocaine during pregnancy inflict serious health consequences, including nervous system disorders, on their fetuses.[40] Drugs of abuse such as cocaine easily cross the placenta and impair fetal growth and development.[41] Infants born to mothers who abuse crack and other forms of cocaine face low birthweight, heartbeat abnormalities, the pain of withdrawal, or even death as they first experience life outside the womb. Some other effects of drugs of abuse on the fetus are listed in the margin.

Environmental Contaminants Infants and young children of pregnant women exposed to environmental contaminants such as lead and mercury show signs of impaired cognitive development. During pregnancy, lead and mercury readily move across the placenta, inflicting severe damage on the developing fetal nervous system.[42]

Unacceptably high concentrations of mercury in fish have prompted the Food and Drug Administration (FDA) and the Environmental Protection Agency (EPA) to issue an advisory to all pregnant women, women who may become pregnant, lactating mothers, and young children, against eating the following large ocean fish: king mackerel, swordfish, shark, and tilefish.[43] The FDA and the EPA further advise the same groups of people to:

- Eat up to 12 ounces a week of a variety of safer fish and shellfish such as canned light tuna, salmon, pollock, catfish, and shrimp; children should be given smaller portions. Albacore "white" tuna has more mercury than canned light tuna and so should be limited to 6 ounces or less per week.
- Check local advisories about the safety of fish caught by family and friends in lakes, rivers, and coastal areas. If no advice is available, eat up to six ounces per week of fish from local waters, but don't eat any other fish during that week.

Foodborne Illness The vomiting and diarrhea caused by many foodborne illnesses can leave a pregnant woman exhausted and dangerously dehydrated. Particularly threatening, however, is **listeriosis,** which can cause miscarriage, stillbirth, or severe brain or other infections to fetuses and newborns. According to the Centers for Disease Control and Prevention, pregnant women are "about 20 times more likely than other healthy adults to get listeriosis."[44] A woman with listeriosis may develop symptoms such as fever, vomiting, and diarrhea in about 12 hours after eating a contaminated food, and serious symptoms may develop a week to six weeks later. A blood test can reliably detect listeriosis, and antibiotics given promptly to the pregnant sufferer can often prevent infection of the fetus or newborn. The margin lists preventive measures pregnant women can take to avoid contracting listeriosis.

Vitamin-Mineral Megadoses Many vitamins are toxic when taken in excess and the minerals even more so. Among vitamins, a single massive dose of preformed vitamin A (100 times the recommended intake) has caused birth defects. Chronic use of lower doses of vitamin A supplements (three to four times the recommended intake) may also cause birth defects. Intakes before the seventh week of pregnancy appear to be the most damaging. For this reason, additional vitamin A is not recommended during pregnancy, and the vitamin is prescribed in the first trimester of pregnancy only upon evidence of deficiency, which is rare.

*Go to **www.eatright.org** and click on "Position Papers," then on "Life Span," and then on "Pregnancy/Breastfeeding."

Fetal effects of abused drugs:

- **Amphetamines:** Suspected nervous system damage; behavioral abnormalities.
- **Barbiturates:** Drug withdrawal symptoms in the newborn, lasting up to six months.
- **Cocaine:** Uncontrolled jerking motions; paralysis; permanent mental and physical damage.
- **Marijuana:** Short-term irritability at birth.
- **Opiates (including heroin):** Drug withdrawal symptoms in the newborn; permanent learning disability (attention deficit/hyperactivity disorder).

To protect their fetuses and newborns from listeriosis, pregnant women should:

- Avoid the following Mexican soft cheeses: queso blanco, queso fresco, queso de hoja, queso de crema, and asadero. Also avoid feta cheese, brie, Camembert, and blue-veined cheeses like Roquefort.
- Use only pasteurized dairy products.
- Eat only thoroughly cooked meat, poultry, and seafood.
- Thoroughly reheat until steaming hot all hot dogs, luncheon meats, and deli meats, including cured meats like salami, before eating them.
- Wash all fruits and vegetables.
- Do not eat refrigerated smoked seafood such as salmon or trout, or any fish labeled "nova-style," "lox," or "kippered," unless it is an ingredient in a cooked dish.
- Do not eat refrigerated pâté and meat spreads. Canned or shelf-stable pâté and meat spreads are safer.

---

listeriosis: a serious foodborne infection that can cause severe brain infection or death in a fetus or newborn; caused by the bacterium *Listeria monocytogenes*, which is found in soil and water.

fetal alcohol spectrum disorders (FASD): a spectrum of physical, behavioral, and cognitive disabilities caused by prenatal alcohol exposure.

fetal alcohol syndrome (FAS): the cluster of symptoms seen in an infant or child whose mother consumed excessive alcohol during her pregnancy. FAS includes, but is not limited to, brain damage, growth retardation, mental retardation, and facial abnormalities.

alcohol-related neurodevelopmental disorder (ARND): a condition caused by prenatal alcohol exposure. ARND is diagnosed when there is a confirmed history of substantial, regular maternal alcohol intake or heavy episodic drinking and behavioral, cognitive, or central nervous system abnormalities known to be associated with alcohol exposure.

alcohol-related birth defects (ARBD): a condition caused by prenatal alcohol exposure. ARBD is diagnosed when there is a history of substantial, regular maternal alcohol intake or heavy episodic drinking and birth defects known to be associated with alcohol exposure.

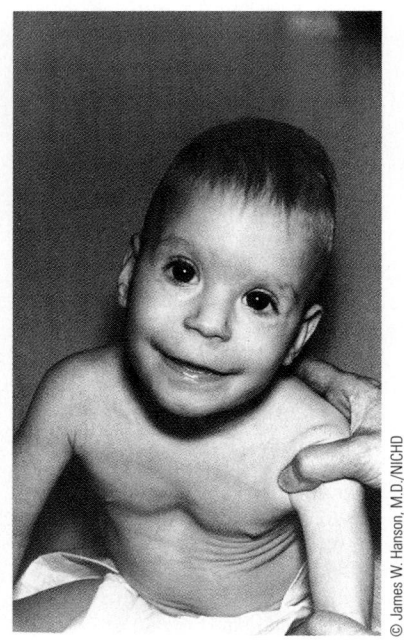

These facial traits are typical of fetal alcohol syndrome, caused by drinking alcohol during pregnancy—low nasal bridge, short eyelid opening, underdeveloped groove in the center of the upper lip, small midface, short nose, and small head circumference.

Dieting Weight-loss dieting, even for short periods, is hazardous during pregnancy. Low-carbohydrate diets or fasts that cause ketosis deprive the growing fetal brain of needed glucose and may impair its development. Such diets are also likely to be deficient in other nutrients vital to fetal growth. Energy restriction during pregnancy is dangerous, regardless of the woman's prepregnancy weight or the amount of weight gained in the previous month.

Sugar Substitutes Artificial sweeteners have been studied extensively and found to be acceptable during pregnancy if used within the FDA's guidelines (see Chapter 3).[45] Still, it would be prudent for pregnant women to use sweeteners in moderation and within an otherwise nutritious and well-balanced diet. Women with phenylketonuria should not use aspartame, as Chapter 3 explains.

Caffeine Caffeine crosses the placenta, and the fetus has only a limited ability to metabolize it. No firm limit for caffeine intake is yet available, but pregnant women who drink more than 3 cups of coffee a day may increase their risk of spontaneous abortion.[46] In light of this evidence, the most sensible course is to limit caffeine consumption to the equivalent of one cup of coffee or two 12-ounce cola beverages a day. Caffeine amounts in food and beverages are listed in Appendix A.

Alcohol Drinking alcohol during pregnancy threatens the fetus with irreversible brain damage, growth retardation, mental retardation, facial abnormalities, vision abnormalities, and many more health problems—a spectrum of symptoms known as **fetal alcohol spectrum disorders** or **FASD.** Children at the most severe end of the spectrum (those with all of the symptoms) are defined as having **fetal alcohol syndrome (FAS).**[47] The fetal brain is extremely vulnerable to a glucose or oxygen deficit, and alcohol causes both by disrupting placental functioning. The lifelong mental retardation and other tragedies of FAS can be prevented by abstaining from drinking alcohol during pregnancy. Once the damage is done, however, the child remains impaired.

The accompanying photo shows the facial abnormalities of FAS, which are easy to depict. A visual picture of the internal harm is impossible, but that damage seals the fate of the child. An estimated 2 to 15 of every 10,000 children are victims of FAS, making it one of the leading known preventable causes of mental retardation in the world.[48]

One out of 10 pregnant women drinks alcohol sometime during her pregnancy; 1 out of 40 pregnant women reports "frequent" drinking (seven or more drinks per week) or binge drinking (five or more drinks on one occasion).[49] Almost half of all pregnancies are unintended, and many are conceived during a binge drinking episode.[50]

For women who know they are pregnant and choose to drink alcohol, the question is how much alcohol is too much. Compared to women who drink less than one drink *per week*, a sizable and significant increase in stillbirths occurs in women who drink five or more drinks per week.[51] Low birthweight is reported among infants born to women who drink 1 ounce (two drinks) of alcohol per day during pregnancy, and FAS is also known to occur with as few as two drinks a day. Birth defects have been reliably observed among the children of women who drink 2 ounces (four drinks) of alcohol daily during pregnancy. The most severe impact is likely to occur in the first two months, before the woman may even be aware that she is pregnant.

Research using animals shows that one-fifth of the amount of alcohol needed to produce major visible defects will produce learning impairment or other defects in the offspring.[52] The cluster of mental problems associated with prenatal alcohol exposure and FASD is known as **alcohol-related neurodevelopmental disorder (ARND),** and the physical malformations are referred to as **alcohol-related birth defects (ARBD).**

For every child diagnosed with full-blown FAS, many more with FASD go undiagnosed until problems develop in the preschool years. Upon reaching adulthood, such children are ill equipped for employment, relationships, and the other facets of life most adults take for granted. Anyone exposed to alcohol before birth may always respond differently to it, and also to certain drugs, than if no exposure had occurred, making addictions likely.

The American Academy of Pediatrics takes the position that women should stop drinking as soon as they *plan* to become pregnant.[53] Researchers have looked for a "safe" alcohol intake limit during pregnancy and have found none. Their conclusion: Abstinence from alcohol is the best policy for pregnant women.

For pregnant women who have already drunk alcohol, the advice is "stop now." A woman who has drunk heavily during the first two-thirds of her pregnancy can still prevent some organ damage by stopping heavy drinking during the third trimester.

REVIEW NOTES

Abstaining from smoking and other drugs, including alcohol, limiting intake of foods known to contain unsafe levels of contaminants such as mercury, taking precautions against foodborne illness, avoiding large doses of nutrients, refraining from dieting, using artificial sweeteners in moderation, and limiting caffeine use are recommended during pregnancy.

ADOLESCENT PREGNANCY

Each year in the United States, an estimated 900,000 adolescent girls become pregnant.[54] Of these, about half choose to continue their pregnancies. A pregnant adolescent presents a special case of intense nutrient needs. Young teenage girls have a hard enough time meeting nutrients needs for their own rapid growth and development, let alone those of pregnancy. Many teens enter pregnancy deficient in vitamins A and D, folate, iron, calcium, and zinc, which places both mother and fetus at risk. Smoking also presents risks, and teens are more likely to smoke while pregnant than older women. Pregnant adolescents have more miscarriages, premature births, stillbirths, and low-birthweight infants than do pregnant adult women.[55] Their greatest risk, though, is death of the infant: mothers under age 16 bear more infants who die within the first year than do women in any other age group. These factors combine to make adolescent pregnancy a major public health problem.

Adequate nutrition is an indispensable component of prenatal care for adolescents and can substantially improve the outlook for both mother and infant. To support the needs of both mother and fetus, a pregnant teenager with a BMI in the normal range is encouraged to gain about 35 pounds to reduce the likelihood of a low-birthweight infant.[56] Pregnant and lactating adolescents would do well to follow the eating pattern presented in Table 1-8, making sure to choose a kcalorie level high enough to support weight gain.

REVIEW NOTES

Proper nutrition and adequate weight gain are especially important in reducing the risk of poor pregnancy outcome in adolescents.

Breastfeeding

The American Academy of Pediatrics (AAP) recommends that infants receive breast milk for at least the first 12 months of life and beyond for as long as mutually desired by mother and child.[57] The American Dietetic Association (ADA)

A woman who decides to breastfeed offers her infant a full array of nutrients and protective factors to support optimal health and development.

advocates breastfeeding for the nutritional health it confers on the infant as well as for the physiological, social, economic, and other benefits it offers the mother.[58] The AAP and the ADA recognize exclusive breastfeeding for 6 months and breastfeeding with complementary foods for at least 12 months as an optimal feeding pattern for infants.[59] Breast milk's unique nutrient composition and protective factors promote optimal infant health and development. The only acceptable alternative to breast milk is iron-fortified formula. Adequate nutrition of the mother supports successful lactation, and without it, lactation is likely to falter or fail.

NUTRITION DURING LACTATION

By continuing to eat nutrient-dense foods, not restricting weight gain unduly, and enjoying ample food and fluid at frequent intervals throughout lactation, the mother who chooses to breastfeed her infant will be nutritionally prepared to do so. An inadequate diet does not support the stamina, patience, and self-confidence that nursing an infant demands. Figure 10-4 (on p. 262) shows how a lactating woman's nutrient needs differ from those of a nonpregnant woman, and Table 10-2 (on p. 263) presents a sample menu that meets those needs.

Energy A nursing woman produces about 25 ounces of milk a day, with considerable variation from woman to woman and in the same woman from time to time, depending primarily on the infant's demand for milk. Producing this milk costs a woman almost 500 kcalories per day above her regular need during the first six months of lactation. To meet this energy need, the woman is advised to eat an extra 330 kcalories of food each day. The other 170 kcalories can be drawn from the fat stores she accumulated during pregnancy. The food energy consumed by the nursing mother should carry with it abundant nutrients. Severe energy restriction hinders milk production and can compromise the mother's health.

Weight Loss A question often raised is whether breastfeeding promotes a more rapid loss of the extra body fat accumulated during pregnancy. Results of studies about the relationship between feeding method and loss of body fat and body weight are inconsistent. When breastfeeding continues for three months or longer, lactation does seem to accelerate a woman's weight loss, but factors such as percentage of body fat and weight gain during pregnancy also play a role.[60] This does not mean that a breastfeeding woman can eat unlimited food and still effortlessly return to her prepregnancy weight. Breastfeeding costs energy, true, but carefully chosen programs of diet and physical activity are still the cornerstones of weight control. Physical activity in particular helps to reduce body fatness and improve fitness while having little effect on a woman's milk production or her infant's weight gain. A gradual weight loss (1 pound per week) is safe and does not reduce milk output.[61] Too large an energy deficit, however, especially soon after birth, will inhibit lactation.

Vitamins and Minerals Another question often raised is whether a mother's milk may lack a nutrient if she fails to get enough in her diet. The answer differs from one nutrient to the next, but in general, nutritional deprivation of the mother reduces the *quantity*, not the *quality*, of her milk. Women can produce milk with adequate protein, carbohydrate, fat, folate, and most minerals, even when their own supplies are limited. For these nutrients, milk quality is maintained at the expense of maternal stores. This is most evident in the case of calcium: dietary calcium has no effect on the calcium concentration of breast milk, but maternal bones lose some of their density during lactation. Such losses are generally made up quickly when lactation ends, and breastfeeding has no long-term harmful effects on women's bones.[62]

Any excess water-soluble vitamins the mother takes in are excreted in the urine; the body does not release them into the milk. The amounts of fat-soluble vitamins in human milk, however, are affected by the mother's excessive or deficient intakes. For example, large doses of vitamin A correspondingly raise the concentration of this vitamin in breast milk. Vitamin supplementation of undernourished women appears to help normalize the vitamin concentrations in their milk and may be beneficial.

Water The volume of breast milk produced depends on how much milk the infant demands, not on how much fluid the mother drinks. The nursing mother is nevertheless advised to drink plenty of liquids each day (about 13 cups) to protect herself from dehydration. To help themselves remember to drink enough liquid, many women make a habit of drinking a glass of milk, juice, or water each time the infant nurses as well as at mealtimes.

Particular Foods Some infants may be sensitive to foods such as cow's milk, onions, or garlic in the mother's diet and become uncomfortable when she eats them. Nursing mothers should not automatically avoid such foods, however. A mother who is breastfeeding her infant is advised to eat whatever nutritious foods she chooses. Then, if a particular food seems to cause the infant discomfort, she can try eliminating that food from her diet for a few days and see if the problem goes away.

CONTRAINDICATIONS TO BREASTFEEDING

Some substances impair maternal milk production or enter the breast milk and interfere with infant development. Some medical conditions prohibit breastfeeding.

Alcohol Alcohol easily enters breast milk and can adversely affect the production, volume, composition, and ejection of breast milk as well as overwhelm an infant's immature alcohol-degrading system.[63] Alcohol concentration in breast milk peaks within one hour after ingestion of even moderate amounts (equivalent to a can of beer). It may alter the taste of the milk to the disapproval of the nursing infant, who may, in protest, drink less milk than normal.

Caffeine Caffeine can make an infant jittery and wakeful. As during pregnancy, caffeine consumption should be moderate.

Cigarette Smoke Health care professionals should actively discourage smoking by lactating women. Research shows that lactating women who smoke produce less milk, and milk with a lower fat content, than mothers who do not smoke. Consequently, their infants gain less weight than infants of nonsmokers.

A lactating woman who smokes not only exposes her infant to nicotine and other chemicals via her breast milk but may also expose the infant to sidestream smoke. Infants who are "smoked over" experience a wide array of health problems—poor growth, hearing impairment, vomiting, breathing difficulties, and even unexplained death.

Medications and Illicit Drugs If a nursing mother must take medication that is secreted in breast milk and is known to affect the infant, then breastfeeding must be put off for the duration of treatment. Meanwhile, the flow of milk can be sustained by pumping the breasts and discarding the milk. Many prescription medications do not reach nursing infants in sufficient quantities to affect them adversely and so have no impact on breastfeeding. Other drugs are not at all compatible with breastfeeding either because they are secreted into the milk and can harm the infant or because they suppress lactation.[64] A nursing mother should

The DRI recommendation for *total* water intake during lactation is 3.8 L/day. This includes 3.1 L or about 13 c as total beverages, including drinking water.

consult with the prescribing physician before taking medicines. Breastfeeding is also contraindicated if the mother uses illicit drugs. Drug addicts, including alcohol abusers, are capable of taking such high doses that their infants can become addicts by way of breast milk.

Many women wonder about using oral contraceptives during lactation. One type that combines the hormones estrogen and progestin seems to suppress milk output, lower the nitrogen content of the milk, and shorten the duration of breastfeeding. In contrast, progestin-only pills have no effect on breast milk or breastfeeding and are considered appropriate for lactating women.

Maternal Illness If a woman has an ordinary cold, she can go on nursing without worry. If susceptible, the infant will catch it from her anyway, and thanks to immunological protection, a breastfed baby may be less susceptible than a formula-fed infant would be. If a woman has active untreated tuberculosis or is receiving therapeutic radioactive isotopes, breastfeeding is contraindicated.[65]

The human immunodeficiency virus (HIV), responsible for causing AIDS, can be passed from an infected mother to her infant during pregnancy, at birth, or through breast milk, especially during the early months of breastfeeding. Thus women in developed countries who have tested positive for HIV should not breastfeed if the infant is not infected. They should choose a safe alternative feeding method, such as breast milk from a milk bank. Milk banks in the United States pasteurize donated human milk and make it available to infants who lack access to milk from their own mothers. Pasteurization destroys harmful organisms, such as HIV, but leaves intact most of the beneficial constituents of the milk.

Throughout the world, breastfeeding prevents millions of infant deaths each year. In developing countries, where the feeding of inappropriate or contaminated formulas causes 1.5 million infant deaths each year, breastfeeding can be critical to infant survival. Thus the question of whether HIV-infected women in developing countries should breastfeed comes down to a delicate balance between risks and benefits. For HIV-positive women in developing countries who are literate, have access to safe water, and have an uninterrupted supply of infant formula, replacement feeding may reduce the risk of infant illness and death by AIDS. For those mothers without safe water and with minimal education, the risk of replacement feeding may be substantial in terms of infant mortality. The World Health Organization and UNICEF (United Nations Children's Fund), in acknowledging the transmission of HIV by way of breast milk, recommend formula feeding for infants of HIV-positive mothers in developing countries if they can be ensured uninterrupted access to safely prepared, nutritionally adequate breast milk substitutes.

> **REVIEW NOTES**
>
> The lactating woman needs enough energy and nutrients to produce about 25 ounces of milk a day. She also needs extra fluid.
>
> Alcohol, caffeine, smoking, and drugs may reduce milk production or enter breast milk and impair infant development.
>
> Some maternal illnesses are incompatible with breastfeeding.

Nutrition of the Infant

Early nutrition affects later development, and early feeding sets the stage for eating habits that will influence nutrition status for a lifetime. Trends change, and experts argue about the fine points, but properly nourishing an infant is relatively simple, overall. Common sense in the selection of infant foods and a nurturing, relaxed environment go far to promote an infant's health and well-being.

In developed countries with well-nourished populations, such as the United States and Canada, the dietary practices that have the most influence on an infant's nutrition status are the type of milk the infant receives and the age at which solid foods are introduced. The remainder of this discussion is devoted to feeding the infant and identifying the nutrients most often deficient in infant diets.

NUTRIENT NEEDS DURING INFANCY

An infant grows faster during the first year than ever again, as Figure 10-9 shows. The growth of infants and children directly reflects their nutritional well-being and is an important parameter in assessing their nutrition status. Health care professionals use growth charts to evaluate the growth and development of children from birth to 20 years of age (see Appendix E).

Nutrients to Support Growth   An infant's birthweight doubles by about four to six months of age and triples by the age of one year. (Consider that if an adult, starting at 120 pounds, were to do this, the person's weight would increase to 360 pounds in a single year.) The infant's length changes more slowly than weight, increasing about 10 inches from birth to one year. By the end of the first year, the growth rate slows considerably. An infant typically gains less than 10 pounds during the second year and grows about 5 inches in height.

Not only do infants grow rapidly, but in proportion to body weight, their basal metabolic rate is remarkably high—about twice that of an adult. The rapid growth and metabolism of the infant demand an ample supply of all the nutrients. Of special importance during infancy are the energy nutrients and the vitamins and minerals critical to the growth process, such as vitamin A, vitamin D, and calcium.

Because they are small, infants need smaller *total* amounts of these nutrients than adults do, but as a percentage of body weight, infants need more than twice as much of most nutrients. Infants require about 100 kcalories per kilogram of body weight per day; most adults require fewer than 40 (see Table 10-6). Figure 10-10 (p. 280) compares a five-month-old infant's needs (per unit of body weight) with those of an adult man. You can see that differences in vitamin D and iodine, for instance, are extraordinary. Around six months of age, energy needs begin to increase less rapidly as the growth rate begins to slow, but some of the energy saved by slower growth is spent on increased activity. When their growth slows, infants spontaneously reduce their energy intakes. Parents should expect their infants to adjust their food intakes downward when appropriate and should not force or coax them to eat more.

Vitamin K nutrition for newborns presents a unique case. A newborn's digestive tract is sterile, and vitamin K–producing bacteria take weeks to establish themselves in the infant's intestines. To prevent uncontrolled bleeding in the newborn, the American Academy of Pediatrics recommends that a single dose of vitamin K be given at birth.[66]

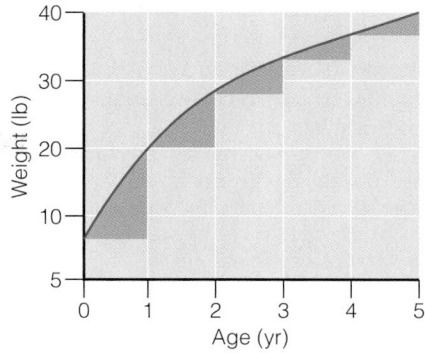

FIGURE 10-9 Weight Gain of Human Infants in Their First Five Years of Life

In the first year, an infant's birthweight may triple, but over the following several years, the rate of weight gain gradually diminishes.

After six months of age, the energy saved by slower growth is spent on increased activity.

TABLE 10-6	Infant and Adult Heart Rate, Respiration Rate, and Energy Needs Compared		
		Infants	Adults
Heart rate (beats/minute)		120 to 140	70 to 80
Respiration rate (breaths/minute)		20 to 40	15 to 20
Energy needs (kcal/body weight)		45/lb (100/kg)	<18/lb (<40/kg)

FIGURE 10-10 Nutrient Recommendations for a Five-Month-Old Infant and an Adult Male Compared on the Basis of Body Weight

Infants may be relatively small and inactive, but they use large amounts of energy and nutrients in proportion to their body size to keep all their metabolic processes going.

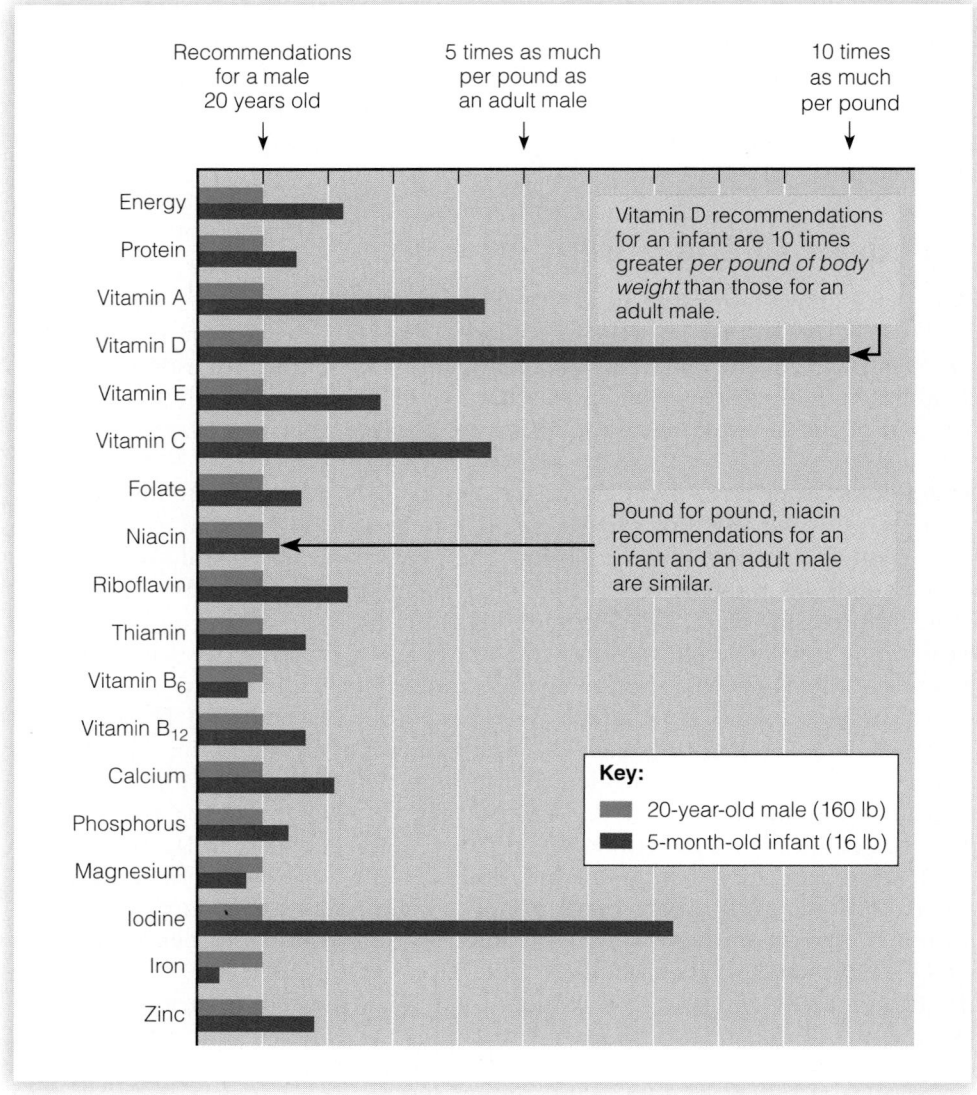

Water   One of the most important nutrients for infants, as for everyone, is water. The younger a child is, the more of its body weight is water. Breast milk or infant formula normally provides enough water to replace fluid losses in a healthy infant. Even in hot, dry climates, neither breastfed nor bottle-fed infants need supplemental water.[67] Because proportionately more of an infant's body water than an adult's is between the cells and in the vascular space, however, this water is easy to lose. Consequently, conditions that cause rapid fluid loss, such as vomiting or diarrhea, require an electrolyte solution designed for infants.

REVIEW NOTES

Infants' rapid growth and development depend on adequate nutrient supplies, including water from breast milk and formula.

BREAST MILK

Breast milk excels as a source of nutrients for the young infant. With the possible exception of vitamin D (discussed later), breast milk provides all the nutrients a healthy infant needs for the first six months of life.[68] It provides many other health benefits as well.

Energy Nutrients The balance of energy nutrients in breast milk differs dramatically from the balance recommended for adults (see Figure 10-11). Yet, for infants, breast milk is the most nearly perfect food, proving that people at different stages of life have different nutrient needs.

The carbohydrate in breast milk (and standard infant formula) is lactose. In addition to being easily digested, lactose enhances calcium absorption.

The lipids in breast milk—and infant formula—provide the main source of energy in the infant's diet. Breast milk contains a generous proportion of the essential fatty acids linoleic acid and linolenic acid, as well as their longer-chain derivatives arachidonic acid and docosahexaenoic acid (DHA). Until recently, infant formula provided only linoleic and linolenic acid. Formula with arachidonic acid and DHA added is now commercially available.[69] Infants can produce some arachidonic acid and DHA from linoleic and linolenic acid, but some infants may need more than they can make.

Arachidonic acid and DHA are found abundantly in the developing brain and the retina of the eye. Research has focused on the visual and mental development of breastfed infants and infants fed standard formula without DHA and arachidonic acid added. Breastfed infants generally score higher on tests of mental development than formula-fed infants do, and researchers are investigating whether this difference can be attributed to DHA and arachidonic acid in breast milk.[70] So far, results are mixed. In one study, researchers found no developmental or visual differences between infants fed standard formula and those fed formula with added DHA and arachidonic acid.[71] In two other studies, infants were given either standard formula or formula with added DHA and arachidonic acid. The infants fed formula fortified with DHA and arachidonic acid had better visual function at one year of age than those who were fed standard formula.[72]

The protein in breast milk is largely **alpha-lactalbumin,** a protein the human infant can easily digest. Another breast milk protein, **lactoferrin,** indirectly benefits the baby's iron nutrition and also acts as an antibacterial agent. Lactoferrin is an iron-gathering compound that helps absorb iron into the infant's bloodstream, keeps intestinal bacteria from getting enough iron to grow out of control, and also works directly to kill some bacteria.[73]

Vitamins and Minerals With the exception of vitamin D, the vitamin content of the breast milk of a well-nourished mother is ample. Even vitamin C, for which cow's milk is a poor source, is supplied generously. The concentration of vitamin D in breast milk is low, however, and vitamin D deficiency impairs bone mineralization. Vitamin D deficiency is most likely in infants who are not exposed to sunlight daily, have darkly pigmented skin, and receive breast milk without vitamin D supplementation.[74] Reports of infants in the United States developing the vitamin D–deficiency disease rickets and recommendations by the American Academy of Pediatrics to keep infants under six months of age out of direct sunlight have prompted updated vitamin D guidelines. The AAP now recommends a vitamin D supplement for all infants who are breastfed exclusively and for any infants who do not receive at least 500 milliliters (15 ounces) per day of vitamin D–fortified formula.[75]

As for minerals, the calcium content of breast milk is ideal for infant bone growth, and the calcium is well absorbed. Breast milk is also low in sodium. The limited amount of iron in breast milk is highly absorbable, and its zinc, too, is absorbed better than from cow's milk, thanks to the presence of a zinc-binding protein.

Supplements for Infants Pediatricians may prescribe supplements containing vitamin D, iron, and fluoride (after six months of age). Table 10-7 (p. 282) offers a schedule of supplements during infancy.

Immunological Protection Breast milk offers the infant unsurpassed protection against infection.[76] Protective factors include antiviral agents, antibacterial agents, and other infection inhibitors.

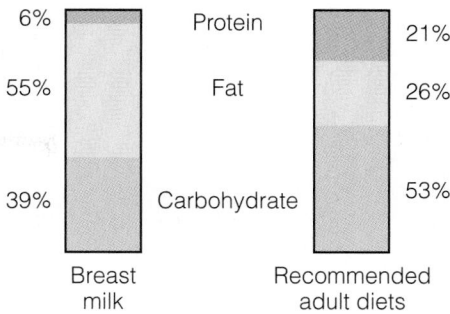

	Breast milk	Recommended adult diets
Protein	6%	21%
Fat	55%	26%
Carbohydrate	39%	53%

[a] The values listed for adults represent approximate midpoints of the acceptable ranges for protein (10 to 35 percent), fat (20 to 35 percent), and carbohydrate (45 to 65 percent).

FIGURE 10-11 Percentages of Energy-Yielding Nutrients in Breast Milk and in Recommended Adult Diets
The proportions of energy-yielding nutrients in human breast milk differ from those recommended for adults.[a]

alpha-lactalbumin (lackt-AL-byoo-min): the chief protein in human breast milk, as **casein** (CAY-seen) is the chief protein in cow's milk.

lactoferrin (lack-toe-FERR-in): a protein in breast milk that binds iron and keeps it from supporting the growth of the infant's intestinal bacteria.

TABLE 10-7	Supplements for Full-Term Infants		
	Vitamin D[a]	Iron[b]	Fluoride[c]
Breastfed infants:			
Birth to six months of age	✓		
Six months to one year	✓	✓	✓
Formula-fed infants:			
Birth to six months of age			
Six months to one year		✓	✓

[a]Vitamin D supplements are recommended for all infants who are exclusively breastfed and for any infants who do not receive at least 500 milliliters (15 ounces) of vitamin D–fortified formula.

[b]Infants four to six months of age need additional iron, preferably in the form of iron-fortified cereal for both breastfed and formula-fed infants and iron-fortified infant formula for formula-fed infants.

[c]At six months of age, breastfed infants and formula-fed infants who receive ready-to-use formulas (these are prepared with water low in fluoride) or formula mixed with water that contains little or no fluoride (less than 0.3 ppm) need supplements.

Source: Adapted from Committee on Nutrition, American Academy of Pediatrics, *Pediatric Nutrition Handbook*, 5th ed., ed. R. E. Kleinman (Elk Grove Village, Ill.: American Academy of Pediatrics, 2004).

During the first two or three days of lactation, the breasts produce **colostrum,** a premilk substance containing antibodies and white cells from the mother's blood. Colostrum is relatively sterile as it leaves the breast, and the infant cannot contract a bacterial infection from it even if the mother has one. Colostrum contains maternal immune factors that inactivate harmful bacteria within the digestive tract. Later, breast milk also delivers immune factors, although not as many as colostrum. Among them are **bifidus factors** and lactoferrin.

Breast milk also contains several enzymes, several hormones (including thyroid hormone and prostaglandins), and lipids, all of which protect the infant against infection. Breastfed babies are less prone to develop stomach and intestinal disorders during the first few months of life and so experience less vomiting and diarrhea than formula-fed infants do. In fact, research shows that breast milk contains not only antibodies against the most common cause of diarrhea in infants and young children but also the protein **lactadherin,** which binds to, and inhibits replication of, the infective agent.*[77] Breastfeeding reduces the severity and duration of symptoms associated with this infection. Breastfeeding also protects against other common illnesses of infancy such as middle ear infection and respiratory illness.[78] Clearly, breast milk is a very special substance. Nutrition in Practice 10 offers suggestions for successful breastfeeding.

Other Potential Benefits Researchers are investigating whether breastfeeding may also help protect against obesity in childhood and later years, but so far results are inconsistent. For example, a well-controlled study of more than 15,000 adolescents and their mothers indicated that those who were mostly breastfed for the first six months of life were less likely to become overweight than those who were fed formula.[79] A study of much younger children (three to five years of age), however, found no clear evidence that breastfeeding influences body weight.[80] A review of more than 60 published studies investigating the relationship between infant feeding and obesity suggests that initial breastfeeding protects

colostrum (co-LAHS-trum): a milklike secretion from the breasts that is rich in protective factors. Colostrum is present during the first day or so after delivery, before milk appears.

bifidus (BIFF-id-us, by-FEED-us) **factors:** factors in colostrum and breast milk that favor the growth of the "friendly" bacterium *Lactobacillus* (lack-toe-ba-SILL-us) *bifidus* in the infant's intestinal tract. These bacteria prevent other, less desirable intestinal inhabitants from flourishing.

lactadherin (lack-tad-HAIR-in): a protein in breast milk that attacks diarrhea-causing viruses.

*The most common cause of diarrhea in the United States is rotavirus. More children are hospitalized for rotavirus infection than for any other single cause.

CASE STUDY   *Woman in Her First Pregnancy*

Ellen Cassidy is a 24-year-old woman who is four months pregnant. This is her first pregnancy, and she is eager to learn how to feed herself during pregnancy as well as her infant after birth. She is 5 feet 3 inches tall and currently weighs 150 pounds. Her prepregnancy weight was 148 pounds. Ellen is very concerned about her 2-pound weight gain.

1. Consult the BMI table (inside back cover), and using the "Healthy Weight" section, find a healthy weight in the middle of the range appropriate for a woman of Ellen's height.
2. Do you think that Ellen's weight at the start of her pregnancy was appropriate for her height? Why or why not?

Should Ellen be concerned about her 2-pound weight gain? Why or why not?

3. What advice should you give Ellen about her weight gain during pregnancy? What other dietary advice would you give her?
4. Discuss methods of infant feeding with Ellen and describe some of the advantages breastfeeding would offer her. What advice will you give Ellen if she decides to breastfeed?
5. What are the advantages of formula feeding? What information should Ellen have about formula feeding?

---

against obesity in later life.[81] The authors cautioned, however, that further review of studies that look at confounding factors such as maternal obesity, smoking, age of assessment, and socioeconomic background in detail is needed.

Breast milk may also offer protection against the development of cardiovascular disease. Compared with formula-fed infants, breastfed infants have lower blood cholesterol as adults.[82]

Breastfeeding may also have a positive effect on later intelligence.[83] In one study, young adults who had been breastfed as long as nine months scored higher on two different intelligence tests than those who had been breastfed less than one month. Many other studies suggest a beneficial effect of breastfeeding on intelligence, but when subjected to strict standards of methodology (for example, large sample size and appropriate intelligence testing), the evidence is less convincing.[84] Nevertheless, the possibility that breastfeeding may positively affect later intelligence is intriguing. It may be that some specific component of breast milk, such as DHA, contributes to brain development or that certain factors associated with the feeding process itself promote intellect.

The Case Study above presents a woman who is four months pregnant. Answering the questions offers practice in thinking through some of the issues related to pregnancy and breastfeeding.

Formula preparation:

- Liquid concentrate (moderately expensive, relatively easy)—mix with equal part water.
- Powdered formula (least expensive, lightest for travel)—read label directions.
- Ready-to-feed (easiest, most expensive)—pour directly into clean bottles.

INFANT FORMULA

Breastfeeding offers many benefits to both mother and infant, and it should be encouraged whenever possible. The mother who has decided to use formula, however, should be supported in her choice just as the breastfeeding mother should be. She can offer the same closeness, warmth, and stimulation during feedings as the breastfeeding mother can.

Many mothers choose to breastfeed at first but wean their children within the first 1 to 12 months. Before infants reach a year of age, mothers must wean them onto *infant formula,* not onto plain cow's milk of any kind—whole, reduced fat, low fat, or fat-free.

Infant Formula Composition Manufacturers can prepare formulas from cow's milk in such a way that they do not differ significantly from human milk in nutrient content. Figure 10-12 (p. 284) illustrates the energy nutrient balance of both. Formulas contain no protective antibodies for human infants, but preventive medical care (vaccinations) and reliable public health measures (clean water) help minimize

The infant thrives on infant formula offered with affection.

© Vic Bider/PhotoEdit

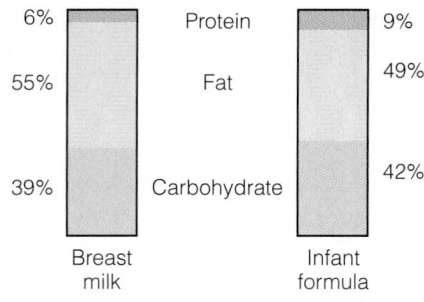

6%	Protein	9%
55%	Fat	49%
39%	Carbohydrate	42%
Breast milk		Infant formula

FIGURE 10-12 Percentages of Energy-Yielding Nutrients in Breast Milk and in Infant Formula
The average proportions of energy-yielding nutrients in human breast milk and formula differ slightly. In contrast, cow's milk provides too much protein and too little carbohydrate.

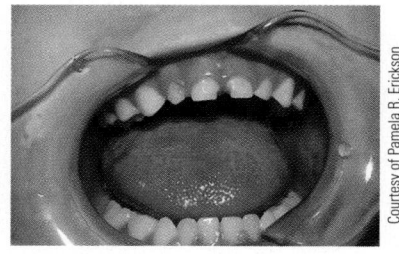

Nursing bottle syndrome in an early stage.

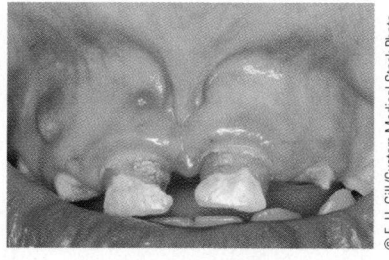

Nursing bottle syndrome, an extreme example. The upper teeth have decayed all the way to the gum line.

hypoallergenic formulas: clinically tested infant formulas that do not provoke reactions in 90% of infants or children with confirmed cow's milk allergy. Like all infant formulas, hypoallergenic formulas must demonstrate nutritional suitability to support infant growth and development. Extensively hydrolyzed and free amino acid–based formulas are examples.

nursing bottle tooth decay: extensive tooth decay due to prolonged tooth contact with formula, milk, fruit juice, or other carbohydrate-rich liquid offered to an infant in a bottle.

this disadvantage. The educated mother whose water supply is reliable can prepare safe, sanitary formulas. Lead-contaminated water, however, is a major source of lead poisoning in infants.

Infant Formula Standards National and international standards have been set for the nutrient contents of infant formulas. U.S. standards are based on AAP recommendations, and the FDA mandates quality control procedures to ensure that these standards are met. All standard formulas are therefore nutritionally similar. Small differences in nutrient content are sometimes confusing but usually unimportant.

Special Formulas Standard formulas are inappropriate for some infants. For example, premature babies require special formulas. Infants allergic to milk protein can drink special **hypoallergenic formulas** or formulas based on soy protein.[85] Soy formulas are lactose-free and so can be used for infants with lactose intolerance as well. They are also useful as an alternative to milk-based formulas for vegetarian families. For infants with other special needs, many other variations are available.

Risks of Formula Feeding In developing countries and in poor areas of the United States, formula may be unavailable, overdiluted in an attempt to save money, or prepared with contaminated water. Overdilution of formula can cause malnutrition and growth failure. Contaminated formula often causes infections leading to diarrhea, dehydration, and failure to absorb nutrients. Wherever sanitation is poor, breastfeeding should take priority over feeding formula. Breast milk is sterile, and its antibodies enhance an infant's resistance to disease.

Iron in Formula The AAP recommends iron-fortified formulas for all formula-fed infants.[86] Low-iron formulas have no role in infant feeding. Use of iron-fortified formulas has risen in recent decades and is credited with the decline of iron-deficiency anemia in U.S. infants.

Nursing Bottle Tooth Decay Dentists advise against putting an infant to bed with a bottle. Salivary flow, which normally cleanses the mouth, diminishes as the infant falls asleep. Sucking for long times pushes the jawline out of shape and causes a bucktoothed profile (protruding upper and receding lower teeth). Furthermore, prolonged sucking on a bottle of formula, milk, or juice bathes the upper teeth in a carbohydrate-rich fluid that nourishes decay-producing bacteria. (The tongue covers and protects most of the lower teeth, but they, too, may be affected.) The result is extensive and rapid tooth decay. To prevent **nursing bottle tooth decay,** no child should be put to bed with a bottle as a pacifier.

THE TRANSITION TO COW'S MILK

The age at which whole cow's milk should be introduced to the infant's diet has long been a source of controversy. The AAP advises that whole cow's milk is not appropriate during the first year.[87] Children one to two years of age should not be given reduced-fat, low-fat, or fat-free milk routinely; they need the fat of whole milk. Between the ages of two and five years, a gradual transition from whole milk to the lower-fat milks can take place, but care should be taken to avoid excessive restriction of dietary fat.

In some infants, particularly those younger than six months of age, whole cow's milk causes intestinal bleeding, which can lead to iron deficiency. Cow's milk is also a poor source of iron. Consequently, it both causes iron loss and fails to replace iron. Furthermore, the bioavailability of iron from infant cereal and other foods is reduced when cow's milk replaces breast milk or iron-fortified formula during the first year. Compared to breast milk or iron-fortified formula, cow's milk is higher in calcium and lower in vitamin C, characteristics that reduce iron absorption. Furthermore, the higher protein concentration of cow's milk can stress the infant's kidneys. In short, cow's milk is a poor choice during the first year of life; infants need breast milk or iron-fortified infant formula.

INTRODUCING FIRST FOODS

Changes in the body organs during the first year affect the infant's readiness to accept solid foods. Until the child is several months old, the immature stomach and intestines can digest milk sugar (lactose), but not starch. This is one of the many reasons why breast milk and formula are such good foods for an infant; they provide simple, easily digested carbohydrate that supplies energy for the infant's growth and activity.

When to Introduce Solid Food The AAP supports exclusive breastfeeding for approximately six months but recognizes that infants are often developmentally ready to accept complementary foods between four and six months of age.[88] Thus foods may be started gradually beginning sometime between four and six months, depending on the infant's readiness. Indications of readiness for solid foods include:

- The infant can sit with support and control head movements.
- The infant is six months old.

Table 10-8 (p. 286) presents a suggested sequence for introducing new foods.

The addition of foods to an infant's diet should be governed by three considerations: the infant's nutrient needs, the infant's physical readiness to handle different forms of foods, and the need to detect and control allergic reactions. With respect to nutrient needs, the nutrient needed earliest is iron, then vitamin C.

Foods to Provide Iron and Vitamin C Iron deficiency is common in young children throughout the world, especially between the ages of six months and three years when they are growing fast and milk, which is a poor source of iron, has a large place in their diets. The iron an infant has stored from before birth typically runs out after the birthweight doubles, long before the end of the first year. Infants can derive adequate iron first from breast milk or formula with iron, then from iron-fortified cereals, and from meat or meat alternates such as legumes. Once infants are consuming iron-fortified cereals, parents or caregivers should begin selecting vitamin C–rich foods to go with meals to enhance iron absorption. The best sources of vitamin C are fruits and vegetables (see p. 210).

Fruit juice is a source of vitamin C, but excessive juice intake can lead to diarrhea in infants and young children.[89] Also, some infants and young children may fail to grow and thrive when they drink so much juice that it displaces other, more nutrient- and energy-dense foods from their diets. The AAP recommends limiting juice consumption for infants and children.[90] Fruit juices should be served in a cup, not a bottle, and should not be offered before the infant is six months of age. Juices should be used moderately (4 to 6 ounces per day), so as not to displace other foods.

Physical Readiness for Solid Foods The ability to swallow solid food develops at around four to six months, and food offered by spoon helps to develop swallowing ability. At eight months to a year, an infant can sit up, can handle finger foods, and begins to teethe. At that time, hard crackers and other hard finger foods may be introduced to

Foods such as iron-fortified cereals and formulas, mashed legumes, and strained meats provide iron.

© Polara Studios, Inc.

TABLE 10-8	Infant Development and Recommended Foods

Note: Because each stage of development builds on the previous stage, the foods from an earlier stage continue to be included in all later stages.

Age (mo)	Feeding Skill	Foods Introduced into the Diet
0–4	Turns head toward any object that brushes cheek. Initially swallows using back of tongue; gradually begins to swallow using front of tongue as well. Strong reflex (extrusion) to push food out during first 2 to 3 months.	Feed breast milk or infant formula.
4–6	Extrusion reflex diminishes, and the ability to swallow nonliquid foods develops. Indicates desire for food by opening mouth and leaning forward. Indicates satiety or disinterest by turning away and leaning back. Sits erect with support at 6 months. Begins chewing action. Brings hand to mouth. Grasps objects with palm of hand.	Begin iron-fortified cereal mixed with breast milk, formula, or water. Begin pureed vegetables and fruits.
6–8	Able to feed self with fingers. Develops pincher (finger to thumb) grasp. Begins to drink from cup.	Begin mashed vegetables and fruits. Begin plain baby food meats. Begin plain, unsweetened fruit juices from cup.
8–10	Begins to hold own bottle. Reaches for and grabs food and spoon. Sits unsupported.	Begin breads and cereals from table. Begin yogurt. Begin pieces of soft, cooked vegetables and fruit from table. Gradually begin finely cut meats, fish, casseroles, cheese, eggs, and legumes.
10–12	Begins to master spoon, but still spills some.	Include breads and cereals from table, in addition to infant cereal; soft or cooked fruits and vegetables; and finely chopped or ground meat, fish, or poultry; eggs; and mashed legumes.[a]

[a] Portions of foods for infants and young children are smaller than those for an adult. For example, a grain serving might be $1/2$ slice of bread instead of 1 slice, or $1/4$ cup rice instead of $1/2$ cup.

Source: Adapted in part from Committee on Nutrition, American Academy of Pediatrics, *Pediatric Nutrition Handbook*, 5th ed., ed. R. E. Kleinman (Elk Grove Village, Ill.: American Academy of Pediatrics, 2004), pp. 103–115.

promote the development of manual dexterity and control of the jaw muscles. These feedings must occur under the watchful eye of an adult because the infant can also choke on such foods.

Some parents want to feed solids at an earlier age, on the theory that "stuffing the baby" at bedtime promotes sleeping through the night. There is no proof for this theory. On average, infants start to sleep through the night at about the same age (three to four months) regardless of when solid foods are introduced.

Allergy-Causing Foods New foods should be introduced singly and at intervals spaced to permit detection of allergies. For example, when cereals are introduced, rice cereal is offered first for several days; it causes allergy least often. Wheat cereal is offered last; it is the most common offender. If a cereal causes an allergic reaction (irritability due to skin rash, digestive upset, or respiratory discomfort), its use should be discontinued before going on to the next food.

Choice of Infant Foods Baby foods commercially prepared in the United States and Canada are safe, and except for mixed dinners and heavily sweetened desserts, they generally have high nutrient density. An alternative for the parent who wants the infant to have family foods is to "blenderize" a small portion of the table food (cooked without salt or sugar) at each meal.

Foods to Omit Sweets of any kind (including baby food "desserts") have no place in an infant's diet. The added food energy conveys few, if any, nutrients to support growth and contributes to obesity. Canned vegetables are also inappropriate for infants; they often contain too much sodium. Honey should never be fed to infants because of the risk of botulism. Infants and even young children cannot safely chew and swallow any of the foods listed in the margin; they can easily choke on these foods, a risk not worth taking. Nonfood items of small size should always be kept out of the infant's reach to prevent choking. Also, an infant's caregiver must be on guard against food poisoning and take precautions against it as described in Chapter 1.

Foods at One Year Whole milk is the best food to supply most of the nutrients the infant needs at one year of age; 2 to 3$\frac{1}{2}$ cups a day meet those needs sufficiently. More milk than this displaces iron-rich foods and can lead to the iron-deficiency anemia known as milk anemia. Other foods—meat and meat alternates, iron-fortified cereal, enriched or whole-grain bread, fruits, and vegetables—should be supplied in variety and in amounts sufficient to round out total energy needs. Ideally, a one-year-old will sit at the table, eat many of the same foods everyone else eats, and drink liquids from a cup—not a bottle. Table 10-9 shows a sample menu that meets a one-year-old's requirements.

LOOKING AHEAD

Probably the most important single measure to undertake during the first year is to encourage eating habits that will support continued normal weight as the child grows. This means introducing a variety of nutritious foods in an inviting way, not forcing the infant to finish the bottle or the baby food jar, avoiding concentrated sweets and empty-kcalorie foods, and encouraging physical activity. Parents

To prevent choking, do not give infants or young children:
- Gum.
- Popcorn.
- Whole grapes.
- Cherries.
- Raw celery.
- Carrots.
- Whole beans.
- Hot dog slices.
- Hard or gel-type candies.
- Marshmallows.
- Nuts.
- Peanut butter.

Keep these nonfood items out of their reach:
- Coins.
- Balloons.
- Small balls.
- Pen tops.
- Other items of similar size.

Milk anemia develops when an excessive milk intake displaces iron-rich foods from the diet.

TABLE 10-9	Sample Menu for a One-Year-Old

Breakfast	**Afternoon snack**
$\frac{1}{2}$ c whole milk	$\frac{1}{2}$ c whole milk
$\frac{1}{2}$ c iron-fortified cereal	$\frac{1}{2}$ slice toast
$\frac{1}{2}$ c orange juice	1 tbs apple butter
Morning snack	**Dinner**
$\frac{1}{2}$ c yogurt	1 c whole milk
$\frac{1}{2}$ c fruit[a]	2 oz chopped meat or well-cooked mashed legumes
Lunch	$\frac{1}{4}$ c potato, rice, or pasta
$\frac{1}{2}$ c whole milk	$\frac{1}{2}$ c vegetables[b]
$\frac{1}{2}$ c vegetables[b]	$\frac{1}{4}$ c fruit[a]
1 egg or $\frac{1}{4}$ c tofu	
$\frac{1}{2}$ c noodles	

[a] Include citrus fruits, melons, and berries.

[b] Include dark green, leafy, and deep yellow vegetables.

Ideally, a one-year-old eats many of the same healthy foods as the rest of the family.

should avoid teaching infants to seek food as a reward, to expect food as comfort for unhappiness, or to associate food deprivation with punishment. If infants cry from thirst, give them water, not milk or juice. Infants seem to have no internal "kcalorie counter," and they stop eating when their stomachs feel full. Nutrient-dense, low-kcalorie foods will satisfy as long as they provide bulk.

Normal dental development is also promoted by supplying nutritious foods, avoiding sweets, and discouraging the association of food with reward or comfort. Dental health is the subject of Nutrition in Practice 3.

The AAP recommends against a fat-modified diet during infancy. The available evidence does not warrant dietary manipulation to lower blood cholesterol in infants.

MEALTIMES

The wise parent of a one-year-old offers nutrition and love together. Both promote growth. Children "fed with love" grow more in both weight and height than children fed the same food in an emotionally negative climate.

The person feeding a one-year-old should be aware that exploring and experimenting are normal and desirable behaviors at this time in a child's life. The child is developing a sense of autonomy that, if allowed to develop, will lay the foundation for later confidence and effectiveness as an individual. The child's impulses, if consistently denied, can turn to shame and self-doubt. In light of the developmental and nutrient needs of one-year-olds, and in the face of their often contrary and willful behavior, a few feeding guidelines may be helpful:

- *Discourage unacceptable behavior (such as standing at the table or throwing food) by removing the child from the table to wait until later to eat.* Be consistent and firm, not punitive. The child will soon learn to sit and eat.
- *Let the child explore and enjoy food.* This may mean the child eats with fingers for a while. Use of the spoon will come in time.
- *Don't force food on children.* Provide children with nutritious foods, and let them choose which ones and how much they will eat. Gradually, they will acquire a taste for different foods. If children refuse milk, provide cheese, cream soups, and yogurt.
- *Limit sweets strictly.* Infants have little room in their 1000-kcalorie daily energy allowance for empty-kcalorie sweets, except occasionally.

These recommendations reflect a spirit of tolerance that serves the best interests of the child emotionally as well as physically. The Nutrition Assessment Checklist helps to identify nutrition-related factors that may help prevent or correct potential problems in pregnant women and infants. Details of nutrition assessment are presented in Chapter 13.

REVIEW NOTES

The primary food for infants during the first 12 months is either breast milk or iron-fortified formula.

In addition to nutrients, breast milk also offers immunological protection.

At about four to six months, infants should gradually begin eating solid foods. By one year, they are drinking from a cup and eating many of the same foods as the rest of the family.

NUTRITION ASSESSMENT CHECKLIST FOR PREGNANT WOMEN AND INFANTS

Medical History

Check the medical record for:

☐ Alcohol or illicit drug abuse

☐ Chronic diseases

☐ Gestational diabetes

☐ History of previous pregnancies (number, intervals, outcomes, multiple births, and gestational age birthweights)

☐ Hypertension

☐ Neural tube defect in an infant born previously

☐ Preeclampsia

Note risk factors for complications during pregnancy, including:

☐ Cigarette smoking

☐ Food faddism

☐ Lactose intolerance

☐ Low socioeconomic status

☐ Significant or prolonged vomiting

☐ Very young or old age

☐ Weight-loss dieting

Note any complaints of:

☐ Constipation

☐ Heartburn

☐ Morning sickness

Medications

For pregnant women who are using drug therapy for medical conditions, note:

☐ Potential for contraindication to breastfeeding

☐ GI tract side effects that might reduce food intake or change nutrient needs

Dietary Intake

For all pregnant women, especially those considered at risk nutritionally, assess the diet for:

☐ Total energy

☐ Protein

☐ Calcium, phosphorus, magnesium, iron, and zinc

☐ Folate and vitamin B_{12}

☐ Vitamin D

For infants, note:

☐ Method of feeding (breastfeeding, formula, or both)

☐ Frequency and duration of breastfeeding

☐ Amount of infant formula

☐ Practice of putting infant to bed with bottle

☐ Solid foods the infant is fed, if any

☐ Amount of food the infant is fed

Anthropometric Data

Measure baseline height and weight:

☐ Prepregnancy weight

☐ Infant birthweight

Reassess weight at each medical checkup and determine whether gains are appropriate. Note:

☐ Weight gain during pregnancy

☐ Gestational age

☐ Weight, length, and head circumference of infants

Laboratory Tests

Monitor the following laboratory tests for pregnant women:

☐ Hemoglobin, hematocrit, or other tests of iron status

☐ Blood glucose

Monitor the following laboratory tests for infants:

☐ Blood glucose of infants born to mothers with gestational diabetes

☐ Results of tests for inborn errors

Physical Signs

Blood pressure measurement is a routine measurement in physical exams, but is especially important for pregnant women. Look for physical signs of:

☐ Iron deficiency

☐ Edema

☐ Protein-energy malnutrition

☐ Folate deficiency

SELF CHECK

1. The most important single predictor of an infant's future health and survival is:
   a. the infant's birthweight.
   b. the infant's iron status at birth.
   c. the mother's weight at delivery.
   d. the mother's prepregnancy weight.

2. A mother's prepregnancy nutrition is important to a healthy pregnancy because it determines the development of:
   a. the largest baby possible.
   b. adequate maternal iron stores.
   c. an adequate fat supply for the mother.
   d. healthy support tissues—the placenta, amniotic sac, umbilical cord, and uterus.

3. A pregnant woman needs an extra 450 calories above the allowance for nonpregnant women during which trimester(s)?
   a. first
   b. second
   c. third
   d. first, second, and third

4. Two nutrients needed in large amount during pregnancy for rapid cell proliferation are:
   a. vitamin B_{12} and vitamin C.
   b. calcium and vitamin B_6.
   c. folate and vitamin B_{12}.
   d. copper and zinc.

5. For a woman who is at the appropriate weight for height and is carrying a single fetus, the recommended weight gain during pregnancy is:
   a. 40 to 60 pounds.
   b. 25 to 35 pounds.
   c. 10 to 20 pounds.
   d. 20 to 40 pounds.

6. Rewards of physical activity during pregnancy may include:
   a. weight loss.
   b. decreased incidence of pica.
   c. relief from morning sickness.
   d. reduced stress and easier labor.

7. Breast milk is recommended for the first 12 months of life because it offers complete nutrition and _____ to the infant.
   a. fluoride
   b. fructose
   c. immunological protection
   d. pica

8. An acceptable substitute for breast milk during the first year is:
   a. low-fat cow's milk.
   b. apple juice.
   c. water.
   d. iron-fortified infant formula.

9. Indications of readiness for solid foods include:
   a. the infant cries a lot.
   b. the infant is two months old.
   c. the infant is not sleeping through the night.
   d. the infant can sit with support and can control head movements.

10. During the first year of life, the most important step to undertake to encourage healthy eating is to:
    a. give food as a reward or for comfort.
    b. give sweets as a reward for eating vegetables.
    c. introduce a variety of nutritious foods in an inviting way.
    d. restrict fat to less than 30 percent of energy intake.

Answers to these questions can be found in Appendix H.

CLINICAL APPLICATIONS

1. Consider the different factors in a pregnant woman's history that can affect her nutrition status and the outcome of her pregnancy. Describe what steps you would take to remedy potential problems for the following clients:

   a. A 15-year-old adolescent of low socioeconomic status is in her first trimester of pregnancy. She began the pregnancy at a normal, healthy weight, but her weight gain during pregnancy so far has been less than expected. Her favorite beverages are soft drinks; her favorite foods are French fries and boxed macaroni and cheese.

   b. A lactose-intolerant, 22-year-old pregnant woman has been eating a vegan diet for the past year or so. She began the pregnancy slightly underweight (BMI = 18.4), but her weight gain has been adequate and consistent during the four months of her pregnancy. She complains of feeling tired all the time.

2. What information would you give to a pregnant woman who is considering breastfeeding her infant, but isn't quite sure why she should?

3. The parents of a two-month-old infant have been told by the child's grandparents that they should introduce solid foods to help the baby sleep through the night. What advice would you give the parents?

NUTRITION ON THE NET

For further study of the topics in this chapter, access these websites.

Find updates and quick links to these and other nutrition-related sites at our website: **www.wadsworth.com/nutrition**

Visit the pregnancy and child health center of the Mayo Clinic: **www.mayohealth.org**

Learn more about having a healthy infant and about birth defects from the March of Dimes: **www.modimes.org**

Learn more about neural tube defects from the Spina Bifida Association of America: **www.sbaa.org**

Search for "birth defects," "pregnancy," "adolescent pregnancy," "maternal and infant health," and "breastfeeding" at the U.S. Government health information site: **www.healthfinder.gov**

Search for "pregnancy" at the American Dietetic Association site: **www.eatright.org**

Learn more about the WIC program search for "WIC" at this site: **www.usda.gov**

Visit the American College of Obstetricians and Gynecologists: **www.acog.org**

Learn more about gestational diabetes from the American Diabetes Association: **www.diabetes.org**

Learn more about breastfeeding from LaLeche League International: **www.lalecheleague.org**

Read *A Woman's Guide to Breastfeeding* at the American Academy of Pediatrics site: **www.aap.org**

NOTES

[1] M. S. Kramer, The epidemiology of adverse pregnancy outcomes: An overview, *Journal of Nutrition* 133 (2003): 1592S–1596S.

[2] L. Adair and D. Dahly, Developmental determinants of blood pressure in adults, *Annual Review of Nutrition* 25 (2005): 407–434; C. M. Law and coauthors, Fetal, infant, and childhood growth and adult blood pressure: A longitudinal study from birth to 22 years of age, *Circulation* 105 (2002): 1088–1092; W. Palinski and C. Napoli, The fetal origins of atherosclerosis: Maternal hypercholesterolemia, and cholesterol-lowering or antioxidant treatment during pregnancy influence in utero programming and postnatal susceptibility to atherogenesis, *The FASEB Journal*, 16 (2002): 1348–1360.

[3] M. Hack and coauthors, Outcomes in young adulthood for very-low-birthweight infants, *New England Journal of Medicine* 346 (2002): 149–157.

[4] J. A. Martin and coauthors, Annual summary of vital statistics—2003, *Pediatrics* 115 (2005): 619–634.

[5] Kramer, 2003.

[6] U. Ramakrishnan, Nutrition and low birth weight: From research to practice, *American Journal of Clinical Nutrition* 79 (2004): 17–21.

[7] M. L. Watkins and coauthors, Maternal obesity and risk for birth defects, *Pediatrics* 111 (2003): 1152–1158.

[8] Position of the American Dietetic Association: Nutrition and lifestyle for a healthy pregnancy outcome, *Journal of the American Dietetic Association* 102 (2002): 1479–1490: J. M. Baeten, E. A. Bukusi, and M. Lambe, Pregnancy complications and outcomes among overweight and obese nulliparous women, *American Journal of Public Health* 91 (2001): 436–440.

[9] Watkins and coauthors, 2003.

[10] M. C. Lacroix and coauthors, Placental growth hormones, *Endocrine* 1 (2002): 73–79.

[11] A. Singhal and coauthors, Programming of lean body mass: A link between birth weight, obesity, and cardiovascular disease? *American Journal of Clinical Nutrition* 77 (2003): 726–730; B. E. Birgisdottir and coauthors, Size at birth and glucose intolerance in a relatively genetically homogeneous, high-birth weight population, *American Journal of Clinical Nutrition* 76 (2002): 399–403; T. W. McDade and coauthors, Prenatal undernutrition, postnatal environments, and antibody response to vaccination in adolescence, *American Journal of Clinical Nutrition* 74 (2001): 543–548.

[12] M. F. Picciano, Pregnancy and lactation: Physiological adjustments, nutritional requirements and the role of dietary supplements, *Journal of Nutrition* 133 (2003): 1997S–2002S.

[13] Standing Committee on the Scientific Evaluation of Dietary Reference Intakes, Food and Nutrition Board, Institute of Medicine, *Dietary Reference Intakes for Energy, Carbohydrate, Fiber, Fat, Fatty Acids, Cholesterol, Protein, and Amino Acids* (Washington, D.C.: National Academies Press, 2005), pp. 185–194.

[14] W. C. Heird and A. Lapillonne, The role of essential fatty acids in development, *Annual Review of Nutrition* 25 (2005): 549–571; S. J. Otto and coauthors, Changes in the maternal essential fatty acid profile during early pregnancy and the relation of the profile to diet, *American Journal of Clinical Nutrition* 73 (2001): 302–307.

[15] T. J. Mathews, Trends in spina bifida and anencephalus in the United States, 1991–2001, *National Center for Health Statistics Health and Stats*, **www.cdc.gov/nchs/products/pubs/pubd/hestats/spine-anen.htm**, site visited February 28, 2006.

[16] Centers for Disease Control and Prevention, Spina bifida and anencephaly before and after folic acid mandate—United States, 1995–1996 and 1999–2000, *Morbidity and Mortality Weekly Report* 53 (2004): 362–365; E. A. Yetley and J. I. Rader, Modeling the level of fortification and post-fortification assessments: U.S. experience, *Nutrition Reviews* 62 (2004): S50–S59.

[17] L. B. Bailey and R. J. Berry, Folic acid supplementation and the occurrence of congenital heart defects, orofacial clefts, multiple births, and miscarriage, *American Journal of Clinical Nutrition* 81 (2005): 1213S–1217S.

[18] K. O'Brien, Maternal and fetal mineral partitioning: Whose needs predominate? *Nutrition Today* 40 (2005): 130–137.

[19] K. O'Brien and coauthors, Maternal iron status influences iron transfer to the fetus during the third trimester of pregnancy, *American Journal of Clinical Nutrition* 77 (2003): 924–930.

[20] T. O. Scholl, Iron status during pregnancy: Setting the stage for mother and infant, *American Journal of Clinical Nutrition* 81 (2005): 1218S–1222S; J. L. Beard, Effectiveness and strategies of iron supplementation during pregnancy, *American Journal of Clinical Nutrition* 71 (2000): 1288S–1294S.

[21] J. C. King, Determinants of maternal zinc status during pregnancy, *American Journal of Clinical Nutrition* 71 (2000): 1334S–1343S.

[22] Position of the American Dietetic Association, 2002.

[23] M. M. Black and coauthors, Special Supplemental Nutrition Program for Women, Infants, and Children participation and infants' growth and health: A multisite surveillance study, *Pediatrics* 114 (2004): 169–176.

[24] V. Oliveira and M. Prell, Sharing the economic burden: Who pays for WIC's infant formula? U.S. Department of Agriculture Economic Research Service, *Amber Waves*, available at **www.ers.usda.gov/AmberWaves/September04/Features/infantformula.htm**.

[25] Black and coauthors, 2004.

[26] Position of the American Dietetic Association, 2002.

[27] R. Artal and M. O'Toole, Guidelines of the American College of Obstetricians and Gynecologists for exercise during pregnancy and the postpartum period, *British Journal of Sports Medicine* 37 (2003): 6–12; American College of Obstetricians and Gynecologists, Exercise during pregnancy and postpartum period, ACOG Committee Opinion no. 267, *Obstetrics and Gynecology* 99 (2002): 171–173.

[28]R. W. Corbett, C. Ryan, and S. P. Weinrich, Pica in pregnancy: Does it affect pregnancy outcomes? *American Journal of Maternal and Child Nursing* 28 (2003): 183–189.

[29]R. J. Kaaja and I. A. Greer, Manifestations of chronic disease during pregnancy, *Journal of the American Medical Association* 294 (2005): 2751–2757; Position statement from the American Diabetes Association: Gestational diabetes mellitus, *Diabetes Care* 26 (2003): S103–S105.

[30]Report of the Expert Committee on the diagnosis and classification of diabetes mellitus, *Diabetes Care* 26 (2003): S5–S20.

[31]Position statement from the American Diabetes Association, 2003.

[32]Report of the Expert Committee on the diagnosis and classification of diabetes mellitus, 2003.

[33]C. G. Solomon and E. W. Seely, Preeclampsia—searching for the cause, *New England Journal of Medicine* 350 (2004): 641–642.

[34]R. A. de la Chica and coauthors, Chromosomal instability in amniocytes from fetuses of mothers who smoke, *Journal of the American Medical Association* 293 (2005): 1212–1222.

[35]G. P. Bogler and L. T. Kozlowski, Differential influence of maternal smoking on infant birth weight: Gene-environment interaction and targeted intervention, *Journal of the American Medical Association* 287 (2002): 241–242.

[36]American Academy of Pediatrics, Task Force on Sudden Infant Death Syndrome, The changing concept of sudden infant death syndrome: Diagnostic coding shifts, controversies regarding the sleeping environment, and new variables to consider in reducing risk, *Pediatrics* 116 (2005): 1245–1255; Centers for Disease Control and Prevention, Smoking during pregnancy—United States, 1990–2002, *Morbidity and Mortality Weekly Report* 53 (2004): 911–915.

[37]J. R. DiFranza, C. A. Aligne, and M. Weitzman, Prenatal and postnatal environmental tobacco smoke exposure and children's health, *Pediatrics* 113 (2004): 1007–1015; K. I. McMartin and coauthors, Lung tissue concentrations of nicotine in sudden infant death syndrome (SIDS), *Journal of Pediatrics* 140 (2002): 205–209.

[38]Centers for Disease Control and Prevention, Smoking during pregnancy—United States, 2004; S. J. Ventura and coauthors, Trends and variations in smoking during pregnancy and low birth weight: Evidence from the birth certificate, 1990–2000, *Pediatrics* 111 (2003): 1176–1180.

[39]Position of the American Dietetic Association, 2002.

[40]M. S. Scher, G. A. Richardson, and N. L. Day, Effects of prenatal cocaine/crack and other drug exposure on electroencephalographic sleep studies at birth and one year, *Pediatrics* 105 (2000): 39–48.

[41]L. T. Singer and coauthors, Cognitive outcomes of preschool children with prenatal cocaine exposure, *Journal of the American Medical Association* 291 (2004): 2448–2456; L. T. Singer and coauthors, Cognitive and motor outcomes of cocaine-exposed infants, *Journal of the American Medical Association* 287 (2002): 1952–1960; E. S. Bandstra and coauthors, Intrauterine growth of full-term infants: Impact of prenatal cocaine exposure, *Pediatrics* 108 (2001): 1309–1319.

[42]P. W. Davidson, G. J. Myers, and B. Weiss, Mercury exposure and child development outcomes, *Pediatrics* 113 (2004): 1023–1029; S. E. Schober and coauthors, Blood mercury levels in US children and women of childbearing age, 1999–2000, *Journal of the American Medical Association* 289 (2003): 1667–1674; A. Gomaa and coauthors, Maternal bone lead as an independent risk factor for fetal neurotoxicity: A prospective study, *Pediatrics* 110 (2002): 110–118.

[43]FDA and EPA announce the revised consumer advisory on methylmercury in fish, *News*, 2004, available from **www.cfsan.fda.gov**.

[44]Center for the Evaluation of Risks to Human Reproduction, **http://cerhr.niehs.nih**.

[45]Position of the American Dietetic Association: Use of nutritive and nonnutritive sweeteners, *Journal of the American Dietetic Association* 104 (2004): 255–275.

[46]S. Cnattingius and coauthors, Caffeine intake and the risk of first-trimester spontaneous abortion, *New England Journal of Medicine* 343 (2000): 1839–1845; M. A. Klebanoff and coauthors, Maternal serum paraxanthine, a caffeine metabolite, and the risk of spontaneous abortion, *New England Journal of Medicine* 341 (1999): 1639–1644.

[47]H. E. Hoyme and coauthors, A practical approach to diagnosis of fetal alcohol spectrum disorders: Clarification of the 1996 Institute of Medicine criteria, *Pediatrics* 115 (2005): 39–47.

[48]Centers for Disease Control, Fetal alcohol syndrome, **www.cdc .gov/ncbddd/fas/fasask.htm,** site visited February 28, 2006.

[49]J. Tsai and R. L. Floyd, Alcohol consumption among women who are pregnant or who might become pregnant—United States, 2002, *Morbidity and Mortality Weekly Report* 53 (2004): 1178–1179.

[50]T. S. Naimi and coauthors, Binge drinking in the preconception period and the risk of unintended pregnancy: Implications for women and their children, *Pediatrics* 111 (2003): 1136–1141.

[51]U. Kesmodel and coauthors, Moderate alcohol intake during pregnancy and the risk of stillbirth and death in the first year of life, *American Journal of Epidemiology* 155 (2002): 305–312.

[52]Committee on Substance Abuse and Committee on Children with Disabilities, American Academy of Pediatrics, Fetal alcohol syndrome and alcohol-related neurodevelopmental disorders, *Pediatrics* 106 (2000): 358–361.

[53]Committee on Substance Abuse and Committee on Children with Disabilities, 2000.

[54]J. D. Klein and the Committee on Adolescence, American Academy of Pediatrics, Adolescent pregnancy: Current trends and issues, *Pediatrics* 116 (2005): 281–286.

[55]D. S. Elfenbein and M. E. Felice, Adolescent pregnancy, *Pediatric Clinics of North America* 50 (2003): 781–800, viii.

[56]D. J. Hunt and coauthors, Effects of nutrition education programs on anthropometric measurements and pregnancy outcomes of adolescents, *Journal of the American Dietetic Association* 102 (2002): S100–S102.

[57]American Academy of Pediatrics, Policy statement: Breastfeeding and the use of human milk, *Pediatrics* 115 (2005): 496–506.

[58]Position of the American Dietetic Association: Promoting and supporting breastfeeding, *Journal of the American Dietetic Association* 105 (2005): 810–818.

[59]American Academy of Pediatrics, 2005; Position of the American Dietetic Association, 2005.

[60]G. Kac and coauthors, Breastfeeding and postpartum weight retention in a cohort of Brazilian women, *American Journal of Clinical Nutrition* 79 (2004): 487–493; L. N. Haiek and coauthors, Postpartum weight loss and infant feeding, *Journal of the American Board of Family Practice* 14 (2001): 85–94.

[61]M. A. McCrory, Does dieting during lactation put infant growth at risk? *Nutrition Reviews* 59 (2001): 18–27; C. A. Lovelady and coauthors, The effect of weight loss in overweight, lactating women on the growth of their infants, *New England Journal of Medicine* 342 (2000): 449–453.

[62]L. M. Paton and coauthors, Pregnancy and lactation have no long-term deleterious effect on measures of bone mineral in healthy women: A twin study, *American Journal of Clinical Nutrition* 79 (2003): 707–714.

[63]American Academy of Pediatrics, 2005.

[64]S. Ito and A. Lee, Drug excretion into breast milk—overview, *Advanced Drug Delivery Reviews* 55 (2003): 617–627; Committee on Drugs, American Academy of Pediatrics, The transfer of drugs and other chemicals into human milk, *Pediatrics* 108 (2001): 776–789.

[65]American Academy of Pediatrics, 2005.

[66]American Academy of Pediatrics, Policy statement, Controversies concerning vitamin K and the newborn, *Pediatrics* 112 (2003): 191–192.

[67]Committee on Nutrition, American Academy of Pediatrics, *Pediatric Nutrition Handbook,* 5th ed., ed. R. E. Kleinman (Elk Grove Village, Ill.: American Academy of Pediatrics, 2004), pp.110–111.

[68]American Academy of Pediatrics, 2005.

[69]J. D. Carver, Advances in nutritional modifications of infant formulas, *American Journal of Clinical Nutrition* 77 (2003): 1550S–1554S.

[70]W. W. Koo, Efficacy and safety of docosahexaenoic acid and arachidonic acid addition to infant formulas: Can one buy better vision and intelligence? *Journal of the American College of Nutrition* 22 (2003): 101–107; E. E. Birch and coauthors, A randomized controlled trial of long-chain polyunsaturated fatty acid supplementation of formula in term infants after weaning at 6 wk of age, *American Journal of Clinical Nutrition* 75 (2002): 570–580.

[71]N. Auestad and coauthors, Growth and development in term infants fed long-chain polyunsaturated fatty acids: A double-masked, randomized, parallel, prospective, multivariate study, *Pediatrics* 108 (2001): 372–381.

[72]E. E. Birch and coauthors, Visual maturation of term infants fed long-chain polyunsaturated fatty acid–supplemented or control formula for 12 mo, *American Journal of Clinical Nutrition* 81 (2005): 871–879; Birch and coauthors, 2002.

[73]B. Lonnerdal, Nutritional and physiologic significance of human milk proteins, *American Journal of Clinical Nutrition* 77 (2003): 1537S–1543S.

[74]L. M. Gartner, F. R. Greer, and the Section on Breastfeeding and Committee on Nutrition, Prevention of rickets and vitamin D deficiency: New guidelines for vitamin D intake, *Pediatrics* 111 (2003): 908–910.

[75]Gartner, Greer, and the Section on Breastfeeding and Committee on Nutrition, 2003.

[76]American Academy of Pediatrics, 2005; Position of the American Dietetic Association, 2005.

[77]D. S. Newburg, G. M. Ruiz-Palacios, and A. L. Morrow, Human milk glycans protect infants against enteric pathogens, *Annual Review of Nutrition* 25 (2005): 37–58.

[78]American Academy of Pediatrics, 2005; Position of the American Dietetic Association, 2005.

[79]M. W. Gillman and coauthors, Risk of overweight among adolescents who were breastfed as infants, *Journal of the American Medical Association* 285 (2001): 2461–2467.

[80]M. L. Hediger and coauthors, Association between infant breastfeeding and overweight in young children, *Journal of the American Medical Association* 285 (2001): 2453–2460.

[81]C. G. Owen and coauthors, Effect of infant feeding on the risk of obesity across the life course: A quantitative review of published evidence, *Pediatrics* 115 (2005): 1367–1377.

[82]R. M. Martin and coauthors, Breastfeeding and atherosclerosis: Intima-media thickness and plaques at 65-year follow-up of the Boyd Orr cohort, *Arteriosclerosis, Thrombosis, and Vascular Biology* 25 (2005): 1482–1488; C. G. Owen and coauthors, Infant feeding and blood cholesterol: A study in adolescents and a systematic review, *Pediatrics* 110 (2002): 597–608.

[83]M. C. Daniels and L. S. Adair, Breastfeeding influences cognitive development in Filipino children, *Journal of Nutrition* 135 (2005): 2589–2595; E. L. Mortensen and coauthors, The association between duration of breastfeeding and adult intelligence, *Journal of the American Medical Association* 287 (2002): 2365–2371.

[84]A. Jain, J. Concato, and J. M. Leventhal, How good is the evidence linking breastfeeding and intelligence? *Pediatrics* 109 (2002): 1044–1053.

[85]L. Seppo and coauthors, A follow-up study of nutrient intake, nutritional status, and growth in infants with cow milk allergy fed either a soy formula or an extensively hydrolyzed whey formula, *American Journal of Clinical Nutrition* 82 (2005): 140–145; Committee on Nutrition, American Academy of Pediatrics, Formula feeding of term infants, in R. E. Kleinman, ed., *Pediatric Nutrition Handbook,* 5th ed. (Elk Grove Village, Ill.: American Academy of Pediatrics, 2004), pp. 87–97; Committee on Nutrition, American Academy of Pediatrics, Hypoallergenic infant formulas, *Pediatrics* 106 (2000): 346–349.

[86]Committee on Nutrition, American Academy of Pediatrics, Formula feeding of term infants, 2004; Committee on Nutrition, American Academy of Pediatrics, Iron fortification of infant formulas, *Pediatrics* 104 (1999): 119–123.

[87]Committee on Nutrition, American Academy of Pediatrics, *Pediatric Nutrition Handbook,* 2004, p. 111.

[88]Committee on Nutrition, American Academy of Pediatrics, *Pediatric Nutrition Handbook,* 2004, p. 105.

[89]Committee on Nutrition, American Academy of Pediatrics, Complementary feeding, in R. E. Kleinman, ed., *Pediatric Nutrition Handbook,* 5th ed. (Elk Grove Village, Ill.: American Academy of Pediatrics, 2004), pp. 103–115.

[90]Committee on Nutrition, American Academy of Pediatrics, Complementary feeding, 2004; J. D. Skinner and B. R. Carruth, A longitudinal study of children's juice intake and growth: The juice controversy revisited, *Journal of the American Dietetic Association* 101 (2001): 432–437; Committee on Nutrition, American Academy of Pediatrics, The use and misuse of fruit juice in pediatrics, *Pediatrics* 107 (2001): 1210–1213.

Encouraging Successful Breastfeeding

As discussed in Chapter 10, breastfeeding offers benefits to both mother and infant. The American Academy of Pediatrics (AAP), the American Dietetic Association, and the U.S. Department of Health and Human Services all advocate breastfeeding as the preferred means of infant feeding.[1] Promotion of breastfeeding is an integral part of the WIC program's nutrition education component. During the late 1980s, breastfeeding was on the decline after reaching a high of about 60 percent of mothers in 1984. National efforts to promote breastfeeding seem to be working, at least to some extent: an encouraging trend of breastfeeding is emerging, with more than 70 percent of mothers initiating breastfeeding in 2003.[2] Nevertheless, only about one in three infants is still being breastfed at six months of age. The AAP and many other health organizations recommend exclusive breastfeeding for the first six months of life.[3] Exclusive breastfeeding is defined as an infant's consumption of human milk with no supplementation of any kind (no water, no other type of milk, no juice, and no other foods) except for vitamins, minerals, and medications. The AAP recommends that breastfeeding continue for at least a year and thereafter for as long as mutually desired.[4] In the United States, only about one in five infants is still breastfeeding at one year of age. Increasing the rates of breastfeeding initiation and duration is one of the goals of *Healthy People 2010*:

> Increase the proportion of mothers who breastfeed immediately after birth, for the first six months, and preferably, through the infant's first year of life. Increase the proportion of mothers who breastfeed exclusively.[5]

Despite the trend toward increasing breastfeeding, the percentage of mothers choosing to breastfeed their infants and continuing to do so still falls short of goals.

Why don't more women choose to breastfeed their infants?

Many experts cite two major deterrents: public advertising of infant formula, and the medical community's failure to encourage breastfeeding. As an example of the medical lack of encouragement, some hospitals routinely separate mother and infant soon after birth. The child's first feeding then comes from the bottle rather than the breast. Furthermore, many hospitals send new mothers home with free samples of infant formula. The World Health Organization opposes this practice because it sends a misleading message that medical authorities favor infant formula over breast milk for infants. Even in hospitals where women are encouraged to breastfeed and are supported in doing so, little, if any, assistance is available after hospital discharge when many breastfeeding women still need assistance. Up to half of mothers who initially breastfeed their infants stop within a month—seemingly due to lack of knowledge.

TABLE NP10-1 Ten Steps to Successful Breastfeeding

To promote breastfeeding, every maternity facility should:
• Develop a written breastfeeding policy that is routinely communicated to all health care staff.
• Train all health care staff in the skills necessary to implement the breastfeeding policy.
• Inform all pregnant women about the benefits and management of breastfeeding.
• Help mothers initiate breastfeeding within half an hour of birth.
• Show mothers how to breastfeed and how to maintain lactation, even if they need to be separated from their infants.
• Give newborn infants no food or drink other than breast milk, unless medically indicated.
• Practice rooming-in, allowing mothers and infants to remain together 24 hours a day.
• Encourage breastfeeding on demand.
• Give no artificial nipples or pacifiers to breastfeeding infants.
• Foster the establishment of breastfeeding support groups and refer mothers to them at discharge from the facility.

Sources: U.S. Department of Health and Human Services, Office on Women's Health, *Breastfeeding: HHS Blueprint for Action on Breastfeeding* (Washington, D.C.: U.S. Department of Health and Human Services, 2000); United Nations Children's Fund and World Health Organization, *The UNICEF/Baby-Friendly Hospital Initiative: Ten Steps to Successful Breastfeeding* (New York: UNICEF, 1992).

Women who receive early and repeated breastfeeding information and support breastfeed their infants longer than other women do. Information and instruction are especially important during the *prenatal* period when most women decide whether to breastfeed or to feed formula. Nurses and other health care professionals can play a crucial role in encouraging successful breastfeeding by offering women adequate, accurate information about breastfeeding that permits them to make informed choices. Table NP10-1 lists ten steps maternity facilities and health care professionals can take to promote successful breastfeeding among new mothers.

If breastfeeding is a natural process, what do mothers need to learn?

Although lactation is an automatic physiological process, breastfeeding requires some learning. This learning is most successful in a supportive environment. It begins with preparatory steps taken before the infant is born.

What are these preparatory steps?

Toward the end of pregnancy and throughout lactation, a woman who intends to breastfeed should stop using soap

FIGURE NP1 0-1 Infant's Grasp on Mother's Breast

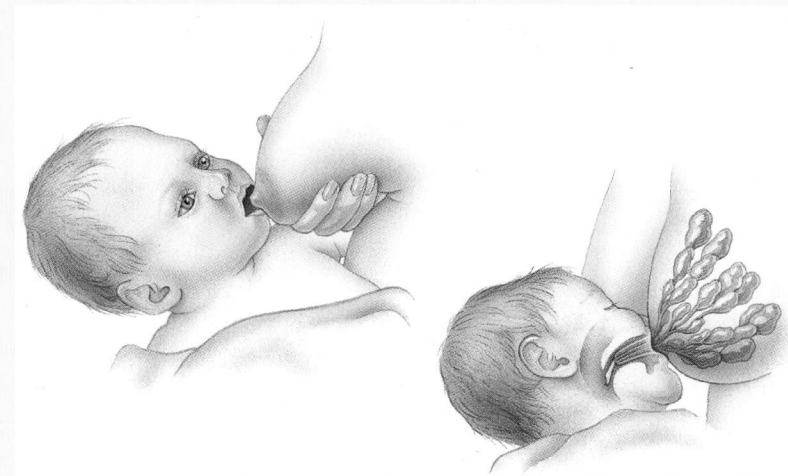

The mother supports the breast with her fingers and thumb behind the areola to present the nipple to the infant. Once the infant latches onto the breast, the infant's lips and gums pump the areola, releasing milk from the mammary glands into the milk ducts that lie beneath the areola.

and lotions on her breasts. The natural secretions of the breasts themselves lubricate the nipple area best. A woman who plans to breastfeed should also acquire at least two nursing bras before her infant is born. The bras should provide good support and have drop-flaps so that either breast can be freed for nursing.

How soon after birth should breastfeeding start?

As soon as possible. Immediately after the delivery, for a short period, the infant is intensely alert and intent on suckling. This is the ideal time for the first breastfeeding and facilitates successful lactation.

What does the new mother need to know in order to continue breastfeeding her infant successfully?

She needs to learn how to relax and position herself so that she and the infant will be comfortable and the infant can breathe freely while nursing. She also needs to understand that infants have a **rooting reflex** that makes them turn toward any touch on the face. (The glossary on p. 296 defines this and other relevant terms.) Consequently, she should touch the infant's cheek to her nipple so that the infant will turn the right way to nurse. The mother can then place four fingers under the breast and her thumb on top to support the breast and present the nipple to the infant. The mother's fingers and thumb should be behind the areola, the colored ring around the nipple so as not to interfere with the infant latching onto the breast (see Figure NP10-1). With the breast supported, the mother tickles the infant's lips with the breast until the infant's mouth opens wide. The mother can then gently bring the infant forward onto the breast. The nipple must rest well back on the infant's tongue so that the infant's

gums will squeeze on the glands that release the milk and swallowing will be effortless. To break the suction, if necessary, the mother can slip a finger between the infant's mouth and her breast.

Does it hurt to have the infant sucking so hard on the breast?

Breastfeeding should not be painful if the infant is positioned correctly. The mother has a **letdown reflex** that forces milk to the front of her breast when the infant begins to nurse, virtually propelling the milk into the infant's mouth. Letdown is necessary for the infant to obtain milk easily, and the mother needs to relax for letdown to occur. The mother who assumes a comfortable position in an environment without interruptions will find it easiest to relax.

How long should the infant be allowed to nurse at each feeding?

Although the infant sucks half the milk from the breast within the first 2 minutes, and 80 to 90 percent of it within 4 minutes, sucking on each breast for 10 to 15 minutes is encouraged. The sucking itself, as well as the complete removal of milk from the breast, stimulates the mammary glands to produce milk for the next nursing session. Successive sessions should start on alternate breasts to ensure that each breast is emptied regularly. This pattern maintains the same supply and demand for each breast and thus prevents either breast from overfilling.

Infants should be fed "on demand" and not be held to a rigid schedule. The breastfed infant may average 8 to 12 feedings per 24-hour period during the first month or so. Once the mother's milk supply is well established and the infant's capacity has increased, the intervals between feedings will become longer.

What if a mother wants to skip one or two feedings daily—for example, because she works outside the home?

The mother can express breast milk into a bottle ahead of time, freeze the breast milk, and, when needed, substitute the expressed breast milk for a nursing session. Breast milk can be kept refrigerated for 48 hours or frozen. Frozen milk can be kept for one month in a freezer attached to a refrigerator or for three to six months in a zero-degree deep freezer.[6]

The mother can hand express her breast milk or use one of several different breast pumps available. The bicycle-horn type of manual breast pump is difficult to keep clean and is not recommended. Cylinder-type manual pumps or electric breast pumps are safer and are also more efficient. Alternatively, a mother can substitute formula for those missed feedings and continue to breastfeed at other times.

What about problems associated with breastfeeding such as sore nipples or infection of the breast?

Most problems associated with breastfeeding can be resolved. Many mothers experience sore nipples during the initial days of breastfeeding. Sore nipples need to be treated kindly, but nursing can continue. Improper feeding position is a frequent cause of sore nipples: the mother should make sure the infant is taking the entire nipple and part of the areola onto the tongue. She should nurse on the less sore breast first to get letdown going while the infant is sucking hardest; then she can switch to the sore breast. Between times, she should expose her nipples to light and air to heal them.

Before lactation is well established, when the schedule changes, or when a feeding is missed, the breasts may become full and hard—an uncomfortable condition known as **engorgement.** The infant cannot grasp an engorged nipple and so cannot provide relief by nursing. A gentle massage or warming the breasts with a cloth soaked in warm water or in a shower helps to initiate letdown and to release some of the accumulated milk; then the mother can pump out some of her milk and allow the infant to nurse.

Infection of the breast, known as **mastitis,** is best managed by continuing to breastfeed. By drawing off the milk, the infant helps to relieve pressure in the infected area. The infant is safe because the infection is between the milk-producing glands, not inside them.

Even if everything is going smoothly, the nursing mother should ideally have enough help and support so that she can rest in bed a few hours each day for the first week or so. Successful breastfeeding requires the support of all those who care. This, plus adequate nutrition, ample fluids, fresh air, and physical activity, will do much to enhance the well-being of mother and infant.

Glossary of Breastfeeding Terms

engorgement: overfilling of the breasts with milk.
letdown reflex: the reflex that forces milk to the front of the breast when the infant begins to nurse.
mastitis: infection of a breast.
rooting reflex: a reflex that causes an infant to turn toward whichever cheek is touched, in search of a nipple.

Notes

[1] American Academy of Pediatrics, Policy statement: Breastfeeding and the use of human milk, *Pediatrics* 115 (2005): 496–506; Position of the American Dietetic Association: Promoting and supporting breastfeeding, *Journal of the American Dietetic Association* 105 (2005): 810–818; U.S. Department of Health and Human Services, Office on Women's Health, *Breastfeeding: HHS Blueprint for Action on Breastfeeding* (Washington, D.C.: U.S. Department of Health and Human Services, 2000).

[2] Centers for Disease Control and Prevention, 2003 National Immunization Survey, available at **www.cdc.gov/breastfeeding/NIS_data,** site visited January 16, 2006.

[3] American Academy of Pediatrics, 2005.

[4] American Academy of Pediatrics, 2005.

[5] *Healthy People 2010: National Health Promotion and Disease Prevention Objectives* (Washington, D.C.: U.S. Department of Health and Human Services, 2000).

[6] Breastfeeding beyond infancy, in American Academy of Pediatrics, *New Mother's Guide to Breastfeeding,* ed. J. Y. Meek and S. Tippins (New York: Bantam Books, 2002), pp. 158–188.

Nutrition through the Life Span: Childhood and Adolescence

CHAPTER 11

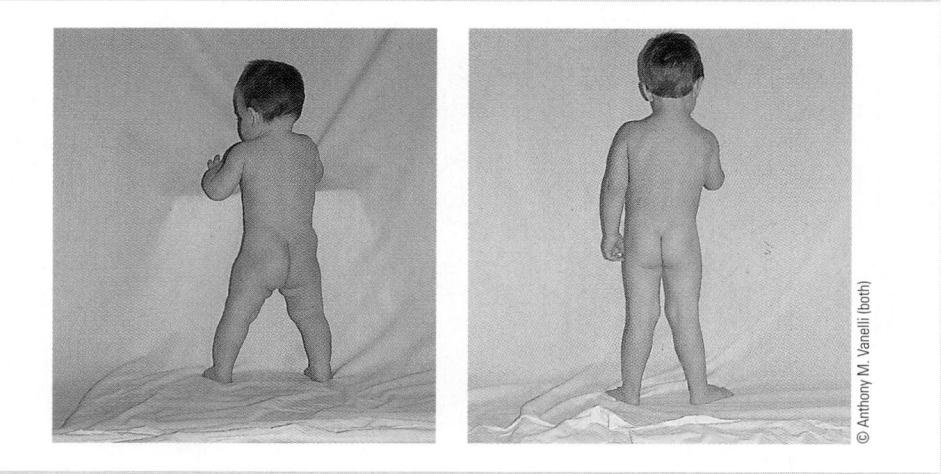

FIGURE 11-1 Body Shape of a One-Year-Old and a Two-Year-Old Compared
The body shape of a one-year-old (left) changes dramatically by age two (right). The two-year-old has lost much of the baby fat; the muscles (especially in the back, buttocks, and legs) have firmed and strengthened; and the leg bones have lengthened.

Nutrient needs change throughout life, depending on rates of growth, activity, and many other factors. Nutrient needs also vary from individual to individual, but generalizations are possible and useful. Sound nutrition throughout childhood promotes normal growth and development; facilitates academic and physical performance; and helps prevent obesity, diabetes, heart disease, cancer, and other degenerative diseases in adulthood. As children enter the teen years, a foundation built by years of eating nutritious foods best prepares them to meet the upcoming demands of rapid growth.

Early and Middle Childhood

After the age of one, growth rate slows, but the body continues to change dramatically (see Figure 11-1). At one, infants have just learned to stand and toddle; by two, they walk confidently and are learning to run, jump, and climb. Nutrition and physical activity have helped them prepare for these new accomplishments by adding to the mass and density of their bone and muscle tissue. Thereafter, their bones continue to grow longer and their muscles to gain size and strength, though unevenly and more slowly, until adolescence.

ENERGY AND NUTRIENT NEEDS

An infant's appetite declines markedly around the first birthday, consistent with the slowed growth rate. Thereafter, the appetite fluctuates. At times children seem to be insatiable, and at other times they seem to live on air and water. Parents and other caregivers need not worry about this—a child will need and demand much more food during periods of rapid growth than during slow periods. The perfect regulation of appetite in children of normal weight guarantees that their food energy intakes will be right for each stage of growth.

Children's Appetites Many people mistakenly believe that they must "make" their children eat the right amounts of food, and children's erratic appetites often reinforce this belief. Although children's food energy intakes vary widely from meal to meal, total daily energy intake remains remarkably constant.[1] If children eat less at one meal, they eat more at the next, and vice versa.

Parents do, however, need to help children choose the right foods, and with overweight children, they may need to help more, as described later. Overweight children may not adjust their energy intakes appropriately and may disregard appetite-regulation signals and eat in response to external cues, such as television commercials.

Energy Individual children's energy needs vary widely, depending on their growth and physical activity. A one-year-old child needs approximately 800 kcalories a day; an active six-year-old needs twice as many kcalories a day. By age ten, an active child needs about 2000 kcalories a day. Total energy needs increase gradually with age, but energy needs per kilogram of body weight actually decline. Physically active children of any age need more energy because they expend more, and inactive children can become obese even when they eat less food than the average. Unfortunately, the prevalence of overweight in preschool children 2 to 5 years of age and adolescents 12 to 19 years of age has more than doubled over the past 30 years, and it has more than tripled for children 6 to 11 years of age.[2] An estimated 9 million U.S. children over the age of six are obese. Obesity poses hazards to the health of children both now and in the future (Nutrition in Practice 11 provides details). Strategies to prevent obesity in children must focus on balancing energy intake and energy expenditure. In other words, factors that influence both eating and physical activity must be addressed.

Some children, notably those adhering to a vegan diet, may have difficulty meeting their energy needs. Grains, vegetables, and fruits provide plenty of fiber, adding bulk, but may provide too few kcalories to support growth. Soy products, other legumes, and nut or seed butters offer more concentrated sources of energy to support optimal growth and development.[3]

Nutrients Steady growth during childhood necessitates a gradual increase in intakes of most nutrients. Nutrient recommendations cluster children into age groupings that reflect similarities in growth rate, biological changes, and hormone status (see inside front cover).

Ideally, children accumulate stores of nutrients before adolescence. Then, when they take off on the adolescent growth spurt and their nutrient intakes cannot keep pace with the demands of rapid growth, they can draw on the nutrient stores accumulated earlier. This is especially true of calcium; the denser the bones are in childhood, the better prepared they will be to support teen growth and still withstand the inevitable bone losses of later life.[4] Consequently, the way children eat influences their nutritional health during childhood, during their teen years, and for the rest of their lives.

Food Patterns for Children To provide all the needed nutrients, a child's meals and snacks should include a variety of foods from each food group—in amounts suited to the child's appetite and needs. Figure 11-2 (p. 300) shows MyPyramid designed for children 6 to 11 years of age and includes the recommended amounts of food for an 1800-kcalorie intake. Table 11-1 (p. 301) lists amounts of food for several kcalorie levels below 1800 kcalories, which are appropriate for most younger children and sedentary older children. Estimated daily kcalorie needs for children are shown in Table 11-2 (p. 301). Table 1-8 in Chapter 1 provides recommended daily amounts of foods from each group for higher kcalorie levels, which are appropriate for active older children.

Based on an assessment tool developed by the U.S. Department of Agriculture (USDA), the great majority of children two to nine years of age consume a diet ranked "poor" or "needs improvement."[5] Among two- to three-year-olds, 60 percent were found to have diets that "need improvement"; among four- to six-year-olds, 76 percent; and among seven- to nine-year-olds, 80 percent. The consequences of such diets may not be evident to the casual observer, but nutritionists

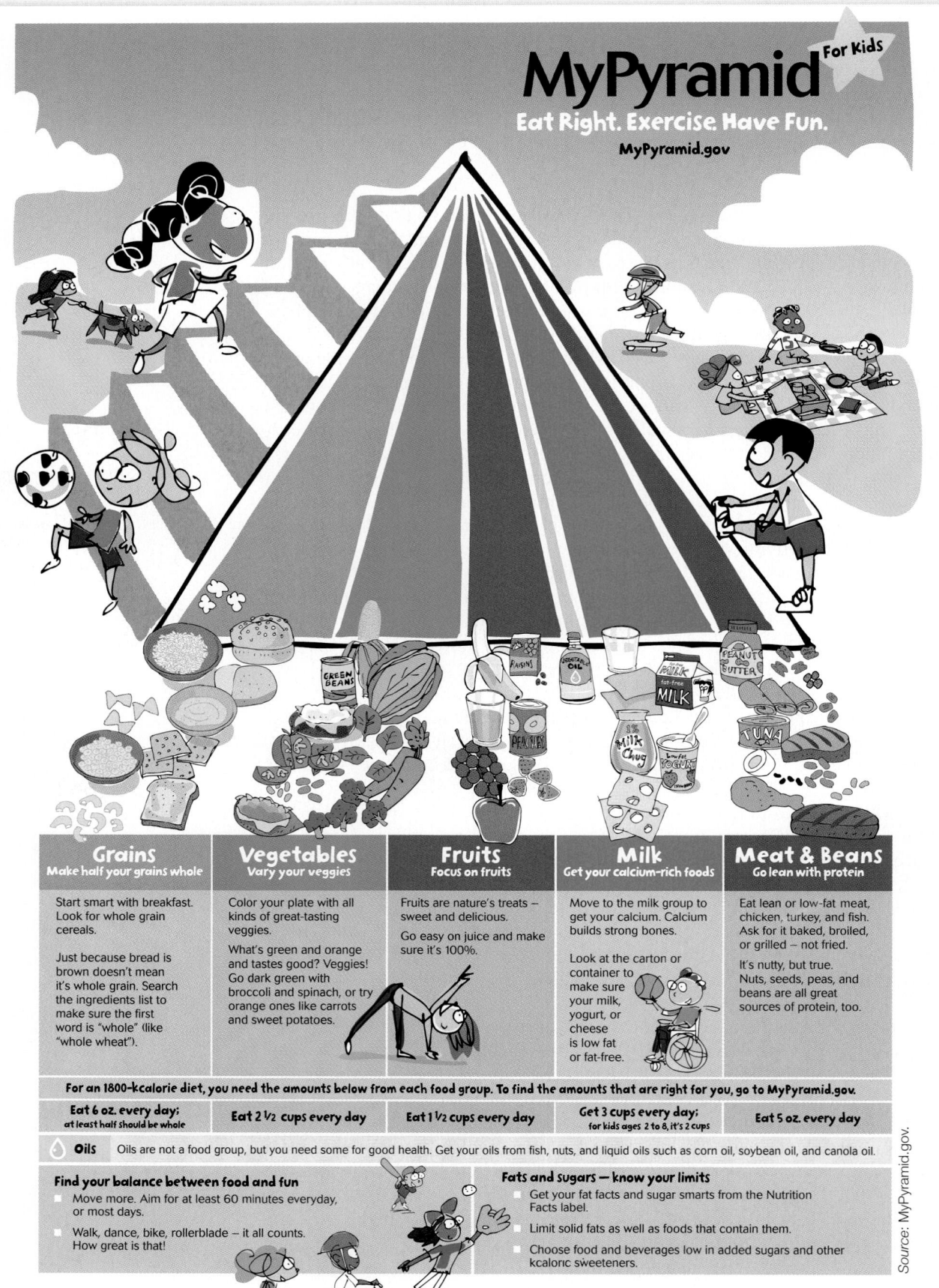

FIGURE 11-2 MyPyramid for Children

TABLE 11-1	Recommended Daily Amounts for Each Food Group (1000 to 1600 kCalories)			
Food Group	1000 kcal	1200 kcal	1400 kcal	1600 kcal
Fruit	1 c	1 c	1½ c	1½ c
Vegetables	1 c	1½ c	1½ c	2 c
Grains	3 oz	4 oz	5 oz	5 oz
Meat and legumes	2 oz	3 oz	4 oz	5 oz
Milk	2 c	2 c	2 c	3 c
Oils	3 tsp	3 tsp	3 tsp	4 tsp

Note: The discretionary kcalorie allowance for these patterns is about 100 kcalories.

TABLE 11-2		
Estimated Daily kCalorie Needs for Children		
Children	Sedentary[a]	Active[b]
2 to 3 yr	1000	1400
Females		
4 to 8 yr	1200	1800
9 to 13 yr	1600	2200
Males		
4 to 8 yr	1400	2000
9 to 13 yr	1800	2600

[a] *Sedentary* describes a lifestyle that includes only the activities typical of day-to-day life.

[b] *Active* describes a lifestyle that includes at least 60 minutes per day of moderate physical activity (equivalent to walking more than 3 miles per day at 3 to 4 miles per hour) in addition to the activities of day-to-day life.

know that nutrient deficiencies during growth often have far-reaching effects on physical and mental development.[6] Among nutrition concerns for U.S. children are inadequate intakes of calcium and fiber and excessive intakes of saturated fat.[7]

Children's Food Choices Parents and other caregivers can do much to foster the development of healthy eating habits in a child. The challenge is to deliver nutrients in the form of meals and snacks that are both nutritious and appealing so that children will learn to enjoy a variety of nutritious foods.

Candy, cola, and other concentrated sweets must be limited in children's diets. If such foods are permitted in large quantities, the only possible outcomes are nutrient deficiencies, obesity, or both. Children can't be trusted to choose nutritious foods on the basis of taste alone; the preference for sweets is innate, and children naturally gravitate to them. Overweight children, especially, need help in selecting nutrient-dense foods that will meet their nutrient needs within their energy allowances. Underweight children or active, normal-weight children can enjoy higher-kcalorie foods, but these should still be nutritious. Examples are ice cream and pudding in the milk group and whole-grain or enriched pancakes and crackers in the bread group.

MALNUTRITION IN CHILDREN

Most children in the United States and Canada are adequately nourished. Hunger and malnutrition are prevalent among some groups, however. Children in very low-income families, for example, are more likely to be hungry and malnourished. An estimated 13 million U.S. children fall under the broad definition of food insecurity—there are times when they do not know where their next meal is coming from or when it will come.[8]

Effects of Hunger Both short-term and long-term hunger exert negative effects on behavior and health. Short-term hunger, such as when a child misses a meal, impairs the child's ability to pay attention and to be productive. Hungry children are irritable, apathetic, and uninterested in their environment. Long-term hunger impairs growth and immune defenses. Food assistance programs such as the WIC program (discussed in Chapter 10) and the School Breakfast and National School Lunch Programs (discussed later in this chapter) are designed to protect against hunger and improve the health of children.

Reminder: *Food insecurity* is having limited or uncertain access to foods of sufficient quality or quantity to sustain a healthy and active life.

© 2001 PhotoDisc, Inc.

Healthy, well-nourished children are alert in the classroom and energetic at play.

Hunger and School Performance Children who eat nutritious breakfasts function better than their peers who do not. A nutritious breakfast is a central feature of a diet that meets the needs of children and supports their healthy growth and development.[9] When a child consistently skips breakfast or is allowed to choose sugary foods (candy or marshmallows) in place of nourishing ones (whole-grain cereals), the child will fail to get enough of several nutrients. Nutrients missed from a skipped breakfast won't be "made up" at lunch and dinner but will be left out completely that day.

Children who eat no breakfast are more likely to be overweight, perform poorly in tasks requiring concentration, have shorter attention spans, score lower on tests, and be tardy or absent more often than their well-fed peers.[10] Common sense dictates that it is unreasonable to expect anyone to learn and perform work when no fuel has been provided. By the late morning, discomfort from hunger may become distracting even if a child has eaten breakfast. Chronically underfed children suffer all the more. Teachers aware of the late-morning slump in their classrooms wisely request that a midmorning snack be provided; it improves classroom performance all the way to lunchtime.

Iron Deficiency and Behavior In U.S. children and adolescents, as in infants after six months, iron deficiency remains common despite iron fortification of foods and other programs to combat this deficiency. The best-known and most widespread effects of iron deficiency are its impacts on behavior. Most people are familiar with the role of iron in carrying oxygen in the blood. Iron also works as part of large molecules to release energy within cells and plays key roles in many molecules of the brain and nervous system. A lack of iron not only causes an energy crisis but also directly affects behavior, mood, attention span, and learning ability.

Iron deficiency is usually diagnosed by a deficit of iron in the *blood*, after the deficiency has progressed all the way to anemia. A child's *brain*, however, is sensitive to slightly lowered iron concentrations long before the blood deficits appear. Iron's effects are hard to distinguish from the effects of other factors in children's lives, but it is likely that iron deficiency manifests itself in a lowering of "motivation to persist in intellectually challenging tasks," a shortening of the attention span, and a reduction of overall intellectual performance.[11] Iron-deficient children perform less well on tests and have more conduct disturbances than their nonanemic classmates.

Preventing Iron Deficiency To avert iron deficiency, children's foods must deliver 7 to 10 milligrams of iron per day. To achieve this goal, milk intakes must be limited after infancy because milk is a poor source of iron. Children should receive enough milk products to ensure adequate calcium and riboflavin intakes, but no more. That means 2 cups of milk per day up to age eight, increasing to 3 cups per day from age nine on. After age two, if reduced-fat milk is used instead of whole milk, the saved kcalories can be invested in iron-rich foods such as lean meats, fish, poultry, eggs, and legumes. Whole-grain or enriched breads and cereals also contribute iron. Table 11-3 lists iron-rich foods children like.

TABLE 11-3	Iron-Rich Foods Children Like[a]

Breads, cereals, and grains

Canned macaroni (1/2 c)
Canned spaghetti (1/2 c)
Cream of wheat (1/4 c)
Fortified dry cereals (1 oz)[b]
Noodles, rice, or barley (1/2 c)
Tortillas (1 flour or whole wheat, 2 corn)
Whole-wheat, enriched, or fortified bread (1 slice)

Vegetables

Baked flavored potato skins (1/2 skin)
Cooked mung bean sprouts or snow peas (1/2 c)[c]
Cooked mushrooms (1/2 c)
Green peas (1/2 c)
Mixed vegetable juice (1 c)

Fruits

Canned plums (3 plums)
Cooked dried apricots (1/4 c)
Dried peaches (4 halves)
Raisins (1 tbs)

Meats and legumes

Bean dip (1/4 c)
Canned pork and beans (1/3 c)
Lean chopped roast beef or cooked ground beef (1 oz)
Liverwurst on crackers (1/2 oz)
Meat casseroles (1/2 c)
Mild chili or other bean/meat dishes (1/4 c)
Peanut butter and jelly sandwich (1/2 sandwich)
Sloppy joes (1/2 sandwich)

[a] Each serving provides at least 1 milligram of iron. Vitamin C–rich foods included with these snacks increase iron absorption.

[b] Some fortified breakfast cereals contain more than 10 milligrams of iron per half-cup serving (read the labels).

[c] Raw sprouts may pose a bacterial hazard to young children.

Other Nutrient Deficiencies Iron is only one of several dozen nutrients that can be displaced by a diet of nutrient-poor foods. Any of the other nutrients may be lacking as well, and the deficiencies of those nutrients may also cause both behavioral and physical symptoms (see Table 11-4). A child with behavioral symptoms of nutrient deficiencies may be irritable, aggressive, disagreeable, or sad and withdrawn. Such a child may be labeled "hyperactive," "depressed," or "unlikable," but in fact these traits may arise from simple, albeit marginal, malnutrition. Parents and medical practitioners often overlook the possibility that malnutrition may account for abnormalities of appearance and behavior. Any departure from normal healthy appearance and behavior is a sign of possible poor nutrition. In any such case, inspection of the child's diet by a registered dietitian or other qualified health care professional is in order. Any suspicion of dietary inadequacies, *no matter what other causes may be implicated*, should prompt steps to correct those inadequacies immediately.

LEAD POISONING IN CHILDREN

The damage caused by malnutrition may be compounded by environmental factors such as lead poisoning. A two-way interaction is typical: lead poisoning can cause an iron deficiency, and an iron deficiency can impair the body's defenses against lead absorption. Adequate calcium may slow lead's absorption or interfere with its toxic effects in the body. Like iron deficiency, mild lead toxicity has

TABLE 11-4	Physical Signs of Malnutrition in Children		
	Well-Nourished	**Malnourished**	**Possible Nutrient Deficiencies**
Hair	Shiny, firm in the scalp	Dull, brittle, dry, loose; falls out	Protein-energy malnutrition (PEM)
Eyes	Bright, clear pink membranes; adjust easily to light	Pale membranes; spots; redness; adjust slowly to darkness	Vitamin A, the B vitamins, zinc, and iron
Teeth and gums	No pain or caries, gums firm, teeth bright	Missing, discolored, decayed teeth. Gums bleed easily and are swollen and spongy.	Minerals and vitamin C
Face	Clear complexion without dryness or scaliness	Off-color, scaly, flaky, cracked skin	PEM, vitamin A, and iron
Glands	No lumps	Swollen at front of neck, cheeks	PEM and iodine
Tongue	Red, bumpy, rough	Sore, smooth, purplish, swollen	B vitamins
Skin	Smooth, firm, good color	Dry, rough, spotty; "sandpaper" feel or sores; lack of fat under skin	PEM, essential fatty acids, vitamin A, B vitamins, and vitamin C
Nails	Firm, pink	Spoon-shaped, brittle, ridged	Iron
Internal systems	Regular heart rhythm, heart rate, and blood pressure; no impairment of digestive function, reflexes, or mental status	Abnormal heart rate, heart rhythm, or blood pressure; enlarged liver, spleen; abnormal digestion; burning, tingling of hands, feet; loss of balance, coordination; mental confusion, irritability, fatigue	PEM and minerals
Muscles and bones	Muscle tone; posture, long bone development appropriate for age	"Wasted" appearance of muscles; swollen bumps on skull or ends of bones; small bumps on ribs; bowed legs or knock-knees	PEM, minerals, and vitamin D

Paint is the primary source of lead in children's lives.

nonspecific effects, including diarrhea, irritability, reduced ability of the blood to carry oxygen, and fatigue. The symptoms may be reversible if exposure stops soon enough. With higher levels of lead, the signs become more pronounced, yet pinpointing a cause may still be difficult. Children lose their general cognitive, verbal, and perceptual abilities and develop learning disabilities and behavior problems. Still more severe lead toxicity can cause irreversible nerve damage, paralysis, mental retardation, and death.

Lead toxicity is most prevalent among children under age six—more than 300,000 may have blood concentrations high enough to cause mental, behavioral, and other health problems.[12] Lead aggressively attacks fetuses, infants, and children because the body absorbs lead most efficiently during times of rapid growth. Blood concentrations of lead generally reach a peak in two-year-old children; age two is the typical time of exploring surroundings "hand to mouth" and ingesting lead-tainted dirt and paint chips. Children's behaviors and activities—putting their hands in their mouths, playing in dirt, and eating nonfood items—favor their chances of exposure to lead.[13]

A ban on leaded gasoline and reductions in other uses of lead have dramatically decreased the amount of lead in the environment.[14] The result has been a gratifying decline in children's blood lead concentrations since the 1970s. The decline follows exactly the reduction in the nation's use of leaded gasoline, leaded house paint, and lead-soldered food cans. A nationwide lead-monitoring system is now in place, and aggressive community programs are testing and treating children for lead poisoning. Despite such safeguards, exposure to lead still remains a threat to about 25 percent of U.S. children, especially those in low-income families who live in older houses that still contain deteriorating lead-based paint.[15] The accompanying "How to" suggests strategies to protect against lead poisoning.

FOOD ALLERGIES

Reminder: *Antigens* are substances foreign to the body that elicit the formation of antibodies or an inflammation reaction from immune system cells. Food antigens are usually glycoproteins (large proteins with glucose molecules attached).

Reminder: *Antibodies* are large proteins that are produced in response to antigens and then inactivate the antigens.

Food allergies are frequently blamed for physical and behavioral abnormalities in children, but just 6 percent of children are diagnosed with true food allergies.[16] Food allergies diminish with age, until in adulthood they affect about 1 or 2 percent of the population.[17] A true food allergy occurs when a whole food protein or other large molecule enters the body and elicits an immunologic response. (Recall that large molecules of food are normally dismantled in the digestive tract to smaller ones before absorption.) The body's immune system reacts to a food protein or other large molecule as it does to an antigen—by producing antibodies or other defensive agents. A problem that results from exposure to food substances, but does not involve the immune system, is known as **food intolerance.**

Asymptomatic and Symptomatic Allergies Allergies may have one or two components. They always involve antibodies; they may or may not involve symptoms. A person may produce antibodies without having any symptoms (known as *asymptomatic allergy*) or may produce antibodies and have symptoms (known as *symptomatic allergy*). A person who experiences symptoms without producing antibodies, however, does not have an allergy. This means that allergies have to be diagnosed by testing for antibodies. Once a food allergy has been diagnosed, therapy requires strict elimination of the offending food. Children with allergies, like all children, need all their nutrients, so it is important to include other foods that offer the same nutrients as the omitted foods.[18]

food allergies: adverse reactions to foods that involve an immune response; also called *food-hypersensitivity reactions*.

food intolerance: an adverse response to a food or food additive that does not involve the immune system.

anaphylactic (AN-ah-feh-LAC-tic) **shock:** a life-threatening whole-body allergic reaction to an offending substance.

Allergy Symptoms A symptomatic allergy will exhibit different symptoms depending on the location of the reaction. In the digestive tract, the allergy may cause nausea or vomiting; in the skin, it may cause rashes; and in the nasal passages and lungs, it may cause inflammation or asthma. A dangerous, generalized, all-systems shock reaction known as **anaphylactic shock** can also occur.

Immediate and Delayed Reactions Allergic reactions to food can occur with different timings, simply classified as immediate and delayed. In both, the antigen interacts immediately with the immune system, but symptoms may appear within minutes or after several (up to 24) hours. Identifying the food that causes an immediate allergic reaction is easy because symptoms correlate closely with the time of eating the food. Identifying the food that caused a delayed reaction is more difficult because the symptoms may not appear until a day after the offending food was eaten; by this time, many other foods will have been eaten, too, complicating the picture. Allergic reactions to single foods are common. Reactions to multiple foods are the exception, not the rule.

Anaphylactic Shock The life-threatening food allergy reaction of anaphylactic shock is most often caused by peanuts, tree nuts, milk, eggs, wheat, soybeans,

---

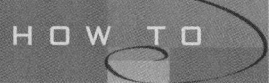

HOW TO   *Protect against Lead Toxicity*

Researchers simultaneously made three major discoveries about lead toxicity: lead poisoning has *subtle* effects, the effects are *permanent*, and they occur at *low levels of exposure*. The amount of lead recognized to cause harm is only 10 micrograms per 100 milliliters of blood. Some research shows that blood lead concentration *below* this amount may adversely affect children's scores on intelligence tests.[a] Consequently, consumers should take ultraconservative measures to protect themselves, and especially their infants and young children, from lead poisoning. The American Academy of Pediatrics and the Centers for Disease Control recommend screening in communities with a substantial number of houses built before 1950 and in those with a substantial number of children with elevated lead levels. In addition to screening children most likely to be exposed, pediatricians should alert all parents to the possible dangers of lead exposure and explain prevention strategies.

Preventive strategies include:

- In contaminated environments, keep small children from putting dirty or old painted objects in their mouths, and make sure children wash their hands before eating. Similarly, keep small children from eating any nonfood items. Lead poisoning has been reported in young children who have eaten crayons or pool cue chalk.
- Wet-mop floors and damp-sponge walls regularly. Children's blood lead levels decline when the homes they live in are cleaned regularly.
- Be aware that other countries do not have the same regulations protecting consumers against lead. Children have been poisoned by eating crayons made in China and drinking fruit juice canned in Mexico.
- Do not use lead-contaminated water to make infant formula.
- Once you have opened canned food, store it in a lead-free container to prevent lead migration into the food.

- Do not store acidic foods or beverages (such as vinegar or orange juice) in ceramic dishware or alcoholic beverages in pewter or crystal decanters.
- Many manufacturers are now making lead-safe products.[b] Old, handmade, or imported ceramic cups and bowls may contain lead and should not be used to heat coffee or tea or acidic foods such as tomato soup.
- U.S. wineries have stopped using lead in their foil seals, but older bottles may still be around and other countries may still use lead; to be safe, wipe the foil-sealed rim of a wine bottle with a clean wet cloth before removing the cork.
- Feed children nutritious meals regularly.
- Before using your newspaper to wrap food, mulch garden plants, or add to your compost, confirm with the publisher that the paper uses no lead in its ink.

The Environmental Protection Agency (EPA) also publishes a booklet, *Lead and Your Drinking Water,* in which the following cautions appear:

- Have the water in your home tested by a competent laboratory.
- Use only cold water for drinking, cooking, and making formula (cold water absorbs less lead).
- When water has been standing in pipes for more than two hours, flush the cold-water pipes by running water through them for 30 seconds before using it for drinking, cooking, or mixing formulas.
- If lead contamination of your water supply seems probable, obtain additional information and advice from the EPA and your local public health agency.

By taking these steps, parents can protect themselves and their children from this preventable danger.[c]

[b] *A Shopper's Guide to Low-Lead China* is available from the Environmental Defense Fund, 257 Park Avenue South, New York, NY 10010; telephone (800) 284-3322.

[c] The National Lead Information Center provides two hotlines: call (800) LEAD-FYI (532-3394) for general information or (800) 424-LEAD (424-5323) with specific questions.

[a] R. L. Canfield and coauthors, Intellectual impairment in children with blood lead concentrations below 10 µg per deciliter, *New England Journal of Medicine* 348 (2003): 1517–1526.

These eight normally wholesome foods—milk, shellfish, fish, peanuts, tree nuts, eggs, wheat, and soybeans (and soy products)—may cause life-threatening symptoms in people with allergies.

Symptoms of impending anaphylactic shock:

- Tingling sensation in mouth.
- Swelling of the tongue and throat.
- Irritated, reddened eyes.
- Difficulty breathing, asthma.
- Hives, swelling, rashes.
- Vomiting, abdominal cramps, diarrhea.
- Drop in blood pressure.
- Loss of consciousness.
- Death.

epinephrine: one of the stress hormones secreted whenever emergency action is needed; prescribed therapeutically to relax the bronchioles during allergy or asthma attacks.

fish, or shellfish. Among these foods, eggs, milk, soy, and peanuts most often cause problems in children. Children are more likely to outgrow allergies to eggs, milk, and soy than allergies to peanuts. Peanuts cause more life-threatening reactions than do all other food allergies combined. Research is currently under way to help those with peanut allergies tolerate small doses, thus saving lives and minimizing reactions.[19] One possible solution depends on finding a natural, hypoallergenic peanut among the 14,000 varieties of peanuts that grow. Families of children with a life-threatening food allergy and school personnel who supervise them must guard them against any exposure to the allergen. The child must learn to identify which foods pose a problem and then learn and use refusal skills for all foods that may contain the allergen.

Parents of allergic children can pack safe foods for lunches and snacks and ask school officials to strictly enforce a "no swapping" policy in the lunchroom. The child must be able to recognize the symptoms of impending anaphylactic shock, such as a tingling of the tongue, throat, or skin, or difficulty breathing. Any person with food allergies severe enough to cause anaphylactic shock should wear a medical alert bracelet or necklace. Finally, the responsible child and the school staff should be prepared to administer injections of **epinephrine,** which prevents anaphylaxis after exposure to the allergen. Many preventable deaths occur each year when people with food allergies accidentally ingest the allergen but have no epinephrine.

Food Labeling As of 2006, food labels must announce the presence of common allergens in plain language, using the names of the eight most common allergy-causing foods.[20] For example, a food containing "textured vegetable protein" must say "soy" on its label. Similarly, "casein" must be identified as "milk," and so forth. Food producers must also prevent cross-contamination during production and clearly label the foods in which it is likely to occur.[21] For example, equipment used for making peanut butter must be scrupulously clean before being used to pulverize cashew nuts for cashew butter to protect unsuspecting cashew butter consumers from peanut allergens.

Technology may soon offer new solutions. Drugs under development may interfere with the immune response that causes allergic reactions.[22] Through genetic engineering, scientists may one day banish allergens from peanuts, soybeans, and other foods to make them safer.

Other Adverse Reactions to Foods Adverse reactions to foods that are not true food allergies include:

- A reaction specific to the flavor enhancer monosodium glutamate, or MSG.
- Reactions to chemicals in foods, such as the natural laxative in prunes.
- Symptoms of digestive diseases, such as hernias and ulcers, that are aggravated by eating any food.
- Enzyme deficiencies, such as lactose intolerance, that cause symptoms superficially indistinguishable from those of food allergy.
- Psychological reactions based on the belief that certain foods cause certain symptoms.

The simple dislike of a food may be a clue to allergy or to any of these reactions.

Food Dislikes Parents are advised to watch for signs of food dislikes and take them seriously. Children's food aversions may be the result of nature's efforts to protect them from allergic or other adverse reactions. Test for allergies, and then apply nutrition knowledge conscientiously in deciding how to alter the diet.

CASE STUDY   *Boy with Disruptive Behavior*

Freddie Willis is a six-year-old boy who seldom sits still, often misbehaves, and is frequently sick. Freddie's eating habits are erratic and poor, as is his appetite. He often misses breakfast because he is too tired to get up in time to eat before school. By midmorning, Freddie is irritable and disruptive in the classroom. At lunchtime he trades the peanut butter and banana sandwich his mother packed in his lunchbox for a piece of cake. After school he hurries home to watch television while he eats his favorite snack—cola and potato chips. At dinner-time Freddie picks at his food because he isn't very hungry. Later on, when it's time for bed, Freddie complains that he's hungry. His parents let him stay up to have a bowl of cereal (the kind with marshmallows) before he finally falls asleep.

1. What factors in Freddie's daily routine might be contributing to his restless behavior?
2. Discuss some changes in diet that might improve Freddie's health and disposition.

HYPERACTIVITY

Hyperactivity affects behavior and learning in about 5 to 10 percent of young school-aged children. Left untreated, it can interfere with a child's social development and ability to learn. Treatment focuses on relieving the symptoms and controlling the associated problems; there is no cure. Physicians often manage hyperactivity through behavior modification, special educational techniques, psychological counseling, and, in some cases, drug therapy.[23]

Parents of hyperactive children sometimes seek help from alternative therapies, including special diets. They mistakenly believe a solution may lie in manipulating the diet—most commonly, by excluding sugar or food additives. Adding carrots or eliminating candy is such a simple solution that many parents eagerly give such diet advice a try. These dietary changes will not solve the problem of true hyperactivity. Studies have consistently found no convincing evidence that sugar causes hyperactivity or worsens behavior. Recommendations to restrict sugar in children's diets to prevent or treat behavior problems are groundless. The above Case Study offers an opportunity to think about these issues in relation to a specific child.

Children can become excitable, rambunctious, and unruly as a result of a desire for attention, lack of sleep, overstimulation, watching too much television or playing too many video games, too much caffeine from colas or chocolate, or a lack of physical activity. Such behaviors may suggest that more consistent care is needed. It helps to insist on regular hours of sleep, regular mealtimes, and regular outdoor activity.

REVIEW NOTES

Children's appetites and nutrient needs reflect their stage of growth.

Long-term hunger and malnutrition impair growth and health.

Short-term hunger exerts more subtle effects on children's health and behavior—such as poor academic performance.

Iron deficiency is widespread and has many physical and behavioral consequences.

Lead toxicity is prevalent among young children and can have irreversible effects on health and behavior.

True food allergies are somewhat rare in children, and children can outgrow some food allergies.

Some allergies, however, can cause dangerous, life-threatening reactions in both children and adults.

"Hyper" behavior is not caused by poor nutrition; misbehavior may reflect inconsistent care.

hyperactivity: inattentive and impulsive behavior that is more frequent and severe than is typical of others of a similar age; professionally called **attention deficit/hyperactivity disorder (ADHD)**.

Food Choices and Eating Habits of Children

The childhood years are the parents' last chance to influence their children's food choices. Parents who want to promote nutritious choices and healthful habits provide access to nutrient-dense, delicious foods and opportunities for active play at home. Food choices and regular physical activity can not only promote healthy growth but, as mentioned earlier, can also help prevent the degenerative diseases of later life. Many experts agree that early childhood is the time to put into effect practices that, until recently, were recommended only for adults. "Childhood Obesity and the Early Development of Chronic Diseases" is the title of the Nutrition in Practice that follows this chapter.

MEALTIMES AT HOME

Feeding children requires not only providing a variety of nutritious foods but also nurturing the children's self-esteem and well-being. Parents face a number of challenges in preparing meals that both appeal to their children's tastes and provide needed nutrients. Because the interactions between parents and children can set the stage for lifelong attitudes and habits, a child's preferences should be treated with respect, even when nutrient needs must take precedence.

Honoring Children's Preferences Researchers attempting to explain children's food preferences encounter many contradictions. Children say they like colorful foods, yet most often reject green and yellow vegetables while favoring brown peanut butter and white potatoes, apple wedges, and bread. They do like raw vegetables better than cooked ones, though, so it is wise to offer vegetables that are raw or slightly undercooked and crunchy and bright in color. They should be warm, not hot, because a child's mouth is much more sensitive than an adult's. The flavor should be mild (a child has more taste buds), and smooth foods such as mashed potatoes or pea soup should have no lumps (a child wonders, with some disgust, what the lumps might be).

Young children like to eat at little tables and to be served little portions of food. They also love to eat with other children and have been observed to stay at the table longer and eat more food when in the company of their peers. Parents who serve food in a relaxed and casual manner, without anxiety, provide an environment in which a child's negative emotions will be minimized.

Avoiding Power Struggles It is not surprising that problems over food often arise during the second or third year, when children begin asserting their independence. Many of these problems stem from the conflict between children's developmental stages and capabilities and parents who, in attempting to do what they think is best for their children, try to control every aspect of eating. Such conflicts can disrupt children's abilities to regulate their own food intakes or to determine their own likes and dislikes. For example, many people share the misconception that children must be persuaded or coerced to try new foods. In fact, the opposite is true. When children are forced to try new foods, especially when offered rewards for eating a particular food (if you eat your vegetables, you can play video games), they are less likely to try those foods again than are children who are left to decide for themselves. Similarly, when chil-

Eating is more fun for children when friends are there.

© Mary Kate Denny/PhotoEdit

dren are restricted from eating their favorite foods, they are more likely to want those foods. As dietitian and family therapist Ellyn Satter notes, the parent is responsible for *what* the child is offered to eat, but the child is responsible for *how much* and even *whether* to eat.

When introducing new foods at the table, parents are advised to offer them one at a time and only in small amounts at first. The more often a food is presented to a young child, the more likely the child will like that food. Between five and ten exposures to a new food are necessary before a toddler shows an enhanced preference for the food. Whenever possible, the new food should be presented at the beginning of the meal, when the child is hungry, but the child should make the decision to accept or reject it. Parents have their own inclinations and dislikes; so do children. It is best never to make an issue of food acceptance. A power struggle almost invariably sets a firm pattern of resistance and permanently closes the child's mind.

Television's Influence Watching television adversely affects children's nutritional health.[24] As Chapter 7 pointed out, watching television contributes to obesity.[25] Children who have television sets in their bedrooms spend more time watching TV and are more likely to be overweight than children who do not have televisions in their rooms.[26] Children who watch television for more than four hours a day, or during meals, are least likely to eat fruits and vegetables and most likely to be obese.[27] Not only are they inactive, but they often snack on the fattening foods that are advertised.

The average child sees about 30,000 commercials a year, and almost all of them urge viewers to purchase sugarcoated breakfast cereals, candy bars, chips, fast foods, and carbonated beverages. Those foods add sugar, fat, and salt to the diet and displace foods that provide needed nutrients. Many parents and pediatricians believe that food ads aimed at children should be banned because they support corporate profits rather than children's health. Alternatively, parents can teach their children how to evaluate food ads and make healthful choices.[28]

Preventing Choking When feeding children, parents must always be alert to the dangers of choking. A choking child is a silent child—an adult should be present whenever a child is eating. Make sure the child sits when eating; choking is more likely when a child is running or falling (see p. 287 for a list of foods and nonfood items most likely to cause choking).

Play First Ideally, each meal is preceded, not followed, by the activity the child looks forward to the most. A number of schools have discovered that children eat a much better lunch if it is served after, rather than before, recess. Otherwise children "hurry up and eat" so that they can go play.

Child Participation Allowing children to help plan and prepare the family's meals provides enjoyable learning experiences and encourages children to eat the foods they have prepared. Vegetables are pretty, especially when fresh, and provide opportunities for children to learn about color, growing things and their seeds, and shapes and textures—all of which are fascinating to young children. Measuring, stirring, decorating, and arranging foods are skills that even a very young child can practice with enjoyment and pride (see Table 11-5).

TABLE 11-5	Food Skills of Preschoolers[a]
Age 1–2 years, when large muscles develop, the child:	
uses short-shanked spoon. helps feed self. lifts and drinks from cup. helps scrub, tear, break, or dip foods.	
Age 3 years, when medium hand muscles develop, the child:	
spears food with fork. feeds self independently. helps wrap, pour, mix, shake, or spread foods. helps crack nuts with supervision.	
Age 4 years, when small finger muscles develop, the child:	
uses all utensils and napkin. helps roll, juice, mash, or peel foods. cracks egg shells.	
Age 5 years, when fine coordination of fingers and hands develops, the child:	
helps measure, grind, grate, and cut (soft foods with a dull knife). uses hand-cranked egg beater with supervision.	

[a]These ages are approximate. Healthy, normal children develop at their own pace.

TABLE 11-6	Healthful Snack Ideas—Think Food Groups, Alone and in Combination

Selecting two or more foods from different food groups adds variety and nutrient balance to snacks. The combinations are endless, so be creative.

Grain Products

Grain products are filling snacks, especially when combined with other foods:

- Cereal with fruit and milk
- Crackers and cheese
- Whole-grain toast with peanut butter
- Popcorn with grated cheese
- Oatmeal raisin cookies with milk

Vegetables

Cut-up fresh, raw vegetables make great snacks alone or in combination with foods from other food groups:

- Celery with peanut butter
- Broccoli, cauliflower, and carrot sticks with a flavored cottage cheese dip

Fruits

Fruits are delicious snacks and can be eaten alone—fresh, dried, or juiced—or combined with other foods:

- Apples and cheese
- Bananas and peanut butter
- Peaches with yogurt
- Raisins mixed with sunflower seeds or nuts

Meats and Meat Alternates

Meat and meat alternates add protein to snacks:

- Refried beans with nachos and cheese
- Tuna on crackers
- Luncheon meat on whole-grain bread

Milk and Milk Products

Milk can be used as a beverage with any snack, and many other milk products, such as yogurt and cheese, can be eaten alone or with other foods as listed above.

Snacks Parents may find that their children often snack so much that they aren't hungry at mealtimes. Instead of teaching children *not* to snack, teach them *how* to snack. Provide snacks that are as nutritious as the foods served at mealtime. Snacks can even be mealtime foods that are served individually over time, instead of all at once on one plate. When providing snacks to children, think of the food groups and offer such snacks as pieces of cheese, sliced strawberries, cooked baby carrots, and egg salad on whole-wheat crackers (see Table 11-6). Snacks that are easy to prepare should be readily available to children, especially if the children arrive home after school before their parents.

Preventing Dental Caries Children frequently snack on sticky, sugary foods that stay on the teeth and provide an ideal environment for the growth of bacteria that cause dental caries. Teach children to brush and floss after meals, to brush or rinse after eating snacks, to avoid sticky foods, and to select crisp or fibrous foods frequently.

Serving as Role Models In an effort to practice these many tips, parents may overlook perhaps the single most important influence on their children's food habits—themselves.[29] Parents who do not eat oranges should not be surprised when their children refuse to eat oranges. Likewise, parents who dislike the smell of brussels sprouts may not be able to persuade children to try them. Children learn much through imitation. Parents, older siblings, and other caregivers set an irresistible example by sitting with younger children, eating the same foods, and having pleasant conversations during mealtime.

While serving and enjoying food, caregivers can promote both physical and emotional health at every stage of a child's life. They can help their children to develop both a positive self-concept and a positive attitude toward food. If the beginnings are right, children will grow without the conflicts and confusions over food that lead to nutrition and health problems.

NUTRITION AT SCHOOL

While parents are doing what they can to establish good eating habits in their children at home, others are preparing and serving foods to their children at day-care centers and schools. In addition, children begin learning about food and nutrition in the classroom. Meeting the nutrition and education needs of children is critical to supporting their healthy growth and development.[30] The American Dietetic Association states that "schools and communities have a shared responsibility to provide all students with access to high-quality foods and school-based nutrition services as an integral part of the total education program."

The U.S. government funds programs to provide nutritious, high-quality meals for children at school. Both the School Breakfast Program and the National School Lunch Program provide meals at a reasonable cost to children from families with the financial means to pay. Meals are available free or at reduced cost to children from low-income families.

School Breakfast The School Breakfast Program is available in more than 80 percent of the nation's schools that offer school lunch, and close to 9 million children participate in it.[31] Nevertheless, for many children who need it, the School Breakfast Program is either unavailable, or the children do not participate in it.[32] The majority of children who eat school breakfasts are from low-income families. As research results continue to emphasize the positive impact breakfast has on school performance and health, vigorous campaigns to expand school breakfast programs are under way.

School Lunch More than 28 million children receive lunches through the National School Lunch Program—more than half of them free or at a reduced price.[33] School lunches are designed to provide at least a third of the recommendation for energy, protein, vitamin A, vitamin C, iron, and calcium. They must also include specified numbers of servings from each food group. In an effort to help reduce cardiovascular disease risk, all government-funded meals served at schools must follow the *Dietary Guidelines for Americans*. Table 11-7 (p. 312) shows school lunch patterns for children of different ages.

Parents often rely on school lunches to meet a significant part of their children's nutrient needs on school days. Indeed, students who regularly eat school lunches have higher intakes of many nutrients and fiber than students who do not.[34] Children don't always like what they are served, however, and school lunch programs must strike a balance between what children want to eat and what will nourish them and guard their health.

Competing Influences at School Serving nutritious lunches is only half the battle; students need to eat them, too. Short lunch periods and long waiting lines prevent some students from eating a school lunch and leave others with too little time to complete their meals.[35] Nutrition efforts at schools are also undermined when students can buy what the USDA labels "competitive foods"—meals from

The school breakfast must contain at a minimum:
- One serving of fluid milk.
- One serving of fruit or vegetable or full-strength juice.
- Two servings of bread or bread alternates; or two servings of meat or meat alternates; or one of each.

The American Dietetic Association has set nutrition standards for child-care programs. Among them, meal plans should:
- Be nutritionally adequate and consistent with the *Dietary Guidelines for Americans*.
- Involve parents in planning.
- Follow recommended meal patterns that balance energy and nutrients with children's ages, appetites, activity levels, and special needs while respecting cultural and ethnic differences.
- Minimize added fat, sugar, and sodium.
- Emphasize fresh fruit, fresh and frozen vegetables, and whole grains.
- Provide furniture and eating utensils that are age appropriate and developmentally suitable to encourage children to accept and enjoy mealtime.

Source: Position of the American Dietetic Association: Benchmarks for nutrition programs in child care settings, *Journal of the American Dietetic Association* 105 (2005): 979–986.

TABLE 11-7   School Lunch Patterns for Different Ages[a]

Food Group	Preschool (Age)		Grade School through High School (Grade)		
	1 to 2	3 to 4	K to 3	4 to 6	7 to 12
Meat or meat alternate					
1 serving:					
Lean meat, poultry, or fish	1 oz	$1^1/_2$ oz	$1^1/_2$ oz	2 oz	3 oz
Cheese	1 oz	$1^1/_2$ oz	$1^1/_2$ oz	2 oz	3 oz
Large egg(s)	$^1/_2$	$^3/_4$	$^3/_4$	1	$1^1/_2$
Cooked dry beans or peas	$^1/_4$ c	$^3/_8$ c	$^3/_8$ c	$^1/_2$ c	$^3/_4$ c
Peanut butter	2 tbs	3 tbs	3 tbs	4 tbs	6 tbs
Yogurt	$^1/_2$ c	$^3/_4$ c	$^3/_4$ c	1 c	$1^1/_2$ c
Peanuts, soynuts, tree nuts, or seeds[b]	$^1/_2$ oz	$^3/_4$ oz	$^3/_4$ oz	1 oz	$1^1/_2$ oz
Vegetable and/or fruit					
2 or more servings, both to total	$^1/_2$ c	$^1/_2$ c	$^1/_2$ c	$^3/_4$ c	$^3/_4$ c
Bread or bread alternate[c]					
Servings	5 per week	8 per week	8 per week	8 per week	10 per week
Milk					
1 serving of fluid milk	$^3/_4$ c	$^3/_4$ c	1 c	1 c	1 c

[a]The quantities listed represent per-lunch minimums for each age and grade except those for the oldest group, which are recommendations. Schools unable to serve the recommended quantities for grades 7 to 12 must provide at least the amount shown for grades 4 to 6.

[b]These meat alternates may be used to meet no more than half of the meat or meat alternate requirement; therefore, they must be used in a meal with another meat or meat alternate.

[c]Schools must serve daily at least $^1/_2$ serving of bread or bread alternate to the youngest age group and at least 1 serving to older children.

Source: U.S. Department of Agriculture, National School Lunch Program Regulations, revised January 1, 1998.

fast-food restaurants or a la carte foods such as pizza or snack foods and carbonated beverages from snack bars, school stores, and vending machines.[36] These foods and beverages compete with nutritious school lunches. Children receive a mixed message when they are left on their own to choose between the health-supporting school lunch and the high-fat, high-salt foods with low nutrient density that their taste buds may prefer. In one study, students who selected competitive foods in addition to, or instead of, school meals consumed more energy and fat and less calcium and vitamin A than those who selected only the school lunch.[37]

Increasingly, school-based nutrition issues are being addressed by legislation. Some states restrict the sale of competitive foods and have higher rates of participation in school meal programs than the national average. Federal legislation mandates that all school districts that participate in the USDA's National School Lunch Program develop and put in place a local wellness policy.[38] Nutrition professionals advocate further legislative measures that would prohibit sales of food and beverages from vending machines or school stores in middle and high schools until 30 minutes after the end of the last meal unless they are part of the school foodservice and meet *Dietary Guidelines* standards.[39] Reducing the prices of nutritious foods also greatly increases the likelihood that students will purchase them.[40]

> **REVIEW NOTES**
>
> Adults at home and at school need to provide children with nutrient-dense foods and teach them how to make healthful choices.
>
> Adults also need to provide ample opportunity for children to be physically active.

The Teen Years

As children pass through **adolescence** on their way to becoming adults, they change in many ways. Their physical changes make their nutrient needs high, and their emotional, intellectual, and social changes make meeting those needs a challenge.

Teenagers make many more choices for themselves than they did as children. They are not fed, they eat. Food choices made during the teen years profoundly affect health, both now and in the future. At the same time, social pressures thrust choices at them: whether to drink alcoholic beverages and whether to develop their bodies to meet extreme ideals of slimness or athletic prowess. Their interest in nutrition derives from personal, immediate experiences. They are concerned with how diet can improve their lives now—they try the latest fad diet in order to fit into a new bathing suit, avoid greasy foods in an effort to clear acne, or eat a plate of pasta to prepare for a big sporting event. In presenting information on the nutrition and health of adolescents, this chapter includes topics of interest to teens.

GROWTH AND DEVELOPMENT DURING ADOLESCENCE

With the onset of adolescence, the steady growth of childhood speeds up abruptly and dramatically, and the growth patterns of females and males become distinct. Hormones direct the intensity and duration of the adolescent growth spurt, profoundly affecting every organ of the body, including the brain. After two to three years of intense growth and a few more at a slower pace, physically mature adults emerge.

In general, a female's adolescent growth spurt begins at age 10 or 11 and a male's, at 12 or 13. The spurt's duration is about two and a half years. Before **puberty,** the differences between male and female body composition are minimal. During the adolescent spurt, gender differences become apparent in the skeletal system, lean body mass, and fat stores. In males, the lean body mass—muscle and bone—becomes much greater, and in females, fat becomes a larger percentage of the total body weight. On average, males grow 8 inches taller, and females, 6 inches taller. Males gain approximately 45 pounds, and females, about 35 pounds.

ENERGY AND NUTRIENT NEEDS

The energy needs of adolescents vary greatly, depending on the current rate of growth, gender, body composition, and physical activity.[41] Boys' energy needs may be especially high; they experience a more intense growth spurt and, as mentioned, develop more lean body mass than girls do. An active teenage boy of 15 may need 3500 kcalories or more a day just to maintain his weight. In general, because girls enter their growth spurts earlier and grow less than boys, their energy needs peak sooner and decline earlier than those of their male peers. An inactive girl of 15 whose growth is nearly at a standstill may need fewer than 1800 kcalories a day if she is to avoid excessive weight gain. Thus teenage girls need to pay special attention to being physically active and selecting foods of high nutrient density so that they will meet their nutrient needs without exceeding their energy needs.

adolescence: the period of growth from the beginning of puberty until full maturity. Timing of adolescence varies from person to person.

puberty: the period in life in which a person becomes physically capable of reproduction.

Obesity The insidious problem of obesity becomes ever more apparent in adolescence and often continues into adulthood. Energy balance is often difficult to regulate in this society—an estimated 15 percent of U.S. children and adolescents 6 to 19 years of age are overweight.[42] The problem is most evident in African-American females, and in Hispanic children of both genders. Without intervention, overweight teens will face numerous physical and socioeconomic consequences for years to come. The consequences of obesity are so dramatic and our society's attitude toward obese people is so negative that even teens of normal weight perceive a need to control their weight. Healthy, normal-weight teenagers are often "on diets" and make all kinds of unhealthy weight-loss attempts—even taking up smoking.[43] Some adolescents may benefit from lower-kcalorie diets that increase fruits, vegetables, fat-free milk, and other nutritious foods while limiting cookies, cakes, soft drinks, fried snacks, and other less healthy choices. Most weight-loss dieting undertaken by adolescents, particularly females, however, is self-prescribed and generally unhealthful and can easily lead to nutrient deficiencies.[44] When taken to extremes, restrictive diets bring dramatic physical consequences of their own, as Nutrition in Practice 7 explains.

Vitamins Recommendations for most vitamins increase during the teen years (see the tables on the inside front cover). Several of the vitamin recommendations for adolescents are similar to those for adults, including the recommendation for vitamin D. During puberty, both the activation of vitamin D and the absorption of calcium are enhanced, thus supporting the intense skeletal growth of the adolescent years without additional vitamin D.

Iron The need for iron increases during adolescence for both females and males, but for different reasons. Iron needs increase for females as they start to menstruate, and for males, as their lean body mass develops. Hence, the RDA increases at age 14 for both males and females. Because menstruation continues throughout a woman's childbearing years, the RDA for iron remains high into late adulthood. For males, the RDA returns to preadolescent values in early adulthood.

In addition, iron needs increase when the adolescent growth spurt begins, whether that occurs before or after age 14.[45] This shifting requirement makes pinpointing an adolescent's need somewhat complicated.

Furthermore, iron recommendations for girls before age 14 do not reflect the iron losses of menstruation. The average age of menarche (first menstruation) in the United States is 12.5 years, however.[46] Therefore, for girls under the age of 14 who have started to menstruate, an additional 2.5 milligrams of iron per day is recommended.[47] Thus the RDA for iron depends not only on age and gender but also on whether the individual is in a growth spurt or has begun to menstruate, as listed in the margin.

Iron intakes often fail to keep pace with increasing needs, especially for females, who typically consume fewer iron-rich foods such as meat and fewer total kcalories than males. Not surprisingly, iron deficiency is most prevalent among adolescent girls. Iron-deficient children and teens score lower on standardized tests than those who are not iron deficient.

Calcium Adolescence is a crucial time for bone development, and the requirement for calcium reaches its peak during these years. Unfortunately, low calcium intakes among adolescents have reached crisis proportions: 85 percent of females and 64 percent of males ages 12 to 19 years have too-low calcium intakes. Paired with physical inactivity, low calcium intakes can compromise the development of peak bone mass, greatly increasing the risk of osteoporosis and other bone disease later on.[48] Increasing milk products in the diet to meet calcium recommendations greatly increases bone density.[49] Once again, however, teenage girls are most vul-

---

Iron RDA for males:

- 9–13 yr (8 mg/day).
- 9–13 yr in growth spurt (10.9 mg/day).
- 14–18 yr (11 mg/day).
- 14–18 yr in growth spurt (13.9 mg/day).

Iron RDA for females:

- 9–13 yr (8 mg/day).
- 9–13 yr in menarche (10.5 mg/day).
- 9–13 yr in menarche and growth spurt (11.6 mg/day).
- 14–18 yr (15 mg/day).
- 14–18 yr in growth spurt (16.1 mg/day)

nerable, for their milk—and therefore their calcium—intakes begin to decline at the time when their calcium needs are greatest.[50] Furthermore, women have much greater bone losses than men in later life. In addition to dietary calcium, bones grow stronger with physical activity, but few high schools require students to attend physical activity classes, so most adolescents must make a point to be physically active during leisure hours.

FOOD CHOICES AND HEALTH HABITS

Teenagers like the freedom to come and go as they choose and eat what they want when they have time. In the face of many demands on their time, including afterschool jobs, social activities, and home responsibilities, they almost inevitably fall into irregular eating habits, relying on quick snacks or fast foods for meals.[51] Only about one-third of adolescents eat evening meals at home with their families, but almost 80 percent say that family meals are important to them.[52] The adolescent who does eat at home with family members consumes more nutritious fruits, vegetables, whole grains, and calcium-rich foods and fewer soft drinks than others.[53]

Many adolescents also begin to skip breakfast on a regular basis, missing out on important nutrients that are not made up at later meals during the day. Compared with those who skip breakfast, teenagers who do eat breakfast have higher intakes of vitamins A, C, and riboflavin, as well as calcium, iron, and zinc.[54] Teenagers who eat breakfast are therefore more likely to meet their nutrient intake recommendations.

Ideally, in light of adolescents' busy schedules and desire for freedom, the adult becomes a **gatekeeper,** controlling the type and availability of food in the teenager's environment. Teenage sons and daughters and their friends should find plenty of nutritious, easy-to-grab food in the refrigerator (meats for sandwiches, low-fat cheeses, fresh, raw vegetables and fruits, fruit juices, and milk) and more in the cupboards (whole-grain breads, peanut butter, nuts, popcorn, and cereal). In many households today, all the adults work outside the home, and teenagers perform some of the gatekeeper's roles, such as shopping for groceries or choosing fast or prepared foods.

Snacks On average, about a fourth of an adolescent's total daily energy intake comes from snacks, which, if chosen carefully, can contribute some of the needed nutrients (see Table 11-6 on p. 310). Often, however, adolescents choose foods that are too high in saturated fat and sodium and too low in fiber to support the future health of their arteries. Their calcium intakes often fall short unless they snack on dairy products, and they may fail to obtain enough iron and vitamin A. For iron and other nutrients, a teenager might snack on iron-containing meat sandwiches or tuna fish, low-fat bran muffins, or tortillas with spicy bean spread along with a glass of orange juice to help maximize the iron's absorption.

Beverages Increasingly, as Figure 11-3 shows, adolescents are drinking soft drinks with lunch, supper, and snacks. About the only time they select fruit juices is at breakfast. When they drink milk, they are more likely to consume it with a meal (especially breakfast) than as a snack. Soft drinks, when chosen as the primary beverage, may affect the density of the bones because they displace milk from the diet.[55] Because of their greater food intakes, boys are more likely to drink enough milk to meet their calcium needs, whereas girls typically fall short of calcium recommendations. Regular soft drink consumption is also linked with overweight in adolescents.[56]

Soft drinks present another problem, when caffeine intake becomes excessive. Caffeine seems to be relatively harmless, however, when used in moderate doses

Nutritious snacks contribute valuable nutrients to an active teen's diet.

gatekeeper: with respect to nutrition, a key person who controls other people's access to foods and thereby exerts a profound impact on their nutrition. Examples are the spouse who buys and cooks the food, the parent who feeds the children, and the caregiver in a day-care center.

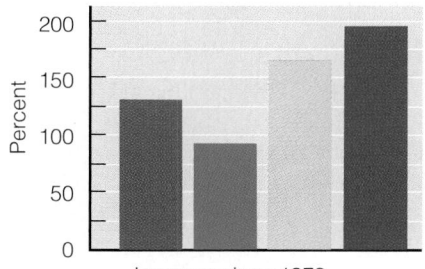

Source: S. A. French, B.-H. Lin, and J. F. Guthrie, National trends in soft drink consumption among children and adolescents age 6–17 years; Prevalence, amounts, and sources, 1977/1978 to 1994/1998, *Journal of the American Dietetic Association* 103 (2003): 1326–1331.

FIGURE 11-3 Increase in Soft Drink Consumption of U.S. Adolescents Average soft drink consumption by adolescents more than doubled between 1978 and the present.

Appendix A provides a table of the caffeine contents of beverages, foods, and medications.

(the equivalent of fewer than, say, three 12-ounce cola beverages a day). In greater amounts, it can cause the symptoms associated with anxiety—sweating, tenseness, and inability to concentrate.

Eating Away from Home Adolescents eat about one-third of their meals away from home, and their nutritional welfare is enhanced or hindered by the choices they make. A lunch of a hamburger, a chocolate shake, and french fries supplies substantial quantities of many nutrients at a kcalorie cost of about 800, an energy intake many adolescents can afford. When they eat this sort of lunch, teens can adjust their breakfast and dinner choices to include fruits and vegetables for vitamin A, vitamin C, folate, and fiber, and lean meats and legumes for iron and zinc. Fortunately, many fast-food restaurants are offering more nutritious choices than the standard hamburger meal.

Peer Influence Teenagers are intensely engaged in day-to-day life with their peers and preparing for their future lives as adults. Adults need to remember that adolescents have the right to make their own decisions—even if those decisions are not in line with the adults' own views. Gatekeepers can set up the environment so that nutritious foods are available and can stand by with reliable nutrition information and advice, but the rest is up to the teenagers. Ultimately, they make the choices.

PROBLEMS ADOLESCENTS FACE

Physical maturity and growing independence present adolescents with new choices to make. The consequences of those choices will influence their nutritional health both today and throughout life. Some teenagers begin using drugs, alcohol, and tobacco; others wisely refrain. Information about the use of these substances is presented here because most people are first exposed to them during adolescence, but it actually applies to people of all ages.

Marijuana Almost half of the high school students in the United States report having at least tried marijuana.[57] When inhaled by smoking, the active chemicals in marijuana are rapidly and almost completely absorbed from the lungs.* They then travel in the blood to the various body tissues that metabolize them. The active ingredients from a single marijuana cigarette can linger in the body's fat a month or more before being excreted in urine.

Marijuana is unique among drugs in that it seems to enhance the enjoyment of eating, especially of sweets, a phenomenon commonly known as "the munchies." Why or how this effect occurs is not known; it may be a social effect induced by suggestibility, or perhaps the drug stimulates appetite. Whatever the reason, prolonged use of the drug does not seem to bring about weight gain.

Cocaine Cocaine stimulates the nervous system and elicits the stress response—constricted blood vessels, raised blood pressure, widened pupils of the eyes, and increased body temperature. It also drives away feelings of fatigue. Cocaine occasionally causes immediate death—usually by heart attack, stroke, or seizure in an already damaged body system.

Weight loss is common, and cocaine abusers often develop eating disorders. Notably, the craving for cocaine replaces hunger; rats given unlimited cocaine will

*The active ingredient of marijuana, which is primarily responsible for its intoxicating effects, is delta-9-tetrahydrocannabinol, or THC.

choose it over food until they starve to death. Thus, unlike marijuana use, cocaine use has major nutritional consequences.

Ecstasy The club drug known as ecstasy has become alarmingly popular in recent years. Ecstasy can damage brain cells, impair memory, increase the heart rate, and dangerously raise body temperature. Furthermore, uncertainties about the sources of ecstasy and other drugs or contaminants added during manufacture exacerbate the dangers of ecstasy use. People who use ecstasy regularly lose weight for many of the same reasons listed in the margin.

Drug Abuse, in General The effects of other addictive drugs vary in degree but are similar to those caused by cocaine. Drug abusers face the multiple nutrition problems listed in the margin. During withdrawal from drugs, an important part of treatment is to identify and correct these nutrition problems.

Alcohol Abuse Sooner or later all teenagers face the decision of whether to drink alcohol. The law forbids the sale of alcohol to people under 21, but most adolescents who seek alcohol can obtain it. By the end of high school, 77 percent of students have tried alcohol, and about half have been drunk at least once.[58]

Nutrition in Practice 19 describes how alcohol affects health and disease. Alcohol provides energy but no nutrients, and it can displace nutritious foods from the diet. Alcohol alters nutrient absorption and metabolism, so imbalances develop.

Smoking The prevalence of cigarette smoking among U.S. adolescents is on the decline, but every day, 3000 young people start smoking. Cigarette smoking is a pervasive health problem, causing thousands of people to suffer from cancer and diseases of the cardiovascular, digestive, and respiratory systems. These effects are beyond the scope of nutrition, but smoking cigarettes does influence hunger, body weight, and nutrient status.

Smoking a cigarette eases feelings of hunger. When smokers receive a hunger signal, they can quiet it with cigarettes instead of food. Such behavior ignores body signals and postpones energy and nutrient intake. In rats, nicotine reduces food intake and increases the rate of energy expenditure, causing weight loss.

Indeed, smokers tend to weigh less than nonsmokers and to gain weight when they stop smoking. Weight gain is often a concern for people contemplating giving up cigarettes. They should know that the average person who quits smoking gains less than 10 pounds. Smokers wanting to quit need to prepare for this possibility and adjust their diet and activity habits so as to maintain weight during and after quitting. Smoking cessation programs need to include strategies for weight management.

Nutrient intakes of smokers and nonsmokers differ. Smokers tend to have lower intakes of dietary fiber, vitamin A, beta-carotene, folate, and vitamin C. The association between smoking and low vitamin C intake may be noteworthy, considering the altered metabolism of vitamin C in smokers and the protective effect of foods rich in this vitamin against some types of cancer. Research shows that compared to nonsmokers, smokers require almost twice as much vitamin C to maintain steady body pools. Oxidants in cigarette smoke accelerate vitamin C metabolism and deplete smokers' body stores of this antioxidant. This depletion is even evident to some degree in nonsmokers who are exposed to passive smoke.

Smokeless Tobacco Like cigarettes, smokeless tobacco use is linked to many health problems, from minor mouth sores to tumors in the nasal cavities, cheeks, gums, and throat. The risk of mouth and throat cancers is even greater than for smoking tobacco. Other drawbacks to tobacco chewing and snuff dipping include bad breath, stained teeth, and blunted senses of smell and taste. Tobacco chewing also

---

Nutrition problems of drug abusers:

- They buy drugs with money that could be spent on food.
- They lose interest in food during "highs."
- They use drugs that depress appetite.
- Their lifestyle fails to promote good eating habits.
- They use intravenous (IV) drugs. They may contract AIDS, hepatitis, or other infectious diseases, which increase their nutrient needs. Hepatitis also causes taste changes and loss of appetite.
- Medicines used to treat drug abuse may alter nutrition status.

The vitamin C recommendation for people who regularly smoke cigarettes is an additional 35 mg/day.

club drug: any of a wide variety of drugs used by young adults at all-night dance parties such as "raves."

ecstasy: a street or slang term used for the drug methylenedioxymethamphetamine (MDMA). Ecstasy is chemically similar to amphetamine (speed) and mescaline (a hallucinogen).

damages the gums, tooth surfaces, and jawbones, making it likely that users will lose their teeth in later life.

> **REVIEW NOTES**
>
> Nutrient needs rise dramatically as children enter the rapid growth phase of the teen years.
>
> The busy lifestyles of teenagers add to the challenge of meeting their nutrient needs—especially for iron and calcium.
>
> In addition to making wise food choices, adolescents need to refrain from using substances that will impair their health—including illicit drugs, alcohol, and tobacco.

Chapter 13 offers details about nutrition assessment.

Assessment of nutrition status in healthy children and adolescents can confirm that development is normal or can catch potential problems early. The Nutrition Assessment Checklist highlights problems to look for when working with children and adolescents.

NUTRITION ASSESSMENT CHECKLIST FOR CHILDREN AND ADOLESCENTS

Medical History

Check the medical record for:

- ☐ Alcohol, tobacco, or illicit drug abuse
- ☐ Attention deficit/hyperactivity disorder (ADHD)
- ☐ Diabetes or other chronic disorders
- ☐ Eating disorders
- ☐ Food allergies
- ☐ Lactose intolerance
- ☐ Obesity
- ☐ Pregnancy

Medications

For children or adolescents being treated with drug therapy for medical conditions, note:

- ☐ Side effects that might reduce food intake or change nutrient needs
- ☐ Proper administration of medication with respect to food intake

Dietary Intake

For all children and adolescents, especially those considered at risk nutritionally, assess the diet for:

- ☐ Total energy
- ☐ Protein
- ☐ Calcium and iron
- ☐ Vitamin A, vitamin C, and folate
- ☐ Fiber

Note the following:

- ☐ Number of days each week a nutritious breakfast is eaten

- ☐ Number of hours the child or teen sleeps each day
- ☐ Number of soft drinks the child or teen drinks each day
- ☐ Number of fast-food meals eaten each day
- ☐ Number and type of snacks eaten each day
- ☐ Type and amount of physical activity
- ☐ Amount of caffeine consumed

Anthropometric Data

Measure baseline height and weight.

- ☐ Reassess height, weight, and growth patterns at each medical checkup.
- ☐ Note significant obesity or underweight and intervention strategies employed.

Laboratory Tests

Monitor the following laboratory tests for children and adolescents:

- ☐ Hemoglobin, hematocrit, or other tests of iron status
- ☐ Blood glucose for children or adolescents with diabetes
- ☐ Blood lead concentrations

Physical Signs

Look for physical signs of:

- ☐ Protein-energy malnutrition
- ☐ Iron deficiency
- ☐ Vitamin A deficiency
- ☐ Vitamin C deficiency
- ☐ Folate deficiency

SELF CHECK

1. Which of the following is a chracteristic of iron deficiency in children?
   a. it rarely develops in those with high intakes of milk.
   b. it affects brain function before anemia sets in.
   c. it is a primary factor in hyperactivity.
   d. mild deficiency enhances mental performance by lowering physical activity level thereby leading to increased attention span.

2. Three symptoms of lead toxicity are:
   a. diarrhea, irritability, and fatigue.
   b. low blood sugar, hair loss, and skin rash.
   c. increased heart rate, hyperactivity, and dry skin.
   d. bleeding gums, brittle fingernails, and swollen glands.

3. Allergic reactions to foods are most often caused by:
   a. corn, rice, or meats.
   b. eggs, peanuts, or milk.
   c. red meats, milk, or MSG.
   d. seafood, dark greens, or lactose.

4. When introducing new foods to children:
   a. reward children as they try new foods.
   b. offer many choices to encourage variety.
   c. offer one new food at the end of the meal.
   d. offer one new food at the beginning of the meal.

5. Which of the following is *not* true? Children who watch a lot of television are likely to:
   a. become obese.
   b. spend less time being physically active.
   c. learn healthy eating tips from programs.
   d. eat the foods most often advertised on television.

6. Which of the following strategies is *not* effective?
   a. Play first, eat later.
   b. Provide small portions.
   c. Encourage children to help prepare meals.
   d. Use dessert as a reward for eating vegetables.

7. During the growth spurt of adolescence:
   a. females gain more weight than males.
   b. males gain more fat, proportionately, than females.
   c. differences in body composition between males and females become apparent.
   d. similarities in body composition between males and females become apparent.

8. Two nutrients that are usually lacking in adolescents' diets are:
   a. zinc and fat.
   b. iron and calcium.
   c. protein and thiamin.
   d. vitamin A and riboflavin.

9. To help teenagers consume a balanced diet, parents can:
   a. monitor the teens' food intake.
   b. give up—parents can't influence teenagers.
   c. keep the cupboards and refrigerator well stocked.
   d. forbid snacking and insist on regular, well-balanced meals.

10. Smoking increases the need for:
    a. iron.
    b. folate.
    c. vitamin C.
    d. vitamin E.

Answers to these questions are in Appendix H.

CLINICAL APPLICATIONS

1. At two and a half years old, Travis is healthy, though slightly underweight, and headstrong. Travis's mother hovers over him at every meal and insists that he take several bites of every food on his plate, even if he dislikes the food or is not familiar with it. Even though Travis is hungry when he sits down to a meal, his mother's constant urging to get him to eat quickly quells any interest he had in eating. Travis simply folds his arms across his chest, closes his mouth tightly, and refuses to eat any more food. After more begging, pleading, and nagging, Travis's mother becomes angry and sends him away from the table. Travis is not allowed to snack between meals because his mother is concerned that snacks will ruin his appetite.

   • What factors might be contributing to Travis's refusal to eat?

   • Travis's mother is concerned about her son's underweight. What strategies would you suggest to help Travis gain weight?

   • What advice would you offer Travis's mother to help her improve mealtimes with her son?

2. Loni is a physically inactive, slightly overweight 15-year-old who started smoking cigarettes at the age of 13 to help her lose weight. She planned to quit smoking as soon as she dropped a few pounds. Smoking reduced her appetite somewhat, but when she did eat, she chose chips, cola drinks, and fast foods such as chicken nuggets and burgers. At 15, Loni is still overweight and is smoking more than a pack of cigarettes a day. She uses her lunch money to fund her cigarette habit, which leaves only a little change for crackers or chips from the vending machine for lunch. She's noticed that her complexion looks dry and off-color, her teeth are less white than they used to be, she's easily fatigued, and she gets sick more often.

   • What would you tell Loni to motivate her to stop smoking?

   • What nutrients might be affected by Loni's smoking, and how might her diet be adjusted to meet these needs?

   • What dietary advice would you suggest to help Loni look and feel healthier?

NUTRITION ON THE NET

For further study of the topics in this chapter, access these websites.

Find updates and quick links to these and other nutrition-related sites at our website: **www.wadsworth.com/nutrition**

Learn how to care for children and adolescents from the American Academy of Pediatrics and the Canadian Paediatric Society: **www.aap.org** and **www.cps.ca**

Download the current growth charts and learn about their most recent revision: **www.cdc.gov/growthcharts**

Get information on MyPyramid for young children from the USDA: **www.mypyramid.gov**

Get tips for feeding children from the American Dietetic Association and the I Am Your Child Program: **www.eatright.org** and **www.iamyourchild.org**

Get tips for keeping children healthy from the Nemours Foundation: **www.kidshealth.org**

Visit the National Center for Education in Maternal and Child Health: **www.ncemch.org**

Learn about child nutrition programs: **www.fns.usda.gov/fns**

Learn how to reduce lead exposure in your home from the U.S. Department of Housing and Urban Development's Office of Lead Hazard Control: **www.hud.gov/lead**

Learn more about food allergies from the American Academy of Allergy, Asthma, and Immunology and the Food Allergy Network: **www.aaaai.org, www.foodallergy.org**

Learn more about hyperactivity from Children and Adults with Attention Deficit Disorders: **www.chadd.org**

Visit the Milk Matters section of the National Institute of Child Health and Development (NICHD): **www.nichd.nih.gov**

Get weight-loss tips for children and adolescents: **www.shapedown.com**

Read the message for parents and teens on the risks of tobacco use from the American Academy of Pediatrics: **www.aap.org**

Visit the Tobacco Information and Prevention Source (TIPS) of the Centers for Disease Control and Prevention: **www.cdc/gov/tobacco**

NOTES

[1] M. K. Fox and coauthors, Relationship between portion size and energy intake among infants and toddlers: Evidence of self-regulation, *Journal of the American Dietetic Association* 106 (2006): S77–S83.

[2] Institute of Medicine, Overview of IOM's childhood obesity prevention study, *Fact Sheet,* September 2004, available at **www.iom.edu**.

[3] V. Messina and A. R. Mangels, Considerations in planning vegan diets: Children, *Journal of the American Dietetic Association* 101 (2001): 661–669.

[4] F. R. Greer, N. F. Krebs, and the Committee on Nutrition, Optimizing bone health and calcium intakes of infants, children, and adolescents, *Pediatrics* 117 (2006): 578–585; H. Vatanparast and coauthors, Positive effects of vegetable and fruit consumption and calcium intake on bone mineral accrual in boys during growth from childhood to adolescence: The University of Saskatchewan Pediatric Bone Mineral Accrual Study, *American Journal of Clinical Nutrition* 82 (2005): 700–706; J. O. Fisher and coauthors, Meeting calcium recommendations during middle childhood reflects mother-daughter beverage choices and predicts bone mineral status, *American Journal of Clinical Nutrition* 79 (2004): 698–706.

[5] A. Carlson and coauthors, U.S. Department of Agriculture Center for Nutrition Policy and Promotion, Report card on the diet quality of children ages 2 to 9, *Nutrition Insights* 25 (2001), available at **www.usda.gov/cnpp/Insights/Insight25.pdf**.

[6] F. E. Viteri and H. Gonzalez, Adverse outcomes of poor micronutrient status in childhood and adolescence, *Nutrition Reviews* 60 (2002): S77–S83.

[7] Position of the American Dietetic Association: Dietary guidance for healthy children ages 2 to 11 years, *Journal of the American Dietetic Association* 104 (2004): 660–677.

[8] Forum on Child and Family Statistics, *America's Children: Key National Indicators of Well-Being, 2005,* available at **www.childstats.gov**.

[9] G. C. Rampersaud and coauthors, Breakfast habits, nutritional status, body weight, and academic performance in children and adolescents, *Journal of the American Dietetic Association* 105 (2005): 743–760; S. G. Affenito and coauthors, Breakfast consumption by African-American and white adolescent girls correlates positively with calcium and fiber intake and negatively with body mass index, *Journal of the American Dietetic Association* 105 (2005): 938–945; Position of the American Dietetic Association, 2004.

[10] Rampersaud and coauthors, 2005; Affenito and coauthors, 2005; Position of the American Dietetic Association 2004; C. S. Berkey and coauthors, Longitudinal study of skipping breakfast and weight change in adolescents, *International Journal of Obesity and Related Metabolic Disorders* 27 (2003): 1258–1266.

[11]J. S. Halterman and coauthors, Iron deficiency and cognitive achievement among school-aged children and adolescents in the United States, *Pediatrics* 107 (2001): 1381–1386.

[12]Centers for Disease Control and Prevention, Blood lead levels—United States, 1999–2002, *Morbidity and Mortality Weekly Report* 54 (2005): 513–616.

[13]Committee on Environmental Health, American Academy of Pediatrics, Policy statement: Lead exposure in children: Prevention, detection, and management, *Pediatrics* 116 (2005): 1036–1046.

[14]Centers for Disease Control and Prevention, 2005.

[15]Committee on Environmental Health, American Academy of Pediatrics, 2005.

[16]U.S. Department of Health and Human Services, National Institutes of Health, National Institute of Allergy and Infectious Diseases, *Food Allergy: An Overview*, NIH publication no. 04-5518 (July 2004), available at **www.niaid.nih.gov**; R. Formanek, Food allergies: When food becomes the enemy, *FDA Consumer*, July/August 2001, pp. 10–16; J. M. Yeung, R. S. Applebaum, and R. Hildwine, Criteria to determine food allergen priority, *Journal of Food Protection* 63 (2000): 982–986.

[17]H. Skolnick and coauthors, The natural history of peanut allergy, *Journal of Allergy and Clinical Immunology* 107 (2001): 367–374; Formanek, 2001.

[18]L. Christie and coauthors, Food allergies in children affect nutrient intake and growth, *Journal of the American Dietetic Association* 102 (2002): 1648–1651.

[19]H. Metzger, Two approaches to peanut allergy, *New England Journal of Medicine* 348 (2003): 1046–1048.

[20]Food Allergen Labeling and Consumer Protection Act of 2004, available at **http://thomas.loc.gov/cgi-bin/query/F?c108:6:./temp/~c108Dz8zuL:e48634.**

[21]Formanek, 2001.

[22]B. Merz, Studying peanut anaphylaxis, *New England Journal of Medicine* 348 (2003): 975–976; Metzger, 2003; X. M. Li and coauthors, Persistent protective effect of heat-killed *Escherichia coli* producing "engineered," recombinant peanut proteins in a murine model of peanut allergy, *Journal of Allergy and Clinical Immunology* 112 (2003): 159–167.

[23]M. L. Wolraich and coauthors, Attention-deficit/hyperactivity disorder among adolescents: A review of the diagnosis, treatment, and clinical implications, *Pediatrics* 115 (2005): 1734–1746.

[24]S. J. Salmon, K. J. Campbell, and D. A. Crawford, Television viewing habits associated with obesity risk factors: A survey of Melbourne schoolchildren, *Medical Journal of Australia* 184 (2006): 64–67; D. M. Matheson and coauthors, Children's food consumption during television viewing, *American Journal of Clinical Nutrition* 79 (2004): 1088–1094.

[25]R. M. Viner and T. J. Cole, Television viewing in early childhood predicts adult body mass index, *Journal of Pediatrics* 147 (2005): 429–435.

[26]B. A. Dennison, T. A. Erb, and P. L. Jenkins, Television viewing and television in bedroom associated with overweight risk among low-income preschool children, *Pediatrics* 109 (2002): 1028–1035.

[27]K. A. Coon and coauthors, Relationships between use of television during meals and children's food consumption patterns, *Pediatrics* 107 (2001): 167 [**www.pediatrics.org/cgi/content/full/107/1/e7**]; C. J. Crespo and coauthors, Television watching, energy intake, and obesity in U.S. children: Results from the Third National Health and Nutrition Examination Survey, 1988–1994, *Archives of Pediatrics and Adolescent Medicine* 155 (2001): 360–365.

[28]T. J. Hindin, I. R. Contento, and J. D. Gussow, A media literacy nutrition education curriculum for head start parents about the effects of television advertising on their children's food requests, *Journal of the American Dietetic Association* 104 (2004): 192–198.

[29]J. Wardle, S. Carnell, and L. Cooke, Parental control over feeding and children's fruit and vegetable intake: How are they related? *Journal of the American Dietetic Association* 105 (2005): 227–232; A.

T. Galloway and coauthors, Parental pressure, dietary patterns, and weight status among girls who are "picky eaters," *Journal of the American Dietetic Association* 105 (2005): 541–548; L. J. Cooke and coauthors, Demographic, familial and trait predictors of fruit and vegetable consumption by pre-school children, *Public Health Nutrition* 2 (2004): 251–252.

[30]Position of the American Dietetic Association: Benchmarks for nutrition programs in child care settings, *Journal of the American Dietetic Association* 105 (2005): 979–986.

[31]Position of the American Dietetic Association: Local support for nutrition integrity in schools, *Journal of the American Dietetic Association* 106 (2006): 122–133.

[32]Position of the American Dietetic Association, 2006.

[33]Position of the American Dietetic Association, 2006.

[34]Position of the American Dietetic Association, 2004; K. W. Cullen and I. Zakeri, Fruits, vegetables, milk, and sweetened beverages consumption and access to a la carte/snack bar meals at school, *American Journal of Public Health* 94 (2004): 463–467; P. M. Gleason and C. W. Suitor, Eating at school: How the National School Lunch Program affects children's diets, *American Journal of Agricultural Economics* 85 (2003): 1047–1051.

[35]Position of the American Dietetic Association, 2006.

[36]Position of the American Dietetic Association, 2006; Committee on School Health, American Academy of Pediatrics, Soft drinks in schools, *Pediatrics* 113 (2004): 152–154; J. L. Kramer-Atwood and coauthors, Fostering healthy food consumption in schools: Focusing on the challenges of competitive foods, *Journal of the American Dietetic Association* 102 (2002): 1228–1233.

[37]S. B. Templeton and coauthors, Competitive foods increase the intake of energy and decrease the intake of certain nutrients by adolescents consuming school lunch, *Journal of the American Dietetic Association* 105 (2005): 215–220.

[38]Position of the American Dietetic Association, 2006.

[39]Position of the American Dietetic Association, Society for Nutrition Education, and American School Food Service Association, Nutrition services: An essential component of comprehensive school health programs, *Journal of the American Dietetic Association* 103 (2003): 505–514.

[40]S. A. French, Pricing effects on food choices, *Journal of Nutrition* 133 (2003): 841S–843S.

[41]Standing Committee on the Scientific Evaluation of Dietary Reference Intakes, Food and Nutrition Board, Institute of Medicine, *Dietary Reference Intakes for Energy, Carbohydrate, Fiber, Fat, Fatty Acids, Cholesterol, Protein, and Amino Acids* (Washington, D.C.: National Academies Press, 2005), pp. 177–182.

[42]Food and Nutrition Board, Institute of Medicine, *Preventing Childhood Obesity: Health in the Balance* (Washington, D.C.: National Academies Press, 2005), p. 56.

[43]S. E. Saarni and coauthors, Intentional weight loss and smoking in young adults, *International Journal of Obesity and Related Metabolic Disorders* 28 (2004): 796–802; J. Cawley, S. Markowitz, and J. Tauras, Lighting up and slimming down: The effects of body weight and cigarette prices on adolescent smoking initiation, *Journal of Health Economics* 23 (2004): 293–311; D. Neumark-Sztainer and coauthors, Weight-control behaviors among adolescent girls and boys: Implications for dietary intake, *Journal of the American Dietetic Association* 104 (2004): 913–920.

[44]Neumark-Sztainer and coauthors, 2004.

[45]Standing Committee on the Scientific Evaluation of Dietary Reference Intakes, Food and Nutrition Board, Institute of Medicine, *Dietary Reference Intakes for Vitamin A, Vitamin K, Arsenic, Boron, Chromium, Copper, Iodine, Iron, Manganese, Molybdenum, Nickel, Silicon, Vanadium, and Zinc* (Washington, D.C.: National Academy Press, 2001), pp. 290–393.

[46]W. C. Chumlea and coauthors, Age at menarche and racial comparisons in US girls, *Pediatrics* 111 (2003): 110–113.

[47]Committee on the Scientific Evaluation of Dietary Reference Intakes, 2001, pp. 290–393.

[48]Federal Update, Milk matters, *Journal of the American Dietetic Association* 102 (2002): 469.

[49]H. J. Kalkwarf, J. C. Khoury, and B. P. Lanphear, Milk intake during childhood and adolescence, adult bone density, and osteoporotic fractures in US women, *American Journal of Clinical Nutrition* 77 (2003): 257–265.

[50]S. A. Bowman, Beverage choices of young females: Changes and impact on nutrient intakes, *Journal of the American Dietetic Association* 102 (2002): 1234–1239.

[51]M. Story, D. Neumark-Sztainer, and S. French, Individual and environmental influences on adolescent eating behaviors, *Journal of the American Dietetic Association* 102 (2002): S40–S51.

[52]Story, Neumark-Sztainer, and French, 2002.

[53]D. Neumark-Sztainer and coauthors, Family meal patterns: Associations with sociodemographic characteristics and improved dietary intake among adolescents, *Journal of the American Dietetic Association* 103 (2003): 317–322.

[54]Rampersaud and coauthors, 2005.

[55]Vatanparast and coauthors, 2005; G. Mrdjenovic and D. A. Levitsky, Nutritional and energetic consequences of sweetened drink consumption in 6- to 13-year-old children, *Journal of Pediatrics* 142 (2003): 604–610.

[56]J. James and coauthors, Preventing childhood obesity by reducing consumption of carbonated drinks: Cluster randomised controlled trial, *British Medical Journal* 328 (2004): 1237.

[57]American Academy of Pediatrics, J. W. King, and Committee on Substance Abuse, Tobacco, alcohol, and other drugs: The role of the pediatrician in prevention, identification, and management of substance abuse, *Pediatrics* 115 (2005): 816–821.

[58]American Academy of Pediatrics, King, and Committee on Substance Abuse, 2005.

Childhood Obesity and the Early Development of Chronic Diseases

When most people think of health problems in children and adolescents, they typically think of ear infections, colds, and acne, not type 2 diabetes and hypertension. Today, however, unprecedented numbers of U.S. children are being diagnosed with obesity and the serious "adult diseases" such as type 2 diabetes that accompany overweight.[1] For children born in the United States in the year 2000, the risk of developing type 2 diabetes sometime in their lives is estimated to be 30 percent for boys and 40 percent for girls.[2] U.S. children are not alone—rapidly rising rates of obesity are threatening the health of an alarming number of children around the globe.[3] Without immediate intervention, some 60 million children are destined to suffer type 2 diabetes and hypertension in childhood followed by **cardiovascular disease (CVD)** in early adulthood.

Over the past three decades, researchers have been observing how changes in body weight, blood lipids, blood pressure, and individual behaviors correlate with the development of CVD over time—from infancy to childhood through adolescence and into young adulthood. Some major findings have emerged from this research:

- Changes inside the arteries—changes predictive of CVD—are evident in childhood.
- Obesity in children affects these changes.
- Behaviors that influence the development of obesity and of CVD are learned and begin early in life. These behaviors include overeating, physical inactivity, and cigarette smoking.

This Nutrition in Practice focuses on efforts to prevent childhood obesity, type 2 diabetes, and CVD (see the glossary on p. 306 for definitions of the relevant terms). The years of childhood are emphasized here, for the earlier in life health-promoting habits become established, the better they will stick.

What about genetics? Don't some people inherit the tendency to become obese or develop diabetes or CVD regardless of the lifestyle habits they adopt?

For obesity, as well as for CVD, hypertension, and type 2 diabetes, genetics does not appear to play a *determining* role; that is, a person is not simply destined at birth to develop them. Instead, genetics appears to play a *permissive* role—the potential is inherited, and will then develop, if given a push by factors in the environment such as poor diet, sedentary lifestyle, and cigarette smoking. Researchers note that the relationship between genes and the environment is a synergistic one—their combined effects are greater than the sum of their individual effects.[4] The Pima Indians exemplify this gene-environment interaction. Their population seems to be genetically predisposed to obesity and type 2 diabetes—they have high rates of both diseases.[5] Those who live in the restrictive, remote environment of the Mexican mountains, however, have a lower prevalence of obesity and type 2 diabetes than those who live in the obesity-promoting environment of Arizona.[6]

What about events that take place during fetal development—malnutrition, for example? Can they affect a person's tendency to develop diseases later in life?

A theory called *fetal programming* or *fetal origins of disease* states that maternal malnutrition or other harmful conditions at a critical period of fetal development may have lifelong effects on an individual's pattern of genetic expression and therefore on the tendency to develop obesity and certain diseases.[7] Blood pressure is the most studied outcome in researching this theory because it is easy to measure and is a known risk factor for CVD.[8] Most studies use infant birthweight as the indicator of fetal nutrition status. For the most part, research shows that lower birthweight increases the risk of adult hypertension.[9] Researchers' efforts to explain why fetal nutrient insufficiency and lower birthweight enhance the risk of adult hypertension center on factors that affect the function of the kidneys, vascular system, nervous system, and other organs or systems involved in the complex process of blood pressure regulation. For example, fetal growth restriction may impair kidney growth and development as well as the function of blood vessels.

Research also suggests that *postnatal* growth influences adult blood pressure.[10] People with relatively low birthweights who grow to be relatively large adults seem to have the highest risk of hypertension as adults.[11] This type of growth pattern—low birthweight followed by rapid "catch-up" growth—may be a risk factor for later hypertension, although more research is needed to confirm or refute these findings.

Why has type 2 diabetes become so prevalent?

Type 2 diabetes, a chronic disease closely linked with obesity, has been on the rise among children and adolescents as the prevalence of obesity in U.S. youth has increased in recent years. Obesity is the most important risk factor for type 2 diabetes in young people—an estimated 85 percent of the children diagnosed with type 2 diabetes are obese.[12] Most are diagnosed during puberty, but as children become more obese and less active, the disease is appearing in younger and younger children. Type 2 diabetes is most likely to occur in those who are obese and sedentary and have a family history of diabetes.

How does type 2 diabetes develop?

In type 2 diabetes, the body's cells become insulin-resistant—that is, the cells become less sensitive to insulin, reducing the amount of glucose entering the cells from the blood. The combination of obesity and insulin resistance produces a cluster of symptoms, including high blood pressure and high blood lipids, which in turn promotes the development of atherosclerosis and the early development of CVD.[13] Other common problems evident by early adulthood include kidney disease, blindness, and miscarriages. The complications of diabetes, especially when encountered at a young age, can shorten life expectancy. Chapter 20 offers a detailed discussion of diabetes.

Prevention and treatment of type 2 diabetes depend on weight management, which can be particularly difficult in a young person's world of food advertising, video games, and pocket money for candy bars. The activity and dietary suggestions to help defend against heart disease discovered later in this discussion apply to type 2 diabetes as well.

How does CVD develop, and when does its development begin?

Most CVD involves **atherosclerosis**—the accumulation of cholesterol and other blood lipids along the walls of the arteries. Frequently, atherosclerosis and its complications interfere with the flow of blood to the heart and can lead to coronary heart disease (CHD), which, in turn, raises the likelihood of a heart attack. When atherosclerosis interferes with blood flow to the brain, a stroke can result. Infants are born with healthy, smooth, clear arteries, but within the first decade of life, **fatty streaks** may begin to appear. During adolescence, these fatty streaks may begin to turn to **plaques** (Figure 21-2 in Chapter 21 shows the formation of plaques in atherosclerosis). By early adulthood, the plaques may begin to calcify and become raised lesions, especially in boys and young men. As the lesions grow more numerous and thicken, the heart disease rate begins to rise, and the rise becomes dramatic at about age 45 in men and 55 in women. From this point on, arterial damage and blockage progress rapidly, and heart attacks and strokes threaten life. In short, the consequences of atherosclerosis, which become apparent only in adulthood, have their beginnings in the first decades of life.[14]

Children with the highest risks of developing heart disease are sedentary and obese, with diabetes, high blood pressure, and high blood cholesterol.[15] In contrast, children with the lowest risks of heart disease are physically active and of normal weight, with low blood pressure and favorable lipid profiles.

Parents don't need to worry about their children's blood cholesterol, do they?

Atherosclerotic lesions reflect blood cholesterol: as blood cholesterol increases, lesion coverage increases. Cholesterol values at birth are similar in all populations; differences emerge in early childhood. Standard values for cholesterol screening in children and adolescents are listed in Table NP11-1.

TABLE NP11-1 Cholesterol Values for Children and Adolescents

Disease Risk	Total Cholesterol (mg/dL)	LDL Cholesterol (mg/dL)
Acceptable	<170	<100
Borderline	170–199	≥100
High	≥200	≥130

Note: Adult values appear in Chapter 21. A deciliter (dL) is one-tenth of a liter or 100 milliliters.

In general, blood cholesterol tends to rise as saturated fat intakes increase. Blood cholesterol also correlates with childhood obesity, especially central obesity.[16] In obese children, the LDL (low-density lipoprotein) cholesterol value is often too high, and the HDL (high-density lipoprotein) value is too low for health. These relationships are apparent throughout childhood, and their magnitude increases with age.

Children who are both overweight and have high blood cholesterol are likely to have parents who developed CVD early.[17] For this reason, screening is recommended for children and adolescents whose parents or grandparents have CVD, those whose parents have elevated blood cholesterol, and those whose family history is unavailable, especially if other risk factors are evident.[18] Because blood cholesterol in children is a good predictor of adult values, some experts recommend universal screening for all children, and particularly for those who are overweight, smoke, are sedentary, or consume diets high in saturated fat. Early—but not advanced—atherosclerotic lesions are reversible, making screening and education a high priority.

Is hypertension a concern for children and adolescents?

Pediatricians routinely monitor blood pressure in children and adolescents. High blood pressure may signal an underlying disease or the early onset of hypertension. Hypertension accelerates the development of atherosclerosis.[19] Hypertension may develop in the first decades of life, especially among obese children, and worsen with time. Children with hypertension can often make dramatic improvements by participating in regular aerobic activity and by losing weight or maintaining their weight as they grow taller. Evidence is needed to clarify whether restricting sodium in children's and adolescents' diets lowers blood pressure.

Why is there an epidemic of childhood obesity in the United States?

Over the past three decades, as the prevalence of childhood obesity throughout the United States has more than doubled for young children and adolescents and tripled for children 6 to 11 years of age, the society our children live in has changed considerably.[20] In many families today, both parents work outside the home, and they work longer hours; more emphasis is placed on convenience foods and foods eaten away from home; meal choices at school are more diverse and often less nutritious; sedentary activities such as watching television and playing video or computer games occupy much of children's free time; and opportunities for physical activity and outdoor play both during school and outside of school have declined. All of these factors and many others influence children's eating and activity patterns.

Children learn food behaviors from their families, and entire families may be eating too much, dieting inappropriately, and exercising too little. Research shows that one in four toddlers (19 to 24 months of age) exceeds estimated energy requirements as a result of eating such foods as candy, pizza, chicken nuggets, sodas, sweet tea, and salty snacks like cheese puffs and chips.[21] Thus, when researchers ask, "Are today's children eating more kcalories than those of 30 years ago?" the answer comes back, "Yes." Some researchers report an increase of 100 to 200 kcalories a day for all age groups, enough to account for significant weight gain.[22]

Research links added sugars, and especially high-fructose corn syrup—the easily consumed, energy-dense liquid sugar added to soft drinks—with excess body fatness in children. Each 12-ounce can of soft drink provides the equivalent of about 10 teaspoons of sugar and 150 kcalories. More than half of children consume at least one soft drink each day at school; adolescent males consume the most—four or more cans daily.[23] According to one estimate, the risk of obesity increases by 60 percent with each sugared soft drink consumed daily.[24]

Although the tremendous increase in soft drink consumption may play some role in the obesity epidemic, much of the epidemic can be explained by lack of physical activity. Children have become more sedentary, and sedentary children are more often overweight. A child who spends more than an hour or two each day in front of a television, computer monitor, or other media can become obese and develop unhealthy blood lipids even while eating fewer kcalories than a more active child.[25] Physically active children have higher HDL, lower LDL, and lower blood pressure than sedentary children, and these positive findings often persist into adulthood.

Just as blood cholesterol and obesity track over the years, so does a person's level of physical activity. Compared with inactive adolescents, those who are physically active weigh less, smoke less, eat diets lower in saturated fats, and have better blood lipid profiles. The message is clear: physical activity offers numerous health benefits, and children and adolescents who are active today are most likely to be active for years to come.

What can concerned adults do to help prevent childhood obesity?

In light of all these findings, parents and teachers are encouraged to make major efforts to prevent childhood obesity. Suggestions include the following: encourage children to eat slowly, to pause and enjoy the company of their table companions, and to stop eating when they are full. Teach them how to select nutrient-dense snacks and to serve themselves appropriate portions. Never force children to clean their plates.

Above all, be sensitive in teaching children nutrition principles that can help to prevent obesity. Children can easily get the idea that their worth is tied to their body weight. Some parents fail to realize that society's ideal of slimness can be perilously close to starvation and that a child encouraged to "diet" cannot obtain the energy and nutrients required for normal growth and development. Even healthy children without eating disorders have been observed to limit their growth through "dieting." Pediatricians warn parents to avoid extremes; they caution that while intentions may be good, excessive food restriction may create nutrient deficiencies and impair growth. Furthermore, parental control over eating may instigate battles and foster attitudes about foods that can lead to inappropriate eating behaviors. Weight gain in truly overweight children can be controlled safely without compromising growth, but the process should be overseen by a health care professional.

Are adult dietary recommendations appropriate for children?

Regardless of family history, all children over age two should eat a variety of foods and maintain desirable weight (see Table NP11-2). Children (4 to 18 years of age) should receive at least 25 percent and no more than 35 percent of total energy from fat, less than 10 percent from saturated fat, and less than 300 milligrams of cholesterol per day.[26]

Recommendations limiting fat and cholesterol are not intended for infants or children under two years old. Infants and toddlers need a higher percentage of fat to support their rapid growth.

Healthy children over age two can begin the transition to eating according to recommendations by eating fewer foods high in saturated fat and selecting more fruits and vegetables

TABLE NP11-2 American Heart Association (AHA) Pediatric Dietary Strategies for Individuals Aged >2 Years

Balance dietary kcalories with physical activity to maintain normal growth.
Engage in 60 minutes of moderate-to-vigorous play or physical activity daily.
Eat vegetables and fruits daily; limit juice intake.
Use vegetable oils and soft margarines low in saturated fat and *trans*-fatty acids instead of butter or most other animal fats in the diet.
Eat whole-grain breads and cereals rather than refined-grain products.
Reduce the intake of sugar-sweetened beverages and foods.
Use fat-free or low-fat milk and dairy products daily.
Eat more fish, especially fatty fish, broiled or baked.
Reduce salt intake, including salt from processed foods.

Source: American Heart Association, Samuel S. Gidding, and coauthors, Dietary recommendations for children and adolescents: A guide for practitioners, *Pediatrics* 117 (2006): 544–559.

TABLE NP11-3 Tips for Parents to Implement AHA Pediatric Dietary Guidelines

Reduce added sugars, including sugar-sweetened drinks and juices.
Use canola, soybean, corn oil, safflower oil, or other unsaturated oils in place of solid fats during food preparation.
Use the portion sizes recommended on food labels when preparing and serving food.
Use fresh, frozen, and canned vegetables and fruits, and serve at every meal; be careful with added sauces and sugar.
Introduce and regularly serve fish as an entrée.
Remove the skin from poultry before eating.
Use only lean cuts of meat and reduced-fat meat products.
Limit high-kcalorie sauces such as Alfredo, cream sauces, cheese sauces, and hollandaise.
Eat whole-grain breads and cereals rather than refined products; read labels and ensure that "whole grain" is the first ingredient on the food label of these products.
Eat more legumes (beans) and tofu in place of meat for some entrées.
Breads, breakfast cereals, and prepared foods, including soups, may be high in salt and/or sugar; read food labels for content and choose high-fiber, low-salt/low-sugar alternatives.

Source: American Heart Association, Samuel S. Gidding, and coauthors, Dietary recommendations for children and adolescents: A guide for practitioners, *Pediatrics* 117 (2006): 544–559.

(see Table NP11-3). Healthy meals can occasionally include moderate amounts of a child's favorite foods, even if they are high in saturated fat such as french fries and ice cream.[27] A steady diet of offerings from some "children's menus" in restaurants, such as chicken nuggets, hot dogs, and french fries, easily exceeds a prudent intake of saturated fat, *trans* fat, and kcalories, however, and invites both nutrient shortages and gains of body fat.[28] Most restaurant chains are changing children's menus to include steamed vegetables and broiled or grilled poultry—additions welcomed by busy parents who often dine out or purchase take-out foods.

Other fatty foods, such as nuts, vegetable oils, and safer varieties of fish, such as light canned tuna or salmon, are important for their essential fatty acids. Low-fat milk and milk products deserve special attention in a child's diet for the needed calcium and other nutrients they supply.

Can parents or caregivers do anything else to help children reduce their risks of CVD?

Even though the focus of this text is nutrition, another risk factor for CVD that starts in childhood and carries over into adulthood must also be addressed—cigarette smoking. Each day 3000 children light up for the first time—typically, in grade school. Among high school students, two out of three

have tried smoking, and one in seven smokes regularly.[29] Approximately 80 percent of all adult smokers began smoking before the age of 18.

Of those adolescents who continue smoking, half will eventually die of smoking-related causes. Efforts to teach children about the dangers of smoking need to be aggressive. Children are not likely to consider the long-term health consequences of tobacco use. They are more likely to be struck by the immediate health consequences, such as shortness of breath when playing sports, or social consequences, such as having bad breath. Whatever the context, the message to all children and teens should be clear: don't start smoking. If you've already started, quit now.

In conclusion, treatment of established obesity is notoriously unsuccessful, making prevention of childhood obesity and the diseases it engenders a high national priority.[30] Education is clearly needed. Classroom lessons that are reinforced by lunchroom offerings and other school policies often help children change their eating habits for the better.[31] When school nutrition policy conflicts with classroom teachings and adults set poor examples, however, children are likely to follow their taste buds, not nutrition teachings.

The best advice to help turn the tide of obesity and related diseases is the easiest to give and perhaps the most difficult to follow: don't smoke, choose a diet in accord with the *Dietary Guidelines for Americans 2005*, follow the USDA Food Guide (Chapter 1), and make it a habit to be physically active each day. Last, but by no means least, parents and other significant adults can help mold children's behaviors by the examples they set.

Glossary

atherosclerosis (ath-er-oh-scler-OH-sis): a type of artery disease characterized by accumulations of lipid-containing material on the inner walls of the arteries.
athero = porridge or soft
scleros = hard
osis = condition
cardiovascular disease (CVD): a general term for all diseases of the heart and blood vessels. Atherosclerosis is the main cause of CVD. When the arteries that carry blood to the heart muscle become occluded, the heart suffers damage known as **coronary heart disease (CHD).**
cardio = heart
vascular = blood vessels
fatty streaks: accumulations of cholesterol and other lipids along the walls of the arteries.
plaques: mounds of lipid material, mixed with smooth muscle cells and calcium, which develop in the artery walls in atherosclerosis.

Notes

1. J. P. Kaplan, C. T. Liverman, and V. I. Kraak, eds., *Preventing Childhood Obesity: Health in the Balance* (Washington, D.C.: National Academies Press, 2005), pp. 1–20; M. L. Cruz and coauthors, Pediatric obesity and insulin resistance: Chronic disease risk and implications for treatment and prevention beyond body weight modification, *Annual Review of Nutrition* 25 (2005): 435–468; T. Lobstein, L. Baur, and R. Uauy, Obesity in children and young people: A crisis in public health, *Obesity Reviews* 5 (2004): 4–85.

2. Kaplan, Liverman, and Kraak, eds., 2005.

3. M. Kohn and M. Booth, The worldwide epidemic of obesity in adolescents, *Adolescent Medicine* 14 (2003): 1–9; L. S. Lieberman, Dietary, evolutionary, and modernizing influences on the prevalence of type 2 diabetes, *Annual Review of Nutrition* 23 (2003): 345–377; Committee on Nutrition, American Academy of Pediatrics, Prevention of pediatric overweight and obesity, *Pediatrics* 112 (2003): 424–430.

4. R. J. F. Loos and T. Rankinen, Gene-diet interactions on body weight changes, *Journal of the American Dietetic Association* 105 (2005): S29–S34.

5. L. F. Baier and R. L. Hanson, Genetic studies of the etiology of type 2 diabetes in Pima indians, *Diabetes* 53 (2004): 1181–1186.

6. Loos and Rankinen, 2005; A. M. Kriska and coauthors, Physical activity, obesity, and the incidence of type 2 diabetes in a high-risk population, *American Journal of Epidemiology* 158 (2003): 669–675; J. Esparza and coauthors, Daily energy expenditure in Mexican and USA Pima Indians: Low physical activity as a possible cause of obesity, *International Journal of Obesity and Related Metabolic Disorders* 24 (2000): 55–59.

7. L. Adair and D. Dahly, Developmental determinants of blood pressure in adults, *Annual Review of Nutrition* 25 (2005): 407–434; G. Wu and coauthors, Maternal nutrition and fetal development, *Journal of Nutrition* 134 (2004): 2169–2172.

8. Adair and Dahly, 2005.

9. Adair and Dahly, 2005.

10. Adair and Dahly, 2005.

11. D. J. Barker and coauthors, Growth and living conditions in childhood and hypertension in adult life: A longitudinal study, *Journal of Hypertension* 20 (2002): 1951–1956; S. M. Robinson and D. J. Barker, Coronary heart disease: A growth disorder, *Proceedings of the Nutrition Society* 61 (2002): 537–542.

12. T. S. Hannon, F. Rao, and S. A. Arslanian, Childhood obesity and type 2 diabetes mellitus, *Pediatrics* 116 (2005): 473–480; D. S. Ludwig and C. B. Ebbeling, Type 2 diabetes mellitus in children: Primary care and public health considerations, *Journal of the American Medical Association* 286 (2001): 1427–1430.

13. G. S. Boyd and coauthors, Effect of obesity and high blood pressure on plasma lipid levels in children and adolescents, *Pediatrics* 116 (2005): 442–446; R. Kohen-Avramoglu, A. Theriault, and K. Adeli, Emergence of the metabolic syndrome in childhood: An epidemiological overview and mechanistic link to dyslipidemia, *Clinical Biochemistry* 36 (2003): 413–420.

14. S. Li and coauthors, Childhood cardiovascular risk factors and carotid vascular changes in adulthood: The Bogalusa Heart Study, *Journal of the American Medical Association* 290 (2003): 2271–2276; K. B. Keller and L. Lemberg, Obesity and the metabolic syndrome, *American Journal of Clinical Care* 12 (2003): 167–170.

15. V. N. Muratova and coauthors, The relation of obesity to cardiovascular risk factors among children: The CARDIAC project, *West Virginia Medical Journal* 98 (2002): 263–267.

16. O. Fiedland and coauthors, Obesity and lipid profiles in children and adolescents, *Journal of Pediatric Endocrinology and Metabolism* 15 (2002): 1011–1016; T. Dwyer and coauthors, Syndrome X in 8-y-old Australian children: Stronger associations with current body fatness than with infant size or growth, *International Journal of Obesity and Related Metabolic Disorders* 26 (2002): 1301–1309.

17. B. Glowinska, M. Urban, and A. Koput, Cardiovascular risk factors in children with obesity, hypertension and diabetes: Lipoprotein(a) levels and body mass index correlate with family history of cardiovascular disease, *European Journal of Pediatrics* 161 (2002): 511–518.

18. A. Wiegman and coauthors, Family history and cardiovascular risk in familial hypercholesterolemia: Data in more than 1000 children, *Circulation* 107 (2003): 1473–1478.

19. National High Blood Pressure Education Program Working Group on High Blood Pressure in Children and Adolescents, The Fourth Report on the Diagnosis, Evaluation, and Treatment of High Blood Pressure in Children and Adolescents, *Pediatrics* 114 (2004): 555S–576S.

20. Kaplan, Liverman, and Kraak, 2005.

21. S. A. Lederman and coauthors, Summary of the presentations at the Conference on Preventing Childhood Obesity, December 8, 2003, *Pediatrics* 114 (2004): 1146–1173.

22. S. Kranz, A. M. Siega-Riz, and A. H. Herring, Changes in diet quality of American preschoolers between 1977 and 1998, *American Journal of Public Health* 94 (2004): 1525–1530; S. J. Nielsen, A. M. Siega-Riz, and B. M. Popkin, Trends in energy intake in U.S. between 1977 and 1996: Similar shifts seen across age groups, *Obesity Research* 10 (2002): 370–378.

23. Committee on School Health, American Academy of Pediatrics, Soft drinks in schools, *Pediatrics* 113 (2004): 152–154.

24. D. S. Ludwig, K. E. Peterson, and L. S. Gortmaker, Relation between consumption of sugar-sweetened drinks and childhood obesity: A prospective, observational analysis, *Lancet* 357 (2001): 505–508.

25. M. H. Proctor and coauthors, Television viewing and change in body fat from preschool to early adolescence: The Framingham Children's Study, *International Journal of Obesity and Related Metabolic Disorders* 27 (2003): 827–833.

26. Standing Committee on the Scientific Evaluation of Dietary Reference Intakes, Food and Nutrition Board, Institute of Medicine, *Dietary Reference Intakes for Energy, Carbohydrate, Fiber, Fat, Fatty Acids, Cholesterol, Protein, and Amino Acids* (Washington, D.C.: National Academies Press, 2005), pp. 769–879.

27. E. Satter, A moderate view on fat restriction for young children, *Journal of the American Dietetic Association* 100 (2000): 32–35.

28. J. Hurley and B. Liebman, Kids' cuisine: "What would you like with your fries?" *Nutrition and Action Healthletter* 31 (2004): 12–15.

29. Trends in cigarette smoking among high school students—United States, 1991–2001, *Morbidity and Mortality Weekly Report* 51 (2002): 409–412.

30. S. Caprio and M. Genel, Confronting the epidemic of childhood obesity, *Pediatrics* 115 (2005): 494–495.

31. S. M. Gross and B. Cinelli, Coordinated school health program and dietetics professionals: Partners in promoting healthful eating, *Journal of the American Dietetic Association* 104 (2004): 793–798.

CONTENTS

Nutrition through the Life Span: Later Adulthood

CHAPTER 12

The last two chapters were devoted to stages of the life cycle that require special nutrition attention: pregnancy, lactation, infancy, childhood, and adolescence. Much of the text before that focused on nutrition to support wellness during adulthood. This chapter describes the special nutrition needs of the later adult years.

The most urgent nutrition need of older people, however, is to have made good food choices in the past! All of life's nutrition choices incur health consequences for the better or for the worse. A single day's intakes of nutrients may exert only a minute effect on body organs and their functions, but over years and decades, the repeated effects accumulate to have major impacts. This being the case, it is of great importance for everyone, of every age, to pay close attention today to nutrition.

The U.S. population is graying. The majority of citizens are now middle-aged, and the ratio of old people to young is increasing, as Figure 12-1 shows. Our society uses the arbitrary age of 65 to define the transition point between middle age and old age, but growing "old" happens day by day, with changes occurring gradually over time. Since 1950 the population of people over 65 has more than doubled, and people over 85 years old are the fastest-growing age group.[1] The U.S. Bureau of the Census projects that by the year 2040 more than a million Americans will be 100 years old or older.

Life expectancy in the United States is 77 years, up from about 47 years in 1900.[2] Women today live about 7 years longer than men. Advances in medical science—antibiotics and other treatments—are largely responsible for almost doubling the life expectancy since 1900. Improved nutrition and an abundant food supply have also contributed to lengthening life expectancy. The human **life span,** currently estimated at 130 years, is the upper limit of human **longevity,** even given optimal nutrition. With work progressing in medical and genetic technologies, however, the human life span may be extended significantly.

The study of the aging process is among the youngest of the scientific disciplines. Not until the twentieth century did human beings achieve a life expectancy worthy of a science devoted to studying it. The idea that nutrition can influence the way the human body ages is particularly appealing because diet is a factor that people can control and change.

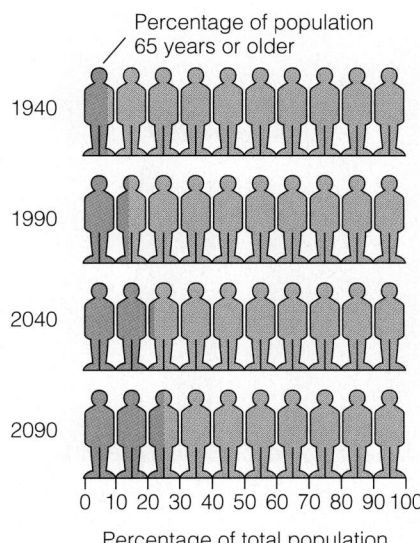

Percentage of population 65 years or older

1940

1990

2040

2090

0 10 20 30 40 50 60 70 80 90 100
Percentage of total population

FIGURE 12-1 The Aging of the U.S. Population
In 1970, 6.8 percent of the population was 65 or older. In 1990, 12.7 percent of us had reached age 65; by 2040, 21.7 percent will have reached age 65; and by 2090, nearly one out of four Americans will be 65 or older.

Nutrition and Longevity

What has been learned so far about the effects of nutrition and environment on longevity provides incentive for researchers to keep asking questions about how and why human beings age. Among their questions are:

- To what extent is aging inevitable, and can it be slowed through changes in lifestyle and environment?
- What roles does nutrition play in aging, and what roles can it play in retarding aging?

With respect to the first question, aging is a natural process, programmed into the genes at conception. People can, however, slow the process within the natural limits set by heredity. They can adopt healthy lifestyle habits such as eating nutritious food and engaging in physical activity. In fact, an estimated 70 to 80 percent of the average person's life expectancy may depend on individual health-related behaviors; genes determine the remaining 20 to 30 percent.[3]

With respect to the second question, good nutrition helps to maintain a healthy body and can therefore ease the aging process in many significant ways.[4] Clearly, nutrition can improve the **quality of life** in the later years.

life expectancy: the average number of years lived by people in a given society.

life span: the maximum number of years of life attainable by a member of a species.

longevity: long duration of life.

quality of life: a person's perceived physical and mental well-being.

Regular physical activity promotes a healthy, independent lifestyle.

SLOWING THE AGING PROCESS

One approach researchers use to search out the secret of long life has been to study older people. Some people are young for their ages, whereas others are old for their ages. What makes the difference?

Healthy Habits Six lifestyle habits seem to have a profound influence on people's health and therefore on their **physiological age:**

- Sleeping regularly and adequately.
- Eating well-balanced meals, including breakfast, regularly.
- Maintaining a healthy body weight.
- Engaging in regular physical activity.
- Not smoking.
- Not using alcohol, or using it in moderation.

Over the years, the effects of these lifestyle choices accumulate—that is, those who follow all of these practices live longer and have fewer disabilities as they age.[5] They are in better health, even if older in **chronological age,** than people who do not adopt these behaviors. Even though people cannot alter the years of their births, they can alter the probable lengths and quality of their lives. Physical activity seems to be most influential in preventing or slowing the many changes that many people seem to accept as an inevitable consequence of old age. In other words, physical activity and long life seem to go together.[6]

Physical Activity The many and remarkable benefits of regular physical activity are not limited to the young. Compared to those who are inactive, older adults who are active weigh less; have greater flexibility, more endurance, better balance and better health; and live longer.[7] They reap additional benefits from various activities as well: aerobic activities improve cardiorespiratory endurance, blood pressure, and blood lipid concentrations; moderate endurance activities improve the quality of sleep; and strength training significantly improves posture and mobility. In fact, regular physical activity is a powerful predictor of a person's mobility in the later years. Physical activity also increases blood flow to the brain, thereby preserving mental ability, alleviating depression, and supporting independence.[8]

Muscle mass and muscle strength tend to decline with aging, making older people vulnerable to falls and immobility. Falls are a major cause of fear, injury, disability, dependence, and even death among older adults. Regular physical activity tones, firms, and strengthens muscles, helping to improve confidence, reduce the risk of falling, and minimize the risk of injury should a fall occur. Strength training, even in frail, elderly people over 85 years of age, has been shown not only to improve balance, muscle strength, and mobility, but also to increase energy expenditure and energy intake. This finding highlights another reason to be physically active: a person spending energy on physical activity can afford to eat more food and with it, more nutrients. People who are committed to an ongoing fitness program have higher energy and nutrient intakes than more sedentary people.

Activities of all kinds are recommended to maintain and promote health. Strength training improves muscle strength, which enhances a person's ability to perform many of life's daily tasks such as climbing stairs and carrying packages.[9] Muscle strength during midlife (between the ages of 45 and 65) predicts health and disability 25 years later. Greater muscle strength during midlife may act as a strength *reserve* later on, protecting older adults from disability even when chronic conditions develop. In short, improving overall strength during early and middle adulthood could potentially lower the risk of later physical disability.

Ideally, physical activity should be part of each day's schedule and should be intense enough to prevent muscle atrophy and to speed up the heartbeat and respiration rate. Although aging affects both speed and endurance to some degree,

physiological age: a person's age as estimated from her or his body's health and probable life expectancy.

chronological age: a person's age in years from his or her date of birth.

TABLE 12-1	Exercise Guidelines for Older Adults			
	Endurance	**Strength**	**Balance**	**Flexibility**
Examples				
Start easy	Be active 5 minutes on most or all days.	Using 0- to 2-pound weights, do one set of 8 repetitions twice a week.	Hold onto table or chair with one hand, then with one finger.	Hold stretch 10 seconds; do each stretch three times.
Progress gradually to goal	Be active 30 minutes (minimum) on most or all days.	Increase weight as able; do two sets of 8–15 repetitions twice a week.	Do not hold onto table or chair; then close eyes.	Hold stretch 30 seconds; do each stretch five times.
Cautions and comments	Stop if you are breathing so hard you can't talk or if you feel dizziness or chest pain.	Breathe out as you contract and in as you relax (do not hold breath); use smooth, steady movements.	Incorporate balance techniques with strength exercises as you progress.	Stretch after strength and endurance exercises for 20 minutes, three times a week; use slow, steady movements; bend joints slightly.

Source: *Exercise: A Guide from the National Institute on Aging,* **www.nia.nih.gov,** accessed May 2005.

older adults can still train and achieve exceptional performances. Healthy older adults who have not been active can ease into a suitable routine. They can start by walking short distances until they are walking at least 10 minutes continuously, and then gradually increase their distance to a 30- to 45-minute walk at least five days a week. With persistence, people can achieve great improvements at any age. Table 12-1 provides exercise guidelines for seniors; people with medical conditions should check with a physician before beginning an exercise routine, as should sedentary men over 40 and women over 50 who want to participate in a vigorous program.

Restriction of kCalories In their efforts to understand longevity, researchers have not only observed people but have also manipulated influencing factors, such as diet, in animals. This research has produced some interesting and suggestive findings. For example, animals live longer and have fewer age-related diseases when their energy intakes are restricted.

Several mechanisms to explain how energy restriction prolongs life in animals have been proposed but not proved. Research suggests that food restriction may extend the life span by preventing damaging lipid oxidation thereby delaying the onset of age-related diseases, such as atherosclerosis.[10] Gene activity also appears to play a key role. Energy restriction in animals prevents alterations in gene expression that are associated with aging.[11] Experiments with food restriction and longevity in animals have *not* suggested any direct applications to human nutrition.

Moderate energy restriction (80 percent of usual intake) in human beings may be valuable. When people restrict energy intake moderately, body weight, body fat, and blood pressure drop, and HDL cholesterol rises—favorable changes for preventing chronic diseases. The reduction in oxidative damage that occurs with energy restriction in animals also occurs in people whose diets include antioxidant nutrients and phytochemicals. Diets such as the Mediterranean diet that include an abundance of fruits, vegetables, olive oil, whole grains, and legumes—with their array of antioxidants and phytochemicals—support good health and long life.[12]

NUTRITION AND DISEASE PREVENTION

Nutrition alone, even if ideal, cannot ensure a long and robust life. Nevertheless, nutrition clearly affects aging and longevity in human beings by way of its role in disease prevention. Among the better-known relationships between nutrition and disease are the following:

- Appropriate energy intake helps prevent *obesity, diabetes,* and related *cardiovascular diseases* such as atherosclerosis and hypertension (Chapters 7, 20, and 21) and may influence the development of some forms of *cancer* (Chapter 23).
- Adequate intakes of essential nutrients prevent *deficiency diseases* such as scurvy, goiter, anemia, and the like (Chapters 8 and 9).
- Variety in food intake, as well as ample intakes of certain fruits and vegetables, may be protective against certain types of *cancer* (Chapter 23).
- Moderation in sugar intake helps prevent *dental caries* (Nutrition in Practice 3).
- Appropriate fiber intakes help prevent disorders of the digestive tract such as *constipation, diverticulosis,* and possibly *colon cancer* (Chapters 3 and 18).
- Moderate sodium intake and adequate intakes of potassium, calcium, and other minerals help prevent *hypertension* (Chapters 9 and 21).
- An adequate calcium intake throughout life helps protect against *osteoporosis* (Chapter 9).

Other, less well-established links between nutrition and disease are being discovered each day. Research that focuses on how life factors affect aging and disease processes is vital to ensuring that more and more people can look forward to long, healthy lives.

> ### REVIEW NOTES
>
> Life expectancy in the United States increased dramatically in the twentieth century.
>
> Factors that enhance longevity include limited or no alcohol use, regular balanced meals, weight control, adequate sleep, abstinence from smoking, and regular physical activity.
>
> Nutrition alone, even if ideal, cannot guarantee a long and robust life. At the very least, however, nutrition—especially when combined with regular physical activity—can influence aging and longevity in human beings by supporting good health and preventing disease.

Nutrition-Related Concerns during Late Adulthood

Nutrition through the prime years may play a greater role than has been realized in preventing many changes once thought to be inevitable consequences of growing older. The following discussions of cataracts and macular degeneration, arthritis, and the aging brain show that nutrition may provide at least some protection against some of the conditions commonly associated with aging.

CATARACTS AND MACULAR DEGENERATION

cataracts: thickenings of the eye lenses that impair vision and can lead to blindness.

Cataracts are age-related thickenings in the lenses of the eye that impair vision. If not surgically removed, they ultimately lead to blindness.

Oxidative stress appears to play a significant role in the development of cataracts, and the antioxidant nutrients and phytochemicals in fruits and vegetables may help minimize the damage.[13] Some studies suggest that a diet providing ample carotenoids, vitamin C, and vitamin E may be especially important for preventing early onset of cataracts.[14] Also, people who follow the *Dietary Guidelines for Americans* are reported to have fewer cataracts, but cataracts can occur even in well-nourished individuals due to exposure to ultraviolet light, oxidative damage, viral infections, toxic substances, genetic disorders, injury, or other trauma.[15] Most cataracts are vaguely called senile cataracts, meaning "caused by aging." In the United States, more than half of all adults age 65 and older have a cataract.

One other diet-related factor may play a role in cataract development: overweight. In a study of more than 100,000 men and women, those with a body mass index (BMI) above 30 had a significantly greater risk of cataracts compared to those with a BMI less than 23.[16] How obesity may influence the development of cataracts is not known, but researchers speculate that conditions that typically accompany obesity such as glucose intolerance and insulin resistance may provide clues.

Another cause of visual loss among older people is **macular degeneration,** a deterioration of the macular region of the eye. Age-related macular degeneration (AMD) is the leading cause of blindness in individuals over 65 years of age in the United States; nearly 2 million people are afflicted with the disease.[17] AMD is more prevalent among white people than among black people, and the prevalence of AMD increases dramatically with age—more than 15 percent of white women over the age of 80 have AMD. As with cataracts, risk factors for AMD include oxidative stress from sunlight; some studies have found that supplements of antioxidant nutrients plus zinc and of the carotenoids lutein and zeaxanthin may help to protect against AMD.[18] Diets high in omega-3 fatty acids from fish may also offer some protection against AMD.[19] The omega-3 fatty acid docosahexaenoic acid (DHA) is a major structural lipid in the membrane of the retina, and DHA also affects signaling pathways in the retina.[20]

macular (MACK-you-lar) **degeneration:** deterioration of the macular area of the eye that can lead to loss of central vision and eventual blindness. The **macula** is a small, oval, yellowish region in the center of the retina that provides the sharp, straight-ahead vision so critical to reading and driving.

arthritis: inflammation of a joint, usually accompanied by pain, swelling, and structural changes.

osteoarthritis: a painful, chronic disease of the joints that occurs when the cushioning cartilage in a joint breaks down; joint structure is usually altered, with loss of function; also called *degenerative arthritis.*

ARTHRITIS

Arthritis is the leading cause of disability among older adults—affecting close to 60 percent of all older people.[21] The most common type of **arthritis** that disables older people is **osteoarthritis,** a painful swelling of the joints. During movement, the ends of bones are normally protected from wear by cartilage and by small sacs of fluid that lubricate the joint. With age, the cartilage sometimes disintegrates, and the joints become malformed and painful to move. Osteoarthritis afflicts millions of people around the world, especially the elderly.

One known connection between osteoarthritis and nutrition is overweight. Weight loss can help overweight people with osteoarthritis, partly because the joints affected are often weight-bearing joints that are stressed and irritated by having to carry excess poundage. Interestingly, though, weight loss often relieves the worst pain of osteoarthritis in the hands as well, even though they are not weight-bearing joints. Jogging and other weight-bearing activities do not worsen osteoarthritis. In fact, both aerobic activity and weight training offer modest improvements in physical performance and pain relief.[22]

Nutrition quackery to treat arthritis is abundant, but no one universally effective diet for arthritis relief is known. Table 12-2 presents some of the many *non*effective dietary

TABLE 12-2	*Non*effective Dietary Strategies for Arthritis
Alfalfa tea	Garlic
Aloe vera liquid	Honey
Amino acid supplements	Inositol
Blackstrap molasses	Kelp
Burdock root	Lecithin
Calcium	Para-amino benzoic acid (PABA)
Celery juice	Raw liver
Cod liver oil	Superoxide dismutase (SOD)
Copper supplements	Vitamin D
Dimethyl sulfoxide (DMSO)	Vitamin megadoses
Fasting	Watercress
Fresh fruit	Yeast

treatments for osteoarthritis. Traditional medical intervention for arthritis includes medication and surgery. Two popular supplements for treating osteoarthritis—glucosamine and chondroitin—may indeed alleviate pain and improve mobility, and they may even slow the progression of osteoarthritis if substantial additional research supports findings to date.[23]

Another type of arthritis, known as **rheumatoid arthritis,** has a possible link to diet through the immune system. In rheumatoid arthritis, the immune system mistakenly attacks the bone coverings as if they were made of foreign tissue. In some individuals, certain foods, notably vegetables and olive oil, may moderate the inflammatory responses and provide some relief.[24]

Another nutrient linked to rheumatoid arthritis is the omega-3 fatty acid found in fish oil, eicosapentaenoic acid (EPA). Research shows that the same diet recommended for heart health—one low in saturated fat from meats and milk products and high in oils from fish—helps prevent or reduce the inflammation in the joints that makes arthritis so painful.[25] Researchers theorize that EPA probably interferes with the action of prostaglandins, compounds involved in inflammation.

rheumatoid arthritis: a disease of the immune system involving painful inflammation of the joints and related structures.

neurons: nerve cells; the structural and functional units of the nervous system. Neurons initiate and conduct nerve transmissions.

cerebral cortex: the outer surface of the cerebrum, which is the largest part of the brain.

senile dementia: the loss of brain function beyond the normal loss of physical adeptness and memory that occurs with aging.

Alzheimer's disease: a progressive, degenerative disease that attacks the brain and impairs thinking, behavior, and memory.

Both foods and mental challenges nourish the brain.

THE AGING BRAIN

The brain, like all of the body's organs, responds to both inherited and environmental factors that can enhance or diminish its amazing capacities. One of the challenges researchers face when studying the aging of the human brain is to distinguish among changes caused by normal, age-related, physiological processes; changes caused by diseases; and changes caused by cumulative, extrinsic factors such as diet.

The brain normally changes in some characteristic ways as it ages. For one thing, its blood supply decreases. For another, the number of **neurons,** the brain cells that specialize in transmitting information, diminishes as people age. When the number of nerve cells in one part of the **cerebral cortex** diminishes, hearing and speech are affected. Losses of neurons in other parts of the cortex can impair memory and cognitive function. When the number of neurons in the hindbrain diminishes, balance and posture are affected. Losses of neurons in other parts of the brain affect still other functions.

Nutrient Deficiencies and Brain Function Clinicians now recognize that much of the cognitive loss and forgetfulness generally attributed to aging is due in part to extrinsic, and therefore controllable, factors such as nutrient deficiencies. The ability of neurons to synthesize specific neurotransmitters depends in part on the availability of precursor nutrients that are obtained from the diet.[26] The neurotransmitter serotonin, for example, derives from the amino acid tryptophan. To function properly, the enzymes involved in neurotransmitter synthesis require vitamins and minerals. Thus nutrient deficiencies may contribute to the loss of memory and cognition that some older adults experience.[27] Such losses may be preventable or at least diminished or delayed through diet. Table 12-3 summarizes some of the better-known connections between brain function and nutrients.

In some instances, the degree of cognitive loss is extensive. Such **senile dementia** may be attributable to a specific disorder such as a brain tumor or Alzheimer's disease.

Alzheimer's Disease In **Alzheimer's disease,** the most prevalent form of senile dementia, brain cell death occurs in the areas of the brain that coordinate memory and cognition. Alzheimer's disease afflicts 4.5 million people in the United States, and that number is expected to almost triple by the year 2050.[28] Diagnosis of Alzheimer's depends on its characteristic symptoms: the victim gradually loses memory and reasoning, the ability to communicate, physical capabilities, and eventually life itself.

Researchers are closing in on the cause of Alzheimer's disease.* Clearly, genetic factors are involved.[29] Free radicals may also be involved.[30] Nerve cells in the brains of people with Alzheimer's disease show evidence of free-radical attack—damage to DNA, cell membranes, and proteins. They also show evidence of the minerals that trigger free-radical attacks—iron, copper, zinc, and aluminum. Some research suggests that the antioxidant nutrients can limit free-radical damage and delay or prevent Alzheimer's disease.[31]

Most people have heard of an association between aluminum and the development of Alzheimer's, although a causal connection seems unlikely. Brain concentrations of aluminum in people with Alzheimer's exceed normal brain concentrations by some 10 to 30 times, but blood and hair aluminum remains normal, indicating that the accumulation is caused by something in the brain itself, not by an overload of aluminum in the body. Thus the high brain aluminum must be at least partly a result, rather than a cause, of Alzheimer's.

Some research suggests a connection between elevated levels of the amino acid homocysteine and Alzheimer's.[32] As noted in Chapter 8, research shows a correlation between elevated blood levels of homocysteine and atherosclerosis. Researchers recognize an association between atherosclerosis and Alzheimer's, which lends support to the hypothesis that elevated blood levels of homocysteine may play a role in Alzheimer's disease. The possible link between elevated homocysteine and Alzheimer's, in turn, raises the question whether adequate amounts of B vitamins such as vitamin B_6, folate, and vitamin B_{12} may help prevent Alzheimer's. Increased intakes (slightly above the RDA) of vitamin B_6, folate, and vitamin B_{12} help to lower homocysteine. Indeed, some research shows that low blood concentrations of folate and vitamin B_{12} may be risk factors for the development of Alzheimer's disease.[33]

Treatment for Alzheimer's disease involves providing care to clients and support to their families. There is no cure for Alzheimer's disease, but medications are used to treat some of the symptoms. The Food and Drug Administration has approved several prescription medications for people with mild-to-moderate Alzheimer's disease.[34] The medications act to increase the level in the brain of acetylcholine, which is essential to memory. The drugs have effects on the symptoms of the disease, but there is no evidence that they slow its progression.

Maintaining appropriate body weight may be the most important nutrition concern for the person with Alzheimer's. Depression and forgetfulness can lead to poor food intake, and restlessness may increase energy needs. Perhaps the best that a caregiver can do nutritionally for a person with Alzheimer's is to supervise food planning and mealtimes. Providing well-liked and well-balanced meals and snacks in a cheerful atmosphere encourages food consumption. To minimize confusion, offer a few ready-to-eat foods, in bite-size pieces, with seasonings and sauces. To avoid mealtime disruptions, control distractions such as television, children, and the telephone.

TABLE 12-3	Summary of Nutrient-Brain Relationships
Brain Function	**Depends on an Adequate Intake of:**
Short-term memory	Vitamin B_{12}, vitamin C, vitamin E
Performance in problem-solving tests	Riboflavin, folate, vitamin B_{12}, vitamin C
Mental health	Thiamin, niacin, zinc, folate
Cognition	Folate, vitamin B_6, vitamin B_{12}, iron, vitamin E
Vision	Essential fatty acids, vitamin A
Neurotransmitter synthesis	Tyrosine, tryptophan, choline

REVIEW NOTES

Cataracts, age-related macular degeneration, and arthritis afflict millions of older adults, while others face senile dementia and other losses of brain function.

Some of these problems may be inevitable, but others are preventable, and good nutrition may play a key preventive role.

*A report on the genetic and other aspects of Alzheimer's is available from Alzheimer's Disease Education and Referral Center, P.O. Box 8250, Silver Springs, MD 20907-8250.

Energy and Nutrient Needs during Late Adulthood

Knowledge about the nutrient needs and nutrition status of older adults has grown considerably in the last decade or so. The Dietary Reference Intakes (DRI) cluster people over the age of 50 into two age categories—51 to 70 years old, and 71 years and older. Research is showing that the nutrition needs of people 50 to 70 years old may be very different from those of people over 70—a pattern that makes sense considering the wide age spans involved.

Setting standards for older people is difficult, though, because individual differences become more pronounced as people grow older. One person may tend to omit vegetables from his diet, and by the time he is old, he will have an associated set of nutrition problems. Another may have omitted milk and milk products all her life—her nutrition problems will be different. Also, as people age, they may suffer different chronic diseases and take different medications—both have impacts on nutrient needs. Table 12-4 lists some changes of aging that can affect nutrition. Even before all this, people start out with different genetic predispositions and ways of handling nutrients, and the effects of these become magnified over the years. Researchers have difficulty even defining "healthy aging," a prerequisite to developing recommendations that are designed to meet the "needs of practically all healthy persons." Still, some generalizations are valid. The next sections give special attention to a few nutrients of concern.

ENERGY AND ENERGY NUTRIENTS

Energy needs decline with advancing age. As a general rule, adult energy needs decline an estimated 5 percent per decade. For one thing, as people age, they usually reduce their physical activity, although they need not do so. For another, lean body mass diminishes, slowing the basal metabolic rate. Loss of muscle mass, known as **sarcopenia,** can be significant in the later years (its prevalence is more than 50 percent among those older than 80), and its consequences, dramatic (see Figure 12-2).[35] As skeletal muscle mass diminishes, people lose their ability to move and to maintain balance, making falls likely. The limitations that accom-

TABLE 12-4	Examples of Physical Changes of Aging That Affect Nutrition
Mouth	Tooth loss, gum disease, and reduced salivary output impede chewing and swallowing. Swallowing disorders and choking may become likely. Discomfort and pain associated with eating may reduce food intake.
Digestive tract	Intestines lose muscle strength resulting in sluggish motility that leads to constipation (see Chapter 18). Stomach inflammation, abnormal bacterial growth, and greatly reduced acid output impair digestion and absorption. Pain may cause food avoidance or reduced intake.
Hormones	For example, the pancreas secretes less insulin and cells become less responsive, causing abnormal glucose metabolism.
Sensory organs	Diminished senses of smell and taste can reduce appetite; diminished sight can make food shopping and preparation difficult.
Body composition	Weight loss and decline in lean body mass lead to lowered energy requirements. May be preventable or reversible through physical activity.
Urinary tract	Increased frequency of urination may limit fluid intake.

sarcopenia (SAR-koh-PEE-nee-ah): loss of skeletal muscle mass, strength, and quality.

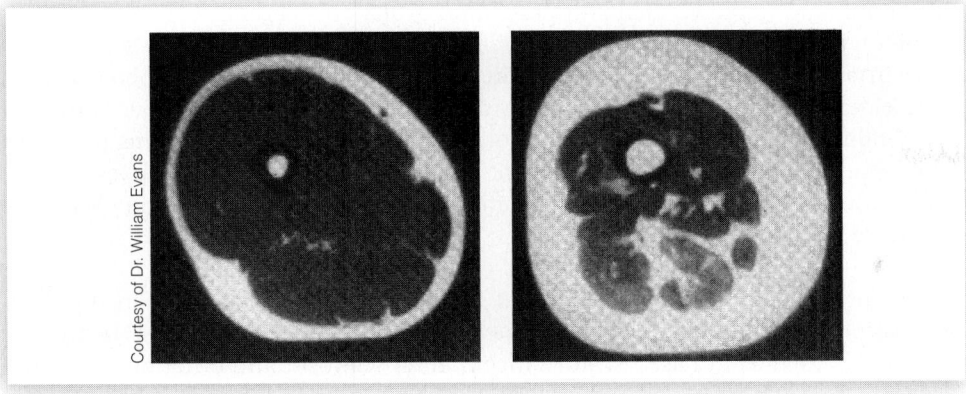

Courtesy of Dr. William Evans

FIGURE 12-2 Sarcopenia

These cross sections of two women's thighs may appear to be about the same size from the outside, but the 20-year-old woman's thigh (left) is dense with muscle tissue. The 64-year-old woman's thigh (right) has lost muscle and gained fat, changes that may be largely preventable with strength-building physical activities.

pany the loss of muscle mass and strength play a key role in the diminishing health that often accompanies aging. To some extent, however, declines in lean body mass and energy needs may not be entirely inevitable. Physical activity may be a key, as may a nutrient-dense diet.[36] Optimal nutrition and regular physical activity can help maintain muscle mass and strength and minimize the changes in body composition associated with aging.[37] Physical activity not only increases energy expenditure but, along with sound nutrition, enhances bone density and supports many body functions as well.

The lower energy expenditures of many older adults require that they eat less food energy to maintain their weights. Accordingly, the energy RDA for adults decreases slightly, beginning at age 51. Energy intakes typically decline in parallel with needs. Still, many older adults are overweight, indicating that their food intakes do not decline enough to compensate for their reduced energy expenditures. Overweight and obesity in older adults increase the risk for many diseases and for disabilities as well.[38]

On limited energy allowances, people must select mostly nutrient-dense foods. There is little leeway for sugars, fats, oils, or alcohol. Older adults can follow the USDA Food Guide (see pp. 18–19), making sure to choose the recommended amounts of food from each food group daily that are appropriate to their energy needs (see Table 1-9 p. 21).

Protein The protein needs of older adults appear to be about the same as those of younger people. Protein is especially important for older adults to support a healthy immune system and to prevent muscle wasting. Since energy needs decrease, however, the protein has to be obtained from low-kcalorie sources of high-quality protein, such as lean meats, poultry, fish, and eggs; fat-free and low-fat milk products; legumes; and grains.

Underweight or malnourished older adults need protein- and energy-dense snacks such as hard-boiled eggs, tuna fish and crackers, peanut butter on graham crackers, and hearty soups. Drinking liquid nutritional formulas between meals can also boost energy and nutrient intakes.

Carbohydrate and Fiber As always, abundant carbohydrate is needed to protect protein from being used as an energy source. The recommendation to obtain ample amounts of mostly whole-grain breads, cereals, rice, and pasta holds true for older people. As for younger people, a steady supply of carbohydrate is essential for optimal brain functioning.[39] With age, fiber takes on extra importance for its role in alleviating constipation, a common complaint among older adults and among nursing home residents in particular. Fruits and vegetables supply viscous (soluble) fibers thought to help ward off diseases of aging, but factors such as transportation problems, limited cooking facilities, and chewing problems limit some elderly people's intakes of fresh fruits and vegetables.[40] Most older adults do not obtain the recommended daily 25 or more grams of fiber (14 grams per 1000 kcalories).[41] When low fiber intakes are combined with low fluid intakes, inadequate physical activity, and constipating medications, constipation becomes inevitable.

Fat As is true for people of all ages, fat intake needs to be moderate in the diets of most older adults—enough to enhance flavors and provide valuable nutrients, but not so much as to raise risks of cancer, atherosclerosis, and other degenerative diseases. The recommendation should not be taken too far; limiting fat too severely may lead to nutrient deficiencies and weight loss—two problems that carry greater health risks in the elderly than overweight.

WATER

Dehydration is a risk for older adults, who may not notice or pay attention to their thirst, or who find it difficult and bothersome to get a drink or to get to a bathroom.[42] Older adults who have lost bladder control may be afraid to drink too much water. Despite real fluid needs, older people do not seem to feel as thirsty or notice mouth dryness as readily as younger people. Many nursing home employees say it is hard to persuade their elderly clients to drink enough water and fruit juices.

Total body water decreases as people age, so even mild stresses such as fever or hot weather can precipitate rapid dehydration in older adults. Dehydrated older adults seem to be more susceptible to urinary tract infections, pneumonia, pressure ulcers, confusion, and disorientation.[43] An intake of 9 cups a day of total beverages, including water, is recommended for women; for men, the recommendation is 13 cups a day of total beverages, including water.[44]

VITAMINS AND MINERALS

As research reveals more about how specific vitamins and minerals influence disease prevention and how age-related physiological changes affect nutrient metabolism, optimal intakes of vitamins and minerals for different groups of older adults are being defined. This section highlights the vitamins and minerals of greatest concern to older adults.

Vitamin D Older adults face a greater risk of vitamin D deficiency than younger people do. Only vitamin D–fortified milk provides significant vitamin D, and many older adults drink little or no milk. Consequently, the vitamin D intakes of many older adults are less than half of recommendations. Further compromising the vitamin D status of many older people, especially those in nursing homes, is their limited exposure to sunlight. Finally, aging reduces the skin's capacity to make vitamin D and the kidneys' ability to convert it to its active form. The margin lists Vitamin D recommendations to prevent bone loss and to maintain vitamin D status in older people, especially those who engage in minimal outdoor activity.

Vitamin D AI during late adulthood:
- 10 μg/day (51–70 yr).
- 15 μg/day (>70 yr).

Vitamin B$_{12}$ The DRI committee recommends that adults aged 51 years and older obtain 2.4 micrograms of vitamin B$_{12}$ daily *and* that vitamin B$_{12}$–fortified foods (such as fortified cereals) or supplements be used to meet much of the DRI recommended intake.[45] The committee's recommendation reflects the finding that between 10 and 30 percent of people older than 50 years lose the ability to produce enough stomach acid to make the protein-bound form of vitamin B$_{12}$ avail-

Vitamin B$_{12}$ RDA during late adulthood:
- 2.4 μg/day.

able for absorption. Synthetic vitamin B_{12} is reliably absorbed, however. Given the poor cognition, anemia, and devastating neurological effects associated with a vitamin B_{12} deficiency, an adequate intake is imperative.

One cause of the malabsorption of protein-bound vitamin B_{12} is a condition known as **atrophic gastritis.** The prevalence of atrophic gastritis among those 60 years of age and older is high.

Folate As is true of vitamin B_{12}, folate intakes of older adults typically fall short of recommendations. The elderly are also more likely to have medical conditions or to take medications that can compromise folate status.

Iron Among the minerals, iron deserves first mention. Iron-deficiency anemia is less common in older adults than in younger people, but it still occurs in some, especially in those with low food energy intakes. Aside from diet, other factors in many older people's lives make iron deficiency likely: chronic blood loss from disease conditions and medicines, and poor iron absorption due to reduced secretion of stomach acid and antacid use. Anyone concerned with older people's nutrition should keep these possibilities in mind.

Zinc Zinc intake is commonly low in older people. Zinc deficiency can depress the appetite and blunt the sense of taste, thereby leading to low food intakes and worsening of zinc status. Many medications that older adults commonly use can impair zinc absorption or enhance its excretion and thus lead to deficiency.

Calcium The importance of abundant dietary calcium throughout life to protect against osteoporosis has been emphasized throughout this book. The calcium intakes of many people, especially women, in the United States are well below the recommended amount. If milk causes stomach discomfort, as many older adults report, then lactose-modified milk or other calcium-rich foods should take its place.

NUTRIENT SUPPLEMENTS FOR OLDER ADULTS

People judge for themselves how to manage their nutrition, and some turn to supplements. Advertisers target older people with appeals to take supplements and eat "health" foods, claiming that these products prevent disease and promote longevity. About 60 percent of all women over 60 years of age take some type of nutrient supplement, and about 40 percent of older men do.[46] Quite often those who take supplements are not deficient in the nutrients being supplemented.

Elderly people often benefit from a balanced low-dose vitamin and mineral supplement, however.[47] Such supplements supply many of the needed minerals along with the vitamins often lacking in older people's diets without providing too much of any one nutrient. Many times, those taking such supplements suffer fewer infectious diseases.[48]

Food is still the best source of nutrients for everybody, however. Supplements are just that—supplements to foods, not substitutes for them. For anyone who is motivated to obtain the best possible health, it is never too late to learn to eat well, become physically active, and adopt other lifestyle changes such as quitting smoking, moderating alcohol use, and the like. Table 12-5 (p. 340) summarizes the nutrient concerns of aging. Table 12-6 (p. 340) offers strategies for growing old healthfully.

THE EFFECTS OF DRUGS ON NUTRIENTS

As people grow older, the use of medicines—from over-the-counter types such as aspirin and laxatives to prescription medications of all kinds—becomes commonplace. Most medications interact with one or more nutrients in several ways, usually resulting in greater-than-normal needs for these nutrients. Chapter 14 offers

Folate RDA during late adulthood:
• 400 µg/day.

Iron RDA during late adulthood:
• 8 mg/day.

Zinc RDA during late adulthood:
• 11 mg/day (men).
• 8 mg/day (women).

Calcium AI during late adulthood:
• 1200 mg/day.

atrophic gastritis (a-TRO-fik gas-TRI-tis): a condition characterized by chronic inflammation of the stomach accompanied by a diminished size and functioning of the mucosa and glands.

TABLE 12-5	Summary of Nutrient Concerns of Aging	

Nutrient	Effect of Aging	Comments
Water	Lack of thirst and decreased total body water make dehydration likely.	Mild dehydration is a common cause of confusion. Difficulty obtaining water or getting to the bathroom may compound the problem.
Energy	Need decreases as muscle mass decreases (sarcopenia).	Physical activity moderates the decline.
Fiber	Likelihood of constipation increases with low intakes and changes in the GI tract.	Inadequate water intakes and lack of physical activity, along with some medications, compound the problem.
Protein	Needs may stay the same or increase slightly.	Low-fat, high-fiber legumes and grains meet both protein and other nutrient needs.
Vitamin B_{12}	Atrophic gastritis is common.	Deficiency causes neurological damage; supplements may be needed.
Vitamin D	Increased likelihood of inadequate intake; skin synthesis declines.	Daily sunlight exposure in moderation or supplements may be beneficial.
Calcium	Intakes may be low; osteoporosis is common.	Stomach discomfort commonly limits milk intake; calcium substitutes or supplements may be needed.
Iron	In women, status improves after menopause; deficiencies are linked to chronic blood losses and low stomach acid output.	Adequate stomach acid is required for absorption; antacid or other medicine use may aggravate iron deficiency; vitamin C and meat increase absorption.

TABLE 12-6	Strategies for Growing Old Healthfully

- Choose nutrient-dense foods.

- Be physically active. Walk, run, dance, swim, bike, or row for aerobic activity. Lift weights, do calisthenics, or pursue some other activity to tone, firm, and strengthen muscles. Practice balancing on one foot or doing simple movements with your eyes closed. Modify activities to suit changing abilities and tastes.

- Maintain appropriate body weight.

- Reduce stress (cultivate self-esteem, maintain a positive attitude, manage time wisely, know your limits, practice assertiveness, release tension, and take action).

- For women, discuss with a physician the risks and benefits of estrogen replacement therapy.

- For people who smoke, discuss with a physician strategies and programs to help you quit.

- Expect to enjoy sex, and learn new ways of enhancing it.

- Use alcohol only moderately, if at all; use drugs only as prescribed.

- Take care to prevent accidents.

- Expect good vision and hearing throughout life; obtain glasses and hearing aids if necessary.

- Take care of your teeth; obtain dentures if necessary.

- Be alert to confusion as a disease symptom, and seek diagnosis.

- Take medications as prescribed; see a physician before self-prescribing medicines or herbal remedies and a registered dietitian before self-prescribing supplements.

- Control depression through activities and friendships; seek professional help if necessary.

- Drink plenty of water every day.

- Practice mental skills. Keep on solving math problems and crossword puzzles, playing cards or other games, reading, writing, imagining, and creating.

- Make financial plans early to ensure security.

- Accept change. Work at recovering from losses; make new friends.

- Cultivate spiritual health. Cherish personal values. Make life meaningful.

- Go outside for sunshine and fresh air as often as possible.

- Be socially active—play bridge, join an exercise or dance group, take a class, teach a class, eat with friends, volunteer time to help others.

- Stay interested in life—pursue a hobby, spend time with grandchildren, take a trip, read, grow a garden, or go to the movies.

- Enjoy life.

a discussion of diet-drug interactions and describes the many reasons why elderly people are vulnerable to such interactions.

The most common drug that can affect nutrition in older people is alcohol. The effects of alcohol on people of all ages are explained in Nutrition in Practice 19.

REVIEW NOTES

Table 12-5 summarizes the nutrient concerns of aging.

The ever-growing number of older people creates an urgent need to know more about how their nutrient requirements differ from those of others and how such knowledge can enhance their health.

Food Choices and Eating Habits of Older Adults

To provide any benefit, strategies and interventions to improve a person's nutrition status must be based on knowledge of food preferences and eating patterns. Menus and feeding programs for older adults must take into consideration not only the food likes and dislikes but also the living conditions, economic status, and medical conditions of this diverse group of people. If nutrition intervention is to be successful, it is essential to know what foods people will eat, in what settings they like to eat these foods, and whether they can buy and prepare meals.

Older people are, for the most part, independent, socially sophisticated, mentally lucid, fully participating members of society who report themselves to be happy and healthy. In fact, chronic disabilities among the elderly have declined dramatically in recent years. Older people spend more money per person on foods to eat at home than other age groups and less money on foods away from home. Manufacturers would be wise to cater to the preferences of older adults by providing good-tasting, nutritious foods in easy-to-open, single-serving packages with labels that are easy to read. Such products enable older adults to maintain their independence; most of them want to take care of themselves and need to feel a sense of control and involvement in their own lives. As discussed earlier, another way older adults can take care of themselves is by remaining or becoming physically active. Physical activity helps preserve one's ability to perform daily tasks and so promotes independence.

Individual Preferences Familiarity, taste, and health beliefs are most influential on older people's food choices. Eating foods that are familiar, especially ethnic foods that recall family meals and pleasant times, can be comforting. Older adults are choosing poultry and fish, low-fat milk and milk products, and high-fiber breads and grains, indicating that they recognize the importance of diet in supporting good health. Few older adults, however, consume the recommended amounts of milk products.

Meal Setting The food choices and eating habits of older adults are also affected by the changes in lifestyle that often accompany aging in this society. Whether people live alone, with others, or in institutions affects the way they eat. For example, men living alone are most likely to be poorly nourished. Older adults who live alone do not make poorer food choices than those who live with companions; rather, they consume too little food: loneliness is directly related to inadequacies, especially of energy intakes.

Depression Another factor affecting food intake and appetite in older people is depression. Loss of appetite and motivation to cook or even to eat frequently

Shared meals can brighten the day and enhance the appetite.

CASE STUDY *Elderly Man with a Poor Diet*

Mr. Brezenoff is 75 years old and lives alone. He has slowly been losing weight since his wife died a year ago. At 5 feet 8 inches tall, he currently weighs 124 pounds. His previous weight was 150 pounds. In talking with Mr. Brezenoff, you realize that he doesn't even like to talk about food, let alone eat it. "My wife always did the cooking before, and I ate well. Now I just don't feel like eating." You manage to find out that he skips breakfast, has soup and bread for lunch, and sometimes eats a cold-cut sandwich or a frozen dinner for supper. He seldom sees friends or relatives. Mr. Brezenoff has also lost several teeth and doesn't eat any raw fruits or vegetables because he finds them hard to chew. He lives on a meager but adequate income.

1. Consult the BMI table (inside back cover), and judge whether Mr. Brezenoff is at a healthy weight. What other assessments might you use to back up your judgment? Is his weight loss significant?
2. What factors are contributing to Mr. Brezenoff's poor food intake? What nutrients are probably deficient in his diet?
3. Look at Mr. Brezenoff as an individual and suggest ways he can improve his diet and his lifestyle.
4. What other aspects of Mr. Brezenoff's physical and mental health should you consider in helping him to improve his food intake?

Risk factors for malnutrition in older adults:

- **Disease.**
- **Eating poorly.**
- **Tooth loss or oral pain.**
- **Economic hardship.**
- **Reduced social contact.**
- **Multiple medications.**
- **Involuntary weight loss or gain.**
- **Needs assistance with self-care.**
- **Elderly person older than 80 years.**

accompanies depression. An overwhelming feeling of grief and sadness at the death of a spouse, friend, or family member may leave many people, particularly elderly people, with a feeling of powerlessness to overcome the depression. The support and companionship of family and friends, especially at mealtimes, can help overcome depression and enhance appetite. The above Case Study presents a man who has several of these problems. Use the suggestions here, and in the last sections of this chapter, to help develop solutions. The Nutrition Assessment Checklist helps to pinpoint nutrition-related factors to look for when working with older adults. To *determine* the risk of malnutrition in older clients, health care providers can keep in mind the characteristics listed in the margin.

Food Assistance Programs Federally funded programs can provide food and nutrition services for older adults.[49] The Older Americans Act (OAA) provides many different services and support to help older adults remain independent. An integral component of the OAA is the OAA Nutrition Program, formerly known as the Elderly Nutrition Program, which offers services that promote health to older adults. Table 12-7 summarizes food assistance programs available to the elderly.

Meals for Singles Many older adults live alone, and singles of all ages face challenges in purchasing, storing, and preparing food. Large packages of meat and vegetables are often intended for families of four or more, and even a head of lettuce can spoil before one person can use it all. Many singles live in small dwellings and have little storage space for foods. A limited income presents additional obstacles. This section offers suggestions that can help to solve some of the problems singles face.

People who have the means to shop and cook for themselves can cut their food bills just by being wise shoppers. Large supermarkets are usually less expensive than convenience stores. A grocery list helps reduce impulse buying, and specials and coupons can save money when the items featured are those that the shopper needs and uses.

Buying the right amount so as not to waste any food is a challenge for people eating alone. They can buy fresh milk in the size best suited for personal needs. Pint-size and even cup-size boxes of milk are available and can be stored unopened on a shelf for up to three months without refrigeration.

Many foods that offer a variety of nutrients for practically pennies have a long shelf life; staples such as rice, pastas, dry powdered milk, and dried legumes can be purchased in bulk and stored for months at room temperature. Other foods

Boxes of milk that can be stored at room temperature have been exposed to temperatures above those of pasteurization just long enough to sterilize the milk—a process called **ultrahigh temperature (UHT).**

TABLE 12-7	Food Assistance Programs for Older Adults

OAA Nutrition Program

Services: Provides congregate and home-delivered meals to improve older people's nutrition status. Includes transportation to congregate meal sites; shopping assistance; information and referral; and, to some extent, nutrition counseling and education.

Impact: Improves the nutrient content of high-risk older adults' diets and offers socialization and recreation. Many of the nutrition programs around the country go above and beyond federal requirements of congregate and home meals by offering lunch clubs, ethnic meals, and meals for older homeless people.

Food Stamp Program

Services: Supplements income for low-income households by means of a card similar to a debit card that can be used to purchase food.

Impact: Serves more as an income supplement for some elderly participants than as a device to improve nutrition status. For other elderly food stamp participants, nutrient intakes are higher than those of nonparticipants with similar incomes.

Meals on Wheels

Services: Delivers meals directly to the homebound elderly; integrated into the meal delivery services provided by the OAA Program.

Impact: Focuses on filling the need for weekend and holiday meals for homebound elderly people, a service that is limited in the OAA Nutrition Program.

Senior Farmers Market Nutrition Program

Services: Provides low-income older adults with coupons that can be exchanged for fresh fruits, vegetables, and herbs at community-supported farmers' markets and roadside stands; administered by the USDA. State agencies may limit sales to specific foods that are locally grown to encourage recipients to support farmers in their own states.

Impact: Increases fresh fruit and vegetable consumption, provides nutrition information, and even reaches the homebound elderly, a group of people who normally do not have access to farmers' markets.

that are usually a good buy include whole pieces of cheese rather than sliced or shredded cheese, fresh produce in season, variety meats such as chicken livers, and cereals that require cooking instead of ready-to-serve cereals.

A person who has ample freezing space can buy large packages of meat, such as pork chops, ground beef, or chicken, when they are on sale. Then the meat can be immediately wrapped in individual servings for the freezer. All the individual servings can be put in a bag marked appropriately with the contents and the date.

Frozen vegetables are more economical in large bags than in small boxes. The amount needed can be taken out, and the bag closed tightly with a twist tie or rubber band. If the package is returned quickly to the freezer each time, the vegetables will stay fresh for a long time.

Finally, breads and cereals usually must be purchased in larger quantities. Again the amount needed for a few days can be taken out and the rest stored in the freezer.

People who do not have freezers can ask the grocer to break open a package of wrapped meat and rewrap the portion needed. Similarly, eggs can be purchased by the half-dozen. Eggs do keep for long periods, though, if stored properly in the refrigerator.

Fresh fruits and vegetables can be purchased individually. A person can buy fresh fruit at various stages of ripeness: a ripe one to eat right away, a semiripe one to eat soon after, and a green one to ripen on the windowsill. If vegetables are packaged in large quantities, the grocer can break open the package so that a smaller amount can be purchased. Small cans of fruits and vegetables, even though they are more expensive per unit, are a reasonable alternative, considering that it is expensive to buy a regular-size can and let the unused portion spoil.

Creative chefs think of various ways to use foods when only large amounts are available. For example, a head of cauliflower can be divided into thirds. Then one-third is cooked and eaten hot. Another third is put into a vinegar and oil marinade for use in a salad. And the last third can be used in a casserole or stew.

A variety of vegetables and meats can be enjoyed stir-fried; inexpensive vegetables such as cabbage, celery, and onion are delicious when crisp cooked in a little oil with herbs or lemon added. Interesting frozen vegetable mixtures are available in larger grocery stores. Cooked, leftover vegetables can be dropped in at the last minute. A bonus of a stir-fried meal is that there is only one pan to wash. Similarly, a microwave oven allows a home chef to use fewer pots and pans. Meals and leftovers can also be frozen or refrigerated in microwavable containers to reheat as needed.

Many frozen dinners or grocery store take-out foods offer nutritious options. Adding a fresh salad, a whole-wheat roll, and a glass of milk can make a nutritionally balanced meal.

Also, single people shouldn't hesitate to invite someone to share meals with them whenever there is enough food. It's likely that the person will return the invitation, and both parties will get to enjoy companionship and a meal prepared by others.

REVIEW NOTES

Food choices of older adults are affected by health status and changed life circumstances.

Older people can benefit from both the nutrients provided and the social interaction available at congregate meals.

Other government programs deliver meals to those who are homebound.

With creativity and careful shopping, those living alone can prepare nutritious, inexpensive meals.

NUTRITION ASSESSMENT CHECKLIST FOR OLDER ADULTS

Medical History

Check the medical record for:

☐ Alcohol abuse

☐ Alzheimer's disease or other dementia or confusion

☐ Arthritis

☐ Cataracts

☐ Chronic diseases (cancer, heart disease, hypertension, diabetes)

☐ Cigarette, cigar, or pipe smoking; use of other tobacco products

☐ Constipation

☐ Dehydration

☐ Dental disease or tooth loss

☐ Depression

☐ Inflammation of the stomach (gastritis)

☐ Swallowing disorders

Medications

For older adults being treated with drug therapy for medical conditions, note:

☐ Use of multiple medications—prescription and/or over-the-counter medications such as laxatives and pain relievers

☐ Side effects that might reduce food intake or change nutrient needs

☐ Proper administration of medication with respect to food intake

☐ Malnutrition—is the person's nutrition status questionable even before considering side effects of medications that worsen nutrition status?

☐ Diminished mental capacity that might interfere with taking correct medications and doses

☐ Dehydration (can alter effects of medications)

Dietary Intake

For all older adults, especially those at risk nutritionally, assess the diet for:

☐ Total energy

☐ Protein

☐ Calcium, iron, and zinc

☐ Vitamin B_6, vitamin B_{12}, folate, and vitamin D

Note the following:

☐ Number of meals eaten each day

☐ Number and ages of people in household

☐ Amount of milk consumed each day

☐ Type and frequency of outdoor activity

☐ Type and frequency of physical activity

☐ Financial resources

☐ Transportation resources

☐ Physical disabilities

☐ Mental alertness

Anthropometric Data

Measure baseline height and weight.

☐ Reassess height and weight at each medical checkup.

☐ Note significant overweight or underweight, which warrants intervention.

☐ Use skinfold measures to reveal altered body composition that may indicate malnutrition and loss of lean tissue.

Laboratory Tests

☐ Hemoglobin, hematocrit, or other tests of iron status

☐ Serum albumin or other measures of protein status

☐ Serum folate

☐ Serum B_{12}

Physical Signs

Look for physical signs of:

☐ Protein-energy malnutrition

☐ Iron and zinc deficiency

☐ Folate deficiency

SELF CHECK

1. The fastest-growing age group in the United States is:
   a. 21 years of age.
   b. 30 to 45 years of age.
   c. 50 to 70 years of age.
   d. over 85 years of age.

2. Which of the following lifestyle habits can enhance the length and quality of people's lives?
   a. moderate smoking
   b. 6 hours of sleep daily
   c. regular physical activity
   d. skipping breakfast

3. Among the better-known relationships between nutrition and disease prevention are:
   a. appropriate fiber intake helps prevent goiter.
   b. moderate sodium intake helps prevent obesity.
   c. moderate sugar intake helps prevent hypertension.
   d. appropriate energy intake helps prevent diabetes and cardiovascular disease.

4. A disease of the immune system that involves painful inflammation of the joints is:
   a. sarcopenia.
   b. osteoarthritis.
   c. senile dementia.
   d. rheumatoid arthritis.

5. Examples of low-kcalorie, high-quality protein foods include:
   a. cottage cheese, sour cream, and eggs.
   b. green and yellow vegetables and citrus fruits.
   c. potatoes, rice, pasta, and whole-grain breads.
   d. lean meats, poultry, fish, legumes, fat-free milk, and eggs.

6. For malnourished and underweight people, protein- and energy-dense snacks include:
   a. fresh fruits and vegetables.
   b. yogurt and cottage cheese.
   c. whole grains and high-fiber legumes.
   d. hard-boiled eggs and peanut butter and graham crackers.

7. Which of the following does not explain why dehydration is a risk for older adults?
   a. They do not seem to feel thirsty.
   b. Total body water increases with age.
   c. They may find it difficult to get a drink.
   d. They may have difficulty swallowing liquids.

8. Inadequate milk intake and limited exposure to sunlight contribute to older adults' risk of:
   a. vitamin A deficiency.
   b. vitamin D deficiency.
   c. riboflavin deficiency.
   d. vitamin B_6 deficiency.

9. Two risk factors for malnutrition in older adults are:
   a. loneliness and multiple medication use.
   b. increased energy needs and lack of fiber.
   c. decreased mineral absorption and antioxidant intake.
   d. high carbohydrate intake and lack of physical activity.

10. Two strategies to improve nutrition status when growing old include:
   a. increase vitamin A intake and exercise 30 minutes daily.
   b. choose nutrient-dense foods and maintain appropriate weight.
   c. avoid high-fiber foods and take a daily vitamin-mineral supplement.
   d. eat at least one big meal per day and drink at least 10 glasses of water daily.

Answers to these questions appear in Appendix H.

CLINICAL APPLICATIONS

1. Ms. Hamilton is an 80-year-old woman in excellent health who lives alone, eats a well-balanced diet, enjoys an active social life, and walks every day. Consider the way Ms. Hamilton's health and nutrition status might be affected by the following situations:

   • Many of Ms. Hamilton's friends pass away or move into extended care facilities.

   • Ms. Hamilton falls and breaks her hip.

   • Ms. Hamilton begins to feel isolated and depressed.

   Describe interventions the health care professional can take to help Ms. Hamilton deal with each situation to prevent her from falling into a downward spiral.

NUTRITION ON THE NET

For further study of the topics in this chapter, access these websites.

Find updates and quick links to these and other nutrition-related sites at our website: **www.wadsworth.com/ nutrition**

Search for "aging," "arthritis," and "Alzheimer's" on the U.S. Government health information site: **www.healthfinder .gov**

Visit the National Aging Information Center of the Administration on Aging: **www.aoa.dhhs.gov/naic**

Visit the American Geriatrics Society: **www.american geriatrics.org**

Visit the National Institution on Aging: **www.nih.gov/nia**

Visit the American Association of Retired Persons: **www .aarp.org**

Get nutrition tips for growing older in good health from the American Dietetic Association: **www.eatright.org**

Learn more about cataracts from the National Eye Institute and the American Society of Cataract and Refractive Surgery: **www.nei.nih.gov** and **www.ascrs.org**

Learn more about arthritis from the Arthritis Foundation and the National Institute of Arthritis and Musculoskeletal and Skin Diseases Information Clearinghouse of the National Institutes of Health: **www.arthritis.org** and **www.nih .gov/niams**

Learn more about Alzheimer's disease from the NIA Alzheimer's Disease Education and Referral Center and the Alzheimer's Association: **www.alzheimers.org** and **www.alz.org**

Learn more about malnutrition in older adults and the Nutrition Screening Initiative from the American Academy of Family Physicians: **www.aafp.org**

Find out about federal government programs designed to help senior citizens maintain good health: **www.seniors.gov**

NOTES

[1] Federal Interagency Forum on Aging-Related Statistics, Older Americans 2004: Key Indicators of Well-Being, November 2004, available at **www.agingstats.gov**; Trends in aging—United States and worldwide, *Morbidity and Mortality Weekly Report* 52 (2003): 101–106.

[2] J. A. Martin and coauthors, Annual summary of vital statistics—2003, *Pediatrics* 115 (2005): 619–634.

[3] T. Perls, Genetic and environmental influences on exceptional longevity and the AGE nomogram, *Annals of the New York Academy of Sciences* 959 (2002): 1–13.

[4]Position of the American Dietetic Association: Nutrition across the spectrum of aging, *Journal of the American Dietetic Association* 105 (2005): 616–633.

[5]D. K. Houston and coauthors, Dairy, fruit, and vegetable intakes and functional limitations and disability in a biracial cohort: The Atherosclerosis Risk in Communities Study, *American Journal of Clinical Nutrition* 81 (2005): 515–522; K. T. B. Knoops and coauthors, Mediterranean diet, lifestyle factors, and 10-year mortality in elderly European men and women, *Journal of the American Medical Association* 292 (2004): 1433–1439; G. E. Fraser and D. J. Shavlik, Ten years of life: Is it a matter of choice? *Archives of Internal Medicine* 161 (2001): 1645–1652.

[6]C. R. Richardson and coauthors, Physical activity and mortality across cardiovascular disease risk groups, *Medicine and Science in Sports and Exercise* 36 (2004): 1923–1929; F. B. Hu and coauthors, Adiposity as compared with physical activity in predicting mortality among women, *New England Journal of Medicine* 351 (2004): 2694–2703; M. E. Cress and coauthors, Physical activity programs and behavior counseling in older adult populations, *Medicine and Science in Sports and Exercise* 36 (2004): 1997–2003.

[7]Position of the American Dietetic Association, 2005; Cress and coauthors, 2004; E. W. Gregg and coauthors, Relationship of changes in physical activity and mortality among older women, *Journal of the American Medical Association* 289 (2003): 2379–2386.

[8]C. H. Hillman and coauthors, Physical activity and executive control: Implications for increased cognitive health during older adulthood, *Research Quarterly for Exercise and Sport* 75 (2004): 176–185; W. J. Strawbridge and coauthors, Physical activity reduces the risk of subsequent depression for older adults, *American Journal of Epidemiology* 156 (2002): 328–334.

[9]K. A. Martin Ginis and coauthors, Weight training to activities of daily living: Helping older adults make a connection, *Medicine and Science in Sports and Exercise* 38 (2006): 116–121; P. S. Ades and coauthors, Resistance training on physical performance in disabled older female cardiac patients, *Medicine and Science in Sports and Exercise* 35 (2003): 1265–1270.

[10]L. K. Heilbronn and E. Ravussin, Calorie restriction and aging: Review of the literature and implications for studies in humans, *American Journal of Clinical Nutrition* 78 (2003): 361–369.

[11]Position of the American Dietetic Association, 2005.

[12]Knoops and coauthors, 2004.

[13]W. G. Christen and coauthors, Fruit and vegetable intake and the risk of cataract in women, *American Journal of Clinical Nutrition* 81 (2005): 1417–1422.

[14]C. Chitchumroonchokchai and coauthors, Xanthophylls and α-tocopherol decrease UVB-induced lipid peroxidation and stress signaling in human lens epithelial cells, *Journal of Nutrition* 134 (2004): 3225–3232; A. Taylor and coauthors, Long-term intake of vitamins and carotenoids and odds of early age-related cortical and posterior subcapsular lens opacities, *American Journal of Clinical Nutrition* 75 (2002): 540–549.

[15]S. M. Moeller and coauthors, Overall adherence to the *Dietary Guidelines for Americans* is associated with reduced prevalence of early age-related nuclear lens opacities in women, *Journal of Nutrition* 134 (2004): 1812–1819.

[16]J. M. Weintraub and coauthors, A prospective study of the relationship between body mass index and cataract extraction among US women and men, *International Journal of Obesity and Related Metabolic Disorders* 26 (2002): 1588–1595.

[17]Eye Diseases Prevalence Research Group, Prevalence of age-related macular degeneration in the United States, *Archives of Ophthalmology* 122 (2004): 564–572.

[18]N. I. Krinsky, J. T. Landrum, and R. A. Bone, Biologic mechanisms of the protective role of lutein and zeaxanthin in the eye, *Annual Review of Nutrition* 23 (2003): 171–201; Age-Related Eye Disease Study Research Group, Antioxidants and zinc to prevent progression of age-related macular degeneration, *Journal of the American Medical Association* 286 (2001): 2466–2468.

[19]B. M. Kuehn, Studies probe diet's role in eye disease, *Journal of the American Medical Association* 294 (2005): 32–33.

[20]J. P. SanGiovanni and E. Y. Chew, The role of omega-3 long-chain polyunsaturated fatty acids in health and disease of the retina, *Progress in Retinal and Eye Research* 24 (2005): 87–138.

[21]Position of the American Dietetic Association, 2005.

[22]S. P. Messier and coauthors, Exercise and dietary weight loss in overweight and obese older adults with knee osteoarthritis: The Arthritis, Diet, and Activity Promotion Trial, *Arthritis and Rheumatism* 50 (2004): 1501–1510.

[23]R. F. Bruyere and coauthors, Structural and symptomatic efficacy of glucosamine and chondroitin in knee osteoarthritis: A comprehensive meta-analysis, *Archives of Internal Medicine* 163 (2003): 1514–1522.

[24]L. Skoldstam, L. Haffors, and G. Johansson, An experimental study of a Mediterranean diet intervention for patients with rheumatoid arthritis, *Annals of the Rheumatic Diseases* 62 (2003): 208–214.

[25]O. Adam, Dietary fatty acids and immune reactions in synovial tissue, *European Journal of Medical Research* 8 (2003): 381–387; L. Cleland, M. James, and S. Proudman, The role of fish oils in the treatment of rheumatoid arthritis, *Drugs* 63 (2003): 845–853.

[26]R. J. Wurtman and coauthors, Effects of normal meals rich in carbohydrates or proteins on plasma tryptophan and tyrosine ratios, *American Journal of Clinical Nutrition* 77 (2003): 128–132.

[27] K. L. Tucker and coauthors, High homocysteine and low B vitamins predict cognitive decline in aging men: The Veterans Affairs Normative Aging Study, *American Journal of Clinical Nutrition* 82 (2005): 627–635; G. Ravaglia and coauthors, Homocysteine and folate as risk factors for dementia and Alzheimer disease, *American Journal of Clinical Nutrition* 82 (2005): 636–643; P. Quadri and coauthors, Homocysteine, folate, and vitamin B-12 in mild cognitive impairment, Alzheimer disease, and vascular dementia, *American Journal of Clinical Nutrition* 80 (2004): 114–122.

[28]L. E. Hebert and coauthors, Alzheimer disease in the US population: Prevalence estimates using the 2000 census, *Archives of Neurology* 60 (2003): 1119–1122.

[29]T. D. Bird, Genetic factors in Alzheimer's Disease, *New England Journal of Medicine* 352 (2005): 862–864.

[30]P. I. Moreira and coauthors, Oxidative stress: The old enemy in Alzheimer's disease pathophysiology, *Current Alzheimer Research* 2 (2005): 403–408.

[31]P. P. Zandi and coauthors, Reduced risk of Alzheimer disease in users of antioxidant vitamin supplements: The Cache County Study, *Archives of Neurology* 61 (2004): 82–88; M. J. Engelhart and coauthors, Dietary intake of antioxidants and risk of Alzheimer disease, *Journal of the American Medical Association* 287 (2002): 3223–3229; M. C. Morris, Dietary intake of antioxidant nutrients and the risk of incident Alzheimer disease in a biracial community study, *Journal of the American Medical Association* 287 (2002): 3230–3237.

[32]Ravaglia and coauthors, 2005; Tucker and coauthors, 2005; Quadri and coauthors, 2004.

[33]Tucker and coauthors, 2005; Ravaglia and coauthors, 2005; Quadri and coauthors, 2004.

[34]L. Bren, Alzheimer's: Searching for a cure, *FDA Consumer*, July/August 2003, pp. 19–25.

[35]H. K. Kamel, Sarcopenia and aging, *Nutrition Reviews* 61 (2003): 157–167; I. Janssen, S. B. Heymsfield, and R. Ross, Low relative skeletal muscle mass (sarcopenia) in older persons is associated with functional impairment and physical disability, *Journal of the American Geriatric Society* 50 (2002): 889–896; C. W. Bales and C. S. Ritchie, Sarcopenia, weight loss, and nutritional frailty in the elderly, *Annual Review of Nutrition* 22 (2002): 309–323.

[36]K. S. Nair, Aging muscle, *American Journal of Clinical Nutrition* 81 (2005): 953–963; D. K. Houston and coauthors, Dairy, fruit, and

vegetable intakes and functional limitations and disability in a biracial cohort: The Atherosclerosis Risk in Communities Study, *American Journal of Clinical Nutrition* 81 (2005): 515–522.

[37]Nair, 2005; Houston and coauthors, 2005; P. Szulc and coauthors, Hormonal and lifestyle determinants of appendicular skeletal muscle mass in men: The MINOS Study, *American Journal of Clinical Nutrition* 80 (2004): 496–503.

[38]Position of the American Dietetic Association, 2005.

[39]D. Benton and S. Nabb, Carbohydrate, memory, and mood, *Nutrition Reviews* 61 (2003): S61–S67.

[40]Position of the American Dietetic Association, 2005; N. R. Sahyoun and E. Krall, Lower dietary quality among older adults with self-perceived ill-fitting dentures, *Journal of the American Dietetic Association* 103 (2003): 1494–1499.

[41]Standing Committee on the Scientific Evaluation of Dietary Reference Intakes, Food and Nutrition Board, Institute of Medicine, *Dietary Reference Intakes for Energy, Carbohydrate, Fiber, Fat, Fatty Acids, Cholesterol, Protein, and Amino Acids* (Washington D.C.: National Academies Press, 2005), pp. 387–389.

[42]Position of the American Dietetic Association, 2005; D. R. Thomas, Dehydration in older adults, *Nutrition and the M.D.*, November 2004, pp. 1–3.

[43]Standing Committee on the Scientific Evaluation of Dietary Reference Intakes, Food and Nutrition Board, Institute of Medicine, *Dietary Reference Intakes for Water, Potassium, Sodium, Chloride, and Sulfate* (Washington, D.C.: National Academies Press, 2005), pp. 118–127; Position of the American Dietetic Association: Liberalization of the diet prescription improves quality of life for older adults in long-term care, *Journal of the American Dietetic Association* 105 (2005): 1955–1965.

[44]Standing Committee on the Scientific Evaluation of Dietary Reference Intakes, *Dietary Reference Intakes for Water, Potassium, Sodium, Chloride, and Sulfate,* 2005: pp. 149–150.

[45]R. Chernoff, Micronutrient requirements in older women, *American Journal of Clinical Nutrition* 81 (2005): 1240S–1245S.

[46]Executive Summary: Conference on dietary supplement use in the elderly—proceedings of the conference held January 14–15, 2003, Natcher Auditorium, National Institutes of Health, Bethesda, MD, *Nutrition Reviews* 62 (2004): 160–175.

[47]Position of the American Dietetic Association: Nutrition across the spectrum of aging, 2005.

[48]M. A. Johnson and K. H. Porter, Micronutrient supplementation and infection in institutionalized elders, *Nutrition Reviews* 55 (1997): 400–404.

[49]Position of the American Dietetic Association, 2005.

Disease Prevention

Earlier chapters described individual nutrients' connections with diseases and may have left the mistaken impression of "one disease–one nutrient" relationships. Indeed, valid links do exist between saturated fat and heart disease, calcium and osteoporosis, and antioxidant vitamins and cancer, but focusing only on these links oversimplifies the story. In reality, each nutrient may have connections with several diseases because its role in the body is not specific to a disease, but to a body function. Furthermore, each of the chronic diseases develops in response to multiple risk factors, including many nondietary factors such as genetics, physical inactivity, and smoking. An integrated and balanced approach to disease prevention therefore includes attention to all of the many factors involved. Figure NP12-1 illustrates some of the relationships between risk factors and chronic diseases.

Notice how many of the diseases have a genetic component. A family history of a certain disease is a powerful indicator of a person's tendency to contract that disease. Still,

lifestyle factors are often pivotal in determining whether that tendency will be expressed. This Nutrition in Practice focuses on one lifestyle factor related to disease prevention—diet.

I'm only 20 years old. Do the foods I choose to eat now really influence whether I will one day have heart disease, cancer, or another chronic disease?

A remark by a former surgeon general is worth repeating: If you do not smoke or drink excessively, "your choice of diet can influence your long-term health prospects more than any other action you might take." Indeed, healthy young adults today are privileged to be the first generation in history with enough knowledge now to lay a foundation for healthy years ahead through a lifetime of proper nutrition. It isn't as

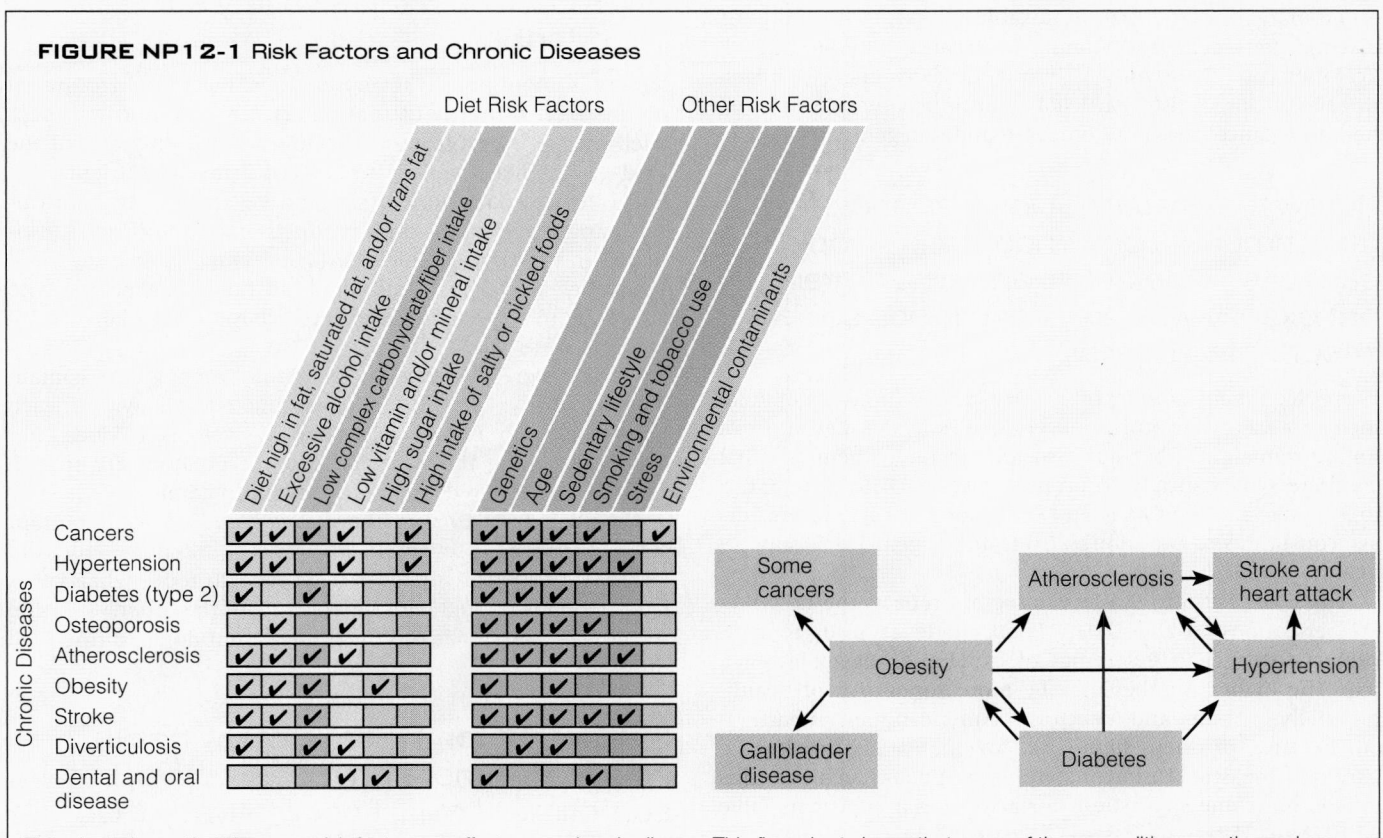

FIGURE NP12-1 Risk Factors and Chronic Diseases

This chart shows that the same risk factor can affect many chronic diseases. Notice, for example, how many diseases have been linked to a sedentary lifestyle. The chart also shows that a particular disease, such as atherosclerosis, may have several risk factors.

This flow chart shows that many of these conditions are themselves risk factors for other chronic diseases. For example, a person with diabetes is likely to develop atherosclerosis and hypertension. These two conditions, in turn, worsen each other and may cause a stroke or heart attack. Notice how all of these chronic diseases are linked to obesity.

complicated as it may seem. In fact, the more detailed our knowledge of nutrition science becomes, the simpler the truth becomes: people who consume the moderate, adequate, balanced, kcalorie-controlled diet recommended by the *Dietary Guidelines for Americans 2005* (see Table NP12-1) enjoy a longer, healthier life than those who do not.[1] The National Heart, Lung, and Blood Institute and the American Cancer Society offer suggestions specifically for disease prevention. Table NP12-1 shows how similar they are to the *Dietary Guidelines*, confirming that it's time to put recommendations into practice in the interest of health for everyone.

What are the specifics for putting recommendations into practice?

Primary among the recommendations is to choose unsaturated fats in place of saturated fat and *trans* fat. To accomplish this requires people to limit such foods as high-fat milk and dairy products, stick margarine, high-fat baked goods and convenience foods, foods with cream or cheese sauces, commercially fried foods, fat-marbled meat cuts, sausages, and fatty ground beef.

Conversely, foods such as nuts, olive oil, canola oil, and liquid margarines are high in unsaturated fats that supply the essential fatty acids and vitamin E. Because they are high in kcalories, use them only within your daily energy budget. Regular meals of fish, particularly fatty fish such as salmon, help to balance intakes of omega-6 and omega-3 fatty acids.

I know the *Dietary Guidelines* and other recommendations encourage people to eat fruits, vegetables, and whole grains, but exactly why are these foods so important to health?

Every legitimate source of dietary advice urges people to include a variety of fruits and vegetables in the diet, not just for nutrients and fiber but also for the phytochemicals that combine synergistically to promote health.[2] The U.S. "Eat 5 to 9 a Day" and the Canadian "Reach for It" programs encourage consumers to eat enough fruits and vegetables to support health.

The U.S. "Eat 5 to 9 a Day" program reflects the new recommendations of the USDA Food Guide as it urges consumers to eat 5 to 9 servings of fruits and vegetables each day. The higher number may be more supportive of health, especially for men, and for African American men in particular because of their high risk for many life-threatening chronic diseases. Unfortunately, few people achieve this intake, but some suggestions for achieving it appear in Table NP12-2 (p. 352).

As for whole grains, most guidelines call for their inclusion, and the *Dietary Guidelines* recommend that whole grains make up at least half of the day's grain intake. Why not aim for more? If whole grains make up the majority of your grain choices, then the grains you consume daily will supply nutrients, fiber, and phytochemicals associated with good health.

Vegetables rich in fiber, phytochemicals, and the antioxidant nutrients (beta-carotene, vitamin C, and vitamin E) help to protect against chronic diseases.

Dietary recommendations emphasize variety—eat foods from each different group, and within the groups, vary each day's choices. Why is variety so important?

Variety is important for several reasons. Each food group represents a set of nutrients that differs from the nutrients supplied by the other groups. Selecting foods from each of the food groups helps ensure a balanced array of nutrients. Eat foods high in potassium (fruits and vegetables), high in calcium and magnesium (milk products and appropriate substitutes), low in fat and high in fiber (whole grains, legumes, vegetables, and fruits), and ample in fluids. For people who are prone to hypertension, experts advise fewer high-sodium foods and less salt.

Furthermore, because foods within each group contain different arrays of nutrients, varying choices within each group helps to ensure adequacy as well as balance. For example, among the vegetables, sweet potatoes are rich in vitamin A, while red peppers are rich in vitamin C.

Finally, advice to vary the diet is based on an important concept in the prevention of cancer initiation—dilution. Whenever a person switches from food to food, whatever is in one food is diluted with whatever is in the others. It is safe to eat *some* salty foods or grilled meats, but don't eat them all the time.

I know that for the sake of my health, I need to be more physically active, but sometimes I'm just not motivated. Do you have any suggestions?

Expend energy in daily physical activity that you enjoy. Take the dog outside and go for a run. Learn to dance or take a bike ride. If the threat of cardiovascular disease doesn't motivate you, then be physically active to improve your self-image, to improve your morale, to control your body weight,

TABLE NP12-1 Dietary Guidelines for Lowering Disease Risks

Dietary Guidelines for Americans 2005 Key Recommendations[a]	National Heart, Lung and Blood Institute's Therapeutic Life Changes (TLC) Diet to Lower LDL Cholesterol[b]	American Cancer Society Recommendations for Nutrition and Physical Activity for Cancer Prevention
Adequate nutrients within energy needs • Consume a variety of nutrient-dense foods and beverages within and among the basic food groups; limit intakes of saturated and *trans* fats, cholesterol, added sugars, salt, and alcohol. • Plan a balanced diet with the USDA Food Guide or the DASH eating plan (see Chapter 21). Weight management • To maintain body weight in a healthy range through life, balance kcalories from foods and beverages with kcalories expended. Physical activity • Engage in regular physical activity and reduce sedentary activities to promote health, psychological well-being, and a healthy body weight. Food groups to encourage • Consume sufficient amounts of a variety of fruits and vegetables, milk and milk products, and whole grains while staying within energy needs. Fats • Keep saturated fat and *trans* fat to less than 10% of kcalories, and cholesterol to less than 300 milligrams. • Keep total fat intake to between 20 and 35% of kcalories from mostly unsaturated fats. Carbohydrates • Choose fiber-rich fruits, vegetables, and whole grains often. • Choose and prepare foods and beverages with little added sugars. Sodium and potassium • Choose and prepare foods with little salt (less than 2300 milligrams sodium or approximately 1 teaspoon salt). • At the same time, consume potassium-rich foods. Alcoholic beverages • Those who drink alcohol should do so sensibly and in moderation. Some people should abstain from alcohol. Food safety • Keep foods safe from microbes that cause illness.	Carbohydrates • 50–60% kcalories from carbohydrates (primarily complex carbohydrates from whole-food sources). • 20–30 g dietary fiber per day. Lipids • 25–35% kcalories from total fats. • Up to 10% kcalories from polyunsaturated fats. • Up to 20% kcalories from monounsaturated fats. • 200 mg or less of cholesterol per day. • Less than 7% of total kcalories as saturated fat. • Keep *trans*-fatty acids low. Protein • Approximately 15% kcalories from protein. These additional dietary changes can further reduce LDL cholesterol in those with higher heart disease risk.[c] • Increase intake of viscous (soluble) fiber. • Replace some animal protein sources with soy protein foods.	Eat a variety of healthful foods, with an emphasis on plant sources. • Eat five or more servings of a variety of vegetables and fruits each day. • Choose whole grains in preference to processed (refined) grains and sugars. • Limit consumption of red meats, especially those high in fat and processed meats. • Choose foods that maintain a healthful weight. Adopt a physically active lifestyle. • Adults: engage in at least moderate activity for 30 minutes or more on five or more days of the week; 45 minutes or more of moderate-to-vigorous activity on five or more days per week may further enhance reductions in the risk of breast and colon cancer. • Children and adolescents: engage in at least 60 minutes per day of moderate-to-vigorous physical activity at least five days per week. Maintain a healthful weight throughout life. • Balance kcaloric intake with physical activity. • Lose weight if currently overweight or obese. If you drink alcoholic beverages, limit consumption.

[a] Abbreviated. A more complete set can be found in Chapter 1 of this text, or online at **www.healthierus.gov/dietaryguidelines/**.

[b] Abbreviated. A more complete version of the TLC diet can be found in Chapter 21.

[c] Also recommended for people with elevated LDL; 2 g per day of sterol or stanol esters.

TABLE NP12-2 Suggestions for Consuming More Fruits and Vegetables

Eat 5 to 9 a Day

Tips for consuming more fruits and vegetables:

- Vow to try a new fruit or vegetable once each month. Read some cookbooks for ideas.

- Eat a rainbow of fruits and vegetables: the more reds, oranges, yellows, greens, blues, and purples the better.

- Use the salad bar to buy ready-to-eat vegetables if you are in a hurry.

- Keep a fruit bowl in plain view, filled with fresh fruit, small raisin or dried cranberry packs, and shelf-stable fruit cups.

- Add dried fruit bits to salads, cereal, or yogurt.

- Place carrot and celery sticks in a glass of water like a bouquet, and keep them in the refrigerator for crisp, healthy snacks.

- Try a grilled mushroom sandwich or "veggie burger" instead of a beef hamburger sandwich.

- Try using soy products in many ways. Soy drinks, ground and patty meat replacers, tofu, and soy snacks count as a vegetable and may offer unique benefits from soy protein, fiber, and phytochemicals.

- Drink 100% fruit or vegetable juices. Choose 100% juice bars for a frozen treat.

- Vegetable sauces count: 1/2 cup tomato spaghetti sauce or salsa counts as 1/2 cup "other vegetables" (see Chapter 1).

- Blend smoothies from bananas, fruit juice, and berries with ice or yogurt.

- Choose larger portions of lower-kcalorie vegetables, such as cooked leafy greens or carrots.

- Add beans to salads, stews, and meat dishes. Beans soak up the flavor of meat and stretch the servings per pound.

to improve your complexion, or to make friends—but do become more physically active. The health benefits of regular physical activity are numerous and well worth the effort it may take to get started.

In the end, people's choices are their own. Whoever you are, take the time to work out ways of making your diet meet the guidelines you know will support your health. If you are healthy and of normal weight, if you are physically active, and if your diet on most days follows the guidelines just provided, then you can indulge occasionally in a cheesy pizza, marbled steak, banana split, or even a greasy fast-food burger and fries without inflicting much damage on your health. Especially, take time to enjoy your meals: the sights, smells, and tastes of good foods are among life's greatest pleasures. Joy, even the simple joy of eating, contributes to health.

Notes

[1] Joint WHO/FAO Expert Consultation, *Diet, Nutrition and the Prevention of Chronic Diseases* (Geneva, Switzerland: World Health Organization, 2003), pp. 30–53.

[2] Phytochemicals and cancer prevention: From epidemiology to mechanism of action, *Journal of the American Dietetic Association* 106 (2006): 20–21; R. H. Liu, Health benefits of fruits and vegetables are from additive and synergistic combinations of phytochemicals, *American Journal of Clinical Nutrition* 78 (2003): 517S–520S.

Nutrition Care and Assessment

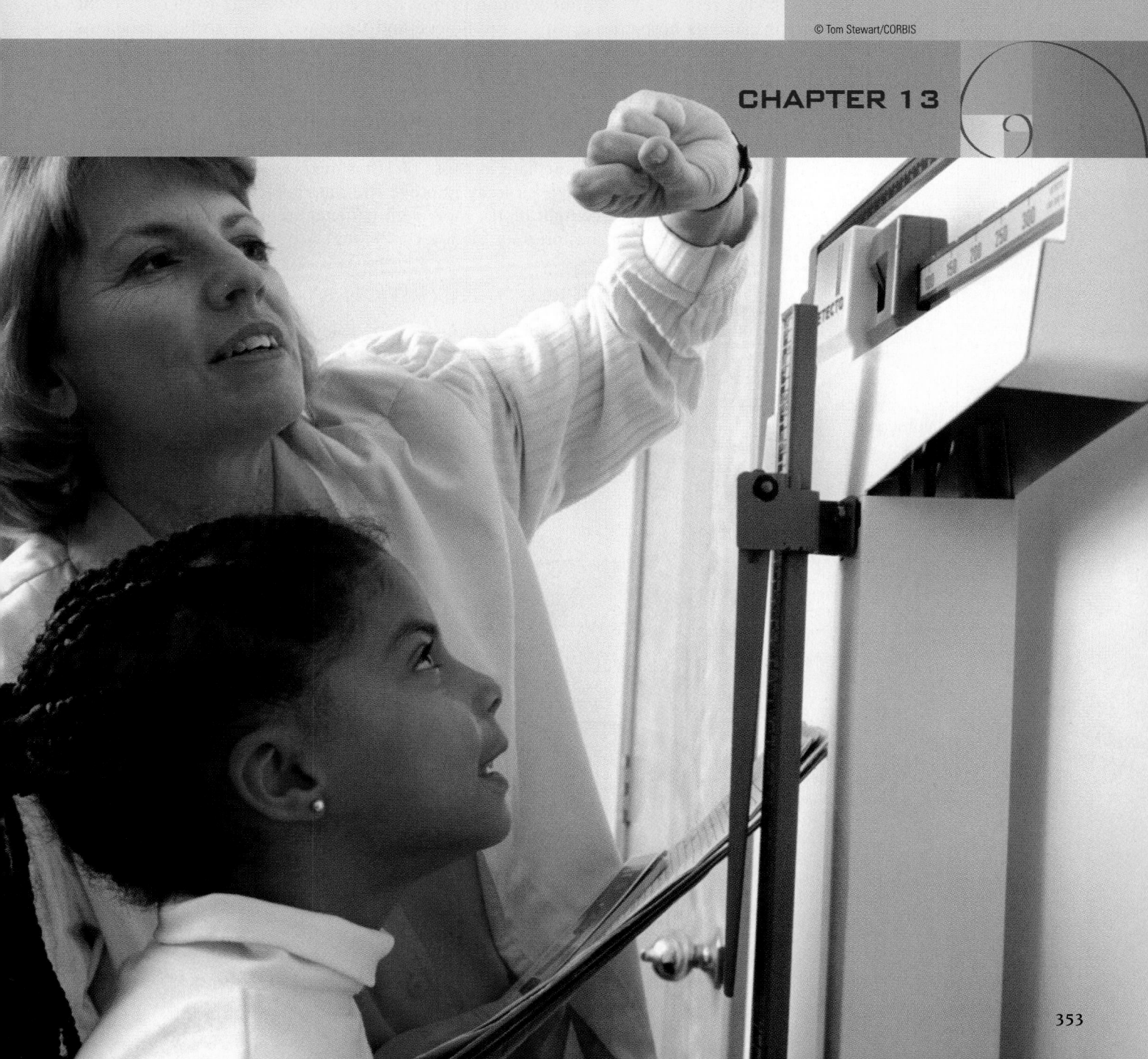

CHAPTER 13

Reminder: Nurses may use either *client* or *patient* when referring to an individual under their care. The clinical chapters emphasize the care of those with serious illnesses; thus the term *patient* is used throughout these chapters.

The first 12 chapters of this book introduced the nutrients and described how the appropriate dietary choices can support good health. Turning now to clinical nutrition, the remaining chapters describe the potential effects of illnesses and their treatments on nutrition status and nutrient needs. Nurses and other health care providers who consider nutrition care in their treatment decisions are best prepared to help patients recover from disease and maintain an optimal quality of life. This chapter focuses on the process used for providing nutrition care and the components of nutrition assessment.

Nutrition in Health Care

Many health problems can alter nutrient needs and lead to malnutrition. Conversely, poor nutrition status can influence both the course of disease and the body's response to treatment. Malnutrition has been reported in 40 to 60 percent of patients hospitalized with acute illness, and those with no nutrition problems often exhibit a decline in nutrition status within three weeks of admission.[1] Poor nutrition status can ultimately weaken immune function and compromise a person's healing ability.

For the busy nurse who has many responsibilities, it may be easy to put a patient's nutrition needs on the back burner, as the benefits of diet therapy are not always as obvious or immediate as those of other medical treatments. Correcting nutritional problems, however, may improve the outcome of medical treatments and help to prevent complications. Moreover, patients are often concerned about the impact their diet has on their disease condition.

HOW ILLNESS AFFECTS NUTRITION STATUS

An illness, its symptoms, and its treatments may lead to malnutrition by reducing food intake, interfering with digestion and absorption, or altering nutrient metabolism and excretion (see Figure 13-1). For example, the nausea caused by some illnesses can diminish appetite and reduce food intake; similarly, inflammation in

FIGURE 13-1 Ways in Which Illness Can Affect Nutrition Status

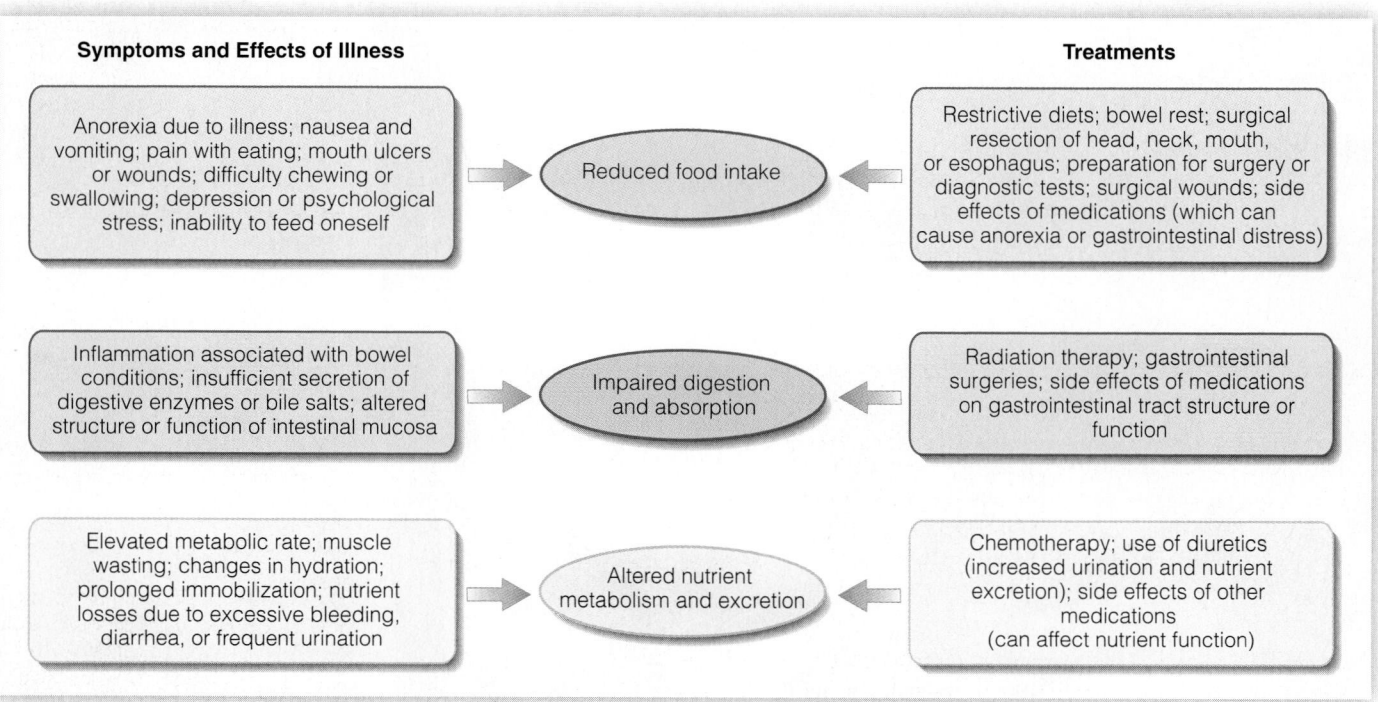

Symptoms and Effects of Illness		Treatments
Anorexia due to illness; nausea and vomiting; pain with eating; mouth ulcers or wounds; difficulty chewing or swallowing; depression or psychological stress; inability to feed oneself	Reduced food intake	Restrictive diets; bowel rest; surgical resection of head, neck, mouth, or esophagus; preparation for surgery or diagnostic tests; surgical wounds; side effects of medications (which can cause anorexia or gastrointestinal distress)
Inflammation associated with bowel conditions; insufficient secretion of digestive enzymes or bile salts; altered structure or function of intestinal mucosa	Impaired digestion and absorption	Radiation therapy; gastrointestinal surgeries; side effects of medications on gastrointestinal tract structure or function
Elevated metabolic rate; muscle wasting; changes in hydration; prolonged immobilization; nutrient losses due to excessive bleeding, diarrhea, or frequent urination	Altered nutrient metabolism and excretion	Chemotherapy; use of diuretics (increased urination and nutrient excretion); side effects of other medications (can affect nutrient function)

the mouth or esophagus may make the physical act of eating uncomfortable. Surgeries of the gastrointestinal tract often require temporary dietary restrictions. Some medications can cause anorexia or gastrointestinal discomfort, or interfere with nutrient function and metabolism. In many cases, the dietary changes required during illness are temporary and can be tailored to accommodate an individual's preferences and lifestyle. Chronic illnesses, however, may require long-term dietary modifications. For example, diabetes treatment requires lifelong changes in diet and lifestyle that some people may have difficulty adhering to. The challenge for nurses and other health professionals is to help their patients appreciate the potential benefits of treatment and accept dietary changes that can improve their health.

In addition to the direct effects of illness on nutrition status, the cost of health care can drain financial resources and limit ability to obtain high-quality foods. Some individuals may lack the space and equipment necessary to store and prepare special meals. Emotional health may also suffer as a result of chronic disease or terminal illness, and this can influence nutrition status as well.

RESPONSIBILITY FOR NUTRITION CARE

Each member of the health care team has a unique role in nutrition care. By working together, they can ensure that nutrition problems are quickly addressed.

Physicians Physicians are responsible for meeting all of a patient's medical needs, including nutrition. They prescribe **diet orders** and other orders related to nutrition care, including those for nutrition assessment and dietary counseling. Physicians rely on nurses, registered dietitians, and other health professionals to alert them to nutrition problems, suggest strategies for handling nutrition problems, and provide nutrition services.

Nurses Nurses interact frequently with patients and thus are in an ideal position to identify people who would benefit from nutrition services. Nurses often screen patients for nutrition problems and may participate in nutrition and dietary assessments. Nurses also provide direct nutrition care, such as encouraging patients to eat, finding practical solutions to food-related problems, recording a patient's food intake, and answering questions about special diets. As members of **nutrition support teams,** nurses are responsible for administering tube and intravenous feedings. In facilities that do not employ registered dietitians, nurses often assume responsibility for much of the nutrition care.

Registered Dietitians Registered dietitians are food and nutrition experts who are qualified to provide **medical nutrition therapy.** They conduct nutrition and dietary assessments; diagnose nutritional problems; develop, implement, and evaluate **nutrition care plans** (described in a later section); plan and approve menus; and provide nutrition education. Registered dietitians may also work as managers of food and cafeteria services in health care institutions.

Registered Dietetic Technicians Registered dietetic technicians work in partnership with registered dietitians and assist in the implementation and monitoring of nutrition services. Depending on their background and experience, they may screen patients for nutrition problems, provide patient education and counseling, develop menus and recipes, ensure appropriate meal delivery, and monitor patients' food choices and intakes. Dietetic technicians sometimes supervise food-service operations and may have roles in purchasing, inventory, quality control, sanitation, or safety.

Other Health Care Professionals Other health care professionals may also assist with nutrition care. Pharmacists, physical therapists, occupational therapists, speech

Reminder: A *registered dietitian* has met the academic and professional requirements to qualify for the RD credential conferred by the American Dietetic Association (or Dietitians of Canada), including a bachelor's degree in nutrition or dietetics, a supervised internship, and the successful completion of a national examination.

diet orders: specific instructions concerning dietary management; also called **diet prescriptions.**

nutrition support teams: health care professionals responsible for the provision of nutrients by tube feeding or intravenous infusion.

medical nutrition therapy: nutrition care provided by a registered dietitian; includes diagnosing nutrition problems, prescribing diet plans, and providing dietary counseling.

nutrition care plans: strategies for meeting an individual's nutritional needs.

therapists, social workers, nursing assistants, and home health care aides can be instrumental in alerting dietitians or nurses to nutrition problems or may share relevant information about a patient's health status or personal needs.

Note that the job descriptions in different institutions vary, and the roles of health professionals may overlap. In some cases, nutrition care is incorporated into the medical care plan developed by the entire health care team. Such plans, known as **critical pathways,** outline coordinated plans of care for specific medical diagnoses, treatments, or procedures.

IDENTIFYING RISK FOR MALNUTRITION

The **Joint Commission on Accreditation of Healthcare Organizations (JCAHO)** recommends that **nutrition screening** be conducted within 24 hours of a patient's admission to a hospital (or other extended-care facility) to identify those who may need more extensive assessment and follow-up care. Nutrition screening should be accurate enough to identify nutritional risk, yet simple enough to be completed within 5 to 15 minutes. Usually, a nurse, nursing assistant, registered dietitian, or dietetic technician performs and documents the screening, which varies according to the patient population, the type of care offered by the health care facility, and the patient's medical problems. The information used in screening includes the admitting diagnosis, information from the medical record, physical measurements and laboratory test results collected during the admission process, and responses given by the patient or caregiver to an interview or questionnaire. A nutrition or health screening often leads to a referral for nutrition care. Table 13-1 lists examples of information collected during screening that help to identify those who have a high risk of developing malnutrition. Nutrition screening is also frequently included in outpatient services and community health programs.

Nursing care plans often include **nursing diagnoses** that have nutritional implications (see Table 13-2). For example, a nursing diagnosis of "impaired swallowing" alerts the nurse to potential nutrition problems and suggests the need for a modified diet. Such a diagnosis can accompany various medical conditions,

critical pathways: interdisciplinary care plans for specific diagnoses, treatments, or procedures that merge the medical and nursing plans with those of other disciplines, such as physical therapy, nutrition, and mental health.

Joint Commission on Accreditation of Healthcare Organizations (JCAHO): a nonprofit organization that sets standards for health care performance and safety and confers accreditation on health care organizations that meet those standards.

nutrition screening: an examination process that identifies patients who require intervention for existing or potential nutritional problems.

nursing diagnoses: clinical judgments about actual or potential health problems that provide the basis for selecting appropriate nursing interventions.

TABLE 13-1	Criteria for Identifying Malnutrition Risk
Age, medical diagnosis, severity of illness	
Height and weight, body mass index (BMI), recent unintentional weight changes	
Results of laboratory tests that indicate poor health status	
Recent changes in appetite or food intake	
Problems or symptoms that make eating difficult (such as chewing or swallowing difficulty, or nausea and vomiting)	
Food allergies or intolerances, or extensive dietary restrictions	
Presence of anemia, tissue wasting, or pressure sores	
History of diabetes, renal disease, or other chronic illness	
Use of medications that can impair nutrition status	
Depression or social isolation	

TABLE 13-2	Nursing Diagnoses with Nutritional Implications
Chronic confusion	
Chronic pain	
Constipation	
Diarrhea	
Disturbed body image	
Feeding self-care deficit	
Imbalanced nutrition: less than body requirements	
Imbalanced nutrition: more than body requirements	
Impaired dentition	
Impaired oral mucous membrane	
Impaired physical mobility	
Impaired swallowing	
Nausea	
Readiness for enhanced nutrition	
Risk for aspiration	
Risk for imbalanced nutrition: more than body requirements	

Source: NANDA International, *Nursing Diagnoses: Definitions and Classification 2005–2006* (Philadelphia: NANDA International, 2005).

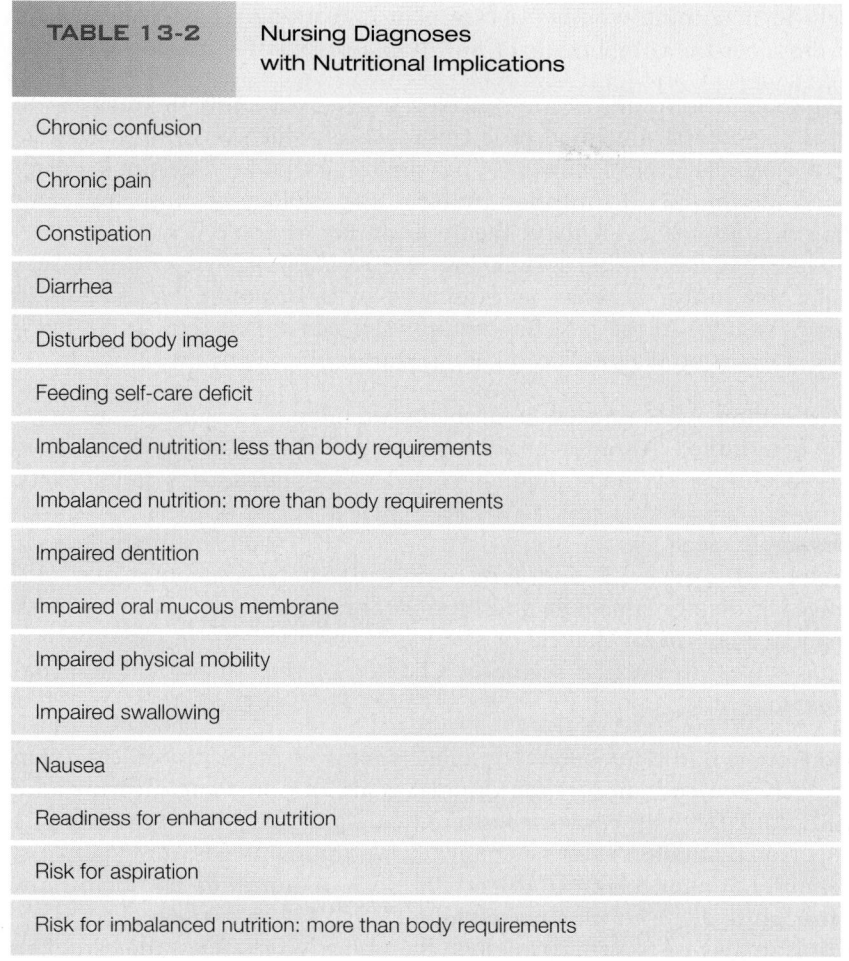

FIGURE 13-2 The Nutrition Care Process

including developmental disabilities, diseases involving the esophagus, and neurological conditions. Examples of nursing diagnoses associated with conditions that frequently require nutritional intervention are provided throughout the clinical nutrition chapters.

THE NUTRITION CARE PROCESS

Registered dietitians use a systematic approach to medical nutrition therapy called the **nutrition care process.** The steps of the nutrition care process include nutrition assessment, nutrition diagnosis, nutrition intervention, and nutrition monitoring and evaluation, as shown in Figure 13-2.[2] When providing nutrition care, these steps are frequently revisited in order to reassess and revise diagnoses and intervention strategies. Each step of the nutrition care process must be documented in the medical record, providing a record for future reference and facilitating communication among members of the health care team.

Nutrition Assessment Nutrition assessment involves the collection of information needed to evaluate a patient's nutrition status and dietary needs. This information can be obtained from medical, social, and diet histories; anthropometric data; biochemical analyses; and physical examination. The assessment data are used to develop a plan of action to prevent or correct nutrient imbalances and

As a comparison, the *nursing process* consists of five steps:

1. Assessment.
2. Nursing diagnosis.
3. Outcome identification and planning.
4. Implementation.
5. Evaluation.

nutrition care process: a problem-solving method that dietetics professionals use to evaluate nutrition-related problems and provide safe and effective nutrition care.

may also help to determine whether a care plan is working. The second half of this chapter describes the components of nutrition assessment in more detail.

Nutrition Diagnosis The results of the nutrition assessment allow the dietitian to identify actual or potential nutrition problems, each of which receives a separate diagnosis.[3] Nutrition diagnoses follow a format similar to that used in nursing diagnoses; they include the specific nutrition problem, the etiology or cause, and signs and symptoms that provide evidence of the problem. For example, a nutrition diagnosis might state, "Unintentional weight loss *(the problem)* related to insufficient kcaloric intake *(the etiology or cause)* as evidenced by a 10-pound weight loss (8 percent of body weight) in the past few months *(the sign or symptom)*." Like nursing diagnoses, a nutrition diagnosis can change during the course of an illness.

Nutrition Intervention After nutritional problems are identified, appropriate treatments can be determined. An intervention may include dietary changes, nutrition education, or a change in medication. It considers an individual's food habits, lifestyle, and other personal factors. Goals are stated in terms of measurable outcomes; for example, goals for an overweight person with diabetes might include changes in blood glucose levels and body weight. Desirable outcomes may also include changes in dietary behaviors and lifestyle; for example, an interview with a diabetes patient may reveal that he or she has learned to control carbohydrate intake and has started to exercise regularly. Chapter 14 gives additional information about nutrition intervention.

Nutrition Monitoring and Evaluation The effectiveness of the nutrition care plan needs to be evaluated periodically: the original goals and outcome measures are reviewed and compared with earlier assessment data and diagnoses. Sometimes, a change in a person's situation causes a change in nutritional needs; for example, a new medication may alter tolerance to certain foods. A nutrition care plan must be flexible enough to adapt to the new situation.

If a patient is unable or unwilling to make the suggested changes, the care plan must be redesigned and should take into account the reasons that the earlier plan was not successful. This new plan may need to include motivational techniques or additional patient education. If the patient remains unwilling to modify behaviors despite the expected benefits, the health care provider can try again at a later time when the patient may be more receptive.

REVIEW NOTES

Illnesses and their treatments can affect food intake and nutrient needs, leading to malnutrition. In turn, poor nutrition can reduce the effectiveness of medical treatment.

The combined efforts of each member of the health care team ensure that patients receive optimal nutrition care.

Nutrition screening identifies individuals who can benefit from nutrition assessment and follow-up care. Nursing diagnoses can also identify potential nutrition problems.

The nutrition care process includes four interrelated steps: nutrition assessment, nutrition diagnosis, nutrition intervention, and nutrition monitoring and evaluation.

Nutrition Assessment

A nutrition assessment provides the information needed to identify nutrition problems and design a nutrition care plan; follow-up assessments help to determine if the care plan is effective. Ideally, the assessment should be sensitive enough to detect subtle nutrition problems and specific enough to identify problem nutrients. The

TABLE 13-3	Historical Information Used in Nutrition Assessment	
Medical History	**Social History**	**Diet History**
Current complaint(s)	Socioeconomic status	Dietary pattern
Past medical condition(s)	Cultural/ethnic identity	Dietary restrictions
Family history of illness	Educational level	Use of alcohol
Surgical history	Living situation	Food allergies and intolerances
Medication history	Shopping arrangements	Chewing and swallowing ability
Use of dietary/herbal supplements	Cooking facilities	Need for feeding assistance

remainder of this chapter discusses the types of information and measures that are most commonly used in a nutrition assessment.

HISTORICAL INFORMATION

Historical information reveals valuable clues about nutrition status and nutrient requirements, as well as the personal preferences that need consideration when developing a nutrition care plan. Table 13-3 summarizes the elements of medical, social, and diet histories that contribute to a nutrition assessment. This information can be obtained from the medical record or by interviewing the patient or caregiver.

Medical History Reviewing the medical history can help the health care provider determine whether food intake is likely to be affected by illness or if specific dietary changes are needed. Factors such as age, gender, and weight are also considered when assessing nutritional risk. In addition, a medical history provides information about the prescription drugs, over-the-counter medications, and dietary supplements used by an individual. Some medications can have detrimental effects on nutrition status, and some dietary components can alter the absorption and metabolism of drugs. Chapter 14 provides examples of the diet-drug interactions that need to be considered when planning nutrition care.

Social History Social factors influence food choices as well as a person's ability to deal with health and nutrition problems. For example, cultural heritage can affect food preferences. Financial concerns can restrict access to health care or place a nutritious diet out of reach. Some individuals may depend on others to prepare or procure food. A person who lives alone or is depressed may eat poorly or be unable to follow complex dietary instructions.

Diet History A **diet history** is a comprehensive record of dietary practices. It includes information about food intakes, meal patterns, and physical problems that influence dietary choices (see Table 13-3). The information may be obtained using various methods including an interview about recent food intake (for example, a *24-hour recall*) and a survey about usual food choices (such as a *food frequency questionnaire*). The following section describes several common methods of gathering food intake information.

DIETARY ASSESSMENT METHODS

Obtaining an accurate account of a person's usual food intake is challenging, as results may vary depending on both the individual's memory and honesty and the

diet history: a comprehensive record of a person's food intake and dietary practices.

assessor's skill and training. In addition, each method has its own strengths and weaknesses, so best results are obtained from using a combination of methods. Table 13-4 summarizes the different methods commonly used and each method's advantages and disadvantages.

The 24-Hour Recall The 24-hour recall is a guided interview in which an individual recounts all of the foods and beverages consumed in the past 24 hours or during the previous day. The interviewer includes questions about the times when meals or snacks were eaten, the amounts consumed, and the ways in which foods were prepared. The assessor may begin by asking, "What is the first thing you ate or drank yesterday morning?" After the first food items are described, the follow-up questions might be, "What time was that?" and "How much did you eat?" Questioning continues until the intake record for the day is

24-hour recall: a record of foods consumed in the previous 24 hours; sometimes modified to include foods consumed in a typical day.

TABLE 13-4	Dietary Assessment Methods		
Method	**Description**	**Advantages**	**Disadvantages**
24-hour recall	Guided interview in which the foods and beverages consumed in a 24-hour period are described in detail.	• Results are not dependent on literacy or educational level of respondent. • Interview occurs after food is consumed, so it does not interfere with food choices. • Relatively easy and quick assessment method.	• Reliant on memory. • Food items that cause embarrassment (alcohol, desserts) may be omitted. • Under- and overestimation of food intakes are common. • Skill of interviewer affects outcome. • Data from a single day cannot represent the respondent's usual intake accurately. • Seasonal variations may not be addressed.
Food frequency questionnaire	Written survey of food consumption during a specific period of time, often a one-year period.	• Examines long-term food intake, so day-to-day and seasonal variability should not affect results. • Completed after food is consumed, so does not interfere with food choices. • Low-cost method.	• Reliant on memory. • Not good for monitoring short-term changes in food intake. • Serving sizes are often difficult for respondents to evaluate without assistance. • Calculated nutrient intakes may not be accurate. • Food lists include common foods only. • Food lists for the general population are of limited value in special populations.
Food record	Written account of food consumed during a specified period, usually several consecutive days. Accuracy is improved by including weights or measures of foods.	• Process does not rely on memory. • Recording foods as they are consumed improves likelihood of obtaining accurate food intake data. • Useful for controlling intake because keeping records can increase awareness of food choices.	• Recording process itself influences food intake. • Time-consuming and burdensome for respondent; requires high degree of motivation. • Underreporting is common. • Requires literacy and the physical ability to write. • Seasonal changes in diet are not taken into account.
Direct observation	Observation of meal trays or shelf inventories before and after eating; possible only in residential facilities.	• Process does not rely on memory. • Does not interfere with person's food intake. • Can be used to evaluate acceptability of prescribed diet.	• Possible only in residential situations. • Labor-intensive.

complete. Food models or measuring cups and spoons can be used to help the individual visualize and describe the amounts consumed. After the day's intake is recounted, the interviewer asks whether the intake that day is fairly typical and, if not, how it varies from the person's usual intake. A recall interview may be conducted on several nonconsecutive days to get a better representation of a person's usual diet.

A recall interview can obtain useful data for developing an acceptable nutrition care plan and identifying food items that may need to be restricted due to illness. It is a poor technique, however, for determining the adequacy of a diet because it does not take into account fluctuations in food intake or seasonal variations. Moreover, food intakes are often underestimated because the process relies on the individual's memory and reporting accuracy. People often forget to mention beverages, condiments, and snack foods unless specifically prompted to do so, and some find it embarrassing to report consumption of foods such as chocolate, butter, and red meat.[4]

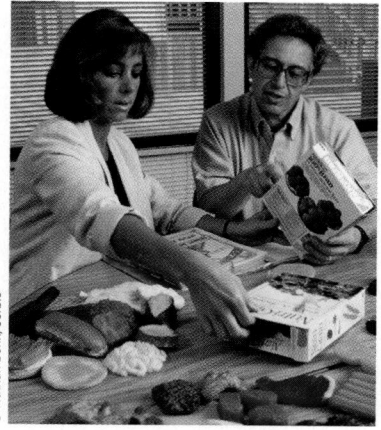

Food models and measuring utensils can help an individual visualize portion sizes.

Food Frequency Questionnaire A **food frequency questionnaire** surveys the foods and beverages regularly consumed during a specific time period. Some questionnaires are qualitative only: the food lists contain commonly eaten foods, categorized by food group, with boxes to check to indicate frequency of consumption. Other types of questionnaires provide semiquantitative information by asking for portion sizes as well. Figure 13-3 shows a sample section of a semiquantitative questionnaire that surveys fruit intake over the previous year. Because the respondent is often asked to estimate food intakes over a one-year period, the results should not be affected by seasonal changes in diet. Conversely, a disadvantage of this method is its inability to determine recent changes in food intake.

Simple versions of food frequency questionnaires focus on food categories relevant to a person's medical condition. For example, a questionnaire designed to evaluate calcium intake may include only milk products, fortified foods, certain fruits and vegetables, and dietary supplements that contain calcium. A computer

food frequency questionnaire: a survey of foods routinely consumed. Some questionnaires ask about the types of food eaten and yield only qualitative information, whereas others include questions about portions consumed and yield semiquantitative data as well.

FIGURE 13-3 Sample Section of a Food Frequency Questionnaire

FRUIT	Never or less than once per month	1 per mon.	2–3 per mon.	1 per week	2 per week	3–4 per week	5–6 per week	Every day	MEDIUM SERVING	YOUR SERVING SIZE S	M	L
EXAMPLE: Bananas	○	○	○	●	○	○	○	○	1 medium	○ 1/2	● 1	○ 2
Bananas	○	○	○	○	○	○	○	○	1 medium	○ 1/2	○ 1	○ 2
Apples, applesauce	○	○	○	○	○	○	○	○	1 medium or 1/2 cup	○ 1/2	○ 1	○ 2
Oranges (not including juice)	○	○	○	○	○	○	○	○	1 medium	○ 1/2	○ 1	○ 2
Grapefruit (not including juice)	○	○	○	○	○	○	○	○	1/2 medium	○ 1/4	○ 1/2	○ 1
Cantaloupe	○	○	○	○	○	○	○	○	1/4 medium	○ 1/8	○ 1/4	○ 1/2
Peaches, apricots (fresh, in season)	○	○	○	○	○	○	○	○	1 medium	○ 1/2	○ 1	○ 2
Peaches, apricots (canned or dried)	○	○	○	○	○	○	○	○	1 medium or 1/2 cup	○ 1/2	○ 1	○ 2
Prunes, or prune juice	○	○	○	○	○	○	○	○	1/2 cup	○ 1/4	○ 1/2	○ 1
Watermelon (in season)	○	○	○	○	○	○	○	○	1 slice	○ 1/2	○ 1	○ 2
Strawberries, other berries (in season)	○	○	○	○	○	○	○	○	1/2 cup	○ 1/4	○ 1/2	○ 1
Any other fruit, including kiwi, fruit cocktail, grapes, raisins, mangoes	○	○	○	○	○	○	○	○	1/2 cup	○ 1/4	○ 1/2	○ 1

HOW OFTEN spans the columns from "Never or less than once per month" through "Every day". *HOW MUCH* spans "MEDIUM SERVING" and "YOUR SERVING SIZE".

analysis can then quickly estimate the individual's calcium intake and compare it to recommendations.

Food Record A **food record** is a written account of foods and beverages consumed during a specified time period, usually several consecutive days. Foods are recorded as they are consumed in order to obtain the most complete and accurate record possible; thus the process does not rely on memory. A detailed food record includes the types and amounts of foods and beverages consumed, times of consumption, and methods of preparation. It can provide valuable information about food intake as well as a person's response to and compliance with medical nutrition therapy. Unfortunately, food records require a great deal of time to complete, and people need to be highly motivated to keep accurate records. Another drawback is that the recording process itself may influence food intake. Furthermore, day-to-day and seasonal variations in food intake make it difficult to obtain accurate estimates of nutrient values in just a few days or even a week.

Direct Observation In facilities that serve meals, food intakes can be directly observed and analyzed. This method can also reveal a person's food preferences, changes in appetite, and any problems with a prescribed diet. Nurses use direct observation to conduct **kcalorie counts** to determine the food energy (and often, protein) consumed by patients during a single day or several consecutive days. To perform a kcalorie count, the nurse estimates food intake by recording the dietary items that a patient is given at meals and subtracting the amounts remaining after meals are completed; the nurse can then make an estimate of the kcaloric content of foods and beverages actually consumed. Although a useful means of discerning patients' intakes, direct observation requires regular and careful documentation and can be labor-intensive and costly.

ANTHROPOMETRIC DATA

Measures of body size, called **anthropometric** measurements, can reveal problems related to overnutrition and protein-energy malnutrition (PEM). The most common anthropometric values used are height (or length) and weight, which help to evaluate growth in children and nutrition status in adults. Other helpful data include skinfold measurements (described in Appendix E) and circumferences of the head, waist, and limbs.

Height (or Length) Poor growth in children can signify malnutrition. In adults, height measurements alone do not reflect current nutrition status but can be used for estimating a person's energy needs or appropriate body weight. Length is measured in infants and children younger than 24 months of age, and height is usually measured in older children and adults. Length can also be measured in adults and children who cannot stand unassisted due to physical or medical reasons. The "How to" on p. 363 describes some standard techniques for measuring length and height.

In adults, height can be estimated from equations that include either knee height or the full arm span, both of which correlate well with height. The measure of knee height is frequently used in bedridden patients; specific formulas are available for different age, gender, and ethnic groups.[5] For children with disabilities that affect stature, alternative measures of linear growth include the full arm span, lower-leg lengths (knee to heel, similar to the knee height measure), and upper-arm lengths (shoulder to elbow), which can be compared with reference percentiles.[6]

Body Weight During clinical care, health care providers monitor body weights carefully. Weight changes can reflect changes in hydration status, and an involun-

---

Reminder: *Protein-energy malnutrition* is a deficiency of protein and food energy, characterized by weight loss and loss of muscle mass.

The difference between length and height is related to the person's body position while the measurement is taken. *Length* is measured while a person is recumbent (lying down), whereas *height* is measured while a person is standing upright.

Knee height, measured in a sitting position, extends from the heel to the top of the knee. Full arm span, measured when the arms are extended horizontally, is the distance from the tip of one middle finger to the other.

food record: a detailed log of food eaten during a specified time period, usually several days. A food record may also include information regarding disease symptoms, physical activity, and medication use; also called a **food diary.**

kcalorie counts: the determination of food energy (and often, protein) consumed by patients during one or more days.

anthropometric (AN-throw-poe-MEH-trik): related to physical measurements of the human body, such as height, weight, body circumferences, and percentage of body fat.

Measure Length and Height

To improve the accuracy of length and height measurements, keep the following in mind:

- Always measure—never ask! Self-reported heights are less accurate than measured heights. If height is not measured, document that the height is self-reported.
- Measure the length of infants and young children by using a measuring board with a fixed headboard and a movable footboard. It generally takes two people to measure length. One person gently holds the infant's head against the headboard; the other straightens the infant's legs and moves the footboard to the bottom of the infant's feet.
- Measure height next to a wall on which a nonstretchable measuring tape or board has been fixed. Ask the person to stand erect without shoes and with heels together. The person's eyes and head should be facing forward, with heels, buttocks, and shoulder blades touching the wall. Place a ruler or other flat, stiff object on the top of the head at a right angle to the wall and carefully note the height measurement.
- Immediately record length and height measurements to the nearest $1/8$ inch or 0.1 centimeter.
- For evaluating growth rate in young children, use the appropriate growth chart (Appendix E) when plotting results. If length is measured, use the growth chart for children between 0 and 36 months; if height is measured, use the chart for individuals between 2 and 20 years.
- Higher values are obtained from supine measurements than from vertical height measurements due to gravity.

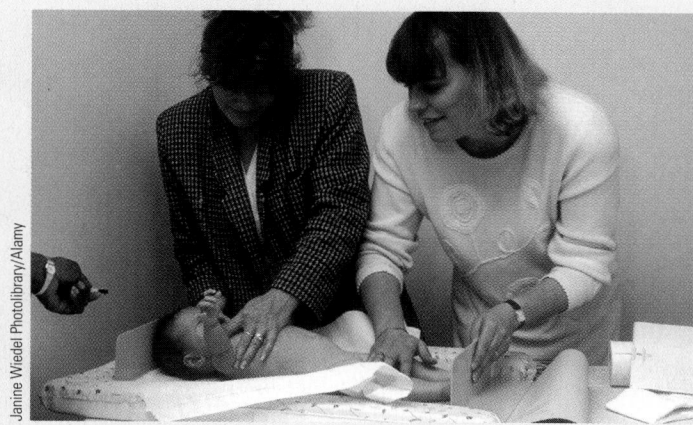

It takes two people to measure the length of an infant.

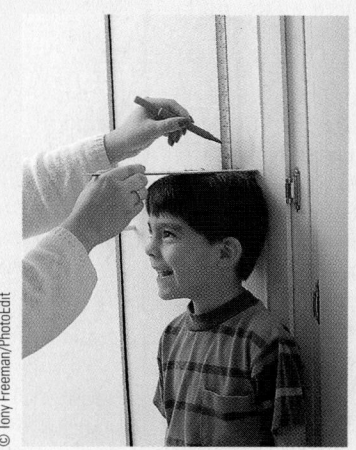

Standing erect allows for an accurate height measurement.

tary loss of body weight may signify PEM. Body weights can be compared with healthy ranges on height-weight tables and growth charts or used to calculate the body mass index (BMI). The "How to" on p. 364 includes suggestions for improving the accuracy of weight measurements. Table 13-5 (p. 364) describes a quick method for estimating desirable weight.

Weight data can be expressed as percent of "ideal body weight" (%IBW) or percent of "usual body weight" (%UBW) in order to assess the degree of nutritional risk associated with illness. The %UBW is more effective for interpreting weight changes that occur in overweight and obese individuals, as %IBW may fail to identify significant weight loss. Conversely, in patients who have been underweight throughout life, %IBW can overstate the degree of weight loss due to illness. The "How to" on p. 365 describes how to estimate %IBW and %UBW, and Table 13-6 (p. 365) shows how to interpret these values.

Head Circumference A head circumference measure helps to assess brain growth and malnutrition in children up to three years of age, although this measure is not necessarily reduced in a malnourished child.[7] Head circumference values can also track brain development in premature and small-for-gestational-age infants. To measure head circumference, the assessor encircles the largest circumference measure of a child's head with a nonstretchable measuring tape: the tape is placed

Reminder:

$$BMI = \frac{weight\ (kg)}{height\ (m)^2}.$$

A healthy BMI typically falls between 18.5 and 25.

HOW TO   *Measure Weight*

Tips for measuring weight include:

- Always measure—never ask! Self-reported weights are often inaccurate. If weight is not measured, document that the weight is self-reported.
- Valid weight measurements require scales that have been carefully maintained, calibrated, and checked for accuracy at regular intervals. Beam balance and electronic scales are the most accurate. Bathroom scales are inaccurate and inappropriate in the clinical setting.
- Measure an infant's weight with a scale that allows the infant to sit or lie down. The tray should be large enough to support an infant or young child up to 40 pounds, and the scale should weigh in $^1/_2$-ounce or 10-gram increments. For accurate results, weigh infants without clothes or diapers. Excessive movement by the infant can reduce accuracy.
- Children who can stand are weighed in the same way as adults, using beam balance or electronic scales with platforms large enough for standing comfortably. If repeated weight measurements are needed, each weighing should take place at the same time of day (preferably before breakfast), in the same amount of clothing, after the person has voided, and on the same scale. Record weights to the nearest $^1/_4$ pound or 0.1 kilogram.
- Special scales and hospital beds with built-in scales are available for weighing people who are bedridden.

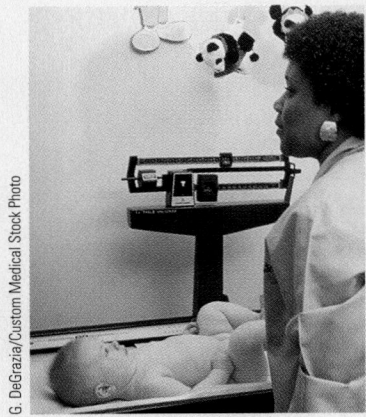

Infants are weighed on scales that allow them to sit or lie down.

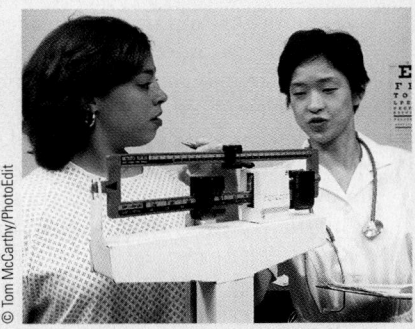

Beam balance scales allow accurate weight measurements for older children and adults.

TABLE 13-5	Quick Estimate of Desirable Body Weight[a]

Men

For first 5 feet, consider 106 pounds a reasonable weight. For each inch over 5 feet, add 6 pounds.

For each inch under 5 feet, subtract 6 pounds.

Add 10% for a large-framed individual; subtract 10% for a small-framed individual.

Example: For a man 5 feet 8 inches tall (medium frame), a desirable weight would be 154 pounds (106 lb + 48 = 154 lb).

Women

For first 5 feet, consider 100 pounds a reasonable weight. For each inch over 5 feet, add 5 pounds.

For each inch under 5 feet, subtract 5 pounds.

Add 10% for a large-framed individual; subtract 10% for a small-framed individual.

Example: For a woman 5 feet 6 inches tall (medium frame), a desirable weight would be 130 pounds (100 lb + 30 = 130 lb).

[a] This method does not account for differences in age or race.

To estimate %IBW, compare an individual's current weight with a reasonable (ideal) weight obtained from height-weight tables or calculated using the quick method in Table 13-5:

$$\%IBW = \frac{current\ weight}{ideal\ weight} \times 100.$$

For example, suppose you wish to calculate %IBW for a man who is 5 feet 8 inches tall and weighs 123 pounds. Using Table 13-5, you estimate that a reasonable weight for this man would be 154 pounds.

$$\%IBW = \frac{123\ lb}{154\ lb} \times 100 = 80\%.$$

The man in this example weighs 80 percent of his ideal body weight. A look at Table 13-6 indicates that at 80 percent of IBW he is mildly underweight.

To estimate %UBW, compare a person's current weight with the weight that the person generally maintains:

$$\%UBW = \frac{current\ weight}{usual\ weight} \times 100.$$

For example, if a woman loses 32 pounds during illness and her usual weight is 145 pounds, her current weight would be 113 pounds. These values can be incorporated into the above equation:

$$\%UBW = \frac{113\ lb}{145\ lb} \times 100 = 78\%.$$

The woman in this example weighs 78 percent of her usual weight. A look at Table 13-6 shows that a person at 78 percent of UBW is moderately underweight.

just above the eyebrows and ears and around the occipital prominence at the back of the head. The measurement is read to the nearest $^1/_8$ inch or 0.1 centimeter.

Circumferences of Waist and Limbs Circumferences of the waist and limbs help to evaluate body fat and muscle mass, respectively. Waist circumference correlates with visceral fat and can help in assessing overnutrition (see Figure 6-9 on p. 143). Circumferences of the mid-upper-arm, mid-thigh, and mid-calf regions can help in assessing the effects of illness, aging, and PEM on skeletal muscle content. For improved accuracy, circumference measurements are often used together with skinfold measurements to correct for the subcutaneous fat in limbs.

Anthropometric Assessment in Infants and Children To evaluate growth patterns, the nurse takes periodic measurements of height (or length), weight, and head circumference and plots them on growth charts, such as those provided in Appendix E. The most commonly used growth charts compare height (or length) to age, weight to age, head circumference to age, weight to length, and BMI to age. Although individual growth patterns vary, a child's growth will generally stay at about the same percentile throughout childhood; a sharp drop in a previously steady growth pattern suggests malnutrition. Growth patterns that fall below the 5th percentile may also be cause for concern, although genetic influences must be considered when interpreting low values. Growth charts with BMI-for-age percentiles can be used to assess risk of underweight and overweight in children over two years of age: the 10th and 85th percentiles are used as cutoffs to identify children who may be malnourished or overweight, respectively.[8]

Anthropometric Assessment in Adults Nurses routinely record and monitor weight and height during treatment. Weight changes must be evaluated carefully: although unintentional weight *loss* can indicate malnutrition, weight *gain* may result from fluid retention rather than overnutrition. Fluid retention often accompanies worsening disease in

TABLE 13-6	Use of Body Weight for Assessing Nutritional Risk	
%IBW	**%UBW**	**Nutritional Risk**
>120	—	Obesity
110–120	—	Overweight
90–109	—	Adequate weight—not at risk
80–89	85–95	Risk of mild malnutrition
70–79	75–84	Risk of moderate malnutrition
<70	<75	Risk of severe malnutrition

patients with heart failure, liver cirrhosis, and kidney failure and can mask the weight loss associated with PEM. In assessing the significance of weight loss, the rate should be considered as well as the amount: a 10 percent involuntary weight loss within a six-month period suggests a risk of PEM. Some types of medications can also lead to weight loss or gain.

Many of the illnesses discussed in later chapters are associated with lean tissue losses that resist nutrition intervention. Losses in both lean tissue and height are common with aging, even though body weights may remain stable. Including anthropometric measures such as skinfold measurements and limb circumferences can help nurses identify changes in body composition that need to be addressed in the treatment plan.

BIOCHEMICAL ANALYSES

Biochemical analyses provide information about protein-energy nutrition, vitamin and mineral status, fluid and electrolyte balance, and organ function. Most tests are based on analyses of blood and urine samples, which contain proteins, nutrients, and metabolites that reflect nutrition status. Table 13-7 lists and describes common blood tests with nutritional implications. Laboratory tests relevant to specific diseases will be discussed in the chapters that follow.

Interpreting laboratory values can be challenging when a number of different factors influence test results. For example, serum protein values can be affected by fluid imbalances, pregnancy, medications, and exercise. Similarly, serum levels of vitamins and minerals are often poor indicators of nutrient deficiency because the values are affected by multiple physiological factors; therefore, a variety of tests are generally needed to diagnose a nutrition problem. Taken together with other assessment data, however, laboratory test results help to present a clearer picture than is possible otherwise.

> Blood test results are reported in terms of either *plasma* or *serum* levels. *Plasma* is the yellow fluid that remains after cells are removed; it still contains clotting factors. *Serum* is the fluid remaining after both cells and clotting factors are removed.

> Fluid retention can result in lab results that are deceptively low. Dehydration may cause lab results to be deceptively high.

Plasma Proteins Plasma protein levels can help in the assessment of protein status, although levels may fluctuate for different reasons. For example, both PEM and liver disease can reduce plasma proteins. Metabolic stress causes the release of hormones that alter plasma protein levels. Values are also influenced by pregnancy, kidney function, zinc status, and some medications. Because plasma proteins are affected by so many factors, their values must be considered along with other data to evaluate nutrition status. The following paragraphs describe several of the plasma proteins commonly measured during illness.

> Plasma proteins are synthesized in the liver, so plasma levels can reflect liver function.

Albumin Albumin is the most abundant plasma protein, and its levels are routinely measured. Although many medical conditions influence albumin, it is slow to reflect changes in nutrition status because of its large body pool and slow rate of degradation. In people with chronic PEM, albumin levels remain normal for long periods of time despite depletion of body proteins; levels fall only after prolonged malnutrition. Likewise, when malnutrition is treated, albumin concentrations increase slowly, so albumin is not a sensitive indicator of effective treatment.

> The term *half-life* defines the length of time that a substance remains in plasma. The albumin in plasma has a 3-week half-life, meaning that half of the amount circulating in plasma is degraded in a 3-week period.

Transferrin Transferrin transports iron, so its concentrations respond to both PEM and iron status. Transferrin breaks down in the body more rapidly than albumin, but it responds relatively slowly to nutrition therapy. In addition, evaluating protein-energy status using transferrin is difficult if an iron deficiency is also present. Transferrin levels rise as iron deficiency worsens and fall as iron status improves.

> Transferrin's half-life in plasma is approximately 8 to 10 days.

Prealbumin and Retinol-Binding Protein Levels of prealbumin (also called transthyretin) and retinol-binding protein decrease rapidly during PEM and respond quickly to changes in protein intake. Thus these proteins are more sensitive than albumin to changes in protein status. Like other plasma proteins, their usefulness in nutrition assessment is limited because they are affected by a number of different

> Half-lives of prealbumin and retinol-binding protein are 2 days and 12 hours, respectively.

TABLE 13-7	Routine Laboratory Tests with Nutritional Implications

This table presents a partial listing of some uses of commonly performed lab tests that have implications for nutritional problems.

Laboratory Test	Acceptable Range	Description
Hematology		
Red blood cell (RBC) count	Male: 4.3–5.7 million/μL Female: 3.8–5.1 million/μL	Number of RBC; aids anemia diagnosis.
Hemoglobin (Hb)	Male: 13.5–17.5 g/dL Female: 12.0–16.0 g/dL	Hemoglobin content of RBC; aids anemia diagnosis.
Hematocrit (Hct)	Male: 39–49% Female: 35–45%	Percentage RBC in total blood volume; aids anemia diagnosis.
Mean corpuscular volume (MCV)	81–99 fL	RBC size, helps to distinguish between microcytic and macrocytic anemias.
Mean corpuscular hemoglobin concentration (MCHC)	31–37% Hb/cell	Hb concentration within RBCs, helps to distinguish iron-deficiency anemia.
White blood cell (WBC) count	4500–11,000 cells/μL	Number of WBC; general assessment of immunity.
Blood Chemistry		
Serum Proteins		
Total protein	6.4–8.3 g/dL	Protein levels are not specific to disease or highly sensitive; they can reflect poor protein intake, illness or infections, changes in hydration or metabolism, pregnancy, or medications.
Albumin	3.4–4.8 g/dL	May reflect illness or PEM; slow to respond to improvement or worsening of disease.
Transferrin	200–400 mg/dL >60 yr: 180–380 mg/dL	May reflect illness, PEM, or iron deficiency; slightly more sensitive to changes than albumin.
Prealbumin (transthyretin)	10–40 mg/dL	May reflect illness or PEM; more responsive to health status changes than albumin or transferrin.
C-reactive protein	68–8200 ng/mL	Indicator of inflammation or disease.
Serum Enzymes		
Creatine kinase (CK)	Male: 38–174 U/L Female: 26–140 U/L	Different forms of CK are found in muscle, brain, and heart. High levels in blood may indicate heart attack, brain tissue damage, or skeletal muscle injury.
Lactate dehydrogenase (LDH)	208–378 U/L	LDH is found in many tissues. Specific types may be elevated after heart attack, lung damage, or liver disease.
Alkaline phosphatase	25–100 U/L	Found in many tissues; often measured to evaluate liver function.
Aspartate aminotransferase (AST, formerly SGOT)	10–30 U/L	Usually monitored to assess liver damage; elevated in most liver diseases. Levels are somewhat increased after muscle injury.
Alanine aminotransferase (ALT, formerly SGPT)	Male: 10–40 U/L Female: 7–35 U/L	Usually monitored to assess liver damage; elevated in most liver diseases. Levels are somewhat increased after muscle injury.
Serum Electrolytes		
Sodium	136–146 mEq/L	Helps to evaluate hydration status or neuromuscular, kidney, and adrenal functions.
Potassium	3.5–5.1 mEq/L	Helps to evaluate acid-base balance and kidney function; can detect potassium imbalances.
Chloride	98–106 mEq/L	Helps to evaluate hydration status and detect acid-base and electrolyte imbalances.
Other		
Glucose	74–106 mg/dL	Detects risk of glucose intolerance, diabetes mellitus, and hypoglycemia; helps to monitor diabetes treatment.
Glycosylated hemoglobin (Hb A$_{1c}$)	5.0–7.5% of Hb	Used to monitor long-term blood glucose control (approximately 1 to 3 months prior).
Blood urea nitrogen (BUN)	6–20 mg/dL	Primarily used to monitor kidney function; value is altered by liver failure, dehydration, or shock.
Uric acid	Male: 3.5–7.2 mg/dL Female: 2.6–6.0 mg/dL	Used for detecting gout or changes in kidney function; levels affected by age and diet; varies among different ethnic groups.
Creatinine (serum or plasma)	Male: 0.7–1.3 mg/dL Female: 0.6–1.1 mg/dL	Used to monitor renal function.

Note: μL = microliter; dL = deciliter; fL = femtoliter; ng = nanogram; U/L = units per liter; mEq = milliequivalents.

Source: L. Goldman and J. C. Bennett, eds., *Cecil Textbook of Medicine* (Philadelphia: Saunders, 2000).

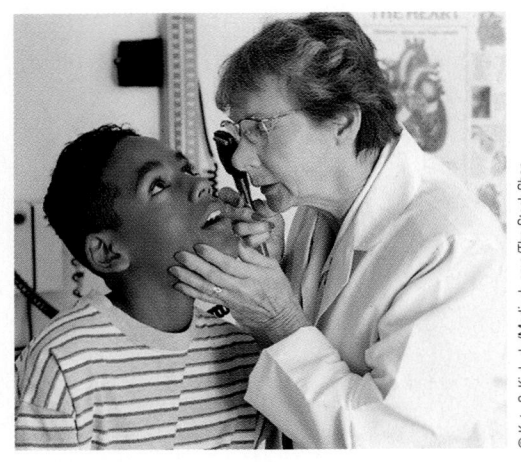

Physical signs of malnutrition are evident in parts of the body where cells are replaced at a rapid rate.

factors, including metabolic stress, zinc deficiency, and various medical conditions. Prealbumin and retinol-binding protein are more expensive to measure than albumin, so they are not routinely included during nutrition assessment.

PHYSICAL EXAMINATIONS

As with the other assessment methods, interpreting physical signs of malnutrition requires skill and clinical judgment. Most physical signs are nonspecific; they can reflect any of several nutrient deficiencies as well as conditions unrelated to nutrition. For example, cracked lips may be caused by several B vitamin deficiencies but may also be caused by sunburn, windburn, or dehydration. Dietary and laboratory data are usually needed as additional evidence to confirm suspected nutrient deficiencies.

Signs of malnutrition appear most often in parts of the body where cell replacement occurs at a rapid rate, such as the hair, skin, and digestive tract (including the mouth and tongue). Table 13-8 lists some clinical signs of nutrient deficiencies. Many of the symptoms listed occur only in advanced stages of deficiency. Chapters 8 and 9 provide additional examples of clinical signs of nutrient imbalances.

Fluid Imbalances Fluid imbalances accompany many illnesses and can also result from use of certain medications. Attention to the physical signs of fluid retention or dehydration can help health care providers correctly interpret the results of blood tests and the body weight measurement.

Fluid retention (also called *edema*) can accompany malnutrition, infection, or injury. It can be caused by impaired blood circulation and is often associated with

TABLE 13-8	Clinical Signs of Nutrient Deficiencies		
Body System	**Acceptable**	**Signs of Malnutrition**	**Other Possible Causes**
Hair	Shiny, firm in scalp	Dull, brittle, dry, loose; falls out (PEM); corkscrew hair (copper)	Excessive hair bleaching; hair loss from aging, chemotherapy, or radiation therapy
Eyes	Bright, clear pink membranes; adjust easily to light	Pale membranes (iron); spots, dryness, nightblindness (vitamin A); redness at corners of eyes (B vitamins)	Anemia, unrelated to nutrition; eye disorders; allergies
Lips	Smooth	Dry, cracked, or with sores in the corner of the lips (B vitamins)	Sunburn, windburn, excessive salivation from ill-fitting dentures or other disorders
Mouth and gums	Red tongue without swelling, normal sense of taste; teeth without caries; gums without bleeding, swelling, or pain	Smooth or magenta tongue (B vitamins), decreased taste sensations (zinc); swollen, bleeding gums (vitamin C)	Medications, periodontal disease (poor oral hygiene)
Skin	Smooth, firm, good color	Poor wound healing (PEM, vitamin C, zinc); dry, rough, lack of fat under skin (essential fatty acids, PEM, B vitamins); bruising, bleeding under skin (vitamins C and K)	Poor skin care, diabetes mellitus, aging, medications
Nails	Smooth, firm, pink	Ridged (PEM); spoon shaped, pale (iron)	
Other		Dementia, peripheral neuropathy (B vitamins); swollen glands at front of neck (PEM, iodine); bowed legs (vitamin D)	Disorders of aging (dementia), diabetes mellitus (peripheral neuropathy)

diseases of the heart, kidney, liver, and lungs. Physical signs include weight gain, facial puffiness, swelling of limbs, abdominal distention, and tight-fitting shoes. Fluid retention can be detected after 5 to 10 pounds of excess fluid accumulate in the body of an average-sized person.[9]

Dehydration can result from fever, sweating, vomiting, diarrhea, excessive urination, and skin injury or burns (due to fluid loss through skin lesions). Symptoms include thirst, dry skin or mouth, and reduced skin tension. The urine may be dark yellow or amber colored, and urine volume may be unusually low. Dehydration risk is greatest in the elderly, who have reduced thirst responses to water deprivation.

Functional Assessment Nutrient deficiencies can impair normal physiological functions; for example, zinc deficiency can depress immunity and slow wound healing. Protein-energy malnutrition and illness can cause **wasting,** the breakdown and loss of body tissues. Functional tests help health practitioners evaluate the changes in physiological functions and losses in body strength that accompany malnutrition or disease. For example, exercise tolerance can be assessed using a treadmill or cycle ergometer. Respiratory muscle strength might be tested by holding a strip of paper 4 inches from the mouth and observing how the paper moves while breathing. Assessment of immunity may include testing the skin response to antigens that cause redness and swelling when immune function is adequate. The chapters that follow include additional examples of functional assessment.

INTEGRATING ASSESSMENT DATA

Combining the results from several methods can improve the accuracy of nutrition assessment. One technique for doing so, known as Subjective Global Assessment (SGA), combines historical information with the results of a physical examination to predict the nutrition status of acute care patients (see Table 13-9).

TABLE 13-9	Elements of Subjective Global Assessment
Medical and diet histories	• Body weight changes in past six months and past two weeks • Change in dietary intake and duration of change • Current diet: whether suboptimal, low-kcalorie, liquid, or starvation diet • Gastrointestinal symptoms: nausea, diarrhea, vomiting, or anorexia • Functional ability: full capacity or suboptimal, walking or bedridden • Current medical diagnosis • Degree of metabolic stress: low, moderate, or high
Physical examination	• Loss of subcutaneous fat: in triceps or chest • Muscle wasting: in quadriceps or deltoids • Ankle edema • Sacral (lower spine) edema • Ascites
SGA rating	• Well-nourished if recent weight gain, mild fat/muscle loss, improvement in histories • Moderate malnutrition suspected if >5% weight loss (not from hydration change), decreased food intake, mild fat/muscle wasting • Severe malnutrition if >10% weight loss (not from hydration change), severe fat/muscle wasting, some edema

Source: A. S. Detsky and coauthors, What is subjective global assessment of nutritional status? *Journal of Parenteral and Enteral Nutrition* 11 (1987): 8–13.

wasting: the gradual atrophy (loss) of body tissues; associated with protein-energy malnutrition or chronic illness.

CASE STUDY   *Nutrition Screening and Assessment*

Joan Ellison is an 85-year-old retired businesswoman who has been a widow for 10 years. She uses a walker and has poorly fitting dentures. She was recently admitted to the hospital with pneumonia and also has congestive heart failure and diabetes. She routinely takes several medications to control blood glucose, hypertension, and heart function, and, in addition to these, the physician ordered antibiotics to treat the pneumonia. During an initial nutrition screening, Mrs. Ellison stated that she had been eating very poorly over the past two weeks. She said that she usually weighs about 125 pounds, a fact that was documented in her medical chart from a previous visit. Although she felt she was losing weight, she didn't know how much weight she might have lost or when she started losing weight. On admission to the hospital, Mrs. Ellison weighed 115 pounds and was 5 feet 2 inches tall. Her serum albumin level was 3.0 g/dL. A physical exam revealed edema, and several other laboratory tests confirmed that she was retaining fluid. As a result of the nutrition screening, Mrs. Ellison was referred to a nurse for a complete nutrition assessment.

1. From the brief description provided, which items in Mrs. Ellison's medical, social, and diet histories might alert the nurse that she is at risk of malnutrition?
2. Identify a healthy body weight for Mrs. Ellison and calculate her %IBW and %UBW. What do the results reveal? What effect does fluid retention have on Mrs. Ellison's weight?
3. How can fluid retention alter Mrs. Ellison's serum protein levels? What physical symptoms may have suggested that she was retaining excess fluid?
4. What tools can be used to estimate Mrs. Ellison's usual food intake? What medical, physical, and social factors are likely to affect her dietary intake?
5. Describe other types of assessment information that may help the nurse determine whether Mrs. Ellison should be referred to a registered dietitian.

Ascites, the abnormal accumulation of fluid in the abdominal cavity, is discussed in Chapter 19.

Elements of SGA may include weight and dietary changes, gastrointestinal symptoms, work capacity, level of metabolic stress, degree of muscle wasting and fat loss, and presence of edema or ascites. SGA is widely used to evaluate nutrition status and has been found to be applicable to different patient populations. The Case Study above can help you review the different components of a nutrition assessment.

REVIEW NOTES

Nutrition assessments include historical information, anthropometric data, biochemical analyses, and physical examinations. Health care providers assess food intake using 24-hour recall interviews, food frequency questionnaires, food records, and direct observation.

Anthropometric measurements help to evaluate growth patterns, overnutrition and undernutrition, and body composition. Plasma proteins such as albumin, transferrin, and prealbumin can help in assessment of protein status but are influenced by various medical conditions. Physical exams can detect signs of nutrient deficiency, fluid imbalances, and functional impairments related to nutritional problems.

By combining the data from different assessment methods, nurses and dietitians can better identify patients who are most likely to develop nutrition problems.

SELF CHECK

1. Mr. Smith experiences loss of appetite, difficulty swallowing, and mouth pain as a consequence of illness. Mr. Smith is at risk of malnutrition due to:
   a. altered metabolism.
   b. reduced food intake.
   c. altered excretion of nutrients.
   d. altered digestion and absorption.

2. The central role of nurses in health care makes them well positioned for:
   a. calculating patients' nutrient needs.
   b. providing medical nutrition therapy.
   c. conducting complete nutrition assessments.
   d. identifying patients at risk for malnutrition.

3. The nutrition care process is a systematic approach for:
   a. identifying the nutrient content of foods.
   b. ordering special diets.
   c. conducting nutrition screening.
   d. meeting the nutrition needs of patients.

4. To conduct complete nutrition assessments, clinicians rely on all of the following sources of information *except:*
   a. nutrition care plans.
   b. body measurements.
   c. medical, social, and diet histories.
   d. biochemical analyses.

5. All of the following factors place a person at risk for malnutrition *except:*
   a. health problems that are frequently associated with PEM.
   b. the use of prescription medications that affect nutrient needs.
   c. a social history that reveals the individual lives with a spouse in a middle-income neighborhood.
   d. a significant reduction in food intake over the past five or more days.

6. Which dietary assessment method does a nurse use to conduct a kcalorie count?
   a. 24-hour recall interview
   b. food frequency questionnaire
   c. food record
   d. direct observation

7. Height and weight measurements:
   a. are both affected by fluid status.
   b. cannot be performed on bedridden clients.
   c. are routine measurements in health care facilities.
   d. require equipment that is not readily available in most health care facilities.

8. The %IBW of a person who weighs 185 pounds and has a healthy body weight of 150 pounds is:
   a. 123 percent.
   b. 150 percent.
   c. 23 percent.
   d. 81 percent.

9. A malnourished patient has just begun to eat after days without significant amounts of food. Which of the following blood tests would change most quickly as the patient's nutrition status improves?
   a. albumin
   b. transferrin
   c. serum electrolytes
   d. retinol-binding protein

10. Which sign of PEM would be unlikely to show up in a physical examination?
    a. low plasma protein levels
    b. dull, brittle hair
    c. poor wound healing
    d. wasted appearance

Answers to these questions appear in Appendix H.

CLINICAL APPLICATIONS

1. Describe the possible nutritional implications of these findings from a patient's medical and social histories: age 77, lives alone, recently lost spouse, uses a walker, no natural teeth or dentures, history of hypertension and diabetes, uses medications that cause frequent urination.

2. Nurses and nurses' aides frequently shoulder much of the responsibility for collecting food intake data for kcalorie counts because they often deliver food trays and snacks and later retrieve them. Why is it important for a nurse or aide to verify and record both what the patient receives (foods and amounts) and the foods that remain uneaten? When might patients be enlisted in the collection of food intake data, and when might such a course be unwise?

3. Calculate the %IBW and %UBW for a man who is 5 feet 11 inches tall with a current weight of 160 pounds and a usual body weight of 180 pounds. Use the method in Table 13-5 (p. 364) to estimate a desirable ("ideal") body weight. What additional information do you need to interpret the implications of this amount of weight loss?

NUTRITION ON THE NET

For further study of the topics in this chapter, access these websites.

Find updates and quick links to these and other nutrition-related sites at our website: **www.wadsworth.com/nutrition**

Calculate your nutrient intake using food composition data from the USDA Nutrient Data Laboratory: **www.nal.usda.gov/fnic/foodcomp/search/**

Visit the website of the Joint Commission on Accreditation of Healthcare Organizations (JCAHO) to learn more about the quality of health care and accreditation: **www.jointcommission.org**

Learn about the Nutrition Screening Initiative from the American Academy of Family Physicians (type "nutrition screening" in the search box): **www.aafp.org**

NOTES

[1] D. R. Thomas and coauthors, Malnutrition in subacute care, *American Journal of Clinical Nutrition* 75 (2002): 308–313.

[2] K. Lacey and E. Pritchett, Nutrition care process and model: ADA adopts road map to quality care and outcomes management, *Journal of the American Dietetic Association* 103 (2003): 1061–1072.

[3] K. Lacey and N. Cross, A problem-based nutrition care model that is diagnostic driven and allows for monitoring and managing outcomes, *Journal of the American Dietetic Association* 102 (2002): 578–589.

[4] L. C. Tapsell, V. Brenninger, and J. Barnard, Applying conversation analysis to foster accurate reporting in the diet history interview, *Journal of the American Dietetic Association* 100 (2000): 818–824.

[5] E. Saltzman and K. M. Mogensen, Physical assessment, in A. M. Coulston, C. L. Rock, and E. R. Monsen, eds., *Nutrition in the Prevention and Treatment of Disease* (San Diego: Academic Press, 2001), pp. 43–58.

[6] V. A. Stallings and E. B. Fung, Clinical nutrition assessment of infants and children, in M. E. Shils and coeditors, *Modern Nutrition in Health and Disease* (Baltimore: Williams & Wilkins, 1999), pp. 885–893.

[7] Stallings and Fung, 1999.

[8] K. M. Flegal, R. Wei, and C. Ogden, Weight-for-stature compared with body mass index-for-age growth charts for the United States from the Centers for Disease Control and Prevention, *American Journal of Clinical Nutrition* 75 (2002): 761–766.

[9] G. A. Modest, Edema, in J. Noble and coeditors, *Textbook of Primary Care Medicine* (St. Louis: Mosby, 2001), pp. 178–182.

Nutritional Genomics

Imagine this situation: a physician scrapes a sample of cells from inside your cheek and submits it to a **genomics** lab. Within an hour, you receive a report that reveals your disease susceptibilities and recommends dietary and lifestyle changes that can maintain health. You may even be given a prescription for a dietary supplement to prevent the diseases that you are most likely to develop. Unlikely? Perhaps, but these possibilities are being explored by scientists working in the new field of **nutritional genomics,** the study of dietary effects on **gene expression.**[1] Recent research suggests that some dietary factors may be more helpful (or more harmful) in people who have particular genetic variations. The promise of nutritional genomics is a custom-designed dietary prescription that fits each person's specific needs. The glossary on pages 375–376 defines genomics and related terms.

What is a genome?

Genetic information is encoded in DNA molecules within the nuclei of almost all of the cells in our bodies. Figure NP13-1 shows how the genetic material is organized within the **genome,** the complete set of genetic information within our cells. The DNA molecules are tightly packed along with associated proteins within the 46 **chromosomes.** Segments of a DNA strand that can eventually be translated into proteins are called **genes.** The sequence of **nucleotides** within each gene encodes the amino acid sequence of a particular protein. Scientists estimate that there are between 20,000 and 25,000 genes in the human genome.[2] Only a small percentage of the genome codes are for proteins, however: most DNA consists of **noncoding sequences,** whose function, if any, is unclear.

When proteins are made, the information in the DNA sequence is first transcribed (copied) to messenger RNA molecules, which carry the genetic information out of the nucleus. Gene expression can be measured by determining the amounts of messenger RNA in a tissue sample. The expression of thousands of genes can be measured simultaneously using **microarray technology** (see photo p. 374).

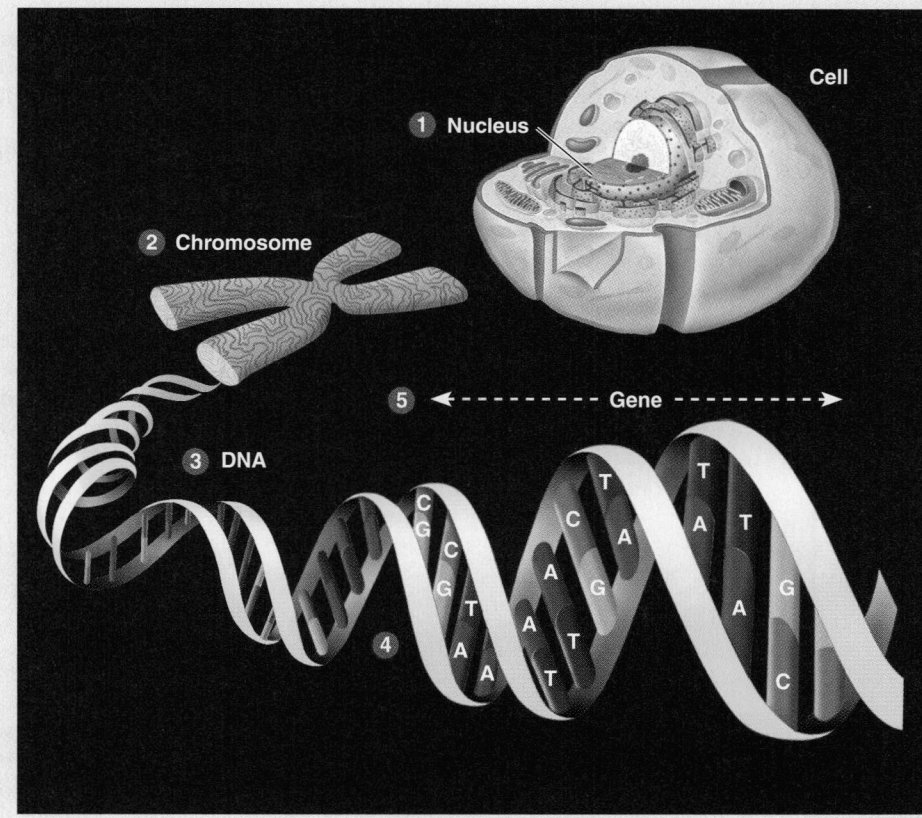

FIGURE NP13-1 The Human Genome

1. The human genome is a complete set of genetic material organized into 46 chromosomes, located within the nucleus of a cell.

2. A chromosome is made of DNA and associated proteins.

3. The double helical structure of a DNA molecule is made up of two long chains of nucleotides. Each nucleotide is composed of a phosphate group, a 5-carbon sugar, and a base.

4. The sequence of nucleotide bases (C, G, A, T) determines the amino acid sequence of proteins. These bases are connected by hydrogen bonding to form base pairs: adenine (A) with thymine (T) and guanine (G) with cytosine (C).

5. A gene is a segment of DNA that includes the information needed to synthesize one or more proteins.

Source: Adapted from "A Primer: From DNA to Life," Human Genome Project, U.S. Department of Energy Office of Science, **www.ornl.gov/sci/techresources/Human_Genome/primer_pic.shtml**.

A DNA microarray allows researchers to monitor the expression of thousands of genes simultaneously.

© Science VU/Visuals Unlimited

How did research in nutritional genomics begin?

The Human Genome Project, an international effort by industry and government scientists to sequence the human genome, was completed in April 2003. The project led to enormous advances in the research technologies needed to study genes and genetic variation. Knowing the DNA sequences within human chromosomes allows researchers to study how alterations in diet and lifestyle can alter the expression of a multitude of genes. The next steps are to identify the individual genes in the genome and the roles of their protein products, the genes and proteins associated with diseases, and the dietary and lifestyle choices that influence the expression of genes involved in disease.

Genetic differences among individuals have been studied for years, as have the specialized dietary therapies used to treat various **inherited disorders.** For example, an individual may inherit a genetic defect that inhibits the normal metabolism of an essential nutrient and may therefore need to consume a diet that contains either more or less of this nutrient. An example of this type of condition is phenylketonuria (PKU), discussed in Nutrition in Practice 17. Genomic research takes this concept further: instead of focusing on alterations in one or two genes, researchers study the expression of *multiple* genes.

How do nutrients alter gene expression?

Some nutrients can switch gene expression on or off.[3] The **promoter** region of a gene (a DNA region involved with gene activation) acts as the master switch. A large variety of proteins known as **transcription factors** bind to areas on the promoter and either enhance or inhibit gene expression. A combination of dietary factors and hormones influences the types of transcription factors that reach the nucleus and

their tendency to bind to DNA. Specific examples of how nutrients can influence gene expression include:

- The transcription factor that enhances gene expression of enzymes required for cholesterol synthesis enters the nucleus only when the cellular cholesterol content is low.
- The transcription factor that inhibits expression of ferritin, an iron-storage protein, changes its affinity for DNA based on the iron content of the cell.

How much genetic variation is there among people?

Except for identical twins, no two individuals are genetically identical. The variation in the genomes of any two persons, however, is only about 0.1 percent, a difference of only one base in every 1000. The most common genetic differences, known as **polymorphisms,** are changes in single nucleotides **(single-nucleotide polymorphisms).** Such variations are significant only if they affect the amino acid sequence of a protein in such a way that protein function is altered.

Genetic variation gives rise to the diversity among human beings—it explains most of the differences in our physical appearances and metabolic characteristics. Along with environmental factors, it also determines our susceptibilities to disease. Diseases affected by a single gene tend to be relatively rare and usually exert their effects early in life. In contrast, common diseases like heart disease and cancer are influenced by many genes and typically develop over several decades or even longer. In these more complex **multigene,** or **polygenic,** disorders, many genes can contribute to disease risk, but no single gene may be sufficient to cause the disease on its own.

What are some examples of single-gene disorders?

Examples of single-gene disorders include PKU, sickle-cell anemia, and the iron-overload disease hemochromatosis. Single-gene disorders may seriously disrupt metabolism and often require significant dietary or medical intervention. Not all single-gene disorders have life-threatening ramifications, however. For example, lactose intolerance can result from an alteration in the promoter of the lactase gene; it may cause gastrointestinal discomfort but is readily managed by simple dietary changes.

How are multigene disorders different from single-gene disorders?

Multigene diseases are usually sensitive to a number of environmental influences, including diet and lifestyle.[4] These environmental factors can directly influence the expression of the genes involved. Also, multigene diseases tend to develop over many years, so determining genetic susceptibility may allow a person to modify diet and lifestyle appropriately and reduce the risk of developing the disease.

Heart disease is an example of a disease with multiple gene influences. Its many risk factors represent the involvement of an assortment of genes, which affect disparate aspects of physiology and metabolism. Consider that the major risk factors for heart disease include elevated blood cholesterol levels, obesity, diabetes, and hypertension. The underlying cause of any of these risk factors is rarely known; currently, clinicians screen for the presence of risk factors, but not the reasons why they occur. Should genomic research prove successful, a future assessment approach might be to identify specific genetic alterations and changes in metabolism that lead to the development of individual risk factors. For example, tests may determine whether blood cholesterol levels are high due to excessive cholesterol absorption, excessive liver production, or reduced cholesterol degradation.[5] This information could then guide health care providers to the most appropriate intervention, allowing a better match between treatment recommendations and a person's genetic profile.

Can genomic research be used to explore the differences in nutrient needs among people?

Even though most people apparently can meet their nutrient needs by consuming nutrients at recommended levels, it would be useful to learn more about genetic variations within healthy populations. The techniques that have emerged from genomic research may provide a means for fine-tuning nutrient recommendations for different individuals.[6] Moreover, ideal indicators of nutrient status are still lacking for several of the minerals, such as zinc, magnesium, and chromium. Scientists hope to eventually produce genomic maps that will indicate how various nutrient deficiencies and combinations of deficiencies affect gene expression. These maps may eventually provide data that can help diagnose nutrient deficiencies.

Will knowledge about the human genome substantially change the manner in which health care is provided?

The enthusiasm surrounding genomic research must be put into perspective in terms of both the status of clinical medicine at present and people's willingness to make difficult lifestyle choices. Critics have questioned whether genetic markers for disease would be more useful than simple and inexpensive clinical measurements, which reflect both genetic *and* environmental influences. In other words, knowing that a person is genetically predisposed toward high cholesterol levels is not necessarily more useful than knowing the person's actual blood cholesterol level.[7] Furthermore, if a disease has many genetic risk factors, each gene that contributes to susceptibility may have little influence on its own, so the benefits of identifying an individual genetic marker would be small. The long-range possibility is that many genetic markers will eventually be identified; the hope is that the combined information will be a more useful and accurate predictor of disease and its effective treatments.

Obtaining additional knowledge about disease risk may not be useful, however, unless people are motivated to make serious lifestyle changes. For example, despite the present abundance of disease prevention recommendations, many people seem unwilling to make the changes known to improve health. Researchers have estimated that heart disease and type 2 diabetes are 80 percent and 90 percent preventable, respectively, by changing one's lifestyle to include an appropriate diet, a healthy body weight, and regular exercise, among other factors.[8] Given the difficulty that people have with current recommendations, it is unlikely that they will enthusiastically adopt an even more detailed list of dietary and lifestyle modifications.

What ethical concerns are raised by having extensive knowledge about an individual's genome?

The ability to obtain detailed genetic information raises several concerns. A primary consideration is confidentiality: Should information about a person's susceptibility to disease be released to others without that person's consent? Because environmental factors play such an important role in disease risk, genetic predisposition usually cannot predict whether a person will develop a particular disease. Nevertheless, future insurers of medical services may attempt to charge higher rates or base acceptance criteria on applicants' disease susceptibilities as evidenced by genetic testing. Another important concern is whether genetic testing is always in the best interest of children. Although early knowledge of a child's predisposition to illnesses may be useful for parents who want to provide optimal care, the release of this information could threaten the child's privacy and increase the potential for "genetic discrimination" in the future.

Although genomic research has the potential to improve our ability to diagnose and treat disease, it is still unclear how knowledge of the genome will be translated into useful medical treatments. Still, health care professionals will need to keep informed of the ethical, legal, and social implications of genomics in the fields of medicine and nutrition as this remarkable research continues.

Glossary

chromosomes: structures within the nucleus of a cell that contain the cell's DNA and associated proteins.

gene expression: the process by which a cell converts the genetic code into RNA and protein.

genes: segments of DNA that contain the information needed to make proteins.

genome (JEE-nome): the full complement of genetic material in the chromosomes of a cell.

genomics (jee-NO-miks): the study of genomes.

inherited disorders: medical conditions resulting from genetic defects.

microarray technology: research technology that monitors the expression of thousands of genes simultaneously.

multigene or **polygenic:** involving a number of genes, rather than a single gene.

noncoding sequences: regions of DNA that do not code for proteins. Some noncoding sequences may have regulatory or structural properties, but most have no known function.

nucleotides: the subunits of DNA and RNA molecules. These compounds—cytosine (C), thymine (T), uracil (U), guanine (G), and adenine (A)—are each composed of a phosphate group, a 5-carbon sugar (ribose), and a nitrogen-containing base. A DNA molecule is made up of two long chains of nucleotides held together by hydrogen bonding between nucleotide bases on opposing strands; each hydrogen-bonded nucleotide couple is called a **base pair.**

nutritional genomics: the study of dietary effects on genetic expression; also known as **nutrigenomics.**

polymorphisms: differences in the DNA sequences among individuals. A **single-nucleotide polymorphism** involves a single nucleotide at a particular area in the DNA strand.
- **poly** = many
- **morph** = form
- **ism** = condition

promoter: a region of DNA involved with gene activation.

transcription factors: proteins that bind DNA at specific sequences to regulate gene expression.

Notes

[1] J. B. German, M.-A. Roberts, and S. M. Watkins, Personal metabolomics as a next generation nutritional assessment, *Journal of Nutrition* 133 (2003): 4260–4266; S. M. Brown, *Essentials of Medical Genomics* (Hoboken, N.J.: Wiley-Liss, 2003).

[2] International Human Genome Sequencing Consortium, Finishing the euchromatic sequence of the human genome, *Nature* 431 (2004): 931–945.

[3] S. D. Clarke, The human genome and nutrition, in B. A. Bowman and R. M. Russell, eds., *Present Knowledge in Nutrition* (Washington, D.C.: ILSI Press, 2001), pp. 750–760.

[4] W. C. Willett, Balancing life-style and genomics research for disease prevention, *Science* 296 (2002): 695–698.

[5] German, Roberts, and Watkins, 2003.

[6] B. N. Ames, H. Elson-Schwab, and E. A. Silver, High-dose vitamin therapy stimulates variant enzymes with decreased coenzyme binding affinity (increased K_m): Relevance to genetic disease and polymorphisms, *American Journal of Clinical Nutrition* 75 (2002): 616–658; R. A. Sunde, Research needs for human nutrition in the post-genome sequencing era, *Journal of Nutrition* 131 (2001): 3319–3323.

[7] Willett, 2002.

[8] Willett, 2002.

Nutrition Intervention and Diet-Drug Interactions

CHAPTER 14

This chapter describes the implementation of nutrition care in clinical practice. Ensuring that nutrient needs are met is a key part of this process, so the chapter also introduces common dietary modifications. The final section discusses diet-drug interactions that need consideration when health practitioners administer nutrition care.

Implementing Nutrition Care

As explained in Chapter 13, nurses are well positioned to identify patients with nutritional problems. In addition, nursing care plans may include nursing diagnoses that suggest the need for nutritional intervention. This section describes how the nurse can incorporate nutrition care into a care plan.

CARE PLANNING

Once the nurse or dietitian has collected assessment information, the next steps of the nursing process can be carried out. Table 14-1 shows an example of how nutrition care might be incorporated into a nursing care plan. The example describes a patient at risk for protein-energy malnutrition and wasting due to his swallowing difficulty and intolerance to physical activity. After diagnosing his nutrition-related problems, the nurse identifies **expected outcomes** and interventions that can correct these problems. Note that some of the interventions may fall within the scope of nursing practice whereas others require the assistance of other health professionals.[1]

APPROACHES TO NUTRITION CARE

As discussed in Chapter 13, nutrition care often involves dietary modification and nutrition education. The care plan should be compatible with the desires and abilities of the person it is designed to help. The challenge is greater if dietary changes are required for extended periods.

Long-Term Dietary Intervention When long-term changes are necessary, a care plan must take into account the person's current food habits, lifestyle, and degree of motivation. Behavior change is a process that occurs in stages; therefore, more than one consultation is usually necessary. The following approaches may be helpful in implementing long-term dietary changes:[2]

- *Determine the individual's readiness for change.* Some people have little desire to change their dietary behaviors, and even those who are willing may not be fully prepared to take the necessary steps. The nurse needs to consider a patient's readiness to adopt new dietary behaviors before attempting to implement an ambitious care plan.
- *Emphasize what to eat, rather than what not to eat.* Emphasizing foods to include in the diet, rather than those to restrict, can make dietary changes more appealing. For example, encouraging additional fruits and vegetables is a more attractive message than telling the patient to restrict butter, cream sauces, and ice cream.
- *Suggest only one or two changes at a time.* People are more likely to adopt a dietary plan that does not deviate too much from their usual diet. If they succeed in adopting one or two changes, they are more likely to stick to the plan and be open to additional suggestions. Stricter plans may yield quicker results but are useful only for highly motivated people.

Nutrition Education Nutrition education allows patients to learn about the dietary factors that affect their particular medical condition. Ideally, this knowledge motivates them to change their diet and lifestyle to improve their health status.

expected outcomes: patient-oriented goals that are derived from nursing diagnoses.

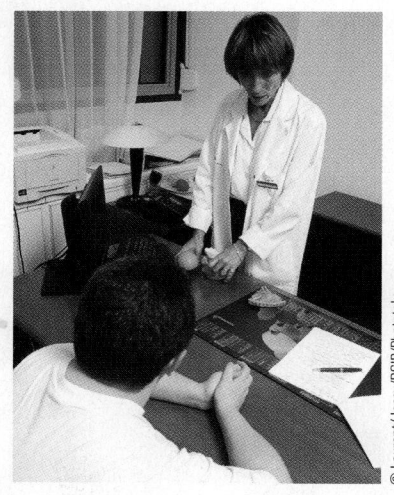

© Laurent/Jessy/BSIP/Phototake

Dietary counseling requires sensitivity to cultural orientation, educational background, and motivation for change.

TABLE 14-1	Incorporating Nutrition Care into the Nursing Care Plan

The following example illustrates how the nurse can address nutritional problems while working through the different phases of the nursing process.

Nursing Process	Nutrition-Related Features
Assessment: The nurse includes nutrition assessment data in the overall assessment database.	The assessment data related to nutrition care are categorized and documented. *Subjective data:* • 74-year-old man with emphysema and dysphagia (difficulty swallowing); lives with wife. • Uses oxygen therapy at home, but mostly while sleeping; frequently feels "out of breath" during the day. • Reports coughing while eating; senses that food gets "stuck" in his throat. • Believes he has lost weight recently because his clothes seem looser than usual. Says usual weight is 160 pounds, which is documented in the medical record. • Physically inactive; mostly watches TV during the day. *Objective data:* Height: 5′9″; weight: 145 lb; BMI: 21.5; %UBW: 90.6%; albumin: 3.6 g/dL; continually needs to clear throat; exhibits dyspnea (labored breathing) with walking.
Diagnosis: The nurse develops diagnoses that suggest a need for nutritional intervention.	Nursing diagnoses with nutritional implications include: • Impaired swallowing: *related to neuromuscular impairment.* • Risk for aspiration: *related to ineffective swallowing reflex.* • Imbalanced nutrition: less than body requirements *related to inability to ingest foods.* • Activity intolerance: *related to imbalance between oxygen supply and demand.*
Outcome identification/planning: The nurse identifies outcomes that are likely to improve nutrition status.	Desirable outcomes that may improve nutrition status include: • Patient will demonstrate ability to consume appropriate foods without coughing or aspiration. • Patient will consume adequate amounts of food and beverages. • Patient and spouse will identify foods and beverages that can be consumed without difficulty. • Patient will show no further evidence of weight loss. • Patient will communicate understanding of the need for a gradual increase in physical activity. • Patient will use oxygen therapy as needed to permit a steady increase in activity.
Implementation: The nurse develops appropriate interventions that the nurse can perform independently or with the help of other health professionals.	Depending on the nurse's background and knowledge, independent nursing interventions may include: • Educating the patient about body positioning and feeding techniques that can improve his ability to eat without coughing or discomfort. • Providing the patient with information about high-kcalorie and high-protein foods to minimize further weight and lean tissue loss. • Explaining the need for daily physical activity, including suggestions that may improve the patient's ability to tolerate additional activity. • Teaching the patient to conserve energy while performing activities of daily living. • Providing information about lightweight, portable, supplemental oxygen that can assist the patient during the day. Interventions that require the assistance of other health professionals include: • Consulting attending physician about the possible need for additional dysphagia testing, to help determine most appropriate dysphagia diet. • Referring the patient and spouse to a dysphagia specialist who can give specific feeding recommendations related to the patient's type of dysphagia. • Referring the patient to a dietitian to review and modify his diet so that the patient can improve energy intake and make appropriate food and beverage selections.

Evaluation: The nurse monitors and evaluates the patient's progress in reaching each outcome. The care plan is revised as needed.

Source: S. S. Ralph and C. M. Taylor, *Nursing Diagnosis Reference Manual* (Philadelphia: Lippincott Williams & Wilkins, 2005).

Mikey is an 11-year-old boy who was admitted to the hospital after he passed out while playing with friends. Tests confirm a diagnosis of type 1 diabetes mellitus. Mikey remains in the hospital for several days until his blood glucose and ketone levels are under control. During this time, he and his family learn about diabetes, the diet Mikey needs to follow, the use of insulin, how to monitor blood glucose levels, and the required coordination of diet, insulin, and physical activity. The details of diabetes mellitus are reserved for Chapter 20, but for now you can consider the steps that are necessary for implementing nutrition care.

1. The nurse handling Mikey's case must identify *expected outcomes* when she writes her care plan. What expected outcomes are appropriate for Mikey's situation?

2. Given the chronic nature of Mikey's illness and his age, what approaches should be used when discussing the required dietary and medical treatments with Mikey and his family?

3. What factors need consideration when designing a nutrition education program for Mikey and his parents?

4. Mikey will need additional care so that he can learn more about diabetes and to make the adjustments that will allow him to cope with his condition. Why is it important to plan follow-up care before Mikey leaves the hospital?

A nutrition education program should be tailored to a person's age, level of literacy, and cultural background. Learning style should also be considered: some people learn best by discussion supplemented with written materials, whereas others prefer visual examples, such as food models and measuring devices.[3] Information can be provided in one-on-one sessions or group discussions. The meeting should also assess the person's understanding of the material and commitment to making changes. Follow-up sessions can reveal whether the person has successfully adopted a dietary plan. A nurse who counsels a woman who is lactose intolerant and hesitant to use milk products can proceed as follows:

- The nurse can provide sample menus of a nutritionally adequate diet that limits milk and milk products. Together the nurse and the woman can design menus that consider her food preferences.
- Using diet analysis software, the nurse can demonstrate how altering food choices changes the calcium content of a meal.
- The nurse can explain how to use the Daily Value on food labels to estimate the calcium content of packaged foods.
- The nurse can provide information about the advantages and disadvantages of different calcium supplements.
- The nurse can assess the woman's understanding by having her identify non-milk products that are high in calcium.

Ideally, the nurse would be able to monitor the woman's progress in a subsequent counseling session. The accompanying Case Study provides an opportunity for you to review the implementation of nutrition care.

REVIEW NOTES

By using the steps of the nursing process, a nurse can incorporate nutrition care into a nursing care plan.

A care plan should be compatible with a person's food preferences and willingness to make dietary changes.

Nutrition education must be individualized to accommodate a patient's needs and learning style.

Modified Diets

During many illnesses, a person can meet energy and nutrient needs by following a **standard diet.** In other cases, a **modified diet** is prescribed. Modified diets are altered either by changing the consistency or nutrient content or by including or eliminating certain foods. This section presents examples of common modified diets and explains their use during illness. Nutrition in Practice 14 discusses how foodservice departments accommodate patients' needs for special diets in hospitals and residential facilities.

THE DIET ORDER

As mentioned in Chapter 13, the physician has the primary responsibility for prescribing an appropriate diet for a patient in a medical facility. Diet orders must be precise to avoid confusion; for example, a "low-sodium diet" should specify the amount of sodium permitted, as "low-sodium" could be interpreted to mean any amount between 500 and 3000 milligrams. The physician often relies on the dietitian or nurse to recommend changes in the diet order when warranted. If a diet order seems inappropriate or outdated, the nurse should alert the physician that a mistake may have been made.

Diet Manuals   For each modified diet, the exact foods to include or exclude are detailed in a **diet manual.** The dietetics staff at a health care institution may compile the manual using resources from the American Dietetic Assocation or other dietetics organizations. Small facilities may adopt the diet manual prepared by another hospital or a dietetics organization.

Diet Progression   A common diet order reads, "progress diet from clear liquids to a regular diet as tolerated." **Diet progression** is a change in diet to adapt to a patient's increased tolerance to foods. For example, after surgery involving the gastrointestinal (GI) tract, a patient might be given clear beverages initially and then gradually be provided with other beverages or solid foods that are unlikely to cause discomfort. In some cases, small, frequent feedings are offered initially, and the patient progresses to larger meals as tolerance allows. The nurse is often responsible for monitoring a patient's tolerance to a diet; symptoms such as nausea, vomiting, diarrhea, and gastrointestinal pain suggest intolerance.

Nothing by Mouth (NPO)   An order to not give a patient anything at all—food, beverages, or medications—is indicated by NPO, an abbreviation for *non per os,* meaning "nothing by mouth." For example, an order may read "NPO for 24 hours" or "NPO until after X-ray." The NPO order is commonly used during certain acute illnesses or diagnostic tests involving the GI tract.

DIETARY MODIFICATIONS

Table 14-2 (p. 382) lists examples of modified diets that are often prescribed during illness.[4] The texture and consistency of foods can be altered for people with chewing or swallowing impairments. Some dietary modifications relieve the symptoms of disease; for example, restricting dietary sodium can help to control fluid accumulation, and eliminating legumes and certain vegetables can reduce flatulence. Increasing the nutrient density of a diet may prevent or reverse malnutrition. Diets are also adjusted to improve nutrition-sensitive risk factors for chronic diseases, such as high blood cholesterol and hypertension. If a patient has more than one of these problems, several aspects of a diet may need to be modified.

The modified diets discussed in this section should be adjusted to satisfy individual preferences and tolerances. They also need to be altered as a patient's condition

standard diet: a diet that includes all foods and meets the nutrient needs of healthy people; sometimes called a **regular diet.**

modified diet: a diet that is adjusted in consistency, in energy or nutrient content, or by the inclusion or elimination of certain foods; sometimes called a **therapeutic diet.**

diet manual: a resource that specifies the foods allowed and restricted on modified diets and provides sample menus.

diet progression: a change in diet as a patient's tolerances permit.

TABLE 14-2   Examples of Modified Diets

Type of Diet	Description of Diet	Appropriate Uses
Modified Texture and Consistency		
Mechanically altered diets	Contain foods that are modified in texture. Pureed diets include only pureed foods; mechanical soft diets may include solid foods that are mashed, minced, ground, or soft.	Pureed diets are used for people with swallowing difficulty, poor lip and tongue control, or oral hypersensitivity. Mechanical soft diets are appropriate for people with limited chewing ability or certain swallowing impairments.
Blenderized liquid diet	Contains fluids and foods that are blenderized to liquid form.	For people who cannot chew, swallow easily, or tolerate solid foods.
Clear liquid diet	Contains clear fluids or foods that are liquid at room temperature and leave minimal residue in the colon.	For preparation for bowel surgery or colonoscopy, for acute GI disturbances (such as after GI surgeries), or as a transition diet after intravenous feeding. For short-term use only.
Therapeutic Diets		
Fat-restricted diet	Restricts fat to low (<50 g/day) or very low (<25 g/day) levels in the diet.	For people who have certain malabsorptive disorders or symptoms of diarrhea, flatulence, or steatorrhea (fecal fat) resulting from dietary fat intolerance.
Fiber-restricted diet	Restricts fiber to low levels in the diet (<10 g/day).	For acute phases of intestinal disorders or to reduce fecal output before surgery. Not recommended for long-term use.
Sodium-restricted diet	Restricts sodium; degree of restriction depends on symptoms and disease severity.	To prevent fluid retention or induce fluid loss; used in hypertension, congestive heart failure, renal disease, and liver disease.
High-kcalorie, high-protein diet	Contains foods that are kcalorie and protein dense.	Used for increased kcalorie and protein requirements (in cancer, AIDS, burns, trauma, and other illnesses); also used to reverse malnutrition, improve nutritional status, or promote weight gain.

Sources: American Dietetic Association, *Nutrition Care Manual* (Chicago: American Dietetic Association, 2005); American Dietetic Association, *Manual of Clinical Dietetics* (Chicago: American Dietetic Association, 2000).

changes. Later chapters include other types of modified diets and additional dietary strategies for treating nutritional problems.

Mechanically Altered Diets Individuals who have difficulty chewing or swallowing may benefit from mechanically altered diets. Impaired swallowing, or **dysphagia**, is associated with a number of diseases and disease treatments. Dysphagia may accompany neurological diseases, surgical procedures involving the head and neck, and physiological or anatomical abnormalities that restrict the movement of food within the throat or esophagus. Dysphagia diets are highly individualized because swallowing difficulties vary in severity, and swallowing ability can fluctuate over time. Diets used for dysphagia may also be recommended for people with limited chewing abilities or dental problems.

Table 14-3 (p. 383) provides examples of foods included in mechanically altered diets. Although the names for these diets vary, a diet may contain mostly pureed foods *(pureed diet)*, whereas a less restrictive diet may include moist, soft-textured foods that easily form a bolus *(mechanical soft diet)*. For people with chewing problems, many foods can be ground or minced. Chapter 17 provides additional information about the diets used for treating dysphagia.

Blenderized Liquid Diet Blenderized diets may be prescribed following oral or facial surgeries (for example, jaw wiring) or used by individuals with chewing problems.

dysphagia: difficulty swallowing.

TABLE 14-3	Foods Included in Mechanically Altered Diets

Depending on the feeding problem, a mechanically altered diet may include foods that are pureed, mashed, ground, minced, or soft textured. Foods vary according to tolerance.

Pureed Diets	Mechanical Soft Diets
Milk products: Milk, smooth yogurt, pudding.	**Milk products:** Milk, yogurt with soft fruit, pudding, cottage cheese.
Fruits: Pureed fruits and juices without pulp, skin, seeds or chunks; well-mashed fresh bananas; applesauce.	**Fruits:** Canned or cooked fruits without seeds or skin, fruit juices with small amounts of pulp, ripe bananas.
Vegetables: Pureed cooked vegetables without seeds or skins, mashed potatoes, pureed potatoes with gravy.	**Vegetables:** Soft, well-cooked vegetables that are not rubbery or fibrous; well-cooked, moist potatoes.
Meats and meat substitutes: Pureed meats (with gravy), pureed casseroles (with broth), hummus or other pureed legume spread.	**Meats and meat substitutes:** Ground, minced, or tender meat, poultry, or fish with gravy or sauce; tofu; well-cooked legumes; scrambled eggs.
Breads and cereals: Smooth cooked cereals such as cream of wheat, slurried breads and pancakes,[a] pureed rice and pasta.	**Breads and cereals:** Cooked cereals or moistened dry cereals with minimal texture, soft pancakes or breads, well-cooked noodles or dumplings in sauce or gravy.

[a] Slurried foods are mixed with liquid until the consistency is appropriate; they may be gelled and shaped to improve their appearance.

Source: American Dietetic Association, *Nutrition Care Manual* (Chicago: American Dietetic Association, 2005).

Soft or tender foods from all food groups can be blenderized (often with added liquid), including cereals and breads; cooked vegetables; fresh or cooked fruits without skins and seeds; cooked, tender meats and fish; and potatoes, rice, and pasta. Foods that do not blend well are excluded; examples include nuts and seeds, dried fruits, sausage and frankfurters, hard cheeses, raw vegetables, and corn.

Clear Liquid Diet Clear liquids require minimal digestion and are easily tolerated by the GI tract. They are often the first foods offered to patients before some GI procedures, after GI surgery, or after intravenous feeding. The **clear liquid diet** consists of clear fluids and foods that are liquid at room temperature and leave little **residue** in the intestine (for this reason, milk products are not included in the diet). Permitted foods include clear or pulp-free fruit juices, clear broths, bouillon, consommé, fruit-flavored or unflavored gelatin, fruit ices made from clear juices, frozen juice bars, and plain hard candy. Although the diet provides fluid and electrolytes, its nutrient and energy contents are extremely limited. If used for longer than a day or two, it should be supplemented with commercially prepared low-residue formulas that provide required nutrients. Figure 14-1 gives an example of a one-day clear liquid menu.

Fat-Restricted Diet Fat restriction is recommended for reducing the symptoms of fat malabsorption that may accompany diseases of the liver, gallbladder, pancreas, lymphatic system, and intestines. Fat restriction may also alleviate the symptoms of heartburn. Although fat intake is occasionally limited to as little as 25 grams daily, it should not be restricted more than necessary

clear liquid diet: a diet that consists of foods that are liquid at room temperature and leave almost no residue (undigested material) in the intestines after digestion and absorption.

residue: material left in the intestine after digestion; includes mostly dietary fiber and undigested starches and proteins.

FIGURE 14-1 Menu—Clear Liquid Diet

SAMPLE *Menu—Clear Liquid Diet*

Breakfast
Strained orange juice
Flavored gelatin
Ginger ale
Coffee or tea, sugar

Lunch
Bouillon or consommé
Flavored gelatin
Frozen juice bars
Apple or grape juice
Coffee or tea, sugar

Supper
Bouillon or consommé
Flavored gelatin
Fruit ice
Cranberry juice
Coffee or tea, sugar

Snacks
Soft drinks
Fruit ices
Hard candy

because fat is an important source of kcalories. Chapter 18 gives additional information about fat-restricted diets.

Most foods included in a fat-restricted diet provide less than 1 gram of fat per serving. The diet includes fat-free milk products, most breads and cooked grains, fat-free broths and soups, vegetables prepared without fats, most fruits, and fat-free candies and sweets (Table 18-4 on p. 475). Restricted foods include low-fat and whole-milk products, baked products with added fat (like muffins), and most prepared desserts. Lean meat and meat substitutes are permitted but may be restricted to 4 to 6 ounces per day, depending on the degree of restriction. Some patients with malabsorptive conditions do not tolerate large amounts of lactose or dietary fiber, so foods that include these substances may also need to be excluded from the diet.

Fiber-Restricted Diet Fiber restriction is recommended during acute phases of intestinal disorders when the presence of fiber may exacerbate intestinal discomfort or cause diarrhea or blockages. Fiber-restricted diets are sometimes used before surgery to minimize fecal volume and after surgery during transition to a regular diet. Long-term fiber restriction is discouraged, however, because it is associated with constipation, diverticulosis, and other problems.

Fiber-restricted diets often eliminate whole-grain breads and cereals, nuts and seeds, raw and dried fruits, berries, dried beans and peas, chunky peanut butter, winter squash, and most raw vegetables. If required, even greater reductions in colonic residue can be achieved by excluding most fruits and vegetables and milk products. Additional information about the fiber content of foods can be found in Chapter 3 (pp. 73–78) and Appendix A.

Reminder: Although the Upper Level for sodium is 2300 mg, intakes in the United States generally exceed this amount.

Sodium-Restricted Diet Sodium restriction can help to prevent or correct fluid retention and may be recommended for treatment of hypertension, congestive heart failure, kidney disease, or liver disease. The degree of restriction depends on the illness, the severity of symptoms, and the drug treatment prescribed. In most cases, sodium is restricted to 2000 or 3000 milligrams daily, although more severe restrictions may be used in the hospital setting.

A sodium-restricted diet requires the person to omit the use of salt when cooking and at the table, avoid most prepared foods and condiments, and limit consumption of milk and milk products (if excessive). Because so many processed foods are high in sodium, people following a low-sodium diet should check food labels and consume only low-sodium products. Sodium restriction is difficult to implement on a long-term basis because many people find low-sodium diets unpalatable and fail to adhere to them. Sodium restriction is discussed further in Chapters 21 and 22.

High-kCalorie, High-Protein Diet The high-kcalorie, high-protein diet is used to increase kcalorie and protein intakes in patients who have unusually high requirements or in those who are eating poorly. High-fat foods are added to increase energy intakes; consequently, the diet may exceed 35 percent kcalories from fat. Consuming small, frequent meals and commercial liquid supplements can also help a patient meet increased energy, protein, and other nutrient needs.

Examples of foods included in high-kcalorie, high-protein diets are listed in Table 14-4. Some of these foods are high in saturated fat, which is limited in heart-healthy diets. These foods are used liberally in malnourished patients to help correct their immediate nutrition problems—weight loss and muscle wasting.

ALTERNATIVE FEEDING ROUTES

In most cases, patients meet their nutrient needs by consuming regular foods. If their nutrient needs are high or their appetites poor, liquid formulas can be added

TABLE 14-4	Foods Included in High-kCalorie, High-Protein Diets
Milk products	Whole milk, half-and-half, cream Cheese Milk shakes, eggnog Ice cream, whipped cream
Fruits	Dried fruit Canned fruit in heavy syrup Avocado
Vegetables	Vegetables prepared with butter, margarine, sour cream, mayonnaise, or salad dressing Cream of vegetable soups

Meats and high-protein foods	All meats, fish, and poultry, including bacon, frankfurters,
	    and luncheon meats; eggs
	All meats, prepared fried or covered in cream sauces and
	    gravies
	Nuts and seeds, peanut and other nut butters, coconut

Breads and cereals	Granola and dry cereals prepared with whole milk or cream and dried fruit Hot cereals with whole milk or cream, or added fat Pasta, rice, and potatoes with added fat Pancakes, waffles, French toast

to their diets to supplement their intakes. Sometimes, however, a person's medical condition makes it difficult to meet nutrient needs orally. Two options remain: **tube feedings** and **intravenous feedings,** described more fully in Chapter 15.

- *Tube feedings.* Nutritionally complete formulas can be delivered through a tube placed directly into the stomach or intestine. Tube feedings are preferred to intravenous feedings if the GI tract is functioning. A person in a coma, for example, is unable to eat but may be able to digest foods and absorb nutrients normally. In this situation, a tube feeding would be the appropriate option.
- *Intravenous feedings.* A person's medical condition sometimes prohibits the use of the GI tract to deliver nutrients. If the person is malnourished and the GI tract cannot be used for a long period of time, intravenous feedings can meet nutritional needs.

IMPROVING FOOD INTAKE

People in hospitals and other medical facilities often lose their appetites as a result of their medical condition, treatment, or emotional distress. In addition, some medications and other treatments can dramatically alter taste perceptions. Patients usually receive meals at specified times, whether they are hungry or not, and often must eat in bed without companionship; under these conditions, eating can be more of a chore than a pleasurable experience. Meals may also be unwelcome if the person is in pain or has been sedated.

Nurses often play a central role in helping patients to eat. If either appetite or the sense of taste is affected by illness, the nurse may be able to work with the patient to identify foods that are most enjoyable. The nurse can also check to see that foods and utensils are arranged attractively when meals are served and help the patient wash up before a meal. The "How to" on page 386 lists additional suggestions that may help to improve food intake at mealtimes.

tube feedings: liquid formulas delivered through a tube placed in the stomach or intestine.

intravenous feedings: the provision of nutrients through a vein, bypassing the intestine. The intravenous provision of nutrients is called *parenteral nutrition.*

How to   Help Hospital Patients Improve Their Food Intakes

1. Empathize with the patient. Show that you understand how difficult eating may be. Imagine feeling too sick to move or too tired to sit up.
2. Motivate. Be sure the patient understands how important nutrition is to recovery.
3. Help patients select the foods they like and mark menus appropriately. When appropriate and permissible, let friends or family members bring favorite foods from outside the hospital.
4. For patients who are weak, suggest foods that require little effort to eat. Eating a roast beef sandwich, for example, requires less effort than cutting and eating a steak. Drinking soup from a cup may be easier than eating it with a spoon.
5. Help patients prepare for meals. Help them get comfortable, either in bed or in a chair. Adjust the extension table to a comfortable distance and height, and make sure it is

clean. Take these steps before the tray arrives, so the meal can be served promptly and at the right temperature.
6. When the food cart arrives, check the patient's tray. Confirm that the patient is receiving the right diet, that the foods on the tray are those selected from the menu, and that the foods look appealing. Order a new tray if foods are not appropriate.
7. Help with eating, if necessary. Help patients to open containers or cut foods, and assist with feeding if patients cannot feed themselves.
8. Try to solve eating problems. Encourage patients with little appetite to eat the most nutritious foods first and to drink liquids between meals.
9. Take a positive attitude toward the hospital's food. Never say something like "I couldn't eat this either." Instead, say, "The foodservice department really tries to make foods appetizing. I'm sure we can find a solution."

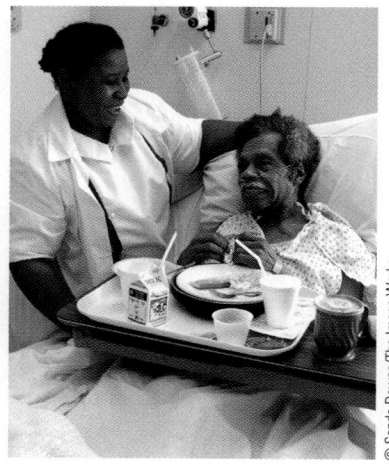

People enjoy eating when they feel comfortable and cared for.

REVIEW NOTES

The physician has the primary responsibility for writing a diet order. Diet manuals specify the foods to include in modified diets.

Diets prescribed during illness can be modified in consistency or in nutrient content.

Mechanically altered diets are used for people with swallowing and chewing difficulties.

Clear liquid diets may be used briefly after acute gastrointestinal disturbances or intravenous feedings.

Some medical conditions may require the restriction of specific nutrients, such as fat, fiber, or sodium.

A high-kcalorie, high-protein diet may help to prevent or reverse malnutrition, improve nutritional status, or promote weight gain.

Diet-Drug Interactions

When counseling patients, nurses should be alert to possible diet-drug interactions, which can raise health care costs and result in serious, and sometimes fatal, complications. With hundreds of diet-drug interactions known and more to be identified in the future, nurses and other health care professionals need to take steps to prevent them. The remaining sections of this chapter describe the main types of diet-drug interactions and their clinical significance.[5] Diet-drug interactions generally fall into the following categories:

- Medications can alter food intake by reducing appetite or by causing complications that make food consumption difficult or unpleasant. Some medications may increase appetite and cause weight gain.
- Medications can alter the absorption, metabolism, and excretion of nutrients. Conversely, nutrients and other food components can alter the absorption, metabolism, and excretion of medications.
- Some interactions between dietary components and medications can be toxic.

TABLE 14-5	Examples of Diet-Drug Interactions

Drugs May Alter Food Intake by:

Altering the appetite (amphetamines suppress appetite; corticosteroids increase appetite).

Interfering with taste or smell (amphetamines change taste perceptions).

Inducing nauses or vomiting (digitallis may do both).

Interfering with oral function (some antidepressants may cause dry mouth).

Causing sores or inflammation in the mouth (methotrexate may cause painful mouth ulcers).

Drugs May Alter Nutrient Absorption by:

Changing the acidity of the digestive tract (antacids may interfere with iron and folate absorption).

Damaging mucosal cells (cancer chemotherapy may damage mucosal cells).

Binding to nutrients (bile acid binders bind to fat-soluble vitamins).

Foods and Nutrients May After Drug Absorption by:

Stimulating secretion of gastric acid (the antifungal agent ketoconazole is absorbed better with meals due to increased acid secretion).

Altering rate of gastric emptying (intestinal absorption of drugs may be delayed when they are taken with food).

Binding to drugs (calcium binds to tetracycline, reducing drug and calcium absorption).

Competing for absorption sites in the intestines (dietary amino acids interfere with levodopa absorption).

Drugs and Nutrients May Interact and Alter Metabolism by:

Acting as structural analogs (as do warfarin and vitamin K).

Using similar enzyme systems (phenobarbital induces liver enzymes that increase metabolism of folate, vitamin D, and vitamin K).

Competing for transport on plasma proteins (fatty acids and drugs may compete for the same sites on the plasma protein albumin).

Drugs May Alter Nutrient Excretion by:

Altering reabsorption in the kidneys (some diuretics increase the excretion of sodium and potassium).

Causing diarrhea or vomiting (diarrhea and vomiting may cause electrolyte losses).

Food May Alter Medication Excretion by:

Inducing activities of liver enzymes that metabolize drugs to allow their excretion (components of charcoal-broiled meats increase metabolism of warfarin, theophylline, and acetominophen).

Toxicity May Occur from Combining Foods and Drugs by:

Increasing side effects of the drug (caffeine in beverages can increase adverse effects of stimulants).

Increasing drug action to excessive levels (grapefruit components may block metabolism of drugs and enhance drugs' actions and side effects).

Specific examples of diet-drug interactions are shown in Table 14-5. The "How to" on p. 388 offers some practical advice about preventing diet-drug interactions.

DRUG EFFECTS ON FOOD INTAKE

Medications can reduce food intake by inducing nausea or vomiting, altering taste sensations, suppressing the appetite, drying the mouth, or causing inflammation or lesions in the mouth or GI tract. Sedatives and medications that cause drowsiness can make a person too tired to eat. Side effects such as abdominal discomfort, constipation, diarrhea, dizziness, or confusion can also interfere with food intake.

Complications that reduce food intake are significant only when they continue for a long period. Although many drugs can cause nausea in some individuals, the

The Joint Commission on Accreditation of Healthcare Organizations has recommended that all patients be educated about potential diet-drug interactions. Nurses can help by informing patients of precautions related to medications and watching for signs of problems that may arise.

To prevent diet-drug interactions, first list the types and amounts of over-the-counter drugs, prescription medications, and dietary supplements that the patient uses on a regular basis. Look up each medication in a drug reference and make a note of:

- The appropriate method of administration (twice daily or at bedtime, for example).
- How the medication should be administered with respect to foods, beverages, and specific nutrients (for example, take on an empty stomach, take with food, do not take with milk, or do not drink alcoholic beverages while using the medication).
- How the medication should be used with respect to other medications.
- The side effects that may affect food intake (nausea and vomiting, constipation or diarrhea, or sedation, for example) or nutrient needs (interference with nutrient absorption or metabolism, for example).

A similar process can be used to review the dietary supplements that a person is taking: a reliable reference may list their appropriate uses, possible side effects, and potential interactions with food and medications.

Patients who take multiple medications may need help learning when to take each medication to avoid drug-drug or diet-drug interactions. The nurse can use information from a patient's diet history (see Chapter 13) to help the patient coordinate meals and drugs so as to avoid interactions.

Some medications have well-known effects on nutritional status. The nurse should remain alert for signs of problems, especially when:

- Nutritional problems are a frequent result of using the medication.
- A patient uses multiple medications.
- The patient is in a high-risk group; for example, a child, a pregnant or lactating woman, an older adult, or a person who is malnourished, abuses alcohol, or has impaired liver or kidney function.
- The patient needs to use the medication for a long period of time.

Check with the pharmacist for additional information about medications and potential interactions.

---

nausea often subsides after the first few doses of the medication and thus has little effect on nutrition status. If side effects persist, other medications are sometimes used to treat them; for example, antinauseants and antiemetics may help to reduce nausea and vomiting and thereby improve food intake.

Unintentional weight gain can result from the use of some antipsychotics, antidepressants, and corticosteroids (for example, prednisone). People using these drugs may be unable to feel satiated and may gain 40 to 60 pounds in just a few months. For some conditions, however, weight gain is desirable. Patients with diseases that cause wasting, such as cancer or AIDS, are sometimes prescribed appetite enhancers such as megestrol acetate, a progesterone analog, or dronabinol, which is derived from the active ingredient in marijuana.

DRUG EFFECTS ON NUTRIENT ABSORPTION

The medications most likely to cause widespread nutrient malabsorption are those that damage the intestinal mucosa. Antineoplastic and antiretroviral drugs are especially detrimental, although nonsteroidal anti-inflammatory drugs (NSAIDs) and some antibiotics can have similar, though milder, effects. This section describes additional ways in which medications may alter nutrient absorption.

Drug-Nutrient Binding Some medications bind nutrients in the GI tract, preventing their absorption. For example, bile acid binders, used to reduce cholesterol levels, may bind to fat-soluble vitamins. Some antibiotics, notably tetracycline and ciprofloxacin, bind to the calcium in foods and supplements, reducing the absorption of both the calcium and the antibiotic.

Altered Stomach Acidity Medications that reduce stomach acid can potentially impair the absorption of vitamin B_{12}, folate, and iron. Examples include antacids, which neutralize stomach acid by acting as weak bases, and antiulcer drugs (proton pump inhibitors and H2 blockers), which interfere with acid secretion.

Direct Inhibition Several drugs impede nutrient absorption by interfering with their intestinal metabolism or transport into mucosal cells. For example, the antibiotics trimethoprim and pyrimethamine compete with folate for absorption into intestinal cells.

DIETARY EFFECTS ON DRUG ABSORPTION

Most drugs are absorbed in the upper small intestine. Major influences on drug absorption include the stomach emptying rate, level of acidity, and direct interactions with dietary components. The drug's formulation may also influence its absorption.

Stomach Emptying Rate Drugs reach the small intestine more quickly when the stomach is empty. Therefore, taking a medication with meals may delay its absorption, although the total amount absorbed may not be affected. For example, aspirin works faster when taken on an empty stomach, but taking it with food is often encouraged to reduce stomach irritation. Slow stomach emptying can sometimes enhance drug absorption because the drug's absorption sites do not become saturated. Slow drug absorption can be a problem, however, if high concentrations are needed for effectiveness, as when a hypnotic is taken to induce sleep.

Stomach Acidity Some drugs are better absorbed in an acid medium, so conditions that are too alkaline may reduce their absorption. Hence, antacid medications may reduce the absorption of other drugs. Drugs that can be damaged by acid are often available in coated forms that resist the stomach's acid environment.

Interactions with Food Components Some dietary substances can bind to drugs and inhibit their absorption. For example, high-fiber diets may decrease the absorption of some tricyclic antidepressants. Phytates in foods can bind to digoxin, a drug prescribed for heart disease. As mentioned earlier, calcium can bind to some antibiotics, reducing absorption of both the calcium and the drug.

Reminder: *Phytates* are compounds found in many plant foods, including whole grains and legumes. Phytates can bind to minerals and reduce their absorption.

DRUG EFFECTS ON NUTRIENT METABOLISM

Drugs and nutrients share similar enzyme systems in the small intestine and liver. Consequently, some drugs may enhance or inhibit the activities of enzymes needed for nutrient metabolism. For example, the anticonvulsants phenobarbital and phenytoin induce the liver enzymes that metabolize folate, vitamin D, and vitamin K; therefore, persons using these drugs require supplements of these vitamins.

The drug methotrexate, used to treat cancer and inflammatory conditions, interferes with folate metabolism. Methotrexate resembles folate in structure (see Figure 14-2, p. 390) and competes with folate for the enzyme that converts folate to its active form. The adverse effects of using methotrexate therefore include symptoms of folate deficiency. These adverse effects can be reduced by using a pre-activated form of folate (called leucovorin) that is often prescribed along with methotrexate to "rescue" the rapidly dividing cells in the body, which have high folate requirements.

Corticosteroids, used as anti-inflammatory agents and immunosuppressants, have actions that mimic those of the hormone cortisol. Long-term corticosteroid use can have broad effects on nutritional health and may cause weight gain, muscle wasting, bone loss, and hyperglycemia, with eventual development of osteoporosis and diabetes.

Cortisol is a steroid hormone secreted by the adrenal cortex as part of the body's stress response.

FIGURE 14-2 Folate and Methotrexate

By competing for the enzyme that activates folate, methotrexate prevents cancer cells from obtaining the folate they need to multiply. In the process, normal cells are also deprived of the folate they need.

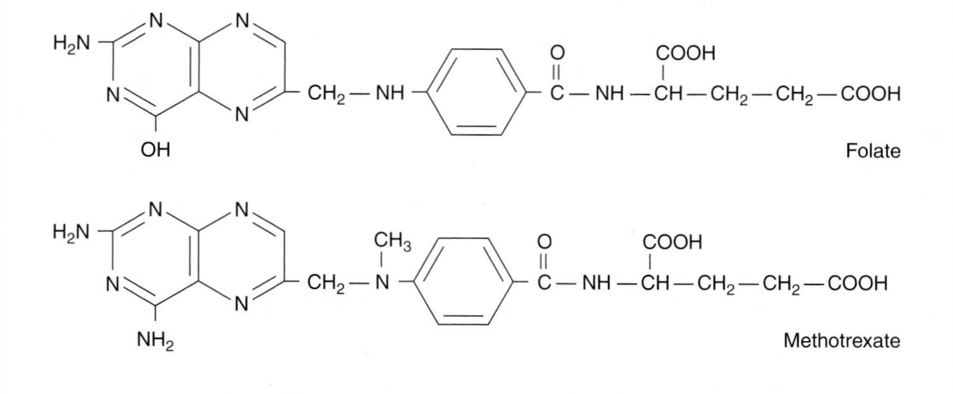

DIETARY EFFECTS ON DRUG METABOLISM

Some food components can affect the activities of enzymes that metabolize drugs or may counteract drug effects in other ways. Compounds in grapefruit juice (and whole grapefruit) have been found to inhibit or inactivate enzymes that metabolize a number of different drugs (see Table 14-6).[6] As a result of the reduced enzyme action, blood concentrations of the drugs increase, leading to stronger

TABLE 14-6	Grapefruit Juice—Drug Interactions—Selected Examples	
Drug Category	**Drugs Affected by Grapefruit Juice**	**Drugs Unaffected by Grapefruit Juice**
Cardiovascular drugs	Felodipine Nicardipine Nifedipine Verapamil	Amlodipine Diltiazem Propafenone Quinidine
Cholesterol-lowering drugs	Atorvastatin Lovastatin Simvastatin	Pravastatin
Central nervous system drugs	Buspirone Carbamazepine Diazepam Triazolam	Clomipramine Haloperidol
Anti-infective drugs	Saquinavir	Clarithromycin Itraconazole
Estrogens	Ethinylestradiol	17-β-estradiol
Anticoagulants		Acenocoumarol Warfarin
Immunosuppressants	Cyclosporine Tacrolimus	Prednisone
Antiasthmatic drugs		Theophylline

Source: D. G. Bailey, M. O. Arnold, and J. D. Spence, Inhibitors in the diet: Grapefruit juice–drug interactions, in R. H. Levy and coeditors, *Metabolic Drug Interactions* (Phildelphia: Lippincott Williams & Wilkins, 2000), pp. 661–669.

physiological effects. The effect of the grapefruit juice lasts for a substantial period after the juice is consumed; in experiments with a drug prescribed for heart disease, the juice's effect had an estimated half-life of 12 hours.

A number of dietary substances affect the activity of the anticoagulant drug warfarin. The most important interaction is with vitamin K, which is structurally similar to warfarin. Warfarin acts by blocking the enzyme that activates vitamin K, thereby preventing the synthesis of blood-clotting factors. The amount of warfarin prescribed is dependent, in part, on how much vitamin K is in the diet. If vitamin K consumption from foods or supplements increases substantially, it can weaken the effect of the drug. Individuals using warfarin are advised to consume similar amounts of vitamin K daily to keep warfarin activity stable. The dietary sources highest in vitamin K are green leafy vegetables.

A number of popular herbs contain natural compounds that may enhance the activity of warfarin and therefore should be avoided during warfarin treatment. These herbs include St. John's wort, ginkgo, garlic, ginseng, dong quai, danshen, and others.[7]

Reminder: the term *half-life* can be used to define the time period of a chemical effect. If the grapefruit effect has a 12-hour half-life, this means that after 12 hours, its biological effect is half of the maximum effect measured.

DRUG EFFECTS ON NUTRIENT EXCRETION

Some medications may alter mineral reabsorption in the kidneys, causing an increase or decrease in urinary losses. For example, some diuretics cause an accelerated excretion of calcium, potassium, and magnesium. Risk of mineral depletion is highest if multiple drugs with the same effect are used, if kidney function is impaired, or if medications are used for a long time. Note that some diuretics can lead to mineral retention, rather than excretion.

A number of drugs can cause an increase in vitamin B_6 excretion. An example is isoniazid (INH), an antituberculosis drug that is similar in structure to vitamin B_6. This drug induces excretion of vitamin B_6 and therefore may lead to a vitamin B_6 deficiency. Because the drug must be taken for at least six months to treat infection, vitamin B_6 supplements are routinely given to prevent deficiency.

When the kidneys reabsorb a substance, they retain it in the blood. Substances that are not reabsorbed are excreted in urine.

DIETARY EFFECTS ON DRUG EXCRETION

Inadequate excretion of medications can cause toxicity, whereas excessive losses may reduce the amount available for therapeutic effect. Some food components can alter drug reabsorption by the kidneys. For example, the amount of the medication lithium reabsorbed by the kidneys is generally similar to the amount of sodium reabsorbed. Consequently, dehydration or sodium depletion, which increase sodium reabsorption, may result in lithium retention. Similarly, a person with a high sodium intake will excrete more sodium in the urine and therefore more lithium. Individuals using lithium are advised to maintain a consistent sodium intake from day to day in order to maintain a stable blood level of lithium.

Urine acidity can affect drug excretion due to the effects of pH on a compound's ionic (chemical) form. The medication quinidine, used to treat arrhythmias, is excreted more readily in acidic urine. Foods or drugs that cause urine to become more alkaline (for example, sodium bicarbonate) may reduce quinidine excretion and raise blood levels.

Lithium is used to prevent mood swings in patients with manic-depressive disorder.

DIET-DRUG INTERACTIONS AND TOXICITY

Interactions between food components and drugs may sometimes cause toxicity or exacerbate a drug's side effects. The combination of tyramine, a compound in some foods (Table 14-7, p. 392), and monoamine oxidase (MAO) inhibitors, which treat depression, can be fatal. MAO inhibitors block an enzyme that normally

TABLE 14-7	Foods Restricted in a Tyramine-Controlled Diet
Beverages	Red wines including chianti, sherry[a]
Cheeses	Aged cheeses, American, camembert, cheddar, gouda, gruyère, mozzarella, parmesan, provolone, romano, roquefort, stilton[b]
Meats	Liver; dried, salted, smoked, or pickled fish; sausage; pepperoni; salami; dried meats
Vegetables	Fava beans; Italian broad beans; sauerkraut; snow peas; fermented pickles and olives
Other	Brewer's yeast;[c] all aged and fermented products; soy sauce in large amounts; cheese-filled breads, crackers, and desserts; salad dressings containing cheese

Note: The tyramine contents of foods vary from product to product depending on the methods used to prepare, process, and store the food. In some cases, as little as 1 ounce of cheese can cause a severe hypertensive reaction in people taking monoamine oxidase inhibitors. In general, the following foods contain small enough amounts of tyramine that they can be consumed in small quantities; ripe avocado, banana, yogurt, sour cream, acidophilus milk, buttermilk, raspberries, and peanuts.

[a]Most wine and domestic beer can be consumed in small quantities.

[b]Unfermented cheeses, such as ricotta, cottage cheese, and cream cheese, are allowed.

[c]Products made with baker's yeast are allowed.

inactivates tyramine as well as the hormones epinephrine and norepinephrine. When people who take MAO inhibitors consume excessive tyramine, the tyramine causes a sudden release of accumulated norepinephrine. This surge in norepinephrine results in severe headaches, rapid heartbeat, and a dangerous increase in blood pressure. For this reason, people taking MAO inhibitors are advised to restrict their intakes of foods rich in tyramine.

Considering the many ways that medications and nutrients can interact and the number of medications and dietary supplements available, it is no wonder that serious side effects are increasingly recognized. Nurses and other health professionals are challenged to understand the mechanisms of diet-drug interactions, identify them when they occur, and prevent them whenever possible.

REVIEW NOTES

Medications can alter food intake and affect the absorption, metabolism, and excretion of nutrients. Components of foods can similarly affect drug activity in the body.

Drugs can alter food intake by reducing appetite, altering the sense of taste, causing gastrointestinal discomfort, or damaging the mucosal lining of the GI tract.

Drugs can affect nutrient absorption by binding to nutrients, changing stomach acidity, or interfering with nutrient transport into intestinal cells. Dietary substances can influence drug absorption by altering the stomach emptying rate, changing stomach acidity, or directly binding to drugs.

Drugs may enhance or inhibit the activities of enzymes needed for normal nutrient metabolism, and conversely, dietary components may enhance or inhibit the activities of enzymes that metabolize drugs.

Diet-drug interactions can cause excessive nutrient losses and alter the urinary excretion of medications.

SELF CHECK

1. A successful nutrition intervention would include a long list of:
   a. dietary changes that the patient should consider making.
   b. foods that the patient should avoid.
   c. appetizing meals and foods that the patient can include in his or her diet.
   d. reasons why the patient should make dietary changes.

2. The most important factor(s) that affect how nutrition education is presented is (are):
   a. the person's nutrient needs and nutrition status.
   b. the person's abilities and motivation.
   c. the person's medical history.
   d. the entries in the medical record.

3. Foods permitted on the clear liquid diet include all of the following *except:*
   a. milk.
   b. fruit ices.
   c. flavored gelatin.
   d. consommé.

4. The modified diet *least* likely to provide adequate nutrients and kcalories is the:
   a. pureed diet.
   b. clear liquid diet.
   c. mechanical soft diet.
   d. high-kcalorie, high-protein diet.

5. Fiber restriction may be recommended:
   a. for patients with dysphagia.
   b. for patients who are unable to absorb fat normally.
   c. for patients with heartburn.
   d. during the acute phases of some intestinal disorders.

6. A nurse notices a food on a patient's tray that she thinks should be excluded from the patient's prescribed diet. An appropriate action for the nurse to take would be to check the:
   a. care plan.
   b. diet order.
   c. diet manual.
   d. medical record.

7. Examples of drug-related symptoms that can significantly limit food intake include:
   a. skin rash and ringing in the ears.
   b. persistent nausea and vomiting.
   c. insomnia and sensitivity to sunlight.
   d. nasal congestion and hair loss.

8. Factors that typically interfere with drug absorption include:
   a. binding between drugs and food components.
   b. use of antacid therapies.
   c. rapid stomach emptying rate.
   d. all of the above.

9. Compounds in grapefruit juice:
   a. bind to antibiotics, reducing absorption.
   b. cause excessive drug excretion.
   c. strengthen effects of certain drugs.
   d. alter acidity in the stomach, impairing drug absorption.

10. People who use MAO inhibitors must limit consumption of:
    a. whole milk and yogurt.
    b. grapefruit juice.
    c. dark green leafy vegetables.
    d. aged cheeses.

Answers to these questions appear in Appendix H.

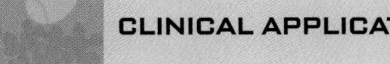

CLINICAL APPLICATIONS

1. A middle-aged man with congestive heart failure, who is taking multiple medications, has just been prescribed a low-sodium diet. The man tells you that the dietitian has talked with him about his diet but that he is totally confused. He confides that diet is the last thing on his mind right now. What actions should you take?

2. An elderly woman in a residential home has just been prescribed a high-kcalorie, high-protein diet. She has been taking several medications to treat both a heart problem and a mild case of bronchitis. You notice that she eats only a few bites at mealtimes and seems disinterested in food. Describe several steps that can be taken to uncover and address problems that the woman might be having with food.

3. A hospital patient who has a dysphagia problem has been receiving a mechanical soft diet. You notice that she coughs repeatedly while eating meals and eats only small portions of the foods on her plate. You suspect that the diet is inappropriate and that she may need a more restrictive dysphagia diet, but her medical chart doesn't specify the severity of the dysphagia.

   - To address these problems in your nursing care plan, which health professionals would you need to consult, and what information would you require from each of them?

   - The next meal is due within the hour, and you suspect the woman will be intolerant to the foods she is scheduled to receive. You haven't yet heard from the health practitioners you consulted while developing your care plan. What options might you consider in handling this immediate problem?

NUTRITION ON THE NET

For further study of the topics in this chapter, access these websites.

Find updates and quick links to these and other nutrition-related sites at our website: **www.wadsworth.com/nutrition**

Read more about nursing diagnoses at the website of NANDA International: **www.nanda.org**

Design patient education materials with information from this government website: **http://medlineplus.gov/**

Visit the Food and Drug Administration's home page for links to drug information: **www.fda.gov**

NOTES

[1] S. S. Ralph and C. M. Taylor, *Nursing Diagnosis Reference Manual* (Philadelphia: Lippincott Williams & Wilkins, 2005).

[2] K. Glanz, Current theoretical bases for nutrition intervention and their uses, in A. M. Coulston, C. L. Rock, and E. R. Monsen, eds., *Nutrition in the Prevention and Treatment of Disease* (San Diego: Academic Press, 2001), pp. 83–93; M. C. Rosal and coauthors, Facilitating dietary change: The patient-centered counseling model, *Journal of the American Dietetic Association* 101 (2001): 332–338, 341.

[3] J. M. Heins and L. Delahanty, Tools and techniques to facilitate eating behavior change, in A. M. Coulston, C. L. Rock, and E. R. Monsen, eds., *Nutrition in the Prevention and Treatment of Disease* (San Diego: Academic Press, 2001), pp. 105–122.

[4] American Dietetic Association, *Nutrition Care Manual* (Chicago: American Dietetic Association, 2005).

[5] Z. M. Pronsky and J. P. Crowe, Food-drug interactions, in L. K. Mahan and S. Escott-Stump, eds., *Krause's Food, Nutrition, and Diet Therapy* (Philadelphia: Saunders, 2004); B. G. Katzung, ed., *Basic and Clinical Pharmacology* (New York: Lange Medical Books/McGraw-Hill, 2001); V. Utermohlen, Diet, nutrition, and drug interactions, in M. E. Shils and coeditors, *Modern Nutrition in Health and Disease* (Baltimore: Williams & Wilkins, 1999), pp. 1619–1641.

[6] D. G. Bailey, Grapefruit juice–drug interaction issues, in J. I. Boullata and V. T. Armenti, eds., *Handbook of Drug-Nutrient Interactions* (Totowa, N.J.: Humana Press, 2004), pp. 175–194.

[7] A. Fugh-Berman and E. Ernst, Herb-drug interactions: Review and assessment of report reliability, *British Journal of Clinical Pharmacology* 52 (2001): 587–596.

Food and Foodservice

Foodservice departments strive to prepare appetizing and nutritious meals and may accommodate dozens of special diets.

The work of a foodservice department can appear deceptively simple, with appropriate meals being delivered to patients who need specific types of diets. Behind the scenes, however, a complex system is at work. A foodservice department faces a daily challenge in planning, producing, and delivering hundreds of nutritious meals and accommodating dozens of special diets and food preferences.

Although this discussion focuses on the foodservice in hospitals, much of the information applies to foodservice in any health care facility, including nursing homes, assisted living centers, rehabilitation centers, and residential mental health care facilities. An important difference between hospitals and long-term health care facilities deserves mention, however. When patients in hospitals eat poorly, they can make up for nutrient deficits by eating well when they return home. Residents of a long-term care facility do not have this option. For this reason, foodservice departments in long-term care facilities must make even greater efforts to ensure that their patients receive and consume nutritious foods. The glossary on page 397 defines some of the terms related to this discussion.

How much freedom do patients have in selecting foods?

Most hospitals provide **selective menus** from which patients can select their meals. A patient who must follow a modified diet receives menus that include only the foods specified in the hospital's diet manual for that particular diet (examples of menus are shown in Figure NP14-1, p. 396). Often, patients need to make menu selections for a day or two in advance so that the foodservice department can estimate the amounts and types of food it needs to prepare. Once food selections have been made and the menus collected, a member of the foodservice staff (usually, a dietetic technician or dietitian) may check them to make sure that the selections are appropriate.

By allowing a choice, this system ensures that patients will receive the foods they prefer and are most likely to eat. An added advantage is that patients can become familiar with the modified diets as they select foods from the appropriate menus. Completed menus can also provide valuable clues about a person's usual eating habits or understanding of a modified diet.

Do all facilities offer selective menus?

Not all. Some facilities provide **nonselective menus** (menus with preselected food items) or menus that include elements of both systems **(semiselective menus).** Nonselective menus have been gaining popularity in hospital foodservice because they simplify operations and may help to cut costs.

If patients often make their own selections, why do they sometimes receive foods they don't like?

If a menu is not marked correctly or is misplaced, the patient may receive a meal selected by the foodservice department. When a facility uses selective menus, other problems may arise:

- Patients may have difficulty seeing, reading, understanding, or physically marking menus.
- Patients may not understand that their selections will be for the next (or another) day.
- Patients may be out of their rooms (for tests, procedures, or exercise) or asleep when the menus arrive and may miss the menu pickup time.
- Patients may be too ill or too disinterested in food to make menu selections.

Problems with menu procedures can often be corrected by explaining the system or by taking the time to help patients mark menus. Foodservice departments often conduct periodic surveys to uncover problems that patients may have with menu selections, food quality, or foodservice. Keep in mind that many foodservice employees do not have a nutrition background, and their ability to interpret diet orders and provide accurate information is limited.

Don't patients make valid complaints about foods sometimes?

Yes, this can happen as well. The hospital may not prepare foods in the same way that the patient does at home—sometimes a considerable problem for a person who must eat three meals a day for many days at the hospital. A patient may be expecting to enjoy a favorite food for dinner, only to

FIGURE NP14-1

SAMPLE Lunch Menu

LOW-FAT/LOW CHOLESTEROL/CARDIAC SUNDAY

✿ Lunch ✿
LF = Low Fat   LSLF = Low Sodium, Low Fat

Meats
LSLF Baked chicken   LSLF Baked fish (cod)

Starchy Vegetables
LSLF Rice   LSLF Boiled potatoes

Vegetables
LSLF Baby carrots   LSLF Green beans

Soup/Salad/Juice	Dressings
LSLF Coleslaw	Diet French
Gelatin	Diet Thousand Island
Tomato soup	Diet Italian
Tossed salad	

Desserts
Pears   Fresh fruit

Breads
LF Dinner roll   Bran bread
White bread   LS Crackers
Wheat bread

Beverages & Condiments
Coffee	Creamer
Decaf. coffee	Sugar
Hot tea	Sugar substitute
Decaf. hot tea	Herb seasoning
Iced tea	Lemon
Buttermilk	Margarine
Fat-free milk	Mustard
	Diet mayonnaise
	Catsup

Name _____ Room ___

LOW SODIUM	SUNDAY

✿ Lunch ✿
LF = Low Fat   LSLF = Low Sodium, Low Fat

Meats
LSLF Baked chicken   LSLF Baked fish (cod)

Starchy Vegetables
LSLF Rice   LSLF Boiled potatoes

Vegetables
LSLF Baby carrots   LSLF Green beans

Soup/Salad/Juice	Dressings
LSLF Coleslaw	Diet French
LS Chicken broth	Diet Thousand Island
Apple juice	Diet Italian
Tossed salad	

Desserts
Pears   Fresh fruit

Breads
Dinner roll   Bran bread
White bread   LS Crackers
Wheat bread

Beverages & Condiments
Coffee	Sugar
Decaf. coffee	Sugar substitute
Hot tea	Creamer
Decaf. hot tea	Lemon
Iced tea	Herb seasoning
Whole milk	Margarine
2% milk	Diet mustard
Fat-free milk	Diet mayonnaise
No salt	Diet catsup

Name _____ Room ___

RENAL	SUNDAY

✿ Lunch ✿
LF = Low Fat   LSLF = Low Sodium, Low Fat

Meats
LSLF Baked chicken   LSLF Baked fish

Starchy Vegetables
LSLF Rice   LSLF Dialyzed potatoes

Vegetables
LSLF Baby carrots   LSLF Green beans

Soup/Salad/Juice	Dressings
Lemonade	Diet French
LSLF Coleslaw	Diet Thousand Island
Tossed salad	Diet Italian
(no tomato)	

Desserts
Pears   Apple pie

Breads
Dinner roll   Bran bread
White bread   LS Crackers
Wheat bread

Beverages & Condiments
Coffee	Sugar
Decaf. coffee	Sugar substitute
Hot tea	Creamer
Decaf. hot tea	Lemon
Iced tea	Margarine
	Diet mustard
	Mayonnaise
No salt	

Name _____ Room ___

---

be disappointed with the way the food is prepared. In addition, meals may not be served at the right temperature by the time the patient is ready to eat, especially if the meal is delivered while the patient is away for a medical procedure. Complaints may have little to do with the meal itself, however, and instead may serve as a way for patients to vent fear, frustration, anger, and physical pain.

How are hospital meals delivered to patients?

Meals may be produced in a central kitchen and delivered directly to patients' rooms, using serving equipment that keeps hot foods hot and cold foods cold. Another popular practice is to produce meals in advance, deliver trays to the nursing unit, and then reheat hot food items in areas close to patients' rooms. Generally, foodservice personnel deliver food carts directly to the nursing unit, and then either nursing or foodservice personnel take trays to patients.

How are food safety issues addressed by foodservice departments?

Each institution has protocols for handling food products based on the identification of potential hazards and critical control points in food preparation, usually referred to as **HACCP (Hazard Analysis and Critical Control Point).**[1] Generally, a HACCP program addresses food handling, cooking, and storage procedures; cleaning and disinfecting of utensils, surfaces, and equipment; and staff sanitation issues. Personnel involved with preparing or delivering meals need to be aware of the specific HACCP systems at their facility.

What can the nurse do to improve patients' enjoyment of meals?

Nurses can make sure that patients have a chance to wash up before meals, and can check to see that foods are arranged attractively. Placing an occasional "surprise" on the tray—a decoration or funny card, for example—may help patients look forward to meals or perk up sagging spirits. Whenever possible, the patient's room should remain calm and quiet during mealtime. Excessive activity, like room maintenance or ward rounds, can distract patients and reduce appetite. If a patient receives the wrong type of meal or receives foods that differ substantially from those requested, the nurse should contact the foodservice department. Chapter 14 provides additional suggestions that may help to make meals more enjoyable.

Patient's experiences with foodservice strongly influence their perception of the overall hospital stay.[2] Good communication between patients and the hospital staff regarding meals and food quality can substantially improve a patient's satisfaction with foodservice procedures.

Glossary

HACCP (Hazard Analysis and Critical Control Point): systems of food or formula preparation that identify food safety hazards and critical control points during foodservice procedures; pronounced *hassip*.

nonselective menus: menus that do not allow choices and list only preselected food items.

selective menus: menus with two or more choices in some or all menu categories.

semiselective menus: menus that combine aspects of both selective and nonselective menus.

Notes

[1] K. W. McClusky, Implementing Hazard Analysis Critical Control Points, *Journal of the American Dietetic Association* 104 (2004): 1699–1700.

[2] C. A. Watters and coauthors, Exploring patient satisfaction with foodservice through focus groups and meal rounds, *Journal of the American Dietetic Association* 103 (2003): 1347–1349.

Enteral and Parenteral Nutrition Support

© Ed Eckstein/Phototake

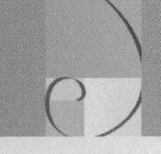

CHAPTER 15

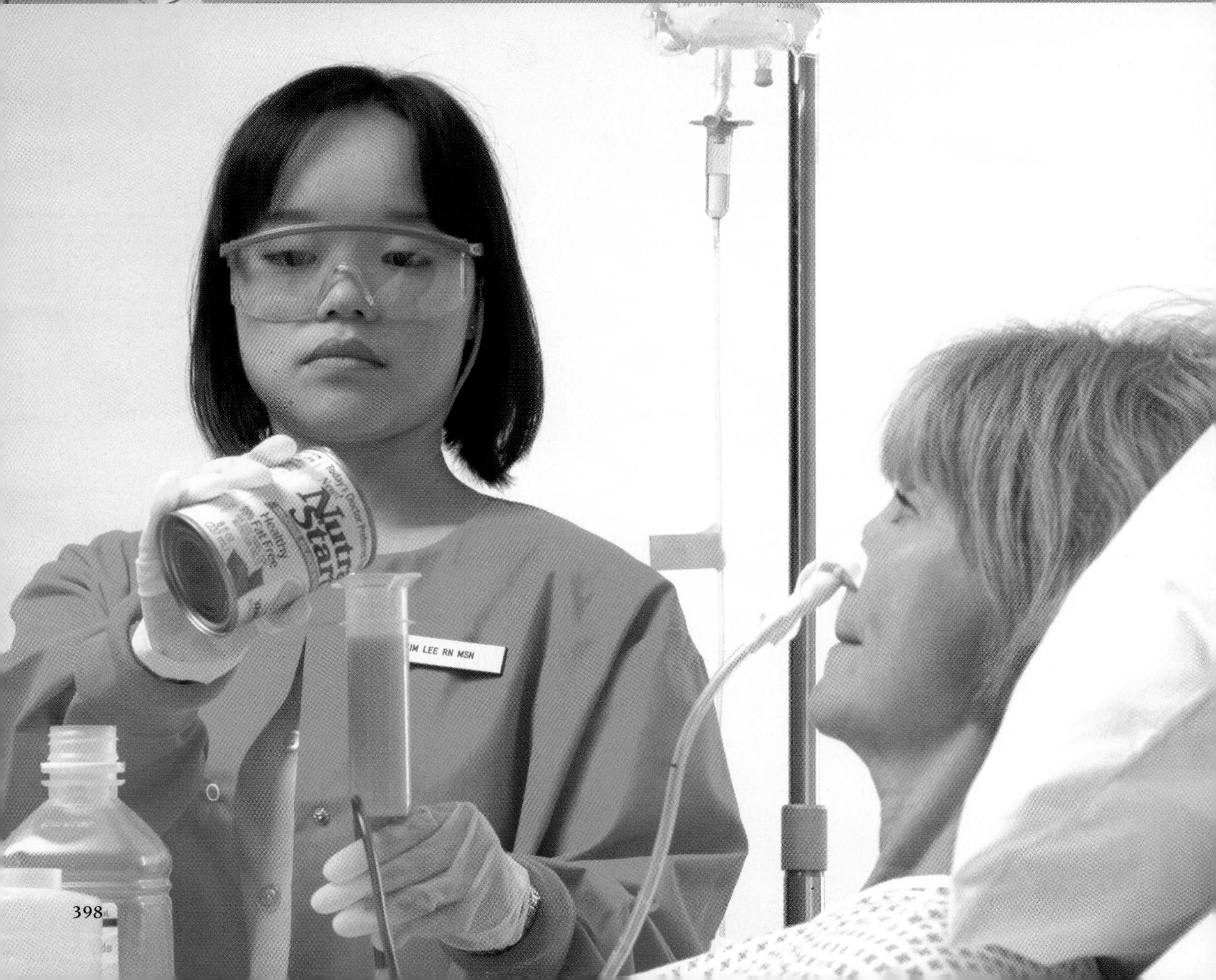

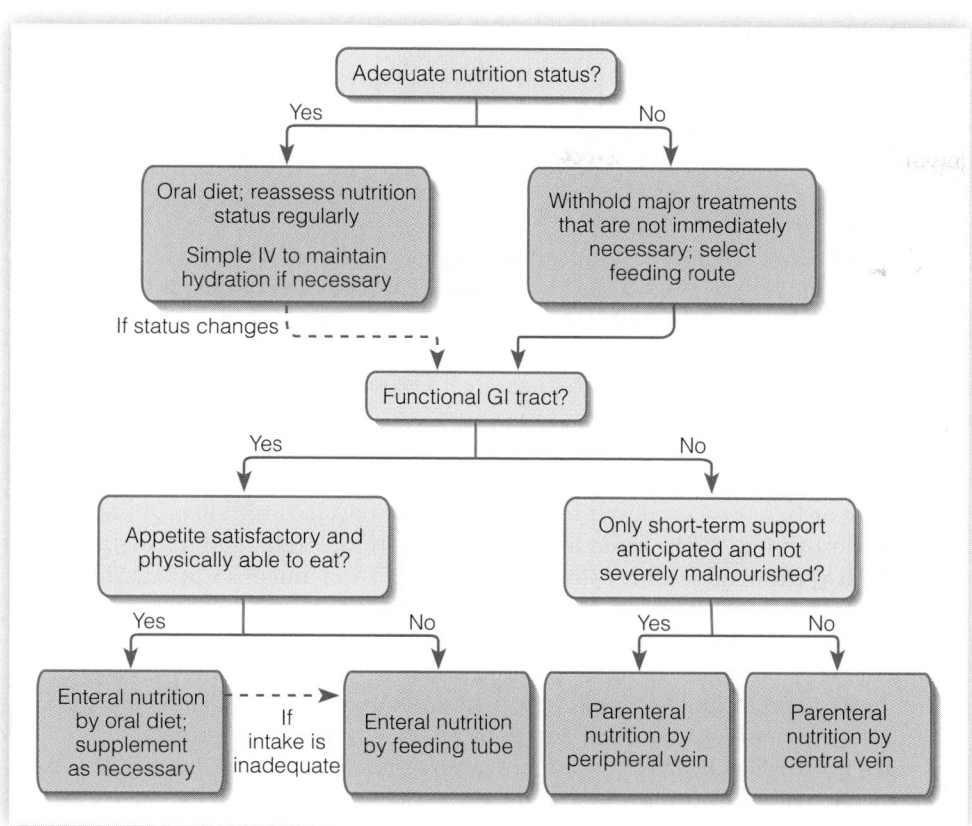

FIGURE 15-1 Selecting a Feeding Route

Patients are often too sick to obtain the energy and nutrients they need by consuming foods. Furthermore, some illnesses may interfere with eating, digestion, or absorption to such a degree that conventional foods cannot supply the necessary nutrients. In such cases, **nutrition support,** the delivery of formulated nutrients, can meet a patient's nutritional needs. **Enteral nutrition** provides nutrients using the gastrointestinal (GI) tract. Enteral nutrition includes oral diets or supplements but often refers to the use of tube feedings, which supply nutrients directly to the stomach or intestine via a thin, flexible tube. **Parenteral nutrition** provides nutrients intravenously to patients who do not have adequate gastrointestinal function to handle enteral feedings. If the GI tract remains functional, enteral nutrition support is preferred, partly to avoid the expense and complications associated with intravenous feedings and partly to preserve healthy GI function. Figure 15-1 summarizes the decision-making process for selecting the most appropriate feeding method.

Enteral Nutrition Support

If gastrointestinal function is normal and a poor appetite is the primary nutrition problem, enteral formulas can be provided as an oral supplement to the usual diet. If patients cannot consume enough food or drink enough formula to meet nutrient needs, tube feedings can deliver the required nutrients.

ENTERAL FORMULAS

A huge number of enteral formulas that are designed to meet a variety of medical and nutritional needs are on the market; some examples are listed in Appendix G. Formulas can be used alone or given along with other foods. Many formulas can supply all of an individual's nutrient requirements when consumed in sufficient

NURSING DIAGNOSIS

imbalanced nutrition: less than body requirements is appropriate for a person who needs to supplement the diet with an enteral formula.

nutrition support: the delivery of formulated nutrients by feeding tube or intravenous infusion.

enteral (EN-ter-al) **nutrition:** the provision of nutrients using the GI tract, including the use of tube feedings and oral diets.

parenteral (par-EN-ter-al) **nutrition:** the intravenous provision of nutrients that bypasses the GI tract.
   par = beside
   entero = intestine

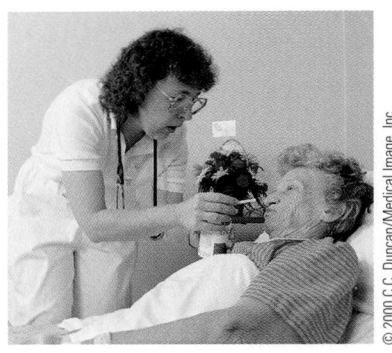

© 2000 C.C. Duncan/Medical Image, Inc.

Patients can drink enteral formulas when they are unable to consume enough food from a conventional diet.

Reminder: The *macronutrients* are carbohydrates, fats, and proteins.

standard formulas: general-purpose enteral formulas that contain mostly intact proteins and polysaccharides; also called **polymeric formulas**.

blenderized formulas: enteral formulas that are made by blenderizing whole foods.

hydrolyzed formulas: enteral formulas that contain macronutrients that have been partially or fully hydrolyzed; also called **monomeric**, or **elemental formulas**.

medium-chain triglycerides (MCT): triglycerides that contain fatty acids that are 8 to 10 carbons in length. MCT do not require digestion and can be absorbed in the absence of lipase or bile.

disease-specific formulas: enteral formulas designed to meet the nutrient needs of patients with specific illnesses.

modular formulas: enteral formulas that contain only one or two macronutrients; used to enhance other formulas, meet specific nutrient needs, or create individualized formulas for people with unique needs.

osmolality (OZ-moe-LAL-ih-tee): the osmotic property of a solution, based on its concentrations of molecules and ionic particles. Osmolality is expressed as milliosmoles (mOsm) per kilogram.

isotonic formula: a formula with an osmolality similar to that of blood serum (300 mOsm/kg).
  iso = equal
  tono = pressure

hypertonic formula: a formula with an osmolality greater than that of blood serum.

volume. Nutritionally complete formulas are essential for a patient who is using a tube feeding or oral liquid diet for more than a few days.

Types of Formulas Formulas are typically classified according to their macronutrient composition. The main types of formulas and their nutrient composition are as follows:

- **Standard formulas** are used for individuals who can digest and absorb nutrients without difficulty. They contain intact proteins isolated from milk or soybeans or a combination of purified proteins. Carbohydrate sources include modified starches, glucose polymers (such as maltodextrin), and sugars. A few formulas, called **blenderized formulas,** are made from whole foods and derive their protein primarily from pureed meat or poultry.
- **Hydrolyzed formulas** are provided to patients who have compromised digestive or absorptive functions. The formulas contain macronutrients that have been partially or fully broken down to fragments that require little (if any) digestion before absorption. Hydrolyzed formulas are often low in fat and may contain **medium-chain triglycerides (MCT)** to ease digestion and absorption. These formulas are generally lactose-free and result in minimal fecal output.
- **Disease-specific formulas** are designed to meet the specific nutrient needs of patients with particular illnesses. Products have been developed for liver, kidney, and lung diseases; glucose intolerance; and metabolic stress. Disease-specific formulas are generally expensive, and their effectiveness is controversial.
- **Modular formulas** usually contain only one or two macronutrients and are used to enhance other formulas. Several modular formulas may be combined with liquid vitamin and mineral preparations to create individualized formulas for patients with unique nutrient needs.

Nutrient and Energy Densities The percentages of protein, carbohydrate, and fat vary considerably in enteral formulas. Protein content ranges from 8 to 29 percent of total kcalories; note that protein needs are high in patients with severe metabolic stress, but protein restrictions are necessary in patients with renal failure. Carbohydrate and fat provide most of the energy in enteral formulas; standard formulas often provide 40 to 50 percent of kcalories from carbohydrate and 30 to 45 percent from fat.

Fiber content often influences the selection of an enteral formula. Fiber may be helpful for normalizing intestinal function, treating diarrhea and constipation, and maintaining blood glucose control. Fiber-containing formulas may be avoided, however, during acute intestinal conditions, pancreatitis, or procedures involving the intestines.

The energy density of enteral formulas ranges from 0.5 to 2.0 kcalories per milliliter of fluid. Standard formulas provide 1.0 to 1.2 kcalories per milliliter and are appropriate for patients with average fluid requirements. Formulas that have higher energy densities can meet energy and nutrient needs in a smaller volume of fluid and thus benefit patients with high nutrient needs or fluid restrictions.

Osmolality Osmolality refers to the osmotic property of a solution, that is, a solution's tendency to shift from one fluid compartment to another across a semipermeable membrane. The osmolality depends on a solution's concentrations of molecules and ionic particles. An enteral formula with an osmolality similar to that of blood serum (about 300 milliosmoles per kilogram) is an **isotonic formula,** whereas a **hypertonic formula** has an osmolality greater than that of blood serum.

Most enteral formulas have osmolalities between 300 and 700 milliosmoles per kilogram; generally, hydrolyzed formulas and nutrient-dense formulas have higher osmolalities than standard formulas. Most people are able to tolerate both isotonic and hypertonic feedings without difficulty.[1] When medications are infused along with enteral feedings, however, the osmotic load increases substantially and may contribute to the diarrhea experienced by many tube-fed patients.

ENTERAL NUTRITION IN MEDICAL CARE

A person with a functioning GI tract who cannot meet nutrient needs with conventional foods alone may be a candidate for enteral nutrition support. Enteral feedings are preferred over intravenous feedings because they help to stimulate or maintain gut function, cause fewer complications, and are less costly.[2] Similarly, oral feedings are preferred to tube feedings if the person is able to drink an enteral formula, as drinking formulas prevents the stress, complications, and expense associated with tube feedings.

Oral Use of Enteral Formulas Enteral formulas can fully meet the nutrient needs of people who can consume only liquids or who require hydrolyzed nutrients. More often, enteral formulas are used to supplement conventional diets when individuals cannot consume enough food to meet their needs. Enteral formulas provide a reliable source of nutrients and can add energy and protein to the diets of malnourished patients. Patients who are weak or debilitated may also find it easier to manage formulas than meals.

When a patient drinks a formula, taste becomes an important consideration. Allowing patients to sample different products and flavors and select the ones they prefer helps to promote acceptance. The "How to" offers additional suggestions for helping patients accept and enjoy oral formulas.

Tube Feedings An individual with a functional GI tract who is unable to consume enough food or formula orally may need to be fed via a feeding tube. A tube feeding delivers a nutritionally complete formula directly to the stomach or intestine. Candidates for tube feedings include:

- People with severe swallowing difficulties.
- People who have little or no appetite for extended periods, especially if malnourished.
- People with gastrointestinal obstructions, some types of fistulas, or impaired motility in the upper GI tract.

HOW TO   *Help Patients Accept Oral Formulas*

People using enteral formulas are often quite ill and have poor appetites. Even when a person enjoys a formula, the taste can become monotonous in time. Hydrolyzed formulas are usually less palatable than standard formulas, and patients may find them difficult to drink. Nurses can help by trying these suggestions:

- Let the patient sample different formulas that are appropriate for his or her needs, and use only those that the patient enjoys.
- Serve formulas attractively and remind patients to drink them. Formulas offered in a glass on an attractive plate may be more appealing than those served from a can with an unfamiliar name.
- If a patient finds the smell of a formula unappealing, it may help to cover the top of the glass with plastic wrap or a lid, leaving just enough room for a straw.

- Provide easy access. Keep the formula close to the patient's bed where it can be reached with little effort and within sight so that the patient is reminded to drink it. Patients who are very ill may lack the motivation to reach for the formula, let alone drink it.
- Try keeping the formula in an ice bath so that it will be cool and refreshing when the patient drinks it. Check with the patient to make sure the colder temperature is suitable.
- For patients with little appetite, offer the formula in smaller amounts that are easy to tolerate, and serve it more frequently during the day.
- If the patient stops enjoying the formula, recommend different flavors or try other formulas.

- People who have undergone intestinal resections and are beginning enteral feedings.
- People who are mentally incapacitated due to confusion, dementia, or neurological difficulties.
- People in a coma.
- People with extremely high nutrient requirements.
- People on mechanical ventilators.

Feeding Routes The feeding route chosen depends on the medical condition, the expected duration of tube feeding, and the potential complications of a particular route. Figure 15-2 illustrates the main feeding routes, and the glossary on p. 403 describes each route. Table 15-1 summarizes the advantages and disadvantages of each route.

When a patient is expected to be tube-fed for less than four weeks, a **nasogastric** or **nasoenteric** route is generally chosen; for these routes, the feeding tube is passed into the GI tract via the nose. The patient is frequently awake during **transnasal** (through the nose) placement of a feeding tube. While the patient is in a slightly upright position with head tilted, the tube is inserted into a nostril and passed into the stomach **(nasogastric)**, duodenum **(nasoduodenal)**, or jejunum **(nasojejunal)**. If the patient is awake and alert, he or she can swallow water to ease the tube's passage. The final position of the feeding tube tip is verified by abdominal X-ray or other means. In infants, **orogastric** placement, in which the feeding tube is passed into the stomach via the mouth, is preferred over transnasal routes; this placement allows the infant to breathe more normally during feedings.

When a patient will be tube-fed for longer than four weeks or if the nasoenteric route is inaccessible due to an obstruction or other medical reasons, a direct

The final location of the feeding tube determines how the feeding route is classified.

FIGURE 15-2 Tube Feeding Routes

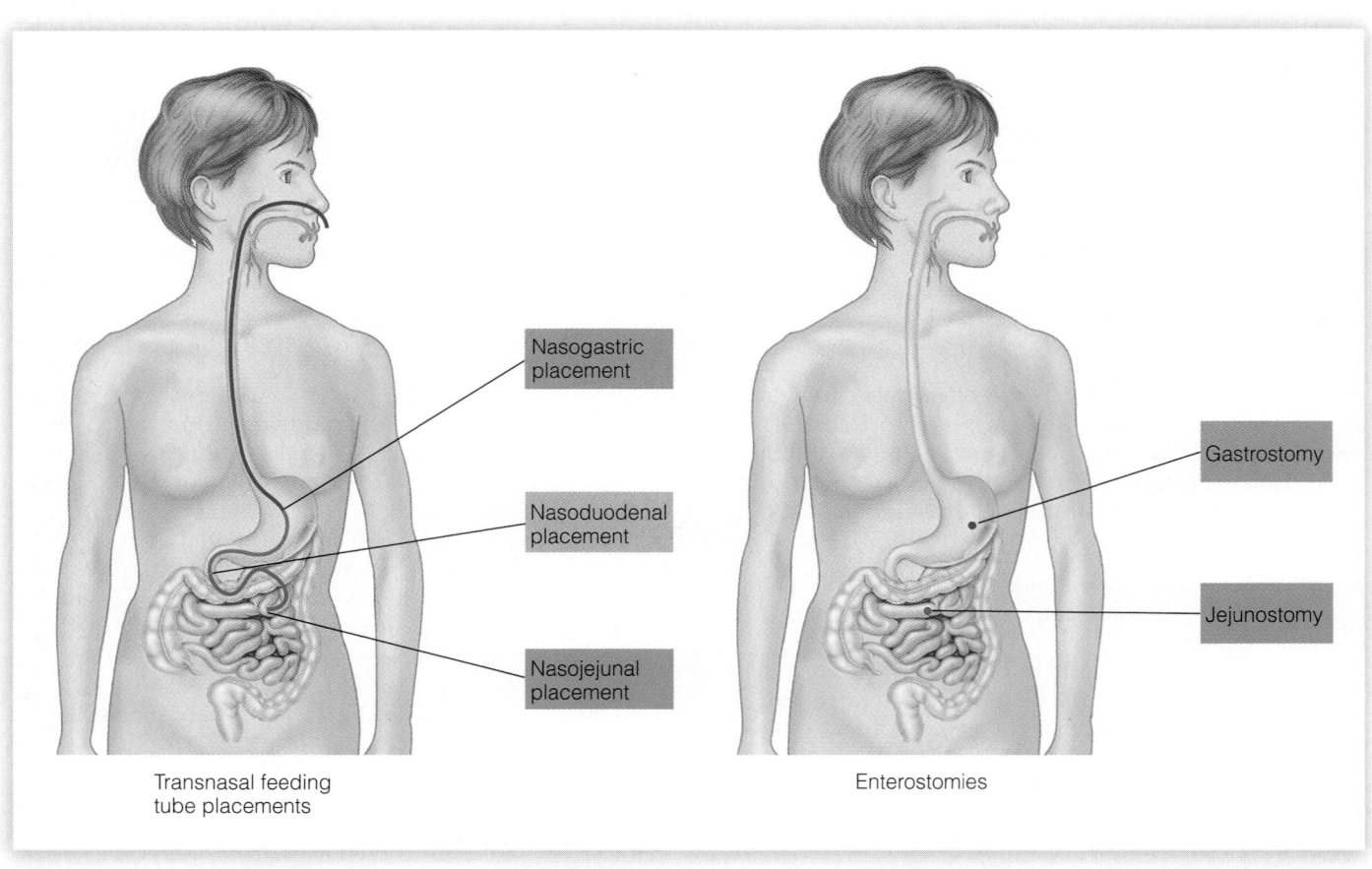

Nasogastric placement

Nasoduodenal placement

Nasojejunal placement

Gastrostomy

Jejunostomy

Transnasal feeding tube placements

Enterostomies

GLOSSARY of Tube Feeding Routes

For each type of tube placement, the terms are listed in order from the upper to lower organs of the digestive system.

transnasal: through the nose. A **transnasal feeding tube** is one that is inserted through the nose.

nasogastric (NG): tube is placed into the stomach via the nose.

nasoenteric: tube is placed into the GI tract via the nose. (*Nasoenteric feedings* usually refer to *nasoduodenal* and *nasojejunal* feedings.)

nasoduodenal (ND): tube is placed into the duodenum via the nose.

nasojejunal (NJ): tube is placed into the jejunum via the nose.

orogastric: tube is placed into the stomach via the mouth. This method is often used to feed infants because a nasogastric tube can hinder the infant's breathing.

enterostomy (EN-ter-AH-stoe-mee): an opening into the GI tract through which a feeding tube can be passed.

gastrostomy (gah-STRAH-stoe-mee): an opening into the stomach through which a feeding tube can be passed. A nonsurgical technique for creating a gastrostomy under local anesthesia is called *percutaneous endoscopic gastrostomy (PEG)*.

jejunostomy (JE-ju-NAH-stoe-mee): an opening in the jejunum through which a feeding tube can be passed. A nonsurgical technique for creating a jejunostomy is called *percutaneous endoscopic jejunostomy (PEJ)*. The tube can either be guided into the jejunum via a gastrostomy or passed directly into the jejunum *(direct PEJ)*.

TABLE 15-1   Comparison of Tube Feeding Routes[a]

Insertion Method and Feeding Site	Advantages	Disadvantages
Transnasal	Does not require surgery or incisions for placement.	Easy to remove by disoriented patients, long-term use may irritate the nasal passages, throat, and esophagus.
Nasogastric	Easiest to insert and confirm placement; feedings can often be given intermittently and without an infusion pump.	Highest risk of aspiration in compromised patients.
Nasoduodenal and nasojejunal	Lower risk of aspiration in compromised patients; allow for enteral nutrition earlier than gastric feedings following severe stress, may allow for enteral feeding when obstruction, fistulas, or other medical conditions prevent gastric feeding.	More difficult to insert and confirm placement; feedings require an infusion pump for administration; may take longer to reach nutrition goals.
Tube enterostomies	Allow lower esophageal sphincter to remain closed, reducing the risk of aspiration; more comfortable than transnasal insertion for long-term use; site is not visible under clothing.	May require general anesthesia for insertion; require incisions; greater risk of complications from the insertion procedure; greater risk of infection; may cause skin irritation around the insertion site.
Gastrostomy	Feedings can often be given intermittently and without a pump; easier to insert than a jejunostomy.	Moderate risk of aspiration in high-risk patients.
Jejunostomy	Lowest risk of aspiration; allows for enteral nutrition earlier following severe stress; may allow for enteral feeding when obstructions, fistulas, or medical conditions prevent gastric feeding.	Most difficult to insert; feedings require an infusion pump for administration; may take longer to reach nutrition goals.

[a] Relative to other tube feeding routes. The actual advantages and disadvantages of different insertion procedures depend on the person's medical condition.

route to the stomach or intestine may be created by passing the tube through an **enterostomy,** an opening in the stomach **(gastrostomy)** or jejunum **(jejunostomy).** An enterostomy can be made by surgical incision or nonsurgically using local anesthesia.

Gastric feedings, such as the nasogastric and gastrostomy routes, are preferred whenever possible. These feedings are more easily tolerated and less complicated to deliver than intestinal feedings because the stomach controls the rate at which nutrients enter the intestine. Gastric feedings are not possible, however, if patients have gastric obstructions or motility disorders that interfere with the stomach's ability to empty. Gastric feedings may also be a problem for patients at high risk of **aspiration,** a common complication in which formula or GI secretions enter the lungs, often from the backflow of stomach contents. Although health practitioners often administer nasoenteric feedings to minimize the possibility of aspiration, studies have not consistently shown that gastric feedings are associated with increased aspiration risk.[3]

Feeding Tubes   Feeding tubes are soft and flexible and come in a variety of lengths and diameters. The tube selected largely depends on the patient's age and size, the feeding route, and the formula's viscosity. Once the appropriate length is determined, the tube selected is often the smallest tube through which the formula will flow without clogging.

The outer diameter of a tube is measured in **French units,** in which each unit equals $^1/_3$ millimeter; thus a "12 French" feeding tube has a 4-millimeter diameter. The inner diameter depends on the thickness of the tubing material. Double-lumen tubes are also available; these allow one tube to be used for both intestinal feedings and **gastric decompression,** a procedure in which the stomach contents of patients with motility disorders is removed by suction.

Formula Selection   The formula is selected after careful assessment of the patient's age, medical problems, nutritional status, and ability to digest and absorb nutrients; some of the considerations are shown in Figure 15-3. Generally, the best formula is one that meets the patient's medical and nutrient needs with the lowest risk of complications and the lowest cost. The vast majority of patients use standard formulas. A person with a functional, but impaired, GI tract may require a hydrolyzed formula. Some nutrition-related factors that influence formula selection include:

- *A patient's energy, protein, and fluid requirements.* Nutrient needs must be met using the volume of formula a patient can tolerate. If fluids need to be restricted, the formula should be able to deliver the necessary nutrients in the volume prescribed.
- *The need for fiber modifications.* The choice of formulas is narrowed if fiber intake needs to be low or high.
- *Individual tolerances (food allergies and sensitivities).* Most formulas are lactose-free because many patients who need enteral formulas have some degree of lactose intolerance. Many formulas are also gluten-free and can accommodate the needs of individuals with celiac disease (gluten sensitivity).

Health care facilities stock a limited number of formulas, so formula selection is limited by availability. Initially, the dietitian or physician may make an educated guess as to the best formula, based on the criteria previously mentioned. The decision can be reappraised based on the patient's response to the formula.

ADMINISTRATION OF TUBE FEEDINGS

After the feeding route and formula have been selected, attention is given to delivering the formula. The methods of tube feeding administration vary somewhat from one health care facility to the next. The procedures presented in the following sections are suggested guidelines.

NURSING DIAGNOSIS

risk for aspiration applies to many people on tube feedings.

Aspiration risk is high in patients with esophageal disorders, neurological diseases, and conditions that reduce consciousness or cause dementia.

1 French = $^1/_3$ mm.

12 French = $12 \times ^1/_3$ mm = 4 mm.

aspiration: drawing in by suction or breathing; a common complication of enteral feedings in which foreign material enters the lungs, often from reflux of stomach contents.

French units: units of measure used to indicate the size of a feeding tube's outer diameter. One French unit equals $^1/_3$ millimeter.

gastric decompression: the removal of the stomach contents (including swallowed saliva, stomach secretions, and gas) of patients who have motility disorders or obstructions that prevent stomach emptying.

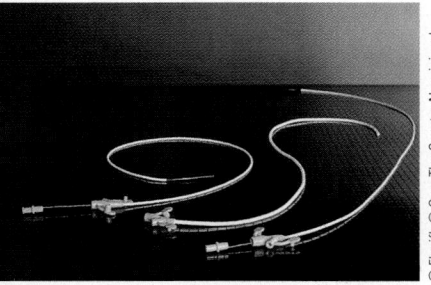

© Flexiflo® Over-The-Counter Nasojejunal Feeding Tube, Courtesy of Ross Products Div., Abbott Laboratories, Columbus OH.

Feeding tubes come in many lengths and diameters. The thin wires protruding from the end of the feeding tubes are stylets, which stiffen the tube to ease insertion and are discarded thereafter. The Y-connector (shown here in orange) provides a port for administering water or medications without disrupting the feeding.

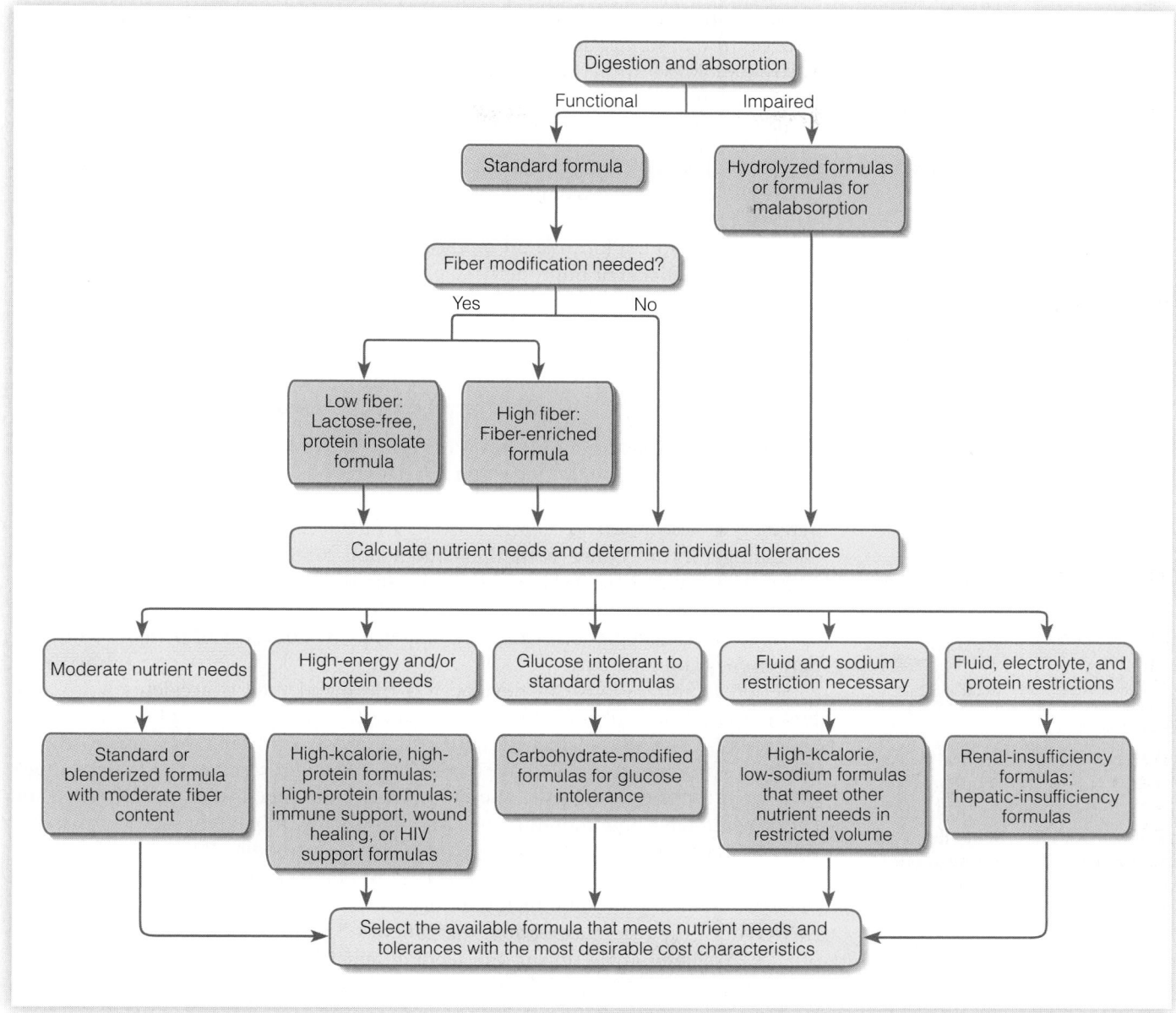

FIGURE 15-3 Selecting a Formula

Safe Handling Individuals who are ill or malnourished often have suppressed immune systems, making them vulnerable to infection from foodborne illness. To prevent contamination, the personnel involved in preparing and delivering formulas should work in clean environments, using clean equipment and clean hands. The foodservice department or pharmacy most often assumes responsibility for preparing formulas.

Formulas are available in **open feeding systems** and **closed feeding systems.** With an *open feeding system*, the formula needs to be transferred from its original packaging to a feeding container. Examples include formulas that are packaged in cans or bottles, concentrates that need to be diluted, and powders that require reconstitution. In a *closed feeding system*, the formula is prepackaged in a container that can be connected directly to a feeding tube. Closed systems are less likely to become contaminated, require less nursing time, and can hang for longer periods of time than open systems. Although closed systems cost more initially, they may be less expensive in the long run by preventing bacterial contamination and thus avoiding the costs of treating infections.

open feeding systems: delivery systems that require formula to be transferred from the original packaging to feeding containers before being administered through feeding tubes.

closed feeding systems: delivery systems in which formula comes prepackaged in containers that are ready to be attached to feeding tubes for administration.

Safety Guidelines After the formula reaches the nursing station, the nursing staff assumes responsibility for its safe handling. Hands should be carefully washed before handling formulas and feeding containers. Some facilities require that non-sterile gloves be worn whenever formulas are handled. The following steps can reduce the risk of formula contamination when using open feeding systems:

- Before opening a can of formula, clean the can opener and the lid. If you do not use the entire can at one feeding, label the can with the time it was opened.
- Store opened cans or mixed formulas in clean, closed containers. Refrigerate the unused portion of formula promptly.
- Discard unlabeled or improperly labeled containers and all opened containers of formula that are not used within 24 hours.
- Hang no more than an 8- to 12-hour supply of formula when using an open feeding system. Discard any formula that remains, rinse out the feeding bag and tubing, and add fresh formula to the feeding bag. Use a new feeding container and tubing (except for the feeding tube itself) every 24 hours.

Closed feeding systems should hang for no longer than 24 to 48 hours; contamination is more likely with longer time periods.

Initiating a Tube Feeding Before starting a tube feeding, nurses can ease fears by discussing the procedure with the patient and family members, who may feel anxious about using a feeding tube. The discussion should address the reasons why tube feeding is appropriate and the benefits and risks of the procedure. The "How to" offers suggestions that may help ease the concerns of patients who may benefit from tube feeding.

Serious complications can develop if a transnasal tube is accidentally inserted into the respiratory tract or if formula or GI secretions are aspirated into the lungs. To minimize the risk of incorrect tube placement, clinicians may use X-rays to verify the position of the feeding tube before a feeding is initiated. After the tube's placement has been confirmed, the nurse secures the tube to the patient's nose and cheek with tape and monitors the position of the tubing throughout the day. Another technique is to test the pH of a sample of bodily fluid drawn into the feeding tube, as the pH of stomach fluid is much lower than the pH of fluid obtained from the intestine or respiratory tract.

A gastric sample usually has a pH between 1 and 4. An intestinal sample should have a pH between 6 and 7.

To reduce the risk of aspiration, the patient's upper body is elevated to at least a 30- to 45-degree angle during the feeding and for 30 minutes after the feeding whenever possible. The addition of blue food coloring to formula is sometimes suggested as a means of identifying aspirated formula in lung secretions; however, this practice is now discouraged because several deaths have been attributed to its use.[4]

Formula Delivery A day's nutrient needs can be met by delivering relatively large amounts of formula several times a day (**intermittent feedings**) or smaller amounts continuously throughout the day (**continuous feedings**). A patient may also start on a continuous feeding and gradually transition to an intermittent feeding. Each method has specific uses, advantages, and disadvantages.

Intermittent feedings are best tolerated when they are delivered into the stomach (not the intestine). Generally, about 250 to 400 milliliters is delivered over 20 to 40 minutes using a gravity drip method or an infusion pump. The volume of formula required to meet a patient's nutrient needs is divided into several daily feedings. Because of the relatively high volume of formula delivered at a time, intermittent feedings may be difficult for some patients to tolerate, and the risk of aspiration may be higher than with continuous feedings. An advantage of intermittent feedings is that they are similar to the usual pattern of eating and allow the patient freedom of movement between meals.

Rapid delivery of a large volume of formula into the stomach (250 to 500 milliliters in less than 15 minutes) is called a **bolus feeding**. This type of feeding may

intermittent feedings: delivery of about 250 to 400 mL of formula over 20 to 40 minutes.

continuous feedings: slow delivery of formula at a constant rate over an 8- to 24-hour period.

bolus (BOH-lus) **feeding:** delivery of about 250 to 500 mL of formula in less than 15 minutes.

Help Patients Cope with Tube Feedings

The thought of being "force-fed" is frightening to many people. Some may envision thick feeding tubes or fear that the procedure will be painful. Others may associate tube feedings with disabling injury or irreversible illness. Patients may be less apprehensive once they understand the insertion procedure, the expected duration of the tube feeding, and the strategic role that nutrition plays in recovery from disease. The pointers that follow can help nurses prepare patients for transnasal tube feedings:

- Allow the patient to see and touch the feeding tube. Seeing firsthand that the tube is soft and narrow (only about half the diameter of a pencil) often alleviates anxiety.
- Show the patient how the feeding apparatus is attached to the feeding tube, and explain how the feeding will work. For young children, use dolls or stuffed toys to demonstrate tube insertion and feeding procedures.
- Explain that the patient remains fully alert during the procedure and helps pass the tube by swallowing. A numbing solution sprayed on the back of the throat minimizes discomfort and prevents gagging during the procedure.
- Tell the patient that once the tube has been inserted, most people become accustomed to its presence within a few hours. In most cases, the patient can easily swallow foods and liquids with the tube in place. If permitted, favorite foods or beverages can still be enjoyed.
- Assure the patient that the tube feeding will be temporary, if such assurance is appropriate.

A tube feeding may be frightening for some patients, but others may be relieved to know that they can receive sound nutrition without any effort. As they feel better and begin to eat again, the volume of the feeding can be reduced and then discontinued when oral intake is adequate.

Tube feedings may cause some patients to feel that they have lost control over an important aspect of their lives. They may also feel self-conscious about how the feeding tube looks or feel awkward when moving around with the equipment. A few measures can help:

- Involve patients in the decision-making and care process whenever possible. Patients can help to arrange their daily feeding schedules and can perform some of the feeding procedures themselves.
- Show patients how to manipulate the feeding equipment so that they can get out of bed and move around.
- Encourage patients to maintain contact with friends and keep busy with the hobbies and activities they enjoy. This measure is especially important for children, teens, and those on long-term feedings.

When caring for infants and children, keep the developmental age of the child in mind and work with parents to ensure that appropriate feeding skills are mastered. Infants can be provided with a pacifier during feedings to help maintain the associations between sucking, swallowing, and fullness. When possible, the formula can be provided by bottle to an infant, or by spoon to a child, to further develop skills.

The more complex the procedure, the easier it becomes for health care professionals to focus on the procedure and disregard a patient's emotional response. No matter how many technicalities you have to keep in mind, remember to stay focused on the person receiving your care.

be given every 3 to 4 hours using a syringe. Bolus feedings can cause abdominal discomfort, nausea, and cramping in some patients, especially when a feeding is initiated. The risk of aspiration is also greater than with other methods of feeding. For these reasons, bolus feedings are used only in patients who are not critically ill.

Continuous feedings are delivered slowly and at a constant rate over a period of 8 to 24 hours. This type of feeding is recommended for critically ill patients because delivering relatively small volumes at a time may reduce nausea, diarrhea, and the risk of aspiration. Continuous feedings are generally used in patients who receive intestinal feedings. An infusion pump is required to ensure accurate and steady flow rates; consequently, the feedings can limit the patient's freedom of movement and are also more costly.

Formula Volume and Strength Formula administration schedules vary among institutions, so protocols should be reviewed carefully before working with patients. In general, almost all patients can receive undiluted formula (either isotonic or hypertonic) at the start of a feeding.[5] Formulas, especially hypertonic formulas, are sometimes started slowly, and the volume gradually increased. In rare cases, formulas may be diluted until tolerated by the patient, although this practice is discouraged.

Intermittent feedings may start at about 120 to 150 milliliters at the initial feeding and be increased by 50 to 100 milliliters at each feeding until the goal volume is reached. Continuous feedings may start at about 30 to 40 milliliters per hour and be raised by 30 milliliters per hour every 6 to 8 hours. For both intermittent and continuous feedings, the rate and amount of increase depend on the patient's tolerance to the formula. If the new rate is not tolerated, the rate of delivery progresses more slowly to give the person additional time to adapt. If a patient on an intermittent feeding cannot tolerate the feeding, a continuous feeding may be a better choice. The "How to" describes several ways to plan tube feeding schedules.

Checking Gastric Residuals When a patient receives a gastric feeding, the nurse regularly measures the **gastric residual volume** (the volume of formula remaining in the stomach after feeding) to ensure that the stomach is emptying properly. The gastric residual is measured by gently withdrawing the gastric contents through the feeding tube using a syringe, usually before each intermittent feeding and every 4 to 6 hours during continuous feedings. Although opinions vary, some experts recommend that an evaluation be conducted if the gastric residual exceeds 200 milliliters, and feedings withheld if it exceeds 500 milliliters.[6] If the tendency to accumulate fluids persists, the physician may recommend intestinal feedings or begin drug therapy to stimulate gastric emptying.

gastric residual volume: the volume of formula remaining in the stomach from a previous feeding.

To estimate fluid requirements in adults and children:

- Adults: allow 30 to 40 mL/kg; 30 mL/kg in older adults
- Children: allow 50 to 60 mL/kg
- Infants: allow 150 mL/kg

Meeting Water Needs Attention must also be paid to the patient's fluid requirements. Many adults require about 2000 milliliters (about 2 quarts) of water daily. Fluids may be restricted in persons with kidney, liver, or heart disease. Additional water is required in those with fever, high urine output, diarrhea, excessive sweating, severe vomiting, fistula drainage, high-output ostomies, blood loss, or open wounds.

The water in formulas can meet a substantial portion of water needs. Standard formulas contain about 85 percent water, or about 850 milliliters of water per liter of formula. Nutrient-dense formulas contain about 69 to 72 percent water; exact amounts can be obtained from the product label or manufacturer's information sheet.

NURSING DIAGNOSIS

deficient fluid volume or *excess fluid volume* apply to patients who are receiving too little or too much water relative to needs.

In addition, water needs are partially met by flushes given via feeding tubes. To prevent clogging, feeding tubes are routinely flushed with water before and after each bolus or intermittent feeding and about every 4 hours during continuous feedings. The water used for routine flushes should be included when estimating fluid intakes.

- For intermittent or bolus feedings, feeding tubes should be flushed with 20 to 60 mL of water before and after formula delivery.
- For continuous feedings, feeding tubes should be flushed with 20 to 60 mL of water every 4 hours.

Transition to Table Foods Once the condition requiring a tube feeding resolves, the volume of formula is tapered off, and the patient gradually shifts to an oral

---

HOW TO   *Plan a Schedule for Administering Tube Feedings*

After selecting a formula that meets the patient's medical and nutrient needs, the clinician determines the volume of formula that meets those needs. Consider a patient who needs 2000 kcalories daily and is using a standard formula that provides 1.0 kcalorie per milliliter. The total volume of formula required would be 2000 milliliters per day:

$$x \text{ mL} \times 1.0 \text{ kcal/mL} = 2000 \text{ kcal.}$$

$$x \text{ mL} = \frac{2000 \text{ kcal}}{1.0 \text{ kcal/mL}} = 2000 \text{ mL.}$$

If the patient is to receive intermittent feedings six times a day, he will need about 330 milliliters of formula at each feeding:

$$2000 \text{ mL} \div 6 \text{ feedings} = 333 \text{ mL/feeding.}$$

Alternatively, if he is to receive intermittent feedings eight times a day, he will need 250 milliliters (or about one can of ready-to-feed formula) at each feeding:

$$2000 \text{ mL} \div 8 \text{ feedings} = 250 \text{ mL/feeding.}$$

He will probably tolerate this volume of formula best if it is delivered over a 30-minute period at each feeding. If the patient is to receive the formula continuously over 24 hours, he will need about 85 milliliters of formula each hour:

$$2000 \text{ mL} \div 24 \text{ hours} = 83 \text{ mL/hr.}$$

diet. Sometimes the patient can begin the transition by drinking the same formula that is delivered by tube. Oral intake should supply about two-thirds of estimated nutrient needs before the tube feeding is discontinued.

DELIVERING MEDICATIONS THROUGH FEEDING TUBES

Patients receiving tube feedings sometimes require one or more medications that need to be delivered through feeding tubes. Because medications can interact with enteral formulas in the same ways that they interact with foods, diet-drug interactions must be considered. Medications can also cause feeding tubes to clog. The "How to" provides some guidelines that may help to prevent complications.

Continuous feedings are ordinarily stopped during the administration of medication so that the components of enteral formulas do not interfere with the medication's absorption. The feeding is typically halted for 15 minutes before and 15 minutes after medication delivery. Some medications may require a longer formula-free interval; for example, feedings need to be stopped for one to two hours before and after administering phenytoin, a medication that controls seizures. In such cases, the delivery rate of formula needs to be increased so that the correct amount of formula can be delivered.

COMPLICATIONS OF TUBE FEEDING

Complications are a frequent occurrence during tube feedings. Common complications include gastrointestinal problems, such as nausea and diarrhea; mechanical problems related to the tube feeding process; and metabolic problems, such as biochemical alterations and nutrient deficiencies. Examples of the complications that may arise and some preventive and corrective measures are summarized in Table 15-2 (p. 410).

Many complications of tube feeding are preventable if the most appropriate feeding route, formula, and delivery method are chosen. Attention to a patient's primary medical condition and medication use is important as well. Nurses routinely

---

HOW TO   *Administer Medications to Patients Receiving Tube Feedings*

The pharmacist is your best resource for learning how and when medications can be administered via feeding tubes, especially when you are dealing with an unfamiliar medication. Check with the pharmacist to learn the following:

- Whether a particular medication is known to be incompatible with formulas.
- The proper timing of medication administration to avoid drug-nutrient interactions.
- For patients using intestinal feedings, whether a medication can be absorbed without exposure to stomach acid.
- Whether a liquid form of a medication is available, and if so, the appropriate dosage of the liquid form.
- If only tablets are available, whether the tablets can be crushed and mixed with water. Enteric-coated and sustained-release medications should not be crushed due to the potential for adverse effects.

In general, it is best to give medications by mouth instead of by tube whenever possible. In some cases, the injectable form of a medication may be the best option. For medications that must be given by feeding tube:

- Do not mix medications with enteral formulas. Do not mix medications together.
- Before administering medications, ensure that the feeding tube is placed correctly, that it is not clogged, and that the gastric residual is not excessive.
- Position the patient in a semi-upright position (30 degrees or higher) to prevent aspiration.
- Flush the feeding tube with 15 to 30 milliliters of warm (body temperature) water before and after administering a medication. When more than one medication is administered, flush the tube with 5 milliliters of water after each medication is given.
- Use liquid forms of medications whenever possible. Dilute very viscous or hypertonic liquid medications with 10 to 30 milliliters of water before administering them through the feeding tube.
- If tablets are used, crush tablets to a fine powder and mix with 10 milliliters of warm water before administering.

TABLE 15-2	Causes and Prevention or Correction of Tube Feeding Complications	

Complications	Possible Causes	Preventive/Corrective Measures
Aspiration of formula	Compromised lower esophageal sphincter, delayed gastric emptying	Use nasoenteric, gastrostomy, or jejunostomy feedings in high-risk patients; check tube placement; elevate head of bed during and for 45 minutes after feeding; check gastric residuals.
Clogged feeding tube	Formula too thick for tube	Select appropriate tube size; flush tubing with water before and after giving formula; use infusion pump to deliver thick formulas. Remedies that may help to unclog feeding tubes include flushes with warm water or solutions that contain pancreatic enzymes or bicarbonate; consult pharmacist for more options.
	Medications delivered through feeding tube	Use oral, liquid, or injectable medications whenever possible; dilute thick or sticky liquid medications with water before administering; crush tablets to a fine powder and mix with water (except enteric-coated or sustained-release medications); flush tubing with water before and after medications are given; give medications individually; do not add medications to the feeding container.
Constipation	Low-fiber formula	Provide additional fluids; use high-fiber formula.
	Lack of exercise	Encourage walking and other activities, if appropriate.
Dehydration and electrolyte imbalance	Excessive diarrhea	See items under *Diarrhea*.
	Inadequate fluid intake	Provide additional fluid.
	Carbohydrate intolerance	Use continuous drip administration of formula; monitor blood glucose; select a formula with a lower amount or different type of carbohydrate; provide a formula with a higher fat content.
	Excessive protein intake	Monitor blood electrolyte levels; reduce protein intake.
Diarrhea, cramps, abdominal distention	Bacterial contamination	Use fresh formula every 24 hours; store opened or mixed formula in a refrigerator; rinse feeding bag and tubing before adding fresh formula; change feeding apparatus every 24 hours; prepare formula with clean hands using clean equipment in a clean environment.
	Lactose intolerance	Use lactose-free formula in patients with current or potential lactose intolerance.
	Hypertonic formula	Use small volume of formula and increase volume gradually.
	Rapid formula administration	Slow administration rate or use continuous drip feedings.
	Malnutrition/low serum albumin	Use small volume of dilute formula and increase volume and concentration gradually.
Hyperglycemia	Diabetes, hypermetabolism, drug therapy	Check blood glucose; slow administration rate; provide adequate fluids; select a formula with a lower amount or different type of carbohydrate; provide a formula with a higher fat content.
Nausea and vomiting	Obstruction	Discontinue tube feeding.
	Delaying gastric emptying	Check gastric residual; slow administration rate, use continuous drip feedings, or discontinue tube feeding.
	Intolerance to concentration or volume of formula	Use small volume of formula and increase volume and concentration gradually; use continuous drip feedings.
	Psychological reaction to tube feeding	Address patient's concerns.
Skin irritation at enterostomy site	Leakage of GI secretions and friction caused by the tube	Keep site clean; inspect area for redness, tenderness, and drainage; use protective skin cream.

Note: Many of the complications presented here can be caused by the patient's primary disorder or drug therapy rather than the tube feeding itself. In such a case, the corrective measure would include treatment of the disorder or a change in drug therapy. Additionally, other corrective measures that require a physician's order are not shown here.

Sharyn Eschler is a 24-year-old graphics designer who suffered multiple fractures when she fell from a cliff while hiking. She has been in the hospital for seven days and has no appetite. Sharyn has lost 8 pounds over the course of her hospitalization. Due to the nature of her injuries, she is in traction and is immobile, although the head of her bed can be elevated 45 degrees. From the diet history, it appears that Sharyn's nutrition status was adequate prior to hospitalization. The health care team agrees that a nasoduodenal tube feeding should be instituted before her nutritional status deteriorates further. The standard formula selected for the feeding is lactose-free, and Sharyn's nutrient requirements can be met with 2200 milliliters of the formula per day.

1. What steps can be taken to prepare Sharyn for tube feeding? Why might nasoduodenal placement of the feeding tube be preferred to nasogastric placement?
2. The physician's orders specify that the feeding should be given continuously over 18 hours. Develop an appropriate tube feeding schedule.
3. After three days of feeding, Sharyn develops diarrhea. Check Table 15-2 to determine the possible causes. What measures can be taken to correct the diarrhea?
4. Describe precautions that should be taken if Sharyn is to receive medications through the feeding tube.

monitor patients' weights, hydration status, and results of laboratory tests to help them detect problems before complications develop. The Case Study above can help you consider the many factors involved in tube feedings.

REVIEW NOTES

Enteral formulas can supplement conventional diets or be used for tube feedings. They differ in their macronutrient composition, energy density, osmolality, and fiber content.

A nasoenteric feeding route is preferred for short-term tube feedings, whereas enterostomies are used for longer-term feedings. Gastric feedings are preferred, although they are often avoided in patients at risk of aspiration.

Depending on the feeding route and the patient's medical condition, the formula can be delivered in bolus feedings, intermittently, or continuously.

Medications should be given separately from formula and accompanied by water flushes to prevent tube clogging.

Parenteral Nutrition Support

The first half of this chapter described how enteral formulas can supplement or replace conventional foods. Because enteral formulas cannot be used if intestinal function is inadequate, the ability to meet nutrient needs by vein is a lifesaving option for critically ill persons. The procedure is costly, however, and is associated with potentially dangerous complications. As previous sections suggested, enteral nutrition support is preferred if the GI tract is functional.

INDICATIONS FOR PARENTERAL SUPPORT

Parenteral nutrition is indicated for patients who do not have functioning GI tracts and who are either malnourished or likely to become so (review Figure 15-1). The following conditions may require use of parenteral nutrition:

• Short-bowel syndrome (part of the small intestine has been removed).
• Severe pancreatitis.
• Malabsorption disorders.
• Intestinal obstructions or fistulas.
• Severe burns or trauma.

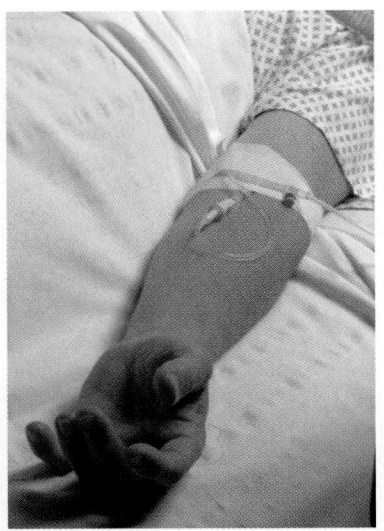

The peripheral veins can provide access to the blood for the delivery of parenteral solutions.

Hypertonic solutions can cause phlebitis (inflammation within the vein), resulting in redness, swelling, and tenderness at the infusion site.

Note that the osmotic property of a solution can be expressed either as *osmolarity* or as *osmolality* (introduced on p. 400).

- *Osmolarity* refers to the number of solutes per liter of solution.
- *Osmolality* refers to the number of solutes per kilogram of solvent.

Due to the measurement techniques used in clinical laboratories, the term *osmolality* is often preferred when referring to biological solutions like blood or urine.

A 10% amino acid solution supplies 10 g of amino acids per 100 mL of solution.

peripheral veins: small-diameter veins that carry blood from the arms and legs.

central veins: large-diameter veins located close to the heart.

peripheral parenteral nutrition (PPN): a type of nutrition support in which intravenous feedings are delivered into peripheral veins.

total parenteral nutrition (TPN): a type of nutrition support in which intravenous feedings are delivered into a central vein.

catheter: a thin tube placed within a narrow lumen (such as a blood vessel) or body cavity; can be used to infuse or withdraw fluids or to keep a passage open.

- Critical illnesses or wasting disorders.
- Bone marrow transplants.
- Being malnourished and having a high risk of aspiration.

VENOUS ACCESS

Once the decision to use parenteral nutrition has been made, the access site must be selected. The access sites for intravenous feedings fall into two main categories: nutrients may be delivered into the **peripheral veins** located in the arms and legs or into the large-diameter **central veins** located near the heart.

Peripheral Parenteral Nutrition In **peripheral parenteral nutrition (PPN)**, nutrient needs are met using only the peripheral veins. Peripheral veins can be damaged by concentrated solutions, so the *osmolarity* of parenteral solutions is usually limited to 900 milliosmoles per liter.[7] Thus PPN can supply only limited amounts of energy and protein. It is used most often in patients who need short-term nutrition support (about 7 to 10 days) and who do not have high nutrient needs or fluid restrictions. PPN is not possible if the peripheral veins are not strong enough to tolerate the procedure. In many cases, it is necessary to rotate venous access sites to prevent inflammation.

Total Parenteral Nutrition (TPN) Most patients meet their nutrient needs using the larger, central veins where blood volume is greater and nutrient concentrations do not need to be limited. Because this method can reliably meet a person's complete nutrient requirements, it is called **total parenteral nutrition (TPN).** Central veins lie close to the heart where the large volume of blood rapidly dilutes parenteral solutions. Therefore, patients with very high nutrient needs or fluid restrictions are able to receive the nutrient-dense solutions they require. TPN is also preferred for patients who require long-term intravenous feedings.

There are several ways to access central veins. The tip of a central venous **catheter** can be placed directly into a large-diameter central vein or threaded into a central vein through a peripheral vein (see Figure 15-4). Although insertion of peripherally inserted central catheters is less invasive and more easily performed than the direct insertion of catheters into central veins, some studies have found higher complication rates associated with peripheral insertion.[8]

PARENTERAL SOLUTIONS

The pharmacies located within health care institutions are often responsible for preparing parenteral solutions. This arrangement is beneficial because the pharmacist can customize formulations to meet patients' nutrient needs and because the solutions have a limited shelf life. Prescriptions for parenteral solutions are highly individualized and may need to be recalculated daily until the patient's condition is stable. Because the nutrients are provided intravenously, they must be given in forms that are safe to inject directly into the bloodstream.

Amino Acids Parenteral solutions contain all of the essential amino acids and different combinations of nonessential amino acids. Amino acid concentrations range from 3.5 to 15 percent; the more concentrated solutions are used only for TPN. Just as in regular foods, the amino acids provide 4 kcalories per gram. Disease-specific amino acid solutions are available for patients with liver failure, kidney failure, and metabolic stress.

Carbohydrate Glucose is usually the main source of energy in parenteral feedings. It is provided in the form dextrose monohydrate, in which each glucose molecule is associated with a single water molecule. Dextrose monohydrate provides 3.4

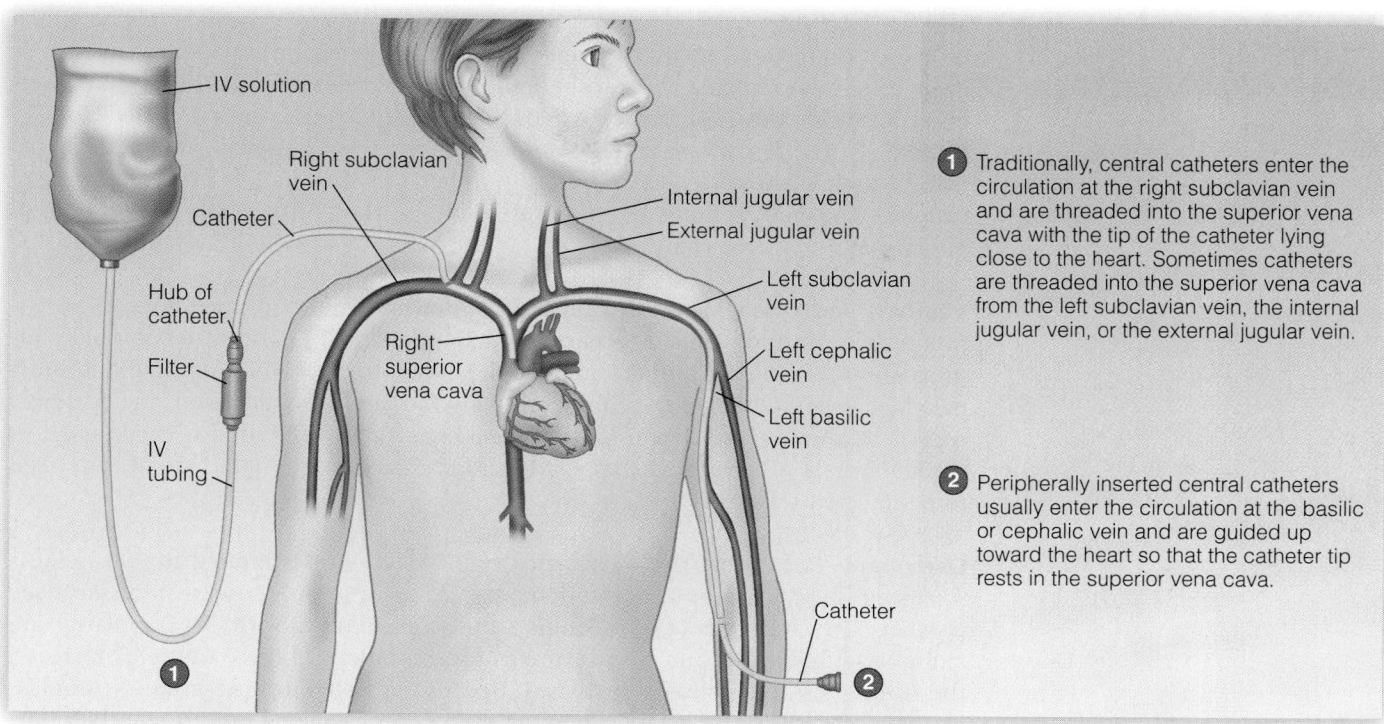

① Traditionally, central catheters enter the circulation at the right subclavian vein and are threaded into the superior vena cava with the tip of the catheter lying close to the heart. Sometimes catheters are threaded into the superior vena cava from the left subclavian vein, the internal jugular vein, or the external jugular vein.

② Peripherally inserted central catheters usually enter the circulation at the basilic or cephalic vein and are guided up toward the heart so that the catheter tip rests in the superior vena cava.

FIGURE 15-4 Accessing Central Veins for Total Parenteral Nutrition

kcalories per gram, slightly less than pure glucose, which provides 4 kcalories per gram. Dextrose solutions are available in concentrations between 2.5 and 70 percent. Concentrations greater than 10 percent are used only in TPN solutions.

In parenteral solutions, the dextrose concentration is indicated by a "D" followed by its concentration in water (W) or normal saline (NS). For example, D5 or D5W indicates that a solution contains 5 percent dextrose in water. Similarly, D5/NS means that a solution contains 5 percent dextrose in normal saline.

Lipids Lipid emulsions supply essential fatty acids and are a significant source of energy. The emulsions usually contain triglycerides from soybean oil and safflower oil, phospholipids to serve as emulsifying agents, and glycerol to make the solutions isotonic. Lipid emulsions are available in 10, 20, and 30 percent solutions, containing 1.1, 2.0, and 3.0 kcalories per milliliter, respectively. Therefore, a 500-milliliter container of 10 percent lipid emulsion would provide 550 kcalories. The same volume of a 20 percent lipid emulsion would provide 1000 kcalories.

Lipid emulsions are often provided daily and may supply 20 to 30 percent of total kcalories. Including lipids as an energy source reduces the need for energy from dextrose and lowers the risk of hyperglycemia in glucose-intolerant patients. Lipid infusions must be restricted in patients with hypertriglyceridemia, however. There is also some concern that lipid emulsions that contain excessive linoleic acid may suppress some aspects of the immune response.

Fluids and Electrolytes Daily fluid needs approximate 30 to 40 milliliters per kilogram of body weight in young adults and 30 milliliters per kilogram of body weight in older adults (averaging between 1500 and 2500 milliliters for most people). The amounts are adjusted according to fluid losses and the results of hydration assessment.

The electrolytes added to parenteral solutions include sodium, potassium, chloride, calcium, magnesium, and phosphorus. The amounts in parenteral solutions differ from DRI values because the nutrients are infused directly into the blood

A 10% dextrose solution provides 10 g of dextrose monohydrate per 100 mL of solution.

- For 500 mL of a 10% lipid emulsion:
  500 mL × 1.1 kcal/mL = 550 kcal.
- For 500 mL of a 20% lipid emulsion:
  500 mL × 2 kcal/mL = 1000 kcal.

The lipid emulsion in a parenteral solution can give it a milky white color.

and are not influenced by absorption as they are when consumed orally. Blood tests are administered daily to monitor electrolyte levels until patients have stabilized. Electrolyte imbalances can be lethal, so electrolyte management by experienced professionals is necessary whenever intravenous therapies are used.

The electrolyte content of parenteral solutions is expressed in milliequivalents (mEq), which are units indicating the number of ionic charges provided by electrolytes. The body's fluids are neutral solutions that contain equal numbers of positive and negative charges.

Vitamins and Trace Minerals Commercial multivitamin and trace mineral preparations are routinely added to parenteral solutions. All of the water-soluble vitamins and vitamins A, D, and E are supplied; vitamin K is often omitted and must be added separately. The trace minerals added to parenteral solutions include zinc, copper, chromium, selenium, and manganese. Iron is excluded because it alters the stability of other ingredients in parenteral mixtures; special forms of iron need to be injected separately.

Osmolarity Recall that the **osmolarity** of PPN solutions is limited to 900 milliosmoles per liter, whereas TPN solutions may be as nutrient dense as necessary (see p. 412). The components of a solution that contribute most to its osmolarity are amino acids, dextrose, and electrolytes: as concentrations of these nutrients increase, the osmolarity of a solution increases. Because lipids contribute little to osmolarity, lipid emulsions are used to increase the energy provided in PPN solutions.

Types of Parenteral Solutions When a parenteral solution contains dextrose, amino acids, and lipids, it is called a **total nutrient admixture (TNA)**, a **3-in-1**, or an **all-in-one** solution. A **2-in-1 solution** excludes lipids, and a lipid emulsion is administered separately, often by using a second port in the catheter. The administration of TNA solutions is simpler because only one infusion pump is required. Lipids are usually administered separately when they are not a major energy source and are used only to provide essential fatty acids. The "How to" describes a method for calculating the macronutrient and energy content of a parenteral solution when the macronutrients are expressed as percentages.

ADMINISTERING PARENTERAL NUTRITION

Providing parenteral nutrition is complex and requires skills from a variety of disciplines. Many hospitals organize nutrition support teams, consisting of physicians, nurses, dietitians, and pharmacists, that specialize in the provision of both intravenous and tube feedings. The nurse, who performs direct patient care, plays a central role in administering and monitoring parenteral feedings.

Insertion and Care of Intravenous Catheters Although a skilled nurse can place catheters into peripheral veins, the insertion of catheters directly into central veins must be performed by a qualified physician. Patients may be awake for the procedure and given local anesthesia. Unnecessary apprehension can be avoided by explaining the procedure to the patient beforehand.

Catheter-related problems frequently cause complications (see Table 15-3). Catheters may be improperly positioned or may dislodge after placement. Air can leak into catheters, obstructing blood flow. Catheters in peripheral veins may cause phlebitis, necessitating reinsertion at an alternate site. A catheter may become clogged from blood clotting or from a buildup of scar tissue around the catheter tip. Catheters are also a leading cause of infection: contamination may be introduced during insertion or may develop at the placement site.

To reduce the risk of complications, nurses use aseptic techniques when inserting catheters, changing tubing, or changing a dressing that covers the catheter site.

Reminder: A *nutrition support team* is a multidisciplinary team of health care professionals who are responsible for the provision of nutrients by tube feeding or intravenous infusion.

NURSING DIAGNOSIS

risk for infection applies to people receiving parenteral feedings.

osmolarity: the concentration of osmotically active particles in a solution, expressed as milliosmoles per liter (mOsm/L). *Osmolality* is an alternative expression of a solution's osmotic properties that is used in clinical practice and uses the units milliosmoles per kilogram (mOsm/kg).

total nutrient admixture (TNA): a parenteral solution that contains dextrose, amino acids, and lipids; also called a **3-in-1** or an **all-in-one** solution.

2-in-1 solution: a parenteral solution that contains dextrose and amino acids but excludes lipids.

HOW TO Calculate the Macronutrient and Energy Content of Parenteral Solutions

Suppose a person is receiving 1800 milliliters of a parenteral solution that contains 16 percent dextrose, 5 percent amino acids, and 2.5 percent lipids. How many grams of protein, carbohydrate, and fat is the person receiving, and what is the energy content of the solution?

Step 1: Determine the macronutrient content of the solution.

$$\textbf{16\% dextrose} = \frac{16 \text{ g dextrose}}{100 \text{ mL}}.$$

$$\frac{16 \text{ g dextrose}}{100 \text{ mL}} \times 1800 \text{ mL} = 288 \text{ g of dextrose}.$$

$$\textbf{5\% amino acids} = \frac{5 \text{ g amino acids}}{100 \text{ mL}}.$$

$$\frac{5 \text{ g amino acids}}{100 \text{ mL}} \times 1800 \text{ mL} = 90 \text{ g of amino acids}.$$

$$\textbf{2.5\% lipids} = \frac{2.5 \text{ g lipids}}{100 \text{ mL}}.$$

$$\frac{2.5 \text{ g lipids}}{100 \text{ mL}} \times 1800 \text{ mL} = 45 \text{ g of lipids}.$$

Remember that a percentage is a fraction in which the denominator is always 100. Therefore, a 10 percent amino acid solution contains 10 grams of amino acids per 100 grams of solution. Note that 1 gram of fluid is equivalent to 1 milliliter.

Step 2: Determine the energy content of the solution. (Remember that dextrose monohydrate provides 3.4 kcalories per gram.)

$$288 \text{ g dextrose} \times 3.4 \text{ kcal/g} = 979 \text{ kcal}.$$

$$90 \text{ g amino acids} \times 4.0 \text{ kcal/g} = 360 \text{ kcal}.$$

$$45 \text{ g lipid} \times 9.0 \text{ kcal/g}^a = 405 \text{ kcal}.$$

$$\textbf{Total} = 979 \text{ kcal} + 360 \text{ kcal} + 405 \text{ kcal} = 1744 \text{ kcal}.$$

(handwritten) $\frac{50}{100} \times 100\,mL = 500 \text{ dextrose}$

(handwritten) $\frac{7}{100} \times 1200 = 84$

a Intravenous lipid actually provides more than 9.0 kcalories per gram, but this value is an acceptable estimate.

(handwritten) $500 \times 3.4 = 1,700$ $84 \times 4.0 = 336$

Unusual bleeding or a wet dressing suggests a problem with catheter placement. A change in infusion rate may indicate a clogged catheter. Infection may be indicated by redness or swelling around the catheter site or by an unexplained fever. Routine inspections of equipment and frequent monitoring of patients' symptoms help to minimize the problems associated with catheter use.

Administration of Parenteral Solutions Infusion protocols vary among institutions. A common approach is to start infusing the solution at a slow rate and gradually increasing the rate over a two- to three-day period. For example, 40 milliliters per hour can be infused during the first 24 hours of administration (supplying 960 milliliters) and the rate increased by one liter per day until the goal rate is reached. Another approach is to give the full volume of a nutrient-dilute solution on the first day and advance nutrient concentrations as tolerated. Some protocols suggest starting solutions at full strength unless there is a risk that the patient will become hyperglycemic.

Parenteral solutions can be infused continuously over 24 hours (**continuous parenteral nutrition**) or during 10- to 16-hour periods only (**cyclic parenteral nutrition**). Continuous feedings are given to critically ill and malnourished patients who cannot receive adequate nutrition during shorter time periods. Cyclic feedings are often provided at night so that patients can participate in routine activities during the day. This method is especially suited to patients who require long-term parenteral support or

continuous parenteral nutrition: continuous administration of parenteral solutions over a 24-hour period.

cyclic parenteral nutrition: administration of a parenteral solution over a 10- to 16-hour period.

TABLE 15-3	Potential Complications of Parenteral Nutrition
Catheter-Related	**Metabolic**
Air embolism	Abnormalities in liver function
Blood clotting at catheter tip	Electrolyte imbalances
Clogging of catheter	Gallbladder disease
Dislodgment of catheter	Hyperglycemia, hypoglycemia
Improper placement	Hypertriglyceridemia
Infection, sepsis	Metabolic bone disease
Phlebitis	Nutrient deficiencies
Tissue injury	Refeeding syndrome

will be infusing parenteral solutions at home. Patients may begin with continuous feedings and transition to cyclic feedings as their condition improves.

Regular monitoring can help to prevent complications. The parenteral solution and tubing are checked daily for signs of contamination. Routine testing of glucose, lipids, and electrolyte levels helps to determine tolerance to solutions. Frequent reassessment of nutritional status may be necessary until a patient has stabilized. Rapid changes in infusion rate are discouraged in some patients due to a risk of developing hyperglycemia or hypoglycemia.[9]

Discontinuing Intravenous Feedings Patients must have adequate GI function before parenteral feedings can be tapered off and enteral feedings begun. During the transition to oral feedings, a combination of feeding methods is often used. By tapering off parenteral feedings at the same time that tube feedings or oral feedings are begun, the two feeding methods together supply the needed nutrients. As described in Chapter 14, clear liquids are often the first foods offered. Small enteral feedings are given initially to determine tolerance. If gastrointestinal symptoms (such as nausea, vomiting, bloating, or diarrhea) develop, oral feedings are limited in size or frequency until the intestines adapt. Once two-thirds to three-fourths of nutrient needs can be provided enterally, the intravenous feedings may be discontinued.

Transitioning to an oral diet is sometimes difficult because a person's appetite remains suppressed for several weeks after parenteral nutrition is terminated. Patients receiving continuous parenteral feedings may have better appetites during the day if they are switched to nocturnal cyclic feedings before beginning oral intakes.

MANAGING METABOLIC COMPLICATIONS

As discussed in a previous section, the catheters used for parenteral nutrition may cause a number of serious complications. This section describes some metabolic complications that may result from parenteral feedings (review Table 15-3).[10]

For most patients receiving parenteral nutrition, blood glucose levels should not exceed 200 mg/dL.

Hyperglycemia Hyperglycemia most often occurs in patients who are glucose intolerant or undergoing severe metabolic stress. It can be prevented by providing insulin along with feedings or by restricting the amount of dextrose in a solution. Dextrose infusions are generally limited to less than 5 milligrams per kilogram of body weight per minute in critically ill adult patients so that the carbohydrate intake does not exceed the maximum glucose oxidation rate.

Hypoglycemia Although uncommon, hypoglycemia sometimes occurs when feedings are interrupted or discontinued. In patients at risk, such as young infants, feedings may be tapered off over several hours before discontinuation.

Hypertriglyceridemia Hypertriglyceridemia may develop in critically ill patients who cannot tolerate the amount of lipid emulsion supplied. It may also result from excessive carbohydrate feedings or severe infection, which impair lipid clearance. If blood triglyceride levels exceed 350 to 400 milligrams per deciliter, lipid infusions should be reduced or stopped.[11]

Refeeding Syndrome Severely malnourished patients who are fed aggressively (parenterally or otherwise) may develop **refeeding syndrome,** characterized by electrolyte and fluid imbalances and hyperglycemia. These effects occur because dextrose infusions raise circulating insulin levels, which promote anabolic processes that quickly remove phosphate, potassium, and magnesium from the blood. The altered electrolyte levels can lead to fluid retention and life-threatening changes in organ systems. To prevent refeeding syndrome, nurses usually start parenteral feedings slowly and carefully monitor electrolyte levels when malnourished patients begin receiving nutrition support.

refeeding syndrome: a condition that sometimes develops when a severely malnourished person is aggressively fed; characterized by electrolyte and fluid imbalances and hyperglycemia.

CASE STUDY   *Geologist Requiring Parenteral Nutrition*

Jerry Huang, a 27-year-old geologist with an inflammatory intestinal disease, underwent a surgical procedure in which a substantial portion of his small intestine was removed. He had received TPN prior to surgery and continued to receive it afterwards. After ten days, tube feeding was begun, which initially delivered very small feedings.

1. List some reasons why the nutrition support team initially chose TPN to provide nutrition support to this patient. How would you explain the need for parenteral feedings to Jerry?
2. Describe the components of a typical TPN solution. Calculate the energy content of 1 liter of a solution that provides 140 grams of dextrose monohydrate, 45 grams of amino acids, and 20 grams of lipids. If Jerry's energy requirement is 2100 kcalories per day, how many liters of solution will he need each day?
3. Why is it important that Jerry begin enteral feedings as soon as possible? Assuming that Jerry eventually tolerates a tube feeding, in what ways can the health care team help Jerry make the transition from parenteral feedings to tube feeding? Consider some of the physiological problems Jerry might face when he begins eating an oral diet.
4. If Jerry is unable to meet nutrient needs orally, he may need to continue tube feeding or TPN at home. As you read through the following sections, consider the factors that would make Jerry a good candidate for a home nutrition support program. Consider both the benefits of a proposed program and the problems he could encounter.

Abnormalities in Liver Function Fatty liver often results from parenteral feedings, but it is usually corrected when the parenteral feedings are discontinued. Long-term parenteral nutrition, however, may result in chronic, irreversible liver disease that may eventually lead to liver failure. The cause of the liver abnormalities is unclear.

Gallbladder Disease When parenteral nutrition continues for more than four weeks, sludge (thickened bile) often builds up in the gallbladder and may eventually lead to gallstone formation. Patients requiring long-term parenteral nutrition may be given cholecystokinin injections (to cause gallbladder contraction and bile release) or have their gallbladders removed surgically.

Metabolic Bone Disease Long-term parenteral nutrition has been associated with lower bone density, which may be related to alterations in calcium, phosphorus, and vitamin D metabolism. Another hypothesis is that parenteral feedings alter the function of the parathyroid gland in some way, which then leads to disrupted bone metabolism.[12] There is no satisfactory remedy for this complication at present.

REVIEW NOTES

Peripheral parenteral nutrition is provided to patients who need short-term nutrition support and who do not have high nutrient needs or fluid restrictions. Total parenteral nutrition can supply nutrient-dense solutions and provide long-term intravenous feedings.

Parenteral solutions include amino acids, dextrose, electrolytes, vitamins, and minerals. Lipid emulsions may be included in the mixture or may be administered separately.

Critically ill patients may require continuous feedings, whereas healthier patients and long-term users may prefer cyclic feedings.

Catheters are frequently the cause of complications, which include improper placement or dislodgment, infection, clotting, embolism, and phlebitis.

Metabolic complications include hyperglycemia and hypoglycemia, hypertriglyceridemia, fluid and electrolyte imbalances, and diseases affecting the liver, gallbladder, and bone. The Case Study can check your understanding of the concepts introduced in this section.

Portable pumps and convenient carrying cases allow people who require home nutrition support to move about freely.

Nutrition Support at Home

Occasionally, a patient must continue to receive nutrition support (tube feedings or parenteral nutrition) after a medical condition has stabilized. For such a person, home nutrition support might be an option.

The use of home nutrition support is rapidly expanding. Current technology allows for the safe administration of nutrition support in home settings, and insurance coverage often pays a substantial portion of the costs. Home health services and home infusion pharmacies can provide the equipment, enteral formulas or parenteral solutions, and services necessary for home nutrition care. Most importantly, patients using these services can continue to receive specialized nutrition care while leading normal lives.

CANDIDATES FOR HOME NUTRITION SUPPORT

Individuals referred for home nutrition support usually need long-term nutrition care for chronic medical conditions. Users of home nutrition services must be intellectually capable of learning the necessary procedures, monitoring the treatment, and managing complications as necessary. The home should be clean and have adequate storage for formulas or solutions and equipment. The costs should be clearly explained to families who cannot get insurance reimbursement. Good candidates for home nutrition support include the following:

- People who have functioning GI tracts and illnesses that prevent food from reaching the digestive tract may benefit from home enteral nutrition. Examples include patients with head and neck cancers and patients with neurological impairments that affect swallowing.
- People who have illnesses that severely impair nutrient absorption or cause motility problems in the stomach or intestines may benefit from home parenteral nutrition. Examples include individuals who have had large portions of their small intestine removed and those with intestinal obstructions or malabsorption conditions.

PLANNING HOME NUTRITION CARE

As with nutrition support provided in health care facilities, planning for home care involves decisions about access sites, formulas, and nutrient delivery methods. Users of home services should be involved in the decision making to ensure long-term compliance and satisfaction.

Home Enteral Nutrition Access to the GI tract is possible using either nasal tubes or enterostomies. People sometimes learn to place nasogastric tubes themselves, which may improve acceptance of the therapy. Active children and adults often prefer low-profile gastrostomy tubes, which allow them to lead a more normal lifestyle. Jejunostomy tubes may be required for some individuals but are less convenient because the frequent feedings required for jejunostomies can interfere with daytime activities.

The choice of formula for home use is inflenced by its cost and availability. Insurance reimbursements do not always include the cost of formula, which is considered to be a "food" product. For this reason, some people choose to prepare simple formulas at home. Blenderizing home-cooked foods is possible, but the foods need to be strained to remove particles and clumps that may obstruct the tube. Closed (ready-to-hang) feeding systems are useful for avoiding contamination risk but are not appropriate for intermittent feedings that require smaller amounts of formula.

The advantages and disadvantages associated with different feeding techniques and administration schedules should be fully discussed with patients. For gastric feedings, bolus infusions are simplest and can be quickly delivered. Gravity drip infusions eliminate the need for an infusion pump, but the delivery rates are less reliable. If intermittent feeding schedules are appropriate, they should be tailored to daily routines. Portable pumps can free individuals from the need to infuse formula at home and can also be used when traveling.

Home Parenteral Nutrition   Although both peripheral parenteral nutrition and total parenteral nutrition (TPN) can be provided at home, long-term therapy requires access to the larger, central veins that are appropriate for TPN. The catheter can be inserted so that the exit site is in an area that is accessible to the patient.

Parenteral solutions need to be sterile and aseptically prepared, and people who mix their own solutions must be carefully trained. Ready-made parenteral solutions require refrigeration and are stable for limited periods; for example, TNA (total nutrient admixture) solutions may be stable for only a week when refrigerated.

Most people prefer cyclic infusions to continuous infusions and transition to cyclic infusions before discharge from the hospital. Because an infusion pump is required for home TPN, sufficient battery backup should be on hand in case electrical service is interrupted. Portable pumps are helpful for individuals who lead an active lifestyle or prefer to infuse during the day.

QUALITY OF LIFE ISSUES

Although home nutrition programs can help to improve health and extend life, consumers of home services and their families may struggle with the lifestyle adjustments required.[13] In addition to the economic impact of nutrition support, home feedings are often time-consuming and inconvenient. Activities and work schedules must be planned around feedings. Extra planning is needed and precautions must be taken when a person wants to travel or participate in sports activities. Explaining one's medical needs to friends and acquaintances may be embarrassing.

Among physical difficulties, people receiving nocturnal feedings often cite disturbed sleep as a major problem. Disruptions may be due to multiple nighttime bathroom visits, noisy infusion pumps, or difficulty finding a comfortable sleeping position when "hooked up." People using parenteral support sometimes prefer infusing solutions during the day to improve their sleeping patterns.

Among social issues, the inability to consume meals with family and friends is often a great concern. Some individuals may be able to eat small amounts of food or may decide to sit at the table to participate in the conversation. Joining friends at restaurants and attending certain types of social events, however, can be problematic.

People who depend on nutrition support face a number of stressful issues that can affect quality of life. Although parenteral and enteral nutrition are life-sustaining therapies, both are associated with serious complications. Many people find that their lifestyles need to be greatly altered to accommodate nutrition therapy and may experience fear, anxiety, and depression. Support groups or counseling resources may help patients cope with ongoing stresses.

The Oley Foundation is an excellent source of outreach services, emotional support, and current information for people who require home nutrition support (**www.oley.org**).

REVIEW NOTES

Candidates for home enteral nutrition services have functional GI tracts but are unable to consume food orally. Parenteral nutrition candidates have illnesses that impair nutrient absorption or cause motility problems.

Patients and caregivers should participate in decisions about access sites, formulas, and nutrient delivery methods. Formulas and solutions can be prepared in the home.

The use of portable pumps may help individuals lead a normal lifestyle. Nevertheless, lifestyle adjustments to nutrition support may be difficult and stressful.

NUTRITION ASSESSMENT CHECKLIST FOR PEOPLE RECEIVING TUBE FEEDINGS

Medical History

Check the medical record for medical conditions that:

☐ Alter nutrient needs and influence the formula selection

☐ Influence the selection of tube placement sites (gastric versus intestinal) and feeding routes

☐ Suggest the length of time that tube feeding will be needed

Monitor the medical record for complications that may influence the formula selection or delivery technique, including:

☐ Aspiration

☐ Constipation

☐ Fluid and electrolyte imbalances

☐ Diarrhea

☐ Hyperglycemia

☐ Nausea and vomiting

☐ Skin irritation

Medications

Check medications for those that can cause side effects similar to those associated with the tube feeding, such as:

☐ Nausea and vomiting

☐ Diarrhea

☐ Constipation

☐ GI discomfort

For medications delivered through the feeding tube, check:

☐ Form of medication and possible alternates

☐ Viscosity of liquid medications

☐ Potential for diet-drug interactions

Dietary Intake

To assess nutritional adequacy, check to see if:

☐ Formula is appropriate for patient's needs

☐ Supplemental water is provided to meet needs

☐ Formula is administered as prescribed

Anthropometric Data

Measure baseline height and weight, and monitor daily weights. If weight is not appropriate:

☐ Determine whether energy needs have been correctly assessed.

☐ Check to see if formula is being delivered as prescribed.

☐ Check for signs of dehydration or overhydration.

Laboratory Tests

Check serum and urine tests for signs of:

☐ Fluid and electrolyte imbalances

☐ Glucose intolerance

☐ Adequacy of protein intake (serum protein levels)

☐ Improvement or deterioration of medical condition

Physical Signs

Look for physical signs of:

☐ Dehydration or overhydration

☐ Delayed gastric emptying (gastric residual volume)

☐ Malnutrition

NUTRITION ASSESSMENT CHECKLIST FOR PEOPLE RECEIVING PARENTERAL NUTRITION

Medical History

Check the medical record for medical conditions that:

☐ Prevent the use of enteral nutrition

☐ Indicate the appropriate feeding route (peripheral versus central)

☐ Suggest the length of time that parenteral nutrition will be required

Monitor the medical record for complications that may influence the parenteral solution formulation or delivery technique, including:

☐ Acid-base imbalances

☐ Fluid and electrolyte imbalances

☐ Hyperglycemia or hypoglycemia

☐ Hypertriglyceridemia

☐ Nutrient deficiencies

☐ Refeeding syndrome

Medications

For medications added to the parenteral solution, determine the:

☐ Medication's compatibility with the parenteral solution

☐ Length of time that the medication can remain stable in solution

For medications infused separately, determine:

☐ Length of time that the feeding may need to be stopped

☐ Adjustments in solution infusion to compensate for the medication delivery

Dietary Intake

To assess nutritional adequacy, check to see if:

☐ Patient's nutrient needs were correctly determined

☐ Solution is administered as prescribed

☐ Infusion pump is operating correctly

Anthropometric Data

Measure baseline height and weight, and monitor daily weights. If weight is not appropriate:

☐ Determine whether energy needs have been correctly assessed.

☐ Check to see if parenteral solution is being delivered as prescribed.

☐ Check for signs of dehydration or overhydration.

Laboratory Tests

Check serum and urine tests for signs of:

☐ Fluid, electrolyte, and acid-base imbalances

☐ Hyperglycemia or hypoglycemia

☐ Hypertriglyceridemia

☐ Nutrient deficiencies

☐ Adequacy of protein intake (serum protein levels)

☐ Improvement or deterioration of medical condition

Physical Signs

Routinely monitor the following:

☐ Catheter insertion site for signs of infection or inflammation

☐ Blood pressure, temperature, pulse, and respiration for signs of fluid, electrolyte, and acid-base imbalances

Look for physical signs of:

☐ Dehydration or overhydration

☐ Protein-energy malnutrition

☐ Malnutrition

SELF CHECK

1. The terms *osmolality* and *osmolarity* refer to:
   a. energy density.
   b. nutrient density.
   c. fiber content.
   d. concentrations of molecules and ionic particles.

2. An important measure that may prevent bacterial contamination in tube feeding formulas is:
   a. nonstop feeding of formula.
   b. using the same feeding bag and tubing each day.
   c. discarding opened containers of formula within 24 hours.
   d. adding formula to the feeding container before it empties completely.

3. Compared to intermittent feedings, continuous feedings:
   a. require an infusion pump.
   b. allow greater freedom of movement.
   c. are more similar to normal patterns of eating.
   d. are associated with more GI side effects.

4. A patient needs 1800 milliliters of formula a day. If the patient is to receive formula intermittently every 4 hours, how many milliliters of formula will he need at each feeding?
   a. 225
   b. 300
   c. 400
   d. 425

5. The term that describes the volume of formula remaining in the stomach from a previous feeding is:
   a. residue.
   b. osmolar load.
   c. gastric residual.
   d. intermittent feeding.

6. The nurse using the feeding tube to deliver medications recognizes that:
   a. medications given by feeding tube generally do not cause GI complaints.
   b. medications can usually be added directly to the feeding container.
   c. enteral formulas do not interact with medications in the same way that foods do.
   d. thick or sticky liquid medications and crushed tablets can clog feeding tubes.

7. TPN is preferred over PPN for a patient who:
   a. does not have high nutrient requirements.
   b. needs long-term parenteral nutrition support.
   c. has strong peripheral veins and moderate nutrient needs.
   d. needs parenteral feedings as a supplement to tube feedings.

8. For a patient receiving central TPN who also receives intravenous lipid emulsions two or three times a week, the lipid emulsions serve primarily as a source of:
   a. essential fatty acids.
   b. cholesterol.
   c. fat-soluble vitamins.
   d. concentrated energy.

9. Refeeding syndrome causes dangerous fluctuations in:
   a. electrolytes.
   b. liver enzymes.
   c. triglycerides.
   d. ketone bodies.

10. Patients using home parenteral nutrition:
    a. are unable to use TNA solutions.
    b. are usually given continuous rather than cyclic infusions.
    c. require infusion pumps for use at home.
    d. are generally unable to work out of the home or travel.

Answers to these questions can be found in Appendix H.

CLINICAL APPLICATIONS

1. The administration of tube feedings and parenteral nutrition require attention to many technical details, making it easy to focus on the procedure rather than the patient. Imagine that your brother, sister, or a parent requires a transnasal tube feeding. How might this person react to the need for a tube feeding? How would you explain the benefits and possible problems associated with the procedure? Think about the ways you would want the staff nurse to help.

2. A liter of a TPN solution contains 500 milliliters of 50 percent dextrose solution and 500 milliliters of 5 percent amino acid solution. Determine the daily energy and protein intakes of a person who receives 2 liters of such a solution. Calculate the average daily energy intake if the person also receives 500 milliliters of a 20 percent fat emulsion three times a week.

3. Consider what it might be like to be using home parenteral nutrition, with no foods allowed by mouth. What would be the advantages of living at home instead of in a hospital or other residential facility? Can you think of some disadvantages?

   Think about how you might manage daily feedings: consider the time, cost, and commitment required to maintain the therapy. If not allowed to consume foods, what possible difficulties might you encounter? How would you handle holidays and special occasions that center around food?

NUTRITION ON THE NET

For further study of the topics in this chapter, access these websites.

Find updates and quick links to these and other nutrition-related sites at our website: **www.wadsworth.com/nutrition**

To find out more about organizations that promote the appropriate use of enteral and parenteral nutrition, visit the sites of the:

• American Society for Parenteral and Enteral Nutrition: **www.clinnutr.org**

• Canadian Parenteral-Enteral Nutrition Association: **www.cpena.ca/home.html**

• British Association for Parenteral and Enteral Nutrition: **www.bapen.org.uk**

• European Society of Parenteral and Enteral Nutrition: **www.espen.org**

To learn about home nutrition support, visit the website of the Oley Foundation, a national, nonprofit organization that provides information, outreach services, and emotional support for consumers of home enteral and parenteral services: **www.oley.org**

NOTES

[1]C. R. Parrish and S. McCray, Enteral feeding: Dispelling myths, *Practical Gastroenterology* 27 (September 2003): 33–50.

[2]L. Gramlich and coauthors, Does enteral nutrition compared to parenteral nutrition result in better outcomes in critically ill adult patients? A systematic review of the literature, *Nutrition* 20 (2004): 843–848.

[3]D. A. Neumann and M. H. DeLegge, Gastric versus small-bowel tube feeding in the intensive care unit: A prospective comparison of efficacy, *Critical Care Medicine* 30 (2002): 1436–1438.

[4]L. Klein, Is blue dye safe as a method of detection for pulmonary aspiration? *Journal of the American Dietetic Association* 104 (2004): 1651–1652; J. P. Maloney and T. A. Ryan, Detection of aspiration

in enterally fed patients: A requiem for bedside monitors of aspiration, *Journal of Parenteral and Enteral Nutrition* 26 (2002): S34–S42.

[5]Parrish and McCray, 2003.

[6]Parrish and McCray, 2003.

[7]L. Matarese, Composite foods and formulas, parenteral and enteral nutrition, in A. M. Coulston, C. L. Rock, and E. R. Monsen, eds., *Nutrition in the Prevention and Treatment of Disease* (San Diego: Academic Press, 2001), pp. 245–260.

[8]L. J. Walshe and coauthors, Complication rates among cancer patients with peripherally inserted central catheters, *Journal of Clinical Oncology* 20 (2002): 3276–3281; C. T. Cowl and coauthors, Complications and cost associated with parenteral nutrition delivered to hospitalized patients through either subclavian or peripherally inserted central catheters, *Clinical Nutrition* 19 (2000): 237–243.

[9]M. M. McMahon, Parenteral nutrition, in L. Goldman and D. Ausiello, eds., *Cecil Textbook of Medicine* (Philadelphia: Saunders, 2004), pp. 1322–1326; A.S.P.E.N. Board of Directors and The Clinical Guidelines Task Force, Guidelines for the use of parenteral and enteral nutrition in adult and pediatric patients, *Journal of Parenteral and Enteral Nutrition* 26 (2002): 1SA–138SA.

[10]A.S.P.E.N. Board of Directors and The Clinical Guidelines Task Force, 2002.

[11]McMahon, 2004.

[12]W. Goodman and coauthors, Altered diurnal regulation of blood ionized calcium and serum parathyroid hormone concentrations during parenteral nutrition, *American Journal of Clinical Nutrition* 71 (2000): 560–568.

[13]B. Ehrenpreis and A. Hilf, Home parenteral nutrition: The consumer's perspective, in *Lifeline Letter*, **http://oley.org/lifeline/LivingHPN.pdf,** site visited April 17, 2006.

Ethical Issues in Nutrition Care

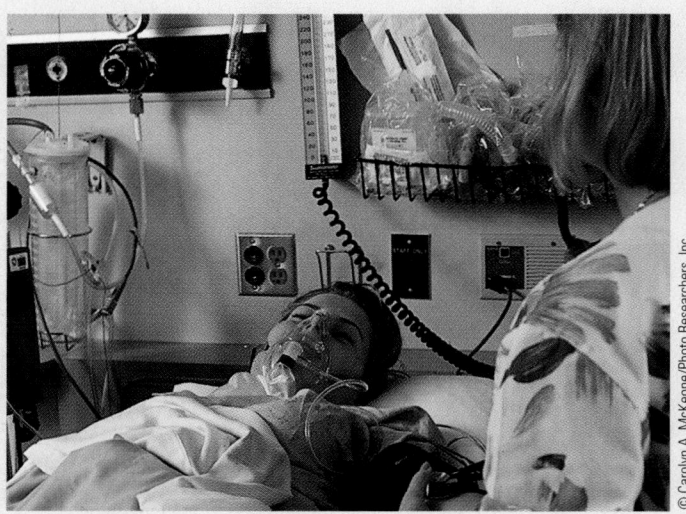

Is it ever morally and legally appropriate to withhold or withdraw nutrition support?

© Carolyn A. McKeone/Photo Researchers, Inc.

As with other medical technologies, the availability of specialized nutrition support forces health care professionals and members of our society to face difficult **ethical** issues. When medical treatments prolong life by merely delaying death, the lifetime that remains may be of extremely low quality. This discussion examines the ethical dilemmas that clinicians must face when dealing with patients in critical care. The accompanying glossary defines the relevant terms.

If providing nutrition care can do little to promote recovery, is it appropriate to withhold or withdraw nutrition support?

In attempting to answer questions such as this one, health care professionals must consider the following ethical principles:[1]

- A patient has the right to make decisions concerning his or her own well-being (**patient autonomy**), even if refusing treatment could result in death. It is generally accepted that a patient's preferences should take precedence over the desires of others.[2]
- A patient should be fully informed of a treatment's benefits and risks in a fair and honest manner (**disclosure**). A patient's acceptance of a treatment that has been adequately disclosed is considered **informed consent**.
- A patient must have the mental capacity to make appropriate health care decisions (**decision-making capacity**). If a patient is mentally incapable of doing so, a person designated by the patient should serve as a **surrogate** decision maker.

- The potential benefits (**beneficence**) of any treatment should outweigh its potential harm (**maleficence**).
- Health care providers must determine whether the provision of health care to one patient would unfairly limit the care of other patients (**distributive justice**).

Although these principles may seem simple and obvious, it is often difficult to determine the appropriate action to take during intensive care.[3] When clinicians and families disagree, the courts may be asked to decide.

What kinds of treatments can help to sustain a patient's life?

Nutrition support and hydration are both considered life-sustaining treatments, as withholding or withdrawing either can result in death. Other life-sustaining treatments include **cardiopulmonary resuscitation (CPR)**, which supplies oxygen and restores a person's ability to breathe and pump blood; **defibrillation**, in which an electronic device shocks the heart and reestablishes normal contractions; **mechanical ventilation**, which substitutes for lung function; and **dialysis**, which substitutes for kidney function.

Do patients have a right to life-sustaining treatments?

Although life-sustaining treatments are readily provided to patients who have a reasonable chance of recovering from illness, it is sometimes difficult to determine the best course of action for patients who are dying or who are unlikely to regain consciousness. Under such circumstances, such treatments may be considered **futile** because they are unable to improve the outcome of disease or increase the patient's comfort and well-being. If patients or caregivers demand treatment that health practitioners have determined to be useless, a legal resolution may be required. Conversely, medical personnel may find it objectionable to withdraw life support knowing that the consequence is the death of a patient.

How have the courts resolved conflicts involving nutrition support?

One of the landmark cases involving nutrition support concerned Nancy Cruzan, who suffered permanent and irreversible brain damage after a car crash in 1983, when she was 26 years of age.[4] After she had been in a **persistent vegetative state** for five years, her parents requested permission to discontinue tube feeding, but hospital staff refused to honor the request and the matter was taken to court. The Missouri Supreme Court determined that Nancy had never definitively stated her "right to die" wishes and that her parents were unable to make such a request for her. The court also stated that preserving life, no matter what its quality, should take precedence over all other considerations. Nancy's parents appealed the ruling, but in 1990 the U.S. Supreme Court upheld the Missouri Supreme Court in a five-to-four decision. Three witnesses were eventually found who could testify that Nancy would not desire life-sustaining treatment under the circumstances, and the court finally granted permission to remove the feeding tube. This case illustrates the importance of having an **advance directive** (discussed in a later section) that clearly indicates one's preferences for medical treatment in the event of incapacitation.

In a more recent case that received widespread media attention, the spouse and parents of a patient in a persistent vegetative state fought a ten-year legal battle over her medical care. In 1990, at the age of 25, Terri Schiavo suffered a full cardiac arrest.[5] She initially fell into a coma, but her condition evolved into a persistent vegetative state that was considered irreversible. Despite the neurologists' diagnosis and a series of CT and MRI scans showing extensive brain atrophy, her parents maintained that she was minimally conscious and would improve with rigorous treatment. Her husband, who was legally responsible for her care, insisted that she would never have wanted to be kept alive in a vegetative state. Like Nancy Cruzan, Terri had never expressed her wishes in an advance directive.

In 1998, Terri's husband filed a petition to have her feeding tube removed, and a Florida court approved the motion in February 2000. Although Terri's parents appealed, an appeals court affirmed the decision, and the Florida Supreme Court declined to review the case. In April 2001, Terri's physicians removed her feeding tube, but within days, a federal circuit court judge ordered it to be reinserted and reopened the case. Eventually, the motions filed by the parents were dismissed, and Terri's feeding tube was removed for the second time in October 2003. Within days, the Florida legislature passed a bill known as "Terri's Law" that gave the governor the authority to intervene, and Governor Jeb Bush ordered her feeding tube restored. A year later, Florida's Supreme Court declared Terri's Law to be unconstitutional. Although the governor appealed the decision, his appeal was rejected in January 2005. Terri's feeding tube was removed for the third time in March 2005. Despite emergency petitions by her parents and an attempt by the U.S. Congress to have her case reconsidered, the courts refused to grant a restraining order, and Terri died 13 days after her feeding tube was removed.

How can people ensure that their wishes will be considered in the event that they become incapacitated?

People can declare their preferences about medical treatments in a **living will**, sometimes called a **medical directive**. Living wills can include detailed instructions about life-sustaining procedures that a person does or does not want. Advance directives are incorporated into the medical record and updated when appropriate. They take effect only if a physician determines that a patient lacks the ability to understand and make decisions about available treatments.

Another important directive is a **durable power of attorney** (sometimes called a **health care proxy**) in which another person (a **health care agent**) is appointed to act as decision maker in the event of incapacitation. The agent should understand one's medical preferences and be absolutely trustworthy. Only one person can be designated, although one or two alternates may also be listed. If an agent is given comprehensive power to supervise care, he or she may make decisions about medical staff, health care facilities, and medical procedures.

Laws regarding advance directives vary from state to state. In some states, nutrition and hydration are not considered life-sustaining treatments, and a person's instructions about them may need to be indicated separately. Some states restrict the use of advance directives to terminal illness or disallow them if a woman is pregnant. Generally, advance directives created in one state are honored in another. If no advance directive is available and a person's preferences are unknown, decisions are based on a patient's best interests as determined by an immediate family member.[6]

How does a "do-not-resuscitate" order differ from other advance directives?

A **do-not-resuscitate (DNR) order** is frequently used to withhold CPR in the event of cardiopulmonary arrest, which occurs too suddenly for deliberate decision making.[7] A DNR order is written in the medical record as are other directives, but it does not exclude the use of other life-prolonging measures. A DNR order is most often used in patients with serious illnesses or advanced age. Some institutions allow a physician to write a DNR order for a patient who has a poor prognosis, but the physician must inform the patient or surrogate if this is done.

Have advance directives changed the way that medical care is provided?

Not really. Only about 20 percent of people in the United States have completed an advance directive.[8] Furthermore, advance directives are often unavailable when intensive care decisions are made: one study found that only 57.5 percent of patient charts indicating the existence of an advance directive actually contained a copy.[9] In addition, physicians often make treatment decisions without first discussing them with patients or caregivers.[10] In many cases, life-sustaining treatments are begun without prior knowledge of patients or their decision makers, or treatments continue even if patients want them stopped.

Advance directives are sometimes too general or vague to guide treatment decisions. Patients who are fully aware of treatment options and clearly state their preferences are more likely to be successful at obtaining the care they desire.

Glossary

advance directive: a written or oral instruction regarding one's preferences for medical treatment to be used in the event of becoming incapacitated.

beneficence (be-NEF-eh-sens): the act of performing beneficial services rather than harmful ones.

cardiopulmonary resuscitation (CPR): life-sustaining treatment that supplies oxygen and restores a person's ability to breathe and pump blood.

decision-making capacity: the ability to understand pertinent information and make appropriate decisions; known as **decision-making competency** within the legal system.

defibrillation: life-sustaining treatment in which an electronic device is used to shock the heart and reestablish a pattern of normal contractions. Defibrillation is used when a heart has arrhythmias or has experienced cardiac arrest.

dialysis: life-sustaining treatment in which a patient's blood is filtered using selective diffusion through a semipermeable membrane; substitutes for kidney function.

disclosure: the act of revealing pertinent information. For example, clinicians should accurately describe proposed tests and procedures, their benefits and risks, and alternative approaches.

distributive justice: the equitable distribution of resources.

do-not-resuscitate (DNR) order: a request by a patient or surrogate to withhold cardiopulmonary resuscitation.

durable power of attorney: a legal document (sometimes called a **health care proxy**) that gives legal authority to another (a *health care agent*) to make medical decisions in the event of incapacitation.

ethical: in accordance with accepted principles of right and wrong.

futile: medical care that will not improve the medical circumstances of a patient.

health care agent: a person given legal authority to make medical decisions for another in the event of incapacitation.

informed consent: a patient's or caregiver's agreement to undergo a treatment that has been adequately disclosed. Persons must be mentally competent in order to make the decision.

living will: a written statement that specifies the medical procedures desired or not desired in the event that a person is unable to communicate or is incapacitated; also called a **medical directive**.

maleficence (mah-LEF-eh-sens): the act of doing evil or harm.

mechanical ventilation: life-sustaining treatment in which a mechanical ventilator is used to substitute for a patient's failing lungs.

patient autonomy: a principle of self-determination, such that patients (or surrogate decision makers) are free to choose the medical interventions that are acceptable to them, even if they choose to refuse interventions that may extend their lives.

persistent vegetative state (PVS): a vegetative mental state resulting from brain injury that persists for at least one month. Individuals lose awareness and the ability to think but retain noncognitive brain functions, such as motor reflexes and normal sleep patterns.

surrogate: a substitute; a person who takes the place of another.

Notes

[1] M. A. Grippi, Ethics in critical care, in A. P. Fishman and coeditors, *Fishman's Manual of Pulmonary Diseases and Disorders* (New York: McGraw-Hill, 2002), pp. 1111–1114.

[2] E. J. Emanuel, Bioethics in the practice of medicine, in L. Goldman and D. Ausiello, eds., *Cecil Textbook of Medicine* (Philadelphia: Saunders, 2004), pp. 5–9.

[3] Emanuel, 2004.

[4] J. O. Maillet, R. L. Potter, and L. Heller, Position of the American Dietetic Association: Ethical and legal issues in nutrition, hydration, and feeding, *Journal of the American Dietetic Association* 102 (2002): 716–726.

[5] R. Cranford, Facts, lies, and videotapes: The permanent vegetative state and the sad case of Terri Schiavo, *Journal of Law, Medicine, and Ethics* 33 (2005): 363–372.

[6] American College of Physicians, Ethics manual, *Annals of Internal Medicine* 128 (1998): 576–594.

[7] American College of Physicians, 1998.

[8] Emanuel, 2004.

[9] Institute of Medicine, *Approaching Death: Improving Care at the End of Life* (Washington, D.C.: National Academy Press, 1997), pp. 202–203.

[10] Emanuel, 2004.

Nutrition in Metabolic and Respiratory Stress

© Stewart Cohen/Getty Images

CHAPTER 16

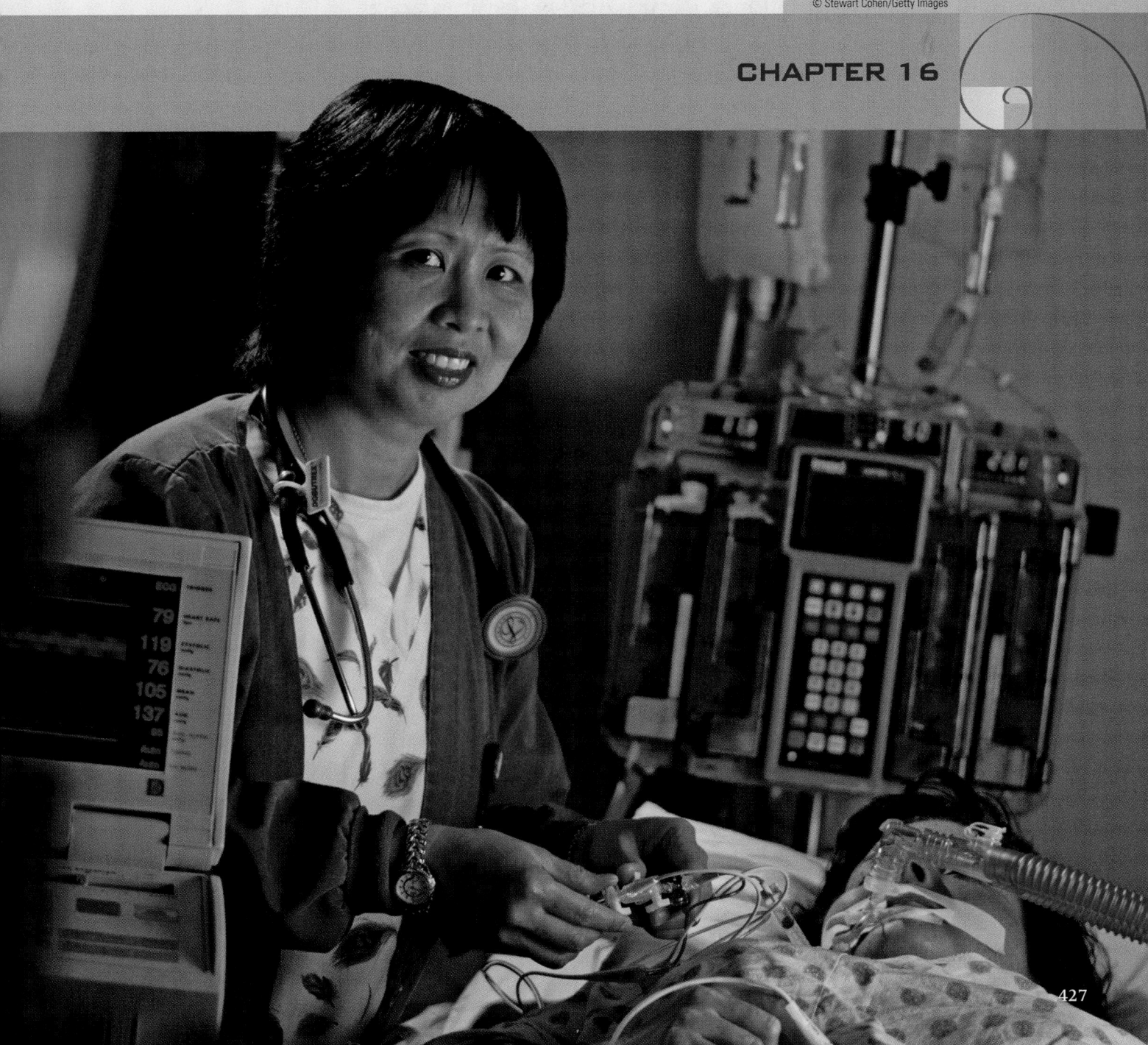

The body's response to severe stress can alter metabolism enough to threaten survival. Many patients require life support measures and intensive monitoring. Stress also raises nutritional needs considerably, increasing the risk of malnutrition even in previously healthy individuals. **Metabolic stress,** a disruption in the body's internal chemical environment, can result from uncontrolled infections or extensive tissue damage, such as deep, penetrating wounds or multiple broken bones. As the first part of this chapter explains, the body's stress response is an attempt to restore balance, but it can have both helpful and harmful effects. The final part of this chapter describes **respiratory stress,** which is characterized by inadequate oxygen and excessive carbon dioxide in the blood and tissues. Both metabolic and respiratory stress can lead to **hypermetabolism** (above-normal metabolic rate), **wasting** (breakdown of muscle mass and loss of strength), and, in severe circumstances, life-threatening complications.

The Body's Responses to Stress and Injury

The **stress response** is the body's *nonspecific* response to a variety of stressors, such as infection, burns, fractures, surgery, and extensive bleeding. During stress, the body's actions focus on immediate survival, and functions of lesser consequence are delayed. Energy is of primary importance, and therefore the energy nutrients are mobilized from storage and made available in the blood. Heart rate and respiration (breathing rate) increase to deliver oxygen and nutrients to cells more quickly, and blood pressure rises. Meanwhile, energy is diverted from processes that are not life sustaining, such as growth, reproduction, and long-term immunity. If stress continues for a long period, interference with these processes begins to cause damage, which can result in growth retardation and illness.

HORMONAL RESPONSES TO STRESS

The stress response is mediated by several hormones, which are released into the blood soon after the onset of injury (see Table 16-1). The catecholamines (epinephrine and norepinephrine), often called the "fight-or-flight" hormones, stimulate heart muscle, alter the rate of blood flow, and raise metabolic rate. Epinephrine also prompts the secretion of glucagon by the pancreas, causing the release of nutrients from storage. The steroid hormone cortisol enhances protein degradation, thereby raising amino acid levels in the blood and making them available for conversion to glucose. All of these hormones result in similar effects on glucose and fat metabolism, causing the breakdown of glycogen, the production of glucose from amino acids, and the breakdown of triglycerides in adipose tissue. Two other hormones induced by stress, aldosterone and antidiuretic hormone, help to maintain blood volume by stimulating the kidneys to reabsorb more sodium and water, respectively.

Cortisol's effects can be detrimental when stress is prolonged. In excess, cortisol causes the depletion of protein in muscle, bone, connective tissue, and skin. It impairs wound healing, so high cortisol levels are especially dangerous for a patient with severe injuries. Because cortisol inhibits protein synthesis, consuming more protein cannot easily reverse tissue losses. Excess cortisol also disrupts calcium metabolism and causes insulin resistance and abnormal fat deposition. In addition, cortisol suppresses immune responses, increasing susceptibility to infection. Note that the pharmaceutical forms of cortisol are common anti-inflammatory medications (such as *cortisone* and *prednisone*); their long-term use can cause undesirable side effects such as muscle wasting, thinning of the skin, diabetes, and early osteoporosis.

NURSING DIAGNOSIS

imbalanced nutrition: less than body requirements often applies to a person undergoing severe stress.

NURSING DIAGNOSIS

impaired tissue integrity and *risk for infection* may apply to a person who undergoes prolonged stress.

The catecholamines, glucagon, and cortisol have actions that oppose those of insulin and are therefore sometimes referred to as *counterregulatory hormones.*

metabolic stress: a disruption in the body's chemical environment due to the effects of disease or injury. Metabolic stress is characterized by changes in metabolic rate, heart rate, blood pressure, hormonal status, and nutrient metabolism.

respiratory stress: inadequate gas exchange between the air and blood, resulting in lower-than-normal oxygen levels and higher-than-normal carbon dioxide levels.

hypermetabolism: a higher-than-normal metabolic rate.

wasting: the breakdown of lean tissue that results from disease or malnutrition.

stress response: the chemical and physical changes that occur within the body during stress.

TABLE 16-1	Metabolic Effects of Hormones Released during the Stress Response
Hormone	**Metabolic Effects**
Catecholamines	• Increase in metabolic rate • Glycogen breakdown in liver and muscle • Glucose production from amino acids • Release of fatty acids from adipose tissue • Glucagon secretion from pancreas
Glucagon	• Glycogen breakdown in liver • Glucose production from amino acids • Release of fatty acids from adipose tissue
Cortisol	• Protein degradation • Enhancement of glucagon's action on liver glycogen • Glucose production from amino acids • Release of fatty acids from adipose tissue
Aldosterone	• Retention of sodium
Antidiuretic hormone	• Retention of water

THE INFLAMMATORY RESPONSE

Cells of the immune system mount a quick, nonspecific response to infection or tissue injury. This so-called **inflammatory response** serves to contain and destroy infectious agents (and their products) and prevent further tissue damage. As in the stress response, there is a delicate balance between a response that protects tissues from further injury and an excessive response that can cause additional damage to tissue.

The Inflammatory Process The inflammatory response begins with the dilation of blood vessels that deliver blood to the site of an injury (arterioles) and the constriction of small blood vessels that carry blood away from an infected area (venules). The capillaries within the damaged tissue become more permeable, allowing the inflowing fluid to accumulate in tissue and cause localized edema. These changes in blood vessels prevent the spread of infection and encourage the entry of immune cells that can destroy foreign agents (see Figure 16-1, p. 430). Among the first cells to arrive are the **phagocytes**, which slip through gaps between the endothelial cells that form the vessel walls. The phagocytes engulf microorganisms and destroy them with hydrolytic enzymes and reactive forms of oxygen. When inflammation becomes chronic, these normally useful products of phagocytes can damage healthy tissue.

Mediators of Inflammation Numerous chemical substances control the inflammatory process. These *mediators* are released from damaged tissue, blood vessel cells, and activated immune cells. Many of them help to regulate more than one step in the process. Histamine, a small molecule similar to an amino acid in structure, is released from granules within **mast cells,** causing vasodilation and capillary permeability. Other proteins that participate in the inflammatory process include **cytokines,** produced by white blood cells, and **eicosanoids,** which are derived from dietary fatty acids. Note that most anti-inflammatory medications, including both steroidal drugs (such as cortisone and prednisone) and nonsteroidal anti-inflammatory drugs (such as aspirin and ibuprofen), act by blocking eicosanoid synthesis.

The classic signs of inflammation that accompany altered blood flow are:

- **Swelling**—from the accumulation of fluid at the site of injury.
- **Redness**—from the dilation of small blood vessels in the injured area.
- **Heat**—from the influx of warm arterial blood.
- **Pain**—from the pressure of edema within damaged tissue and the actions of certain chemical mediators on pain receptors.

Antihistamines are medications taken to reduce the effects of histamine.

inflammatory response: the metabolic responses of the immune system to infection or injury.

phagocytes (FAG-oh-sites): white blood cells (neutrophils and macrophages) that have the ability to engulf and destroy antigens.
   phagein = to eat

mast cells: cells within connective tissue (close to blood vessels) that produce and release histamine.

cytokines (SIGH-toe-kynes): proteins produced by white blood cells that regulate immune cell development and immune responses.

eicosanoids (eye-KO-sah-noyds): 20-carbon molecules derived from dietary fatty acids that help to regulate blood pressure, blood clotting, and other body functions.
   eicosa = twenty

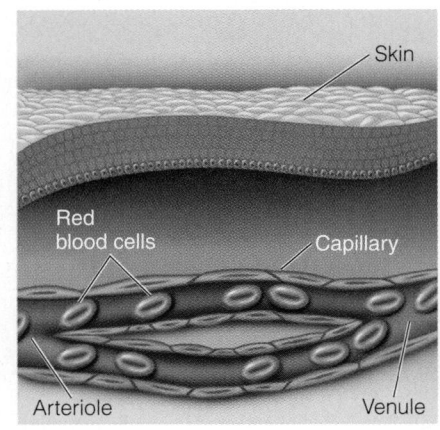
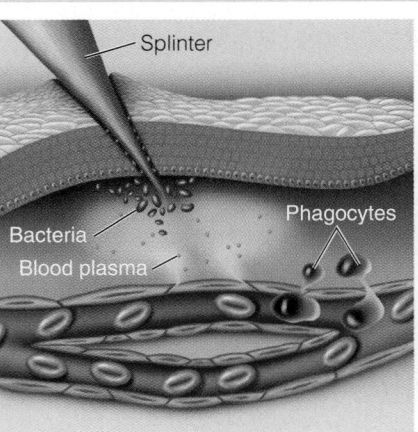
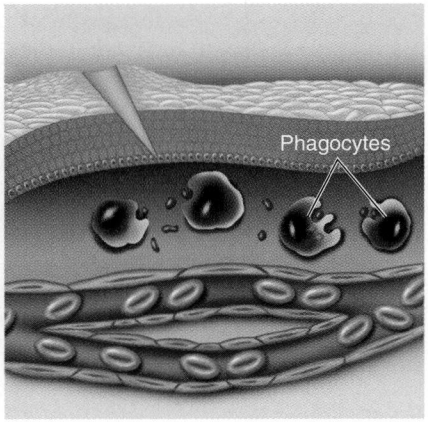

Skin

Red blood cells

Capillary

Arteriole

Venule

Splinter

Bacteria

Blood plasma

Phagocytes

Phagocytes

Cells lining the blood vessels lie close together, and normally do not allow the contents to cross into tissue.

When tissues are damaged, immune cells release histamine, which dilates some blood vessels, increasing blood flow to the damaged area. Fluid leaks out of capillaries (causing swelling), and phagocytes escape between the small gaps in the blood vessel walls.

Phagocytes engulf bacteria and disable them with hydrolytic enzymes and reactive forms of oxygen.

FIGURE 16-1 The Inflammatory Process

C-reactive protein, the best clinical indicator of the acute-phase response, also becomes elevated during many chronic diseases.

systemic (sih-STEM-ic): relating to the entire body.

acute-phase response: changes in body chemistry resulting from infection, inflammation, or injury; characterized by alterations in plasma proteins.

C-reactive protein: an acute-phase protein released from the liver during acute inflammation or stress.

complement: a group of plasma proteins that assist the activities of antibodies.

Changing dietary fat sources can have subtle effects on the inflammatory process.[1] The major precursor for the eicosanoids is arachidonic acid, which is a product of the omega-6 fatty acids in vegetable oils. Some omega-3 fatty acids compete with arachidonic acid and inhibit the production of the most powerful inflammatory mediators. Partially replacing vegetable oils rich in omega-6 fatty acids with sources high in omega-3 fatty acids (such as fish oil) helps to suppress inflammation, but it is not a reliable treatment.

Systemic Effects of Inflammation The inflammatory process leads to **systemic** effects as well as the localized effects described earlier. Within hours (or, in some cases, days) after inflammation, infection, or severe injury, the liver steps up its production of certain proteins in an effort known as the **acute-phase response.**[2] These acute-phase proteins include **C-reactive protein, complement,** blood-clotting proteins such as fibrinogen and prothrombin, and others. At the same time, plasma concentrations of albumin, iron, and zinc fall (recall from Chapter 13 that albumin levels are often measured to assess health and nutrition status). The acute-phase response is accompanied by muscle catabolism to make amino acids available for glucose production, tissue repair, and immune protein synthesis; consequently, negative nitrogen balance (and wasting) frequently results. Other clinical features include an elevated metabolic rate, increased neutrophil numbers, lethargy, anorexia, and often, fever.

Severe inflammation that persists for more than a few days can lead to a life-threatening condition: the **systemic inflammatory response syndrome (SIRS).** SIRS is a whole-body response to unresolved inflammation and is diagnosed when a patient's condition becomes severe enough to raise heart rate, respiratory rate, white blood cell levels, and/or body temperature to critical levels. If identical symptoms result from infection, the condition is called **sepsis.** Complications associated with SIRS or sepsis include excessive fluid retention and tissue edema, low blood pressure, and impaired blood flow. The inability to supply tissues with sufficient blood may even-

Although severe metabolic stress can have damaging consequences, the healthy body handles minor stresses quickly and efficiently.

tually cause symptoms of **shock,** a syndrome that threatens the functioning of multiple organs. Shock and multiple organ failure are discussed further in Nutrition in Practice 16.

REVIEW NOTES

The stress response and inflammatory response are nonspecific responses to stressors that cause infection and injury.

The stress response is mediated by the catecholamine hormones, cortisol, and glucagon, which together raise nutrient levels in blood, stimulate heart rate, and constrict blood vessels. Aldosterone and antidiuretic hormone help to maintain adequate blood volume.

The inflammatory process can also cause systemic effects that alter metabolism, heart rate, blood pressure, body temperature, and immune cell functions.

Nutrition Treatment of Acute Stress

As mentioned earlier, an excessive response to metabolic stress can worsen illness and even threaten survival. Therefore, a critical care unit must manage both the acute medical condition that initiated stress and the complications that arise due to the stress and inflammatory responses. Immediate concerns during severe stress are to restore lost fluids and electrolytes and to remove underlying stressors. Thus initial treatments include administering intravenous solutions to correct fluid and electrolyte imbalances, treating infections, repairing wounds, draining **abscesses** (pus), and removing dead tissue **(debridement).** After stabilization, nutrient needs can be estimated and medical nutrition therapy provided.

DETERMINING NUTRITIONAL REQUIREMENTS

The most notable metabolic changes in patients undergoing metabolic stress include hypermetabolism, negative nitrogen balance, hyperglycemia, and insulin resistance.[3] Hypermetabolism and negative nitrogen balance may result in wasting, which can worsen organ function and delay recovery. Hyperglycemia increases the risk of infection, a dangerous problem during critical illness. Therefore, the principal nutritional goals are to provide a diet that preserves lean tissue content, maintains immune defenses, and promotes healing.

Feeding an acutely stressed patient is challenging. Overfeeding increases the risks of refeeding syndrome and its associated hyperglycemia. Underfeeding worsens nitrogen balance and may increase lean tissue losses. Assessing nutritional needs can be complicated because fluid imbalances prevent accurate measurements of weight, and laboratory data reflect the metabolic alterations of illness rather than the person's nutritional status.

The amounts of protein and energy to provide during acute illness are controversial and still under investigation. Research results have been mixed, in part because experimentation is difficult in patients who are critically ill. Also, the wide assortment of acute medical conditions that cause metabolic stress makes each patient's situation somewhat unique. The guidelines presented here are subject to change as new findings help to resolve the complex issues related to nutrient intakes and delivery methods. In all cases, clinicians need to closely observe patients' responses to feedings and readjust nutrient intakes as necessary.

Estimation of Energy Needs A common method for determining the energy needs of acutely stressed individuals is to estimate basal energy expenditure (BEE) using the **Harris-Benedict equation** and then multiply the result by a stress factor to account for the increased energy requirements of stress and healing. This method

Reminder: *Refeeding syndrome* may develop when a severely malnourished person is aggressively fed; it is associated with fluid and electrolyte imbalances and hyperglycemia.

systemic inflammatory response syndrome (SIRS): a whole-body response to acute inflammation; characterized by raised heart and respiratory rates, abnormal white blood cell counts, and altered body temperature.

sepsis: an acute inflammatory response caused by infection; characterized by symptoms similar to those of SIRS.

shock: a dangerous physiological response to injury, bleeding, or infection resulting from an insufficient blood supply; associated with reduced blood pressure, raised heart and respiratory rates, and muscle weakness.

abscesses (AB-sess-es): accumulated pus that is surrounded by inflamed tissue.

debridement: the surgical removal of dead, damaged, or contaminated tissue resulting from burns or wounds; helps to prevent infection and hasten healing.

Harris-Benedict equation: an equation used to estimate basal energy expenditure.

TABLE 16-2	Estimating Energy Needs for Acute Metabolic Stress

Step 1. Estimate energy needs to support basal energy expenditure (BEE) using the Harris-Benedict equation.[a]

Women: BEE = 655.1 + [9.563 × weight (kg)] + [1.85 × height (cm)] − [4.676 × age (years)].
Men: BEE = 66.5 + [13.75 × weight (kg)] + [5.003 × height (cm)] − [6.775 × age (years)].

Step 2. Multiply BEE by an appropriate stress factor for acute illness. For example:[b]

Postoperative (no complications): 1.00 to 1.05
Peritonitis: 1.05 to 1.25
Cancer: 1.10 to 1.45
Long bone fracture: 1.25 to 1.30
Severe infection: 1.3 to 1.55
Multiple trauma: 1.3 to 1.55
Burns (over 40% body surface): 2.0

Adjustment for obesity:

For persons with BMI > 30, use an adjusted body weight in the BEE equation. One suggestion is to use an adjusted weight based on ideal body weight (IBW):[c]

Adjusted weight = IBW + [0.5 × (actual body weight − IBW)].

[a]L.J. Hoffer, Protein and energy provision in critical illness, *American Journal of Clinical Nutrition* 78 (2003): 906–911.

[b]W. W. Souba and D. Wilmore, Diet and nutrition in the care of the patient with surgery, trauma, and sepsis, *Modern Nutrition in Health and Disease* (Baltimore: Williams & Wilkins, 1999), p. 1593.

[c]N. Barak and coauthors, Evaluation of stress factors and body weight adjustments currently used to estimate energy expenditure in hospitalized patients, *Journal of Parenteral and Enteral Nutrition* 26 (2002): 231–238.

is described in Table 16-2, and the "How to" presents an example. Although a direct measurement of basal metabolism using indirect calorimetry is much preferred, many facilities do not have the proper equipment, personnel, or time to make routine measurements in critically ill patients. Stress factors vary according to the severity of the illness and the patient's nutritional status, and the particular values that are used vary among institutions; those listed in Table 16-2 are given as examples. The energy expenditure of critical care patients may be raised further due to fever, mechanical ventilation, restlessness, or the presence of open wounds. Note that patients who are critically ill are usually bedridden and inactive, so the energy needed for physical activity is minimal.

Another popular method of estimating energy needs is to multiply a person's body weight by a factor appropriate for the medical condition. For example, energy needs for patients with sepsis have been estimated to be 25 to 30 kcalories per kilogram body weight daily;[4] a patient weighing 160 pounds (72.7 kilograms) may therefore require between 1818 and 2181 kcalories per day. The energy intake can be started within this range and then adjusted as the patient's body weight and other determinants of nutrition status change.

Reminder: The protein RDA for adults is 0.8 g per kilogram body weight.

Estimation of Protein Needs To help preserve lean tissue, the protein intakes recommended during acute stress are higher than DRI values. Even with adequate protein, however, negative nitrogen balance cannot be prevented because metabolic processes during stress encourage protein catabolism. Protein recommendations vary according to the severity of a condition and are usually estimated to be between 1.0 and 2.0 grams per kilogram body weight for most conditions. The "How to" includes a sample calculation of protein requirements.

The amino acids glutamine and arginine are sometimes added to the diets of acutely stressed and immune-compromised patients. Several studies have suggested that glutamine supplementation is associated with fewer infections, shorter

HOW TO   Estimate Energy and Protein Needs during Acute Stress

Grace is a 39-year-old female who is 5 feet 3 inches tall and weighs 130 pounds. She recently broke a leg and an arm while snowboarding. Her energy needs can be estimated using the Harris-Benedict equation and a stress factor as follows:

Weight in kilograms = 130 lb ÷ 2.2 lb/kg = 59 kg.

Height in centimeters = 63 in × 2.54 cm/in = 160 cm.

BEE = 655.1 + [9.563 × weight (kg)] + [1.85 × height (cm)] − [4.676 × age (years)]

= 655.1 + (9.563 × 59) + (1.85 × 160) − (4.676 × 39) = 1333 kcal.

Stress factor (see Table 16-2): 1.25.

BEE × stress factor = 1333 × 1.25 = 1666 kcal.

Grace's energy needs are about 1666 kcalories. Her weight can be monitored to determine if her actual needs are higher or lower. In addition, her energy needs are likely to change as stress resolves.

Grace's protein requirements can be estimated using an appropriate protein factor between 1.0 and 2.0 grams per kilogram body weight (g/kg). Assuming that Grace could obtain adequate protein using the factor 1.0 g/kg:

Protein intake: 1.0 g/kg × 59 kg = 59 g protein.

Using the protein factor 1.5 g/kg instead:

Protein intake: 1.5 g/kg × 59 kg = 88.5 g protein.

To calculate the amount of fat and carbohydrate to provide, clinicians usually subtract the protein kcalories from the total energy needs and then supply the remaining kcalories from carbohydrates and fat. Using the second protein prescription in the example, Grace would be given 1312 kcalories from carbohydrate and fat:

Carbohydrate and fat kcalories = 1666 kcal − (88.5 g protein × 4 kcal/g).

1666 kcal (total) − 354 kcal (protein) = 1312 kcal (carbohydrate and fat).

hospital stays, and reduced mortality rates in critically ill patients.[5] Arginine supplementation has been shown to have beneficial effects on the immune responses of postoperative patients.[6] Although glutamine and arginine are often added to enteral formulas promoted for wound healing and enhanced immunity, their use remains controversial.

Carbohydrates and Lipids The bulk of energy needs are supplied from carbohydrate and fat. Carbohydrate is usually the main source of energy and may provide up to 70 percent of kcalories depending on a patient's condition. When parenteral feedings are necessary, dextrose is usually provided to critically ill patients at no more than 5 milligrams per kilogram body weight per minute (see p. 416). Fat provides both energy and essential fatty acids and may supply up to 40 percent of kcalories.

Micronutrients Acutely stressed patients may have increased micronutrient needs, but specific requirements remain unknown. In hypermetabolic patients, the need for B vitamins may be higher to support the increase in energy metabolism. A number of micronutrients, such as zinc, vitamin C, and vitamin A, have critical roles in immunity and wound healing, and their supplementation may speed recovery in certain circumstances. Patients with burns and tissue injuries may have increased requirements for trace minerals due to tissue losses; in several studies, supplementation of zinc, copper, and selenium reduced infection rates in burn patients during recovery.[7]

Plasma levels of micronutrients are often altered during critical illness. The acute-phase response causes a redistribution of some micronutrients (such as zinc and iron) that lowers their blood levels; therefore, micronutrient status is sometimes difficult to interpret. Blood concentrations of trace minerals should be monitored in patients receiving parenteral nutrition support to ensure that excessive amounts are not given intravenously.

CASE STUDY   *Mortgage Broker with a Severe Burn*

David Bray, a 42-year-old mortgage broker, has been admitted to intensive care. He suffered a severe burn covering over 40 percent of his body when he was trapped inside a burning building. His wife told the nurse that Mr. Bray's height is 6 feet and that he weighs about 175 pounds. The physician ordered lab work, including serum protein concentrations, but the results have not yet been received.

1. Identify Mr. Bray's immediate needs after the injury. Describe the initial concerns of the health care team and the measures they might take soon after Mr. Bray's arrival at the hospital.

2. Considering Mr. Bray's condition, what problems might the health care team encounter when they attempt to obtain information that can help them assess his nutritional status?

3. Calculate Mr. Bray's energy and protein needs (use a protein factor of 2 grams/kilogram). Discuss other nutrients that may be of concern during recovery.

4. Due to complications that developed during tube feeding, Mr. Bray was able to obtain only 65 percent of his energy requirements. What other feeding options can be considered?

APPROACHES TO NUTRITION CARE

As mentioned previously, the initial care following acute stress focuses on maintaining fluid and electrolyte balances. Simple intravenous solutions often contain dextrose, providing minimal kcalories. Once feedings begin, patients may require a combination of methods to meet their nutritional needs. If the medical problem, a medical procedure (such as mechanical ventilation), or poor appetite interferes with food intake, nutrition support may be required.

As Chapter 15 explained, enteral nutrition support is preferred over parenteral nutrition in patients with normal intestinal function. Parenteral nutrition support may be needed, however, if patients cannot achieve adequate nutrient intakes from enteral feedings. In one study of critically ill patients, only 53 percent could meet nutrient needs from enteral feedings alone, and energy intakes averaged 77 percent of the amounts prescribed.[8] Consequently, parenteral nutrition is sometimes used to supplement enteral feedings; for patients who are likely to become malnourished during critical illness, it may be the main source of nutrients.

Once patients transition to oral feedings, meeting protein and energy needs may be difficult, and enteral formulas are often given to supplement the diet. Many such formulas are of high nutrient density, and some contain extra amounts of nutrients believed to promote healing, such as vitamin A, zinc, and the amino acids arginine and glutamine.[9] A high-kcalorie, high-protein diet is often prescribed, although care must be taken not to overfeed patients who are at risk of developing refeeding syndrome or hyperglycemia. Nutrient needs should be reassessed frequently until the patient's condition stabilizes. The accompanying Case Study reviews the nutrition care of a patient undergoing acute metabolic stress.

REVIEW NOTES

Severe metabolic stress causes hypermetabolism and negative nitrogen balance and may result in wasting.

The nutrition objectives during stress are to provide a diet that can preserve muscle tissue, maintain immune defenses, and promote healing.

Energy intake should sustain nitrogen balance but should not result in overfeeding. Protein recommendations are increased to help prevent tissue losses and allow healing of damaged tissue.

Enteral and parenteral feedings are sometimes needed to meet the high nutrient requirements of acutely stressed patients.

Nutrition and Respiratory Stress

Some medical conditions upset the gas exchange process between the air and blood and lead to respiratory stress, which is characterized by a reduction in the blood's oxygen supply and an increase in carbon dioxide levels. Excessive carbon dioxide in the blood disrupts the breathing pattern and can interfere with food intake in some individuals. Moreover, the labored breathing caused by respiratory disorders entails a higher energy cost than normal breathing does, raising energy needs and increasing carbon dioxide production further. Lung diseases can make physical activity difficult and may eventually result in muscle wasting. Weight loss and malnutrition can therefore become dangerous outcomes of some types of respiratory illnesses.

CHRONIC OBSTRUCTIVE PULMONARY DISEASE

Chronic obstructive pulmonary disease (COPD) refers to a group of conditions characterized by persistent obstruction of airflow through the lungs. The two major types of COPD are **chronic bronchitis** and **emphysema,** and many patients display features of both conditions. Figure 16-2 illustrates the main airways (**bronchi** and **bronchioles**) and air sacs (**alveoli**) of the normal respiratory system, and Figure 16-3 (p. 436) shows how they are altered in COPD. In chronic bronchitis, excessive secretions of mucus clog the lungs' airways, which ultimately thicken and become too narrow for adequate mucus clearance. A chronic, productive cough is the main symptom. Emphysema is characterized by the breakdown of the lungs' elastic structure and the destruction of the walls of the bronchioles and alveoli, changes that significantly reduce the surface area involved in respiration. Both chronic bronchitis and emphysema reduce the capacity of the lungs to maintain

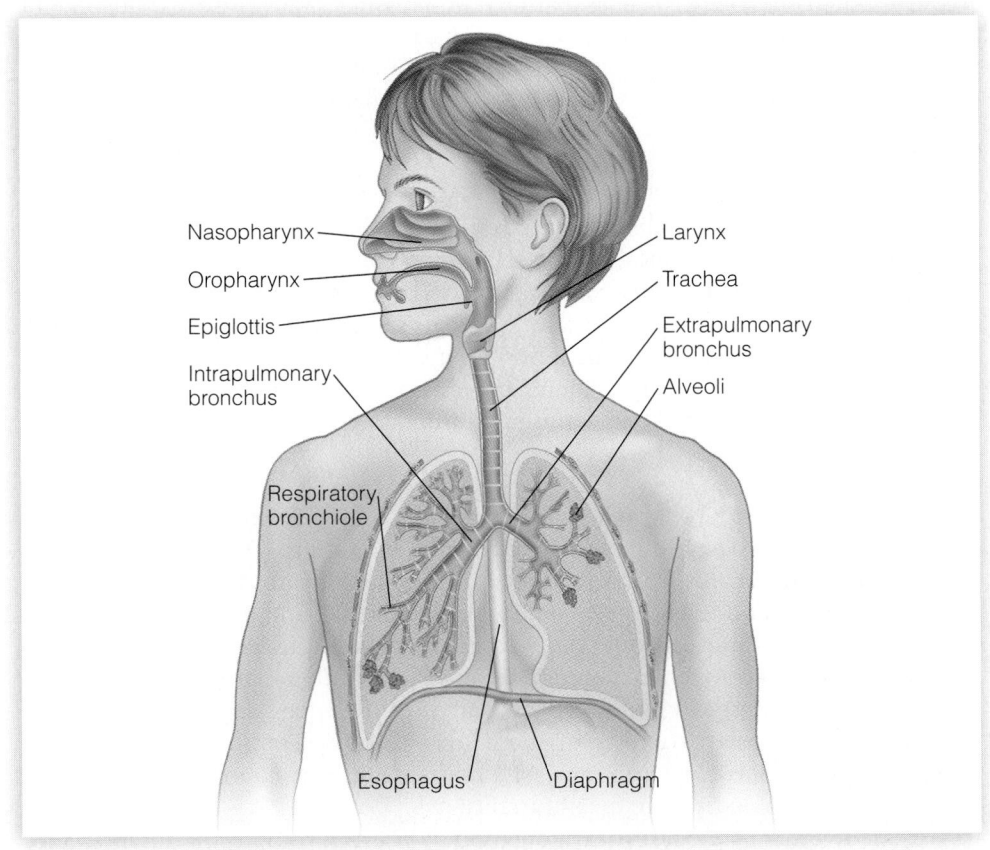

Based on a drawing in Carol Mattson Porth, *Pathophysiology,* 5th ed. (Lippincott Williams and Wilkins, 1998).

FIGURE 16-2 The Respiratory System
Inhaled air travels via the trachea to the bronchi and bronchioles, the major airways of the lungs. Oxygen and carbon dioxide are exchanged across the thin-walled alveoli, which are surrounded by capillaries.

chronic obstructive pulmonary disease (COPD): a group of lung diseases characterized by persistent obstructed airflow through the lungs and airways; include chronic bronchitis and emphysema.

chronic bronchitis (bron-KYE-tis): persistent inflammation of the mucous membranes lining the main airways of the lungs. Chronic inflammation leads to narrower airways and difficulty with breathing.

emphysema (EM-fih-ZEE-mah): disease characterized by progressive damage to the alveoli (air sacs) in the lungs; causes difficulty with breathing.

bronchi, bronchioles: the main airways of the lungs. The singular form of bronchi is *bronchus.*

alveoli (al-VEE-oh-lie): air sacs in the lungs. One sac is an *alveolus.*

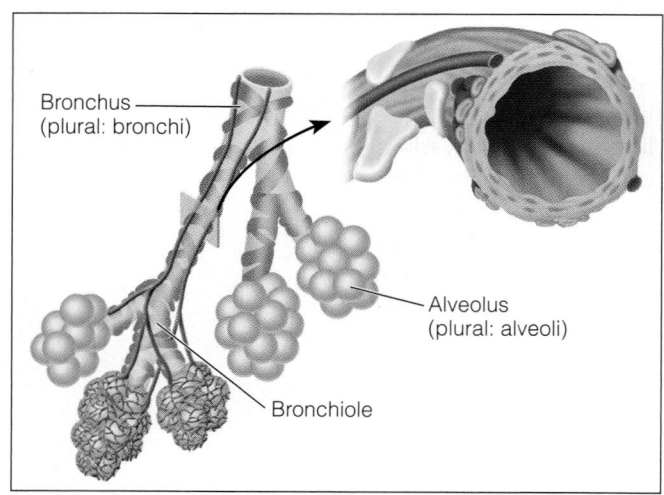

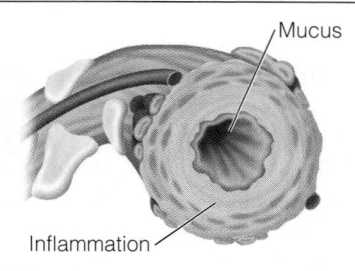

Chronic bronchitis causes inflammation, excessive secretion of mucus, and narrowing of bronchi, factors that reduce normal airflow.

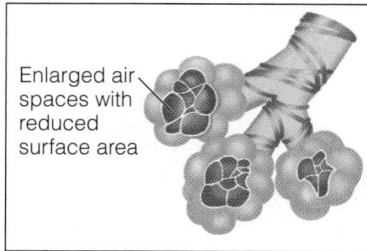

Emphysema causes gradual destruction of the walls separating alveoli and reduces lung elasticity.

Healthy bronchi provide an open passageway for air. Healthy alveoli permit gas exhange between the air and blood.

Panels 1 and 2 adapted from "Medical Encyclopedia: Bronchitis and Normal Condition in Tertiary Bronchus," Medline Plus. Copyright 2005, A.D.A.M., Inc. **http://www.nlm.nih.gov/medlineplus/ency/imagepages/19357.htm.** Panel 3 adapted from "Causes and Risk Factors for Emphysema," Emphysema-Symptoms.com. Copyright 2005 Emphysema-Symptoms.com. All rights reserved. **http://www.emphysema-symptoms.com/html/emphysemacauses.php3.**

FIGURE 16-3 Chronic Obstructive Pulmonary Disease

NURSING DIAGNOSIS

that generally apply to people with COPD include *impaired gas exchange, ineffective breathing pattern, ineffective airway clearance, activity intolerance, adult failure to thrive,* and *fatigue.*

Note that corticosteroids promote catabolic processes and can exacerbate the muscle loss that often accompanies COPD.

dyspnea (DISP-nee-ah): shortness of breath.

Patients who need supplemental oxygen can use lightweight, portable equipment that allows them to move about freely.

normal oxygen and carbon dioxide levels in the blood; shortness of breath (**dyspnea**) results and may eventually cause respiratory or heart failure. COPD ranks as the fourth leading cause of death in the United States.

COPD is a debilitating condition. Generally, dyspnea worsens as the condition progresses, resulting in dramatic reductions in physical activity and quality of life. Activities of daily living such as bathing or dressing may cause exhaustion or breathlessness. Muscle wasting often occurs and is only partially reversible by exercise. As with other chronic illnesses, anxiety and depression are a concern, and psychological distress may reduce a COPD patient's ability to cope with the demands of treatment.

Causes of COPD Smoking tobacco is the primary risk factor for COPD and is especially damaging when combined with respiratory infections or an occupational exposure to dusts or chemicals. Only a minority of smokers (15 to 20 percent) develop COPD, however; thus genetic susceptibility also plays a role in its development. Genetic factors are especially likely in patients with early-onset COPD. Alpha-1 antitrypsin deficiency, an inherited disorder, occurs in 1 to 2 percent of patients with COPD. These individuals have inadequate blood levels of a plasma protein (alpha-1 antitrypsin) that normally inhibits enzymatic breakdown of the lungs' connective tissue.

Treatment of COPD The primary objectives of COPD treatment are to prevent the disease from progressing and relieve major symptoms (dyspnea and coughing). People with COPD are encouraged to quit smoking to prevent the disease from progressing, and to get vaccinated against influenza and pneumonia to avoid complications. The most frequently prescribed medications are bronchodilators, which improve airflow, and corticosteroids (anti-inflammatory medications), which help to prevent symptom recurrence.[10] (See the Diet-Drug Interactions feature for nutrition-related effects of these medications.) For people with severe COPD, supplemental oxygen therapy (12 hours daily) can maintain normal oxygen levels in the blood and reduce mortality risk.

The main goals of medical nutrition therapy are to promote the maintenance of a healthy body weight and prevent muscle loss. Researchers have found a cor-

DIET-DRUG INTERACTIONS Check this table for notable nutrition-related effects of the medications discussed in this chapter.

	Gastrointestinal Effects	Interactions with Dietary Substance	Metabolic Effects
Bronchodilators (theophylline, dyphylline)	Increased gastric acid secretion, acid reflux.	Caffeine may enhance drug effects.	
Corticosteroids (prednisone)			Glucose intolerance, sodium retention, negative nitrogen balance, appetite stimulation, weight gain, growth suppression in children.

relation between low body weights and increased mortality.[11] Rehabilitation programs often include exercise training to improve exercise tolerance, thus reducing the likelihood that patients will become sedentary and lose additional muscle mass.

Improving Food Intake Food intake often declines in severe cases of COPD, but for different reasons in each person. Dyspnea can sometimes interfere with chewing or swallowing. Appetite may be affected by medications, depression or anxiety, or changes in taste perception. Physical changes in the diaphragm and lungs may reduce abdominal volume, leading to early satiety. Some patients may become too disabled to shop or prepare food or may lack adequate support at home. The dietitian or nurse must assess the unique needs of a COPD patient before proposing a nutrition care plan.

Some patients may benefit by eating frequent, small meals rather than two or three large ones.[12] The lower energy content of small meals reduces the carbon dioxide load,[13] plus abdominal discomfort and dyspnea may be reduced when less food is consumed. If bloating is a problem, foods that increase gas formation should be avoided. Some individuals may eat better if oxygen is provided at mealtimes. Consuming adequate fluids should be encouraged to help prevent the secretion of overly thick mucus; however, some patients should consume liquids between meals so as not to interfere with food intake. For undernourished patients, a high-kcalorie, high-protein diet may be helpful, but excessive energy intakes may increase carbon dioxide output and increase respiratory stress. Liquid supplements are sometimes given between meals to improve weight gain or exercise endurance, but high-energy feedings (more than 250 kcalories) may induce satiety and reduce energy intake at mealtime.[14]

Some patients with COPD may be overweight or obese, which puts an additional strain on the respiratory system. An energy-restricted diet to promote gradual weight loss is encouraged for these individuals.

Specialized Formulas Some enteral formulas available for use in pulmonary disease provide more kcalories from fat and fewer from carbohydrate than standard formulas. The ratio of carbon dioxide production to oxygen consumption in cells is lower when fat is consumed, so theoretically these formulas should lower respiratory requirements. Clinical studies, however, have not confirmed that reduced-carbohydrate formulas decrease respiratory stress in COPD patients, and they currently are not recommended as a component of treatment.[15]

Incorporating an Exercise Program Loss of muscle can be more readily prevented or reversed if the treatment plan includes a carefully designed exercise program.[16] With exercise, patients are likely to see improvements in their endurance and become less fearful of their physical limitations. For some patients, the combination of an exercise plan and oral supplement may be better for maintaining weight

The altered sense of taste in patients with COPD may be due to chronic mouth breathing, which dries the mouth. Taste is also affected by the use of certain medications, including some bronchodilators.

John Norback is an 82-year-old man who has emphysema that severely affects both lungs. He is 5 feet 9 inches tall and currently weighs 150 pounds, about 20 pounds less than his weight in earlier years. He lives with a daughter and son-in-law and eats meals with their family. He becomes breathless when eating and when walking around the house, and he feels tired all the time. A medical clinic recently ordered oxygen therapy for home use, but supplies have not yet arrived. Mr. Norback's daughter is concerned about her father's recent weight loss and breathlessness.

1. Assess Mr. Norback's risk of malnutrition, using information from Table 13-6 in Chapter 13 (p. 365). What factors may have contributed to his weight loss?
2. What are possible reasons for Mr. Norback's difficulty with eating? List some dietary suggestions that may help to improve his appetite and food intake. How might the use of oxygen therapy help?
3. Based on the history given, what factors may account for Mr. Norback's tiredness? What suggestions would you give Mr. Norback and his daughter regarding physical activity?

acute respiratory distress syndrome (ARDS): respiratory failure triggered by acute lung injury; a medical emergency that causes dyspnea and pulmonary edema and usually requires assisted (mechanical) ventilation.

hypoxemia (high-pox-SEE-me-ah): a low level of oxygen in the blood.

hypercapnia (high-per-CAP-nee-ah): excessive carbon dioxide in the blood.

hypoxia (high-POX-see-ah): a low amount of oxygen in body tissues.

acidosis: acid accumulation in body tissues; depresses the central nervous system and can lead to disorientation and, eventually, coma.

cyanosis (sigh-ah-NOH-sis): a bluish cast in skin due to the color of deoxygenated hemoglobin. Cyanosis is most evident in individuals with lighter, thinner skin; it is mostly seen on lips, cheeks, and ears and under nails.

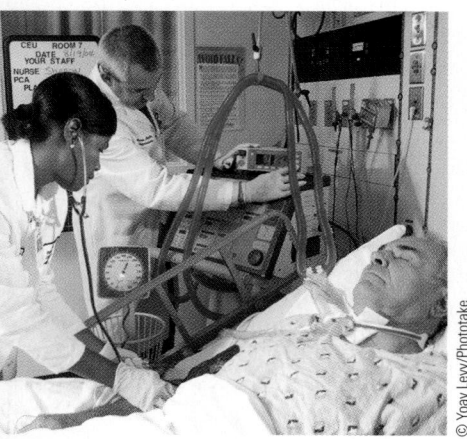

Mechanical ventilation controls the rate and amount of oxygen supplied to a person's airways.

and improving muscle status than either component of treatment alone.[17] The accompanying Case Study allows you to review the nutrition care for a patient with COPD.

RESPIRATORY FAILURE

In respiratory failure, the gas exchange between the air and the circulating blood is greatly impaired. This condition can develop from chronic disease (such as COPD) or may arise suddenly (*acute* respiratory failure). Various factors affecting lung function may be the cause. Respiratory failure may result from obstruction in the upper airways or from weakness or paralysis of the muscles involved in respiration. An embolus lodged within the lungs may prevent blood flow, or toxic substances may damage lung tissue. Surgery sometimes results in respiratory failure due to the depressive effects of anesthesia or because some abdominal procedures can affect breathing.[18] Severe trauma and infection are common triggers of **acute respiratory distress syndrome (ARDS),** an acute form of respiratory failure marked by extensive lung damage. ARDS is a life-threatening condition that usually requires the use of mechanical ventilation to restore normal oxygen and carbon dioxide levels.

Consequences of Respiratory Failure Impaired gas exchange results in **hypoxemia** (low blood levels of oxygen) and **hypercapnia** (excessive carbon dioxide in the blood). An inadequate oxygen supply within tissues **(hypoxia)** inhibits cell function and can ultimately cause cell death. Hypercapnia can lead to **acidosis,** which interferes with the functions of the central nervous system and the heart. To compensate for respiratory failure, a person breathes more rapidly, and heart rate quickens. The skin may become sweaty and develop a bluish cast **(cyanosis).** Headache, confusion, and drowsiness may occur. Severe cases of respiratory failure can cause heart arrhythmias and, ultimately, coma.

Treatment of Respiratory Failure Treatment of respiratory failure focuses on supporting lung function and correcting the underlying disorder. Because respiratory failure can be caused by a number of different conditions, the treatment plan can vary greatly. In individuals with chronic lung disease, providing oxygen therapy via a face mask or nasal tubing can relieve symptoms, but patients with ARDS usually require mechanical ventilation until they are able to breathe independently. Fluids may need careful monitoring to maintain fluid balance and prevent overload; diuretics are sometimes prescribed to mobilize the fluid that has accumulated in lung tissue. Medications may be needed to treat infections, keep airways open, or relieve inflammation. Complications are common in ARDS and must be forestalled to prevent multiple organ dysfunction.

Nutrition Care during Acute Respiratory Failure Dietary recommendations are individualized according to the patient's condition. The primary concern is to supply enough energy and protein to support lung function without overtaxing the compromised respiratory system.[19] Fluid restrictions may be necessary to help reverse pulmonary edema. As usual, when nutrition support is necessary, enteral nutrition is preferred over parenteral nutrition.

- *Energy.* Energy needs can be estimated by using the Harris-Benedict equation to determine BEE and adjusting the result with an appropriate stress factor (review Table 16-2). Body weight may need to be corrected for edema, which is often present in patients with acute respiratory failure. A stress factor of 1.2 may be used initially, although needs may be higher in malnourished patients or if fever or infections are present. Energy intakes greater than 1.5 times BEE are not recommended; excessive energy intakes generate extra carbon dioxide and may increase the risk for complications.
- *Fluids.* Dehydration may develop due to a low fluid intake, an increase in bronchial mucus secretions, or diuretic therapy. Although dehydration may impede the clearance of lung secretions, pulmonary edema is often present, and fluid restriction is required to prevent the accumulation of additional fluids in lung tissue. The edema may make it difficult to assess whether a critically ill patient is maintaining weight.

Nutrition Support Patients with acute respiratory failure are often unable to eat meals and may need nutrition support. Tube feedings are usually used if the intestine is functional, and intestinal feedings are preferred over gastric feedings because they reduce the risk of aspiration. Nutrient-dense formulas (2 kcalories per milliliter) can be used in patients with fluid restrictions. Pulmonary formulas that provide less carbohydrate and more fat are available (under the assumption that reduced carbohydrate will reduce carbon dioxide production), but these have not been shown to improve outcomes. If risk of aspiration is too high to continue enteral feedings, parenteral nutrition support may be considered.

REVIEW NOTES

Chronic obstructive pulmonary disease (COPD) is a debilitating, progressive illness that can lead to malnutrition, muscle wasting, and activity intolerance. Depending on individual needs, the goals of nutrition therapy are to improve food intake, maintain proper weight, preserve muscle mass, and improve exercise endurance.

Acute respiratory failure can develop from COPD or arise in persons with no history of lung disease. It often follows acute lung injury due to infection, trauma, or inhalation of toxic substances. Medical nutrition therapy may include nutrition support and fluid restrictions.

NUTRITION ASSESSMENT CHECKLIST

FOR PEOPLE UNDERGOING METABOLIC OR RESPIRATORY STRESS

Medical History

Check the medical record to determine:

- ☐ Cause of stress
- ☐ Severity of stress
- ☐ Route of feeding (oral, tube feeding, or parenteral)

- ☐ Whether any organ system is compromised

For patients with COPD, check to determine:

- ☐ Degree of breathing difficulty
- ☐ Use of oxygen therapy
- ☐ Activity tolerance

Review the medical record for complications related to under-feeding or overfeeding, such as:

☐ Dehydration or fluid overload

☐ Electrolyte imbalances

☐ Fatty liver

☐ Hyperglycemia

☐ Hypertriglyceridemia

Medications

Record all medications and note:

☐ Side effects that may alter food intake or nutrition status

☐ Use of theophylline, in patients who may need to avoid caffeine

Dietary Intake

If the patient is not meeting nutrition goals:

☐ Monitor intakes to ensure that the patient is receiving the diet prescribed.

☐ Investigate appetite problems or difficulties with eating.

☐ Consider interventions to improve food intake.

☐ Consider need for supplementation.

☐ In patients with COPD, consider problems that may hamper the patient's ability to prepare or consume foods.

Anthropometric Data

Measure baseline height and weight and monitor daily weights. Remember that body weight can fluctuate in acutely ill patients who undergo fluid resuscitation. Once a patient's weight has stabilized:

☐ Reevaluate protein and energy needs.

☐ Consider the need to alter the energy prescription to meet weight goals.

Laboratory Tests

Laboratory tests that may be affected by stress and therefore require careful interpretation include:

☐ Albumin

☐ Transferrin

☐ Prealbumin

☐ C-reactive protein

☐ Serum iron and zinc

☐ Total lymphocyte count (white blood cell counts are often elevated)

Monitor laboratory tests for signs of:

☐ Dehydration or fluid overload

☐ Electrolyte and acid-base imbalances

☐ Hyperglycemia

☐ Hypertriglyceridemia

☐ Nutrient deficiencies

☐ Negative nitrogen balance

☐ Organ dysfunction or organ function that has normalized

Physical Signs

Regularly assess vital signs including:

☐ Blood pressure

☐ Pulse

☐ Body temperature

☐ Respiration

Look for physical signs of:

☐ Protein-energy malnutrition

☐ Dehydration or fluid overload

☐ Nutrient deficiencies and excesses

SELF CHECK

1. Which of the following metabolic changes accompanies acute stress?
   a. reduced plasma concentrations of glucose and fatty acids
   b. lower blood volume and blood pressure
   c. increased insulin action
   d. catabolism of protein in skeletal muscle and connective tissue

2. Tissue injury is followed by:
   a. fluid accumulation in damaged tissue.
   b. reduced blood flow to injured tissue.
   c. reduced capillary permeability.
   d. decreased body temperature.

3. What is a possible effect of replacing vegetable oils rich in omega-6 fatty acids with oils rich in omega-3 fatty acids?
   a. improvement in blood circulation
   b. suppression of inflammation
   c. protection against sepsis
   d. hypertriglyceridemia

4. The acute-phase response results in increased plasma concentrations of:
   a. albumin.
   b. iron.
   c. C-reactive protein.
   d. zinc.

5. Which of the following statements concerning protein and energy recommendations during acute metabolic stress is true?
   a. Protein and energy recommendations are similar to those for healthy people.
   b. Protein and energy recommendations are reduced because a stressed individual cannot metabolize nutrients normally.

c. Acutely stressed individuals can benefit from as much protein and energy as can be provided.

d. Protein and energy recommendations are high in order to minimize muscle tissue losses.

6. The amount of protein recommended for an acutely stressed individual who weighs 150 pounds ranges from about ____ grams of protein per day.
   a. 55 to 85
   b. 68 to 136
   c. 126 to 158
   d. 150 to 300

7. The primary risk factor for COPD is:
   a. alpha-1 antitrypsin deficiency.
   b. occupational exposure to dusts or chemicals.
   c. smoking tobacco.
   d. respiratory infections.

8. A primary feature of emphysema is:
   a. obstruction within the bronchi.
   b. obstruction within the bronchioles.
   c. destruction of the walls separating the alveoli.
   d. excessive lung elasticity.

9. The weight loss and wasting that often occur in COPD can be caused by:
   a. reduced food intake.
   b. increased metabolic rate.
   c. reduced exercise tolerance.
   d. all of the above.

10. Medical nutrition therapy for a person undergoing respiratory failure includes:
    a. careful attention to providing enough, but not too much, energy.
    b. a generous fluid intake to facilitate mucus clearance.
    c. a high-fat intake to prevent weight loss.
    d. a high-carbohydrate intake to limit carbon dioxide production.

Answers to these questions appear in Appendix H.

CLINICAL APPLICATIONS

1. James is a 49-year-old male who is 6 feet 2 inches tall and has a usual body weight of 180 pounds. He was severely injured by an explosion in the chemistry lab where he works and is now in the intensive care unit. Using the method described in the "How to" on p. 433, estimate his energy requirement (use the factor for multiple trauma) and protein requirement (use 1.5 gram/kilogram body weight).

2. Turning again to the case described in item 1, assume that James requires a tube feeding and can tolerate a standard enteral formula. Check Appendix G to find at least three formulas that the nutrition support team might select for tube feeding. Determine the volume of each formula that would be needed to meet James's energy and protein needs. Would this volume also meet the recommendations for vitamins and minerals?

3. Ayla is a 23-year-old law student admitted to the hospital following an automobile accident in which she broke several bones and ruptured part of her small intestine. She has been in the hospital for several weeks and has just begun eating table foods. Her brother, who was driving the vehicle, was also seriously injured and nearly lost his life. Aside from the increased nutritional needs imposed by the stress of the accident, describe how the following factors might interfere with Ayla's ability to improve nutrition status:

- Ayla's injuries are painful.
- Ayla's medications cause drowsiness.
- Ayla is depressed.
- Ayla is often out of her room for X-rays and other diagnostic tests when the menus and food trays arrive.
- Ayla's food intake is sometimes restricted due to the procedures she is undergoing.

How might these problems be resolved to improve Ayla's food intake?

NUTRITION ON THE NET

For further study of the topics in this chapter, access these websites.

Find updates and quick links to these and other nutrition-related sites at our website: **www.wadsworth.com/nutrition.**

To find additional information relevant to critical care, visit these sites:

- American Association of Critical Care Nurses: **www.aacn.org**
- American Society for Parenteral and Enteral Nutrition: **www.clinnutr.org**

This nonprofit group provides comprehensive, up-to-date information for burn care professionals: **www.burnsurgery.org**

To learn more about lung diseases, visit these sites:

- American Lung Association: **www.lungusa.org**
- Canadian Lung Association: **www.lung.ca**
- National Heart, Lung, and Blood Institute: **www.nhlbi.nih.gov**

NOTES

[1] E. Lopez-Garcia and coauthors, Consumption of (n-3) fatty acids is related to plasma biomarkers of inflammation and endothelial activation in women, *Journal of Nutrition* 134 (2004): 1806–1811; T. Pischon and coauthors, Habitual dietary intake of n-3 and n-6 fatty acids in relation to inflammatory markers among US men and women, *Circulation* 108 (2003): 155–160.

[2] C. A. Dinarello and R. Porat, The acute phase response, in L. Goldman and D. Ausiello, eds., *Cecil Textbook of Medicine* (Philadelphia: Saunders, 2004), pp. 1733–1735.

[3] A.S.P.E.N. Board of Directors and The Clinical Guidelines Task Force, Guidelines for the use of parenteral and enteral nutrition in adult and pediatric patients, *Journal of Parenteral and Enteral Nutrition* 26 (2002): 1SA–138SA.

[4] A.S.P.E.N. Board of Directors and The Clinical Guidelines Task Force, 2002.

[5] F. Novak and coauthors, Glutamine supplementation in serious illness: A systematic review of the evidence, *Critical Care Medicine* 30 (2002): 2022–2029.

[6] K. C. McCowen and B. R. Bistrian, Immunonutrition: Problematic or problem solving? *American Journal of Clinical Nutrition* 77 (2003): 764–770.

[7] A. Shenkin, Micronutrients and outcome, *Nutrition* 13 (1997): 825–828.

[8] A.S.P.E.N. Board of Directors and The Clinical Guidelines Task Force, 2002.

[9] American Dietetic Association, *Nutrition Care Manual* (Chicago: American Dietetic Association, 2005.)

[10] N. Anthonisen, Chronic obstructive pulmonary disease, in L. Goldman and D. Ausiello, eds., *Cecil Textbook of Medicine* (Philadelphia: Saunders, 2004), pp. 509–515; H. R. Gosker and coauthors, Skeletal muscle dysfunction in chronic obstructive pulmonary disease and chronic heart failure: Underlying mechanisms and therapy perspectives, *American Journal of Clinical Nutrition* 71 (2000): 1033–1047.

[11] A. M. Schols and E. F. Wouters, Nutritional assessment and support of the stable COPD patient, in T. Similowski, W. A. Whitelaw, and J.-P. Derenne, eds., *Clinical Management of Chronic Obstructive Pulmonary Disease* (New York: Marcel Dekker, 2002), pp. 686–687.

[12] S. Escott-Stump, *Nutrition and Diagnosis-Related Care* (Baltimore: Lippincott Williams & Wilkins, 2002), pp. 620–622; American Dietetic Association, *Manual of Clinical Dietetics* (Chicago: American Dietetic Association, 2000).

[13] A. M. Malone, Enteral formula selection: A review of selected product categories, *Practical Gastroenterology* 29 (June 2005): 44–74.

[14] M. A. P. Vermeeren and coauthors, Acute effects of different nutritional supplements on symptoms and functional capacity in patients with chronic obstructive pulmonary disease, *American Journal of Clinical Nutrition* 73 (2001): 295–301.

[15] Malone, 2005; Schols and Wouters, 2002; Vermeeren and coauthors, 2001.

[16] C. F. Donner and A. Patessio, Exercise in stable COPD, in T. Similowski, W. A. Whitelaw, and J.-P. Derenne, eds., *Clinical Management of Chronic Obstructive Pulmonary Disease* (New York: Marcel Dekker, 2002), pp. 731–758; R. M. Senior, Chronic obstructive pulmonary disease: Epidemiology, pathophysiology, pathogenesis, clinical course, management, and rehabilitation, in A. P. Fishman and coeditors, *Fishman's Manual of Pulmonary Diseases and Disorders* (New York: McGraw-Hill, 2002), pp. 118–141.

[17] M. C. Steiner and coauthors, Nutritional enhancement of exercise performance in chronic obstructive pulmonary disease: A randomised controlled trial, *Thorax* 58 (2003): 745–751.

[18] M. A. Grippi, Acute respiratory failure in the surgical patient, in A. P. Fishman and coeditors, *Fishman's Manual of Pulmonary Diseases and Disorders* (New York: McGraw-Hill, 2002), pp. 1034–1043.

[19] L. M. Bellini, Nutrition in acute respiratory failure, in A. P. Fishman and coeditors, *Fishman's Manual of Pulmonary Diseases and Disorders* (New York: McGraw-Hill, 2002), pp. 1082–1089; American Dietetic Association, 2000.

Multiple Organ Failure

Multiple organ failure is the cause of death in up to one-half of intensive care patients.[1] Described as a failure of two or more of the body's organ systems, multiple organ failure most frequently involves the lungs, liver, kidneys, and gastrointestinal (GI) tract. Involvement of three or more organ systems is associated with a fatality rate of nearly 100 percent. Multiple organ failure is not a disease per se, but rather a late stage of severe illness or injury that results from a severe inflammatory response (discussed in the preceding chapter). Multiple organ failure can be initiated by a number of very different critical illnesses, including acute respiratory failure, trauma, sepsis, burn injuries, extensive surgery, and pancreatitis. This Nutrition in Practice discusses how multiple organ failure develops, the manner in which it is treated, and the importance of prevention.[2]

How long has multiple organ failure been a major clinical problem?

As a clinical entity, multiple organ failure was first recognized only after World War II. Prior to the mid-twentieth century, patients with severe illnesses or multiple injuries frequently died of shock or circulatory failure. After fluid replacement and blood transfusions became standard treatments, the kidneys became the organs at highest risk and kidney failure the most common cause of death. Eventually, physicians learned to better support kidney function by providing appropriate electrolyte solutions and improving urine output. With improved kidney care, the lungs became the most vulnerable organ after severe injury. Improved treatment of respiratory failure eventually led to the current situation: advances in critical care allow patients to survive severe illnesses and injuries, but the body's defenses often overburden organs that were not originally injured.

Why does critical illness lead to multiple organ failure?

As discussed in this chapter, injury and infection cause the release of chemical mediators that have systemic (whole-body) effects. A severe, persistent inflammatory response may cause systemic inflammatory response syndrome (SIRS), which is associated with a constellation of symptoms including fever, raised heart and respiratory rates, and abnormal white blood cell counts. SIRS is a normal adaptive response to a severe insult, but if not reversed quickly enough, it can progress to shock, which is characterized by extremely low blood pressure and an inadequate blood supply for the tissues and organs of the body.

As might be expected from a systemic reduction in blood availability, shock can impair numerous organ systems. The abnormal delivery of oxygen and nutrients to tissues and insufficient removal of wastes result in irreversible injury to cells and tissues. Although each organ system is affected differently, ultimately one or more organs may begin to fail. The

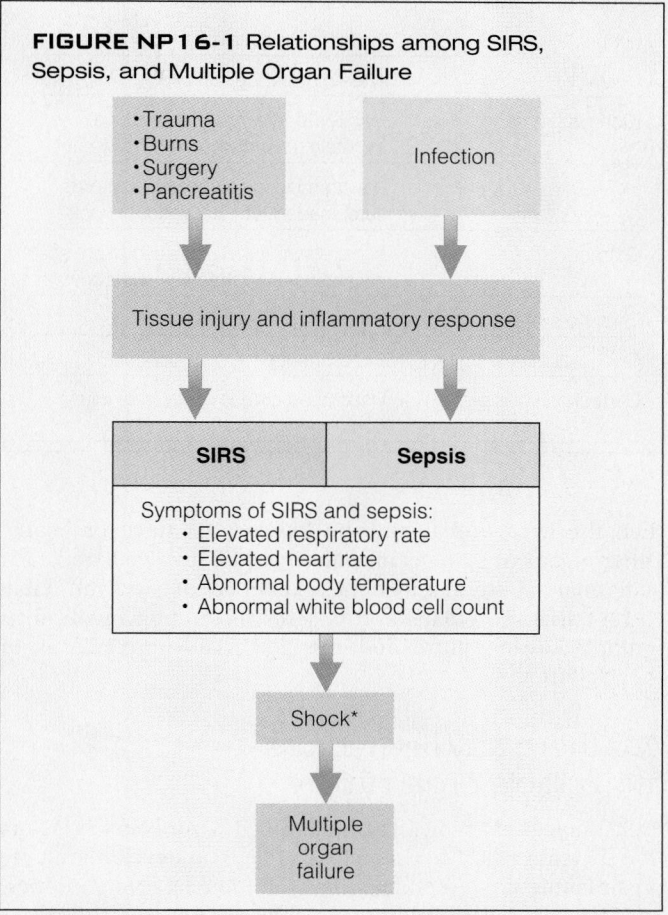

FIGURE NP16-1 Relationships among SIRS, Sepsis, and Multiple Organ Failure

- Trauma
- Burns
- Surgery
- Pancreatitis

Infection

Tissue injury and inflammatory response

SIRS	Sepsis

Symptoms of SIRS and sepsis:
- Elevated respiratory rate
- Elevated heart rate
- Abnormal body temperature
- Abnormal white blood cell count

Shock*

Multiple organ failure

*After critical injury, shock may sometimes precede and be the cause of SIRS.

failure of one organ may place excessive demands on another, causing the second to fail as well. The progression of SIRS to multiple organ failure reflects the inability of the body's defenses and medical treatments to counter the detrimental effects of a sustained and potent inflammatory response.

The specific pathophysiology of multiple organ failure is poorly understood. Although early reports attempted to link the development of multiple organ failure directly to sepsis, sepsis is not present in all cases. Infection often results from impaired immune function and therefore is a frequent consequence of multiple organ failure, but it may not necessarily be the underlying trigger of organ dysfunction. Recall from this chapter that sepsis gives rise to the identical symptoms seen in SIRS. Figure NP16-1 illustrates the relationships among SIRS, infection, sepsis, and multiple organ failure.

Do organs fail in a specific pattern?

Although the clinical course differs greatly among patient populations, the sequence of multiple organ failure often follows a similar pattern among patients: first the lungs fail,

TABLE NP16-1 Physiological Effects of Organ or System Failure

Organ or System	Effects of Failure
Lungs	Inability to maintain gas exchange
Liver	Altered metabolic processes
Kidneys	Inability to regulate blood volume, maintain electrolytes, remove wastes
Heart	Low cardiac output, low blood pressure, inadequate circulation, shock
GI tract	Impaired digestion and absorption, abnormal bleeding, bacterial translocation
Immune system	Infection, sepsis
Coagulation system	Excessive bleeding or coagulation
Central nervous system	Decreased perceptions, brain injury, coma

TABLE NP16-2 Factors That Influence Risk of Multiple Organ Failure

Age over 55 years
Prior chronic disease
Persistent SIRS
Major infection
Blood transfusions
Severity of tissue injury
Length of time between injury and arrival at hospital
Malnutrition

then the liver, and finally the kidneys, GI tract, or heart.[3] Other organs or systems may also become involved, and each additional failure reduces the likelihood of survival. Table NP16-1 lists the organs and systems most often involved in multiple organ failure and the potential consequences of their failure.

Are there any risk factors for multiple organ failure?

Epidemiological studies have identified a number of factors that increase risk. For example, people who develop multiple organ failure are often older, have multiple or severe injuries, and develop severe infections. Table NP16-2 lists the major risk factors associated with multiple organ failure; some of these are discussed below:

- *Age.* Patients over 55 years old are several times more likely to develop multiple organ failure than are younger patients. In elderly patients, the increased risk may be due to the presence of chronic illnesses that directly affect organ function, such as heart disease, lung disease, diabetes, or liver damage. Aging also decreases the functional reserve of organs, thereby reducing an older patient's ability to deal with the additional stress that arises during critical illness.
- *Severity of SIRS.* The length of time that SIRS persists is related to the development of multiple organ failure. In one study, patients who had SIRS that persisted for more than three days were more likely to develop multiple organ failure than patients who had SIRS for less than two days.[4]
- *Infection.* Prolonged SIRS can suppress immune function and increase the risk of developing an infection. During hospital stays, critically ill patients often contract pneumonia—the principal infection associated with multiple organ

failure. The risks of infection and sepsis greatly increase with the use of invasive catheters, which are frequently needed during intensive care to provide oxygen support, intravenous fluid resuscitation, nutrition support, and urine clearance.
- *Blood transfusions.* Blood transfusions are immunosuppressive and may increase a patient's risks of developing infection or sepsis. Blood transfusions frequently have adverse effects that can add further stress; they may cause acute lung injury, allergic reactions, red blood cell hemolysis (breakdown), and other complications.

What is the treatment for multiple organ failure?

Once multiple organ failure has developed, extensive medical support is needed until the inflammatory response has abated. Unfortunately, aggressive treatments can have damaging effects of their own and may cause further injury to organs that are already weakened by illness. Health practitioners must be aware of the adverse effects of aggressive therapies and alert to a patient's responses to treatments. Examples of therapies that are often used to manage organ failure include:

- *Lung support.* Mechanical ventilation is used to assist injured lungs and sustain gas exchange.
- *Fluid resuscitation.* Fluids and electrolytes are supplied to restore blood volume and maintain electrolyte balance.
- *Heart and blood vessel function.* Medications help to sustain or increase cardiac output and maintain adequate blood pressure.
- *Kidney support.* Hemofiltration or dialysis helps to prevent the buildup of toxic metabolites in blood.
- *Infection.* Antibiotic therapy may reverse or prevent infections.
- *Nutrition support.* Enteral and parenteral nutrition support provide nutrients, help to prevent excessive wasting, and promote recovery.

What can be done to reduce the incidence of multiple organ failure?

Because mortality rates for multiple organ failure are so high, prevention must be considered at the earliest stages of injury and treatment before an excessive inflammatory response can cause further damage. Health practitioners have learned to identify the conditions that may increase organ stress whether they are due to a disease process, an inflammatory response, or an aggressive treatment that is intended to provide organ support. Although improvements in care over the past few decades have reduced some of the complications that arise during intensive care, rates of mortality from multiple organ failure have not changed. Thus a focus on prevention is critical until a better understanding of the pathophysiology of multiple organ failure is achieved, which may lead to additional therapeutic options.

Notes

[1] D. Johnson and I. Mayers, Multiple organ dysfunction syndrome: A narrative review, *Canadian Journal of Anesthesia* 48 (2001): 502–509.

[2] J. Parrillo, Approach to the patient with shock, in L. Goldman and D. Ausiello, eds., *Cecil Textbook of Medicine* (Philadelphia: Saunders, 2004), pp. 608–615; Johnson and Mayers, 2001; A. E. Baue, E. Faist, and D. E. Fry, eds., *Multiple Organ Failure: Pathophysiology, Prevention, and Therapy* (New York: Springer-Verlag, 2000); T. W. Evans and M. Smithies, Organ dysfunction, *British Medical Journal* 318 (1999): 1606–1608.

[3] P. J. Offner and E. E. Moore, Risk factors for MOF and pattern of organ failure following severe trauma, in A. E. Baue, E. Faist, and D. E. Fry, eds., *Multiple Organ Failure: Pathophysiology, Prevention, and Therapy* (New York: Springer-Verlag, 2000).

[4] Offner and Moore, 2000.

CONTENTS

Nutrition and Upper Gastrointestinal Disorders

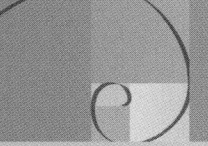
© Getty Images

CHAPTER 17

Gastrointestinal illnesses account for a significant fraction of hospital admissions and visits to health practitioners each year. Diagnosis is not always straightforward, however, as a substantial number of patients with gastrointestinal complaints exhibit no physical abnormalities. Evaluation therefore requires a detailed review of a patient's symptoms and responses to dietary adjustments or earlier treatments. Because gastrointestinal complications frequently accompany other illnesses, the medical history can sometimes uncover the underlying source of distress.

Conditions Affecting the Mouth and Esophagus

This section describes how people with dry mouth can relieve discomfort and examines the causes and treatments of the two most common problems affecting the esophagus: dysphagia (difficulty swallowing) and gastroesophageal reflux disease. Recall that Chapter 14 described several types of mechanically altered diets that can be used by people who have difficulty chewing and swallowing (pp. 381–383 and Tables 14-2 and 14-3).

DRY MOUTH

Dry mouth **(xerostomia),** caused by reduced salivary flow, is a side effect of many medications and is associated with a number of diseases and disease treatments. Antihistamines, antihypertensive agents, antidepressants, decongestants, and other medications can cause dry mouth. Poorly controlled diabetes is often associated with dry mouth, as are conditions that directly affect salivary gland function, such as **Sjögren's syndrome.** Salivary glands can be damaged, sometimes permanently, by the radiation therapy that is used to treat head and neck cancers. Mouth breathing is also a common cause of dry mouth.

The consequences of dry mouth can impair both health and quality of life. Dry mouth is associated with increased plaque, tooth decay, and gum disease. Mouth infections are more common. Dry mouth can interfere with speech and cause bad breath. Chewing and swallowing are more difficult, and taste sensation is diminished. Dentures may be uncomfortable to wear, and ulcerations may develop where they contact the mouth. Dry mouth may cause a person to reduce food intake and thereby increase malnutrition risk. Table 17-1 (p. 448) lists suggestions for managing dry mouth.

DYSPHAGIA

The act of swallowing is complex. First, food is chewed, mixed with saliva, and propelled into the pharynx using the tongue. At the same time, the soft palate and larynx close to prevent regurgitation of food material through the nose, and the epiglottis closes over the trachea to prevent aspiration of food and saliva into the lungs. Next, the bolus of food travels through the esophagus, and the lower esophageal sphincter relaxes to allow passage into the stomach. Due to the multiple tasks involved in swallowing, dysphagia can result from many different physical or neurological conditions; it can be categorized according to the phase of swallowing that is impaired (see Table 17-2, p. 448).

Oropharyngeal dysphagia, which involves the transfer of food from the mouth and pharynx to the esophagus, is often due to a problem affecting the tongue, other oral tissues, or the swallowing reflex. Symptoms include an inability to initiate swallowing, coughing during or after swallowing (due to aspiration), and nasal regurgitation. Other signs include bad breath, a gurgling noise after swallowing, a hoarse or "wet" voice, or a speech disorder. Oropharyngeal dysphagia occurs frequently in elderly people and is often caused by stroke.[1]

NURSING DIAGNOSIS

such as *impaired oral mucous membrane* is appropriate for a person with dry mouth.

xerostomia (ZEE-roh-STOE-me-ah): dry mouth caused by reduced salivary flow.
xero = dry
stomia = mouth

Sjögren's (SHOW-grenz) **syndrome:** an autoimmune disease characterized by the destruction of secretory glands, resulting in dry mouth and dry eyes.

oropharyngeal dysphagia (or-oh-fah-ren-JEE-al diss-FAY-jee-ah): an inability to transfer food from the mouth and pharynx to the esophagus; usually caused by a neurological or muscular disorder.

TABLE 17-1	Suggestions for Managing Dry Mouth

Take frequent sips of water or a sugarless beverage.

Use sugarless candy or gum to help stimulate salivary flow.

Suck on ice cubes or frozen fruit juice bars (unless their coldness causes discomfort).

Avoid citrus juices and spicy or salty foods if they cause mouth irritation.

Avoid dry foods like toast, chips, and crackers.

Avoid caffeine, alcohol, and smoking, which may dry the mouth.

Consume foods that have a high fluid content such as soups, stews, sauces and gravies, yogurt, and pureed fruits.

Try over-the-counter saliva substitutes (available as gels, sprays, and tablets), especially just before meals and at bedtime.

Try rinsing the mouth with small amounts of vegetable oil or softened margarine.

Use a humidifier during the night.

Pay strict attention to oral hygiene, brushing and flossing at least twice daily. Try to brush immediately after each meal.

Avoid alcohol- and detergent-containing mouthwashes that may dry and irritate the mouth.

If dry mouth is caused by a medication, ask your physician about possible alternatives.

Ask your physician if using a medication to stimulate saliva secretion can be of benefit; examples include nicotinic acid tablets and pilocarpine.

TABLE 17-2	Causes of Dysphagia

Oropharyngeal Dysphagia

Brain stem tumors
Developmental disabilities
Lou Gehrig's disease (amyotrophic lateral sclerosis)
Multiple sclerosis
Myasthenia gravis
Obstructive structural anomalies
Obstructive tumors
Parkinson's disease
Poliomyelitis
Severe inflammation
Stroke
Thyroid enlargement

Esophageal Dysphagia

Achalasia
AIDS
Esophageal spasm
Scleroderma
Strictures (induced by inflammation, medication, radiation)
Tumors

Esophageal dysphagia involves the passage of materials through the esophageal lumen and into the stomach. People with this condition often complain of food "sticking" in the esophagus after it is swallowed. The problem is usually an obstruction in the esophagus or a motility disorder. An obstruction may be due to a **stricture** (abnormal narrowing), tumor, or compression of the esophagus by surrounding tissues. Whereas an obstruction usually affects the passage of solid foods (not liquids), a motility disorder hinders the passage of both solids and liquids. **Achalasia,** the most common motility disorder, is a degenerative nerve condition affecting the esophagus; it is characterized by impaired peristalsis and incomplete relaxation of the lower esophageal sphincter when swallowing.[2]

Complications of Dysphagia Nurses should be alert to signs of dysphagia, as it sometimes goes unnoticed. A serious and potentially life-threatening complication associated with dysphagia is aspiration, which may cause airway obstruction, choking, or respiratory infections, including pneumonia. If a dysphagia problem reduces food consumption, malnutrition and weight loss may occur. People who cannot swallow liquids are at increased risk of dehydration.

Dietary Interventions Because a wide variety of defects can cause dysphagia, finding the best diet is often a challenge. Even after careful assessment of a person's swallowing abilities, the most appropriate foods may be determined only by trial and error. Modifying the physical properties of foods and beverages and using alternative feeding methods may help to compensate for swallowing difficulties. A person's swallowing abilities may fluctuate over time, so the dietary plan needs frequent reassessment.

In recent years, there has been an attempt to standardize dysphagia diets to make dietetic practice and terminology more similar among institutions in the United States. In 2002, the American Dietetic Association published the National Dysphagia Diet, developed by a panel of dietitians, speech and language therapists, and a food scientist.[3] Table 17-3 presents brief descriptions of the different levels of the diet and sample meals. After a diet is selected, it must be adjusted to suit the person's swallowing abilities and tolerances. A consultation with a swallowing expert, such as a speech and language therapist, is often necessary.

TABLE 17-3   National Dysphagia Diet

Level 1: Dysphagia Pureed

Foods should be pureed, homogeneous, and cohesive. This diet is for patients with moderate-to-severe dysphagia and poor oral or chewing ability.

Sample meals:

Breakfast: Cream of wheat, pureed pancakes, mashed banana, fruit juice without pulp (thickened as needed), coffee or tea (if thin liquids are acceptable).

Lunch or dinner: Pureed soup, pureed chicken, mashed potatoes with gravy, pureed carrots, broccoli soufflé, applesauce, chocolate pudding.

Level 2: Dysphagia Mechanically Altered

Foods should be moist and soft textured and should easily form a bolus. This diet is for patients with mild-to-moderate dysphagia; some chewing ability is required.

Sample meals:

Breakfast: Scrambled eggs, slightly moistened dry cereals (only those with a simple texture, such as puffed rice cereal or corn flakes), cooked fruit without skin or seeds, fruit juice (thickened as needed), coffee or tea (if thin liquids are allowed).

Lunch or dinner: Soup with easy-to-chew meat and vegetables; well-cooked pasta with moist meatballs and meat sauce; soft, tender vegetables (not fibrous or rubbery); soft fruit pie (with bottom crust only).

Level 3: Dysphagia Advanced

Foods should be moist and be in bite-sized pieces when swallowed. Individuals using this diet need to tolerate mixed textures. This diet is for patients with mild dysphagia.

Sample meals:

Breakfast: Muffin with margarine or butter, cereal and milk (except coarse or dry; milk is restricted if thin liquids are not tolerated), soft peeled fresh fruit or berries, coffee or tea (if thin liquids are tolerated).

Lunch or dinner: Clam chowder; thin-sliced tender meat (can be in sandwich form); rice; cooked, tender vegetables or shredded lettuce with dressing; fresh melon; chocolate chip cookie (without nuts).

Liquid Consistencies (only those that are tolerated are allowed in the diet)

Thin: Watery fluids; may include milk, coffee, tea, juices, carbonated beverages.

Nectarlike: Fluids thicker than water that can be sipped through a straw; may include buttermilk, eggnog, tomato juice.

Honeylike: Fluids that can be eaten with a spoon but do not hold their shape; may include honey, tomato sauce, yogurt.

Spoon-thick: Thick fluids that must be eaten with a spoon and can hold their shape; may include milk pudding, thickened applesauce.

---

esophageal dysphagia: an inability to move food through the esophagus; usually caused by an obstruction or a motility disorder.

stricture: abnormal narrowing of a passageway due to inflammation, scarring, or other structural changes.

achalasia (ack-ah-LAY-zhah): an esophageal disorder characterized by weakened peristalsis and impaired relaxation of the lower esophageal sphincter.

  a = without
  chalasia = relaxation

Diamond Crystal Specialty Foods

Can you tell that the foods in this photo are pureed foods shaped with commercial thickeners?

Food Properties and Preparation Foods included in dysphagia diets should have easy-to-manage textures and consistencies. Soft, cohesive foods are easier to handle than hard or crumbly foods. Moist foods are preferred over dry foods. Some foods within a category may be acceptable, and others may not; for example, some cookies are soft and tender whereas others are hard and brittle. Sticky or gummy foods, such as peanut butter and cream cheese, may be difficult to clear from the mouth and throat. Some patients find more viscous beverages like milk shakes easier to manage than thin liquids such as water or juice.

The textures of foods can be altered to make them easier to swallow. Foods are often pureed, mashed, ground, or minced (review Table 14-3 on p. 383). Foods that have more than one texture, such as vegetable soup or cereal with milk, are harder to handle, so ingredients may be blended to a single consistency and items such as nuts and seeds omitted. Commercial starch thickeners or baby cereals can be used to thicken watery liquids.

Consuming foods that have a similar consistency can quickly become monotonous, however. By using commercial thickeners, pureed foods can be formed into attractive shapes. Including a variety of flavors and colors can also make a meal more appealing. The "How to" offers additional suggestions that can improve the acceptance of pureed and other mechanically altered diets.

Feeding Strategies Depending on the nature of the swallowing problem, some patients can learn new techniques to help them compensate for their disability. For example, people with oropharyngeal dysphagia can do exercises that strengthen the jaws, tongue, or larynx or learn new methods of swallowing that allow them to consume a normal diet. Changing the position of the head and neck while eating can also minimize some swallowing problems.[4] Speech and language therapists are often responsible for teaching patients these techniques.

GASTROESOPHAGEAL REFLUX DISEASE

Gastroesophageal reflux disease (GERD) is a condition of gastric reflux that causes frequent discomfort and, sometimes, tissue damage. Reflux of the stomach's acidic contents can irritate the esophagus, and small amounts may enter the mouth. People who suffer from GERD often refer to these symptoms as *heartburn* or *acid indigestion.* Reflux does not necessarily cause symptoms or injury—it occurs occasionally in healthy people and is a problem only if it creates complications and requires lifestyle changes or medical treatment.

HOW TO   Improve Acceptance of Mechanically Altered Foods

Take a moment to think about a meal of pureed or ground foods. A typical dinner of baked chicken, potatoes, carrots, and green beans can look like mounds of differently colored mush. The foods may taste great, but the person may have little incentive to try the first bite. To improve appetite, be creative when preparing and serving meals:

- Prepare a person's favorite foods and foods that have pleasant smells. The smell and thought of one's favorite foods can help to stimulate the appetite.
- Prepare foods that have strong flavors, for example, curries and chili. Seasonings and spices can enliven food flavors.
- Consider colors and shapes when planning meals and arranging foods on a plate. Substitute brightly colored veg-

etables for white vegetables; for example, replace mashed potatoes with mashed sweet potatoes. Arranging foods attractively on a plate with colorful garnishes can also add color and eye appeal.
- Try layering ingredients so that an entrée looks like a fancy casserole. For example, recipes can resemble such popular entrées as lasagna, shepherd's pie, and moussaka.

Efforts to improve the visual appearance of foods can go a long way toward helping people to eat nourishing meals and maintain a healthy weight.

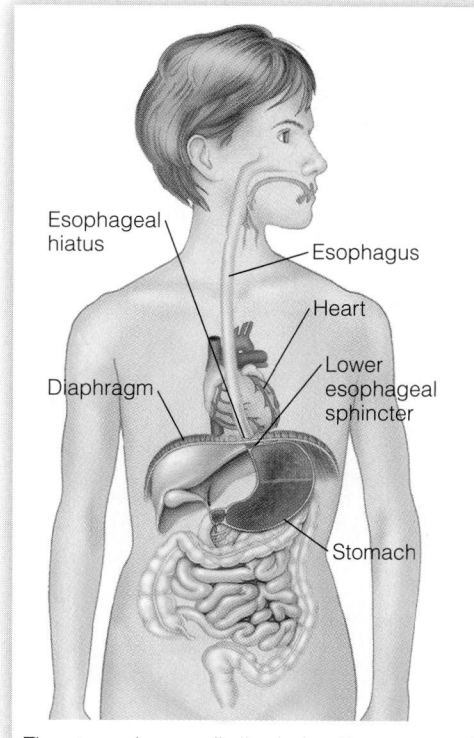
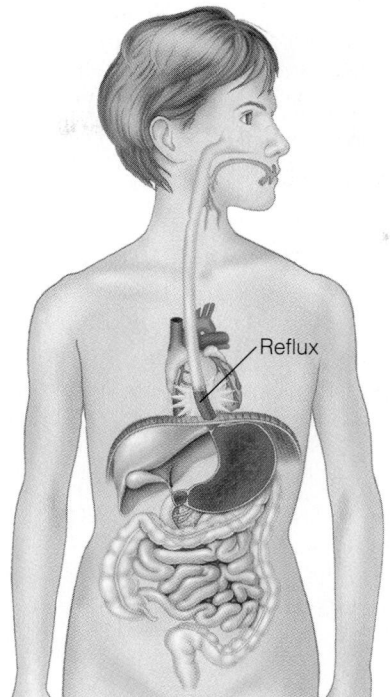
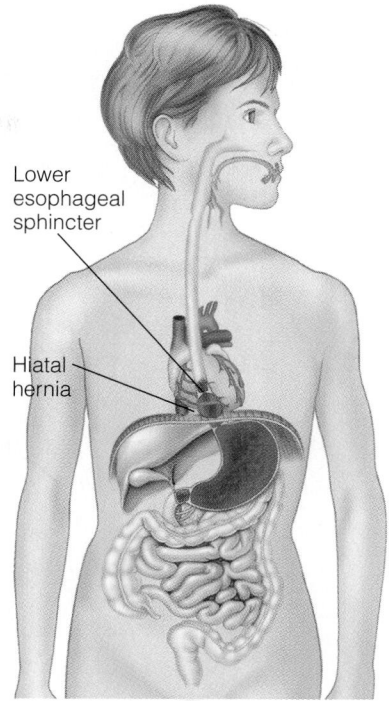

The stomach normally lies below the diaphragm, and the esophagus passes through the esophageal hiatus. The lower esophageal sphincter prevents reflux of stomach contents.

Whenever the pressure in the stomach exceeds the pressure in the esophagus, as can occur with overeating and overdrinking, the chance of reflux increases. The resulting "heartburn" is so-named because it is felt in the area of the heart.

Risk of acid reflux may increase as a consequence of a hiatal hernia. A "sliding" hiatal hernia occurs when part of the stomach, along with the lower esophageal sphincter, rises above the diaphragm.

FIGURE 17-1 The Upper GI Tract, Acid Reflux, and Hiatal Hernia

Causes of GERD The lower esophageal sphincter is the main barrier to gastric reflux; GERD can result if the sphincter muscle is weak or relaxes inappropriately. Medical conditions that interfere with the sphincter's mechanism or prevent rapid clearance of acid from the esophagus can also predispose a person to GERD.

Conditions associated with high rates of GERD include pregnancy, asthma, and **hiatal hernia,** a condition in which a portion of the stomach protrudes above the diaphragm (see Figure 17-1). Pregnancy is the most common predisposing condition; as many as two-thirds of pregnant women report heartburn, which often begins in the first trimester.[5] Various other conditions or substances can exacerbate GERD by either weakening the sphincter or raising pressure within the stomach (see Table 17-4, p. 452 for examples). A number of medications increase the risk of reflux, as does the use of nasogastric tubes in tube feedings.

Consequences of GERD If gastric acid remains in the esophagus long enough to damage the esophageal lining, the resulting inflammation is called **reflux esophagitis.** Severe and chronic inflammation may lead to esophageal ulcers, with consequent bleeding. Healing and scarring of ulcerated tissue may narrow the inner diameter of the esophagus, causing esophageal stricture. A slowly progressive dysphagia for solid foods sometimes results, and swallowing occasionally becomes painful. Pulmonary disease can develop if gastric contents are aspirated into the lungs. Chronic reflux is also associated with **Barrett's esophagus,** a condition in which damaged esophageal cells are gradually replaced by cells that resemble those in gastric or intestinal tissue; such cellular changes increase the risk of developing esophageal cancer.

NURSING DIAGNOSIS

acute pain and *risk for aspiration* may be appropriate for a person suffering from GERD.

hiatal hernia: a condition in which the upper portion of the stomach protrudes above the diaphragm. Most cases are asymptomatic.

reflux esophagitis: inflammation in the esophagus related to the reflux of acidic stomach contents.

Barrett's esophagus: a condition in which esophageal cells damaged by chronic exposure to stomach acid are replaced by cells that resemble those in the stomach or small intestine, sometimes becoming cancerous.

TABLE 17-4	Conditions and Substances Associated with Esophageal Reflux	
Conditions That Raise the Likelihood of Reflux	**Substances That Weaken Lower Esophageal Sphincter Pressure**	
Ascites (accumulation of fluid in the abdomen)	Alcohol	
Delayed gastric emptying	Anticholinergic agents	
Eating large meals	Caffeine	
Lying flat after eating	Calcium channel blockers	
Obesity	Chocolate	
Pregnancy	Cigarette smoking	
Wearing clothes that fit tightly across the waist or abdomen	Diazepam	
	Garlic	
	High-fat foods	
	Meperidine	
	Onions	
	Peppermint and spearmint oils	
	Progesterone	
	Theophylline	

proton-pump inhibitors: a class of drugs that inhibits the enzyme that pumps hydrogen ions (protons) into the stomach. Example include omeprazole (Prilosec) and lansoprazole (Prevacid).

histamine-2 receptor blockers: a class of drugs that suppresses acid secretion by inhibiting receptors on acid-producing cells (commonly called H2 blockers). Examples include cimetidine (Tagamet), ranitidine (Zantac), and famotidine (Pepcid).

Treatment of GERD Treatment objectives are to alleviate symptoms and facilitate the healing of damaged tissue. Severe ulcerative disease may require immediate acid-suppressing medication, whereas a mild case may be managed with dietary and lifestyle modifications. The "How to" below offers suggestions that may help to prevent recurrence of gastrointestinal reflux.

Medications that suppress gastric acid secretion help the healing process by reducing the damaging effects of acid on esophageal tissue. **Proton-pump inhibitors** are the most effective of the antisecretory agents and are used both for rapid healing of esophagitis and as a maintenance treatment. Other antisecretory drugs include **histamine-2 receptor blockers** (often referred to as H2 blockers) and antacids,

---

HOW TO   *Manage Gastrointestinal Reflux Disease*

Management of GERD often requires lifestyle changes to help minimize discomfort and reduce recurrence of acid reflux. Recommendations generally include the following:

- Avoid eating bedtime snacks or lying down after meals. Meals should be consumed at least two to three hours before bedtime.
- Reduce nighttime reflux by elevating the head of the bed on 6-inch blocks, inserting a foam wedge under the mattress, or propping pillows under the head and upper torso.
- Consume only small meals and drink liquids between meals so that the stomach does not become overly distended, which can exert pressure on the lower esophageal sphincter.
- Limit foods that weaken lower esophageal sphincter pressure or increase gastric acid secretion; these include chocolate, fatty foods, spearmint and peppermint, coffee (both caffeinated and decaffeinated), and tea.
- Avoid cigarettes and alcohol; both relax the lower esophageal sphincter.

- Avoid bending over and wearing tight-fitting garments; both can cause pressure in the stomach to increase, heightening the risk of reflux.
- Advise obese individuals to lose weight if they are adequately motivated. Obesity can increase abdominal pressure.
- During periods of esophagitis, avoid foods and beverages that may irritate the esophagus, such as citrus fruits and juices, tomato products, pepper, spicy foods, carbonated beverages, and very hot or very cold foods (depending on individual tolerances).
- Avoid using nonsteroidal anti-inflammatory drugs (NSAIDs) such as aspirin, naproxen, and ibuprofen, which can damage the esophageal mucosa.

Food tolerances among people with GERD can vary markedly. Nurses can help patients pinpoint food intolerances by advising them to keep a record of the foods and beverages consumed as well as any resulting symptoms.

CASE STUDY   *Accountant with GERD*

Lisa Rinaldi is a 49-year-old accountant who is 5 feet 4 inches tall and weighs 165 pounds. She recently underwent a complete physical examination. She told her physician that she had been feeling fairly well until she began experiencing heartburn, which has progressively become more frequent and painful. The heartburn often occurs after she eats a large meal and is particularly bad after she goes to bed at night. By direct examination of the esophageal lumen using an endoscope (a thin, flexible tube equipped with an optical device), the physician found evidence of reflux esophagitis and a slight narrowing throughout the length of the esophagus.

Mrs. Rinaldi's medical history does not indicate any significant health problems. During her last physical, her physician advised her to stop smoking cigarettes and to lose 20 pounds, but she has not attempted either. The nutrition assessment

reveals that Mrs. Rinaldi is feeling stressed because it is the middle of the tax season. She usually has little time for breakfast, eats a lunch of fast foods while continuing to work at her desk, and eats a large dinner at around 8:00 P.M. She generally has wine with dinner and another alcoholic beverage later in the evening.

1. Explain to Mrs. Rinaldi the meaning of the medical diagnoses "reflux esophagitis" and "esophageal stricture."
2. From the brief history provided, list the factors and behaviors that increase Mrs. Rinaldi's risks of experiencing reflux. What recommendations can you make to help her change these behaviors?
3. What medications might the physician prescribe and why?

which neutralize gastric acid. Although antacids are frequently used to relieve occasional heartburn, they are not necessarily appropriate for GERD because they have only short-term effects, are associated with gastrointestinal side effects, and may cause some nutrient deficiencies when used long term. The accompanying Case Study includes questions that review the usual treatments for a patient with GERD.

REVIEW NOTES

Dry mouth can increase the risk of developing dental problems, diminish taste sensation, and lead to reduced food intake. It can be managed with oral hygiene, dietary changes, and saliva substitutes.

Dysphagia may interfere with food intake and increase the risk of aspiration. Treatment includes dietary adjustments, strengthening exercises, and using different swallowing techniques.

Gastroesophageal reflux disease (GERD) can lead to esophageal ulcers, inflammation, bleeding, and stricture. Treatment includes the use of acid-suppressing drugs and lifestyle changes.

Conditions Affecting the Stomach

Stomach disorders range from occasional bouts of discomfort to severe conditions that require surgery. This section begins with a discussion of **dyspepsia** (often called "indigestion"), the feeling of pain or discomfort in the upper abdomen that occurs after food consumption. More serious stomach conditions that may benefit from dietary adjustments include **gastritis** and **peptic ulcers,** which often result from bacterial infection or from the use of medications that damage the stomach lining.[6]

DYSPEPSIA

Dyspepsia refers to the general discomfort in the upper abdominal area that people may experience after eating certain foods. Symptoms include stomach pain, heartburn, fullness, nausea, and bloating. These symptoms sometimes indicate the presence of more serious illnesses, including GERD or peptic ulcer disease. Although a

NURSING DIAGNOSIS

acute pain is often appropriate for a person with dyspepsia.

dyspepsia: a feeling of pain, bloating, or discomfort in the upper abdominal area, often called "indigestion"; a symptom of illness rather than a disease itself.
  dys = bad; impaired
  pepsia = refers to digestion

gastritis: inflammation of stomach tissue.

peptic ulcers: ulcers in the gastrointestinal mucosa resulting from exposure to gastric secretions; may develop in the esophagus, stomach, or duodenum.
  peptic = related to digestion

majority of the population experiences occasional dyspepsia, most are unlikely to seek medical attention.

Causes of Dyspepsia Abdominal pain can be difficult to diagnose, and the cause is not always evident from a physical examination. Symptoms can be caused by various medical conditions, including peptic ulcers, GERD, motility disorders, malabsorptive disorders (discussed in Chapter 18), gallbladder disease, and tumors in the abdominal region. Medications and dietary supplements can sometimes cause gastrointestinal distress. Intestinal conditions such as irritable bowel syndrome or lactose intolerance can mimic dyspepsia. Although pinpointing the cause of gastric symptoms can be difficult, a complete examination is in order if symptoms include weight loss, persistent vomiting, dysphagia, anemia, or bleeding, which suggest the presence of serious illness.

The feeling of bloating may be caused by excessive gas in the stomach, which accumulates when air is swallowed. Swallowing air often accompanies gum chewing, smoking, rapid eating, drinking carbonated beverages, and using a straw. Omitting these practices generally helps to correct the problem.

Potential Food Intolerances Although people may attribute their symptoms to overeating or eating certain foods or spices, controlled studies have been unable to find associations between specific foods and dyspepsia. Coffee (including decaffeinated) can cause symptoms in many people who complain of dyspepsia and may also increase acid reflux and cause heartburn.[7] Spicy foods may cause some injury to the mucosal lining and exacerbate the pain from a preexisting ulcer. High-fat meals can slow gastric emptying and thereby exacerbate dyspepsia. To minimize symptoms, people with dyspepsia are sometimes advised to consume small meals with well-cooked foods that are not overly seasoned and to consume meals in a relaxed atmosphere.[8]

NAUSEA AND VOMITING

Nausea and vomiting accompany many illnesses and are common side effects of medications. Although occasional vomiting is not dangerous, prolonged vomiting can cause fluid and electrolyte imbalances and may require medical care. Chronic vomiting can also reduce food intake and lead to malnutrition and nutrient deficiencies.

The timing of vomiting gives clues to its cause. Vomiting that occurs within an hour after a meal suggests peptic ulcer or a psychological cause. If it occurs more than one hour after a meal, possible causes include food poisoning, an obstruction that prevents stomach emptying, or a stomach motility disorder.

Treatment of Nausea and Vomiting The main goal of treatment is to find and correct the underlying disorder. Most cases are short-lived and require no treatment. Restoring hydration may be necessary in some cases. If a medication is the cause, taking it with food may help. If the cause is unknown or the underlying disorder cannot be corrected, medications that suppress nausea and vomiting can be prescribed. People with **intractable vomiting** —vomiting that is not easily controlled—may require intravenous nutrition support.

Dietary Interventions Sometimes nausea can be prevented or improved with dietary measures. Eating and drinking slowly may be helpful, as may eating small meals that do not distend the stomach. Drinking clear, cold beverages such as carbonated drinks or fruit juices may ease symptoms. Foods that may help to reduce nausea include dry, salty foods like crackers or pretzels. Fried or spicy foods and foods with strong odors should be avoided. Foods that are cold or at room temperature may be better tolerated than hot meals. Individuals sometimes have strong food aversions when nauseated, and tolerances vary greatly.

intractable vomiting: vomiting that is not easily managed or controlled.

TABLE 17-5	Potential Causes of Gastritis

Infection	Internal (bodily) Causes
Bacterial: *Helicobacter pylori, Actinomyces israelii*	Autoimmune
Fungal: *Candida albicans*	Bile reflux
Parasitic: Cryptosporidiosis, nematode infection	Stress
Viral: Cytomegalovirus	Systemic illness/sepsis
Chemical Substances	**Miscellaneous**
Alcohol	High salt intake
Cocaine	Food sensitivity (allergy)
Drugs (especially aspirin and other NSAIDs)	Foreign bodies
Ingestion of corrosive materials	Radiation therapy

GASTRITIS

Gastritis is a general term referring to inflammation of the stomach mucosa. As shown in Table 17-5, gastritis can result from infection, irritating substances, and diseases and treatments that damage the stomach lining. Most often, gastritis results from **Helicobacter pylori** infection or the use of nonsteroidal anti-inflammatory drugs (NSAIDs), both primary causes of peptic ulcer disease as well.

If the gastric mucosa shows signs of **hemorrhage** (severe bleeding), tissue erosion, or ulcers, the condition may be called **acute erosive gastritis,** even if inflammation is not present. Gastritis that becomes chronic and is associated with tissue destruction is known as atrophic gastritis; it is especially prevalent among older adults, affecting as many as 30 percent of people over 60 years of age.[9]

Complications of Gastritis The extensive tissue damage that sometimes develops in chronic gastritis can disrupt gastric secretory functions. If hydrochloric acid secretions become abnormally low **(hypochlorhydria)** or absent **(achlorhydria),** absorption of nonheme iron and vitamin B_{12} can be impaired, and the risk of deficiencies increases.[10] Pernicious anemia, a condition characterized by the destruction of stomach cells that produce intrinsic factor, is a late complication of atrophic gastritis and a primary cause of vitamin B_{12} deficiency (see p. 206).

Dietary Interventions Dietary recommendations depend on an individual's symptoms. If gastritis is asymptomatic, no dietary adjustments are needed. If pain or discomfort is present, irritating foods and beverages should be avoided; these usually include alcohol, coffee (including decaffeinated), tea, cola beverages, spicy foods, and fatty or greasy foods. If food consumption increases pain or causes nausea and vomiting, food intake should be avoided for 24 to 48 hours to rest the stomach. Nutrition support may be necessary if food is not tolerated for a prolonged period. If gastritis results in hypochlorhydria or achlorhydria, supplementation of iron and vitamin B_{12} may be warranted.

PEPTIC ULCER DISEASE

A peptic ulcer is an ulceration that develops in the gastrointestinal mucosa as a result of the destructive effects of hydrochloric acid and pepsin. It is most often caused by *Helicobacter pylori* infection, which has been implicated in more than 60 percent of **gastric ulcers** and approximately 80 percent of **duodenal ulcers.**[11] A second major cause of ulcers is the use of NSAIDs, which have both topical and systemic effects that can damage mucosal tissue. Less frequently, ulcers may develop

The suffix *-itis* refers to the presence of inflammation in an organ or tissue.

Reminder: *Atrophic gastritis* is characterized by the progressive destruction of stomach tissue (see p. 339).

The specific reasons that ulcers develop are not known; fewer than 20% of people with chronic *H. pylori* infection actually develop a peptic ulcer.

NURSING DIAGNOSIS

acute pain is often appropriate for a person with gastritis. In chronic gastritis *imbalanced nutrition: less than body requirements* may apply.

Helicobacter pylori: a type of bacterium that colonizes gastric mucosa; a major cause of gastritis and peptic ulcer disease.

hemorrhage: severe bleeding; a copious flow of blood from blood vessels.

acute erosive gastritis: a condition in which the gastric mucosa is acutely injured, often by the toxic effects of chemical substances or radiation treatment. Damage may include hemorrhaging, tissue erosion, and ulcers.

hypochlorhydria (HIGH-poe-clor-HIGH-dree-ah): a reduction in gastric acid secretion.

achlorhydria (AY-clor-HIGH-dree-ah): absence of gastric acid secretion.

gastric ulcers: peptic ulcers that develop in stomach tissue. Approximately 15% of ulcers occur in the stomach.

duodenal ulcers: peptic ulcers that develop in the duodenum. Approximately 85% of ulcers occur in the duodenum.

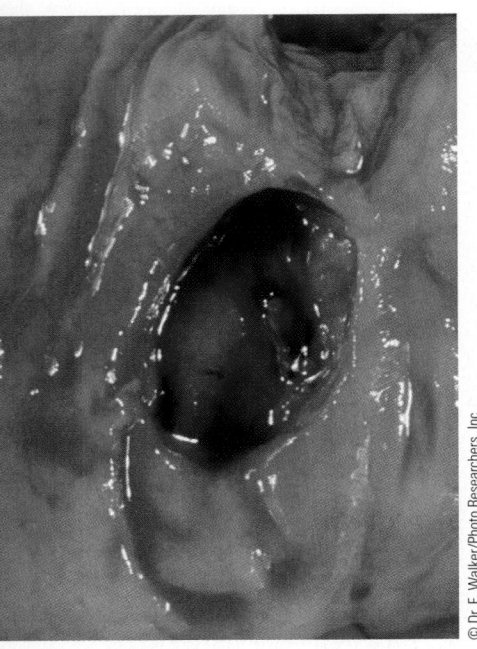

A peptic ulcer, such as the gastric ulcer shown here, damages mucosal tissue and may cause pain and bleeding.

NURSING DIAGNOSIS

for peptic ulcers, may include *acute pain* and *nausea*.

for bleeding ulcers, *deficient fluid volume* and *imbalanced nutrition: less than body requirements* may apply.

from conditions that cause excessive acid secretion. Ulcer risk can be increased by cigarette smoking, emotional stress, and genetic factors.

Effects of Emotional Stress Although most ulcers result from either *H. pylori* infection or NSAID use, an estimated 10 to 20 percent of ulcers develop in people who have no exposure to either.[12] Emotional stress is not believed to cause ulcers per se, but it has effects on physiological processes and behaviors that may increase a person's vulnerability. Physiological effects of stress vary among individuals but may include rapid stomach emptying (which increases the acid load in the duodenum), hormonal changes that impair wound healing, and increases in acid and pepsin secretions. Stress may also lead to behavioral changes including increased use of alcohol, tobacco, and NSAIDs—all potential risk factors. Although the role of stress in ulcer development is not fully understood, current evidence suggests that it may play a contributory role.

Signs and Symptoms Peptic ulcer symptoms vary. Some people are asymptomatic or experience only mild discomfort. Ulcer "pain" may be experienced as a hunger pain, a sensation of gnawing, or a burning pain in the stomach region. The pain or discomfort of ulcers may be relieved by food and recur several hours after a meal, especially if the ulcer is duodenal. Gastric ulcers may sometimes be aggravated by food and can cause loss of appetite and eventual weight loss. Ulcer symptoms tend to go into remission regularly and recur every few weeks or months.

Complications of Peptic Ulcers Peptic ulcers are a major cause of gastrointestinal bleeding, which is the first sign of an ulcer in about 10 to 15 percent of cases. Bleeding is suspected if a person feels weak or fatigued or shows other signs of anemia. Severe bleeding or hemorrhage is evidenced by black, tarry stool samples or, occasionally, vomit that resembles coffee grounds. Other serious complications of ulcers include perforations of the stomach or duodenum and gastric outlet obstruction.

Drug Therapy for Ulcers The goals of ulcer treatment are to relieve pain, promote healing, and prevent recurrence. Treatment often requires a combination of antibiotics to eradicate *H. pylori* infection and discontinuing the use of aspirin and other NSAIDs, which irritate the gastric mucosa and may delay healing. Antisecretory drugs may be prescribed to relieve pain and allow healing; these include proton-pump inhibitors, H2 blockers, or antacids (as used in GERD; see the earlier discussion on p. 452). Bismuth preparations (such as Pepto-Bismol) or sucralfate may help by coating the gastrointestinal lining and preventing further tissue erosion. See the Diet-Drug Interactions feature for nutrition-related effects of these medications.

Dietary Considerations Alterations in diet are advised only if symptoms are affected by food consumption. As with gastritis, dietary recommendations are individualized to personal tolerances. The patient should avoid foods that may irritate the gastrointestinal lining such as alcohol, coffee and caffeine-containing beverages, and spicy foods. Large meals should be avoided so that gastric secretions do not persist for long periods. There is no evidence that dietary adjustments alter the rate of healing.

REVIEW NOTES

Abdominal pain, nausea, and vomiting can be caused by many different medical conditions, and the primary treatment is to remove any underlying causes.

The main cause of gastritis and peptic ulcer disease is *Helicobacter pylori* infection, which may be eradicated by antibiotic therapy. The second most common cause of these conditions is NSAID use, which can directly damage the mucosal lining.

Extensive damage to the mucosa can reduce gastric secretions and increase risks of iron and vitamin B_{12} deficiencies.

DIET-DRUG INTERACTIONS Check this table for notable nutrition-related effects of the medications discussed in this chapter.

	Gastrointestinal Effects	Interactions with Dietary Substances	Metabolic Effects
Antacids (aluminum hydroxide, magnesium hydroxide, calcium carbonate)	Constipation (aluminum- or calcium-containing antacids), diarrhea (magnesium-containing antacids).	May decrease iron, folate, or vitamin B_{12} absorption.	Hypophosphatemia (from excess aluminum, magnesium, or calcium).
Antibiotics (for *Helicobacter pylori* infection; may include amoxicillin, metronidazole, tetracycline)	Diarrhea (amoxicillin, tetracycline), nausea and vomiting (tetracycline), altered taste sensation (metronidazole).	Avoid alcohol with metronidazole; tetracycline decreases iron absorption and binds calcium in the GI tract, reducing absorption of both the tetracycline and the calcium.	
Antisecretory agents (proton-pump inhibitors, H2 blockers)	Constipation, nausea and vomiting, abdominal pain (proton-pump inhibitors).	May decrease iron, folate, and vitamin B_{12} absorption.	
Coating agents (bismuth preparations, sucralfate)	Constipation, diarrhea.		Hypophosphatemia, calcium retention (sucralfate).

Gastric Surgery

In recent years, gastric surgery has become a popular treatment for severe obesity. Less frequently, surgery is used to treat peptic ulcers that are resistant to drug therapy or to correct ulcer complications. In addition, gastric surgery may be necessary for treating stomach cancer. As gastric surgery interferes with stomach function either temporarily or permanently, dietary adjustments are generally needed after surgery.

GASTRECTOMY

Although popular in the past, surgical treatment of ulcers is now rarely necessary due to the efficacy of current drug therapies. Peptic ulcer disease is only occasionally treated by **gastrectomy** (see Figure 17-2, p. 458), a surgical procedure in which diseased portions of the stomach are removed. Other types of gastric **resection** operations can treat ulcer complications, such as obstruction of the pyloric sphincter. In a **vagotomy** procedure, the **vagus nerve** is severed in order to suppress gastric acid secretion. Because this procedure may impair gastric motility, it is sometimes followed by a **pyloroplasty,** which widens the pyloric sphincter to ensure drainage from the stomach to the duodenum.

The Postgastrectomy Diet Following surgery, fluids and foods are withheld until some healing has occurred.[13] Initially, fluids are supplied intravenously and fluid balance is carefully monitored. Ice chips or small sips of water (cold or warm) may be allowed 24 to 48 hours after surgery. Patients are given liquids for the first few meals and can usually tolerate solid foods by the fourth or fifth day after surgery. Tube feedings may be provided if complications prevent a normal progression to solid foods.[14]

Dietary adjustments after gastrectomy are influenced by the size of the remaining stomach and the more rapid gastric emptying that results. A smaller stomach not only limits meal size but also affects food tolerances because of the potential for **dumping syndrome,** as described in the next section. Initially, the patient is offered several small meals and snacks that include only one or two food items

gastrectomy (gah-STREK-ta-mee): the surgical removal of part of the stomach (partial gastrectomy) or the entire stomach (total gastrectomy).

resection: the surgical removal of part of an organ or body structure.

vagotomy (vay-GOT-oh-mee): surgery that severs the vagus nerve in order to suppress gastric acid secretion. This surgery may require a follow-up *pyloroplasty* procedure to allow stomach drainage.

vagus nerve: the cranial nerve that regulates hydrochloric acid secretion and peristalsis. Effects elsewhere in the body include regulation of heart rate and bronchiole constriction.

pyloroplasty (pye-LORE-oh-PLAS-tee): surgery that enlarges the pyloric sphincter.

dumping syndrome: symptoms that result from the rapid emptying of an osmotic load from the stomach into the small intestine. Early symptoms include nausea, abdominal cramps, weakness, and diarrhea; later symptoms are those of hypoglycemia.

FIGURE 17-2 Typical Gastrectomy Procedures
In a gastrectomy, part or all of the stomach is surgically removed. The dashed lines show the removed section.

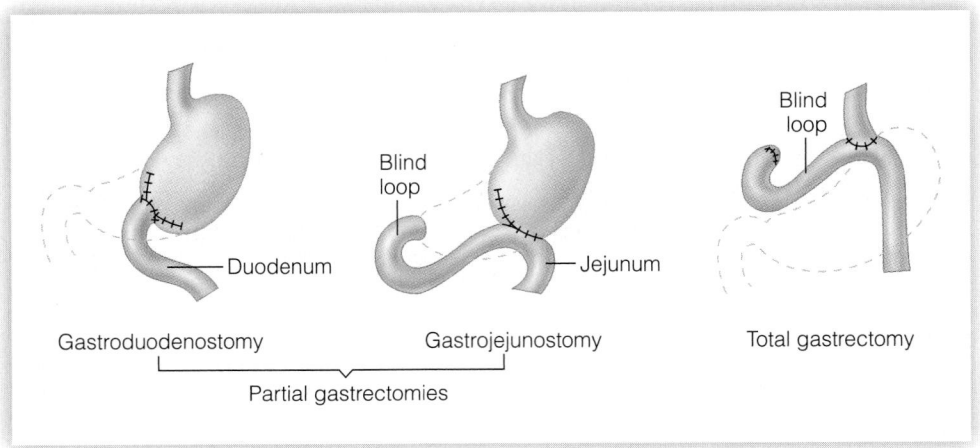

providing protein (fish, lean meats, eggs), fat, and complex carbohydrates (bread, potatoes, vegetables); if tolerated, the diet is slowly progressed to include five or six meals per day. Sweets and sugars should be avoided because they increase osmolarity in the small intestine and potentiate the dumping syndrome. Some patients may need to avoid milk products due to lactose intolerance. Fiber may be added to meals to help delay stomach emptying and reduce diarrhea. Although tolerances vary, some patients may have difficulty with fatty foods, highly spiced foods, carbonated drinks, caffeine-containing beverages, alcohol, extremely hot or cold foods, peppermint, and chocolate. Liquids are restricted during meals due to the limited stomach capacity and because liquids can speed the emptying rate. Table 17-6 lists foods that are often permitted and those that are limited in postgastrectomy diets.

TABLE 17-6   Postgastrectomy Diet

	Foods Recommended (as tolerated)	Foods to Limit (unless tolerated)
Meat and meat alternates	Lean tender meats, fish, poultry, shellfish, eggs, peanut butter	Fried, highly seasoned, and spicy foods
Milk and milk products	Milk, plain yogurt, mild cheeses	Milk shakes, chocolate milk, heavy cream, fruit yogurt
Breads and cereals	Whole-grain or enriched breads, crackers, bagels, low-fat muffins, dry cereal, oatmeal and other cooked cereals, rice, noodles, pasta	Bread or other baked goods that contain dried fruit, nuts, or seeds; fruitcake; frosted cereals; granola; pastries; doughnuts
Vegetables	Any, as tolerated	Candied yams or sweet potatoes, fried vegetables
Fruit	Fresh fruits, unsweetened canned fruits and fruit juices	Dried fruits, sweetened juices, fruits canned in heavy syrup
Desserts	Plain cake, cookies, custard, artificially sweetened pudding and gelatin desserts	Desserts made with chocolate or dried fruit, sweetened gelatin desserts, ice cream and ice milk, candy, marshmallows
Beverages	Milk, diluted or unsweetened fruit drinks, unsweetened carbonated drinks, tea	Coffee, alcoholic beverages, sweetened fruit drinks, sweetened carbonated drinks
Sweeteners and condiments	Sugar substitutes, salt, pepper, mildly flavored sauces and gravies, spices as tolerated	Sugar, syrup, honey, jam, molasses

Source: Adapted from American Dietetic Association, *Manual of Clinical Dietetics* (Chicago: American Dietetic Association, 2000), Chapter 25: Gastric Surgery.

TABLE 17-7	Symptoms of Dumping Syndrome

Early Dumping Syndrome	Late Dumping Syndrome
Symptoms may begin within 30 minutes after eating.	**Symptoms may begin 1 to 3 hours after eating.**
Abdominal fullness, cramps	Anxiety
Diarrhea	Confusion, difficulty thinking
Dizziness	Headache
Flushing, sweating	Hunger
Nausea and vomiting	Palpitations
Rapid heartbeat	Sweating
Weakness, feeling faint	Weakness, feeling faint

Dumping Syndrome The dumping syndrome is a common complication of both gastrectomy and gastric bypass surgery; it involves a group of symptoms resulting from abnormally rapid gastric emptying. Ordinarily, the pyloric sphincter controls the rate of flow from the stomach into the duodenum. After some types of stomach surgery, the hypertonic gastric contents are no longer regulated and can rush into the small intestine more quickly after meals, causing a number of unpleasant effects. Early symptoms can occur within 30 minutes and may include nausea, vomiting, abdominal cramping, diarrhea, lightheadedness, rapid heartbeat, and others (see Table 17-7). These symptoms may be due to intestinal distention (causing release of an excess amount of vasoactive chemicals), a shift of fluid from blood vessels to the intestine that lowers blood volume, and an increase in peristaltic activity. Several hours later, symptoms of hypoglycemia may occur (see Table 17-7) because the unusually large spike in blood glucose following the meal (due to rapid nutrient influx and absorption) can result in an excessive insulin response.

Dietary adjustments can greatly minimize or prevent dumping syndrome. The goals are to limit the amount of food material that reaches the intestine, slow the rate of gastric emptying, and reduce foods that increase hypertonicity. Therefore, meal size is limited, fluids are restricted during meals, and sugars (including milk sugar) are restricted. The "How to" (p. 460) lists practical suggestions for reducing the occurrence of dumping syndrome. In some cases, drugs that inhibit gastrointestinal motility may help. The accompanying Case Study provides the opportunity to design a menu for a gastrectomy patient who is at risk for dumping syndrome.

CASE STUDY   *Biology Teacher Requiring Gastric Surgery*

Karl Toebe, a 58-year-old biology teacher, was admitted to the hospital for gastric surgery after numerous medical treatments failed to manage his severe peptic ulcer disease. A gastrojejunostomy was performed, and after about 24 hours, Mr. Toebe was able to take small sips of warm water. The health care team anticipates multiple nutrition-related problems and is taking measures to prevent them.

1. Review Figure 17-2 to better understand Mr. Toebe's surgical procedure. Consider the possibilities that he might experience the following symptoms: early satiety, nausea and vomiting, weight loss, dumping syndrome, fat malabsorption, anemia, and bone disease. Explain why each of these conditions may occur.

2. What type of diet will the physician prescribe for Mr. Toebe after he begins eating solid foods? Create a day's worth of menus, using foods from Table 17-6.

3. What advice can you give Mr. Toebe that will help to prevent dumping syndrome? List several foods from each major food group that may cause dumping symptoms.

Dietary adjustments can minimize or prevent symptoms of dumping syndrome. The following suggestions often help:

- Eat smaller meals to fit the reduced capacity of the stomach. Increase the number of meals consumed daily so that energy intake is adequate.
- Eat in a relaxed setting. Eat slowly and chew food thoroughly.
- Limit the amount of fluid taken with meals. Avoid consuming beverages within 45 minutes before and after meals, but be sure to include adequate fluid intake during the day to avoid dehydration.
- Avoid juices and sweetened beverages and foods that contain high amounts of sugar. Avoid carbonated beverages if they cause bloating.
- Use artificial sweeteners to sweeten beverages and desserts.
- Avoid foods and beverages that are very hot or very cold, unless tolerated.

- Include fiber-rich foods in each meal. Sometimes adding soluble fibers like pectin or guar gum to meals can help to control symptoms.
- Avoid milk and most milk products, which are high in lactose. Enzyme-treated milk should also be avoided because the breakdown products of lactose (glucose and galactose) can also cause symptoms. Cheese may be better tolerated because its lactose content is low. Make an effort to consume nonmilk calcium sources such as green leafy vegetables, tofu, and fish with bones.
- If symptoms of hypoglycemia continue, try including a protein-rich food in each meal.
- Lie down for 20 to 30 minutes (or longer) after eating to help slow the transit of food to the small intestine. While eating a meal, sit upright.

Fat malabsorption reduces calcium absorption because the negatively charged fatty acids combine with calcium (which is positively charged) and prevent its absorption.

Postsurgical Complications and Nutrition Status Substantial weight loss can sometimes be an unintended consequence of gastrectomy.[15] The patient may need time to learn how much food can be consumed without causing discomfort. The symptoms associated with meals may lead to food avoidance, weight loss, and, eventually, malnutrition. Other nutrition problems that may occur after gastrectomy include the following:

- *Fat malabsorption.* Fat digestion may be impaired due to the accelerated transit of food material, which prevents normal mixing of fat with enzymes and bile. If the duodenum has been removed or bypassed, less lipase is secreted into the intestine for fat digestion. The resulting fat malabsorption can lead to deficiencies of fat-soluble vitamins and some minerals. Supplemental pancreatic enzymes are sometimes provided to improve fat digestion.
- *Bone disease.* Fat malabsorption can cause malabsorption of both calcium and vitamin D. Because patients may also avoid milk products to minimize dumping, they are at extremely high risk of calcium and vitamin D deficiencies. Bone density should be monitored during the years following surgery, and supplementation of calcium and vitamin D is often recommended.
- *Anemia.* Reduced gastric secretions can impair absorption of both iron and vitamin B_{12}. If the duodenum has been removed or is bypassed, iron absorption is further reduced because the duodenum is a major site of iron absorption. Iron and vitamin B_{12} deficiencies can eventually lead to anemia, which may take several years to develop. Supplementation of both iron and vitamin B_{12} is often warranted.

bariatric (BAH-ree-AH-trik) **surgery:** surgery that treats severe obesity.

baros = weight

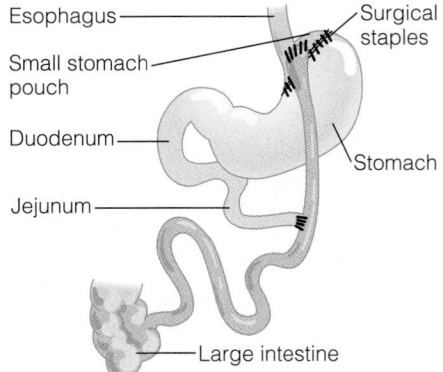

In a gastric bypass, the surgeon constructs a small gastric pouch and creates an outlet directly to the jejunum. The dark pink area highlights the flow of food through the GI tract. The pale pink area indicates the sections that have been bypassed.

FIGURE 17-3 Gastric Bypass Surgery for Severe Obesity

BARIATRIC SURGERY

Bariatric surgery, the type of surgery that treats severe obesity, was introduced in Chapter 7 (p. 165). Figure 17-3 illustrates the gastric bypass procedure (called a Roux-en-Y gastric bypass), one of the more popular surgical options for weight reduction. By creating a small gastric pouch, this procedure reduces gastric capacity and thus restricts meal size. In addition, the digestive route created by the surgery bypasses part of the small intestine, restricting absorptive capacity. The long-term (5- to 14-year) weight loss achieved following a gastric bypass averages between 49 and 62 percent of excess weight.[16] Weight loss is most rapid in the first 6 months after surgery, and weight stabilizes after about 18 to 24 months.

Although bariatric surgeries are effective treatments for morbid obesity, patients should have realistic expectations about the amount of weight they are likely to lose, the diet they need to follow, and the complications that may ensue. Bariatric surgery can dramatically affect health and nutrition status, and patients require lifelong management.[17]

Dietary Guidelines after Bariatric Surgery The gastric pouch created by surgery eventually expands to hold about a cup of food, but its initial capacity is only a few tablespoons. As with the postgastrectomy diet, only ice chips and small sips of water are given during the first day or two after bariatric surgery.[18] A liquid diet is then given for one to two weeks (in small, frequent meals at first); the diet progresses to pureed foods for a similar time period, then to soft foods, and, finally, to regular foods. Some foods may be difficult to manage, especially dry, doughy, or fibrous foods, which may cause pain or vomiting. Fluids need to be consumed separately from meals to avoid excessive distention of the gastric pouch.

Foods that can be problematic after bariatric surgery include chicken, red meat, rice, and white bread.

Patient education and counseling are critical for weight loss and management. Food portions must be carefully controlled to avoid symptoms of dumping syndrome and, later, to maintain weight loss. Patients should learn the elements of a healthy diet as well as the foods that may cause abdominal discomfort, vomiting, or dumping. Dietary supplements need to be taken regularly to avoid nutrient deficiencies. The "How to" includes additional dietary suggestions for patients who have undergone bariatric surgery.

Postsurgical Concerns in Bariatric Surgery The complications that may arise after bariatric surgery are similar to those for gastrectomy patients and may include dumping syndrome, fat malabsorption, and multiple nutrient deficiencies. Rapid weight loss also increases a person's risk of developing gallbladder disease; patients at especially high risk sometimes have their gallbladders removed while undergoing bariatric surgery. After weight loss, plastic surgery may be necessary to remove extra skin, especially on the abdomen, buttocks, hips, and thighs.

HOW TO   Alter Dietary Habits to Achieve and Maintain Weight Loss after Bariatric Surgery

Patients need to learn new dietary habits after bariatric surgery. The following recommendations may help:

- Chew food thoroughly and consume only small amounts. Use a small spoon, and take small bites. Relax and enjoy the meal, taking at least 20 minutes to eat.
- Understand that, at first, the appropriate portion of each food served at mealtime may be only a few spoonfuls. Learn to recognize the sensations that occur when the gastric pouch is full. Signs of fullness may include pressure in the stomach region, a slight feeling of nausea, or pain in the upper chest or shoulder.
- Learn to recognize foods that cause problems. Foods that are dry, sticky, or fibrous may be difficult to tolerate during the weeks after surgery.
- To control vomiting, try eating smaller volumes of food, eating more slowly, and avoiding foods that are known to

cause difficulty. Continued vomiting may be a sign that appropriate food behaviors are not being maintained.
- Eat only at mealtimes. Snacking throughout the day can become a bad habit that causes weight to be regained.
- Avoid consuming liquids within 45 minutes of mealtime. Consume liquids between meals only. Avoid high-kcalorie drinks like soda, alcoholic beverages, and milk shakes. Carbonated beverages increase stomach gas and may cause bloating.
- Drink adequate fluids between meals to avoid dehydration. Most people meet a substantial portion of their fluid needs by eating foods, but a patient who has had bariatric surgery cannot do this. Therefore, fluid intakes need to be increased after surgery.
- Engage in regular physical activity. Activity is a valuable aid to weight maintenance and can help to maintain lean tissue while weight is being lost.

REVIEW NOTES

Gastric surgeries, used to treat peptic ulcer disease and obesity, require dietary adjust-ments after surgery and are associated with multiple complications that affect nutri-tion status.

Dumping syndrome, which results from the rapid influx of nutrients from the stom-ach into the small intestine, is a common complication of gastric surgeries.

The gastric bypass procedure, effective for treating severe obesity, restricts both meal size and absorptive capacity.

NUTRITION ASSESSMENT CHECKLIST FOR PEOPLE WITH UPPER GI TRACT DISORDERS

Medical History

Check the medical history to uncover conditions or treat-ments that may:

- ☐ Lead to dry mouth
- ☐ Interfere with chewing or swallowing
- ☐ Lead to dyspepsia, nausea, or vomiting

Check for a medical diagnosis of:

- ☐ Gastritis or peptic ulcer
- ☐ GERD
- ☐ Hiatal hernia
- ☐ Pernicious anemia

For a patient who has undergone gastric surgery, check for the following complications:

- ☐ Anemia
- ☐ Bone disease
- ☐ Dumping syndrome
- ☐ Fat malabsorption

Medications

Record all medications and note:

- ☐ Medications that may cause dry mouth
- ☐ Medications that may cause nausea and vomiting
- ☐ Aspirin or NSAID use in patients with gastritis or peptic ulcer disease

Note that many medications can cause nausea, especially the first few doses. Suggest that medications be taken with food, when possible, to help alleviate nausea.

Dietary Intake

To devise an acceptable meal plan, obtain:

- ☐ An accurate and thorough record of food intake
- ☐ A record of foods that provoke symptoms of dyspepsia, nausea, GERD, gastritis, peptic ulcers, or dumping syndrome

For patients on long-term dysphagia diets, monitor:

- ☐ Appetite
- ☐ Tolerances to foods
- ☐ Variety of foods offered and regularly consumed

Anthropometric Data

Measure baseline height and weight. Address weight loss early to prevent malnutrition for patients with:

- ☐ Dysphagia or difficulty chewing
- ☐ Dyspepsia or nausea of long duration
- ☐ Malabsorption
- ☐ Dumping syndrome

Laboratory Tests

Check laboratory tests for signs of dehydration for patients with:

- ☐ Persistent vomiting
- ☐ Dumping syndrome

Check laboratory tests for nutrition-related anemia in patients with:

- ☐ Gastritis
- ☐ Previous gastric surgeries
- ☐ Conditions that require long-term use of antisecretory medications

Physical Signs

Look for physical signs of:

- ☐ Dehydration—in patients with persistent vomiting or dumping syndrome
- ☐ Iron and vitamin B_{12} deficiencies—in patients with hypochlorhydria or achlorhydria

SELF CHECK

1. If a patient with dysphagia has difficulty swallowing solids but can easily swallow liquids:
   a. the problem is probably a motility disorder.
   b. the patient most likely has achalasia.
   c. the problem is probably an esophageal obstruction.
   d. the patient may also develop oropharyngeal dysphagia.

2. The nurse working with a patient with dysphagia should recognize that:
   a. only pureed foods should be given to minimize the risk of aspiration.
   b. the patient can have any food that can be comfortably and safely chewed and swallowed.
   c. highly seasoned foods are often restricted.
   d. conventional diets are unable to meet total nutrient needs and supplements are always necessary.

3. Gastroesophageal reflux disease (GERD) is:
   a. characterized by frequent backflow of the stomach's gastric secretions into the esophagus.
   b. a protuberance of a portion of the stomach above the lower esophageal sphincter.
   c. an erosion of the lining of the stomach caused by excess acid in gastric secretions.
   d. an obstruction of the lower esophagus that results in dysphagia.

4. Conditions associated with an increased risk of developing GERD include:
   a. hiatal hernia.
   b. asthma.
   c. pregnancy.
   d. all of the above.

5. For the patient with persistent vomiting, the major nutrition-related concern(s) is/are:
   a. dehydration and malnutrition.
   b. reflux esophagitis.
   c. dyspepsia.
   d. peptic ulcers.

6. Chronic gastritis frequently leads to:
   a. dumping syndrome.
   b. bone disease.
   c. iron and vitamin B_{12} deficiencies.
   d. excessive hydrochloric acid secretion.

7. The primary cause of most peptic ulcers is:
   a. consumption of spicy foods.
   b. hypochlorhydria.
   c. smoking cigarettes.
   d. *Helicobacter pylori* infection.

8. Foods discouraged for patients with gastritis or active ulcers include those that:
   a. are high in fiber.
   b. irritate the gastric mucosa.
   c. are easy to swallow.
   d. contain simple sugars.

9. People at risk of dumping syndrome should generally avoid:
   a. high-fiber foods.
   b. sweets and sugars.
   c. beverages.
   d. bread and potatoes.

10. The nurse assessing a patient who underwent a gastrectomy several years ago should be alert to signs of:
    a. dysphagia.
    b. GERD.
    c. anemia.
    d. gastritis.

Answers to these questions appear in Appendix H.

CLINICAL APPLICATIONS

1. Many of the diets described in this chapter are highly individualized: a particular food may cause discomfort for one person and have no effect on another. Describe some practical ways to keep track of food intolerances.

2. Although some individuals may require a mechanically altered diet for just a few weeks, others may have medical problems that require long-term use of such diets. Consider the difference between working with a person who has had a swallowing problem for years and a person who recently had mouth surgery and is just beginning to eat again.

- Explain how the needs of these individuals may differ. What nutrition-related problems may develop if a person has been following a restrictive dysphagia diet for several years?

- Using Table 17-3 and the "How to" on p. 450, create a day's worth of menus for a person who requires long-term use of a pureed dysphagia diet and tolerates only liquids that have a "honeylike" consistency.

NUTRITION ON THE NET

For further study of the topics in this chapter, access these websites.

Find updates and quick links to these and other nutrition-related sites at our website: **www.wadsworth.com/ nutrition**

Visit the websites of these organizations to find information that is helpful both for health practitioners and patients with gastrointestinal problems:

- American College of Gastroenterology: **www.acg.gi.org**
- American Gastroenterological Association: **www.gastro.org**

- National Institute of Diabetes and Digestive and Kidney Diseases, which is a division of the National Institutes of Health: **www.niddk.nih.gov**
- South Denver Gastroenterology: **www.gutfeelings.com**

Find more information about dysphagia at the Dysphagia Resource Center: **www.dysphagia.com**

Find out more about GERD from the GERD Information Resource Center: **www.gerd.com**

Learn more about *Helicobacter pylori* from the Helicobacter Foundation: **www.helico.com**

NOTES

[1] M. R. Spieker, Evaluating dysphagia, *American Family Physician* 61 (2000): 3639–3648.

[2] R. C. Orlando, Diseases of the esophagus, in L. Goldman and D. Ausiello, eds., *Cecil Textbook of Medicine* (Philadelphia: Saunders, 2004), pp. 814–823.

[3] S. L. McCallum, The National Dysphagia Diet: Implementation at a regional rehabilitation center and hospital system, *Journal of the American Dietetic Association* 103 (2003): 381–384; G. McCullough, C. Pelletier, and C. Steele, National Dysphagia Diet: What to swallow? *The ASHA Leader,* November 4, 2003, pp. 16, 27; The National Dysphagia Diet Task Force, *The National Dysphagia Diet: Standardization for Optimal Care* (Chicago: American Dietetic Association, 2002).

[4] V. Singh and coauthors, Multidisciplinary management of dysphagia: The first 100 cases, *Journal of Laryngology and Otology* 109 (1995): 419–424.

[5] J. E. Richter, Gastroesophageal reflux disease during pregnancy, *Gastroenterology Clinics of North America* 32 (2003): 235–261.

[6] S. L. Friedman, K. R. McQuaid, and J. H. Grendell, eds., *Current Diagnosis and Treatment in Gastroenterology* (New York: Lange Medical Books/McGraw-Hill, 2003); M. Feldman, L. S. Friedman, and M. H. Sleisenger, eds., *Sleisenger and Fordtran's Gastrointestinal and Liver Disease* (Philadelphia: Saunders, 2002).

[7] N. J. Talley, Functional gastrointestinal disorders: Irritable bowel syndrome, nonulcer dyspepsia, and noncardiac chest pain, in L. Goldman and D. Ausiello, eds., *Cecil Textbook of Medicine* (Philadelphia: Saunders, 2004), pp. 806–814.

[8] S. Escott-Stump, *Nutrition and Diagnosis-Related Care* (Baltimore: Lippincott Williams & Wilkins, 2002).

[9] R. M. Russell, The aging process as a modifier of metabolism, *American Journal of Clinical Nutrition* 72 (2000): 529S–532S.

[10] Russell, 2000; Committee on Dietary Reference Intakes, *Dietary Reference Intakes for Vitamin A, Vitamin K, Arsenic, Boron, Chromium, Copper, Iodine, Iron, Manganese, Molybdenum, Nickel, Silicon, Vanadium, and Zinc* (Washington, D.C.: National Academy Press, 2000).

[11] Feldman, Friedman, and Sleisenger, 2002.

[12] S. Levenstein, The very model of a modern etiology: A biopsychosocial view of peptic ulcer, *Psychosomatic Medicine* 62 (2000): 176–185.

[13] P. L. Bayer, Medical nutrition therapy for upper gastrointestinal tract disorders, in L. K. Mahan and S. Escott-Stump, eds., *Krause's Food, Nutrition, and Diet Therapy* (Philadelphia: Saunders, 2004), pp. 686–704; American Dietetic Association, *Manual of Clinical Dietetics* (Chicago: American Dietetic Association, 2000).

[14] American Dietetic Association, *Nutrition Care Manual* (Chicago: American Dietetic Association, 2005).

[15] C. R. Parrish, Post-gastrectomy: Managing the nutrition fall-out, *Nutrition Issues in Gastroenterology* 18 (2004): 63–75; American Dietetic Association, 2000.

[16] D. R. Ferraro, Management of the bariatric surgery patient: Lifelong postoperative care—Board Review, *Clinician Reviews* 14 (2004): 73–79.

[17] W. Marcason, What are the dietary guidelines following bariatric surgery? *Journal of the American Dietetic Association* 104 (2004): 487–488; Ferraro, 2004.

[18] Marcason, 2004.

Inborn Errors of Metabolism

An **inborn error of metabolism** is an inherited trait, caused by a genetic **mutation,** that results in the absence, deficiency, or malfunction of a protein that has a critical metabolic role.[1] The severity of the inborn error is related to the degree of impairment caused by the altered or missing protein. This Nutrition in Practice describes several inborn errors of metabolism and discusses the role of diet in two of these disorders: phenylketonuria and galactosemia. The accompanying glossary defines the related terms.

What problems can result from inborn errors of metabolism?

The protein affected by an inborn error may function as an enzyme, receptor, transport protein, or structural protein. When the body fails to make a protein, body functions that depend on that protein are impaired. For example, when an enzyme is missing or malfunctioning in a metabolic pathway that typically converts compound A to compound B, compound A will accumulate and compound B will not be made. The excess of compound A and the lack of compound B may have harmful effects. Furthermore, the imbalances in one pathway may affect other pathways and ultimately cause a number of metabolic and physiologic disturbances.

What role can diet play in treating inborn errors of metabolism?

Medical nutrition therapy is the primary treatment for many inborn errors that involve nutrient metabolism. Once the biochemical pathway affected by a mutation is identified, a health practitioner may be able to manipulate elements of the diet to compensate for deficiencies and excesses. Dietary intervention generally involves restricting substances that cannot be properly metabolized and supplying substances that cannot be produced. Thus dietary changes may be able to improve outcomes of some inborn errors by:

- Preventing the accumulation of toxic metabolites.
- Replacing nutrients that are deficient as a result of a defective metabolic pathway.
- Providing a diet that supports normal growth and development and maintains health.

Successful treatment of an inborn error depends on the ability to screen newborns and diagnose metabolic diseases before irreversible damage can occur. After a genetic disorder is diagnosed, family members undergo **genetic counseling** to evaluate the likelihood that their future offspring may inherit the disorder. During counseling, couples may learn about reproductive options such as artificial insemination, *in vitro* fertilization, or prenatal monitoring after conception.

Are there treatments for inborn errors that don't involve dietary changes?

Some nondietary therapies are used to treat inborn errors, although the options are somewhat limited. In some cases, the missing protein is infused; this is the primary means of treating **hemophilia,** caused by deficiency of one of the plasma proteins needed for clotting blood. Drug therapy is the main treatment for some inborn errors, including **cystic fibrosis** (discussed in Chapter 18), which is characterized by a defect that prevents normal chloride transport across cell membranes.[2] Future approaches to inborn errors may include **gene therapy,** a treatment that introduces DNA sequences into the chromosomes of affected cells, prompting the cells to express the protein needed to correct the abnormality.

What is an example of an inborn error that benefits from dietary treatment?

A classic example is **phenylketonuria (PKU),** a metabolic disorder that affects amino acid metabolism. PKU occurs in approximately 1 out of every 10,000 births in the United States each year. In PKU, the missing or defective protein is a liver enzyme that converts the essential amino acid phenylalanine to the amino acid tyrosine (see Figure NP17-1, p. 466). Without this enzyme, phenylalanine and its **metabolites** (metabolic products) accumulate and damage the developing nervous system. The impairment in the metabolic pathway also prevents liver synthesis of tyrosine and tyrosine-derived compounds (such as the neurotransmitter epinephrine). Under these conditions, tyrosine becomes essential: because the body cannot make tyrosine, the diet must supply it. Although PKU's most debilitating effect is on brain development, other symptoms may manifest if the condition is untreated. Infants with PKU may have poor appetites and grow slowly. They may be irritable or have tremors or seizures. Their bodies and urine may have a musty odor. Their skin coloring may be unusually pale, and they may develop skin rashes.

How is PKU diagnosed?

Although PKU is not evident at birth, it must be diagnosed in the first few days of life so that early treatment can prevent its devastating effects. For this reason, newborns are routinely screened for PKU in all fifty states.[3] A standard blood test for phenylalanine is typically conducted by heel puncture after the infant has consumed several meals containing protein (usually after 24 hours). Abnormal results require further testing. The screening of newborns for PKU is one of the most common genetic tests in the United States and many other countries. Before newborn screening, infants with PKU

FIGURE NP17-1 Biochemical Alterations in PKU

Normal:

Normally, the amino acid phenylalanine follows two pathways, one in the liver, the other in the kidneys. In the liver, the enzyme phenylalanine hydroxylase adds a hydroxyl group (OH) to produce the amino acid tyrosine. Tyrosine, in turn, produces melanin, the pigmented compound found in skin and brain cells; the neurotransmitters epinephrine and norepinephrine; and the hormone thyroxin. In the kidneys, enzymes convert phenylalanine to by-products that are excreted.

In the liver:

Phenylalanine → (Phenylalanine hydroxylase) → Tyrosine → Melanin / Epinephrine / Norepinephrine / Thyroxin

In the kidneys:

Phenylalanine → Phenylpyruvic acid (a ketone body) → Other phenyl acids (excreted)

In PKU:

Individuals with PKU lack the liver enzyme phenylalanine hydroxylase, impairing conversion of phenylalanine to tyrosine. Phenylalanine accumulates in the liver and blood, reaching the kidneys in abnormally high concentrations. In the kidneys, an aminotransferase enzyme converts phenylalanine to the ketone body phenylpyruvic acid, which spills into the urine—thus the name phenylketonuria.

In the liver:

Phenylalanine (accumulates) → (Phenylalanine hydroxylase (deficient)) → Tyrosine (deficient)

In the kidneys:

Phenylalanine (accumulates) → Phenylpyruvic acid (accumulates) → Other phenyl acids (accumulate)

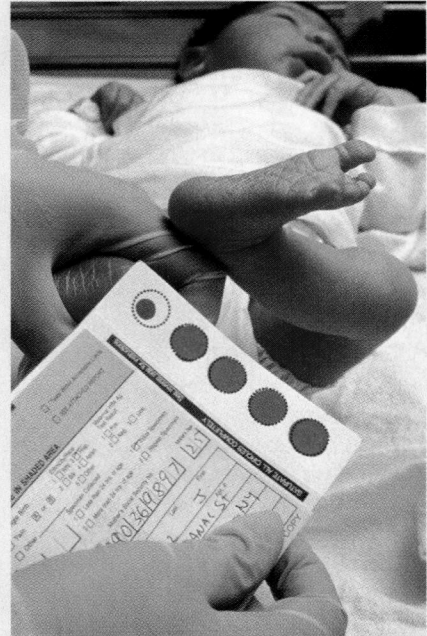

A simple blood test screens newborns for PKU—a common inborn error of metabolism.

demonstrated developmental delays (for example, inability to crawl) by six to nine months of age. By the time parents recognized the problem, the damage was irreversible.

What is the treatment for PKU?

The only current treatment for PKU is a diet that restricts phenylalanine and supplies tyrosine so that the blood levels of these amino acids are maintained within safe ranges. Because phenylalanine is an essential amino acid, the diet cannot exclude it completely. Children with PKU need phenylalanine to grow, but they cannot handle excesses without detrimental effects. Therefore, their diets must provide enough phenylalanine to support growth and health but not so much as to cause harm. The diet must also provide tyrosine, which is an essential nutrient for PKU children. To ensure that blood concentrations of phenylalanine and tyrosine are close to normal, blood tests are performed periodically and diets adjusted when necessary. If the dietary treatment is conscientiously followed, it can prevent the symptoms described earlier.

How do people affected by PKU manage such a strict diet?

Central to the PKU diet is the use of an enteral formula that is phenylalanine-free yet supplies energy, amino acids, vitamins, and minerals.[4] Some formulas supply small amounts of phenylalanine and are useful for infants who must meet most of their nutrient needs by consuming formula. Formula requirements need to be recalculated periodically to accommodate the growing infant's shifting needs for protein, phenylalanine, tyrosine, and energy.

Once food consumption begins, a phenylalanine-free formula supplies the needed amino acids, and foods that contain phenylalanine are carefully monitored. All proteins contain some phenylalanine; therefore, high-protein foods such as meat, fish, poultry, milk, cheese, legumes, and nuts (including peanut butter) are omitted. Fruits, vegetables, and cereals also contain phenylalanine, so only limited amounts are allowed. Low-protein flours and mixes are available for making low-phenylalanine breads, pasta, cakes, and cookies. Foods that do not contain phenylalanine, such as jams, jellies, and most sweeteners, can be used freely. Growth rates and nutrition status are monitored to ensure that the diet is adequate.

Parents and children may need to develop creative ways to make the diet enjoyable. The formula can be flavored or combined with fruits or juices to make smoothies or frozen juice bars. Sandwiches can include low-phenylalanine breads and fillings such as mashed bananas or avocados, shredded carrots and olives, or tomato slices with mayonnaise. Children often enjoy creating special recipes with permitted foods to make their choices more varied and to share meals with friends.

How long should a person with PKU continue the dietary treatment?

Lifelong adherence to a phenylalanine-restricted diet is currently recommended for all individuals with PKU.[5] Elevated phenylalanine levels can adversely affect cognitive function at any age. Case studies have suggested that PKU patients who discontinue dietary management may have problems with attention span, concentration, and memory. It is especially important that women with PKU maintain safe phenylalanine concentrations during pregnancy. Elevated phenylalanine levels, especially during the first trimester, have been associated with birth defects, congenital heart disease, and mental retardation in the offspring of PKU mothers who have discontinued dietary treatment.[6]

What is an another example of an inborn error that requires dietary changes?

Galactosemia is an example of an inborn error of carbohydrate metabolism. Individuals with galactosemia are deficient in one of the enzymes needed to metabolize galactose, a sugar that is primarily found in milk products (recall that each lactose molecule contains a molecule of galactose). An accumulation of galactose can cause damage in multiple tissues. Infants with galactosemia who are given milk react with severe vomiting and liver jaundice within days of the initial feeding. Serious liver damage can develop and progress to symptomatic cirrhosis. Other complications may include kidney failure, cataracts, and brain damage. Treatment in the first weeks of life can prevent the most detrimental effects of galactose accumulation, but if treatment is delayed, the damage to the brain is irreversible.[7]

What is the dietary treatment for galactosemia?

The diet for galactosemia is much simpler than the diet for PKU. For one thing, galactose is not an essential nutrient. The galactosemia diet essentially eliminates galactose from the diet and does not need to provide a carefully determined amount of any nutrient, as the PKU diet does. In addition, dietary galactose is primarily obtained from lactose (the milk sugar), so the main focus of dietary treatment is the exclusion of milk and milk products. A number of other foods that contain galactose in substantial amounts, such as organ meats and some legumes, fruits, and vegetables, must also be avoided or restricted. Patients receive food lists that identify the galactose content of common foods.

Infants diagnosed with galactosemia are given lactose-free formulas to meet their nutrient needs. Once a child can consume adequate amounts of regular foods, special formulas are unnecessary. However, care must be taken to ensure that the diet supplies adequate calcium.

How effective is the dietary treatment for galactosemia?

Although the early introduction of a galactose-restricted diet can eliminate the acute toxic effects of galactosemia, complications of the disease may develop despite an individual's compliance with diet therapy. For example, most patients experience delays in speech and language development. Ovarian failure occurs in up to 85 percent of women who have galactosemia.[8] In addition, some evidence suggests that IQ declines as a person with galactosemia ages. The reasons for these long-term complications are not fully understood.

As our scientific understanding of human genetics and biochemistry increases, more inborn errors are being recognized. Mainstays of treatment for these diseases include effective diagnosis, early treatment, and control of environmental factors that cause toxicity. In some cases, dietary changes are central to treatment and can prevent serious complications. Not all inborn errors are easily treated, however. Future developments in biotechnology may someday allow medical practitioners to correct genetic errors using gene therapy.

Glossary

cystic fibrosis: an inherited disorder that affects the transport of chloride across epithelial cell membranes; primarily affects the gastrointestinal and respiratory systems.

galactosemia (ga-LACK-toe-SEE-me-ah): an inherited disorder that affects galactose metabolism. Accumulated galactose causes damage to the liver, kidney, and brain in untreated patients.

gene therapy: treatment for inherited disorders, in which DNA sequences are introduced into the chromosomes of affected cells, prompting the cells to express the protein needed to correct the disease.

genetic counseling: support for families at risk of genetic disorders; involves diagnosis of disease, identification of inheritance patterns within the family, and review of reproductive options.

hemophilia (HE-moh-FEEL-ee-ah): inherited bleeding disorders characterized by deficiency or malfunction of plasma proteins needed for clotting blood.

inborn error of metabolism: an inherited trait (present at birth) that causes the absence, deficiency, or malfunction of a protein that has a critical metabolic role.

metabolites: products of metabolism; the compounds produced by a biochemical pathway.

mutation: an inheritable alteration in the DNA sequence of a gene.

phenylketonuria (FEN-il-KEY-toe-NU-ree-ah) or **PKU:** an inherited disorder that affects the conversion of the essential amino acid phenylalanine to the amino acid tyrosine.

Notes

[1] L. J. Elsas II, Inborn errors of metabolism, in L. Goldman and D. Ausiello, eds., *Cecil Textbook of Medicine* (Philadelphia: Saunders, 2004), pp. 185–191.

[2] L. M. Bellini and M. A. Grippi, Cystic fibrosis, in A. P. Fishman and coeditors, *Fishman's Manual of Pulmonary Disease and Disorders* (New York: McGraw-Hill, 2002), pp. 176–186.

[3] C. M. Trahms, Inborn errors of metabolism, in A. M. Coulston, C. L. Rock, and E. R. Monsen, eds., *Nutrition in the Prevention and Treatment of Disease* (San Diego, Academic Press, 2001), pp. 209–225.

[4] S. Escott-Stump, *Nutrition and Diagnosis-Related Care* (Baltimore: Lippincott Williams & Wilkins, 2002).

[5] S. D. Cederbaum and C. R. Scriver, Disorders of phenylalanine and tyrosine metabolism, in L. Goldman and D. Ausiello, eds., *Cecil Textbook of Medicine* (Philadelphia: Saunders, 2004), pp. 1282–1285.

[6] Cederbaum and Scriver, 2004.

[7] L. J. Elsas II, Galactosemia, in L. Goldman and D. Ausiello, eds., *Cecil Textbook of Medicine* (Philadelphia: Saunders, 2004), pp. 1268–1270.

[8] Elsas, Galactosemia, 2004.

Nutrition and Lower Gastrointestinal Disorders

© Clouds Hill Imaging Ltd./Corbis

CHAPTER 18

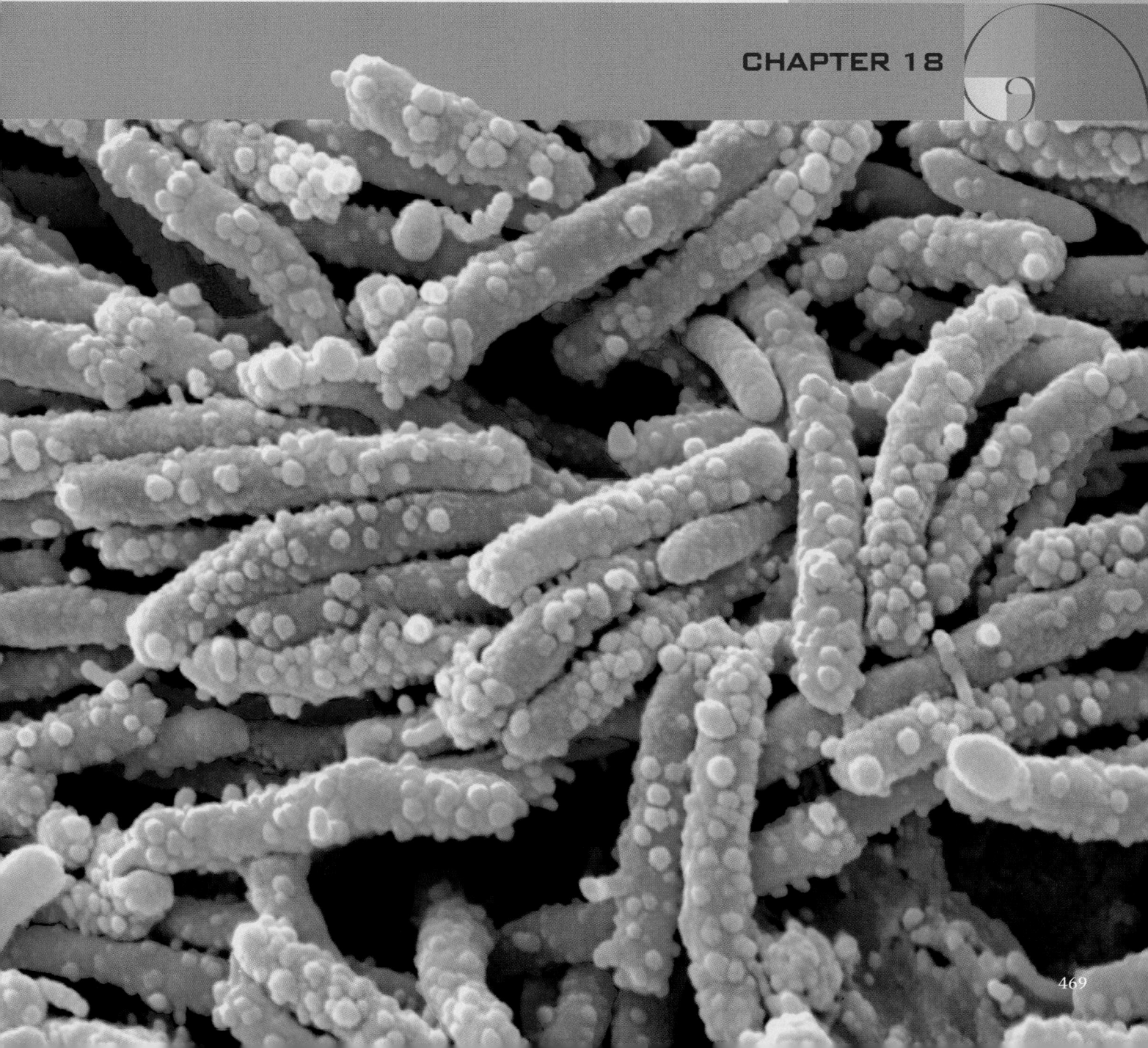

As you may recall from Chapter 2, most digestive and absorptive processes occur in the lower gastrointestinal (GI) tract. Consequently, lower GI disorders can interfere substantially with a person's diet and lifestyle. Some of the diets required for these disorders are complicated and difficult to follow, and foods that are tolerated can vary considerably. In visits with patients, nurses can make sure that the dietary prescription is understood and can suggest ways to make restrictive diets more acceptable.

Common Problems of the Lower Intestine

Intestinal symptoms may result from disorders affecting the lower GI tract or may be due to illnesses and medical treatments that have systemic effects. The most common intestinal symptoms and their causes and treatments are discussed in this section.

CONSTIPATION

A person who complains of constipation is usually concerned about difficulty passing stools or infrequent bowel movements. Sometimes, however, people who complain of constipation have a mistaken notion of what constitutes "normal" bowel habits. Constipation may be diagnosed when a person has fewer than three bowel movements per week. Constipation is much more prevalent among women than men and increases somewhat with aging. During pregnancy, 40 percent of women report some degree of constipation.

Causes of Constipation In Western societies, constipation generally correlates with low food intake, low-fiber diets, and inactivity. All of these factors can extend the time that wastes take to travel through the colon, leading to increased water removal and dry, hard stools that are difficult to pass. Medical conditions often associated with constipation include hypothyroidism, diabetes mellitus, and chronic renal failure. Neurological conditions such as Parkinson's disease, spinal cord lesions, and multiple sclerosis may cause motor problems that lead to constipation. Constipation is also a common side effect of several classes of medications and some dietary supplements, including opiate-containing analgesics, tricyclic antidepressants, antihistamines, calcium channel blockers, aluminum-containing antacids, and iron and calcium supplements.

Treatment of Constipation The main treatment for constipation is a gradual increase in fiber intake. High-fiber diets increase stool weights and promote a more rapid transit of materials through the colon. Foods that increase stool weight the most are wheat bran, fruits, and vegetables.[1] Bran intake can be increased by adding bran cereals and whole-wheat bread to the diet or by mixing bran powder with beverages or foods. A transition to a high-fiber diet may be difficult for some people because it can increase intestinal gas; therefore, high-fiber foods should be added gradually, as tolerated. Fiber supplements like methylcellulose (Citrucel), psyllium (Metamucil, Fiberall), and polycarbophil (a synthetic fiber) are also effective (see Table 18-1, p. 471). Unlike other fibers, methylcellulose and polycarbophil do not increase intestinal gas.

Some other measures may also help constipation. Consuming adequate fluid prevents dehydration, which can draw water from the colon to increase hydration in the rest of the body. Adding prunes or prune juice to the diet is often recommended because prunes contain compounds that have a mild laxative effect. Increasing daily exercise can help to stimulate peristalsis.

Reminder: Fiber-containing foods are described in Figure 3-4 (p. 77), and Appendix A includes the fiber content of common foods. The fiber DRI for women and men aged 19 to 50 years are 25 and 38 g, respectively.

Andrew McClenaghan/Photo Researchers, Inc.

High-fiber foods promote regular bowel movements.

TABLE 18-1	Laxatives and Bulk-Forming Agents			
Laxative Type	**Active Ingredients**	**Product Examples**	**Method of Action**	**Cautions**
Fiber (bulk formers)	Methylcellulose, polycarbophil, psyllium, malt soup extract	Metamucil, Citrucel, FiberLax	Fiber supplements increase stool weight and aid in formation of soft, bulky stools. Similar effects are achieved by adding bran to the diet. For mild constipation. Safe for long-term use.	Some fiber supplements may increase flatulence. Psyllium may cause an allergic reaction.
Emollient (stool softeners)	Docusate sodium	Colace, Surfak	Detergent action promotes the mixing of water with stools. Prevents formation of dry, hard stools.	Does not increase stool weight. Limited effectiveness.
Nonabsorbable sugars (osmotic laxatives)	Lactulose, sorbitol, mannitol	Chronulac, Cephulac	Unabsorbed sugars attract water to large intestine and promote softer stools. Must be used for several days to take effect. Safe for long-term use.	May cause flatulence and cramps. Can lose effectiveness over time.
Saline laxatives (osmotic laxatives)	Magnesium hydroxide, magnesium citrate, sodium sulfate	Milk of magnesia	Unabsorbed salts attract and retain water in large intestine and stimulate contractions.	May cause bloating and watery stools or diarrhea. Should be used with caution. Avoid using in renal patients and children.
Stimulant or irritant laxatives	Senna, bisacodyl, cascara, castor oil, aloe	Ex-Lax, Correctol, Dulcolax	Act as local irritants to colonic tissue; stimulate peristalsis and mucosal secretions. For moderate-to-severe constipation. Long-term use is discouraged.	Usually given only after milder treatments fail. May alter fluid and electrolyte balances. May lead to laxative dependency.

Laxatives Many laxatives can be purchased without prescription. They work by increasing stool weight, increasing the water content of the stool, or stimulating peristaltic contractions. Table 18-1 includes examples of popular laxatives and describes their modes of action. Enemas and suppositories (chemicals introduced into the rectum) are also used to promote defecation; they work by distending and stimulating the rectum or by lubricating the stool.

Medical Interventions Patients with severe constipation who do not respond to dietary or laxative treatments may require more aggressive therapies. Some prescription drugs (for example, prucalopride) work by targeting the neural receptors involved with colonic peristalsis. Surgical interventions are a last resort and include colonic resections and colostomy operations (see p. 487).

INTESTINAL GAS

As mentioned, increased intestinal gas (**flatulence**) is sometimes an unpleasant side effect of consuming a high-fiber diet. Undigested fibers pass into the colon where they are fermented by bacteria, which produce gas as a by-product. Similar effects occur when other carbohydrates are incompletely digested or absorbed; these include fructose, sugar alcohols (sorbitol, mannitol, maltitol), the undigestible carbohydrates in beans (raffinose, stachyose), and some forms of resistant starch, found in grain products and potatoes. Table 18-2 (p. 472) lists some foods commonly associated with excessive gas production.[2] Conditions that cause malabsorption (discussed

flatulence: the condition of having excessive intestinal gas, which causes abdominal discomfort.

TABLE 18-2

Dietary Substances That May Increase Intestinal Gas

Apples

Beer

Broccoli

Brussels sprouts

Cabbage

Carbonated beverages

Cauliflower

Dried beans and peas

Fruit juices

Milk products (if lactose intolerant)

Onions

Pears

Potatoes

Turnips

An oral rehydration solution can be mixed from the following ingredients:

- 2.6 g sodium chloride (table salt).
- 1.5 g potassium chloride.
- 2.9 g sodium bicarbonate (baking soda).
- 13.5 g glucose.
- 1 liter water.

intractable: not easily managed or controlled.

bacterial overgrowth: excessive bacterial colonization of the stomach and small intestine; may be caused by low gastric acidity, altered gastrointestinal motility, mucosal damage, or contamination; interferes with normal digestion and absorption.

later in this chapter) also cause flatulence because the undigested nutrients can be metabolized by intestinal bacteria. Swallowed air also contributes to intestinal gas. Note that people often attribute abdominal bloating or pain to excessive gas, but these symptoms do not correlate well with increased intestinal gas.[3]

DIARRHEA

Diarrhea is characterized by the passage of frequent, watery stools. In most cases, it lasts for only a day or two and subsides without complication. Severe or persistent diarrhea, however, can cause dehydration and electrolyte imbalances. Serious cases of diarrhea are often accompanied by other symptoms, such as fever, cramps, dyspepsia, or bleeding, that help in diagnosing its cause.

Causes of Diarrhea Diarrhea can be classified according to the cause of fluid loss.[4] *Osmotic diarrhea* often results from nutrient malabsorption, as when poorly absorbed sugars like sorbitol or fructose attract water to the colon, increasing fecal water content. *Motility disorders* can accelerate the entry of fluids into the colon, where they are inadequately reabsorbed. In *secretory diarrhea,* the intestines are stimulated to secrete excessive fluid into the colon. Secretory diarrhea is often due to bacterial food poisoning but may also be caused by conditions that cause severe intestinal inflammation.

Acute diarrhea starts abruptly and may persist for several weeks; it is frequently caused by infections or occurs as a side effect of medication use. Chronic diarrhea persists for a month or longer and can result from altered GI tract motility, intestinal inflammation, malabsorptive and endocrine disorders, infectious diseases, radiation treatment, and many other conditions.

Treatment of Diarrhea Correcting the underlying medical condition is the first step in treating diarrhea. Antibiotics may be given to treat infections. If a medication causes diarrhea, an alternate drug can be prescribed. If certain foods are responsible, they can be omitted from the diet. Bulk-forming agents such as psyllium (Metamucil) can help to reduce the liquidity of the stool. If chronic diarrhea does not respond to treatment, antidiarrheal drugs may be prescribed to slow GI motility or reduce intestinal secretions. People with severe, **intractable** diarrhea sometimes require total parenteral nutrition.

Table 18-3 lists foods that may either aggravate or alleviate diarrhea. Generally, foods and beverages that contain fructose, sugar alcohols, or lactose worsen symptoms, although individual tolerances vary.

Rehydration Therapy Severe diarrhea requires the replacement of lost fluid and electrolytes. Oral rehydration solutions can be purchased or easily mixed using water, salts, and glucose or sucrose (see the recipe in the margin). The addition of carbohydrate to the rehydration solution facilitates sodium and water absorption. Commercial sports drinks are not recommended as oral rehydration solutions because their sodium contents are too low to replace losses that result from severe diarrhea. If diarrhea results in extreme dehydration, intravenous solutions can be used to quickly replenish fluid and electrolytes.

BACTERIAL OVERGROWTH

Ordinarily, the stomach and small intestine are protected from **bacterial overgrowth** by gastric acid, which destroys bacteria, and peristalsis, which flushes bacteria through the small intestine before they multiply. When bacterial overgrowth does occur, it disrupts fat digestion and absorption, as excessive bacteria in the small intestine can dismantle the bile acids needed for fat emulsification. In addition, deficiencies of the fat-soluble vitamins may develop. The bacteria also compete for vitamin B_{12}, impairing its absorption and increasing the risk of vitamin

TABLE 18-3	Foods That May Affect Diarrhea

Foods That May Worsen Diarrhea	Foods That May Lessen Diarrhea
Apple juice	Applesauce
Caffeine-containg beverages	Bananas
Coffee	Barley
Dates	Cheese
Fried foods	Oat bran
Fructose-sweetened drinks	Oatmeal
Grapes	Peanut butter (smooth)
Honey	Potatoes
Milk and milk products	Rice (boiled)
Pear juice	Soda crackers
Prune juice	Tapioca
Sugar-free candies	Yogurt

Note: Individual tolerances vary.

Sources: American Dietetic Association, *Manual of Clinical Dietetics* (Chicago: American Dietetic Association, 2000), p. 423; M. H. Beers and R. Berkow, eds., *The Merck Manual of Diagnosis and Therapy* (Whitehouse Station, N.J.: Merck Research Laboratories, 1999), pp. 275–278.

B_{12} deficiency. Typical symptoms of bacterial overgrowth include chronic diarrhea, abdominal discomfort, bloating, weakness, and weight loss.

Causes of Bacterial Overgrowth Conditions that impair intestinal motility and allow material to stagnate can predispose a person to bacterial overgrowth. For example, in some types of gastric surgery, a portion of the small intestine is bypassed, preventing the flow of material in the bypassed region and allowing bacteria to flourish (see the "blind loop" shown in Figure 17-2 on p. 458). Intestinal motility can also be reduced by strictures, obstructions, and diverticula in the small intestine (discussed later in the chapter).

Reduced gastric acid secretions can also lead to bacterial overgrowth. Possible causes include atrophic gastritis, acid-suppressing medications, and acid-reducing surgery (vagotomy) for peptic ulcer disease.

Treatment for Bacterial Overgrowth Treatment may include antibiotics to suppress bacterial growth and surgical correction of the anatomical defects that contribute to stasis. Dietary supplements are provided to reverse nutrient deficiencies, especially for fat-soluble vitamins, calcium (which combines with malabsorbed fatty acids), and vitamin B_{12}. **Medium-chain triglycerides (MCT),** which do not require bile for digestion and absorption, can be used as an alternate source of dietary fat.

STEATORRHEA

Steatorrhea refers to excessive fat in the stools, which results from fat maldigestion or malabsorption. Fat maldigestion may occur in medical conditions that reduce the availability of bile or pancreatic lipase. For example, liver diseases and

medium-chain triglycerides (MCT): triglycerides that contain fatty acids that are 8 to 10 carbons in length. MCT do not require digestion and can be absorbed in the absence of lipase or bile.

steatorrhea (stee-AH-tor-REE-ah): excessive fat in the stools resulting from fat malabsorption; characterized by stools that are loose, frothy, and foul smelling due to a high fat content.
> **steat** = fat
> **rheo** = flow

bacterial overgrowth can interfere with bile's availability, whereas pancreatitis and cystic fibrosis can alter pancreatic secretions, which include lipase. Fat malabsorption may result from inflammatory conditions in the intestine or radiation treatment for cancer. Motility disorders that cause rapid gastric emptying or rapid intestinal transit can also lead to steatorrhea because they prevent the normal mixing of dietary fat with lipase and bile.

Consequences of Fat Malabsorption Fat malabsorption is associated with losses of food energy, essential fatty acids, fat-soluble vitamins, and some minerals (see Figure 18-1). Weight loss may occur unless alternate energy sources are provided. Deficiencies of fat-soluble vitamins and essential fatty acids are common in chronic conditions. Absorption of some minerals, including calcium, magnesium, and zinc, may be impaired because they can form **soaps** with unabsorbed fatty acids and bile acids. Calcium deficiency can lead to bone loss, which is further aggravated by the vitamin D deficiency that is common in steatorrhea.

Another complication of steatorrhea is an increased risk of kidney stones, which are most often composed of calcium oxalate. The oxalates in foods ordinarily bind to calcium in the small intestine and are excreted in the stool. If calcium instead binds to fatty acids or bile acids, the oxalate is free to be absorbed into the blood and is ultimately excreted in urine. The risk of developing oxalate stones increases when urinary oxalate levels are high. Kidney stones are discussed further in Chapter 22.

Dietary Adjustments If steatorrhea does not improve, a fat-restricted diet may be recommended (see Table 18-4). The main objectives of fat restriction are to relieve abdominal symptoms that are aggravated by fat intake (usually diarrhea and flatulence) and to reduce vitamin and mineral losses. Fat should not be restricted more than necessary because fat is an important source of energy. Although MCT oil is sometimes used as a replacement for fat, it does not provide essential fatty acids. The "How to" on p. 476 offers suggestions for following a fat-restricted diet and for using MCT oil.

NURSING DIAGNOSIS

imbalanced nutrition: less than body requirements often applies to people with steatorrhea.

Oxalates are plant compounds that bind with some minerals to form complexes that the body cannot absorb. They are present in green leafy vegetables such as beet greens and spinach.

soaps: chemical compounds formed between positively charged minerals and fatty acids.

FIGURE 18-1 The Consequences of Fat Malabsorption

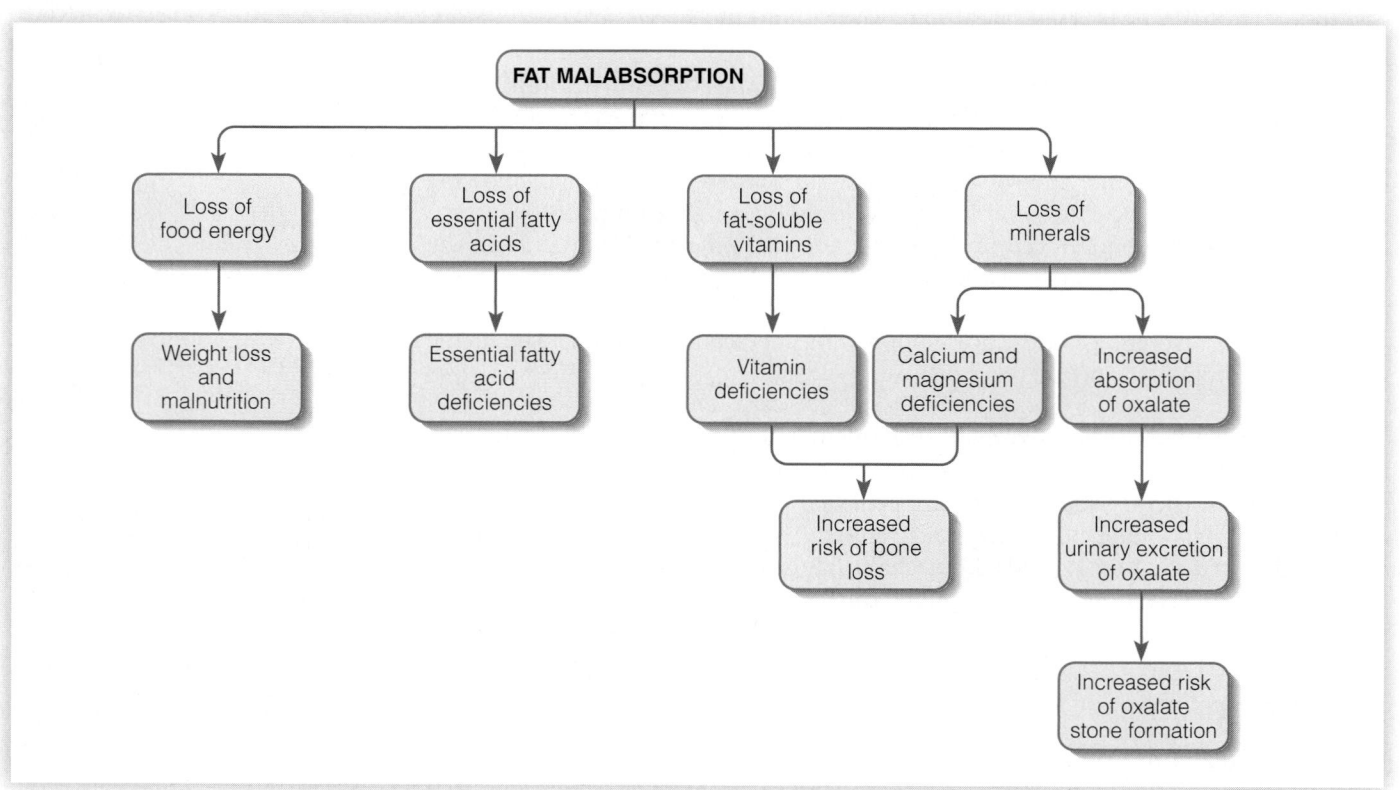

TABLE 18-4	Fat-Restricted Diet
General guidelines	*For fat restriction of 25 grams per day:* Limit meat and meat alternates to 4 ounces daily (cooked weight); limit fat equivalents to 1 per day (see Fats section below). *For fat restriction of 50 grams per day:* Limit meat and meat alternates to 6 ounces daily (cooked weight); limit fat equivalents to 3 to 5 per day. *To raise fat content:* Allow additional servings of meat or fat. *To lower fat content:* Restrict servings of meat or fat.
Meat and meat alternates	*Recommended:* Choose lean meat, fish, and poultry only. Preparation methods can include broiling, roasting, grilling, or boiling. Trim visible fat and remove poultry skin before consuming. For sandwiches, select turkey breast or other low-fat luncheon meats. Limit eggs to 2 per week or use low-fat egg substitutes. Meat alternates include tofu and dried beans and peas. *Avoid:* Pork and beans, sausage, bacon, frankfurters, spareribs, duck, goose, tuna packed in oil, fried meats.
Milk and milk products	Choose milk products that contain less than 1 gram fat per serving. *Recommended:* Fat-free milk, fat-free yogurt, fat-free sour cream substitutes, fat-free Half-and-Half and cream substitutes, fat-free cheeses. *Avoid:* Milk products that are not fat-free.
Breads, cereals, rice, and pasta	Choose breads, cereals, rice, and pasta dishes that contain less than 1 gram fat per serving. *Recommended:* Whole-grain breads, soda crackers, cooked cereals and most cold cereals, plain tortillas, bagels, English muffins, fat-free muffins, graham crackers, plain rice, plain noodles and pasta. *Avoid:* Biscuits, pancakes, waffles, doughnuts, granola, snack crackers that contain fat, corn chips, cornbread, fried rice, pasta sauces with added fats.
Vegetables	Choose vegetables that contain less than 1 gram fat per serving. *Recommended:* All vegetables prepared without added fats. *Avoid:* Buttered or fried vegetables, creamed vegetables, au gratin style, french-fried potatoes, olives, sauces with added fats.
Fruit	Choose fruits that contain less than 1 gram fat per serving. *Recommended:* All fruits prepared without added fats. *Avoid:* Avocado, fruit dips made with fat or coconut.
Desserts	Choose desserts that contain less than 1 gram fat per serving. *Recommended:* Sherbet, fruit ices, fruit whips, flavored gelatin, angel food cake, meringues, fat-free puddings, fat-free baked products, fat-free ice cream or frozen yogurt, fat-free candies (marshmallows, jelly beans, hard candy). *Avoid:* Cakes, cookies, pies, and pastries made with fat; puddings made with whole milk or eggs, ice cream, candies made with fat (caramel, chocolates).
Fats	Choose 1 fat equivalent daily if fat restriction is 25 grams per day, and 3 to 5 fat equivalents daily if fat restriction is 50 grams per day. *One fat equivalent is equal to:* Vegetable oil: 1 tsp. Butter, margarine: 1 tsp, or 1 tbs diet margarine. Mayonnaise: 1 tsp, or 1 tbs reduced-kcalorie mayonnaise. Salad dressing: 1 tbs, or 2 tbs reduced-kcalorie dressing. Nuts: 6 almonds or cashews, 10 peanuts, 2 tsp peanut butter, 4 halves walnuts or pecans.
Beverages	Choose beverages that contain less than 1 gram fat per serving. *Recommended:* Coffee, tea, soft drinks, juices, fat-free milk, coffee substitutes. *Avoid:* Beverages made with milk (unless fat-free milk) or added cream, chocolate milk, eggnog, milk shakes.

Source: Adapted from American Dietetic Association, *Manual of Clinical Dietetics* (Chicago: American Dietetic Association, 2000), Chapter 58, Fat-restricted diet.

Fat-restricted diets can be difficult to follow. Fats add flavors, aromas, and textures to foods—characteristics that make foods more enjoyable. Unlike some diets that can be introduced gradually, a fat-restricted diet is often implemented immediately, allowing little time for adaptation. These suggestions may help:

- Fat is better tolerated if provided in small portions. Divide the day's allotment into several servings that can be consumed throughout the day.
- Use variety to enhance enjoyment of meals: vary flavors, textures, colors, and seasonings.
- Look for fat-free items when grocery shopping. Incorporate fat-free ingredients when preparing favorite recipes.
- Try fat-free and low-fat condiments to improve the diet's palatability. Experiment with herbs and spices. Instead of butter, use fruit butters on toast. Use butter-flavored granules on vegetables. Replace the mayonnaise in sandwiches with spicy mustard. Replace salad dressings with flavored vinegars.

- Avoid products that contain the fat substitute olestra, which may aggravate GI symptoms.

If patients are interested in using MCT oil:

- Explain that MCT products are expensive, but that the cost is sometimes covered by medical insurance.
- Advise patients to add MCT oil to the diet gradually. Diarrhea and abdominal cramps may result if too much is used at once. Tolerance to MCT oil may improve in time.
- Advise patients that MCT oil may have an unpleasant taste when used alone. Suggest using MCT oil in recipes as a substitute for regular oil. MCT oil can replace oil in salad dressings, be incorporated into sauces, and be used in cooking or baking. It can also be added to fat-free milk products to make milk shakes.
- Point out that MCT oil should not be used to fry foods because it decomposes at lower temperatures than most cooking oils.

REVIEW NOTES

Disorders of the lower GI tract can cause intestinal symptoms such as constipation, diarrhea, intestinal gas, bacterial overgrowth, and steatorrhea.

Constipation accompanies many different illnesses and is often improved by a high-fiber diet or fiber supplements. Diarrhea may result from food poisoning, malabsorption, and motility disorders and may lead to fluid and electrolyte imbalances.

Bacterial overgrowth may result from conditions that reduce gastric acidity or intestinal motility. Steatorrhea can be caused by reduced availability of bile or lipase or by fat malabsorption.

Malabsorption Syndromes

Malabsorption syndromes can lead to nutrient deficiencies and weight loss and cause serious complications. Malabsorption is most often caused by either a lack of digestive enzymes or inflammatory conditions that damage the intestinal mucosa. Malabsorption can also result when the absorptive capacity of the small intestine is reduced by surgical resection.

LACTOSE INTOLERANCE

Approximately 75 percent of people worldwide have some degree of lactose intolerance, caused by the loss or reduction of lactase, the intestinal enzyme that digests the lactose in milk products. Lactose intolerance has a genetic component and is especially prevalent among individuals of certain ethnic groups, including Asians, African Americans, Native Americans, Ashkenazi Jews, and Latinos.[5] It can also result from GI conditions that damage the small intestinal mucosa. The primary symptoms of lactose intolerance are diarrhea and increased intestinal gas.

Lactose intolerance is generally managed by dietary adjustments. Although people are sometimes reluctant to consume milk products, clinical studies have found that individuals with lactose intolerance can tolerate up to 2 cups of milk daily without significant symptoms.[6] People who avoid milk for fear of intestinal discomfort can be urged to gradually increase consumption of lactose-containing foods as tolerated. They may more readily tolerate milk when intake is divided throughout the day and the milk is taken with food.[7] Some people tolerate chocolate milk better than plain milk. Most aged cheeses are well tolerated because they contain little lactose. Yogurts that contain live bacterial cultures are usually acceptable, as the bacteria contain lactase that may aid in lactose digestion. Other options are to add a lactase preparation to milk or to take enzyme tablets before consuming lactose-containing foods. Lactose-free milk is also commercially available.

People who develop lactose intolerance as a result of intestinal illness are advised to temporarily restrict milk and milk products. Foods that contain lactose can be reintroduced in small amounts once the condition improves. All individuals who restrict milk products should be encouraged to consume alternate food sources of calcium and vitamin D; dietary sources of these nutrients are discussed in Chapters 8 and 9 (see pp. 193–195 and 231–235).

Most people with lactose intolerance can drink milk, especially if they drink it along with other foods and limit the amount they consume at any one time.

PANCREATITIS

Pancreatitis develops when digestive enzymes within the pancreas are prematurely activated and destroy pancreatic tissue. Mild cases can subside in a few days, but some cases persist for weeks or months. Chronic pancreatitis can cause irreversible damage to pancreatic tissue and permanent loss of function.

Acute Pancreatitis Acute pancreatitis is most often caused by gallstones or excessive alcohol use, but it may also result from high blood triglyceride levels or exposure to toxins.[8] Severe abdominal pain, nausea and vomiting, and abdominal distention are common symptoms. In most patients, the condition resolves within a week with no complications. More severe cases may lead to renal failure, sepsis, and other complications and often require prolonged hospitalization.

Medical Nutrition Therapy for Acute Pancreatitis Oral fluids and food are withheld until pain and tenderness have subsided, and fluids and electrolytes are supplied intravenously. After three to seven days, small amounts ($^1/_2$ to 1 cup) of fluids may be offered; if tolerated, the diet progresses to soft foods and finally to solids, given in small, frequent feedings.[9] As fat has a greater stimulating effect on pancreatic secretions than do carbohydrates or protein, low-fat diets may be better tolerated initially.[10] Severe cases of pancreatitis may require an elemental formula delivered by jejunal tube feeding; protein and energy needs are often high due to the catabolic and hypermetabolic effects of severe inflammation.

Chronic Pancreatitis Chronic pancreatitis is characterized by permanent damage to pancreatic tissue, resulting in impaired secretion of digestive enzymes. About 70 percent of chronic pancreatitis cases are caused by excessive alcohol consumption.[11]

In chronic pancreatitis, abdominal pain is often severe and unrelenting and worsens with eating. Fat maldigestion develops sooner than maldigestion of protein or carbohydrate, and steatorrhea is common in advanced cases. Food avoidance (due to pain associated with eating) and malabsorption may lead to weight loss and malnutrition. Advanced cases are associated with reductions in both insulin and glucagon secretions, and diabetes eventually develops in up to 30 percent of patients.

Medical Nutrition Therapy for Chronic Pancreatitis Protein and energy needs are high in patients who have lost weight or are malnourished. Dietary supplements may be needed to correct nutrient deficiencies, which may be due to malabsorption or to the alcohol abuse that caused the disease. Patients should avoid alcohol completely, as it can worsen pancreatic function.

Reminder: Pancreatic digestive secretions include bicarbonate, which neutralizes the acidic gastric contents, and digestive enzymes, which break down protein, carbohydrate, and fat.

NURSING DIAGNOSIS

for people with acute pancreatitis may include *acute pain, nausea, diarrhea,* and *deficient fluid volume.*

Reminder: An *elemental formula* contains hydrolyzed nutrients that require minimal digestion and are easily absorbed.

NURSING DIAGNOSIS

imbalanced nutrition: less than body requirements is often appropriate for those with chronic pancreatitis.

Reminder: *Exocrine* refers to "outside the body." Exocrine secretions include digestive juices, lubricants like mucus and tears, and sweat.

The nutrition problems associated with chronic obstructive lung diseases were described in Chapter 16.

enteric-coated: refers to medications or enzyme preparations that can withstand gastric acidity and dissolve only at a higher pH.

cystic fibrosis: an inherited disease characterized by the production of abnormally viscous exocrine secretions; often leads to respiratory illness and pancreatic insufficiency.

celiac (SEE-lee-ack) **disease:** a condition characterized by an abnormal immune reaction to wheat gluten that causes severe intestinal damage and nutrient malabsorption; also called **gluten-sensitive enteropathy** or **celiac sprue.**

wheat gluten (GLU-ten): a family of water-insoluble proteins in wheat. The protein in wheat gluten that has toxic effects in celiac disease is called **gliadin** (GLY-ah-din).

Steatorrhea is usually treated with pancreatic enzyme replacement. Pancreatic enzymes are often **enteric-coated** to resist the acidity of the stomach and do not dissolve until the pH is above 5.5. If non-enteric-coated preparations are used, acid-suppressing drugs may be required. Fecal fat concentrations are monitored to determine if the enzyme treatment has been effective.

CYSTIC FIBROSIS

Cystic fibrosis is the most common life-threatening genetic disorder among Caucasians. It is caused by a mutation that results in unusually viscous exocrine secretions, leading to a broad range of serious complications. Until a few decades ago, few infants born with cystic fibrosis survived to adulthood. Now, with early detection and advances in medical treatment, the average life span has increased to approximately 30 years.[12]

Consequences of Cystic Fibrosis The major complications of cystic fibrosis involve the lungs, pancreas, and sweat glands. Individuals with cystic fibrosis develop persistent respiratory infections, which cause inflammation of the bronchial tissues and progressive airway obstruction. Thickened pancreatic secretions obstruct the pancreatic ducts, causing digestive enzymes to accumulate in the pancreas and destroy pancreatic tissue. Fewer pancreatic enzymes reach the small intestine, leading to malabsorption of protein, fat, and fat-soluble vitamins. Salt losses in sweat are usually excessive, increasing the risk of dehydration. Children with cystic fibrosis are chronically undernourished, grow poorly, and have difficulty maintaining normal body weight. Complications that may develop over time include pancreatitis, hyperglycemia (due to destruction of insulin-producing cells), and diabetes.

Medical Nutrition Therapy for Cystic Fibrosis Energy needs are high in patients with cystic fibrosis due to increased requirements, nutrient malabsorption, and reduced food consumption. Children may need to consume between 120 and 150 percent of recommended energy intakes to achieve normal growth and maintain optimal nutrition status.[13] A patient with cystic fibrosis is encouraged to eat high-kcalorie and high-fat foods, eat frequent meals and snacks, and supplement meals with milk shakes or liquid dietary supplements. Supplemental tube feedings can help to improve nutrition status if energy intakes are inadequate.

Pancreatic enzyme replacement therapy is a central feature of cystic fibrosis treatment. Supplemental enzymes must be included with every meal or snack. For infants and small children, the contents of capsules are mixed in small amounts of liquid or a soft food (like applesauce) and fed with a spoon. Enzyme dosages may need to be adjusted if malabsorption continues, as evidenced by poor growth or GI symptoms such as steatorrhea, intestinal gas, or abdominal pain.

The risk of nutrient deficiency depends on the degree of malabsorption; nutrients of greatest concern include fat-soluble vitamins, essential fatty acids, and calcium. Multivitamin and fat-soluble vitamin supplements are routinely recommended. The liberal use of table salt and salty foods is encouraged to make up for losses of sodium in sweat.

CELIAC DISEASE

Celiac disease is characterized by an abnormal immune response to a protein fraction in **wheat gluten** and to related proteins in barley and rye. The reaction to gluten causes severe damage to the intestinal mucosa and subse-

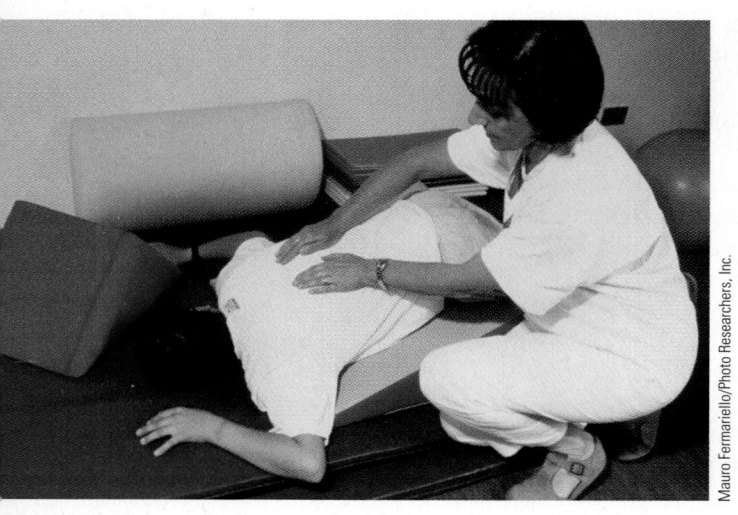

Postural drainage, a type of physical therapy used in cystic fibrosis, helps to clear the thick, sticky secretions that block airways and increase infection risk.

Mauro Fermariello/Photo Researchers, Inc.

quent malabsorption. Symptoms include GI disturbances like diarrhea, steatorrhea, and flatulence. Celiac disease may affect as many as 1 in every 133 persons in the United States.[14]

Consequences of Celiac Disease The immune reaction to gluten can cause striking changes in intestinal tissue. In affected areas, the absorptive surface appears flattened, and villi may be shortened or absent. The reduction in mucosal surface area—and, hence, in intestinal digestive enzymes—may be substantial. The damage may be restricted to the duodenum or may involve the full length of the small intestine. People with severe cases may malabsorb all nutrients to some degree, especially the macronutrients, fat-soluble vitamins, electrolytes, calcium, magnesium, zinc, iron, folate, and vitamin B_{12}. Because lactase deficiency can result from mucosal damage, GI symptoms may be exacerbated by milk products. Children with celiac disease often have stunted growth and are severely underweight. As a result of nutrient deficiencies, iron-deficiency anemia is common, and bone mineral density is usually low.

Some gluten-sensitive individuals may have few GI symptoms but react to gluten by developing a severe rash. This condition is called **dermatitis herpetiformis** and requires dietary adjustments similar to those for celiac disease.

Medical Nutrition Therapy for Celiac Disease The treatment for celiac disease is lifelong adherence to a gluten-free diet. Improvement in symptoms is often evident within several weeks, although mucosal healing can sometimes take years. If lactase deficiency is suspected, lactose-containing foods should be avoided until the intestine has recovered. Dietary supplements can be used to meet micronutrient needs and reverse deficiencies.

The gluten-free diet eliminates foods that contain wheat, barley, and rye (see Table 18-5, p. 480). Because many foods contain ingredients derived from these grains, foods that are problematic are not always obvious. Even small amounts of gluten may cause symptoms in some persons, so ingredient lists on food labels should be checked carefully. Special gluten-free products can be purchased to replace common food items such as bread, pasta, and cereals. Although somewhat expensive, these foods increase food choices and allow celiac patients to enjoy foods that would otherwise be forbidden.

Although most people with celiac disease can safely consume moderate amounts of oats, oats grown in the United States may be contaminated with wheat, barley, or rye.[15] Oats are often grown in rotation with other grains and may become contaminated during harvesting or processing. Although some oat millers have developed procedures that remove other grains from oats, no oats available in the United States are guaranteed to be gluten-free. Individuals who wish to try oats can be advised to limit their intakes to the amounts found to be safe (about $1/2$ cup of dry rolled oats per day) and to contact millers or product manufacturers to determine whether contamination with gluten-containing products is likely.[16]

INFLAMMATORY BOWEL DISEASES

Inflammatory bowel diseases are chronic inflammatory conditions involving the GI tract. Both genetic and environmental factors are believed to contribute to the development of these diseases, but the exact triggers are unknown. Table 18-6 (p. 481) compares the two major forms of inflammatory bowel disease, **Crohn's disease** and **ulcerative colitis**. Crohn's disease usually involves the small intestine and may lead to nutrient malabsorption, whereas ulcerative colitis affects the colon, which is past the absorptive areas. Both diseases may cause nutrient losses due to tissue damage, bleeding, and diarrhea.

Complications of Crohn's Disease Crohn's disease may occur in any region of the GI tract, but most cases involve the ileum and/or colon. Lesions may develop

NURSING DIAGNOSIS

for people with celiac disease may include *diarrhea, risk for disproportionate growth,* and *imbalanced nutrition: less than body requirements.*

dermatitis herpetiformis (DERM-ah-TYE-tis HER-peh-tih-FOR-mis): a gluten-sensitive disorder characterized by a severe skin rash. Gastrointestinal symptoms may be mild or absent.

Crohn's disease: an inflammatory bowel disease that usually occurs in the lower portion of the small intestine and the colon. Inflammation may pervade the entire intestinal wall.

ulcerative colitis (ko-LYE-tis): an inflammatory bowel disease that involves the colon. Inflammation affects the mucosa and submucosa.

Gluten-free products help people with celiac disease enjoy a wider variety of foods.

TABLE 18-5	Gluten-Free Diet
Meat and meat alternates	*Recommended:* Fresh, frozen, salted, and smoked meats (unless processed meats contain any prohibited grains); products made with hydrolyzed vegetable protein (HVP) or hydrolyzed plant protein (HPP); eggs; dried beans and peas; tofu. *Questionable:* Luncheon meats, sandwich spreads, meat loaf, frozen burgers, sausage, imitation meat products, meat extenders, egg substitutes, dried egg products, dry roasted nuts, peanut butter. *Avoid:* Products that are breaded or prepared in cream sauces, gravies.
Milk and milk products	*Recommended:* Milk, buttermilk, plain yogurt, cheese. *Questionable:* Milk shakes, cheese spreads, flavored yogurt, frozen yogurt, chocolate milk. *Avoid:* Malted milk and malted milk powders.
Breads, cereals, rice, and pasta	*Recommended:* Breads, baked products, and cereals made with corn, rice, soy, potato starch, potato flour, hominy, buckwheat, millet, teff, sorghum, amaranth, quinoa, arrowroot, and tapioca; pasta and noodles made with grains or starches listed above; corn tacos and corn tortillas. *Questionable:* Oatmeal and oat bran; rice crackers, rice cakes, and corn cakes. *Avoid:* Breads, baked products, cereals, tortillas, or pastas made with wheat, rye, barley, triticale, spelt, kamut, wheat germ, wheat bran, graham flour, durum flour, wheat starch, bulgur, farina, or semolina from wheat; commercially prepared mixes for biscuits, cornbread, muffins, pancakes, or waffles; malt and malt flavoring; pretzels; matzo.
Fruits and vegetables	*Recommended:* Any unprocessed fruits or vegetables. *Questionable:* French fries, especially in fast-food restaurants; commercial salad dressings; fruit pie fillings; dried fruits. *Avoid:* Scalloped potatoes (with wheat flour), creamed vegetables, vegetables dipped in batters.
Desserts	*Recommended:* Ice cream, sherbet, egg custards, or gelatin desserts that do not contain gluten; pure baking chocolate; chocolate chips; hard candy. *Questionable:* Icing, powdered sugar, candies, chocolate bars, marshmallows. *Avoid:* Puddings thickened with wheat flour; ice cream or sherbets that contain gluten stabilizers; baked products or doughnuts made with wheat, rye, or barley; ice cream cones; licorice.
Beverages	*Recommended:* Coffee; tea; cocoa; soft drinks; distilled alcoholic beverages such as rum, gin, whiskey, and vodka; wine. *Questionable:* Instant tea or coffee, coffee substitutes, chocolate drinks, hot cocoa mixes. *Avoid:* Beer, ale, lager, malted beverages, cereal beverages (Postum), beverages that contain nondairy cream substitutes.

Source: Adapted from American Dietetic Association, *Manual of Clinical Dietetics* (Chicago: American Dietetic Association, 2000), Chapter 11, Celiac disease.

in different areas in the intestine, with normal tissue separating affected regions (called "skip" lesions). The inflammation may extend deeply into intestinal tissue and be accompanied by ulcerations, fissures, and **fistulas** (abnormal passages between tissues). The resultant scar tissue thickens and stiffens the intestine, narrowing the lumen and sometimes causing obstructions. About 60 to 70 percent of patients require surgical resections during the course of illness, although disease often recurs in the remaining intestine. Patients with Crohn's disease are at increased risk of developing intestinal cancers.

Malnutrition may result from the reduced food intake, malabsorption, and surgical resections that shorten the bowel. If the ileum is affected, bile acids may become depleted, causing malabsorption of fat, fat-soluble vitamins, calcium, magnesium, and zinc; similarly, vitamin B_{12} deficiency can develop unless supplements

fistulas (FIST-you-luz): abnormal passages between body tissues; may lead from one hollow organ to another or to an organ surface.

TABLE 18-6	Comparison of Crohn's Disease and Ulcerative Colitis	
	Crohn's Disease	**Ulcerative Colitis**
Location of inflammation	Approximately 40% of cases involve the ileum and cecum, 30% are in the small intestine only, and 20% are in the colon.	Inflammation is confined to the rectum and colon; it begins at the rectum and spreads into the colon.
Pattern of inflammation	Discrete areas separated by normal tissue ("skip" lesions)	Continuous inflammation that begins at the rectum and ends abruptly within the colon
Depth of damage	Damage throughout all layers of tissue; causes deep fissures that give intestinal tissue a "cobblestone" appearance	Damage primarily in the mucosa and submucosa
Fistulas	Common	Usually do not occur
Cancer risk	Increased	Greatly increased

are taken. Anemia may result from nutrient malabsorption, bleeding, or the metabolic effects of chronic illness.

Complications of Ulcerative Colitis Ulcerative colitis always involves the rectum and usually extends into the colon. Inflammation is continuous along the length of intestine affected, ending abruptly at the area where healthy tissue begins. Tissue erosion or ulceration develops primarily in the mucosa and submucosa (the top two layers of intestinal tissue). During active episodes, patients have frequent, urgent bowel movements that are small in volume. Stools are often streaked with blood and contain mucus.

Although mild disease may cause few complications, weight loss, fever, and weakness are common when most of the colon is involved. Severe disease is often associated with anemia (due to blood loss), dehydration, and electrolyte imbalances. Protein losses from inflamed tissue can be substantial. A **colectomy** (removal

Most of the bile used during digestion is eventually reabsorbed in the ileum and returned to the liver.

Vitamin B_{12} is absorbed in the ileum. Some patients with Crohn's disease require regular vitamin B_{12} injections.

colectomy: removal of a portion or all of the colon.

NURSING DIAGNOSIS

for people with inflammatory bowel diseases may include *acute pain, nausea, diarrhea, risk for deficient fluid volume,* and *imbalanced nutrition: less than body requirements.*

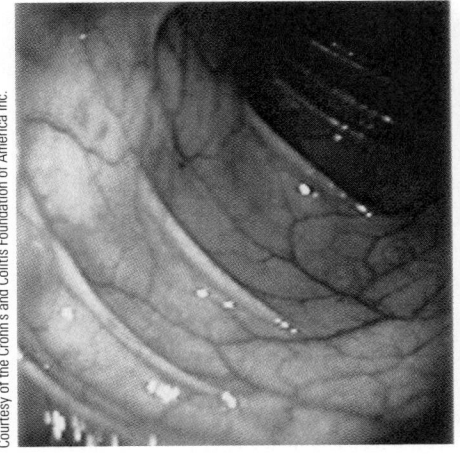

The healthy colon has a smooth surface with a visible pattern of fine blood vessels.

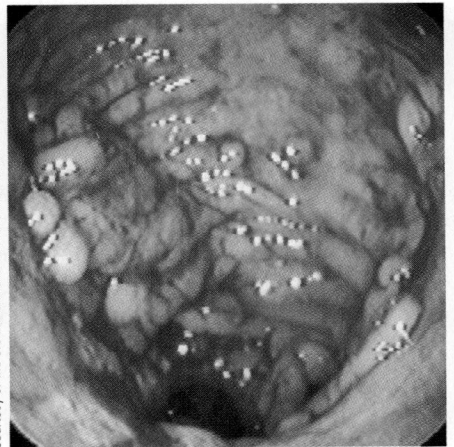

In Crohn's disease, the mucosa has a "cobblestone" appearance due to deep fissuring in the inflamed mucosal tissue.

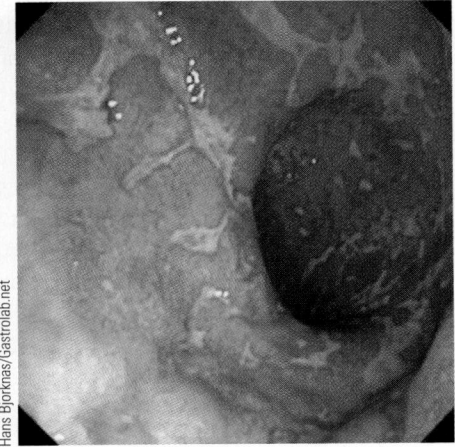

In ulcerative colitis, the colon appears inflamed and reddened, and ulcers are visible.

DIET-DRUG INTERACTIONS Check this table for notable nutrition-related effects of the medications discussed in this chapter.

	Gastrointestinal Effects	Interactions with Dietary Substances	Metabolic Effects
Antidiarrheals	Constipation		
Anti-inflammatory drugs (sulfasalazine, corticosteroids)	Nausea, heartburn (sulfasalazine)	Sulfasalazine may decrease folate absorption; supplementation is recommended.	Anemia (sulfasalazine); fluid retention, hyperglycemia, hypocalcemia, hypokalemia, hypophosphatemia, increased appetite, protein catabolism (corticosteroids).
Laxatives	Diarrhea	Mineral oil may decrease absorption of fat-soluble vitamins, but is not often used.	Fluid and electrolyte imbalances, laxative dependency.
Pancreatic enzyme replacements	Diarrhea, nausea, stomach cramps, irritation to GI mucosa		

of the colon) is performed in 20 to 25 percent of patients and prevents future recurrence.[17] Colon cancer risk is substantially increased in ulcerative colitis patients.

Drug Treatment of Inflammatory Bowel Diseases Drug therapies help to control symptoms, reduce inflammation, and minimize complications. The medications prescribed include antidiarrheal agents, immunosuppressants, and anti-inflammatory drugs (usually corticosteroids and salicylates). Although these medications help in achieving and maintaining remission, some may cause side effects that are detrimental to nutrition status (check the Diet-Drug Interactions feature).

Medical Nutrition Therapy for Inflammatory Bowel Diseases Crohn's disease often requires aggressive dietary management, as it can lead to growth failure in children, protein-energy malnutrition (PEM), and nutrient deficiencies. Nutrition care of patients is highly variable, however; dietary measures depend on the symptoms and complications that develop (see Table 18-7). Some people require high-kcalorie, high-protein diets to prevent or treat malnutrition. Liquid supplements can help to increase energy intakes and improve weight gain. Tube feedings can be used to supplement the diet or may be the sole means of providing nutrients; elemental formulas are most easily tolerated. Some individuals may tolerate meals better if high-fiber and lactose-containing foods are restricted. Vitamin and mineral supplements are generally needed, especially if malabsorption is present. Table 18-7 includes examples of other adjustments that may be beneficial.

The diet for ulcerative colitis may require few adjustments. During severe illness, dietary goals are to restore fluid and electrolyte balances and correct deficiencies that result from protein and blood losses. A low-fiber diet may reduce irritation by minimizing fecal volume. Sometimes food must be withheld to allow bowel rest. Fluids and electrolytes may need to be replaced intravenously, and some patients require parenteral nutrition. As in Crohn's disease, the symptoms and complications that develop can be managed with specific dietary measures (see Table 18-7).

SHORT-BOWEL SYNDROME

The treatment of Crohn's disease, cancer of the small intestine, and other intestinal disorders may include the surgical removal (resection) of a major portion of

TABLE 18-7	Management of Symptoms and Complications in Crohn's Disease
Symptom or Complication	**Possible Dietary Measures**
Growth failure/weight loss	High-kCalorie diet
	Enteral supplements
	Elemental tube feedings
Anorexia/pain with eating	Small, frequent meals
	Enteral supplements
	If long-term (> 5 to 7 days): elemental tube feedings
Malabsorption	High-kCalorie diet
	Nutrient supplementation
Steatorrhea	Fat restriction
	Medium-chain triglycerides
	Nutrient supplementation
Diarrhea	Fluid and electrolyte replacement
	Nutrient supplementation
Lactose intolerance	Avoidance of lactose-containing foods
Nutrient deficiencies	Nutrient-dense diet
	Nutrient supplementation
Intestinal recovery	High-protein diet
	Glutamine supplementation
Strictures/fistulas	Low-fiber diet
Severe bowel obstruction/high-output fistulas/severe exacerbations of disease	Total parenteral nutrition

the small intestine. **Short-bowel syndrome** is the malabsorption syndrome that results when the absorptive capacity of the remaining intestine is insufficient for meeting nutritional needs. Without appropriate dietary adjustments, short-bowel syndrome can result in fluid and electrolyte imbalances and multiple nutrient deficiencies. Symptoms include diarrhea, steatorrhea, dehydration, weight loss, and growth impairment in children.

Figure 18-2 (p. 484) reviews nutrient absorption in the GI tract and describes how absorption is affected by surgical resections. Generally, up to 50 percent of the small intestine can be resected without serious nutritional consequences. More extensive resections lead to generalized malabsorption, and patients may need lifelong parenteral nutrition to supplement oral intakes.

Intestinal Adaptation After a resection, the remaining intestine undergoes **intestinal adaptation,** an adaptive response that dramatically improves its absorptive efficiency. Adaptation begins soon after surgery and continues for several years. The ileum has a greater capacity for adaptation than the jejunum; thus removal of the ileum ultimately has more severe consequences than removal of the jejunum.

short-bowel syndrome: a malabsorption syndrome that follows resection of the small intestine which causes insufficient absorptive capacity in the remaining intestine.

intestinal adaptation: after resection, the process of intestinal recovery that leads to improved absorptive capacity.

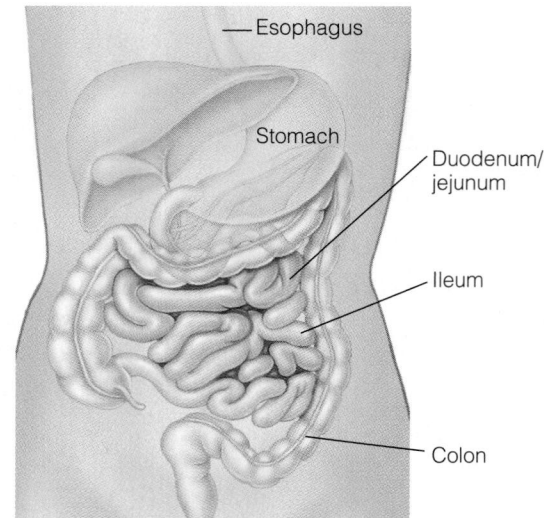

WHAT IS ABSORBED

Duodenum/ jejunum
- Simple carbohydrates
- Fats
- Amino acids
- Vitamins*
- Minerals*
- Water

Ileum
- Bile salts
- Vitamin B$_{12}$
- Water
(Assumes absorptive function of duodenum and jejunum with adaptation)

Colon
- Water
- Electrolytes
- Short-chain fatty acids

POSSIBLE CONSEQUENCES OF RESECTION

Duodenum/ jejunum
- Minimal consequences if the ileum remains intact
- Calcium and iron malabsorption if duodenum resected

Ileum
- Fat malabsorption
- Protein malabsorption
- Malabsorption of fat-soluble vitamins and vitamin B$_{12}$
- Reduced calcium, magnesium, and zinc absorption
- Fluid losses
- Diarrhea/steatorrhea

Colon
- Fluid and electrolyte losses
- Diarrhea
(Losses are compounded if ileum is also resected)

Labels on diagram: Esophagus, Stomach, Duodenum/jejunum, Ileum, Colon

*The absorption of vitamins and minerals begins in the duodenum and continues throughout the length of the small intestine.

FIGURE 18-2 Nutrient Absorption and Consequences of Intestinal Surgeries
About 90 to 95 percent of nutrient absorption takes place in the first half of the small intestine. After a resection, nutrient absorption may be reduced.

Resection of the ileum can have permanent consequences on both vitamin B$_{12}$ nutrition and bile acid reabsorption, and depletion of bile acids worsens both fat malabsorption and diarrhea. If the sphincter between the ileum and cecum is removed, colonic bacteria can infiltrate the small intestine and cause bacterial overgrowth. A functional colon is helpful because its resident bacteria metabolize unabsorbed nutrients and produce some usable nutrients. An intact colon also helps to reduce losses of fluids and electrolytes.

Medical Nutrition Therapy for Short-Bowel Syndrome Immediately after resection, fluids and electrolytes must be supplied intravenously. In the first few weeks after surgery, fluid losses from diarrhea can be substantial (sometimes exceeding 2 liters per day), so appropriate rehydration is critical to recovery. Diarrhea gradually lessens as intestinal adaptation progresses.

Total parenteral nutrition meets nutrient needs after surgery and is gradually reduced as oral feedings increase. To promote intestinal adaptation, oral feedings are started as soon as possible, in as little as five days after surgery in some cases. Initial oral intakes may consist of occasional sips of liquid formulas, progressing to larger amounts, and then to solid foods, as tolerated. Very small, frequent feedings can utilize the remaining intestine most efficiently. Low-fat diets are sometimes suggested to reduce steatorrhea, and a high-carbohydrate diet can help to stimulate production of short-chain fatty acids in the colon. Lactose-containing foods are sometimes poorly tolerated. Vitamin and mineral supplements assist in meeting nutrient needs during the period of adapatation. Patients with short-bowel syndrome are at high risk of developing kidney stones due to calcium malabsorption (see p. 474), so a low-oxalate diet is often suggested. Use the accompanying Case Study to review the material on Crohn's disease and short-bowel syndrome.

REVIEW NOTES

Widespread malabsorption can be a consequence of diseases that disrupt normal digestion and absorption.

CASE STUDY   Economist with Short-Bowel Syndrome

Judi Morel is a 28-year-old economist with an 8-year history of Crohn's disease. Judi is 5 feet 7 inches tall. Three years ago, she underwent a small bowel resection and remained free of active disease for two years. During that time, her symptoms subsided; she was able to tolerate most foods without any problem and gained weight. Ten months ago, Judi experienced a severe flare-up of her Crohn's disease. Since that time, she has lost 15 pounds and currently weighs 118 pounds. She has experienced severe abdominal pain and fatigue that have persisted despite aggressive medical management that included intravenous nutrition. Five days ago, Judi finally underwent another resection, which left her with 40 percent of healthy small intestine. Her colon is intact. She is experiencing extensive diarrhea.

1. Describe the manifestations of Crohn's disease, and explain why surgery is sometimes performed as part of the treatment. Describe the complications that may affect nutrient needs in people with Crohn's disease.
2. What nutrition concerns are suggested by Judi's recent weight loss? What other nutritional problems did Judi probably experience as a consequence of Crohn's disease?
3. Discuss the complications that may follow an extensive intestinal resection. What factors may affect a person's ability to meet nutrient needs with an oral diet?
4. Discuss the dietary progression that is recommended following an intestinal resection. After Judi is able to eat solid foods, what factors may affect the type of diet that is recommended for her?

Enzyme deficiencies result from chronic pancreatitis and cystic fibrosis, and pancreatic enzyme replacement is central to treatment.

Celiac disease and Crohn's disease cause damage to the intestinal mucosa and impair its absorptive function.

Ulcerative colitis is an inflammatory bowel condition that affects the colon only, and severe cases can be cured by colectomy.

Short-bowel syndrome can be a consequence of resections of the small intestine, but intestinal adaptation can significantly improve absorptive capacity.

Disorders of the Large Intestine

The large intestine moves undigested materials to the rectum and helps to maintain fluid and electrolyte balances. Its resident bacteria ferment undigested nutrients and produce short-chain fatty acids and some vitamins, which our bodies can absorb and use. This section describes several conditions that may disrupt the normal functioning of the large intestine.

IRRITABLE BOWEL SYNDROME

People with **irritable bowel syndrome** experience chronic and recurring intestinal symptoms that cannot be explained by specific physical abnormalities. Symptoms include both diarrhea and constipation, as well as flatulence, bloating, and distention. In some patients, symptoms may be mild; in others, defecation disturbances can interfere with work and social activities and dramatically alter a person's lifestyle and sense of well-being. Irritable bowel syndrome accounts for about 12 percent of primary care visits and is more common in women than in men. About 30 percent of people with the disorder eventually become asymptomatic.[18]

Although the causes of the disorder remain elusive, people with irritable bowel syndrome tend to have excessive colonic responses to meals, GI hormones, and stress.[19] Many persons exhibit hypersensitivity to intestinal distention; for example, they may have an exaggerated response to normal meal transit and intestinal gas. Intestinal motility after meals may be excessive, leading to diarrhea, or reduced, causing constipation. Symptoms often worsen during periods of psychological stress, and in some cases, stressful life events may trigger the illness. Some evidence

NURSING DIAGNOSIS

for people with irritable bowel syndrome may include *diarrhea, constipation, acute pain, chronic pain, ineffective health maintenance,* and *ineffective therapeutic regimen management.*

irritable bowel syndrome: an intestinal disorder of unknown cause that affects the functioning of the lower bowel. Symptoms include abdominal pain, flatulence, diarrhea, and constipation.

CASE STUDY   New College Graduate with Irritable Bowel Syndrome

Marcy Hudson is a 22-year-old recent college graduate who began her first professional job in a bank one month ago. As a college student, she occasionally experienced abdominal pain and cramping after eating. She also had frequent bouts of diarrhea and felt somewhat better after bowel movements. Once Marcy began her new job, her symptoms occurred more frequently. At first she attributed her symptoms to job stress, but when the symptoms continued for several months, she decided to see her physician. After taking a careful history and conducting tests to rule out other bowel disorders, the physician diagnosed irritable bowel syndrome. The physician prescribed bulk-forming agents and advised Marcy to keep a record of her food intake and symptoms for one week. Marcy was then referred to a dietitian for a review of her dietary record. The dietitian noticed that Marcy routinely drank several cups of coffee in the morning and had large meals for

lunch and dinner. Marcy often ate out in Mexican restaurants and favored highly spiced foods and refried beans. Between meals, she snacked on low-carb foods containing sugar alcohols and drank several cans of soda daily. Her dietary fiber intake, however, totaled only about 15 grams daily.

1. Describe the characteristics of irritable bowel syndrome to Marcy, and indicate the role that stress might play in her illness.
2. Explain how the record of food intake and symptoms might be helpful in devising an appropriate diet plan for Marcy. Are any of the foods in Marcy's diet likely to be aggravating her symptoms?
3. What dietary measures may benefit individuals with irritable bowel syndrome? What problems might the dietary changes cause?

---

suggests that an infection may be the cause of the initial GI disturbance and that tissue sensitization persists after the infection has healed.

Treatment of Irritable Bowel Syndrome Medical treatment of irritable bowel syndrome often includes dietary adjustments, stress management, and behavioral therapies. Medications may be prescribed to manage symptoms although they are not always helpful. The drugs prescribed may include antidiarrheal agents, anticholinergics, antidepressants, and laxatives.

Medical Nutrition Therapy for Irritable Bowel Syndrome Although dietary changes may be useful, measures that help one symptom can sometimes make another worse. The usual dietary advice is to increase fiber intake, which helps to reduce constipation and improve stool bulk. To minimize discomfort from intestinal gas, fiber-containing foods should be added gradually. Other foods that produce gas should be avoided unless well tolerated (review Table 18-2). If diarrhea persists, a bulking agent (psyllium) may be effective. Avoidance of milk products may benefit those who are lactose intolerant. Caffeine and alcohol can exacerbate symptoms. The placebo effect has a strong influence on food tolerance, so foods that are perceived to be problematic should be discussed with patients so that the diet is not restricted unnecessarily.

The diet history may reveal dietary behaviors that worsen symptoms. Generally, small, frequent meals are better tolerated than larger ones. Eating quickly should be discouraged as it increases the amount of air that is swallowed. Some patients may find a low-fat diet easier to tolerate. Fluid intake should be assessed for adequacy. A careful evaluation of the dietary patterns that exacerbate symptoms may uncover the foods and habits most closely associated with intestinal discomfort. Review the accompanying Case Study to apply your knowledge about irritable bowel syndrome to a clinical situation.

diverticulosis (DYE-ver-tic-you-LOH-sis): a disorder characterized by the presence of small outpockets (called diverticula) in the intestinal wall, which often develop by middle age.

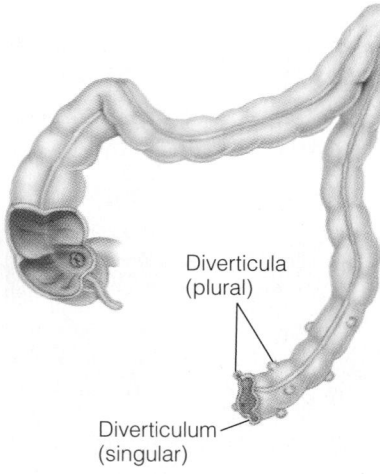

Diverticula (plural)

Diverticulum (singular)

FIGURE 18-3 Diverticula in the Colon
Diverticula are small pouches that develop in weakened areas of the intestinal wall. The condition of having diverticula is known as diverticulosis.

DIVERTICULAR DISEASE OF THE COLON

Diverticulosis refers to the presence of pebble-sized outpockets in the intestinal wall, called diverticula (see Figure 18-3). The prevalence of diverticulosis increases with age, occurring in about half of adults over 60 years of age.[20] Most people with diverticulosis are symptom-free and remain unaware of the condition until a complication develops.

Dietary fiber influences the development of diverticulosis. By increasing stool weight and bulk, fiber reduces the workload of the circular muscles that move wastes through the colon. Low-fiber diets require more vigorous muscle contractions, increasing the pressure within segments immediately adjacent to the circular muscles. This increased pressure induces small areas of intestinal tissue to balloon outward over time.[21]

Diverticulitis Inflammation or infection sometimes develops in the area around a diverticulum. This condition, called **diverticulitis**, is the most common complication of diverticulosis. It is thought to result from hardened fecal matter that abrades the mucosal lining, causing inflammation and possibly a microperforation that leads to subsequent infection. If the infection spreads to adjacent organs, fistulas may develop. More rarely, the infection spreads to the peritoneal cavity, causing life-threatening illness. Symptoms of diverticulitis include persistent abdominal pain, fever, and alternating constipation and diarrhea.

Treatment for Diverticular Disease Treatment for diverticulosis is necessary only if symptoms develop; the treatment focuses on reducing pain and alleviating constipation. Increasing dietary fiber may help to prevent disease progression, so patients are sometimes advised to add wheat bran to meals or use bulk-forming agents such as psyllium. Although dietary advice sometimes includes a recommendation to avoid nuts, popcorn, and foods that contain seeds, there is no evidence that these restrictions can reduce complications.[22]

Patients with diverticulitis may need antibiotics to treat infections and, possibly, pain-control medications. In mild cases, a clear liquid diet may be advised initially, with progression to solid foods as tolerated. In more severe cases, bowel rest is necessary (oral fluids and food are withheld), and fluids are given intravenously; oral intakes are gradually reintroduced as the condition improves. Surgical interventions are sometimes necessary to treat complications of diverticulitis and may include removal of the affected portion of colon.

OSTOMIES

An ostomy is a surgically created opening (called a **stoma**) in the abdominal wall through which dietary waste can be eliminated. A temporary ostomy is sometimes constructed so that part or all of the colon can be bypassed after injury or extensive surgery. A permanent ostomy is necessary after a partial or total colectomy. To construct the stoma, the cut end of the remaining segment of functioning intestine is brought through an opening in the abdominal wall and stitched in place so that it empties to the exterior. The stoma can be formed from a section of the colon (**colostomy**) or ileum (**ileostomy**), as shown in Figure 18-4 (p. 488).

To collect waste, a disposable bag is affixed to the skin around the stoma and emptied during the day, as needed. Alternatively, an interior pouch can be surgically created behind the stoma using intestinal tissue and emptied with a catheter when convenient. Stool consistency varies according to the length of colon that is functional. If a small portion of the colon is absent or bypassed, the stools may continue to be semisolid. If the entire colon has been removed or is by bypassed, absorption of fluid and electrolytes into the body is reduced substantially, and the output is liquid.

Medical Nutrition Therapy for Ostomies Following surgery, the diet gradually progresses from clear liquids (low in sugars) to a normal meal plan that contains low-fiber foods. Small, frequent meals may be better tolerated at first. Questionable foods should be added back to the diet one at a time and in small amounts to assess their effects. A food that causes problems can be tried again at a later time.

NURSING DIAGNOSIS

for people with diverticulitis may include *acute pain, diarrhea, constipation,* and *risk for deficient fluid volume* (due to diarrhea).

NURSING DIAGNOSIS

appropriate for people with ostomies include *disturbed body image, risk for diarrhea, risk for deficient fluid volume, ineffective therapeutic regimen management,* and *risk for impaired skin integrity.*

diverticulitis (DYE-ver-tic-you-LYE-tis): an inflammation or infection that involves diverticula.

stoma (STOE-ma): a surgical opening made in the abdominal wall.

colostomy (co-LAHS-toe-me): a surgical procedure that creates a stoma using a section of the colon.

ileostomy (ill-ee-OS-toe-me): a surgical procedure that creates a stoma using the ileum.

FIGURE 18-4 Colostomy and Ileostomy

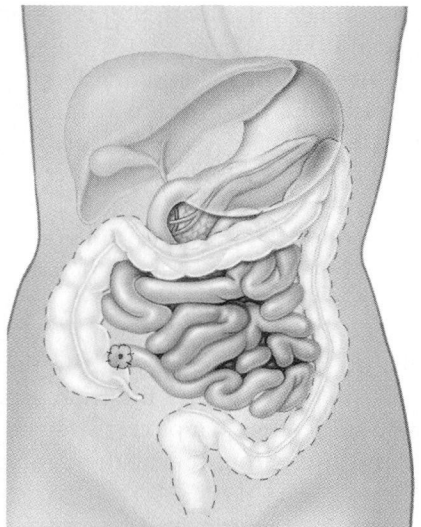

Colostomy	Ileostomy
In a colostomy, a portion of the colon is removed or bypassed, and the stoma is formed from the remaining section of functional colon.	In an ileostomy, the entire colon is removed or bypassed, and the stoma is formed from the ileum.

People with ileostomies need to chew thoroughly to ensure that foods can be adequately digested and to prevent obstructions, which are a common complication. Foods high in insoluble fibers are sometimes avoided because they reduce intestinal transit time and may increase output. Because the colon is no longer available for reabsorbing water, the diet should provide at least 8 cups of liquid daily to prevent dehydration.

Dietary concerns after colostomies depend on the length of colon removed. If a large portion is removed, recommendations may be similar to those given to ileostomy patients. In other cases, a high-fiber diet is often recommended to improve stool consistency and promote regularity.

Obstructions Foods that are incompletely digested can cause obstructions, a primary concern of ileostomy patients. Although almost any food can be consumed if cut into small pieces and carefully chewed, the following foods may cause difficulty: corn, cabbage, celery, coconut, dried fruit, green peppers, lettuce, nuts, peas, pineapple, popcorn, spinach, and turnip greens.[23]

Reducing Gas and Odors Persons with ostomies are often concerned about foods that may increase gas production or cause strong odors. Foods that may cause excessive gas include those listed in Table 18-2 on p. 472; practices that increase gas formation include smoking, gum chewing, using drinking straws, and eating quickly. Foods that sometimes produce unpleasant odors include fish, eggs, dried beans and peas, onions, garlic, asparagus, brussels sprouts, green peppers, and beer. Foods that may help to reduce odors include buttermilk, cranberry juice, parsley, and yogurt.[24]

Diarrhea Examples of foods that may either aggravate or reduce diarrhea were listed in Table 18-3 on p. 473. What works may differ for each individual, however, and is best determined by trial and error.

REVIEW NOTES

Irritable bowel syndrome is associated with abdominal pain and alternating diarrhea and constipation. Diverticulosis is often asymptomatic until complications develop. Both irritable bowel syndrome and diverticulosis may benefit from a high-fiber diet.

Colostomies and ileostomies are surgical procedures in which stomas are created in the abdominal wall using the colon or ileum. Fluid and electrolyte requirements are greater after an ostomy because colon function is reduced or absent.

Although foods that are poorly digested may cause obstructions in people with ostomies, thorough chewing can reduce risk. Foods can be avoided if they provoke diarrhea or cause excessive gas or strong odors, although individual tolerances differ.

NUTRITION ASSESSMENT CHECKLIST FOR PEOPLE WITH LOWER GI TRACT DISORDERS

Medical History

Check the medical record for diseases that:

- ☐ Interfere with pancreatic enzyme secretion, such as chronic pancreatitis or cystic fibrosis
- ☐ Interfere with nutrient absorption, such as celiac disease or Crohn's disease
- ☐ Cause chronic GI symptoms, such as irritable bowel syndrome or ulcerative colitis

Check for surgical procedures involving the lower GI tract, such as:

- ☐ Intestinal resection
- ☐ Ileostomy
- ☐ Colostomy

Check for the following symptoms or complications:

- ☐ Anemia
- ☐ Bacterial overgrowth
- ☐ Bone disease
- ☐ Constipation
- ☐ Diarrhea, dehydration
- ☐ Fistulas
- ☐ Lactose intolerance
- ☐ Nutrient deficiencies
- ☐ Obstructions
- ☐ Oxalate kidney stones
- ☐ Poor growth, in children
- ☐ Steatorrhea

Medications

Check for medications or dietary supplements that may:

- ☐ Cause constipation or diarrhea
- ☐ Interfere with food intake by causing nausea, vomiting, cramps, dry mouth, or drowsiness
- ☐ Alter appetite or nutrient needs

Dietary Intake

Note the following conditions and contact the dietitian if you suspect a problem with:

- ☐ Poor appetite or food intake
- ☐ Food intolerances
- ☐ Inadequate fiber intake, in those with constipation
- ☐ Lactose intolerance, in those with diarrhea
- ☐ Fluid intake

Anthropometric Data

Measure baseline height and weight. Address weight loss early to prevent malnutrition in patients with:

- ☐ Severe or persistent diarrhea
- ☐ Malabsorption

Laboratory Tests

Check laboratory tests for signs of dehydration, electrolyte imbalances, nutrient deficiencies, and nutrition-related anemias in patients with:

- ☐ Severe or persistent diarrhea
- ☐ Malabsorption
- ☐ Intestinal resections

Physical Signs

Look for physical signs of:

- ☐ Dehydration
- ☐ Protein-energy malnutrition
- ☐ Essential fatty acid and fat-soluble vitamin deficiencies
- ☐ Folate and vitamin B_{12} deficiencies
- ☐ Mineral deficiencies

SELF CHECK

1. The nurse advising an elderly patient with constipation encourages the patient to:
   a. consume a low-fat diet rich in potassium.
   b. consume a high-protein diet rich in calcium.
   c. gradually add high-fiber foods to the diet.
   d. eliminate gas-forming foods from the diet.

2. Osmotic diarrhea often results from:
   a. excessive motility of fluids within the colon.
   b. excessive fluid secretion by the intestines.
   c. viral, bacterial, or protozoal infections.
   d. nutrient malabsorption.

3. Common nutrition problems associated with bacterial overgrowth in the stomach and small intestine include:
   a. sensitivity to gluten.
   b. fat malabsorption and vitamin B_{12} deficiency.
   c. constipation.
   d. permanent loss of digestive enzymes.

4. Nutrition problems that may result from fat malabsorption include all of the following, *except:*
   a. weight loss.
   b. essential amino acid deficiencies.
   c. bone loss.
   d. oxalate kidney stones.

5. The majority of acute pancreatitis cases can be attributed to:
   a. bacterial and viral infections.
   b. cystic fibrosis.
   c. excessive alcohol use and gallstones.
   d. elevated triglyceride levels.

6. The most appropriate diet for a person with cystic fibrosis is a:
   a. high-kcalorie diet.
   b. high-fiber diet.
   c. gluten-free diet.
   d. fat-restricted diet.

7. A person on a gluten-free diet must avoid products containing:
   a. wheat, corn, and rice.
   b. barley, soybeans, and corn.
   c. wheat, barley, and rye.
   d. buckwheat, rice, and millet.

8. A patient with Crohn's disease may develop all of the following nutrition problems, *except:*
   a. fat malabsorption.
   b. dumping syndrome.
   c. vitamin B_{12} deficiency.
   d. anemia.

9. If 60 percent of the small intestine remains after a jejunal resection:
   a. lifelong parenteral nutrition is the only option available.
   b. bile salts must be taken orally.
   c. pancreatic enzyme replacement is required.
   d. oral diets may eventually be able to meet nutrient needs.

10. After an ileostomy, the most serious concern is that:
   a. the diet is too restrictive to meet nutrient needs.
   b. waste disposal causes frequent daily interruptions.
   c. incompletely digested foods may cause obstructions.
   d. fluid restrictions prevent patients from drinking beverages freely.

Answers to these questions appear in Appendix H.

CLINICAL APPLICATIONS

1. Using Table 18-4 on p. 475 as a guide, plan a day's menus for a diet containing 50 grams of fat. Take care to make the menus both palatable and nutritious. How can these menus be improved using the suggestions in the box on p. 476?

2. As stated in this chapter, treatment of celiac disease is deceptively simple—eliminate wheat, barley, and rye, and possibly oats. Remaining on a gluten-free diet is more challenging than it appears, however.

   • Randomly select ten of your favorite snack and convenience foods. Take a trip to the grocery store and check the labels of the products you selected to see if they would be allowed on a gluten-free diet. Keep in mind that the labels may not list all offending ingredients.

   • Find acceptable substitutes for the products that are not allowed, either by substituting other foods or by checking for gluten-free products in the grocery store. If you have access to the Internet, you may want to investigate websites that advertise gluten-free products to get an idea of what's available.

NUTRITION ON THE NET

For further study of the topics in this chapter, access these websites.

Find updates and quick links to these and other nutrition-related sites at our website: **www.wadsworth.com/nutrition**

Visit the websites of these organizations to find information that is helpful for both health practitioners and patients with gastrointestinal problems:

- American College of Gastroenterology: **www.acg.gi.org**
- American Gastroenterological Association: **www.gastro.org**
- National Institute of Diabetes and Digestive and Kidney Diseases, which is a division of the National Institutes of Health: **www.niddk.nih.gov**

- South Denver Gastroenterology: **www.gutfeelings.com**

Find more information about celiac disease by visiting these websites:

- Celiac Disease Foundation: **www.celiac.org**
- Canadian Celiac Association: **www.celiac.ca**
- Celiac Sprue Association: **www.csaceliacs.org**
- Gluten Intolerance Group: **www.gluten.net**

Learn more about inflammatory bowel diseases at the website of the Crohn's and Colitis Foundation of America: **www.ccfa.org**

Find additional information about cystic fibrosis at the website of the Cystic Fibrosis Foundation: **www.cff.org/home**

NOTES

[1] Standing Committee on the Scientific Evaluation of Dietary Reference Intakes, Food and Nutrition Board, Institute of Medicine, *Dietary Reference Intakes for Energy, Carbohydrate, Fiber, Fat, Fatty Acids, Cholesterol, Protein, and Amino Acids* (Washington, D.C.: National Academies Press, 2002).

[2] F. L. Suarez and M. D. Levitt, Intestinal gas, in M. Feldman, L. S. Friedman, and M. H. Sleisenger, eds., *Sleisenger and Fordtran's Gastrointestinal and Liver Disease: Pathophysiology, Diagnosis, Management* (Philadelphia: Saunders, 2002); L. J. Cheskin and D. L. Miller, Nutrition in the prevention and treatment of common gastrointestinal symptoms, in A. M. Coulston, C. L. Rock, and E. R. Monsen, eds., *Nutrition in the Prevention and Treatment of Disease* (San Diego: Academic Press, 2001), pp. 549–562; American Dietetic Association, *Manual of Clinical Dietetics* (Chicago: American Dietetic Association, 2000), p. 423.

[3] S. L. Friedman, K. R. McQuaid, and J. H. Grendell, eds., *Current Diagnosis and Treatment in Gastroenterology* (New York: Lange Medical Books/McGraw-Hill, 2003); M. Feldman, L. S. Friedman, and M. H. Sleisenger, eds., *Sleisenger and Fordtran's Gastrointestinal and Liver Disease: Pathophysiology, Diagnosis, Management* (Philadelphia: Saunders, 2002).

[4] Friedman, McQuaid, and Grendell, 2003; Feldman, Friedman, and Sleisenger, 2002.

[5] D. L. Swagerty, Jr., A. D. Walling, and R. M. Klein, Lactose intolerance, *American Family Physician* 65 (2002): 1845–1850; J. R. Saltzman and coauthors, A randomized trial of *Lactobacillus acidophilus* BG2FO4 to treat lactose intolerance, *American Journal of Clinical Nutrition* 69 (1999): 140–146.

[6] F. L. Suarez and coauthors, Lactose maldigestion is not an impediment to the intake of 1500 mg calcium daily as dairy products, *American Journal of Clinical Nutrition* 68 (1998): 1118–1122.

[7] L. D. McBean and G. D. Miller, Allaying fears and fallacies about lactose intolerance, *Journal of the American Dietetic Association* 98 (1998): 671–676.

[8] C. Owyang, Pancreatitis, in L. Goldman and D. Ausiello, eds., *Cecil Textbook of Medicine* (Philadelphia: Saunders, 2004), pp. 879–886.

[9] C. Dervenis, Enteral nutrition in severe acute pancreatitis: Future development, *Journal of the Pancreas (Online)* 5 (2004): 60–66; E. P. DiMagno and S. Chari, Acute pancreatitis, in M. Feldman, L. S. Friedman, and M. H. Sleisenger, eds., *Sleisenger and Fordtran's Gastrointestinal and Liver Disease: Pathophysiology, Diagnosis, Management* (Philadelphia: Saunders, 2002).

[10] C. Baum, D. Moxon, and M. Scott, Gastrointestinal disease, in B. A. Bowman and R. M. Russell, eds., *Present Knowledge in Nutrition* (Washington, D.C.: ILSI Press, 2001), pp. 472–482.

[11] Owyang, 2004.

[12] M. J. Welsh, Cystic fibrosis, in L. Goldman and D. Ausiello, eds., *Cecil Textbook of Medicine* (Philadelphia: Saunders, 2004), pp. 515–519.

[13] S. W. Powers and S. R. Patton, A comparison of nutrient intake between infants and toddlers with and without cystic fibrosis, *Journal of the American Dietetic Association* 103 (2003): 1620–1625.

[14] A. Fasano and coauthors, Prevalence of celiac disease in at-risk and not-at-risk groups in the United States: A large multicenter study, *Archives of Internal Medicine* 163 (2003): 286–292.

[15] American Dietetic Association, *Nutrition Care Manual* (Chicago: American Dietetic Association, 2005); L. Hogberg and coauthors, Oats to children with newly diagnosed coeliac disease: A randomised double blind study, *Gut* 53 (2004): 649–654; E. K. Janatuinen and coauthors, No harm from five year ingestion of oats in coeliac disease, *Gut* 50 (2002): 332–335.

[16] T. Thompson, Oats and the gluten-free diet, *Journal of the American Dietetic Association* 103 (2003): 376–379.

[17] W. F. Stenson, Inflammatory bowel disease, in L. Goldman and D. Ausiello, eds., *Cecil Textbook of Medicine* (Philadelphia: Saunders, 2004), pp. 861–868.

[18] N. J. Talley, Functional gastrointestinal disorders: Irritable bowel syndrome, nonulcer dyspepsia, and noncardiac chest pain, in L. Goldman and D. Ausiello, eds., *Cecil Textbook of Medicine* (Philadelphia: Saunders, 2004), pp. 806–814.

[19] Talley, 2004.

[20] C. L. Simmang and G. T. Shires, Diverticular disease of the colon, in M. Feldman, L. S. Friedman, and M. H. Sleisenger, eds., *Sleisenger and Fordtran's Gastrointestinal and Liver Disease: Pathophysiology, Diagnosis, Management* (Philadelphia: Saunders, 2002), pp. 2100–2112.

[21] B. E. Stabile and T. D. Arnell, Diverticular disease of the colon, in S. L. Friedman, K. R. McQuaid, and J. H. Grendell, eds., *Current Diagnosis and Treatment in Gastroenterology* (New York: Lange Medical Books/McGraw-Hill, 2003), pp. 436–451; Simmang and Shires, 2002.

[22] Stabile and Arnell, 2003.

[23] American Dietetic Association, 2005.

[24] American Dietetic Association, 2005.

Mental Health and Nutrition Status

Mentally healthy individuals have the capacity to feed themselves well and maintain healthy lifestyles. People with mental and emotional problems, however, often have poor diets and lifestyle behaviors that contribute to ill health. Many of the conditions discussed in this book can lead to emotional distress and loss of hope, causing a person to lose interest in sustaining healthy behaviors. Some illnesses may result in **dementia,** the loss of intellectual function. Psychiatric illnesses can also have detrimental effects on diet and health. The nurse who recognizes the relationship between physical illness and mental health is in a better position to offer effective care. The accompanying glossary defines the relevant terms.

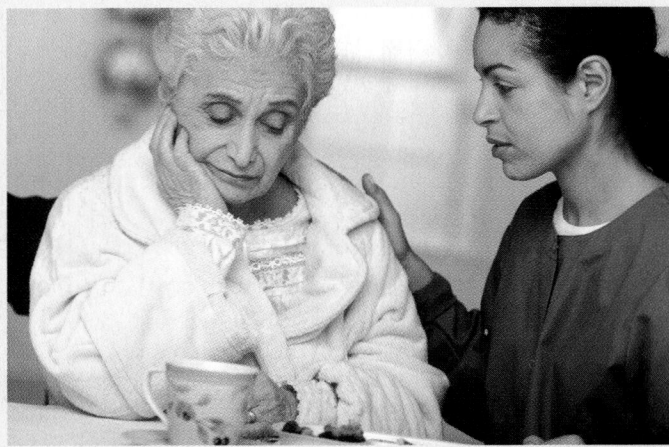

Depression and loneliness can profoundly affect nutrition status.

How can emotional and mental health affect what a person eats?

To understand the connection between mental health and nutrition, consider how depression and anxiety affect your own eating habits. Do you lose your appetite? Overeat? Eat "junk" foods instead of balanced meals? Although transient emotional stress may have little impact on nutrition status, prolonged emotional difficulties can lead to underweight, overweight, or nutrient imbalances.

Mental health problems that are difficult to overcome, such as depression, are likely to lead to nutritional problems. Quite often people who are depressed lose interest in caring for themselves and in participating in usual activities such as eating, socializing, or pursuing hobbies. When these individuals cut themselves off from pleasurable activities and friendships, depression deepens and becomes a self-aggravating condition. Thus people who are depressed may have little interest in preparing and eating food.

What factors may contribute to depression?

Factors contributing to depression include pain, loss of physical independence, and economic hardships imposed by serious illness. Terminal illnesses entail especially difficult emotional adjustments. Furthermore, when people become ill and lose significant amounts of weight, they may develop depression as they lose strength and become unable to perform routine tasks. The anxiety, pessimism, and sleeping difficulties that often accompany illness may worsen depression.[1] Table NP 18-1 lists examples of nursing diagnoses that may be associated with depression.

Medical conditions that are strongly linked with depression include gastrointestinal disorders, genitourinary conditions, musculoskeletal conditions, cardiovascular diseases, back problems, and asthma.[2] Depression is most common in hospitalized patients, affecting an estimated 10 to 14 percent.[3] Both treatment outcomes and death rates are worsened by depression.[4]

Disease treatments are sometimes overlooked as causes of depression. Medications linked to depression include some anticonvulsants, antihistamines, antihypertensives, antibiotics, and immunosuppressant agents.[5] Patients undergoing invasive treatments, recurring pain, and chronic disability are also at risk.

Caring nurses should remain alert for signs of depression. If it is recognized early, treatment can be initiated before health status markedly deteriorates. Therapeutic strategies may include support groups, an exercise program, psychotherapy, or, if needed, antidepressant medications.

TABLE NP18-1 Nursing Diagnoses That May Be Associated with Depression

- Adult failure to thrive
- Caregiver role strain
- Chronic low self-esteem
- Chronic pain
- Chronic sorrow
- Disturbed body image
- Disturbed sleep pattern
- Dysfunctional grieving
- Fatigue
- Hopelessness
- Ineffective coping
- Post-trauma syndrome
- Powerlessness
- Risk for loneliness
- Risk for self-directed violence
- Self-care deficit
- Sexual dysfunction
- Social isolation

Is aging associated with an increased risk of depression?

Depression is not a normal consequence of aging; in fact, major depression actually becomes less common as people age.[6] When it does occur, disability and the onset of new medical illnesses are likely causes: approximately 12 to 13 percent of elderly people who are hospitalized or receiving home health care experience depression.[7]

In addition to being associated with illness, depression in the elderly often results from loneliness due to social isolation and the loss of loved ones, mobility, or a sense of purpose. Many authorities believe that loneliness is particularly relevant to malnutrition among the elderly. For many individuals, eating is as much a social and psychological event as a biological one. Without companionship, appetite diminishes. Approximately 30 percent of older adults live alone.[8] Their most pressing need seems to be for companionship; food takes second place. Social interaction is important to mental health, and elderly people of all classes in our society, both the financially secure and the poverty-stricken, tend to become isolated. Jack Weinberg, professor of psychiatry at the University of Illinois, wrote perceptively of this problem:

> In our efforts to provide the aged with a proper diet, we often fail to perceive it is not what the older person eats but with whom that will be the deciding factor in proper care for him. The oft-repeated complaint of the older patient that he has little incentive to prepare food for only himself is not merely a statement of fact but also a rebuke to the questioner for failing to perceive his isolation and aloneness and to realize that food . . . for one's self lacks the condiment of another's presence which can transform the simplest fare to the ceremonial act with all its shared meaning.[9]

How does malnutrition contribute to depression?

A spiral of sadness can develop when a lonely person neglects to eat well. Malnutrition worsens the apathy that is felt due to loneliness—and then the person has even less energy with which to procure and prepare food. Be alert for signs of this problem in elderly people who live alone or who have recently lost a spouse or other loved one.

© Thinkstock/SuperStock

Elderly individuals may find companionship to be a more pressing need than food.

Nurses, counselors, and social workers can help their elderly patients work through depression and find solutions to their loneliness. Dietitians and diet technicians can help them understand how depression affects food intake and health and how eating a well-balanced diet can prevent additional health and nutrition problems. Meal plans that include easy-to-prepare foods and nutrition supplements can help some people meet their nutrient needs when they lack the motivation to eat. Encouraging patients to eat with family or friends or at congregate meal sites can help combat loneliness.

What types of psychiatric illnesses may eventually lead to malnutrition?

Individuals with psychiatric illnesses are likely to develop malnutrition and ill health if their illness affects their ability to take care of themselves. Psychiatric illnesses are often characterized by dementia, illogical thinking, **paranoia, delusions,** depression, or anxiety—any of which may lead to inappropriate eating habits and interfere with nutrition status. Some of these disorders include **schizophrenia,** Alzheimer's disease (see Chapter 12), **mood disorders,** and substance abuse. People with mental illnesses characterized by illogical thinking or dementia may need assistance in taking care of routine daily activities, including purchasing, preparing, and consuming food. Those who are paranoid may believe that foods are being used to poison them. People suffering from delusions may attribute magical powers to certain foods and insist on eating only those particular foods. Medications used in the treatment of mental illnesses can also cause weight gain or loss, interact with nutrients, and alter nutrition status.

Nutrition affects the brain and the mind, and the brain and the mind affect the way people eat. All are interrelated, and the wise health care provider keeps these interrelationships in focus.

Glossary

delusions (deh-LOO-zhuns): false beliefs that are firmly maintained despite lack of proof or evidence to the contrary.

dementia (deh-MEN-she-ah): irreversible loss of intellectual function.

mood disorders: mental illness characterized by episodes of severe depression or excessive excitement (mania) or both.

paranoia (PAHR-ah-NOY-ah): mental illness characterized by irrational distrust of others and delusions of persecution.

schizophrenia (SKITZ-oh-FREN-ee-ah): mental illness characterized by an altered concept of reality and, in some cases, delusions and hallucinations.

Reminder: Alzheimer's disease is a degenerative disease of the brain involving memory loss and major structural changes in neural networks.

Notes

[1] K. H. Ladwig and coauthors, Gender differences in emotional disability and negative health perception in cardiac patients 6 months after stent implantation, *Journal of Psychosomatic Research* 48 (2000): 501–508.

[2] L. M. Gagnon and S. B. Patten, Major depression and its association with long-term medical conditions, *Canadian Journal of Psychiatry,* March 2002, available at **http://www.cpa-apc.org/Publications/Archives/CJP/2002/march/orMajorDepression.asp** (site visited August 31, 2005).

[3] W. Katon and M. D. Sullivan, Depression and chronic medical illness, *Journal of Clinical Psychiatry* 51 (1990): 3S–11S.

[4] V. Pignay-Demaria and coauthors, Depression and anxiety and outcomes of coronary artery bypass surgery, *Annals of Thoracic Surgery* 75 (2003): 314–321; S. von Ammon and coauthors, Medical illness, past depression, and present depression: A predictive triad for in-hospital mortality, *American Journal of Psychiatry* 158 (2001): 43–48.

[5] C. Ryan and M. E. Shea, Recognizing depression in older adults: The role of the dietitian, *Journal of the American Dietetic Association* 96 (1996): 1042–1044.

[6] Gagnon and Patten, 2002; Ryan and Shea, 1996.

[7] M. G. Cole and N. Dendukuri, Risk factors for depression among elderly community subjects: A systematic review and meta-analysis, *American Journal of Psychiatry* 160 (2003): 1147–1156; M. L. Bruce and coauthors, Major depression in elderly home health care patients, **www.ncoa.org/content.cfm?sectionID=106**, site *American Journal of Psychiatry* 159 (2002): 1367–1374.

[8] National Council on Aging, Facts about older Americans, March 2005, visited September 10, 2005.

[9] J. Weinberg, Psychological implications of the nutritional needs of the elderly, *Journal of the American Dietetic Association* 60 (1972): 293–296.

Nutrition and Liver Diseases

© Getty Images

CHAPTER 19

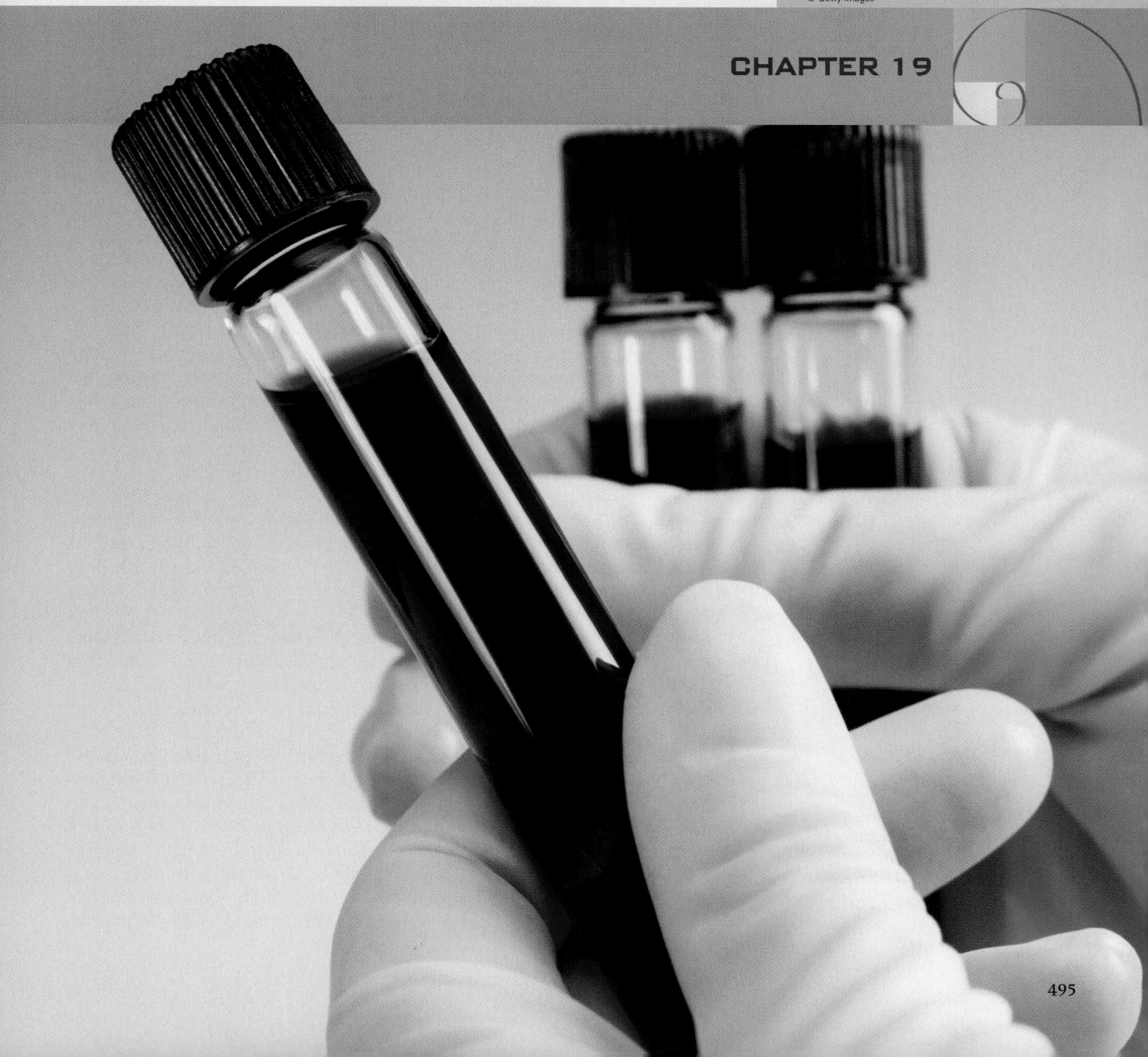

The liver is the most active organ in the body. It plays a central role in processing, storing, and redistributing the nutrients provided by the meals we eat. The liver also produces bile, which emulsifies fat during digestion. In addition, the liver synthesizes most of the proteins that circulate in plasma, including albumin, clotting proteins, and transport proteins. The liver also detoxifies drugs and alcohol and processes excess nitrogen so that it can be safely excreted as urea. If the liver's numerous roles are upset by liver damage or disease, the effects on health and nutritional status can be profound.

Liver disease progresses slowly. Its symptoms are sometimes so mild that complications may develop before liver disease is diagnosed. Once liver disease is recognized, preserving remaining liver function becomes a primary concern because the liver can eventually regenerate some healthy tissue, improving prognosis. Preventing additional damage is the principal means of avoiding liver failure or transplantation.

Fatty Liver and Hepatitis

Fatty liver and hepatitis are the two most common disorders affecting the liver. Although both conditions may be mild and are usually reversible, each may progress to more serious illness and eventually cause liver damage.

FATTY LIVER

Fatty liver is an accumulation of fat in liver tissue. Ordinarily, the liver's triglycerides are packaged into very-low-density lipoproteins (VLDL) and exported to the bloodstream. Although the exact reason why fat accumulates is unknown, fatty liver represents an imbalance between the amount of fat synthesized or picked up from the blood and the amount exported to the blood via VLDL.

Fatty liver is a clinical finding that is common to many conditions. It is present in the majority of patients who have alcoholic liver disease (discussed in Nutrition in Practice 19) and can also result from exposure to drugs and toxic metals. It is associated with obesity, diabetes mellitus, and diseases of malnutrition, including kwashiorkor and marasmus. Fatty liver may follow gastrointestinal bypass surgery or long-term total parenteral nutrition. Its causes are not always clear, however, as it occurs in as many as 14 percent of adults in the United States.[1]

Consequences of Fatty Liver In many individuals, fatty liver is asymptomatic and causes no harm. It is often accompanied by other symptoms, however, such as liver enlargement (**hepatomegaly**), inflammation, and fatigue. If fatty liver is drug induced or associated with certain metabolic disorders, it may progress quickly and result in liver damage or even liver failure.[2]

The liver enzymes ALT (alanine aminotransferase) and AST (aspartate aminotransferase) are involved in amino acid catabolism.

Fatty liver is a frequent cause of abnormal levels of liver enzymes in the blood. Laboratory findings may include elevated blood concentrations of the liver enzymes ALT and AST, as well as increased levels of triglycerides, cholesterol, and glucose. Table 19-2 on p. 499 provides normal ranges for these liver enzymes.

Treatment of Fatty Liver The usual treatment for fatty liver is to eliminate the factors that cause it. For example, if fatty liver is due to alcohol abuse or drug treatment, it may improve after the substances are discontinued. In patients with elevated blood lipids, fatty liver may improve after blood lipid levels are lowered. An appropriate treatment for obese or diabetic patients might be weight reduction or blood glucose control. Rapid weight loss should be discouraged, however, as it may accelerate the progression of liver disease.[3] It should be noted that lifestyle modifications are not always successful in reversing fatty liver, especially in patients who lack the usual risk factors.

fatty liver: an accumulation of triglycerides in the liver; also called **hepatic steatosis** (STEE-ah-TOE-sis).

   hepatic = pertaining to the liver
   steato = fat

hepatomegaly (HEP-ah-toe-MEG-ah-lee): enlargement of the liver.

TABLE 19-1	Features of Hepatitis Viruses			
Hepatitis Virus	% of Viral Cases	Major Mode of Transmission	Chronic Disease Rate (% of cases)	Vaccination Available
A	48%	Fecal-oral	None	Yes
B	34%	Bloodborne; sexual transmission	<10% of adult cases >90% of infant cases	Yes
C	15%	Bloodborne	80–90%	No

Note: Although not listed here, a small fraction of viral hepatitis cases are caused by hepatitis viruses D and E.

HEPATITIS

Hepatitis, a condition of liver inflammation, can result from any factor that causes damage to liver tissue. It is most often caused by infection with specific viruses, which are designated by the letters A, B, and C. These viruses are primarily spread by blood contact with infected persons or by ingesting contaminated food or water. Hepatitis can also be caused by excessive alcohol intake or exposure to certain drugs and toxic chemicals. A number of herbal remedies are reported to cause hepatitis; these include chaparral, germander, ma huang, saw palmetto, and jin bu huan.[4] Less common causes of hepatitis include infection with other viruses and autoimmune disease.

Viral Hepatitis Hepatitis A virus (HAV) is extremely contagious and is the most common cause of acute viral hepatitis (see Table 19-1). It is usually spread by the fecal-oral route, which occurs when foods and beverages become contaminated with fecal material. Outbreaks of hepatitis A are often associated with floods and other natural disasters, when inadequately treated sewage may contaminate water supplies with fecal matter. Less frequently, HAV is spread by consuming under-cooked shellfish obtained from contaminated waters. HAV infection usually resolves within a few months and does not cause chronic illness or permanent liver damage.

About half of viral hepatitis cases are caused by infection with hepatitis B and C viruses. Hepatitis B virus (HBV) is transmitted by infected blood or needles or by sexual contact and is carried by approximately 1.25 million persons in the United States. As a precautionary measure, HBV vaccinations are recommended for health care workers, sexually active adults, and newborn infants and children. Hepatitis C virus (HCV) is transmitted by blood contact as well, but is less efficiently spread by sexual contact. HCV is a major cause of chronic hepatitis and is currently carried by 2.7 million individuals in the United States. No vaccine is currently available to protect against HCV infection. Chronic cases of either HBV or HCV infection may lead to cirrhosis (discussed in a later section) and liver cancer.

Symptoms of Hepatitis The effects of hepatitis depend on the cause and severity of the disease. Both mild and chronic cases of hepatitis are often asymptomatic. The onset of acute hepatitis may be accompanied by fatigue, nausea, anorexia, and pain in the liver area. The liver is often slightly enlarged. **Jaundice** (yellow coloration of tissues) can develop, causing discoloration of the skin, urine, and the whites of the eyes. Other symptoms of hepatitis may include fever, headache, muscle weakness, and skin rashes. Serum levels of the liver enzymes ALT and AST are typically elevated. Chronic hepatitis can be associated with complications that are typical of liver cirrhosis (see pp. 498–501).

There are fewer new cases of HCV than of HBV each year, but more HCV cases become chronic. Therefore, there are more HCV carriers than HBV carriers.

Jaundice results when liver dysfunction impairs metabolism of bilirubin, a breakdown product of hemoglobin that is normally eliminated in bile; bilirubin therefore accumulates in the blood and other tissues.

NURSING DIAGNOSIS

for people with hepatitis include *fatigue, acute pain, activity intolerance, risk for deficient fluid volume* (due to vomiting and diarrhea), and *imbalanced nutrition: less than body requirements*.

hepatitis (hep-ah-TYE-tis): inflammation of the liver.

jaundice (JAWN-dis): yellow discoloration of the skin and mucous membranes due to an accumulation of bilirubin, a breakdown product of hemoglobin that exits the body via bile secretions.

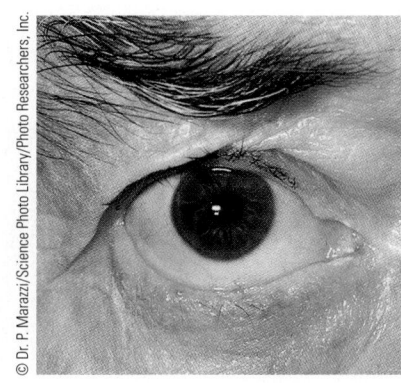

Jaundice is a yellow discoloration of tissues that is most easily seen in the whites of the eyes.

Treatment of Hepatitis Hepatitis is treated with supportive care, including bed rest and an appropriate diet. Hepatitis patients should avoid substances that aggravate the liver, such as alcohol or drugs that can cause liver damage. Hepatitis A typically resolves without medications, whereas antiviral agents may be provided for HBV and HCV infections. Nonviral forms of hepatitis may be treated with anti-inflammatory and immunosuppressant drugs.

Medical nutrition therapy is individualized according to a patient's symptoms and nutrition status. Some individuals require no dietary changes. Those who are malnourished or have lost weight due to illness may require a high-kcalorie, high-protein diet to replenish nutrient stores. A person with anorexia or gastrointestinal discomfort may find small, frequent meals easier to tolerate. Liquid supplements can help to improve nutrient intakes. Fluid and electrolyte replacement may be necessary for patients with persistent vomiting.

> **REVIEW NOTES**
>
> Fatty liver can be a consequence of excessive alcohol intake, drug toxicity, and chronic diseases such as diabetes and obesity. Hepatitis is frequently caused by viral infection, but may also result from alcohol abuse and drug toxicity.
>
> Although fatty liver is often benign, hepatitis may become chronic and lead to cirrhosis and liver cancer.
>
> Treatment of hepatitis may include bed rest, medications, and dietary measures that improve nutrition status.

Cirrhosis

Cirrhosis is the end-stage condition that results from long-term liver disease. Liver disease gradually destroys liver tissue, leading to scarring (fibrosis) in some regions and small areas of regenerated, healthy tissue in others. As the disease progresses, the scarring becomes more extensive, leaving fewer areas of healthy tissue. A cirrhotic liver is often shrunken in size and has an irregular, nodular appearance (see the photo). Cirrhosis impairs liver function and can eventually lead to liver failure. It is estimated to affect approximately three million people in the United States.[5]

The most common cause of cirrhosis in the United States is hepatitis C infection, which is responsible for 26 percent of cases.[6] Alcoholic liver disease, which develops in about 10 to 20 percent of chronic alcoholics,[7] is the second most common cause. Bile duct blockages can lead to cirrhosis by impeding bile flow; the resulting accumulation of toxic bile acids in the liver causes liver injury. All types of chronic hepatitis, if untreated, can eventually lead to cirrhosis. Other causes of cirrhosis include drug-induced liver injury and inherited metabolic disorders that cause toxic substances to accumulate in the liver.

cirrhosis (sih-ROE-sis): an advanced stage of liver disease in which extensive scarring replaces healthy liver tissue, causing impaired liver function and liver failure.

Normal liver tissue is smooth and has a regular texture.

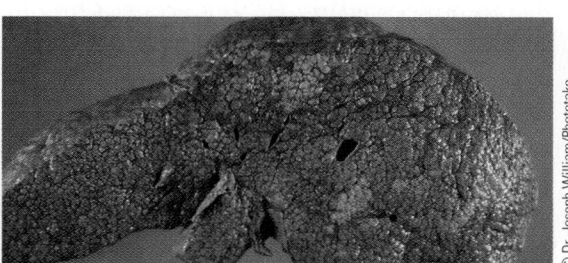

A cirrhotic liver has an irregular, nodular appearance. The nodules represent clusters of regenerating cells within the damaged liver tissue.

CONSEQUENCES OF CIRRHOSIS

Up to 40 percent of people with cirrhosis are asymptomatic.[8] Because liver disease progresses slowly, the effects of cirrhosis may be minimal at first. Initial symptoms may include fatigue, weakness, anorexia, and weight loss. Later, the decline in liver function can lead to anemia, impaired blood clotting, and greater susceptibility to infections. If bile obstruction occurs, jaundice and fat malabsorption can result. The physical changes in liver tissue may interfere with blood flow, causing fluid to accumulate in blood vessels and body tissues. Advanced cirrhosis can disrupt kidney and lung function. Figure 19-1 illustrates some of the clinical effects of liver cirrhosis, and later sections describe some of these complications in more detail.

Table 19-2 shows the laboratory tests that are generally used to monitor the extent of liver damage. Because liver disease injures liver tissue, liver enzymes spill into the bloodstream. Levels of bilirubin may be elevated if the liver is too damaged to process it or if bile ducts are blocked and prevent its excretion. Reduced synthesis of plasma proteins by the liver lowers albumin levels and extends blood-clotting time. Liver damage also impairs the conversion of ammonia to urea, causing ammonia levels in the blood to rise.

Portal Hypertension A large volume of blood normally flows through the liver. The portal vein and hepatic artery supply approximately 1500 milliliters (about 1.5 quarts) of blood each minute to the extensive network of vessels in the liver. The scarred tissue of a cirrhotic liver impedes this blood flow, which is mostly supplied by the portal vein. The resistance to blood flow within the liver causes a rise in pressure within the portal vein called **portal hypertension**.

Collaterals and Gastroesophageal Varices When blood flow through the portal vein is impeded, the blood is diverted to the smaller blood vessels surrounding the liver. These **collaterals** develop throughout the gastrointestinal (GI) tract and in regions near the abdominal wall. As pressure builds, the collateral vessels become enlarged and engorged, forming **varices**. Esophageal and gastric varices are vulnerable to rupture, as they have thin walls and often

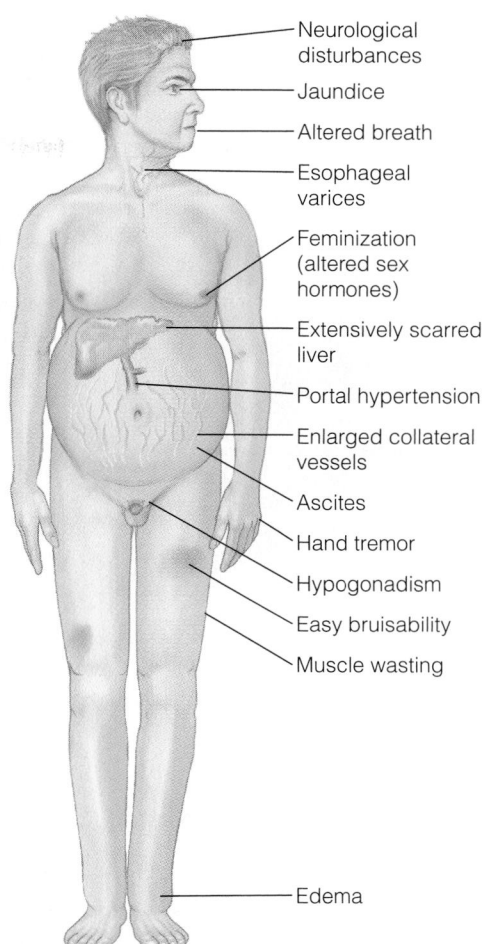

Neurological disturbances
Jaundice
Altered breath
Esophageal varices
Feminization (altered sex hormones)
Extensively scarred liver
Portal hypertension
Enlarged collateral vessels
Ascites
Hand tremor
Hypogonadism
Easy bruisability
Muscle wasting
Edema

FIGURE 19-1 Clinical Effects of Liver Cirrhosis

TABLE 19-2	Laboratory Tests for Evaluation of Liver Disease	
Laboratory Test	Normal Ranges (serum)	Values in Liver Disease
Alanine aminotransferase (ALT)	Male: 10–40 U/L Female: 7–35 U/L	Elevated
Albumin	3.4–4.8 g/dL	Decreased
Alkaline phosphatase	25–100 U/L	Normal or elevated
Ammonia	15–45 μg N/dL	Elevated
Aspartate aminotransferase (AST)	10–30 U/L	Elevated
Bilirubin (total)	0.3–1.2 mg/dL	Elevated
Blood urea nitrogen (BUN)	6–20 mg/dL	Normal or decreased
Prothrombin time[a]	10–13 seconds	Prolonged

[a]The test for prothrombin time evaluates the clotting ability of blood.

NURSING DIAGNOSIS

for people with cirrhosis include *chronic pain, fatigue, nausea, risk for impaired skin integrity, ineffective protection,* and *imbalanced nutrition: less than body requirements.*

Reminder: The *portal vein* is a large blood vessel that carries nutrient-rich blood from the digestive tract to the liver.

portal hypertension: elevated blood pressure in the portal vein; may be caused by obstructed blood flow through the liver.

collaterals: blood vessels that enlarge in order to allow an alternative pathway for diverted blood.

varices (VAH-rih-seez): abnormally dilated blood vessels (singular: *varix*).

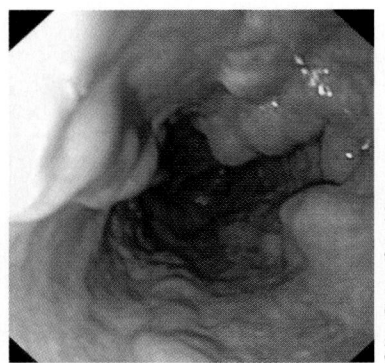

Esophageal varices, such as the one shown here, may protrude into the lumen and be vulnerable to rupture and bleeding.

The aromatic amino acids—phenylalanine, tyrosine, and tryptophan—have carbon rings in their side groups. The branched-chain amino acids include leucine, isoleucine, and valine; their side groups have a branched structure.

ascites (ah-SIGH-teez): the abnormal accumulation of fluid in the abdominal cavity.

sinusoids: the small capillary-like passages that carry blood through liver tissue.

hepatic encephalopathy (en-SEF-ah-LOP-ah-thee): a condition in advanced liver disease that is characterized by altered neurological functioning, including personality changes, reduced mental abilities, and disturbances in motor function.
  encephalo = brain
  pathy = disease

hepatic coma: loss of consciousness resulting from severe liver disease.

bulge into the lumen. If ruptured, they can cause massive bleeding that is sometimes fatal. The blood loss is exacerbated by the liver's reduced production of blood-clotting factors.

Ascites Within ten years of disease onset, about 50 percent of cirrhosis patients develop **ascites,** an accumulation of fluid in the abdominal cavity. The development of ascites indicates that liver damage has reached a critical stage, as half of patients with ascites die within two years.[9]

Ascites is thought to be a consequence of portal hypertension, reduced albumin synthesis by the diseased liver, and altered kidney function. Portal hypertension raises pressure in the liver's **sinusoids,** forcing fluid to leak from the blood into the peritoneal (abdominal) cavity. The fluid accumulation is exacerbated by low plasma albumin levels, as albumin helps to retain fluid in blood vessels. The increased pressure in the portal vein triggers the release of factors (such as nitric oxide) that dilate blood vessels, thereby lowering the pressure within vessels elsewhere; this activates sodium and water retention by the kidneys and leads to additional water pooling.[10] Ascites can cause abdominal discomfort and early satiety, which contribute to malnutrition. As ascites can raise body weight considerably, however, weight changes may be difficult to interpret.

Hepatic Encephalopathy Advanced liver disease sometimes leads to **hepatic encephalopathy,** a disorder characterized by abnormal neurological functioning. Symptoms of hepatic encephalopathy include changes in personality, mental abilities, and motor functions (see Table 19-3). At worst, amnesia, seizures, and **hepatic coma** may develop. Although reversible, hepatic encephalopathy is associated with a one-year survival rate of only 40 percent.[11]

The causes of hepatic encephalopathy remain elusive, although elevated blood ammonia levels may be partly responsible due to ammonia's neurotoxicity. Other compounds that are potentially toxic to brain tissue, such as sulfur compounds, short-chain fatty acids, and GABA (a neurotransmitter), may accumulate in brain cells and alter neurotransmitter activity.[12] Another theory is that neurotransmitter functioning is altered by an increased ratio of aromatic amino acids to branched-chain amino acids in brain tissue, a result of altered amino acid metabolism in the liver. Most likely, a combination of metabolic abnormalities contributes to disrupted neurological functioning.[13]

Elevated Blood Ammonia Levels Much of the body's free ammonia is produced by bacterial action on unabsorbed dietary protein in the colon. One of the liver's functions is to extract ammonia from portal blood and convert it to urea, which is then excreted by the kidneys. In advanced liver disease, ammonia-laden portal blood bypasses the liver by way of collateral vessels and reaches the general blood

TABLE 19-3	Symptoms of Hepatic Encephalopathy	
Early Stages	**Middle Stages**	**Later Stages**
Lack of attention	Poor memory	Disorientation
Irritability, depression	Drowsiness	Amnesia
Impaired judgment	Slurred speech	Muscular rigidity
Lack of coordination	Jerking movements	Abnormal reflexes
Tremor	Sleep disorders	Delirium, stupor

TABLE 19-4	Possible Causes of Malnutrition in Liver Disease

Mechanism	Examples
Reduced nutrient intake	Anorexia, early satiety (due to ascites), nausea and vomiting, restrictive diets, effects of medications (including gastrointestinal disturbances and taste changes), abdominal pain, fatigue, fasting for medical procedures
Malabsorption/nutrient losses	Fat malabsorption (due to reduced bile flow), vomiting, diarrhea, gastrointestinal bleeding, effects of medications (including malabsorption and nutrient losses from diuretic use)
Altered metabolism/increased nutrient needs	Hypermetabolism, catabolism, infections/inflammation, inadequate protein synthesis, reduced nutrient storage and metabolism in the liver

circulation, causing a substantial increase in the ammonia that reaches brain tissue. Although there is strong evidence that ammonia contributes to hepatic encephalopathy, blood levels of ammonia do not correlate well with the degree of neurological impairment.

Malnutrition and Wasting Most patients with cirrhosis develop protein-energy malnutrition (PEM) and experience some degree of wasting.[14] Malnutrition is usually caused by a combination of factors (see Table 19-4). Patients often consume less food due to reduced appetite, gastrointestinal symptoms, or fatigue. If the diet is restricted in sodium (to treat ascites), meals may seem monotonous or unpalatable. Fat malabsorption is common, and diarrhea, vomiting, and gastrointestinal bleeding all contribute to nutrient losses. If cirrhosis is a consequence of alcohol abuse, multiple nutrient deficiencies may be present.

TREATMENT OF CIRRHOSIS

Treatment of cirrhosis is individualized according to the severity of the illness and the complications that develop. Supportive care, including an appropriate diet and avoidance of liver toxins, promotes recovery and helps to prevent further damage. Abstinence from alcohol is critical for preserving liver function and extending survival. Patients who have developed cirrhosis due to viral infection may benefit from the use of antiviral medications. Liver transplantation may be necessary in advanced cirrhosis.

Drug Therapy Drugs are prescribed to treat the symptoms and complications associated with cirrhosis. Medications for portal hypertension and varices may either dilate blood vessels (for example, nitroglycerin) or constrict them (vasopressin and octreotide). Diuretics may help to control portal hypertension and ascites; common examples include spironolactone and furosemide. To stimulate the appetite and promote weight gain, megestrol acetate or dronabinol is sometimes prescribed. Drug treatment of hepatic encephalopathy focuses on controlling blood ammonia levels. Lactulose, a nonabsorbable disaccharide that is often used as a laxative, helps to reduce ammonia production and absorption in the colon. The antibiotic neomycin is an alternative treatment for elevated ammonia that works by altering bacterial populations. The Diet-Drug Interactions feature lists potential nutritional problems associated with these medications.

DIET-DRUG INTERACTIONS Check this table for notable nutrition-related effects of the medications discussed in this chapter.

	Gastrointestinal Effects	Interactions with Dietary Substances	Metabolic Effects
Appetite stimulants (megestrol acetate, dronabinol)	Nausea, vomiting, diarrhea		Hyperglycemia (megestrol acetate)
Diuretics (furosemide, spironolactone[a])	Dry mouth, anorexia, decreased taste perception	Furosemide's bioavailability is reduced when taken with food.	Fluid and electrolyte imbalances,[a] hyperglycemia (spironolactone), hyperlipidemia (spironolactone), thiamin and zinc deficiencies
Immunosuppressants (cyclosporine, tacrolimus)	Nausea, vomiting, diarrhea, anorexia (tacrolimus)	Grapefruit juice can raise serum concentrations of these drugs to toxic levels. Cyclosporine potentiates the effects of alcohol. Tacrolimus's bioavailability is reduced when taken with food.	Electrolyte imbalances, hypertension, hyperglycemia, hyperlipidemia
Lactulose	Diarrhea		Fluid and electrolyte imbalances
Nitroglycerin	Decreased taste perception	Potentiates effects of alcohol.	

[a]*Furosemide* is a "potassium-wasting" diuretic; patients should increase intakes of potassium-rich foods. *Spironolactone* is a "potassium-sparing" diuretic; patients should avoid supplemental potassium and potassium-containing salt substitutes.

MEDICAL NUTRITION THERAPY FOR CIRRHOSIS

Medical nutrition therapy for patients with cirrhosis is customized to each person's needs, which vary considerably and depend on accompanying complications. In all cases, dietary substances that can cause further liver injury should be avoided; examples include alcohol, herbal supplements, and vitamin or mineral megadoses. Table 19-5 lists the general dietary guidelines for cirrhosis, which are discussed in the sections that follow.

Energy To estimate energy requirements, health practitioners can calculate basal energy expenditure (BEE) and then apply a stress factor, as described in Table 16-2 on p. 432. For most patients with cirrhosis, the stress factor 1.2 can be used initially and adjusted as necessary.[15] If possible, indirect calorimetry should be used to determine the basal metabolic rate. For patients with ascites, BEE calculations should use either the patient's desirable weight or an estimated dry weight (weight without ascites). A value for dry weight can be obtained after diuretic therapy or after a medical procedure that directly removes excess abdominal fluid.

Many patients with cirrhosis have difficulty consuming enough food to achieve good nutritional status. Four to six daily feedings may be better tolerated than three meals per day. Liquid dietary supplements can help to improve energy intakes. The "How to" on p. 503 offers additional suggestions that can help a patient meet energy needs.

Protein Patients with cirrhosis risk developing PEM and wasting; therefore, protein intakes should be high enough to maintain nitrogen balance. Protein intakes should range between 0.8 and 1.2 grams per kilogram body weight per day (based on desirable weight or dry weight). To prevent protein catabolism, adequate energy must be consumed.

Although it was formerly believed that high-protein diets could harm patients with hepatic encephalopathy, protein restrictions are no longer considered helpful

Reminder: The protein RDA for healthy adults is 0.8 g/kg.

TABLE 19-5	Dietary Guidelines for Liver Cirrhosis
Energy	• Energy needs may be approximately 20% above basal energy expenditure (BEE); calculate or measure BEE and apply the stress factor 1.2 initially (see Table 16-2 on p. 432). • Energy requirements may be higher in patients with infection or malnutrition. Energy requirements may be lower in patients who would benefit from weight loss.
Protein	• Provide 0.8 to 1.2 grams of protein per kilogram body weight per day to maintain nitrogen balance and prevent wasting.
Carbohydrate	• No carbohydrate restrictions unless the patient has insulin resistance or diabetes. • For persons with insulin resistance or diabetes, provide up to 50% to 60% of kcalories from carbohydrates (mainly complex carbohydrates); carbohydrate intake should be consistent from day to day and at each meal and snack.
Fat	• No fat restrictions unless fat malabsorption is present. • If fat is malabsorbed, restrict fat as necessary to control steatorrhea (see Chapter 18); use medium-chain triglycerides (MCT) to increase kcalories.
Sodium and fluid	• Restrict sodium as necessary to control ascites; limiting sodium to no more than 2000 milligrams per day is adequate restriction in most cases. • For difficult-to-manage ascites, restrict fluids to 1200 to 1500 milliliters per day.
Vitamins and minerals	• Ensure adequate intake from diet or supplements based on individual needs.

HOW TO   Help the Person with Cirrhosis Eat Enough Food

Individuals with cirrhosis often have difficulty consuming enough food to prevent malnutrition and its consequences. Ascites and gastrointestinal symptoms such as nausea and vomiting may interfere with food intake. Fatigue may cause disinterest in food preparation. Sodium restrictions may make foods unpalatable. To improve food intake:

• If nutrient restrictions are necessary, make sure the patient fully understands how to modify the diet so that food intake isn't restricted unnecessarily. Provide lists of acceptable foods and menus. Explain how recipes can be altered so that favorite foods can still be incorporated into the diet.

• Suggest between-meal snacks during the day and a snack at bedtime. An oral supplement like Ensure can substitute for a snack and requires no preparation. Snacks should not be consumed within two hours of meals, or they may reduce appetite at mealtime.

• If the patient has little appetite or is quickly satiated, suggest foods that are higher in food energy, such as whole milk instead of reduced-fat milk or canned fruit that is packed in heavy syrup instead of fruit juice.

• Recommend energy boosters. Cream sauces and gravies can add kcalories to entrées. Fruit juices and fruit nectars can substitute for drinking water. The following additions can boost the energy content of meals:
  • Sour cream and butter—on vegetables and potatoes.
  • Mayonnaise—in sandwiches and salads.
  • Half-and-Half and light cream—in soups and on cereals.

• Hard-boiled egg—in casseroles and meat loaf.
• Cheese—in salads and casseroles and melted on steamed vegetables.
• Peanut and nut butters and cream cheese—on crackers or celery and in milk shakes.
• Chopped nuts—in salads, cooked cereals, and bakery products.

Low-sodium diets are recommended for treating ascites and other medical conditions, including kidney and heart disorders. The "How to" feature on p. 554 offers suggestions to help patients implement sodium restrictions. To improve the palatability of low-sodium meals:

• Suggest that patients replace the salt they use for cooking and seasoning with strong-flavored herbs and spices like coriander, chili powder, cumin, curry powder, garlic, ginger, mint, lemon, and parsley.

• Advise patients to check food labels to learn the sodium content of the foods they eat. Similar products may be available that are lower in sodium. (Persons using potassium-sparing diuretics should be cautioned to avoid salt substitutes that replace sodium with potassium.)

Offer support and encouragement to the patient with cirrhosis. Severe weight loss is less likely to occur if dietary advice is provided before problems progress.

and may worsen malnutrition and wasting.[16] In an attempt to normalize altered amino acid ratios in brain tissue and improve mental status, some health care providers prescribe enteral formulas with added branched-chain amino acids and reduced aromatic amino acids. Clinical studies testing the use of these formulas have yielded mixed results, however, and their routine use is not currently recommended.[17]

Carbohydrate and Fat   Carbohydrate provides a substantial proportion of energy needs. Many patients with cirrhosis have insulin resistance, however, and require medications or insulin to manage their hyperglycemia. These individuals should follow the dietary guidelines for diabetes: consume mostly complex carbohydrates, and consume them at regular intervals throughout the day. A few small studies have suggested that high-fiber, low glycemic index diets can improve glucose tolerance in patients with cirrhosis, but larger studies are needed to confirm the benefit.[18]

If steatorrhea is present, fat intake is restricted, and medium-chain triglycerides (MCT) may be used to provide additional energy. Essential fatty acids cannot be obtained from MCT oils and may need to be supplemented. Severe steatorrhea also warrants supplementation of fat-soluble vitamins, calcium, magnesium, and zinc.

Sodium and Fluid   Patients with ascites are generally advised to restrict sodium. Ascites is partly caused by the kidneys' reabsorption of sodium into the blood, which results in sodium and water retention. Therefore, treatment usually combines moderate sodium restriction (no more than 2 grams per day) and diuretic therapy to produce a fluid loss of approximately one pound daily.[19] Potassium intake should be monitored if a potassium-wasting diuretic (such as furosemide) is used.

Many patients find low-sodium diets unpalatable, so some health practitioners may allow a more liberal sodium intake and depend on diuretics to mobilize excess fluids. If patients do not respond to sodium restriction and diuretic therapy, fluid may be removed directly by surgical puncture (**paracentesis**) or may be diverted to the bloodstream using a catheter (**peritoneovenous shunt**). Patients who have ascites that is difficult to manage may be advised to restrict fluids to 1200 to 1500 milliliters daily.[20]

Vitamins and Minerals   Vitamin and mineral deficiencies are common in cirrhosis patients due to the effects of illness, complications of disease, or alcoholism that may have induced the liver disease. Multivitamin supplementation is often necessary. If steatorrhea is present, fat-soluble nutrients can be provided in water-soluble forms. Patients with esophageal varices may find it easier to ingest supplements in liquid form.

Enteral and Parenteral Nutrition Support   If a person with cirrhosis is unable to consume enough food, tube feedings or intravenous feedings may be warranted. Specialized enteral (tube feeding) products are sometimes preferred; these include formulas that are high in kcalories, low in electrolytes, or high in branched-chain amino acids. If a patient has esophageal varices, the feeding tube should be as narrow and flexible as possible to prevent rupture and bleeding.

Parenteral nutrition support should be considered in patients who are unable to tolerate enteral feedings due to intestinal obstruction, gastrointestinal bleeding, or uncontrollable vomiting. For those with hyperglycemia, the rate of dextrose infusion should be limited to the amount that can be oxidized within the body, about 5 milligrams per kilogram body weight per minute. To avoid excessive fluid delivery, concentrated nutrient formulas are recommended for patients with ascites, and central veins are used for feedings. The Case Study on p. 505 allows you to apply your understanding of cirrhosis to a clinical situation.

Reminder: Hyperglycemia is a common complication of both parenteral feedings and liver disease.

paracentesis (pahr-ah-sen-TEE-sis): a surgical puncture of a body cavity with an aspirator to draw out excess fluid.

peritoneovenous (PEH-rih-toe-nee-oh-VEE-nus) **shunt:** a surgical passage created between the peritoneum and the jugular vein to divert fluid and relieve ascites. The peritoneum is the membrane that surrounds the abdominal cavity.

CASE STUDY   *Carpenter with Cirrhosis*

Stu Hamilton, a 49-year-old carpenter, has just been diagnosed with cirrhosis, which is a consequence of alcohol abuse over the past 25 years. Although he recognizes that he has an alcohol problem and recently entered an alcohol rehabilitation program, he is still drinking. At 5 feet 7 inches tall, Mr. Hamilton, who formerly weighed 155 pounds, now weighs 125 pounds. Although he is still living at home, he is showing signs of mental deterioration. He is jaundiced and appears thin, although his abdomen is distended with ascites. Laboratory findings indicate elevated serum concentrations of AST, ALT, and ammonia; reduced albumin levels; and hyperglycemia.

1. Do Mr. Hamilton's laboratory values suggest liver disease? Compare the results of his laboratory tests with the values shown in Table 19-2.
2. From the limited information available, evaluate Mr. Hamilton's nutrition status. What medical problem makes it diffi-cult to interpret his present weight? Describe the development of ascites in liver disease and explain how the diet is usually adjusted for a patient with ascites.
3. Calculate Mr. Hamilton's energy and protein needs. Describe the general diet you might recommend for him. What suggestions do you have for increasing his energy intake?
4. Explain the significance of Mr. Hamilton's elevated blood ammonia levels. What are some signs that would indicate that he is undergoing mental decline?
5. Describe each of the following complications of liver disease: portal hypertension, jaundice, gastroesophageal varices. What complication may result if esophageal varices are not treated?

REVIEW NOTES

Liver cirrhosis is characterized by fibrosis and permanent liver dysfunction. The primary causes of cirrhosis in the United States are hepatitis C infection and alcohol abuse.

Symptoms of cirrhosis include fatigue, gastrointestinal disturbances, anorexia, and weight loss. Complications include portal hypertension, gastroesophageal varices, ascites, and hepatic encephalopathy.

Treatment of cirrhosis is highly individualized and depends on the accompanying symptoms and complications. Both drug therapies and dietary adjustments are usually necessary. If warranted, the diet may need to be restricted in fat, sodium, or fluids.

Liver Transplantation

Acute or chronic liver disease can potentially lead to liver failure, in which case liver transplantation is the only remaining treatment option. The most common illnesses that precede liver transplantation are chronic hepatitis C infection and alcoholic liver disease, which account for about 40 percent of liver transplant cases.[21] The five-year survival rate among transplant recipients is 70 to 75 percent, although complications such as ascites and hepatic encephalopathy may worsen the prognosis.[22]

Nutrition Status of Transplant Patients   As mentioned earlier, advanced liver disease is usually associated with malnutrition, which can increase the risk of complications following a liver transplant. Evaluating nutrition status in transplant candidates can be difficult because liver dysfunction and malnutrition often have similar metabolic effects. In addition, fluid retention can mask weight loss and alter anthropometric and laboratory values. Correcting malnutrition prior to surgery can help speed recovery after surgery. Depending on the cause of liver disease or the complications that develop during illness, transplant patients may be deficient in vitamins B_6, B_{12}, and C, fat-soluble vitamins, calcium, copper, iron, magnesium, phosphorus, potassium, and zinc.[23]

Post-Transplantation Concerns The immediate concerns following a transplant are organ rejection and infection. Immunosuppressive drugs, including prednisone, cyclosporine, and tacrolimus, help to reduce the immune responses that cause rejection, but they also raise the risk of infection. Infections are the most common cause of death following a liver transplant; therefore, antibiotics and antiviral medications are prescribed to reduce infection risk.[24]

Immunosuppressive drugs can affect nutrition status in numerous ways. Gastrointestinal side effects include nausea, vomiting, diarrhea, abdominal pain, and mouth sores. Some of the drugs may cause hyperglycemia or outright diabetes, which may need to be controlled with insulin. Electrolyte and fluid imbalances are common, and some medications may alter appetite and taste perception. Other possible effects include hypertension, hyperlipidemias, protein catabolism, and increased osteoporosis risk.

Protein and energy requirements are increased after transplantation due to the stress of surgery. High-kcalorie, high-protein snacks and enteral supplements can help the transplant patient meet postsurgical needs. Vitamin and mineral supplementation is also an integral part of nutrition care. To help transplant patients avoid developing foodborne illnesses, nurses can provide information about food safety measures, such as cooking meats adequately, washing fresh produce, and avoiding foods that may be contaminated.[25] Chapter 1 (pp. 30–31) provides additional information about food safety.

> **NURSING DIAGNOSIS**
>
> for people with liver transplants include *ineffective protection, risk for infection,* and *imbalanced nutrition: less than body requirements.*

> **REVIEW NOTES**
>
> Liver transplantation has improved the long-term outlook for patients with advanced liver disease. Transplant patients are usually malnourished, however, and may have medical problems that affect transplant success.
>
> Due to the potential for organ rejection, immunosuppressive drugs are prescribed following surgery. Use of these drugs increases the risk of infection, and the drugs have side effects that can impair nutrition status and general health.

NUTRITION ASSESSMENT CHECKLIST FOR PEOPLE WITH LIVER DISORDERS

Medical History

Check the medical record to determine:

☐ Type of liver disorder

☐ Cause of liver disorder

☐ If the patient has received a liver transplant

Review the medical record for complications that may alter nutritional needs including:

☐ Anemia

☐ Ascites

☐ Esophageal varices

☐ Hepatic encephalopathy

☐ Impaired kidney or lung function

☐ Infections

☐ Insulin resistance or diabetes mellitus

☐ Malabsorption

☐ Malnutrition

Medications

In patients with liver dysfunction, the risk of diet-drug interactions is high because most drugs are metabolized in the liver. Risk of interactions is intensified for patients with:

☐ Ascites (medications may take a long time to reach the liver)

☐ Renal failure (medications are often metabolized further in the kidneys and excreted in the urine)

☐ Malnutrition

☐ Multiple prescriptions

☐ Long-term medication use

Dietary Intake

For patients with fatty liver, pay special attention to:

☐ Energy intake, if the patient is overweight or malnourished, has diabetes, or is receiving total parenteral nutrition

☐ Carbohydrate, if the patient has diabetes or is receiving total parenteral nutrition

☐ Alcohol use

For patients with hepatitis, cirrhosis, or ascites:

☐ Check appetite.

☐ Ensure that energy and nutrient intakes are adequate.

☐ Determine alcohol consumption.

☐ Determine whether sodium or fluid restriction is warranted.

☐ Base energy needs on desirable weight or estimated dry weight to avoid overfeeding.

Anthropometric Data

Take baseline height and weight measurements and monitor weight regularly. For patients with ascites and edema:

☐ Monitor weight changes to evaluate the degree of fluid retention.

☐ Remember that the patient may be malnourished, and weight deceptively high.

Laboratory Tests

Note that albumin and serum proteins are often reduced in people with liver disease and cannot always be used as indicators of nutrition status. Review the following laboratory test results to assess liver function:

☐ Albumin

☐ Alkaline phosphatase

☐ ALT and AST

☐ Ammonia

☐ Bilirubin

☐ Prothrombin time

Check laboratory test results for complications associated with liver failure including:

☐ Anemia

☐ Fluid retention

☐ Hyperglycemia

☐ Decreased renal function

Physical Signs

Look for physical signs of:

☐ Fluid retention (ascites and edema)

☐ PEM (muscle wasting and unintentional weight loss)

☐ Nutrient deficiencies

SELF CHECK

1. Which of the following dietary strategies would be most appropriate for reversing fatty liver associated with diabetes mellitus?
   a. low-protein diet
   b. fat-restricted diet
   c. fluid- and sodium-restricted diet
   d. modifying energy to achieve a desirable weight and modifying carbohydrates to attain blood glucose control

2. Which of the following statements about hepatitis is true?
   a. Chronic hepatitis can progress to cirrhosis.
   b. Whatever the cause of hepatitis, symptoms are typically severe.
   c. People with hepatitis require high-kcalorie, high-protein diets.
   d. HCV infections can be spread through contaminated foods and water.

3. The most common cause of liver cirrhosis in the United States is:
   a. hepatitis A infection.
   b. hepatitis B infection.
   c. hepatitis C infection.
   d. alcohol abuse.

4. Esophageal varices are a dangerous complication of liver disease primarily because they:
   a. interfere with food intake.
   b. can lead to massive bleeding.
   c. divert blood flow from the GI tract.
   d. cause portal hypertension and collateral development.

5. A complication of liver disease that contributes to the development of ascites is:
   a. portal hypertension.
   b. rising blood ammonia levels.
   c. elevated serum albumin levels.
   d. insulin resistance.

6. A patient with cirrhosis may develop personality changes and motor dysfunction, which are signs of:
   a. jaundice.
   b. hepatic encephalopathy.
   c. hyperammonemia.
   d. hepatic coma.

7. With respect to protein intake, patients with cirrhosis should:
   a. consume no more than the protein RDA.
   b. restrict protein intake to 0.6 gram per kilogram body weight.
   c. use formulas with modified amino acids to meet their protein needs.
   d. maintain nitrogen balance by consuming 0.8 to 1.2 grams of protein per kilogram body weight per day.

8. People with ascites must often restrict dietary intake of:
   a. fat.
   b. protein.
   c. sugars.
   d. sodium.

9. After a liver transplant, a primary health concern is:
   a. jaundice.
   b. gallstones.
   c. infection.
   d. hepatic coma.

10. Dietary concerns after a liver transplant include all of the following *except:*
    a. the severe protein restrictions are difficult to adhere to.
    b. patients are at increased risk of foodborne illness.
    c. medications may cause gastrointestinal side effects.
    d. medications may reduce appetite and alter taste perception.

Answers to these questions appear in Appendix H.

CLINICAL APPLICATIONS

1. Vijaya Reddy is a college student who visited relatives near her parents' birthplace in Anantapur, India, during summer vacation. Although her relatives provided boiled or purified water at their home, they occasionally took Vijaya to local restaurants, where tap water was served. Several weeks after Vijaya returned home, she developed flu-like symptoms and started feeling extremely tired. She also experienced upper abdominal pain and felt nauseous after meals. After her roommate told her that her eyes and skin appeared yellow, she knew something was definitely wrong. A physician at the student health center diagnosed hepatitis.

   • Which type of hepatitis did Vijaya most likely have? What additional symptoms can develop? Is Vijaya's condition likely to become chronic?

   • What medical treatment is suggested for hepatitis? Describe the dietary modifications that may be necessary in some cases.

2. As discussed in the section on cirrhosis, many patients develop protein-energy malnutrition and wasting during the course of illness. Review Table 19-4 to find examples of problems that may lead to malnutrition. Select three nutrition or medical problems (from the *Examples* column) and discuss complications of liver disease that can cause the problems you selected. What dietary or medical treatments may help in managing the problems?

NUTRITION ON THE NET

For further study of the topics in this chapter, access these websites.

Find updates and quick links to these and other nutrition-related sites at our website: **www.wadsworth.com/nutrition**

Visit the websites of the American Liver Foundation and Canadian Liver Foundation to find information that is helpful both for health practitioners and patients with liver diseases: **www.liverfoundation.org** and **www.liver.ca**

Find resources and support for children with liver diseases by visiting Liverkids: **www.kinsey.id.au/liverkids**

Learn more about hepatitis by visiting the Hepatitis Foundation International: **www.hepfi.org**

Uncover more information about liver transplants at the Center Span Transplant News Network: **www.centerspan.org**

NOTES

[1] A. M. Diehl, Alcoholic and nonalcoholic steatohepatitis, in L. Goldman and D. Ausiello, eds., *Cecil Textbook of Medicine* (Philadelphia: Saunders, 2004), pp. 933–936.

[2] A. M. Diehl and F. Poordad, Nonalcoholic fatty liver disease, in M. Feldman, L. S. Friedman, and M. H. Sleisenger, eds., *Sleisenger and*

Fordtran's Gastrointestinal and Liver Disease: Pathophysiology, Diagnosis, Management (Philadelphia: Saunders, 2002), pp. 1393–1401.

[3] Diehl and Poordad, 2002.

[4] G. C. Farrell, Liver disease caused by drugs, anesthetics, and toxins, in M. Feldman, L. S. Friedman, and M. H. Sleisenger, eds., *Sleisen-*

ger and Fordtran's Gastrointestinal and Liver Disease: Pathophysiology, Diagnosis, Management (Philadelphia: Saunders, 2002), pp. 1403–1447.

[5]T. D. Schiano and H. C. Bodenheimer, Complications of chronic liver disease, in S. L. Friedman, K. R. McQuaid, and J. H. Grendell, eds., *Current Diagnosis and Treatment in Gastroenterology* (New York: Lange Medical Books/McGraw-Hill, 2003), pp. 639–663.

[6]D. C. Wolf, Cirrhosis, available at **www.emedicine.com/med/topic3183.htm,** site visited September 17, 2005.

[7]S. H. Rigby and K. B. Schwarz, Nutrition and liver disease, in A. M. Coulston, C. L. Rock, and E. R. Monsen, eds., *Nutrition in the Prevention and Treatment of Disease* (San Diego: Academic Press, 2001), pp. 601–613.

[8]S. L. Friedman and T. D. Schiano, Cirrhosis and its sequelae, in L. Goldman and D. Ausiello, eds., *Cecil Textbook of Medicine* (Philadelphia: Saunders, 2004), pp. 936–944.

[9]Friedman and Schiano, 2004.

[10]B. A. Runyon, Ascites and spontaneous bacterial peritonitis, in M. Feldman, L. S. Friedman, and M. H. Sleisenger, eds., *Sleisenger and Fordtran's Gastrointestinal and Liver Disease: Pathophysiology, Diagnosis, Management* (Philadelphia: Saunders, 2002), pp. 1517–1542.

[11]Schiano and Bodenheimer, 2003.

[12]J. G. Fitz, Hepatic encephalopathy, hepatopulmonary syndromes, hepatorenal syndrome, coagulopathy, and endocrine complications of liver disease, in M. Feldman, L. S. Friedman, and M. H. Sleisenger, eds., *Sleisenger and Fordtran's Gastrointestinal and Liver Disease: Pathophysiology, Diagnosis, Management* (Philadelphia: Saunders, 2002), pp. 1543–1565.

[13]Fitz, 2002.

[14]Schiano and Bodenheimer, 2003.

[15]American Dietetic Association, *Nutrition Care Manual* (Chicago: American Dietetic Association, 2005).

[16]S. Escott-Stump, *Nutrition and Diagnosis-Related Care* (Baltimore: Lippincott Williams & Wilkins, 2002).

[17]B. Als-Nielsen and coauthors, Branched-chain amino acids for hepatic encephalopathy (Cochrane Review), *The Cochrane Library* Issue 3 (2004).

[18]H. Barkoukis, K. M. Fiedler, and E. Lerner, A combined high-fiber, low-glycemic index diet normalizes glucose tolerance and reduces hyperglycemia and hyperinsulinemia in adults with hepatic cirrhosis, *Journal of the American Dietetic Association* 102 (2002): 1503–1507; D. J. Jenkins and coauthors, Low glycemic index foods and reduced glucose, amino acid, and endocrine responses in cirrhosis, *American Journal of Gastroenterology* 84 (1989): 732–739.

[19]Schiano and Bodenheimer, 2003.

[20]American Dietetic Association, 2005.

[21]E. B. Keeffe, Hepatic failure and liver transplantation, in L. Goldman and D. Ausiello, eds., *Cecil Textbook of Medicine* (Philadelphia: Saunders, 2004), pp. 944–949.

[22]J. R. Lake, Liver transplantation, in S. L. Friedman, K. R. McQuaid, and J. H. Grendell, eds., *Current Diagnosis and Treatment in Gastroenterology* (New York: Lange Medical Books/McGraw-Hill, 2003), pp. 813–834.

[23]American Dietetic Association, 2005.

[24]Lake, 2003.

[25]D. Shattuck, The dietitian's role on the transplantation team, *Journal of the American Dietetic Association* 102 (2002): 902–903.

Alcohol in Health and Disease

As Chapter 19 described, excessive alcohol consumption is a primary cause of liver disease. Alcohol can be toxic to other organs as well, such as the brain, gastrointestinal tract, and pancreas. In addition, **alcohol abuse** can lead to a number of nutrient deficiencies. Moderate use of alcohol, however, has been shown to reduce deaths from coronary heart disease in middle-aged and older adults. This Nutrition in Practice discusses current recommendations concerning alcohol and the health problems and benefits associated with its use. The glossary defines the relevant terms.

What are the current dietary guidelines for alcohol consumption?

For those who choose to drink alcoholic beverages, the *Dietary Guidelines for Americans 2005* specify that women and men should limit their alcohol intakes to one drink and two drinks per day, respectively. One **drink** is equivalent to 12 ounces of beer, 5 ounces of wine, 10 ounces of wine cooler, or 1½ ounces of 80 proof distilled spirits such as gin, rum, vodka, and whiskey. Some individuals should not consume alcohol at all. These include pregnant and lactating women, women of childbearing age who may become pregnant, children and adolescents, people using medications that can interact with alcohol, and people who are unable to voluntarily restrict their alcohol intake. Alcohol should also be avoided by anyone who is involved in an activity that requires attention or coordination, such as driving or operating machinery.[1]

About 70 percent of men and 56 percent of women in the United States drink alcoholic beverages.[2] Most alcohol drinkers (about 69 percent) are light drinkers, consuming three drinks or fewer per week. Only 8 percent of drinkers exceed the recommendations given in the *Dietary Guidelines*.

12oz beer

10oz wine cooler

1½oz hard liquor (80 proof whiskey, gin, brandy, rum, vodka)

5 oz wine

© Polara Studios, Inc.

Each of the amounts shown is equivalent to one drink.

What happens to alcohol in the body?

Recall that alcohol is a source of food energy, providing 7 kcalories per gram. A small amount of alcohol is metabolized in the stomach, but most of the alcohol consumed is quickly absorbed in the stomach and small intestine and passes readily into the body's cells, where alcohol concentrations reach levels similar to those in the blood. The liver is the site of most alcohol metabolism; as there is no storage pool for alcohol and it can be toxic to cells, its metabolism in the liver takes priority over that of other substances. The main product of alcohol metabolism is **acetate,** which can be used as a source of energy by most tissues.[3]

Because alcohol receives priority, its metabolism interferes with the metabolism of other substances. Alcohol suppresses both the storage of glycogen and the availability of glucose between meals. Heavy drinkers are at risk of developing hypoglycemia, an effect that is accentuated in people with diabetes who use insulin or medication to reduce insulin levels. Alcohol also suppresses the breakdown of fat for energy, leading to fat accumulation in the liver and increased production of triglyceride-carrying lipoproteins (VLDL). Alcohol can also inhibit protein synthesis.

The liver's ability to metabolize alcohol is limited, however; the average adult metabolizes only 7 to 10 grams of alcohol per hour,[4] an amount equivalent to three-fourths of a drink. The alcohol that remains in the blood has ready access to the body's cells, where it can alter functioning or cause damage to tissues. Alcohol levels in the brain rise quickly because the brain receives a large fraction of the circulating blood.

How is alcohol toxic to cells?

Alcohol alters the structure of cell membranes, increases their permeability, and interferes with the actions of cell membrane proteins. Under certain conditions, alcohol exposure can induce cell death. A metabolite of alcohol—**acetaldehyde**—causes numerous adverse effects; it binds to proteins and interferes with their functions, prevents formation of microtubules within cells, alters immune function, inhibits DNA repair, and causes oxidative damage.

How does alcohol affect brain function?

Alcohol acts as a central nervous system depressant; it can cause sedation, slow reaction times, and relieve anxiety. In excess it impairs judgment, reduces inhibitions, and impairs speech and motor function. Extremely high blood alcohol levels can lead to coma, respiratory depression, and death.

Chronic heavy drinking can lead to certain types of neurological damage.[5] Symptoms include tingling in the hands and feet, lack of muscular coordination, and changes in a person's manner of walking. Visual impairments, such as blurred vision and optic nerve degeneration, can also occur.

What are some other long-term consequences of drinking too much alcohol?

Alcoholic liver disease is the most common complication of alcohol abuse, occurring in about 10 to 20 percent of chronic heavy drinkers.[6] The liver disorders that develop are similar to those described in Chapter 19: fatty liver, hepatitis, and, eventually, cirrhosis and liver failure. In addition to liver damage, alcohol can cause damage to the GI tract, pancreas, and heart (see Figure NP19-1). Damage to GI mucosal tissue increases the risks of gastritis, ulcers, and GI cancers and may lead to nutrient malabsorption and diarrhea. Damage to the pancreas can alter pancreatic secretions and eventually result in pancreatitis and diabetes. Excessive alcohol consumption can raise the risk of heart attacks by elevating blood pressure and weakening heart muscle's contractibility.

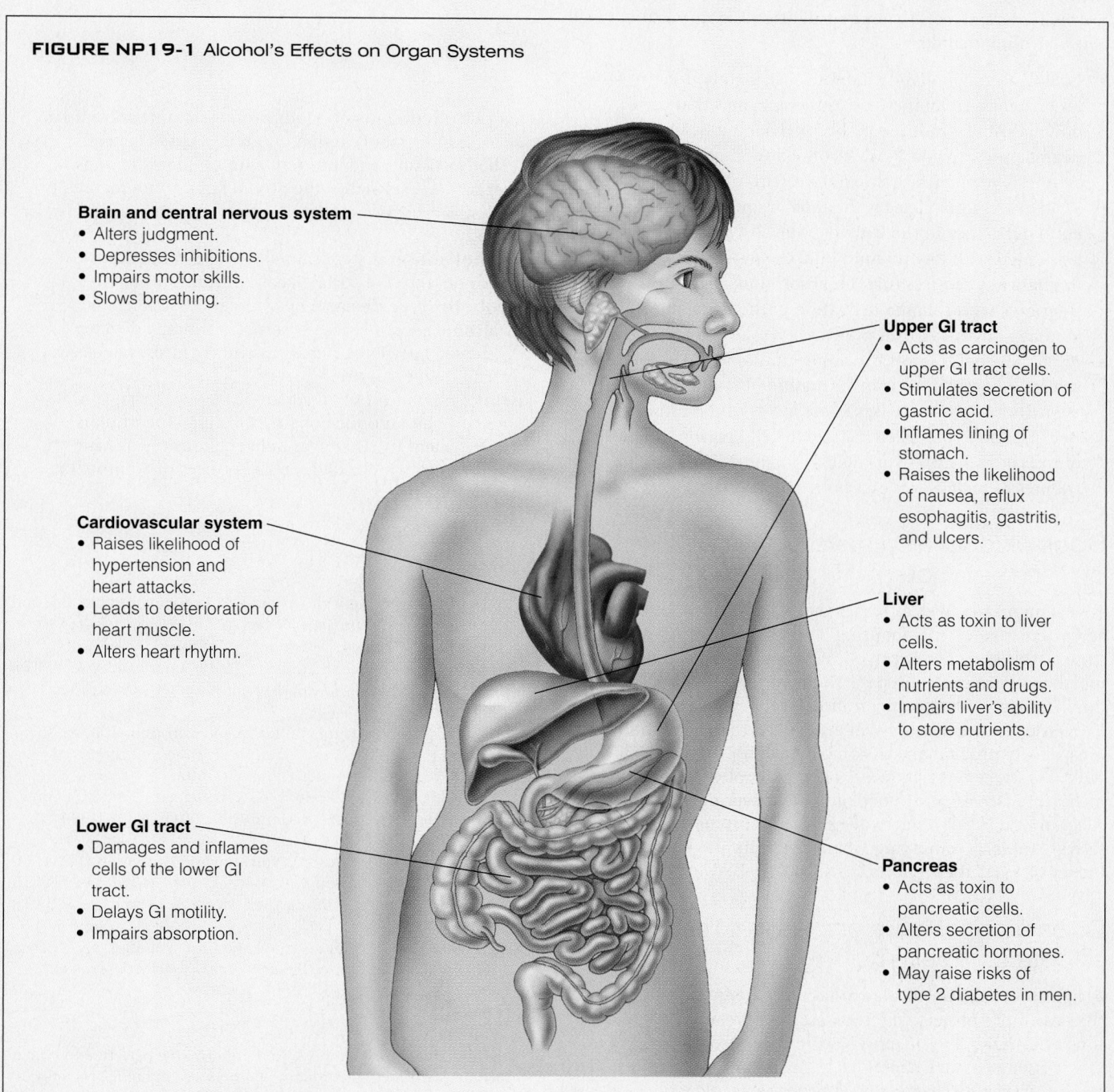

FIGURE NP19-1 Alcohol's Effects on Organ Systems

Brain and central nervous system
- Alters judgment.
- Depresses inhibitions.
- Impairs motor skills.
- Slows breathing.

Cardiovascular system
- Raises likelihood of hypertension and heart attacks.
- Leads to deterioration of heart muscle.
- Alters heart rhythm.

Lower GI tract
- Damages and inflames cells of the lower GI tract.
- Delays GI motility.
- Impairs absorption.

Upper GI tract
- Acts as carcinogen to upper GI tract cells.
- Stimulates secretion of gastric acid.
- Inflames lining of stomach.
- Raises the likelihood of nausea, reflux esophagitis, gastritis, and ulcers.

Liver
- Acts as toxin to liver cells.
- Alters metabolism of nutrients and drugs.
- Impairs liver's ability to store nutrients.

Pancreas
- Acts as toxin to pancreatic cells.
- Alters secretion of pancreatic hormones.
- May raise risks of type 2 diabetes in men.

What are the effects of excessive alcohol consumption on nutrition status?

An excessive intake of alcohol causes nutrient deficiences for several reasons. It can upset food intake: because alcohol supplies 7 kcalories per gram, it displaces other energy sources along with the essential nutrients such foods would provide. As mentioned earlier, it can cause widespread malabsorption as a result of direct damage to the GI mucosa. Alcohol also interferes with the way the body processes nutrients. Examples of common deficiencies in persons who abuse alcohol include:[7]

- *Vitamin A.* Alcohol and vitamin A are metabolized by similar enzymes, so an increase in the enzymes that break down alcohol—induced by heavy drinking—can increase degradation of vitamin A. Alcohol also interferes with the synthesis of vitamin A's carrier protein in the blood.
- *Thiamin.* Alcohol abuse is the most frequent cause of thiamin deficiency in the United States. Heavy drinking is associated with low thiamin intakes, and alcohol ingestion dramatically reduces thiamin absorption. Alcohol also interferes with thiamin activation in the body and increases thiamin losses in urine.
- *Folate.* Alcohol reduces the absorption of folate in the small intestine, inhibits the transformation of folate to its coenzyme forms, and increases folate losses in the urine. Because folate has a central role in cell division (and cells of the GI tract turn over rapidly), folate deficiency contributes to malabsorption of other nutrients as well.

Does alcohol also have disruptive effects on the metabolism of medications?

Yes. Some of the liver enzymes that degrade alcohol have the additional role of metabolizing drugs. When alcohol is consumed, the enzymes involved in alcohol degradation cannot metabolize drugs (to prepare them for excretion), so heavy drinking can increase medication potency. Thus it may be dangerous to use alcohol while taking certain medications. It is especially problematic to combine alcohol and drugs if they have similar effects in the body; examples of such drugs include sedatives and blood glucose–lowering medications.[8] Alcohol can also reduce drug absorption because alcohol slows stomach emptying; hence, medications may take longer to exert their effects.[9]

Does alcohol have any beneficial effects on health?

A light-to-moderate alcohol intake can reduce heart disease risk. Alcohol's protective effects are seen mainly in older persons who have one or more classic risk factors for heart disease (discussed in Chapter 21). Although the reasons that alcohol is beneficial are not entirely clear, alcohol increases levels of HDL cholesterol (protecting against further development of atherosclerosis) and also reduces the tendency for blood clotting. These effects occur when any type of alcoholic beverage is consumed and are seen most frequently when alcohol is consumed at least three times per week.[10]

The benefits or harm associated with alcohol consumption depend on a person's health status, age, and the amount consumed. Most people in the United States who consume alcoholic beverages do not drink more alcohol than is recommended; however, nurses should be alert to those who may be abusing alcohol and potentially endangering their health.

Glossary

acetaldehyde (ah-set-AL-duh-hide): an intermediate in alcohol metabolism that, in excess, can damage the body's tissues and interfere with cellular functions.

acetate (AH-seh-tate): the product of alcohol metabolism that is used as a source of energy by many tissues in the body.

alcohol abuse: the continued use of alcohol despite the development of social, legal, or health problems.

alcoholic liver disease: liver disease that is related to alcohol consumption. Disorders that may develop include fatty liver, hepatitis, and cirrhosis, which may lead to liver failure.

drink: an amount of an alcoholic beverage that contains about half an ounce of pure alcohol. One drink is equivalent to 12 ounces of beer, 5 ounces of wine, 10 ounces of wine cooler, or 1^1/$_2$ ounces of 80 proof distilled spirits.

Notes

[1] U.S. Department of Agriculture and U.S. Department of Health and Human Services, *Nutrition and Your Health: Dietary Guidelines for Americans 2005*, 6th ed. (Washington, D.C.: 2005).

[2] National Center for Health Statistics, *Health, United States, 2004: With Chartbook on Trends in the Health of Americans* (Hyattsville, Md.: 2004), pp. 235–237.

[3] P. M. Suter, Alcohol: Its role in health and nutrition, in B. A. Bowman and R. M. Russell, eds., *Present Knowledge in Nutrition* (Washington, D.C.: ILSI Press, 2001), pp. 497–507.

[4] S. B. Masters, The alcohols, in B. G. Katzung, ed., *Basic and Clinical Pharmacology* (New York: McGraw-Hill, 2001), pp. 382–394.

[5] Masters, 2001.

[6] S. H. Rigby and K. B. Schwarz, Nutrition and liver disease, in A. M. Coulston, C. L. Rock, and E. R. Monsen, eds., *Nutrition in the Prevention and Treatment of Disease* (San Diego: Academic Press, 2001), pp. 601–613.

[7] K. E. Light and R. Hakkak, Alcohol and nutrition, in B. J. McCabe, E. H. Frankel, and J. J. Wolfe, eds., *Handbook of Food-Drug Interactions* (Boca Raton, Fl.: CRC Press, 2003), pp. 167–189; Suter, 2001.

[8] Masters, 2001.

[9] Light and Hakkak, 2003.

[10] K. J. Mukamal and coauthors, Roles of drinking pattern and type of alcohol consumed in coronary heart disease in men, *New England Journal of Medicine* 348 (2003): 109–118.

Nutrition and Diabetes Mellitus

CHAPTER 20

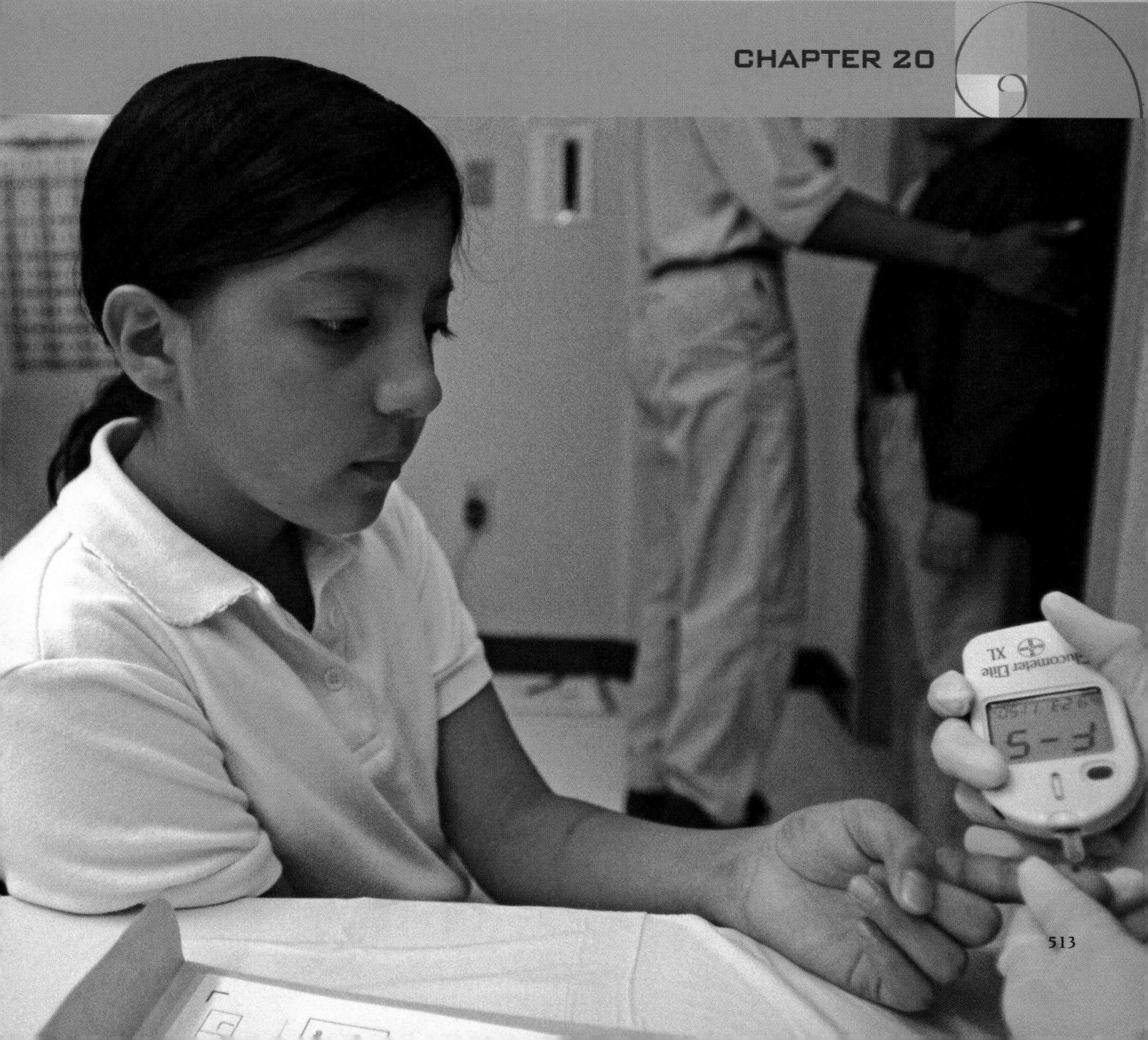

513

Reminder: *Insulin* is a pancreatic hormone that regulates blood glucose concentrations. Its many metabolic actions are countered mainly by the hormone *glucagon.*

Reminder: *Osmolarity* refers to the concentration of osmotically active particles in solution. Hyperglycemia causes the body's fluids to become *hyerposmolar,* meaning that they have an abnormally high osmolarity.

Normal fasting plasma glucose levels are approximately 75 to 106 mg/dL (published values vary).

NURSING DIAGNOSIS

for people with diabetes may include *impaired urinary elimination, risk for deficient fluid volume, disturbed sensory perception, risk for infection,* and *adult failure to thrive.*

diabetes: (DYE-ah-BEE-teez) **mellitus:** a group of metabolic disorders characterized by hyperglycemia and disordered insulin metabolism.

> **diabetes** = siphon (in Greek), referring to the excessive passage of urine that is characteristic of untreated diabetes
> **mellitus** = honey-sweet

hyperglycemia: elevated blood glucose concentrations.

renal threshold: the point at which the blood concentration of a substance exceeds the kidneys' capacity for reabsorption, causing the substance to spill into the urine.

glycosuria: (GLY-koh-SOOR-ee-ah): an abnormal amount of glucose in urine.

polyuria (pah-lee-YOOR-ee-ah): excessive urine secretion.

polydipsia (pah-lee-DIP-see-ah): excessive thirst.

polyphagia (pah-lee-FAY-jee-ah): excessive appetite or eating.

oral glucose tolerance test: a test that evaluates a person's ability to tolerate a glucose load. A common protocol for diabetes diagnosis is ingestion of a 75 g glucose load followed by measurement of plasma glucose after a two-hour interval.

prediabetes: the condition in which blood glucose levels are higher than normal but not high enough to be diagnosed as diabetes; considered a major risk factor for future diabetes and cardiovascular diseases. Formerly called **impaired glucose tolerance.**

The incidence of **diabetes mellitus** is steadily increasing in the United States and in many other countries. It now affects an estimated 6.6 percent of the U.S. population, or more than 18 million people.[1] About 29 percent of persons with diabetes do not know that they have it,[2] a danger because damage to the body often occurs before symptoms develop. Diabetes ranks sixth among the leading causes of death in the United States. It also contributes to the development of other life-threatening diseases, including heart disease and kidney failure, which are discussed in the two chapters that follow.

Overview of Diabetes Mellitus

The term *diabetes mellitus* refers to metabolic disorders characterized by elevated blood glucose concentrations and disordered insulin metabolism. People with diabetes may have impaired insulin secretion or cells that do not respond to insulin normally. The result is **hyperglycemia,** a marked elevation in blood glucose levels that can ultimately cause damage to blood vessels, nerves, and tissues.

SYMPTOMS OF DIABETES

Symptoms of diabetes are usually related to the degree of hyperglycemia present (see Table 20-1). When the plasma glucose concentration rises above about 200 milligrams per deciliter (mg/dL), it exceeds the **renal threshold;** that is, the kidneys are unable to reabsorb all of the glucose back into the blood, and some glucose spills into the urine **(glycosuria).** High glucose levels in urine draw additional water out of the blood, increasing the amount of urine produced. Thus the symptoms that arise in diabetes typically include frequent urination **(polyuria),** dehydration, and increased thirst **(polydipsia).** Hyperglycemia can also cause blurred vision due to the exposure of eye tissues to hyperosmolar fluids. Increased infections are common in diabetes, a possible consequence of hyperglycemia, impaired circulation, or depressed immune responses. Some people lose weight and have an increased appetite **(polyphagia)** as a result of the nutrient depletion that occurs when insulin is deficient. In some cases, constant fatigue is the only symptom and may be related to altered energy metabolism, dehydration, or other effects of the disease.

DIAGNOSIS OF DIABETES

The diagnosis of diabetes is based primarily on plasma glucose levels, which can be measured under fasting conditions or at random times during the day. In some cases, an **oral glucose tolerance test** is given: the patient ingests a 50- or 75-gram glucose load, and plasma glucose is measured at one or more time intervals following glucose ingestion. The presence of symptoms helps to confirm the diagnosis. The following criteria are currently used to diagnose diabetes:

- The plasma glucose concentration of a blood sample obtained at a random time during the day (without regard to food intake) is 200 mg/dL or greater, and classic symptoms of diabetes (such as polyuria, polydipsia, and unexplained weight loss) are present.
- The plasma glucose concentration is 126 mg/dL or greater after a fast of at least eight hours.
- The plasma glucose concentration measured two hours after a 75-gram glucose load is 200 mg/dL or greater.
- Diagnosis is confirmed if a subsequent test yields similar results.

The term **prediabetes** is used to refer to blood glucose levels between normal and diabetic, that is, between 100 and 125 mg/dL when fasting or between 140 and 200 mg/dL when measured two hours after a 75-gram glucose load. A person with prediabetes is at risk of developing diabetes and cardiovascular diseases.[3]

TYPES OF DIABETES

Table 20-2 lists the features of the two main types of diabetes, type 1 and type 2 diabetes. Pregnancy can lead to abnormal glucose tolerance and the condition known as *gestational diabetes*, which often resolves after pregnancy but is a risk factor for type 2 diabetes. Diabetes can also be caused by medical conditions that either damage the pancreas or interfere with insulin function.

Type 1 Diabetes **Type 1 diabetes** accounts for about 5 to 10 percent of diabetes cases. It is usually caused by **autoimmune** destruction of the pancreatic beta cells, which produce and secrete insulin. By the time symptoms develop, the damage to the beta cells has usually progressed so far that insulin must be supplied exogenously, usually by injection. Although the trigger for the autoimmune attack is unknown, both inherited and environmental factors are likely to be involved.[4] Individuals who develop type 1 diabetes are at increased risk of developing other autoimmune disorders.

Type 1 diabetes usually develops during childhood or adolescence, and symptoms may appear abruptly in previously healthy children.[5] Classic symptoms are frequent urination, weight loss, and increased thirst. **Ketoacidosis**—acidosis due to excessive production of ketone bodies—is sometimes the first sign of disease.[6] Disease onset tends to be more gradual in individuals who develop type 1 diabetes in later years. Blood tests that detect antibodies to insulin, pancreatic islet cells, and pancreatic enzymes can confirm the diagnosis and help to predict development of the disease in close relatives.

TABLE 20-1	Symptoms of Diabetes Mellitus
Frequent urination (polyuria)
Dehydration, dry mouth
Increased thirst (polydipsia)
Blurred vision
Increased infections
Weight loss
Increased hunger (polyphagia)
Fatigue

Gestational diabetes was introduced in Chapter 10 and is discussed later in this chapter.

type 1 diabetes: the type of diabetes that accounts for 5 to 10% of diabetes cases and usually results from autoimmune destruction of pancreatic beta cells. In this type of diabetes, the pancreas produces little or no insulin.

autoimmune: an immune response directed against the body's own tissues.
auto = self

ketoacidosis (KEY-to-ah-sih-DOE-sis): lowering of pH in blood and tissues due to excessive production of ketone bodies.

TABLE 20-2	Features of Type 1 and Type 2 Diabetes	
	Type 1	Type 2
Prevalence in diabetic population	5 to 10% of cases	90 to 95% of cases
Age of onset	<30 years	>45 years[a]
Associated conditions	Autoimmune diseases, viral infection, inherited factors	Obesity, aging, inherited factors
Major defect	Destruction of pancreatic beta cells; insulin deficiency	Insulin resistance; insulin deficiency (relative to needs)
Insulin secretion	Little or none	Varies; may be normal, increased, or decreased
Requirement for insulin therapy	Always	Sometimes
Other names	Juvenile-onset diabetes	Adult-onset diabetes
	Insulin-dependent diabetes mellitus (DDM)	Noninsulin-dependent diabetes mellitus (NIDDM)
	Ketosis-prone diabetes	Ketosis-resistant diabetes

[a] Incidence of type 2 diabetes is increasing in children and adolescents; in over 90 percent of these cases, it is associated with overweight or obesity and a family history of type 2 diabetes.

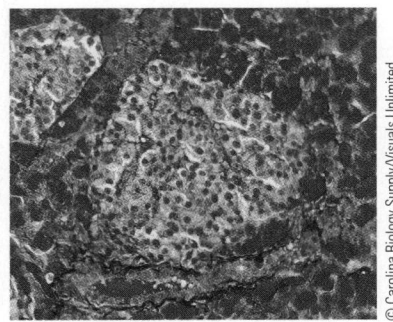

Cross sections of the pancreas reveal distinct areas known as the islets of Langerhans, which contain the beta cells that produce insulin.

Type 2 Diabetes **Type 2 diabetes** is the most prevalent form of diabetes, accounting for 90 to 95 percent of cases, and is often asymptomatic. The primary defect in type 2 diabetes is **insulin resistance,** a reduced sensitivity to insulin in muscle, adipose, and liver cells. To compensate, the pancreas secretes larger amounts of insulin, and plasma insulin concentrations can rise to abnormally high levels (hyperinsulinemia). Over time, the pancreas becomes less able to compensate for the cells' reduced sensitivity to insulin, and hyperglycemia worsens. The high demand for insulin can eventually exhaust the beta cells of the pancreas and lead to impaired insulin secretion and reduced plasma insulin concentrations. Type 2 diabetes is therefore associated both with insulin resistance and with relative insulin deficiency; that is, the amount of insulin is insufficient to compensate for its diminished effect in the cells.

Although the actual causes of type 2 diabetes are unknown, the risk is substantially increased by obesity (especially abdominal obesity), aging, and physical inactivity. An estimated 80 to 90 percent of individuals with type 2 diabetes are obese, and obesity itself can directly cause some degree of insulin resistance.[7] The prevalence of type 2 diabetes increases with age and approaches 18 percent in persons between 65 and 74 years of age; many of these cases, however, remain undiagnosed.[8] Inherited factors strongly influence risk, and type 2 diabetes is more common in certain ethnic populations, including Native Americans, Hispanic Americans, Mexican Americans, African Americans, Asian Americans, and Pacific Islanders.

Type 2 Diabetes in Children and Adolescents Although most cases of type 2 diabetes are diagnosed in individuals over 45 years old, children and adolescents who are overweight or have a family history of diabetes are at increased risk. Because type 2 diabetes is frequently asymptomatic, it is generally detected in children only when high-risk groups are screened for the disease. For example, when 167 obese children of different ethnic groups were screened, prediabetes was detected in 25 percent of children between 4 and 10 years of age and in 21 percent of adolescents between 11 and 18 years of age.[9] Among the Pima Indians in Arizona, a population with one of the highest rates of type 2 diabetes in the world, overt type 2 diabetes was reported in 2.2 percent of 10- to 14-year-old children and in 5 percent of 15- to 19-year-old teens.[10] Routine screening and prevention programs that target food intake and activity patterns can be important safeguards for preventing diabetes in children at risk.

ACUTE COMPLICATIONS OF DIABETES

Untreated diabetes can result in life-threatening complications. Insulin deficiency can cause severe disturbances in energy metabolism, and hyperglycemia can lead to fluid and electrolyte imbalances. In treated diabetes, hypoglycemia is a possible complication of inappropriate disease management.

Diabetic Ketoacidosis in Type 1 Diabetes A severe lack of insulin causes diabetic ketoacidosis. The insulin deficiency results in unrestrained breakdown of the triglycerides in adipose tissue and excessive release of fatty acids into the blood. This promotes a substantial increase in the liver's production of ketone bodies (ketosis). Ketone bodies are acidic and can reach extremely high levels in the bloodstream (ketoacidosis) and spill into the urine (ketonuria). Blood pH typically falls below 7.3. Blood glucose concentrations usually exceed 250 mg/dL and may rise above 1000 mg/dL in severe cases.[11] The main features of diabetic ketoacidosis thus include ketosis, acidosis, and hyperglycemia.

Patients with ketoacidosis may exhibit symptoms of both acidosis and dehydration. Acidosis is partially corrected by exhalation of carbon dioxide, so hyperventilation or deep breathing is characteristic. Ketone accumulation is sometimes evident by a fruity odor on a person's breath (**acetone breath**). The fluid loss (polyuria) that

Reminder: *Ketone bodies* are products of fat metabolism that are produced in the liver; they accumulate in tissues when fatty acids are released in abnormally high amounts from adipose tissue.

Normal blood pH ranges from 7.35 to 7.45.

Bicarbonate, a buffer in the bloodstream, can correct acidosis. The acid (H^+) and bicarbonate (HCO_3^-) combine to form carbonic acid (H_2CO_3), which breaks down to water (H_2O) and carbon dioxide (CO_2). The carbon dioxide is then exhaled.

type 2 diabetes: the type of diabetes that accounts for 90 to 95% of diabetes cases and usually results from insulin resistance coupled with insufficient insulin secretion. Obesity is present in 80 to 90% of cases.

insulin resistance: reduced sensitivity to insulin in muscle, adipose, and liver cells.

acetone breath: distinctive fruity odor on the breath of a person with ketosis.

accompanies hyperglycemia lowers blood volume and blood pressure and depletes electrolytes. Mental state may vary from alertness to comatose (**diabetic coma**).[12]

Diabetic ketoacidosis may result from inappropriate treatment (such as missed insulin injections), illness or infection, alcohol abuse, and other physiological stressors.[13] The condition usually develops quickly, within hours or a few days. Although diabetic ketoacidosis can occur in type 2 diabetes as well, it develops only rarely because even relatively low insulin concentrations can prevent excessive ketone body production. When ketosis does occur in type 2 diabetes, it is usually associated with metabolic stress due to infection or serious illness.

Hyperosmolar Hyperglycemic State in Type 2 Diabetes The **hyperosmolar hyperglycemic state** is a condition of severe hyperglycemia that usually develops in the absence of significant ketosis. It often evolves slowly, over several days or weeks, and is most often associated with type 2 diabetes. Blood glucose levels typically exceed 600 mg/dL and may rise above 2000 mg/dL. The extreme hyperglycemia causes substantial fluid losses, leading to depleted blood volume and electrolyte imbalances. Blood plasma may become so hyperosmolar as to cause neurological abnormalities, such as abnormal reflexes, motor impairments, reduced verbal ability, and seizures; about 10 percent of patients lapse into coma. The hyperosmolar hyperglycemic state often develops because patients are unable to recognize thirst or adequately replace fluid losses due to age, illness, sedation, or incapacity.[14]

Hypoglycemia **Hypoglycemia,** or low blood glucose, arises from the inappropriate management of diabetes rather than from the disease itself. It can result from using excessive amounts of insulin or antidiabetic drugs, prolonged exercise, skipped or delayed meals, inadequate food intake, or consuming alcohol without food. Hypoglycemia most often occurs in type 1 diabetes and accounts for about 3 to 4 percent of deaths in insulin-treated patients.[15]

Symptoms of hypoglycemia include hunger, sweating, shakiness, heart palpitations, slurred speech, and confusion. Mental confusion may prevent a person from recognizing the problem and taking such corrective action as ingesting glucose tablets, juice, or candy. If hypoglycemia occurs during the night, patients may be completely unaware of its presence.

CHRONIC COMPLICATIONS OF DIABETES

Prolonged exposure to high glucose concentrations can destroy cells and tissues. Glucose and glucose fragments react with proteins, forming compounds that accumulate and cause damage to cells and blood vessels. Chronic complications of diabetes typically affect the large blood vessels, smaller vessels such as arterioles and capillaries, and the nervous system.

Disorders of the Large Blood Vessels The damage caused by diabetes accelerates the development of atherosclerosis in the coronary arteries and arteries in the limbs. Cardiovascular diseases are the leading cause of death in people with diabetes, accounting for 75 percent of deaths.[16] Type 2 diabetes is often accompanied by multiple risk factors for coronary heart disease, including hypertension, abnormal blood lipids, and obesity. In addition, people with diabetes have increased tendencies for thrombosis (blood clot formation) and abnormal ventricle function, both of which can worsen the clinical course of heart disease.[17]

Impaired blood flow in the arteries of limbs increases the risk of **claudication** (pain while walking) and contributes to the development of foot ulcers (see the photo). Left untreated, foot ulcers can lead to **gangrene** (tissue death), and some patients require foot amputation, a major cause of disability in diabetes. About 15 to 20 percent of persons with diabetes are hospitalized with foot complications during the course of illness.[18]

Diabetic coma was a frequent cause of death before insulin was routinely used to manage diabetes.

NURSING DIAGNOSIS

for people with diabetic ketoacidosis or the hyperosmolar hyperglycemic state may include *deficient fluid volume, disturbed thought processes,* and *ineffective therapeutic regimen management.*

People with type 2 diabetes frequently develop the *metabolic syndrome,* a cluster of symptoms associated with insulin resistance (including hyperglycemia, hypertension, and altered blood lipids) that substantially increase heart disease risk (see Nutrition in Practice 20).

diabetic coma: a coma that occurs in uncontrolled diabetes; may be due to diabetic ketoacidosis, the hyperosmolar hyperglycemic state, or excessive doses of insulin or certain antidiabetic drugs.

hyperosmolar hyperglycemic state: extreme hyperglycemia that is associated with hyperosmolar blood, dehydration, and altered mental status; formerly called *hyperglycemic hyperosmolar nonketotic coma.*

hypoglycemia: abnormally low concentrations of blood glucose.

claudication (CLAW-dih-KAY-shun): pain in the legs while walking; usually due to an inadequate supply of blood to muscles.

gangrene: death of tissue due to a deficient blood supply and/or infection.

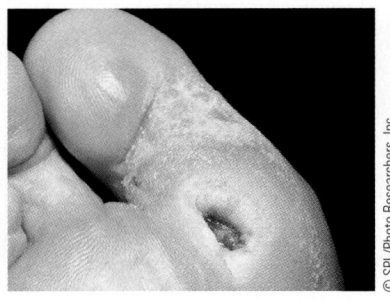

© SPL/Photo Researchers, Inc.

Foot ulcers are a common complication of diabetes because blood circulation is impaired, which slows healing, and nerve damage dampens foot pain, delaying recognition and treatment of cuts and bruises.

NURSING DIAGNOSIS

for persons with chronic complications of diabetes may include *ineffective tissue perfusion, risk for injury, risk for infection, risk for impaired skin integrity,* and *disturbed sensory perception.*

Studies that evaluated the benefits of intensive diabetes treatment include the *Diabetes Control and Complications Trial* and the *United Kingdom Prospective Diabetes Study.*

gastroparesis (GAS-troe-pah-REE-sis): delayed stomach emptying.

glycemic (gly-SEE-mic): pertaining to blood glucose.

Disorders of the Small Blood Vessels Long-term diabetes can cause progressive damage to the small blood vessels of the retina (retinopathy), resulting in visual impairments and blindness. Damage to the small blood vessels of the kidneys (nephropathy) frequently occurs in the later stages of diabetes and may lead to renal failure. Retinopathy and nephropathy progress most rapidly when diabetes is poorly controlled, and intensive management substantially reduces the risks of developing these conditions.[19]

Nerve Damage Nerve degeneration (neuropathy) occurs in about 50 percent of diabetes cases. Symptoms of neuropathy vary and may be experienced as pain or burning, numbness and tingling in the hands and feet, or loss of sensation. Pain and cramping, especially in the legs, are often severe during the night and may interrupt sleep. Neuropathy also contributes to the development of foot ulcers because cuts and bruises may go unnoticed until wounds are severe. Other manifestations of neuropathy include sweating abnormalities, sexual dysfunction, constipation, and delayed stomach emptying (gastroparesis).[20]

REVIEW NOTES

In type 1 diabetes, the pancreas secretes little or no insulin, and insulin therapy is necessary for survival. Type 2 diabetes is characterized by insulin resistance coupled with relative insulin deficiency.

In diabetic ketoacidosis, which is more common in type 1 diabetes, hyperglycemia is accompanied by ketosis and acidosis. The hyperosmolar hyperglycemic state is usually a consequence of type 2 diabetes; it is associated with severe hyperglycemia and depleted blood volume and may lead to mental impairment and coma.

Chronic complications of diabetes include cardiovascular diseases, impaired blood circulation, retinopathy, nephropathy, and neuropathy.

Treatment of Diabetes Mellitus

Diabetes is a chronic and progressive illness that requires lifelong treatment. Managing blood glucose levels is a delicate balancing act that involves meal planning, proper timing of medications, and physical exercise; frequent adjustments in treatment are often necessary to establish good **glycemic** control. Individuals with type 1 diabetes require insulin therapy for survival. Type 2 diabetes is initially treated with diet therapy and exercise, but most patients are eventually treated with oral medications or insulin. Although the health care team must determine the appropriate therapy, an individual with diabetes ultimately assumes much of the responsibility for treatment and therefore requires education in self-management of the disease.

TREATMENT GOALS

The main goal of diabetes treatment is to maintain blood glucose levels within a desirable range to prevent or reduce the risk of complications. Several multicenter clinical trials have shown that intensive diabetes treatment, which keeps blood glucose levels tightly controlled, can reduce the incidence and severity of nephropathy, retinopathy, and neuropathy.[21] Therefore, maintenance of near-normal glucose levels has become the fundamental objective of all diabetes care plans. Other goals of treatment include maintaining healthy blood lipid concentrations, controlling blood pressure, and managing weight—measures that can help to prevent or delay diabetes complications as well. Table 20-3 provides examples of the major differences between conventional and intensive therapies for type 1 diabetes. Although intensive therapy is also associated with some risks, including an increased risk of hypoglycemia, its benefits outweigh these disadvantages.

TABLE 20-3	Comparison of Conventional and Intensive Therapies for Type 1 Diabetes	
	Conventional Therapy	Intensive Therapy
Blood glucose monitoring	Monitored daily	Monitored at least three times daily
Insulin therapy	One or two daily injections; no daily adjustments	Three or more daily injections or use of external insulin pump; dosage adjusted according to results of glucose monitoring and expected carbohydrate intake
Advantages	Fewer incidences of severe hypoglycemia; less weight gain	Delayed progression of retinopathy, nephropathy, and neuropathy
Disadvantages	More rapid progression of retinopathy, nephropathy, and neuropathy	Two- to threefold increase in severe hypoglycemia; weight gain; increased risk of becoming overweight

Newly diagnosed patients and their families have much to learn about diabetes and its management. Diabetes education provides an individual with the knowledge and skills necessary to implement treatment. The primary instructor is often a **Certified Diabetes Educator (CDE),** a health care professional (often a nurse or dietitian) who has specialized knowledge about diabetes treatment and the health education process. To manage diabetes, patients need to learn about appropriate meal planning, medication administration, blood glucose monitoring, weight management, appropriate physical activity, and prevention of complications.

EVALUATING DIABETES TREATMENT

Diabetes treatment is largely evaluated by monitoring glycemic status. Good glycemic control requires frequent home monitoring of blood glucose using a glucose meter, referred to as **self-monitoring of blood glucose.** Glucose testing provides valuable feedback when the patient adjusts food intake, medications, and physical activity and is helpful for preventing hypoglycemia. Ideally, patients with type 1 diabetes should monitor blood glucose three or more times daily—and more frequently when therapy is adjusted. Self-monitoring of blood glucose is also useful in type 2 diabetes, although the recommended frequency depends on the specific needs of individual patients.[22]

Long-Term Glycemic Control Health care providers periodically evaluate long-term glycemic control by measuring **glycated hemoglobin** (abbreviated **HbA_{1c}**). The glucose in blood freely enters red blood cells and attaches to hemoglobin molecules in direct proportion to the amount of glucose present. Because the life span of red blood cells averages 120 days, the percentage of HbA_{1c} is a measure of glycemic control during the preceding two to three months. In people without diabetes, HbA_{1c} is typically less than 6 percent of total hemoglobin. The goal of diabetes treatment is an HbA_{1c} value under 7 percent, but it is often markedly higher even in people with diabetes who are maintaining near-normal blood glucose levels.[23]

Monitoring for Long-Term Complications Individuals with diabetes are routinely monitored for signs of long-term complications. Blood pressure is measured at each checkup. Lipid screening is suggested annually for most adult patients. Routine checks for urinary protein **(microalbuminuria)** help to determine if nephropathy has developed. Physical examinations generally screen for signs of retinopathy, neuropathy, and foot problems.[24]

Ketone Testing Ketone testing is used to check for the development of ketoacidosis if symptoms are present or if risk has increased due to acute illness, stress, or

Certified Diabetes Educator (CDE): a health care professional who specializes in diabetes management education. Certification is obtained from the National Certification Board for Diabetes Educators.

self-monitoring of blood glucose: home monitoring of blood glucose levels using a glucose meter.

glycated hemoglobin (HbA_{1c}): hemoglobin molecules to which glucose is attached. The percentage of such molecules is used to evaluate long-term glycemic control. Also called **glycosylated hemoglobin.**

microalbuminuria: the presence of small amounts of albumin (protein) in the urine, a sign of diabetic nephropathy.

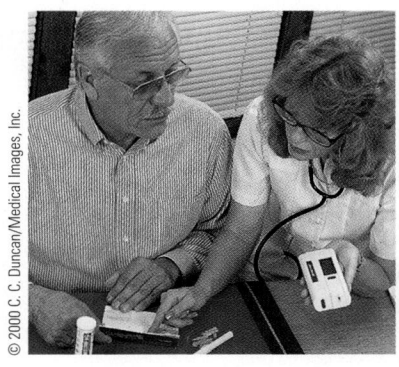

© 2000 C. C. Duncan/Medical Images, Inc.

Self-monitoring of blood glucose can help persons with diabetes learn how to maintain blood glucose levels within a desirable range.

pregnancy. Both blood and urine tests are available for home use, although the blood tests are currently more reliable. Ketone testing is most useful for patients who have type 1 diabetes or gestational diabetes.

BODY WEIGHT CONCERNS

Whereas individuals with newly diagnosed type 1 diabetes are likely to be thin, most people with type 2 diabetes are overweight or obese. Weight and growth patterns are routinely monitored to evaluate whether energy intakes are appropriate.

Body Weight in Type 1 Diabetes In general, people with type 1 diabetes are less likely to be overweight than those in the general population. In some individuals, however, excessive weight gain is an unwanted side effect of improved glycemic control, especially in those undergoing intensive insulin therapy. Although the cause of weight gain is unclear, it is possibly related to the insulin treatment, which may increase appetite or reduce metabolic rate.[25] Although efforts should be made to prevent excessive weight gain, concerns about weight should not discourage the use of intensive therapy, which is associated with longer life expectancy and fewer complications than occur with conventional therapy (review Table 20-3). It is also important to ensure that growing children receive sufficient energy for normal growth and development.

Body Weight in Type 2 Diabetes Because excessive body fat can worsen insulin resistance, weight loss is often recommended for those who are overweight or obese. Even moderate weight loss (10 to 20 pounds) can help to improve glycemic control, blood lipid levels, and blood pressure. Weight loss is most beneficial early in the course of diabetes, before insulin secretion has diminished.

Not all persons with type 2 diabetes are overweight or obese. Older adults and those in long-term care facilities are often underweight and may need to gain weight. Low body weight increases risks of morbidity and mortality in these individuals.

MEDICAL NUTRITION THERAPY: NUTRIENT RECOMMENDATIONS

Medical nutrition therapy has a considerable influence on diabetes outcome, as the appropriate dietary choices can improve blood glucose levels and slow the progression of diabetic complications. As always, the nutrition care plan must take personal preferences and lifestyle habits into account. Dietary intakes need to be modified to accommodate growth, lifestyle changes, aging, and any complications that develop. Although all members of the diabetes care team should understand the principles of dietary treatment, a registered dietitian is best suited for designing and implementing the medical nutrition therapy provided to diabetes patients. This section presents the nutrient recommendations for diabetes. A later section describes meal-planning strategies.

Total Carbohydrate Intake The amount of carbohydrate ingested has the greatest influence on blood glucose levels after meals—the more grams of carbohydrate ingested, the greater the glycemic response. The carbohydrate recommendation is based in part on the person's metabolic needs (that is, the type of diabetes or degree of glucose tolerance) and individual preferences. In general, it is recommended that 60 to 70 percent of energy intake should derive from a combination of carbohydrate and monounsaturated fat, suggesting a carbohydrate intake equivalent to about 50 percent of total kcalories.[26]

Carbohydrate Sources Different carbohydrate-containing foods have varying effects on blood glucose levels; for example, consuming a portion of white rice may cause blood glucose to rise more than would a similar portion of barley. This *glycemic effect* of foods is influenced by a food's fiber content, the preparation method, the other foods included in a meal, and individual tolerances. At present, the glycemic effect of individual foods is not a primary consideration when treating diabetes, as the evidence thus far is insufficient to conclude that consuming only foods with a low glycemic effect offers a long-term benefit.[27] High-fiber, whole-grain products, however, which have more moderate effects on blood glucose than do highly processed starchy foods, are among the foods frequently recommended for persons with diabetes.

Fiber Fiber recommendations for diabetes are similar to those for the general population. Although some studies have suggested that very high intakes of fiber (50 grams or more per day) may improve glycemic control, the benefits have not been consistent across studies, and many individuals may have difficulty tolerating such large amounts of fiber.

The fiber DRI for adult women and men ranges from 21 to 38 g. Check the DRI table on the inside front cover for the specific DRI values for different age and gender groups.

Sugars A common misperception is that people with diabetes need to avoid sugar and sugar-containing foods. In reality, however, table sugar (sucrose), made up of glucose and fructose, has a lower glycemic effect than that of starch. Because moderate consumption of sugar has not been shown to adversely affect glycemic control,[28] sugar recommendations for people with diabetes are similar to those for the general population, which suggest minimizing foods and beverages that contain added sugars. Sugars and sugary foods must be counted as part of the daily carbohydrate allowance, however.

Although fructose has a minimal glycemic effect, its use as an added sweetener is not advised because excessive dietary fructose may adversely affect blood lipid levels (it is not necessary to avoid the naturally occurring fructose in fruits and vegetables, however). Sugar alcohols (such as sorbitol and maltitol) have lower glycemic effects than glucose, fructose, or sucrose, but their use has not been found to significantly improve long-term glycemic control. Artificial sweeteners (such as aspartame, saccharin, and sucralose) contain no digestible carbohydrate and can be safely used in place of sugar.

Dietary Fat As mentioned earlier, people with diabetes are at high risk of developing cardiovascular diseases. Guidelines for dietary fat are similar to those for other persons at risk: saturated fat intake should be limited to less than 10 percent of kcalories and cholesterol intake to less than 300 milligrams daily. If an individual's LDL cholesterol levels are elevated, saturated fat should be limited to 7 percent of kcalories and cholesterol intake to 200 milligrams daily. Dietary strategies for cardiovascular disease are discussed further in Chapter 21.

Protein Some evidence suggests that people with diabetes have higher protein requirements than those for the general population due to the increased protein turnover associated with hyperglycemia.[29] However, protein intakes in the United States generally range from 15 to 20 percent of total kcalories—enough to cover any increased needs in diabetes. Although protein intakes within this range have not been found to hasten the development of diabetic nephropathy, higher protein intakes are discouraged because they may be detrimental to kidney function.

Alcohol Use in Diabetes Alcohol can be used in moderation by adults with diabetes. Guidelines are similar to those for the general population, which advise a daily limit of one drink for women and two drinks for men. Those using insulin or medications that promote insulin secretion, however, should consume food when they use alcoholic beverages to avoid hypoglycemia. Alcohol can cause hypoglycemia by interfering with glucose production in the liver. Conversely, an excessive alcohol intake can worsen hyperglycemia, and it can also raise triglyceride levels in

Reminder: One drink is equivalent to 12 ounces of beer, 5 ounces of wine, 10 ounces of wine cooler, or 1¹/₂ ounces of 80 proof distilled spirits such as gin, rum, vodka, and whiskey.

susceptible persons.[30] People who should avoid alcohol include pregnant women and individuals with pancreatitis, advanced neuropathy, abnormally high triglyceride levels, or a history of alcohol abuse.

Micronutrients Micronutrient recommendations for people with diabetes are the same as for the general population. Vitamin and mineral supplementation is not recommended unless nutrient deficiencies develop. Exceptions include the use of folate supplements during pregnancy to prevent neural tube defects and calcium supplements to reduce osteoporosis risk in older adults. Although some studies have suggested that supplemental chromium can improve glycemic control in type 2 diabetes, results have not been consistent. At present, chromium supplementation is not recommended for those with type 2 diabetes.[31]

MEDICAL NUTRITION THERAPY: MEAL-PLANNING STRATEGIES

Dietitians use a number of meal-planning strategies to help people with diabetes maintain glycemic control. Emphasis is usually given to controlling carbohydrate intake and portion sizes. A regular eating pattern, with meals spaced throughout the day, is typically recommended. Providing sample menus can help to illustrate general principles. Persons using intensive insulin therapy must coordinate insulin injections with meals and adjust insulin dosages to carbohydrate intake, as discussed in a later section.

Carbohydrate Counting Carbohydrate counting techniques are simpler and more flexible than other menu-planning approaches and are widely used for planning diabetes diets. Carbohydrate counting works as follows: after a dietitian determines a person's nutrient and energy needs, the individual is given a daily carbohydrate allowance, often divided into a pattern of meals and snacks according to individual preferences. The carbohydrate allowance can be expressed in grams or as the number of carbohydrate portions allowed per meal (see Table 20-4). The user of the plan need only be concerned about meeting carbohydrate goals and can select from any of the carbohydrate-containing food groups when planning meals (see Table 20-5 and Figure 20-1). Although encouraged to make healthy food choices, the individual has the freedom to choose the foods desired at each meal without risking loss of glycemic control. Some people may also need guidance about noncarbohydrate foods to help them choose a healthy diet that improves blood lipids or energy intakes. The "How to" on pp. 523–524 explains how to implement carbohydrate counting in clinical practice.

    Carbohydrate counting is taught at different levels of complexity depending on a person's needs and abilities. The basic carbohydrate counting method just described can be helpful for most people, although it requires a consistent carbohydrate intake to match the medication or insulin regimen. Advanced carbohydrate counting allows more flexibility but is best suited for patients using intensive insulin therapy. With this method, a person can determine the specific dosage of insulin needed to cover the amount of carbohydrate consumed at a meal. The person is then free to choose the types and portions of food desired without sacrificing glycemic control. Advanced carbohydrate counting requires some training and should be attempted only after more basic methods are mastered.

Exchange Lists for Meal Planning The exchange list system is still taught by some dietitians, although it is more complex and difficult for patients to learn than carbohydrate counting. The exchange system sorts foods according to their proportions of carbohydrate, fat, and protein so that each item in a food group (or "exchange list") has a similar macronutrient and energy content (see pp. C-1 to C-2). Thus any food on a list can be exchanged, or traded, for any other food on

Use Basic Carbohydrate Counting in Clinical Practice

1. The first step in basic carbohydrate counting is to determine an appropriate carbohydrate intake and suitable distribution pattern; an example is shown in Table 20-4. A nutrition assessment can help to estimate a person's usual energy and carbohydrate intakes. The carbohydrate level should be acceptable to the person using the plan. Frequent monitoring of blood glucose levels can help determine whether additional carbohydrate restriction would be helpful.

The example given in Table 20-4 illustrates a meal pattern for a person consuming 2000 kcalories daily with a carbohydrate allowance of 50 percent of kcalories. This is calculated as follows:

$$50\% \times 2000 \text{ kcal} = 1000 \text{ kcal of carbohydrate.}$$

$$\frac{1000 \text{ kcal carbohydrate}}{4 \text{ kcal/g carbohydrate}} = 250 \text{ g carbohydrate/day.}$$

$$\frac{250 \text{ g carbohydrate}}{15 \text{ g/1 carbohydrate portion}} = \frac{16.7 \text{ carbohydrate}}{\text{portions/day.}}$$

2. The distribution of carbohydrates among meals and snacks is based on both individual preferences and metabolic needs. In type 1 diabetes, the insulin regimen must coordinate with the individual's dietary and lifestyle choices. People using conventional insulin therapy must have a consistent carbohydrate intake from day to day to match their particular insulin prescription, whereas those using intensive therapy can alter insulin dosages when carbohydrate intakes change. People with type 2 diabetes are encouraged to develop dietary patterns that suit their lifestyle and medication schedules. For all types of diabetes, the carbohydrate recommendation may need to be altered periodically to improve blood glucose control.

3. Carbohydrate counting can be done in one of two ways:

- Count the grams of carbohydrate provided by foods.
- Count carbohydrate portions, expressed in terms of servings that contain approximately 15 grams each.

Success with carbohydrate counting requires knowledge about the food sources of carbohydrates and an understanding of portion control. As shown in Table 20-5, food selections that contain about 15 grams of carbohydrate are interchangeable. The portions of foods that contain 15 grams may vary substantially,

TABLE 20-5	Portion Sizes of Carbohydrate-Containing Foods

Food Groups with Sample Portion Sizes

Bread, cereal, rice, and pasta: 1 portion = 15 g carbohydrate.

1 slice of bread or 1 tortilla
1/2 English muffin
3/4 c unsweetened, ready-to-eat cereal
1/2 c cooked oatmeal
1/3 c cooked rice or pasta

Fruit: 1 portion = 15 g carbohydrate.

1 medium apple, orange, or peach
1 small banana
3/4 c blueberries or chopped pineapple
1/2 c apple or orange juice

Milk products: 1 portion = 12 g carbohydrate; may be rounded up to 15 g for ease in counting carbohydrate portions.

1 c milk (whole, low-fat, or fat-free)
1 c buttermilk
6 oz plain yogurt

Starchy vegetables: 1 portion = 15 g carbohydrate.

1 small (3 oz) potato
1/2 c canned or frozen corn
1/3 c baked beans
1 c winter squash, cubed

Sweets and desserts: Considerable variation in carbohydrate content; portions listed contain approximately 15 g.

1/2 c ice cream
2 sandwich cookies (with cream filling)
1/2 frosted cupcake
1 granola bar (1 oz)
1 tbs honey

Nonstarchy vegetables: 1 portion = 3 to 6 g carbohydrate; 3 servings are equivalent to 1 carbohydrate portion; can be disregarded if less than 3 servings are consumed.

1 c cooked cauliflower
1/2 c cooked cabbage, collards, or kale
1/2 c cooked okra
1/2 c diced or raw tomatoes

TABLE 20-4	Sample Carbohydrate Distribution for a 2000-kCalorie Diet

	Carbohydrate Allowance	
Meals	**Grams**	**Portions[a]**
Breakfast	60	4
Lunch	60	4
Afternoon snack	30	2
Dinner	75	5
Evening snack	30	2
Totals	**255 g**	**17**

Note: The carbohydrate allowance in this example is approximately 50 percent of total kcalories.

[a]1 portion = 15 g carbohydrate = 1 portion of starchy food, milk, or fruit.

Note: Unprocessed meats, fish, and poultry contain negligible amounts of carbohydrate.

however, even among foods in a single food group. Accurate carbohydrate counting often requires instruction and practice in portion control using measuring cups, spoons, and a food scale. Food lists that indicate the carbohydrate contents of common foods are available from the American Diabetes Association and the American Dietetic Association; these are helpful resources for learning carbohydrate counting methods.

When using packaged foods, individuals should check the Nutrition Facts panel of food labels to find the carbohydrate content of a serving. If the fiber content is greater than 5 grams per serving, it should be subtracted from the *Total Carbohydrate* value, as fiber does not contribute to blood glucose. If the sugar alcohol content is greater than 5 grams per serving, half of the grams of sugar alcohol can be subtracted from the *Total Carbohydrate* value.

4. Once they have learned the basic carbohydrate counting method, individuals can select whatever foods they wish, as long as they do not exceed their carbohydrate goals. Figure 20-1 shows a day's menu that follows the dietary plan shown in Table 20-4. Although carbohydrate counting focuses on a single macronutrient, people using this technique should be encouraged to follow a healthy eating plan that meets other dietary objectives as well.

FIGURE 20-1

SAMPLE Menu—Translating Carbohydrate Portions into a Day's Meals

	Carbohydrate Portions			Carbohydrate Portions
Breakfast:		**Afternoon snack:**		
Carbohydrate goal = 4 portions or 60 g.		**Carbohydrate goal = 2 portions or 30 g.**		
3/4 c unsweetened, ready-to-eat cereal	1	2 sandwich cookies		1
1/2 c low-fat milk	1/2	1 c low-fat millk		1
1 scrambled egg	—	**Dinner:**		
1 slice whole-wheat toast (with margarine or butter)	1	**Carbohydrate goal = 5 portions or 75 g.**		
6 oz orange juice	1 1/2	4 oz grilled steak		—
Coffee (without milk or sugar)	—	1 small baked potato (with margarine or butter)		1
		Corn on cob, 1 large ear		2
Lunch:		1/2 c steamed collard greens[a]	⎤	
Carbohydrate goal = 4 portions or 60 g.		1 c sliced, raw tomatoes[a]	⎦	1
1 tuna salad sandwich (includes 2 slices whole-grain bread, mayonnaise)	2	1/2 c ice cream		1
6 oz yogurt (plain) with 3/4 c blueberries and artificial sweetener	2	**Evening snack:**		
		Carbohydrate goal = 2 portions or 30 g.		
Diet cola	—	1 medium apple		1
		1 oz granola bar		1

[a] Three servings of nonstarchy vegetables are equivalent to 1 carbohydrate portion.

the same list without affecting the macronutrient balance in a day's meals. Although the exchange list system may be helpful for individuals who want a structured dietary plan that provides specific percentages of protein, carbohydrate, and fat, it offers no advantages for maintaining glycemic control and is less flexible than carbohydrate counting. This system of meal planning is described further in Appendix C (Appendix B for Canadians).

The exchange lists can be helpful resources for individuals using carbohydrate counting methods because the portions in the exchange lists are interchangeable with the portions used in carbohydrate counting. Foods listed in the starch, fruit, and milk exchange lists, for example, are equivalent to carbohydrate "portions," as each item contains approximately 15 grams of carbohydrate (see pp. C-2 and C-3; note that the carbohydrate in milk exchanges can be rounded up to 15 grams). The list labeled "Sweets, Desserts, and Other Carbohydrates" (p. C-0) indicates the number of carbohydrate portions per serving in the far-right column.

TABLE 20-6	Insulin Preparations			
Form of Insulin	Common Preparations	Onset of Action	Peak Activity	Duration of Action
Rapid-acting	Lispro Aspart	15 min	30 min to 2 hr	3 to 5 hr
Short-acting	Regular	30 min	2 to 4 hr	5 to 8 hr
Intermediate-acting	Lente NPH	1 to 3 hr	5 to 10 hr	18 to 24 hr
Long-acting	Glargine Ultralente	2 to 4 hr 4 to 6 hr	Steady effects 8 to 12 hr	24 hr Over 30 hr
Insulin mixtures (with sample ratios)	NPH/regular (70:30) NPH/regular (50:50)	Variable; depends on formulation	Variable; depends on formulation	Variable; depends on formulation

INSULIN THERAPY

Insulin therapy is necessary for people who cannot produce enough insulin to meet their metabolic needs. It is therefore required by individuals with type 1 diabetes and those with type 2 diabetes who cannot maintain glycemic control with oral medications, diet, and exercise. The pancreas normally secretes insulin in relatively low amounts between meals and during the night (called *basal insulin*) and in much higher amounts when meals are ingested. Ideally, the insulin treatment should reproduce the natural pattern of insulin secretion as closely as possible.

Insulin Preparations The forms of insulin available differ by their onset of activity, timing of peak activity, and duration of effects. Table 20-6 and Figure 20-2 show how insulin preparations are classified: they may be rapid-acting (lispro and aspart), short-acting (regular), intermediate-acting (lente and NPH), or long-acting (glargine and ultralente), thereby allowing substantial flexibility in establishing a suitable insulin regimen. Mixtures of several types of insulin can produce greater glycemic control than any one type alone. Several premixed formulations are also available (see Table 20-6).

FIGURE 20-2 Effects of Insulin Preparations

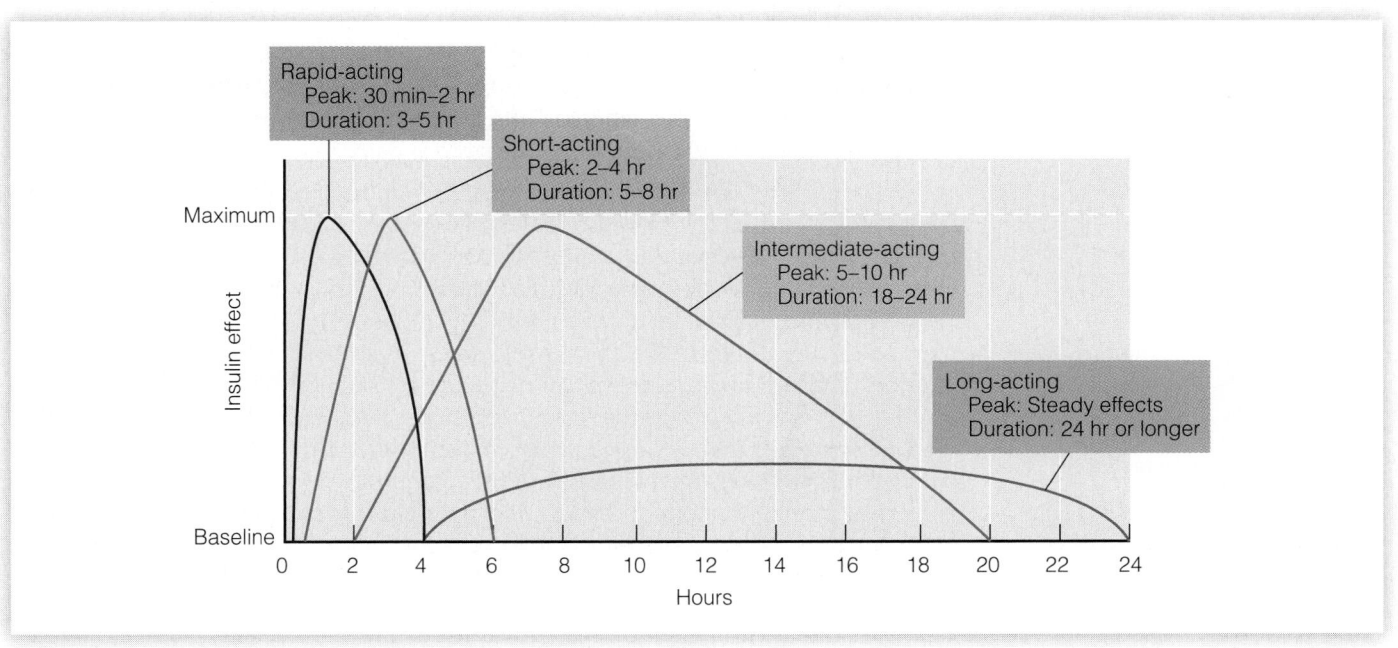

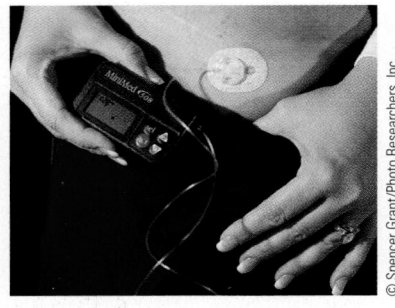

External insulin pumps deliver insulin continuously through thin, flexible tubing inserted into the skin.

Because insulin is a protein, it would be destroyed by digestive processes if taken orally.

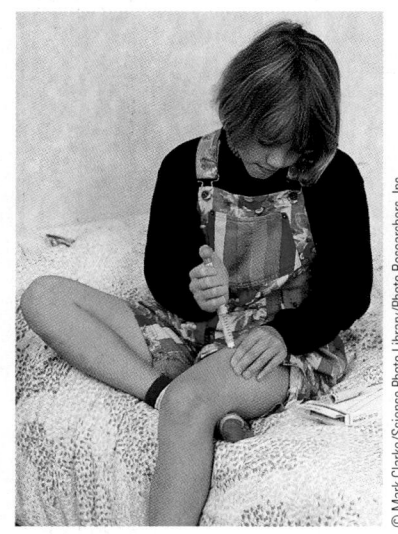

Children often become adept at administering the insulin they require.

Rapid-acting insulin begins working in only 15 minutes, so it can be injected right before a meal. Short-acting insulin requires a half-hour wait before the meal can begin.

carbohydrate-to-insulin ratio: the amount of carbohydrate that can be handled per unit of insulin. On average, every 15 g of carbohydrate requires about 1 unit of rapid- or short-acting insulin.

Insulin Delivery Insulin is most often administered by injection, either self-administered or provided by caregivers. Although individual syringes are most commonly used, several other options are available. Injection ports, inserted through the skin, can remain in place for several days, eliminating the need for multiple punctures. Another option is to use an insulin pump, a computerized device that can be programmed to deliver basal insulin continuously and bolus doses at mealtimes (see the photo). The pump infuses insulin through thin, flexible tubing that remains in the skin. The pump can be worn under clothes, attached to a belt, or kept in a pocket.

Insulin Regimen for Type 1 Diabetes Type 1 diabetes is best managed with intensive insulin therapy, which involves multiple daily injections of several types of insulin or use of an insulin pump (review Table 20-3). Usually, intermediate- or long-acting insulin meets basal insulin needs, and rapid- or short-acting insulin is injected before meals. Three or more daily injections are required for good glycemic control. Simpler regimens involve twice-daily injections of a mixture of intermediate- and short-acting insulin. Regimens that include three or more injections allow for greater flexibility in meal timing. With fewer injections, the timing of both meals and injections must be similar from day to day to avoid periods of insulin deficiency or excess.[32]

A person using intensive therapy must learn to accurately determine the amount of insulin to inject before each meal. The amount required depends on premeal blood glucose levels, the carbohydrate content of the meal, and the person's body weight and sensitivity to insulin. To determine insulin sensitivity, a person keeps careful records of food intake, insulin dosages, and blood glucose levels. Eventually, these records are analyzed to determine the appropriate **carbohydrate-to-insulin ratio** for that individual, which assists in calculating insulin dosages at mealtime.[33]

After insulin therapy is initiated, persons with type 1 diabetes may experience a temporary remission of disease symptoms and a reduced need for insulin, known as the "honeymoon phase." The remission is due to a temporary improvement in pancreatic function and may last for several weeks or months. In all cases, diabetes eventually returns, and full insulin treatment must be reinstated.

Insulin Regimen for Type 2 Diabetes Approximately 30 percent of people diagnosed with type 2 diabetes are treated with insulin therapy.[34] Although initial treatment of type 2 diabetes may involve diet therapy, physical activity, and oral antidiabetic medications, long-term results with these treatments are often disappointing. As the disease progresses, pancreatic function worsens, and many individuals require insulin therapy to maintain glycemic control.

Various regimens can be used to control type 2 diabetes. Some persons may be treated with insulin alone, whereas others may use insulin in combination with oral antidiabetic agents. Often, only one or two daily injections are needed. Some regimens involve a mixture of rapid- and intermediate-acting insulin in the morning and an injection of intermediate- or long-acting insulin at dinner or before bedtime. In other cases, only a single injection of intermediate- or long-acting insulin may be needed at bedtime. Dosages and timing are adjusted according to the results of blood glucose self-monitoring.[35]

Insulin Therapy and Hypoglycemia Hypoglycemia is the most common complication of insulin therapy, although it may also result from use of some oral antidiabetic drugs. Hypoglycemia most often results from intensive insulin therapy, as the attempt to attain near-normal blood glucose levels increases the risk of overtreatment. Hypoglycemia can be corrected with the immediate intake of glucose or a carbohydrate-containing food. Usually, 15 to 20 grams of carbohydrate can relieve hypoglycemia within 10 to 20 minutes, although blood glucose levels should be reevaluated within an hour in case additional treatment is necessary.[36] Foods that provide pure glucose yield a better response than foods that contain other sugars, such as sucrose or fructose. People using insulin are usually advised to carry glucose tablets or a source of carbohydrate that can be readily ingested.

Fasting Hyperglycemia Insulin therapy must sometimes be adjusted to prevent fasting hyperglycemia, which has three possible causes. The usual cause is a waning of insulin action during the night due to insufficient insulin. A second possibility, known as the **dawn phenomenon,** occurs when blood glucose levels increase in the morning due to the early morning secretion of growth hormone, which counteracts insulin's actions. Less frequently, fasting hyperglycemia develops as a result of nighttime hypoglycemia, which causes hormonal responses that stimulate glucose production; the resulting condition is known as **rebound hyperglycemia.** Whatever the cause, fasting hyperglycemia can be treated by adjusting the dosage or formulation of insulin administered in the evening.

ORAL ANTIDIABETIC AGENTS

Treatment of type 2 diabetes often requires the use of oral medications. These drugs can improve hyperglycemia by four modes of action: they can improve insulin secretion, reduce glucose production in the liver, improve use of glucose by the tissues, or delay carbohydrate absorption. Treatment may involve the use of a single medication (monotherapy) or a combination of several (combination therapy). By utilizing several mechanisms at once, combination therapy achieves more rapid and sustained glycemic control than is possible with monotherapy.[37] Table 20-7 lists examples of oral antidiabetic agents, and the Diet-Drug Interactions feature lists their nutrition-related effects. As medications cannot replace the benefits offered by dietary modifications and physical activity, persons with diabetes should be advised to continue both.

PHYSICAL ACTIVITY AND DIABETES MANAGEMENT

Regular physical activity is a central feature of disease management in type 2 diabetes, as it can substantially improve glycemic control. A regular exercise program improves insulin sensitivity and thereby reduces insulin requirements. It also helps

Each of the following sources provides approximately 15 g of carbohydrate:

- Glucose tablets: 2 to 3 tablets.
- Table sugar: 4 tsp.
- Honey: 1 tbs.
- Jelly beans: 15 small.
- Grape juice, unsweetened: $1/2$ c.
- Orange juice, canned: $1/2$ c.

TABLE 20-7	Oral Antidiabetic Agents	
Mode of Action	**Drug Category**	**Common Examples**
Stimulate insulin secretion by pancreas	Sulfonylureas	Chlorpropamide Tolbutamide Glyburide Glipizide
	Meglitinides	Repaglinide Nateglinide
Inhibit liver glucose production	Biguanides	Metformin
Increase insulin sensitivity	Thiazolidinediones	Pioglitazone Rosiglitazone
Delay carbohydrate absorption	Alpha-glucosidase inhibitors	Acarbose Miglitol

dawn phenomenon: morning hyperglycemia that is caused by the early morning release of growth hormone, which counteracts insulin's glucose-lowering effects.

rebound hyperglycemia: hyperglycemia that results from the release of counterregulatory hormones following nighttime hypoglycemia; also called the **Somogyi phenomenon.**

DIET-DRUG INTERACTIONS Check this table for notable nutrition-related effects of the medications discussed in this chapter.

	Gastrointestinal Effects	Interactions with Dietary Substances	Metabolic Effects
Sulfonylureas	Nausea, vomiting, cramps, diarrhea	Avoid using with alcohol due to a toxic reaction that causes flushing, throbbing head and neck pain, shortness of breath, palpitations, and sweating. Avoid using with dietary supplements that contain ginseng, garlic, fenugreek, coriander, and celery, as they may increase risk of hypoglycemia.	Hypoglycemia, weight gain, allergic skin reactions
Biguanides (metformin)	Abdominal pain, nausea, vomiting, gas, cramps, diarrhea, metallic taste, anorexia		Asymptomatic vitamin B_{12} deficiency
Thiazolidinediones			Weight gain, fluid retention, edema, anemia
Alpha-glucosidase inhibitors	Abdominal pain, nausea, gas, cramps, diarrhea		Elevated liver enzymes, hyperbilirubinemia

to improve blood lipid levels, lower blood pressure, and promote weight loss. Although exercise is less effective for glycemic control in type 1 diabetes, it helps to maintain other aspects of health, including cardiovascular conditioning.

Physical Activity in Insulin Therapy People with type 1 diabetes must carefully adjust food intake and insulin therapy to prevent hypoglycemia during physical activity. Insulin dosages that precede exercise often need to be reduced substantially. Blood glucose levels should be checked both before and after an activity. If blood glucose is below 100 mg/dL, carbohydrate should be consumed before exercise begins. Additional carbohydrate may be needed during or after prolonged activity, or even several hours after the activity is completed. Physical activity should be avoided if blood glucose levels exceed 250 mg/dL with ketosis, or 300 mg/dL without ketosis, as hyperglycemia may worsen.[38]

Physical Activity in Type 2 Diabetes People with type 2 diabetes are often overweight and sedentary, and many develop complications during the course of disease. Before an exercise program is planned, a medical evaluation should screen for problems that may be aggravated by certain activities. Complications involving the heart and blood vessels, eyes, kidneys, feet, and nervous system may limit the types of activity recommended.[39]

Only mild or moderate exercise may be prescribed at first. For obese, inactive persons, a short walk at a comfortable pace may be the first activity suggested. Persons with retinopathy should avoid heavy lifting or straining, which may raise blood pressure and damage eye tissue. Peripheral neuropathy precludes repetitive weight-bearing exercises such as jogging and step exercises, as these activities may lead to foot ulcerations. Proper hydration should be encouraged before and during exercise, as dehydration can adversely affect blood glucose levels and heart function.

CASE STUDY   *Child with Type 1 Diabetes*

Nora is a 12-year-old girl who was diagnosed with type 1 diabetes two years ago. She practices intensive therapy and has had the support of her parents and an excellent diabetes management team. With their help, Nora has been able to assume the bulk of the responsibility for her diabetes care and has managed to control her blood glucose remarkably well. In the last few months, however, Nora has been complaining bitterly about the impositions diabetes has placed on her life and her interactions with friends. Sometimes she refuses to monitor her blood glucose levels, and she has skipped insulin injections a few times. Recently, Nora was admitted to the emergency room complaining of fever, nausea, vomiting, and intense thirst. The physician noted that Nora was confused and lethargic. A urine test was positive for ketones, and her blood glucose levels were 400 mg/dL. The diagnosis was diabetic ketoacidosis.

1. Describe the metabolic events that lead to ketoacidosis. Were Nora's symptoms and laboratory tests consistent with the diagnosis?
2. Review Table 20-3 and consider the advantages and disadvantages that intensive therapy might have for Nora.
3. Discuss how Nora's age might influence her ability to cope with and manage her diabetes. Why might she feel that diabetes is disrupting her life? What suggestions may help?
4. Review the list of complications associated with long-term diabetes. How might you explain the importance of glycemic control to a 12-year-old girl?

SICK-DAY MANAGEMENT

Illness, infection, or injury can cause hormonal changes that raise blood glucose levels. In people with type 1 diabetes, illness also increases the risk of diabetic ketoacidosis. A patient with type 1 diabetes should be advised to test blood glucose frequently during illness and to test blood or urine for ketones. Some carbohydrate should be ingested regularly to avoid ketosis; a daily intake of 150 to 200 grams of carbohydrate (about 45 to 50 grams every three to four hours) is recommended. If appetite is poor, carbohydrate-sweetened beverages or frozen juice bars may be easier to consume than solid foods. Fluid intake should be monitored to prevent dehydration.[40]

REVIEW NOTES

Diabetes treatment often includes medical nutrition therapy, insulin or oral medications, and appropriate physical activity. Glycemic control is evaluated by monitoring blood glucose levels and glycated hemoglobin.

Carbohydrate intake is the main factor that influences blood glucose levels after meals. The total amount of carbohydrate ingested is more important than the type of carbohydrate consumed.

Carbohydrate counting is widely used in menu planning and can be taught at different levels of complexity depending on individual needs.

Insulin therapy is required for patients who are unable to produce sufficient insulin and may be used in both type 1 and type 2 diabetes. Oral medications can improve insulin secretion and sensitivity, lower glucose production by the liver, and delay carbohydrate absorption.

Physical activity can improve glycemic control in type 2 diabetes and also enhances various aspects of general health. The Case Study provides an opportunity to review the factors that influence treatment for a 12-year-old with type 1 diabetes.

Diabetes Management in Pregnancy

Women with diabetes face new challenges during pregnancy. Due to hormonal changes, pregnancy increases insulin resistance and the need for insulin, so maintaining glycemic control may be more difficult. In addition, up to 7 percent of

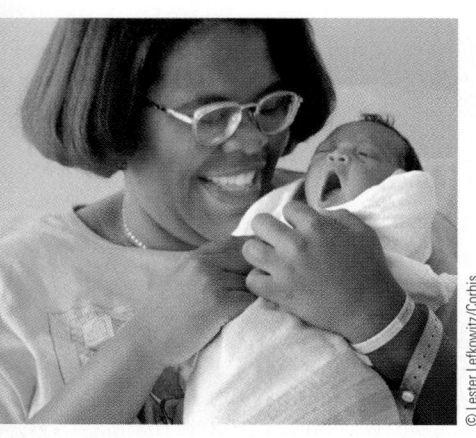

© Lester Lefkowitz/Corbis

Glycemic control during pregnancy offers the best chance of a safe delivery and a healthy infant.

NURSING DIAGNOSIS

for diabetic pregnancies may include *impaired fetal nutrition: more than body requirements, risk for delayed development: fetal, risk for disproportionate growth: fetal, risk for impaired tissue integrity: fetal,* and *risk for impaired tissue integrity: maternal.*

nondiabetic women develop gestational diabetes and require treatment during pregnancy.[41] Women with gestational diabetes are at greater risk of developing type 2 diabetes later in life, and their children are at increased risk of developing obesity and type 2 diabetes as they enter into adulthood.

A pregnancy complicated by diabetes increases health risks for both mother and fetus. Uncontrolled diabetes is linked with an increased rate of miscarriages. Incidences of birth defects and fetal deaths are higher than normal. Newborns are more likely to suffer from respiratory distress and to develop metabolic problems such as hypoglycemia, jaundice, and hypocalcemia. Women with type 2 diabetes and gestational diabetes often deliver babies with **macrosomia** (abnormally large bodies), which makes delivery more difficult and can result in birth trauma or a cesarean section.[42]

PREGNANCY IN TYPE 1 OR TYPE 2 DIABETES

Women with diabetes who achieve glycemic control at conception and during the first trimester can substantially reduce the risks of birth defects and spontaneous abortion during pregnancy. For this reason, it is recommended that women contemplating pregnancy receive preconception care to avoid the complications associated with poorly controlled diabetes.[43] Maintaining glycemic control during the second and third trimesters can minimize the risks of macrosomia and morbidity in newborn infants.

Nutrient requirements during pregnancy are generally similar for women with and without diabetes. The dietary adjustments suggested for improving glycemic control should be based on a woman's dietary habits and the results of blood glucose monitoring. Regular meals and snacks help to avoid hypoglycemia, which is more likely to occur during pregnancy because glucose is continuously supplied to the fetus. An evening snack is usually required to prevent overnight hypoglycemia and ketosis. Insulin and medication changes are often needed during pregnancy, and the woman may have to adjust her dietary habits further as a result.

GESTATIONAL DIABETES

Risk of gestational diabetes is highest in women who have a family history of diabetes; are obese; are members of certain ethnic groups, including Hispanic Americans, Native Americans, Asian Americans, African Americans, and Pacific Islanders; or have given birth to infants weighing over 9 pounds. To ensure that problems are dealt with promptly, physicians routinely test all women for gestational diabetes between 24 and 28 weeks of gestation. In high-risk women, screening should begin prior to pregnancy or soon after conception. Even mild hyperglycemia can have adverse effects on a developing fetus and may lead to complications during pregnancy.[44]

Women with gestational diabetes who are overweight or obese may need to adjust their energy intakes during pregnancy. Although adequate energy is needed for fetal development, a modest kcaloric reduction (about 30 percent less than total energy needs) may improve glycemic control without increasing the risk of ketosis. Restricting carbohydrate to 40 to 45 percent of total energy intake can improve blood glucose levels after meals. Because carbohydrate may be poorly tolerated in the morning, reducing carbohydrate at breakfast can be helpful. The remaining carbohydrate intake should be spaced throughout the day into several meals and snacks, including an evening snack to prevent ketosis during the night. Regular aerobic activity is often recommended, as it can help to improve glycemic control. Women who fail to achieve glycemic goals by diet and exercise alone may need insulin therapy, which is given to approximately 20 to 25 percent of women with gestational diabetes.[45] Oral antidiabetic drugs generally are not prescribed during pregnancy because they may be toxic to the developing fetus. The Case Study on p. 531 reviews the connections between gestational diabetes and type 2 diabetes.

macrosomia (MAK-roh-SOH-mee-ah): the condition of having an abnormally large body. In infants, refers to birthweights of 4000 g (8 lb 13 oz) and above.

CASE STUDY   School Counselor with Type 2 Diabetes

Alicia Cordova is a 41-year-old Mexican American woman recently diagnosed with type 2 diabetes. Mrs. Cordova developed gestational diabetes while she was pregnant with her second child. Her blood glucose levels returned to normal following pregnancy, and she was advised to get regular checkups, maintain a desirable weight, and engage in regular physical activity. Although she reports that she does not overeat and that she exercises regularly, she has been unable to maintain a healthy weight. At 5 feet 3 inches tall, Mrs. Cordova currently weighs 155 pounds. She has decided to lose weight and join a gym because she is concerned about the long-term effects of diabetes and the possibility that she may need insulin injections. She is also concerned about her husband and children because they are overweight and not very active. The physician refers Mrs. Cordova to a dietitian to help her plan a diet.

1. What factors in Mrs. Cordova's medical history increase her risk for diabetes? Are her husband and children also at risk?
2. Describe the general characteristics of a diet and exercise program that would be appropriate for Mrs. Cordova. How might weight loss and physical activity benefit her diabetes?
3. If Mrs. Cordova is unable to control her blood glucose with diet and physical activity, what treatment might be suggested? Can you explain to Mrs. Cordova why she would probably not require insulin at this time?
4. What dietary and lifestyle changes may help to prevent diabetes in Mrs. Cordova's husband and children?

REVIEW NOTES

Careful management of blood glucose levels before and during pregnancy may reduce complications in mother and infant. Most nutrient requirements during pregnancy are similar for women with and without diabetes.

Carbohydrate intake should be distributed into several meals and snacks, including an evening snack to prevent overnight ketosis. Carbohydrate restriction may be recommended, especially in women with gestational diabetes.

Moderate energy restriction may help to improve glycemic control in overweight and obese women.

NUTRITION ASSESSMENT CHECKLIST FOR PEOPLE WITH DIABETES

Medical History

Check the medical record to determine:

☐ Type of diabetes
☐ Duration of diabetes
☐ Acute and chronic complications
☐ Conditions, including pregnancy, that may alter treatment

Medications

For people with preexisting diabetes who use oral antidiabetic drugs, insulin, or both, note:

☐ Type of medication
☐ Administration schedule

Check for use of other medications, including:

☐ Medications that affect blood glucose levels
☐ Cholesterol- and triglyceride-lowering medications
☐ Antihypertensive medications

Dietary Intake

To devise an acceptable meal plan and coordinate medications, obtain:

☐ An accurate and thorough record of food intake and meal patterns
☐ An account of usual physical activities

At medical checkups, reassess the person's ability to:

☐ Maintain appropriate carbohydrate intake
☐ Maintain appropriate energy intake
☐ Monitor blood glucose levels at home
☐ Adjust insulin and diet to accommodate sick days
☐ Use appropriate foods to treat hypoglycemia

Anthropometric Data

Take accurate baseline height and weight measurements as a basis for:

☐ Appropriate energy intake
☐ Initial insulin therapy

Periodically reassess height and weight for children and weight for adults and pregnant women to ensure that the meal plan provides an appropriate energy intake.

Laboratory Tests

Monitor the success of diabetes treatment using these tests:

☐ Glycated hemoglobin

☐ Blood lipid concentrations

☐ Blood or urinary ketones

☐ Urinary protein (microalbuminuria)

Physical Signs

Look for signs of:

☐ Nerve damage

☐ Vision problems

☐ Foot ulcers

☐ Dehydration, especially in older adults

SELF CHECK

1. Which of the following is characteristic of type 1 diabetes?
   a. It frequently goes undiagnosed.
   b. The pancreas makes little or no insulin.
   c. It is the predominant form of diabetes.
   d. It often arises during pregnancy.

2. Which of the following describes type 2 diabetes?
   a. It is usually an autoimmune disease.
   b. The pancreas makes little or no insulin.
   c. Diabetic ketoacidosis is a common complication.
   d. Chronic complications may develop before it is diagnosed.

3. The chronic complications associated with all types of diabetes result from:
   a. altered kidney function.
   b. infections that deplete nutrient reserves.
   c. weight gain and hypertension.
   d. damage to blood vessels and nerves.

4. Long-term glycemic control is usually evaluated by:
   a. self-monitoring of blood glucose.
   b. testing urinary ketone levels.
   c. measuring glycated hemoglobin.
   d. testing urinary protein levels (microalbuminuria).

5. Regarding dietary carbohydrate, a patient with diabetes should be most concerned about:
   a. consuming the correct quantity of carbohydrate at each meal or snack.
   b. consuming the correct proportion of sugars, starches, and fiber in meals.
   c. avoiding added sugars and kcaloric sweeteners.
   d. choosing meals with ideal proportions of protein, carbohydrate, and fat.

6. Which of the following is true regarding the general use of alcohol in diabetes?
   a. A serving of alcohol is considered part of the carbohydrate allowance.
   b. Alcohol contributes to hyperglycemia and should be avoided completely.
   c. Alcohol can cause hypoglycemia and should therefore be consumed with food if patients use insulin or medications that stimulate insulin secretion.
   d. Patients can use alcohol in unlimited quantities unless they are pregnant.

7. The most ideal meal-planning strategy for people with diabetes is:
   a. carbohydrate counting.
   b. the exchange list system.
   c. following menus and recipes provided by a registered dietitian.
   d. any approach that best helps the patient control blood glucose levels.

8. A patient using intensive insulin therapy is likely to use a regimen that involves:
   a. twice-daily injections that combine short-, intermediate-, and long-acting insulin in each injection.
   b. a mixture of intermediate- and long-acting insulin injected between meals.
   c. multiple daily injections that supply basal insulin and precise insulin dosages for each meal.
   d. use of both insulin and oral antidiabetic agents.

9. Sudden hyperglycemia in a person who has previously maintained good glycemic control can be precipitated by:
   a. infections or illnesses.
   b. chronic alcohol ingestion.
   c. undertreatment of hypoglycemia.
   d. prolonged exercise.

10. Women with pregnancies complicated by diabetes:
    a. generally benefit from larger meals and a snack at bedtime.
    b. often need less carbohydrate at breakfast.
    c. need more carbohydrate than women with diabetes who are not pregnant.
    d. need more kcalories to support the pregnancy than women without diabetes.

Answers to these questions appear in Appendix H.

CLINICAL APPLICATIONS

1. Using the carbohydrate counting method described in the "How to" on pp. 523–524, determine an appropriate carbohydrate intake (in both grams and portions) for a man with type 2 diabetes who requires approximately 2600 kcalories daily. Assume he would benefit from a carbohydrate allowance that is 50 percent of his energy intake. Using information from Tables 20-4 and 20-5, develop a one-day sample menu that is likely to meet his carbohydrate goals. Use the exchange lists in Appendix C to find additional examples of foods to include in your menu.

2. Take a trip to a pharmacy or use information from an online drugstore to price these items: blood glucose meter, test strips for the glucose meter selected, lancets, insulin, and syringes. Determine the approximate cost of insulin and syringes for a person who uses 14 units of short-acting insulin (regular) and 26 units of intermediate-acting insulin (lente or NPH) in three injections daily. Also estimate the cost of testing blood glucose four times daily. Approximately how much would these supplies cost per month?

NUTRITION ON THE NET

For further study of the topics in this chapter, access these websites.

Find updates and quick links to these and other nutrition-related sites at our website: **www.wadsworth.com/nutrition**

Visit the American Diabetes Association and the Joslin Diabetes Center to find information on a wide range of topics related to diabetes: **www.diabetes.org** and **www.joslin.org**

Find comprehensive and reliable information about diabetes, suitable for both health practitioners and consumers, at the websites of the National Institute of Diabetes and Digestive and Kidney Diseases and the Centers for Disease Control: **www.niddk.nih.gov** and **www.cdc.gov/diabetes**

Find out how to become a diabetes educator by visiting the American Association of Diabetes Educators site: **www.aadenet.org**

NOTES

1. M. Lethbridge-Çejku and J. Vickerie, Summary health statistics for U.S. adults: National Health Interview Survey, 2003, National Center for Health Statistics, *Vital Health Statistics* 10 (July 2005).
2. Centers for Disease Control and Prevention, Prevalence of diabetes and impaired fasting glucose in adults—United States, 1999–2000, *MMWR Morbidity and Mortality Weekly Reports* 52 (2003): 833–837.
3. American Diabetes Association, Diagnosis and classification of diabetes mellitus, *Diabetes Care* 27 (2004): S5–S10.
4. K. N. Frayn, *Metabolic Regulation: A Human Perspective* (Oxford, U.K.: Blackwell Science, 2003).
5. R. S. Sherwin, Diabetes mellitus, in L. Goldman and D. Ausiello, eds., *Cecil Textbook of Medicine* (Philadelphia: Saunders, 2004), pp. 1424–1452.
6. A. Peters Harmel and R. Mathur, eds., *Davidson's Diabetes Mellitus: Diagnosis and Treatment* (Philadelphia: Saunders, 2004).
7. Peters Harmel and Mathur, 2004; American Diabetes Association, 2004.
8. Centers for Disease Control and Prevention, 2003.
9. F. R. Kaufman, Diabetes management in children and adolescents, in A. Peters Harmel and R. Mathur, eds., *Davidson's Diabetes Mellitus: Diagnosis and Treatment* (Philadelphia: Saunders, 2004), pp. 299–321.
10. Kaufman, 2004; P. A. Tartaranni and C. Bogardus, Obesity and diabetes mellitus, in D. Porte, Jr., R. S. Sherwin, and A. Baron, eds., *Ellenberg and Rifkin's Diabetes Mellitus* (New York: McGraw-Hill, 2003), pp. 401–413.
11. Sherwin, 2004.
12. Peters Harmel and Mathur, 2004.
13. Sherwin, 2004.
14. R. Matz, Hyperglycemic hyperosmolar syndrome, in D. Porte, Jr., R. S. Sherwin, and A. Baron, eds., *Ellenberg and Rifkin's Diabetes Mellitus* (New York: McGraw-Hill, 2003), pp. 587–599.
15. Sherwin, 2004.
16. D. M. Kendall, Reducing cardiovascular risk in type 2 diabetes and the metabolic syndrome: The emerging role of insulin resistance, in A. Peters Harmel and R. Mathur, eds., *Davidson's Diabetes Mellitus: Diagnosis and Treatment* (Philadelphia: Saunders, 2004), pp. 239–257.
17. L. H. Young and D. A. Chyun, Heart disease in patients with diabetes, in D. Porte, Jr., R. S. Sherwin, and A. Baron, eds., *Ellenberg and Rifkin's Diabetes Mellitus* (New York: McGraw-Hill, 2003), pp. 823–844.
18. R. G. Frykberg, Diabetic foot ulcers: Pathogenesis and management, *American Family Physician* 66 (2002): 1655–1662.
19. Peters Harmel and Mathur, 2004.
20. Peters Harmel and Mathur, 2004.
21. American Diabetes Association, Implications of the United Kingdom Prospective Diabetes Study, *Diabetes Care* 21 (1998): 2180–2184;

Diabetes Control and Complications Trial Research Group, The effect of intensive treatment of diabetes on the development and progression of long-term complications in insulin-dependent diabetes mellitus, *New England Journal of Medicine* 329 (1993): 977–986.

[22]American Diabetes Association, Tests of glycemia in diabetes, *Diabetes Care* 27 (2004): S91–S93.

[23]American Diabetes Association, Tests of glycemia in diabetes, 2004.

[24]American Diabetes Association, Standards of medical care in diabetes, *Diabetes Care* 27 (2004): S15–S35.

[25]K. V. Williams and coauthors, Improved glycemic control reduces the impact of weight gain on cardiovascular risk factors in type 1 diabetes: The Epidemiology of Diabetes Complications Study, *Diabetes Care* 22 (1999): 1084–1091.

[26]C. A. Beebe, Nutrition and physical activity in diabetes, in A. Peters Harmel and R. Mathur, eds., *Davidson's Diabetes Mellitus: Diagnosis and Treatment* (Philadelphia: Saunders, 2004), pp. 49–69; American Diabetes Association, Nutrition principles and recommendations in diabetes, *Diabetes Care* 27 (2004): S36–S46.

[27]American Diabetes Association, Nutrition principles and recommendations in diabetes, 2004.

[28]Beebe, 2004.

[29]American Diabetes Association, Nutrition principles and recommendations in diabetes, 2004.

[30]American Diabetes Association, Nutrition principles and recommendations in diabetes, 2004; Beebe, 2004; G. Ben and coauthors, Effects of chronic alcohol intake on carbohydrate and lipid metabolism in subjects with type II (non-insulin-dependent) diabetes, *American Journal of Medicine* 90 (1991): 70–76.

[31]F. Guerrero-Romero and M. Rodriguez-Moran, Complementary therapies for diabetes: The case for chromium, magnesium, and antioxidants, *Archives of Medical Research* 36 (2005): 250–257; G. Y.

Yeh and coauthors, Systematic review of herbs and dietary supplements for glycemic control in diabetes, *Diabetes Care* 26 (2003): 1277–1294.

[32]Peters Harmel and Mathur, 2004; S. M. Strowig and P. Raskin, Intensive management of type 1 diabetes mellitus, in D. Porte, Jr., R. S. Sherwin, and A. Baron, eds., *Ellenberg and Rifkin's Diabetes Mellitus* (New York: McGraw-Hill, 2003), pp. 501–514.

[33]Strowig and Raskin, 2003.

[34]D. M. Nathan, Insulin treatment of type 2 diabetes mellitus, in D. Porte, Jr., R. S. Sherwin, and A. Baron, eds., *Ellenberg and Rifkin's Diabetes Mellitus* (New York: McGraw-Hill, 2003), pp. 515–522.

[35]Nathan, 2003.

[36]M. J. Franz and coauthors, Evidence-based nutrition principles and recommendations for the treatment and prevention of diabetes and related complications, *Diabetes Care* 25 (2002): 148–198.

[37]S. Mudaliar and R. R. Henry, The oral antidiabetic agents, in D. Porte, Jr., R. S. Sherwin, and A. Baron, eds., *Ellenberg and Rifkin's Diabetes Mellitus* (New York: McGraw-Hill, 2003), pp. 531–564.

[38]American Diabetes Association, Physical activity/exercise and diabetes, *Diabetes Care* 27 (2004): S58–S62.

[39]American Diabetes Association, Physical activity/exercise and diabetes, 2004.

[40]Franz and coauthors, 2002.

[41]American Diabetes Association, Gestational diabetes mellitus, *Diabetes Care* 27 (2004): S88–S90.

[42]Peters Harmel and Mathur, 2004.

[43]American Diabetes Association, Preconception care of women with diabetes, *Diabetes Care* 27 (2004): S76–S78.

[44]Peters Harmel and Mathur, 2004; Sherwin, 2004.

[45]American Diabetes Association, Nutrition principles and recommendations in diabetes, 2004.

Metabolic Syndrome

As explained in Chapter 20, insulin resistance—a reduced sensitivity to insulin in muscle, adipose, and liver cells—can lead to hyperglycemia and hyperinsulinemia and, eventually, to type 2 diabetes. Insulin resistance is also a central feature of the **metabolic syndrome,** a group of disorders that substantially increases the risk of developing cardiovascular disease (CVD). The metabolic syndrome is a cluster of at least three of the following: insulin resistance, obesity, **hypertriglyceridemia** (elevated blood triglycerides), reduced HDL cholesterol levels, and hypertension' (high blood pressure). This Nutrition in Practice describes how the metabolic syndrome is diagnosed, how and why it might develop, its consequences, and current treatment approaches. The accompanying glossary defines the relevant terms.

How is the metabolic syndrome diagnosed, and how many people in the United States does it affect?

Table NP20-1 lists the laboratory values used to identify metabolic syndrome, which is currently estimated to affect 24 percent of the adult population in the United States.[1] As Figure NP20-1 shows, the prevalence of metabolic syndrome increases with age. Risk also varies among ethnic groups: Hispanic Americans have the highest incidence in the United States, with an overall prevalence of 36 percent.[2]

What causes the metabolic syndrome?

Although the precise cause of the metabolic syndrome is not known, the close relationship between abdominal obesity and insulin resistance suggests that the current obesity crisis in the United States may be partly responsible for its high prevalence. Excessive abdominal fat can induce metabolic changes that lead to insulin resistance, which then leads to other abnormalities.

What kinds of metabolic changes are induced by excess abdominal fat?

Adipose (fat) cells that reside in the abdominal area are more metabolically active than adipose cells elsewhere.[3] Hence,

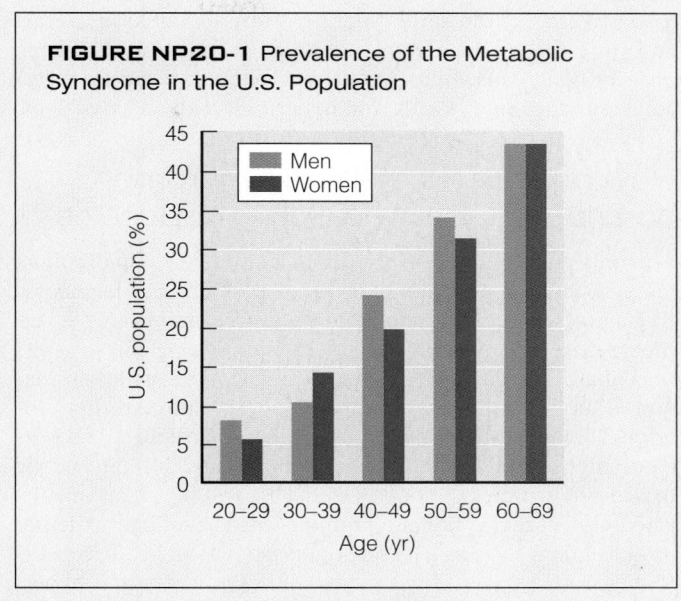

FIGURE NP20-1 Prevalence of the Metabolic Syndrome in the U.S. Population

Source: Data from E. S. Ford, W. H. Giles, and W. H. Dietz, Prevalence of the metabolic syndrome among U.S. adults: Findings from the Third National Health and Nutrition Examination Survey, *Journal of the American Medical Association* 287 (2002): 356–359.

triglycerides break down more rapidly, increasing fatty acid levels in the blood. These higher fatty acid concentrations inhibit the actions of insulin receptors, the proteins that recognize and bind insulin at cell surfaces.[4] Unless the pancreas can secrete enough insulin to compensate, glucose uptake from the blood is reduced, contributing to hyperglycemia.

Obesity can also alter production of the hormones and proteins made in adipose cells. For example, it causes reduced secretion of **adiponectin,** a hormone that improves insulin sensitivity.[5] Conversely, **resistin,** a hormone that contributes to insulin resistance, is released in greater amounts. Enlarged adipose cells also boost their production of certain cytokines that induce the synthesis of liver proteins that promote inflammation and blood coagulation.[6] People who are obese often have elevated levels of C-reactive protein,[7] a marker of inflammation linked to an increased risk of CVD.

Can obesity cause other problems related to the metabolic syndrome?

Obesity increases the risk of developing high blood pressure, a common component of the metabolic syndrome. Both the insulin resistance and hyperinsulinemia mentioned earlier may be implicated in raising blood pressure.[8] Insulin resistance interferes with the normal relaxation and dilation of blood vessels. Hyperinsulinemia promotes reabsorption of sodium by the kidneys, resulting in fluid retention and increased blood volume. These effects can contribute to increased blood pressure.

Abdominal obesity is often associated with blood lipid abnormalities.[9] Increases in body weight are linked with

TABLE NP20-1 Features of the Metabolic Syndrome

Metabolic syndrome is diagnosed when a person has three or more of the following symptoms.	
Symptom	**Diagnostic Criteria**
Hyperglycemia	Fasting plasma glucose ≥110 mg/dL
Abdominal obesity	Waist circumference >40" in men, >35" in women
Hypertriglyceridemia	≥150 mg/dL
Reduced HDL cholesterol	<40 mg/dL in men, <50 mg/dL in women
Hypertension	≥130/85 mm Hg

higher triglyceride and LDL cholesterol levels and lower HDL cholesterol levels. As a result of obesity, adipose cells are less responsive to insulin and release more fatty acids into the bloodstream. At the same time, they are less able to extract and store triglycerides from chylomicrons and VLDL. To keep up with the greater influx of fatty acids, the liver must accelerate its production of VLDL, and hypertriglyceridemia develops.

How does the metabolic syndrome contribute to cardiovascular disease risk?

A number of factors are responsible, as obesity, lipid abnormalities, and hypertension are all independent risk factors for CVD. Both insulin resistance and elevated lipoprotein levels can cause damage to blood vessels, accelerating the progression of atherosclerosis.[10] The resulting blood vessel inflammation induces liver secretion of **fibrinogen,** a protein that promotes blood clot formation. C-reactive protein, which is elevated by both inflammation and obesity, inhibits **nitric oxide** production by blood vessel cells, an effect that impairs blood vessel activity and also promotes blood clotting.[11] Another pro-coagulant factor—**plasminogen activator inhibitor-1**—is overproduced as a consequence of both obesity and hyperinsulinemia. The combined effect of these multiple abnormalities can worsen atherosclerosis and increase the risks of developing heart attack and stroke. Individuals with insulin resistance are also at increased risk of developing diabetes, another major risk factor for CVD.

What is the treatment for the metabolic syndrome?

The metabolic syndrome is primarily treated with dietary and lifestyle changes, with the goal of correcting abnormalities that increase CVD risk.[12] In most individuals, a combination of weight loss and physical activity can improve insulin resistance, blood pressure, and blood lipid levels. Even a small weight loss (10 to 20 pounds) can improve symptoms, although many people find this difficult to achieve. Additional dietary strategies depend on a patient's specific symptoms. If dietary and lifestyle changes are not successful, medications may be prescribed. Because effective treatment requires lifelong commitment, health care providers should work with patients to develop a treatment plan that they are willing to adopt.

What dietary strategies, other than weight loss, are suggested for people with the metabolic syndrome?

In individuals with hypertriglyceridemia, the general recommendation is to reduce intake of added sugars and refined grain products (soda, juices, white bread, sweetened cereal, desserts) and increase servings of whole grains and foods high in fiber (whole-wheat bread, oatmeal, legumes, fruits, vegetables).[13] In some people, carbohydrate restriction may help to reduce blood triglyceride levels and improve hyperglycemia.[14] Including fish in the diet each week may also improve triglyceride levels. Individuals with hypertension are encouraged to reduce sodium intake and increase consumption of fruits and vegetables and low-fat milk products. A diet low in saturated fat, *trans* fats, and cholesterol can help to reduce LDL cholesterol levels. Chapter 21 includes additional information about dietary modifications that can reduce CVD risk.

Why is physical activity recommended for people with the metabolic syndrome?

Regular physical activity helps with weight management and may also improve blood lipid concentrations, hypertension, and insulin resistance—all changes that can reduce the risk of developing CVD. A regular exercise program can also prevent or delay the onset of diabetes in persons at risk.[15] A program that includes both aerobic exercise and strength training is best. A minimum of 30 minutes of moderate aerobic activity (brisk walking, jogging, cycling) daily is suggested, although longer periods (one hour daily) are recommended for weight control.[16] A sedentary lifestyle can worsen the progression of metabolic syndrome and should be discouraged.

© Rolf Bruderer/Corbis

Regular exercise can reduce the risks of developing the metabolic syndrome, cardiovascular diseases, and type 2 diabetes.

Are medications used to treat the metabolic syndrome?

If dietary and lifestyle changes are unsuccessful, medications may be prescribed to correct hypertriglyceridemia and hypertension (Chapter 21 provides details). Insulin resistance is not routinely treated with drug therapy in nondiabetic patients due to insufficient evidence that the medications can benefit individuals with the metabolic syndrome.[17]

As explained in this Nutrition in Practice, the metabolic syndrome consists of a cluster of related disorders that increase the risk for developing CVD. Whereas the common features of the metabolic syndrome are independent risk factors for CVD, in combination they may raise risk two- to threefold. Treatment of the metabolic syndrome emphasizes dietary and lifestyle changes. The following chapter provides additional information about lifestyle changes that can reduce CVD risk.

GLOSSARY

adiponectin (AH-dih-poe-NECK-tin): a hormone produced by adipose cells that improves insulin sensitivity.

fibrinogen (fye-BRIN-oh-jen): a liver protein that promotes blood clot formation.

hypertriglyceridemia (HYE-per-try-GLISS-er-eye-DEEM-ee-ah): elevated blood triglyceride levels.

metabolic syndrome: a cluster of interrelated clinical symptoms, including obesity, insulin resistance, high blood pressure, and abnormal blood lipids, which together increase cardiovascular disease risk two- to threefold; also called **syndrome X** or **insulin resistance syndrome.**

nitric oxide: a compound produced by blood vessel cells that helps to regulate blood vessel activity, including dilation and constriction.

plasminogen activator inhibitor-1: a protein that promotes blood clotting by inhibiting blood clot degradation within blood vessels.

resistin (re-ZIST-in): a hormone produced by adipose cells that induces insulin resistance.

Notes

[1] D. E. Moller and K. D. Kaufman, Metabolic syndrome: A clinical and molecular perspective, *Annual Review of Medicine* 56 (2005): 45–62.

[2] Z. T. Bloomgarden, American Association of Clinical Endocrinologists (AACE) consensus conference on the insulin resistance syndrome, *Diabetes Care* 26 (2003): 1297–1303.

[3] Moller and Kaufman, 2005.

[4] G. A. Bray and C. M. Champagne, Obesity and the metabolic syndrome: Implications for dietetics practitioners, *Journal of the American Dietetic Association* 104 (2004): 86–89.

[5] Moller and Kaufman, 2005; P. A. Kern and coauthors, Adiponectin expression from human adipose tissue, *Diabetes* 52 (2003): 1779–1785.

[6] Bray and Champagne, 2004; S. M. Grundy, Inflammation, hypertension, and the metabolic syndrome, *Journal of the American Medical Association* 290 (2003): 3000–3002.

[7] Grundy, 2003.

[8] Grundy, 2003.

[9] A. Tiengo and A. Avogaro, Cardiovascular disease, in P. Bjorntorp, ed., *International Textbook of Obesity* (West Sussex, U.K.: John Wiley & Sons, 2001), pp. 365–377.

[10] Moller and Kaufman, 2005.

[11] S. M. Grundy and coauthors, Clinical management of metabolic syndrome: Report of the American Heart Association/National Heart, Lung, and Blood Institute/American Diabetes Association Conference on Scientific Issues Related to Management, *Circulation* 109 (2004): 551–556.

[12] Grundy and coauthors, 2004; D. Deen, Metabolic syndrome: Time for action, *American Family Physician* 69 (2004): 2875–2882.

[13] Grundy and coauthors, 2004.

[14] D. M. Kendall, Reducing cardiovascular risk in type 2 diabetes and the metabolic syndrome: The emerging role of insulin resistance, in A. Peters Harmel and R. Mathur, eds., *Davidson's Diabetes Mellitus: Diagnosis and Treatment* (Philadelphia: Saunders, 2004), pp. 239–257.

[15] D. H. Wasserman, Z.-Q. Shi, and M. Vranic, Metabolic implications of exercise and physical fitness in physiology and diabetes, in D. Porte, Jr., R. S. Sherwin, and A. Baron, eds., *Ellenberg and Rifkin's Diabetes Mellitus* (New York: McGraw-Hill, 2003), pp. 453–480.

[16] Grundy and coauthors, 2004; Deen, 2004.

[17] Grundy and coauthors, 2004.

CONTENTS

Nutrition and Disorders of the Heart and Blood Vessels

CHAPTER 21

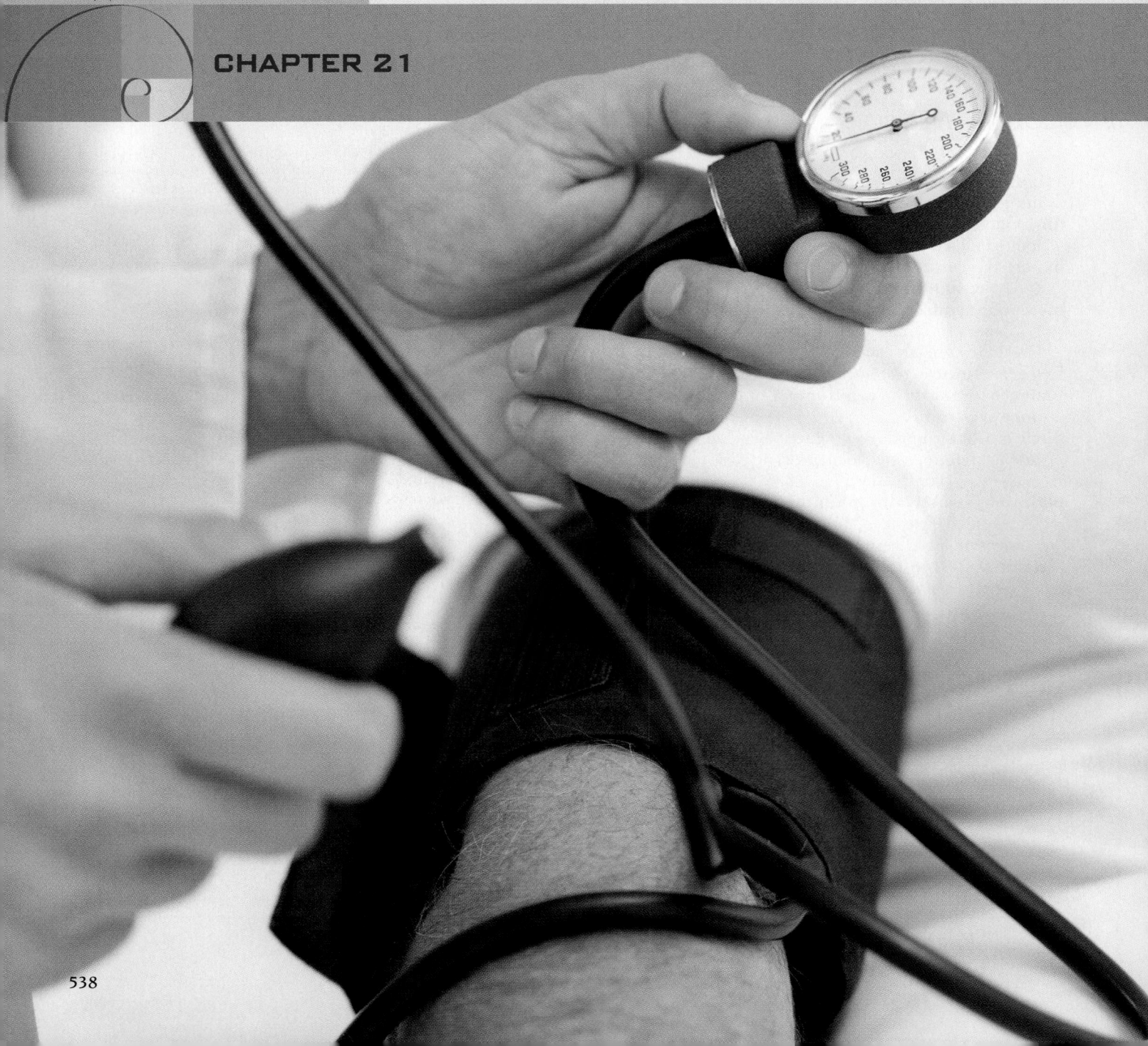

Cardiovascular disease (CVD) is a general term describing diseases of the heart and blood vessels. **Coronary heart disease (CHD),** the most common form of CVD, is caused by **atherosclerosis** in the coronary arteries that supply blood to the heart muscle. If atherosclerosis restricts blood flow in these arteries, the resulting deprivation of oxygen and nutrients can destroy heart tissue and cause a **myocardial infarction (MI)**—a **heart attack.** When the blood supply to brain tissue is blocked, a **stroke** occurs. Both heart attack and stroke may result in disablement or death. This chapter describes these and other cardiovascular disorders. Figure 21-1 shows the percentages of deaths resulting from all types of CVD. The accompanying glossary defines some common terms related to CVD.

Cardiovascular disease is responsible for approximately 38 percent of deaths in the United States, claiming more lives than the next five leading causes of death combined.[1] Although many people assume that heart disorders are men's diseases, more women than men die each year from the various types of CVD.[2] Furthermore, CVD is a global health issue; it is the leading cause of death in Europe and contributes to one-third of deaths worldwide.[3]

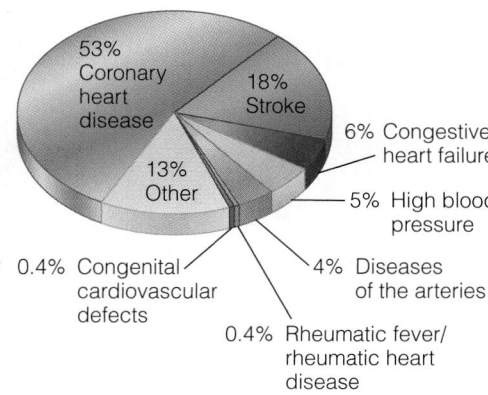

Source: Based on preliminary data for 2002 from the Centers for Disease Control/National Center for Health Statistics and the National Heart, Lung, and Blood Institute.

FIGURE 21-1 Percentage Breakdown of Deaths from Cardiovascular Diseases in the United States, 2002

Atherosclerosis

In atherosclerosis, the artery walls become progressively thickened due to an accumulation of fatty deposits, smooth muscle cells, and fibrous connective tissue, collectively known as **plaque.** Atherosclerosis initially arises in response to minimal but chronic injuries that damage the inner arterial wall. The first lesions tend to develop at regions where arteries branch or bend because the blood flow is disturbed

Atherosclerosis is the most common form of *arteriosclerosis,* a more general term for arterial diseases that are characterized by abnormally thickened walls and lost elasticity.

GLOSSARY of Terms Related to Cardiovascular Diseases

aneurysm: (AN-you-rih-zum): an abnormal enlargement or bulging of a blood vessel (usually an artery) caused by damage to or weakness in the blood vessel wall.

angina (an-JYE-nah or AN-ji-nah) **pectoris:** a condition caused by ischemia in the heart muscle that results in discomfort or dull pain in the chest region. The pain often radiates to the left shoulder and arm, or to the back, neck, and lower jaw.

atherosclerosis (ATH-er-oh-scler-OH-sis): a type of artery disease characterized by accumulations of fatty material on the inner walls of arteries.

cardiovascular disease (CVD): a general term describing diseases of the heart and blood vessels.
   cardio = heart
   vascular = blood vessels

coronary heart disease (CHD): a condition characterized by reduced blood flow in the coronary arteries that can eventually cause damage to heart tissue; also called *coronary artery disease.*

embolism (EM-boh-lizm): the obstruction of a blood vessel by an embolus, causing sudden tissue death.
   embol = to insert, plug

embolus (EM-boh-lus): an abnormal particle, such as a blood clot or air bubble, that travels in the blood.

ischemia (is-KEY-mee-a): inadequate blood supply within tissues due to obstructed blood flow in the arteries.

myocardial (MY-oh-CAR-dee-al) **infarction** (in-FARK-shun) or **MI:** death of heart muscle caused by a sudden reduction in coronary blood flow; also called a **heart attack** or *cardiac arrest.*
   myo = muscle
   cardial = heart
   infarct = tissue death

plaque (PLACK): an abnormal accumulation of fatty deposits, smooth muscle cells, and fibrous connective tissue in blood vessels.

stroke: a sudden injury to brain tissue resulting from disturbed blood flow through an artery that supplies blood to the brain; also called a *cerebrovascular accident.*
   cerebro = brain

thrombosis (throm-BOH-sis): the formation or presence of a blood clot in blood vessels. A *coronary thrombosis* occurs in a coronary artery, and a *cerebral thrombosis* occurs in an artery that supplies blood to the brain.
   thrombo = clot

thrombus: a blood clot formed within a blood vessel that remains attached to its place of origin.

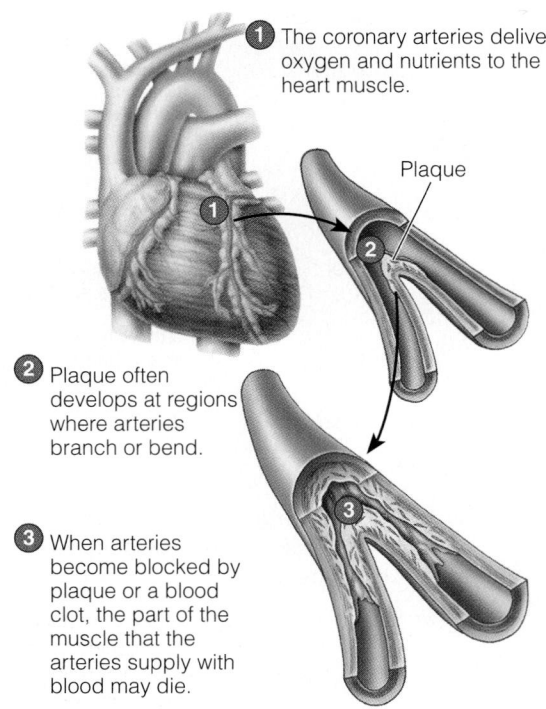

① The coronary arteries deliver oxygen and nutrients to the heart muscle.

Plaque

② Plaque often develops at regions where arteries branch or bend.

③ When arteries become blocked by plaque or a blood clot, the part of the muscle that the arteries supply with blood may die.

FIGURE 21-2 Plaque Formation in Atherosclerosis

in those areas (see Figure 21-2). Eventually, the plaque thickens and hardens as additional lipids and connective tissue accumulate.

CONSEQUENCES OF ATHEROSCLEROSIS

As atherosclerosis worsens, it can eventually narrow the lumen of an artery and interfere with blood flow. Some types of plaque are highly susceptible to rupture, which promotes the formation of a blood clot within the artery **(thrombosis).** A blood clot **(thrombus)** can enlarge in time and ultimately obstruct blood flow. A portion of a clot can also break free **(embolus)** and travel through the circulatory system until it lodges in a narrowed artery and shuts off blood flow to the surrounding tissue **(embolism).** Most complications of atherosclerosis result from the deficiency of blood and oxygen within the tissue served by an artery **(ischemia).**

Atherosclerosis can affect almost any organ or tissue in the body and, accordingly, is a major cause of disablement or death. Obstructed blood flow in the coronary arteries can cause pain or discomfort in the chest and surrounding regions **(angina pectoris)** or lead to a heart attack. As mentioned earlier, a stroke may result from impaired blood flow to the brain. Inadequate blood circulation in the legs can cause fatigue and pain while walking. Blockage of the arteries that supply the kidneys can result in kidney disease or even acute kidney failure.

Atherosclerosis is the most common cause of an **aneurysm**—a saclike distention of the blood vessel wall. Plaque can weaken the blood vessel, allowing it to expand and balloon out. Aneurysms can rupture and lead to massive bleeding and death, particularly when a large vessel such as the aorta is affected. In the arteries of the brain, an aneurysm may lead to bleeding within the brain, coma, or stroke.

CAUSES OF ATHEROSCLEROSIS

Atherosclerosis begins to develop as early as childhood or adolescence and typically progresses over several decades before symptoms develop.[4] The factors that initiate atherosclerosis either cause direct damage to the artery wall or allow lipid materials to penetrate its surface. Atherosclerosis is generally worsened by factors that induce plaque rupture or blood coagulation.[5]

C-reactive protein, an acute-phase protein secreted during the inflammatory response, is associated with increased heart disease risk (see Chapter 16).

Inflammation and Infection Plaque formation is an inflammatory response to an injury on the artery wall, and therefore the body's immune system is directly involved in its development.[6] There is also evidence that a persistent infection within the body may contribute to plaque formation. A number of different bacterial and viral antigens have been identified in plaque.[7]

Hypertension The stress of blood flow along the artery wall can cause physical damage to arteries. Hypertension (high blood pressure) intensifies the stress of blood flow on arterial tissue. Hypertension is discussed in a later section.

Smoking Smoking has multiple effects that contribute to plaque development and subsequent complications. Some components in smoke impair the normal functions of arterial cells and induce vasoconstriction; other substances in smoke increase oxidative stress and promote blood clotting. Passive smoking has similar effects.[8]

Elevated LDL and VLDL High blood levels of low-density lipoproteins (LDL) and very-low-density lipoproteins (VLDL) promote atherosclerosis, particularly if the lipoproteins have been oxidized. Oxidized LDL and VLDL are actively taken up and retained in the artery wall.[9] Oxidation may result from free-radical generation by white blood cells and blood vessel cells or from various enzyme reactions. High-density lipoproteins (HDL) help to prevent the oxidation of LDL, so low levels of HDL may contribute to the development of atherosclerosis. It is not yet known whether the oxidation of dietary lipids significantly contributes to the levels of oxidized LDL and VLDL in blood.[10]

Elevated concentrations of a variant form of LDL called *lipoprotein(a)* have been found to speed the progression of advanced atherosclerosis and to double the risk of CHD.[11] Abnormally high levels are genetically determined and have been associated with premature development of heart disease.

Diabetes Mellitus Diabetes increases risks of both early and advanced atherosclerosis. Chronic hyperglycemia leads to the production of chemical substances that damage blood vessels and thereby worsen atherosclerosis. Other effects of diabetes contribute to plaque's progression by promoting the formation of blood clots.[12]

Aging As a person ages, arterial cells tend to degenerate, and CVD risk factors accumulate. Risk of atherosclerosis increases substantially in men who are age 45 or older and in women who are age 55 or older. Premenopausal women are at lower risk of atherosclerosis due to estrogen's protective effect on arterial function. Levels of the amino acid homocysteine (see p. 203), which may damage artery walls and increase oxidative stress, rise with age and are generally higher in men; however, researchers have not determined whether the damage is caused by homocysteine itself or by a factor associated with it.[13]

Disorders in which blood lipids are abnormally high are called *hyperlipidemias* or *dyslipidemias*.

Oral estrogen therapy, given after menopause, has both positive and negative effects on atherosclerosis risk: it can improve blood lipid levels, but can also increase blood-clotting activity.

REVIEW NOTES

Atherosclerosis is characterized by a buildup of plaque on an artery's wall. Rupture of plaque can lead to thrombosis and obstruction of blood flow.

Atherosclerosis can lead to complications such as angina pectoris, heart attack, stroke, pain while walking, kidney disease, and aneurysms.

Plaque is initiated by factors that damage the artery wall and advanced by rupture of the plaque and blood coagulation. Factors that cause plaque formation and progression include inflammation, hypertension, smoking, hyperlipidemias, and diabetes.

Coronary Heart Disease (CHD)

Coronary heart disease (CHD) is the most common type of cardiovascular disease and the leading cause of death in the United States.[14] As discussed earlier, CHD is usually caused by atherosclerosis in the large and medium-sized arteries that supply the heart muscle with oxygen and nutrients. Because atherosclerosis develops over many years, prevention should begin well before signs and symptoms appear.

EVALUATING RISK FOR CORONARY HEART DISEASE

For most people, prevention of CHD begins by reducing risk. Population studies have suggested that about 90 percent of people with CHD have at least one of the four classic risk factors: smoking, high LDL cholesterol, high blood pressure, and diabetes.[15] These and other major risk factors that can be modified by changes in

NURSING DIAGNOSIS

health-seeking behaviors applies to people who wish to alter their health habits to reduce risk of CHD.

diet and lifestyle are listed in Table 21-1; only age, gender, and family history cannot be modified.

TABLE 21-1	Risk Factors for CHD

Major Risk Factors for CHD (not modifiable)

- Increasing age.
- Male gender.
- Family history of premature heart disease.

Major Risk Factors for CHD (modifiable)

- High blood LDL cholesterol.
- Low blood HDL cholesterol.
- High blood pressure (hypertension).
- Diabetes.
- Obesity (especially abdominal obesity).
- Physical inactivity.
- Cigarette smoking.
- An "atherogenic" diet (high in saturated fats and low in vegetables, fruits, and whole grains).

Note: Risk factors highlighted in yellow have relationships with diet.

Source: Expert Panel on Detection, Evaluation, and Treatment of High Blood Cholesterol in Adults (Adult Treatment Panel III), *Third Report of the National Cholesterol Education Program (NCEP)*, NIH publication no. 02-5215 (Bethesda, Md.: National Heart, Lung, and Blood Institute, 2002), pp. II-15 to II-20.

A *lipoprotein profile* provides laboratory values for the different types of lipoproteins. It is sometimes called a *blood lipid profile*.

Abdominal obesity is suggested by a waist circumference of >40 inches for men and >35 inches for women.

HDL protect against atherosclerosis by removing cholesterol from circulation and helping to prevent LDL oxidation.

CHD Risk Assessment Risk assessment requires several key laboratory measures (see Table 21-2) and a medical history. A complete lipoprotein profile, which includes measures of total cholesterol, LDL and HDL cholesterol, and triglycerides, should be obtained every 5 years starting at 20 years of age. Sometimes the ratio of total cholesterol to HDL cholesterol is used to predict CHD risk: a high total cholesterol value suggests elevated LDL cholesterol levels, and a low HDL value is often linked with the multiple risk factors of the metabolic syndrome. Overweight and obesity predispose to CHD, particularly when abdominal obesity is present. High blood pressure and cigarette smoking strongly contribute to CHD risk. Finally, certain medical conditions, including diabetes mellitus and the metabolic syndrome, confer high risk for CHD. The "How to" on p. 543 presents a screening method for assessing a person's ten-year risk of developing CHD that includes some of these risk factors.

Blood Cholesterol Levels and CHD Risk Once a person's risks have been identified, treatment focuses on lowering LDL cholesterol. Elevated LDL cholesterol levels are directly related to the development of atherosclerosis, and clinical studies have confirmed that LDL-lowering treatments can successfully reduce CHD mortality rates. CHD is seldom seen in populations that maintain desirable LDL levels.

As mentioned earlier, HDL help to protect against atherosclerosis, and therefore a low HDL value is highly predictive of CHD risk. In addition, low HDL levels often coexist with other risk factors, such as high triglyceride (VLDL) levels. Moreover, some factors that increase CHD risk—obesity, smoking, inactivity, and male gender—also reduce HDL. It is not known whether raising HDL will help to reduce

TABLE 21-2	Standards for CHD Risk Assessment		
Clinical Measures	**Desirable**	**Borderline Risk**	**High Risk**
Total blood cholesterol (mg/dL)	<200	200–239	≥240
LDL cholesterol (mg/dL)	<100[a]	130–159	160–189[b]
HDL cholesterol (mg/dL)	≥60	59–40	<40
Triglycerides, fasting (mg/dL)	<150	150–199	200–499[c]
Body mass index (BMI)[d]	18.5–24.9	25–29.9	≥30
Blood pressure (systolic and/or diastolic pressure)	<120/<80	120–139/80–89[e]	≥140/≥90[f]

[a] 100–129 mg/dL LDL indicates a near or above optimal level. <70 mg/dL is a desirable goal for very high-risk persons.

[b] ≥190 mg/dL LDL indicates a very high risk.

[c] ≥500 mg/dL triglycerides indicates a very high risk.

[d] Body mass index (BMI) was defined in Chapter 6; BMI standards are found on the inside back cover.

[e] These values indicate prehypertension.

[f] These values indicate stage one hypertension; ≥160/≥100 indicates stage two hypertension. Physicians use these classifications to determine medical treatment.

HOW TO *Assess a Person's Risk of Heart Disease*

This assessment estimates a person's ten-year risk for developing a major coronary event associated with CHD, such as a heart attack.* A high score does not mean that the person *will* develop a heart attack, but it warns of the possibility and suggests the need to consult a physician. To use this algorithm, you need to know a person's age, total and HDL cholesterol levels, and blood pressure.

Age (years)

	Men	Women
20–34	−9	−7
35–39	−4	−3
40–44	0	0
45–49	3	3
50–54	6	6
55–59	8	8
60–64	10	10
65–69	11	12
70–74	12	14
75–79	13	16

HDL (mg/dL)

	Men	Women
≥60	−1	−1
50–59	0	0
40–49	1	1
<40	2	2

Systolic Blood Pressure (mm Hg)

	Untreated		Treated	
	Men	Women	Men	Women
<120	0	0	0	0
120–129	0	1	1	3
130–139	1	2	2	4
140–159	1	3	2	5
≥160	2	4	3	6

Total Cholesterol (mg/dL)

	Age 20–39		Age 40–49		Age 50–59		Age 60–69		Age 70–79	
	Men	Women	Men	Women	Men	Women	Men	Women	Men	Women
<160	0	0	0	0	0	0	0	0	0	0
160–199	4	4	3	3	2	2	1	1	0	1
200–239	7	8	5	6	3	4	1	2	0	1
240–279	9	11	6	8	4	5	2	3	1	2
≥280	11	13	8	10	5	7	3	4	1	2

Smoking (any cigarette smoking in the past month)

Smoker	8	9	5	7	3	4	1	2	1	1
Nonsmoker	0	0	0	0	0	0	0	0	0	0

Scoring Heart Disease Risk

Add up the total points: _____ . Using the table at the right, find the total in the first column and check the second column to learn the percentage risk of developing severe CHD within the next ten years: a person's risk for an acute coronary event can be identified as being less than 10 percent (low risk), 10 to 20 percent (moderate risk), or over 20 percent (high risk). Treatment strategies vary according to a person's risk category.

Men		Women	
Total	Risk	Total	Risk
<0	<1%	<9	<1%
0–4	1%	9–12	1%
5–6	2%	13–14	2%
7	3%	15	3%
8	4%	16	4%
9	5%	17	5%
10	6%	18	6%
11	8%	19	8%
12	10%	20	11%
13	12%	21	14%
14	16%	22	17%
15	20%	23	22%
16	25%	24	27%
≥17	≥30%	≥25	≥30%

*An electronic version of this assessment is available on the ATP III page of the National Heart, Lung, and Blood Institute's website (**www.nhlbi.nih.gov/guidelines/cholesterol**). Another risk inventory is available from the American Heart Association (**www.americanheart.org**). *Source*: Adapted from Expert Panel on Detection, Evaluation, and Treatment of High Blood Cholesterol in Adults (Adult Treatment Panel III), *Third Report of the National Cholesterol Education Program (NCEP)*, NIH publication no. 02-5215 (Bethesda, Md.: National Heart, Lung, and Blood Institute, 2002), section III.

CHD risk, but weight loss, physical activity, and smoking cessation can all independently help to lower risk.

THERAPEUTIC LIFESTYLE CHANGES FOR LOWERING CHD RISK

People who have CHD or multiple risk factors for CHD are often advised to make dietary and lifestyle changes before considering drug therapy. An approach to risk reduction promoted by the National Cholesterol Education Program, known as Therapeutic Lifestyle Changes (TLC), is summarized in Table 21-3.[16] The main features of TLC include a cholesterol-lowering diet, weight reduction, and regular physical activity. If TLC is followed carefully, substantial progress may be seen after six weeks. People with a high risk of CHD should try to lower LDL cholesterol with at least a three-month trial of TLC before considering drug therapy. This section describes the elements of TLC in detail.

Saturated Fat Of the dietary lipids, saturated fat has the strongest effect on blood cholesterol levels, and replacing saturated fat with monounsaturated or polyun-

The National Cholesterol Education Program was developed by a division of the National Institutes of Health and is implemented by medical and health professional associations, voluntary health organizations, community programs, and governmental agencies.

TABLE 21-3	Reducing Risk of CHD with Therapeutic Lifestyle Changes

Dietary Strategies

- Limit saturated fat to less than 7% of total kcalories and cholesterol to less than 200 milligrams a day. Maintaining a fat intake that is 25 to 35% of total kcalories may help with this goal.

- Replace saturated fats with carbohydrates from whole grains, legumes, fruits, and vegetables or with unsaturated fats from fish, vegetable oils, and nuts.

- Avoid food products that contain *trans*-fatty acids. The *trans* fat content in packaged foods is shown on the Nutrition Facts panel of food labels.

- Choose foods high in soluble fibers, including oats, barley, beans, and fruit. Psyllium seed husks can be used as a food supplement to help lower LDL cholesterol levels.

- Regularly consume food products that contain added plant sterols or stanols.

- Regularly consume foods that contain soy protein to replace those that contain animal fat.

- To reduce blood pressure, choose a diet that is high in fruits and vegetables, low-fat milk products, nuts, and whole grains. Limit sodium intake to 2400 milligrams per day.*

- Fish can be consumed regularly as part of a CHD risk-reduction diet.

- If alcohol is consumed, it should be limited to one drink daily for women and two drinks daily for men.

Lifestyle Choices

- Physical activity: At least 30 minutes of moderate-intensity endurance activity should be undertaken on most days of the week. The eventual goal should be an expenditure of at least 2000 kcalories weekly.

- Smoking cessation: Exposure to any form of tobacco smoke should be minimized.

Weight Reduction

- Weight reduction may improve other CHD risk factors. The general goal of a weight-management program should be to prevent weight gain, reduce body weight, and maintain a lower body weight over the long term. The initial goal of a weight-loss program should be to lose no more than 10% of original body weight.

*According to DRI recommendations, sodium intake should be limited to 2300 milligrams daily.

saturated fat can generally lower LDL levels. The TLC recommendation is to consume less than 7 percent of total kcalories as saturated fat. On average, the American diet provides about 11 percent of total kcalories from saturated fat.

Cutting down on saturated fat involves more than just switching from butter to vegetable oil—the main sources of saturated fat in most diets are whole-milk products, high-fat meats, and baked goods. Choosing lean meats or fish, using fat-free or low-fat milk products, and avoiding certain types of bakery products are usually more effective ways of reducing saturated fat.

Replacing saturated fat with carbohydrate can also lower LDL cholesterol, but such a change may lower HDL cholesterol as well and also raise blood triglyceride levels. This effect may be offset somewhat by limiting added sugars and including fiber-rich foods; ideally, the diet should include generous amounts of whole grains, legumes, fruits, and vegetables. The TLC diet recommends a carbohydrate intake in the range of 50 to 60 percent of total kcalories.

> Diets high in carbohydrate—especially those that are high in added sugars—can raise blood triglyceride levels in some people.

Total Fat To help reduce saturated fat, the TLC recommendation for total fat is 25 to 35 percent of kcalories. People with the metabolic syndrome, however, typically have elevated blood triglycerides and may benefit from a fat intake of 30 to 35 percent—which would reduce their carbohydrate intake. A fat intake higher than this may promote weight gain in some people.

Dietary Cholesterol A high cholesterol intake can raise LDL levels, and reducing dietary cholesterol lowers LDL cholesterol in most people. The TLC recommendation is a cholesterol intake of less than 200 milligrams per day. Currently, the daily cholesterol intake in the United States averages 256 milligrams, although it is higher in men (331 milligrams) than in women (213 milligrams). Eggs contribute about one-third of the cholesterol in the American diet, followed by meats, milk, and cheese.

Trans Fats *Trans*-fatty acids raise LDL cholesterol levels, and when they replace saturated fats in the diet (as when stick margarine replaces butter), they may also cause a decline in HDL cholesterol. Most sources of *trans* fats are products made with partially hydrogenated oils, including baked goods, such as crackers, cookies, and doughnuts, and fried foods like french fries and fried chicken. The TLC recommendation is to keep *trans* fats as low as possible.

> Many food manufacturers have been reducing their use of hydrogenated oils and reformulating products so that they contain minimal amounts of *trans* fats.

Soluble Fibers Soluble, viscous fibers inhibit the absorption of cholesterol and bile by binding them in the intestinal tract and may also influence the liver's production of cholesterol by other means. Dietary sources of soluble fibers include oats, barley, legumes, and fruits. The soluble fiber from psyllium seed husks, frequently used to treat constipation, is effective for lowering cholesterol levels when used as a dietary supplement.

Plant Sterols and Stanols Foods or supplements that contain significant amounts of plant sterols or plant stanols can help to lower LDL cholesterol levels. Plant sterols and stanols are added to food products, such as margarines or cheeses, or supplied in dietary supplements. These plant compounds work by interfering with cholesterol and bile absorption.

> Plant sterols are extracted from soybeans and pine-tree oils and are hydrogenated to produce the plant stanols that are added to commercial products.

Soy Diets that are low in saturated fat and cholesterol and high in soy protein can reduce LDL cholesterol levels, especially when the soy protein replaces foods that contain animal fats. Approximately 25 grams of soy protein daily—about four servings of soy milk or tofu—are needed for significant benefit.[17] Whether the LDL-lowering effect is due to the soy protein alone or to other components of soy, such as *isoflavones* or *saponins*, remains unknown.

Sodium and Potassium Intakes Sodium can raise blood pressure in some people, whereas potassium has blood pressure–lowering effects. A low-sodium diet that contains generous amounts of fruits and vegetables, low-fat milk products, nuts, and whole grains has been found to substantially reduce blood pressure, largely due to the diet's content of potassium and several other minerals with blood pressure–lowering effects. This diet (the *DASH Eating Plan*) and other factors that influence blood pressure are discussed in a later section.

Reminder: The American Heart Association recommends consuming two servings of fish per week, with an emphasis on fatty fish.

Fish and Omega-3 Fatty Acids The omega-3 fatty acids in fatty fish, known as EPA and DHA, may benefit people who previously have had a heart attack by suppressing the inflammatory response, reducing blood clotting, stabilizing heart rhythm, and lowering triglyceride levels. In addition, fish can be beneficial in a diet for reducing CHD risk because fish is low in saturated fat and often replaces entrées that contain animal fats. Chapter 4 provides additional information about omega-3 fatty acids, as well as a discussion about the use of fish oil supplements.

Alcohol Moderate consumption of alcohol—from beer, wine, or liquor—has favorable effects on HDL cholesterol levels, atherosclerosis, inflammation, and blood-clotting activity.[18] These benefits are most apparent in men and women who are at least 45 and 55 years old, respectively. Of note, only low or moderate amounts of alcohol—no more than one drink daily for women and two for men—have been found to lower CHD risk, and higher intakes are associated with higher mortality rates. One "drink" is equivalent to 12 ounces of beer, 5 ounces of wine, 10 ounces of wine cooler, or $1^{1}/_{2}$ ounces of 80 proof distilled spirits such as gin, rum, vodka, and whiskey.

Regular Physical Activity Regular physical activity reverses a number of risk factors for CHD. It can lower triglycerides, raise HDL, lower blood pressure, promote weight loss, improve insulin sensitivity, strengthen heart muscle, and increase coronary artery size and tone. The American Heart Association recommends at least 30 minutes of moderate-intensity endurance exercise most days of the week.[19] Aerobic activities that use large muscle groups help the heart the most: such activities include brisk walking, running, swimming, cycling, stair-stepping, and cross-country skiing. Some people may find it easier to schedule several short exercise sessions each day, but each should be at least 15 minutes long.[20]

Cigar and pipe smoking can also increase the risk of CHD, but the risk may not be quite as great because the smoke is less likely to be inhaled.

Smoking Cessation Smoking is a major risk factor for CHD as well as other types of cardiovascular disease. Compounds in cigarette smoke damage blood vessel cells, decrease the oxygen-carrying capacity of the blood (contributing to ischemia), promote blood coagulation, and raise heart rate and blood pressure.[21] Quitting smoking improves CHD risk almost immediately, and eventually the damage caused by smoking is reversed.

Weight Reduction In persons who are obese, weight reduction can improve such CHD risk factors as high blood pressure, elevated blood triglycerides, and low HDL cholesterol. Individuals should focus on weight reduction *after* they have adopted other dietary measures to lower LDL, however. This approach ensures that LDL reduction is given priority and that the individual does not receive a multitude of dietary suggestions at one time. For some, avoiding additional weight gain may be a desirable starting point.

Regular aerobic exercise can strengthen the cardiovascular system, promote weight loss, reduce blood pressure, and improve blood glucose and lipid levels.

© Ronnie Kaufman/Corbis

Successful Adherence to Lifestyle Changes Altering one's lifestyle is challenging. Nurses can help to motivate patients by explaining the reasons for each change, setting obtainable goals, and providing practical suggestions. In some individuals, high LDL cholesterol levels may persist despite adherence to a TLC program; drug therapy may be the only effective treatment for such people. Review Table 21-3 (p. 544) for a summary of the Therapeutic Lifestyle Changes discussed in this section. The "How to" box on p. 547 offers suggestions for implementing a heart-healthy diet.

HOW TO *Implement a Heart-Healthy Diet*

Following a heart-healthy diet can require major changes in dietary choices. Patients may find it easier to adopt a new diet if only a few changes are made at a time. Discussing positive choices (what to eat) first, rather than negative ones (what not to eat), may also aid compliance. These suggestions can help patients implement their diet.

Breads, Cereals, and Pasta

- Choose whole-grain breads and cereals. Make sure the first ingredient on bread and cereal labels is "whole wheat" rather than "enriched wheat flour."
- Bakery products often contain *trans*-fatty acids. Choose foods whose labels *do not* list any *trans* fat in the Nutrition Facts panel or "hydrogenated oil" in the ingredients list. Crackers, chips, cookies, and doughnuts often include *trans* fats.
- Avoid products that contain tropical oils (coconut, palm, and palm kernel oil), which are high in saturated fat.

Fruits and Vegetables

- Consume fruits and vegetables frequently. Keeping the refrigerator stocked with a variety of colorful fruits and vegetables (baby carrots, grapes, blueberries, melon) makes it easier to choose healthy foods when the urge to nibble arises.
- Incorporate at least one or two servings of fruits and vegetables into each meal. People who rarely eat fruits or vegetables may start by adding at least one of their favorites to each meal.
- Choose canned products carefully. Canned vegetables (especially tomato-based products) may be high in sodium. Fruits that are canned in juice are higher in nutrient density than those canned in syrup.
- Restrict high-sodium foods such as pickles, olives, sauerkraut, and kimchee.
- Avoid french fries from fast-food restaurants, which are often loaded with *trans* fats.

Lunch and Dinner Entrées

- Limit meat, fish, and poultry servings to a maximum intake of 5 ounces per day.
- Select lean cuts of beef, such as sirloin tip, round steak, and arm roast, and lean cuts of pork, such as center-cut ham, loin chops, and tenderloin. Trim visible fat before cooking.
- Select extra-lean ground meat and drain well after cooking. Use lean ground turkey, without skin added, in place of ground beef.
- Limit cholesterol-rich organ meats (liver, brain, sweetbreads) and shrimp.
- Limit egg yolks to no more than two per week because the yolks are high in cholesterol (about 215 milligrams per yolk). Replace whole eggs in recipes with egg whites or commercial egg substitutes or similar reduced-cholesterol products.
- Include more vegetarian entrées or legume dishes to boost soluble fiber and soy protein intakes. Pasta and stir-fry recipes can help to reduce meat intake and increase vegetables in the diet.

- Restrict these high-sodium foods:
  - Cured or smoked meats such as beef jerky, bologna, corned or chipped beef, frankfurters, ham, luncheon meats, salt pork, and sausage.
  - Salty or smoked fish, such as anchovies, caviar, salted or dried cod, herring, sardines, and smoked salmon.
  - Packaged, canned, or frozen soups, sauces, and entrées.

Milk Products

- Milk products can be good sources of protein, calcium, vitamin D, and potassium. To obtain two to three servings daily, include a portion of fat-free or low-fat milk, yogurt, or cottage cheese in each meal.
- Use yogurt or fat-free sour cream to make dips or salad dressings. Substitute evaporated fat-free milk for heavy cream.
- Restrict foods high in saturated fat or sodium, such as cheese, processed cheeses, ice cream, and many other milk-based desserts.

Fats and Oils

- Add nuts (not salted) and avocados to meals to increase monounsaturated fat intakes and make meals more appetizing.
- Include vegetable oils in salad dressings and recipes, such as canola, corn, olive, peanut, safflower, sesame, soybean, and sunflower oils.
- Use margarines with added plant sterols or stanols regularly to lower LDL cholesterol levels.
- Select soft margarines in tubs or liquid form; they are low in *trans* fats. Avoid stick margarines and solid vegetable shortenings.
- Avoid products that contain tropical oils (coconut, palm, and palm kernel oil), which are high in saturated fat.

Spices and Seasonings

- Use salt only at the end of cooking, and you will need to add much less. Use salt substitutes at the table.
- Spices and herbs improve the flavor of foods without adding sodium. Try using more garlic, ginger, basil, curry or chili powder, cumin, pepper, lemon, mint, oregano, rosemary, and thyme.
- Check the sodium content on labels. Flavorings and sauces that are usually high in sodium include bouillon cubes, soy sauce, steak and barbecue sauces, relishes, mustard, and catsup.

Snacks and Desserts

- Select low-sodium and low–saturated fat choices such as unsalted pretzels and nuts, plain popcorn, and unsalted chips and crackers.
- Choose canned or dried fruits and some raw vegetables to boost fruit and vegetable intake.
- Enjoy angel food cake, which is made without egg yolks and added fat.
- Select low-fat frozen desserts such as sherbet, sorbet, fruit bars, and some low-fat ice creams.

Blood triglycerides:
- Borderline high: 150–199 mg/dL.
- High: ≥200 mg/dL.

LIFESTYLE CHANGES FOR HYPERTRIGLYCERIDEMIA

Hypertriglyceridemia (elevated blood triglycerides) is common in people with the metabolic syndrome and diabetes mellitus and may also result from other disorders. Severe hypertriglyceridemia can cause serious complications, including fatty deposits in the liver and skin and acute pancreatitis. Whereas diet and lifestyle contribute to mild hypertriglyceridemia ("borderline-high" triglycerides), genetic factors are usually responsible for severe cases ("high" and "very high" triglyceride levels).

Mild Hypertriglyceridemia Dietary and lifestyle changes can improve mild hypertriglyceridemia. Overweight and obesity, a sedentary lifestyle, and cigarette smoking all may raise triglyceride levels. The dietary factors that influence triglycerides the most are high intakes of carbohydrate (60 percent or more of total kcalories) and alcohol. Thus controlling body weight, becoming physically active, quitting smoking, avoiding a high carbohydrate intake, and restricting alcohol are basic treatments for hypertriglyceridemia. As mentioned earlier, high triglycerides are often associated with low HDL, and the lifestyle changes listed here are likely to improve HDL levels as well.

Severe Hypertriglyceridemia Medications are usually necessary for those with severe hypertriglyceridemia. Weight reduction and physical activity are still emphasized, and a very low-fat diet, providing less than 15 percent of kcalories from fat, may be required in extreme cases to prevent dangerous complications. If the dietary fat restriction is severe, medium-chain triglycerides (see p. 400 and p. 476) can be useful for replacing fats and oils in the diet.

VITAMIN SUPPLEMENTATION AND CHD RISK

People often ask about the potential benefits of certain types of vitamin supplements for reducing CHD risk. There is ongoing research interest in the supplements described in this section.

B Vitamin Supplements and Homocysteine As mentioned earlier, elevated blood homocysteine is a known risk factor for CHD, but whether homocysteine itself is directly damaging or is simply an indicator of other abnormalities remains unknown. Possibly, homocysteine has harmful effects on the artery wall or heightens blood-clotting activity that worsens atherosclerosis.[22] Although increased intakes of folate, vitamin B_6, and vitamin B_{12} can lower homocysteine levels, it is not known whether reducing homocysteine with diet or supplements will reduce the risk of CHD.

Antioxidant Vitamin Supplements Because oxidized LDL promote atherosclerosis, researchers have hypothesized that antioxidant supplementation may help to reduce CHD risk. Several epidemiological studies have suggested that antioxidant-rich diets may protect against CHD, but because persons who consume such diets usually also have a healthy lifestyle and body weight, it has been difficult to determine whether the antioxidants are responsible for the effect. Most studies that tested supplementation with single antioxidants (like vitamins C and E), combinations, or multivitamins have produced weak or inconsistent results,[23] and several studies have suggested possible harm.[24] Until more data are available, there is no recommendation to use antioxidant supplements for heart disease prevention.

DRUG THERAPIES FOR CHD PREVENTION

Individuals who cannot reach LDL goals with dietary and lifestyle changes alone may be prescribed one or more medications. The most common LDL-lowering drugs reduce cholesterol synthesis in the liver (statins) or reduce cholesterol and

DIET-DRUG INTERACTIONS Check this table for notable nutrition-related effects of the medications discussed in this chapter.

	Gastrointestinal Effects	Interactions with Dietary Substances	Metabolic Effects
Anticoagulants (warfarin)		Requires consistent vitamin K intake to maintain effectiveness. Enhanced drug effects with supplementation of vitamin E, dong quai, danshen, fish oils, garlic, and ginkgo. Reduced drug effects with coenzyme Q, ginseng, and green tea. Avoid alcohol.	
Antihypertensives			
Beta-blockers			Elevated serum potassium levels, hypoglycemia
Calcium channel blockers	Nausea, constipation	Avoid herbal supplements that contain natural licorice. Avoid grapefruit juice, which may enhance drug effects.	
ACE inhibitors	Reduced taste sensation	Avoid herbal supplements that contain natural licorice. Avoid potassium supplements and salt substitutes containing potassium.	Elevated serum potassium levels
Antilipimics			
Statins	Constipation, flatulence, GI discomfort	Avoid grapefruit juice, which may enhance drug effects.	Elevated serum liver enzymes
Bile acid sequestrants	Flatulence, diarrhea, constipation	Cause reduced absorption of fat-soluble vitamins.	Electrolyte imbalances, iron deficiency
Nicotinic acid	GI discomfort	Avoid alcoholic beverages, coffee, and tea, which may increase side effects.	Elevated serum liver enzymes and uric acid levels, hyperglycemia, low blood pressure
Digoxin	Anorexia, nausea, stomach cramps, diarrhea	High-fiber foods and magnesium supplements can reduce drug absorption. St. John's wort may reduce drug efficacy.	Elevated serum potassium and reduced serum magnesium levels. Drug toxicity can develop if body potassium levels are low.
Diuretics (furosemide, spironolactone[a])	Dry mouth, anorexia, decreased taste perception	Furosemide bioavailability is reduced when taken with food.	Fluid and electrolyte imbalances,[a] hyperglycemia (spironolactone), hyperlipidemia (spironolactone), thiamin and zinc deficiencies
Nitroglycerin	Decreased taste perception	Increases effects of alcohol.	

[a] *Furosemide* is a "potassium-wasting" diuretic; patients should increase intakes of potassium-rich foods. *Spironolactone* is a "potassium-sparing" diuretic; patients should avoid supplemental potassium and potassium-containing salt substitutes.

bile absorption in the small intestine (bile acid sequestrants). The vitamin niacin (nicotinic acid), taken in high amounts, lowers blood triglycerides and raises HDL levels. Individuals using lipid-lowering medications should continue the TLC program so that the lowest-possible dosages are used.

In addition to lipid-lowering medications, some people may benefit from drugs that suppress blood clotting (anticoagulants and aspirin) or reduce blood pressure. Nitroglycerin may be given to alleviate angina as needed. Some medications may affect nutritional status or food intake (see the Diet-Drug Interactions feature on this page); the interactions can be even more complicated when multiple medications are used.

The dietary strategies that a cardiac patient should adopt are similar to the Therapeutic Lifestyle Changes (TLC) described earlier.

TREATMENT FOR HEART ATTACK

As explained earlier, a heart attack may result when a blood clot blocks a coronary artery and cuts off the supply of blood and oxygen to heart muscle. Drug therapies given immediately after a heart attack may include thrombolytic drugs (clot-busting drugs), anticoagulants, aspirin, painkillers, and medications that regulate heart rhythm and reduce blood pressure. Patients are not given food or liquids, except for sips of water, until their condition stabilizes. Once food is permitted, patients are initially offered small portions of soft foods that are low in sodium, saturated fat, and cholesterol. The sodium restriction helps to limit fluid retention, but may be lifted after several days if the patient shows no signs of heart failure.

A heart attack patient needs to regain strength and learn strategies that can reduce the risk of a future heart attack. Cardiac rehabilitation programs, found in hospitals and outpatient clinics, often last for several months and include exercise therapy, help with smoking cessation, dietary instruction, stress management, and medication counseling. Home-based programs are also beneficial, but they are more limited in scope and lack the benefit of group interaction.

REVIEW NOTES

Long-term CHD management emphasizes risk reduction. Modifiable risk factors include elevated LDL and triglyceride levels, high blood pressure, cigarette smoking, diabetes, obesity, sedentary lifestyle, and an atherogenic diet.

The Therapeutic Lifestyle Changes (TLC) approach includes dietary and lifestyle modifications that can help to reduce LDL levels and eliminate other risk factors. Dietary recommendations are to reduce saturated fat, *trans* fats, and cholesterol; increase soluble fiber; and incorporate plant sterols and stanols into the diet.

Treatment recommendations for mild hypertriglyceridemia emphasize weight control, regular physical activity, smoking cessation, avoiding a high carbohydrate intake, and restricting alcohol. Severe hypertriglyceridemia may require drug therapies and dietary fat restriction.

Medications given after a heart attack suppress blood clotting and regulate heart rhythm, and patients are initially offered heart-healthy, soft foods. To reduce the risk of a future heart attack, patients must learn strategies similar to the TLC approach.

Hypertension

Blood pressure is measured both when heart muscle contracts (*systolic* blood pressure) and when it relaxes (*diastolic* blood pressure). Measurements are expressed as millimeters of mercury (mm Hg).

	Systolic	Diastolic
Desirable blood pressure	<120	<80
Prehypertension	120–139	80–89
Hypertension	≥140	≥90

Hypertension (high blood pressure) affects nearly one-third of adults in the United States.[25] Prevalence is especially high among African Americans, who develop hypertension earlier in life and sustain higher average blood pressures throughout their lives than other ethnic groups. An estimated 30 percent of people with hypertension are unaware that they have it.[26]

Although people cannot feel the physical effects of hypertension, it is a primary risk factor for atherosclerosis and cardiovascular diseases. High blood pressure causes the heart to work harder to eject blood into the arteries; this effort weakens heart muscle and increases the risk of developing heart arrhythmias, congestive heart failure, and even sudden death. Hypertension is also a primary cause of stroke and kidney failure. Reducing blood pressure can dramatically reduce the incidence of these diseases.

FACTORS THAT INFLUENCE BLOOD PRESSURE

Although the underlying causes of most cases of hypertension are not fully understood, much is known about the physiological factors that affect blood pressure. As shown in Figure 21-3, blood pressure arises from the heart muscle con-

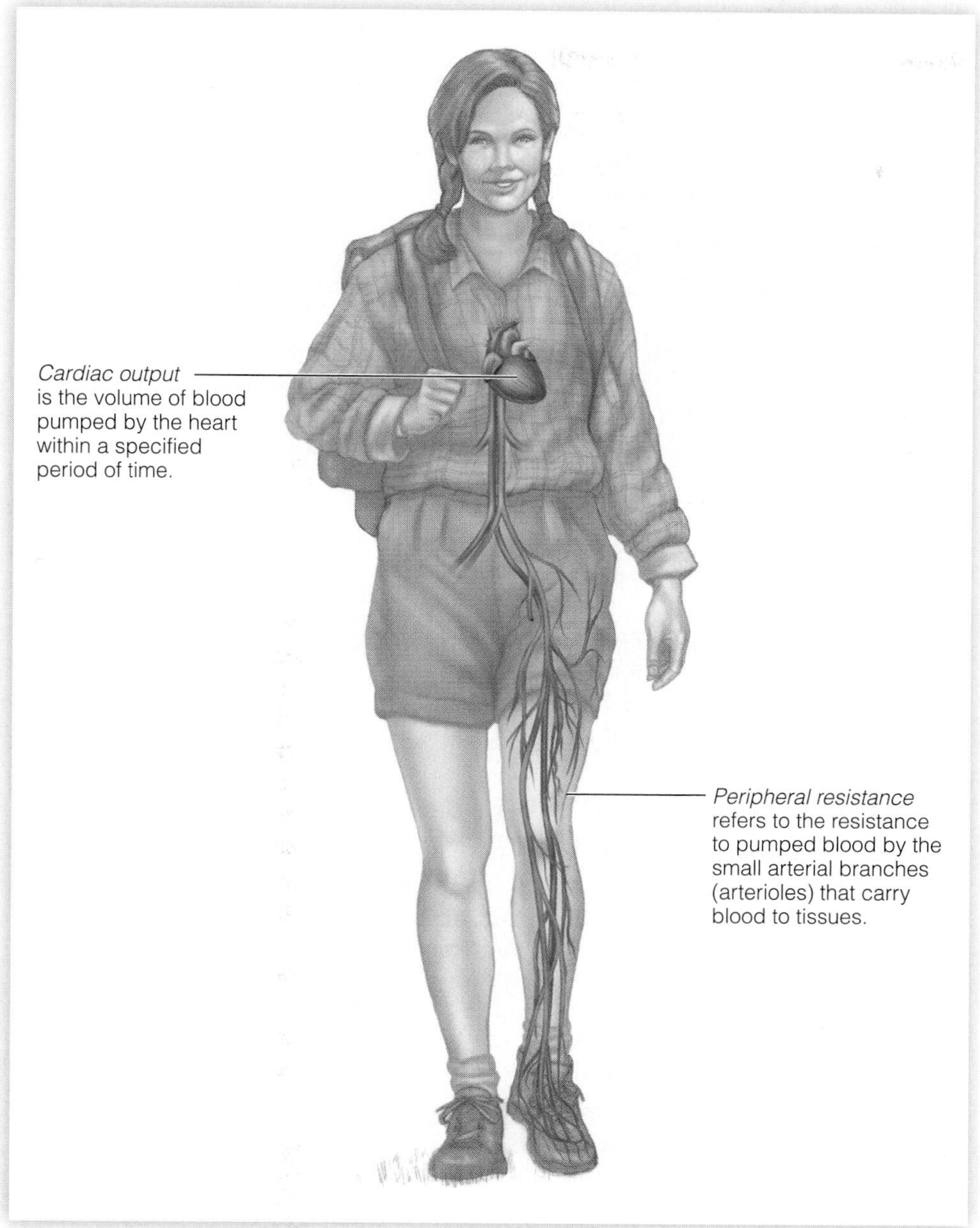

FIGURE 21-3 Determinants of Blood Pressure

Cardiac output is the volume of blood pumped by the heart within a specified period of time.

Peripheral resistance refers to the resistance to pumped blood by the small arterial branches (arterioles) that carry blood to tissues.

The equation describing this relationship is blood pressure (BP) = cardiac output (CO) × peripheral resistance (PR).

cardiac output: the volume of blood pumped by the heart within a specified period of time.

peripheral resistance: the resistance to pumped blood in the small arterial branches (arterioles) that carry blood to tissues.

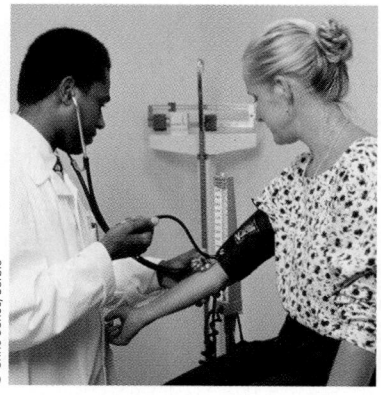

Screening people for hypertension is a first step toward early detection and prevention of complications.

tractions that pump blood away from the heart (**cardiac output**) and the resistance the blood encounters in the arterioles (**peripheral resistance**). When either cardiac output or peripheral resistance increases, blood pressure rises. Cardiac output is raised when heart rate or blood volume increases; peripheral resistance is affected mostly by the diameters of the arterioles. Blood pressure is therefore influenced by the nervous system, which regulates heart muscle contractions and the arteriole's diameters, and hormonal signals, which may cause fluid retention or blood vessel constriction. The kidneys also play a role in regulating blood pressure by controlling the secretion of the hormones involved in vasoconstriction and retention of sodium and water.

CONTRIBUTING FACTORS FOR HYPERTENSION

In 95 percent of cases, the cause of hypertension is unknown.[27] In other cases, hypertension may be caused by a physical abnormality, such as a hormone imbalance, kidney disorder, or atherosclerosis in the renal arteries. Stiffness and thickening of the aorta or arteries (arteriosclerosis) due to age, diabetes, or other reasons can increase resistance to blood flow, thereby raising blood pressure. A number of risk factors have been strongly implicated in the development of hypertension:

- *Aging.* Hypertension risk increases with age. Individuals who have normal blood pressure at age 55 still have a 90 percent risk of developing high blood pressure during their lifetimes.[28]
- *Genetic.* Hypertension risk is similar among family members. It is also more prevalent and severe in certain ethnic groups: the prevalence is about 38 percent in African Americans, compared with 23 percent in both whites and Mexican Americans.[29]
- *Obesity.* Most people with hypertension—an estimated 60 percent—are obese.[30] Obesity raises blood pressure, in part, by altering kidney function and promoting fluid retention.[31]
- *Salt sensitivity.* Among those with hypertension, approximately 30 to 50 percent have blood pressure that is sensitive to salt and can benefit by reducing salt in their diets.[32]
- *Alcohol.* Heavy alcohol consumption (defined as three or more drinks daily) is strongly associated with hypertension. Alcohol's specific role in blood pressure is unclear, however.[33]
- *Diet.* A diet containing generous amounts of fruits, vegetables, nuts, and low-fat milk products can lower blood pressure. These foods provide the major minerals potassium, calcium, and magnesium, which help to reduce blood pressure when included in substantial amounts in the diet.

NURSING DIAGNOSIS

for persons with hypertension may include *imbalanced nutrition: more than body requirements* and *ineffective health maintenance.*

TREATMENT OF HYPERTENSION

The goal of hypertension treatment is to reduce blood pressure to <140/<90 mm Hg (or <130/<80 mm Hg in people with diabetes or kidney disease).

Both lifestyle modifications and drug therapies are used to treat hypertension. Lifestyle changes that lower blood pressure include weight reduction if overweight or obese; a diet low in sodium and rich in potassium, calcium, and magnesium; regular physical activity; and a moderate alcohol intake. Combining two or more lifestyle modifications can enhance results. Table 21-4 describes the lifestyle changes that reduce blood pressure and the expected reduction in systolic blood pressure for each change.

TABLE 21-4	Lifestyle Modifications for Blood Pressure Reduction	
Modification	**Recommendation**	**Expected Reduction in Systolic Blood Pressure**
Weight reduction	Maintain healthy body weight (BMI below 25).	5–20 mm Hg/10 kg lost
DASH eating plan	Adopt a diet rich in fruits, vegetables, and low-fat milk products with reduced saturated fat intake.	8–14 mm Hg
Sodium restriction	Reduce dietary sodium intake to less than 2400 milligrams sodium (less than 6 grams salt) per day.*	2–8 mm Hg
Physical activity	Perform aerobic physical activity for at least 30 minutes per day, most days of the week.	4–9 mm Hg
Moderate alcohol consumption	Men: Limit to 2 drinks per day. Women and lighter-weight men: Limit to 1 drink per day.	2–4 mm Hg

*According to DRI recommendations, sodium intake should be limited to 2300 milligrams daily.

Source: Adapted from *Reference Card from the Seventh Report of the Joint National Committee on Prevention, Detection, Evaluation, and Treatment of High Blood Pressure (JNC 7)*, NIH publication no. 03-5231 (Bethesda, Md.: National Institutes of Health, National Heart, Lung, and Blood Institute, and National High Blood Pressure Education Program, May 2003).

TABLE 21-5	The DASH Eating Plan			
Food Group	**Recommended Servings for Different Energy Intakes** (servings per day except as noted)			
	1600 kcal	2000 kcal	2600 kcal	3100 kcal
Grains and grain products[a] (1 serving = 1 slice bread, 1 oz dry cereal,[b] or $^1/_2$ c cooked rice, pasta, or cereal)	6	7–8	10–11	12–13
Vegetables (1 serving = $^1/_2$ c cooked vegetables, 1 c raw leafy vegetables, or 6 oz vegetable juice)	3–4	4–5	5–6	6
Fruits (1 serving = 1 medium fruit; $^1/_2$ c fresh, frozen, or canned fruit; $^1/_4$ c dried fruit; or 6 oz fruit juice)	4	4–5	5–6	6
Milk products (low-fat or fat-free) (1 serving = 8 oz milk, 1 yogurt, or 1$^1/_2$ oz cheese)	2–3	2–3	3	3–4
Meat, poultry, and fish (1 serving = 3 oz cooked meat, poultry, or fish)	1–2	1–2	2	2–3
Nuts, seeds, and legumes (1 serving = $^1/_3$ c nuts, 2 tbs seeds, or $^1/_2$ c cooked dry beans or peas)	3–4 per week	4–5 per week	1	1
Fats and oils (1 serving = 1 tsp vegetable oil or soft margarine, 1 tbs low-fat mayonnaise, or 2 tbs light salad dressing)	2	2–3	3	4
Sweets (1 serving = 1 tbs sugar, jelly, or jam; $^1/_2$ oz jelly beans; or 8 oz lemonade)	0	5 per week	2	2

[a]Whole grains are recommended for most servings consumed.

[b]One ounce of dry cereal may be equivalent to $^1/_2$ to 1$^1/_4$ cups, depending on the cereal. Check the food label for the portion size.

Source: Dietary Guidelines for Americans, 2005, available at **www.healthierus.gov/dietaryguidelines**.

Dietary Approaches for Reducing Blood Pressure Several research studies have shown that a significant reduction in blood pressure can be achieved by following a diet that emphasizes fruits, vegetables, and low-fat dairy products and includes whole grains, poultry, fish, and nuts.[34] The diet tested in these studies, now known as the *DASH Eating Plan,* provides more fiber, potassium, magnesium, and calcium than the typical American diet. The diet also limits red meat, sweets, sugar-containing beverages, saturated fat (to 7 percent of kcalories), and cholesterol (to 150 milligrams per day), so it is beneficial for reducing CHD risk as well. The DASH Eating Plan, shown in Table 21-5, is a dietary pattern that meets the goals specified in the *Dietary Guidelines for Americans 2005.*[35]

The DASH Eating Plan is even more effective when accompanied by a low sodium intake. In a research study that tested the blood pressure–lowering effects of the DASH dietary pattern in combination with sodium restriction, the best results were achieved when sodium was reduced to 1500 milligrams daily—a lower level than the 2400 milligrams usually recommended for people with hypertension.[36] Although some individuals may find low-sodium diets difficult to adhere to, those

The DASH Eating Plan was the test diet used in a study called "Dietary Approaches to Stop Hypertension."

with hypertension should be encouraged to reduce sodium to whatever extent possible, as even a mild reduction may help to improve blood pressure. The "How to" on this page lists practical suggestions for reducing sodium intake.

Weight Reduction Weight reduction can reduce blood pressure considerably (review Table 21-4). In controlled studies, participants who lost 22 pounds (10 kilograms) lowered systolic blood pressure by an average of 7.0 mm Hg, and greater weight loss was associated with greater reductions in blood pressure.[37] The improvement persisted for at least a year and a half: the prevalence of hypertension among participants was found to be 20 to 50 percent lower among those who lost weight.

Drug Therapies for Reducing Blood Pressure Most people with hypertension use two or more medications to meet their blood pressure goals. Using a combination of drugs with different modes of action can reduce the dosage of each drug needed and minimize side effects. Most treatments include diuretics, which lower blood pressure by reducing blood volume. Other drugs commonly prescribed are similar to those used to treat various heart conditions (see the Diet-Drug Interactions feature on p. 549). Drug dosages may need to be adjusted in follow-up visits until the blood pressure goal is reached.

REVIEW NOTES

Nearly one in three persons in the United States has hypertension, which increases the risk of developing heart disease, heart failure, stroke, and kidney failure.

Blood pressure is elevated by factors that increase blood volume, heart rate, and resistance to blood flow. Although the underlying cause of most cases of hypertension is unknown, major risk factors include aging, family history, ethnicity, obesity, and certain dietary choices.

Treatment usually includes a combination of lifestyle modifications and drug therapies. The case study on p. 555 provides an opportunity to review the risk factors and treatments for CHD and hypertension.

HOW TO   *Reduce Sodium Intake*

- Select fresh, unprocessed foods. Packaged foods, canned goods, and frozen meals are often high in sodium.
- Use unsalted snack foods including tortilla chips, popcorn, and nuts. Some snack foods may need to be avoided.
- Do not use salt at the table or while cooking. Salt substitutes may be useful for some people. Salt substitutes often contain potassium, however; these are not appropriate for persons using diuretics that allow potassium to accumulate in the blood.
- Avoid eating in fast-food restaurants; most choices are very high in sodium.
- Check food labels. The labeling term "low sodium" is a better guide than the terms "reduced sodium" (contains 25 percent less sodium than the regular product) or "light in sodium" (contains 50 percent less sodium). To be labeled "low sodium," a food product must contain less than 140 milligrams of sodium per serving. Keep your goal sodium level (about 2400 milligrams) in mind when you read labels.

- Recognize the high-salt or high-sodium foods in each food category, including:
  - Bakery products made with baking powder or baking soda (sodium bicarbonate); read labels to make sure.
  - Processed meats: ham, corned beef, bologna, salami, sausage, bacon, frankfurters, and pastrami.
  - Fish: salted fish, shellfish, and canned fish.
  - Tomato-based products: tomato sauce, tomato juice, pizza, canned tomatoes, and catsup.
  - Canned soups and broths: even reduced-sodium varieties may contain excessive sodium.
  - Cheese: cottage cheese, American cheese, parmesan, and most other hard cheeses.
  - Condiments and relishes: bouillon cubes, olives, and pickled vegetables.
  - Flavoring sauces: soy sauce, barbecue sauce, and steak sauce.
- Check for the word *sodium* on medication labels. Sodium is often an ingredient in some types of antacids and laxatives.

Computer Programmer with Cardiovascular Disease

Mr. Reid, a 48-year-old African American computer programmer, is 5 feet 9 inches tall and weighs 240 pounds. He sits for long hours at work and is too tired to exercise when he gets home at night. His meals usually include fatty meats, eggs, and cheese, and he likes dairy desserts such as pudding and ice cream. He has a family history of CHD and hypertension. His recent laboratory tests show that his blood pressure is 160/100 mm Hg, his LDL cholesterol is 160 mg/dL, and his HDL level is 35 mg/dL. He smokes a pack of cigarettes a day and usually has two glasses of wine at both lunch and dinner.

1. Name Mr. Reid's major risk factors for CHD and hypertension. Which can be modified? What complications might occur if he doesn't seek treatment for his blood lipids and blood pressure?

2. What dietary changes would you recommend that could help to improve Mr. Reid's blood pressure and his LDL and HDL cholesterol? Explain the rationale for each dietary change. Prepare a day's menus for Mr. Reid using the DASH diet as an outline for your choices.

3. What other laboratory tests or measurements would you need to better assess Mr. Reid's condition? Why?

4. Describe several benefits that Mr. Reid might obtain from a program that includes weight reduction and regular physical activity. Explain why the use of alcohol can be both a protective and a damaging lifestyle habit.

5. Assuming that Mr. Reid does not make any changes in his diet and lifestyle and suffers a heart attack, discuss the elements of a cardiac rehabilitation program that would be critical for his long-term survival.

Congestive Heart Failure

Congestive heart failure (CHF) is a condition characterized by the heart's inability to pump adequate blood, causing a buildup of fluids in the veins and tissues. It is often a consequence of disorders that create extra work for the heart muscle, such as CHD and hypertension. The extra workload causes the heart to enlarge or pump faster or harder, and eventually the overburdened heart may fail completely. CHF develops mostly in the elderly: approximately 75 percent of persons with CHF in the United States are age 65 or older.[38]

CONSEQUENCES OF CONGESTIVE HEART FAILURE

In CHF, fluid may accumulate in the liver, abdomen, and lower extremities, causing chest pain, difficulty with digestion and absorption, and swelling in the legs, ankles, and feet. Fluid can sometimes build up in the lungs, causing shortness of breath and limited tolerance for activity; in severe cases, CHF can lead to acute respiratory failure. With inadequate blood flow, the functions of other organs, such as the liver and kidneys, may be impaired.

Heart failure can lead to reduced food intake as well. If abdominal bloating and liver enlargement develop, pain and discomfort may worsen with meals. End-stage heart failure is often accompanied by **cardiac cachexia,** a condition of malnutrition brought about by changes in body chemistry and worsened by reduced appetite and food intake. Cardiac cachexia is characterized by severe weight loss and tissue wasting. The resultant weakness can cause substantial disability.

congestive heart failure (CHF): a condition in which the heart is unable to pump adequate blood, resulting in fluid congestion in tissues and in the veins leading to the heart.

cardiac cachexia: the severe muscle wasting and weight loss that accompany congestive heart failure

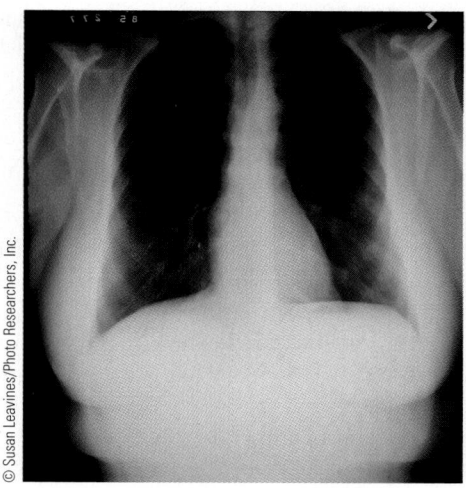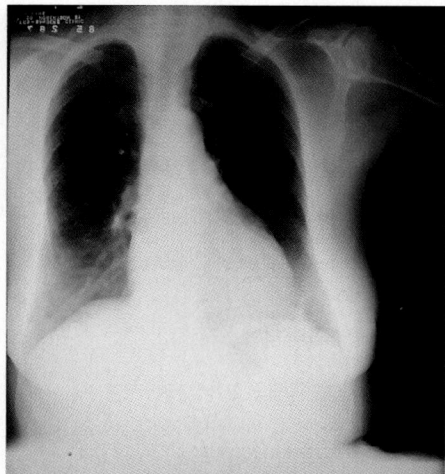

© Susan Leavines/Photo Researchers, Inc.

An overburdened heart enlarges in an effort to supply blood to the body's tissues.

MEDICAL MANAGEMENT OF CONGESTIVE HEART FAILURE

There is no cure for CHF. It is a chronic, progressive disease that may require frequent hospitalizations. Many patients face a combination of debilitating symptoms, restrictive treatments, and an uncertain outcome. Important goals of medical therapy are to enhance the patient's quality of life and to slow disease progression.

The specific treatment for CHF depends on the nature and severity of the illness.[39] In general, medications help to manage congestion and improve heart function. Diuretics are given to reverse or prevent fluid retention. Dietary sodium and fluid restrictions may help to prevent fluid accumulation. Vaccinations for influenza and pneumonia can reduce the risk of developing respiratory infections. CHF patients are also encouraged to participate in exercise programs to avoid becoming physically disabled and to improve endurance.

The recommendation of 2000 to 3000 mg of sodium is not very restrictive. The current DRI recommendation is to limit sodium intake to 2300 mg daily.

Medical Nutrition Therapy The main dietary recommendation for CHF is a moderate sodium intake of 2000 to 3000 milligrams daily, depending on the severity of the illness. Sodium restriction reduces the likelihood of fluid retention. Patients with more severe cases of CHF may need to reduce sodium intake to 2000 milligrams per day (review the "How to" on p. 554.) Under some circumstances, fluid restriction may also be helpful. For patients who have difficulty eating due to abdominal or chest pain, small, frequent meals may be better tolerated than larger meals.

Other Dietary Recommendations Patients with CHF may be prone to constipation due to diuretic use and reduced physical activity. Maintaining an adequate fiber intake can help to minimize constipation problems. Because of alcohol's deleterious effects on blood pressure, heart function, and blood lipids, alcoholic beverages should be avoided.[40]

There are no known therapies that can reverse cardiac cachexia, and prognosis is poor. For some patients, liquid supplements, tube feedings, or parenteral nutrition support can be supportive additions to treatment.

> **REVIEW NOTES**
>
> In congestive heart failure, the heart is unable to pump adequate blood to tissues. Consequences may include congestion in the veins, lungs, and other organs and impaired organ function.
>
> CHF is usually a chronic, progressive heart condition that results from other cardiovascular illnesses.
>
> Treatment of CHF includes drug therapies that reduce congestion and strengthen heart function. The main dietary recommendation is moderate sodium restriction.

NURSING DIAGNOSIS

for stroke patients include *adult failure to thrive, constipation, disturbed sensory perception, impaired physical mobility, impaired swallowing, impaired verbal communication, risk for injury,* and *feeding self-care deficit.*

Stroke

Stroke is the third leading cause of death in the United States after heart disease and cancer. It is the second most common cause of neurological disability after Alzheimer's disease. Most strokes—about 88 percent—are **ischemic strokes** and result from the obstruction of blood flow to brain tissue. **Hemorrhagic strokes** occur in 9 percent of cases and result from bleeding within the brain, which damages brain tissue. Most strokes are a consequence of atherosclerosis or hypertension, or both.

Strokes that occur suddenly and are short-lived (lasting 2 to 30 minutes) are called **transient ischemic attacks.** These brief strokes are a warning sign that a more severe stroke may follow. They are usually treated with aspirin and other drugs that inhibit blood clotting.

ischemic strokes: strokes that result from the obstruction of blood flow to brain tissue.

hemorrhagic strokes: strokes that result from bleeding within the brain, which destroys or compresses brain tissue.

transient ischemic attacks: brief ischemic episodes that cause short-term neurological symptoms, such as blurred vision, slurred speech, numbness, paralysis, or difficulty speaking.

STROKE PREVENTION

Stroke is largely preventable by recognizing its risk factors and making lifestyle choices that reduce risk. Many of the risk factors are similar to those for heart disease, such as hypertension, cigarette smoking, diabetes mellitus, elevated LDL cholesterol, and a history of cardiovascular disease. Medications that suppress blood clotting help to reduce the risk of ischemic stroke, especially in people who have suffered a first stroke or a transient ischemic attack. Such drugs include aspirin, antiplatelet drugs, and anticoagulants (usually warfarin). Anticoagulant therapy requires regular follow-up and occasional adjustments in dosage to prevent excessive bleeding.

> Warfarin acts by interfering with vitamin K's blood-clotting function (see Chapter 8).

STROKE MANAGEMENT

The effects of a stroke vary according to the area of the brain that has been injured. Body movements, senses, and speech are often impaired. One side of the body may be weakened or paralyzed. Early diagnosis and treatment are necessary to preserve brain tissue and minimize long-term disability. Ideally, thrombolytic drugs are used within the first few hours after an ischemic stroke to prevent further brain damage.[41]

The main nutritional goals are to maintain nutrition status and overall health despite a patient's disabilities. Dysphagia (difficulty swallowing) is a frequent complication of stroke and is associated with poorer prognosis. Difficulty with speech prevents patients from describing the problems they may be having with eating and from communicating their food preferences. Coordination problems can make it hard for them to grasp utensils or bring food from table to mouth. In some cases, tube feedings may be necessary until the patient has regained these skills. Nutrition in Practice 21 gives additional information about feeding people with disabilities such as those that follow stroke.

REVIEW NOTES

The two major types of strokes, ischemic and hemorrhagic stroke, may be a consequence of atherosclerosis, hypertension, or both. Transient ischemic attacks are short-lived "mini-strokes" and are a warning sign that a more severe stroke may occur.

Strokes are largely preventable by reversing modifiable risk factors, which include hypertension, cigarette smoking, diabetes mellitus, and elevated LDL cholesterol.

Treatment includes the use of anticlotting drugs such as aspirin, antiplatelet drugs, and anticoagulants. A patient who has had a major stroke may have problems eating normally due to lack of coordination and difficulty swallowing.

NUTRITION ASSESSMENT CHECKLIST FOR PEOPLE WITH CARDIOVASCULAR DISEASES

Medical History

Check the medical record for a diagnosis of:

☐ Coronary heart disease

☐ Hypertension

☐ Congestive heart failure

☐ Strokes

Review the medical record for complications related to cardiovascular diseases:

☐ Heart attacks

☐ Transient ischemic attacks

☐ Cardiac cachexia

Note risk factors for CHD related to diet, including:

☐ Elevated LDL or triglyceride levels

☐ Obesity or overweight

☐ Diabetes

☐ Hypertension

Medications

For patients using drug treatments for cardiovascular diseases, note:

☐ Side effects that may alter food intake

☐ Medications that may interact with grapefruit juice

☐ Use of warfarin, which influences vitamin K intake

☐ Use of diuretics associated with potassium imbalances

☐ Potential diet-drug or herb-drug interactions

Dietary Intake

For patients with CHD or hypertension, assess the diet for:

☐ Energy intake

☐ Saturated fat, *trans* fat, cholesterol, and sodium content

☐ Soluble fiber and soy protein content

☐ Intake of whole grains, fruits, vegetables, legumes, and nuts

☐ Alcohol

For patients with complications related to cardiovascular diseases:

☐ Check adequacy of food intake in patients with congestive heart failure.

☐ Check physical disabilities that may interfere with food preparation or consumption following a stroke.

Anthropometric Data

Measure baseline height and weight and reassess weight at each medical checkup. Note whether patients are meeting weight goals, including:

☐ Weight loss or maintenance in patients who are overweight

☐ Weight maintenance in patients with advanced congestive heart failure

Remember that weight may be deceptively high in people who are retaining fluids, especially those with congestive heart failure.

Laboratory Tests

Monitor the following laboratory tests in people with cardiovascular diseases:

☐ LDL cholesterol, triglycerides, and HDL cholesterol

☐ Blood glucose in those with diabetes

☐ Blood potassium in those using diuretics or cardiac glycosides

☐ Indicators of fluid retention in those with congestive heart failure

☐ Blood-clotting times in those using anticoagulants

Physical Signs

Blood pressure measurement is routine in physical exams, but is especially important for people who:

☐ Have cardiovascular diseases

☐ Have experienced a heart attack or stroke

☐ Have risk factors for CHD or hypertension

Look for signs of:

☐ Potassium imbalances (muscle weakness, numbness and tingling, irregular heartbeat) in those using diuretics or cardiac glycosides

☐ Fluid overload in patients with congestive heart failure

SELF CHECK

1. Ischemia in the coronary arteries is a frequent cause of:
   a. angina pectoris.
   b. hemorrhagic stroke.
   c. aneurysm.
   d. hypertension.

2. Risk factors for atherosclerosis include all of the following, *except:*
   a. smoking.
   b. hypertension.
   c. diabetes mellitus.
   d. elevated HDL cholesterol.

3. The dietary lipids that have the strongest LDL cholesterol–raising effects are:
   a. monounsaturated fats.
   b. cholesterol.
   c. saturated fats.
   d. plant sterols.

4. The omega-3 fatty acids EPA and DHA, which may improve some risk factors for heart disease, are obtained by consuming:
   a. fatty fish.
   b. soy products.
   c. egg yolks and organ meats.
   d. nuts and seeds.

5. Moderate alcohol consumption can improve heart disease risk because it:
   a. lowers blood pressure.
   b. improves nutrition status.
   c. offsets the damage from smoking.
   d. increases HDL cholesterol levels.

6. Patients with hypertriglyceridemia may benefit from:
   a. reducing sodium intake.
   b. consuming moderate amounts of alcohol.
   c. avoiding a high carbohydrate intake.
   d. reducing cholesterol intake.

7. In most cases of hypertension, the cause is:
   a. excessive alcohol use.
   b. atherosclerosis.
   c. hormonal imbalances.
   d. unknown.
8. Hypertensive patients can benefit from all of the following dietary and lifestyle modifications, *except:*
   a. including fat-free or low-fat milk products in the diet.
   b. reducing total fat intake.
   c. consuming generous amounts of fruits, vegetables, legumes, and nuts.
   d. reducing sodium intake.

9. Medical nutrition therapy for a patient with congestive heart failure usually includes:
   a. weight loss.
   b. reducing total fat intake.
   c. sodium restriction.
   d. cholesterol restriction.
10. Hemorrhagic stroke:
   a. is the most common type of stroke.
   b. results from obstructed blood flow within brain tissue.
   c. comes on suddenly and usually lasts for up to 30 minutes.
   d. results from bleeding within the brain, which damages brain tissue.

Answers to these questions appear in Appendix H.

CLINICAL APPLICATIONS

1. List the risk factors for CHD. Find possible interrelationships among the factors. For example, a woman over age 55 is also at risk for diabetes; a person with diabetes is more likely to have hypertension.
2. Review the DASH Eating Plan shown in Table 21-5. As the chapter described, the DASH dietary pattern is helpful for lowering blood pressure and for reducing CHD risk as well.
   • List elements of the DASH diet that are consistent with TLC dietary strategies.

• Suggest ways in which a person following the DASH diet might accomplish these additional dietary modifications:
   • Consume a higher percentage of fat from monounsaturated sources.
   • Reduce intake of *trans*-fatty acids.
   • Include EPA and DHA and soy protein.

NUTRITION ON THE NET

For further study of the topics in this chapter, access these websites.

Find updates and quick links to these and other nutrition-related sites at our website: **www.wadworth.com/nutrition**

To search for additional information about cardiovascular diseases, obtain the American Heart Association dietary guidelines, or find links to other relevant materials, visit the website of the American Heart Association: **www.americanheart.org**

Information about cardiovascular diseases, the DASH Eating Plan, and implementation of heart-healthy diets is available at the websites of the National Heart, Lung, and Blood Institute and the Heart and Stroke Foundation of Canada: **www.nhlbi.nih.gov** and **www.heartandstroke.ca**

To learn about improving health care and life expectancy of ethnic minority populations at high risk for cardiovascular diseases, visit the website of the International Society on Hypertension in Blacks: **www.ishib.org**

NOTES

[1]American Heart Association, *Heart Disease and Stroke Statistics—2005 Update* (2005), p. 4, **www.americanheart.org/downloadable/heart/1105390918119HDSStats2005Update.pdf**, site visited October 17, 2005.

[2]American Heart Association, *Heart Disease and Stroke Statistics—2005 Update,* 2005.

[3]American Heart Association, *International Cardiovascular Disease Statistics* (2005), p. 1, **www.americanheart.org/downloadable/heart/1107369401339FS06INTSrev0119.pdf**, site visited October 17, 2005.

[4]D. M. Davidson and coauthors, Children and adolescents, in J. Foody, ed., *Preventive Cardiology* (Totowa, N.J.: Humana Press, 2001), pp. 423–444.

5 A. P. Burke and coauthors, Healed plaque ruptures and sudden coronary death: Evidence that subclinical rupture has a role in plaque progression, *Circulation* 103 (2001): 934–940; P. Libby, Changing concepts of atherogenesis, *Journal of Internal Medicine* 247 (2000): 349–358; J. Willeit and coauthors, Distinct risk profiles of early and advanced atherosclerosis, *Arteriosclerosis, Thrombosis, and Vascular Biology* 20 (2000): 529–537.

6 T. A. Pearson and coauthors, Markers of inflammation and cardiovascular disease, *Circulation* 107 (2003): 499–511.

7 M. A. Lauer, Inflammation and infection in coronary artery disease, in J. Foody, ed., *Preventive Cardiology* (Totowa, N.J.: Humana Press, 2001).

8 R. B. Buchsbaum and J. C. Buchsbaum, Tobacco as a cardiovascular risk factor, in J. Foody, ed., *Preventive Cardiology* (Totowa, N.J.: Humana Press, 2001), pp. 178–181.

9 J. S. Cohn, C. Marcoux, and J. Davignon, Detection, quantification, and characterization of potentially atherogenic triglyceride-rich remnant lipoproteins, *Arteriosclerosis, Thrombosis, and Vascular Biology* 19 (1999): 2474–2486; S. C. Whitman and coauthors, Uptake of type III hypertriglyceridemic VLDL by macrophages is enhanced by oxidation, especially after remnant formation, *Arteriosclerosis, Thrombosis, and Vascular Biology* 17 (1997): 1707–1715.

10 I. Staprans and coauthors, Oxidized cholesterol in the diet is a source of oxidized lipoproteins in human serum, *Lipid Research* 44 (2003): 705–715; A. Mertens and P. Holvoet, Oxidized LDL and HDL: Antagonists in atherothrombosis, *FASEB Journal* 15 (2001): 2073–2084.

11 E. J. Schaefer, Lipoproteins, nutrition, and heart disease, *American Journal of Clinical Nutrition* 75 (2002): 191–212; Willeit and coauthors, 2000.

12 K. C. B. Tan and coauthors, Advanced glycation end products and endothelial dysfunction in type 2 diabetes, *Diabetes Care* 25 (2002): 1055–1059; A. Chait and J. D. Brunzell, Diabetes mellitus, lipids, and atherosclerosis, in D. LeRoith, S. I. Taylor, and J. M. Olefsky, eds., *Diabetes Mellitus: A Fundamental and Clinical Text*, 2nd ed. (Philadelphia: Lippincott Williams & Wilkins, 2000), pp. 934–939.

13 H. O'Grady and coauthors, Homocysteine and occlusive arterial disease, *British Journal of Surgery* 89 (2002): 838–844.

14 American Heart Association, *Heart Disease and Stroke Statistics—2005 Update*, 2005.

15 American Heart Association, *Heart Disease and Stroke Statistics—2005 Update*, 2005.

16 Expert Panel on Detection, Evaluation, and Treatment of High Blood Cholesterol in Adults (Adult Treatment Panel III), *Third Report of the National Cholesterol Education Program (NCEP)*, NIH publication no. 02-5215 (Bethesda, Md.: National Heart, Lung, and Blood Institute, 2002), **www.nhlbi.nih.gov/guidelines/cholesterol/atp3full.pdf**, site visited October 22, 2005.

17 U.S. Department of Health and Human Services and Food and Drug Administration, Food labeling: Health claims; soy protein and coronary heart disease: Final rule, *Federal Register* 64 (1999): 699–733.

18 K. J. Mukamai and E. B. Rimm, Alcohol's effects on the risk for coronary heart disease, *Alcohol Research and Health* 25 (2001): 255–261.

19 P. D. Thompson and coauthors, Exercise and physical activity in the prevention and treatment of atherosclerotic cardiovascular disease, *Circulation* 107 (2003): 3109–3116.

20 G. G. Blackburn, Exercise in the prevention of coronary artery disease, in J. Foody, ed., *Preventive Cardiology* (Totowa, N.J.: Humana Press, 2001), pp. 144–155.

21 Buchsbaum and Buchsbaum, 2001.

22 S. Nader and K. Robinson, A recognized risk factor: Homocysteine and coronary artery disease, in J. Foody, ed., *Preventive Cardiology* (Totowa, N.J.: Humana Press, 2001), pp. 223–226.

23 C. D. Morris and S. Carson, Routine vitamin supplementation to prevent cardiovascular disease: A summary of the evidence for the U.S. preventive services task force, *Annals of Internal Medicine* 139 (2003): 56–57.

24 E. R. Miller and coauthors, Meta-analysis: High-dosage vitamin E supplementation may increase all-cause mortality, *Annals of Internal Medicine* 142 (2005): 37–46; D. H. Lee and coauthors, Does supplemental vitamin C increase cardiovascular disease risk in women with diabetes? *American Journal of Clinical Nutrition* 80 (2004): 1194–1200.

25 L. E. Fields and coauthors, The burden of adult hypertension in the United States 1999 to 2000: A rising tide, *Hypertension* 44 (2004): 398–404.

26 American Heart Association, *Heart Disease and Stroke Statistics—2005 Update*, 2005.

27 R. Victor, Arterial hypertension, in L. Goldman and D. Ausiello, eds., *Cecil Textbook of Medicine* (Philadelphia: Saunders, 2004), pp. 346–363.

28 W. F. Ganong, Cardiovascular disorders: Vascular disease, in S. J. McPhee and coeditors, *Pathophysiology of Disease*, 2nd ed. (Stamford, Conn.: Appleton & Lange, 1997), p. 271.

29 American Heart Association, *Heart Disease and Stroke Statistics—2005 Update*, 2005.

30 T. A. Kotchen and J. M. Kotchen, Nutrition, diet, and hypertension, in M. E. Shils and coeditors, *Modern Nutrition in Health and Disease*, 10th ed. (Philadelphia: Lippincott Williams & Wilkins, 2006), pp. 1095–1107.

31 M. R. Wofford and J. E. Hall, Pathophysiology and treatment of obesity hypertension, *Current Pharmaceutical Design* 10 (2004): 3621–3637.

32 Kotchen and Kotchen, 2006.

33 X. Xin and coauthors, Effects of alcohol reduction on blood pressure: A meta-analysis of randomized controlled trials, *Hypertension* 38 (2001): 1112.

34 F. M. Sacks and coauthors, Effects on blood pressure of reduced dietary sodium and the Dietary Approaches to Stop Hypertension (DASH) Diet, *New England Journal of Medicine* 344 (2001): 3–10; L. J. Appel and coauthors, A clinical trial on the effects of dietary patterns on blood pressure, *New England Journal of Medicine* 336 (1997): 1117–1124.

35 U.S. Department of Agriculture and U.S. Department of Health and Human Services, *Dietary Guidelines for Americans, 2005* (Washington, D.C.: Government Printing Office, January 2005).

36 Sacks and coauthors, 2001.

37 Obesity Education Initiative, *Clinical Guidelines on the Identification, Evaluation, and Treatment of Overweight and Obesity in Adults: The Evidence Report*, NIH publication no. 98-4083 (Bethesda, Md.: National Heart, Lung, and Blood Institute, 1998).

38 B. M. Massie, Heart failure: Pathophysiology and diagnosis, in L. Goldman and D. Ausiello, eds., *Cecil Textbook of Medicine* (Philadelphia: Saunders, 2004), pp. 291–299.

39 S. A. Hunt and coauthors, ACC/AHA Guidelines for the Evaluation and Management of Chronic Heart Failure in the Adult: Executive Summary, *Journal of the American College of Cardiology* 38 (2001): 2101–2113.

40 S. Lutton and N. Anzlovar, Nutrition and congestive heart failure, in A. M. Coulston, C. L. Rock, and E. R. Monsen, *Nutrition in the Prevention and Treatment of Disease* (San Diego: Academic Press, 2001), pp. 325–333.

41 J. A. Zivin, Ischemic cerebrovascular disease, in L. Goldman and D. Ausiello, eds., *Cecil Textbook of Medicine* (Philadelphia: Saunders, 2004), pp. 2287–2298.

Helping People with Feeding Disabilities

Chapter 21 referred to difficulties following a stroke that can interfere with the ability to eat independently. This Nutrition in Practice discusses a broader problem faced by individuals who must cope with disabilities that interfere with the process of eating, such as those that interfere with chewing and swallowing. These obstacles can arise at any time during a person's life and from any number of causes. An infant may be born with a physical impairment such as cleft palate; an adolescent may lose motor control following injuries sustained in an automobile accident; an older adult may struggle with the pain of arthritis or the mental deterioration of dementia. Table NP21-1 lists some of the conditions that may lead to feeding problems.

In what ways can disabilities impair a person's ability to eat?

Eating and drinking require a substantial number of individual coordinated motions. Consider an infant learning the skills required for feeding: each step—sitting, grasping cups and utensils, bringing food to the mouth, biting, chewing, and swallowing—requires coordinated movements. An injury or disability that interferes with any of these movements can lead to feeding problems and inadequate food intake. Total food intake has been found to be significantly reduced when people with inefficient motor functions take a prolonged time to eat.[1] Difficulties that affect procurement of food, such as the inability to drive or walk or carry groceries, can also lower food intake and lead to malnutrition and weight loss.

How may disease symptoms or medications alter food intake?

Examples of disease symptoms that can alter food intake include nausea, frequent coughing or choking, difficulty breathing, and gastroesophageal reflux. Individuals with speech and hearing problems may have a difficult time communicating with caregivers about thirst and hunger. Mobility problems can influence nutrition status by causing bone demineralization and pressure sores.

Conditions that require the use of multiple medications can also have a significant impact on nutrition status.[2] Medications may increase or decrease appetite, interfere with nutrient metabolism, or have gastrointestinal effects that cause pain or discomfort with eating.

Can disabilities alter a person's energy needs?

Yes, certain disabilities can either increase or decrease energy requirements. Disabilities that affect muscle tension and mobility can reduce physical activity and, consequently, energy requirements. Other disabilities, such as certain forms of cerebral palsy, cause involuntary muscle activity that raises energy requirements.[3] Loss of a limb due to amputation reduces energy needs in proportion to the weight and metabolism represented by the missing limb, but can result in greater energy needs if an individual increases activity to compensate for the loss, such as by propelling a wheelchair. Because the effects of disabilities are often unpredictable, the health care practitioner may find it difficult to assess energy requirements until weight gain or loss has occurred.

Overweight and obesity often accompany conditions that limit mobility or result in short stature; examples include Down syndrome and spina bifida. Obesity may also develop because the family or caregiver provides an inappropriate amount of food, sometimes out of sympathy for the disabled individual.[4] In these cases, the health care provider may need to counsel the family or caregiver about appropriate food choices and portion sizes.

Which health professionals typically work with people who have feeding problems?

Evaluating and treating feeding problems often involve the joint efforts of health care professionals from a variety of disciplines, including nurses, dietitians, occupational and physical therapists, speech-language pathologists, and dentists. Together, these professionals evaluate each patient's dietary needs and assess abilities to chew, sip, swallow, grasp utensils, use utensils to pick up foods, and bring foods from the plate to the mouth. A speech-language pathologist most often evaluates chewing and swallowing abilities and trains patients to use lips, tongue, and throat for eating and speaking. An occupational therapist can demonstrate alternative feeding strategies, including changes in body position that improve feeding, techniques for handling utensils and food, and use of special feeding devices.

Direct observation of a patient during mealtimes allows health professionals to assess current eating behaviors,

TABLE NP21-1 Conditions That May Lead to Feeding Problems

The following conditions may lead to feeding problems by interfering with a person's ability to suck, bite, chew, swallow, or coordinate hand-to-mouth movements.

Accidents	Language, visual, or hearing impairment
Amputations	Microcephalia
Arthritis	Multiple sclerosis
Birth defects	Muscle weakness
Cerebral palsy	Muscular dystrophy
Cleft palate	Neuromotor dysfunction
Down syndrome	Parkinson's disease
Head injuries	Polio
Huntington's chorea	Spinal cord injuries
Hydrocephalia	Stroke

demonstrate feeding techniques, monitor the patient's and caregiver's understanding of the techniques, and evaluate how well the care plan is working. To illustrate, consider an example of a child with a feeding problem caused by hypersensitivity to oral stimulation. The health care professional may start by teaching the caregiver to gently and playfully stroke the child's face with a hand, wash cloth, or soft toy. Once the child tolerates touch on less sensitive areas of the face, the health care professional may encourage the caregiver to slowly begin to rub the child's lips, gums, palate, and tongue. With time, the child may be better able to tolerate the presence of food in the mouth. Examples of other strategies that can help feeding problems are listed in Table NP21-2.

Can special equipment be used to help people with certain feeding difficulties?

Yes. Figure NP21-1 shows a few of the many special feeding devices that are available and describes their uses. These devices can make a remarkable difference in a person's ability to eat independently. Other examples of adaptive equipment include specialized chairs to improve posture, bolsters inserted under arms to improve elbow stability, and raised trays or eating surfaces to simplify hand-to-mouth movements.[5]

Sometimes, despite the best efforts of all involved, a patient is unable to consume enough food by mouth. In these cases, tube feedings can help to improve nutrition status. Tube feedings are also recommended for patients who have severe dysphagia (difficulty swallowing) or aspiration pneumonia.[6]

In what ways can feeding difficulties affect family life?

Mealtimes are a critical time for social interaction, and individuals with feeding problems may encounter emotional and social problems if unable to participate. Children may fail to develop social skills, whereas adults may miss the social stimulation that mealtimes provide. Individuals should be encouraged to sit with family and friends during meals so that they are not deprived of the social and cultural aspects of eating.

The responsibility of caring for a person with a feeding problem can frequently overwhelm the caregiver. Caring for a disabled person requires time and patience—often many new therapies must be learned and administered. The caregiver may spend many hours preparing special foods, monitoring use of adaptive feeding equipment, and helping with feedings. Moreover, a disabled person may need help with other tasks as well, and all may require a considerable amount of time. In many cases, a caregiver may receive little or no assistance. These conditions may lead to strained interactions between caregiver and patient and cause frustration and depression.[7] Psychologists can offer counseling to patients or caregivers to help them adjust, and all members of the health care team can offer emotional support and practical suggestions to ease caregivers' responsibilities and frustrations.

TABLE NP21-2 Interventions for Feeding-Related Problems

Inability to suck

- Use squeeze bottles, which do not require sucking, to express liquids into the mouth.
- Place a spoon on the center of the tongue and apply downward pressure to stimulate sucking.
- Apply rhythmic, slow strokes on the tongue to alter tongue position and improve sucking response.

Inability to chew

- Place foods between gums and teeth to promote chewing.
- Improve chewing skills with different textured foods; for example, fruit leathers stimulate jaw movements but dissolve quickly enough to minimize choking.
- Provide soft foods that require minimal chewing or are easily chewed.

Inability to swallow

- Provide thickened liquids, pureed foods, and moist foods that form boluses easily.
- Provide cold formulas, frozen fruit juice bars, and ice; cold substances promote swallowing movements by the tongue and soft palate.
- Make sure the patient's jaw and lips are closed to facilitate swallowing action.
- Correct posture and head position if they interfere with swallowing ability.

Inability to grasp or coordinate movements

- Provide utensils that have modified handles, or are smaller or larger as necessary.
- Encourage use of hands for feeding if utensils are difficult to maneuver.
- Provide plates with food guards to prevent spilling.
- Supply clothing protection.

Impaired vision

- Place foods (meats, vegetables) in similar locations on the plate at meals.
- Provide plates with food guards to prevent spilling.

Sources: J. Case-Smith and R. Humphry, Feeding and oral motor skills, in J. Case-Smith, A. S. Allen, and P. N. Pratt, eds., *Occupational Therapy for Children* (St. Louis: Mosby–Year Book, 1996), pp. 430–460; S. Escott-Stump, *Nutrition and Diagnosis-Related Care* (Baltimore: Lippincott Williams & Wilkins, 2002), pp. 64–65.

Successful therapy for people with feeding disabilities requires the involvement of many health care professionals and depends on accurate identification of impaired feeding skills and determination of appropriate interventions. Ideally, with training, people with disabilities attain total independence—they are able to prepare, serve, and eat nutritionally adequate food daily without help. In some cases, these goals can be met with

FIGURE NP21-1 Examples of Adaptive Feeding Devices

Utensils

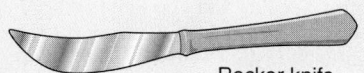

Rocker knife

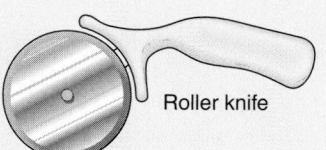

Roller knife

People with only one arm or hand may have difficulty cutting foods and may appreciate using a *rocker knife* or a *roller knife*.

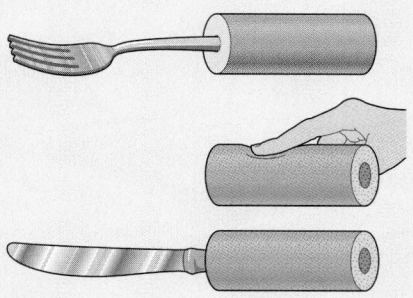

People with a limited range of motion can feed themselves better when they use *flatware with built-up handles*.

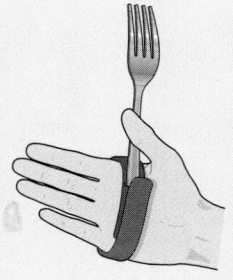

People with extreme muscle weakness may be able to eat with a *utensil holder*.

For people with tremors, spasticity, and uneven jerky movements, *weighted utensils* can aid the feeding process.

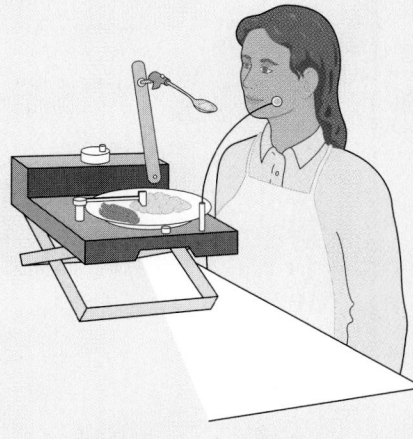
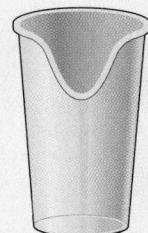

Battery-powered feeding machines enable people with severe limitations to eat with less assistance from others.

Plates

People who have limited dexterity and difficulty maneuvering food find *scoop dishes* or *food guards* useful.

People with uncontrolled or excessive movements might move dishes around while eating and may benefit from using *unbreakable dishes with suction cups*.

Cups

People with limited neck motion can use a *cutout plastic cup*.

Two-handed cups enable people with moderate muscle weakness to lift a cup with two hands.

People with uncontrolled or excessive movements might prefer to drink liquids from a *covered cup* or glass with a *slotted opening* or *spout*.

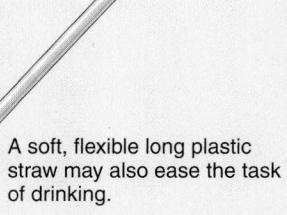

A soft, flexible long plastic straw may also ease the task of drinking.

the help of caregivers. The combined efforts of the health care team can support both patients and caregivers in enhancing quality of life and in achieving independence to the greatest degree possible.

Notes

[1] E. B. Fung and coauthors, Feeding dysfunction is associated with poor growth and health status in children with cerebral palsy, *Journal of the American Dietetic Association* 102 (2002): 361–368.

[2] Position of the American Dietetic Association: Providing nutrition services for infants, children, and adults with developmental disabilities and special health care needs, *Journal of the American Dietetic Association* 104 (2004): 97–107.

[3] Position of the American Dietetic Association, 2004.

[4] H. H. Cloud, Expanding roles for dietitians working with persons with developmental disabilities, *Journal of the American Dietetic Association* 97 (1997): 129–130.

[5] J. Case-Smith and R. Humphry, Feeding and oral motor skills, in J. Case-Smith, A. S. Allen, and P. N. Pratt, eds., *Occupational Therapy for Children* (St. Louis: Mosby–Year Book, 1996), pp. 430–460.

[6] Position of the American Dietetic Association, 2004.

[7] Fung and coauthors, 2002.

Nutrition and Renal Diseases

© age fotostock/SuperStock

CHAPTER 22

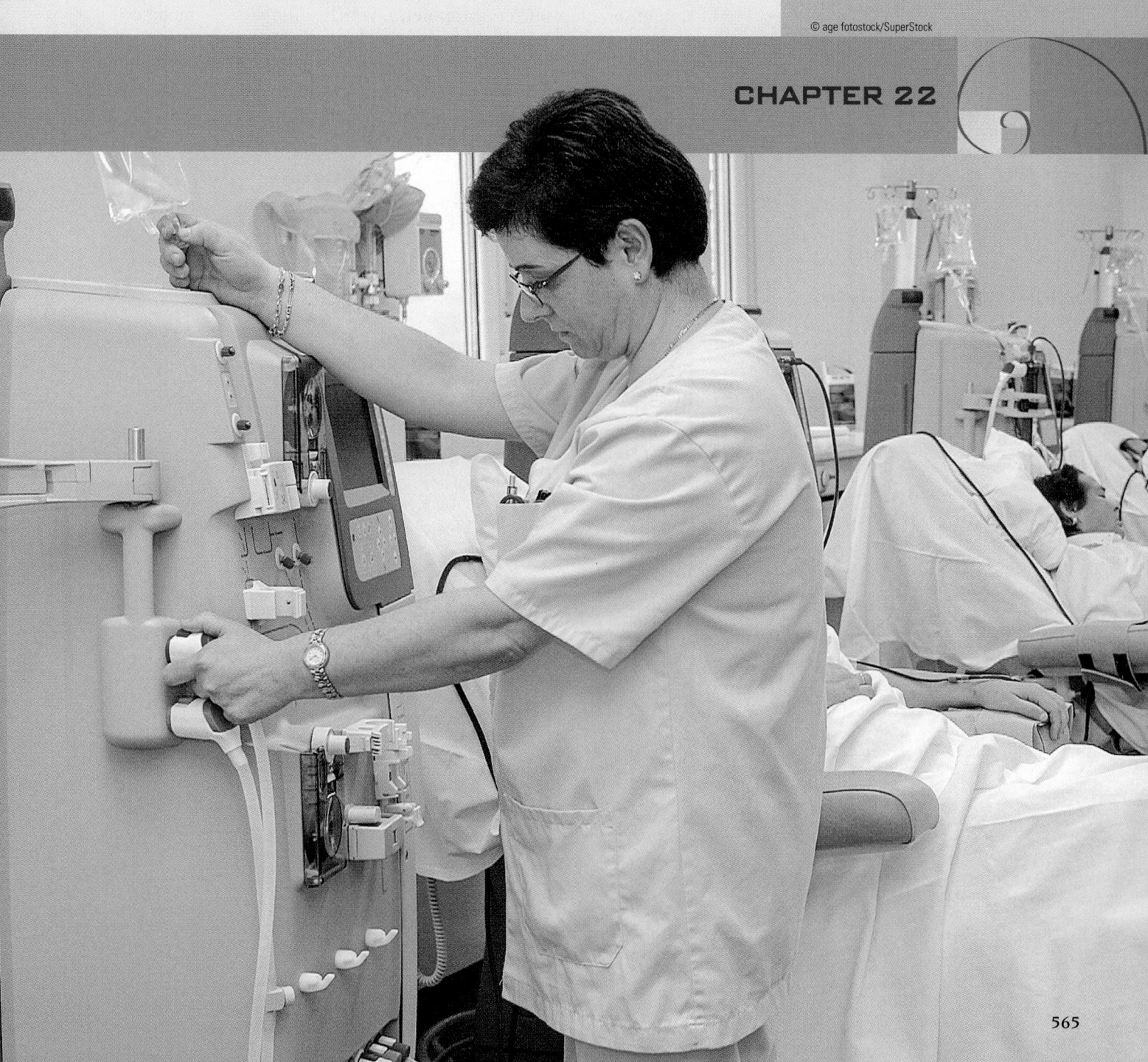

nephron (NEF-ron): the functional unit of the kidneys, consisting of a glomerulus and tubules.

glomerulus (gloh-MEHR-yoo-lus): a tuft of capillaries within the nephron that filters water and solutes from blood as urine production begins (plural: *glomeruli*).

tubules: tubelike structures of the nephron that process filtrate during urine production. The tubules are surrounded by capillaries that reabsorb substances retained by tubule cells.

erythropoietin (eh-RITH-ro-POY-eh-tin): a hormone made by the kidneys that stimulates red blood cell production.

nephrotic (neh-FROT-ik) **syndrome:** a kidney disorder characterized by urinary protein losses exceeding 3.5 g per day. Accompanying symptoms often include low serum albumin, elevated blood lipids, and edema.

proteinuria (PRO-teen-NEW-ree-ah): loss of protein, especially albumin, in the urine; also known as *albuminuria*.

The kidneys sit just above the waist on each side of the spinal column. As part of the urinary system, they are responsible for filtering blood and removing excess fluid and wastes for elimination in urine. Figure 22-1 shows the kidneys' placement and structure and one of their functional units, the **nephron.** Within each nephron, the **glomerulus,** a ball-shaped tuft of capillaries, functions like a sieve: it retains blood cells and plasma proteins in blood while allowing fluid and small solutes to enter the nephron's system of **tubules.** As the material passes through the tubules, some components are reabsorbed and returned to the body, while the remaining substances form the final urine product. In this manner, the kidneys regulate the extracellular fluid's volume and osmolarity, electrolyte concentrations, and acid-base balance. They also excrete metabolic waste products like urea and creatinine, as well as various drugs and toxins. Other roles of the kidneys include the following:

- Secretion of the enzyme renin, which helps to regulate blood pressure.
- Production of the hormone **erythropoietin,** which stimulates red blood cell production.
- Conversion of vitamin D to its active form, thereby helping to maintain bone tissue.

As this chapter explains, renal diseases that interfere with the kidneys' various functions can severely disrupt health.

The Nephrotic Syndrome

The **nephrotic syndrome** is not a specific disease; rather, the term refers to any kidney disorder that results in urinary protein losses **(proteinuria)** exceeding 3.5 grams

FIGURE 22-1 The Kidneys and Nephron Function

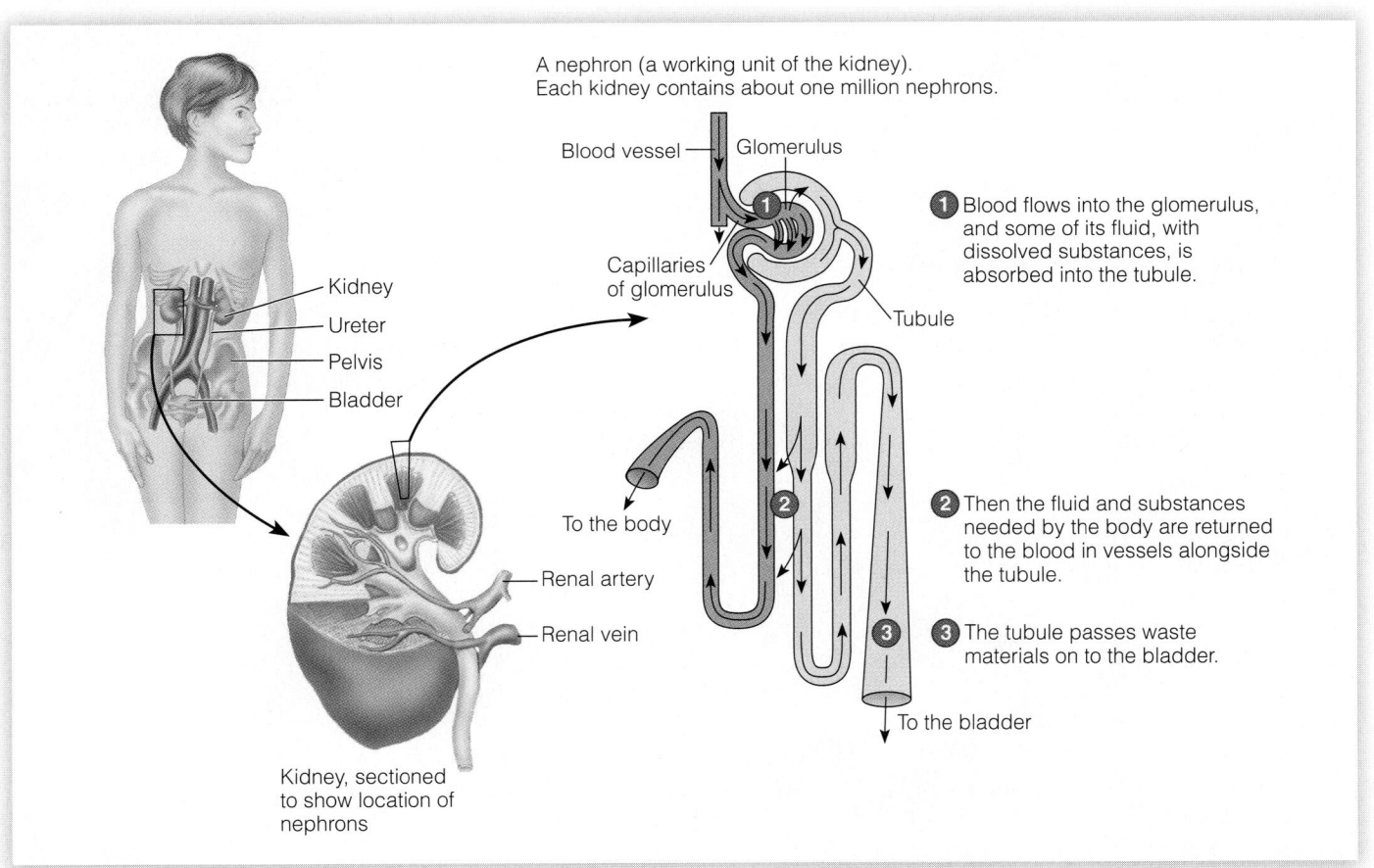

per day. The nephrotic syndrome arises when damage to the glomeruli increases their permeability to plasma proteins, allowing protein to escape into the urine. Possible causes include infections, chemical damage, immunological and hereditary disorders, diabetes mellitus, and various disorders that directly involve the glomeruli. Along with proteinuria, clinical findings often include low serum albumin levels, edema, elevated blood lipids, and blood coagulation disorders. The nephrotic syndrome can sometimes progress to renal failure.

CONSEQUENCES OF THE NEPHROTIC SYNDROME

In the nephrotic syndrome, urinary protein losses average about 8 grams daily, but additional protein is lost due to protein catabolism within the kidney tubules.[1] The liver attempts to compensate for these losses by increasing its synthesis of some plasma proteins, but this results in higher concentrations of some proteins and lower levels of others. Consequently, the complications of the nephrotic syndrome are exacerbated by disturbances in protein metabolism.[2]

Edema Urinary losses of albumin, the most abundant plasma protein, contribute to the fluid shift from blood plasma to the interstitial spaces, causing edema. In addition, disordered kidney function leads to sodium retention, causing additional fluid retention within the body.

Reminder: Plasma proteins, such as albumin, help to maintain fluid balance within the blood.

Risk of Cardiovascular Disease People with the nephrotic syndrome frequently have elevated levels of low-density lipoproteins (LDL), very-low-density lipoproteins (VLDL), and the more damaging LDL variant known as lipoprotein(a). Furthermore, risk of blood clotting is higher due to urinary losses of proteins that inhibit blood clotting and elevated levels of plasma proteins that favor clotting. These factors raise the risks of developing heart disease and stroke.

Other Effects of the Nephrotic Syndrome The proteins lost in urine include antibodies and vitamin D–binding protein. Depletion of antibodies increases susceptibility to infection. Loss of vitamin D–binding protein results in lower vitamin D and calcium levels and increases the risk of rickets in children. If proteinuria persists, protein-energy malnutrition (PEM) and muscle wasting may develop. Figure 22-2 summarizes the effects of urinary protein losses in the nephrotic syndrome.

NURSING DIAGNOSIS

excess fluid volume, risk for infection, and *imbalanced nutrition: less than body requirements* may apply to people with the nephrotic syndrome.

TREATMENT OF THE NEPHROTIC SYNDROME

Treatment goals are to relieve symptoms and prevent kidney damage. Any underlying disorder, if diagnosed, should be corrected. The drugs most often prescribed include anti-inflammatory drugs (usually corticosteroids), ACE inhibitors (which reduce protein losses), diuretics, antihypertensives, immunosuppressants, and lipid-lowering medications. Medical nutrition therapy helps to prevent PEM and alleviate edema.

Protein and Energy Meeting protein and energy needs helps to minimize losses of muscle tissue. High-protein diets are not advised, however, because they can exacerbate urinary protein losses.[3] Instead, protein intake should fall between 0.8 and 1.0 gram per kilogram body weight per day. An adequate energy intake (about 35 kcalories per kilogram body weight daily) sustains weight and spares protein. Weight loss or infections signal the need for additional kcalories.

Fat As Chapter 21 explained, a diet low in saturated fat, cholesterol, and refined sugars helps to control elevated blood lipids. People with the nephrotic syndrome are often unable to control blood lipids by diet alone, however, and physicians may need to prescribe lipid-lowering medications.

FIGURE 22-2 Consequences of Urinary Protein Losses in the Nephrotic Syndrome

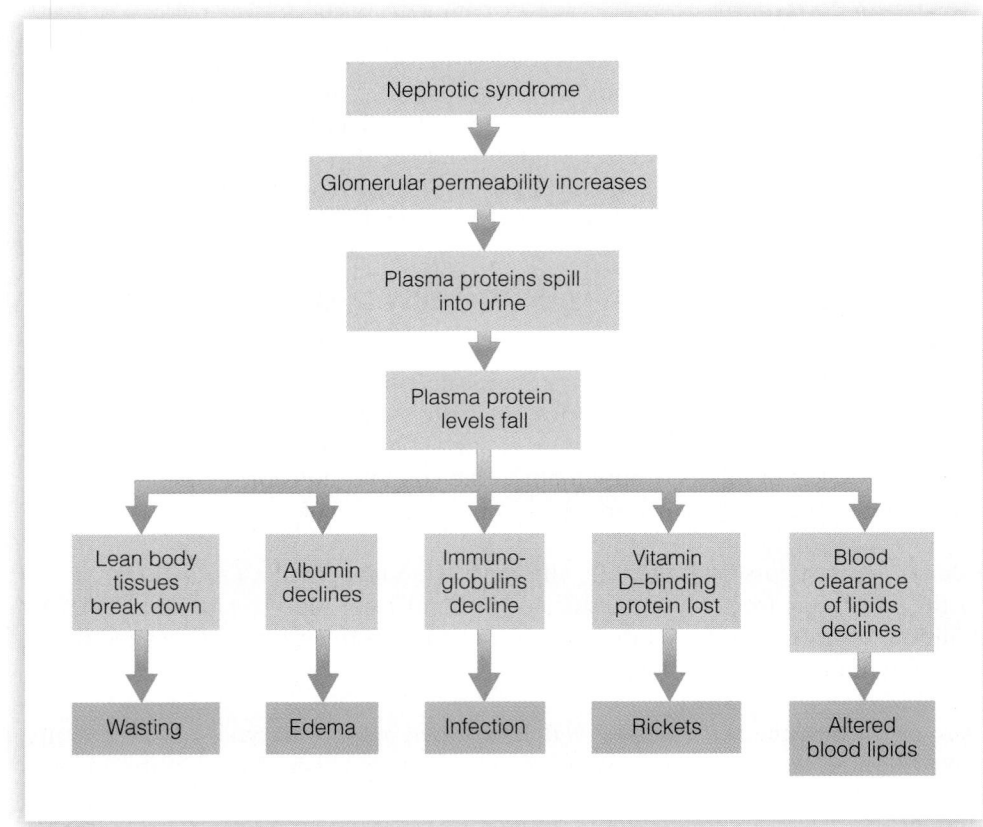

Sodium Sodium restriction helps to control edema; an intake of no more than 2 to 3 grams daily is often suggested. Table 22-1 provides guidelines for following a diet restricted to 2 grams of sodium. If diuretics prescribed for edema cause potassium wasting, patients are encouraged to select foods rich in potassium.

Nutrient deficiencies may develop if carrier proteins for nutrients are lost in urine.

Vitamins and Minerals Patients with the nephrotic syndrome may require vitamin D and calcium supplementation to help prevent bone loss and rickets. Multivitamin supplementation is often advised to help avoid additional nutrient deficiencies.

> **REVIEW NOTES**
>
> The nephrotic syndrome is characterized by urinary protein losses exceeding 3.5 grams per day. Complications include edema, lipid abnormalities, blood coagulation disorders, reduced immunity, rickets, and PEM.
>
> The diet should provide sufficient protein and energy to maintain health, but excess protein should be avoided. Other dietary adjustments may be needed to correct edema, lipid disorders, and nutrient deficiencies.

Acute Renal Failure

In **acute renal failure,** kidney function deteriorates rapidly, over hours or days. The loss of kidney function reduces urine output and allows nitrogenous wastes to build up in blood. The degree of renal dysfunction varies from mild to severe. With prompt treatment, acute renal failure is often reversible, although mortality rates are high, ranging from 35 to 65 percent.[4]

CAUSES OF ACUTE RENAL FAILURE

Many different disorders can lead to acute renal failure, and it often develops as a consequence of severe illness, injury, or surgery. To aid in diagnosis and treatment,

acute renal failure: abrupt loss of kidney function over a period of hours or days.

TABLE 22-1	Sodium-Restricted Diet

General Guidelines

About 75% of sodium in a typical diet comes from processed foods, about 10% from unprocessed natural foods, and about 15% from table salt. With this in mind:

- Choose fresh foods and foods frozen or canned without added salt.
- Avoid adding salt to foods while cooking.
- Avoid adding salt to foods at the table.
- When eating out, ask that meals be prepared without salt.

Sodium in Foods

All foods contain sodium, but some contain more than others. Use the information about the average sodium contents of foods to tailor the diet to the individual's preferences.

Food Group	Serving Size	Sodium (mg) per Serving
Fresh meats, poultry, freshwater fish; low-sodium canned meats and fish; low-sodium peanut butter and cheese; unsalted cottage cheese, soybeans, and textured vegetable protein	1 oz	20
Regular fat-free, low-fat, and whole milk and yogurt	8 oz	120
Eggs	1	60
Fresh artichokes, beets, carrots, and celery; beet, collard, dandelion, mustard, and turnip greens	1/2 c	50
Regular canned vegetables	1/2 c	300
Regular white and whole-grain bread	1 slice	150
Butter and margarine	1 tsp	50
Salt	1/2 tsp	1000

Other Foods

These foods can be used freely or with some limits with respect to sodium, although some of these foods may need to be restricted for weight or blood lipid control:

- All fruits and fruit juices.
- Low-sodium canned or frozen vegetables without added salt except those listed above; low-sodium vegetable juices.
- Low-sodium bread and bread products; puffed rice and wheat and shredded wheat cereals; rice; pasta.
- Soups, casseroles, and recipes made with allowed foods and ingredients.
- Unsalted butter, margarine, nuts, and gravy; low-sodium mayonnaise and salad dressing; shortening.
- Low-sodium catsup, mustard, tabasco sauce, and other condiments; low-sodium baking powder.

These foods and dishes prepared with them are high in sodium and should be avoided:

- Cured, canned, salted, or smoked meats, poultry, and fish such as bacon, luncheon meats, corned beef, kosher meats, and canned tuna and salmon; imitation fish products, salted textured vegetable protein, peanut butter, and nuts.
- Buttermilk, regular cheeses.
- Maraschino cherries; crystallized or glazed fruits; dried fruits with sodium sulfite added.
- Pickles, pickled vegetables, sauerkraut, and regular vegetable juices.
- Instant and quick-cooking hot cereals; commercial bread products made from self-rising flour or cornmeal; salted snack foods.
- Salt pork and bacon; commercial salad dressing; olives; regular gravy; catsup, baking powder, soy sauce, bouillon.

A Sample Diet Restricted to Two Grams Sodium

Using the information above, many diet plans that meet individual needs are possible. Using the guidelines for a heart-healthy diet, a typical plan for a day might look like this:

Food Group	Sodium (mg)
Meat, 6 oz (6 × 20 mg)	120
Milk, 3 c (3 × 120 mg)	360
Fruit, 3 servings	negligible
Vegetables, 1/2 c vegetables with some sodium (1 × 50)	50
Vegetables, other vegetables and legumes, 2 servings	negligible
Whole-grain bread, 4 slices (4 × 150 mg)	600
Salted margarine, 6 servings (6 × 50 mg)	300
Total	1430

Individuals can use the remainder of the sodium allowance for whatever foods they choose. The sodium content of other foods can be determined by reading food labels or using food composition tables. An individual may choose to use some (1/4 teaspoon) table salt or a favorite food that contains sodium.

TABLE 22-2	Causes of Acute Renal Failure	
Prerenal Factors (60 to 70% of cases)	**Intrarenal Factors** (25 to 40% of cases)	**Postrenal Factors** (5 to 10% of cases)
• **Low blood volume or pressure:** Hemorrhage, burns, sepsis or shock, anaphylactic reactions, nephrotic syndrome, gastrointestinal losses, diuretics, antihypertensive medications • **Renal artery disorders:** Blood clots or emboli, stenosis, aneurysm, trauma • **Heart disorders:** Congestive heart failure, heart attack, arrhythmias	• **Vascular disorders:** Sickle-cell disease, diabetes mellitus, transfusion reactions • **Obstructions (within kidney):** Inflammation, tumors, stones, scar tissue • **Renal injury:** Infections, environmental contaminants, drugs, medications, *E. coli* food poisoning	• **Obstructions (ureter or bladder):** Strictures, tumors, stones, trauma • **Prostate disorders:** Cancer or hyperplasia • **Renal vein thrombosis** • **Bladder disorders:** Neurological conditions, bladder rupture • **Pregnancy**

NURSING DIAGNOSIS

for people with acute renal failure include *excess fluid volume, imbalanced nutrition: less than body requirements, impaired urinary elimination, risk for infection, risk for injury, decreased cardiac output, fatigue,* and *risk for impaired oral mucous membrane.*

To measure serum phosphate, the phosphorus content of the blood is analyzed; thus the terms *serum phosphate* and *serum phosphorus* are often used interchangeably.

Normal urine volume exceeds 800 mL per day, which is equivalent to about 27 fluid ounces or 3.4 cups.

A progressive rise in BUN or creatinine suggests the presence of acute renal failure. Refer to Table 13-7 on p. 367 for normal laboratory values.

oliguria (OL-lih-GOO-ree-ah): an abnormally low amount of urine, often less than 400 mL per day.

hyperkalemia (HIGH-per-ka-LEE-me-ah): elevated serum potassium levels.

hyperphosphatemia (HIGH-per-fos-fa-TEE-me-ah): elevated serum phosphate levels.

uremia (you-REE-me-ah): abnormal accumulation of nitrogen-containing substances, especially urea, in the blood; also called **azotemia** (AZE-oh-TEE-me-ah).

its causes are commonly classified as prerenal, intrarenal, or postrenal.[5] Factors classified as *prerenal* cause a sudden reduction in blood flow to the kidneys and often involve a severe stressor such as heart failure, shock, or blood loss. Factors that damage kidney tissue, such as infections, toxins, drugs, or direct trauma, are classified as *intrarenal* causes of failure. Factors that prevent excretion of urine due to urinary tract obstructions are classified as *postrenal*. Table 22-2 provides examples of specific disorders that may cause acute renal failure.

CONSEQUENCES OF ACUTE RENAL FAILURE

A decline in renal function alters the composition of blood. The kidneys may become unable to produce urine or regulate the levels of electrolytes, acid, and nitrogenous wastes in blood. Diagnosis is often a complex task, however, because the clinical effects can be subtle and vary according to the underlying cause of disease.

Fluid and Electrolyte Imbalances About half of patients experience **oliguria**, producing less than 400 milliliters of urine per day.[6] This impairs the excretion of both water and electrolytes, resulting in sodium retention and elevated levels of potassium, phosphate, and magnesium in the blood. Elevated potassium **(hyperkalemia)** is of particular concern because potassium imbalances can alter heart rate and lead to heart failure. Elevated serum phosphate levels **(hyperphosphatemia)** promote excessive secretion of parathyroid hormone and a reduction in blood calcium levels. Due to the sodium retention and reduced urine production, edema is a common symptom of acute renal failure and may be apparent as puffiness in the face and hands and swelling of the feet and ankles.

Uremia As a result of impaired kidney function, nitrogen-containing waste products—blood urea nitrogen (BUN), creatinine, and uric acid—accumulate in the blood. Furthermore, the tissue catabolism that accompanies illness generates more nitrogenous wastes than usual. The clinical outcome, called **uremia**, includes symptoms such as fatigue, lethargy, confusion, headache, anorexia, a metallic taste in the mouth, nausea and vomiting, and diarrhea. In more serious cases, elevated blood pressure, rapid heartbeat, seizures, and delirium or coma may occur. It is sometimes difficult to distinguish the symptoms of uremia from those of an underlying illness.[7]

TREATMENT OF ACUTE RENAL FAILURE

Treatment of acute renal failure involves a combination of drug therapy, **dialysis**, and medical nutrition therapy to restore fluid and electrolyte balances and minimize blood concentrations of toxic waste products. Correcting the underlying illness is necessary to prevent further damage to the kidneys.

In oliguric patients (those unable to produce urine), recovery from renal failure sometimes begins with a period of **diuresis**, in which large amounts of fluid are excreted. Because tubular function is minimal at this stage, electrolytes may not be sufficiently reabsorbed; consequently, both fluid depletion and electrolyte imbalances become a concern. Patients with this pattern of recovery (generally those whose renal failure is due to tubular injury) require close monitoring in case they require fluid and electrolyte replacement.

Drug Therapy The degree of kidney function affects the dosages of drugs administered because many drugs are eliminated in the urine. Edema is treated with diuretics to mobilize fluids; furosemide (Lasix) is the usual choice. Patients with hyperkalemia are given potassium-exchange resins that bind potassium ions in the gastrointestinal (GI) tract, ensuring potassium excretion in the stool. Rapid correction of hyperkalemia requires the use of insulin, which causes a temporary shift of extracellular potassium into the cells. (Glucose must be supplied along with insulin to prevent hypoglycemia.) If acidosis is present, bicarbonate may be administered orally or intravenously.[8]

Protein and Energy Although its effects are highly variable, acute renal failure is typically a catabolic condition associated with hypermetabolism and muscle wasting. Thus sufficient protein and energy must be ingested to preserve muscle mass. Initially, the patient can be provided with 35 kcalories per kilogram body weight per day, while body weight is monitored to ensure that energy intake is adequate. If available, indirect calorimetry provides the best estimate of energy needs.

Protein contributes nitrogen, increasing the kidneys' workload, but intake should be sufficient to prevent negative nitrogen balance and additional wasting. Protein recommendations are influenced by kidney function, the degree of catabolism, and the use of dialysis (dialysis removes nitrogenous wastes). Protein restriction to about 0.6 to 0.8 gram per kilogram body weight per day may be necessary for patients with limited kidney function who are not treated with dialysis. Higher intakes (about 1.2 to 1.3 grams per kilogram daily) may be recommended if kidney function improves or the treatment includes dialysis. Patients who are catabolic or septic may need additional protein but require dialysis to accommodate the additional nitrogen load.[9]

Fluids Health practitioners can assess fluid status by monitoring weight fluctuations, blood pressure, pulse rates, and appearance of the skin and mucous membranes. Another method is to measure serum sodium concentrations: a low level of sodium often indicates excessive fluid intake, and a high level suggests inadequate intake.

Fluid balance must be restored in patients who are either overhydrated or dehydrated. Thereafter, fluid needs can be estimated by measuring urine output and adding about 500 milliliters to account for the water lost from skin, lungs, and perspiration. An individual with fever, vomiting, or diarrhea requires additional fluid. Patients undergoing dialysis can ingest fluids more freely: 1.5 to 2.0 liters per day may be permitted, depending on hydration state.

Electrolytes Serum electrolyte levels are monitored closely to determine appropriate electrolyte intakes. Generally, potassium and phosphorus need to be restricted. Sodium restriction may be necessary to prevent fluid retention and hypertension: patients with oliguria may need to limit sodium intakes to 2 to 3 grams daily,

Nutrition in Practice 22 describes common dialysis procedures, including continuous renal replacement therapy, which is the approach usually used for treating acute renal failure.

dialysis (dye-AH-lih-sis): a procedure for removing wastes and excess fluid from the blood after the kidneys have stopped functioning. The two main types are *hemodialysis* and *peritoneal dialysis* (see Nutrition in Practice 22).

diuresis (DYE-uh-REE-sis): increased urine production.

CASE STUDY   *Store Manager with Acute Renal Failure*

Catherine Garber is a 42-year-old store manager admitted to the hospital's intensive care unit. She was first seen in the emergency room with severe edema, headache, nausea and vomiting, and a rapid heart rate. She reported an inability to pass more than minimal amounts of urine in the past two days. Her son, who drove her to the emergency room, reported that she had missed work for several days and seemed confused and unusually tired. Laboratory tests revealed elevated serum creatinine, BUN, and potassium levels. After learning from her medical history that Mrs. Garber had begun taking penicillin earlier in the week, the physician diagnosed acute renal failure, probably caused by a reaction to the medication. Mrs. Garber is 5 feet 3 inches tall and weighs 125 pounds.

1. Describe the probable reason for Mrs. Garber's inability to produce urine. Is her reaction to penicillin considered a prerenal, intrarenal, or postrenal cause of renal failure? What other problems can cause acute renal failure?
2. What medications can the physician prescribe to treat Mrs. Garber's edema and hyperkalemia? What recommendation is likely concerning her continued use of penicillin?
3. What factors should be kept in mind when determining Mrs. Garber's energy, protein, fluid, and electrolyte needs during acute renal failure? How would diuresis alter these recommendations? As you read through the discussion of chronic renal failure, consider how Mrs. Garber's diet would change if her renal failure were to become chronic.

although lower intakes are sometimes necessary. As mentioned previously, oliguric patients who experience diuresis at the beginning of the recovery period may need electrolyte replacement to compensate for urinary losses.

Enteral and Parenteral Nutrition   Some patients need enteral or parenteral nutrition support to obtain adequate energy. Enteral support (tube feeding) is generally preferred over parenteral nutrition because it is less likely to cause infection and sepsis. Enteral formulas for renal failure are more kcalorically dense and have lower protein and electrolyte concentrations than standard formulas. Total parenteral nutrition is necessary only if patients are severely malnourished or cannot consume food for more than 14 days.[10]

> **REVIEW NOTES**
>
> Acute renal failure is characterized by a rapid loss in kidney function, causing a buildup of fluid, electrolytes, and nitrogenous wastes in blood. If hyperkalemia develops, it can have potentially serious consequences on heart function.
>
> Acute renal failure is treated with medications, dietary modifications, and dialysis. The accompanying Case Study allows you to check your understanding of acute renal failure.

Chronic Renal Failure

The kidneys' ability to function despite loss of nephrons is referred to as *renal reserve.*

Unlike acute renal failure, in which kidney function declines suddenly and rapidly, chronic renal failure is characterized by gradual and irreversible deterioration. Because the kidneys have a large functional reserve, the disease typically progresses over many years without causing symptoms. Patients are typically diagnosed late in the course of illness, after over 75 percent of kidney function has been lost.[11]

The most common causes of chronic renal failure are diabetes mellitus and hypertension, which are estimated to cause 43 and 26 percent of cases, respectively.[12] Other conditions that lead to renal failure include inflammatory, immunological, or hereditary diseases that directly involve the kidneys. In a few cases, chronic renal failure follows acute renal failure.

CONSEQUENCES OF CHRONIC RENAL FAILURE

In the early stages of chronic renal failure, the nephrons compensate by enlarging so that they can handle the extra workload. As the nephrons deteriorate, however, there is additional work for the remaining nephrons. The overburdened nephrons continue to degenerate, until finally the kidneys are unable to function adequately. Once the extent of kidney damage necessitates active treatment—either dialysis or a kidney transplant—the condition is classified as **end-stage renal disease (ESRD)**. Table 22-3 lists the common clinical effects of the early and advanced stages of chronic renal failure.

Renal failure is evaluated using the **glomerular filtration rate (GFR)**, the rate at which the kidneys form filtrate. GFR can be estimated using predictive equations that are based on serum creatinine levels, age, gender, race, and body size. Table 22-4 shows how kidney disease is classified according to estimated GFR. Other laboratory measures used to assess kidney function include urinary protein levels, BUN, and the ratio of albumin to creatinine in a urine sample.

Altered Electrolytes and Hormones Fluid and electrolyte disturbances usually do not develop until the final stage of renal failure. As GFR falls, the increased activity by the remaining nephrons is sufficient to maintain electrolyte excretion. A number of hormonal adaptations also help to regulate electrolyte levels, but these changes may cause complications of their own. Increased secretion of aldosterone helps to prevent increases in serum potassium, but contributes to the development of hypertension (in patients who were not previously hypertensive). Increased secretion of parathyroid hormone helps to prevent elevations in serum phosphorus, but contributes to bone loss and development of **renal osteodystrophy**, a bone disorder common in renal patients. Electrolyte imbalances may develop when GFR becomes extremely low (less than 5 milliliters

TABLE 22-3	Clinical Effects of Chronic Renal Failure

Early Stages

- Anorexia
- Fatigue
- Headache
- Hypertension
- Itching
- Kidney inflammation or nephrotic syndrome
- Nausea and vomiting
- Proteinuria, hematuria (blood in urine)

Advanced Stages

- Anemia, bleeding tendency
- Cardiovascular disease
- Confusion, mental impairments
- Electrolyte abnormalities
- Fluid retention
- Hormonal abnormalities
- Metabolic acidosis
- Peripheral neuropathy
- Protein-energy malnutrition
- Reduced immunity
- Renal osteodystrophy

Reminder: *Creatinine* is a waste product of creatine, a nitrogen-containing compound in muscle cells.

Reminder: *Aldosterone* promotes sodium (and therefore water) retention and potassium excretion.

Reminder: *Parathyroid hormone* helps to regulate serum concentrations of calcium and phosphorus.

end-stage renal disease (ESRD): an advanced stage of chronic renal failure in which dialysis or a kidney transplant is necessary to sustain life.

glomerular filtration rate (GFR): the rate at which filtrate is formed within the kidneys, normally approximately 125 mL/min.

renal osteodystrophy: a bone disorder that develops in patients with chronic renal failure as a consequence of increased secretion of parathyroid hormone (which stimulates calcium release from bone), reduced serum calcium, acidosis, and impaired vitamin D activation by the kidneys.

TABLE 22-4	Evaluation of Chronic Renal Disease	

Stage of Disease	Description	GFR[a] (mL/min per 1.73 m²)
1	Kidney damage with normal or increased GFR	≥90
2	Kidney damage with mildly decreased GFR	60–89
3	Moderately decreased GFR	30–59
4	Severely decreased GFR	15–29
5	Kidney failure	<15 (or undergoing dialysis)

[a] GFR is estimated from the Modification of Diet in Renal Disease study equation and is based on age, gender, race, and calibration for serum creatinine. Normal GFR is approximately 125 milliliters per minute.

Source: A. S. Levey and coauthors, National Kidney Foundation practice guidelines for chronic kidney disease: Evaluation, classification, and stratification, *Annals of Internal Medicine* 139 (2003): 137–147.

per minute), when hormonal adaptations are inadequate, or when intakes of water and electrolytes are either very restricted or excessive.

Because the kidneys are responsible for maintaining acid-base balance, acidosis often develops in chronic renal failure. Although usually mild, the acidosis worsens renal bone disease because compounds in bone (protein and phosphates, for example) are released to buffer the acid in blood.

Uremic Syndrome Uremia develops during the final stages of chronic renal failure, when GFR is below 15 milliliters per minute and BUN exceeds 60 milligrams per deciliter.[13] The various symptoms and complications that develop during this stage of illness are collectively known as the **uremic syndrome.** The uremia itself can cause subtle mental dysfunctions and neuromuscular changes such as muscle cramping, twitching, and restless leg syndrome. Other complications include:

- *Impaired hormone synthesis.* Diseased kidneys are unable to produce erythropoietin, causing anemia. Reduced production of active vitamin D contributes to bone disease.
- *Impaired hormone degradation.* Imbalances develop in hormones involved in growth, reproduction, fluid balance, blood glucose regulation, and nutrient metabolism.
- *Bleeding abnormalities.* Defects in platelet function and clotting factors prolong bleeding time and contribute to bruising, gastrointestinal bleeding, and anemia.
- *Increased cardiovascular disease risk.* Risk factors include hypertension, increased insulin resistance, and abnormal blood lipids.
- *Reduced immunity.* Patients with uremia have poor immune responses and are at high risk of developing infections, which are a frequent cause of death.

Protein-Energy Malnutrition Patients with chronic renal disease often develop PEM and wasting. Studies suggest that renal patients have inadequate protein and energy intakes, even during early stages of the disease.[14] Anorexia is thought to contribute to poor food intake and may result from hormonal disturbances, nausea and vomiting, restrictive diets, uremia, and medications. Nutrient losses also contribute to malnutrition and may be a consequence of vomiting, diarrhea, gastrointestinal bleeding, concurrent catabolic illness, and dialysis.

TREATMENT OF CHRONIC RENAL FAILURE

The goals of treatment for chronic renal failure are to slow disease progression and prevent or alleviate symptoms. Dietary measures help to prevent PEM and weight loss. Once renal failure reaches the final stages, dialysis or a kidney transplant is necessary to sustain life.

Drug Therapy Drug therapies help to control some of the complications associated with chronic renal failure. Treatment of hypertension is critical for slowing disease progression and reducing cardiovascular disease risk; thus antihypertensive drugs are often prescribed (see Chapter 21). Anemia is usually treated by injection or intravenous administration of erythropoietin (epoetin). Other common drug treatments include phosphate binders (taken with food) to reduce serum phosphorus levels, sodium bicarbonate to reverse acidosis, and cholesterol-lowering medications. Supplementation with active vitamin D (also called calcitriol) helps to raise serum calcium and reduce parathyroid hormone levels.

Dialysis Dialysis replaces kidney function by removing excess fluid and wastes from blood. In **hemodialysis,** the blood is circulated through a **dialyzer** (artificial kidney), where it is bathed by **dialysate,** a solution that selectively removes fluid and wastes. In **peritoneal dialysis,** dialysate is infused into a person's peritoneal cavity, and blood is filtered by the peritoneum (the membrane that surrounds the

---

NURSING DIAGNOSIS

excess fluid volume, imbalanced nutrition: less than body requirements, impaired urinary elimination, risk for infection, risk for injury, decreased cardiac output, fatigue, and *risk for impaired oral mucous membrane* may apply to people with chronic renal failure.

uremic syndrome: the cluster of symptoms associated with a GFR below 15 mL/min, including uremia, anemia, bone disease, hormonal imbalances, bleeding impairment, increased cardiovascular disease risk, and reduced immunity.

hemodialysis (HE-moh-dye-AL-ih-sis): removal of fluids and wastes from blood by passing the blood through a dialyzer.

dialyzer (DYE-ah-LYE-zer): a machine used for hemodialysis; also called an *artificial kidney.*

dialysate (dye-AL-ih-sate): the solution used in dialysis to draw wastes and fluids from the blood.

peritoneal (PEH-rih-toe-NEE-al) **dialysis:** removal of fluids and wastes by using the peritoneal membrane to filter blood.

TABLE 22-5	Dietary Recommendations for Chronic Renal Failure		
Nutrient	Predialysis	Hemodialysis	Peritoneal Dialysis
Energy[a] (kcal/kg)	<60 years old: 35 ≥60 years old: 30–35	<60 years old: 35 ≥60 years old: 30–35	<60 years old: 35 ≥60 years old: 30–35 (total kcalories should include those absorbed from dialysate)
Protein (g/kg) (50% from high-quality proteins)	0.6–0.75	1.2	1.2–1.3
Fat	As needed to maintain a healthy lipid profile	As needed to maintain a healthy lipid profile	As needed to maintain a healthy lipid profile
Fluid (mL/day)	Unrestricted if urine output is normal	1000 plus urine output	1500–2000; monitor closely
Sodium (mg/day)	2000	2000	2000
Potassium (mg/day)	Individualized according to laboratory values	2000–3000	3000–4000
Calcium (mg/day)	1200	≤2000 from diet and medications	≤2000 from diet and medications
Phosphorus (mg/day)	Individualized according to laboratory values	800–1000	800–1000

[a] Values listed apply to adults; recommendations for children should not fall below RDA levels.

Source: Reprinted from *Journal of the American Dietetic Association* 104 (2004): 404–409, J. A. Beto and V. K. Bansal, Medical nutrition therapy in chronic kidney failure: Integrating clinical practice guidelines. © 2004, with permission from the American Dietetic Association.

abdominal cavity); after several hours, the dialysate is drained, removing unneeded fluid and wastes. Dialysis is discussed in more detail in Nutrition in Practice 22.

Medical Nutrition Therapy The patient's diet strongly influences disease progression and the development of complications. As the renal diet is complex and nutrient needs change frequently during the course of illness, a dietitian who specializes in renal disease is best suited to provide medical nutrition therapy. Table 22-5 summarizes the general dietary guidelines for different stages of renal disease. Note that hemodialysis and peritoneal dialysis affect nutrient needs differently. Because patients' needs can vary considerably, actual recommendations should be based on the results of a nutrition assessment.

Energy Energy intake should be high enough to maintain a healthy weight and prevent wasting. To help prevent malnutrition, patients are encouraged to consume high-kcalorie foods and beverages when possible, including hard candy, honey, sugar, jam, salad dressings, and soft margarine. Malnourished patients may require oral supplements or tube feedings to maintain weight.

Patients undergoing peritoneal dialysis can absorb a substantial amount of glucose from the dialysate, which can contribute as many as 800 kcalories daily. These kcalories must be included when estimating energy intake. Weight gain is sometimes a problem when peritoneal dialysis continues for a long period.

The dialysate contains glucose in order to draw fluid from the blood to the peritoneal cavity by osmosis.

Protein A low-protein diet is often prescribed to help slow the progression of renal failure; guidelines suggest intakes between 0.6 to 0.75 grams per kilogram body weight per day.[15] Low-protein diets produce fewer nitrogenous wastes and therefore reduce the risk of uremia. In addition, low-protein diets supply less phosphorus than high-protein diets, reducing the risks associated with hyperphosphatemia.

The protein RDA for adults is 0.8 g/kg.

In a renal diet, half of the protein consumed should be from high-quality sources such as eggs, milk, meat, poultry, and fish.

Reminder: High-quality protein can be obtained from eggs, milk products, meat, poultry, fish, and soybeans.

hypokalemia (HIGH-po-ka-LEE-me-ah): low serum potassium levels.

Because renal patients often develop PEM, however, the diet must provide enough protein to meet needs and prevent wasting. To ensure that adequate amounts of essential amino acids are consumed, about 50 percent of the protein should be from high-quality protein sources. Low-protein breads, pastas, and other grain-based products are commercially available and can help renal patients improve energy intake without increasing protein consumption.

Because low-protein diets are difficult to adhere to and patients are at high risk of wasting, dietitians may suggest consuming higher amounts of protein to preserve health.[16] Once dialysis is begun, protein restrictions can be relaxed, as dialysis removes nitrogenous wastes and causes some amino acid losses as well.

Lipids To control elevated blood lipids and reduce heart disease risk, patients with chronic renal disease are advised to restrict saturated fat and cholesterol. As renal diets typically include high-fat foods to improve caloric intakes, patients should be instructed to select foods that provide mostly unsaturated fats, such as nuts and seeds, salad dressings and mayonnaise, avocados, and soybean products.

Fluids and Sodium As renal failure progresses, the patient excretes less urine and cannot handle normal amounts of sodium and fluids. Suggested intakes are based on total urine output, changes in body weight and blood pressure, and serum sodium levels. A rise in body weight and blood pressure suggests that the person is retaining sodium and fluid; conversely, declines in these measurements indicate fluid loss. Most persons with renal disease tend to retain sodium and may benefit from mild restriction; less frequently, a patient may have a salt-wasting condition that requires additional dietary sodium.

Fluids are not restricted until urine output decreases. For a person who is neither dehydrated nor overhydrated, the daily fluid intake should match the daily urine output. (The water obtained from solid foods is roughly equivalent to water losses from skin and lungs.) In dialysis patients, sodium and fluid intakes should be controlled so that only about 2 pounds of water weight are gained daily—this excess fluid is then removed during the next dialysis treatment. Patients should be advised that foods like flavored gelatin, soups, fruit ices, frozen fruit juice bars, and ice milk contribute to fluid intakes.

Potassium Before dialysis treatments become necessary, most renal patients can handle typical intakes of potassium. For some individuals, the amount allowed is adjusted according to serum potassium levels. Those using potassium-wasting diuretics may require potassium supplementation.

Dialysis patients need to control potassium intakes to prevent hyperkalemia or, more rarely, **hypokalemia.** Restriction is usually necessary for persons treated with hemodialysis, whereas those undergoing peritoneal dialysis can consume potassium more freely. Recommended intakes are based on serum potassium levels, renal function, medications, and the dialysis procedure used.

All fresh foods provide potassium, but some fruits and vegetables contain such large amounts that some patients need to restrict intakes; examples include avocados, bananas, Brussels sprouts, cantaloupe, dried fruit, dried beans and peas, prune juice, spinach, tomatoes, and winter squash. Other high-potassium choices include milk, molasses, and nuts. Salt substitutes that contain potassium chloride should be avoided on a potassium-restricted diet.

Calcium, Phosphorus, and Vitamin D To prevent bone disease, calcium and phosphorus intakes may need adjustment, even during the early stages of renal failure. As mentioned earlier, the kidneys are unable to produce active vitamin D, which greatly impairs calcium absorption. Elevated serum phosphorus levels indicate the need for dietary phosphorus restriction; in many patients, phosphate binders (taken with meals) are used to control serum levels. Once dialysis treat-

People on a renal diet can consume most fruits and vegetables in limited amounts.

ments begin, phosphorus restrictions and calcium and vitamin D supplementation become standard treatment. Serum calcium levels must be monitored to guard against **hypercalcemia,** which can develop in response to simultaneous vitamin D and calcium supplementation.

High-protein foods are also high in phosphorus, so protein-restricted diets curb phosphorus intakes as well. After dialysis treatments begin and protein intakes are liberalized, phosphate binders become essential for phosphorus control. Because foods that are rich in calcium (such as milk and milk products) are usually high in phosphorus and are therefore restricted, patients must rely on calcium supplements to meet their calcium needs.

Vitamins and Minerals The restrictive renal diet interferes with vitamin and mineral intakes, increasing the risk of deficiencies. In addition, patients treated with dialysis lose water-soluble vitamins and some trace minerals into the dialysate. Dietary supplements for renal patients typically supply generous amounts of folate and vitamin B_6, 1 milligram and 10 milligrams per day, respectively, along with recommended amounts of the other water-soluble vitamins. Supplemental vitamin C should be limited to 100 milligrams per day because excessive intakes can contribute to kidney stone formation in those at risk (see p. 580). Supplements containing vitamins A and E are not recommended because these vitamins sometimes accumulate in patients with renal failure.[17]

Iron deficiency is common in hemodialysis patients and may be due to gastrointestinal bleeding, reduced iron absorption, iron losses into the dialysate, and blood losses associated with the dialysis treatment.[18] Intravenous administration of iron, in conjunction with erythropoietin therapy, is more effective than oral iron supplementation for improving iron status.

Enteral and Parenteral Nutrition Enteral and parenteral nutrition support are helpful for renal patients who cannot eat adequate amounts of food. As in acute renal failure, formulas that are suitable for chronic renal failure are more kcalorically dense and have lower protein and electrolyte concentrations than standard formulas. **Intradialytic parenteral nutrition** is an option for supplying supplemental nutrients: this technique combines parenteral feedings with hemodialysis treatments. An advantage of this approach is that the volume of parenteral solution infused can be simultaneously removed (recall that fluid intake is controlled in dialysis patients). Intradialytic parenteral nutrition has been successful at increasing weight gain, muscle mass, and body fat in malnourished hemodialysis patients.[19]

Dietary Compliance Adhering to a renal diet is probably the most difficult aspect of treatment for renal disease patients. Patients often require extensive counseling once multiple dietary restrictions become necessary. Depending on the stage of illness and the patient's laboratory values, the renal diet may limit protein, fluids, sodium, potassium, and phosphorus, thereby affecting food selections from all major food groups. Because these diets have so many restrictions, patient compliance is often a problem.[20] The "How to" provides suggestions to help patients comply with renal diets. The accompanying Case Study can help you apply your knowledge about chronic renal failure and hemodialysis.

KIDNEY TRANSPLANTS

A preferred alternative to dialysis in end-stage renal disease is kidney transplantation.[21] A successful kidney transplant restores kidney function, allows a more liberal diet, and frees the patient from routine dialysis. Given the choice, many patients would prefer transplants, but the demand for suitable kidneys far exceeds the supply. Other barriers to transplantation include age, poor health, financial

hypercalcemia (HIGH-per-kal-SEE-me-ah): elevated serum calcium levels.

intradialytic parenteral nutrition: the infusion of nutrients during hemodialysis, often providing amino acids, dextrose, lipids, and some trace minerals.

Patients with renal disease and their caregivers face considerable challenges as they learn to manage a renal diet. The following suggestions may help:

1. *To keep track of fluid intake:*
   - Fill a container with an amount of water equal to your total fluid allowance. Each time you use a liquid food or beverage, discard an equivalent amount of water from the container. The amount remaining in the container will show you how much fluid you have left for the day.
   - Be sure to save enough fluid to take medications.
2. *To help control thirst:*
   - Chew gum or suck hard candy.
   - Freeze beverages to a semisolid state so that they take longer to consume.
   - Add lemon juice or crumpled mint leaves to water to make it more refreshing.
   - Gargle with refrigerated mouthwash.
3. *To increase the energy content of meals:*
   - Add extra margarine or butter to rice, noodles, breads, crackers, and cooked vegetables. Add extra salad dressing or oil to salads.
   - Add nondairy whipped toppings to desserts.
   - Include fried foods in your diet.

4. *To include more of your favorite vegetables in meals:*
   - Consult your nurse or dietitian to learn whether you can safely use the process of leaching to remove some of the potassium from vegetables.
   - To leach potassium from vegetables: Cut the vegetables into $1/8$-inch slices and rinse. Soak the vegetables in a large amount of warm water for two hours—about ten parts of water to one part of vegetables. Cook the vegetables using five parts of water to one part of vegetables.
5. *To prevent the diet from becoming monotonous:*
   - Experiment with new combinations of allowed foods.
   - Substitute nondairy products for milk products. Nondairy products, which are lower in protein, phosphorus, and potassium, can substitute for milk and add energy to the diet.
   - Add flavor to foods by seasoning with garlic, onion, chili powder, curry powder, oregano, mint, basil, parsley, pepper, or lemon juice.
   - Consult a nurse or dietitian when you want to eat restricted foods. Many restricted foods can be used occasionally and in small amounts if the menu is carefully adjusted.

NURSING DIAGNOSIS

risk for infection and *ineffective protection* apply to persons with kidney transplants.

difficulties, and abnormalities of the urinary tract. Fewer than 20 percent of patients who develop end-stage renal disease receive a kidney transplant.[22]

Immunosuppressive Drug Therapy Following transplant surgery, patients require high doses of immunosuppressive drugs to prevent tissue rejection. These drugs have multiple effects that can alter nutrition status. Side effects may include nausea, vomiting, diarrhea, glucose intolerance, altered blood lipids, fluid retention, hyper-

CASE STUDY   Banker with Chronic Renal Failure

Thomas Stone is a 55-year-old banker who developed chronic renal failure as a result of hypertension. His condition was discovered several years ago when routine laboratory tests revealed elevated serum creatinine and BUN levels. Since then, he has been taking antihypertensive medications and restricting dietary sodium, although he reported difficulty following the low-protein diet that was also prescribed. Mr. Stone recently visited his doctor with complaints of low urine output and reduced sensation in his hands and feet. He also reported feeling drowsy at work and mentioned that he was bruising more than usual. The examination revealed a 9-pound weight gain since his last visit and swelling in his ankles and feet. Tests revealed that his GFR had fallen to 10 milliliters per minute. Mr. Stone is 5 feet 8 inches tall and normally weighs 160 pounds.

1. Explain how chronic renal failure progresses. What happens to GFR, serum creatinine levels, and BUN as renal function declines?
2. What clinical effects would you expect during the final stage of kidney failure? Explain the significance of each of Mr. Stone's physical complaints.
3. Explain why a low-sodium, low-protein diet was prescribed for Mr. Stone at a former visit. What energy and protein intakes were probably recommended?
4. The physician determines that Mr. Stone's kidney failure has reached the final stage and prescribes hemodialysis. How will dialysis alter Mr. Stone's diet? Calculate his new energy and protein needs and compare them to his former diet. What other nutrients will need consideration?

DIET-DRUG INTERACTIONS Check this table for notable nutrition-related effects of the medications discussed in this chapter.

	Gastrointestinal Effects	Interactions with Dietary Substances	Metabolic Effects
Anti-inflammatory/immunosuppressive drugs			
Corticosteroids			Glucose intolerance, sodium retention, negative nitrogen balance, appetite stimulation, weight gain, growth suppression in children
Cyclosporine		Grapefruit juice raises drug levels and increases risk of toxicity. Avoid potassium supplements and salt substitutes with potassium. St. John's wort can alter drug efficacy.	Hyperkalemia, hypomagnesia, hyperglycemia, hyperlipidemia
Phosphate binders (calcium-containing)	Constipation		Electrolyte imbalances
Potassium-exchange resins (sodium polystyrene sulfonate)			Fluid retention, hypokalemia, hypocalcemia
Potassium citrate	Nausea, vomiting, stomach pain, diarrhea		Hyperkalemia

tension, and infection. The Diet-Drug Interactions feature summarizes the nutrition-related effects of these drugs and others mentioned in this chapter. Because immunosuppressive drug therapy increases the risk of foodborne infection, food safety guidelines should be provided to patients and caregivers.

Medical Nutrition Therapy Protein and energy requirements increase after surgery due to stress and the catabolic effects of drug therapy. Once recovery is under way, side effects from drugs can strongly influence dietary treatment. Typical dietary modifications are shown in Table 22-6. Hyperglycemia (due to drug treatment) may be improved by controlling carbohydrate intake, although oral medications or insulin therapy may also be required. Because blood lipids are frequently elevated, patients should limit saturated fat and cholesterol intakes. Sodium, potassium, and phosphorus intakes are often liberalized following a transplant, but serum electrolyte levels must be monitored closely because some drug therapies can cause hyperkalemia or hypophosphatemia. Calcium supplementation is advised due to urinary calcium losses associated with corticosteroids (often used as immunosuppressants). Patients should avoid foods that can potentially cause foodborne illness, such as raw or undercooked meat, fish, and eggs; unpasteurized milk and juices; cheese made with unpasteurized milk; and fresh bean sprouts.[23]

REVIEW NOTES

Chronic renal failure causes gradual loss of kidney function and most often results from long-standing diabetes mellitus or hypertension.

Depending on the stage of disease, complications of chronic renal failure may include fluid and electrolyte disturbances, hypertension, renal osteodystrophy, mental impairments, bleeding abnormalities, anemia, reduced immunity, abnormal blood lipids, and insulin resistance.

TABLE 22-6	Dietary Guidelines Following a Kidney Transplant

- **Energy:** Initially, 30–35 kcalories per kilogram body weight per day; reduce to 25-30 kcalories/kilogram body weight per day to maintain healthy body weight, and adjust as necessary.

- **Protein:** Initially, 1.3 to 1.5 grams per kilogram body weight per day; reduce to 1.0 gram/kilogram per day after 6–8 weeks.

- **Carbohydrate:** For hyperglycemia, maintain consistent carbohydrate intake each day.

- **Fat:** Limit saturated fat (to <10% of total kcalories) and cholesterol (to <300 milligrams/day) to help control serum lipids.

- **Sodium:** Generally unrestricted. Restrict to 2–4 grams/day if fluid retention and hypertension are present.

- **Potassium:** Generally unrestricted. Adjust according to serum potassium levels if necessary.

- **Calcium:** 1200 milligrams/day to minimize bone loss associated with drug therapy.

- **Phosphorus:** Unrestricted.

- **Fluid:** Unrestricted.

Sources: J. A. Beto and V. K. Bansal, Medical nutrition therapy in chronic kidney failure: Integrating clinical practice guidelines, *Journal of the American Dietetic Association* 104 (2004): 404–409; American Dietetic Association, *Manual of Clinical Dietetics* (Chicago: American Dietetic Association, 2000).

> Treatment includes drug therapies, medical nutrition therapy, and regular dialysis. Dietary measures usually include a low-protein diet, controlled fluid and sodium intakes, potassium and phosphorus restrictions, and calcium and vitamin D supplementation.
>
> Kidney transplantation can restore renal function and liberalize dietary restrictions.

Kidney Stones

kidney stones: crystalline masses that form in the urinary tract; also called *renal calculi* and *nephrolithiasis*.

hypercalciuria (HIGH-per-kal-see-YOO-ree-ah): elevated urinary calcium levels.

hyperoxaluria (HIGH-per-ox-ah-LOO-ree-ah): elevated urinary oxalate levels.

Approximately 12 percent of men and 5 percent of women in the United States develop one or more **kidney stones** by the age of 70.[24] A kidney stone is a crystalline mass that forms within the urinary tract. Although often asymptomatic, a stone's passage can cause severe pain or block the urinary tract. Stones tend to recur but can be prevented with dietary measures and medical treatment.

FORMATION OF KIDNEY STONES

Kidney stones develop when stone constituents become concentrated in urine, allowing crystals to form and grow. About 75 percent of kidney stones are composed of calcium oxalate. Less commonly, stones are made from uric acid, the amino acid cystine, or magnesium ammonium phosphate (the latter are known as *struvite* stones). Stone formation is promoted by factors that reduce urine volume, block urine flow, or increase concentrations of stone-forming substances.

Calcium Oxalate Stones The most common abnormality in people with calcium oxalate stones is **hypercalciuria** (elevated urinary calcium). Hypercalciuria can result from excessive calcium absorption, impaired calcium reabsorption in kidney tubules, or elevated serum levels of parathyroid hormone or vitamin D. Some people with calcium oxalate stones excrete normal amounts of calcium in the urine, however, and the reason they form stones is unknown.

Elevated urinary oxalate levels, or **hyperoxaluria**, also promote the formation of calcium oxalate crystals. Hyperoxaluria may reflect an increase in the body's synthesis of oxalate or increased absorption from dietary sources. People who

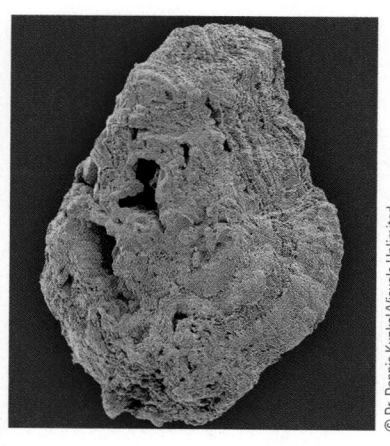

The most common type of kidney stone is composed of calcium oxalate crystals, as shown here.

© Dr. Dennis Kunkel/Visuals Unlimited

TABLE 22-7	Foods High in Oxalate	
Vegetables	**Fruits**	**Other**
Beans, green and wax	Blackberries	Chocolate and chocolate beverages*
Beets*	Blueberries	Cocoa
Celery	Currants, red	Coffee
Chard, Swiss	Gooseberries	Draft beer
Collard greens	Grapes, Concord	Fruit cake
Dandelion greens	Lemon peel	Grits
Eggplant	Lime peel	Nuts, nut butters*
Endive	Orange peel	Peanut butter*
Escarole	Raspberries	Pepper
Leeks	Rhubarb*	Soybean crackers
Legumes	Strawberries*	Tea*
Okra		Tofu
Parsley		Wheat bran*
Potatoes, sweet		Wheat germ
Spinach*		
Squash, summer		

Note: The oxalate content of many foods has not been analyzed and even fewer studies have been conducted to determine which foods raise urinary oxalate. *The foods marked with an asterisk have been documented to raise urinary oxalate and should be avoided by people who form calcium stones.

form calcium oxalate stones are advised to reduce their dietary intake of oxalate (see Table 22-7) and to avoid supplementation with vitamin C, which degrades to oxalate in the body.[25]

Uric Acid Stones Uric acid stones develop when the urine is abnormally acidic, contains excessive uric acid, or both. These stones are frequently associated with **gout,** a metabolic disorder characterized by elevated uric acid levels in the blood and urine. Other diseases that increase the risk of uric acid stones include leukemia and lymphoma, which are associated with overproduction of uric acid in the body. A diet rich in **purines** also contributes to high uric acid levels (see Table 22-8, p. 582).

Cystine and Struvite Stones Cystine stones can form in people with the inherited disorder **cystinuria,** in which the renal tubules are unable to reabsorb the amino acid cystine. This results in abnormally high concentrations of cystine in the urine, leading to subsequent crystallization and stone formation.

Struvite stones, composed primarily of magnesium ammonium phosphate, form in alkaline urine as a result of bacterial degradation of urea to ammonia. Struvite stones can accompany chronic urinary infections or disorders that interfere with urinary flow.

Reminder: Fat malabsorption promotes oxalate absorption, thereby increasing the risk of forming calcium oxalate stones (see Chapter 18).

gout: a metabolic disorder characterized by elevated uric acid in the blood and urine and the deposition of uric acid in and around the joints, causing acute joint inflammation.

purine (PYOO-reen): a product of nucleotide metabolism that degrades to uric acid.

cystinuria (SIS-tin-NOO-ree-ah): an inherited disorder characterized by elevated urinary excretion of several amino acids, including cystine.

struvite (STROO-vite): crystals of magnesium ammonium phosphate.

TABLE 22-8	Foods High in Purines		
Organ Meats	**Meat and Meat Products**	**Seafood**	
Brains	Game meat	Anchovies	
Kidney	Gravies	Herring	
Liver	Meat extracts	Mackerel	
Sweetbreads		Sardines	
		Scallops	

Source: J. A. T. Pennington, *Bowes and Church's Food Values of Portions Commonly Used* (Philadelphia: J. B. Lippincott, 1994), p. 387.

CONSEQUENCES OF KIDNEY STONES

NURSING DIAGNOSIS

acute pain, risk for infection, and *impaired urinary elimination* may apply to people with kidney stones.

In most cases, kidney stones do not pose serious medical problems. Small stones can readily pass through the ureters and out of the body with minimal treatment.

Renal Colic A stone passing through the ureter can produce severe, continuous pain, called **renal colic.** Generally, the pain begins in the back and intensifies as the stone travels toward the bladder (review Figure 22-1 on p. 566). The pain can be severe enough to cause nausea and vomiting and sometimes requires medication. When the stone reaches the bladder, the pain abruptly stops. Blood may appear in the urine (called **hematuria**) as a result of damage to the kidney or ureter lining.

Urinary Tract Complications Depending on the location of the stone, symptoms may include urination urgency, frequent urination, or inability to urinate. Stones that are unable to pass through the ureter can cause a urinary tract obstruction and possibly lead to infection.

PREVENTION AND TREATMENT OF KIDNEY STONES

Solutes are less likely to crystallize and form stones in dilute urine. Therefore, people who form kidney stones are advised to drink 12 to 16 cups of fluids daily in order to maintain urine volumes of least $2^1/2$ liters per day.[26] Acceptable fluid sources include tea, coffee, wine, and beer, but apple and grapefruit juices should be limited because they may increase the risk of stones.[27]

Calcium Oxalate Stones Dietary measures and drug treatments aim to reduce urinary calcium and oxalate levels. In addition, urinary uric acid levels must be kept low because uric acid crystals contribute to stone formation.[28] Thiazide diuretics can help to reduce urinary calcium by enhancing calcium reabsorption in the kidney tubules. Other medications include cholestyramine, which reduces oxalate absorption, and allopurinol, which reduces uric acid production in the body. Potassium citrate can inhibit the formation and growth of crystals but may cause stomach upset and diarrhea.[29]

Medical nutrition therapy includes adjustments in calcium, oxalate, protein, and sodium intakes. Patients should consume adequate calcium from food sources (800 to 1200 milligrams per day) because dietary calcium combines with oxalate in the intestines, reducing oxalate absorption and helping to control hyperoxaluria. Conversely, low-calcium diets promote oxalate absorption and higher urinary oxalate levels. Foods high in oxalate should be restricted because dietary

Because calcium supplements can elevate urinary calcium levels, they are not as helpful as food sources of calcium.

renal colic: the intense pain that occurs when a kidney stone passes through the ureter.

hematuria (HE-mah-TOO-ree-ah): blood in the urine.

oxalate contributes to urinary oxalate content (review Table 22-7).[30] As high protein and sodium intakes increase urinary calcium excretion, moderate protein consumption and sodium restriction are advised.

Uric Acid Stones Drug treatments for uric acid stones include allopurinol to reduce urinary uric acid and potassium citrate to reduce urine acidity. Diets restricted in purines may also help to control urinary uric acid content (review Table 22-8). Because all meats, poultry, fish, and shellfish contain considerable amounts of purines, strict dietary control over a long period may be difficult to achieve. In addition, the benefits of purine restriction are unknown.

Cystine and Struvite Stones High fluid intakes may prevent the formation of cystine stones in some patients, whereas others require drug therapy to reduce cystine production in the body. Medications frequently prescribed include penicillamine and tiopronin, which reduce cystine levels, and potassium citrate, which reduces urine acidity. Preventing urinary tract infections is an important strategy for preventing struvite stones. Patients with these stones may require antibiotic therapy to prevent further stone formation.

Drinking plenty of water throughout the day is the most important measure for preventing kidney stones.

REVIEW NOTES

Kidney stones form when stone constituents—calcium oxalate, uric acid, cystine, or magnesium ammonium phosphate—crystallize in urine. Complications include renal colic, difficulty with urination, and obstruction.

Kidney stones may be prevented by maintaining urine volumes of at least $2^{1}/_{2}$ liters daily. Other dietary measures include consuming enough calcium to control oxalate absorption, dietary oxalate and purine restrictions, moderate protein intake, and sodium restriction.

NUTRITION ASSESSMENT CHECKLIST FOR PEOPLE WITH RENAL DISORDERS

Medical History

Check the medical record to determine:

- [ ] Degree of renal function
- [ ] Cause of nephrotic syndrome or renal failure
- [ ] Type of dialysis, if appropriate
- [ ] If the patient has received a kidney transplant
- [ ] Type of kidney stone

Review the medical record for complications that may alter nutritional needs:

- [ ] Anemia
- [ ] Diabetes mellitus
- [ ] Edema or oliguria
- [ ] Hyperlipidemia
- [ ] Hypertension
- [ ] Metabolic stress or infection
- [ ] Protein-energy malnutrition

Medications

Assess risks for medication-related malnutrition related to:

- [ ] Long-term use of medications

- [ ] Multiple medication use, especially if medications affect nutrition status

For all patients with renal diseases, note:

- [ ] Whether medications or supplements contain electrolytes that must be controlled
- [ ] Use of drugs or herbs that may be toxic to the kidneys

Dietary Intake

For patients with the nephrotic syndrome, renal failure, or kidney transplants, assess intakes of:

- [ ] Protein and energy
- [ ] Fluid
- [ ] Vitamins, especially vitamin D
- [ ] Minerals, especially calcium, phosphorus, iron, and electrolytes

For patients with kidney stones or a history of kidney stones:

- [ ] Stress the need to drink plenty of fluids throughout the day.
- [ ] Assess intake of calcium, oxalate, sodium, protein, purines, or vitamin C, as appropriate for the type of stone.

Anthropometric Data

Take accurate baseline height and weight measurements. Keep in mind that:

☐ Fluid retention due to the nephrotic syndrome or renal failure can mask malnutrition.

☐ For dialysis patients, the weight measured immediately after the dialysis treatment (called the *dry weight*) most accurately reflects the person's true weight. Rapid weight gain between dialysis treatments reflects fluid retention. If fluid retention is excessive, review fluid intake to determine if the patient understands and is complying with diet recommendations.

Laboratory Tests

Note that serum protein levels are often low in patients with the nephrotic syndrome or renal failure. Review the following laboratory test results to assess the degree of renal function and response to treatments:

☐ Glomerular filtration rate (GFR)

☐ Creatinine

☐ Urinary protein

☐ Blood urea nitrogen (BUN)

☐ Serum electrolytes

Check laboratory test results for complications associated with renal disease, including:

☐ Anemia

☐ Hyperglycemia

☐ Hyperlipidemia

☐ Hyperparathyroidism (related to bone disease)

Physical Signs

For patients with the nephrotic syndrome or renal failure, look for physical signs of:

☐ Dehydration or fluid retention

☐ Iron deficiency

☐ Uremia

☐ Bone disease

☐ Hyperkalemia

SELF CHECK

1. Which of the following is not a function of the kidneys?
   a. activation of vitamin K
   b. maintenance of acid-base balance
   c. elimination of metabolic waste products
   d. maintenance of fluid and electrolyte balances

2. The nephrotic syndrome frequently results in:
   a. the uremic syndrome.
   b. oliguria.
   c. edema.
   d. renal colic.

3. Dietary recommendations for the nephrotic syndrome include:
   a. a high-protein intake.
   b. sodium restriction.
   c. potassium and phosphorus restrictions.
   d. fluid restriction.

4. Hyperkalemia is often treated by:
   a. eliminating potassium from the diet.
   b. using diuretics to increase potassium losses.
   c. increasing fluid consumption.
   d. using potassium-exchange resins, which bind potassium in the GI tract.

5. Fluid requirements for oliguric patients are estimated by adding about ____ milliliters to the volume of urine output.
   a. 100
   b. 300
   c. 500
   d. 750

6. The most common cause of chronic renal failure is:
   a. diabetes mellitus.
   b. hypertension.
   c. autoimmune disease.
   d. exposure to toxins.

7. A person with chronic renal failure, who has been following a renal diet for several years, begins hemodialysis treatment. An appropriate dietary adjustment would be to:
   a. reduce protein intake.
   b. consume protein more liberally.
   c. increase intakes of sodium and water.
   d. consume potassium and phosphorus more liberally.

8. Which of the following nutrients may be unintentionally restricted when a patient follows a renal diet?
   a. fluid
   b. calcium
   c. potassium
   d. phosphorus

9. Most kidney stones are made primarily from:
   a. struvite.
   b. uric acid.
   c. calcium oxalate.
   d. cystine.

10. Treatment for all kidney stones includes:
    a. dietary oxalate restriction.
    b. dietary protein restriction.
    c. vitamin C supplementation.
    d. a fluid intake that maintains a urine volume of at least $2^1/_2$ liters a day.

Answers to these questions appear in Appendix H.

CLINICAL APPLICATIONS

1. A person with chronic renal failure may need multiple medications to control disease progression and treat symptoms and complications. For people with diabetes and hyperlipidemias who develop renal failure, medications might include insulin, oral hypoglycemic drugs, antihypertensives, diuretics, lipid-lowering medications, and phosphate binders. Review the nutrition-related side effects of these medications. Describe the ways in which these medications may make it harder for people to maintain nutrition status.

2. Because the diet for chronic renal failure is so restrictive, patients find the diet difficult to manage and maintain over the long term. Review the suggestions in the "How to" on p. 578. Can you think of additional suggestions that may help? Give suggestions for helping people adjust to the different aspects of their renal diets.

NUTRITION ON THE NET

For further study of the topics in this chapter, access these websites.

Find updates and quick links to these and other nutrition-related sites at our website: **www.wadsworth.com/nutrition**

To find information related to kidney diseases and dialysis, visit these sites:

National Institute of Diabetes and Digestive and Kidney Diseases: **www.niddk.nih.gov**

American Association of Kidney Patients: **www.aakp.org**

National Kidney Foundation: **www.kidney.org**

Kidney Foundation of Canada: **www.kidney.ca**

To find materials for kidney transplant patients and their families, visit these sites:

Renalnet Kidney Information Clearinghouse: **www.renalnet.org**

The Transplant Living website developed by the United Network of Organ Sharing: **www.transplantliving.org**

To find more information about kidney stones, visit the Oxalosis and Hyperoxaluria Foundation: **www.ohf.org**

To see photographs of kidney stones, visit the website of the Louis C. Herring Laboratory: **www.herringlab.com**

NOTES

[1] G. B. Appel, Glomerular disorders, in L. Goldman and D. Ausiello, eds., *Cecil Textbook of Medicine* (Philadelphia: Saunders, 2004), pp. 726–733.

[2] G. A. Kaysen, Proteinuria and the nephrotic syndrome, in R. W. Schrier, ed., *Renal and Electrolyte Disorders* (Philadelphia: Lippincott Williams & Wilkins, 2003), pp. 580–622.

[3] Kaysen, 2003; S. Escott-Stump, *Nutrition and Diagnosis-Related Care* (Baltimore: Lippincott Williams & Wilkins, 2002), pp. 665–666.

[4] W. E. Mitch, Acute renal failure, in L. Goldman and D. Ausiello, eds., *Cecil Textbook of Medicine* (Philadelphia: Saunders, 2004), pp. 703–708.

[5] Mitch, 2004; C. J. Richard, Renal disorders, in L. E. Lancaster, ed., *Core Curriculum for Nephrology Nursing* (Pitman, N.J.: Anthony J. Jannetti, 2001), pp. 85–115.

[6] C. L. Edelstein and R. W. Schrier, Acute renal failure: Pathogenesis, diagnosis, and management, in R. W. Schrier, ed., *Renal and Electrolyte Disorders* (Philadelphia: Lippincott Williams & Wilkins, 2003), pp. 401–455.

[7] F. N. Hutchison, Management of acute renal failure, in A. Greenberg, ed., *Primer on Kidney Diseases* (San Diego: National Kidney Foundation and Academic Press, 2001), pp. 275–280.

[8] Mitch, 2004.

[9] J. Stover and G. Morrison, Renal disease, in L. Hark and G. Morrison, eds., *Medical Nutrition and Disease: A Case-Based Approach* (Malden, Mass.: Blackwell Science, 2003), pp. 328–349; Escott-Stump, 2002.

[10] Edelstein and Schrier, 2003.

[11] R. G. Luke, Chronic renal failure, in L. Goldman and D. Ausiello, eds., *Cecil Textbook of Medicine* (Philadelphia: Saunders, 2004), pp. 708–716.

[12] W. Wang and L. Chan, Chronic renal failure: Manifestations and pathogenesis, in R. W. Schrier, ed., *Renal and Electrolyte Disorders* (Philadelphia: Lippincott Williams & Wilkins, 2003), pp. 456–497.

[13] Luke, 2004.

[14] R. Mehrotra and J. D. Kopple, Nutritional management of maintenance dialysis patients: Why aren't we doing better? *Annual Reviews of Nutrition* 21 (2001): 343–379.

[15] J. A. Beto and V. K. Bansal, Medical nutrition therapy in chronic kidney failure: Integrating clinical practice guidelines, *Journal of the American Dietetic Association* 104 (2004): 404–409.

[16] Beto and Bansal, 2004.

[17] Beto and Bansal, 2004.

[18]N. Tolkoff-Rubin and N. Goes, Treatment of irreversible renal failure, in L. Goldman and D. Ausiello, eds., *Cecil Textbook of Medicine* (Philadelphia: Saunders, 2004), pp. 716–726.

[19]A. K. Mortelmans and coauthors, Intradialytic parenteral nutrition in malnourished hemodialysis patients: A prospective long-term study, *Journal of Parenteral and Enteral Nutrition* 23 (1999): 90–95; K. Hiroshige and coauthors, Prolonged use of intradialysis parenteral nutrition in elderly malnourished chronic haemodialysis patients, *Nephrology Dialysis Transplantation* 13 (1998): 2081–2087.

[20]C. L. Durose and coauthors, Knowledge of dietary restrictions and the medical consequences of noncompliance by patients on hemodialysis are not predictive of dietary compliance, *Journal of the American Dietetic Association* 104 (2004): 35–41.

[21]Tolkoff-Rubin and Goes, 2004.

[22]Wang and Chan, 2003.

[23]American Dietetic Association, *Nutrition Care Manual* (Chicago: American Dietetic Association, 2005).

[24]I. Juknevicius and K. A. Hruska, Renal calculi (nephrolithiasis), in L. Goldman and D. Ausiello, eds., *Cecil Textbook of Medicine* (Philadelphia: Saunders, 2004), pp. 761–767.

[25]D. G. Assimos, Vitamin C supplementation and urinary oxalate excretion, *Reviews in Urology* 6 (2004): 167; M. S. Parmar, Kidney stones, *British Medical Journal* 328 (2004): 1420–1424.

[26]American Dietetic Association, 2005.

[27]Escott-Stump, 2002.

[28]Juknevicius and Hruska, 2004.

[29]D. S. Goldfarb, Reconsideration of the 1988 NIH Consensus Statement on prevention and treatment of kidney stones: Are the recommendations out of date? *Reviews in Urology* 4 (2002): 53–60.

[30]M. S. Krishnamurthy, K. A. Hruska, and P. S. Chandhoke, The urinary response to an oral oxalate load in recurrent calcium stone formers, *Journal of Urology* 169 (2003): 2030–2033.

Dialysis

Although there is no perfect substitute for one's own kidneys, dialysis offers a life-sustaining treatment option for people with chronic renal failure. Dialysis can serve as a permanent treatment or as a temporary measure to sustain life until a suitable kidney donor can be found. Dialysis can also restore fluid and electrolyte balances in patients with acute renal failure. Clinicians who routinely work with renal patients should understand how dialysis procedures work. This Nutrition in Practice describes the process of dialysis and outlines the different types of procedures used. The accompanying glossary defines the relevant terms.

How does dialysis work?

Dialysis removes excess fluids and wastes from the blood by employing the processes of **diffusion, osmosis,** and **ultrafiltration** (see Figure NP22-1). The dialysate, a solution similar in composition to normal blood plasma, is delivered to a compartment beside a **semipermeable membrane;** the person's blood flows along the other side of the membrane. The semipermeable membrane acts like a filter: small molecules like urea and glucose can pass through microscopic pores in the membrane, whereas large molecules are unable to cross.

In *hemodialysis,* the tiny tubes that carry blood through the dialyzer are made of materials that serve as semipermeable membranes. In *peritoneal dialysis,* the body's peritoneal membrane, rich with blood vessels, is used to filter the blood.

How are solutes separated from the blood in dialysis?

The chemical composition of the dialysate affects the movement of solutes across the semipermeable membrane. When the concentration of a substance is lower in the dialysate than in the blood, the substance, provided it can cross the membrane, will diffuse out of the blood. To maximally remove waste products like urea from the blood, the dialysate contains no urea. For many other solutes, the dialysate is adjusted so that only excesses will be removed. Potassium can be removed from the blood, for example, by providing a dialysate with a lower concentration of potassium than the person's blood. The dialysate must contain some potassium, however; otherwise the blood potassium would fall too low.

The dialysate can also be used to add needed components back into the blood. For a person with acidosis, for example, bases such as bicarbonate are added to the dialysate and move by diffusion into the blood to alleviate acidosis.

How is fluid removed from the blood?

Because albumin and other plasma proteins are so adept at retaining fluids in blood, osmosis alone is not an efficient process for removing fluid. In hemodialysis, a **pressure gradient** is created between the blood and the dialysate. Most modern dialyzers produce *positive* pressure in the blood compartment and *negative* pressure in the dialysate compartment,[1] establishing a pressure gradient that "pushes" water (and accompanying solutes) through the pores of the membrane. This process, called ultrafiltration, relies on pumps to establish an appropriate flow rate between the blood and the dialysate.

How does the health practitioner know if the dialysis treatment has been effective?

A number of methods have been devised for gauging the adequacy of dialysis treatment. The most common method is **urea kinetic modeling,** a technique that evaluates the amount of

---

FIGURE NP22-1 Diffusion, Osmosis, and Ultrafiltration

Diffusion

Osmosis

Ultrafiltration

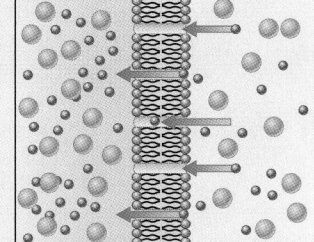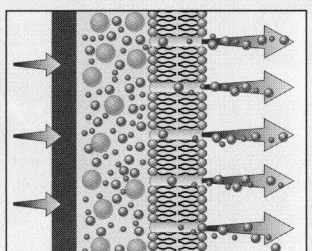

Small molecules (electrolytes and waste products) move from an area of high concentration to an area of low concentration by diffusion.

Water moves from an area of high water concentration to an area of low water concentration. In other words, water moves toward the side where solutes are more concentrated.

Pressure squeezes water and small molecules through the pores of a semipermeable membrane during ultrafiltration.

urea cleared from the blood. The formula used most often is Kt/V, where K is the amount of urea cleared, t is the time spent on dialysis, and V is the blood volume. The value obtained indicates whether the patient has undergone sufficient dialysis; the goal is a Kt/V result of approximately 1.2. Because technical data need to be incorporated into the calculation (such as dialyzer clearance data, blood flow rate, and dialysate flow rate), the computation is usually done by computer analysis. Current treatment guidelines recommend that hemodialysis adequacy be evaluated at least monthly or more often if problems develop or if the patient is noncompliant.[2]

How long does a hemodialysis treatment last?

As described previously, hemodialysis utilizes a dialyzer to cleanse the patient's blood. Although dialyzers vary in efficiency, the treatment usually lasts 3 to 4 hours and is required at least three times weekly. Some studies suggest that patients undergoing daily hemodialysis for briefer periods (2 to 2.5 hours) may tolerate dialysis treatment better and have fewer complications, but this approach has not been widely adopted.[3] Most patients visit dialysis centers to obtain treatment: home hemodialysis programs are available, but only about 2 percent of patients use them.

Are any complications associated with hemodialysis?

Yes. Although lifesaving, hemodialysis is associated with a substantial number of complications.[4] Problems at the vascular access site include infections and blood clotting. Hypotension can develop while blood is circulated through the dialyzer. Muscle cramping often occurs during the procedure, especially in the hands, legs, and feet. Blood losses can worsen anemia, which is already severe in two-thirds of patients beginning hemodialysis treatment. Patients may also experience headaches, weakness, nausea, vomiting, restlessness, and agitation.[5]

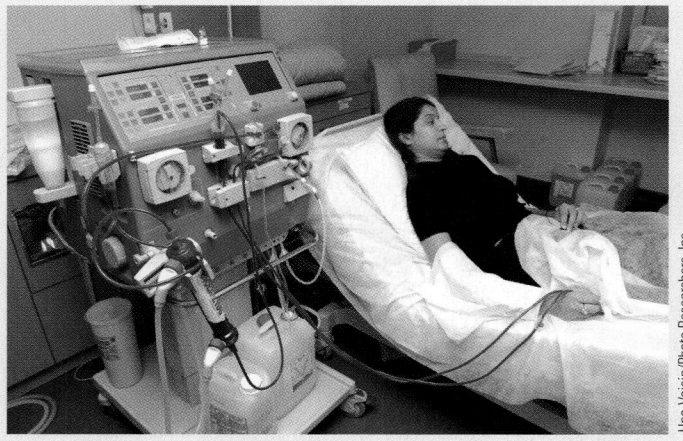

Hpa-Voisin/Photo Researchers, Inc.

During hemodialysis, blood passes through a dialyzer where wastes are extracted, and the cleansed blood is returned to the body.

How does peritoneal dialysis work?

In peritoneal dialysis, the peritoneal membrane surrounding the abdominal organs serves as a semipermeable membrane. The dialysate is infused into a catheter that empties into the peritoneal space—the space within the abdomen near the intestines (see Figure NP22-2). In the most common procedure, **continuous ambulatory peritoneal dialysis (CAPD),** the dialysate remains in the peritoneal cavity for 4 to 6 hours, after which it is drained and replaced with fresh dialysate (about 2 to 3 liters in adults). Generally, the dialysate solution is exchanged four times daily and requires only about 30 minutes to drain and replace.

Because a pressure gradient cannot be created in the peritoneal cavity as it can in a dialyzer, the glucose concentration in the dialysate must be high enough to create enough **oncotic pressure** to draw fluid from the blood. As indicated in Chapter 22, a substantial amount of glucose can be absorbed into the patient's blood and may contribute to weight gain over time. The high glucose load may also cause hyperglycemia and hypertriglyceridemia in some patients.

What are the advantages and disadvantages of peritoneal dialysis?

Peritoneal dialysis offers a number of advantages over hemodialysis: vascular access is not required, dietary restrictions are fewer, and the procedure can be scheduled when convenient. The most common complication is infection, which can occur at the catheter site or in the peritoneal cavity **(peritonitis).** Other problems that may arise include blood clotting in the catheter, catheter migration, and abdominal hernia due to the dialysate volume.[6]

What are the features of continuous renal replacement therapy?

In people with acute renal failure, **continuous renal replacement therapy (CRRT)** removes fluids and wastes. CRRT utilizes the process of **hemofiltration,** in which blood is gently pumped across a filtration membrane over a prolonged time period. (This differs from dialysis treatments that rely on the diffusion of wastes across a membrane into the dialysate.) Either a pump or the patient's own blood pressure may move the blood across the membrane. The procedure can be used to remove fluids, solutes, or both. Some patients require fluid replacement during the procedure to maintain adequate blood volume, so hydration status must be closely monitored.

The use of CRRT is advantageous in acute care situations because it corrects imbalances without causing sudden shifts in blood volume, which are poorly tolerated in acute care patients.[7] In addition, replacement fluids can include parenteral feedings without upsetting fluid balance. Complications include clotting problems, damage to arteries, and inadequate blood flow rates in hypotensive patients.

Dialysis and CRRT help to remove the wastes and fluids that are normally removed by healthy kidneys. Although these pro-

FIGURE NP22-2 Peritoneal Dialysis

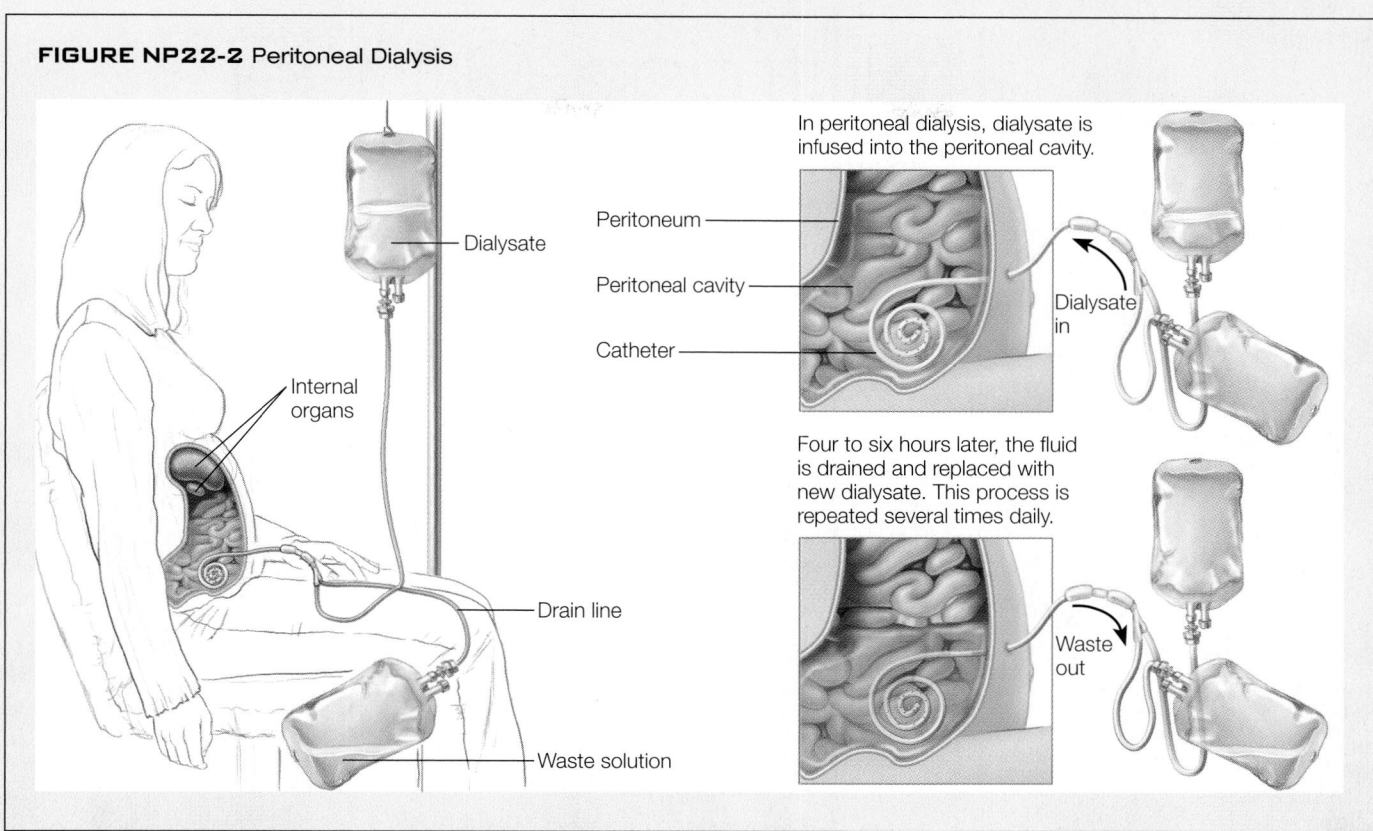

In peritoneal dialysis, dialysate is infused into the peritoneal cavity.

Peritoneum

Peritoneal cavity

Catheter

Dialysate in

Four to six hours later, the fluid is drained and replaced with new dialysate. This process is repeated several times daily.

Waste out

Dialysate

Internal organs

Drain line

Waste solution

cedures cannot restore the kidneys' hormonal functions, they provide a lifesaving means for alleviating symptoms of uremia, hypertension, and edema.

Glossary

continuous ambulatory peritoneal dialysis (CAPD): the most common method of peritoneal dialysis; involves frequent exchanges of dialysate, which remains in the peritoneal cavity throughout the day.

continuous renal replacement therapy (CRRT): a slow, continuous method of removing solutes and fluid from blood by gently pumping blood across a filtration membrane over a prolonged time period.

diffusion: movement of solutes from an area of high concentration to one of low concentration.

hemofiltration: removal of fluid and solutes by pumping blood across a membrane. No osmotic gradients are created during the process.

oncotic pressure: the pressure exerted by fluid on one side of a membrane as a result of osmosis.

osmosis: movement of water across a membrane toward the side where solutes are more concentrated.

peritonitis: inflammation of the peritoneal membrane.

pressure gradient: the change in pressure over a given distance. In dialysis, a pressure gradient is created between the blood and the dialysate.

semipermeable membrane: a membrane that allows some particles to pass through, but not others.

ultrafiltration: removal of fluids and solutes from blood by using pressure to transfer the blood across a semipermeable membrane.

urea kinetic modeling: a method of determining the adequacy of dialysis treatment by calculating urea clearance from blood.

Notes

[1]C. F. Gutch, Principles of hemodialysis, in C. F. Gutch, M. H. Stoner, and A. L. Corea, eds., *Review of Hemodialysis for Nurses and Dialysis Personnel* (St. Louis: Mosby, 1999), pp. 35–45.

[2]National Kidney Foundation, K/DOQI Clinical practice guidelines for hemodialysis adequacy: Update 2000, **www.kidney.org/ professionals/kdoqi/guidelines_updates/doqiuphd_intro .html**, visited November 6, 2005.

[3]A. Pierratos, New approaches to hemodialysis, *Annual Review of Medicine* 55 (2004): 179–189.

[4]N. Tolkoff-Rubin and N. Goes, Treatment of irreversible renal failure, in L. Goldman and D. Ausiello, eds., *Cecil Textbook of Medicine* (Philadelphia: Saunders, 2004), pp. 716–726.

[5]Gutch, 1999.

[6]Tolkoff-Rubin and Goes, 2004.

[7]M. Rolston and coauthors, Dialyzers, dialysate, and delivery systems, in C. F. Gutch, M. H. Stoner, and A. L. Corea, eds., *Review of Hemodialysis for Nurses and Dialysis Personnel* (St. Louis: Mosby, 1999), pp. 46–71.

CONTENTS

Nutrition, Cancer, and HIV Infection

CHAPTER 23

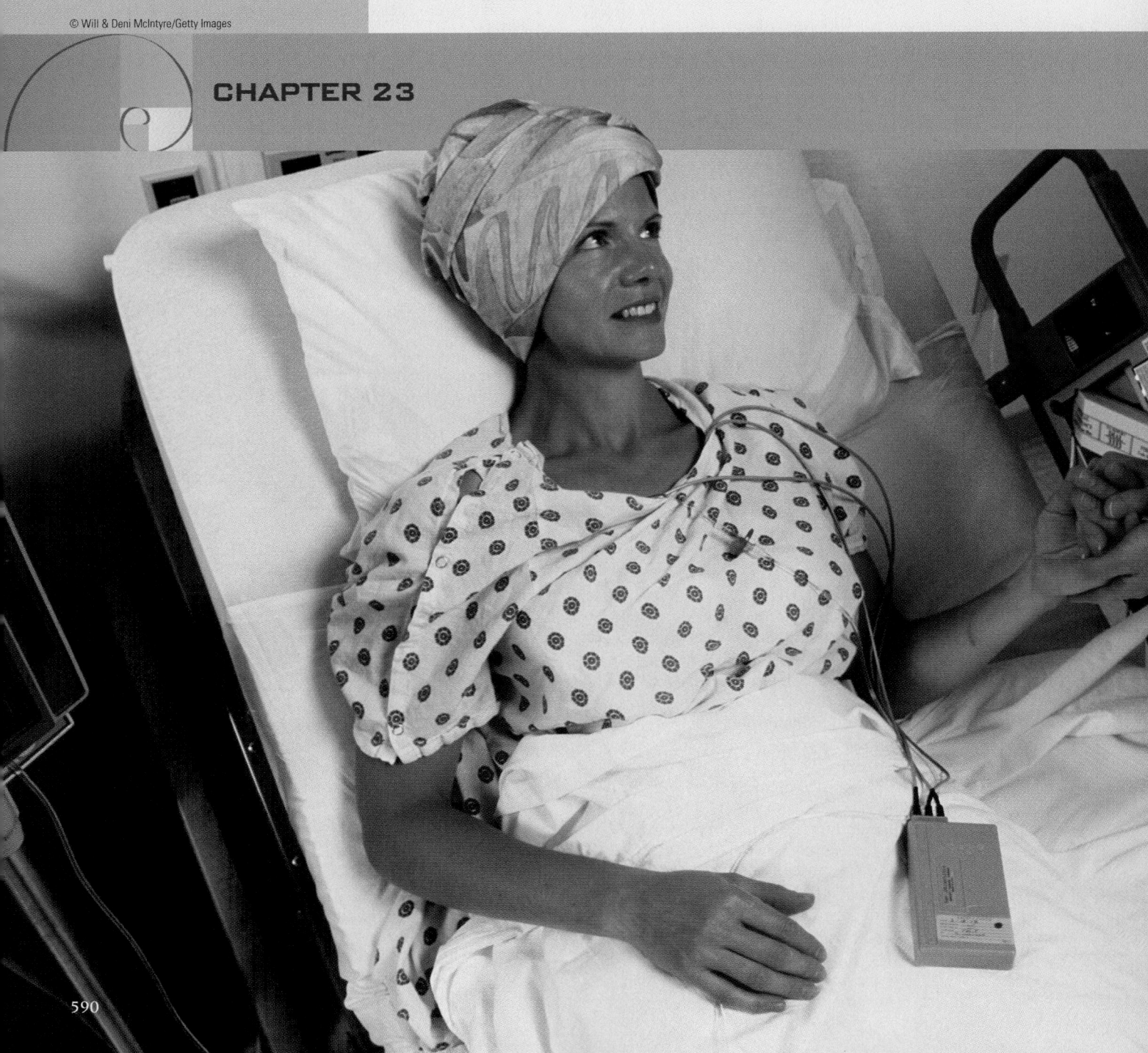

Although **cancers** and **HIV (human immunodeficiency virus)** infections are distinct disorders, from a nutritional standpoint, they share some similarities. Both disorders have debilitating effects that influence nutritional needs, and both can lead to severe wasting in advanced cases. These illnesses require medical nutrition therapy that is highly individualized based on the symptoms manifested and the organ systems involved.

Cancer

Cancer, the growth of **malignant** tissue, ranks just below cardiovascular disease as a cause of death in the United States. Cancer is not a single disorder, however; there are many cancers, that is, many different kinds of malignant growths. The different types of cancer have different characteristics, occur in different locations in the body, take different courses, and require different treatments. Whereas an isolated, nonspreading type of skin cancer may be removed in a physician's office with no effect on nutrition status, advanced cancers, especially those of the gastrointestinal (GI) tract and pancreas, can seriously impair nutrition status.

HOW CANCER DEVELOPS

The development of cancer, called **carcinogenesis,** often proceeds slowly and continues for several decades. A cancer arises from mutations in the genes that control cell division in a single cell. These mutations may promote cellular growth, interfere with growth restraint, or prevent cellular death.[1] The affected cell thereby loses its built-in capacity for halting cell division and produces daughter cells with the same genetic defects. As the abnormal mass of cells, called a **tumor,** grows, blood vessels form to supply the tumor with the nutrients it needs to support its growth. The tumor can disrupt the functioning of the normal tissue around it, and some tumor cells may **metastasize,** spreading to another region in the body. In leukemia (cancer affecting the white blood cells), the cells do not form a tumor, but rather accumulate in blood and other tissues. Figure 23-1 illustrates the steps in cancer development.

The reasons that cancers develop are numerous and varied. Vulnerability to cancer is sometimes inherited, as when a person is born with a genetic defect that alters DNA structure, function, or repair. Certain metabolic processes may initiate carcinogenesis, as when phagocytes (immune cells) produce oxidants that cause DNA damage, or chronic inflammation increases the rate of cell division, increasing the risk of a damaging mutation.[2] More often, cancers are caused by interactions between a person's genes and environmental agents. Exposure to cancer-causing substances, or **carcinogens,** may either induce genetic mutations that lead to cancer or promote proliferation of cancerous cells. Table 23-1 (p. 592) provides examples of environmental factors that increase cancer risk.

NUTRITION AND CANCER RISK

Like other environmental factors, diet and lifestyle strongly influence cancer risk. Certain food components may directly damage DNA, alter the metabolism of carcinogens by liver enzymes, or inhibit the formation of carcinogens in the body.[3] In addition, energy balance and growth rates affect the rate of cell division and consequently influence the rates at which mutations form and are replicated. Table 23-2 (p. 593) lists examples of nutrition-related factors that may increase or decrease the risk of developing cancer.

Nutrition and Increased Cancer Risk As shown in Table 23-2, obesity is a risk factor for a number of different cancers, including some relatively common cancers

---

Cancers are classified by the tissues or cells from which they develop:

- *Adenomas* (ADD-eh-NO-muz) arise from glandular tissues.
- *Carcinomas* (CAR-sih-NO-muz) arise from epithelial tissues.
- *Gliomas* (gly-OH-muz) arise from glial cells of the central nervous system.
- *Leukemias* (loo-KEY-mee-uz) arise from white blood cell precursors.
- *Lymphomas* (lim-FOE-muz) arise from lymph tissue.
- *Melanomas* (MEL-ah-NO-muz) arise from pigmented skin cells.
- *Sarcomas* (sar-KO-muz) arise from connective tissues, such as muscle or bone.

An abnormal mass of cells that is noncancerous is called a *benign* tumor.

cancers: malignant growths or tumors that result from abnormal and uncontrolled cell division.

HIV (human immunodeficiency virus): the virus that causes acquired immune deficiency syndrome (AIDS). HIV destroys immune cells and progressively impedes the body's ability to fight infections and certain cancers.

malignant (ma-LIG-nant): describes a cancerous cell or tumor, which can injure healthy tissue and spread cancer to other regions of the body.

carcinogenesis (CAR-sin-oh-JEN-eh-sis): the process of cancer development.

tumor: an abnormal tissue mass that has no physiological function; also called a *neoplasm* (NEE-oh-plazm).

metastasize (meh-TAS-tah-size): the spread of cancer cells from one part of the body to another.

carcinogens (CAR-sin-oh-jenz or car-SIN-oh-jenz): substances that can cause cancer (the adjective is *carcinogenic*).

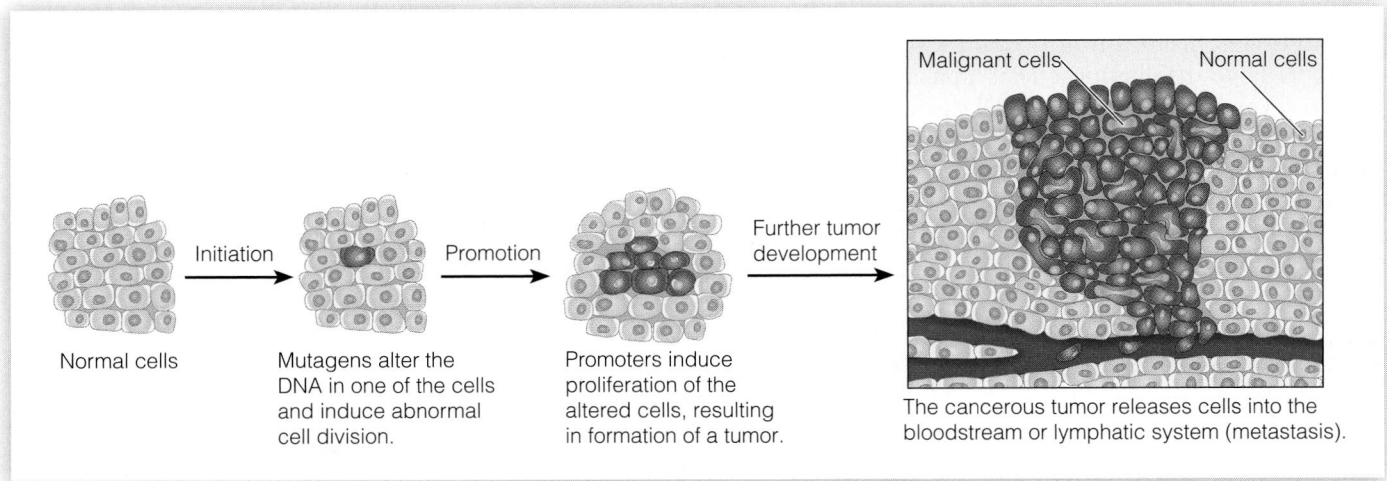

FIGURE 23-1 Cancer Development

like colon cancer and postmenopausal breast cancer. Obesity increases cancer risk by altering levels of hormones that influence cell growth, such as the sex hormones, insulin, and several kinds of growth factors. In the case of breast cancer in postmenopausal women, for example, the hormone estrogen is likely involved: obese women have higher estrogen levels than do lean women, because adipose tissue produces estrogen.

Although studies in animals have suggested that high-fat diets can promote tumor growth, studies of humans have not proved that fat's effects are indepen-

TABLE 23-1	Environmental Factors That Increase Cancer Risk
Environmental Factor	**Cancer Site(s)**
Aflatoxins (toxins in moldy peanuts or grains)	Liver
Alcohol[a]	Oral cavity, pharynx, esophagus, larynx, liver, colon, rectum, breast
Asbestos[b]	Lung, pleura, peritoneum
Chromium (hexavalent) compounds	Lung
Estrogen-progesterone replacement therapy	Breast
Immunosuppressive medications	Lymphoid tissues
Infection with *Helicobacter pylori*	Stomach
Infection with hepatitis B and hepatitis C viruses	Liver
Infection with human papillomavirus (HPV)	Cervix
Radiation	White blood cells (leukemia), breast, thyroid, lung
Sun exposure (ultraviolet radiation)	Skin
Tobacco[a]	Lung, oral cavity, pharynx, esophagus, larynx, renal pelvis, pancreas, bladder, kidney

[a]A combined exposure to alcohol and tobacco multiplies risks of developing cancers of the oral cavity, pharynx, esophagus, and larynx.

[b]Risk is greatly increased in cigarette smokers.

Source: W. J. Blot, Epidemiology of cancer, in L. Goldman and D. Ausiello, eds., *Cecil Textbook of Medicine* (Philadelphia: Saunders, 2004), pp. 1116–1120.

TABLE 23-2	Nutrition-Related Factors That Influence Cancer Risk	
	Nutrition-Related Factor[a]	**Cancer Site(s)**
Factors that increase cancer risk	Obesity	Colon, kidney, pancreas, esophagus, endometrium, gallbladder, breast (in postmenopausal women)
	Total fat[b]	Colon, prostate
	Red meat	Prostate, colon, rectum
	Calcium (over 1500 mg daily)	Prostate
	Salted and salt-preserved foods	Stomach
	Low level of physical activity	Colon
Factors that decrease cancer risk	Tomato products	Prostate
	Cruciferous vegetables (broccoli, cauliflower, brussels sprouts)	Prostate, bladder, lung
	Allium vegetables (onion, garlic)	Stomach
	Citrus fruits	Lung
	Folate-containing foods and supplements	Colon, esophagus, breast, white blood cells (leukemia)
	Calcium (up to 1000 mg daily)	Colon, rectum
	High level of physical activity	Colon

[a]Altered cancer risk is associated with high intakes of the dietary substances listed. The risks associated with alcohol are included in Table 23-1.

[b]The effect of fat may be due to its direct association with higher energy intakes (obesity) or with higher intakes of red meat.

Source: W. C. Willett and E. Giovannucci, Epidemiology of diet and cancer risk, in M. E. Shils and coeditors, *Modern Nutrition in Health and Disease* (Philadelphia: Lippincott Williams & Wilkins, 2006), pp. 1267–1279.

dent from those of energy intake and physical activity.[4] Evidence from population studies is mixed: high-fat diets often, but not always, correlate with high cancer rates. Within single populations, cancer rates do not reliably reflect fat intakes. In addition, the type of fat consumed may be critical: studies of prostate cancer implicate animal fats but not vegetable fat, while consuming fatty fish may be protective.[5]

Food preparation methods are responsible for producing certain types of carcinogens in foods. Cooking meat, poultry, and fish at high temperatures causes carcinogens to form on food surfaces.[6] Carcinogens also accompany the smoke that adheres to foods during grilling and are present in the charred surfaces of meat and fish. The cancer risk from eating such foods is unclear, however, as the biological actions of these carcinogens are modulated by other dietary components, including compounds in vegetables and other plant foods. In several population studies, consumption of well-cooked meats has been linked to cancers of the colon, breast, and stomach.[7]

Cruciferous vegetables, such as cauliflower, broccoli, and brussels sprouts, contain nutrients and phytochemicals that inhibit cancer development.

NURSING DIAGNOSIS

health-seeking behaviors applies to people who attempt to reduce cancer risk by modifying their diet and lifestyle.

NURSING DIAGNOSIS

imbalanced nutrition: less than body requirements, adult failure to thrive, and *ineffective protection* apply to those with cancer cachexia.

Reminder: The immune responses to tissue injury include the release of *cytokines* into the bloodstream (described in Chapter 16).

Reminder: *Protein turnover* refers to the continuous degradation and synthesis of the body's proteins.

cancer cachexia (ka-KEK-see-ah): a wasting syndrome, associated with cancer, that is characterized by anorexia, muscle wasting, weight loss, and fatigue.

Nutrition and Decreased Cancer Risk A considerable number of human studies have found a link between the consumption of certain fruits and vegetables and reduced incidences of cancers (review Table 23-2). Fruits and vegetables contain both nutrients and phytochemicals with antioxidant activity, and these substances may prevent or reduce the oxidative reactions in cells that cause DNA damage. Phytochemicals may also help to inhibit carcinogen production in the body, enhance immune functions that protect against cancer development, and promote enzyme reactions that inactivate carcinogens.[8] In addition, certain fruits and vegetables provide the B vitamin folate, which plays roles in DNA synthesis and repair; thus inadequate folate intakes may allow DNA damage to accumulate.

Although research reports in the 1970s and 1980s suggested that a fiber-rich diet could protect against colon cancer, recent studies have cast doubt on the earlier analyses.[9] The earlier studies depended on the ability of colon cancer patients to recall the foods they had consumed during the preceding years, whereas more recent studies—considered more reliable—tracked the subjects' health behaviors and cancer outcomes for extended periods (10 to 20 years). Moreover, some studies that had found fiber to be protective did not analyze factors such as physical activity, smoking, or folate intake, all of which can influence cancer outcome. A fiber-rich diet may be protective, in part, because high-fiber foods usually contain high levels of nutrients and phytochemicals that are protective against cancer. Table 23-3 summarizes dietary and lifestyle recommendations for reducing cancer risk.

CONSEQUENCES OF CANCER

Once cancer develops, its consequences depend on the location of the tumor, its severity, and the treatment. The complications that develop are often related to the impingement of the tumor on surrounding tissues. Nonspecific effects of cancer include anorexia, lethargy, weight loss, night sweats, and fever.[10] The likelihood of effective treatment is highest with early detection and intervention, yet during the early stages many cancers produce no symptoms and the person may be unaware of the threat to health.

Wasting Associated with Cancer Anorexia, muscle wasting, weight loss, and fatigue typify **cancer cachexia,** which occurs in as many as 80 percent of people with cancer.[11] Weight loss is often evident at the time that cancer is diagnosed, and severe malnutrition, often seen in the later stages of cancer, is the ultimate cause of death in many cases. Without adequate energy and nutrients, the body is poorly equipped to maintain organ function, support immune defenses, and mend damaged tissues. An involuntary weight loss of more than 10 percent, which indicates significant malnutrition, is cause for concern.[12]

Many factors play a role in the wasting associated with cancer. Cytokines, released by both tumor cells and immune cells, induce a hypermetabolic, catabolic state. The combined effects of a poor appetite, accelerated and abnormal metabolism, and diversion of nutrients to support tumor growth result in a lower supply of energy and nutrients at a time when demands are high. Appetite and food intake are further disturbed by the effects of treatments and medications prescribed for cancer patients.

Metabolic Changes The metabolic changes that arise in cancer exacerbate the wasting described in the previous section. Cancer patients exhibit an increased rate of protein turnover, but reduced muscle protein synthesis.[13] Muscle contributes amino acids for glucose production, further straining the body's supply of protein. Triglyceride breakdown increases, elevating serum lipids. Many patients develop insulin resistance. These metabolic abnormalities help to explain why people with cancer fail to regain lean tissue or maintain healthy body weights even when they are consuming adequate energy and nutrients.

TABLE 23-3	Recommendations for Reducing Cancer Risk[a]

Choose a diet rich in a variety of plant-based foods.

- Plant foods, such as vegetables, fruits, whole grains, and beans, should cover two-thirds or more of the plate.
- Fish, poultry, meat, or low-fat milk products should cover one-third or less of the plate.
- Limit consumption of processed foods and refined sugar.

Eat plenty of vegetables and fruits.

- Consume five or more servings of a variety of vegetables and fruits each day.
- Include vegetables that are dark green and leafy, as well as those that are deep orange in color.
- Include citrus fruits and other foods high in vitamin C.

Maintain a healthy weight and be physically active.

- Avoid being underweight or overweight, and limit weight gain during adulthood to less than 11 pounds (5 kilograms).
- If occupational activity is low or moderate, take an hour's brisk walk or participate in a similar exercise daily.
- Exercise vigorously for at least one hour each week.

Drink alcohol in moderation, if at all.

- Avoid alcohol consumption.
- If alcohol is consumed, limit it to less than two drinks a day for men and one for women.

Select foods low in fat and salt.

- Limit consumption of fatty foods, particularly those of animal origin. If red meat is eaten, limit intake to less than 3 ounces daily.
- Choose modest amounts of vegetable oils.
- Limit consumption of salted foods and use of cooking and table salt.
- Use herbs and spices to season foods.

Prepare and store foods safely.

- Use refrigeration and other appropriate methods to preserve perishable foods as purchased and at home.
- Do not eat charred food.
- Consume meat and fish grilled in direct flame and cured and smoked meats only occasionally.

Most importantly, do not smoke or use tobacco in any form.

[a] American Institute for Cancer Research, *The New American Plate,* revised ed. (2004), available at http://www.aicr.org/site/PageServer? (site visited) April 30, 2006; American Institute for Cancer Research, *Food, Nutrition and the Prevention of Cancer: A Global Perspective* (1997).

Anorexia and Reduced Food Intake   Anorexia is a major contributor to the wasting associated with cancer. Some factors that contribute to anorexia or otherwise reduce food intake include:

- *Chronic nausea and early satiety.* People with cancer frequently experience nausea and a premature feeling of fullness after eating small amounts of food.
- *Fatigue.* People with cancer often tire easily and lack the energy to prepare and eat meals. Once cachexia develops, these tasks become even more difficult to handle.

chemotherapy: the use of drugs to arrest or destroy cancer cells. Such drugs are called *antineoplastic agents.*

radiation therapy: the use of X-rays, gamma rays, or atomic particles to destroy cancer cells.

- *Pain.* People in pain may have little interest in eating, particularly if eating makes the pain worse.
- *Mental stress.* A cancer diagnosis can cause distress, anxiety, and depression, all of which may reduce appetite. Facing and undergoing cancer treatments causes additional psychological stress.
- *Effects of cancer therapies.* Therapies for cancer (including medications, chemotherapy, radiation therapy, surgery, and bone marrow transplants) can affect food intake by causing nausea, vomiting, altered taste perceptions, food aversions, inflammation of the mouth and esophagus, dry mouth, mouth sores, difficulty swallowing, intestinal cramping, diarrhea, and constipation.
- *Obstructions.* A tumor may partially or completely obstruct a portion of the GI tract, causing complications such as nausea and vomiting, early satiety, delayed gastric emptying, and bacterial overgrowth. Some patients with obstructions are unable to tolerate oral diets.

TREATMENTS FOR CANCER

The primary medical treatments for cancer—surgery, chemotherapy, radiation therapy, or any combination of the three—aim to remove cancer cells, prevent further tumor growth, and alleviate symptoms. As treatment decisions are difficult and cancer therapies have considerable side effects, patients rely on health care providers to help them make informed decisions.

Surgery Surgery is performed to remove tumors, determine the extent of cancer, and discern whether nearby tissues are involved.[14] Often, surgery must be followed by other cancer treatments to prevent growth of new tumors. The acute metabolic stress caused by surgery raises protein and energy needs and can exacerbate wasting. Surgery also contributes to pain, fatigue, and anorexia, all of which can reduce food intake at a time when nutritional needs are substantial. Blood loss contributes to nutrient losses and further exacerbates malnutrition. Some surgeries can have long-term effects on nutritional status (see Table 23-4).

Chemotherapy Chemotherapy relies on the use of drugs to inhibit tumor growth. Some of these drugs interfere with the process of cell division; others sterilize cells that are in a resting phase and not actively dividing. Ideally, chemotherapy would wipe out cancer cells without destroying healthy ones. Unfortunately, most of these drugs have toxic effects on normal cells as well and are especially damaging to rapidly dividing cells, such as those of the GI tract, skin, and bone marrow. Some of the newer drugs are able to target properties specific to cancer cells and are better tolerated by the body's tissues.[15] Table 23-5 includes a summary of the nutrition-related side effects that may result from chemotherapy.

Radiation Therapy Radiation therapy treats cancer by bombarding cancer cells with X-rays, gamma rays, or various atomic particles, which damage DNA and lead to cell death. Newer techniques are able to focus radiation directly at tumors and minimize damage to nearby tissues. An advantage of radiation therapy over surgery is that it can

TABLE 23-4	Nutrition-Related Side Effects of Cancer Surgeries

Head and Neck Surgeries

Difficulty in chewing/swallowing
Inability to chew/swallow

Esophageal Resection

Diarrhea
Fistula formation
Reduced gastric acid secretion
Reduced gastric motility
Steatorrhea (fat malabsorption)
Stenosis (constriction)

Gastric Resection

Dumping syndrome
General malabsorption
Hypoglycemia
Lack of gastric acid
Vitamin B_{12} malabsorption

Intestinal Resection

Blind loop syndrome
Diarrhea
Fluid and electrolyte imbalances
Hyperoxaluria
Malabsorption
Steatorrhea

Pancreatic Resection

Diabetes mellitus
Malabsorption

TABLE 23-5	Nutrition-Related Side Effects of Chemotherapy and Radiation		
	Reduced Nutrient Intake	**Accelerated Nutrient Losses**	**Altered Metabolism**
Chemotherapy	Abdominal pain	Diarrhea	Fluid and electrolyte imbalances
	Anorexia	Intestinal ulcers	Hyperglycemia
	Mouth ulcers	Malabsorption	Interference with vitamins or other metabolites
	Nausea	Vomiting	Negative nitrogen and calcium balances
	Taste alterations		Secondary effects of malnutrition, infection, or
	Vomiting		tissue damage (inflammation)
Radiation	Anorexia	Blood loss from intestine and bladder	Fluid and electrolyte imbalances as a conse-
	Damage to teeth and jaws	Diarrhea	quence of vomiting, diarrhea, or malabsorption
	Dysphagia	Fistulas	Secondary effects of malnutrition,
	Esophagitis	Intestinal obstructions	infection, or tissue damage (inflammation)
	Mouth ulcers	Malabsorption	
	Nausea	Radiation enteritis	
	Reduced salivary secretions	Vomiting	
	Taste alterations		
	Thick salivary secretions		
	Vomiting		

shrink tumors while preserving organ structure and function. Compared with chemotherapy, radiation therapy is better able to target specific regions of the body, rather than involving all body cells. Nonetheless, radiation therapy can damage healthy tissues and sometimes has long-term effects on nutritional status (see Table 23-5). Radiation to the head and neck area can damage the salivary glands and taste buds, causing inflammation, dry mouth, and a reduced sense of taste; in severe cases, damage may be permanent. Radiation treatment in the lower abdominal area can cause **radiation enteritis,** an inflammatory condition of the small intestine that causes nausea, vomiting, malabsorption, and diarrhea.

Bone Marrow Transplants Bone marrow transplants replace bone marrow that has been destroyed by chemotherapy or radiation therapy; they are also one of the primary treatments for leukemia. If possible, bone marrow cells are collected from the patient before chemotherapy or radiation treatment begins so that it is not necessary to find a separate donor.[16] If another person's cells are used, the patient must take immunosuppressant drugs to prevent **tissue rejection.**

The procedures used in transplant patients can have a substantial impact on food intake and nutrition status. The intensive chemotherapy, radiation treatments, and immunosuppressant drugs can all impair immune function substantially, increasing the risk of foodborne illness. Other common complications include anorexia, dry mouth, altered taste sensations, inflamed mucous membranes, malabsorption, nausea, vomiting, and diarrhea. Patients are often unable to consume adequate food and may require nutrition support, as described in a later section.

Medications to Combat Anorexia and Wasting To combat anorexia, medications may be prescribed to stimulate the appetite and promote weight gain. One of the most effective medications, megestrol acetate, is a synthetic compound similar in structure to the hormone progesterone.[17] Dronabinol, which resembles the psychoactive ingredient in marijuana, stimulates the appetite at doses that have minimal mental effects. Under investigation are medications that may help to restore

One drug that inhibits cell division is *methotrexate*, which closely resembles the B vitamin folate (see Figure 14-2 on p. 390). Folate is required for synthesis of DNA. Methotrexate inhibits cancer cell division by blocking activity of the enzyme that converts folate to its active form.

NURSING DIAGNOSIS

for people with bone marrow transplants include *imbalanced nutrition: less than body requirements, ineffective protection, risk for infection, impaired oral mucous membrane, disturbed sensory perception, nausea,* and *diarrhea.*

radiation enteritis: inflammation of intestinal tissue caused by exposure to radiation.

bone marrow transplants: procedures that replace bone marrow that has been destroyed by cancer treatments; also used to treat certain types of cancers and blood disorders.

tissue rejection: destruction of donor tissue by the recipient's immune system, which recognizes the donor cells as foreign.

lean body mass, such as anabolic steroids, growth hormone, and insulin-like growth factor.

The term *complementary and alternative medicine (CAM)* refers to health care practices that have not been proved to be effective and therefore, are not included as part of conventional treatment. Nutrition in Practice 23 provides additional information about CAM.

Alternative Therapies   Many patients turn to *complementary and alternative medicine (CAM)* to assist them with their fight against cancer. Although few abandon conventional medicine, up to 80 percent of cancer patients combine one or more CAM approaches with standard treatment.[18] Few patients discuss their use of CAM with physicians.

Dietary supplements and herbal remedies are among the most frequently used CAM therapies. Although many supplements can be used without risk, some may have adverse effects or interfere with conventional treatments. Use of the herbal remedy St. John's wort, for example, can reduce the effectiveness of some anticancer drugs.[19] As another example, some studies suggest that antioxidant supplements interfere with chemotherapy and radiation treatments.[20] Clinical trials of several popular supplements are in progress to learn more about their potential effects and interactions with treatments.

MEDICAL NUTRITION THERAPY FOR CANCER

The objectives of medical nutrition therapy for cancer patients are to minimize loss of weight and muscle tissue, correct nutrient deficiencies, and provide a diet that can be tolerated and enjoyed despite the complications of illness. Appropriate nutrition care helps patients preserve their strength and improves recovery after stressful cancer treatments. Moreover, malnourished cancer patients develop more complications and have shorter survival times than patients who maintain good nutrition status.[21]

Nutritional needs among cancer patients vary considerably, as there are many forms of cancer and a variety of potential treatments. Furthermore, a person's needs may change at different stages of illness. Patients should be screened for malnutrition when cancer is diagnosed and reassessed during the treatment and recovery periods.

Protein and Energy   For patients at risk of weight loss and wasting, protein and energy needs are considerable. Daily protein requirements may be 1.0 to 1.5 grams per kilogram body weight to maintain lean body mass and 1.5 to 2.0 grams per kilogram to repair and rebuild tissues.[22] Energy needs may reach 145 percent of

HOW TO   *Increase kCalories and Protein in Meals*

To increase the energy content of a meal, try these suggestions:

- *Butter or margarine*. Melt on pasta, potatoes, rice, and cooked vegetables. Add to hot cereals, casseroles, and soups. Spread liberally on bread, crackers, and rolls.
- *Mayonnaise*. Add to pasta, tuna, and potato salads. Use as a dressing for raw or cooked vegetables.
- *Cream cheese*. Spread on raw vegetables, toast, and crackers. Mix into chopped fruit. Use as a spread in sandwiches made with luncheon meats.
- *Half-and-Half and cream*. Replace milk or water with Half-and-Half or cream in soups, sauces, hot chocolate, desserts, mashed potatoes, and cold and cooked cereals.
- *Nuts*. Add chopped nuts to pasta dishes, stir-fried vegetables, fruit salads, and green salads. Use nut meals in baked products.
- *Beverages*. Replace water and nonkcaloric beverages with sweetened drinks, fruit juices, and milk shakes.

These suggestions can help to add protein to a meal:

- *Powdered milk (use full-fat milk powder if available)*. Add to recipes that include milk. Dissolve extra milk powder into milk-containing beverages. Stir into hot cereals, potato dishes, casseroles, and sauces. Add to scrambled eggs, hamburger, and meat loaf.
- *Cheese*. Melt on burgers, meat loaf, cooked vegetables, scrambled eggs, casseroles, and potatoes. Add cottage cheese to casseroles, egg dishes, pasta recipes, and salad dressings. Grate hard cheeses and sprinkle on soups, salads, and cooked vegetable dishes.
- *Eggs*. Add raw eggs when preparing casseroles, meatballs, and hamburgers. Add chopped hard-cooked eggs to salads, vegetable dishes, sandwich fillings, and pasta and potato salads.
- *Meats*. Add meat pieces to soups, egg dishes, casseroles, bean dishes, and pasta sauces. Add minced meats to vegetable dishes. Add chunks of cooked chicken or turkey to salads.

basal energy expenditure (see Table 16-2 on p. 432). Health practitioners should regularly monitor patients' weight changes and adjust intake recommendations as necessary. Patients who cannot eat adequate food may be able to meet their needs by supplementing the diet with nutrient-dense formulas. The "How to" provides suggestions that can help to increase the energy and protein content of meals.

Although weight loss is a problem for many cancer patients, breast cancer patients often gain weight. In one survey, 63 percent of women with breast cancer reported weight gains, ranging from 5 to 27 pounds.[23] Weight gain occurs more often in premenopausal women and in those undergoing extensive chemotherapy. By discussing weight maintenance soon after diagnosis and encouraging physical activity, health practitioners can help patients avoid unncecessary weight gain.[24]

Managing Symptoms and Complications A thorough nutrition assessment may uncover specific problems that interfere with food consumption. Table 23-6 lists dietary considerations related to cancers affecting different sites in the body. The "How to" on p. 601 outlines a variety of dietary measures that may improve food intake and alleviate symptoms. Patients' responses to these strategies may vary considerably, and in some cases, a number of adjustments may be necessary.

Enteral and Parenteral Nutrition Support Nutrition support is used in limited situations during cancer treatment. Generally, tube feedings and parenteral nutrition are provided to patients who have long-term or permanent gastrointestinal impairment or are experiencing complications that interfere with food intake.[25] Many patients undergoing radiation treatment for head and neck cancers, for example, require long-term tube feeding and often need to continue tube feedings at home. Parenteral nutrition is reserved for patients who have inadequate GI function, such as individuals with chronic radiation enteritis. Whenever possible, enteral nutrition is strongly preferred over parenteral nutrition, to preserve GI function and avoid infection.

Reminder: The high-kcalorie, high-protein diet was described in Chapter 14.

Radiation to the head and neck regions often causes dysphagia and mouth sores.

TABLE 23-6	Dietary Considerations for Specific Cancers

Cancer Sites	Common Complications[a]	Possible Dietary Measures
Brain and nervous system	Chewing and swallowing difficulties, difficulty feeding oneself	Mechanically altered diet, use of adaptive feeding devices (see Nutrition in Practice 21)
Head and neck[b]	Swallowing difficulty, aspiration, inflamed mucosa, dry mouth, altered taste sensation	Tube feeding, mechanically altered diet
Esophagus	Swallowing difficulty, obstruction, acid reflux, inflamed mucosa	Tube feeding, mechanically altered diet
Stomach	Anorexia, delayed stomach emptying, early satiety, dumping syndrome, malabsorption	Tube feeding (for obstruction or unmanageable dumping syndrome), postgastrectomy diet, small frequent meals, limited sugars and insoluble fibers (see Chapter 17)
Intestine	Fluid and electrolyte imbalances, altered bowel function, malabsorption, lactose intolerance, inflamed mucosa, bacterial overgrowth, short bowel syndrome (if resected), obstruction	Tube feeding or total parenteral nutrition for obstruction, enteritis, or short bowel syndrome; fat- and lactose-restricted diet (see Chapter 18)
Pancreas	Malabsorption, bile insufficiency, hyperglycemia	Fat-restricted diet, enzyme replacement (see Chapter 18), small frequent meals, carbohydrate-controlled diet (Chapter 20)

[a] Actual complications depend on the specific methods used for treating the cancer.

[b] Includes cancers of the pharynx, larynx, salivary glands, and oral and nasal cavities.

CASE STUDY   *Public Relations Consultant with Cancer*

Jane Woodhouse is a 58-year-old public relations consultant who was recently diagnosed with colon cancer after a routine colonoscopy, a procedure in which the colon is examined using a flexible tube attached to an optical device. Mrs. Woodhouse is scheduled to have surgery to remove the segment of colon that contains the tumor and to determine if the cancer has spread to the surrounding lymph nodes and, possibly, other organs. The nurse completing the nutrition assessment finds that Mrs. Woodhouse is 5 feet 5 inches tall and weighs 178 pounds. She spends most of the day sitting and has little time to engage in recreational exercise. Her diet is high in fat and typically includes red meat at both lunch and dinner. She eats two or three servings of fruits and vegetables each day, although she doesn't like green leafy vegetables very much. She rarely drinks milk or consumes milk products.

1. Review Table 23-2 on p. 593 and describe the factors in Mrs. Woodhouse's diet and lifestyle that may have contributed to the development of colon cancer.

2. What symptoms and complications can arise after colon surgery and impair nutrition status? If the cancer team decides that Mrs. Woodhouse needs follow-up chemotherapy, how might the chemotherapy affect her nutrition status?

3. If Mrs. Woodhouse is unresponsive to treatment and her cancer progresses, she may develop cancer cachexia. Describe this syndrome and its causes. What are the benefits of preventing or correcting the wasting associated with cancer?

4. Provide suggestions that may help Mrs. Woodhouse handle these problems should they develop: poor appetite, fatigue, taste alterations, nausea and vomiting, chewing and swallowing difficulties, mouth sores, dry mouth, diarrhea, and weight loss.

Patients who undergo bone marrow transplants may require total parenteral nutrition (TPN) before and after the transplant, as the GI tract is often severely damaged by the preparatory procedure (which may include high-dose chemotherapy or radiation treatment). When GI function returns, the patient can begin consuming small amounts of food along with TPN. As oral intake improves, TPN is gradually tapered. Because recipients of bone marrow transplants are severely immunocompromised, they should be instructed to follow safe food-handling practices to minimize the risk of foodborne illness (see Chapter 1). In addition, they need to avoid foods that are likely to contain unsafe levels of bacteria, such as fresh fruits and vegetables and undercooked meat, poultry, and eggs.

> ### REVIEW NOTES
>
> Cancer arises from mutations in the genes that control cell division. Some dietary substances promote carcinogenesis, whereas others may help to prevent cancer.
>
> Cancer's effects on nutritional status depend on the type of cancer, its severity, and the methods of treatment. Cancer cachexia is a frequent complication of cancer and may be related to anorexia, altered metabolism, and responses to treatment.
>
> Medical nutrition therapy for cancer patients aims to minimize weight loss and wasting, correct deficiencies, and manage complications that impair food intake. The accompanying Case Study allows you to apply information about nutrition and cancer to a clinical situation.

HIV Infection

acquired immune deficiency syndrome (AIDS): the late stage of illness caused by infection with the human immunodeficiency virus (HIV); characterized by severe damage to immune function.

A diagnosis of HIV infection, which leads to **acquired immune deficiency syndrome (AIDS),** can be devastating. HIV (human immunodeficiency virus) attacks the immune system and disables a person's defenses against infections and certain cancers. Patients may expect an ever-worsening course of illness and, possibly,

 **Help Patients Handle Food-Related Problems**

In people with cancer or HIV infections, various complications can interfere with eating. Nurses can try to identify the specific problems that patients are having and offer appropriate solutions. Not every suggestion will work for each patient; encourage patients to experiment and find the strategies that work best.

- **I just don't have an appetite.**
    - Eat small meals and snacks at regular times each day.
    - Eat the largest meal at the time of day when you feel the best.
    - Include nutrient-dense foods in meals, and consume them before other foods.
    - Indulge in favorite foods throughout the day. Serve foods attractively.
    - Avoid drinking large amounts of liquids before or with meals.
    - Eat in a pleasant and relaxed environment. Eat with family and friends when possible.
    - Listen to your favorite music or enjoy a program on TV while you eat.
    - Take a walk before you eat.

- **I am too tired to fix meals and eat.**
    - Let family members and friends prepare food for you.
    - Obtain foods that are easy to prepare and easy to eat, like sandwiches, frozen dinners, take-out meals from restaurants, instant breakfast drinks, liquid formulas, and energy bars.

- **Foods just don't taste right.**
    - Brush your teeth or use mouthwash before you eat.
    - Consume foods chilled or at room temperature.
    - Choose eggs, fish, poultry, and milk products instead of meats.
    - Experiment with sauces, seasonings, herbs, and spices to improve food flavor.
    - Use plastic, rather than metal, eating utensils.
    - Save your favorite foods for times when you are not feeling nauseated.

- **I am nauseated a lot of the time, and sometimes I need to vomit.**
    - Consume liquids throughout the day to replace fluids.
    - If you become nauseated from chemotherapy treatments, avoid eating for at least two hours before treatments.
    - Consume smaller meals, and eat slowly.
    - Avoid foods and meals that have strong odors or are fatty, greasy, or gas forming.

- **I am having problems chewing and swallowing food.**
    - Experiment with food consistencies to find the ones you can manage best. Thin liquids, dry foods, and sticky foods (like peanut butter) are often difficult to swallow.
    - Add sauces and gravies to dry foods.
    - Drink fluids with meals to ease chewing and swallowing.
    - Try using a straw to drink liquids.

- Tilt your head forward and backward to see if you can swallow more easily when your head is positioned differently.

- **I have sores in my mouth and they hurt when I eat.**
    - Use cold or frozen foods; they are often soothing.
    - Try soft foods like ice cream, milk shakes, bananas, applesauce, mashed potatoes, cottage cheese, and macaroni and cheese.
    - Avoid foods that irritate mouth sores like citrus fruits and juices, tomatoes and tomato-based products, spicy foods, foods that are very salty, foods with seeds (like poppy seeds and sesame seeds) that can scrape the sore, and coarse foods like raw vegetables and toast.
    - Ask your doctor about using a local anesthetic solution like lidocaine before eating to reduce pain.
    - Use a straw for drinking liquids, in order to bypass the sores.

- **My mouth is really dry.**
    - Rinse your mouth with warm salt water or mouthwash frequently. Avoid using mouthwash that contains alcohol.
    - Drink small amounts of liquid frequently between meals.
    - Ask your doctor or pharmacist about medications that can help dry mouth.
    - Use sour candy or gum to stimulate the flow of saliva.
    - Add broth, sauces, gravies, mayonnaise, butter, or margarine to dry foods.
    - Make sure you brush your teeth and floss regularly to prevent cavities and oral infections.

- **I am having trouble with diarrhea.**
    - Drink plenty of fluids. Salty broths and soups, diluted fruit juices, and sports drinks are good choices. For severe diarrhea, try oral rehydration formulas that are commercially prepared.
    - Avoid foods and beverages that increase gas, such as legumes, onions, vegetables of the cabbage family, foods that contain sorbitol or mannitol, and carbonated beverages.
    - Try using lactase enzyme replacements when you use milk products in case you are experiencing lactose intolerance. Yogurt and aged cheeses may be easier to tolerate than milk and fresh cheeses.
    - Avoid high-fat foods if you are fat intolerant.
    - Avoid caffeine.
    - Eat smaller meals, and eat more frequently.
    - Check with your doctor about using digestive enzyme replacements if you have had diarrhea for a long time.

- **I am having trouble with constipation.**
    - Drink plenty of fluids. Try warm fluids, especially in the morning.
    - Eat whole-grain breads and cereals, nuts, fresh fruits and vegetables, prunes, and prune juice. Avoid refined carbohydrate foods like white bread, white rice, and pasta.
    - Engage in physical activity regularly.

TABLE 23-7	HIV and AIDS Epidemic at a Glance, 2005	
	World	United States
Living with HIV or AIDS	40,300,000	1,000,000
Newly infected with HIV	4,900,000	40,000
AIDS deaths	3,100,000	18,000

Source: UNAIDS/WHO, *AIDS Epidemic Update: December 2005*, **http://www**
.unaids.org/epi/2005/doc/report_pdf.asp, site visited May 1, 2006.

TABLE 23-8	Risk Factors for HIV Infection

- History of receiving blood transfusions or clotting factors between 1978 and 1985.

- Infant born to mother with HIV infection.

- Intravenous drug use in which syringes are shared among users.

- Sexual contact with multiple partners.

- Sexual contact with intravenous drug users, prostitutes, or individuals with a history of HIV or other sexually transmitted diseases.

- Unsafe sexual practices.

NURSING DIAGNOSIS

for those with HIV infection may include *ineffective protection, risk for infection, risk for impaired oral mucous membrane, risk for impaired skin integrity, chronic pain, diarrhea, risk for deficient fluid volume, fatigue, imbalanced nutrition: less than body requirements, hopelessness*, and *death anxiety*.

T cells are lymphocytes that develop in the thymus gland. The other lymphocytes are the *B cells* (which develop in bone marrow) and *natural killer cells*.

helper T cells: lymphocytes that have a specific protein called CD4 on their surfaces and therefore are also known as *CD4+ T cells;* the cells most affected in HIV infection.

opportunistic infections: infections from microorganisms that normally do not cause disease in healthy people but are damaging to persons with compromised immune function.

death. In recent years, however, treatment options have expanded, and patients have benefited by vast improvements in quality of life.

The HIV/AIDS epidemic continues to sweep across countries, especially in sub-Saharan Africa. Table 23-7 shows its impact worldwide and in the United States. For many years, the destructive effects of HIV infection seemed unstoppable, but in the mid-to-late 1990s, the death rate from AIDS began to decline in the United States, and the progression from HIV to AIDS slowed dramatically. The disease still has no cure, but remarkable progress has been made in understanding and treating HIV infection. Without a cure, the best course is prevention. HIV is most often sexually transmitted and can be spread by direct contact with contaminated body fluids, such as blood, semen, vaginal secretions, and breast milk.

Because many people remain symptom-free during the early stages of infection, they may not realize that they can pass the infection to others. To reduce the spread of HIV infection, those at risk are encouraged to undergo testing (see Table 23-8). A blood test can usually detect HIV antibodies within several months after exposure and, often, after one or two weeks. An estimated 25 percent of persons in the United States who have HIV infection are unaware that they are infected.[26]

CONSEQUENCES OF HIV INFECTION

HIV infection destroys immune cells that have a protein called CD4 on their surfaces. The cells most affected are the **helper T cells,** also called *CD4+ T cells* because the presence of CD4 is a primary characteristic. Early symptoms of infection are nonspecific and may include fever, sore throat, malaise, skin rashes, nausea, muscle and joint pain, and diarrhea. Afterwards, many people remain symptom-free for five to ten years or even longer. If the HIV infection is not treated, however, the depletion of T cells eventually increases the person's susceptibility to **opportunistic infections,** caused by microorganisms that normally do not cause disease in healthy individuals.

The term *AIDS* applies to the advanced stages of HIV infection, in which the inability to fight illness allows a number of serious diseases and complications to develop; such **AIDS-defining illnesses** include severe infections, certain cancers, and wasting. About half of untreated persons with HIV infection develop AIDS within ten years,[27] although the period varies depending on such factors as genetic susceptibility and health status. Health practitioners evaluate the progression of the disease by measuring the concentrations of helper T cells and circulating virus (called the *viral load*) and by monitoring clinical symptoms. Although current drug therapies can dramatically slow the progression of HIV infection, the drugs' side effects make it difficult for patients to adhere to treatments, as discussed in the following sections.

Lipodystrophy Many of the drug treatments that suppress HIV infection cause abnormalities in glucose and fat metabolism in an estimated 25 to 50 percent of patients. These complications, collectively known as the **HIV-lipodystrophy syndrome,** include body fat redistribution, abnormal lipid levels, and insulin resistance.[28] Patients tend to accumulate abdominal fat and lose fat from the face, arms, and legs. Thus they appear to be thin except for a "pot belly." Also observed are breast enlargement (in both men and women), fat accumulation at the base of the neck (called **buffalo hump**), and benign growths composed of fat tissue (called **lipomas**). The changes in body composition are often disfiguring and may cause physical discomfort; moreover, patients often develop hypertriglyceridemia, low HDL (high-density lipoprotein) cholesterol levels, glucose intolerance, and hyperinsulinemia. The reasons for the development of lipodystrophy are unknown.

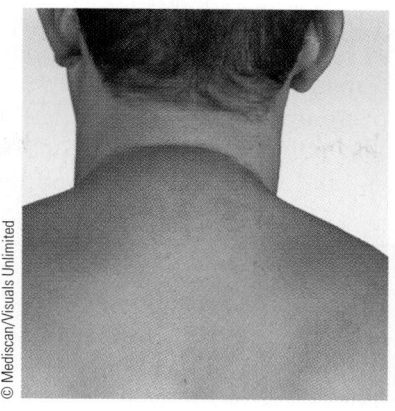

HIV-lipodystrophy syndrome is sometimes evident by the accumulation of fatty tissue at the base of the neck, referred to as *buffalo hump.*

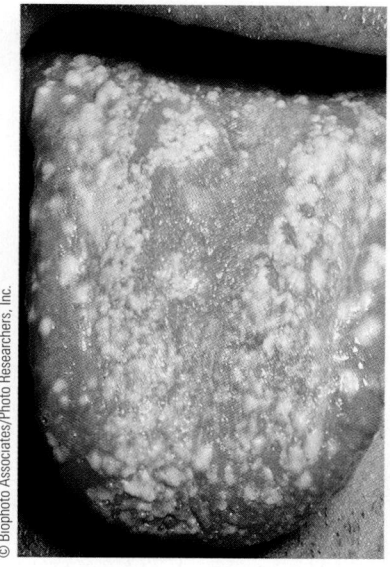

The oral infection thrush is easily identified by the characteristic milky white patches that appear on the tongue.

Weight Loss and Wasting Even with effective treatment of HIV infection, weight loss and wasting are ongoing problems for HIV-infected patients.[29] Wasting has been linked with accelerated disease progression, reduced strength, and fatigue. In the later stages of AIDS, wasting is severe and increases the risk of death. Much as in cancer, the wasting associated with HIV infection has many causes: anorexia and inadequate food intake, altered metabolism, malabsorption, chronic diarrhea, and food-drug interactions.

Anorexia and Reduced Food Intake Poor food intake is a key factor in the development of wasting. Anorexia and reduced food intake may result from various factors, including the following:

- *Emotional distress, pain, and fatigue.* The physical and social problems that accompany chronic illness may cause fear, anxiety, and depression, which contribute to anorexia. Pain and fatigue, which may be associated with some disease complications, can cause anorexia and difficulty with eating.
- *Oral infections.* The oral infections associated with HIV infection can cause discomfort and interfere with food consumption. Common infections include thrush and herpes simplex virus. **Thrush** can cause mouth pain, dysphagia, and altered taste sensation, and **herpes simplex virus** may cause painful lesions around the lips and in the mouth.
- *Respiratory disorders.* Respiratory infections, including pneumonia and tuberculosis, are common in people with HIV infection. Symptoms often include chest pain, shortness of breath, and cough, which interfere with eating and contribute to anorexia.
- *Cancer.* As described earlier in this chapter, cancer leads to anorexia for numerous reasons. In addition, **Kaposi's sarcoma,** a type of cancer frequently associated with HIV infection, can cause lesions in the mouth and throat that make eating painful.
- *Medications.* The medications given to treat HIV infection, other infections, and cancer often cause anorexia, nausea and vomiting, altered taste sensations, food aversions, and diarrhea.

GI Tract Complications Complications involving the GI tract may result from opportunistic infections, medications, or the HIV infection itself.[30] In addition to the oral infections described previously, infections commonly develop in the stomach and intestines. Advanced AIDS is often accompanied by characteristic changes in the lining of the small intestine, likely caused by GI infection: the villi appear shortened and flattened, and the absorptive area is substantially reduced. These changes contribute to malabsorption, steatorrhea, and diarrhea.

The Centers for Disease Control defines *AIDS-related wasting syndrome* as a 10% weight loss within a six-month period accompanied by diarrhea or fever for more than 30 days without a known cause.

The AIDS-related abnormalities in the intestinal mucosa are sometimes referred to as *HIV enteropathy* (EN-ter-OP-ah-thy).

AIDS-defining illnesses: diseases and complications associated with the later stages of an HIV infection; include wasting, recurrent bacterial pneumonia, opportunistic infections, and certain cancers.

HIV-lipodystrophy (LIP-oh-DIS-tro-fee) **syndrome:** a collection of abnormalities in fat and glucose metabolism that result from drug treatments for HIV; includes body fat redistribution, abnormal lipid levels, and insulin resistance. The accumulation of abdominal fat is sometimes called *protease paunch.*

buffalo hump: the accumulation of fatty tissue at the base of the neck.

lipomas (lih-POE-muz): benign tumors composed of fatty tissue.

thrush: a fungal infection of the mouth and throat, most often caused by *Candida albicans.*

herpes simplex virus: a common virus that can cause blisterlike lesions on the lips and in the mouth.

Kaposi's (cap-OH-seez) **sarcoma:** a common cancer in HIV-infected persons that is characterized by lesions on the skin, in the lungs, and in the GI tract.

As described earlier, many patients are unable to tolerate the medications used to suppress HIV and develop nausea, vomiting, and diarrhea. Furthermore, medications that treat GI viral, parasitic, and fungal infections contribute to bacterial overgrowth. Thus HIV-infected patients face an extremely high risk of malnutrition due to the combination of intestinal malabsorption, bacterial overgrowth, and nutrient losses from vomiting and diarrhea.

TREATMENTS FOR HIV INFECTION

Although there is no cure for HIV infection, treatments can help to slow its progression, reduce complications, and alleviate pain. The standard treatment for suppressing HIV infection, called *highly active antiretroviral therapy (HAART),* combines three or more antiretroviral drugs. Table 23-9 lists the major drug categories included in antiretroviral therapy and describes the drugs' modes of action.[31] These antiretroviral agents have multiple adverse effects that make their long-term use difficult to tolerate. In addition to the GI effects discussed previously, side effects include skin rashes, headache, anemia, tingling and numbness, hepatitis, pancreatitis, and kidney stones. Thus, although HAART has improved life span and quality of life for many patients, the drug regimens are difficult to adhere to and cause complications that require continual management. The Diet-Drug Interactions feature summarizes the nutrition-related effects of antiretroviral agents and other drugs mentioned in this chapter.

Control of Anorexia and Wasting Appetite stimulants, physical activity, and anabolic hormones have been successful in reversing weight loss and increasing lean body mass in HIV-infected patients.[32] The medications megestrol acetate and dronabinol (described on p. 597) are sometimes prescribed to stimulate appetite and help with weight gain. Testosterone and human growth hormone have demonstrated positive effects on nitrogen balance and lean tissue mass, especially in combination with resistance training.

Control of Lipodystrophy Treatment strategies for lipodystrophy are under investigation. Both aerobic activity and resistance training may help to reduce abdominal fat, although some patients opt for cosmetic surgery.[33] Patients may be given alternative antiretroviral drugs to alleviate symptoms. Medications may be prescribed to treat abnormal blood lipids and insulin resistance.

Alternative Therapies Like cancer patients, people with HIV infection and AIDS are frequently tempted to try unconventional methods of treatment. Although many

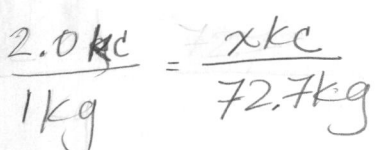

TABLE 23-9	Antiretroviral Drugs for Treatment of HIV Infection	
Category	**Example(s)**	**Mode of Action**
Nucleoside reverse transcriptase inhibitors (NRTI)	Zidovudine (AZT) Didanosine Lamivudine	As analogs of the nucleosides needed for DNA synthesis, NRTI impair the ability of HIV's *reverse transcriptase* enzyme to produce usable copies of DNA.
Non-nucleoside reverse transcriptase inhibitors (NNRTI)	Nevirapine Delavirdine Efavirenz	NNRTI bind active sites on HIV's *reverse transcriptase* enzyme, blocking the ability of HIV to produce DNA copies of its genetic material.
Protease inhibitors (PI)	Saquinavir Ritonavir Indinavir	PI inhibit HIV's *protease enzyme*, which cleaves HIV's gene products into usable structural proteins.
Fusion Inhibitors	Enfuvirtide	Fusion inhibitors prevent HIV from entering cells.

DIET-DRUG INTERACTIONS Check this table for notable nutrition-related effects of the medications discussed in this chapter.

	Gastrointestinal Effects	Interactions with Dietary Substances	Metabolic Effects
Appetite stimulants (megestrol acetate, dronabinol)	Nausea, vomiting, diarrhea		Hyperglycemia (megestrol acetate)
Didanosine	Nausea, vomiting, dry mouth, altered taste perception, anorexia, constipation	Avoid alcohol and aluminum- and magnesium-containing antacids.	Pancreatitis
Enfuvirtide	Nausea, vomiting, anorexia, diarrhea, constipation		Pancreatitis, increased blood triglycerides
Methotrexate	Nausea, vomiting, diarrhea, reduced absorption of vitamin B_{12} and calcium	Milk may reduce methotrexate absorption if they are ingested together.	Increased serum uric acid levels, anemia, liver toxicity
Ritonavir	Nausea, vomiting, altered taste perception, anorexia, diarrhea		Pancreatitis, diabetes, reduced blood levels of copper and zinc; increased levels of triglycerides, liver enzymes, creatine kinase, and uric acid
Zidovudine (AZT)	Nausea, vomiting, altered taste perception, anorexia, mouth sores, constipation		Anemia, reduced blood levels of copper and zinc

Note: Other antiretroviral drugs that treat HIV infection have gastrointestinal and metabolic side effects; only a few are listed here as examples.

alternative therapies are harmless, they can be expensive at a time when financial security is of concern. Monitoring patients' use of dietary supplements is essential to reduce the possibility of drug-nutrient and drug-herb interactions.

MEDICAL NUTRITION THERAPY FOR HIV INFECTION

The initial nutrition assessment provides baseline data with which to monitor progress throughout the course of the disease. The assessment should include an evaluation of body composition. Follow-up measurements may indicate the need to adjust dietary recommendations and drug therapies.

Weight Maintenance A primary objective of nutrition therapy is to maintain weight and lean body mass.[34] Daily energy requirements may be 35 to 45 kcalories per kilogram body weight, and protein needs may be as high as 2.0 to 2.5 grams per kilogram. A regular program of resistance training can improve muscle mass and strength and correct some of the metabolic abnormalities (altered blood lipids and insulin resistance) that are common in HIV-infected patients.

Health practitioners should attempt to determine the dietary and lifestyle factors that may interfere with patients' food intake, appetite, and physical activity, and provide suggestions that may prevent future weight problems. If food consumption is difficult, small, frequent feedings may be better tolerated than several large meals. The addition of nutrient-dense snacks, protein or energy bars, and oral supplements can improve intakes. Liquid formulas may be especially useful for the person who is too tired to eat or prepare meals. The "How to" on p. 598 provides suggestions for adding energy and protein to the diet, and the "How to" on p. 601 includes strategies for improving appetite and dealing with complications that interfere with eating.

Resistance training can help a person with HIV infection maintain muscle mass and strength.

Vitamins and Minerals Vitamin and mineral needs of people with HIV infections are highly variable, and little information is available concerning specific needs. Because nutrient deficiencies are likely to result from reduced food intake, malabsorption, diet-drug interactions, and nutrient losses, multivitamin-mineral supplements are usually recommended.

Additional suggestions for managing insulin resistance and hyperlipidemias are available in Chapters 20 and 21, respectively.

Metabolic Complications HIV patients using antiretroviral drugs frequently develop insulin resistance and elevated triglyceride and LDL (low-density lipoprotein) cholesterol levels, and dietary adjustments should be attempted before medications are prescribed.[35] Patients should be advised to achieve or maintain desirable weight, replace saturated fats with monounsaturated and polyunsaturated fats, and limit intakes of *trans*-fatty acids and cholesterol. Complex carbohydrates are preferred over foods high in sugars. Regular physical activity is recommended for improving both blood lipid levels and insulin resistance. If problems persist, alternative antiretroviral medications are sometimes attempted.

Food Safety The depressed immunity of people with HIV infections places them at extremely high risk of developing foodborne infections. Patients should be cautioned about their high susceptibility for foodborne illness and given detailed instructions about the safe handling and preparation of foods. Water can also be a source of foodborne illness and is a common cause of **cryptosporidiosis** in HIV-infected individuals. Because water quality varies throughout the United States, local health departments should be consulted to determine if the local tap water is safe for patients to drink. If not, or to take additional safety measures, water used for drinking and making ice cubes should be boiled for one minute. Some types of filtered and bottled waters are also safe, but not all.

Enteral and Parenteral Nutrition Support In later stages of illness, people with HIV infections may need aggressive nutrition support if they are unable to consume enough food. Tube feedings are preferred whenever the GI tract is functional; they can be given at night to supplement oral diets during the day. Preventing bacterial contamination of the formula is particularly important. Parenteral nutrition is reserved for patients who are unable to tolerate enteral nutrition, such as those with GI obstructions that prevent food intake. For individuals with severe malabsorption, orally administered hydrolyzed formulas containing medium-chain triglycerides may be as effective as parenteral nutrition for reversing weight loss and wasting.

cryptosporidiosis (KRIP-toe-spo-rid-ee-OH-sis): a foodborne illness caused by the parasite *Cryptosporidium parvum*.

CASE STUDY   Film Producer with HIV Infection

Three years ago, Darrell Meckler, a 34-year-old film producer, sought medical help when he began feeling run-down and developed a painful white fungal infection over his mouth and tongue. The presence of thrush, recent weight loss, and anemia alerted Mr. Meckler's physician to the possibility of an HIV infection. When Mr. Meckler tested positive for HIV, he and his family and friends were devastated by the news, but those close to him have remained supportive. During the three years since Mr. Meckler began antiretroviral drug therapy, he has maintained his weight but has also developed lipodystrophy and hypertriglyceridemia. Mr. Meckler is 6 feet tall and currently weighs 185 pounds. He occasionally develops diarrhea and sometimes anorexia.

1. Describe lipodystrophy, and discuss its typical pattern in people who have an HIV infection. What adjustments in treatment and lifestyle may be helpful?
2. Describe an appropriate diet for Mr. Meckler. What strategies may improve his problems with diarrhea and anorexia? Suggest reasons why people with an HIV infection may develop diarrhea and anorexia.
3. Explain why an HIV infection can lead to wasting as the disease progresses to the later stages.

REVIEW NOTES

By attacking immune cells, HIV causes progressive damage to immune function and may eventually lead to AIDS.

Improved drug therapies have slowed the progression of HIV infection; however, these drugs may promote the HIV-lipodystrophy syndrome, characterized by body fat redistribution, abnormal lipid levels, and insulin resistance.

HIV infection is often associated with weight loss and wasting, anorexia, and various complications that affect food intake. Dietary adjustments, resistance training, and medications can help patients maintain their weight and prevent wasting.

People with HIV infections must pay strict attention to food safety guidelines to prevent foodborne infections. The Case Study provides an opportunity to review the nutrition concerns of a person with HIV infection.

NUTRITION ASSESSMENT CHECKLIST FOR PEOPLE WITH CANCER OR HIV INFECTIONS

Medical History

Check the medical record to determine:

☐ Type and stage of cancer

☐ Stage of HIV infection

Review the medical record for complications that may alter medical nutrition therapy including:

☐ Altered organ function

☐ Altered taste perception

☐ Anorexia

☐ Dry mouth and oral infections

☐ GI symptoms and infections

☐ Hyperlipidemias

☐ Insulin resistance

☐ Malnutrition and wasting

Medications

For patients with cancer or HIV infections:

☐ Check medications to identify potential diet-drug interactions.

☐ Recommend using antinauseants at mealtime, if needed.

☐ Ask about use of dietary supplements, including herbal remedies.

For cancer patients who require chemotherapy:

☐ Recommend strategies to prevent food aversions.

☐ Offer suggestions for managing drug-related complications.

For HIV-infected patients using antiretroviral drug therapy:

☐ Remind patients that some drugs are better absorbed with foods and others must be taken on an empty stomach.

☐ Help patients work out a medication schedule that suits their lifestyle and is timed appropriately in regard to food intake.

☐ Offer suggestions for managing drug-related complications.

Dietary Intake

For patients with poor food intakes and weight loss:

☐ Determine the reasons for reduced food intake.

☐ Offer appropriate suggestions to improve food intake.

☐ Provide interventions before weight loss progresses too far.

For patients with HIV infections who experience weight gain, elevated triglyceride or LDL cholesterol levels, or hyperglycemia:

☐ Assess the diet for energy, total fat, types of fat, carbohydrates, fiber, and sugars.

☐ For hyperlipidemias, recommend a diet low in saturated fat, *trans*-fatty acids, and sugars.

☐ For hyperglycemia, recommend a consistent carbohydrate intake that emphasizes complex carbohydrates.

☐ Recommend regular physical activity for weight control and for improving blood lipid levels and insulin resistance.

Anthropometric Data

Take baseline height and weight measurements, monitor weight regularly, and suggest dietary adjustments for weight maintenance, if necessary. Perform baseline and periodic body composition measurements in HIV-infected patients who are using antiretroviral drug therapy.

Laboratory Tests

Note that albumin and other serum proteins may be reduced in patients with cancer or HIV infections, especially in those experiencing wasting. Check laboratory tests for indications of:

☐ Anemia

☐ Dehydration

☐ Elevated triglyceride levels

☐ Elevated LDL cholesterol levels

☐ Hyperglycemia

For patients with HIV infections, evaluate disease progression by checking:

☐ Helper T cell counts

☐ Viral load

Physical Signs

Look for physical signs of:

☐ Wasting and protein-energy malnutrition

☐ Dehydration (especially for those with fever, vomiting, or diarrhea)

☐ Oral infections

☐ Kaposi's sarcoma

SELF CHECK

1. Which dietary substances may help to protect against cancer?
   a. alcohol
   b. well-cooked meats, poultry, and fish
   c. animal fats
   d. phytochemicals from fruits and vegetables

2. The metabolic changes that often result from cancer include:
   a. increased fat synthesis.
   b. increased protein turnover.
   c. increased muscle protein synthesis.
   d. reduced serum lipids.

3. An advantage of radiation therapy over chemotherapy is that:
   a. radiation is not damaging to rapidly dividing cells.
   b. radiation's side effects do not include malnutrition.
   c. radiation can be directed toward the regions affected by cancer.
   d. the radiation used is too weak to damage GI tissues.

4. Although many cancer patients lose weight, which type of cancer is often associated with weight *gain*?
   a. kidney cancer
   b. breast cancer
   c. colon cancer
   d. Kaposi's sarcoma

5. Oral diets after bone marrow transplants may restrict:
   a. fiber.
   b. carbohydrates.
   c. high-protein foods.
   d. raw fruits and vegetables.

6. The immune cells most seriously damaged by HIV are:
   a. B cells.
   b. helper T cells.
   c. natural killer cells.
   d. neutrophils.

7. HIV-lipodystrophy syndrome is characterized by these changes in body composition:
   a. increased central and peripheral fat.
   b. decreased central and peripheral fat.
   c. increased central and decreased peripheral fat.
   d. decreased central and increased peripheral fat.

8. Mouth sores in people with HIV infections are most frequently due to:
   a. oral infections.
   b. dehydration.
   c. nutrient deficiency.
   d. foodborne illnesses.

9. The medications megestrol acetate and dronabinol:
   a. are used to promote weight gain.
   b. are protease inhibitors that fight HIV infection.
   c. treat common opportunistic infections that develop in AIDS patients.
   d. treat HIV-lipodystrophy syndrome.

10. To prevent cryptosporidiosis, a person with HIV infection may need to:
    a. cook meat, poultry, and fish to an appropriate internal temperature.
    b. avoid consuming undercooked eggs.
    c. avoid foods prepared by people who are sick or who have skin infections.
    d. boil drinking water for one minute.

Answers to these questions appear in Appendix H.

CLINICAL APPLICATIONS

1. Consider the nutrition problems that may develop in a 36-year-old woman with a malignant brain tumor that affects her ability to move the right side of her body (including the tongue) and to speak coherently. She is taking a pain medication that makes her nauseated and sleepy. Her expected survival time is only about six months.

   • If she is right-handed, how might her impairment interfere with eating? What suggestions do you have for overcoming this problem?

   • How might her nutrition status be affected by her inability to communicate effectively? What suggestions may help?

   • Describe ways in which the pain medication she is taking can affect her nutrition status.

2. Various types of chronic conditions can lead to weight loss and wasting. For some of these conditions, such as celiac disease or Crohn's disease (Chapter 18), diet is a cornerstone of treatment. For others, such as cancer and HIV infection, nutrition plays a supportive role. What determines whether nutrition plays a primary role or a supportive role in the treatment of disease?

NUTRITION ON THE NET

For further study of the topics in this chapter, access these websites.

Find updates and quick links to these and other nutrition-related sites at our website:
www.wadsworth.com/nutrition

To learn more about cancer, including risk factors, prevention, screening, detection, treatments (including nutrition), and support networks, visit these sites:

   American Cancer Society: **www.cancer.org**

   National Cancer Institute: **www.cancer.gov**

   CancerSource: **www.cancersource.com**

American Institute for Cancer Research: **www.aicr.org**

American Association for Cancer Research: **www.aacr.org**

To find additional information about HIV infection and AIDS, visit these sites:

   The Body: **www.thebody.com**

   AIDS Education Global Information System: **www.aegis.com**

   UCSF Center for HIV Information: **hivinsite.ucsf.edu**

To review information about safe food handling, visit the Food and Drug Administration's Center for Food Safety and Applied Nutrition: **vm.cfsan.fda.gov**

NOTES

[1] E. T. Liu, Oncogenes and suppressor genes: Genetic control of cancer, in L. Goldman and D. Ausiello, eds., *Cecil Textbook of Medicine* (Philadelphia: Saunders, 2004), pp. 1108–1116.

[2] S. Christen and coauthors, Chronic inflammation, mutation, and cancer, in J. Parsonnet and S. Hornig, eds., *Microbes and Malignancy: Infection as a Cause of Cancer* (New York: Oxford University Press, 1999), pp. 35–88.

[3] W. C. Willett and E. Giovannucci, Epidemiology of diet and cancer risk, in M. E. Shils and coeditors, *Modern Nutrition in Health and Disease* (Philadelphia: Lippincott Williams & Wilkins, 2006), pp. 1267–1279.

[4] Willett and Giovannucci, 2006.

[5] P. D. Terry and coauthors, Intakes of fish and marine fatty acids and the risks of cancers of the breast and prostate and of other hormone-related cancers: A review of the epidemiologic evidence, *American Journal of Clinical Nutrition* 77 (2003): 532–543.

[6] T. Sugimura and coauthors, Heterocyclic amines: Mutagens/carcinogens produced during cooking of meat and fish, *Cancer Science* 95 (2004): 290–299; J. S. Felton and coauthors, Impact of environmental exposures on the mutagenicity/carcinogenicity of heterocyclic amines, *Toxicology* 198 (2004): 135–145; P. Jakszyn and coauthors, Development of a food database of nitrosamines, heterocyclic amines, and polycyclic aromatic hydrocarbons, *Journal of Nutrition* 134 (2004): 2011–2014.

[7]G. N. Wogan and coauthors, Environmental and chemical carcinogenesis, *Seminars in Cancer Biology* 14 (2004): 473–486.

[8]R. H. Liu, Potential synergy of phytochemicals in cancer prevention: Mechanism of action, *Journal of Nutrition* 134 (2004): 3479S–3485S.

[9]Willett and Giovannucci, 2006.

[10]H. S. Rugo, Paraneoplastic syndromes and other non-neoplastic effects of cancer, in L. Goldman and D. Ausiello, eds., *Cecil Textbook of Medicine* (Philadelphia: Saunders, 2004), pp. 1124–1131.

[11]Rugo, 2004.

[12]M. Schattner and M. Shike, Nutrition support of the patient with cancer, in M. E. Shils and coeditors, *Modern Nutrition in Health and Disease* (Philadelphia: Lippincott Williams & Wilkins, 2006), pp. 1290–1313.

[13]B. Eldridge and coauthors, Nutrition and the patient with cancer, in A. M. Coulston, C. L. Rock, and E. R. Monsen, eds., *Nutrition in the Prevention and Treatment of Disease* (San Diego: Academic Press, 2001), pp. 397–412.

[14]J. R. Bertino and W. Hait, Principles of cancer therapy, in L. Goldman and D. Ausiello, eds., *Cecil Textbook of Medicine* (Philadelphia: Saunders, 2004), pp. 1137–1150.

[15]Bertino and Hait, 2004.

[16]S. Z. Pavletic and J. M. Vose, Hematopoietic stem cell transplantation, in L. Goldman and D. Ausiello, eds., *Cecil Textbook of Medicine* (Philadelphia: Saunders, 2004), pp. 999–1003.

[17]M. Tomiska, Palliative treatment of cancer anorexia with oral suspension of megestrol acetate, *Neoplasma* 50 (2003): 227–233.

[18]B. R. Cassileth and G. Deng, Complementary and alternative therapies for cancer, *The Oncologist* 9 (2004): 80–89; M. A. Richardson and coauthors, Complementary/alternative medicine use in a comprehensive cancer center and the implications for oncology, *Journal of Clinical Oncology* 18 (2000): 2505–2514.

[19]M. Markman, Safety issues in using complementary and alternative medicine, *Journal of Clinical Oncology* 20 (2002): 39S–41S.

[20]B. Bruemmer and coauthors, The association between vitamin C and vitamin E supplement use before hematopoietic stem cell transplant and outcomes to two years, *Journal of the American Dietetic Association* 103 (2003): 982–990; H. E. Seifried and coauthors, The antioxidant conundrum in cancer, *Cancer Research* 63 (2003): 4295–4298.

[21]Schattner and Shike, 2006.

[22]S. Escott-Stump, *Nutrition and Diagnosis-Related Care* (Baltimore: Lippincott Williams & Wilkins, 2002), pp. 525–531.

[23]J. A. McInnes and M. T. Knobf, Weight gain and quality of life in women treated with adjuvant chemotherapy for early-stage breast cancer, *Oncology Nursing Forum* 28 (2001): 675–684.

[24]A. L. Schwartz, Exercise and weight gain in breast cancer patients receiving chemotherapy, *Cancer Practice* 8 (2000): 231–237.

[25]Schattner and Shike, 2006.

[26]UNAIDS/WHO, *AIDS Epidemic Update: December 2005*, **http://data .unaids.org/epi/2005/doc/report_pdf.asp**, site visited May 1, 2006.

[27]UNAIDS/WHO, 2005.

[28]P. Koutkia and S. Grinspoon, HIV-associated lipodystrophy: Pathogenesis, prognosis, treatment, and controversies, *Annual Review of Medicine* 55 (2004): 303–317.

[29]A. M. Tang and coauthors, Weight loss and survival in HIV-positive patients in the era of highly active antiretroviral therapy, *Journal of Acquired Immune Deficiency Syndromes* 31 (2002): 230–236.

[30]J. G. Bartlett, Gastrointestinal manifestations of AIDS, in L. Goldman and D. Ausiello, eds., *Cecil Textbook of Medicine* (Philadelphia: Saunders, 2004), pp. 2168–2170.

[31]Panel on Clinical Practices for Treatment of HIV Infection, *Guidelines for the Use of Antiretroviral Agents in HIV-1-Infected Adults and Adolescents*, October 6, 2005, available from **http://AIDSinfo.nih.gov** (site visited November 28, 2005); S. Safrin, Antiviral agents, in B. G. Katzung, ed., *Basic and Clinical Pharmacology* (New York: Lange Medical Books/McGraw-Hill, 2001), pp. 823–844.

[32]S. Grinspoon and K. Mulligan, Weight loss and wasting in patients infected with human immunodeficiency virus, *Clinical Infectious Diseases* 36 (2003): S69–S78.

[33]Koutkia and Grinspoon, 2004.

[34]J. Nerad and coauthors, General nutrition management in patients infected with human immunodeficiency virus, *Clinical Infectious Diseases* 36 (2003): S52–S62; Escott-Stump, 2002, pp. 614–617.

[35]M. Dube and M. Fenton, Lipid abnormalities, *Clinical Infectious Diseases* 36 (2003): S79–S83; M. C. Gelato, Insulin and carbohydrate dysregulation, *Clinical Infectious Diseases* 36 (2003): S91–S95.

Alternative Therapies

The medical treatments described in the clinical chapters are based on our current scientific understanding of human physiology and biochemistry and are generally supported by well-conducted clinical research. This Nutrition in Practice examines therapies that have *not* been scientifically validated and therefore are not currently accepted by conventional medical professionals; they fall into a category called *complementary and alternative medicine (CAM)*.

How popular is CAM in the United States?

In 2002, an estimated 36 percent of adults in the United States used some form of CAM (excluding the use of prayer).[1] CAM is most prevalent among people with chronic, debilitating diseases; for example, 84 percent of AIDS patients reportedly use CAM.[2] CAM therapies remain popular despite the dearth of evidence demonstrating their effectiveness. Reasons for their popularity include consumers' growing interest in self-help measures, the noninvasive nature of many CAM therapies, and the positive interactions consumers have with CAM practitioners.[3]

In response to the enormous popularity of CAM in the United States, in 1998 Congress established the National Center for Complementary and Alternative Medicine (NCCAM), which is now one of the 27 institutes that make up the National Institutes of Health. NCCAM's missions are to investigate complementary and alternative therapies by funding well-designed scientific studies and to provide authoritative information for consumers and health professionals.

Should mainstream health professionals learn more about CAM?

Yes, health care professionals often need to be familiar with CAM therapies so that they can better communicate with patients regarding health care and advise them when an alternative approach conflicts with standard therapy or presents a danger to health. To provide medical students with objective information about CAM, about half of U.S. medical schools now offer elective courses about alternative forms of treatment.[4] The term *alternative* may be somewhat misleading, however, because it implies that unproven methods of treatment are valid alternatives to conventional treatments.

What kinds of practices are considered CAM therapies?

CAM includes any and all therapies that are not usually part of conventional medicine. Consequently, the list of CAM approaches includes hundreds of advertised therapies purchased and used by consumers. Unfortunately, CAM has become a marketing buzzword and is used by unscrupulous sellers of worthless treatments. NCCAM categorizes CAM

therapies as shown in Table NP23-1 and defined in the glossary of alternative therapies on p. 615. The most common alternative practice is the therapeutic use of natural products, such as dietary supplements and herbal remedies.

How effective are herbal remedies in the treatment of disease?

As herbs contain naturally occurring compounds that exert physiological effects, they have been used for centuries to treat medical conditions. Only a limited number of clinical studies support the traditional uses, however. NCCAM is currently funding large, controlled trials of several popular herbal treatments in an effort to obtain reliable efficacy and safety data. Table NP23-2 (p. 612) lists examples of popular herbs, their common uses, and potential adverse effects associated with their use.

TABLE NP23-1 Examples of Complementary and Alternative Medicine

Alternative Medical Systems
- Naturopathic medicine
- Homeopathic medicine
- Traditional Chinese medicine
- Ayurveda

Mind-Body Interventions
- Meditation
- Faith healing (prayer)
- Mental healing (including hypnotherapy)
- Music, art, and dance therapy

Biologically Based Therapies
- Dietary supplements
- Foods and special diets
- Herbal products
- Hormones
- Aromatherapy

Manipulative and Body-Based Methods
- Chiropractic
- Massage therapy
- Osteopathic manipulation
- Reflexology

Energy Therapies
- Biofield therapies (including therapeutic touch, acupuncture, and qi gong)
- Bioelectrical fields (including electrical and magnetic fields)

TABLE NP23-2 Popular Herbs, Their Common Uses, and Adverse Effects

Herb	Scientific Name	Common Uses	Adverse Effects
Black cohosh	*Cimicifuga racemosa*	Relief of menopausal symptoms	Rare; occasional stomach upset, headache, weight gain
Chaparral	*Larrea tridentata*	General tonic, treatment of infection, cancer, and arthritis	Hepatitis, liver failure
Comfrey	*Symphytum officinale*	Wound healing (topical use), treatment of lung and GI disorders	Liver damage
Echinacea	*Echinacea augustifolia, E. pallida, E. purpurea*	Prevention and treatment of upper respiratory infections	Rare; occasional allergic reactions
Feverfew	*Tanacetum parthenium*	Prevention of migraine headache	Mouth and tongue sores, swelling of lips, GI upset
Garlic	*Allium sativum*	Reduction of blood clotting, atherosclerosis, blood pressure, and blood cholesterol	Halitosis (bad breath), body odor; occasional dyspepsia, flatulence, excessive bleeding, anorexia, allergic reactions
Ginger	*Zingiber officinale*	Prevention and treatment of nausea and motion sickness	Rare; occasional heartburn
Ginkgo	*Ginkgo biloba*	Treatment of dementia, memory defects, and circulatory impairment	Rare; occasional stomach upset, headache, skin hypersensitivity, excessive bleeding
Ginseng	*Panax ginseng, P. quinquefolius*	General tonic, reduction of blood glucose levels	Rare
Kava	*Piper methysticum*	Treatment of anxiety, stress, insomnia	Dyspepsia, restlessness, drowsiness, tremor, headache, dermatitis (with heavy use), occasional hepatitis and liver failure
St. John's wort	*Hypericum perforatum*	Treatment of mild-to-moderate depression	Rare; occasional stomach upset, fatigue, dizziness, headache, dry mouth, dermatitis, skin photosensitivity
Saw palmetto	*Serenoa repens*	Reduction of symptoms associated with enlarged prostate	Rare
Valerian	*Valeriana officinalis*	Sedation, treatment of insomnia	Rare
Yohimbe	*Pausinystalia yohimbe*	Treatment of erectile dysfunction	Anxiety, headache, dizziness, nausea, rapid heartbeat, hypertension, increased urinary frequency; isolated reports of renal failure, blood disorders, and airway constriction

Source: M. Rotblatt and I. Ziment, *Evidence-Based Herbal Medicine* (Philadelphia: Hanley & Belfus, Inc., 2002).

Determining the effectiveness of herbal remedies can be complicated, however. Herbs contain many different compounds, and it is often unclear which, if any, might produce the implied beneficial effect. Because the compounds in herbs vary among species and are affected by a plant's growing conditions, different samples of an herb can have different chemical compositions. Furthermore, even when the active ingredients in an herbal preparation have been shown to be effective, the dosage suggested on the label may not provide the amount of active ingredients found to be effective. For example, a consumer group (ConsumerLab.com) tested 13 ginkgo biloba products and found that 7 of the products, when consumed at the recommended dose, lacked adequate levels of one or more compounds believed to be helpful.[5]

How safe are the herbal products that are available in the marketplace?

Consumers of herbal supplements often assume that because plants are "natural," herbal products must be harmless. Many herbal remedies do have toxic effects, however. The most common adverse effects of herbs include diarrhea, nausea, and vomiting.[6] The herbs kava, chaparral, and comfrey have caused liver damage. The use of yohimbe has been linked to hypertension, heart palpitations, and renal failure. In 2004, the Food and Drug Administration (FDA) removed the herb ephedra (also known as *ma huang*) from the market, advising that its side effects (which include elevated blood pressure and rapid heartbeat) could cause heart attack or stroke. The adverse effects of herbs are rarely listed on supplement labels.

Contamination of herbal products is another safety concern. When 251 products imported from Asia were analyzed, 10 percent were found to contain lead, 14 percent arsenic, and 14 percent mercury in excessive amounts.[7] Other contaminants occasionally found in herbal products include molds, bacteria, and pesticides that have been banned for use on food crops. Adulteration of imported products has been a serious concern: a Taiwanese study found that 24 percent of the 2609 herbal products tested contained synthetic drugs that were not declared on the label, and 53 percent of the adulterated products contained two or more added drugs.[8] There have also been reports of serious illnesses and fatalities occurring from the intentional or accidental substitution of one plant species for another.[9]

Is there any risk that herbs may interact with medications?

Yes. Like drugs, herbs may either potentiate or interfere with the effects of medications. Information about herb-drug interactions is limited, however, and much of what is known has been obtained from case studies rather than from controlled clinical trials. An herb may either increase or decrease the effects of medications, or it may raise the risk of toxicity. For example, garlic, ginkgo, and ginseng may increase the risk of bleeding when used with anticoagulant drugs. St. John's wort has been found to inhibit the actions of oral contraceptives, anticoagulants, and other drugs. Individuals may be more susceptible to the adverse effects of a drug if the herb they are using has a similar effect; for example, ginger and ginseng contain compounds that raise blood pressure and may increase the toxicity of drugs that have a similar side effect.[10]

In what ways do alternative medical systems differ from conventional medicine?

Alternative medical systems (review Table NP23-1) are based on beliefs that lack the scientific basis of the theories underlying conventional medicine. Virtually all of these alternative systems were developed well over 100 years ago, before our bodies' biochemical and physiological processes were well understood. The alternative forms of diagnoses and treatments may appeal to consumers because the interventions are nontechnical and seem nonthreatening. In general, however, alternative theories and practices have not been updated to include our current knowledge. Examples of alternative medical systems include the following:

- **Naturopathic medicine** proposes that ill health results from an internal disruption rather than from external disease-causing agents. Naturopathic therapies aim to enhance the natural healing powers of the body and may

include special diets or fasting, herbal remedies and other dietary supplements, **acupuncture**, homeopathy, massage, and various other interventions.

- **Homeopathic medicine** theorizes that "like cures like," meaning that a substance that causes a particular set of symptoms can be used to cure a disease that has similar symptoms. Homeopathic remedies are generally diluted to such an extent that the original substance essentially is no longer present. Homeopaths believe that these remedies have powerful healing effects because the water structure is somehow altered during the dilution process.

- **Traditional Chinese medicine (TCM)** includes a large number of folk practices that originated in China. TCM promotes the theory that the body has pathways (called *meridians*) that conduct energy (called *qi*; pronounced *chee*). The interrupted flow of qi is believed to cause illness. TCM practices allegedly improve the flow of qi and include acupuncture, **qi gong**, herbal remedies, dietary practices, and massage.

What is the theory underlying mind-body interventions?

Mind-body therapies attempt to improve a person's sense of psychological or spiritual well-being despite the presence of illness. The treatments are also used in the hope of reducing stress, dealing with pain, or lowering blood pressure. Some of these therapies have been incorporated into mainstream medicine for stress reduction or relaxation. For example, **biofeedback** training, in which individuals learn to monitor skin temperature, muscle tension, or brain wave activity while practicing relaxation techniques, is frequently taught by behavioral medicine specialists to help patients reduce stress or anxiety. Other techniques to reduce stress and promote relaxation include **meditation,** art and music therapy, and prayer.

The clinical applications of other mind-body therapies are far more questionable. An example is guided **imagery,** in

Biofeedback training is a stress reduction and relaxation technique.

which a person tries to reverse the disease process (for example, shrink a tumor) by using mental pictures. Another example is the use of **faith healing** to cure disease in place of proven conventional treatments.

Which alternative practices involve physical manipulation, and how do they work?

Manipulative interventions include physical touch, forceful movement of different parts of the body, and the application of pressure. Some practitioners maintain that special energy fields are also manipulated during the physical treatment and that proper energy flow induces healing. The most popular practices include the following:

- **Chiropractic** theory proposes that keeping the nervous system free from obstruction allows the body to heal itself because the process of healing is conducted via the spinal cord and nerves. The main treatment is the "adjustment," a manual manipulation that is said to correct a "subluxation" (a pinched nerve or misaligned vertebra) and restore the body's natural healing ability. Although spinal manipulation has mainly been found to be helpful for improving back pain, most chiropractors still assert that chiropractic can cure disease rather than simply relieve symptoms.[11] For example, many promote spinal manipulation to treat infectious diseases, prevent cancer, and regulate menstrual periods, even though the nervous system and spinal alignment do not play roles in the pathology of these conditions.
- **Massage therapy** is the manipulation of muscle and connective tissue to improve muscle function, reduce pain, or promote relaxation. Massage therapists may also apply heat or cold and give advice about exercises that may improve muscle tone and range of motion. Massage is often integrated into conventional physical therapy, although some massage therapists may incorrectly suggest that massage is a valid treatment for a wide range of medical conditions.

What are the alleged effects of "energy" therapies?

Two categories of therapies involve the alleged curative power of "energy":

- **Biofield therapies** are said to influence the energy that surrounds or pervades the human body. Proponents claim that an energy therapy can strengthen or restore a person's "energy flow" and induce healing. Acupuncture, qi gong, and **therapeutic touch** can be included in this category. Note that CAM adherents often use the term *energy* unscientifically and that there is no objective evidence of this sort of energy flow.
- **Bioelectrical** or **bioelectromagnetic therapies** use electric or magnetic fields to allegedly promote healing; for example, magnets have been marketed with claims that they can improve circulation, reduce inflammation, and speed recovery from injuries.

Why do so many consumers choose to use CAM, given that the therapies have not been proved to be beneficial?

Surveys suggest that consumers perceive their visits to CAM therapists as far more pleasant than visits to conventional health practitioners. CAM therapists spend more time with patients, are more attentive, and use less invasive interventions.[12] Self-help measures are encouraged, so the consumer has more control over the treatment. The therapies appear to be more "natural" and to have fewer side effects. In addition, many consumers seem satisfied that these treatments "work." Possible explanations for "cures" include:

- A person may seem "cured" because of misdiagnosis; that is, the condition diagnosed by the CAM practitioner may not have actually existed.
- The condition may have been self-limiting, or it may have gone into temporary remission after the treatment.
- Undue credit may be inappropriately assigned to the CAM therapy when the improvement was actually due to a previous or concurrent conventional treatment.
- The placebo effect may have had an influence on the course of disease.

The central question remains: Do the CAM therapies merely make people *feel* better, or do they really *get* better? This question can be answered only by well-controlled research studies.

Are any potential dangers associated with the use of CAM?

One of the attractions of alternative therapies is the assumption that they are safe. Recall, however, the concerns associated with the use of herbal products discussed earlier, which include the possible toxicity of herbal ingredients, product contamination or adulteration, and interactions with conventional medications. Between 1990 and 1999, the FDA recalled more than 100 dietary supplements due to hazards associated with their use.[13]

Another concern is that use of CAM therapies may delay the use of reliable treatments that have demonstrable benefits. Various reports have described how people with treatable medical conditions suffered permanent disability or death when they were misdiagnosed or improperly treated by CAM practitioners. For example, a rare but well-known risk of spinal cord injury or stroke is associated with a type of cervical manipulation performed by chiropractors.[14] Unfortunately, because most CAM therapies are not regulated or monitored, there are no accurate estimates of their adverse effects.

What should nurses do if they think their patients are using CAM?

Nurses should be aware when their patients are using CAM therapies that may have consequences for the course of their disease and its treatment. It is important to routinely inquire about the use of CAM therapies and to educate patients about the hazards of postponing or stopping conventional treatment. Patients should also be told about potential interactions between conventional treatments and CAM therapies. Some patients may want to learn about differences between evidence-based medical practices and untested CAM theories and may be interested in the **integrative medicine** options available.

All alternative therapies have one characteristic in common: their effectiveness is, for the most part, unproven. As mentioned previously, patients often choose alternative therapies because of their positive interactions with CAM practitioners. Empathizing with patients may go a long way toward winning their trust and improving their compliance with conventional therapy. In addition, health practitioners need to regularly update their knowledge about unconventional practices, using reliable, objective resources, so that they can knowledgeably discuss these options with patients.

Glossary of Alternative Therapies

acupuncture (AK-you-PUNK-cher): a therapy that involves inserting thin needles into the skin at specific anatomical points, allegedly to correct disruptions in the flow of energy within the body.

aromatherapy: inhalation of oil extracts from plants to cure illness or enhance health.

ayurveda: a traditional medical system from India that promotes the use of diet, herbs, meditation, massage, and yoga for preventing and treating illness.

bioelectrical or **bioelectromagnetic therapies:** therapies that involve the unconventional use of electric or magnetic fields to cure illness.

biofeedback: a technique in which individuals are trained to gain voluntary control of certain physiological processes, such as skin temperature or brain wave activity, to help reduce stress and anxiety.

biofield therapies: healing methods based on the belief that illnesses can be healed by manipulating energy fields that purportedly surround and penetrate the body. Examples include *acupuncture, qi gong,* and *therapeutic touch.*

chiropractic (KYE-roh-PRAK-tic): an alternative medical system based on the unproven theory that spinal manipulation can restore health.

- A *subluxation* is a misaligned vertebra or other spinal alteration that may cause illness.

- *Adjustment* is the manipulative therapy practiced by chiropractors.

faith healing: the use of prayer or belief in divine intervention to promote healing.

homeopathic (HO-mee-oh-PATH-ic) **medicine:** a practice based on the theory that "like cures like." Substances believed to cause certain symptoms are prescribed for curing the same symptoms, but are given in extremely diluted amounts.

hypnotherapy: a technique that uses hypnosis and the power of suggestion to improve health behaviors, relieve pain, and promote healing.

imagery: the use of mental images of things or events to aid relaxation or promote self-healing.

integrative medicine: an approach to medical care that combines conventional therapies and CAM therapies that have been shown to be safe and potentially effective.

massage therapy: manual manipulation of muscles to reduce tension, increase blood circulation, improve joint mobility, and promote healing of injuries.

meditation: a self-directed technique of calming the mind and relaxing the body.

naturopathic (NAY-chur-oh-PATH-ic) **medicine:** an approach to medical care using practices alleged to enhance the body's natural healing abilities. Treatments may include a variety of alternative therapies including dietary supplements, herbal remedies, exercise, and homeopathy.

osteopathic (OS-tee-oh-PATH-ic) **manipulation:** a manipulative technique performed by osteopaths that includes deep tissue massage and manipulation of joints, spine, and soft tissues. Doctors of Osteopathic Medicine (DOs) are fully trained and licensed medical physicians.

qi gong (chee GUNG): a Chinese system that combines movement, meditation, and breathing techniques and allegedly cures illness by enhancing the flow of "qi" energy within the body.

reflexology: a technique that applies pressure or massage on areas of the hands or feet to allegedly cure disease or relieve pain in other areas of the body; sometimes called *zone therapy.*

therapeutic touch: a technique of passing hands over a patient to purportedly identify energy imbalances and transfer healing power from therapist to patient; also called *laying on of hands.*

traditional Chinese medicine (TCM): an approach to medical care based on the concept that illness can be cured by enhancing the flow of "qi" energy within a person's body. Treatments may include herbal therapies, physical exercises, meditation, acupuncture, and remedial massage.

Notes

[1] P. M. Barnes and E. Powell-Griner, Complementary and alternative medicine use among adults: United States, 2002, *Advance Data from Vital and Health Statistics* 343 (2004): 1–19.

[2] J. D. Berman and S. E. Straus, Implementing a research agenda for complementary and alternative medicine, *Annual Review of Medicine* 55 (2004): 239–254.

[3] E. Ernst, The role of complementary and alternative medicine, *British Medical Journal* 321 (2000): 1133–1135.

[4] B. Barzansky, H. S. Jonas, and S. I. Etzel, Educational programs in US medical schools, 1999–2000, *Journal of the American Medical Association* 284 (2000): 1114–1120.

[5] ConsumerLab.com, LLC, Product review: Memory enhancement supplements (ginkgo, huperzine A, phosphatidylserine, and acetyl-L-carnitine), **http://consumerlab.com/results/ginkgobiloba.asp**, posted November 30, 2005 (site visited December 5, 2005).

[6] C. H. Halsted, Dietary supplements and functional foods: 2 sides of a coin? *American Journal of Clinical Nutrition* 77 (2003): 1001S–1007S.

[7] Halsted, 2003.

[8] W. F. Huang, K. C. Wen, and M. L. Hsiao, Adulteration by synthetic therapeutic substances of traditional Chinese medicines in Taiwan, *Journal of Clinical Pharmacology* 37 (1997): 344–350.

[9] J. Barnes, L. A. Anderson, and J. D. Phillipson, *Herbal Medicines: A Guide for Healthcare Professionals* (Chicago: Pharmaceutical Press, 2002).

[10] Barnes, Anderson, and Phillipson, 2002.

[11] American Medical Association, *Alternative Medicine (Report 12 of the Council on Scientific Affairs, A-97)* (American Medical Association, 1997), **www.ama-assn.org/ama/pub/category/13638.html,** site visited December 5, 2005.

[12] American Medical Association, 1997.

[13] Berman and Straus, 2004.

[14] A. Vickers and C. Zollman, The manipulative therapies: Osteopathy and chiropractic, *British Medical Journal* 319 (1999): 1176–1179.

Appendixes

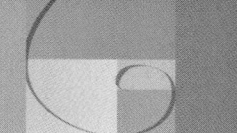

CONTENTS

Table of Food Composition

CONTENTS

This edition of the table of food composition has been updated to reflect current nutrient data for foods, to remove outdated foods, and to add foods that are new to the marketplace.* The nutrient database for this appendix is compiled from a variety of sources, including the USDA Standard Release database (Release 17) and manufacturers' data. The USDA database provides data for a wider variety of foods and nutrients than other sources. Because laboratory analysis for each nutrient can be quite costly, manufacturers tend to provide data only for those nutrients mandated on food labels. Consequently, data for their foods are often incomplete; any missing information on this table is designated as a dash. Keep in mind that a dash means only that the information is unknown and should not be interpreted as a zero. A zero means that the nutrient is not present in the food.

Whenever using nutrient data, remember that many factors influence the nutrient contents of foods. These factors include the mineral content of the soil, the diet fed to the animal or the fertilizer used on the plant, the season of harvest, the method of processing, the length and method of storage, the method of cooking, the method of analysis, and the moisture content of the sample analyzed. With so many influencing factors, users should view nutrient data as a close approximation of the actual amount.

For updates, corrections, and a list of 8,000 additional foods and codes found in the diet analysis software that accompanies this text, visit **www.thomsonedu.com/nutrition** and click on *Diet Analysis Plus*.

- *Fats* Total fats, as well as the breakdown of total fats to saturated, monounsaturated, and polyunsaturated fats, are listed in the table. The fatty acids seldom add up to the total in part due to rounding but also because values are derived from a variety of laboratories and other fatty acid components.
- *Trans Fats* *Trans* fat data has been listed in the table. Because food manufacturers have only been required to report *trans* fats on food labels since January 1 of 2006, much of the data is incomplete. Missing *trans* fat data is designated with a dash. As additional *trans* fat data becomes available, the table will be updated.
- *Vitamin A and Vitamin E* In keeping with the 2001 RDA for vitamin A, this appendix presents data for vitamin A in micrograms (µg) RAE. Similarly, because the 2000 RDA for vitamin E is based only on the alpha-tocopherol form of vitamin E, this appendix reports vitamin E data in milligrams (mg) alpha-tocopherol, listed on the table as Vit E (mgα).
- *Bioavailability* Keep in mind that the availability of nutrients from foods depends not only on the quantity provided by a food, but also on the amount absorbed and used by the body—the bioavailability. The bioavailability of folate from fortified foods, for example, is greater than from naturally occurring sources. Similarly, the body can make niacin from the amino acid tryptophan, but niacin values in this table (and most databases) report preformed niacin only.
- *Using the Table* The foods and beverages in this table have been organized into several categories, which are listed at the head of each right-hand page. Page numbers have been provided, and each group has been color-coded to make it easier to find individual foods.
- *Caffeine Sources* Caffeine occurs in several plants, including the familiar coffee bean, the tea leaf, and the cocoa bean from which chocolate is made. Most human societies use caffeine regularly, most often in beverages, for its stimulant effect and flavor. Caffeine contents of beverages vary depending on the plants they are made from, the climates and soils where the plants are grown, the grind or cut size, the method and duration of brewing, and the amounts served. The accompanying chart shows that in general, a cup of coffee contains the most caffeine; a cup of tea, less than half as much; and cocoa or chocolate, less still. As for cola beverages, they are made from kola nuts, which contain caffeine, but most of their caffeine is added, using the purified compound obtained from decaffeinated coffee beans. A table of caffeine sources appears on page A-80.

*This food composition table has been prepared by Wadsworth Publishing Company. The nutritional data are supplied by Axxya Systems.

TABLE A–1
Food Composition

(Computer code number is for Wadsworth Diet Analysis program)　　(For purposes of calculations, use "0" for t, <1, <.1, <.01, etc.)

DA + Code	Food Description	Quantity	Measure	Wt (g)	H₂O (g)	Ener (kcal)	Prot (g)	Carb (g)	Dietary Fiber (g)	Fat (g)	Sat	Mono	Poly	Trans
												Fat Breakdown (g)		

BREADS, BAKED GOODS, CAKES, COOKIES, CRACKERS, CHIPS, PIES

DA + Code	Food Description	Quantity	Measure	Wt (g)	H₂O (g)	Ener (kcal)	Prot (g)	Carb (g)	Dietary Fiber (g)	Fat (g)	Sat	Mono	Poly	Trans
	Bagels													
8534	Cinnamon & raisin	1	item(s)	71	23	195	7	39	2	1	0.19	0.12	0.48	—
4910	Enriched, all varieties	1	item(s)	71	23	195	7	38	2	1	0.16	0.09	0.49	0
4911	Plain, enriched, toasted	1	item(s)	66	18	195	7	38	2	1	0.16	0.09	0.49	0
8538	Oat bran	1	item(s)	71	23	181	8	38	3	1	0.14	0.18	0.35	—
12079	Whole grain	1	item(s)	85	—	170	9	35	6	2.5	0	—	—	0
	Biscuits													
25008	Biscuits	1	item(s)	41	16	121	3	16	1	5	1.40	1.41	1.82	0
16729	Scone	1	item(s)	42	11	149	4	19	1	6	2.01	2.55	1.26	—
25166	Wheat biscuits	1	item(s)	55	21	162	4	22	1	7	1.90	1.92	2.51	0
	Bread													
325	Boston brown, canned	1	slice(s)	45	21	88	2	19	2	1	0.13	0.09	0.25	—
8716	Bread sticks, plain	4	item(s)	24	1	99	3	16	1	2	0.34	0.86	0.87	—
25176	Cornbread	1	piece(s)	55	26	141	5	18	1	5	2.09	1.44	1.50	0
327	Cracked wheat	1	slice(s)	25	9	65	2	12	1	1	0.23	0.48	0.17	—
9079	Croutons, plain	¼	cup(s)	8	<1	31	1	6	<1	<1	0.11	0.23	0.10	—
8582	Egg	1	slice(s)	40	14	115	4	19	1	2	0.64	0.92	0.44	—
8585	Egg, toasted	1	slice(s)	37	10	117	4	19	1	2	0.60	1.11	0.43	—
329	French	1	slice(s)	25	9	69	2	13	1	1	0.16	0.30	0.17	—
8591	French, toasted	1	slice(s)	23	7	69	2	13	1	1	0.16	0.30	0.17	—
8597	Indian fry	1	item(s)	90	24	296	6	48	2	9	2.08	3.59	2.33	—
332	Italian	1	slice(s)	30	11	81	3	15	1	1	0.26	0.24	0.42	—
1393	Mixed grain	1	slice(s)	26	10	65	3	12	2	1	0.21	0.40	0.24	—
8604	Mixed grain, toasted	1	slice(s)	24	8	65	3	12	2	1	0.21	0.40	0.24	—
8605	Oat bran	1	slice(s)	30	13	71	3	12	1	1	0.21	0.48	0.51	—
8608	Oat bran, toasted	1	slice(s)	27	10	70	3	12	1	1	0.21	0.47	0.50	—
8609	Oatmeal	1	slice(s)	27	10	73	2	13	1	1	0.19	0.43	0.46	—
8613	Oatmeal, toasted	1	slice(s)	25	8	73	2	13	1	1	0.19	0.43	0.46	—
1409	Pita	1	item(s)	60	19	165	5	33	1	1	0.10	0.06	0.32	—
7905	Pita, whole wheat	1	item(s)	64	20	170	6	35	5	2	0.26	0.22	0.68	—
338	Pumpernickel	1	slice(s)	32	12	80	3	15	2	1	0.14	0.30	0.40	—
334	Raisin, enriched	1	slice(s)	26	9	71	2	14	1	1	0.28	0.60	0.18	—
8625	Raisin, toasted	1	slice(s)	24	7	71	2	14	1	1	0.28	0.60	0.18	—
10168	Rice, white	1	slice(s)	42	—	140	1	21	1	6	0.50	—	—	0
8653	Rye	1	slice(s)	32	12	83	3	15	2	1	0.20	0.42	0.26	—
8654	Rye, toasted	1	slice(s)	29	9	82	3	15	2	1	0.20	0.42	0.25	—
336	Rye, light	1	slice(s)	25	9	65	2	12	2	1	0.20	0.30	0.30	—
8588	Sourdough	1	slice(s)	25	9	69	2	13	1	1	0.16	0.30	0.17	—
8592	Sourdough, toasted	1	slice(s)	23	7	69	2	13	1	1	0.16	0.30	0.17	—
491	Submarine or hoagie roll	1	item(s)	135	41	400	11	72	4	8	1.80	3.00	2.20	—
8596	Vienna, toasted	1	slice(s)	23	7	69	2	13	1	1	0.16	0.30	0.17	—
8670	Wheat	1	slice(s)	25	9	65	2	12	1	1	0.22	0.43	0.23	—
8671	Wheat, toasted	1	slice(s)	23	7	65	2	12	1	1	0.22	0.43	0.23	—
340	White	1	slice(s)	25	9	67	2	13	1	1	0.18	0.17	0.34	—
1395	Whole wheat	1	slice(s)	46	15	128	4	24	3	2	0.37	0.53	1.35	—
	Cakes													
386	Angel food, from mix	1	slice(s)	50	16	129	3	29	<1	<1	0.02	0.01	0.06	—
8772	Butter pound, ready to eat, commercially prepared	1	slice(s)	75	18	291	4	37	<1	15	8.67	4.43	0.80	—
8737	Carrot, cream cheese frosting, from mix	1	slice(s)	111	23	484	5	52	1	29	5.43	7.24	15.10	—
4931	Chocolate, chocolate icing, commercially prepared	1	slice(s)	64	15	235	3	35	2	10	3.05	5.61	1.18	—
8756	Chocolate, from mix	1	slice(s)	95	23	340	5	51	2	14	5.16	5.74	2.62	—
393	Devil's food cupcake, chocolate frosting	1	item(s)	35	8	120	2	20	1	4	1.80	1.60	0.60	—
8757	Fruitcake, ready to eat, commercially prepared	1	piece(s)	43	11	139	1	26	2	4	0.45	1.81	1.43	—
1397	Pineapple upside down, from mix	1	slice(s)	115	37	367	4	58	1	14	3.35	5.97	3.77	—
411	Sponge, from mix	1	slice(s)	63	19	187	5	36	<1	3	0.82	0.99	0.41	—
8817	White, coconut frosting, from mix	1	slice(s)	112	23	399	5	71	1	12	4.36	4.14	2.42	—
8819	Yellow, chocolate frosting, ready to eat, commercially prepared	1	slice(s)	64	14	243	2	35	1	11	2.98	6.14	1.35	—
8822	Yellow, vanilla frosting, ready to eat, commercially prepared	1	slice(s)	64	14	239	2	38	<1	9	1.52	3.91	3.30	—
	Snack cakes													
8791	Chocolate snack cake, creme filled, w/frosting	1	item(s)	50	10	188	2	30	<1	7	1.43	2.85	2.62	—
25010	Cinnamon coffee cake	1	piece(s)	72	23	231	4	36	1	8	2.19	2.65	2.99	0

PAGE KEY: A–4 = Breads/Baked Goods A–8 = Cereal/Rice/Pasta A–12 = Fruit A–16 = Vegetables/Legumes A–26 = Nuts/Seeds A–28 = Vegetarian A–30 = Dairy A–36 = Eggs A–36 = Seafood A–38 = Meats A–42 = Poultry A–42 = Processed meats A–44 = Beverages A–48 = Fats/Oils A–50 = Sweets A–52 = Spices/Condiments/Sauces A–54 = Mixed Foods/Soups/Sandwiches A–60 = Fast food A–76 = Convenience A–78 = Baby foods

Chol (mg)	Calc (mg)	Iron (mg)	Magn (mg)	Pota (mg)	Sodi (mg)	Zinc (mg)	Vit A (RAE) (µg)	Thia (mg)	Vit E (mg α)	Ribo (mg)	Niac (mg)	Vit B_6 (mg)	Fola (µg)	Vit C (mg)	Vit B_{12} (µg)	Sele (µg)
0	13	2.70	20	105	229	0.80	15	0.27	0.22	0.20	2.19	0.04	79	<1	0	22
0	53	2.53	21	72	379	0.62	0	0.38	0.07	0.22	3.24	0.04	75	0	0	23
0	53	2.52	20	72	379	0.62	0	0.31	0.08	0.20	2.91	0.03	64	0	0	23
0	9	2.19	22	82	360	0.64	1	0.24	0.23	0.24	2.10	0.03	70	<1	0	24
0	200	1.08	120	0	200	4.5	0	0.44	—	0.5	8	0.6	—	0	1.79	0
<1	33	1.01	6	37	205	0.27	9	0.13	0.01	0.12	1.08	0.01	26	0	<.1	7
49	80	1.31	7	48	288	0.29	—	0.15	0.43	0.16	1.20	0.03	8	<.1	<1	—
<1	57	1.22	16	81	321	0.42	12	0.16	0.01	0.13	1.49	0.03	29	<.1	<.1	12
<1	32	0.95	28	143	284	0.23	11	0.01	0.14	0.05	0.50	0.04	5	0	<.1	10
0	5	1.03	8	30	158	0.21	0	0.14	0.24	0.13	1.27	0.02	39	0	0	9
21	88	1.01	10	59	209	0.57	38	0.13	0.33	0.16	0.98	0.04	36	2	<1	6
0	11	0.70	13	44	135	0.31	0	0.09	—	0.06	0.92	0.08	15	0	<.1	6
0	6	0.31	2	9	52	0.07	0	0.05	—	0.02	0.41	0.00	10	0	0	3
20	37	1.22	8	46	197	0.32	25	0.18	0.10	0.17	1.94	0.03	42	0	<.1	12
21	38	1.24	8	47	200	0.32	26	0.14	0.11	0.16	1.77	0.02	36	0	<.1	12
0	19	0.63	7	28	152	0.22	0	0.13	0.08	0.08	1.19	0.01	37	0	0	8
0	19	0.63	7	28	152	0.22	0	0.10	0.07	0.07	1.07	0.01	22	0	0	8
0	210	3.24	14	67	626	0.45	0	0.39	—	0.27	3.27	0.02	67	0	0	21
0	23	0.88	8	33	175	0.26	0	0.14	0.09	0.09	1.31	0.01	57	0	0	8
0	24	0.90	14	53	127	0.33	0	0.11	0.09	0.09	1.13	0.09	31	<.1	<.1	8
0	24	0.90	14	53	127	0.33	0	0.08	0.08	0.08	1.02	0.08	28	<.1	<.1	8
0	20	0.94	11	44	122	0.27	1	0.15	0.13	0.10	1.45	0.02	24	0	0	9
0	19	0.93	9	33	121	0.28	1	0.12	0.13	0.09	1.29	0.01	19	0	0	9
0	18	0.73	10	38	162	0.28	1	0.11	0.13	0.06	0.85	0.02	17	0	<.1	7
0	18	0.74	10	39	163	0.28	1	0.09	0.13	0.06	0.77	0.02	13	<.1	<.1	7
0	52	1.57	16	72	322	0.50	0	0.36	0.18	0.20	2.78	0.02	64	0	0	16
0	10	1.96	44	109	340	0.97	0	0.22	0.39	0.05	1.82	0.17	22	0	0	28
0	22	0.92	17	67	215	0.47	0	0.10	0.13	0.10	0.99	0.04	30	0	0	8
0	17	0.75	7	59	101	0.19	0	0.09	0.07	0.10	0.90	0.02	28	<.1	0	5
0	17	0.76	7	59	102	0.19	0	0.07	0.07	0.09	0.81	0.02	24	<.1	0	5
0	40	1.08	—	45	160	—	0	0.23	—	0.14	1.20	—	40	0	—	—
0	23	0.91	13	53	211	0.36	0	0.14	0.11	0.11	1.22	0.02	35	<1	0	10
0	23	0.90	12	53	210	0.36	0	0.11	0.11	0.10	1.09	0.02	30	<.1	0	10
0	20	0.70	4	51	175	0.18	0	0.10	—	0.08	0.80	0.01	5	0	<.1	8
0	19	0.63	7	28	152	0.22	0	0.13	0.08	0.08	1.19	0.01	37	0	0	8
0	19	0.63	7	28	152	0.22	0	0.10	0.07	0.07	1.07	0.01	22	0	0	8
0	100	3.80	—	128	683	—	0	0.54	—	0.33	4.50	0.05	—	0	—	42
0	19	0.63	7	28	152	0.22	0	0.10	0.07	0.07	1.07	0.01	22	0	0	8
0	26	0.83	12	50	133	0.26	0	0.10	0.07	0.07	1.03	0.02	23	0	0	8
0	26	0.83	12	50	132	0.26	0	0.08	0.07	0.06	0.93	0.02	19	0	0	8
0	38	0.94	6	25	170	0.19	0	0.11	0.05	0.08	1.10	0.02	28	0	0	4
0	15	1.43	37	144	159	0.69	0	0.14	0.35	0.10	1.83	0.09	30	0	0	18
0	42	0.12	4	68	255	0.07	0	0.05	0.00	0.10	0.09	0.00	10	0	<.1	8
166	26	1.04	8	89	299	0.35	112	0.10	—	0.98		0.03	31	0	<1	7
60	28	1.39	20	124	273	0.54	—	0.15	—	0.17	1.13	0.08	13	1	<1	—
27	28	1.41	22	128	214	0.44	—	0.02	—	0.09	0.37	0.03	11	<.1	<.1	2
55	57	1.53	30	133	299	0.66	38	0.13	—	0.20	1.08	0.04	26	<1	<1	11
19	21	0.70	—	46	92	—	—	0.04	—	0.05	0.30	—	2	0	—	2
2	14	0.89	7	66	116	0.12	3	0.02	0.39	0.04	0.34	0.02	9	<1	<.1	1
25	138	1.70	15	129	367	0.36	71	0.18	—	0.18	1.37	0.04	30	1	<.1	11
107	26	1.00	6	89	144	0.37	49	0.10	—	0.19	0.76	0.04	25	0	<1	12
1	101	1.30	13	111	318	0.37	13	0.14	0.13	0.21	1.19	0.03	35	<1	<.1	12
35	24	1.33	19	114	216	0.40	21	0.08	—	0.10	0.80	0.02	14	0	<1	2
35	40	0.68	4	34	220	0.16	12	0.06	—	0.04	0.32	0.02	17	0	<1	4
9	37	1.68	21	61	213	0.26	3	0.11	1.09	0.15	1.21	0.01	20	0	<.1	1
26	50	1.46	10	81	277	0.38	35	0.14	0.23	0.16	1.17	0.02	30	<.1	<1	10

TABLE A–1

Food Composition

(Computer code number is for Wadsworth Diet Analysis program)   (For purposes of calculations, use "0" for t, <1, <.1, <.01, etc.)

DA + Code	Food Description	Quantity	Measure	Wt (g)	H₂O (g)	Ener (kcal)	Prot (g)	Carb (g)	Dietary Fiber (g)	Fat (g)	Sat	Mono	Poly	Trans
	BREADS, BAKED GOODS, CAKES, COOKIES, CRACKERS, CHIPS, PIES—Continued													
16777	Funnel cake	1	item(s)	90	37	278	7	29	1	14	2.77	4.46	6.33	—
8794	Sponge snack cake, creme filled	1	item(s)	43	9	155	1	27	<1	5	1.09	1.73	1.40	—
	Snacks, chips, pretzels													
29428	Bagel chips, plain	3	item(s)	29	—	130	3	19	1	5	0.50	—	—	—
29429	Bagel chips, toasted onion	3	item(s)	29	—	130	4	20	1	5	0.50	—	—	—
38192	Chex traditional snack mix	1	cup(s)	46	—	198	3	33	2	6	0.76	—	—	—
654	Potato chips, salted	20	item(s)	28	1	152	2	15	1	10	3.11	2.79	3.46	—
8816	Potato chips, unsalted	20	item(s)	28	1	152	2	15	1	10	3.11	2.79	3.46	—
4641	Tortilla chips, plain	6	item(s)	28	1	142	2	18	2	7	1.43	4.39	1.03	—
5096	Pretzels, plain, hard, twists	5	item(s)	30	1	114	3	24	1	1	0.23	0.41	0.37	—
4632	Pretzels, whole wheat	1	ounce(s)	28	1	103	3	23	2	1	0.16	0.29	0.24	—
	Cookies													
8859	Animal crackers	12	piece(s)	30	0	134	2	22	<1	4	1.03	2.29	0.56	—
8876	Brownie, prepared from mix	1	item(s)	24	3	112	1	12	1	7	1.76	2.60	2.26	—
25207	Chocolate chip cookies	1	item(s)	30	4	140	2	16	1	8	2.09	3.26	2.09	0
8915	Chocolate sandwich cookie, extra creme filling	1	item(s)	13	<1	65	<1	9	<1	3	0.50	1.39	1.22	1.10
14145	Fig Newtons	1	item(s)	16	—	55	1	10	1	1	0.50	0.50	0.00	0.50
8920	Fortune cookie	1	item(s)	8	1	30	<1	7	<1	<1	0.05	0.11	0.04	—
25208	Oatmeal cookies	1	item(s)	69	12	234	6	45	3	4	0.70	1.28	1.85	0
25213	Peanut butter cookies	1	item(s)	35	4	163	4	17	1	9	1.65	4.72	2.43	0
33095	Sugar cookies	1	item(s)	16	4	61	1	7	<1	3	0.63	1.27	0.87	0
9002	Vanilla sandwich cookie, creme filling	1	item(s)	10	<1	48	<1	7	<1	2	0.30	0.84	0.76	—
	Crackers													
9008	Cheese crackers (mini)	30	item(s)	30	1	151	3	17	1	8	2.81	3.63	0.74	—
9010	Cheese crackers (mini), low salt	30	item(s)	30	1	151	3	17	1	8	2.82	2.70	1.44	—
9012	Cheese cracker sandwich w/peanut butter	4	item(s)	28	1	139	3	16	1	7	1.23	3.64	1.43	—
8928	Honey graham crackers	4	item(s)	28	1	118	2	22	1	3	0.43	1.14	1.07	—
9016	Matzo crackers, plain	1	item(s)	28	1	112	3	24	1	<1	0.06	0.04	0.17	—
9024	Melba toast	3	item(s)	15	1	59	2	11	1	<1	0.07	0.12	0.19	—
14189	Ritz crackers	5	item(s)	16	<1	80	1	10	1	4	0.50	1.50	0.00	—
9014	Rye crispbread crackers	1	item(s)	10	1	37	1	8	2	<1	0.01	0.02	0.06	—
9028	Rye melba toast	3	item(s)	15	1	58	2	12	1	1	0.07	0.14	0.20	—
9040	Rye wafer	1	item(s)	11	1	37	1	9	3	<.1	0.01	0.02	0.04	—
432	Saltine crackers	5	item(s)	15	1	65	1	11	<1	2	0.44	0.96	0.25	0.54
9046	Saltine crackers, low salt	5	item(s)	15	1	65	1	11	<1	2	0.44	0.96	0.25	—
9048	Snack crackers, round	10	item(s)	30	1	151	2	18	<1	8	1.13	3.19	2.86	—
9050	Snack crackers, round, low salt	10	item(s)	30	1	151	2	18	<1	8	1.13	3.19	2.86	—
9052	Snack cracker sandwich, cheese filling	4	item(s)	28	1	134	3	17	1	6	1.72	3.15	0.72	—
9054	Snack cracker sandwich, peanut butter filling	4	item(s)	28	1	138	3	16	1	7	1.38	3.86	1.30	—
9044	Soda crackers	5	item(s)	15	1	65	1	11	<1	2	0.44	0.96	0.25	0.54
9055	Wheat crackers	10	item(s)	30	1	142	3	19	1	6	1.55	3.43	0.84	—
9057	Wheat crackers, low salt	10	item(s)	30	1	142	3	19	1	6	1.55	3.43	0.84	—
9059	Wheat cracker sandwich, cheese filling	4	item(s)	28	1	139	3	16	1	7	1.16	2.90	2.57	—
9061	Wheat cracker sandwich, peanut butter filling	4	item(s)	28	1	139	4	15	1	7	1.29	3.29	2.48	—
9022	Whole wheat crackers	7	item(s)	28	1	124	2	19	3	5	0.95	1.65	1.85	—
	Pastry													
16754	Apple fritter	1	item(s)	17	6	62	1	6	<1	4	0.87	1.69	1.13	—
5118	Cinnamon sweet roll w/icing, from refrigerator dough	1	item(s)	30	7	109	2	17	1	4	1.00	2.23	0.52	—
4945	Croissant, butter	1	item(s)	57	13	231	5	26	1	12	6.59	3.15	0.62	—
9096	Danish pastry, nut	1	item(s)	65	13	280	5	30	1	16	3.78	8.90	2.78	—
4947	Doughnut, cake	1	item(s)	47	10	198	2	23	1	11	1.70	4.37	3.70	—
9105	Doughnut, cake, chocolate glazed	1	item(s)	42	7	175	2	24	1	8	2.16	4.74	1.04	—
9115	Doughnut, creme filling	1	item(s)	85	32	307	5	26	1	21	4.62	10.27	2.62	—
437	Doughnut, glazed	1	item(s)	60	15	242	4	27	1	14	3.49	7.72	1.74	—
9117	Doughnut, jelly filling	1	item(s)	85	30	289	5	33	1	16	4.12	8.69	2.02	—
10617	Toaster pastry, brown sugar cinnamon	1	item(s)	50	5	210	3	35	1	6	1.00	4.00	1.00	—
30928	Toaster pastry, cream cheese	1	item(s)	54	—	200	3	23	1	11	3.50	—	—	—
	Muffins													
25015	Blueberry	1	item(s)	63	30	160	3	23	1	6	0.87	1.48	3.25	0
4997	Bran, from mix	1	item(s)	50	18	138	3	23	2	5	1.18	2.34	0.72	—
9189	Corn, ready to eat	1	item(s)	57	19	174	3	29	2	5	0.77	1.20	1.83	—

PAGE KEY: A–4 = Breads/Baked Goods   A–8 = Cereal/Rice/Pasta   A–12 = Fruit   A–16 = Vegetables/Legumes   A–26 = Nuts/Seeds   A–28 = Vegetarian
A–30 = Dairy   A–36 = Eggs   A–36 = Seafood   A–38 = Meats   A–42 = Poultry   A–42 = Processed meats   A–44 = Beverages   A–48 = Fats/Oils
A–50 = Sweets   A–52 = Spices/Condiments/Sauces   A–54 = Mixed Foods/Soups/Sandwiches   A–60 = Fast food   A–76 = Convenience   A–78 = Baby foods

Chol (mg)	Calc (mg)	Iron (mg)	Magn (mg)	Pota (mg)	Sodi (mg)	Zinc (mg)	Vit A (RAE) (µg)	Thia (mg)	Vit E (mg α)	Ribo (mg)	Niac (mg)	Vit B6 (mg)	Fola (µg)	Vit C (mg)	Vit B12 (µg)	Sele (µg)
63	128	1.86	18	154	273	0.64	—	0.24	1.55	0.32	1.86	0.05	14	<1	<1	—
7	19	0.55	3	37	155	0.12	2	0.07	0.50	0.06	0.52	0.01	17	<.1	<.1	1
0	0	0.72	—	45	70	—	0	—	—	—	—	—	—	0	0	—
0	0	0.72	—	50	300	—	0	—	—	—	—	—	—	0	0	—
0	0	0.55	0	76	623	0.00	0	0.09	—	0.05	1.22	0.00	12	0	—	—
0	7	0.46	19	362	169	0.31	0	0.05	1.91	0.06	1.09	0.19	13	9	0	2
0	7	0.46	19	362	2	0.31	0	0.05	2.59	0.06	1.09	0.19	13	9	0	2
0	44	0.43	25	56	150	0.43	1	0.02	1	0.05	0.36	0.08	3	0	0	2
0	11	1.30	11	44	515	0.26	0	0.14	—	0.19	1.58	0.03	51	0	0	2
0	8	0.76	9	122	58	0.18	0	0.12	—	0.08	1.86	0.05	15	<1	0	—
0	13	0.82	5	30	1118	0.19	—	0.10	0.04	0.09	1.04	0.00	50	0	<.1	—
18	14	0.44	13	42	82	0.23	42	0.03	—	0.05	0.24	0.02	7	<.1	<.1	3
13	11	0.70	12	62	109	0.24	27	0.07	0.54	0.06	0.82	0.02	16	<.1	<.1	4
0	3	0.37	4	16	64	0.08	0	0.01	0.25	0.02	0.20	0.00	6	0	<.1	<1
0	5	0.36	—	40	60	—	4	0.03	—	0.04	0.22	—	—	<1	—	<1
<1	1	0.12	1	3	22	0.01	<.1	0.01	0.00	0.01	0.15	0.00	5	0	<.1	<1
<.1	26	1.94	49	177	311	1.43	48	0.23	0.23	0.12	1.24	0.09	30	<1	<.1	17
13	28	0.67	22	104	157	0.46	51	0.08	0.74	0.09	1.81	0.05	21	<.1	<.1	5
18	5	0.32	2	13	50	0.08	31	0.04	0.28	0.04	0.28	0.01	8	<.1	<.1	3
0	3	0.22	1	9	35	0.04	0	0.03	0.16	0.02	0.27	0.00	5	0	0	<1
4	45	1.43	11	44	299	0.34	9	0.17	0.66	0.13	1.40	0.17	46	0	<1	3
4	45	1.44	11	32	137	0.33	—	0.18	—	0.12	1.41	0.18	8	0	<1	—
0	14	0.76	16	61	199	0.29	0	0.15	0.16	0.08	1.63	0.04	26	0	<.1	2
0	7	1.04	8	38	169	0.23	0	0.06	0.09	0.09	1.15	0.02	13	0	0	3
0	4	0.90	7	32	1	0.19	0	0.11	0.02	0.08	1.11	0.03	5	0	0	10
0	14	0.56	9	30	124	0.30	0	0.06	0.06	0.04	0.62	0.01	19	0	0	5
0	20	0.72	3	10	135	0.23	—	0.07	—	0.04	0.45	0.01	10	1	0	—
0	3	0.24	8	32	26	0.24	0	0.02	0.08	0.01	0.10	0.02	5	0	0	4
0	12	0.55	6	29	135	0.20	0	0.07	—	0.04	0.71	0.01	13	0	0	6
0	4	0.65	13	54	87	0.31	0	0.05	0.09	0.03	0.17	0.03	5	<.1	0	3
0	18	0.81	4	19	195	0.12	0	0.08	0.15	0.07	0.79	0.01	19	0	0	2
0	18	0.81	4	109	95	0.12	0	0.08	0.02	0.07	0.79	0.01	19	0	0	3
0	36	1.08	8	40	254	0.20	0	0.12	0.61	0.10	1.21	0.02	27	0	0	2
0	36	1.08	8	107	112	0.20	0	0.12	0.61	0.10	1.21	0.02	27	0	0	2
1	72	0.67	10	120	392	0.17	5	0.12	0.06	0.19	1.05	0.01	28	<.1	<.1	6
0	23	0.78	15	60	201	0.32	0	0.14	0.58	0.08	1.71	0.04	24	0	<.1	3
0	18	0.81	4	19	195	0.12	0	0.08	0.15	0.07	0.79	0.01	19	0	0	2
0	15	1.32	19	55	239	0.48	0	0.15	0.15	0.10	1.49	0.04	35	0	0	2
0	15	1.32	19	61	85	0.48	0	0.15	0.15	0.10	1.49	0.04	15	0	0	10
2	57	0.73	15	86	256	0.24	5	0.10	—	0.12	0.89	0.07	18	<1	<.1	7
0	48	0.75	11	83	226	0.23	0	0.11	—	0.08	1.65	0.04	20	0	0	6
0	14	0.86	28	83	185	0.60	0	0.06	0.24	0.03	1.27	0.05	8	0	0	4
14	9	0.25	2	24	7	0.09	—	0.03	0.07	0.04	0.23	0.01	2	<1	<.1	—
0	10	0.80	4	19	250	0.10	—	0.12	—	0.07	1.09	0.01	14	<1	<.1	—
38	21	1.16	9	67	424	0.43	101	0.22	—	0.14	1.25	0.03	35	<1	<.1	13
30	61	1.17	21	62	236	0.57	6	0.14	0.53	0.16	1.50	0.07	54	1	<1	9
17	21	0.92	9	60	257	0.26	—	0.10	—	0.11	0.87	0.03	22	<.1	<1	0
24	89	0.95	14	45	143	0.24	5	0.02	0.09	0.03	0.20	0.01	19	<.1	<.1	2
20	21	1.56	17	68	263	0.68	9	0.29	0.25	0.13	1.91	0.06	60	0	<1	9
4	26	0.36	13	65	205	0.46	2	0.53	—	0.04	0.39	0.03	13	<.1	<.1	5
22	21	1.50	17	67	249	0.64	14	0.27	0.37	0.12	1.82	0.09	58	0	<1	11
0	0	1.80	—	70	190	—	—	0.15	—	0.17	2.00	0.20	40	0	0	—
15	0	1.08	—		230	—	—	—	—	—	—	—	—	0	—	—
20	50	1.15	7	56	288	0.39	20	0.14	0.76	0.15	1.14	0.03	29	<1	<1	9
34	16	1.27	29	74	234	0.57	—	0.10	—	0.12	1.44	0.09	33	0	<.1	—
15	42	1.60	18	39	297	0.31	30	0.16	0.46	0.19	1.16	0.05	46	0	0	9

TABLE A–1
Food Composition

(Computer code number is for Wadsworth Diet Analysis program)   (For purposes of calculations, use "0" for t, <1, <.1, <.01, etc.)

DA + Code	Food Description	Quantity	Measure	Wt (g)	H₂O (g)	Ener (kcal)	Prot (g)	Carb (g)	Dietary Fiber (g)	Fat (g)	Sat	Mono	Poly	Trans
												Fat Breakdown (g)		

BREADS, BAKED GOODS, CAKES, COOKIES, CRACKERS, CHIPS, PIES—Continued

DA + Code	Food Description	Quantity	Measure	Wt (g)	H₂O (g)	Ener (kcal)	Prot (g)	Carb (g)	Dietary Fiber (g)	Fat (g)	Sat	Mono	Poly	Trans
9121	English muffin, plain, enriched	1	item(s)	57	24	134	4	26	2	1	0.15	0.17	0.51	—
29582	English, toasted	1	item(s)	50	19	128	4	25	1	1	0.14	0.16	0.48	—
9145	English, wheat	1	item(s)	57	24	127	5	26	3	1	0.16	0.16	0.48	—
	Granola bars													
38161	Kudos milk chocolate w/fruit & nuts	1	item(s)	28	—	90	2	15	1	3	1.00	—	—	—
38196	Nature Valley banana nut crunchy	1	item(s)	21	—	95	2	14	1	4	0.50	—	—	—
38187	Nature Valley fruit n nut trail mix	1	item(s)	35	—	140	3	25	2	4	0.50	—	—	—
1383	Plain, hard	1	item(s)	25	1	115	2	16	1	5	0.58	1.07	2.95	—
4606	Plain, soft	1	item(s)	28	2	126	2	19	1	5	2.06	1.08	1.51	—
	Pies													
454	Apple pie, from home recipe	1	slice(s)	155	73	411	4	58	2	19	4.73	8.36	5.17	—
470	Pecan pie, from home recipe	1	slice(s)	122	24	503	6	64	0	27	4.87	13.64	6.97	—
472	Pumpkin pie, from home recipe	1	slice(s)	155	91	316	7	41	0	14	4.92	5.73	2.81	—
9007	Pie crust, frozen, ready to bake, enriched, baked	1	slice(s)	16	2	82	1	8	<1	5	1.69	2.51	0.65	—
5052	Pie crust, prepared w/water, baked	1	slice(s)	20	2	100	1	10	<1	6	1.54	3.46	0.77	—
	Rolls													
8555	Crescent dinner roll	1	item(s)	28	10	80	2	14	1	1	0.34	0.70	0.25	—
489	Hamburger roll or bun, plain	1	item(s)	43	15	120	4	21	1	2	0.47	0.48	0.85	—
490	Hard roll	1	item(s)	57	18	167	6	30	1	2	0.35	0.65	0.98	—
5127	Kaiser roll	1	item(s)	57	18	167	6	30	1	2	0.35	0.65	0.98	—
5130	Whole wheat roll or bun	1	item(s)	28	9	76	2	15	2	1	0.24	0.34	0.62	—
	Sport bars													
37026	Balance original chocolate	1	item(s)	50	—	200	14	22	1	6	3.50	—	—	—
37024	Balance original peanut butter	1	item(s)	50	—	200	14	22	1	6	2.50	—	—	—
36580	Clif Bar chocolate brownie energy bar	1	item(s)	68	—	240	10	41	6	4	1.00	—	—	—
36583	Clif Bar crunchy peanut butter energy bar	1	item(s)	68	—	240	12	39	5	5	0.50	—	—	—
36584	Clif Luna tropical crisp energy bar	1	item(s)	48	—	180	10	24	2	5	3.50	0.00	0.00	—
12005	Powerbar apple cinnamon	1	item(s)	65	—	230	10	45	3	3	0.50	1.50	0.50	—
16078	Powerbar banana	1	item(s)	65	—	230	9	45	3	2	0.50	1.00	0.50	—
16080	Powerbar chocolate	1	item(s)	65	—	230	10	45	3	2	0.50	0.50	1.00	—
16079	Powerbar mocha	1	item(s)	65	—	230	10	45	3	3	1.00	1.00	0.50	—
	Tortillas													
1391	Corn tortillas, soft	1	item(s)	26	11	58	1	12	1	1	0.09	0.17	0.29	—
1669	Flour tortilla	1	item(s)	32	9	104	3	18	1	2	0.56	1.21	0.34	—
1390	Taco shells, hard	1	item(s)	13	1	62	1	8	1	3	0.43	1.19	1.13	—
	Pancakes, waffles													
8926	Pancakes, blueberry, from recipe	3	item(s)	114	61	253	7	33	1	10	2.26	2.64	4.74	—
5037	Pancakes, from mix w/egg & milk	3	item(s)	114	60	249	9	33	2	9	2.33	2.36	3.33	—
9219	Waffle, plain, frozen, toasted	2	item(s)	66	28	174	4	27	2	5	0.95	2.12	1.84	—
500	Waffle, plain, from recipe	1	item(s)	75	<.1	218	6	25	2	11	2.14	2.64	5.08	—
30311	Waffle, 100% whole grain	1	item(s)	75	32	201	7	25	2	8	2.35	3.38	2.06	—

CEREAL, FLOUR, GRAIN, PASTA, NOODLES, POPCORN

DA + Code	Food Description	Quantity	Measure	Wt (g)	H₂O (g)	Ener (kcal)	Prot (g)	Carb (g)	Dietary Fiber (g)	Fat (g)	Sat	Mono	Poly	Trans
	Grain													
2861	Amaranth, dry	½	cup(s)	98	10	365	14	65	15	6	1.62	1.40	2.82	—
1953	Barley, pearled, cooked	½	cup(s)	79	54	97	2	22	3	<1	0.07	0.04	0.17	—
1956	Buckwheat groats, cooked, roasted	½	cup(s)	84	64	77	3	17	2	1	0.11	0.16	0.16	—
1957	Bulgur, cooked	½	cup(s)	91	71	76	3	17	4	<1	0.04	0.03	0.09	—
1963	Couscous, cooked	½	cup(s)	79	57	88	3	18	1	<1	0.02	0.02	0.05	—
1967	Millet, cooked	½	cup(s)	120	86	143	4	28	2	1	0.21	0.22	0.61	—
1969	Oat bran, dry	½	cup(s)	47	3	116	8	31	7	3	0.62	1.12	1.30	—
1972	Quinoa, dry	½	cup(s)	85	8	318	11	59	5	5	0.50	1.30	1.99	—
	Rice													
129	Brown, long grain, cooked	½	cup(s)	98	71	108	3	22	2	1	0.18	0.32	0.31	—
2863	Brown, medium grain, cooked	½	cup(s)	97.5	0.07	109.19	2.26	22.92	1.75	0.8	0.16	0.29	0.28	—
37488	Jasmine, saffroned, cooked	½	cup(s)	280	—	340	8	78	0	0	0.00	—	—	0
30280	Pilaf, cooked	½	cup(s)	103	74	129	2	22	1	3	0.67	1.61	0.95	—
28066	Spanish, cooked	½	cup(s)	120	3	25	2	1	<1	<1	0.33	0.07	18.31	0
2867	White glutinous, cooked	½	cup(s)	87	67	84	2	18	1	<1	0.03	0.06	0.06	—
482	White, instant long grain, enriched, boiled	½	cup(s)	83	63	81	2	18	<1	<1	0.04	0.04	0.04	—
484	White, long grain, boiled	½	cup(s)	79	54	103	2	22	<1	<1	0.06	0.07	0.06	—
486	White, long grain, enriched, parboiled, cooked	½	cup(s)	88	63	100	2	22	<1	<1	0.06	0.07	0.06	—
1194	Wild brown, cooked	½	cup(s)	82	0.06	82.81	3.27	17.49	1.47	0.27	0.04	0.04	0.17	—

PAGE KEY: A–4 = Breads/Baked Goods A–8 = Cereal/Rice/Pasta A–12 = Fruit A–16 = Vegetables/Legumes A–26 = Nuts/Seeds A–28 = Vegetarian A–30 = Dairy A–36 = Eggs A–36 = Seafood A–38 = Meats A–42 = Poultry A–42 = Processed meats A–44 = Beverages A–48 = Fats/Oils A–50 = Sweets A–52 = Spices/Condiments/Sauces A–54 = Mixed Foods/Soups/Sandwiches A–60 = Fast food A–76 = Convenience A–78 = Baby foods

Chol (mg)	Calc (mg)	Iron (mg)	Magn (mg)	Pota (mg)	Sodi (mg)	Zinc (mg)	Vit A (RAE) (µg)	Thia (mg)	Vit E (mg α)	Ribo (mg)	Niac (mg)	Vit B$_6$ (mg)	Fola (µg)	Vit C (mg)	Vit B$_{12}$ (µg)	Sele (µg)
0	30	1.43	12	75	264	0.40	0	0.25	—	0.16	2.21	0.02	42	0	<.1	—
0	95	1.36	11	72	252	0.38	0	0.19	0.17	0.14	1.90	0.02	15	<.1	<.1	—
0	101	1.64	21	106	218	0.61	0	0.25	0.26	0.17	1.91	0.05	36	0	0	17
0	200	0.36	—	—	60	—	0	—	—	—	—	—	—	0	0	—
0	10	0.54	—	60	80	—	0	—	—	—	—	—	—	0	—	—
0	0	0.00	—	—	95	—	0	—	—	—	—	—	—	0	—	—
0	15	0.72	24	82	72	0.50	2	0.06	—	0.03	0.39	0.02	6	<1	0	4
<1	30	0.73	21	92	79	0.43	0	0.08	—	0.05	0.15	0.03	7	0	<1	5
0	11	1.74	11	122	327	0.29	17	0.23	—	0.17	1.91	0.05	37	3	0	12
106	39	1.81	32	162	320	1.24	100	0.23	—	0.22	1.03	0.07	32	<1	<1	15
65	146	1.97	29	288	349	0.71	660	0.14	—	0.31	1.21	0.07	33	3	<1	11
0	3	0.36	3	18	104	0.05	0	0.04	0.42	0.06	0.39	0.01	9	0	<.1	<1
0	12	0.43	3	12	146	0.08	0	0.06	—	0.04	0.47	0.01	20	0	0	—
0	39	0.89	6	39	157	0.17	0	0.14	0.02	0.09	1.10	0.01	—	0	<.1	—
0	59	1.43	9	40	206	0.28	0	0.17	0.03	0.14	1.79	0.03	48	0	<.1	8
0	54	1.87	15	62	310	0.54	0	0.27	0.24	0.19	2.42	0.02	54	0	0	22
0	54	1.87	15	62	310	0.54	0	0.27	—	0.19	2.42	0.02	54	0	0	22
0	30	0.69	24	78	136	0.57	0	0.07	—	0.04	1.05	0.06	9	0	0	14
3	100	4.50	40	160	180	3.75	—	0.38	—	0.43	5.00	0.50	100	60	2	18
3	100	4.50	40	130	230	3.75	—	0.38	—	0.43	5.00	0.50	100	60	2	18
0	250	5.40	120	260	150	3.75	—	0.38	—	0.26	4.00	0.40	80	60	1	18
0	250	5.40	120	300	290	3.75	—	0.38	—	0.34	6.00	0.40	100	60	1	14
0	350	6.30	140	120	135	5.25	—	1.50	—	1.70	20.00	2.00	400	60	6	25
0	300	6.30	140	110	90	5.25	0	1.50	—	1.70	20.00	2.00	400	60	6	—
0	300	6.30	140	200	90	5.25	0	1.50	—	1.70	20.00	2.00	400	60	6	—
0	300	6.30	140	150	90	5.25	0	1.50	—	1.70	20.00	2.00	400	60	6	—
0	300	6.30	140	150	90	5.25	0	1.50	—	1.70	20.00	2.00	400	60	6	—
0	46	0.36	17	40	42	0.24	0	0.03	0.07	0.02	0.39	0.06	26	0	0	1
0	40	1.06	8	42	153	0.23	0	0.17	0.06	0.09	1.14	0.02	33	0	0	7
0	21	0.33	14	24	49	0.19	0	0.03	0.22	0.01	0.18	0.04	17	0	0	2
64	235	1.96	18	157	470	0.62	57	0.22	—	0.31	1.74	0.06	41	3	<1	16
81	245	1.48	25	227	576	0.86	82	0.23	—	0.36	1.40	0.12	105	1	<1	—
16	153	2.95	15	84	519	0.38	253	0.25	0.65	0.31	2.93	0.59	36	0	2	11
52	191	1.73	14	119	383	0.50	49	0.19	—	0.26	1.55	0.04	51	<1	<1	35
71	196	1.56	30	173	374	0.85	—	0.15	0.32	0.25	1.47	0.09	14	<1	<1	—
0	149	7.40	259	357	20	3.10	0	0.08	—	0.20	1.25	0.22	48	4	0	—
0	9	1.04	17	73	2	0.64	0	0.07	0.01	0.05	1.62	0.09	13	0	0	7
0	6	0.67	43	74	3	0.51	0	0.03	0.08	0.03	0.79	0.06	12	0	0	2
0	9	0.87	29	62	5	0.52	0	0.05	0.01	0.03	0.91	0.08	16	0	0	1
0	6	0.30	6	46	4	0.20	0	0.05	0.10	0.02	0.77	0.04	12	0	0	22
0	4	0.76	53	74	2	1.09	0	0.13	0.02	0.10	1.60	0.13	23	0	0	1
0	27	2.54	110	266	2	1.46	0	0.55	0.47	0.10	0.44	0.08	24	0	0	21
0	51	7.86	179	629	18	2.81	0	0.17	—	0.34	2.49	0.19	42	0	0	—
0	10	0.41	42	42	5	0.61	0	0.09	0.03	0.02	1.49	0.14	4	0	0	10
0	9.75	0.51	42.9	77.02	0.97	0.6	0	0.09	—	0.01	1.29	0.14	3.9	0	0	38
0	—	2.16	—	—	780	—	—	—	—	—	—	—	—	—	—	—
0	13	1.16	9	55	403	0.38	—	0.13	0.28	0.02	1.24	0.06	4	<1	<.1	—
1	47	0.78	48	1	13	0.13	<1	0.03	0.06	0.19	8.71	0.14	<.1	7	<.1	9
0	2	0.12	4	9	4	0.36	0	0.02	0.03	0.01	0.25	0.02	1	0	0	5
0	7	0.52	4	3	2	0.20	0	0.06	0.01	0.04	0.73	0.01	58	0	0	3
0	8	0.95	9	28	1	0.39	0	0.13	0.03	0.01	1.17	0.07	46	0	0	6
0	17	0.99	11	32	3	0.27	0	0.22	0.01	0.02	1.23	0.02	67	0	0	7
0	2.46	0.49	26.23	82.81	2.46	1.09	0	0.04	—	0.07	1.05	0.11	21.31	0	0	0.65

TABLE A–1
Food Composition

(Computer code number is for Wadsworth Diet Analysis program)   (For purposes of calculations, use "0" for t, <1, <.1, <.01, etc.)

DA + Code	Food Description	Quantity	Measure	Wt (g)	H₂O (g)	Ener (kcal)	Prot (g)	Carb (g)	Dietary Fiber (g)	Fat (g)	Sat	Mono	Poly	Trans
												Fat Breakdown (g)		

CEREAL, FLOUR, GRAIN, PASTA, NOODLES, POPCORN—Continued

Flour & grain fractions

DA + Code	Food Description	Quantity	Measure	Wt (g)	H₂O (g)	Ener (kcal)	Prot (g)	Carb (g)	Dietary Fiber (g)	Fat (g)	Sat	Mono	Poly	Trans
505	All purpose flour, self rising, enriched	½	cup(s)	63	7	221	6	46	2	1	0.10	0.05	0.26	—
503	All purpose flour, white, bleached, enriched	½	cup(s)	63	7	228	6	48	2	1	0.10	0.05	0.26	—
1643	Barley flour	½	cup(s)	56	6	198	4	45	2	1	0.16	0.10	0.38	—
383	Buckwheat flour, whole groat	½	cup(s)	60	7	201	8	42	6	2	0.41	0.57	0.57	—
504	Cake wheat flour, enriched	½	cup(s)	55	7	197	4	43	1	<1	0.07	0.04	0.21	—
426	Cornmeal, degermed, enriched	½	cup(s)	69	8	253	6	54	5	1	0.16	0.28	0.49	—
424	Cornmeal, yellow whole grain	½	cup(s)	61	6	221	5	47	4	2	0.31	0.58	1.00	—
1644	Masa corn flour, enriched	½	cup(s)	57	5	208	5	43	5	2	0.30	0.57	0.98	—
1976	Rice flour, brown	½	cup(s)	79	9	287	6	60	4	2	0.44	0.80	0.79	—
1645	Rice flour, white	½	cup(s)	79	9	289	5	63	2	1	0.30	0.35	0.30	—
1978	Rye flour, dark	½	cup(s)	64	7	207	9	44	14	2	0.20	0.21	0.77	—
1980	Semolina, enriched	½	cup(s)	84	11	301	11	61	3	1	0.13	0.10	0.36	—
2827	Soy flour, raw	½	cup(s)	43	2	186	15	15	4	9	1.27	1.94	4.96	—
1990	Wheat germ, crude	2	tablespoon(s)	14	2	52	3	7	2	1	0.24	0.20	0.86	—
506	Whole wheat flour	½	cup(s)	60	6	203	8	44	7	1	0.19	0.14	0.47	—

Breakfast bars

DA + Code	Food Description	Quantity	Measure	Wt (g)	H₂O (g)	Ener (kcal)	Prot (g)	Carb (g)	Dietary Fiber (g)	Fat (g)	Sat	Mono	Poly	Trans
39230	Atkins Morning Start apple crisp	1	item(s)	37	—	170	11	12	6	9	4.00	—	—	—
10574	Health Valley fat free apple	1	item(s)	38	—	110	2	26	3	0	0.00	0.00	0.00	0
10647	Nutri-Grain blueberry cereal bar	1	item(s)	37	5	140	2	27	1	3	0.50	2.00	0.50	—
10648	Nutri-Grain raspberry cereal bar	1	item(s)	37	5	140	2	27	1	3	0.50	2.00	0.50	—
10649	Nutri-Grain strawberry cereal bar	1	item(s)	37	5	140	2	27	1	3	0.50	2.00	0.50	—

Breakfast cereals, hot

DA + Code	Food Description	Quantity	Measure	Wt (g)	H₂O (g)	Ener (kcal)	Prot (g)	Carb (g)	Dietary Fiber (g)	Fat (g)	Sat	Mono	Poly	Trans
363	Corn grits, white, regular & quick, enriched, cooked w/water & salt	½	cup(s)	121	103	71	2	16	<1	<1	0.03	0.06	0.10	—
8636	Corn grits, yellow, regular & quick, enriched, cooked w/salt	½	cup(s)	121	103	71	2	16	<1	<1	0.03	0.06	0.10	—
1260	Cream of Wheat, instant, prepared	½	cup(s)	121	106	61	2	13	<1	<.1	0.01	0.01	0.04	0
365	Farina, enriched, cooked w/water & salt	½	cup(s)	117	102	56	2	12	<1	<.1	0.01	0.01	0.03	—
8657	Oatmeal, cooked w/water	½	cup(s)	117	100	74	3	13	2	1	0.19	0.37	0.44	—
5500	Oatmeal, maple & brown sugar, instant, prepared	1	item(s)	198	150	200	5	40	2	2	0.42	0.74	0.85	—
5510	Oatmeal, ready to serve, packet	1	item(s)	186	158	112	4	20	3	2	0.38	0.66	0.76	—

Breakfast cereals, ready to eat

DA + Code	Food Description	Quantity	Measure	Wt (g)	H₂O (g)	Ener (kcal)	Prot (g)	Carb (g)	Dietary Fiber (g)	Fat (g)	Sat	Mono	Poly	Trans
1197	All-Bran	1	cup(s)	62	2	160	8	46	20	2	0.00	0.00	1.00	0
1200	All-Bran Buds	1	cup(s)	91	3	212	6	73	42	3	—	—	—	0
1199	Apple Jacks	1	cup(s)	33	1	130	1	30	1	1	—	—	—	0
13633	Bran Flakes, Post	1	cup(s)	40	1	133	4	32	7	1	0.00	0.00	0.71	—
1204	Cap'n Crunch	1	cup(s)	36	1	144	2	30	1	2	0.53	0.39	0.27	—
1205	Cap'n Crunch Crunchberries w/wildberry colors	1	cup(s)	35	1	139	2	29	1	2	0.49	0.39	0.28	—
1206	Cheerios	1	cup(s)	30	1	110	3	22	3	2	0.00	0.50	0.50	—
3415	Cocoa Puffs	1	cup(s)	30	1	120	1	26	0	1	—	—	—	—
1207	Cocoa Rice Krispies	1	cup(s)	41	1	160	1	36	1	1	0.67	0.00	0.00	—
5522	Complete wheat bran flakes	1	cup(s)	39	1	120	4	31	7	1	—	—	—	0
1211	Corn Flakes	1	cup(s)	28	1	100	2	24	1	0	0.00	0.00	0.00	0
1247	Corn Pops	1	cup(s)	31	1	120	1	28	0	0	0.00	0.00	0.00	0
1937	Cracklin' Oat Bran	1	cup(s)	65	0	266	5	47	7	9	2.70	4.70	1.33	0
1220	Froot Loops	1	cup(s)	32	1	120	1	28	1	1	0.50	0.00	0.00	—
38214	Frosted Cheerios	1	cup(s)	30	—	120	2	25	1	1	0.00	0.00	0.00	—
372	Frosted Flakes	1	cup(s)	41	1	160	1	37	1	0	0.00	0.00	0.00	0
38215	Frosted Mini Chex	1	cup(s)	40	—	146	1	36	0	0	0.00	0.00	0.00	0
10268	Frosted Mini-Wheats	5	item(s)	51	3	180	5	41	5	1	0.00	0.00	0.50	0
38216	Frosted Wheaties	1	cup(s)	40	—	146	1	36	<1	0	0.00	0.00	0.00	0
1223	Granola, prepared	½	cup(s)	61	0	299	9	32	5	15	2.76	4.7	6.53	—
13334	Granola, Quaker 100% natural, oats & honey	½	cup(s)	48	0	219	5	31	3	9	3.83	4.0	1.19	—
13335	Granola, Quaker 100% natural, oats, honey & raisins	½	cup(s)	51	0	225	5	34	3	9	3.57	3.80	1.10	—
2415	Honey Bunches of Oats honey roasted	1	cup(s)	40	1	160	3	33	1	2	0.67	1.20	0.13	—
1227	Honey Nut Cheerios	1	cup(s)	30	1	120	3	24	2	2	0.00	0.50	0.00	—
2424	Honeycomb	1	cup(s)	22	<1	83	2	20	<1	<1	0.00	—	—	—
10286	Kashi puffed	1	cup(s)	25	—	70	3	13	2	1	0.00	—	—	—
1231	Kix	1	cup(s)	23	<1	90	2	20	1	<1	0.00	0.00	0.00	—
30569	Life	1	cup(s)	43	2	160	4	33	3	2	0.35	0.64	0.61	—
1233	Lucky Charms	1	cup(s)	30	1	120	2	25	1	1	0.00	0.00	0.00	—

PAGE KEY: A–4 = Breads/Baked Goods   A–8 = Cereal/Rice/Pasta   A–12 = Fruit   A–16 = Vegetables/Legumes   A–26 = Nuts/Seeds   A–28 = Vegetarian
A–30 = Dairy   A–36 = Eggs   A–36 = Seafood   A–38 = Meats   A–42 = Poultry   A–42 = Processed meats   A–44 = Beverages   A–48 = Fats/Oils
A–50 = Sweets   A–52 = Spices/Condiments/Sauces   A–54 = Mixed Foods/Soups/Sandwiches   A–60 = Fast food   A–76 = Convenience   A–78 = Baby foods

Chol (mg)	Calc (mg)	Iron (mg)	Magn (mg)	Pota (mg)	Sodi (mg)	Zinc (mg)	Vit A (RAE) (µg)	Thia (mg)	Vit E (mg α)	Ribo (mg)	Niac (mg)	Vit B$_6$ (mg)	Fola (µg)	Vit C (mg)	Vit B$_{12}$ (µg)	Sele (µg)
0	211	2.92	12	78	794	0.39	0	0.42	0.03	0.26	3.65	0.03	123	0	0	22
0	9	2.90	14	67	1	0.44	0	0.49	0.04	0.31	3.69	0.03	114	0	0	21
0	16	0.71	45	186	4	1.05	0	0.07	—	0.03	2.57	0.16	13	0	0	2
0	25	2.44	151	346	7	1.87	0	0.25	0.19	0.11	3.69	0.35	32	0	0	3
0	8	3.99	9	57	1	0.34	0	0.49	0.01	0.23	3.70	0.02	101	0	0	3
0	3	2.85	28	112	2	0.50	8	0.49	0.10	0.28	3.47	0.18	161	0	0	5
0	4	2.10	77	175	21	1.11	7	0.23	0.26	0.12	2.22	0.19	15	0	0	9
0	80	4.11	63	170	3	1.01	0	0.81	0.09	0.43	5.61	0.21	133	0	0	9
0	9	1.56	88	228	6	1.94	0	0.35	0.95	0.06	5.01	0.58	13	0	0	—
0	8	0.28	28	60	0	0.63	0	0.11	0.09	0.02	2.05	0.34	3	0	0	12
0	36	4.13	159	467	1	3.60	1	0.20	0.90	0.16	2.73	0.28	38	0	0	23
0	14	3.64	39	155	1	0.88	0	0.68	0.22	0.48	5.00	0.09	153	0	0	75
0	88	2.71	183	1070	6	1.67	3	0.25	0.93	0.49	1.84	0.20	147	0	0	3
0	6	0.90	34	128	2	1.77	0	0.27	—	0.07	0.98	0.19	40	0	0	11
0	20	2.33	83	243	3	1.76	0	0.27	0.49	0.13	3.82	0.20	26	0	0	42
0	200	—	—	90	70	—	—	0.23	—	0.26	3.00	—	—	9	—	—
0	0	0.72	—	160	25	—	—	0.09	—	0.03	0.40	—	—	1	—	—
0	200	1.80	8	75	110	1.50	—	0.38	—	0.43	5.00	0.50	40	0	0	—
0	200	1.80	8	70	110	1.50	—	0.38	—	0.43	5.00	0.50	40	0	0	—
0	200	1.80	8	55	110	1.50	—	0.38	—	0.43	5.00	0.50	40	0	0	—
0	4	0.73	6	25	270	0.08	0	0.10	0.02	0.07	0.87	0.03	40	0	0	4
0	4	0.73	6	25	270	0.08	2	0.10	0.02	0.07	0.87	0.03	40	0	0	3
0	27	8.60	2	17	1	0.10	0	0.07	—	0.04	0.60	0.01	357	0	0	—
0	5	0.58	2	15	383	0.09	0	0.07	0.01	0.05	0.57	0.01	40	0	0	11
0	9	0.80	28	66	1	0.57	0	0.13	0.12	0.02	0.15	0.02	5	0	0	9
0	26	6.84	50	126	404	1.04	0	1.02	—	0.05	1.57	0.31	30	0	0	11
0	21	3.96	45	112	241	0.93	0	0.60	—	0.05	0.78	0.19	19	0	0	4
0	300	9.00	200	700	160	3.00	300	0.75	—	0.85	10.00	4.00	800	12	12	6
0	0	13.64	182	909	606	4.55	455	1.14	—	1.29	15.15	6.06	1212	18	18	26
0	0	4.50	8	35	150	1.50	150	0.38	—	0.43	5.00	0.50	100	15	2	2
0	0	10.77	80	253	293	2.00	—	0.50	—	0.57	6.65	0.67	133	0	2	—
0	5	6.00	20	72	269	4.99	3	0.51	—	0.57	6.66	0.67	133	0	0	7
<.1	7	6.14	19	71	242	5.12	2	0.51	—	0.57	6.66	0.67	133	<.1	0	7
0	100	8.10	40	95	280	3.75	150	0.38	—	0.43	5.00	0.50	200	6	2	11
0	100	4.50	8	50	170	3.75	0	0.38	—	0.43	5.00	0.50	100	6	2	2
0	53	5.99	11	67	253	2.00	200	0.50	—	0.57	6.65	0.67	133	20	2	6
0	0	23.94	53	226	279	19.95	299	2.00	—	2.26	26.60	2.66	532	80	8	4
0	0	8.10	3	25	200	0.17	150	0.38	—	0.43	5.00	0.50	100	6	2	1
0	0	1.80	2	25	120	1.50	150	0.38	—	0.43	5.00	0.50	100	6	2	2
0	27	2.38	80	293	186	2.00	299	0.49	—	0.56	6.65	0.67	218	20	2	14
0	0	4.50	8	35	150	1.50	150	0.38	—	0.43	5.00	0.50	100	15	2	2
0	100	4.50	16	55	210	3.75	—	0.38	—	0.43	5.00	0.50	100	6	2	—
0	0	5.99	4	27	200	0.21	200	0.50	—	0.57	6.65	0.67	133	8	2	2
0	133	11.97	—	33	266	3.99	—	0.50	—	0.57	6.65	0.67	266	8	2	—
0	0	15.30	60	170	5	1.50	0	0.38	—	0.43	5.00	0.50	100	0	2	2
0	133	10.77	0	47	266	9.98	—	1.00	—	1.13	13.30	1.33	532	8	4	—
0	48	2.59	107	328	13	2.5	2	0.44	3.59	0.17	1.29	0.18	51	1	0	16.95
1	61	1.21	51	225	20	1.04	1	0.12	—	0.11	0.81	0.07	17	<1	0.1	8.3
1	59	1.24	49	250	19	0.99	<1	0.12	—	0.11	0.8	0.07	16	<1	0.1	8.82
0	0	3.59	21	67	253	0.40	—	0.50	—	0.57	6.65	0.67	133	0	2	—
0	100	4.50	24	95	270	3.75	—	0.38	—	0.43	5.00	0.50	200	6	2	7
0	0	2.03	6	26	165	1.13	—	0.28	—	0.32	3.74	0.37	75	0	1	—
0	0	0.72	—	35	0	—	0	0.03	—	0.03	0.80	0.00	—	0	—	—
0	113	6.08	6	26	203	2.81	113	0.28	—	0.32	3.75	0.38	150	5	1	5
0	124	11.92	41	121	218	5.32	1	0.53	—	0.60	7.10	0.70	142	0	0	11
0	100	4.50	16	60	210	3.75	—	0.38	—	0.43	5.00	0.50	200	6	2	6

TABLE A–1
Food Composition

(Computer code number is for Wadsworth Diet Analysis program) (For purposes of calculations, use "0" for t, <1, <.1, <.01, etc.)

DA + Code	Food Description	Quantity	Measure	Wt (g)	H₂O (g)	Ener (kcal)	Prot (g)	Carb (g)	Dietary Fiber (g)	Fat (g)	Sat	Mono	Poly	Trans
	CEREAL, FLOUR, GRAIN, PASTA, NOODLES, POPCORN—Continued													
1201	Multi-Bran Chex	1	cup(s)	58	1	200	4	49	7	2	0.00	0.00	0.00	0
38220	Multi Grain Cheerios	1	cup(s)	30	—	110	3	24	3	1	0.00	0.00	0.00	—
1238	Nutri-Grain golden wheat	1	cup(s)	40	—	133	4	31	5	1	0.00	0.00	0.67	—
1241	Product 19	1	cup(s)	30	1	100	2	25	1	0	0.00	0.00	0.00	0
32432	Puffed rice, fortified	1	cup(s)	14	<1	56	1	13	<1	<.1	0.02	—	—	—
32433	Puffed wheat, fortified	1	cup(s)	12	0	43.68	1.76	9.55	0.52	0.14	0.02	—	—	—
2420	Raisin Bran	1	cup(s)	59	5	190	4	47	8	1	0.00	0.10	0.36	—
1244	Rice Chex	1	cup(s)	25	1	96	2	22	<1	0	0.00	0.00	0.00	0
1245	Rice Krispies	1	cup(s)	26	1	96	2	23	0	0	0.00	0.00	0.00	0
5593	Shredded Wheat	1	cup(s)	25	1	88	3	20	3	1	0.04	0.01	0.10	—
1248	Smacks	1	cup(s)	36	1	133	3	32	1	1	0.00	0.00	0.00	—
1246	Special K	1	cup(s)	31	1	110	7	22	1	0	0.00	0.00	0.00	0
3428	Total, corn flakes	1	cup(s)	23	1	83	2	18	1	0	0.00	0.00	0.00	0
1253	Total whole grain	1	cup(s)	40	1	146	3	31	4	1	0.00	0.00	0.00	—
1254	Trix	1	cup(s)	30	1	120	1	27	1	1	0.00	0.00	0.00	—
382	Wheat germ, toasted	2	tablespoon(s)	14	0	53.95	4.11	7	2.13	1.51	0.25	0.21	0.93	—
1257	Wheaties	1	cup(s)	30	1	110	3	24	3	1	0.00	0.00	0.00	—
	Pasta, noodles													
449	Chinese chow mein noodles, cooked	½	cup(s)	23	<1	119	2	13	1	7	0.99	1.73	3.90	—
1995	Corn pasta, cooked	½	cup(s)	70	48	88	2	20	3	1	0.07	0.13	0.23	—
448	Egg noodles, enriched, cooked	½	cup(s)	80	55	106	4	20	1	1	0.25	0.34	0.33	0.02
440	Macaroni, enriched, cooked	½	cup(s)	70	46	99	3	20	1	<1	0.07	0.06	0.19	—
1996	Pasta, plain, fresh-refrigerated, cooked	½	cup(s)	64	44	84	3	16	0	1	0.10	0.08	0.27	—
1725	Ramen noodles, cooked	½	cup(s)	114	95	104	3	15	1	4	0.19	0.22	0.21	—
2878	Soba noodles, cooked	½	cup(s)	95	69	94	5	20	0	<.1	0.02	0.02	0.03	—
2879	Somen noodles, cooked	½	cup(s)	88	60	115	4	24	0	<1	0.02	0.02	0.06	—
493	Spaghetti, al dente, cooked	½	cup(s)	65	42	95	4	20	1	1	0.05	0.05	0.15	—
2884	Spaghetti, whole wheat, cooked	½	cup(s)	70	47	87	4	19	3	<1	0.07	0.05	0.15	—
1563	Spinach egg noodles, enriched, cooked	½	cup(s)	80	55	105	4	19	2	1	0.29	0.39	0.28	—
2000	Tricolor vegetable macaroni, enriched, cooked	½	cup(s)	67	46	86	3	18	3	<.1	0.01	0.01	0.03	—
	Popcorn													
476	Air popped	1	cup(s)	8	<1	31	1	6	1	<1	0.05	0.09	0.15	—
4619	Caramel	1	cup(s)	35	1	152	1	28	2	5	1.27	1.01	1.58	—
4620	Cheese flavored	1	cup(s)	37	1	196	3	19	4	12	2.38	3.61	5.72	—
477	Popped in oil	1	cup(s)	33	1	165	3	19	3	9	1.61	2.70	4.43	—
	FRUIT AND FRUIT JUICES													
	Apples													
223	Raw medium, w/peel	1	item(s)	138	118	72	<1	19	3	<1	0.04	0.01	0.07	—
224	Slices	½	cup(s)	55	47	29	<1	8	1	<.1	0.02	0.00	0.03	—
946	Slices w/o skin, boiled	½	cup(s)	85	73	45	<1	12	2	<1	0.05	0.01	0.09	—
948	Dried, sulfured	½	cup(s)	22	7	52	<1	14	2	<.1	0.01	0.00	0.02	—
952	Juice, from frozen concentrate	½	cup(s)	120	105	56	<1	14	<1	<1	0.02	0.00	0.04	—
225	Juice, unsweetened, canned	½	cup(s)	124	109	58	<.1	14	<1	<1	0.02	0.01	0.04	—
226	Applesauce, sweetened, canned	½	cup(s)	128	101	97	<1	25	2	<1	0.04	0.01	0.07	—
227	Applesauce, unsweetened, canned	½	cup(s)	122	108	52	<1	14	1	<.1	0.01	0.00	0.02	—
38492	Crabapples	1	item(s)	35	28	27	<1	7	1	<1	0.02	0.00	0.03	—
	Apricot													
228	Fresh w/o pits	4	item(s)	140	121	67	2	16	3	1	0.04	0.24	0.11	—
230	Halves, dried, sulfured	¼	cup(s)	33	10	79	1	21	2	<1	0.01	0.02	0.02	—
229	Halves w/skin, canned in heavy syrup	½	cup(s)	129	100	107	1	28	2	<1	0.01	0.04	0.02	—
	Avocado													
233	California, whole, w/o skin or pit	1	item(s)	170	<1	284	3	15	12	26	3.59	16.61	3.42	—
234	Florida, whole, w/o skin or pit	1	item(s)	304	<1	365	7	24	17	31	5.90	16.70	5.00	—
2998	Pureed	⅛	cup(s)	29	21	46	1	2	2	4	0.61	2.82	0.52	—
	Banana													
235	Fresh whole, w/o peel	1	item(s)	118	88	105	1	27	3	<1	0.13	0.04	0.09	—
4580	Dried chips	¼	cup(s)	55	2	287	1	32	4	19	16.00	1.08	0.35	—
	Blackberries													
237	Raw	½	cup(s)	72	63	31	1	7	4	<1	0.01	0.03	0.20	—
958	Unsweetened, frozen	½	cup(s)	76	62	48	1	12	4	<1	0.01	0.03	0.18	—
	Blueberries													
238	Raw	½	cup(s)	72	61	41	1	10	2	<1	0.02	0.03	0.11	—
959	Canned in heavy syrup	½	cup(s)	128	98	113	1	28	2	<1	0.03	0.06	0.18	—
960	Unsweetened, frozen	½	cup(s)	78	67	40	1	10	2	1	0.04	0.07	0.22	—

PAGE KEY: A–4 = Breads/Baked Goods   A–8 = Cereal/Rice/Pasta   A–12 = Fruit   A–16 = Vegetables/Legumes   A–26 = Nuts/Seeds   A–28 = Vegetarian   A–30 = Dairy   A–36 = Eggs   A–36 = Seafood   A–38 = Meats   A–42 = Poultry   A–42 = Processed meats   A–44 = Beverages   A–48 = Fats/Oils   A–50 = Sweets   A–52 = Spices/Condiments/Sauces   A–54 = Mixed Foods/Soups/Sandwiches   A–60 = Fast food   A–76 = Convenience   A–78 = Baby foods

Chol (mg)	Calc (mg)	Iron (mg)	Magn (mg)	Pota (mg)	Sodi (mg)	Zinc (mg)	Vit A (RAE) (µg)	Thia (mg)	Vit E (mg α)	Ribo (mg)	Niac (mg)	Vit B$_6$ (mg)	Fola (µg)	Vit C (mg)	Vit B$_{12}$ (µg)	Sele (µg)
0	100	16.20	60	220	390	3.75	158	0.38	—	0.03	5.00	0.50	100	6	2	5
0	100	18.00	24	85	200	15.00	—	1.50	—	1.70	20.00	2.00	400	15	6	—
0	0	1.46	32	146	279	4.99	0	0.50	—	0.57	6.65	0.67	133	20	2	9
0	0	18.00	16	50	210	15.00	225	1.50	—	1.70	20.00	2.00	400	60	6	4
0	1	4.44	4	16	<1	0.14	0	0.36	—	0.25	4.94	0.01	3	0	0	1
0	3.35	3.8	17.39	41.75	0.47	0.28	0	0.31	—	0.21	4.23	0.02	3.83	0	0	14.77
0	20	10.80	80	340	300	2.25	—	0.53	—	0.60	7.00	0.70	140	0	2	—
0	80	7.20	7	28	232	3.00	—	0.30	—	0.34	4.00	0.40	160	5	1	1
0	0	1.44	13	32	256	0.48	120	0.30	—	0.34	4.80	0.40	80	5	1	4
0	10	1.08	31	92	2	0.70	0	0.07	—	0.06	1.77	0.10	12	0	0	1
0	0	0.48	11	53	67	0.40	200	0.50	—	0.57	6.65	0.67	133	8	2	17
0	0	8.70	16	60	220	0.90	225	0.53	—	0.60	7.00	2.00	400	15	6	7
0	750	13.50	0	23	158	11.25	113	1.13	22.50	1.28	15.00	1.50	300	45	5	1
0	1330	23.94	32	120	253	19.95	200	2.00	31.24	2.26	26.60	2.66	532	80	8	2
0	100	4.50	0	15	190	3.75	150	0.38	—	0.43	5.00	0.50	100	6	2	6
0	6.35	1.28	45.2	133.76	0.56	2.35	0	0.23	—	0.11	0.78	0.13	49.72	0.84	0	9.18
0	0	8.10	32	110	220	7.50	150	0.75	2.26	0.85	10.00	1.00	200	6	3	1
0	5	1.06	12	27	99	0.32	0	0.13	—	0.09	1.34	0.02	20	0	0	10
0	1	0.18	25	22	0	0.44	2	0.04	0.78	0.02	0.39	0.04	4	0	0	2
26	10	1.27	15	22	6	0.49	5	0.15	0.14	0.07	1.19	0.03	51	0	<.1	17
0	5	0.98	13	22	1	0.37	0	0.14	0.04	0.07	1.17	0.02	54	0	0	15
21	4	0.73	12	15	4	0.36	4	0.13	—	0.10	0.63	0.02	41	0	<.1	—
18	9	0.89	9	34	415	0.31	—	0.08	—	0.05	0.71	0.03	4	<.1	<.1	—
0	4	0.45	9	33	57	0.11	0	0.09	—	0.02	0.48	0.04	7	0	0	—
0	7	0.46	2	25	141	0.19	0	0.02	—	0.03	0.09	0.01	2	0	0	—
0	7	1.00	12	52	1	0.35	0	0.12	0.04	0.07	0.90	0.04	8	0	0	40
0	11	0.74	21	31	2	0.57	0	0.08	0.21	0.03	0.49	0.06	4	0	0	18
26	15	0.87	19	30	10	0.50	4	0.20	0.46	0.10	1.18	0.09	51	0	<1	17
0	7	0.33	13	21	4	0.29	3	0.08	0.06	0.04	0.72	0.02	44	0	0	13
0	1	0.22	11	24	<1	0.28	1	0.02	0.02	0.02	0.16	0.02	2	0	0	1
2	15	0.61	12	38	73	0.20	1	0.02	0.42	0.02	0.77	0.01	2	0	<.1	1
4	42	0.83	34	97	331	0.75	14	0.05	—	0.09	0.54	0.09	4	<1	<1	4
0	3	0.92	36	74	292	0.87	3	0.04	—	0.04	0.51	0.07	6	<.1	0	2
0	8	0.17	7	148	1	0.06	4	0.02	—	0.04	0.13	0.06	4	6	0	0
0	3	0.07	3	59	1	0.02	2	0.01	—	0.01	0.05	0.02	2	3	0	0
0	4	0.16	3	75	1	0.03	2	0.01	0.04	0.01	0.08	0.04	1	<1	0	<1
0	3	0.30	3	97	19	0.04	0	0.00	0.11	0.03	0.20	0.03	0	1	0	<1
0	7	0.31	6	151	8	0.05	0	0.00	0.01	0.02	0.05	0.04	0	1	0	<1
0	9	0.46	4	148	4	0.04	0	0.03	0.01	0.02	0.12	0.04	0	1	0	<1
0	5	0.45	4	78	4	0.05	1	0.02	0.27	0.04	0.24	0.03	1	2	0	<1
0	4	0.15	4	92	2	0.04	1	0.02	0.26	0.03	0.23	0.03	1	1	0	<1
0	6	0.13	2	68	<1	—	0	0.01	—	0.01	0.04	—	2	3	0	—
0	18	0.55	14	363	1	0.28	134	0.04	1.25	0.06	0.84	0.08	13	14	0	<1
0	18	0.88	11	383	3	0.13	59	0.00	1.43	0.02	0.85	0.05	3	<1	0	1
0	12	0.39	9	181	5	0.14	80	0.03	0.77	0.03	0.49	0.07	3	4	0	<1
0	22	1.00	49	861	14	1.12	104	0.12	3.35	0.24	3.24	0.47	105	15	0	1
0	30	0.50	73	1067	6	1.20	185	0.00	0.09	0.10	2.00	0.20	106	53	0	0
0	3	0.16	8	139	2	0.18	2	0.02	0.60	0.04	0.50	0.07	17	3	0	<1
0	6	0.31	32	422	1	0.18	4	0.04	0.12	0.09	0.78	0.43	24	10	0	1
0	10	0.69	42	296	3	0.41	2	0.05	0.13	0.01	0.39	0.14	8	3	0	1
0	21	0.45	14	117	1	0.38	8	0.01	0.84	0.02	0.47	0.02	18	15	0	<1
0	22	0.60	17	106	1	0.19	5	0.02	0.88	0.03	0.91	0.05	26	2	0	<1
0	4	0.20	4	55	1	0.12	2	0.03	0.41	0.03	0.30	0.04	4	7	0	<.1
0	6	0.42	5	51	4	0.09	3	0.04	0.49	0.07	0.14	0.05	3	1	0	<1
0	6	0.14	4	42	1	0.06	2	0.03	0.37	0.03	0.41	0.05	6	2	0	0

TABLE A–1

Food Composition

(Computer code number is for Wadsworth Diet Analysis program)   (For purposes of calculations, use "0" for t, <1, <.1, <.01, etc.)

DA + Code	Food Description	Quantity	Measure	Wt (g)	H₂O (g)	Ener (kcal)	Prot (g)	Carb (g)	Dietary Fiber (g)	Fat (g)	Fat Breakdown (g)			
											Sat	Mono	Poly	Trans
	FRUIT AND FRUIT JUICES—Continued													
	Boysenberries													
961	Canned in heavy syrup	½	cup(s)	128	98	113	1	29	3	<1	0.01	0.02	0.09	—
962	Unsweetened, frozen	½	cup(s)	66	57	33	1	8	3	<1	0.01	0.02	0.10	—
35576	**Breadfruit**	1	item(s)	384	271	396	4	104	17	1	0.00	0.00	0.00	—
	Cherries													
3000	Sour red, raw	½	cup(s)	78	67	39	1	9	1	<1	0.05	0.06	0.07	—
967	Sour red, canned in water	½	cup(s)	122	110	44	1	11	1	<1	0.03	0.03	0.04	—
240	Sweet, raw	½	cup(s)	73	60	46	1	12	2	<1	0.03	0.03	0.04	—
3004	Sweet, canned in heavy syrup	½	cup(s)	127	98	105	1	27	2	<1	0.04	0.05	0.06	—
969	Sweet, canned in water	½	cup(s)	124	108	57	1	15	2	<1	0.03	0.04	0.05	—
	Cranberries													
3007	Chopped, raw	½	cup(s)	55	48	25	<1	7	3	<.1	0.01	0.01	0.03	—
1638	Cranberry juice cocktail	½	cup(s)	127	108	72	0	18	<1	<1	0.01	0.02	0.06	—
241	Cranberry juice cocktail, low calorie, w/saccharin	½	cup(s)	127	120	24	<.1	6	0	<.1	0.00	0.00	0.00	—
1717	Cranberry apple juice drink	½	cup(s)	123	100	87	<.1	22	<1	<.1	0.00	0.00	0.00	—
242	Cranberry sauce, sweetened, canned	¼	cup(s)	69	42	105	<1	27	1	<1	0.01	0.01	0.05	—
	Dates													
244	Domestic, chopped	¼	cup(s)	44.5	0	126	1	33	4	<1	0.01	0.01	0	—
243	Domestic, whole	¼	cup(s)	44.5	0	126	1	33	4	<1	0.01	0.01	0	—
	Figs													
973	Raw, medium	2	item(s)	101	80	74	1	19	3	<1	0.06	0.07	0.14	—
975	Canned in heavy syrup	½	cup(s)	130	99	114	<1	30	3	<1	0.03	0.03	0.06	—
974	Canned in water	½	cup(s)	124	106	66	<1	17	3	<1	0.02	0.03	0.06	—
	Fruit cocktail & salad													
245	Fruit cocktail, canned in heavy syrup	½	cup(s)	124	100	91	<1	23	1	<.1	0.01	0.02	0.04	—
978	Fruit cocktail, canned in juice	½	cup(s)	119	104	55	1	14	1	<.1	0.00	0.00	0.00	—
977	Fruit cocktail, canned in water	½	cup(s)	119	108	38	<1	10	1	<.1	0.01	0.01	0.02	—
979	Fruit salad, canned in water	½	cup(s)	123	112	37	<1	10	1	<.1	0.01	0.02	0.03	—
	Gooseberries													
981	Raw	½	cup(s)	75	66	33	1	8	3	<1	0.03	0.04	0.24	—
982	Canned in light syrup	½	cup(s)	126	101	92	1	24	3	<1	0.02	0.02	0.14	—
	Grapefruit													
3022	Raw, pink or red	½	cup(s)	115	<.1	48	1	12	2	<1	0.02	0.02	0.04	—
247	Raw, white	½	item(s)	118	107	39	1	10	1	<1	0.02	0.02	0.03	—
251	Juice, pink, sweetened, canned	½	cup(s)	125	109	58	1	14	<1	<1	0.02	0.02	0.03	—
249	Juice, white	½	cup(s)	124	111	48	1	11	<1	<1	0.02	0.02	0.03	—
248	Sections, canned in light syrup	½	cup(s)	127	106	76	1	20	1	<1	0.02	0.02	0.03	—
983	Sections, canned in water	½	cup(s)	122	<.1	44	1	11	<1	<1	0.02	0.02	0.03	—
	Grapes													
255	American, slip skin	½	cup(s)	46	37	31	<1	8	<1	<1	0.05	0.01	0.05	—
256	European, red or green, adherent skin	½	cup(s)	80	<.1	55	1	14	1	<1	0.04	0.01	0.04	—
259	Juice, sweetened, added vitamin C, from frozen concentrate	½	cup(s)	125	109	64	<1	16	<1	<1	0.04	0.01	0.03	—
3159	Juice drink, canned	½	cup(s)	125	109	63	<1	16	0	0	0.00	Mono	0.00	—
3060	Raisins, seeded, packed	¼	cup(s)	41	7	122	1	32	3	<1	0.07	0.01	0.07	—
987	**Guava, raw**	1	item(s)	90	77	46	1	11	5	1	0.15	0.05	0.23	—
35593	**Guava, strawberry**	1	item(s)	6	5	4	<.1	1	<1	<.1	0.01	0.00	0.02	—
3027	**Jackfruit**	½	cup(s)	83	61	78	1	20	1	<1	0.05	0.04	0.07	—
8458	**Kiwi fruit**	1	item(s)	77	63	53	1	11	3	1	0.02	0.03	0.19	—
	Lemon													
992	Raw	1	item(s)	108	94	22	1	12	5	<1	0.04	0.01	0.10	—
262	Juice	1	tablespoon(s)	15	14	4	<.1	1	<.1	0	0.00	0.00	0.00	—
993	Peel	1	teaspoon(s)	2	2	1	<.1	<1	<1	<.1	0.00	0.00	0.00	—
	Lime													
994	Raw	1	item(s)	67	61	15	<1	6	2	<.1	0.01	0.01	0.02	—
269	Juice	1	tablespoon(s)	15	14	4	<.1	1	<.1	<.1	0.00	0.00	0.00	—
995	**Loganberries, frozen**	½	cup(s)	74	62	40	1	10	4	<1	0.01	0.02	0.13	—
	Mandarin orange													
1038	Canned in juice	½	cup(s)	125	111	46	1	12	1	<.1	0.00	0.01	0.01	—
1039	Canned in light syrup	½	cup(s)	126	105	77	1	20	1	<1	0.02	0.02	0.03	—
999	**Mango**	½	item(s)	104	85	67	1	18	2	<1	0.07	0.10	0.05	—
1005	**Nectarine, raw, sliced**	½	cup(s)	69	60	30	1	7	1	<1	0.02	0.06	0.08	—
	Melons													
271	Cantaloupe	½	cup(s)	80	72	27	1	7	1	<1	0.04	0.00	0.07	—
1000	Casaba melon	½	cup(s)	85	78	24	1	6	1	<.1	0.02	0.00	0.03	—

PAGE KEY: A–4 = Breads/Baked Goods   A–8 = Cereal/Rice/Pasta   A–12 = Fruit   A–16 = Vegetables/Legumes   A–26 = Nuts/Seeds   A–28 = Vegetarian
A–30 = Dairy   A–36 = Eggs   A–36 = Seafood   A–38 = Meats   A–42 = Poultry   A–42 = Processed meats   A–44 = Beverages   A–48 = Fats/Oils
A–50 = Sweets   A–52 = Spices/Condiments/Sauces   A–54 = Mixed Foods/Soups/Sandwiches   A–60 = Fast food   A–76 = Convenience   A–78 = Baby foods

Chol (mg)	Calc (mg)	Iron (mg)	Magn (mg)	Pota (mg)	Sodi (mg)	Zinc (mg)	Vit A (RAE) (µg)	Thia (mg)	Vit E (mg α)	Ribo (mg)	Niac (mg)	Vit B$_6$ (mg)	Fola (µg)	Vit C (mg)	Vit B$_{12}$ (µg)	Sele (µg)
0	23	0.55	14	115	4	0.24	3	0.03	—	0.04	0.29	0.05	44	8	0	1
0	18	0.56	11	92	1	0.15	2	0.03	0.57	0.02	0.51	0.04	42	2	0	<1
0	65	2.07	96	1882	8	0.46	8	0.42	—	0.12	3.46	0.00	54	111	0	2
0	12	0.25	7	134	2	0.08	50	0.02	0.05	0.03	0.31	0.03	6	8	0	0
0	13	1.67	7	120	9	0.09	46	0.02	0.28	0.05	0.22	0.05	10	3	0	0
0	9	0.26	8	161	0	0.05	2	0.02	0.05	0.02	0.11	0.04	3	5	0	0
0	11	0.44	11	183	4	0.13	10	0.03	0.29	0.05	0.50	0.04	5	5	0	0
0	14	0.45	11	162	1	0.10	10	0.03	0.29	0.05	0.51	0.04	5	3	0	0
0	4	0.14	3	47	1	0.06	2	0.01	0.66	0.01	0.06	0.03	1	7	0	<.1
0	4	0.19	3	23	3	0.09	0	0.01	0.28	0.01	0.04	0.02	0	45	0	<1
0	11	0.05	3	32	4	0.03	0	0.00	0.06	0.00	0.01	0.00	0	41	0	0
0	6	0.15	2	34	9	0.22	0	0.01	0.15	0.02	0.07	0.03	0	39	0	0
0	3	0.15	2	18	20	0.03	1	0.01	0.57	0.01	0.07	0.01	1	1	0	<1
0	17	0.45	19	292	1	0.12	1	0.02	0.02	0.02	0.56	0.07	9	<1	0	1
0	17	0.45	19	292	1	0.12	1	0.02	0.02	0.02	0.56	0.07	9	<1	0	1
0	35	0.37	17	233	1	0.15	7	0.06	0.11	0.05	0.40	0.11	6	2	0	<1
0	35	0.36	13	128	1	0.14	3	0.03	0.16	0.05	0.55	0.09	3	1	0	<1
0	35	0.36	12	128	1	0.15	2	0.03	0.10	0.05	0.55	0.09	2	1	0	<1
0	7	0.36	6	109	7	0.10	12	0.02	0.50	0.02	0.46	0.06	4	2	0	1
0	9	0.25	8	113	5	0.11	18	0.01	0.47	0.02	0.48	0.06	4	3	0	1
0	6	0.30	8	111	5	0.11	15	0.02	0.47	0.01	0.43	0.06	4	2	0	1
0	9	0.37	6	96	4	0.10	27	0.02	—	0.03	0.46	0.04	4	2	0	1
0	19	0.23	8	149	1	0.09	11	0.03	0.28	0.02	0.23	0.06	5	21	0	<1
0	20	0.42	8	97	3	0.14	9	0.03	—	0.07	0.19	0.02	4	13	0	1
0	25	0.09	10	155	0	0.08	30	0.05	0.15	0.03	0.23	0.06	15	36	0	<1
0	14	0.07	11	175	0	0.08	2	0.04	0.15	0.02	0.32	0.05	12	39	0	2
0	10	0.45	13	203	3	0.08	0	0.05	0.05	0.03	0.40	0.03	13	34	0	<1
0	11	0.25	15	200	1	0.06	2	0.05	0.27	0.02	0.25	0.05	12	47	0	<1
0	18	0.51	13	164	3	0.10	0	0.05	0.11	0.03	0.31	0.03	11	27	0	1
0	18	0.50	12	161	2	0.11	0	0.05	0.11	0.03	0.30	0.02	11	27	0	1
0	6	0.13	2	88	1	0.02	2	0.04	0.09	0.03	0.14	0.05	2	2	0	<.1
0	8	0.29	6	153	2	0.06	6	0.06	0.15	0.06	0.15	0.07	2	9	0	<.1
0	5	0.13	5	26	3	0.05	0	0.02	0.00	0.03	0.16	0.05	1	30	0	<1
0	4	0.13	4	41	1	0.03	0	0.01	0.00	0.02	0.09	0.02	1	20	0	<1
0	12	1.07	12	340	12	0.07	0	0.05	—	0.08	0.46	0.08	1	2	0	<1
0	18	0.28	9	256	3	0.21	28	0.05	0.66	0.05	1.08	0.13	13	165	0	1
0	1	0.01	1	18	2	—	—	0.00	—	0.00	0.04	0.00	—	2	0	—
0	28	0.50	31	251	2	0.35	12	0.02	—	0.09	0.33	0.09	12	6	0	<1
0	30	0.38	14	251	2	0.10	4	—	—	0.02	0.25	0.05	<.1	74	0	—
0	66	0.76	13	157	3	0.11	2	0.05	—	0.04	0.22	0.12	—	83	0	1
0	1	0.00	1	19	<1	0.01	<1	0.00	0.02	0.00	0.02	0.01	2	7	0	<.1
0	3	0.02	<1	3	<1	0.01	<.1	0.00	0.00	0.00	0.01	0.00	<1	3	0	<.1
0	9	0.06	5	78	1	0.05	1	0.02	0.15	0.01	0.10	0.03	7	20	0	<.1
0	1	0.00	1	17	<1	0.01	<1	0.00	0.03	0.00	0.02	0.01	1	5	0	<.1
0	19	0.47	15	107	1	0.25	1	0.04	0.64	0.02	0.62	0.05	19	11	0	<1
0	14	0.34	14	166	6	0.63	54	0.10	0.12	0.04	0.55	0.05	6	43	0	<1
0	9	0.47	10	98	8	0.30	53	0.07	0.13	0.06	0.56	0.05	6	25	0	1
0	10	0.13	9	161	2	0.04	39	0.06	1.16	0.06	0.60	0.14	14	29	0	1
0	4	0.19	6	139	0	0.12	12	0.02	0.53	0.02	0.78	0.02	3	4	0	0
0	7	0.17	10	215	13	0.14	136	0.03	0.04	0.02	0.59	0.06	17	30	0	<1
0	9	0.29	9	155	8	0.06	0	0.01	0.04	0.03	0.20	0.14	7	19	0	<1

TABLE A–1

Food Composition

(Computer code number is for Wadsworth Diet Analysis program)   *(For purposes of calculations, use "0" for t, <1, <.1, <.01, etc.)*

DA + Code	Food Description	Quantity	Measure	Wt (g)	H₂O (g)	Ener (kcal)	Prot (g)	Carb (g)	Dietary Fiber (g)	Fat (g)	Fat Breakdown (g) Sat	Mono	Poly	Trans
	FRUIT AND FRUIT JUICES—Continued													
272	Honeydew	½	cup(s)	89	80	32	<1	8	1	<1	0.03	0.00	0.05	—
318	Watermelon	½	cup(s)	77	71	23	<1	6	<1	<1	0.01	0.03	0.04	—
	Orange													
273	Raw	1	item(s)	131	114	62	1	15	3	<1	0.02	0.03	0.03	—
3040	Peel	1	teaspoon(s)	2	1	2	<.1	1	<1	<.1	0.00	0.00	0.00	—
274	Sections	½	cup(s)	90	78	43	1	11	2	<1	0.01	0.02	0.02	—
275	Juice	½	cup(s)	124	109	56	1	13	<1	<1	0.03	0.04	0.05	—
29630	Juice, fresh squeezed	½	cup(s)	124	109	56	1	13	<1	<1	0.03	0.04	0.05	—
14414	Juice w/calcium & extra vitamin C	½	cup(s)	125	55	1	13	<1	0	0.00	0.00	0.00	0	—
278	Juice, unsweetened, from frozen concentrate	½	cup(s)	125	110	56	1	13	<1	<.1	0.01	0.01	0.01	—
	Papaya													
282	Raw	½	cup(s)	70	62	27	<1	7	1	<.1	0.03	0.03	0.02	—
16830	Dried, strips	2	item(s)	46	12	119	2	30	5	<1	0.13	0.12	0.09	—
35640	**Passion fruit, purple**	1	item(s)	18	13	17	<1	4	3	<1	0.00	0.00	0.00	—
	Peach													
283	Raw, medium	1	item(s)	98	87	38	1	9	1	<1	0.02	0.07	0.08	—
285	Halves, canned in heavy syrup	½	cup(s)	131	104	97	1	26	2	<1	0.01	0.05	0.06	—
286	Halves, canned in water	½	cup(s)	122	114	29	1	7	2	<.1	0.01	0.03	0.03	—
290	Slices, sweetened, frozen	½	cup(s)	125	93	118	1	30	2	<1	0.02	0.06	0.08	—
	Pear													
291	Raw	1	item(s)	166	139	96	1	26	5	<1	0.01	0.04	0.05	—
8672	Asian	1	item(s)	122	108	51	1	13	4	<1	0.01	0.06	0.07	—
293	Danjou	1	item(s)	200	168	120	1	30	5	1	0.00	0.20	0.20	—
294	Halves, canned in heavy syrup	½	cup(s)	133	107	98	<1	25	2	<1	0.01	0.04	0.04	—
1012	Halves, canned in juice	½	cup(s)	124	107	62	<1	16	2	<.1	0.00	0.02	0.02	—
1017	**Persimmon**	1	item(s)	25	16	32	<1	8	0	<1	0.01	0.02	0.02	—
	Pineapple													
295	Raw, diced	½	cup(s)	78	67	37	<1	10	1	<.1	0.01	0.01	0.03	—
3053	Canned in extra heavy syrup	½	cup(s)	130	101	108	<1	28	1	<1	0.01	0.02	0.05	—
1019	Canned in juice	½	cup(s)	125	104	75	1	20	1	<.1	0.01	0.01	0.04	—
296	Canned in light syrup	½	cup(s)	126	108	66	<1	17	1	<1	0.01	0.02	0.05	—
1018	Canned in water	½	cup(s)	123	112	39	1	10	1	<1	0.01	0.01	0.04	—
299	Juice, unsweetened, canned	½	cup(s)	125	107	70	<1	17	<1	<1	0.01	0.01	0.04	—
1024	**Plantain, cooked**	½	cup(s)	77	52	89	1	24	2	<1	0.05	0.01	0.03	—
300	**Plum, raw, large**	1	item(s)	83	72	38	1	9	1	<1	0.01	0.11	0.04	—
1027	**Pomegranate**	1	item(s)	154	125	105	1	26	1	<1	0.06	0.07	0.10	—
	Prunes													
5644	Dried	2	item(s)	17	5	40	<1	11	1	<.1	0.01	0.06	0.02	—
305	Dried, stewed	½	cup(s)	119	<.1	128	1	33	4	<1	0.00	0.15	0.04	—
306	Juice, canned	1	cup(s)	256	208	182	2	45	3	<.1	0.01	0.05	0.02	—
	Raisins, *see* grapes													
	Raspberries													
309	Raw	½	cup(s)	62	53	32	1	7	4	<1	0.01	0.04	0.23	—
310	Red, sweetened, frozen	½	cup(s)	125	91	129	1	33	6	<1	0.01	0.02	0.11	—
311	**Rhubarb, cooked with sugar**	½	cup(s)	120	82	140	1	38	3	<.1	0.00	0.00	0.05	—
	Strawberries													
313	Raw	½	cup(s)	72	65	23	<1	6	1	<1	0.01	0.03	0.11	—
315	Sweetened, frozen, thawed	½	cup(s)	128	100	99	1	27	2	<1	0.01	0.02	0.09	—
16828	**Tangelo**	1	item(s)	95	82	45	1	11	2	<1	0.01	0.02	0.02	—
	Tangerine													
316	Raw	1	item(s)	84	74	37	1	9	2	<1	0.02	0.03	0.03	—
1040	Juice	½	cup(s)	124	110	53	1	12	<1	<1	0.03	0.04	0.05	—
	VEGETABLES, LEGUMES													
	Amaranth													
1042	Leaves, raw	1	cup(s)	28	26	6	1	1	0	<.1	0.03	0.02	0.04	—
1043	Leaves, boiled, drained	½	cup(s)	66	60	14	1	3	0	<1	0.03	0.03	0.05	—
8683	**Arugula leaves, raw**	1	cup(s)	20	18	5	1	1	<1	<1	0.02	0.01	0.06	—
	Artichoke													
1044	Boiled, drained	1	item(s)	120	101	60	4	13	6	<1	0.04	0.01	0.08	—
2885	Hearts, boiled, drained	½	cup(s)	84	71	42	3	9	5	<1	0.03	0.00	0.06	—
	Asparagus													
566	Boiled, drained	½	cup(s)	90	0.08	20	2	4	2	0.19	0.06	0	0.12	—
568	Canned, drained	½	cup(s)	121	114	23	3	3	2	1	0.18	0.03	0.34	—
565	Tips, frozen, boiled, drained	½	cup(s)	90	82	25	3	4	1	<1	0.09	0.01	0.17	—

Chol (mg)	Calc (mg)	Iron (mg)	Magn (mg)	Pota (mg)	Sodi (mg)	Zinc (mg)	Vit A (RAE) (µg)	Thia (mg)	Vit E (mg α)	Ribo (mg)	Niac (mg)	Vit B$_6$ (mg)	Fola (µg)	Vit C (mg)	Vit B$_{12}$ (µg)	Sele (µg)
0	5	0.15	9	203	16	0.08	3	0.03	0.02	0.01	0.37	0.08	17	16	0	1
0	5	0.19	8	86	1	0.08	22	0.03	0.04	0.02	0.14	0.03	2	6	0	<1
0	52	0.13	13	237	0	0.09	14	0.11	0.24	0.05	0.37	0.08	39	70	0	1
0	3	0.02	<1	4	<.1	0.01	<1	0.00	0.00	0.00	0.02	0.00	1	3	0	<.1
0	36	0.09	9	164	0	0.06	10	0.08	0.16	0.04	0.26	0.05	27	48	0	<1
0	14	0.25	14	248	1	0.06	12	0.11	0.05	0.04	0.50	0.05	37	62	0	<1
0	14	0.25	14	248	1	0.06	—	0.11	0.05	0.04	0.50	0.05	38	62	0	—
—	175	—	—	225	0	—	5	0.08	—	—	0.40	0.06	30	54	0	—
0	11	0.12	12	237	1	0.06	6	0.10	0.25	0.02	0.25	0.05	55	48	0	<1
0	17	0.07	7	180	2	0.05	39	0.02	0.51	0.02	0.24	0.01	27	43	0	<1
0	73	0.30	30	783	9	0.21	—	0.06	2.22	0.09	0.93	0.05	58	38	0	—
0	2	0.29	5	63	5	—	—	0.00	—	0.02	0.27	—	3	5	0	<1
0	6	0.25	9	186	0	0.17	16	0.02	0.72	0.03	0.79	0.02	4	6	0	<.1
0	4	0.35	7	121	8	0.12	22	0.01	0.64	0.03	0.80	0.02	4	4	0	<1
0	2	0.39	6	121	4	0.11	33	0.01	0.60	0.02	0.64	0.02	4	4	0	<1
0	4	0.46	6	163	8	0.06	18	0.02	0.77	0.04	0.82	0.02	4	118	0	1
0	15	0.28	12	198	2	0.17	2	0.02	0.20	0.04	0.26	0.05	12	7	0	<1
0	5	0.00	10	148	0	0.02	0	0.01	0.15	0.01	0.27	0.03	10	5	0	<1
0	22	0.50	12	250	0	0.24	—	0.04	1.00	0.08	0.20	0.04	15	8	0	1
0	7	0.29	5	86	7	0.11	0	0.01	0.11	0.03	0.32	0.02	1	1	0	0
0	11	0.36	9	119	5	0.11	0	0.01	0.10	0.01	0.25	0.02	1	2	0	0
0	7	0.63	—	78	<1	—	—	—	—	—	—	—	—	17	0	0
0	10	0.22	9	89	1	0.08	2	0.06	0.02	0.02	0.38	0.09	12	28	0	<.1
0	18	0.49	20	133	1	0.14	1	0.12	—	0.03	0.37	0.10	7	9	0	—
0	17	0.35	17	152	1	0.12	2	0.12	0.01	0.02	0.35	0.09	6	12	0	<1
0	18	0.49	20	132	1	0.15	3	0.11	0.01	0.03	0.37	0.09	6	9	0	1
0	18	0.49	22	156	1	0.15	2	0.11	0.01	0.03	0.37	0.09	6	9	0	<1
0	21	0.33	16	168	1	0.14	0	0.07	0.03	0.03	0.32	0.12	29	13	0	<1
0	2	0.45	25	358	4	0.10	35	0.04	0.10	0.04	0.58	0.18	20	8	0	1
0	5	0.14	6	130	0	0.08	14	0.02	0.21	0.02	0.34	0.02	4	8	0	0
0	5	0.46	5	399	5	0.18	8	0.05	0.92	0.05	0.46	0.16	9	9	0	1
0	9	0.42	8	125	1	0.09	17	0.01	0.00	0.03	0.33	0.04	1	1	0	<1
0	23	0.46	21	383	1	0.19	37	0.00	0.23	0.12	0.85	0.23	0	3	0	<1
0	31	3.02	36	707	10	0.54	0	0.04	0.31	0.18	2.01	0.56	0	10	0	2
0	15	0.42	14	93	1	0.26	1	0.02	0.54	0.02	0.37	0.03	13	16	0	<1
0	19	0.81	16	143	1	0.23	4	0.02	0.90	0.06	0.29	0.04	33	21	0	<1
0	174	0.25	16	115	1	—	—	0.02	—	0.03	0.25	—	—	4	0	—
0	12	0.30	9	110	1	0.10	1	0.02	0.21	0.02	0.28	0.03	17	42	0	<1
0	14	0.60	8	125	1	0.06	1	0.02	0.31	0.10	0.37	0.04	5	50	0	1
0	38	0.10	10	172	0	0.07	—	0.08	0.17	0.04	0.27	0.06	29	51	0	—
0	12	0.08	10	132	1	0.20	29	0.09	0.17	0.02	0.13	0.06	17	26	0	<1
0	22	0.25	10	220	1	0.04	16	0.07	0.16	0.02	0.12	0.05	6	38	0	<1
0	60	0.65	15	171	6	0.25	0	0.01	—	0.04	0.18	0.05	24	12	0	<1
0	138	1.49	36	423	14	0.58	92	0.01	—	0.09	0.37	0.12	38	27	0	1
0	32	0.29	9	74	5	0.09	24	0.01	0.09	0.02	0.06	0.01	19	3	0	<.1
0	54	1.55	72	425	114	0.59	11	0.08	0.23	0.08	1.20	0.13	61	12	0	<1
0	38	1.08	50	297	80	0.41	8	0.05	0.16	0.06	0.84	0.09	43	8	0	<1
0	20.7	0.81	12.6	201.6	12.6	0.54	48.59	0.14	1.35	0.12	0.97	0.07	134.1	6.92	0	5.48
0	19	0.73	12	208	347	0.48	50	0.07	0.38	0.12	1.15	0.13	116	22	0	2
0	21	0.58	12	196	4	0.50	—	0.06	1.08	0.09	0.93	0.02	121	22	0	4

TABLE A–1
Food Composition
(Computer code number is for Wadsworth Diet Analysis program)   (For purposes of calculations, use "0" for t, <1, <.1, <.01, etc.)

DA + Code	Food Description	Quantity	Measure	Wt (g)	H₂O (g)	Ener (kcal)	Prot (g)	Carb (g)	Dietary Fiber (g)	Fat (g)	Fat Breakdown (g) Sat	Mono	Poly	Trans
	VEGETABLES, LEGUMES —Continued													
	Bamboo shoots													
1048	Boiled, drained	½	cup(s)	60	58	7	1	1	1	<1	0.03	0.00	0.06	—
1049	Canned, drained	½	cup(s)	65	62	12	1	2	1	<1	0.06	0.01	0.12	—
	Beans													
1801	Adzuki beans, boiled	½	cup(s)	115	76	147	9	28	8	<1	0.04	—	—	—
511	Baked beans w/franks, canned	½	cup(s)	129	89	182	9	20	9	8	3.02	3.64	1.07	
512	Baked beans w/pork in tomato sauce, canned	½	cup(s)	127	92	124	7	25	6	1	0.50	0.56	0.17	—
513	Baked beans w/pork in sweet sauce, canned	½	cup(s)	127	89	140	7	27	7	2	0.71	0.80	0.24	0
1805	Black beans, boiled	½	cup(s)	86	57	114	8	20	7	<1	0.12	0.04	0.20	
14597	Chickpeas, garbanzo beans, or bengal gram, boiled	½	cup(s)	82	49	134	7	22	6	2	0.22	0.48	0.95	
569	Fordhook lima beans, frozen, boiled, drained	½	cup(s)	85	62	88	5	16	5	<1	0.07	0.02	0.14	—
1806	French beans, boiled	½	cup(s)	89	59	114	6	21	8	1	0.07	0.05	0.40	—
2773	Great northern beans, boiled	½	cup(s)	89	61	104	7	19	6	<1	0.12	0.02	0.17	—
2736	Hyacinth beans, boiled, drained	½	cup(s)	44	38	22	1	4	0	<1	0.05	0.06	0.00	—
515	Lima beans, boiled, drained	½	cup(s)	85	57	105	6	20	5	<1	0.06	0.02	0.13	—
570	Lima beans, baby, frozen, boiled, drained	½	cup(s)	90	65	95	6	18	5	<1	0.06	0.02	0.13	—
579	Mung beans, sprouted, boiled, drained	½	cup(s)	62	<.1	13	1	3	<1	<.1	0.02	0.00	0.02	—
510	Navy beans, boiled	½	cup(s)	91	57	129	8	24	6	1	0.13	0.05	0.22	0
32816	Pinto beans, boiled, drained, no salt added	½	cup(s)	114	106	25	2	5	0	<1	0.04	0.03	0.21	
1052	Pinto beans, frozen, boiled, drained	½	cup(s)	47	27	76	4	15	4	<1	0.03	0.02	0.13	—
514	Red kidney beans, canned	½	cup(s)	128	99	109	7	20	8	<1	0.06	0.03	0.24	—
1810	Refried beans, canned	½	cup(s)	127	96	119	7	20	7	2	0.60	0.71	0.19	—
1053	Shell beans, canned	½	cup(s)	123	111	37	2	8	4	<1	0.03	0.02	0.13	—
1670	Soybeans, boiled	½	cup(s)	86	54	149	14	9	5	8	1.12	1.70	4.36	—
1108	Soybeans, green, boiled, drained	½	cup(s)	90	62	127	11	10	4	6	0.67	1.09	2.71	—
1807	White beans, small, boiled	½	cup(s)	90	57	127	8	23	9	1	0.15	0.05	0.25	—
574	Green string beans, canned, fat added in cooking	½	cup(s)	93	<.1	41	1	4	2	3	0.51	1.23	0.75	—
575	Yellow snap, string or wax beans, boiled, drained	½	cup(s)	62	<.1	22	1	5	2	<1	0.04	0.00	0.09	—
576	Yellow snap, string or wax beans, frozen, boiled, drained	½	cup(s)	68	<.1	19	1	4	2	<1	0.02	0.00	0.05	—
	Beets													
580	Whole, boiled, drained	2	item(s)	100	87	44	2	10	2	<1	0.03	0.04	0.06	—
581	Sliced, boiled, drained	½	cup(s)	85	74	37	1	8	2	<1	0.02	0.03	0.05	—
583	Sliced, canned, drained	½	cup(s)	85	77	26	1	6	1	<1	0.02	0.02	0.04	—
2730	Pickled, canned with liquid	½	cup(s)	114	93	74	1	18	3	<.1	0.01	0.02	0.03	—
584	Beet greens, boiled, drained	½	cup(s)	72	64	19	2	4	2	<1	0.02	0.03	0.05	—
585	**Cowpeas or black-eyed peas, boiled, drained**	½	cup(s)	83	0.06	80	2.61	16.76	4.12	0.31	0.07	0.02	0.13	
	Broccoli													
587	Raw, chopped	½	cup(s)	44	39	15	1	3	1	<1	0.02	0.00	0.02	—
588	Chopped, boiled, drained	½	cup(s)	78	70	27	2	6	3	<1	0.06	0.03	0.13	—
590	Frozen, chopped, boiled, drained	½	cup(s)	92	83	26	3	5	3	<1	0.02	0.01	0.05	—
16848	**Broccoflower, raw, chopped**	½	cup(s)	32	29	10	1	2	1	<.1	0.01	0.01	0.04	—
	Brussels sprouts													
591	Boiled, drained	½	cup(s)	78	69	28	2	6	2	<1	0.08	0.03	0.20	—
592	Frozen, boiled, drained	½	cup(s)	78	67	33	3	6	3	<1	0.06	0.02	0.16	—
	Cabbage													
594	Raw, shredded	1	cup(s)	70	65	17	1	4	2	<.1	0.01	0.01	0.04	—
595	Boiled, drained, no salt added	1	cup(s)	150	140	33	2	7	3	1	0.08	0.05	0.29	—
35611	Chinese (pak choi or bok choy), boiled w/salt, drained	1	cup(s)	170	162	20	3	3	2	<1	0.04	0.02	0.13	—
16869	Kim chee	1	cup(s)	150	138	31	2	6	2	<1	0.04	0.02	0.15	—
596	Red, shredded, raw	1	cup(s)	70	63	22	1	5	1	<1	0.02	0.01	0.09	—
597	Savoy, shredded, raw	1	cup(s)	70	64	19	1	4	2	<.1	0.01	0.00	0.03	—
11710	**Capers**	1	teaspoon(s)	5	—	0	0	0	0	0	0.00	0.00	0.00	0
	Carrots													
600	Raw	½	cup(s)	61	54	25	1	6	2	<1	0.02	0.01	0.06	0
8691	Raw, baby	8	item(s)	80	72	28	1	7	1	<1	0.02	0.01	0.05	0
601	Grated	½	cup(s)	55	49	23	1	5	2	<1	0.02	0.01	0.06	0
602	Sliced, boiled, drained	½	cup(s)	78	0.07	27.29	0.59	6.41	2.33	0.14	0.02	0	0.08	—

PAGE KEY: A–4 = Breads/Baked Goods A–8 = Cereal/Rice/Pasta A–12 = Fruit A–16 = Vegetables/Legumes A–26 = Nuts/Seeds A–28 = Vegetarian A–30 = Dairy A–36 = Eggs A–36 = Seafood A–38 = Meats A–42 = Poultry A–42 = Processed meats A–44 = Beverages A–48 = Fats/Oils A–50 = Sweets A–52 = Spices/Condiments/Sauces A–54 = Mixed Foods/Soups/Sandwiches A–60 = Fast food A–76 = Convenience A–78 = Baby foods

Chol (mg)	Calc (mg)	Iron (mg)	Magn (mg)	Pota (mg)	Sodi (mg)	Zinc (mg)	Vit A (RAE) (µg)	Thia (mg)	Vit E (mg α)	Ribo (mg)	Niac (mg)	Vit B₆ (mg)	Fola (µg)	Vit C (mg)	Vit B₁₂ (µg)	Sele (µg)
0	7	0.14	2	320	2	0.28	0	0.01	—	0.03	0.18	0.06	1	0	0	<1
0	5	0.21	3	52	5	0.43	1	0.02	0.41	0.02	0.09	0.09	2	1	0	<1
0	32	2.30	60	612	9	2.04	0	0.13	—	0.07	0.82	0.11	139	0	0	1
8	62	2.22	36	302	553	2.40	5	0.07	0.59	0.07	1.16	0.06	39	3	0	8
9	71	4.15	44	380	557	7.41	5	0.07	0.13	0.06	0.63	0.09	29	4	0	6
9	77	2.10	43	336	425	1.90	1	0.06	0.04	0.08	0.44	0.11	47	4	0	6
0	23	1.81	60	305	1	0.96	0	0.21	—	0.05	0.43	0.06	128	0	0	1
0	40	2.37	39	239	6	1.25	1	0.10	0.29	0.05	0.43	0.11	141	1	0	3
0	26	1.55	36	258	59	0.63	9	0.06	0.25	0.05	0.91	0.10	18	11	0	1
0	56	0.96	50	327	5	0.57	0	0.12	—	0.05	0.48	0.09	66	1	0	1
0	60	1.89	44	346	2	0.78	0	0.14	—	0.05	0.60	0.10	90	1	0	4
0	18	0.33	18	114	1	0.17	3	0.02	—	0.04	0.21	0.01	20	2	0	1
0	27	2.08	63	485	14	0.67	16	0.12	0.12	0.08	0.88	0.16	22	9	0	2
0	25	1.76	50	370	26	0.50	7	0.06	0.58	0.05	0.69	0.10	14	5	0	2
0	7	0.40	9	63	6	0.29	1	0.03	0.04	0.06	0.50	0.03	18	7	0	<1
0	64	2.26	54	335	1	0.96	0	0.18	0.01	0.06	0.48	0.15	127	1	0	5
0	17	0.75	20	111	58	0.19	0	0.08	—	0.07	0.82	0.06	146	7	0	1
0	24	1.27	25	304	39	0.32	0	0.13	—	0.05	0.30	0.09	16	<1	0	1
0	31	1.61	36	329	436	0.70	0	0.13	0.77	0.11	0.58	0.03	65	1	0	2
10	44	2.10	42	338	378	1.48	0	0.03	0.00	0.02	0.40	0.18	14	8	0	2
0	36	1.21	18	134	409	0.33	13	0.04	0.04	0.07	0.25	0.06	22	4	0	1
0	88	4.42	74	443	1	0.99	0	0.13	0.30	0.25	0.34	0.20	46	1	0	6
0	131	2.25	54	485	13	0.82	7	0.23	—	0.14	1.13	0.05	100	15	0	1
0	65	2.54	61	414	2	0.98	0	0.21	—	0.05	0.24	0.11	123	0	0	1
0	24	0.81	12	100	266	0.26	129	0.01	0.40	0.05	0.18	0.03	—	4	0.00	—
0	29	0.80	16	187	2	0.22	5	0.05	0.28	0.06	0.38	0.03	21	6	0	<1
0	33	0.59	16	85	6	0.32	7	0.02	0.24	0.06	0.26	0.04	16	3	0	<1
0	16	0.79	23	305	77	0.35	2	0.03	0.04	0.04	0.33	0.07	80	4	0	1
0	14	0.67	20	259	65	0.30	2	0.02	0.03	0.03	0.28	0.06	68	3	0	1
0	13	1.55	14	126	165	0.18	1	0.01	0.03	0.03	0.13	0.05	26	3	0	<1
0	12	0.47	17	168	300	0.30	1	0.01	—	0.05	0.28	0.06	31	3	0	1
0	82	1.37	49	654	174	0.36	276	0.08	1.30	0.21	0.36	0.10	10	18	0	1
0	105.59	0.92	42.9	344.85	3.29	0.84	65.17	0.08	0.18	0.12	1.15	0.05	104.77	1.81	0	2.06
0	21	0.32	9	139	15	0.18	15	0.03	0.34	0.05	0.28	0.08	28	39	0	1
0	31	0.52	16	229	32	0.35	76	0.05	1.13	0.10	0.43	0.16	84	51	0	1
0	30	0.56	12	131	10	0.26	52	0.05	1.21	0.07	0.42	0.12	52	37	0	1
0	11	0.23	6	96	7	0.20	0	0.03	0.01	0.03	0.23	0.07	18	28	0	—
0	28	0.94	16	247	16	0.26	30	0.08	0.34	0.06	0.47	0.14	47	48	0	1
0	20	0.37	14	225	12	0.19	36	0.08	0.40	0.09	0.42	0.22	78	35	0	<1
0	33	0.41	11	172	13	0.13	6	0.04	0.10	0.03	0.21	0.07	30	23	0	1
0	47	0.26	12	146	12	0.14	11	0.09	0.18	0.08	0.42	0.17	30	30	0	1
0	158	1.77	19	631	459	0.29	360	0.05	0.15	0.11	0.73	0.28	70	44	0	1
0	145	1.28	27	375	995	0.36	—	0.07	0.08	0.10	0.75	0.34	88	80	0	—
0	32	0.56	11	170	19	0.15	39	0.04	0.12	0.05	0.29	0.15	13	40	0	<1
0	25	0.28	20	161	20	0.19	35	0.05	—	0.02	0.21	0.13	56	22	0	1
0	—	—	—	—	105	—	—	—	—	—	—	—	—	—	0	—
0	20	0.18	7	195	42	0.15	367	0.04	0.40	0.04	0.60	0.08	12	4	0	<.1
0	26	0.71	8	190	62	0.14	552	0.02	—	0.03	0.44	0.08	26	7	0	1
0	18	0.17	7	177	38	0.13	333	0.04	0.36	0.03	0.54	0.08	11	3	0	<.1
0	23.39	0.26	7.8	183.3	45.24	0.15	1914.9	0.05	0.80	0.03	0.5	0.11	10.92	2.8	0	0.54

TABLE A–1
Food Composition

(Computer code number is for Wadsworth Diet Analysis program)   (For purposes of calculations, use "0" for t, <1, <.1, <.01, etc.)

DA + Code	Food Description	Quantity	Measure	Wt (g)	H₂O (g)	Ener (kcal)	Prot (g)	Carb (g)	Dietary Fiber (g)	Fat (g)	Sat	Mono	Poly	Trans
	VEGETABLES, LEGUMES—Continued													
1055	Juice, canned	½	cup(s)	123	109	49	1	11	1	<1	0.03	0.01	0.09	—
32725	**Cassava or manioc**	½	cup(s)	103	61	165	1	39	2	<1	0.08	0.08	0.05	—
	Cauliflower													
605	Raw, chopped,	½	cup(s)	50	46	13	1	3	1	<1	0.02	0.01	0.05	—
606	Boiled, drained	½	cup(s)	62	58	14	1	3	2	<1	0.04	0.02	0.13	—
607	Frozen, boiled, drained	½	cup(s)	90	85	17	1	3	2	<1	0.03	0.01	0.09	—
	Celery													
609	Diced	½	cup(s)	60	58	8	<1	2	1	<1	0.03	0.02	0.05	—
608	Stalk	2	item(s)	80	76	11	1	2	1	<1	0.03	0.03	0.06	—
	Chard													
1056	Swiss chard, raw	1	cup(s)	36	33	7	1	1	1	<.1	0.01	0.01	0.03	—
1057	Swiss chard, boiled, drained	½	cup(s)	88	81	18	2	4	2	<.1	0.01	0.01	0.02	—
	Collard greens													
610	Boiled, drained	½	cup(s)	95	87	25	2	5	3	<1	0.04	0.02	0.16	—
611	Frozen, chopped, boiled, drained	½	cup(s)	85	75	31	3	6	2	<1	0.05	0.02	0.18	—
	Corn													
29614	Yellow corn, fresh, cooked	1	item(s)	100	0.06	107.37	3.3	24.96	2.78	1.27	0.19	0.37	0.59	—
612	Yellow sweet corn, boiled, drained	½	cup(s)	82	57	89	3	21	2	1	0.16	0.31	0.49	—
614	Yellow sweet corn, frozen, boiled, drained	½	cup(s)	82	63	66	2	16	2	1	0.08	0.16	0.26	—
615	Yellow creamed sweet corn, canned	½	cup(s)	128	101	92	2	23	2	1	0.08	0.16	0.25	—
618	**Cucumber**	¼	item(s)	75	72	11	<1	3	<1	<.1	0.03	0.00	0.04	—
16870	**Cucumber, kim chee**	½	cup(s)	75	68	16	1	4	1	<.1	0.02	0.00	0.03	—
	Dandelion greens													
2734	Raw	1	cup(s)	55	47	25	1	5	2	<1	0.09	0.01	0.17	—
620	Chopped, boiled, drained	½	cup(s)	53	47	17	1	3	2	<1	0.08	0.01	0.14	—
1066	**Eggplant, boiled, drained**	½	cup(s)	48	43	17	<1	4	1	<1	0.02	0.01	0.04	—
621	**Endive or escarole, chopped, raw**	1	cup(s)	53	49	9	1	2	2	<1	0.03	0.00	0.05	—
8784	**Jicama or yambean**	½	cup(s)	65	59	25	<1	6	3	<.1	0.01	0.00	0.03	—
	Kale													
29313	Raw	1	cup(s)	67	57	34	2	7	1	<1	0.06	0.03	0.23	—
623	Frozen, chopped, boiled, drained	½	cup(s)	65	59	20	2	3	1	<1	0.04	0.02	0.15	—
	Kohlrabi													
1071	Raw	1	cup(s)	135	123	36	2	8	5	<1	0.02	0.01	0.06	—
1072	Boiled, drained	½	cup(s)	83	74	24	1	6	1	<.1	0.01	0.01	0.04	—
	Leeks													
1073	Raw	1	cup(s)	89	74	54	1	13	2	<1	0.04	0.00	0.15	—
1074	Boiled, drained	½	cup(s)	52	47	16	<1	4	1	<1	0.01	0.00	0.06	—
	Lentils													
522	Boiled	½	cup(s)	99	69	115	9	20	8	<1	0.05	0.06	0.17	—
1075	Sprouted	1	cup(s)	77	52	82	7	17	0	<1	0.04	0.08	0.17	—
	Lettuce													
624	Butterhead, boston, or bibb	1	cup(s)	55	53	7	1	1	1	<1	0.02	0.00	0.06	—
625	Butterhead leaves	11	piece(s)	83	79	11	1	2	1	<1	0.02	0.01	0.10	—
626	Iceberg	1	cup(s)	55	53	6	<1	1	1	<.1	0.01	0.00	0.03	—
628	Iceberg, chopped	1	cup(s)	55	53	6	<1	1	1	<.1	0.01	0.00	0.03	—
629	Looseleaf	1	cup(s)	56	54	8	1	2	1	<.1	0.01	0.00	0.05	—
1665	Romaine, shredded	1	cup(s)	56	53	10	1	2	1	<1	0.02	0.01	0.09	—
	Mushrooms													
15585	Crimini (about 6)	3	ounce(s)	85	28	4	3	2	0	0.00	0.00	0.00	0	0
8700	Enoki	30	item(s)	90	80	31	2	6	2	<1	0.04	0.01	0.14	—
630	Mushrooms, raw	½	cup(s)	35	32	8	1	1	<1	<1	0.02	0.00	0.05	—
1079	Mushrooms, boiled, drained	½	cup(s)	78	71	22	2	4	2	<1	0.05	0.01	0.14	—
1080	Mushrooms, canned, drained	½	cup(s)	78	71	20	1	4	2	<1	0.03	0.00	0.09	—
15587	Portobello, raw	1	item(s)	85	30	3	4	3	0	0.00	0.00	0.00	0	0
2743	Shiitake, cooked	½	cup(s)	73	61	40	1	10	2	<1	0.04	0.05	0.02	—
	Mustard greens													
29319	Raw	1	cup(s)	56	51	15	2	3	2	<1	0.01	0.05	0.02	—
2744	Frozen, boiled, drained	½	cup(s)	75	70	14	2	2	2	<1	0.01	0.08	0.04	—
	Okra													
632	Sliced, boiled, drained	½	cup(s)	80	74	18	1	4	2	<1	0.04	0.02	0.04	—
32742	Frozen, boiled, drained, no salt added	½	cup(s)	92	84	26	2	5	3	<1	0.07	0.05	0.07	—
16866	Batter coated, fried	11	piece(s)	83	55	160	2	13	2	11	1.50	2.80	6.37	—
	Onions													
633	Raw, chopped	½	cup(s)	80	71	34	1	8	1	<.1	0.02	0.02	0.05	—
635	Chopped, boiled, drained	½	cup(s)	106	93	47	1	11	1	<1	0.03	0.03	0.08	—

PAGE KEY: A–4 = Breads/Baked Goods   A–8 = Cereal/Rice/Pasta   A–12 = Fruit   A–16 = Vegetables/Legumes   A–26 = Nuts/Seeds   A–28 = Vegetarian
A–30 = Dairy   A–36 = Eggs   A–36 = Seafood   A–38 = Meats   A–42 = Poultry   A–42 = Processed meats   A–44 = Beverages   A–48 = Fats/Oils
A–50 = Sweets   A–52 = Spices/Condiments/Sauces   A–54 = Mixed Foods/Soups/Sandwiches   A–60 = Fast food   A–76 = Convenience   A–78 = Baby foods

Chol (mg)	Calc (mg)	Iron (mg)	Magn (mg)	Pota (mg)	Sodi (mg)	Zinc (mg)	Vit A (RAE) (µg)	Thia (mg)	Vit E (mg α)	Ribo (mg)	Niac (mg)	Vit B_6 (mg)	Fola (µg)	Vit C (mg)	Vit B_{12} (µg)	Sele (µg)
0	30	0.57	17	359	36	0.22	1176	0.11	1.43	0.07	0.47	0.27	5	10	0	1
0	16	0.28	22	279	14	0.35	1	0.09	0.20	0.05	0.88	0.09	28	21	0	1
0	11	0.22	8	152	15	0.14	1	0.03	0.04	0.03	0.26	0.11	29	23	0	<1
0	10	0.20	6	88	9	0.11	1	0.03	0.04	0.03	0.25	0.11	27	27	0	<1
0	15	0.37	8	125	16	0.12	0	0.03	0.05	0.05	0.28	0.08	37	28	0	1
0	24	0.12	7	157	48	0.08	13	0.01	0.16	0.03	0.19	0.04	22	2	0	<1
0	32	0.16	9	208	64	0.10	18	0.02	0.22	0.05	0.26	0.06	29	2	0	<1
0	18	0.65	29	136	77	0.13	110	0.01	0.68	0.03	0.14	0.04	5	11	0	<1
0	51	1.98	75	480	157	0.29	268	0.03	1.65	0.08	0.32	0.07	8	16	0	1
0	133	1.10	19	110	15	0.22	386	0.04	0.84	0.10	0.55	0.12	88	17	0	<1
0	179	0.95	26	213	43	0.23	489	0.04	1.06	0.10	0.54	0.10	65	22	0	1
0	2.12	0.6	31.81	247.59	242.45	0.47	21.87	0.21	0.09	0.07	1.6	0.05	—	6.16	0	—
0	2	0.50	26	204	14	0.39	11	0.18	0.07	0.06	1.32	0.05	38	5	0	<1
0	2	0.39	23	191	1	0.52	8	0.02	0.06	0.05	1.08	0.08	29	3	0	1
0	4	0.49	22	172	365	0.68	5	0.03	0.09	0.07	1.23	0.08	58	6	0	1
0	12	0.21	10	111	2	0.15	4	0.02	0.02	0.02	0.07	0.03	5	2	0	<1
0	7	3.62	6	88	766	0.38	—	0.02	0.36	0.02	0.35	0.08	17	3	0	—
0	103	1.71	20	219	42	0.23	137	0.11	2.65	0.14	0.45	0.14	15	19	0	<1
0	74	0.95	13	122	23	0.15	260	0.07	1.79	0.09	0.27	0.08	7	9	0	<1
0	3	0.12	5	59	<1	0.06	1	0.04	0.20	0.01	0.29	0.04	7	1	0	<.1
0	27	0.44	8	165	12	0.41	57	0.04	0.23	0.04	0.21	0.01	75	3	0	<1
0	8	0.39	8	98	3	0.10	1	0.01	0.30	0.02	0.13	0.03	8	13	0	<1
0	90	1.14	23	299	29	0.29	515	0.07	—	0.09	0.67	0.18	19	80	0	1
0	90	0.61	12	209	10	0.12	478	0.03	0.60	0.07	0.44	0.06	9	16	0	1
0	32	0.54	26	473	27	0.04	3	0.07	0.65	0.03	0.54	0.20	22	84	0	1
0	21	0.33	16	281	17	0.26	2	0.03	0.43	0.02	0.32	0.13	10	45	0	1
0	53	1.87	25	160	18	0.11	74	0.05	0.82	0.03	0.36	0.21	57	11	0	1
0	16	0.57	7	45	5	0.03	1	0.01	—	0.01	0.10	0.06	13	2	0	<1
0	19	3.30	36	365	2	1.26	0	0.17	0.11	0.07	1.05	0.18	179	1	0	3
0	19	2.47	28	248	8	1.16	2	0.18	—	0.10	0.87	0.15	77	13	0	<1
0	19	0.69	7	132	3	0.11	92	0.03	0.10	0.03	0.20	0.05	40	2	0	<1
0	29	1.02	11	196	4	0.17	137	0.05	0.15	0.05	0.29	0.07	60	3	0	<1
0	11	0.19	4	84	5	0.09	9	0.02	0.10	0.01	0.07	0.03	31	2	0	<1
0	11	0.19	4	84	5	0.09	9	0.02	0.10	0.01	0.07	0.03	31	2	0	<1
0	20	0.48	7	109	16	0.10	208	0.04	0.16	0.05	0.21	0.05	21	10	0	<1
0	19	0.55	8	139	5	0.13	163	0.04	0.07	0.04	0.18	0.04	77	14	0	<1
0	0.67	—	—	33	—	0	—	—	—	—	—	—	—	0	0	—
0	1	0.80	14	343	3	0.51	0	0.08	0.01	0.09	3.28	0.04	27	11	0	14
0	1	0.18	3	110	1	0.18	0	0.03	0.00	0.15	1.35	0.04	6	1	<.1	3
0	5	1.36	9	278	2	0.68	0	0.06	0.01	0.23	3.48	0.07	14	3	0	9
0	9	0.62	12	101	332	0.56	0	0.07	0.01	0.02	1.24	0.05	9	0	0	3
40	0.36	—	—	10	—	0	—	—	—	—	—	—	—	0	0	—
0	2	0.32	10	85	3	0.96	0	0.03	0.01	0.12	1.09	0.12	15	<1	0	18
0	58	0.82	18	199	14	0.11	295	0.05	1.13	0.06	0.45	0.10	105	39	0	1
0	76	0.84	10	104	19	0.15	266	0.03	1.01	0.04	0.19	0.08	53	10	0	<1
0	62	0.22	29	108	5	0.34	11	0.11	0.22	0.04	0.70	0.15	37	13	0	<1
0	88	0.62	47	215	3	0.57	16	0.09	0.29	0.11	0.72	0.04	134	11	0	1
2	54	1.13	32	170	110	0.44	—	0.16	1.51	0.13	1.29	0.11	34	9	<.1	—
0	18	0.15	8	115	2	0.13	0	0.04	0.02	0.02	0.07	0.12	15	5	0	<1
0	23	0.26	12	177	3	0.22	0	0.04	0.02	0.02	0.18	0.14	16	6	0	1

TABLE A–1
Food Composition
(Computer code number is for Wadsworth Diet Analysis program)   (For purposes of calculations, use "0" for t, <1, <.1, <.01, etc.)

DA + Code	Food Description	Quantity	Measure	Wt (g)	H₂O (g)	Ener (kcal)	Prot (g)	Carb (g)	Dietary Fiber (g)	Fat (g)	Sat	Mono	Poly	Trans
	VEGETABLES, LEGUMES—Continued													
2748	Frozen, boiled, drained	½	cup(s)	106	98	30	1	7	2	<1	0.02	0.01	0.04	—
16850	Red onions, sliced, raw	½	cup(s)	58	52	22	1	5	1	<.1	0.02	0.01	0.04	—
636	Scallions, green or spring onions	2	item(s)	30	27	10	1	2	1	<.1	0.01	0.01	0.02	—
1081	Onion rings, breaded & pan fried, frozen, heated	11	item(s)	78	22	318	4	30	1	21	6.70	8.49	3.99	—
16860	**Palm hearts, cooked**	½	cup(s)	73	51	75	2	19	1	<1	0.03	0.00	0.07	—
637	**Parsley, chopped**	1	tablespoon(s)	4	3	1	<1	<1	<1	<.1	0.01	0.01	0.00	—
638	**Parsnips, sliced, boiled, drained**	½	cup(s)	78	63	55	1	13	3	<1	0.04	0.09	0.04	—
	Peas													
639	Green peas, canned, drained	½	cup(s)	85	69	59	4	11	3	<1	0.05	0.03	0.14	—
641	Green peas, frozen, boiled, drained	½	cup(s)	80	64	62	4	11	4	<1	0.04	0.02	0.10	—
35694	Pea pods, boiled w/salt, drained	½	cup(s)	80	71	34	3	6	2	<1	0.04	0.02	0.08	—
1082	Peas & carrots, canned w/liquid	½	cup(s)	128	112	48	3	11	3	<1	0.06	0.03	0.16	—
1083	Peas & carrots, frozen, boiled, drained	½	cup(s)	80	69	38	2	8	2	<1	0.06	0.03	0.16	—
640	Snow or sugar peas, raw	½	cup(s)	32	28	13	1	2	1	<.1	0.01	0.01	0.03	—
2750	Snow or sugar peas, frozen, boiled, drained	½	cup(s)	80	69	42	3	7	2	<1	0.06	0.03	0.13	—
29324	Split peas, sprouted	½	cup(s)	60	37	77	5	17	0	<1	0.07	0.04	0.20	—
	Peppers													
643	Green bell or sweet, raw	½	cup(s)	75	70	15	1	3	1	<1	0.04	0.01	0.05	—
644	Green bell or sweet, boiled, drained	½	cup(s)	68	62	19	1	5	1	<1	0.02	0.01	0.07	—
1664	Green hot chili	1	item(s)	45	39	18	1	4	1	<.1	0.01	0.00	0.05	—
1663	Green hot chili, canned w/liquid	½	cup(s)	68	63	14	1	3	1	<.1	0.01	0.00	0.04	—
1086	Jalapeno, canned w/liquid	½	cup(s)	68	60	18	1	3	2	1	0.07	0.04	0.35	—
8703	Yellow bell or sweet	1	item(s)	186	171	50	2	12	2	<1	0.06	0.03	0.21	—
1087	**Poi**	½	cup(s)	122	87	136	<1	33	<1	<1	0.04	0.01	0.07	—
	Potatoes													
5791	Baked, flesh & skin	1	item(s)	202	144	220	5	51	4	<1	0.05	0.00	0.09	—
645	Baked, flesh only	½	cup(s)	61	46	57	1	13	1	<.1	0.02	0.00	0.03	—
1088	Baked, skin only	1	item(s)	58	27	115	2	27	5	<.1	0.02	0.00	0.02	—
5794	Boiled, drained, skin & flesh	1	item(s)	150	116	129	3	30	2	<1	0.04	0.00	0.06	—
647	Boiled, flesh only	½	cup(s)	78	60	67	1	16	1	<.1	0.02	0.00	0.03	—
5795	Boiled in skin, drained, flesh only	1	item(s)	136	105	118	3	27	2	<1	0.04	0.00	0.06	—
2759	Microwaved	1	item(s)	202	146	212	5	49	5	<1	0.05	0.00	0.09	—
5804	Microwaved, skin only	1	item(s)	58	37	77	3	17	4	<.1	0.02	0.00	0.02	—
2760	Microwaved in skin, flesh only	½	cup(s)	78	57	78	2	18	1	<.1	0.02	0.00	0.03	—
1089	Au gratin, prepared w/butter	½	cup(s)	123	91	162	6	14	2	9	5.80	2.63	0.34	—
1090	Au gratin mix, prepared w/water, whole milk, & butter	½	cup(s)	114	90	106	3	15	1	5	2.94	1.34	0.15	—
648	French fried, deep fried, prepared from raw	14	item(s)	70	32	190	3	24	2	10	1.93	4.21	2.97	—
649	French fried, frozen, heated	14	item(s)	70	40	140	2	22	2	5	0.88	3.33	0.55	—
1091	Hashed brown	½	cup(s)	78	37	207	2	27	2	10	1.11	3.13	2.78	—
653	Mashed, from dehydrated granules w/milk, water, & margarine	½	cup(s)	105	80	122	2	17	1	5	1.27	2.05	1.41	—
652	Mashed, w/margarine & whole milk	½	cup(s)	105	79	119	2	18	2	4	1.05	1.83	1.27	—
1097	Potato puffs, frozen, heated	½	cup(s)	64	34	142	2	20	2	7	3.26	2.79	0.51	—
1093	Scalloped, prepared w/butter	½	cup(s)	123	99	105	4	13	2	5	2.76	1.27	0.20	—
1094	Scalloped mix, prepared w/water, whole milk, & butter	½	cup(s)	114	90	106	2	15	1	5	2.99	1.38	0.22	—
	Pumpkin													
1773	Boiled, drained	½	cup(s)	123	115	25	1	6	1	<.1	0.05	0.01	0.00	—
656	Canned	½	cup(s)	123	110	42	1	10	4	<1	0.18	0.05	0.02	—
	Radicchio									<.1				
2498	Raw	1	cup(s)	40	37	9	1	2	<1	<1	0.02	0.00	0.04	—
8731	Raw, leaves	10	item(s)	80	75	18	1	4	1	<1	0.05	0.01	0.09	—
657	**Radishes**	6	item(s)	27	26	4	<1	1	<1	<.1	0.01	0.00	0.01	—
1099	**Rutabaga, boiled, drained**	½	cup(s)	85	76	33	1	7	2	<1	0.02	0.02	0.08	—
658	**Sauerkraut, canned**	½	cup(s)	114	105	22	1	5	3	<1	0.04	0.01	0.07	—
	Seaweed													
1102	Kelp	½	cup(s)	41	33	17	1	4	1	<1	0.10	0.04	0.02	—
1104	Spirulina, dried	½	cup(s)	8	<1	22	4	2	<1	1	0.20	0.05	0.16	—
1106	**Shallots**	3	tablespoon(s)	30	24	22	1	5	0	<.1	0.01	0.00	0.01	—
	Soybeans													
1670	Boiled	½	cup(s)	86	0.05	148.77	14.31	8.53	5.15	7.71	1.11	1.7	4.35	—
2825	Dry roasted	½	cup(s)	86	1	388	34	28	7	19	2.69	4.11	10.50	—
2824	Roasted, salted	½	cup(s)	86	2	405	30	29	15	22	3.16	4.82	12.33	—

PAGE KEY: A–4 = Breads/Baked Goods A–8 = Cereal/Rice/Pasta A–12 = Fruit A–16 = Vegetables/Legumes A–26 = Nuts/Seeds A–28 = Vegetarian
A–30 = Dairy A–36 = Eggs A–36 = Seafood A–38 = Meats A–42 = Poultry A–42 = Processed meats A–44 = Beverages A–48 = Fats/Oils
A–50 = Sweets A–52 = Spices/Condiments/Sauces A–54 = Mixed Foods/Soups/Sandwiches A–60 = Fast food A–76 = Convenience A–78 = Baby foods

Chol (mg)	Calc (mg)	Iron (mg)	Magn (mg)	Pota (mg)	Sodi (mg)	Zinc (mg)	Vit A (RAE) (µg)	Thia (mg)	Vit E (mg α)	Ribo (mg)	Niac (mg)	Vit B$_6$ (mg)	Fola (µg)	Vit C (mg)	Vit B$_{12}$ (µg)	Sele (µg)
0	17	0.32	6	115	13	0.07	0	0.02	0.01	0.03	0.15	0.07	14	3	0	<1
0	11	0.13	6	90	2	0.11	0	0.02	0.01	0.01	0.09	0.07	11	4	0	—
0	22	0.44	6	83	5	0.12	15	0.02	0.17	0.02	0.16	0.02	19	6	0	<1
0	24	1.32	15	101	293	0.33	9	0.22	—	0.11	2.82	0.06	52	1	0	3
0	13	1.23	7	1318	10	2.72	—	0.03	0.37	0.13	0.62	0.53	15	5	0	—
0	5	0.24	2	21	2	0.04	16	0.00	0.03	0.00	0.05	0.00	6	5	0	<.1
0	29	0.45	23	286	8	0.20	0	0.06	0.78	0.04	0.56	0.07	45	10	0	1
0	17	0.81	14	147	214	0.60	23	0.10	0.03	0.07	0.62	0.05	37	8	0	1
0	19	1.22	18	88	58	0.54	84	0.23	0.02	0.08	1.18	0.09	47	8	0	1
0	34	1.58	21	192	192	0.30	43	0.10	0.31	0.06	0.43	0.12	23	38	0	1
0	29	0.96	18	128	332	0.74	368	0.09	—	0.07	0.74	0.11	23	8	0	1
0	18	0.75	13	126	54	0.36	374	0.18	0.42	0.05	0.92	0.07	21	6	0	1
0	14	0.66	8	63	1	0.09	17	0.05	0.12	0.03	0.19	0.05	13	19	0	<1
0	47	1.92	22	174	4	0.39	53	0.05	0.38	0.10	0.45	0.14	28	18	0	1
0	22	1.36	34	229	12	0.63	5	0.14	—	0.09	1.85	0.16	86	6	0	<1
0	7	0.25	7	130	2	0.10	13	0.04	0.28	0.02	0.36	0.17	8	60	0	0
0	6	0.31	7	113	1	0.08	10	0.04	0.36	0.02	0.32	0.16	11	51	0	<1
0	8	0.54	11	153	3	0.14	27	0.04	0.31	0.04	0.43	0.13	10	109	0	<1
0	5	0.34	10	127	798	0.12	24	0.01	0.47	0.03	0.54	0.10	7	46	0	<1
0	16	1.28	10	131	1136	0.23	58	0.03	0.47	0.05	0.27	0.13	10	7	0	<1
0	20	0.86	22	394	4	0.32	19	0.05	—	0.05	1.66	0.31	48	341	0	1
0	19	1.07	29	223	15	0.27	4	0.16	2.80	0.05	1.34	0.33	26	5	0	1
0	20	2.75	55	844	16	0.65	0	0.22	—	0.07	3.32	0.70	22	26	0	2
0	3	0.21	15	239	3	0.18	0	0.06	0.02	0.01	0.85	0.18	5	8	0	<1
0	20	4.08	25	332	12	0.28	1	0.07	0.02	0.06	1.78	0.36	13	8	0	<1
0	13	1.27	34	572	7	0.47	0	0.15	—	0.03	2.13	0.44	15	18	0	—
0	6	0.24	16	256	4	0.21	0	0.08	0.01	0.01	1.02	0.21	7	6	0	<1
0	7	0.42	30	515	5	0.41	0	0.14	—	0.03	1.96	0.41	14	18	0	<1
0	22	2.50	55	903	16	0.73	0	0.24	—	0.06	3.46	0.69	24	31	0	1
0	27	3.45	21	377	9	0.30	0	0.04	—	0.04	1.29	0.29	10	9	0	<1
0	4	0.32	20	321	5	0.26	0	0.10	—	0.02	1.27	0.25	9	12	0	<1
28	146	0.78	25	485	530	0.85	78	0.08	—	0.14	1.22	0.21	13	12	0	3
17	94	0.36	17	249	499	0.27	59	0.02	—	0.09	1.07	0.05	8	4	0	3
0	9	1.02	28	731	8	0.53	0	0.10	0.09	0.05	1.90	0.33	13	21	0	—
0	6	0.87	15	293	21	0.28	0	0.08	0.08	0.02	1.46	0.22	8	7	0	<1
0	11	0.43	27	449	267	0.37	0	0.13	0.01	0.03	1.80	0.37	12	10	0	<1
2	34	0.22	21	163	181	0.25	49	0.09	0.54	0.09	0.91	0.17	8	7	<1	6
1	21	0.27	20	342	350	0.32	43	0.10	0.44	0.05	1.23	0.26	9	11	<.1	1
0	19	1.00	12	243	477	0.19	0	0.13	0.15	0.05	1.38	0.15	11	4	0	<1
15	70	0.70	23	463	410	0.49	39	0.08	—	0.11	1.29	0.22	13	13	0	2
13	41	0.43	16	231	388	0.28	40	0.02	—	0.06	1.17	0.05	11	4	0	2
0	18	0.70	11	282	1	0.28	306	0.04	0.98	0.10	0.51	0.05	11	6	0	<1
0	32	1.70	28	252	6	0.21	953	0.03	1.30	0.07	0.45	0.07	15	5	0	<1
0	8	0.23	5	121	9	0.25	<1	0.01	0.90	0.01	0.10	0.02	24	3	0	<1
0	15	0.46	10	242	18	0.50	1	0.01	1.81	0.02	0.20	0.05	48	6	0	1
0	7	0.09	3	63	11	0.08	0	0.00	0.00	0.01	0.07	0.02	7	4	0	<1
0	41	0.45	20	277	17	0.30	0	0.07	0.27	0.03	0.61	0.09	13	16	0	1
0	34	1.67	15	193	751	0.22	1	0.02	0.11	0.02	0.16	0.15	27	17	0	1
0	68	1.16	49	36	94	0.50	2	0.02	0.35	0.06	0.19	0.00	73	1	0	<1
0	9	2.14	15	102	79	0.15	2	0.18	0.38	0.28	0.96	0.03	7	1	0	1
0	11	0.36	6	100	4	0.12	18	0.02	—	0.01	0.06	0.10	10	2	0	<1
0	87.72	4.42	73.95	442.89	0.86	0.98	0.86	0.13	0.30	0.24	0.34	0.2	46.43	1.46	0	6.27
0	120	3.40	196	1173	2	4.10	0	0.37	—	0.65	0.91	0.19	176	4	0	17
0	119	3.35	125	1264	140	2.70	9	0.09	0.78	0.12	1.21	0.18	181	2	0	16

TABLE A–1
Food Composition

(Computer code number is for Wadsworth Diet Analysis program)   (For purposes of calculations, use "0" for t, <1, <.1, <.01, etc.)

DA + Code	Food Description	Quantity	Measure	Wt (g)	H₂O (g)	Ener (kcal)	Prot (g)	Carb (g)	Dietary Fiber (g)	Fat (g)	Sat	Mono	Poly	Trans
	VEGETABLES, LEGUMES—Continued													
30282	Soup (miso)	1	cup(s)	240	218	85	6	8	2	3	0.59	1.05	1.47	—
8739	Sprouted, stir fried	3	ounce(s)	85	57	106	11	8	1	6	0.84	1.37	3.41	—
	Soy products													
1813	Soy milk	1	cup(s)	240	214	118	9	11	3	5	0.51	0.78	2.00	—
2838	Tofu, dried, frozen (koyadofu)	3	ounce(s)	85	5	408	41	12	6	26	3.73	5.70	14.57	—
13844	Tofu, extra firm	3	ounce(s)	79	—	80	8	2	1	4	0.50	0.87	2.60	—
13843	Tofu, firm	3	ounce(s)	79	—	80	8	2	1	4	0.50	0.87	2.17	—
1816	Tofu, firm, w/calcium sulfate & magnesium chloride (nigari)	3	ounce(s)	85	0.07	65.48	6.83	2.52	0.34	3.79	0.54	0.83	2.14	—
1817	Tofu, fried	3	ounce(s)	85	43	230	15	9	3	17	2.48	3.79	9.69	—
13841	Tofu, silken	3	ounce(s)	91	—	30	6	0	1	1	0.50	0.51	1.52	—
13842	Tofu, soft	3	ounce(s)	91	—	30	6	1	1	1	0.50	1.00	2.00	—
1671	Tofu, soft, w/calcium sulfate & magnesium chloride (nigari)	3	ounce(s)	85	0.07	51.88	5.57	1.53	0.17	3.13	0.45	0.69	1.76	—
	Spinach													
659	Raw, chopped	1	cup(s)	30	27	7	1	1	1	<1	0.02	0.00	0.05	—
663	Canned, drained	½	cup(s)	108	100	25	3	4	3	1	0.09	0.02	0.23	—
660	Chopped, boiled, drained	½	cup(s)	90	82	21	3	3	2	<1	0.04	0.01	0.10	—
661	Chopped, frozen, boiled, drained	½	cup(s)	95	84	30	4	5	4	<1	0.09	0.00	0.20	—
662	Leaf, frozen, boiled, drained	½	cup(s)	95	84	30	4	5	4	<1	0.09	0.00	0.20	—
8470	Trimmed leaves	1	cup(s)	32	27	3	1	<.1	3	<.1	—	—	—	—
	Squash													
1662	Acorn, baked	½	cup(s)	103	85	57	1	15	5	<1	0.03	0.01	0.06	—
29702	Acorn, boiled, mashed	½	cup(s)	123	110	42	1	11	3	<.1	0.02	0.01	0.04	—
1661	Butternut, baked	½	cup(s)	103	90	41	1	11	3	<.1	0.02	0.01	0.04	—
29451	Butternut, frozen, boiled	½	cup(s)	132	116	51	2	13	2	<.1	0.02	0.01	0.04	—
32773	Butternut, frozen, boiled, mashed, no salt added	½	cup(s)	122	<1	47	1	12	0	<.1	0.02	0.00	0.03	—
29700	Crookneck & straightneck, boiled, drained	½	cup(s)	90	0.08	18	0.81	3.87	1.25	0.27	0.05	0.02	0.11	—
29703	Hubbard, baked	v	cup(s)	103	87	51	3	11	0	1	0.13	0.05	0.27	—
1660	Hubbard, boiled, mashed	½	cup(s)	118	107	35	2	8	3	<1	0.09	0.03	0.18	—
29704	Spaghetti, boiled, drained, or baked	½	cup(s)	78	72	21	1	5	1	<1	0.05	0.02	0.10	—
664	Summer, all varieties, sliced, boiled, drained	½	cup(s)	90	84	18	1	4	1	<1	0.06	0.02	0.12	—
665	Winter, all varieties, baked, mashed	½	cup(s)	103	91	38	1	9	3	<1	0.13	0.05	0.27	—
1112	Zucchini, boiled, drained	½	cup(s)	90	85	14	1	4	1	<.1	0.01	0.00	0.02	—
1113	Zucchini, frozen, boiled, drained	½	cup(s)	113	107	19	1	4	1	<1	0.03	0.01	0.06	—
	Sweet potatoes													
666	Baked, peeled	½	cup(s)	100	76	90	2	21	3	<1	0.03	0.00	0.06	—
667	Boiled, mashed	½	cup(s)	166	133	126	2	29	4	<1	0.05	0.00	0.10	—
668	Candied, home recipe	½	cup(s)	84	56	115	1	23	2	3	1.13	0.53	0.12	—
670	Canned, vacuum pack	½	cup(s)	100	76	91	2	21	2	<1	0.04	0.01	0.09	—
2765	Frozen, baked	½	cup(s)	88	65	88	2	21	2	<1	0.02	0.00	0.05	—
1136	Yams, baked or boiled, drained	½	cup(s)	68	48	79	1	19	3	<.1	0.02	0.00	0.04	—
32785	**Taro shoots, cooked, no salt added**	½	cup(s)	70	67	10	1	2	0	<.1	0.01	0.00	0.02	—
	Tomatillo													
8774	Raw	2	item(s)	68	62	22	1	4	1	1	0.09	0.11	0.28	—
8777	Raw, chopped	½	cup(s)	66	60	21	1	4	1	1	0.09	0.10	0.28	—
	Tomato													
671	Fresh, ripe, red	1	item(s)	123	0.11	22.13	1.08	4.82	1.47	0.24	0.05	0.06	0.16	—
16846	Fresh, cherry	5	item(s)	85	0.07	17.85	0.72	3.94	0.93	0.28	0.03	0.04	0.11	—
3952	Diced, red	½	cup(s)	90	84	16	1	4	1	<1	0.04	0.05	0.12	—
1118	Boiled, red	½	cup(s)	120	113	22	1	5	1	<1	0.02	0.02	0.05	—
675	Juice, canned	½	cup(s)	122	115	21	1	5	<1	<.1	0.01	0.01	0.03	—
75	Juice, no salt added	½	cup(s)	122	115	21	1	5	<1	<.1	0.01	0.01	0.03	—
1699	Paste, canned	2	tablespoon(s)	33	24	27	1	6	1	<1	0.04	0.03	0.07	—
1700	Puree, canned	¼	cup(s)	63	55	24	1	6	1	<1	0.02	0.02	0.05	—
1125	Sauce, canned	¼	cup(s)	61	55	20	1	5	1	<1	0.02	0.02	0.06	—
1120	Stewed, canned, red	½	cup(s)	128	117	33	1	8	1	<1	0.03	0.04	0.10	—
8778	Sun dried	½	cup(s)	27	4	70	4	15	3	1	0.12	0.13	0.30	—
8783	Sun dried in oil, drained	¼	cup(s)	28	15	59	1	6	2	4	0.52	2.38	0.57	—
	Turnips													
677	Turnips, cubed, boiled, drained	½	cup(s)	78	73	17	1	4	2	<.1	0.01	0.00	0.03	—
678	Turnip greens, chopped, boiled, drained	½	cup(s)	72	67	14	1	3	3	<1	0.04	0.01	0.07	—
679	Turnip greens, frozen, chopped, boiled, drained	½	cup(s)	82	74	24	3	4	3	<1	0.08	0.02	0.14	—

PAGE KEY: A–4 = Breads/Baked Goods A–8 = Cereal/Rice/Pasta A–12 = Fruit A–16 = Vegetables/Legumes A–26 = Nuts/Seeds A–28 = Vegetarian A–30 = Dairy A–36 = Eggs A–36 = Seafood A–38 = Meats A–42 = Poultry A–42 = Processed meats A–44 = Beverages A–48 = Fats/Oils A–50 = Sweets A–52 = Spices/Condiments/Sauces A–54 = Mixed Foods/Soups/Sandwiches A–60 = Fast food A–76 = Convenience A–78 = Baby foods

Chol (mg)	Calc (mg)	Iron (mg)	Magn (mg)	Pota (mg)	Sodi (mg)	Zinc (mg)	Vit A (RAE) (µg)	Thia (mg)	Vit E (mg α)	Ribo (mg)	Niac (mg)	Vit B$_6$ (mg)	Fola (µg)	Vit C (mg)	Vit B$_{12}$ (µg)	Sele (µg)
0	64	1.89	37	361	988	0.87	—	0.06	0.96	0.16	2.61	0.17	57	4	<1	—
0	70	0.34	82	482	12	1.79	1	0.36	—	0.16	0.94	0.14	108	10	0	1
0	10	1.39	46	338	29	0.55	5	0.39	3.24	0.17	0.35	0.10	5	0	0	3
0	310	8.28	50	17	5	4.17	22	0.42	—	0.27	1.01	0.24	78	1	0	46
0	60	1.08	78	—	0	—	0	—	0.03	—	—	—	—	0	0	—
0	60	1.08	52	—	0	—	0	—	—	—	—	—	—	0	0	—
0	137.78	1.23	39.12	149.68	6.8	0.85	0.85	0.07	—	0.08	0	0.05	28.06	0.17	0	7.99
0	316	4.14	51	124	14	1.69	1	0.14	0.03	0.04	0.09	0.08	23	0	0	24
0	300	0.73	35	—	65	—	0	—	—	—	—	—	—	0	2	—
0	300	0.72	33	—	65	—	0	—	—	—	—	—	—	0	2	—
0	94.4	0.94	22.96	102.05	6.8	0.54	0.85	0.03	0.01	0.03	0.45	0.04	37.42	0.17	0	7.56
0	30	0.81	24	167	24	0.16	141	0.02	0.61	0.06	0.22	0.06	58	8	0	<1
0	138	2.49	82	375	29	0.50	531	0.02	2.10	0.15	0.42	0.11	106	16	0	2
0	122	3.21	78	419	63	0.68	472	0.09	1.87	0.21	0.44	0.22	131	9	0	1
0	145	1.86	78	287	92	0.47	573	0.07	3.36	0.17	0.42	0.13	115	2	0	5
0	145	1.86	78	287	92	0.47	573	0.07	3.36	0.17	0.42	0.13	115	2	0	5
0	25	2.13	25	134	38	0.18	—	0.03	—	0.06	0.18	0.07	<.1	8	0	—
0	45	0.95	44	448	4	0.17	22	0.17	—	0.01	0.90	0.20	19	11	0	1
0	32	0.69	32	322	4	0.13	50	0.12	—	0.01	0.65	0.14	13	8	0	<1
0	42	0.62	30	291	4	0.13	572	0.07	1.32	0.02	0.99	0.13	19	15	0	1
0	25	0.77	12	176	3	0.16	—	0.07	—	0.05	0.61	0.09	22	5	0	1
0	23	0.70	11	162	2	0.14	406	0.05	—	0.05	0.56	0.08	19	4	0	1
0	18.2	0.41	18.2	183.73	1.73	0.25	29.46	0.04	—	0.03	0.39	0.09	19.93	7.28	0	0.17
0	17	0.48	23	367	8	0.15	310	0.08	—	0.05	0.57	0.18	16	10	0	1
0	12	0.33	15	253	6	0.12	236	0.05	0.14	0.03	0.39	0.12	12	8	0	<1
0	16	0.26	9	91	14	0.16	5	0.03	0.09	0.02	0.63	0.08	6	3	0	<1
0	24	0.32	22	173	1	0.35	10	0.04	0.13	0.04	0.46	0.06	18	5	0	<1
0	23	0.45	13	448	1	0.23	268	0.02	0.12	0.07	0.51	0.17	21	10	0	<1
0	12	0.32	20	228	3	0.16	50	0.04	0.11	0.04	0.39	0.07	15	4	0	<1
0	19	0.54	15	219	2	0.23	11	0.05	0.14	0.05	0.44	0.05	9	4	0	<1
0	38	0.69	27	475	36	0.32	961	1.45	0.71	0.11	1.49	0.29	6	20	0	<1
0	45	1.20	30	382	45	0.33	1310	0.09	1.56	0.08	0.89	0.27	10	21	0	<1
7	22	0.95	9	159	59	0.13	176	0.02	—	0.04	0.33	0.03	9	6	0	1
0	22	0.89	22	312	53	0.18	399	0.04	1.00	0.06	0.74	0.19	17	26	0	1
0	31	0.48	18	332	7	0.26	722	0.06	0.68	0.05	0.49	0.16	19	8	0	1
0	10	0.36	12	458	5	0.14	4	0.06	0.26	0.02	0.38	0.16	11	8	0	<1
0	10	0.29	6	241	1	0.38	2	0.03	—	0.04	0.57	0.08	2	13	0	1
0	5	0.42	14	182	1	0.15	4	0.03	0.26	0.02	1.26	0.04	5	8	0	<1
0	5	0.41	13	177	1	0.15	4	0.03	0.25	0.02	1.22	0.04	5	8	0	<1
0	12.3	0.33	13.52	291.51	6.15	0.2	76.26	0.04	0.66	0.02	0.73	0.09	18.45	15.62	0	0
0	4.25	0.37	9.35	188.69	7.65	0.07	52.7	0.05	0.46	0.03	0.53	0.07	—	16.23	0	—
0	9	0.24	10	213	5	0.15	38	0.03	0.49	0.02	0.53	0.07	14	11	0	0
0	13	0.82	11	262	13	0.17	29	0.04	0.67	0.03	0.64	0.09	16	27	0	1
0	12	0.52	13	279	328	0.18	28	0.06	0.39	0.04	0.82	0.14	24	22	0	<1
0	12	0.52	13	279	12	0.18	28	0.06	0.39	0.04	0.82	0.14	24	22	0	<1
0	12	0.98	14	333	259	0.21	25	0.02	1.41	0.05	1.01	0.07	4	7	0	2
0	11	1.11	14	274	249	0.23	16	0.02	1.23	0.05	0.92	0.08	7	7	0	3
0	8	0.62	10	203	321	0.12	10	0.01	1.27	0.04	0.60	0.06	6	4	0	<1
0	43	1.70	15	264	282	0.22	11	0.06	1.06	0.04	0.91	0.02	6	10	0	1
0	30	2.45	52	925	566	0.54	12	0.14	0.00	0.13	2.44	0.09	18	11	0	1
0	13	0.74	22	430	73	0.21	18	0.05	—	0.11	1.00	0.09	6	28	0	1
0	26	0.14	7	138	12	0.09	0	0.02	0.02	0.02	0.23	0.05	7	9	0	<1
0	99	0.58	16	146	21	0.10	274	0.03	1.35	0.05	0.30	0.13	85	20	0	1
0	125	1.59	21	184	12	0.34	441	0.04	2.18	0.06	0.38	0.05	32	18	0	1

TABLE A–1

Food Composition

(Computer code number is for Wadsworth Diet Analysis program)   (For purposes of calculations, use "0" for t, <1, <.1, <.01, etc.)

DA + Code	Food Description	Quantity	Measure	Wt (g)	H₂O (g)	Ener (kcal)	Prot (g)	Carb (g)	Dietary Fiber (g)	Fat (g)	Sat	Mono	Poly	Trans
	VEGETABLES, LEGUMES—Continued													
	Vegetables, mixed													
1132	Canned, drained	½	cup(s)	82	71	40	2	8	2	<1	0.04	0.01	0.10	—
680	Frozen, boiled, drained	½	cup(s)	91	76	59	3	12	4	<1	0.03	0.01	0.07	—
7489	Vegetable juice, V8 100%	½	cup(s)	120	113	25	1	5	1	0	0.00	0.00	0.00	0
7490	Vegetable juice, V8 low sodium	½	cup(s)	120	113	25	0	7	1	0	0.00	0.00	0.00	0
7491	Vegetable juice, V8 spicy hot	½	cup(s)	120	113	25	1	5	1	0	0.00	0.00	0.00	0
	Water chestnuts													
31073	Sliced, drained	½	cup(s)	75	70	20	<1	5	1	0	0.00	0.00	0.00	0
31087	Whole	½	cup(s)	75	70	20	<1	5	1	0	0.00	0.00	0.00	0
1135	**Watercress**	1	cup(s)	34	32	4	1	<1	<1	<.1	0.01	0.00	0.01	—
	NUTS, SEEDS, AND PRODUCTS													
	Almonds													
32886	Blanched	¼	cup(s)	36	2	211	8	7	4	18	1.41	11.70	4.37	—
32887	Dry roasted, no salt added	¼	cup(s)	35	1	206	8	7	4	18	1.40	11.61	4.36	—
29724	Dry roasted, salted	¼	cup(s)	35	1	206	8	7	4	18	1.40	11.61	4.36	—
29725	Oil roasted, salted	¼	cup(s)	39	1	238	8	7	4	22	1.65	13.66	5.31	—
508	Slivered	¼	cup(s)	34	2	195	7	7	4	17	1.31	10.85	4.12	—
1137	Almond butter, no salt added	1	tablespoon(s)	16	<1	101	2	3	1	9	0.90	6.14	1.98	—
32940	Almond butter, salt added	1	tablespoon(s)	16	<1	101	2	3	1	9	0.90	6.14	1.98	—
1138	**Beechnuts, dried**	¼	cup(s)	57	4	327	4	19	5	28	3.25	12.43	11.41	—
517	**Brazil nuts, unblanched, dried**	¼	cup(s)	35	1	230	5	4	3	23	5.30	8.59	7.20	—
1166	**Breadfruit seeds, roasted**	¼	cup(s)	57	28	118	4	23	3	2	0.41	0.20	0.82	—
1139	**Butternuts, dried**	¼	cup(s)	30	1	184	7	4	1	17	0.39	3.13	12.82	—
	Cashews													
1140	Dry roasted	¼	cup(s)	34	1	197	5	11	1	16	3.14	9.36	2.68	—
518	Oil roasted	¼	cup(s)	33	1	189	5	10	1	16	2.76	8.42	2.78	—
32889	Cashew butter, no salt added	1	tablespoon(s)	16	<1	94	3	4	<1	8	1.56	4.66	1.34	—
32931	Cashew butter, salt added	1	tablespoon(s)	16	<1	94	3	4	<1	8	1.56	4.66	1.34	—
	Coconut													
32896	Dried, not sweetened	¼	cup(s)	60	2	393	4	14	10	38	34.06	1.63	0.42	—
1153	Dried, shredded, sweetened	¼	cup(s)	24	3	122	1	12	1	9	7.68	0.37	0.09	—
520	Shredded	¼	cup(s)	21	10	75	1	3	2	7	6.27	0.30	0.08	—
	Chestnuts													
1152	Chinese, roasted	¼	cup(s)	57	23	136	3	30	0	1	0.10	0.35	0.17	—
32895	European, boiled & steamed	¼	cup(s)	57	39	74	1	16	0	1	0.15	0.27	0.31	—
32911	European, roasted	¼	cup(s)	57	23	139	2	30	3	1	0.23	0.43	0.49	—
32922	Japanese, boiled & steamed	¼	cup(s)	57	49	32	<1	7	0	<1	0.02	0.06	0.03	—
32923	Japanese, roasted	¼	cup(s)	57	28	114	2	26	0	<1	0.07	0.24	0.12	—
4958	**Flaxseeds or linseeds**	¼	cup(s)	57	5	276	11	19	16	19	1.79	3.85	12.54	—
32904	**Ginkgo nuts, dried**	¼	cup(s)	57	7	197	6	41	0	1	0.22	0.42	0.42	—
	Hazelnuts or filberts													
32901	Blanched	¼	cup(s)	57	3	357	8	10	6	35	2.65	27.32	3.15	—
32902	Dry roasted, no salt added	¼	cup(s)	57	1	366	9	10	5	35	2.56	26.43	4.80	—
1156	**Hickorynuts, dried**	¼	cup(s)	30	1	197	4	5	2	19	2.11	9.78	6.57	—
	Macadamias													
1157	Raw	¼	cup(s)	34	<1	241	3	5	3	25	4.04	19.72	0.50	—
32905	Dry roasted, no salt added	¼	cup(s)	34	1	241	3	4	3	25	4.00	19.86	0.50	—
32932	Dry roasted, salt added	¼	cup(s)	34	1	240	3	4	3	25	4.00	19.86	0.50	—
	Mixed nuts													
1159	With peanuts, dry roasted	¼	cup(s)	34	1	203	6	9	3	18	2.36	10.75	3.69	—
32933	With peanuts, dry roasted, salt added	¼	cup(s)	34	1	203	6	9	3	18	2.36	10.75	3.69	—
32906	Without peanuts, oil roasted, no salt added	¼	cup(s)	36	1	221	6	8	2	20	3.27	11.93	4.12	—
	Peanuts													
2807	Dry roasted	¼	cup(s)	37	0	214	9	8	3	18	2.51	8.99	5.72	—
2806	Dry roasted, salted	¼	cup(s)	37	0	214	9	8	3	18	2.51	8.99	5.72	—
1763	Oil roasted, salted	¼	cup(s)	36	0	216	10	5	3	19	3.12	9.33	5.49	—
2804	Raw	¼	cup(s)	37	2	207	9	6	3	18	2.49	8.92	5.68	—
1884	Peanut butter, chunky	1	tablespoon(s)	16	<1	94	4	3	1	8	1.53	3.77	2.27	—
30303	Peanut butter, low sodium	1	tablespoon(s)	16	<1	95	4	3	1	8	1.66	3.88	2.21	—
30305	Peanut butter, reduced fat	1	tablespoon(s)	18	<1	94	5	6	1	6	1.33	2.91	1.85	—
524	Peanut butter, smooth	1	tablespoon(s)	16	<1	96	4	3	1	8	1.60	3.96	2.38	—
	Pecans													
32907	Dry roasted, no salt added	¼	cup(s)	57	1	403	5	8	5	42	3.56	24.92	11.66	

PAGE KEY: A–4 = Breads/Baked Goods   A–8 = Cereal/Rice/Pasta   A–12 = Fruit   A–16 = Vegetables/Legumes   A–26 = Nuts/Seeds   A–28 = Vegetarian
A–30 = Dairy   A–36 = Eggs   A–36 = Seafood   A–38 = Meats   A–42 = Poultry   A–42 = Processed meats   A–44 = Beverages   A–48 = Fats/Oils
A–50 = Sweets   A–52 = Spices/Condiments/Sauces   A–54 = Mixed Foods/Soups/Sandwiches   A–60 = Fast food   A–76 = Convenience   A–78 = Baby foods

Chol (mg)	Calc (mg)	Iron (mg)	Magn (mg)	Pota (mg)	Sodi (mg)	Zinc (mg)	Vit A (RAE) (µg)	Thia (mg)	Vit E (mg α)	Ribo (mg)	Niac (mg)	Vit B$_6$ (mg)	Fola (µg)	Vit C (mg)	Vit B$_{12}$ (µg)	Sele (µg)
0	22	0.86	13	237	121	0.33	474	0.04	0.28	0.04	0.47	0.06	20	4	0	<1
0	23	0.75	20	154	32	0.45	195	0.06	0.40	0.11	0.77	0.07	17	3	0	<1
0	20	0.54	13	270	310	0.24	50	0.05	—	0.03	0.87	0.17	—	30	0	—
0	20	0.36	—	420	70	—	63	0.02	—	0.02	0.75	—	—	30	0	—
0	20	0.36	13	255	370	0.24	50	0.05	—	0.03	0.88	0.17	—	18	0	—
0	7	0.23	—	—	6	—	0	—	—	—	—	—	—	2	—	—
0	7	0.23	—	—	6	—	0	—	—	—	—	—	—	2	—	—
0	41	0.07	7	112	14	0.04	80	0.03	0.34	0.04	0.07	0.04	3	15	0	<1
0	78	1.35	100	249	10	1.13	0	0.07	8.96	0.20	1.33	0.04	11	0	0	1
0	92	1.56	99	257	<1	1.22	0	0.03	8.97	0.30	1.33	0.04	11	0	0	1
0	92	1.56	99	257	117	1.22	0	0.03	8.97	0.30	1.33	0.04	11	0	0	1
0	114	1.44	108	274	133	1.20	0	0.04	10.19	0.31	1.44	0.05	11	0	0	1
0	84	1.45	93	246	<1	1.13	0	0.08	8.73	0.27	1.32	0.04	10	0	0	1
0	43	0.59	48	121	2	0.49	0	0.02	—	0.10	0.46	0.01	10	<1	0	—
0	43	0.59	48	121	72	0.49	0	0.02	—	0.10	0.46	0.01	10	<1	0	1
0	1	1.40	0	578	22	0.20	0	0.17	—	0.21	0.50	0.39	64	9	0	4
0	56	0.85	132	231	1	1.42	0	0.22	2.01	0.01	0.10	0.04	8	<1	0	671
0	49	0.51	35	615	16	0.59	9	0.23	—	0.14	4.20	0.24	34	4	0	8
0	16	1.21	71	126	<1	0.94	2	0.11	—	0.04	0.31	0.17	20	1	0	5
0	15	2.06	89	194	5	1.92	0	0.07	0.32	0.07	0.48	0.09	24	0	0	4
0	14	1.97	89	205	4	1.74	0	0.12	0.30	0.07	0.56	0.10	8	<.1	0	7
0	7	0.80	41	87	2	0.83	0	0.05	—	0.03	0.26	0.04	11	0	0	2
0	7	0.80	41	87	98	0.83	0	0.05	0.15	0.03	0.26	0.04	11	0	0	2
0	15	1.98	54	323	22	1.20	0	0.04	0.26	0.06	0.36	0.18	5	1	0	11
0	4	0.47	12	82	64	0.44	0	0.01	0.10	0.00	0.12	0.07	2	<1	0	4
0	3	0.51	7	75	4	0.23	0	0.01	0.05	0.00	0.11	0.01	5	1	0	2
0	11	0.85	51	271	2	0.53	0	0.09	—	0.05	0.85	0.25	41	22	0	4
0	26	0.98	31	405	15	0.14	1	0.08	—	0.06	0.41	0.13	22	15	0	—
0	16	0.52	19	336	1	0.32	1	0.14	0.28	0.10	0.76	0.28	40	15	0	1
0	6	0.30	10	67	3	0.23	1	0.07	—	0.03	0.31	0.06	10	5	0	—
0	20	1.19	36	242	11	0.81	2	0.26	—	0.40		0.24	33	16	0	—
0	111	3.48	203	381	19	2.34	0	0.10	—	0.09	0.78	0.52	156	1	0	3
0	11	0.91	30	566	7	0.38	31	0.24	—	0.10	6.65	0.36	60	17	0	—
0	84	1.87	91	373	0	1.25	1	0.27	9.92	0.06	0.88	0.33	44	1	0	2
0	70	2.48	98	428	0	1.42	2	0.19	8.66	0.07	1.16	0.35	50	2	0	2
0	18	0.64	52	131	<1	1.29	2	0.26	—	0.04	0.27	0.06	12	1	0	2
0	28	1.24	44	123	2	0.44	0	0.40	0.18	0.05	0.83	0.09	4	<1	0	1
0	23	0.89	40	122	1	0.43	0	0.24	0.19	0.03	0.76	0.12	3	<1	0	1
0	23	0.89	40	122	89	0.43	0	0.24	0.19	0.03	0.76	0.12	3	<1	0	4
0	24	1.27	77	204	4	1.30	<1	0.07	—	0.07	1.61	0.10	17	<1	0	1
0	24	1.27	77	204	229	1.30	0	0.07	3.75	0.07	1.61	0.10	17	<1	0	3
0	38	0.93	90	196	4	1.68	<1	0.18	—	0.17	0.71	0.06	20	<1	0	—
0	20	0.82	64	240	2	1.20	0	0.15	2.56	0.03	4.93	0.09	53	0	0	3
0	20	0.82	64	240	297	1.20	0	0.15	2.89	0.03	4.93	0.09	53	0	0	3
0	22	0.54	63	261	115	1.18	0	0.03	2.50	0.03	4.97	0.16	43	<1	0	1
0	34	1.67	61	257	7	1.19	0	0.23	3.04	0.05	4.40	0.13	88	0	0	3
0	8	0.33	31	101	75	0.52	0	0.02	1.01	0.02	2.19	0.07	15	0	0	1
0	6	0.29	25	107	3	0.47	0	0.01	1.23	0.02	2.14	0.07	12	0	0	—
0	6	0.34	31	120	97	0.50	0	0.05	1.20	0.01	2.63	0.06	11	0	0	—
0	8	0.30	28	88	80	0.47	0	0.01	1.44	0.02	2.14	0.07	12	0	0	1
0	41	1.59	75	240	1	2.87	4	0.26	0.74	0.06	0.66	0.11	9	<1	0	2

TABLE A–1
Food Composition

(Computer code number is for Wadsworth Diet Analysis program)   (For purposes of calculations, use "0" for t, <1, <.1, <.01, etc.)

DA + Code	Food Description	Quantity	Measure	Wt (g)	H₂O (g)	Ener (kcal)	Prot (g)	Carb (g)	Dietary Fiber (g)	Fat (g)	Sat	Mono	Poly	Trans
	NUTS, SEEDS, AND PRODUCTS—Continued													
32936	Dry roasted, salt added	¼	cup(s)	57	1	403	5	8	5	42	3.56	24.92	11.66	—
1162	Halves, oil roasted	¼	cup(s)	28	<1	197	3	4	3	21	1.99	11.27	6.49	—
526	Raw	¼	cup(s)	27	1	187	2	4	3	19	1.67	11.02	5.84	—
12973	**Pine nuts or pignolia, dried**	1	tablespoon(s)	9	<1	58	1	1	<1	6	0.42	1.61	2.93	—
	Pistachios													
1164	Dry roasted	¼	cup(s)	32	1	183	7	9	3	15	1.78	7.75	4.45	—
32938	Dry roasted, salt added	¼	cup(s)	32	1	182	7	9	3	15	1.78	7.75	4.45	—
1167	**Pumpkin or squash seeds, roasted**	¼	cup(s)	57	4	296	19	8	2	24	4.52	7.43	10.90	—
	Sesame													
1169	Sesame seeds, whole, roasted, toasted	3	teaspoon(s)	9	<1	51	2	2	1	4	0.60	1.63	1.89	—
32912	Sesame butter paste	1	tablespoon(s)	16	<1	95	3	4	1	8	1.14	3.07	3.57	—
32941	Tahini or sesame butter	1	tablespoon(s)	15	<1	89	3	3	1	8	1.11	3.00	3.48	—
	Soy nuts													
34173	Deep sea salted	¼	cup(s)	56	—	240	24	18	10	8	2.00	—	—	—
34174	Unsalted	¼	cup(s)	56	—	240	24	18	10	8	2.00	—	—	—
	Sunflower seeds													
528	Kernels, dried	¼	cup(s)	36	2	205	8	7	4	18	1.87	3.41	11.78	—
29721	Kernels, dry roasted, salted	¼	cup(s)	32	<1	186	6	8	3	16	1.67	3.04	10.52	—
29723	Kernels, toasted, salted	¼	cup(s)	34	<1	207	6	7	4	19	1.99	3.63	12.56	—
32928	Sunflower seed butter, salt added	1	tablespoon(s)	16	<1	93	3	4	0	8	0.80	1.46	5.04	—
	Trail mix													
4646	Trail mix	¼	cup(s)	38	3	173	5	17	2	11	2.08	4.70	3.62	—
4647	Trail mix with chocolate chips	¼	cup(s)	38	2	182	5	17	0	12	2.29	5.08	4.23	—
4648	Tropical trail mix	¼	cup(s)	35	3	142	2	23	0	6	2.97	0.87	1.81	—
	Walnuts													
529	Dried black, chopped	¼	cup(s)	31	1	193	8	3	2	18	1.05	4.69	10.96	—
531	English or persian	¼	cup(s)	30	1	196	5	4	2	20	1.84	2.68	14.15	—
	VEGETARIAN FOODS													
	Prepared													
34222	Brown rice & tofu stir-fry (vegan)	8	ounce(s)	227	183	228	12	13	3	16	1.25	4.03	9.54	0
34368	Cheese enchilada casserole (lacto)	8	ounce(s)	227	86	410	18	41	4	19	10.06	6.54	1.24	0
34247	Five bean casserole (vegan)	8	ounce(s)	228	178	178	6	26	5	6	1.11	2.49	1.96	0
34261	Lentil stew (vegan)	8	ounce(s)	228	152	125	8	24	7	<1	0.08	0.07	0.21	0
34397	Macaroni & cheese (lacto)	8	ounce(s)	226	163	181	8	17	<1	9	4.37	2.88	0.89	0
34238	Steamed rice & vegetables (vegan)	8	ounce(s)	228	100	265	5	40	3	10	1.84	3.91	4.07	0
34308	Tofu rice burgers (ovo-lacto)	1	piece(s)	218	78	435	22	68	6	8	1.69	2.39	3.52	—
34276	Vegan spinach enchiladas (vegan)	1	piece(s)	82	59	93	5	15	2	2	0.34	0.55	1.27	—
34243	Vegetable chow mein (vegan)	8	ounce(s)	227	163	166	6	22	2	6	0.65	2.66	2.47	0
34454	Vegetable lasagna (lacto)	8	ounce(s)	225	154	177	12	25	2	4	1.92	0.93	0.34	0
34339	Vegetable marinara (vegan)	8	ounce(s)	229	182	94	3	15	1	3	0.36	1.32	0.92	0
34356	Vegetable rice casserole (lacto)	8	ounce(s)	227	172	230	9	24	4	12	4.67	3.48	2.96	0
34311	Vegetable strudel (ovo-lacto)	8	ounce(s)	227	100	756	19	51	4	54	18.24	26.38	6.17	0
34371	Vegetable taco (lacto)	1	item(s)	227	147	365	13	43	9	17	6.45	5.81	4.02	—
34282	Vegetarian chili (vegan)	8	ounce(s)	227	196	116	6	21	7	2	0.24	0.29	0.74	0
34367	Vegetarian vegetable soup (vegan)	8	ounce(s)	226	204	92	3	14	2	4	0.77	1.67	1.30	0
	Boca burger													
32067	All American flamed grilled patty	1	item(s)	71	—	110	14	6	4	4	1.00	—	—	0
32070	Bigger chef max's favorite	1	item(s)	99	—	130	18	11	5	4	1.00	1.00	1.50	—
32069	Bigger vegan	1	item(s)	99	—	120	18	11	6	0	0.00	0.00	0.00	—
32074	Boca chik'n nuggets	4	item(s)	87	—	190	16	16	2	7	2.00	—	—	0
32075	Boca meatless ground burger	½	cup(s)	57	—	70	11	7	4	1	0.00	—	—	0
32073	Boca tenders	1	item(s)	85	—	140	20	9	3	3	0.00	2.00	1.00	—
32072	Breakfast links	2	item(s)	45	—	100	10	6	5	4	0.00	—	—	0
32071	Breakfast patties	1	item(s)	38	—	80	8	5	3	4	0.00	—	—	0
32068	Roasted garlic patty	1	item(s)	71	—	100	14	7	5	2	0.50	—	—	0
32066	Vegan original patty	1	item(s)	71	—	90	13	4	0	1	0.00	—	—	0
	Gardenburger													
37810	Bbq chik'n with sauce	1	item(s)	142	—	250	14	30	5	8	1.00	—	—	0
39661	Black bean burger	1	item(s)	71	—	80	8	11	4	2	0.00	—	—	0
39666	Buffalo chick'n wing	3	item(s)	95	—	180	9	8	5	12	1.50	—	—	0
37808	Chik'n grill	1	item(s)	71	—	100	13	5	3	3	0.00	—	—	0
39665	Country fried chicken w/creamy pepper gravy	1	item(s)	142	—	190	9	16	2	9	1.00	—	—	0
37805	Crispy nuggets	6	item(s)	82	—	180	4	22	3	9	1.50	—	—	0
39663	Homestyle classic burger	1	item(s)	71	—	110	12	6	4	5	0.50	—	—	0

Chol (mg)	Calc (mg)	Iron (mg)	Magn (mg)	Pota (mg)	Sodi (mg)	Zinc (mg)	Vit A (RAE) (µg)	Thia (mg)	Vit E (mg α)	Ribo (mg)	Niac (mg)	Vit B6 (mg)	Fola (µg)	Vit C (mg)	Vit B12 (µg)	Sele (µg)
0	41	1.59	75	240	217	2.87	4	0.26	0.74	0.06	0.66	0.11	9	<1	0	2
0	18	0.68	33	108	<1	1.23	1	0.13	0.70	0.03	0.33	0.05	4	<1	0	2
0	19	0.68	33	111	0	1.22	1	0.18	0.38	0.04	0.32	0.06	6	<1	0	1
0	1	0.48	22	51	<1	0.55	<.1	0.03	0.80	0.02	0.38	0.01	6	<.1	0	<.1
				<1												
0	35	1.34	38	333	<1	0.74	4	0.27	0.62	0.05	0.46	0.41	16	1	0	3
0	35	1.34	38	333	<1	0.74	4	0.27	0.62	0.05	0.46	0.41	16	1	0	3
0	24	8.48	303	457	10	4.22	11	0.12	0.00	0.18	0.99	0.05	32	1	0	3
0	89	1.33	32	43	1	0.64	0	0.07	—	0.02	0.41	0.07	9	0	0	1
0	154	3.07	58	93	2	1.17	<1	0.04	—	0.03	1.07	0.13	16	0	0	1
0	21	0.66	14	69	5	0.69	<1	0.24	—	0.02	0.85	0.02	15	1	0	<1
0	120	2.16	—	—	300	—	0	—	—	—	—	—	—	0	—	—
0	120	2.16	—	—	20	—	0	—	—	—	—	—	—	0	—	—
0	42	2.44	127	248	1	1.82	1	0.82	12.42	0.09	1.62	0.28	82	1	0	21
0	22	1.22	41	272	250	1.69	<1	0.03	8.35	0.08	2.25	0.26	76	<1	0	25
0	19	2.28	43	164	205	1.78	0	0.11	—	0.10	1.41	0.27	80	<1	0	21
0	20	0.76	59	12	83	0.85	<1	0.05	—	0.05	0.85	0.13	38	<1	0	—
0	29	1.14	59	257	86	1.21	<1	0.17	—	0.07	1.77	0.11	27	1	0	—
2	41	1.27	60	243	45	1.18	1	0.15	—	0.08	1.65	0.10	24	<1	0	—
0	20	0.92	34	248	4	0.41	1	0.16	—	0.04	0.52	0.11	15	3	0	—
0	19	0.98	63	163	1	1.05	1	0.02	0.56	0.04	0.15	0.18	10	1	0	5
0	29	0.87	47	132	1	0.93	<1	0.10	0.21	0.05	0.34	0.16	29	<1	0	1
0	266	4.73	88	375	112	1.51	121	0.14	0.07	0.12	1.08	0.28	32	18	0	11
42	468	2.58	37	204	1219	1.96	107	0.33	0.06	0.38	2.38	0.11	77	22	<1	22
0	48	1.71	42	367	618	0.60	54	0.09	0.53	0.08	0.93	0.11	33	8	<.1	4
0	23	2.35	31	380	289	0.87	18	0.14	0.14	0.10	1.50	0.16	61	13	0	9
22	187	0.77	20	120	768	1.11	82	0.15	0.29	0.24	1.02	0.04	39	<.1	<1	16
0	41	1.43	68	358	1403	0.91	86	0.16	3.05	0.12	2.76	0.30	28	13	<.1	8
51	468	4.78	90	455	2454	2.07	82	0.27	0.12	0.27	3.43	0.30	99	2	<1	43
0	117	1.13	40	168	134	0.68	26	0.07	—	0.07	0.54	0.11	46	1	0	5
0	190	3.65	28	302	371	0.74	8	0.13	0.06	0.12	1.43	0.15	47	7	0	6
10	144	1.91	33	393	637	1.06	31	0.20	0.05	0.27	2.07	0.21	64	15	<1	19
0	15	0.85	17	180	378	0.35	18	0.13	0.50	0.08	1.25	0.11	41	20	0	10
16	176	1.72	28	395	609	1.19	121	0.16	0.35	0.29	1.93	0.18	92	54	<1	6
46	318	3.36	39	299	813	1.98	288	0.45	0.21	0.50	4.52	0.16	123	27	<1	31
21	231	2.58	83	550	893	1.80	81	0.23	0.10	0.18	1.48	0.25	132	12	<1	10
<1	68	2.42	41	532	383	0.78	46	0.13	0.15	0.13	1.26	0.18	58	16	0	5
0	37	1.32	28	443	503	0.44	109	0.11	0.55	0.08	1.54	0.22	38	24	<.1	1
3	150	1.80	—	—	370	—	0	—	—	—	—	—	—	0	—	—
5	150	2.70	—	—	400	—	—	—	—	—	—	—	—	0	—	—
0	60	1.80	—	—	380	—	0	—	—	—	—	—	—	2	—	—
0	80	1.80	—	220	570	—	0	—	—	—	—	—	—	0	—	—
0	80	1.44	—	—	220	—	0	—	—	—	—	—	—	0	—	—
0	80	1.08	—	—	440	—	0	—	—	—	—	—	—	0	—	—
0	60	1.44	—	—	330	—	0	—	—	—	—	—	—	0	—	—
0	60	1.44	—	—	260	—	0	—	—	—	—	—	—	0	—	—
3	100	1.80	—	—	400	—	0	—	—	—	—	—	—	1	—	—
0	80	1.80	—	—	350	—	0	—	—	—	—	—	—	1	—	—
0	150	1.08	—	—	890	—	—	—	—	—	—	—	—	0	—	—
0	40	1.44	—	—	330	—	—	—	—	—	—	—	—	0	—	—
0	40	0.72	—	—	1000	—	—	—	—	—	—	—	—	0	—	—
0	60	3.60	—	—	360	—	—	—	—	—	—	—	—	0	—	—
5	40	1.44	—	—	550	—	—	—	—	—	—	—	—	0	—	—
5	60	0.72	—	—	570	—	—	—	—	—	—	—	—	5	—	—
0	80	1.44	—	—	380	—	—	—	—	—	—	—	—	0	—	—

TABLE A–1
Food Composition

(Computer code number is for Wadsworth Diet Analysis program)   (For purposes of calculations, use "0" for t, <1, <.1, <.01, etc.)

DA + Code	Food Description	Quantity	Measure	Wt (g)	H₂O (g)	Ener (kcal)	Prot (g)	Carb (g)	Dietary Fiber (g)	Fat (g)	Sat	Mono	Poly	Trans
	VEGETARIAN FOODS—Continued													
37807	Meatless breakfast sausage	1	item(s)	43	—	50	5	2	2	4	0.00	—	—	0
37809	Meatless meatballs	6	item(s)	85	—	110	12	8	4	5	1.00	—	—	0
37806	Meatless riblets w/sauce	1	item(s)	142	—	210	17	11	4	5	0.00	—	—	0
29913	Original	3	ounce(s)	85	—	132	7	19	4	4	1.80	1.80	0.60	0
31707	Santa Fe	3	ounce(s)	85	—	156	—	24	5	3	1.20	—	—	—
29915	Veggie medley	3	ounce(s)	85	—	108	7	22	4	0	0.00	0.00	0.00	0
	Loma Linda													
9311	Big franks	1	item(s)	51	30	110	10	2	2	7	1.00	2.00	4.00	0
9315	Chik'n nuggets	5	item(s)	85	40	240	14	13	4	15	2.00	4.50	8.00	0
9317	Corn dogs	1	item(s)	71	31	150	7	22	3	4	0.50	1.00	2.50	0
9323	Fried chik'n with gravy	2	piece(s)	80	46	150	12	5	2	10	1.50	2.50	5.00	0
9326	Linketts, canned	1	item(s)	35	21	70	7	1	1	5	0.50	1.00	2.50	0
9336	Redi-Burger patties, canned	1	slice(s)	85	50	120	18	7	4	3	0.50	0.50	1.50	0
9354	Tender Rounds meatball substitute, canned in gravy	6	piece(s)	80	54	120	13	6	1	5	0.50	1.00	2.50	0
	Morningstar Farms													
33707	America's Original Veggie Dog links	1	item(s)	57	—	80	11	6	1	1	0.00	0.00	0.00	0
9362	Better n Eggs egg substitute	¼	cup(s)	57	50	20	5	0	0	0	0.00	0.00	0.00	0
9368	Breakfast links	2	item(s)	45	27	80	9	3	2	3	0.50	0.50	2.00	0
9371	Breakfast strips	2	item(s)	16	7	60	2	2	1	5	0.50	1.00	3.00	0
33705	Chik Nuggets	4	piece(s)	86	—	180	13	17	5	6	0.50	1.50	4.00	—
11587	Chik Patties	1	item(s)	71	36	150	9	16	2	6	1.00	1.50	2.50	—
2531	Garden veggie patties	1	item(s)	67	40	100	10	9	4	3	0.50	0.50	1.50	0
9412	Natural Touch low fat vegetarian chili, canned	1	cup(s)	230	173	170	18	21	11	1	—	—	—	0
33702	Spicy black bean veggie burger	1	item(s)	78	47	150	11	16	5	5	0.50	1.50	2.50	0
	Worthington													
9422	Chik Stiks	1	item(s)	47	27	110	10	4	2	6	1.00	1.00	3.00	0
9424	Chili, canned	1	cup(s)	230	167	290	19	21	9	15	2.50	3.50	9.00	0
9432	Crispychik patties	1	item(s)	71	37	150	9	16	2	6	1.00	1.50	3.50	0
9440	Dinner roast, frozen	1	slice(s)	85	53	180	12	5	3	12	1.50	5.00	5.00	0
9442	Fillets, frozen	2	piece(s)	85	48	180	16	8	4	9	1.00	3.50	4.50	0
9478	Meatless smoked beef, sliced	6	slice(s)	57	—	130	11	7	1	7	1.00	2.00	4.00	0
9480	Meatless smoked turkey, sliced	3	slice(s)	57	—	140	10	5	0	9	1.00	2.50	5.00	—
9462	Prosage links	2	item(s)	45	27	80	9	3	2	3	0.50	0.50	2.00	0
9486	Stripples bacon substitute	2	item(s)	16	7	60	2	2	1	5	0.50	1.00	2.50	0
9496	Vegetable Skallops	½	cup(s)	85	65	90	15	3	3	2	0.50	0.50	0.00	0
9434	Vegetarian cutlets	1	slice(s)	61	43	70	11	3	2	1	—	—	—	—
	DAIRY													
	Butter: *see* Fats & Oils													
	Cheese													
1433	Blue, crumbled	1	ounce(s)	28	12	100	6	1	0	8	5.29	2.21	0.23	—
884	Brick	1	ounce(s)	28	12	104	7	1	0	8	5.25	2.41	0.22	—
885	Brie	1	ounce(s)	28	14	94	6	<1	0	8	4.87	2.24	0.23	—
34821	Camembert	1	ounce(s)	29	15	87	6	<1	0	7	4.43	2.04	0.21	—
888	Cheddar or colby	1	ounce(s)	28	11	110	7	1	0	9	5.66	2.60	0.27	—
32096	Cheddar or colby, low fat	1	ounce(s)	28	18	49	7	1	0	2	1.23	0.59	0.06	—
5	Cheddar, shredded	¼	cup(s)	28	10	114	7	<1	0	9	5.96	2.65	0.27	—
889	Edam	1	ounce(s)	28	12	100	7	<1	0	8	4.92	2.28	0.19	—
890	Feta	1	ounce(s)	28	15	74	4	1	0	6	4.18	1.29	0.17	—
891	Fontina	1	ounce(s)	28	11	109	7	<1	0	9	5.37	2.43	0.46	—
8527	Goat, soft	1	ounce(s)	28	17	76	5	<1	0	6	4.14	1.37	0.14	—
893	Gouda	1	ounce(s)	28	12	100	7	1	0	8	4.93	2.17	0.18	—
894	Gruyere	1	ounce(s)	28	9	116	8	<1	0	9	5.30	2.81	0.49	—
895	Limburger	1	ounce(s)	28	14	92	6	<1	0	8	4.69	2.41	0.14	—
896	Monterey jack	1	ounce(s)	28	11	104	7	<1	0	8	5.34	2.45	0.25	—
13	Mozzarella, part skim milk	1	ounce(s)	28	15	71	7	1	0	4	2.83	1.26	0.13	—
12	Mozzarella, whole milk	1	ounce(s)	28	14	84	6	1	0	6	3.68	1.84	0.21	—
897	Muenster	1	ounce(s)	28	12	103	7	<1	0	8	5.35	2.44	0.19	—
898	Neufchatel	1	ounce(s)	28	17	73	3	1	0	7	4.14	1.90	0.18	—
14	Parmesan, grated	1	tablespoon(s)	5	1	22	2	<1	0	1	0.87	0.42	0.06	—
17	Provolone	1	ounce(s)	28	11	98	7	1	0	7	4.78	2.07	0.22	—
19	Ricotta, part skim milk	¼	cup(s)	62	46	85	7	3	0	5	3.03	1.42	0.16	—
18	Ricotta, whole milk	¼	cup(s)	62	44	107	7	2	0	8	5.10	2.23	0.24	—
20	Romano	1	tablespoon(s)	5	2	19	2	<1	0	1	0.86	0.39	0.03	—

PAGE KEY: A–4 = Breads/Baked Goods   A–8 = Cereal/Rice/Pasta   A–12 = Fruit   A–16 = Vegetables/Legumes   A–26 = Nuts/Seeds   A–28 = Vegetarian
A–30 = Dairy   A–36 = Eggs   A–36 = Seafood   A–38 = Meats   A–42 = Poultry   A–42 = Processed meats   A–44 = Beverages   A–48 = Fats/Oils
A–50 = Sweets   A–52 = Spices/Condiments/Sauces   A–54 = Mixed Foods/Soups/Sandwiches   A–60 = Fast food   A–76 = Convenience   A–78 = Baby foods

Chol (mg)	Calc (mg)	Iron (mg)	Magn (mg)	Pota (mg)	Sodi (mg)	Zinc (mg)	Vit A (RAE) (µg)	Thia (mg)	Vit E (mg α)	Ribo (mg)	Niac (mg)	Vit B_6 (mg)	Fola (µg)	Vit C (mg)	Vit B_{12} (µg)	Sele (µg)
0	20	0.72	—	—	120	—	—	—	—	—	—	—	—	0	—	—
0	60	1.80	—	—	400	—	—	—	—	—	—	—	—	0	—	—
0	60	1.80	—	—	720	—	—	—	—	—	—	—	—	4	—	—
24	72	0.00	37	232	672	1.07	0	0.12	—	0.18	1.30	0.10	12	0	<1	8
24	96	0.00	—	—	336	—	0	—	—	—	—	—	—	0	—	0
0	48	0.00	32	218	336	0.55	—	0.08	—	0.10	1.08	0.11	13	0	<.1	5
0	0	0.77	—	50	240	0.89	0	0.23	—	0.43	1.60	0.04	—	0	1	—
0	20	1.44	—	210	410	0.43	0	0.75	—	0.51	6.00	0.90	—	0	3	—
0	0	1.08	—	60	500	0.43	0	0.72	—	0.61	1.47	0.87	—	0	2	—
0	20	1.80	—	70	430	0.34	0	1.05	—	0.34	4.00	0.30	—	0	2	—
0	0	0.36	—	15	160	0.46	0	0.12	—	0.20	0.40	0.20	—	0	1	—
0	0	1.06	—	140	450	1.11	0	0.23	—	0.34	6.00	0.40	—	0	2	—
0	20	1.08	—	80	340	0.66	0	0.75	—	0.17	2.00	0.16	—	0	1	—
0	0	0.72	—	60	580	—	0	—	—	—	—	—	—	0	—	—
0	20	0.63	—	75	90	0.60	75	0.03	—	0.34	0.00	0.08	24	0	1	—
0	0	1.44	—	50	320	0.36	0	1.80	—	0.17	2.00	0.30	—	0	3	—
0	0	0.27	—	15	220	0.05	0	0.75	—	0.04	0.40	0.07	—	0	<1	—
0	40	3.60	—	330	590	—	0	1.20	—	0.26	5.00	0.40	—	0	3	—
0	0	1.80	—	210	540	0.31	0	1.80	—	0.17	2.00	0.20	—	0	1	—
0	40	0.72	—	180	350	0.58	—	6.47	—	0.10	0.00	0.00	—	0	0	—
0	40	1.80	—	480	870	1.36	—	0.60	—	0.21	0.00	0.30	—	0	0	—
0	40	1.80	44	320	470	0.93	0	—	—	0.14	0.00	0.21	—	0	<.1	—
0	20	1.80	—	100	300	0.31	0	0.60	—	0.17	6.00	0.40	—	0	2	—
0	40	3.60	—	420	1130	1.24	0	0.06	—	0.07	2.00	0.70	—	0	2	—
0	0	1.80	—	170	440	0.33	0	1.80	—	0.17	2.00	0.20	—	0	1	—
3	40	0.36	—	55	580	0.64	0	1.80	—	0.26	6.00	0.60	—	0	2	—
0	0	1.80	—	130	750	0.92	0	0.68	—	0.14	0.80	0.40	—	0	3	—
0	20	1.80	—	180	510	0.14	0	1.80	—	0.17	6.00	0.40	—	0	2	—
0	100	2.70	—	60	490	0.23	0	1.80	—	0.17	6.00	0.40	—	0	3	—
0	0	1.44	—	50	320	0.36	0	1.80	—	0.17	2.00	0.30	—	0	3	—
0	0	0.36	—	15	220	0.05	0	0.75	—	0.03	0.40	0.08	—	0	<1	—
0	0	0.72	—	10	410	0.67	0	0.03	—	0.03	0.00	0.01	—	0	0	—
0	0	0.00	—	30	340	0.43	0	0.03	—	0.04	0.00	0.04	—	0	0	—
21	150	0.09	7	73	395	0.75	56	0.01	0.07	0.11	0.29	0.05	10	0	<1	4
26	189	0.12	7	38	157	0.73	82	0.00	0.07	0.10	0.03	0.02	6	0	<1	4
28	52	0.14	6	43	176	0.67	49	0.02	0.07	0.15	0.11	0.07	18	0	<1	4
21	112	0.10	6	54	244	0.69	—	0.01	—	0.14	0.18	0.07	18	0	<1	4
27	192	0.21	7	36	169	0.86	74	0.00	0.08	0.11	0.03	0.02	5	0	<1	4
6	118	0.12	5	19	174	0.52	17	0.00	0.02	0.06	0.01	0.01	3	0	<1	4
30	204	0.19	8	28	175	0.88	75	0.01	0.08	0.11	0.02	0.02	5	0	<1	4
25	205	0.12	8	53	270	1.05	68	0.01	0.07	0.11	0.02	0.02	4	0	<1	4
25	138	0.18	5	17	312	0.81	35	0.04	0.05	0.24	0.28	0.12	9	0	<1	4
32	154	0.06	4	18	224	0.98	73	0.01	0.08	0.06	0.04	0.02	2	0	<1	4
13	40	0.54	5	7	105	0.26	82	0.02	0.05	0.11	0.12	0.07	3	0	<.1	1
32	196	0.07	8	34	229	1.09	46	0.01	0.07	0.09	0.02	0.02	6	0	<1	4
31	283	0.05	10	23	94	1.09	76	0.02	0.08	0.08	0.03	0.02	3	0	<1	4
25	139	0.04	6	36	224	0.59	95	0.02	0.06	0.14	0.04	0.02	16	0	<1	4
25	209	0.20	8	23	150	0.84	55	0.00	0.07	0.11	0.03	0.02	5	0	<1	4
18	219	0.06	6	24	173	0.77	36	0.01	0.04	0.08	0.03	0.02	3	0	<1	4
22	141	0.12	6	21	176	0.82	50	0.01	0.05	0.08	0.03	0.01	2	0	1	5
27	201	0.11	8	38	176	0.79	83	0.00	0.07	0.09	0.03	0.02	3	0	<1	4
21	21	0.08	2	32	112	0.15	83	0.00	—	0.05	0.04	0.01	3	0	<.1	1
4	55	0.05	2	6	76	0.19	6	0.00	0.01	0.02	0.01	0.00	1	0	<1	1
19	212	0.15	8	39	245	0.90	66	0.01	0.06	0.09	0.04	0.02	3	0	<1	4
19	167	0.27	9	77	77	0.82	66	0.01	0.04	0.11	0.05	0.01	8	0	<1	10
31	127	0.23	7	65	52	0.71	74	0.01	0.07	0.12	0.06	0.03	7	0	<1	9
5	53	0.04	2	4	60	0.13	5	0.00	0.01	0.02	0.00	0.00	<1	0	<.1	1

TABLE A–1
Food Composition

(Computer code number is for Wadsworth Diet Analysis program)   (For purposes of calculations, use "0" for t, <1, <.1, <.01, etc.)

DA + Code	Food Description	Quantity	Measure	Wt (g)	H₂O (g)	Ener (kcal)	Prot (g)	Carb (g)	Dietary Fiber (g)	Fat (g)	Sat	Mono	Poly	Trans
	DAIRY—Continued													
900	Roquefort	1	ounce(s)	28	11	103	6	1	0	9	5.39	2.37	0.37	—
21	Swiss	1	ounce(s)	28	10	106	8	2	0	8	4.98	2.04	0.27	—
	Imitation cheese													
7998	Shredded imitation cheddar	¼	cup(s)	28	—	90	5	2	0	7	1.50	—	—	—
8028	Shredded imitation mozzarella	¼	cup(s)	28	—	80	6	1	0	6	1.00	—	—	—
	Cottage Cheese													
9	Low fat, 1% fat	½	cup(s)	113	93	81	14	3	0	1	0.73	0.33	0.04	—
8	Low fat, 2% fat	½	cup(s)	113	90	102	16	4	0	2	1.38	0.62	0.07	—
	Cream cheese													
11	Cream cheese	2	tablespoon(s)	29	16	101	2	1	0	10	6.37	2.85	0.37	—
17366	Fat free cream cheese	2	tablespoon(s)	30	23	29	4	2	0	<1	0.27	0.10	0.02	—
10438	Tofutti Better Than Cream Cheese	2	tablespoon(s)	30	—	80	1	1	0	8	2.00	—	6.00	—
	Processed cheese													
22	American cheese, processed	1	ounce(s)	28	11	106	6	<1	0	9	5.58	2.54	0.28	—
24	American cheese food, processed	1	ounce(s)	28	12	94	5	2	0	7	4.23	2.05	0.31	—
25	American cheese spread, processed	1	ounce(s)	28	14	82	5	2	0	6	3.78	1.77	0.18	—
9110	Kraft deluxe singles pasteurized process American cheese	1	ounce(s)	28	—	110	5	1	0	9	6.00	—	—	—
23	Swiss cheese, processed	1	ounce(s)	28	12	95	7	1	0	7	4.55	2.00	0.18	—
	Soy cheese													
10430	Nu Tofu cheddar flavored cheese alternative	1	ounce(s)	28	—	70	6	1	0	4	0.50	2.50	1.00	—
10435	Nu Tofu mozzarella flavored cheese alternative	1	ounce(s)	28	—	70	6	2	0	4	0.50	2.50	1.00	—
	Cream													
26	Half & half	1	tablespoon(s)	15	12	20	<1	1	0	2	1.07	0.50	0.06	—
28	Light coffee or table, liquid	1	tablespoon(s)	15	11	29	<1	1	0	3	1.80	0.84	0.11	—
30	Light whipping cream, liquid	1	tablespoon(s)	15	10	44	<1	<1	0	5	2.90	1.36	0.13	—
32	Heavy whipping cream, liquid	1	tablespoon(s)	15	9	52	<1	<1	0	6	3.45	1.60	0.21	—
34	Whipped cream topping, pressurized	1	tablespoon(s)	4	2	10	<1	<1	0	1	0.52	0.24	0.03	—
	Sour cream													
36	Sour cream	2	tablespoon(s)	24	17	51	1	1	0	5	3.13	1.45	0.19	—
30556	Fat free sour cream	2	tablespoon(s)	32	26	24	1	5	0	0	0.00	0.00	0.00	0
	Imitation cream													
3659	Coffeemate nondairy creamer, liquid	1	tablespoon(s)	16	—	20	0	2	0	1	0.00	0.50	0.00	—
40	Cream substitute, powder	1	teaspoon(s)	2	<.1	11	<.1	1	0	1	0.65	0.02	0.00	—
35972	Nondairy coffee whitener, liquid, frozen	1	tablespoon(s)	16	12	22	<1	2	0	2	0.31	1.20	0.00	—
35975	Nondairy dessert topping, pressurized	1	tablespoon(s)	5	3	12	<.1	1	0	1	0.88	0.09	0.01	—
35976	Nondairy dessert topping, frozen	1	tablespoon(s)	5	3	16	<.1	1	0	1	1.09	0.08	0.03	—
904	Imitation sour cream	2	tablespoon(s)	24	17	50	1	2	0	5	4.27	0.14	0.01	—
	Fluid milk													
57	Fat free, nonfat, or skim	1	cup(s)	245	223	83	8	12	0	<1	0.29	0.12	0.02	—
58	Fat free, nonfat, or skim, w/nonfat milk solids	1	cup(s)	245	221	91	9	12	0	1	0.40	0.16	0.02	—
54	Low fat, 1%	1	cup(s)	244	219	102	8	12	0	2	1.54	0.68	0.09	—
55	Low fat, 1%, w/nonfat milk solids	1	cup(s)	245	220	105	9	12	0	2	1.48	0.69	0.09	—
60	Low fat buttermilk	1	cup(s)	245	221	98	8	12	0	2	1.34	0.62	0.08	—
51	Reduced fat, 2%	1	cup(s)	244	218	122	8	11	0	5	2.35	2.04	0.17	—
52	Reduced fat, 2%, w/nonfat milk solids	1	cup(s)	245	218	125	9	12	0	5	2.93	1.36	0.17	—
50	Whole, 3.3%	1	cup(s)	244	216	146	8	11	0	8	4.55	1.98	0.48	—
	Canned													
61	Whole evaporated	2	tablespoon(s)	32	23	42	2	3	0	2	1.45	0.74	0.08	—
62	Fat free, nonfat, or skim evaporated	2	tablespoon(s)	32	25	25	2	4	0	<.1	0.04	0.02	0.00	—
63	Sweetened condensed	2	tablespoon(s)	38	10	123	3	21	0	3	2.10	0.93	0.13	—
	Dried Milk													
64	Dried buttermilk	¼	cup(s)	30	1	118	10	15	0	2	1.09	0.51	0.07	—
65	Instant nonfat dry milk w/added vitamin A	¼	cup(s)	17	1	63	6	9	0	<1	0.08	0.03	0.00	—
5234	Skim milk powder	¼	cup(s)	18	1	64	6	9	0	<1	0.08	0.03	0.01	—
907	Whole dry milk	¼	cup(s)	32	1	161	9	12	0	9	5.43	2.57	0.22	—
909	**Goat milk**	1	cup(s)	244	212	168	9	11	0	10	6.51	2.71	0.36	—
	Chocolate milk													
69	Low fat	1	cup(s)	250	211	158	8	26	1	3	1.54	0.75	0.09	—
68	Reduced fat	1	cup(s)	250	209	180	8	26	1	5	3.10	1.47	0.18	—
67	Whole milk	1	cup(s)	250	206	208	8	26	2	8	5.26	2.48	0.31	—
33156	Chocolate syrup, fortified, prepared w/milk	1	cup(s)	263	220	197	8	24	<1	8	5.22	2.44	0.31	—

PAGE KEY: A–4 = Breads/Baked Goods A–8 = Cereal/Rice/Pasta A–12 = Fruit A–16 = Vegetables/Legumes A–26 = Nuts/Seeds A–28 = Vegetarian A–30 = Dairy A–36 = Eggs A–36 = Seafood A–38 = Meats A–42 = Poultry A–42 = Processed meats A–44 = Beverages A–48 = Fats/Oils A–50 = Sweets A–52 = Spices/Condiments/Sauces A–54 = Mixed Foods/Soups/Sandwiches A–60 = Fast food A–76 = Convenience A–78 = Baby foods

Chol (mg)	Calc (mg)	Iron (mg)	Magn (mg)	Pota (mg)	Sodi (mg)	Zinc (mg)	Vit A (RAE) (µg)	Thia (mg)	Vit E (mg α)	Ribo (mg)	Niac (mg)	Vit B_6 (mg)	Fola (µg)	Vit C (mg)	Vit B_{12} (µg)	Sele (µg)
25	185	0.16	8	25	507	0.58	82	0.01	—	0.16	0.21	0.03	14	0	<1	4
26	221	0.06	11	22	54	1.22	62	0.02	0.11	0.08	0.03	0.02	2	0	1	5
0	150	0.00	—	—	420	—	—	—	—	—	—	—	—	0	—	—
0	150	0.00	8	—	320	1.20	—	0.00	—	0.26	0.00	0.00	40	0	<1	—
5	69	0.16	6	97	459	0.43	12	0.02	0.01	0.19	0.14	0.08	14	0	1	10
9	78	0.18	7	108	459	0.47	24	0.03	0.02	0.21	0.16	0.09	15	0	1	12
32	23	0.35	2	35	86	0.16	106	0.00	0.09	0.06	0.03	0.01	4	0	<1	1
2	56	0.05	4	49	164	0.26	84	0.02	0.00	0.05	0.05	0.02	11	0	<1	1
0	0	0.00	—	—	135	—	0	—	—	—	—	—	—	0	—	—
27	156	0.05	8	48	422	0.81	72	0.01	0.08	0.10	0.02	0.02	2	0	<1	4
23	162	0.16	9	83	359	0.91	57	0.02	0.06	0.15	0.05	0.02	2	0	<1	5
16	160	0.09	8	69	382	0.74	49	0.01	0.05	0.12	0.04	0.03	2	0	<1	3
25	150	0.00	0	25	450	0.90	84	—	—	0.10	—	—	—	0	<1	—
24	219	0.17	8	61	388	1.02	56	0.00	0.10	0.08	0.01	0.01	2	0	<1	5
0	200	0.36	—	—	190	—	—	—	—	—	—	—	—	0	—	—
0	150	0.36	—	—	190	—	—	—	—	—	—	—	—	0	—	—
6	16	0.01	2	20	6	0.08	15	0.01	0.05	0.02	0.01	0.01	<1	<1	<.1	<1
10	14	0.01	1	18	6	0.04	27	0.00	0.08	0.02	0.01	0.00	<1	<1	<.1	<.1
17	10	0.00	1	15	5	0.04	42	0.00	0.13	0.02	0.01	0.00	1	<.1	<.1	<.1
21	10	0.00	1	11	6	0.03	62	0.00	0.16	0.02	0.01	0.00	1	<.1	<.1	<.1
3	4	0.00	<1	6	5	0.01	7	0.00	0.02	0.00	0.00	0.00	<1	<.1	<.1	<.1
11	28	0.01	3	35	13	0.06	42	0.01	0.14	0.04	0.02	0.00	3	<1	<.1	1
3	40	0.00	3	41	45	0.16	—	0.01	0.00	0.05	0.02	0.01	4	0	<.1	—
0	0	0.00	—	30	0	—	0	0.02	—	0.02	0.20	—	—	0	—	—
0	<1	0.02	<.1	16	4	0.01	<.1	0.00	0.01	0.00	0.00	0.00	0	0	0	<.1
0	1	0.00	<.1	30	13	0.00	—	0.00	—	0.00	0.00	0.00	0	0	0	<1
0	<1	0.00	<.1	1	3	0.00	—	0.00	—	0.00	0.00	0.00	0	0	0	<.1
0	<1	0.01	<.1	1	1	0.00	—	0.00	—	0.00	0.00	0.00	0	0	0	<1
0	1	0.09	1	39	24	0.28	0	0.00	0.18	0.00	0.00	0.00	0	0	0	1
5	223	1.23	22	238	108	2.08	149	0.11	0.02	0.45	0.23	0.09	12	0	1	8
5	316	0.12	37	419	130	1.00	149	0.10	0.00	0.43	0.22	0.11	12	2	1	5
12	264	0.85	27	290	122	2.12	142	0.05	0.02	0.45	0.23	0.09	12	0	1	8
10	314	0.12	34	397	127	0.98	145	0.10	—	0.42	0.22	0.11	12	2	1	6
10	284	0.12	27	370	257	1.03	17	0.08	0.12	0.38	0.14	0.08	12	2	1	5
20	271	0.24	27	342	115	1.17	134	0.10	0.07	0.45	0.22	0.09	12	<1	1	6
20	314	0.12	34	397	127	0.98	137	0.10	—	0.42	0.22	0.11	12	2	1	6
24	246	0.07	24	325	105	0.93	68	0.11	0.15	0.45	0.26	0.09	12	0	1	9
9	82	0.06	8	95	33	0.24	20	0.01	0.04	0.10	0.06	0.02	3	1	<.1	1
1	93	0.09	9	106	37	0.29	38	0.01	0.00	0.10	0.06	0.02	3	<1	<.1	1
13	109	0.07	10	142	49	0.36	28	0.03	0.06	0.16	0.08	0.02	4	1	<1	6
21	360	0.09	33	484	157	1.22	15	0.12	0.03	0.48	0.27	0.10	14	2	1	6
3	215	0.05	20	298	96	0.77	124	0.07	0.00	0.30	0.16	0.06	9	1	1	5
3	222	0.06	21	307	99	0.79	0	0.07	—	0.31	0.16	0.06	9	1	1	5
31	296	0.15	28	431	120	1.08	83	0.09	0.16	0.39	0.21	0.10	12	3	1	5
27	327	0.12	34	498	122	0.73	139	0.12	0.17	0.34	0.68	0.11	2	3	<1	3
8	288	0.60	33	425	153	1.03	145	0.10	0.05	0.42	0.32	0.10	13	2	1	5
18	285	0.60	33	423	150	1.03	138	0.09	0.10	0.41	0.32	0.10	13	2	1	5
30	280	0.60	33	418	150	1.03	65	0.09	0.15	0.41	0.31	0.10	13	2	1	5
34	292	2.68	32	460	147	0.92	—	0.09	—	0.55	6.53	0.11	13	2	1	5

TABLE A–1
Food Composition
(Computer code number is for Wadsworth Diet Analysis program)   (For purposes of calculations, use "0" for t, <1, <.1, <.01, etc.)

DA + Code	Food Description	Quantity	Measure	Wt (g)	H₂O (g)	Ener (kcal)	Prot (g)	Carb (g)	Dietary Fiber (g)	Fat (g)	Sat	Mono	Poly	Trans
	DAIRY—Continued													
908	Cocoa, hot, prepared w/milk	1	cup(s)	250	206	193	9	27	3	6	3.58	1.69	0.09	0.18
33184	Cocoa mix with aspartame, added sodium & vitamin A, no added calcium or phosphorus, prepared with water	1	cup(s)	192	177	56	2	10	1	<1	0.00	0.15	0.01	—
70	**Eggnog**	1	cup(s)	254	189	343	10	34	0	19	11.29	5.67	0.86	—
	Breakfast drinks													
10093	Carnation Instant Breakfast classic chocolate malt, prepared w/skim milk, no sugar added	1	cup(s)	243	—	142	11	21	<1	1	0.89	—	—	—
10091	Carnation Instant Breakfast strawberry creme, prepared w/skim milk	1	cup(s)	273	—	220	13	39	0	<1	0.40	—	—	—
10094	Carnation Instant Breakfast strawberry creme, prepared w/skim milk, no sugar added	1	cup(s)	243	—	134	12	21	0	<1	0.45	—	—	—
10092	Carnation Instant Breakfast vanilla creme, prepared w/skim milk, no sugar added	1	cup(s)	273	—	220	13	39	0	<1	0.40	—	—	—
1417	Ovaltine rich chocolate flavor, prepared w/skim milk	1	cup(s)	243	—	134	12	21	0	<1	0.45	—	—	—
8539	**Malted milk, chocolate mix, fortified, prepared w/milk**	1	cup(s)	265	216	223	9	29	1	9	4.95	2.17	0.54	—
	Milkshakes													
73	Chocolate	1	cup(s)	227	164	270	7	48	1	6	3.81	1.77	0.23	—
74	Vanilla	1	cup(s)	227	169	254	9	40	0	7	4.28	1.98	0.26	—
	Ice cream													
4776	Chocolate	½	cup(s)	66	37	143	3	19	1	7	4.49	2.12	0.27	—
16514	Chocolate, soft serve	½	cup(s)	87	50	177	3	24	1	8	5.17	2.43	0.31	—
12137	Chocolate fudge, fat free no sugar added	½	cup(s)	71	—	100	4	22	0	0	0.00	0.00	0.00	0
82	Light vanilla	½	cup(s)	66	42	109	4	18	<1	3	1.71	0.57	0.10	—
78	Light vanilla, soft serve	½	cup(s)	86	60	108	4	19	0	2	1.40	0.65	0.09	—
16523	Sherbet, all flavors	½	cup(s)	97	64	133	1	29	<1	2	1.12	0.51	0.08	—
4778	Strawberry	½	cup(s)	66	40	127	2	18	1	6	3.43	—	—	—
76	Vanilla	½	cup(s)	66	40	133	2	16	<1	7	4.48	1.96	0.30	—
12146	Vanilla chocolate swirl, fat free, no sugar added	½	cup(s)	71	—	100	4	20	0	0	0.00	0.00	0.00	0
	Soy desserts													
10694	Tofutti low fat vanilla fudge nondairy frozen dessert	½	cup(s)	70	—	120	2	24	0	2	1.00	—	—	—
15721	Tofutti premium chocolate supreme nondairy frozen dessert	½	cup(s)	60	—	180	3	18	0	11	2.00	—	—	—
15720	Tofutti premium vanilla nondairy frozen dessert	½	cup(s)	60	—	190	2	20	0	11	2.00	—	—	—
	Ice milk													
16516	Flavored, not chocolate	½	cup(s)	66	45	91	2	15	0	3	1.72	0.81	0.11	—
16517	Chocolate	½	cup(s)	66	43	95	3	17	<1	2	1.29	0.61	0.08	—
	Pudding													
25032	Chocolate	½	cup(s)	144	110	154	5	23	1	5	2.78	1.94	0.23	0
1923	Chocolate, sugar free, prepared w/2% milk	½	cup(s)	133	—	100	5	14	<1	3	1.50	—	—	—
1722	Rice	½	cup(s)	113	73	175	6	26	1	6	1.99	2.14	0.88	—
4747	Tapioca, ready to eat	1	item(s)	142	105	169	3	28	<1	5	0.85	2.24	1.93	—
25031	Vanilla	½	cup(s)	136	110	116	5	17	<.1	3	1.31	1.21	0.16	0
1924	Vanilla, sugar free, prepared w/2% milk	½	cup(s)	133	90	4	12	<1	2	1.50	10	150	0.00	—
	Frozen yogurt													
4785	Chocolate, soft serve	½	cup(s)	72	46	115	3	18	2	4	2.61	1.26	0.16	—
1747	Fruit varieties	½	cup(s)	113	80	144	3	24	0	4	2.63	1.11	0.11	—
4786	Vanilla, soft serve	½	cup(s)	72	47	117	3	17	0	4	2.46	1.14	0.15	—
	Milk substitutes													
	Lactose free													
16081	Fat free calcium fortified milk	1	cup(s)	240	—	90	9	13	0	0	0.00	—	—	0
36486	Low fat milk	1	cup(s)	240	—	110	8	13	0	3	1.50	—	—	—
36487	Reduced fat milk	1	cup(s)	240	—	130	8	13	0	5	3.00	—	—	—
36488	Whole milk	1	cup(s)	240	—	160	8	12	0	9	5.00	—	—	—
	Rice													
10083	Rice Dream carob rice beverage	1	cup(s)	240	—	150	1	32	0	3	0.00	—	—	—
10087	Rice Dream vanilla enriched rice beverage	1	cup(s)	240	—	130	1	28	0	2	0.00	—	—	—
17089	Rice Dream original rice beverage, enriched	1	cup(s)	240	—	120	1	25	0	2	0.00	—	—	—

PAGE KEY: A–4 = Breads/Baked Goods   A–8 = Cereal/Rice/Pasta   A–12 = Fruit   A–16 = Vegetables/Legumes   A–26 = Nuts/Seeds   A–28 = Vegetarian
A–30 = Dairy   A–36 = Eggs   A–36 = Seafood   A–38 = Meats   A–42 = Poultry   A–42 = Processed meats   A–44 = Beverages   A–48 = Fats/Oils
A–50 = Sweets   A–52 = Spices/Condiments/Sauces   A–54 = Mixed Foods/Soups/Sandwiches   A–60 = Fast food   A–76 = Convenience   A–78 = Baby foods

Chol (mg)	Calc (mg)	Iron (mg)	Magn (mg)	Pota (mg)	Sodi (mg)	Zinc (mg)	Vit A (RAE) (µg)	Thia (mg)	Vit E (mg α)	Ribo (mg)	Niac (mg)	Vit B$_6$ (mg)	Fola (µg)	Vit C (mg)	Vit B$_{12}$ (µg)	Sele (µg)
20	263	1.20	58	493	110	1.58	128	0.10	0.08	0.46	0.33	0.10	13	1	1	7
<1	90	0.75	33	405	171	0.52	27	0.04	0.06	0.21	0.16	0.05	2	<1	<1	2
150	330	0.51	48	419	137	1.17	114	0.09	0.51	0.48	0.27	0.13	3	4	1	11
9	445	4.01	89	632	196	3.38	—	0.35	—	0.45	4.45	0.45	4	27	1	8
9	500	4.47	100	638	360	3.75	—	0.38	—	0.51	5.08	0.48	100	30	1	9
9	445	4.01	89	570	187	3.38	—	0.33	—	0.45	4.45	0.45	89	27	1	8
9	500	4.50	100	630	240	3.75	—	0.38	—	0.51	5.00	0.50	100	30	2	9
9	445	4.01	89	570	187	3.38	—	0.33	—	0.45	4.45	0.45	89	27	1	8
27	339	3.76	45	578	231	1.17	904	0.76	0.16	1.32	11.08	1.01	19	32	1	12
25	299	0.70	36	508	252	1.09	41	0.11	0.11	0.50	0.28	0.06	11	0	1	4
27	331	0.23	27	415	215	0.88	57	0.07	0.11	0.44	0.33	0.10	16	0	1	5
22	72	0.61	19	164	50	0.38	78	0.03	0.20	0.13	0.15	0.04	11	<1	<1	2
22	103	0.33	19	192	44	0.48	—	0.04	0.22	0.13	0.11	0.03	5	1	<1	—
0	80	0.36	—	—	60	—	—	—	—	—	—	—	—	—	0	—	—
17	77	0.05	9	137	49	0.48	91	0.02	0.08	0.11	0.06	0.02	3	<1	<1	1
10	135	0.05	12	190	60	0.46	25	0.04	0.05	0.17	0.10	0.04	5	1	<1	3
5	52	0.14	8	93	44	0.46	—	0.02	0.03	0.07	0.09	0.03	4	4	<1	—
19	79	0.14	9	124	40	0.22	63	0.03	—	0.17	0.11	0.03	8	5	<1	1
29	84	0.06	9	131	53	0.46	78	0.03	0.20	0.16	0.08	0.03	3	<1	<1	1
0	80	0.00	50	0				—								
0	0	0.00	—	8	90	—	0	—	—	—	—	—	—	—	0	—	—
0	0	0.00	—	7	180	—	0	—	—	—	—	—	—	—	0	—	—
0	0	0.00	—	2	210	—	0	—	—	—	—	—	—	—	0	—	—
9	91	0.07	10	138	56	0.29	—	0.04	0.06	0.17	0.06	0.04	4	1	<1	—
6	94	0.17	13	155	41	0.38	—	0.03	0.05	0.12	0.09	0.03	4	<1	<1	—
35	138	1.04	29	211	135	1.07	73	0.04	0.00	0.25	0.18	0.03	7	<1	<1	5
10	150	0.72	—	330	310	—	—	0.06	—	0.26	—	—	—	0	—	—
71	130	1.21	21	250	253	0.61	—	0.10	0.06	0.26	0.73	0.08	14	1	<1	—
1	119	0.33	11	136	226	0.38	0	0.03	0.43	0.14	0.44	0.03	4	1	<1	2
35	133	0.25	14	173	134	0.63	73	0.03	0.00	0.24	0.11	0.03	6	<1	<1	5
—	190	380	—	—	0.03	—	—	0.17	—	—	—		0			
4	106	0.90	19	188	71	0.35	32	0.03	—	0.15	0.22	0.05	8	<1	<1	2
15	113	0.52	11	176	71	0.32	—	0.05	0.10	0.20	0.08	0.05	5	1	<.1	—
1	103	0.22	10	152	63	0.30	42	0.03	0.08	0.16	0.21	0.06	4	1	<1	2
3	500	0.00	—	—	130	—	100	—	—	—	—	—	—	0	—	—
15	300	0.00	—	—	125	—	100	—	—	—	—	—	—	0	—	—
20	300	0.00	—	—	125	—	98	—	—	—	—	—	—	0	—	—
35	300	0.00	—	—	125	—	58	—	—	—	—	—	—	0	—	—
0	20	0.72	—	—	100	—	—	—	—	—	—	—	—	1	—	—
0	300	0.00	—	—	90	—	—	—	—	—	—	—	—	0	2	—
0	300	0.00	13	60	90	0.24	—	0.07	—	0.00	0.84	0.08	—	0	2	—

TABLE A–1
Food Composition

(Computer code number is for Wadsworth Diet Analysis program) (For purposes of calculations, use "0" for t, <1, <.1, <.01, etc.)

DA + Code	Food Description	Quantity	Measure	Wt (g)	H₂O (g)	Ener (kcal)	Prot (g)	Carb (g)	Dietary Fiber (g)	Fat (g)	Fat Breakdown (g)			
											Sat	Mono	Poly	Trans
	DAIRY—Continued													
	Soy													
34750	Soy Dream chocolate enriched soy beverage	1	cup(s)	240	—	210	7	37	1	4	0.50	—	—	—
34749	Soy Dream vanilla enriched soy beverage	1	cup(s)	240	—	150	7	22	0	4	0.50	—	—	—
13840	Vitasoy light chocolate soymilk	1	cup(s)	237	—	100	4	17	0	2	0.50	0.50	1.00	—
13839	Vitasoy light vanilla soymilk	1	cup(s)	237	—	70	4	10	0	2	0.50	0.50	1.00	—
13836	Vitasoy rich chocolate soymilk	1	cup(s)	237	—	160	7	24	1	4	0.50	1.00	2.50	—
13835	Vitasoy vanilla delite soymilk	1	cup(s)	237	—	120	8	13	1	4	0.50	1.00	2.50	—
	Yogurt													
3615	Custard style, fruit flavors	6	ounce(s)	170	127	190	7	32	0	4	2.00	—	—	—
3617	Custard style, vanilla	6	ounce(s)	170	134	190	7	32	0	4	2.00	0.94	0.10	—
32101	Fruit, low fat	1	cup(s)	245	184	243	10	46	0	3	1.82	0.77	0.08	—
29638	Fruit, nonfat, sweetened w/low calorie sweetener	1	cup(s)	241	208	122	11	19	1	<1	0.21	0.10	0.04	—
93	Plain, low fat	1	cup(s)	245	208	154	13	17	0	4	2.45	1.04	0.11	—
94	Plain, nonfat	1	cup(s)	245	209	137	14	19	0	<1	0.28	0.12	0.01	—
32100	Vanilla, low fat	1	cup(s)	245	194	208	12	34	0	3	1.97	0.84	0.09	—
5242	Yogurt beverage	1	cup(s)	245	200	172	6	33	0	2	1.39	0.59	0.06	—
38202	Yogurt smoothie, nonfat, all flavors	1	item(s)	325	—	290	10	60	6	0	0.00	0.00	0.00	0
	Soy yogurt													
10453	White Wave plain silk cultured	8	ounce(s)	227	—	120	5	22	1	3	0.00	—	—	0
34616	Stonyfield Farm Osoy chocolate-vanilla pack organic cultured	1	serving(s)	113	—	90	4	15	3	2	0.00	—	—	—
34617	Stonyfield Farm Osoy strawberry-peach pack organic cultured	1	serving(s)	113	—	90	4	15	3	2	0.00	—	—	—
	EGGS													
96	Raw, whole	1	item(s)	50	38	74	6	<1	0	5	1.55	1.91	0.68	—
97	Raw, white	1	item(s)	33	29	17	4	<1	0	<.1	0.00	0.00	0.00	—
98	Raw, yolk	1	item(s)	17	9	53	3	1	0	4	1.59	1.95	0.70	—
99	Fried	1	item(s)	46	32	92	6	<1	0	7	1.98	2.92	1.22	—
100	Hard boiled	1	item(s)	50	37	78	6	1	0	5	1.63	2.04	0.71	—
101	Poached	1	item(s)	50	38	74	6	<1	0	5	1.54	1.90	0.68	—
102	Scrambled, prepared w/milk & butter	2	item(s)	122	89	203	14	3	0	15	4.49	5.82	2.62	—
	Egg Substitute													
920	Frozen	¼	cup(s)	60	44	96	7	2	0	7	1.16	1.46	3.74	—
918	Liquid	¼	cup(s)	63	52	53	8	<1	0	2	0.41	0.56	1.01	—
4028	Egg Beaters	¼	cup(s)	61	—	30	6	1	0	0	0.00	0.00	0.00	0
	SEAFOOD													
	Fish													
	Cod													
6040	Atlantic cod or scrod, baked or broiled	3	ounce(s)	44	34	46	10	0	0	<1	0.07	0.05	0.13	—
1573	Atlantic cod, cooked, dry heat	3	ounce(s)	85	65	89	19	0	0	1	0.14	0.11	0.25	—
2905	Eel, raw	3	ounce(s)	85	58	156	16	0	0	10	2.01	6.12	0.81	—
	Fish fillets													
25079	Baked	3	ounce(s)	84	80	99	22	0	0	1	0.08	0.07	0.26	—
8615	Batter coated or breaded, fried	3	ounce(s)	85	0.04	197.19	12.46	14.42	0.42	10.44	2.39	2.19	5.32	—
25082	Broiled fish steaks	3	ounce(s)	86	69	129	24	0	0	3	0.37	0.87	0.84	—
25083	Poached fish steaks	3	ounce(s)	86	68	112	21	0	0	2	0.33	0.76	0.74	—
25084	Steamed fish fillets	3	ounce(s)	86	73	80	17	0	0	1	0.12	0.08	0.22	—
25089	**Flounder, baked**	3	ounce(s)	85	65	114	15	<1	<.1	6	1.15	2.17	1.44	0
1825	**Grouper, cooked, dry heat**	3	ounce(s)	85	62	100	21	0	0	1	0.25	0.23	0.34	—
	Haddock													
6049	Baked or broiled	3	ounce(s)	44	33	50	11	0	0	<1	0.07	0.07	0.14	—
1578	Cooked, dry heat	3	ounce(s)	85	63	95	21	0	0	1	0.14	0.13	0.26	—
1886	**Halibut, Atlantic & Pacific, cooked, dry heat**	3	ounce(s)	85	61	119	23	0	0	2	0.35	0.82	0.80	—
1582	**Herring, Atlantic, pickled**	4	piece(s)	60	33	157	9	6	0	11	1.43	7.17	1.01	—
1587	**Jack mackerel, solids, canned, drained**	2	ounce(s)	57	39	88	13	0	0	4	1.05	1.26	0.94	—
8580	**Octopus, common, cooked, moist heat**	3	ounce(s)	85	51	139	25	4	0	2	0.39	0.28	0.41	—
1831	**Perch, mixed species, cooked, dry heat**	3	ounce(s)	85	62	99	21	0	0	1	0.20	0.17	0.40	—
1592	**Pacific rockfish, cooked, dry heat**	3	ounce(s)	85	62	103	20	0	0	2	0.40	0.38	0.50	—
	Salmon													
29727	Smoked chinook (lox)	2	ounce(s)	57	<.1	66	10	0	0	2	0.52	1.14	0.56	—

PAGE KEY: A–4 = Breads/Baked Goods A–8 = Cereal/Rice/Pasta A–12 = Fruit A–16 = Vegetables/Legumes A–26 = Nuts/Seeds A–28 = Vegetarian A–30 = Dairy A–36 = Eggs A–36 = Seafood A–38 = Meats A–42 = Poultry A–42 = Processed meats A–44 = Beverages A–48 = Fats/Oils A–50 = Sweets A–52 = Spices/Condiments/Sauces A–54 = Mixed Foods/Soups/Sandwiches A–60 = Fast food A–76 = Convenience A–78 = Baby foods

Chol (mg)	Calc (mg)	Iron (mg)	Magn (mg)	Pota (mg)	Sodi (mg)	Zinc (mg)	Vit A (RAE) (µg)	Thia (mg)	Vit E (mg α)	Ribo (mg)	Niac (mg)	Vit B_6 (mg)	Fola (µg)	Vit C (mg)	Vit B_{12} (µg)	Sele (µg)
0	300	1.80	60	350	160	0.60	33	0.15	—	0.07	0.80	0.12	60	0	3	—
0	300	1.80	40	260	140	0.60	33	0.15	—	0.07	0.80	0.12	60	0	3	—
0	300	0.72	24	200	140	0.90	0	0.09	—	0.34	—	—	24	0	1	—
0	300	0.72	24	200	110	0.90	0	0.09	—	0.34	—	—	24	0	1	—
0	300	1.08	40	320	150	0.90	0	0.15	—	0.34	—	—	60	0	1	—
0	40	0.72	—	320	115	—	0	—	—	—	—	—	—	0	—	—
15	200	0.00	16	310	90	—	0	—	—	0.26	—	—	—	0	—	—
15	200	0.00	16	300	90	—	0	—	—	0.26	—	—	—	0	—	—
12	338	0.15	32	434	130	1.64	27	0.08	0.05	0.40	0.21	0.09	22	1	1	7
3	370	0.62	41	550	139	1.83	0.10	0.17	0.17	0.50	0.11	0.09	26	1		
15	448	0.20	42	573	172	2.18	34	0.11	0.05	0.52	0.28	0.12	27	2	1	8
5	488	0.22	47	625	189	2.38	5	0.12	0.07	0.57	0.30	0.13	29	2	1	9
12	419	0.17	39	537	162	2.03	29	0.10	0.00	0.49	0.26	0.11	27	2	1	12
13	260	0.22	39	399	98	1.10	—	0.11	0.05	0.51	0.30	0.15	29	2	2	—
5	300	2.70	100	580	290	2.25	—	0.38	—	0.43	5.00	0.50	100	15	2	—
0	700	0.90	—		30									0	0	
0	100	0.72	—	—	20	—	0	—	—	—	—	—	—	0	—	—
0	100	0.72	—	—	20	—	0	—	—	—	—	—	—	0	—	—
212	27	0.92	6	67	70	0.56	70	0.03	0.49	0.24	0.04	0.07	24	0	1	16
0	2	0.03	4	54	55	0.01	0	0.00	0.00	0.15	0.04	0.00	1	0	<.1	7
205	21	0.45	1	18	8	0.38	63	0.03	0.43	0.09	0.00	0.06	24	0	<1	9
210	27	0.91	6	68	94	0.55	91	0.03	0.56	0.24	0.04	0.07	23	0	1	16
212	25	0.60	5	63	62	0.53	85	0.03	0.51	0.26	0.03	0.06	22	0	1	15
211	27	0.92	6	67	147	0.55	70	0.03	0.48	0.24	0.04	0.07	24	0	1	16
429	87	1.46	15	168	342	1.22	174	0.06	1.04	0.53	0.10	0.14	37	<1	1	27
1	44	1.19	9	128	119	0.59	7	0.07	0.95	0.23	0.08	0.08	10	<1	<1	25
1	33	1.32	6	207	111	0.82	11	0.07	0.17	0.19	0.07	0.00	9	0	<1	16
0	20	1.08	4	85	115	0.60	113	0.15	—	0.85	0.20	0.08	60	0	1	—
24	6	0.22	19	108	35	0.26	—	0.04	—	0.03	1.11	0.13	5	<1	<1	17
47	12	0.42	36	207	66	0.49	12	0.07	0.69	0.07	2.14	0.24	7	1	1	32
107	17	0.43	17	231	43	1.38	887	0.13	3.40	0.03	2.98	0.06	13	2	3	6
44	8	0.32	29	489	86	0.49	10	0.03	—	0.05	2.48	0.46	8	3	1	44
28.89	15.3	1.79	20.39	272	452.2	0.37	10.19	0.09	—	0.09	1.78	0.08	17	0	0.94	7.73
37	55	0.99	98	529	64	0.49	55	0.06	—	0.08	6.88	0.36	13	0	1	43
33	48	0.86	85	460	55	0.43	48	0.06	—	0.08	5.97	0.33	12	0	1	37
42	13	0.30	25	323	42	0.35	12	0.07	—	0.06	1.92	0.22	6	1	1	32
44	19	0.35	47	225	281	0.21	39	0.06	0.41	0.08	2.03	0.19	7	3	2	34
40	18	0.97	31	404	45	0.43	43	0.07	—	0.01	0.32	0.30	9	0	1	40
33	19	0.60	22	177	39	0.21	—	0.02	—	0.02	2.05	0.15	4	0	1	18
63	36	1.15	43	339	74	0.41	16	0.03	—	0.04	3.94	0.29	11	0	1	34
35	51	0.91	91	490	59	0.45	46	0.06	—	0.08	6.05	0.34	12	0	1	40
8	46	0.73	5	41	522	0.32	155	0.02	1.03	0.08	1.98	0.10	1	0	3	35
45	137	1.16	21	110	215	0.58	74	0.02	0.58	0.12	3.50	0.12	3	1	4	21
82	90	8.11	51	536	391	2.86	77	0.05	1.02	0.06	3.21	0.55	20	7	31	76
98	87	0.99	32	292	67	1.22	9	0.07	—	0.10	1.62	0.12	5	1	2	14
37	10	0.45	29	442	65	0.45	60	0.04	1.33	0.07	3.33	0.23	9	0	1	40
13	6	0.48	10	99	1134	0.17	15	0.01	—	0.05	2.67	0.15	1	0	2	22

TABLE A–1
Food Composition

(Computer code number is for Wadsworth Diet Analysis program)   (For purposes of calculations, use "0" for t, <1, <.1, <.01, etc.)

DA + Code	Food Description	Quantity	Measure	Wt (g)	H₂O (g)	Ener (kcal)	Prot (g)	Carb (g)	Dietary Fiber (g)	Fat (g)	Sat	Mono	Poly	Trans
	SEAFOOD—Continued													
1594	Broiled or baked w/butter	3	ounce(s)	85	54	155	23	0	0	6	1.16	2.29	2.33	—
2938	Coho, farmed, raw	3	ounce(s)	85	60	136	18	0	0	7	1.54	2.83	1.58	—
154	**Sardines, Atlantic, with bones, canned in oil**	2	item(s)	24	<.1	50	6	0	0	3	0.36	0.92	1.23	—
	Scallops													
155	Mixed species, breaded, fried	3	item(s)	47	<.1	100	8	5	0	5	1.24	2.09	1.32	—
1599	Steamed	3	ounce(s)	85	65	90	14	2	0	3	—	—	—	—
1839	**Snapper, mixed species, cooked, dry heat**	3	ounce(s)	85	60	109	22	0	0	1	0.31	0.27	0.50	—
	Squid													
1868	Mixed species, fried	3	ounce(s)	85	55	149	15	7	0	6	1.60	2.34	1.82	—
16617	Steamed or boiled	3	ounce(s)	85	63	90	15	3	0	1	0.35	0.11	0.51	—
1570	**Striped bass, cooked, dry heat**	3	ounce(s)	85	62	105	19	0	0	3	0.55	0.72	0.85	—
1601	**Sturgeon, steamed**	3	ounce(s)	85	59	111	17	0	0	4	0.97	2.04	0.73	—
1840	**Surimi, formed**	3	ounce(s)	85	65	84	13	6	0	1	0.16	0.13	0.38	—
1842	**Swordfish, cooked, dry heat**	3	ounce(s)	85	58	132	22	0	0	4	1.20	1.68	1.00	—
1846	**Tuna, yellowfin or ahi, raw**	3	ounce(s)	85	60	92	20	0	0	1	0.20	0.13	0.24	—
	Tuna, canned													
159	Light, canned in oil, drained	2	ounce(s)	57	34	113	17	0	0	5	0.87	1.68	1.64	—
355	Light, canned in water, drained	2	ounce(s)	57	42	66	14	0	0	<1	0.13	0.09	0.19	—
33211	Light, no salt, canned in oil, drained	2	ounce(s)	57	34	112	17	0	0	5	0.87	1.67	1.64	—
33212	Light, no salt, canned in water, drained	2	ounce(s)	57	43	66	14	0	0	<1	0.13	0.09	0.19	—
2961	White, canned in oil, drained	2	ounce(s)	57	36	105	15	0	0	5	0.73	1.85	1.69	—
351	White, canned in water, drained	2	ounce(s)	57	41	73	13	0	0	2	0.45	0.44	0.63	—
33213	White, no salt, canned in oil, drained	2	ounce(s)	57	36	105	15	0	0	5	0.94	1.41	1.92	—
33214	White, no salt, canned in water, drained	2	ounce(s)	57	42	73	13	0	0	2	0.45	0.44	0.63	—
	Yellowtail													
2970	Mixed species, raw	2	ounce(s)	57	42	83	13	0	0	3	0.73	1.13	0.81	—
8548	Mixed species, cooked, dry heat	3	ounce(s)	85	0.05	158.94	25.21	0	0	5.71	1.44	2.21	1.52	—
	Shellfish, meat only													
1857	Abalone, mixed species, fried	3	ounce(s)	85	51	161	17	9	0	6	1.40	2.33	1.42	—
16618	Abalone, steamed or poached	3	ounce(s)	85	41	177	29	10	0	1	0.25	0.18	0.18	—
	Crab													
1851	Blue crab, canned	2	ounce(s)	57	43	56	12	0	0	1	0.14	0.12	0.25	—
1852	Blue crab, cooked, moist heat	3	ounce(s)	85	66	87	17	0	0	2	0.19	0.24	0.58	—
8562	Dungeness crab, cooked, moist heat	3	ounce(s)	85	62	94	19	1	0	1	0.14	0.18	0.35	—
1860	**Clams, cooked, moist heat**	3	ounce(s)	85	54	126	22	4	0	2	0.16	0.15	0.47	—
1853	**Crayfish, farmed, cooked, moist heat**	3	ounce(s)	85	69	74	15	0	0	1	0.18	0.21	0.35	—
	Oysters													
8720	Baked or broiled	3	ounce(s)	85	69	90	6	3	0	6	1.38	2.18	1.88	—
152	Eastern, farmed, raw	3	ounce(s)	85	73	50	4	5	0	1	0.38	0.13	0.50	—
8715	Eastern, wild, cooked, moist heat	3	ounce(s)	85	60	116	12	7	0	4	1.31	0.53	1.65	—
8584	Pacific, cooked, moist heat	3	ounce(s)	85	55	139	16	8	0	4	0.87	0.66	1.52	—
1865	Pacific, raw	3	ounce(s)	85	70	69	8	4	0	2	0.43	0.30	0.76	—
1854	**Lobster, northern, cooked, moist heat**	3	ounce(s)	85	65	83	17	1	0	1	0.09	0.14	0.08	—
1862	**Mussels, blue, cooked, moist heat**	3	ounce(s)	85	52	146	20	6	0	4	0.72	0.86	1.03	—
	Shrimp													
1855	Mixed species, cooked, moist heat	3	ounce(s)	85	66	84	18	0	0	1	0.25	0.17	0.37	—
158	Mixed species, breaded, fried	3	ounce(s)	85	0.04	205.69	18.18	9.74	0.34	10.43	1.77	3.24	4.32	—
	BEEF, LAMB, PORK													
	Beef													
4450	Breakfast strips, cooked	2	slice(s)	23	0	101.47	7.07	0.31	0	7.77	3.24	3.8	0.35	—
174	Corned, canned	3	ounce(s)	85	49	213	23	0	0	13	5.25	5.07	0.54	—
33147	Cured, thin sliced	2	ounce(s)	57	31	87	18	2	0	1	0.54	0.48	0.04	—
4581	Jerky	1	ounce(s)	28	0	116.44	9.42	3.12	0.51	7.27	3.08	3.21	0.28	—
	Ground													
4411	Extra lean, broiled, well	3	ounce(s)	85	46	225	24	0	0	13	5.28	5.88	0.50	—
4417	Lean, broiled, medium	3	ounce(s)	85	47	231	21	0	0	16	6.16	6.87	0.59	—
4418	Lean, broiled, well	3	ounce(s)	85	45	238	24	0	0	15	5.89	6.56	0.56	—
4423	Regular, broiled, medium	3	ounce(s)	85	46	246	20	0	0	18	6.91	7.70	0.65	—
	Rib													
4183	Rib, whole, lean & fat, ¼" fat, roasted	3	ounce(s)	85	39	320	19	0	0	27	10.71	11.42	0.94	—
	Roast													
4264	Bottom round, lean & fat, ¼" fat, braised	3	ounce(s)	85	44	241	24	0	0	15	5.71	6.63	0.58	—

PAGE KEY: A–4 = Breads/Baked Goods   A–8 = Cereal/Rice/Pasta   A–12 = Fruit   A–16 = Vegetables/Legumes   A–26 = Nuts/Seeds   A–28 = Vegetarian
A–30 = Dairy   A–36 = Eggs   A–36 = Seafood   A–38 = Meats   A–42 = Poultry   A–42 = Processed meats   A–44 = Beverages   A–48 = Fats/Oils
A–50 = Sweets   A–52 = Spices/Condiments/Sauces   A–54 = Mixed Foods/Soups/Sandwiches   A–60 = Fast food   A–76 = Convenience   A–78 = Baby foods

Chol (mg)	Calc (mg)	Iron (mg)	Magn (mg)	Pota (mg)	Sodi (mg)	Zinc (mg)	Vit A (RAE) (µg)	Thia (mg)	Vit E (mg α)	Ribo (mg)	Niac (mg)	Vit B_6 (mg)	Fola (µg)	Vit C (mg)	Vit B_{12} (µg)	Sele (µg)
40	15	1.02	27	377	99	0.56	—	0.14	1.15	0.05	8.33	0.19	4	2	2	41
43	10	0.29	26	383	40	0.37	48	0.08	—	0.09	5.79	0.56	11	1	2	11
34	108	0.70	9	95	121	0.31	16	0.01	0.49	0.05	1.25	0.04	3	0	2	13
28	20	0.38	27	155	216	0.49	10	0.01	—	0.05	0.69	0.06	23	1	1	13
27	21	0.22	—	238	366	—	—	—	0.16	—	—	—	—	2	—	—
40	34	0.20	31	444	48	0.37	30	0.05	—	0.00	0.29	0.39	5	1	3	42
221	33	0.86	32	237	260	1.48	9	0.05	—	0.39	2.21	0.05	12	4	1	44
227	31	0.63	29	192	356	1.49	—	0.02	1.17	0.32	1.70	0.04	4	3	1	—
88	16	0.92	43	279	75	0.43	26	0.10	—	0.03	2.17	0.29	9	0	4	40
63	11	0.59	30	239	389	0.36	—	0.07	0.53	0.07	8.31	0.19	14	0	2	—
26	8	0.22	37	95	122	0.28	17	0.02	0.54	0.02	0.19	0.03	2	0	1	24
43	5	0.88	29	314	98	1.25	35	0.04	—	0.10	10.02	0.32	2	1	2	52
38	14	0.62	43	378	31	0.44	15	0.37	0.43	0.04	8.33	0.77	2	1	<1	31
10	7	0.79	18	118	202	0.51	13	0.02	0.50	0.07	7.06	0.06	3	0	1	43
17	6	0.87	15	134	192	0.44	10	0.02	0.19	0.04	7.53	0.20	2	0	2	46
10	7	0.79	18	117	28	0.51	13	0.02	—	0.07	7.03	0.06	3	0	1	43
17	6	0.87	15	134	28	0.44	10	0.02	—	0.04	7.53	0.20	2	0	2	46
18	2	0.37	19	189	225	0.27	3	0.01	1.30	0.04	6.63	0.24	3	0	1	34
24	8	0.55	19	134	214	0.27	3	0.00	0.48	0.02	3.29	0.12	1	0	1	37
18	2	0.37	19	189	28	0.27	14	0.01	—	0.04	6.63	0.24	3	0	1	34
24	8	0.55	19	134	28	0.27	3	0.00	—	0.02	3.29	0.12	1	0	1	37
31	13	0.28	17	238	22	0.29	16	0.08	—	0.02	3.86	0.09	2	2	1	21
60.34	24.64	0.53	32.29	457.29	42.5	0.56	26.35	0.14	—	0.04	7.41	0.15	3.4	2.46	1.06	39.77
80	31	3.23	48	241	502	0.81	2	0.19	—	0.11	1.62	0.13	12	2	1	44
143	50	4.85	69	295	980	1.38	—	0.29	6.74	0.13	1.90	0.22	6	3	1	—
50	57	0.48	22	212	189	2.28	1	0.05	1.04	0.05	0.78	0.09	24	2	<1	18
85	88	0.77	28	275	237	3.59	2	0.09	1.56	0.04	2.81	0.15	43	3	6	34
65	50	0.37	49	347	321	4.65	26	0.05	—	0.17	3.08	0.15	36	3	9	40
57	78	23.77	15	534	95	2.32	145	0.13	—	0.36	2.85	0.09	25	19	84	54
116	43	0.94	28	202	82	1.26	13	0.04	—	0.07	1.42	0.11	9	<1	3	29
42	37	5.30	38	126	418	72.22	60	0.07	0.99	0.06	1.04	0.05	8	3	15	—
21	37	4.91	28	105	151	32.23	7	0.09	—	0.06	1.08	0.05	15	4	14	54
89	77	10.19	81	239	359	154.37	46	0.16	—	0.15	2.11	0.10	12	5	30	61
85	14	7.82	37	257	180	28.25	124	0.11	0.72	0.38	3.08	0.08	13	11	24	131
43	7	4.35	19	143	90	14.14	69	0.06	—	0.20	1.71	0.04	9	7	14	65
61	52	0.33	30	299	323	2.48	22	0.01	0.85	0.06	0.91	0.07	9	0	3	36
48	28	5.71	31	228	314	2.27	77	0.26	—	0.36	2.55	0.09	65	12	20	76
166	33	2.63	29	155	190	1.33	58	0.03	1.17	0.03	2.20	0.11	3	2	1	34
150.44	56.95	1.07	34	191.25	292.39	1.17	47.59	0.1	—	0.11	2.6	0.08	20.39	1.27	1.58	35.44
26.89	2.03	0.7	6.1	93.11	509.17	1.43	0	0.02	0.07	0.05	1.46	0.07	1.8	0	0.77	6.05
73	10	1.77	12	116	855	3.03	0	0.02	0.13	0.12	2.07	0.11	8	0	1	36
45	3	1.58	11	140	1582	2.49	0	0.03	0.00	0.12	1.85	0.16	5	0	1	13
13.63	5.67	1.53	14.48	169.54	628.49	2.3	0	0.04	0.14	0.04	0.49	0.05	38.05	0	0.28	3.03
84	8	2.35	21	314	70	5.47	0	0.06	—	0.27	4.97	0.27	9	0	2	19
74	9	1.79	18	256	65	4.56	0	0.04	—	0.18	4.39	0.22	8	0	2	25
86	10	2.08	20	297	76	5.27	0	0.05	—	0.20	5.07	0.26	9	0	2	22
77	9	2.07	17	248	71	4.40	0	0.03	—	0.16	4.90	0.23	8	0	2	16
72	9	1.96	16	252	54	4.45	0	0.06	—	0.14	2.86	0.20	6	0	2	19
82	5	2.65	19	240	43	4.17	0	0.06	0.17	0.20	3.17	0.28	9	0	2	27

TABLE A–1
Food Composition

Computer code number is for Wadsworth Diet Analysis program   (For purposes of calculations, use "0" for t, <1, <.1, <.01, etc.)

DA + Code	Food Description	Quantity	Measure	Wt (g)	H₂O (g)	Ener (kcal)	Prot (g)	Carb (g)	Dietary Fiber (g)	Fat (g)	Fat Breakdown (g)			
											Sat	Mono	Poly	Trans
	BEEF, LAMB, PORK—Continued													
169	Bottom round, separable lean, ¼" fat, roasted	3	ounce(s)	85	0.05	160.64	24.45	0	0	6.26	2.13	2.83	0.24	—
4147	Chuck, arm pot roast, lean & fat, ¼" fat, braised	3	ounce(s)	85	41	282	23	0	0	20	7.97	8.68	0.77	—
4161	Chuck, blade roast, lean & fat, ¼" fat, braised	3	ounce(s)	85	40	293	23	0	0	22	8.70	9.44	0.78	—
5853	Chuck, blade roast, separable lean, ¼" trim, pot roasted	3	ounce(s)	85	0.04	209.1	27.45	0	0	10.15	3.94	4.37	0.33	—
4295	Eye of round, lean, ¼" fat, roasted	3	ounce(s)	85	55	149	25	0	0	5	1.76	2.06	0.15	—
4285	Eye of round, lean & fat, ¼" fat, roasted	3	ounce(s)	85	51	195	23	0	0	11	4.23	4.66	0.39	—
	Steak													
1757	Rib, small end, lean, ¼" fat, broiled	3	ounce(s)	85	49	188	24	0	0	10	3.84	4.01	0.27	—
4349	Short loin, T-bone steak, lean, ¼" fat, broiled	3	ounce(s)	85	52	174	23	0	0	9	3.05	4.23	0.26	—
4348	Short loin, T-bone steak, lean & fat, ¼" fat, broiled	3	ounce(s)	85	43	274	19	0	0	21	8.29	9.58	0.75	—
4360	Top loin, prime, lean & fat, ¼" fat, broiled	3	ounce(s)	85	43	275	22	0	0	20	8.16	8.61	0.73	—
	Variety													
188	Liver, pan fried	3	ounce(s)	85	53	149	23	4	0	4	1.27	0.56	0.49	0.17
4447	Tongue, simmered	3	ounce(s)	85	49	236	16	0	0	19	6.91	8.59	0.56	0.71
	Lamb													
	Chop													
3275	Loin, domestic, lean & fat, ¼" fat, broiled	3	ounce(s)	85	44	269	21	0	0	20	8.36	8.25	1.43	—
3287	Shoulder, arm, domestic, lean & fat, ¼" fat, braised	3	ounce(s)	85	38	294	26	0	0	20	8.39	8.65	1.45	—
3290	Shoulder, arm, domestic, lean, ¼" fat, braised	3	ounce(s)	85	42	237	30	0	0	12	4.28	5.24	0.78	—
	Leg													
3264	Domestic, lean & fat, ¼" fat, cooked	3	ounce(s)	85	46	250	21	0	0	18	7.51	7.50	1.28	—
	Rib													
183	Domestic, lean, ¼" fat, broiled	3	ounce(s)	85	50	200	24	0	0	11	3.95	4.43	1.00	—
182	Domestic, lean & fat, ¼" fat, broiled	3	ounce(s)	85	40	307	19	0	0	25	10.80	10.30	2.01	—
	Shoulder													
187	Arm & blade, domestic, choice, lean, ¼" fat, roasted	3	ounce(s)	85	54	173	21	0	0	9	3.47	3.71	0.81	—
186	Arm & blade, domestic, choice, lean & fat, ¼" fat, roasted	3	ounce(s)	85	48	235	19	0	0	17	7.17	6.94	1.38	—
	Variety													
3375	Brain, pan fried	3	ounce(s)	85	52	232	14	0	0	19	4.82	3.42	1.94	—
3406	Tongue, braised	3	ounce(s)	85	49	234	18	0	0	17	6.66	8.50	1.06	—
	Pork													
	Cured													
161	Bacon, cured, broiled, pan fried or roasted	2	slice(s)	13	2	68	5	<1	0	5	1.73	2.33	0.57	0
29229	Bacon, Canadian style, cured	2	ounce(s)	57	38	89	12	1	0	4	1.26	1.79	0.36	—
35422	Breakfast strips, cured, cooked	3	slice(s)	34	9	156	10	<1	0	12	4.34	5.58	1.92	—
16561	Ham, smoked or cured, lean, cooked	1	slice(s)	42	28	66	11	0	0	2	0.77	1.06	0.27	—
189	Ham, cured, boneless, 11% fat, roasted	3	ounce(s)	85	55	151	19	0	0	8	2.65	3.77	1.20	—
1316	Ham, cured, extra lean, 5% fat, roasted	3	ounce(s)	85	58	123	18	1	0	5	1.54	2.23	0.46	—
29215	Ham, cured, extra lean, 4% fat, canned	2	ounce(s)	57	42	68	10	0	0	3	0.86	1.25	0.22	—
	Chop													
32671	Loin, blade, lean & fat, pan fried	3	ounce(s)	85	42	291	18	0	0	24	8.65	9.97	2.64	—
32672	Loin, center cut, lean & fat, pan fried	3	ounce(s)	85	45	236	25	0	0	14	5.11	6.00	1.62	—
32682	Loin, center rib, boneless, lean & fat, braised	3	ounce(s)	85	49	217	22	0	0	13	5.21	6.13	1.12	—
32603	Loin, center rib, lean, broiled	3	ounce(s)	85	48	186	26	0	0	8	2.94	3.78	0.53	—
32481	Loin, whole, lean, braised	3	ounce(s)	85	52	174	24	0	0	8	2.87	3.54	0.60	—
32478	Loin, whole, lean & fat, braised	3	ounce(s)	85	50	203	23	0	0	12	4.35	5.15	1.00	—
	Leg or ham													
32471	Rump portion, lean & fat, roasted	3	ounce(s)	85	48	214	25	0	0	12	4.47	5.42	1.17	—
32468	Whole, lean & fat, roasted	3	ounce(s)	85	47	232	23	0	0	15	5.50	6.70	1.43	—
	Ribs													
32696	Loin, country style, lean, roasted	3	ounce(s)	85	49	210	23	0	0	13	4.52	5.49	0.94	—
32693	Loin, country style, lean & fat, roasted	3	ounce(s)	85	43	279	20	0	0	22	7.83	9.36	1.71	—
	Shoulder													
32629	Arm picnic, lean, roasted	3	ounce(s)	85	51	194	23	0	0	11	3.66	5.09	1.02	—

PAGE KEY: A–4 = Breads/Baked Goods A–8 = Cereal/Rice/Pasta A–12 = Fruit A–16 = Vegetables/Legumes A–26 = Nuts/Seeds A–28 = Vegetarian A–30 = Dairy A–36 = Eggs A–36 = Seafood A–38 = Meats A–42 = Poultry A–42 = Processed meats A–44 = Beverages A–48 = Fats/Oils A–50 = Sweets A–52 = Spices/Condiments/Sauces A–54 = Mixed Foods/Soups/Sandwiches A–60 = Fast food A–76 = Convenience A–78 = Baby foods

Chol (mg)	Calc (mg)	Iron (mg)	Magn (mg)	Pota (mg)	Sodi (mg)	Zinc (mg)	Vit A (RAE) (µg)	Thia (mg)	Vit E (mg α)	Ribo (mg)	Niac (mg)	Vit B$_6$ (mg)	Fola (µg)	Vit C (mg)	Vit B$_{12}$ (µg)	Sele (µg)
66.3	4.25	2.66	23.79	332.35	56.09	3.92	0	0.06	—	0.2	3.45	0.31	10.19	0	2.29	23.29
84	9	2.64	16	209	51	5.81	0	0.06	0.19	0.20	2.70	0.24	8	0	3	21
88	11	2.64	16	196	54	7.07	0	0.06	0.15	0.20	2.06	0.22	4	0	2	21
73.94	11.05	3.12	19.54	223.55	60.34	8.72	0	0.06	—	0.23	0	0.24	—	0	2.09	22.69
59	4	1.66	23	336	53	4.03	0	0.08	—	0.14	3.19	0.32	6	0	2	23
61	5	1.56	20	308	50	3.69	0	0.07	0.15	0.14	2.97	0.30	6	0	2	22
68	11	2.18	23	335	59	5.94	0	0.09	0.12	0.19	4.08	0.34	7	0	3	19
50	5	3.11	22	278	65	4.34	0	0.09	0.12	0.21	3.94	0.33	7	0	2	9
58	7	2.56	18	234	58	3.56	0	0.08	0.19	0.18	3.29	0.28	6	0	2	10
67	8	1.89	20	294	54	3.85	0	0.07	—	0.15	3.96	0.31	6	0	2	19
324	5	5.24	19	298	65	4.45	6582	0.15	0.39	2.91	14.85	0.87	221	1	71	28
112	4	2.22	13	156	55	34.77	0	0.02	0.25	0.25	2.97	0.13	6	1	3	11
85	17	1.54	20	278	65	2.96	0	0.09	0.11	0.21	6.04	0.11	15	0	2	23
102	21	2.03	22	260	61	5.17	0	0.06	0.13	0.21	5.66	0.09	15	0	2	32
103	22	2.30	25	287	65	6.21	0	0.06	0.15	0.23	5.38	0.11	19	0	2	32
82	14	1.60	20	264	61	3.79	0	0.09	0.12	0.21	5.66	0.11	15	0	2	22
77	14	1.88	25	266	72	4.48	0	0.09	0.15	0.21	5.57	0.13	18	0	2	26
84	16	1.60	20	230	65	3.40	0	0.08	0.10	0.19	5.95	0.09	12	0	2	20
74	16	1.81	21	225	58	5.13	0	0.08	0.15	0.22	4.90	0.13	21	0	2	24
78	17	1.67	20	213	56	4.45	0	0.08	0.12	0.20	5.23	0.11	18	0	2	22
2128	18	1.73	19	304	133	1.70	0	0.14	—	0.31	3.87	0.20	6	20	20	10
161	9	2.24	14	134	57	2.54	0	0.07	—	0.36	3.14	0.14	3	6	5	24
14	1	0.18	4	71	291	0.44	1	0.05	0.04	0.03	1.40	0.04	<1	0	<1	8
28	5	0.39	10	195	799	0.79	0	0.43	0.12	0.10	3.53	0.22	2	0	<1	14
36	5	0.67	9	158	714	1.25	0	0.25	0.09	0.13	2.58	0.12	1	0	1	8
23	3	0.40	9	133	557	1.08	0	0.29	0.11	0.11	2.11	0.20	2	0	<1	—
50	7	1.14	19	348	1275	2.10	0	0.62	0.26	0.28	5.23	0.26	3	0	1	17
45	7	1.26	12	244	1023	2.45	0	0.64	0.21	0.17	3.42	0.34	3	0	1	17
22	3	0.53	10	206	712	1.09	0	0.47	0.10	0.13	3.01	0.26	3	0	<1	8
72	26	0.75	18	282	57	2.71	3	0.53	0.17	0.25	3.36	0.29	3	1	1	30
78	23	0.77	25	361	68	1.96	2	0.97	0.21	0.26	4.76	0.40	5	1	1	33
62	4	0.78	14	329	34	1.76	2	0.45	0.21	0.21	3.67	0.26	3	<1	<1	28
69	26	0.70	24	357	55	2.02	2	0.95	0.25	0.28	5.25	0.40	3	<1	1	40
67	15	0.96	17	329	43	2.11	2	0.56	0.18	0.23	3.90	0.33	3	1	<1	41
68	18	0.91	16	318	41	2.02	2	0.54	0.20	0.22	3.76	0.31	3	1	<1	39
82	10	0.89	23	318	53	2.40	3	0.64	0.19	0.28	3.96	0.27	3	<1	1	40
80	12	0.86	19	299	51	2.52	3	0.54	0.19	0.27	3.89	0.34	9	<1	1	39
79	25	1.10	20	297	25	3.24	2	0.49	—	0.29	3.97	0.37	4	<1	1	36
78	21	0.90	20	293	44	2.01	3	0.76	—	0.29	3.67	0.38	4	<1	1	32
81	8	1.21	17	299	68	3.46	2	0.49	—	0.30	3.67	0.35	4	<1	1	33

TABLE A–1
Food Composition

(Computer code number is for Wadsworth Diet Analysis program)   (For purposes of calculations, use "0" for t, <1, <.1, <.01, etc.)

DA + Code	Food Description	Quantity	Measure	Wt (g)	H₂O (g)	Ener (kcal)	Prot (g)	Carb (g)	Dietary Fiber (g)	Fat (g)	Sat	Mono	Poly	Trans
	BEEF, LAMB, PORK—Continued													
32626	Arm picnic, lean & fat, roasted	3	ounce(s)	85	44	270	20	0	0	20	7.47	9.12	2.00	—
	Rabbit													
3366	Domesticated, roasted	3	ounce(s)	85	52	167	25	0	0	7	2.04	1.84	1.33	—
3367	Domesticated, stewed	3	ounce(s)	85	50	175	26	0	0	7	2.13	1.93	1.39	—
	Veal													
3391	Liver, braised	3	ounce(s)	85	51	163	24	3	0	5	1.69	0.97	0.88	0.26
3319	Rib, lean only, roasted	3	ounce(s)	85	55	150	22	0	0	6	1.77	2.26	0.57	—
1732	**Deer or venison, roasted**	3	ounce(s)	85	55	134	26	0	0	3	1.06	0.75	0.53	—
	POULTRY													
	Chicken													
29562	Flaked, canned	2	ounce(s)	57	0.03	97.47	10.37	0.05	0	5.87	1.62	2.32	1.29	—
	Fried													
29632	Breast, meat only, breaded, baked or fried	3	ounce(s)	85	44	193	25	7	<1	7	1.62	2.66	1.73	—
35327	Broiler breast, meat only, fried	3	ounce(s)	85	51	159	28	<1	0	4	1.10	1.46	0.91	—
36413	Broiler breast, meat & skin, flour coated, fried	3	ounce(s)	85	48	189	27	1	<.1	8	2.08	2.98	1.67	—
35389	Broiler drumstick, meat only, fried	3	ounce(s)	85	53	166	24	0	0	7	1.81	2.50	1.68	—
36414	Broiler drumstick, meat & skin, flour coated, fried	3	ounce(s)	85	48	208	23	1	<.1	12	3.11	4.61	2.75	—
35406	Broiler leg, meat only, fried	3	ounce(s)	85	52	177	24	1	0	8	2.12	2.92	1.89	—
35484	Broiler wing, meat only, fried	3	ounce(s)	85	51	179	26	0	0	8	2.13	2.62	1.76	—
29580	Patty, fillet, or tenders, breaded, cooked	3	ounce(s)	85	42	241	14	13	<1	15	4.62	7.25	1.87	—
	Roasted, meat only													
35409	Broiler chicken leg	3	ounce(s)	85	55	162	23	0	0	7	1.95	2.59	1.68	—
35486	Broiler chicken wing	3	ounce(s)	85	53	173	26	0	0	7	1.92	2.22	1.51	—
35138	Roasting chicken, dark meat	3	ounce(s)	85	57	151	20	0	0	7	2.07	2.82	1.70	—
35136	Roasting chicken, light meat	3	ounce(s)	85	58	130	23	0	0	3	0.92	1.29	0.79	—
35132	Roasting chicken	3	ounce(s)	85	57	142	21	0	0	6	1.54	2.13	1.28	—
	Stewed													
3174	Meat only, stewed	3	ounce(s)	85	0.05	150.44	23.19	0	0	5.7	1.56	2.03	1.3	—
1268	Gizzard, simmered	3	ounce(s)	85	58	124	26	0	0	2	0.57	0.45	0.30	0.11
1270	Liver, simmered	3	ounce(s)	85	57	142	21	1	0	6	1.75	1.20	1.08	0.08
	Duck													
1286	Domesticated, meat & skin, roasted	3	ounce(s)	85	44	286	16	0	0	24	8.22	10.97	3.10	—
1287	Domesticated, meat only, roasted	3	ounce(s)	85	55	171	20	0	0	10	3.54	3.15	1.22	—
	Goose													
35507	Domesticated, meat & skin, roasted	3	ounce(s)	85	44	259	21	0	0	19	5.84	8.72	2.14	—
35524	Domesticated, meat only, roasted	3	ounce(s)	85	49	202	25	0	0	11	3.88	3.69	1.31	—
1297	Liver pâté, smoked, canned	4	tablespoon(s)	52	19	240	6	2	0	23	7.51	13.32	0.44	—
	Turkey													
3256	Ground turkey, cooked	3	ounce(s)	85	51	200	23	0	0	11	2.88	4.16	2.75	—
222	Roasted, fryer roaster breast, meat only	3	ounce(s)	85	58	115	26	0	0	1	0.20	0.11	0.17	—
219	Roasted, dark meat, meat only	3	ounce(s)	85	54	159	24	0	0	6	2.06	1.39	1.84	—
220	Roasted, light meat, meat only	3	ounce(s)	85	56	133	25	0	0	3	0.88	0.48	0.73	—
3263	Patty, batter coated, breaded, fried	1	item(s)	94	47	266	13	15	<1	17	4.41	7.02	4.43	—
1302	Turkey roll, light meat	2	slice(s)	57	41	83	11	<1	0	4	1.15	1.42	0.99	—
1303	Turkey roll, light & dark meat	2	slice(s)	57	40	84	10	1	0	4	1.16	1.30	1.01	—
	PROCESSED MEATS													
	Beef													
1331	Corned beef loaf, jellied, sliced	2	slice(s)	57	39	87	13	0	0	3	1.47	1.52	0.18	—
	Bologna													
13458	Made w/chicken, pork, & beef	1	slice(s)	28	15	90	3	1	0	8	3.00	4.05	1.10	—
13461	Light, made w/pork, chicken, & beef	1	slice(s)	28	18	60	3	2	0	4	1.50	2.04	0.43	—
13459	Beef	1	slice(s)	28	15	90	3	1	0	8	3.50	4.26	0.31	—
13565	Turkey	1	slice(s)	28	19	50	3	1	0	4	1.00	1.09	0.98	—
	Chicken													
13562	Oven roasted white chicken	1	slice(s)	28	20	40	4	1	0	3	0.50	—	—	—
	Ham													
13581	Honey glazed, traditional carved	2	slice(s)	45	—	50	8	1	0	2	0.50	0.68	0.18	—
13777	Deli sliced cooked	1	slice(s)	28	—	30	5	1	0	1	0.50	0.39	0.11	—
13778	Deli sliced honey	1	slice(s)	28	—	35	5	1	0	1	0.50	0.39	0.11	—
8614	**Pork & beef mortadella, sliced**	2	slice(s)	46	24	143	8	1	0	12	4.37	5.23	1.44	—

PAGE KEY: A–4 = Breads/Baked Goods A–8 = Cereal/Rice/Pasta A–12 = Fruit A–16 = Vegetables/Legumes A–26 = Nuts/Seeds A–28 = Vegetarian A–30 = Dairy A–36 = Eggs A–36 = Seafood A–38 = Meats A–42 = Poultry A–42 = Processed meats A–44 = Beverages A–48 = Fats/Oils A–50 = Sweets A–52 = Spices/Condiments/Sauces A–54 = Mixed Foods/Soups/Sandwiches A–60 = Fast food A–76 = Convenience A–78 = Baby foods

Chol (mg)	Calc (mg)	Iron (mg)	Magn (mg)	Pota (mg)	Sodi (mg)	Zinc (mg)	Vit A (RAE) (µg)	Thia (mg)	Vit E (mg α)	Ribo (mg)	Niac (mg)	Vit B_6 (mg)	Fola (µg)	Vit C (mg)	Vit B_{12} (µg)	Sele (µg)
80	16	1.00	14	276	60	2.93	2	0.44	—	0.26	3.33	0.30	3	<1	1	29
70	16	1.93	18	326	40	1.93	0	0.08	—	0.18	7.17	0.40	9	0	7	33
73	17	2.01	17	255	31	2.01	0	0.05	0.37	0.14	6.09	0.29	8	0	6	33
434	5	4.34	17	280	66	9.55	17973	0.15	0.58	2.43	11.18	0.78	281	1	72	16
98	10	0.82	20	264	82	3.82	0	0.05	0.31	0.25	6.38	0.23	12	0	1	9
95	6	3.80	20	285	46	2.34	0	0.15	—	0.51	5.70	—	—	0	—	11
35.34	7.98	0.9	6.84	148.19	410.39	0.8	19.37	0	—	0.07	3.6	0.19	—	0	0.16	—
67	19	1.05	25	223	450	0.84	—	0.08	—	0.10	10.98	0.47	4	0	<1	—
77	14	0.97	26	235	67	0.92	—	0.07	—	0.11	12.57	0.54	3	0	<1	22
76	14	1.01	26	220	65	0.94	—	0.07	—	0.11	11.69	0.49	5	0	<1	20
80	10	1.12	20	212	82	2.74	—	0.07	—	0.20	5.23	0.33	8	0	<1	17
77	10	1.14	20	195	76	2.46	—	0.07	—	0.19	5.13	0.30	9	0	<1	16
84	11	1.19	21	216	82	2.53	—	0.07	—	0.21	5.69	0.33	8	0	<1	16
71	13	0.97	18	177	77	1.80	—	0.04	—	0.11	6.16	0.50	3	0	<1	22
51	14	1.06	17	209	452	0.88	—	0.08	—	0.12	5.71	0.26	9	<1	<1	—
80	10	1.11	20	206	77	2.43	—	0.06	—	0.20	5.37	0.32	7	0	<1	19
72	14	0.99	18	179	78	1.82	—	0.04	—	0.11	6.22	0.50	3	0	<1	21
64	9	1.13	17	191	81	1.81	14	0.05	—	0.16	4.88	0.26	6	0	<1	17
64	11	0.92	20	201	43	0.66	7	0.05	0.23	0.08	8.90	0.46	3	0	<1	22
64	10	1.03	18	195	64	1.29	10	0.05	—	0.13	6.70	0.35	4	0	<1	21
70.55	11.89	0.99	17.85	153	59.5	1.69	12.75	0.04	0.23	0.13	5.19	0.22	5.09	0	0.18	17.76
315	14	2.71	3	152	48	3.76	0	0.02	0.17	0.18	2.65	0.06	4	0	1	35
479	9	9.89	21	224	65	3.38	3384	0.25	0.70	1.69	9.39	0.64	491	24	14	70
71	9	2.30	14	173	50	1.58	54	0.15	0.59	0.23	4.10	0.15	5	0	<1	17
76	10	2.30	17	214	55	2.21	20	0.22	0.59	0.40	4.34	0.21	9	0	<1	19
77	11	2.41	19	280	60	2.23	18	0.07	—	0.28	3.55	0.32	2	0	<1	19
82	12	2.44	21	330	65	2.70	10	0.08	—	0.33	3.47	0.40	10	0	<1	22
78	36	2.86	7	72	362	0.48	521	0.05	—	0.16	1.31	0.03	31	0	5	23
87	21	1.64	20	230	91	2.43	0	0.05	0.29	0.14	4.10	0.33	6	0	<1	32
71	10	1.30	25	248	44	1.48	0	0.04	0.08	0.11	6.37	0.48	5	0	<1	27
72	27	1.98	20	247	67	3.79	0	0.05	0.54	0.21	3.10	0.31	8	0	<1	35
59	16	1.15	24	259	54	1.73	0	0.05	0.08	0.11	5.81	0.46	5	0	<1	27
58	13	2.07	14	259	752	1.35	10	0.09	1.18	0.18	2.16	0.19	26	0	<1	19
24	23	0.73	9	142	277	0.88	0	0.05	0.07	0.13	3.97	0.18	2	0	<1	13
31	18	0.77	10	153	332	1.13	0	0.05	0.19	0.16	2.72	0.15	3	0	<1	17
27	6	1.16	6	57	540	2.32	0	0.00	—	0.06	1.00	0.07	5	0	1	10
30	0	0.36	6	43	290	0.40	0	—	—	—	—	—	—	0	—	—
15	0	0.36	6	46	310	0.45	0	—	—	—	—	—	—	0	—	—
20	0	0.36	4	47	310	0.57	0	0.01	—	0.03	0.68	0.05	4	0	<1	—
20	40	0.36	6	43	270	0.52	0	—	—	—	—	—	—	0	—	—
15	0	0.36	7	85	350	0.32	0	—	—	—	—	—	—	0	—	—
25	0	0.72	—	—	560	—	0	—	—	—	—	—	—	0	—	—
15	0	0.00	—	—	240	—	0	—	—	—	—	—	—	0	—	—
15	0	0.00	—	—	240	—	0	—	—	—	—	—	—	0	—	—
26	8	0.64	5	75	573	0.97	0	0.05	0.10	0.07	1.23	0.06	1	0	1	10

TABLE A–1
Food Composition

(Computer code number is for Wadsworth Diet Analysis program)   (For purposes of calculations, use "0" for t, <1, <.1, <.01, etc.)

DA + Code	Food Description	Quantity	Measure	Wt (g)	H₂O (g)	Ener (kcal)	Prot (g)	Carb (g)	Dietary Fiber (g)	Fat (g)	Sat	Mono	Poly	Trans
												Fat Breakdown (g)		

DA + Code	Food Description	Quantity	Measure	Wt (g)	H₂O (g)	Ener (kcal)	Prot (g)	Carb (g)	Dietary Fiber (g)	Fat (g)	Sat	Mono	Poly	Trans
	PROCESSED MEATS—Continued													
1323	**Pork olive loaf**	2	slice(s)	57	33	133	7	5	0	9	3.32	4.47	1.10	—
1324	**Pork pickle & pimento loaf**	2	slice(s)	57	32	149	7	3	0	12	4.45	5.45	1.47	—
	Sausages & frankfurters													
37296	Beerwurst beef beer salami (bierwurst)	1	slice(s)	29	17	74	4	1	0	6	2.50	2.69	0.21	—
37257	Beerwurst pork beer salami	1	slice(s)	21	13	50	3	<1	0	4	1.32	1.89	0.50	—
35338	Berliner, pork & beef	1	ounce(s)	28	17	65	4	1	0	5	1.72	2.27	0.45	—
37299	Braunschweiger pork liver sausage	1	slice(s)	15	0	51.34	1.97	0.34	0	4.48	1.52	2.08	0.52	—
37298	Bratwurst pork, cooked	1	piece(s)	74	42	181	10	2	0	14	5.15	6.73	1.51	—
1329	Cheesefurter or cheese smokie, beef & pork	1	item(s)	43	23	141	6	1	0	12	4.52	5.89	1.30	—
1330	Chorizo, beef & pork	2	ounce(s)	57	18	258	14	1	0	22	8.15	10.43	1.96	—
8600	Frankfurter, beef	1	item(s)	45	23	149	5	2	0	13	5.26	6.44	0.53	—
202	Frankfurter, beef & pork	1	item(s)	57	32	174	7	1	1	16	6.14	7.79	1.56	—
1293	Frankfurter, chicken	1	item(s)	45	26	116	6	3	0	9	2.49	3.82	1.82	—
3261	Frankfurter, turkey	1	item(s)	45	28	102	6	1	0	8	2.65	2.51	2.25	—
37275	Italian sausage, pork, cooked	1	item(s)	68	34	220	14	1	0	17	6.14	8.13	2.23	—
37307	Kielbasa, kolbassa, pork & beef	2⅛	ounce(s)	61	37	135	10	2	0	9	3.40	4.44	1.06	—
1333	Knockwurst or knackwurst, beef & pork	2	ounce(s)	57	31	174	6	2	0	16	5.79	7.26	1.66	—
37285	Pepperoni, beef & pork	1	slice(s)	11	3	55	2	<1	0	5	1.77	2.32	0.48	—
37313	Polish sausage, pork	2	slice(s)	57	31	163	8	2	—	14	4.91	6.42	1.46	—
206	Salami, beef, cooked, sliced	2	slice(s)	46	28	119	6	1	0	10	4.54	4.90	0.48	—
37272	Salami, pork, dry or hard	1	slice(s)	13	5	52	3	<1	0	4	1.52	2.05	0.48	—
3262	Salami, turkey	2	slice(s)	57	31	125	8	11	<.1	5	1.98	1.80	1.43	0
7162	Sausage, breakfast, turkey	2½	ounce(s)	100	67	190	17	<1	0	13	3.90	6.23	3.33	0
8620	Smoked sausage, beef & pork	2	ounce(s)	57	31	181	7	1	0	16	5.54	6.94	2.23	0
8619	Smoked, sausage, pork	2	ounce(s)	57	22	221	13	1	0	18	6.42	8.30	2.13	—
37273	Smoked, sausage, pork link	1	piece(s)	76	30	295	17	2	—	24	8.58	11.09	2.85	—
1336	Summer sausage, thuringer, or cervelat, beef & pork	2	ounce(s)	57	29	190	9	<1	0	17	6.82	7.35	0.68	—
37294	Vienna sausage, cocktail, beef & pork, canned	1	piece(s)	16	10	45	2	<1	0	4	1.49	2.01	0.27	—
	Spreads													
32419	Pork & beef sandwich spread	4	tablespoon(s)	60	36	141	5	7	<1	10	3.59	4.57	1.54	—
1318	Ham salad spread	¼	cup(s)	60	38	130	5	6	0	9	3.04	4.32	1.62	—
	Turkey													
16049	Breast, hickory smoked, slices	1	slice(s)	56	—	50	11	1	0	0	0.00	0.00	0.00	0
13606	Breast, hickory smoked fat free	1	slice(s)	28	—	25	4	1	0	0	0.00	0.00	0.00	0
16047	Breast, honey roasted, slices	1	slice(s)	56	—	60	11	2	0	0	0.00	0.00	0.00	0
16048	Breast, oven roasted, slices	1	slice(s)	56	—	50	11	1	0	0	0.00	0.00	0.00	0
13583	Breast, traditional carved	2	slice(s)	45	—	40	9	0	0	1	0.00	0.07	0.14	—
13604	Breast, oven roasted, fat free	1	slice(s)	28	—	25	4	1	0	0	0.00	0.00	0.00	0
13567	Turkey ham, 10% water added	1	slice(s)	28	20	35	5	0	0	1	0.00	0.22	0.31	—
13596	Turkey pastrami	2	ounce(s)	56	—	70	11	1	0	2	1.00	—	—	—
13597	Turkey salami	2	ounce(s)	56	—	120	8	1	0	9	2.50	2.92	2.30	—
	BEVERAGES													
	Alcoholic													
	Beer													
866	Ale, mild	12	fluid ounce(s)	360	332	148	1	13	1	0	0.00	0.00	0.00	—
686	Beer	12	fluid ounce(s)	356	336	118	1	6	<1	<1	0.00	0.00	0.00	0
869	Beer, light	12	fluid ounce(s)	354	337	99	1	5	0	0	0.00	0.00	0.00	—
16886	Beer, nonalcoholic	12	fluid ounce(s)	360	353	32	1	5	0	0	0.00	0.00	0.00	—
31608	Budweiser beer	12	fluid ounce(s)	355	328	143	1	11	0	0	0.00	0.00	0.00	—
31609	Bud Light beer	12	fluid ounce(s)	355	335	110	1	7	0	0	0.00	0.00	0.00	—
31613	Michelob Beer	12	fluid ounce(s)	355	323	155	1	13	0	0	0.00	0.00	0.00	—
31614	Michelob Light beer	12	fluid ounce(s)	355	330	134	1	12	0	0	0.00	0.00	0.00	—
	Gin, rum, vodka, whiskey													
687	Distilled alcohol, 80 proof	1	fluid ounce(s)	28	19	64	0	0	0	0	0.00	0.00	0.00	0
688	Distilled alcohol, 86 proof	1	fluid ounce(s)	28	18	70	0	<.1	0	0	0.00	0.00	0.00	0
689	Distilled alcohol, 90 proof	1	fluid ounce(s)	28	17	73	0	0	0	0	0.00	0.00	0.00	0
856	Distilled alcohol, 94 proof	1	fluid ounce(s)	28	17	76	0	0	0	0	0.00	0.00	0.00	0
857	Distilled alcohol, 100 proof	1	fluid ounce(s)	28	16	82	0	0	0	0	0.00	0.00	0.00	0
	Liqueurs													
3142	Coffee liqueur, 63 proof	1	fluid ounce(s)	35	14	107	<.1	11	0	<1	0.04	0.01	0.04	—
33187	Coffee liqueur, 53 proof	1	fluid ounce(s)	35	11	117	<.1	16	0	<1	0.04	0.01	0.04	—

PAGE KEY: A–4 = Breads/Baked Goods A–8 = Cereal/Rice/Pasta A–12 = Fruit A–16 = Vegetables/Legumes A–26 = Nuts/Seeds A–28 = Vegetarian A–30 = Dairy A–36 = Eggs A–36 = Seafood A–38 = Meats A–42 = Poultry A–42 = Processed meats A–44 = Beverages A–48 = Fats/Oils A–50 = Sweets A–52 = Spices/Condiments/Sauces A–54 = Mixed Foods/Soups/Sandwiches A–60 = Fast food A–76 = Convenience A–78 = Baby foods

Chol (mg)	Calc (mg)	Iron (mg)	Magn (mg)	Pota (mg)	Sodi (mg)	Zinc (mg)	Vit A (RAE) (µg)	Thia (mg)	Vit E (mg α)	Ribo (mg)	Niac (mg)	Vit B$_6$ (mg)	Fola (µg)	Vit C (mg)	Vit B$_{12}$ (µg)	Sele (µg)
22	62	0.31	11	169	843	0.78	34	0.17	0.14	0.15	1.04	0.13	1	0	1	9
21	54	0.58	10	193	789	0.80	12	0.17	0.24	0.14	1.17	0.11	3	0	1	8
18	3	0.44	4	67	265	0.71	0	0.02	—	0.04	0.99	0.05	1	0	1	5
12	2	0.16	3	53	261	0.36	0	0.12	—	0.04	0.69	0.07	1	0	<1	—
13	3	0.33	4	80	368	0.70	0	0.11	—	0.06	0.88	0.06	1	0	1	4
23.69	1.36	1.42	1.67	27.49	131.54	0.42	641.01	0.03	—	0.23	1.27	0.05	—	0	3.05	8.81
44	33	0.96	11	157	412	1.70	0	0.37	—	0.14	2.37	0.16	1	1	1	16
29	25	0.46	6	89	465	0.97	20	0.11	0.00	0.07	1.25	0.06	1	0	1	7
50	5	0.90	10	226	700	1.93	0	0.36	0.12	0.17	2.91	0.30	1	0	1	12
24	6	0.68	6	70	513	1.11	0	0.02	0.09	0.07	1.07	0.04	2	0	1	4
29	6	0.66	6	95	638	1.05	10	0.11	0.14	0.07	1.50	0.07	2	0	1	8
45	43	0.90	5	38	617	0.47	18	0.03	0.10	0.05	1.39	0.14	2	0	<1	8
48	48	0.83	6	81	642	1.40	0	0.02	0.28	0.08	1.86	0.10	4	0	<1	7
53	16	1.02	12	207	627	1.62	0	0.42	—	0.16	2.83	0.22	3	1	1	15
41	27	0.88	10	169	566	1.23	0	0.14	—	0.13	1.75	0.11	3	0	1	11
34	6	0.37	6	113	527	0.94	0	0.19	—	0.08	1.55	0.10	1	0	1	8
9	1	0.15	2	38	224	0.28	0	0.04	—	0.03	0.55	0.03	<1	0	<1	—
40	7	0.82	8	102	546	1.10	0	0.29	—	0.08	1.96	0.11	1	1	1	10
33	3	1.01	6	86	524	0.81	0	0.05	0.09	0.09	1.49	0.08	1	0	1	7
10	2	0.17	3	48	289	0.54	0	0.12	—	0.04	0.72	0.07	<1	0	<1	3
45	42	0.87	15	225	616	1.76	1	0.24	0.14	0.17	2.26	0.24	6	12	1	11
92	57	2.20	18	188	665	2.07	0	0.04	0.00	0.12	3.55	0.29	5	1	<1	—
33	7	0.43	7	101	517	0.71	7	0.11	0.07	0.06	1.67	0.09	1	0	<1	0
39	17	0.66	11	191	851	1.60	0	0.40	0.14	0.15	2.57	0.20	3	1	1	12
52	23	0.88	14	255	1137	2.14	0	0.53	—	0.20	3.43	0.27	4	0	1	16
43	7	1.44	8	154	704	1.45	0	0.09	0.12	0.19	2.44	0.15	1	0	3	12
8	2	0.14	1	16	152	0.26	0	0.01	—	0.02	0.26	0.02	1	0	<1	3
23	7	0.47	5	66	608	0.61	16	0.10	1.04	0.08	1.04	0.07	1	0	1	6
22	5	0.35	6	90	547	0.66	0	0.26	1.04	0.07	1.26	0.09	1	0	<1	11
25	0	0.72	—	—	730	—	0	—	—	—	—	—	—	0	—	—
10	0	0.00	—	—	300	—	0	—	—	—	—	—	—	0	—	—
20	0	0.72	—	—	640	—	0	—	—	—	—	—	—	0	—	—
20	0	0.72	—	—	620	—	0	—	—	—	—	—	—	0	—	—
20	0	0.72	—	—	540	—	0	—	—	—	—	—	—	0	—	—
10	0	0.00	—	—	330	—	0	—	—	—	—	—	—	0	—	—
20	0	0.36	6	81	310	0.73	—	—	—	—	—	—	—	0	—	—
40	0	0.72	—	—	590	—	0	—	—	—	—	—	—	0	—	—
50	40	0.72	—	—	500	—	0	—	—	—	—	—	—	0	—	—
0	18	0.11	—	—	18	—	0	0.02	0.00	0.10	1.63	—	—	0	<.1	—
0	18	0.07	21	89	14	0.04	0	0.02	0.00	0.09	1.61	0.18	21	0	<.1	2
0	18	0.14	18	64	11	0.11	0	0.03	0.00	0.11	1.39	0.12	14	0	<.1	2
0	25	0.04	32	90	18	0.04	—	0.02	0.00	0.10	1.63	0.18	22	0	<.1	—
0	18	0.11	21	89	9	0.07	0	0.02	0.00	0.09	1.61	0.18	21	0	<.1	4
0	18	0.14	18	64	9	0.11	0	0.03	0.00	0.11	1.39	0.12	15	0	<.1	4
0	18	0.11	21	89	9	0.07	0	0.02	0.00	0.09	1.61	0.18	21	0	<.1	4
0	18	0.14	18	64	9	0.11	0	0.03	0.00	0.11	1.39	0.12	15	0	<.1	4
0	0	0.01	0	1	<1	0.01	0	0.00	0.00	0.00	0.00	0.00	0	0	0	0
0	0	0.01	0	1	<1	0.01	0	0.00	0.00	0.00	0.00	0.00	0	0	0	0
0	0	0.01	0	1	<1	0.01	0	0.00	0.00	0.00	0.00	0.00	0	0	0	0
0	0	0.01	0	1	<1	0.01	0	0.00	0.00	0.00	0.00	0.00	0	0	0	0
0	0	0.01	0	1	<1	0.01	0	0.00	0.00	0.00	0.00	0.00	0	0	0	0
0	<1	0.02	1	10	3	0.01	0	0.00	—	0.00	0.05	0.00	0	0	0	<1
0	<1	0.02	1	10	3	0.01	0	0.00	0.00	0.00	0.05	0.00	0	0	0	<1

TABLE A–1
Food Composition

(Computer code number is for Wadsworth Diet Analysis program)   (For purposes of calculations, use "0" for t, <1, <.1, <.01, etc.)

DA + Code	Food Description	Quantity	Measure	Wt (g)	H₂O (g)	Ener (kcal)	Prot (g)	Carb (g)	Dietary Fiber (g)	Fat (g)	Fat Breakdown (g)			
											Sat	Mono	Poly	Trans
	BEVERAGES—Continued													
736	Cordials, 54 proof	1	fluid ounce(s)	30	9	106	<.1	13	0	<.1	0.02	0.01	0.04	—
	Wine													
858	Champagne, domestic	5	fluid ounce(s)	150	—	105	<1	4	0	0	0.00	0.00	0.00	0
861	Red wine, California	5	fluid ounce(s)	150	133	125	<1	4	0	0	0.00	0.00	0.00	0
690	Sweet dessert wine	5	fluid ounce(s)	150	106	240	<1	21	0	0	0.00	0.00	0.00	0
1481	White wine	5	fluid ounce(s)	148	132	100	<1	1	0	0	0.00	0.00	0.00	0
1811	Wine cooler	10	fluid ounce(s)	300	270	150	<1	18	<.1	<.1	0.01	0.00	0.02	—
	Carbonated													
692	Club soda	12	fluid ounce(s)	355	355	0	0	0	0	0	0.00	0.00	0.00	0
12010	Coca-Cola Classic cola soda	12	fluid ounce(s)	360	—	146	0	41	0	0	0.00	0.00	0.00	0
12031	Coke diet cola soda	12	fluid ounce(s)	360	—	2	0	<1	0	0	0.00	0.00	0.00	0
693	Cola	12	fluid ounce(s)	426	380	179	<1	46	0	0	0.00	0.00	0.00	—
9522	Cola soda, decaffeinated	12	fluid ounce(s)	372	331	156	<1	40	0	0	0.00	0.00	0.00	0
1415	Cola, low calorie w/aspartame	12	fluid ounce(s)	355	354	4	<1	<1	0	0	0.00	0.00	0.00	0
9524	Cola, decaffeinated, low calorie w/aspartame	12	fluid ounce(s)	355	354	4	<1	<1	0	0	0.00	0.00	0.00	0
1412	Cream soda	12	fluid ounce(s)	371	321	189	0	49	0	0	0.00	0.00	0.00	0
31899	Diet 7 Up	12	fluid ounce(s)	360	—	0	0	0	0	0	0.00	0.00	0.00	0
695	Ginger ale	12	fluid ounce(s)	366	334	124	0	32	0	0	0.00	0.00	0.00	0
694	Grape soda	12	fluid ounce(s)	372	330	160	0	42	0	0	0.00	0.00	0.00	0
1876	Lemon lime soda	12	fluid ounce(s)	368	330	147	0	38	0	0	0.00	0.00	0.00	—
29392	Mountain Dew diet soda	12	fluid ounce(s)	360	—	0	0	0	0	0	0.00	0.00	0.00	0
29391	Mountain Dew soda	12	fluid ounce(s)	360	—	170	0	46	0	0	0.00	0.00	0.00	0
3145	Orange soda	12	fluid ounce(s)	372	326	179	0	46	0	0	0.00	0.00	0.00	0
1414	Pepper-type soda	12	fluid ounce(s)	368	329	151	0	38	0	<1	0.26	0.00	0.00	—
2391	Pepper-type or cola soda, low calorie w/saccharin	12	fluid ounce(s)	355	354	0	0	<1	0	0	0.00	0.00	0.00	0
29389	Pepsi diet cola soda	12	fluid ounce(s)	360	—	0	0	0	0	0	0.00	0.00	0.00	0
29388	Pepsi regular cola soda	12	fluid ounce(s)	360	—	150	0	41	0	0	0.00	0.00	0.00	0
696	Root beer	12	fluid ounce(s)	370	330	152	0	39	0	0	0.00	0.00	0.00	0
31898	7 Up	12	fluid ounce(s)	360	—	240	0	59	0	0	0.00	0.00	0.00	0
12034	Sprite diet soda	12	fluid ounce(s)	360	—	4	0	0	0	0	0.00	0.00	0.00	0
12044	Sprite soda	12	fluid ounce(s)	360	—	144	0	39	0	0	0.00	0.00	0.00	0
	Coffee													
731	Brewed	8	fluid ounce(s)	237	236	9	<1	0	0	0	0.00	0.00	0.00	0
9520	Brewed, decaffeinated	8	fluid ounce(s)	237	235	5	<1	1	0	0	0.00	0.00	0.00	0
16882	Cappuccino	8	fluid ounce(s)	240	224	78	4	6	<1	4	2.53	1.18	0.15	—
16883	Cappuccino, decaffeinated	8	fluid ounce(s)	240	224	78	4	6	<1	4	2.53	1.18	0.15	—
16880	Espresso	8	fluid ounce(s)	237	235	5	<1	1	0	0	0.00	0.00	0.00	0
16881	Espresso, decaffeinated	8	fluid ounce(s)	237	235	5	<1	1	0	0	0.00	0.00	0.00	0
732	Instant, prepared	8	fluid ounce(s)	239	237	5	<1	1	0	0	0.00	0.00	0.00	0
	Fruit drinks													
29357	Crystal Light low calorie lemonade drink	8	fluid ounce(s)	240	—	5	0	0	0	0	0.00	0.00	0.00	0
6012	Fruit punch drink w/added vitamin C, canned	8	fluid ounce(s)	276	242	129	0	33	<1	<.1	0.01	0.01	0.01	0
260	Grape drink, canned	8	fluid ounce(s)	250	221	113	<.1	29	0	0	0.00	0.00	0.00	0
266	Lemonade, from frozen concentrate	8	fluid ounce(s)	248	213	131	<1	34	<1	<1	0.02	0.00	0.04	—
268	Limeade, from frozen concentrate	8	fluid ounce(s)	247	220	104	<.1	26	0	<.1	0.00	0.00	0.00	—
31143	Gatorade Thirst Quencher, all flavors	8	fluid ounce(s)	240	—	50	0	14	0	0	0.00	0.00	0.00	0
17372	Kool-Aid (lemonade/punch/fruit drink)	8	fluid ounce(s)	248	220	108	<1	28	<1	<.1	0.01	0.01	0.02	—
17225	Kool-Aid sugar free, low calorie tropical punch mix, prepared	8	fluid ounce(s)	240	—	5	0	0	0	0	0.00	0.00	0.00	0
14266	Odwalla strawberry 'c' monster fruit drink	8	fluid ounce(s)	240	—	150	2	34	1	1	0.00	—	—	0
10080	Odwalla strawberry lemonade quencher	8	fluid ounce(s)	240	—	120	1	28	1	0	0.00	0.00	0.00	0
10099	Snapple fruit punch	8	fluid ounce(s)	240	—	110	0	29	0	0	0.00	0.00	0.00	0
10096	Snapple kiwi strawberry	8	fluid ounce(s)	240	211	110	0	28	0	0	0.00	0.00	0.00	0
	Slim Fast ready to drink shake													
16056	Dark chocolate fudge	11	fluid ounce(s)	325	—	220	10	42	5	3	1.00	1.50	0.50	—
16054	French vanilla	11	fluid ounce(s)	325	—	220	10	40	5	3	0.50	1.50	0.50	—
16055	Strawberries n cream	11	fluid ounce(s)	325	—	220	10	40	5	3	0.50	1.50	0.50	—
	Tea													
733	Tea, prepared	8	fluid ounce(s)	237	236	2	0	1	0	0	0.00	0.00	0.01	0
33179	Decaffeinated, prepared	8	fluid ounce(s)	237	236	2	0	1	0	0	0.00	0.00	0.01	0
1877	Herbal, prepared	8	fluid ounce(s)	237	236	2	0	<1	0	0	0.00	0.00	0.01	0
734	Instant tea mix, unsweetened, prepared	8	fluid ounce(s)	237	236	2	<.1	<1	0	0	0.00	0.00	0.00	0

PAGE KEY: A–4 = Breads/Baked Goods   A–8 = Cereal/Rice/Pasta   A–12 = Fruit   A–16 = Vegetables/Legumes   A–26 = Nuts/Seeds   A–28 = Vegetarian
A–30 = Dairy   A–36 = Eggs   A–36 = Seafood   A–38 = Meats   A–42 = Poultry   A–42 = Processed meats   A–44 = Beverages   A–48 = Fats/Oils
A–50 = Sweets   A–52 = Spices/Condiments/Sauces   A–54 = Mixed Foods/Soups/Sandwiches   A–60 = Fast food   A–76 = Convenience   A–78 = Baby foods

Chol (mg)	Calc (mg)	Iron (mg)	Magn (mg)	Pota (mg)	Sodi (mg)	Zinc (mg)	Vit A (RAE) (µg)	Thia (mg)	Vit E (mg α)	Ribo (mg)	Niac (mg)	Vit B6 (mg)	Fola (µg)	Vit C (mg)	Vit B12 (µg)	Sele (µg)
0	<1	0.02	<1	5	2	0.01	0	0.00	0.00	0.00	0.02	0.00	0	0	0	—
0	—	—	—	—	—	—	—	—	—	—	0.00	—	—	—	0	—
0	12	1.43	16	171	15	0.15	0	0.02	0.00	0.04	0.12	0.05	1	0	<.1	—
0	12	0.36	14	138	14	0.11	0	0.03	0.00	0.03	0.32	0.00	0	0	0	1
0	13	0.47	15	118	7	0.10	0	0.01	—	0.01	0.10	0.02	0	0	0	<1
0	17	0.81	16	135	25	0.17		0.01	0.03	0.02	0.13	0.04	4	5	<.1	—
0	18	0.04	4	7	75	0.36	0	0.00	0.00	0.00	0.00	0.00	0	0	0	0
0	—	—	—	0	50	—	0	—	—	—	—	—	—	0	—	—
0	—	—	—	18	42	—	0	—	—	—	—	—	—	0	—	—
0	13	0.09	4	4	17	0.04	0	0.00	0.00	0.00	0.00	0.00	0	0	0	<1
0	11	0.07	4	4	15	0.04	0	0.00	0.00	0.00	0.00	0.00	0	0	0	<1
0	11	0.11	4	21	18	0.00	0	0.02	0.00	0.08	0.00	0.00	0	0	0	0
0	14	0.11	4	0	21	0.28	0	0.02	0.00	0.08	0.00	0.00	0	0	0	<1
0	19	0.19	4	4	44	0.26	0	0.00	0.00	0.00	0.00	0.00	0	0	0	0
0	—	—	—	116	53	—	—	—	0.00	—	—	—	—	—	0	—
0	11	0.66	4	4	26	0.18	0	0.00	0.00	0.00	0.00	0.00	0	0	0	<1
0	11	0.30	4	4	56	0.26	0	0.00	0.00	0.00	0.00	0.00	0	0	0	0
0	7	0.26	4	4	41	0.18	0	0.00	0.00	0.00	0.06	0.00	0	0	0	0
0	—	—	—	70	35	—	—	—	—	—	—	—	—	—	0	—
0	—	—	—	0	70	—	—	—	—	—	—	—	—	—	0	—
0	19	0.22	4	7	45	0.37	0	0.00	—	0.00	0.00	0.00	0	0	0	0
0	11	0.15	0	4	37	0.15	0	0.00	—	0.00	0.00	0	0	0	<1	
0	14	0.07	4	14	57	0.11	0	0.00	0.00	0.00	0.00	0.00	0	0	0	<1
0	—	—	—	30	35	—	—	—	—	—	—	—	—	—	—	—
0	—	—	—	0	35	—	—	—	—	—	—	—	—	—	—	—
0	18	0.18	4	4	48	0.26	0	0.00	0.00	0.00	0.00	0.00	0	0	0	<1
0	—	—	—	0	113	—	—	—	—	—	—	—	—	—	—	—
0	—	—	—	110	36	—	0	—	—	—	—	—	—	0	—	—
0	—	—	—	0	71	—	0	—	—	—	—	—	—	0	—	—
0	2	0.02	5	114	2	0.02	0	0.00	0.02	0.12	0.00	0.00	5	0	0	0
0	5	0.12	12	128	5	0.05	0	0.00	0.00	0.00	0.53	0.00	<1	0	0	0
17	152	0.26	22	250	62	0.50	—	0.04	0.10	0.20	0.37	0.05	5	1	<1	—
17	152	0.26	22	250	62	0.50	—	0.04	0.10	0.20	0.37	0.05	5	1	<1	—
0	5	0.12	12	128	5	0.05	0	0.00	0.05	0.00	0.53	0.00	<1	0	0	—
0	5	0.12	12	128	5	0.05	0	0.00	0.05	0.00	0.53	0.00	<1	0	0	—
0	10	0.10	7	72	5	0.02	0	0.00	0.00	0.00	0.56	0.00	0	0	0	<1
0	0	0.00	—	160	20	—	0	—	—	—	—	—	—	0	—	—
0	22	0.58	6	69	61	0.33	—	0.06	0.00	0.06	0.06	0.00	4	99	0	0
0	5	0.45	3	30	15	0.30	—	0.00	0.00	0.01	0.03	0.01	0	85	0	<1
0	10	0.52	5	50	7	0.07	0	0.02	0.02	0.07	0.05	0.02	2	13	0	<1
0	7	0.02	2	22	5	0.02	0	0.00	0.00	0.01	0.02	0.01	2	6	0	<1
0	10	0.18	—	30	110	—	—	—	—	—	—	—	—	1	—	—
0	14	0.46	5	50	31	0.20	—	0.04	—	0.05	0.05	0.01	4	42	0	1
0	0	0.00	—	10	10	—	—	—	—	—	—	—	—	6	—	—
0	20	1.44	—	330	40	—	—	—	—	—	—	—	—	600	0	—
0	20	0.00	—	70	30	—	0	—	—	—	—	—	—	60	0	—
0	0	0.00	—	20	10	—	0	—	—	—	—	—	—	0	0	—
0	0	0.00	—	40	10	—	0	—	—	—	—	—	—	0	0	—
5	400	2.70	140	600	220	2.25	—	0.53	—	0.60	7.00	0.70	120	60	2	18
5	400	2.70	140	600	220	2.25	—	0.53	—	0.60	7.00	0.70	120	60	2	18
5	400	2.70	140	600	220	2.25	—	0.53	—	0.60	7.00	0.70	120	60	2	18
0	0	0.05	7	88	7	0.05	0	0.00	0.00	0.03	0.00	0.00	12	0	0	0
0	0	0.05	7	88	7	0.05	0	0.00	0.00	0.03	0.00	0.00	12	0	0	0
0	5	0.19	2	21	2	0.09	0	0.02	0.00	0.01	0.00	0.00	2	0	0	0
0	7	0.05	5	47	7	0.02	0	0.00	0.00	0.00	0.09	0.00	0	0	0	0

TABLE A–1
Food Composition
(Computer code number is for Wadsworth Diet Analysis program)   (For purposes of calculations, use "0" for t, <1, <.1, <.01, etc.)

DA + Code	Food Description	Quantity	Measure	Wt (g)	H₂O (g)	Ener (kcal)	Prot (g)	Carb (g)	Dietary Fiber (g)	Fat (g)	Sat	Mono	Poly	Trans
	BEVERAGES—Continued													
735	Instant lemon flavored tea mix w/sugar, prepared	8	fluid ounce(s)	259	236	88	<1	22	0	<.1	0.01	0.00	0.02	—
	Water													
1413	Mineral water, carbonated	8	fluid ounce(s)	237	237	0	0	0	0	0	0.00	0.00	0.00	0
33183	Poland spring water, bottled	8	fluid ounce(s)	237	237	0	0	0	0	0	0.00	0.00	0.00	0
1	Tap water	8	fluid ounce(s)	237	237	0	0	0	0	0	0.00	0.00	0.00	—
1879	Tonic water	8	fluid ounce(s)	244	222	83	0	21	0	0	0.00	0.00	0.00	0
	FATS AND OILS													
	Butter													
104	Butter	1	tablespoon(s)	15	2	108	<1	<.1	0	12	6.13	5.00	0.43	—
921	Unsalted	1	tablespoon(s)	15	3	108	<1	<.1	0	12	7.71	3.15	0.46	—
107	Whipped	1	tablespoon(s)	11	2	82	<.1	<.1	0	9	5.76	2.67	0.34	—
944	Whipped, unsalted	1	tablespoon(s)	11	2	82	<.1	<.1	0	9	5.76	2.67	0.34	—
2522	Butter Buds, dry butter substitute	1	teaspoon(s)	2	—	8	0	2	0	0	0.00	0.00	0.00	0
	Fats, cooking													
2671	Beef tallow, semisolid	1	tablespoon(s)	13	0	115	0	0	0	13	6.37	5.35	0.51	—
922	Chicken fat	1	tablespoon(s)	13	<.1	115	0	0	0	13	3.81	5.72	2.68	—
5454	Household shortening w/vegetable oil	1	tablespoon(s)	13	0	115	0	0	0	13	3.39	5.56	2.75	2.20
111	Lard	1	tablespoon(s)	13	0	114	0	0	0	13	4.94	5.68	1.41	—
	Margarine													
114	Margarine	1	tablespoon(s)	14	2	101	<1	<1	0	11	2.23	5.05	3.58	—
116	Soft	1	tablespoon(s)	14	2	101	<1	<.1	0	11	1.95	4.02	4.88	—
117	Soft, unsalted	1	tablespoon(s)	14	3	101	<1	<1	0	11	1.95	5.26	3.62	—
928	Unsalted	1	tablespoon(s)	14	3	101	<.1	<.1	0	11	2.12	5.17	3.53	—
119	Whipped	1	tablespoon(s)	9	1	64	<.1	<.1	0	7	1.17	3.25	2.51	—
	Spreads													
16164	I Can't Believe It's Not Butter! whipped spread	1	tablespoon(s)	14	4	60	0	0	0	7	1.50	1.50	2.50	—
16157	Promise vegetable oil spread, stick	1	tablespoon(s)	14	4	90	0	0	0	10	2.50	2.00	4.00	—
	Oils													
2681	Canola	1	tablespoon(s)	14	0	120	0	0	0	14	0.97	8.01	4.03	—
120	Corn	1	tablespoon(s)	14	0	120	0	0	0	14	1.73	3.29	7.98	0.04
122	Olive	1	tablespoon(s)	14	0	119	0	0	0	14	1.82	9.98	1.35	—
124	Peanut	1	tablespoon(s)	14	0	119	0	0	0	14	2.28	6.24	4.32	—
2693	Safflower	1	tablespoon(s)	14	0	120	0	0	0	14	0.84	10.15	1.95	—
923	Sesame	1	tablespoon(s)	14	0	120	0	0	0	14	1.93	5.40	5.67	—
130	Soybean w/cottonseed oil	1	tablespoon(s)	14	0	120	0	0	0	14	2.45	4.01	6.54	—
128	Soybean, hydrogenated	1	tablespoon(s)	14	0	120	0	0	0	14	2.03	5.85	5.11	—
2700	Sunflower	1	tablespoon(s)	14	0	120	0	0	0	14	1.77	6.28	4.95	—
357	**Pam original no stick cooking spray**	1	serving(s)	0	—	0	0	0	0	0	0.00	0.00	0.00	—
	Salad dressing													
132	Blue cheese	2	tablespoon(s)	31	10	154	1	2	0	16	3.03	3.76	8.51	—
133	Blue cheese, low calorie	2	tablespoon(s)	32	25	32	2	1	0	2	0.82	0.57	0.78	—
1764	Caesar	2	tablespoon(s)	30	10	158	<1	1	<.1	17	2.64	4.05	9.86	—
29654	Creamy, reduced calorie, fat free, cholesterol free, sour cream and/or buttermilk & oil	2	tablespoon(s)	32	24	34	<1	6	0	1	0.16	0.21	0.46	—
29617	Creamy, reduced calorie, sour cream and/or buttermilk & oil	2	tablespoon(s)	30	22	48	<1	2	0	4	0.63	0.98	2.40	—
134	French	2	tablespoon(s)	31	11	143	<1	5	0	14	1.76	2.63	6.56	—
135	French, low fat	2	tablespoon(s)	33	18	76	<1	10	<1	4	0.36	1.92	1.64	—
136	Italian	2	tablespoon(s)	29	17	86	<1	3	0	8	1.32	1.86	3.80	—
137	Italian, diet	2	tablespoon(s)	30	25	23	<1	1	0	2	0.14	0.66	0.51	—
139	Mayonnaise type	2	tablespoon(s)	29	12	115	<1	7	0	10	1.44	2.65	5.29	—
942	Oil & vinegar	2	tablespoon(s)	31	15	140	0	1	0	16	2.84	4.62	7.52	—
1765	Ranch	2	tablespoon(s)	30	12	146	<1	2	<.1	16	2.32	3.85	8.92	—
3666	Ranch, reduced calorie	2	tablespoon(s)	30	21	62	<1	2	<.1	6	1.13	1.79	2.89	—
940	Russian	2	tablespoon(s)	31	11	151	<1	3	0	16	2.23	3.61	9.00	—
939	Russian, low calorie	2	tablespoon(s)	33	21	46	<1	9	<.1	1	0.20	0.29	0.75	—
941	Sesame seed	2	tablespoon(s)	31	12	136	1	3	<1	14	1.90	3.64	7.68	—
142	Thousand island	2	tablespoon(s)	31	15	115	<1	5	<1	11	1.59	2.46	5.68	—
143	Thousand island, low calorie	2	tablespoon(s)	31	19	62	<1	7	<1	4	0.23	1.98	0.82	—
	Sandwich spreads													
138	Mayonnaise w/soybean oil	1	tablespoon(s)	14	2	99	<1	1	0	11	1.64	2.70	5.89	0.04

PAGE KEY: A–4 = Breads/Baked Goods A–8 = Cereal/Rice/Pasta A–12 = Fruit A–16 = Vegetables/Legumes A–26 = Nuts/Seeds A–28 = Vegetarian A–30 = Dairy A–36 = Eggs A–36 = Seafood A–38 = Meats A–42 = Poultry A–42 = Processed meats A–44 = Beverages A–48 = Fats/Oils A–50 = Sweets A–52 = Spices/Condiments/Sauces A–54 = Mixed Foods/Soups/Sandwiches A–60 = Fast food A–76 = Convenience A–78 = Baby foods

Chol (mg)	Calc (mg)	Iron (mg)	Magn (mg)	Pota (mg)	Sodi (mg)	Zinc (mg)	Vit A (RAE) (µg)	Thia (mg)	Vit E (mg α)	Ribo (mg)	Niac (mg)	Vit B_6 (mg)	Fola (µg)	Vit C (mg)	Vit B_{12} (µg)	Sele (µg)
0	5	0.05	5	49	8	0.03	0	0.00	0.00	0.04	0.09	0.01	0	<1	0	<1
0	33	0.00	0	0	2	0.00	0	0.00	—	0.00	0.00	0.00	0	0	0	0
0	2	0.02	2	0	2	0.00	0	0.00	—	0.00	0.00	0.00	0	0	0	0
0	4.74	0.00	2.37	0	4.74	0	0	0	0.57	0	0	0	0	0	0	0
0	2	0.02	0	0	10	0.24	0	0.00	0.00	0.00	0.00	0.00	0	0	0	0
32	4	0.00	<1	4	86	0.01	103	0.00	0.35	0.01	0.01	0.00	<1	0	<.1	<1
32	4	0.00	<1	4	2	0.01	103	0.00	0.35	0.01	0.01	0.00	<1	0	<.1	<1
25	3	0.02	<1	3	94	0.01	78	0.00	0.26	0.00	0.00	0.00	<1	0	<.1	<1
25	3	0.02	<1	3	1	0.01	—	0.00	0.26	0.00	0.01	0.00	<1	0	<.1	—
0	0	0.00	0	2	70	0.00	0	0.00	0.00	0.00	0.00	0.00	<1	0	0	—
14	0	0.00	0	0	0	0.00	0	0.00	0.35	0.00	0.00	0.00	0	0	0	<.1
11	0	0.00	0	0	0	0.00	0	0.00	0.35	0.00	0.00	0.00	0	0	0	<.1
0	0	0.00	0	0	0	0.00	0	0.00	—	0.00	0.00	0.00	0	0	0	—
12	0	0.00	0	0	0	0.01	0	0.00	0.08	0.00	0.00	0.00	0	0	0	<.1
0	4	0.01	<1	6	133	0.00	115	0.00	1.27	0.01	0.00	0.00	<1	<.1	<.1	0
0	4	0.00	<1	5	152	0.00	103	0.00	0.99	0.00	0.00	0.00	<1	<.1	<.1	0
0	4	0.00	<1	5	4	0.00	103	0.00	1.23	0.00	0.00	0.00	<1	<.1	<.1	0
0	2	0.00	<1	4	<1	0.00	115	0.00	1.80	0.00	0.00	0.00	<1	<.1	<.1	0
0	2	0.00	<1	3	97	0.00	—	0.00	0.45	0.00	0.00	0.00	<.1	<.1	<.1	—
0	10	0.18	—	4	70	—	—	1.65	0.00	0.00	0.00	—	—	1	—	—
0	10	0.18	—	9	90	—	—	0.00	—	0.00	0.00	—	—	1	—	—
0	0	0.00	0	0	0	0.00	0	0.00	2.33	0.00	0.00	0.00	0	0	0	0
0	0	0.00	0	0	0	0.00	0	0.00	1.94	0.00	0.00	0.00	0	0	0	0
0	<1	0.09	0	<1	<1	0.00	0	0.00	1.94	0.00	0.00	0.00	0	0	0	0
0	0	0.00	0	0	0	0.00	0	0.00	2.12	0.00	0.00	0.00	0	0	0	0
0	0	0.00	0	0	0	0.00	0	0.00	4.64	0.00	0.00	0.00	0	0	0	0
0	0	0.00	0	0	0	0.00	0	0.00	0.19	0.00	0.00	0.00	0	0	0	0
0	0	0.00	0	0	0	0.00	0	0.00	1.65	0.00	0.00	0.00	0	0	0	0
0	0	0.00	0	0	0	0.00	0	0.00	1.10	0.00	0.00	0.00	0	0	0	0
0	0	0.00	0	0	0	0.00	0	0.00	—	0.00	0.00	0.00	0	0	0	0
0	0	0.00	—	0	0	—	0	—	0.00	—	—	—	0	0	—	—
5	25	0.06	0	11	335	0.08	21	0.00	1.84	0.03	0.03	0.01	9	1	<.1	<1
<1	28	0.16	2	2	384	0.08	—	0.01	0.08	0.03	0.02	0.01	1	<.1	<.1	—
1	7	0.05	1	9	323	0.03	—	0.00	1.57	0.00	0.01	0.00	1	0	<.1	—
0	12	0.08	2	43	320	0.06	0	0.00	0.21	0.02	0.01	0.01	1	0	0	—
0	2	0.04	1	11	307	0.01	—	0.00	0.72	0.00	0.01	0.01	4	<1	<.1	—
0	7	0.25	2	21	261	0.09	7	0.01	1.56	0.02	0.06	0.00	0	0	<.1	0
0	4	0.28	3	35	262	0.07	9	0.01	0.10	0.02	0.15	0.02	1	0	0	1
0	2	0.19	1	14	486	0.04	1	0.00	1.47	0.01	0.00	0.02	0	0	0	1
2	3	0.20	1	26	410	0.06	<1	0.00	0.06	0.00	0.00	0.02	0	0	0	2
8	4	0.06	1	3	209	0.05	19	0.00	0.61	0.01	0.00	0.00	2	0	<.1	<1
0	0	0.00	0	2	<1	0.00	0	0.00	1.44	0.00	0.00	0.00	0	0	0	0
1	4	0.03	1	8	354	0.01	—	0.00	1.85	0.00	0.00	0.00	<1	<.1	<.1	—
<1	5	0.01	1	8	414	0.02	—	0.00	0.73	0.01	0.01	0.00	<1	<1	<.1	—
6	6	0.18	1	48	266	0.13	5	0.02	1.02	0.02	0.18	0.01	3	2	<.1	<1
2	6	0.20	0	51	283	0.03	1	0.00	0.13	0.00	0.00	0.00	1	2	<.1	1
0	6	0.18	0	48	306	0.03	1	0.00	1.53	0.00	0.00	0.00	0	0	0	<1
8	5	0.37	2	33	269	0.08	3	0.45	1.25	0.02	0.13	0.00	0	0	0	<1
<1	5	0.28	2	62	254	0.06	5	0.01	0.31	0.01	0.13	0.00	0	0	0	0
5	2	0.07	<1	5	78	0.02	12	0.00	0.72	0.00	0.00	0.08	1	0	<.1	<1

TABLE A-1

Food Composition

(Computer code number is for Wadsworth Diet Analysis program)   (For purposes of calculations, use "0" for t, <1, <.1, <.01, etc.)

DA + Code	Food Description	Quantity	Measure	Wt (g)	H₂O (g)	Ener (kcal)	Prot (g)	Carb (g)	Dietary Fiber (g)	Fat (g)	Sat	Mono	Poly	Trans
	FATS AND OILS—Continued													
2708	Mayonnaise w/soybean & safflower oils	1	tablespoon(s)	14	0	98.94	0.15	0.37	0	10.95	1.18	1.79	7.59	—
140	Mayonnaise, low calorie	1	tablespoon(s)	16	10	37	<.1	3	0	3	0.53	0.72	1.70	—
141	Tartar sauce	2	tablespoon(s)	28	9	144	<1	4	<.1	14	2.14	4.13	7.57	—
	SWEETS													
4799	**Butterscotch or caramel topping**	2	tablespoon(s)	41	13	103	1	27	<1	<.1	0.05	0.01	0.00	—
	Candy													
1786	Almond Joy candy bar	1	item(s)	49	5	240	2	29	2	13	9.00	3.63	0.74	0
1785	Bit-o-Honey candy	6	item(s)	40	2	170	1	34	0	3	2.00	0	20	—
33375	Butterscotch candy	2	piece(s)	12	1	47	<.1	11	0	<1	0.25	0.10	0.01	—
1701	Chewing gum, stick	1	item(s)	3	<.1	7	0	2	<.1	<.1	0.00	0.00	0.00	—
33378	Chocolate fudge w/nuts, prepared	2	piece(s)	38	3	175	2	26	1	7	2.29	1.41	2.81	—
1787	Jelly beans	15	item(s)	43	3	159	0	40	<.1	<.1	0.00	0.00	0.00	—
1784	Kit Kat wafer bar	1	item(s)	42	1	220	3	27	1	11	7.00	3.53	0.34	0
4674	Krackel candy bar	1	item(s)	41	1	220	3	26	1	11	6.00	3.94	0.37	0
4934	Licorice	4	piece(s)	44	7	147	1	34	1	1	0.18	0.07	0.00	—
1780	Life Savers candy	1	item(s)	2	—	8	0	2	0	<.1	0.00	—	—	0
1790	Lollipop	1	item(s)	28	—	108	0	28	0	0	0.00	0.00	0.00	0
4679	M & Ms peanut chocolate candy, small bag	1	item(s)	49	1	250	5	30	2	13	5.00	5.42	2.07	—
1781	M & Ms plain chocolate candy, small bag	1	item(s)	48	1	240	2	34	1	10	6.00	3.30	0.30	—
4673	Milk chocolate bar	1	item(s)	91	1	483	8	53	2	28	16.69	7.20	0.63	—
1783	Milky Way bar	1	item(s)	58	4	270	2	41	1	10	5.00	3.50	0.35	—
1788	Peanut brittle	1½	ounce(s)	43	<1	206	3	30	1	8	1.76	3.43	1.94	—
1789	Reese's peanut butter cups	2	piece(s)	45	1	250	5	25	1	14	5.00	6.17	2.34	0
4689	Reese's pieces candy, small bag	1	item(s)	46	1	230	6	26	1	11	7.00	0.97	0.46	0
33399	Semisweet chocolate candy, made w/butter	½	ounce(s)	14	<.1	68	1	9	1	4	2.49	1.41	0.13	—
1782	Snickers bar	1	item(s)	59	3	280	4	35	1	14	5.00	6.13	2.89	—
4694	Special Dark chocolate bar	1	item(s)	41	<1	220	2	24	3	13	8.00	4.59	0.41	0
4695	Starburst fruit chews, original fruits	1	package	59	4	240	0	48	0	5	1.00	2.10	1.83	—
4698	Taffy	3	piece(s)	45	2	169	<.1	41	0	1	0.92	0.43	0.05	—
4699	Three Musketeers bar	1	item(s)	60	4	260	2	46	1	8	4.50	2.59	0.27	—
4702	Twix caramel cookie bars	2	item(s)	58	2	280	3	37	1	14	5.00	7.75	0.49	—
4705	York peppermint pattie	1	item(s)	42	4	170	1	34	1	3	2.00	1.32	0.12	0
	Frosting, icing													
4760	Chocolate frosting, ready to eat	2	tablespoon(s)	28	5	112	<1	18	<1	5	1.55	2.54	0.60	—
4771	Creamy vanilla frosting, ready to eat	2	tablespoon(s)	28	4	118	0	19	<.1	5	0.84	1.37	2.24	0
17291	Dec-a-Cake variety pack candy decoration	1	teaspoon(s)	4	—	15	0	3	0	1	0.00	—	—	—
536	White icing	2	tablespoon(s)	40	3	163	<1	32	0	4	0.86	2.07	1.19	—
	Gelatin													
13697	Gelatin snack, all flavors	1	item(s)	99	97	70	1	17	0	0	0.00	0.00	0.00	0
2616	Mixed fruit gelatin mix, sugar free, low calorie, prepared	½	cup(s)	121	—	10	1	0	0	0	0.00	0.00	0.00	0
548	**Honey**	1	tablespoon(s)	21	4	64	<.1	17	<.1	0	0.00	0.00	0.00	0
	Jams, Jellies													
23054	Jams, jellies, preserves, all flavors	1	tablespoon(s)	20	<.1	56	<.1	14	<1	<.1	0.00	0.01	0.00	—
23278	Jams, jellies, preserves, all flavors, low sugar	1	tablespoon(s)	18	<.1	25	<.1	6	<1	<.1	0.00	0.01	0.02	—
545	**Marshmallows**	4	item(s)	29	5	92	1	23	<.1	<.1	0.02	0.02	0.01	—
4800	**Marshmallow cream topping**	2	tablespoon(s)	28	6	91	<1	22	<.1	<.1	0.02	0.02	0.01	—
555	**Molasses**	1	tablespoon(s)	20	4	58	0	15	0	<.1	0.00	0.01	0.01	—
4780	**Popsicle or ice pop**	1	item(s)	59	47	42	0	11	0	0	0.00	0.00	0.00	—
	Sugar													
559	Brown, packed	1	teaspoon(s)	5	<.1	17	0	4	0	0	0.00	0.00	0.00	0
563	Powdered, sifted	⅓	cup(s)	33	<.1	130	0	33	0	<.1	0.01	0.01	0.02	—
561	White granulated	1	teaspoon(s)	4	<.1	15	0	4	0	0	0.00	0.00	0.00	—
	Sugar Substitute													
1760	Equal sweetener, packet	1	item(s)	1	<.1	4	<.1	1	0	0	0.00	0.00	0.00	—
13029	Splenda granular no calorie sweetener	1	teaspoon(s)	1	—	2	0	1	0	0	0.00	0.00	0.00	—
1759	Sweet n Low sugar substitute, packet	1	item(s)	1	<.1	4	0	1	0	0	0.00	0.00	0.00	—
	Syrup													
3148	Chocolate	2	tablespoon(s)	38	12	105	1	24	1	<1	0.19	0.11	0.01	—
29676	Maple	¼	cup(s)	80	26	209	0	54	0	<1	0.03	0.05	0.08	—
4795	Pancake	¼	cup(s)	80	30	187	0	49	1	0	0.00	0.00	0.00	0

PAGE KEY: A–4 = Breads/Baked Goods   A–8 = Cereal/Rice/Pasta   A–12 = Fruit   A–16 = Vegetables/Legumes   A–26 = Nuts/Seeds   A–28 = Vegetarian
A–30 = Dairy   A–36 = Eggs   A–36 = Seafood   A–38 = Meats   A–42 = Poultry   A–42 = Processed meats   A–44 = Beverages   A–48 = Fats/Oils
A–50 = Sweets   A–52 = Spices/Condiments/Sauces   A–54 = Mixed Foods/Soups/Sandwiches   A–60 = Fast food   A–76 = Convenience   A–78 = Baby foods

Chol (mg)	Calc (mg)	Iron (mg)	Magn (mg)	Pota (mg)	Sodi (mg)	Zinc (mg)	Vit A (RAE) (µg)	Thia (mg)	Vit E (mg α)	Ribo (mg)	Niac (mg)	Vit B$_6$ (mg)	Fola (µg)	Vit C (mg)	Vit B$_{12}$ (µg)	Sele (µg)
8.14	2.48	0.06	0.13	4.69	78.38	0.01	11.59	0	3.04	0	0	0.07	1.1	0	0.03	0.22
4	<.1	0.00	<.1	2	80	0.02	0	0.00	0.32	0.00	0.00	0.00	0	0	0	—
11	6	0.21	1	10	200	0.05	—	0.00	0.97	0.00	0.01	0.08	2	<1	<.1	—
<1	22	0.08	3	34	143	0.08	11	0.00	—	0.04	0.02	0.01	1	<1	<.1	0
3	20	0.36	33	138	70	0.40	0	0.02	—	0.08	0.24	—	—	0	—	—
0.00	—	—	85	—	0	—	—	—	—	—	—	0	—	—		
1	<1	0.00	<1	<1	47	0.00	3	0.00	0.01	0.00	0.00	0.00	0	0	0	<.1
0	0	0.00	0	<.1	<.1	0.00	0	0.00	0.00	0.00	0.00	0.00	0	0	0	<.1
5	21	0.75	21	68	16	0.54	14	0.03	0.10	0.04	0.12	0.03	6	<.1	<.1	1
0	1	0.06	1	16	21	0.02	0	0.00	0.00	0.00	0.00	0.00	0	0	0	<1
3	40	0.36	16	126	25	0.52	8	0.07	—	0.23	1.07	0.05	60	0	<.1	2
3	60	0.37	—	169	80	—	0	—	—	—	—	—	—	0	—	—
0	3	0.13	3	28	109	0.07	0	0.01	0.08	0.02	0.04	0.00	0	0	0	—
0	<1	0.04	—	0	1	—	0	0.00	—	0.00	0.00	—	—	0	—	0
0	0	0.00	—	—	11	—	0	0.00	—	0.00	0.00	—	—	0	—	1
5	40	0.36	36	171	25	1.13	15	0.03	—	0.07	1.60	0.04	17	1	<.1	2
5	40	0.36	20	127	30	0.46	15	0.03	—	0.07	0.11	0.01	3	1	<1	1
22	228	0.83	61	399	92	1.00	20	0.06	—	0.26	0.15	0.10	11	2	<1	—
5	60	0.18	20	140	95	0.41	15	0.02	—	0.07	0.20	0.03	6	1	<1	3
5	11	0.52	18	71	189	0.37	17	0.06	1.09	0.02	1.13	0.03	20	0	<.1	1
3	20	0.36	40	233	140	0.82	7	0.11	—	0.08	2.08	0.07	25	0	<.1	2
0	40	0.00	20	182	90	0.35	25	0.04	—	0.07	1.31	0.03	13	0	<.1	1
3	5	0.44	16	52	2	0.23	<1	0.01	—	0.01	0.06	0.01	<1	0	0	<1
5	40	0.36	42	—	140	1.38	15	0.03	—	0.07	1.60	0.05	23	1	<.1	3
3	0	0.72	46	136	0	0.60	0	0.01	—	0.03	0.16	0.01	1	0	0	1
0	10	0.18	1	1	0	0.00	—	0.00	—	0.00	0.00	0.00	0	30	0	<1
4	1	0.03	<1	2	40	0.02	—	0.00	—	0.01	0.01	0.00	0	0	<.1	—
5	20	0.36	18	80	110	0.33	14	0.02	—	0.03	0.20	0.01	0	1	<1	2
5	40	0.36	18	117	115	0.45	15	0.09	—	0.13	0.69	0.02	14	1	<1	1
0	0	0.36	25	71	10	0.31	0	0.01	—	0.04	0.34	0.01	2	0	<.1	—
0	2	0.40	6	55	51	0.08	0	0.00	0.44	0.00	0.03	0.00	<1	0	0	<1
0	1	0.04	<1	10	52	0.02	0	0.00	0.43	0.08	0.06	0.00	2	0	0	<.1
0	0	0.00	—	—	15	—	0	—	—	—	—	—	—	0	—	—
<1	5	0.02	—	7	92	—	—	0.00	0.33	0.01	0.00	—	—	<.1	—	—
0	0	0.00	—	0	40	—	0	—	—	—	—	—	—	0	—	—
0	0	0.00	0	0	50	0.00	0	0.00	0.00	0.00	0.00	0.00	0	0	0	0
0	1	0.09	<1	11	1	0.05	0	0.00	0.00	0.01	0.03	0.01	<1	<1	0	<1
0	4	0.10	1	15	6	0.01	0.00	0.00	0.00	0.02	0.01	0.00	2.20	1.76	0.00	—
0	2	0.05	1	19	<1	0.02	0.76	0.00	0.01	0.01	0.03	0.01	—	4.93	0.00	—
0	1	0.07	1	1	23	0.01	0	0.00	0.00	0.00	0.02	0.00	<1	0	0	<1
0	1	0.06	1	1	23	0.01	0	0.00	0.00	0.00	0.02	0.00	<1	0	0	1
0	41	0.94	48	293	7	0.06	0	0.01	0.00	0.00	0.19	0.13	0	0	0	4
0	0	0.00	1	2	7	0.01	0	0.00	0.00	0.00	0.00	0.00	0	0	0	0
0	4	0.09	1	16	2	0.01	0	0.00	0.00	0.00	0.00	0.00	<.1	0	0	<.1
0	<1	0.01	0	1	<1	0.00	0	0.00	0.00	0.01	0.00	0.00	0	0	0	<1
0	<.1	0.00	0	<.1	0	0.00	0	0.00	0.00	0.00	0.00	0.00	0	0	0	<.1
0	0	0.00	0	0	0	0.00	0	0.00	0.00	0.00	0.00	0.00	0	0	0	0
0	10	0.18	—	—	<1	—	—	0.02	0.00	0.02	0.20	—	—	1	0	—
0	0	0.00	0	—	0	0.00	0	0.00	0.00	0.00	0.00	0.00	0	0	0	0
0	5	0.79	24	84	27	0.27	0	0.00	0.00	0.02	0.12	0.00	1	<.1	0	1
0	54	0.96	11	163	7	3.33	0	0.00	0.00	0.01	0.02	0.00	0	0	0	<1
0	2	0.02	2	12	66	0.06	0	0.00	0.00	0.01	0.01	0.00	0	0	0	0

TABLE A–1
Food Composition

(Computer code number is for Wadsworth Diet Analysis program) (For purposes of calculations, use "0" for t, <1, <.1, <.01, etc.)

DA + Code	Food Description	Quantity	Measure	Wt (g)	H₂O (g)	Ener (kcal)	Prot (g)	Carb (g)	Dietary Fiber (g)	Fat (g)	Sat	Mono	Poly	Trans
	SPICES, CONDIMENTS, SAUCES													
	Spices													
807	Allspice, ground	1	teaspoon(s)	2	<1	5	<1	1	<1	<1	0.05	0.01	0.04	—
1171	Anise seeds	1	teaspoon(s)	2	<1	7	<1	1	<1	<1	0.01	0.21	0.07	—
729	Baker's yeast active	1	teaspoon(s)	4	<1	12	2	2	1	<1	0.02	0.10	0.00	—
683	Baking powder, double acting, w/phosphate	1	teaspoon(s)	5	<1	2	<.1	1	<.1	0	0.00	0.00	0.00	0
1611	Baking soda	1	teaspoon(s)	5	<.1	0	0	0	0	0	0.00	0.00	0.00	0
8552	Basil	1	teaspoon(s)	1	1	<1	<.1	<.1	<.1	<.1	0.00	0.00	0.00	—
34959	Basil, fresh	1	piece(s)	1	<1	<1	<.1	<.1	<.1	<.1	0.00	0.00	0.00	—
808	Basil, ground	1	teaspoon(s)	1	<.1	4	<1	1	1	<.1	0.00	0.01	0.03	—
809	Bay leaf	1	teaspoon(s)	1	<.1	2	<.1	<1	<1	<.1	0.01	0.01	0.01	—
11720	Betel leaves	1	ounce(s)	28	—	17	2	2	0	<.1	—	—	—	—
818	Black pepper	1	teaspoon(s)	2	<1	5	<1	1	1	<1	0.02	0.02	0.02	—
730	Brewer's yeast	1	teaspoon(s)	3	<1	8	1	1	1	0	0.00	0.00	0.00	0
35417	Capers	1	teaspoon(s)	4	—	2	0	0	0	0	0.00	0.00	0.00	—
1172	Caraway seeds	1	teaspoon(s)	2	<1	7	<1	1	1	<1	0.01	0.15	0.07	—
819	Cayenne pepper	1	teaspoon(s)	2	<1	6	<1	1	<1	<1	0.06	0.05	0.15	—
1173	Celery seeds	1	teaspoon(s)	2	<1	8	<1	1	<1	1	0.04	0.32	0.07	—
1174	Chervil, dried	1	teaspoon(s)	1	<.1	1	<1	<1	<.1	<.1	0.00	0.01	0.01	—
810	Chili powder	1	teaspoon(s)	3	<1	8	<1	1	1	<1	0.08	0.09	0.19	—
8553	Chives, chopped	1	teaspoon(s)	1	1	<1	<.1	<.1	<.1	<.1	0.00	0.00	0.00	—
8556	Cilantro	1	teaspoon(s)	2	1	<1	<.1	<.1	<.1	<.1	0.00	0.00	0.00	—
811	Cinnamon, ground	1	teaspoon(s)	2	<1	6	<.1	2	1	<.1	0.01	0.01	0.01	—
812	Cloves, ground	1	teaspoon(s)	2	<1	7	<1	1	1	<1	0.11	0.03	0.15	—
1175	Coriander leaf, dried	1	teaspoon(s)	1	<.1	2	<1	<1	<.1	<.1	0.00	0.01	0.00	—
1176	Coriander seeds	1	teaspoon(s)	2	<1	5	<1	1	1	<1	0.02	0.24	0.03	—
1706	Cornstarch	1	tablespoon(s)	8	1	30	<.1	7	<.1	<.1	0.00	0.00	0.00	—
11729	Cumin, ground	1	teaspoon(s)	5	—	11	<1	1	1	<1	—	—	—	—
1177	Cumin seeds	1	teaspoon(s)	2	<1	8	<1	1	<1	<1	0.03	0.29	0.07	—
1178	Curry powder	1	teaspoon(s)	2	<1	7	<1	1	1	<1	0.04	0.11	0.05	—
1179	Dill seeds	1	teaspoon(s)	2	<1	6	<1	1	<1	<1	0.02	0.20	0.02	—
1180	Dill weed, dried	1	teaspoon(s)	1	<.1	3	<1	1	<1	<1	0.01	0.01	0.00	—
34949	Dill weed, fresh	5	piece(s)	1	1	<1	<.1	<.1	<.1	<.1	0.00	0.01	0.00	—
4949	Fennel leaves, fresh	1	teaspoon(s)	1	1	<1	<.1	<.1	0	<.1	0.00	0.00	0.00	—
1181	Fennel seeds	1	teaspoon(s)	2	<1	7	<1	1	1	<1	0.01	0.20	0.03	—
1182	Fenugreek seeds	1	teaspoon(s)	4	<1	12	1	2	1	<1	0.05	—	—	—
11733	Garam masala, powder	1	ounce(s)	28	—	107	4	13	0	4	—	—	—	—
1067	Garlic clove	1	item(s)	3	2	4	<1	1	<.1	<.1	0.00	0.00	0.01	—
813	Garlic powder	1	teaspoon(s)	3	<1	9	<1	2	<1	<.1	0.00	0.00	0.01	—
1183	Ginger, ground	1	teaspoon(s)	2	<1	6	<1	1	<1	<1	0.03	0.02	0.02	—
1068	Ginger root	2	teaspoon(s)	4	3	3	<.1	1	<.1	<.1	0.01	0.01	0.01	—
35497	Leeks, bulb & lower leaf, freeze-dried	¼	cup(s)	1	<.1	3	<1	1	<.1	<.1	0.00	0.00	0.01	—
1184	Mace, ground	1	teaspoon(s)	2	<1	8	<1	1	<1	1	0.16	0.19	0.07	—
1185	Marjoram, dried	1	teaspoon(s)	1	<.1	2	<.1	<1	<1	<.1	0.00	0.01	0.03	—
1186	Mustard seeds, yellow	1	teaspoon(s)	3	<1	15	1	1	<1	1	0.05	0.65	0.18	—
814	Nutmeg, ground	1	teaspoon(s)	2	<1	12	<1	1	<1	1	0.57	0.07	0.01	—
2747	Onion flakes, dehydrated	1	teaspoon(s)	2	<.1	6	<1	1	<1	<.1	0.00	0.00	0.01	—
1187	Onion powder	1	teaspoon(s)	2	<1	7	<1	2	<1	<.1	0.00	0.00	0.01	—
815	Oregano, ground	1	teaspoon(s)	2	<1	5	<1	1	1	<1	0.04	0.01	0.08	—
816	Paprika	1	teaspoon(s)	2	<1	6	<1	1	1	<1	0.04	0.03	0.17	—
817	Parsley, dried	1	teaspoon(s)	0	<.1	1	<.1	<1	<.1	<.1	0.00	0.01	0.00	—
1189	Poppy seeds	1	teaspoon(s)	3	<1	15	1	1	<1	1	0.14	0.18	0.86	—
1190	Poultry seasoning	1	teaspoon(s)	2	<1	5	<1	1	<1	<1	0.05	0.02	0.03	—
1191	Pumpkin pie spice, powder	1	teaspoon(s)	2	<1	6	<.1	1	<1	<1	0.11	0.02	0.01	—
1192	Rosemary, dried	1	teaspoon(s)	1	<1	4	<.1	1	1	<1	0.09	0.04	0.03	—
11723	Rosemary, fresh	1	teaspoon(s)	1	<1	1	<.1	<1	<.1	<.1	0.02	0.01	0.01	—
2722	Saffron powder	1	teaspoon(s)	1	<.1	2	<.1	<1	<.1	<.1	0.01	0.00	0.01	—
11724	Sage	1	ounce(s)	28	—	34	1	4	0	1	—	—	—	—
1193	Sage, ground	1	teaspoon(s)	1	<.1	2	<.1	<1	<1	<.1	0.05	0.01	0.01	—
822	Salt, table	¼	teaspoon(s)	2	<.1	0	0	0	0	0	0.00	0.00	0.00	0
30189	Salt substitute	¼	teaspoon(s)	1	—	<.1	0	<.1	0	0	0.00	0.00	0.00	0
30190	Salt substitute, seasoned	¼	teaspoon(s)	1	—	1	<.1	<1	0	<.1	0.00	—	—	—
1194	Savory, ground	1	teaspoon(s)	1	<1	4	<.1	1	1	<1	0.05	—	—	—
820	Sesame seed kernels, toasted	1	teaspoon(s)	3	<1	15	<1	1	<1	1	0.18	0.49	0.57	—
11725	Sorrel	1	tablespoon(s)	9	—	2	<1	<1	<.1	<.1	0.00	—	—	—

Chol (mg)	Calc (mg)	Iron (mg)	Magn (mg)	Pota (mg)	Sodi (mg)	Zinc (mg)	Vit A (RAE) (µg)	Thia (mg)	Vit E (mg α)	Ribo (mg)	Niac (mg)	Vit B6 (mg)	Fola (µg)	Vit C (mg)	Vit B12 (µg)	Sele (µg)
0	13	0.13	3	20	1	0.02	1	0.00	—	0.00	0.05	0.00	1	1	0	<.1
0	14	0.78	4	30	<1	0.11	<1	0.01	—	0.01	0.06	0.01	<1	<1	0	<1
0	3	0.66	4	80	2	0.26	0	0.09	0.00	0.22	1.59	0.06	94	<.1	<.1	1
0	339	0.52	2	<1	363	0.00	0	0.00	0.00	0.00	0.00	0.00	0	0	0	<.1
0	0	0.00	0	0	1259	0.00	0	0.00	0.00	0.00	0.00	0.00	0	0	0	<.1
0	1	0.03	1	4	<.1	0.01	2	0.00	—	0.00	0.01	0.00	1	<1	0	<.1
0	1	—	<1	2	<.1	0.00	—	0.00	—	0.00	0.01	0.00	<1	—	0	<.1
0	30	0.59	6	48	<1	0.08	7	0.00	0.10	0.00	0.10	0.03	4	1	0	<.1
0	5	0.26	1	3	<1	0.02	2	0.00	—	0.00	0.01	0.01	1	<1	0	<.1
0	110	2.29	—	156	2	—	—	0.04	—	0.07	0.20	—	—	1	0	—
0	9	0.61	4	26	1	0.03	<1	0.02	—	0.01	0.02	0.01	<1	<1	0	<.1
0	6	0.47	6	51	3	0.21	0	0.42	—	0.11	1.00	0.07	104	0	0	0
0	0	0.00	—	—	140	—	0	—	—	—	—	—	—	0	—	—
0	14	0.34	5	28	<1	0.12	<1	0.01	0.05	0.01	0.08	0.01	<1	<1	0	<1
0	3	0.14	3	36	1	0.04	37	0.01	0.54	0.02	0.16	0.04	2	1	0	<1
0	35	0.90	9	28	3	0.14	<.1	0.00	0.02	0.01	0.06	0.02	<1	<1	0	<1
0	8	0.19	1	28	<1	0.05	2	0.00	—	0.00	0.03	0.01	2	<1	0	<1
0	7	0.37	4	50	26	0.07	39	0.01	—	0.02	0.21	0.10	3	2	0	<1
0	1	0.02	<1	3	<.1	0.01	2	0.00	0.76	0.00	0.01	0.00	1	1	0	<.1
0	1	0.03	<1	8	1	0.00	—	0.00	—	0.00	0.02	0.00	1	1	0	<.1
0	28	0.88	1	12	1	0.05	<1	0.00	0.02	0.00	0.03	0.01	1	1	0	<.1
0	14	0.18	6	23	5	0.02	1	0.00	0.18	0.01	0.03	0.00	2	2	0	<1
0	7	0.25	4	27	1	0.03	2	0.01	—	0.01	0.06	0.00	2	3	0	<1
0	13	0.29	6	23	1	0.08	0	0.00	—	0.01	0.04	—	0	<1	0	<1
0	<1	0.04	<1	<1	1	0.00	0	0.00	0.00	0.00	0.00	0.00	0	0	0	<1
0	20	—	—	44	5	—	—	—	—	—	—	—	—	—	—	—
0	20	1.39	8	38	4	0.10	1	0.01	0.07	0.01	0.10	0.01	<1	<1	0	<1
0	10	0.59	5	31	1	0.08	1	0.01	0.44	0.01	0.07	0.02	3	<1	0	<1
0	32	0.34	5	25	<1	0.11	<.1	0.01	—	0.00	0.06	0.01	<1	<1	0	<1
0	18	0.49	5	33	2	0.03	3	0.00	—	0.00	0.03	0.02	2	1	0	0
0	2	—	1	7	1	0.01	—	0.00	—	0.00	0.02	0.00	2	—	0	—
0	1	0.03	—	4	<.1	—	—	0.00	—	0.00	0.01	0.00	—	<1	0	—
0	24	0.37	8	34	2	0.07	<1	0.01	—	0.01	0.12	0.01	—	<1	0	0
0	7	1.24	7	28	2	0.09	<1	0.01	—	0.01	0.06	0.02	2	<1	0	<1
0	215	9.25	94	411	28	1.07	—	0.10	—	0.09	0.71	—	0	0	0	—
0	5	0.05	1	12	1	0.03	0	0.01	0.00	0.00	0.02	0.04	<.1	1	0	<1
0	2	0.08	2	31	1	0.07	0	0.01	0.02	0.00	0.02	0.08	<.1	1	0	1
0	2	0.21	3	24	1	0.08	<1	0.00	0.32	0.00	0.09	0.02	1	<1	0	1
0	1	0.02	2	17	1	0.01	0	0.00	0.01	0.00	0.03	0.01	<1	<1	0	<.1
0	3	0.06	1	19	<1	0.01	<1	0.01	—	0.00	0.03	0.01	3	1	0	<.1
0	4	0.24	3	8	1	0.04	1	0.01	—	0.01	0.02	0.00	1	<1	0	<.1
0	12	0.50	2	9	<1	0.02	2	0.00	0.01	0.00	0.02	0.01	2	<1	0	<.1
0	17	0.33	10	23	<1	0.19	<.1	0.02	0.10	0.01	0.26	0.01	3	<.1	0	4
0	4	0.07	4	8	<1	0.05	<1	0.01	0.00	0.00	0.03	0.00	2	<1	0	<.1
0	4	0.03	2	27	<1	0.03	<.1	0.01	0.00	0.00	0.02	0.03	3	1	0	<.1
0	8	0.05	3	20	1	0.05	0	0.01	0.01	0.00	0.01	0.03	3	<1	0	<.1
0	24	0.66	4	25	<1	0.07	5	0.01	0.28	0.00	0.09	0.02	4	1	0	<.1
0	4	0.50	4	49	1	0.09	55	0.01	0.63	0.04	0.32	0.08	2	1	0	<.1
0	4	0.29	1	11	1	0.01	2	0.00	0.02	0.00	0.02	0.00	1	<1	0	<.1
0	41	0.26	9	20	1	0.29	0	0.02	0.03	0.00	0.03	0.01	2	1	0	<.1
0	15	0.53	3	10	<1	0.05	1	0.00	0.03	0.00	0.04	0.02	2	1	0	<.1
0	12	0.34	2	11	1	0.04	<1	0.00	0.02	0.00	0.04	0.01	1	<1	0	<.1
0	15	0.35	3	11	1	0.04	2	0.01	—	0.01	0.01	0.02	4	1	0	<.1
0	2	0.05	1	5	<1	0.01	1	0.00	—	0.00	0.01	0.00	1	<1	0	—
0	1	0.08	2	12	1	0.01	<1	0.00	—	0.00	0.01	0.01	1	1	0	<.1
0	170	—	45	110	1	0.48	—	0.03	—	—	—	—	—	—	0	—
0	12	0.20	3	7	<.1	0.03	2	0.01	0.05	0.00	0.04	0.02	2	<1	0	<.1
0	<1	0.00	<.1	<1	581	0.00	0	0.00	0.00	0.00	0.00	0.00	0	0	0	<.1
0	7	0.00	<.1	604	<.1	—	0	—	—	—	—	—	—	0	—	—
0	0	0	476	<1	—	0	—	—	—	—	—	—	0	—	—	—
0	30	0.53	5	15	<1	0.06	4	0.01	—	—	0.06	0.03	—	1	0	<.1
0	4	0.21	9	11	1	0.28	<.1	0.03	0.01	0.01	0.15	0.00	3	0	0	<.1
0	—	—	—	—	<1	—	—	—	—	—	—	—	—	—	—	—

TABLE A–1
Food Composition

(Computer code number is for Wadsworth Diet Analysis program)   (For purposes of calculations, use "0" for t, <1, <.1, <.01, etc.)

DA + Code	Food Description	Quantity	Measure	Wt (g)	H₂O (g)	Ener (kcal)	Prot (g)	Carb (g)	Dietary Fiber (g)	Fat (g)	Fat Breakdown (g)			
											Sat	Mono	Poly	Trans
	SPICES, CONDIMENTS, SAUCES—Continued													
11721	Spearmint	1	teaspoon(s)	2	2	1	<.1	<1	<1	<.1	0.00	0.00	0.01	—
35498	Sweet green peppers, freeze-dried	¼	cup(s)	2	<.1	5	<1	1	<1	<.1	0.01	0.00	0.03	—
11726	Tamarind leaves	1	ounce(s)	28	—	33	2	5	0	1	—	—	—	—
11727	Tarragon	1	ounce(s)	28	—	14	1	2	0	<1	—	—	—	—
1195	Tarragon, ground	1	teaspoon(s)	2	<1	5	<1	1	<1	<1	0.03	0.01	0.06	—
11728	Thyme, fresh	1	teaspoon(s)	1	1	1	<.1	<1	<1	<.1	0.00	0.00	0.00	—
821	Thyme, ground	1	teaspoon(s)	1	<1	4	<1	1	1	<1	0.04	0.01	0.02	—
1196	Turmeric, ground	1	teaspoon(s)	2	<1	8	<1	1	<1	<1	0.07	0.04	0.05	—
11995	Wasabi	1	tablespoon(s)	14	11	11	1	2	<1	<.1	—	—	—	—
1188	White pepper	1	teaspoon(s)	2	<1	7	<1	2	1	<.1	0.02	0.02	0.01	—
	Condiments													
674	Catsup or ketchup	1	tablespoon(s)	15	11	14	<1	4	<1	<.1	0.01	0.01	0.04	—
703	Dill pickle	1	ounce(s)	28	26	5	<1	1	<1	<.1	0.01	0.00	0.02	—
1641	Horseradish sauce, prepared	1	teaspoon(s)	5	3	10	<1	<1	<.1	1	0.59	0.28	0.04	—
140	Mayonnaise, low calorie	1	tablespoon(s)	16	10	37	<.1	3	0	3	0.53	0.72	1.70	—
138	Mayonnaise w/soybean oil	1	tablespoon(s)	14	2	99	<1	1	0	11	1.64	2.70	5.89	0.04
1682	Mustard, brown	1	teaspoon(s)	5	4	5	<1	<1	<.1	<1	—	—	—	—
700	Mustard, yellow	1	teaspoon(s)	5	4	3	<1	<1	<1	<1	0.01	0.11	0.03	—
706	Sweet pickle relish	1	tablespoon(s)	15	9	20	<.1	5	<1	<.1	0.01	0.03	0.02	—
141	Tartar sauce	2	tablespoon(s)	28	9	144	<1	4	<.1	14	2.14	4.13	7.57	—
	Sauces													
685	Barbecue sauce	2	tablespoon(s)	31	25	23	1	4	<1	1	0.08	0.24	0.21	—
834	Cheese sauce	¼	cup(s)	70	49	121	5	5	<1	9	4.19	2.67	1.81	—
32123	Chili enchilada sauce, green	2	tablespoon(s)	57	53	15	1	3	1	<1	0.04	0.04	0.13	0
32122	Chili enchilada sauce, red	2	tablespoon(s)	32	24	27	1	5	2	1	0.08	0.05	0.43	0
29688	Hoisin sauce	1	tablespoon(s)	16	7	35	1	7	<1	1	0.09	0.15	0.27	—
16670	Mole poblano sauce	½	cup(s)	133	103	155	5	11	2	11	2.67	5.15	2.91	—
29689	Oyster sauce	1	tablespoon(s)	16	13	8	<1	2	<.1	<.1	0.01	0.01	0.01	—
1655	Pepper sauce or tabasco	1	teaspoon(s)	5	5	1	<.1	<.1	<.1	<.1	0.01	0.00	0.02	—
347	Salsa	2	tablespoon(s)	16	14	4	<1	1	<1	<.1	0.00	0.00	0.02	—
841	Soy sauce	1	tablespoon(s)	18	13	10	1	2	0	<.1	0.00	0.00	0.01	—
839	Sweet & sour sauce	2	tablespoon(s)	39	30	37	<.1	9	<.1	<.1	0.00	0.00	0.00	—
1613	Teriyaki sauce	1	tablespoon(s)	18	12	15	1	3	<.1	0	0.00	0.00	0.00	0
25294	Tomato sauce	½	cup(s)	112	100	46	2	8	2	1	0.18	0.29	0.72	0
728	White sauce, medium	¼	cup(s)	63	47	92	2	6	<1	7	1.78	2.78	1.79	—
1654	Worcestershire sauce	1	teaspoon(s)	6	4	4	0	1	0	0	0.00	0.00	0.00	0
	Vinegar													
30853	Balsamic	1	tablespoon(s)	15	—	10	0	2	0	0	0.00	0.00	0.00	0
727	Cider	1	tablespoon(s)	15	14	2	0	1	0	0	0.00	0.00	0.00	0
1673	Distilled	1	tablespoon(s)	15	14	2	0	1	0	0	0.00	0.00	0.00	0
15439	Tarragon	1	tablespoon(s)	16	—	0	0	0	0	0	0.00	0.00	0.00	0
	MIXED FOODS, SOUPS, SANDWICHES													
	Mixed Dishes													
16652	Almond chicken	1	cup(s)	242	186	280	22	16	3	15	1.91	6.07	5.62	—
25224	Barbecued chicken	2	piece(s)	177	100	325	27	15	<1	17	4.63	6.78	3.71	0
25227	Bean burrito	1	item(s)	149	82	327	17	33	6	15	8.30	4.73	0.85	0
9516	Beef & vegetable fajita	1	item(s)	223	144	397	23	35	3	18	5.50	7.53	3.45	—
16796	Beef or pork egg roll	2	item(s)	128	85	227	10	19	1	12	2.88	5.96	2.64	—
177	Beef stew w/vegetables, prepared	1	cup(s)	245	201	220	16	15	3	11	4.40	4.50	0.50	—
30233	Beef stroganoff w/noodles	1	cup(s)	256	190	343	20	23	2	19	7.37	5.62	4.47	—
16651	Cashew chicken	1	cup(s)	242	131	644	43	17	3	46	7.75	20.83	14.47	—
475	Cheese pizza	2	slice(s)	126	60	281	15	41	0	6	3.08	1.98	0.98	—
30330	Cheese quesadilla	1	item(s)	54	19	183	6	18	1	10	3.49	3.42	2.16	—
215	Chicken & noodles, prepared	1	cup(s)	240	170	365	22	26	1	18	5.10	7.10	3.90	—
30239	Chicken & vegetables w/broccoli, onion, bamboo shoots in soy based sauce	1	cup(s)	162	112	287	22	6	1	19	5.13	7.65	4.68	—
25093	Chicken cacciatore	1	cup(s)	230	166	266	28	5	1	14	3.98	5.78	3.11	0
28020	Chicken fried turkey steak	3	ounce(s)	85	48	122	13	12	1	2	0.59	0.37	0.78	—
218	Chicken pot pie	1	cup(s)	252	154	542	23	42	3	31	9.79	12.52	7.03	—
30240	Chicken teriyaki	1	cup(s)	244	163	339	51	13	1	7	1.78	2.03	1.71	—
25119	Chicken waldorf salad	½	cup(s)	100	68	178	14	6	1	11	1.76	3.18	5.05	0
25099	Chili con carne	¾	cup(s)	215	175	197	14	21	7	7	2.55	2.83	0.54	0
1062	Coleslaw	¾	cup(s)	90	73	62	1	11	1	2	0.35	0.64	1.22	—
1896	Combination pizza, w/meat & vegetables	2	slice(s)	158	75	368	26	43	5	11	3.07	5.09	1.83	—

PAGE KEY: A–4 = Breads/Baked Goods A–8 = Cereal/Rice/Pasta A–12 = Fruit A–16 = Vegetables/Legumes A–26 = Nuts/Seeds A–28 = Vegetarian A–30 = Dairy A–36 = Eggs A–36 = Seafood A–38 = Meats A–42 = Poultry A–42 = Processed meats A–44 = Beverages A–48 = Fats/Oils A–50 = Sweets A–52 = Spices/Condiments/Sauces A–54 = Mixed Foods/Soups/Sandwiches A–60 = Fast food A–76 = Convenience A–78 = Baby foods

Chol (mg)	Calc (mg)	Iron (mg)	Magn (mg)	Pota (mg)	Sodi (mg)	Zinc (mg)	Vit A (RAE) (µg)	Thia (mg)	Vit E (mg α)	Ribo (mg)	Niac (mg)	Vit B$_6$ (mg)	Fola (µg)	Vit C (mg)	Vit B$_{12}$ (µg)	Sele (µg)
0	4	0.23	1	9	1	0.02	4	0.00	—	0.00	0.02	0.00	2	<1	0	—
0	2	0.17	3	51	3	0.04	3	0.02	0.06	0.02	0.12	0.04	4	30	0	<.1
0	85	1.48	20	—	—	—	—	0.07	—	0.03	1.16	—	—	1	0	—
0	48	—	14	128	3	0.17	—	0.04	—	—	—	—	—	1	0	—
0	18	0.52	6	48	1	0.06	3	0.00	0.10	0.02	0.14	0.04	4	1	0	<.1
0	3	0.14	1	5	<.1	0.01	2	0.00	—	0.00	0.01	0.00	<1	1	0	—
0	26	1.73	3	11	1	0.09	3	0.01	—	0.01	0.07	0.01	4	1	0	<.1
0	4	0.91	4	56	1	0.10	0	0.00	—	0.01	0.11	0.04	1	1	0	<.1
0	13	0.11	—	—	—	—	—	0.02	—	0.01	0.07	—	—	11	0	—
0	6	0.34	2	2	<1	0.03	0	0.00	0.10	0.00	0.01	0.00	<1	1	0	<.1
0	3	0.08	3	57	167	0.04	7	0.00	0.22	0.07	0.23	0.02	2	2	0	<.1
0	3	0.15	3	33	363	0.04	3	0.00	0.03	0.01	0.02	0.00	<1	1	0	0
2	5	0.00	1	7	15	0.01	—	0.00	0.03	0.01	0.00	0.00	1	<.1	<.1	—
4	<.1	0.00	<.1	2	80	0.02	0	0.00	0.32	0.00	0.00	0.00	0	0	0	—
5	2	0.07	<1	5	78	0.02	12	0.00	0.72	0.00	0.00	0.08	1	0	<.1	<1
0	6	0.09	1	7	68	0.02	0	0.00	0.09	0.00	0.01	0.00	<1	<.1	0	—
0	4	0.09	2	8	56	0.03	<1	0.00	0.01	0.00	0.02	0.00	<1	<1	0	2
0	<1	0.13	1	4	122	0.02	1	0.00	0.06	0.00	0.03	0.00	<1	<1	0	0
11	6	0.21	1	10	200	0.05	—	0.00	0.97	0.00	0.01	0.08	2	<1	<.1	—
0	6	0.28	6	54	255	0.06	<1	0.01	0.01	0.01	0.28	0.02	1	2	0	<1
20	128	0.15	6	21	578	0.68	56	0.00	—	0.08	0.02	0.01	3	<1	<.1	2
0	5	0.36	9	126	62	0.11	—	0.03	0.00	0.02	0.63	0.06	6	44	0	0
0	7	1.05	11	231	114	0.15	—	0.02	0.00	0.22	0.61	0.34	7	<1	0	<1
<1	5	0.16	4	19	258	0.05	0	0.00	0.04	0.03	0.19	0.01	4	<.1	0	<1
1	37	1.51	57	283	305	0.95	—	0.07	1.72	0.09	1.82	0.09	14	5	<.1	—
0	5	0.03	1	9	437	0.01	0	0.00	0.00	0.02	0.24	0.00	2	<.1	<.1	1
0	1	0.06	1	6	32	0.01	4	0.00	—	0.00	0.01	0.01	<.1	<1	0	—
0	5	0.16	2	34	69	0.04	5	0.01	0.19	0.01	0.13	0.02	3	2	0	<.1
0	3	0.36	6	32	1029	0.07	0	0.01	0.00	0.02	0.61	0.03	3	0	0	—
0	5	0.20	1	8	98	0.01	0	0.00	—	0.01	0.12	0.04	<1	0	0	—
0	5	0.31	11	41	690	0.02	0	0.01	0.00	0.01	0.23	0.02	4	0	0	<1
0	21	1.08	19	431	199	0.30	48	0.05	0.39	0.05	1.18	0.13	15	15	0	1
4	74	0.21	9	98	221	0.26	—	0.04	—	0.12	0.25	0.03	3	1	<1	—
0	6	0.30	1	45	56	0.01	—	0.00	0.00	0.01	0.04	0.00	0	1	0	—
0	0	0.00	—	—	0	—	0	—	—	—	—	—	—	—	0	—
0	1	0.09	3	15	<1	0.00	—	0.00	0.00	0.00	0.00	0.00	0	0	0	<.1
0	1	0.09	0	2	<1	0.00	0	0.00	0.00	0.00	0.00	0.00	0	0	0	5
0	0	0.00	—	—	0	0	0	—	—	—	—	—	—	—	0	—
40	69	1.97	60	549	526	1.62	—	0.09	4.11	0.20	9.48	0.44	26	7	<1	—
120	26	1.64	31	387	477	2.69	69	0.07	0.01	0.37	6.92	0.39	15	5	<1	19
38	331	2.95	45	384	514	1.92	119	0.24	0.01	0.29	1.82	0.15	115	4	<1	18
45	84	3.74	37	476	757	3.51	—	0.39	0.80	0.30	5.37	0.38	23	27	2	—
74	30	1.66	20	248	547	0.91	—	0.32	1.28	0.25	2.55	0.19	20	4	<1	—
71	29	2.90	—	613	292	—	—	0.15	0.51	0.17	4.70	—	—	17	<.1	15
74	70	3.26	37	393	818	3.66	—	0.21	1.25	0.31	3.80	0.21	17	1	2	—
96	74	2.92	94	640	1355	2.24	—	0.23	4.11	0.22	19.76	0.88	64	11	<1	—
19	233	1.16	32	219	672	1.63	147	0.37	—	0.33	4.96	0.09	69	3	<1	27
13	132	1.21	13	77	230	0.64	—	0.13	0.43	0.14	1.09	0.04	6	15	<.1	—
103	26	2.20	—	149	600	—	—	0.05	—	0.17	4.30	—	—	0	—	29
84	22	1.38	29	344	962	1.70	—	0.08	1.12	0.17	7.90	0.32	13	8	<1	—
103	45	2.21	37	444	451	2.01	53	0.10	0.00	0.21	9.20	0.54	15	8	<1	22
27	69	1.34	19	197	139	1.08	5	0.15	0.00	0.18	3.46	0.22	21	<1	<1	16
69	64	3.38	38	393	651	1.93	607	0.40	1.06	0.40	7.24	0.24	31	11	<1	—
157	52	3.27	67	589	3209	3.75	—	0.15	0.59	0.37	16.69	0.89	23	6	1	—
42	20	0.78	24	197	246	1.13	21	0.04	0.62	0.10	4.05	0.25	15	2	<1	11
27	43	3.16	50	646	865	2.44	25	0.13	0.02	0.23	3.01	0.18	56	10	1	10
7	41	0.53	9	163	21	0.06	48	0.06	—	0.06	0.24	0.11	24	29	0	1
41	202	3.07	36	357	765	2.23	117	0.43	—	0.35	3.92	0.19	65	3	1	22

TABLE A–1
Food Composition

(Computer code number is for Wadsworth Diet Analysis program)   (For purposes of calculations, use "0" for t, <1, <.1, <.01, etc.)

DA + Code	Food Description	Quantity	Measure	Wt (g)	H₂O (g)	Ener (kcal)	Prot (g)	Carb (g)	Dietary Fiber (g)	Fat (g)	Fat Breakdown (g)			
											Sat	Mono	Poly	Trans
	MIXED FOODS, SOUPS, SANDWICHES—Continued													
1574	Crab cakes, from blue crab	1	item(s)	60	43	93	12	<1	0	5	0.89	1.69	1.36	—
32144	Enchiladas w/green chili sauce (enchiladas verdes)	1	item(s)	144	104	207	9	18	3	12	6.35	3.65	0.96	0
2793	Falafel patty	3	item(s)	51	18	170	7	16	0	9	1.22	5.19	2.12	—
28546	Fettuccine alfredo	1	cup(s)	222	81	247	11	42	1	3	1.61	0.79	0.43	0
32146	Flautas	3	item(s)	162	78	438	25	36	4	22	8.22	8.80	2.29	—
29629	Fried rice w/meat or poultry	1	cup(s)	198	129	329	12	41	1	12	2.27	3.53	5.69	—
16649	General tso chicken	1	cup(s)	146	91	293	19	16	1	17	3.98	6.27	5.27	—
1826	Green salad	¾	cup(s)	104	99	17	1	3	2	<.1	0.01	0.00	0.04	—
1814	Hummus	½	cup(s)	123	80	218	6	25	5	11	1.38	6.04	2.56	—
16650	Kung pao chicken	1	cup(s)	162	88	431	29	11	2	31	5.19	13.95	9.69	—
16622	Lamb curry	1	cup(s)	236	188	256	28	3	1	14	3.93	4.92	3.35	—
25253	Lasagna w/ground beef	1	cup(s)	237	157	288	18	22	2	15	7.47	4.84	0.84	0
442	Macaroni & cheese	1	cup(s)	200	122	393	15	40	1	19	8.18	6.72	2.66	—
25105	Meat loaf	1	slice(s)	115	85	244	17	7	<1	16	6.15	6.89	0.83	0
16646	Moo shi pork	1	cup(s)	151	77	512	19	5	1	46	6.84	14.80	22.07	—
16788	Nachos w/beef, beans, cheese, tomatoes, & onions	7	item(s)	551	284	1496	40	119	19	99	22.34	40.19	30.69	—
1668	Pepperoni pizza	2	slice(s)	142	66	362	20	40	1	14	4.47	6.28	2.33	—
655	Potato salad	½	cup(s)	125	95	179	3	14	2	10	1.79	3.10	4.67	—
29637	Ravioli, meat filled, w/tomato or meat sauce, canned	1	cup(s)	251	196	220	9	38	2	4	1.58	1.49	0.41	—
25109	Salisbury steaks w/mushroom sauce	1	serving(s)	135	102	251	17	9	1	15	5.98	6.67	0.76	0
16637	Shrimp creole w/rice	1	cup(s)	243	176	311	27	28	1	9	1.83	3.79	2.88	—
497	Spaghetti & meat balls w/tomato sauce, prepared	1	cup(s)	248	174	330	19	39	3	12	3.90	4.40	2.20	—
28585	Spicy thai noodles (pad thai)	8	ounce(s)	231	74	222	9	36	3	6	0.83	3.33	1.83	0
33073	Stir fried pork & vegetables w/rice	1	cup(s)	235	173	349	15	34	2	16	5.55	6.87	2.62	0
28588	Stuffed shells	2½	item(s)	299	189	292	18	33	3	10	3.81	3.57	1.62	0
16821	Sushi w/egg in seaweed	6	piece(s)	156	117	190	9	20	<1	8	2.09	3.02	1.55	—
16819	Sushi w/vegetables & fish	6	piece(s)	156	102	217	8	44	2	1	0.16	0.14	0.20	—
16820	Sushi w/vegetables in seaweed	6	piece(s)	156	110	182	3	41	1	<1	0.10	0.11	0.11	—
25266	Sweet & sour pork	¾	cup(s)	249	206	264	29	17	1	8	2.59	3.51	1.48	0
16824	Tabouli, tabbouleh, or tabuli	1	cup(s)	160	124	199	3	16	4	15	2.04	10.83	1.37	—
25276	Three bean salad	½	cup(s)	99	82	95	2	10	3	6	0.76	1.41	3.48	0
160	Tuna salad	½	cup(s)	103	65	192	16	10	0	9	1.58	2.96	4.23	0
25241	Turkey & noodles	1	cup(s)	319	228	271	24	21	1	9	2.39	3.48	2.27	0
16794	Vegetable egg roll	2	item(s)	128	90	202	5	20	2	12	2.46	5.71	2.65	—
16818	Vegetable sushi, no fish	6	piece(s)	156	99	225	5	50	2	<1	0.11	0.10	0.14	—
	Sandwiches													
1744	Bacon, lettuce & tomato w/mayonnaise	1	item(s)	164	97	349	11	34	2	19	4.54	7.22	6.07	—
30287	Bologna & cheese w/margarine	1	item(s)	111	46	350	13	28	1	20	8.55	8.40	2.28	—
30286	Bologna w/margarine	1	item(s)	83	34	256	7	26	1	13	4.08	6.31	2.07	—
16546	Cheese	1	item(s)	83	31	262	10	27	1	13	5.59	4.77	1.67	—
8789	Cheeseburger, large, plain	1	item(s)	185	72	609	30	47	0	33	14.84	12.74	2.44	—
8624	Cheeseburger, large, w/bacon, vegetables, & condiments	1	item(s)	195	85	608	32	37	2	37	16.24	14.49	2.71	—
1745	Club w/bacon, chicken, tomato, lettuce, & mayonnaise	1	item(s)	246	137	555	31	48	3	26	5.94	—	—	—
1908	Cold cut submarine w/cheese & vegetables	1	item(s)	228	132	456	22	51	2	19	6.81	8.23	2.28	—
30247	Corned beef	1	item(s)	130	75	268	19	25	2	10	3.75	3.96	0.80	—
25283	Egg salad	1	item(s)	126	72	278	10	29	1	13	2.96	3.97	4.79	—
16686	Fried egg	1	item(s)	96	50	226	10	26	1	9	2.29	3.51	1.64	—
16547	Grilled cheese	1	item(s)	83	27	292	10	27	1	16	6.22	6.29	2.54	—
16659	Gyro w/onion & tomato	1	item(s)	105	67	170	12	21	1	4	1.53	1.41	0.43	—
1906	Ham & cheese	1	item(s)	146	74	352	21	33	2	15	6.44	6.74	1.38	—
31890	Ham w/mayonnaise	1	item(s)	112	55	282	14	27	1	13	3.06	5.04	3.79	—
756	Hamburger, double patty, large, w/condiments & vegetables	1	item(s)	226	121	540	34	40	0	27	10.52	10.33	2.80	—
8793	Hamburger, large, plain	1	item(s)	137	58	426	23	32	2	23	8.38	9.88	2.14	—
8795	Hamburger, large, w/vegetables & condiments	1	item(s)	218	121	512	26	40	3	27	10.42	11.42	2.20	—
25134	Hot chicken salad	1	item(s)	98	49	239	16	23	1	9	2.83	2.61	2.76	0
1411	Hot dog w/bun, plain	1	item(s)	98	53	242	10	18	2	15	5.11	6.85	1.71	—
25133	Hot turkey salad	1	item(s)	98	50	221	16	23	1	7	2.23	1.76	2.28	0
30249	Pastrami	1	item(s)	134	71	331	14	27	2	18	6.18	8.74	1.02	—

PAGE KEY: A–4 = Breads/Baked Goods   A–8 = Cereal/Rice/Pasta   A–12 = Fruit   A–16 = Vegetables/Legumes   A–26 = Nuts/Seeds   A–28 = Vegetarian
A–30 = Dairy   A–36 = Eggs   A–36 = Seafood   A–38 = Meats   A–42 = Poultry   A–42 = Processed meats   A–44 = Beverages   A–48 = Fats/Oils
A–50 = Sweets   A–52 = Spices/Condiments/Sauces   A–54 = Mixed Foods/Soups/Sandwiches   A–60 = Fast food   A–76 = Convenience   A–78 = Baby foods

Chol (mg)	Calc (mg)	Iron (mg)	Magn (mg)	Pota (mg)	Sodi (mg)	Zinc (mg)	Vit A (RAE) (µg)	Thia (mg)	Vit E (mg α)	Ribo (mg)	Niac (mg)	Vit B$_6$ (mg)	Fola (µg)	Vit C (mg)	Vit B$_{12}$ (µg)	Sele (µg)
90	63	0.65	20	194	198	2.45	34	0.05	—	0.05	1.74	0.10	32	2	4	24
27	266	1.08	38	251	276	1.27	—	0.07	0.03	0.16	1.28	0.18	45	59	<1	6
0	28	1.74	42	298	150	0.77	1	0.07	—	0.08	0.53	0.06	47	1	0	1
9	153	1.88	32	123	386	1.48	51	0.35	0.00	0.34	2.60	0.06	103	1	<1	35
73	146	2.66	61	223	886	3.44	0	0.10	0.10	0.17	3.00	0.27	96	0	1	37
102	36	2.66	31	182	821	1.42	—	0.30	1.60	0.19	3.51	0.24	24	3	<1	—
65	27	1.49	24	250	906	1.40	—	0.10	1.62	0.19	6.28	0.28	17	12	<1	—
0	13	0.65	11	178	27	0.22	59	0.03	—	0.05	0.57	0.08	38	24	0	<1
0	60	1.93	36	213	298	1.34	0	0.11	0.92	0.06	0.49	0.49	73	10	0	3
64	49	1.96	63	428	907	1.50	—	0.15	4.32	0.15	13.23	0.59	43	8	<1	—
89	36	2.97	40	495	495	6.62	—	0.09	1.30	0.28	8.05	0.20	27	1	3	—
68	222	2.33	40	437	493	2.81	108	0.19	0.22	0.29	3.02	0.20	50	10	1	22
30	323	2.26	42	263	800	1.95	327	0.25	0.72	0.40	2.18	0.10	12	<1	<1	—
85	54	2.09	21	278	423	3.55	27	0.08	0.00	0.29	3.77	0.13	20	<1	2	17
172	30	1.45	26	330	1078	1.83	—	0.50	5.39	0.38	2.90	0.31	22	8	1	—
82	699	6.71	205	1067	1611	7.55	—	0.31	7.71	0.50	5.62	0.85	59	14	1	—
28	129	1.87	17	305	534	1.04	105	0.27	—	0.47	6.09	0.11	74	3	<1	26
85	24	0.81	19	318	661	0.39	40	0.10	—	0.08	1.11	0.18	9	13	0	5
17	28	2.04	23	337	1354	1.19	—	0.22	0.70	0.20	2.88	0.14	17	22	<1	—
60	64	2.21	23	282	370	3.66	27	0.11	0.00	0.30	4.00	0.13	22	<1	2	17
181	101	4.44	64	439	381	1.73	—	0.29	2.07	0.40	4.77	0.22	12	18	1	—
89	124	3.70	—	665	1009	—	82	0.25	—	0.30	4.00	—	—	22	—	22
37	32	1.58	50	187	598	1.08	38	0.18	0.36	0.13	1.88	0.17	44	22	<.1	3
46	39	2.65	32	394	574	2.07	80	0.51	0.38	0.20	5.07	0.30	102	18	<1	23
35	241	3.18	63	462	543	1.68	280	0.32	0.00	0.36	4.64	0.30	109	15	<1	36
217	42	1.63	18	128	527	0.98	—	0.12	0.67	0.29	1.33	0.13	29	2	<1	—
11	24	2.18	25	204	340	0.79	—	0.26	0.25	0.07	2.77	0.15	14	4	<1	—
0	20	1.54	20	99	153	0.70	—	0.20	0.12	0.04	1.86	0.14	10	2	0	—
74	41	1.78	35	622	624	2.53	64	0.80	0.20	0.37	6.69	0.65	14	10	1	50
0	29	1.25	36	246	799	0.48	—	0.08	2.43	0.05	1.14	0.11	31	29	0	—
0	26	0.96	15	144	224	0.31	12	0.04	0.89	0.06	0.26	0.06	31	9	0	3
13	17	1.03	19	182	412	0.57	25	0.03	0.00	0.07	6.87	0.08	8	2	1	42
77	60	2.69	33	379	576	2.64	108	0.23	0.29	0.32	6.40	0.30	60	1	1	34
60	29	1.61	18	193	548	0.51	—	0.16	1.28	0.21	1.59	0.10	27	6	<1	—
0	23	2.40	23	158	369	0.84	—	0.28	0.16	0.06	2.44	0.13	15	4	0	—
20	76	2.54	27	328	837	0.98	—	0.39	1.16	0.27	3.81	0.20	31	15	<1	—
35	221	2.18	24	185	940	1.68	—	0.30	0.56	0.33	2.77	0.12	21	<.1	1	—
16	60	1.96	15	112	598	0.85	—	0.29	0.50	0.21	2.73	0.08	19	<.1	<1	—
19	216	1.75	20	135	655	1.14	—	0.25	0.47	0.29	2.04	0.07	19	<.1	<1	—
96	91	5.46	39	644	1589	5.55	185	0.48	—	0.57	11.17	0.28	74	0	3	39
111	162	4.74	45	332	1043	6.83	82	0.31	—	0.41	6.63	0.31	86	2	2	33
72	116	4.05	47	463	855	1.65	—	0.61	1.53	0.44	11.92	0.59	48	9	1	—
36	189	2.51	68	394	1651	2.58	71	1.00	—	0.80	5.49	0.14	87	12	1	31
46	67	2.67	20	187	1177	2.24	—	0.24	0.21	0.25	3.23	0.10	22	2	1	—
217	107	2.60	18	147	494	0.94	94	0.26	0.13	0.43	2.27	0.16	82	1	1	24
207	80	2.25	17	120	433	0.85	—	0.27	0.66	0.41	2.06	0.10	34	0	<1	—
19	219	1.76	21	137	696	1.15	—	0.19	0.72	0.28	1.86	0.06	13	<.1	<1	—
34	46	1.85	21	209	272	2.30	—	0.24	0.26	0.21	3.14	0.13	18	4	1	—
58	130	3.24	16	291	771	1.37	96	0.31	0.29	0.48	2.69	0.20	76	3	1	23
36	59	2.10	23	245	1033	1.50	—	0.71	0.50	0.31	4.89	0.26	19	0	<1	—
122	102	5.85	50	570	791	5.67	5	0.36	—	0.38	7.57	0.54	77	1	4	26
71	74	3.58	27	267	474	4.11	0	0.29	—	0.29	6.25	0.23	60	0	2	27
87	96	4.93	44	480	824	4.88	24	0.41	—	0.37	7.28	0.33	83	3	2	34
39	114	1.93	20	150	470	1.22	28	0.20	0.28	0.23	4.93	0.20	54	<1	<1	17
44	24	2.31	13	143	670	1.98	0	0.24	—	0.27	3.65	0.05	48	<.1	1	26
37	113	2.04	22	167	459	1.09	23	0.19	0.29	0.21	4.36	0.23	54	<1	<1	20
51	68	2.64	23	243	1335	2.69	—	0.29	0.27	0.27	4.77	0.13	21	2	1	—

TABLE A–1
Food Composition

(Computer code number is for Wadsworth Diet Analysis program)   (For purposes of calculations, use "0" for t, <1, <.1, <.01, etc.)

DA + Code	Food Description	Quantity	Measure	Wt (g)	H₂O (g)	Ener (kcal)	Prot (g)	Carb (g)	Dietary Fiber (g)	Fat (g)	Sat	Mono	Poly	Trans
	MIXED FOODS, SOUPS, SANDWICHES—Continued													
16701	Peanut butter	1	item(s)	93	24	344	13	37	3	17	3.55	8.16	4.58	—
30306	Peanut butter & jelly	1	item(s)	93	24	330	11	42	3	15	3.00	6.87	3.82	—
1910	Roast beef, plain	1	item(s)	139	68	346	22	33	1	14	3.61	6.80	1.71	—
1909	Roast beef submarine w/mayonnaise & vegetables	1	item(s)	216	127	410	29	44	—	13	7.09	1.84	2.61	—
1907	Steak w/mayonnaise & vegetables	1	item(s)	204	104	459	30	52	2	14	3.81	5.34	3.35	—
25288	Tuna salad	1	item(s)	179	102	414	24	29	2	22	3.61	5.46	11.43	—
31891	Turkey w/mayonnaise	1	item(s)	143	75	330	29	26	1	11	2.61	3.25	4.40	—
30283	Turkey submarine w/cheese, lettuce, tomato, & mayonnaise	1	item(s)	277	156	583	37	51	3	25	7.15	8.03	7.81	—
	Soups													
25296	Bean	1	cup(s)	301	253	191	14	29	6	2	0.67	0.83	0.53	0
711	Bean with pork, condensed, prepared w/water	1	cup(s)	265	223	180	8	24	9	6	1.59	2.28	1.91	—
713	Beef noodle, condensed, prepared w/water	1	cup(s)	244	224	83	5	9	1	3	1.15	1.24	0.49	—
825	Cheese, condensed, prepared w/milk	1	cup(s)	251	207	231	9	16	1	15	9.11	4.09	0.45	—
826	Chicken broth, condensed, prepared w/water	1	cup(s)	244	234	39	5	1	0	1	0.39	0.59	0.27	—
25297	Chicken noodle	1	cup(s)	286	258	117	11	11	1	3	0.78	1.10	0.66	—
827	Chicken noodle, condensed, prepared w/water	1	cup(s)	241	222	75	4	9	1	2	0.65	1.11	0.55	—
724	Chicken noodle, dehydrated, prepared w/water	1	cup(s)	252	237	58	2	9	<1	1	0.31	0.52	0.39	—
823	Cream of asparagus, condensed, prepared w/milk	1	cup(s)	248	213	161	6	16	1	8	3.32	2.08	2.23	—
824	Cream of celery, condensed, prepared w/milk	1	cup(s)	248	214	164	6	15	1	10	3.94	2.46	2.65	—
708	Cream of chicken, condensed, prepared w/milk	1	cup(s)	248	210	191	7	15	<1	11	4.64	4.46	1.64	—
715	Cream of chicken, condensed, prepared w/water	1	cup(s)	244	221	117	3	9	<1	7	2.07	3.27	1.49	—
709	Cream of mushroom, condensed, prepared w/milk	1	cup(s)	248	210	203	6	15	<1	14	5.13	2.98	4.61	—
716	Cream of mushroom, condensed, prepared w/water	1	cup(s)	244	220	129	2	9	<1	9	2.44	1.71	4.22	—
25298	Cream of vegetable	1	cup(s)	285	251	165	7	15	2	9	1.56	4.62	1.92	—
16689	Egg drop	1	cup(s)	244	229	73	8	1	0	4	1.15	1.52	0.59	—
25138	Golden squash	1	cup(s)	258	224	144	8	21	2	4	0.84	2.18	0.88	0
16663	Hot & sour	1	cup(s)	244	210	161	15	5	1	8	2.72	3.40	1.20	—
28054	Lentil chowder	1	cup(s)	229	188	150	11	27	12	<1	0.09	0.08	0.22	0
28560	Macaroni & bean	1	cup(s)	229	129	136	6	21	5	3	0.48	2.06	0.59	0
714	Manhattan clam chowder, condensed, prepared w/water	1	cup(s)	244	224	78	2	12	1	2	0.38	0.38	1.29	—
28561	Minestrone	1	cup(s)	230	177	99	4	16	5	2	0.32	1.30	0.43	0
717	Minestrone, condensed, prepared w/water	1	cup(s)	241	220	82	4	11	1	3	0.55	0.70	1.11	—
28038	Mushroom & wild rice	1	cup(s)	230	188	81	4	12	2	<1	0.05	0.02	0.15	0
828	New England clam chowder, condensed, prepared w/milk	1	cup(s)	248	211	164	9	17	1	7	2.95	2.26	1.09	—
28036	New England style clam chowder	1	cup(s)	229	207	83	3	15	2	<1	0.08	0.03	0.05	0
28566	Old country pasta	1	cup(s)	228	164	135	6	20	3	3	1.17	1.60	0.63	0
725	Onion, dehydrated, prepared w/water	1	cup(s)	246	237	27	1	5	1	1	0.12	0.32	0.07	—
16667	Shrimp gumbo	1	cup(s)	244	206	171	10	19	3	7	1.34	3.02	2.05	—
28037	Southwestern corn chowder	1	cup(s)	229	202	102	5	18	2	<1	0.12	0.12	0.20	0
25140	Split pea	1	cup(s)	165	117	85	4	19	2	<1	0.07	0.03	0.18	0
718	Split pea with ham, condensed, prepared w/water	1	cup(s)	253	207	190	10	28	2	4	1.77	1.80	0.63	—
710	Tomato, condensed, prepared w/milk	1	cup(s)	248	210	161	6	22	3	6	2.90	1.61	1.12	—
719	Tomato, condensed, prepared w/water	1	cup(s)	244	220	85	2	17	<1	2	0.37	0.44	0.95	—
726	Tomato vegetable, dehydrated, prepared w/water	1	cup(s)	253	237	56	2	10	1	1	0.38	0.30	0.08	—
28595	Turkey noodle	1	cup(s)	228	203	106	8	14	2	2	0.27	1.06	0.67	0
28051	Turkey vegetable	1	cup(s)	227	203	98	11	8	2	1	0.32	0.17	0.30	0
720	Vegetable beef, condensed, prepared w/water	1	cup(s)	244	224	78	6	10	<1	2	0.85	0.81	0.12	—
28598	Vegetable gumbo	1	cup(s)	229	168	153	4	26	3	4	0.61	2.93	0.56	0
25141	Vegetable	1	cup(s)	252	225	96	5	20	4	—	0.06	0.04	0.16	0
721	Vegetarian vegetable, condensed, prepared w/water	1	cup(s)	241	223	72	2	12	—	2	0.29	0.82	0.72	—

PAGE KEY: A–4 = Breads/Baked Goods A–8 = Cereal/Rice/Pasta A–12 = Fruit A–16 = Vegetables/Legumes A–26 = Nuts/Seeds A–28 = Vegetarian A–30 = Dairy A–36 = Eggs A–36 = Seafood A–38 = Meats A–42 = Poultry A–42 = Processed meats A–44 = Beverages A–48 = Fats/Oils A–50 = Sweets A–52 = Spices/Condiments/Sauces A–54 = Mixed Foods/Soups/Sandwiches A–60 = Fast food A–76 = Convenience A–78 = Baby foods

Chol (mg)	Calc (mg)	Iron (mg)	Magn (mg)	Pota (mg)	Sodi (mg)	Zinc (mg)	Vit A (RAE) (µg)	Thia (mg)	Vit E (mg α)	Ribo (mg)	Niac (mg)	Vit B$_6$ (mg)	Fola (µg)	Vit C (mg)	Vit B$_{12}$ (µg)	Sele (µg)
1	80	2.47	62	272	479	1.25	0	0.33	2.39	0.25	6.46	0.17	43	0	<.1	—
1	68	2.11	53	239	409	1.06	—	0.27	2.02	0.21	5.45	0.15	37	<1	<.1	—
51	54	4.23	31	316	792	3.39	11	0.38	—	0.31	5.87	0.26	57	2	1	29
73	41	2.81	67	330	845	4.38	30	0.41	—	0.41	5.96	0.32	71	6	2	26
73	92	5.16	49	524	798	4.53	20	0.41	—	0.37	7.30	0.37	90	6	2	42
53	100	3.29	35	302	795	1.08	46	0.26	0.35	0.26	12.29	0.48	70	1	2	71
69	78	3.10	34	315	490	2.94	—	0.30	0.74	0.33	6.64	0.46	24	0	<1	—
70	324	3.88	51	552	2408	2.66	—	0.53	1.19	0.49	12.50	0.54	46	5	2	—
5	80	3.08	61	590	690	1.41	26	0.27	0.03	0.15	3.61	0.23	139	3	<1	8
3	85	2.15	48	421	996	1.09	48	0.09	0.80	0.04	0.59	0.04	34	2	<.1	8
5	15	1.10	5	100	952	1.54	7	0.07	0.68	0.06	1.07	0.04	20	<1	<1	7
48	289	0.80	20	341	1019	0.68	359	0.06	—	0.33	0.50	0.08	10	1	<1	7
0	10	0.51	2	210	776	0.24	0	0.01	0.05	0.07	3.35	0.02	5	0	<1	0
24	26	1.34	16	335	776	0.77	49	0.15	0.02	0.16	5.57	0.13	40	1	<1	10
7	17	0.77	5	55	1106	0.39	36	0.05	0.10	0.06	1.39	0.03	22	<1	<1	6
10	5	0.50	8	33	577	0.20	3	0.20	0.13	0.08	1.09	0.03	18	0	<.1	10
22	174	0.87	20	360	1042	0.92	62	0.10	—	0.28	0.88	0.06	30	4	<1	8
32	186	0.69	22	310	1009	0.20	114	0.07	—	0.25	0.44	0.06	7	1	<1	5
27	181	0.67	17	273	1047	0.67	179	0.07	—	0.26	0.92	0.07	7	1	1	8
10	34	0.61	2	88	986	0.63	163	0.03	—	0.06	0.82	0.02	2	<1	<.1	7
20	179	0.60	20	270	918	0.64	35	0.08	1.24	0.28	0.91	0.06	10	2	<1	4
2	46	0.51	5	100	881	0.59	15	0.05	0.95	0.09	0.72	0.01	5	1	<.1	1
1	68	1.38	17	312	784	0.74	100	0.12	1.06	0.20	3.27	0.12	37	10	<1	5
103	21	0.75	5	220	729	0.48	—	0.02	0.29	0.19	3.03	0.05	15	0	<1	—
4	203	1.63	39	412	500	1.72	454	0.17	0.53	0.38	1.15	0.15	32	10	1	8
34	29	1.89	29	382	1561	1.51	—	0.27	0.12	0.25	4.97	0.20	13	1	<1	—
<1	47	4.07	55	590	26	1.44	163	0.21	0.06	0.12	1.69	0.30	164	13	0	3
<1	64	1.86	32	254	489	0.46	174	0.15	0.35	0.13	1.36	0.09	59	7	0	9
2	27	1.63	12	188	578	0.98	56	0.03	0.34	0.04	0.82	0.10	10	4	4	9
0	68	1.76	31	273	423	0.38	138	0.10	0.23	0.10	0.69	0.07	47	12	0	4
2	34	0.92	7	313	911	0.75	118	0.05	—	0.04	0.94	0.10	36	1	0	8
0	27	1.08	26	332	267	0.87	4	0.06	0.07	0.21	2.97	0.14	18	4	<.1	4
22	186	1.49	22	300	992	0.79	57	0.07	0.45	0.24	1.03	0.13	10	3	10	13
2	69	1.29	26	430	236	0.66	34	0.07	0.02	0.12	1.02	0.20	17	12	3	4
6	51	2.32	47	434	319	0.69	114	0.20	0.01	0.15	2.42	0.23	65	17	<.1	9
0	12	0.15	5	64	849	0.05	0	0.03	0.00	0.06	0.48	0.00	2	<1	0	2
51	99	2.34	51	515	515	0.93	—	0.19	1.90	0.10	2.54	0.19	59	26	<1	—
1	65	1.10	24	374	200	0.73	46	0.08	0.09	0.14	1.65	0.22	27	37	<1	2
0	30	1.25	33	352	608	0.57	112	0.12	0.00	0.09	1.67	0.21	61	9	0	<1
8	23	2.28	48	400	1007	1.32	23	0.15	—	0.08	1.47	0.07	3	2	<1	8
17	159	1.81	22	449	744	0.30	64	0.13	1.24	0.25	1.52	0.16	17	68	<1	2
0	12	1.76	7	264	695	0.24	29	0.09	2.32	0.05	1.42	0.11	15	66	0	<1
0	8	0.63	20	104	1146	0.18	10	0.06	0.35	0.05	0.79	0.05	10	6	0	5
24	27	1.40	22	200	372	0.67	81	0.20	0.02	0.11	2.68	0.15	45	5	<1	13
20	36	1.30	22	383	328	0.90	110	0.08	0.01	0.09	3.33	0.27	21	10	<1	9
5	17	1.12	5	173	791	1.54	95	0.04	0.37	0.05	1.03	0.08	10	2	<1	4
0	52	1.90	35	313	471	0.56	15	0.17	0.58	0.07	1.59	0.16	51	18	0	4
0	41	2.45	38	688	674	0.78	118	0.12	0.00	0.13	2.37	0.27	33	23	0	5
0	22	1.08	7	210	822	0.46	116	0.05	—	0.05	0.92	0.06	10	1	0	4

TABLE A–1

Food Composition

(Computer code number is for Wadsworth Diet Analysis program)   (For purposes of calculations, use "0" for t, <1, <.1, <.01, etc.)

DA + Code	Food Description	Quantity	Measure	Wt (g)	H₂O (g)	Ener (kcal)	Prot (g)	Carb (g)	Dietary Fiber (g)	Fat (g)	Fat Breakdown (g)			
											Sat	Mono	Poly	Trans
	FAST FOOD													
	Arby's													
36094	Au jus sauce	1	serving(s)	85	—	5	<1	1	<.1	<.1	0.02	—	—	—
751	Beef 'n cheddar sandwich	1	item(s)	198	—	480	23	43	2	24	8.00	—	—	—
9279	Cheddar curly fries	1	serving(s)	170	—	460	6	54	4	24	6.00	—	—	—
36131	Chocolate shake	1	serving(s)	397	—	480	10	84	0	16	8.00	—	—	—
36045	Curly fries, large	1	serving(s)	198	—	620	8	78	7	30	7.00	—	—	—
36044	Curly fries, medium	1	serving(s)	128	—	400	5	50	4	20	5.00	—	—	—
9265	Fish fillet sandwich	1	item(s)	220	—	529	23	50	2	27	7.00	9.20	10.60	—
752	Ham 'n cheese sandwich	1	item(s)	170	—	340	23	35	1	13	4.50	—	—	—
36048	Homestyle fries, large	1	serving(s)	213	—	560	6	79	6	24	6.00	—	—	—
36047	Homestyle fries, medium	1	serving(s)	142	—	370	4	53	4	16	4.00	—	—	—
33465	Homestyle fries, small	1	serving(s)	113	—	300	3	42	3	13	3.50	—	—	—
9267	Italian sub sandwich	1	item(s)	312	—	780	29	49	3	53	15.00	—	—	—
36041	Market Fresh grilled chicken caesar salad w/o dressing	1	serving(s)	338	—	230	33	8	3	8	3.50	—	—	—
9291	Roast beef deluxe sandwich, light	1	item(s)	182	—	296	18	33	6	10	3.00	5.00	2.00	—
9251	Roast beef sandwich, giant	1	item(s)	228	—	480	32	41	3	23	10.00	—	—	—
9249	Roast beef sandwich, junior	1	item(s)	129	—	310	16	34	2	13	4.50	—	—	—
750	Roast beef sandwich, regular	1	item(s)	157	—	350	21	34	2	16	6.00	—	—	—
2009	Roast beef sandwich, super	1	item(s)	245	—	470	22	47	3	23	7.00	—	—	—
9269	Roast beef sub sandwich	1	item(s)	334	—	760	35	47	3	48	16.00	—	—	—
9295	Roast chicken deluxe sandwich, light	1	item(s)	194	—	260	23	33	3	5	1.00	—	—	—
9293	Roast turkey deluxe sandwich, light	1	item(s)	194	—	260	23	33	3	5	0.50	—	—	—
36132	Strawberry shake	1	serving(s)	397	—	500	11	87	0	13	8.00	—	—	—
9273	Turkey sub sandwich	1	item(s)	306	—	630	26	51	2	37	9.00	—	—	—
36130	Vanilla shake	1	serving(s)	397	—	470	10	83	0	15	7.00	—	—	—
	Auntie Anne's													
35371	Cheese dipping sauce	1	serving(s)	35	—	100	3	4	0	8	4.00	—	—	—
35353	Cinnamon sugar soft pretzel	1	item(s)	120	—	350	9	74	2	2	0.00	—	—	—
35354	Cinnamon sugar soft pretzel w/butter	1	item(s)	120	—	450	8	83	3	9	5.00	—	—	—
35372	Marinara dipping sauce	1	serving(s)	35	—	10	0	4	0	0	0.00	0.00	0.00	0
35357	Original soft pretzel	1	item(s)	120	—	340	10	72	3	1	0.00	—	—	—
35358	Original soft pretzel w/butter	1	item(s)	120	—	370	10	72	3	4	2.00	—	—	—
35359	Parmesan herb soft pretzel	1	item(s)	120	—	390	11	74	4	5	2.50	—	—	—
35360	Parmesan herb soft pretzel w/butter	1	item(s)	120	—	440	10	72	9	13	7.00	—	—	—
35361	Sesame soft pretzel	1	item(s)	120	—	350	11	63	3	6	1.00	—	—	—
35362	Sesame soft pretzel w/butter	1	item(s)	120	—	410	12	64	7	12	4.00	—	—	—
35364	Sour cream & onion soft pretzel	1	item(s)	120	—	310	9	66	2	1	0.00	—	—	—
35366	Sour cream & onion soft pretzel w/butter	1	item(s)	120	—	340	9	66	2	5	3.00	—	—	—
35373	Sweet mustard dipping sauce	1	serving(s)	35	—	60	1	8	0	2	1.00	—	—	—
35367	Whole wheat soft pretzel	1	item(s)	120	—	350	11	72	7	2	0.00	—	—	—
35368	Whole wheat soft pretzel w/butter	1	item(s)	120	—	370	11	72	7	5	1.50	—	—	—
	Boston Market													
34975	Bbq baked beans	¾	cup(s)	201	—	270	8	48	12	5	2.00	—	—	—
34976	Black beans & rice	1	cup(s)	227	—	300	8	45	5	10	1.50	—	—	—
34978	Butternut squash	¾	cup(s)	193	—	150	2	25	6	6	4.00	—	—	—
35006	Caesar side salad	1	serving(s)	119	—	300	5	13	1	26	4.50	—	—	—
34979	Chicken gravy	1	ounce(s)	28	—	15	0	2	0	1	0.00	—	—	—
34973	Chicken pot pie	1	item(s)	425	—	750	26	57	2	46	14.00	—	—	—
35007	Cole slaw	¾	cup(s)	184	—	300	2	30	3	19	3.00	—	—	—
35057	Cornbread	1	item(s)	68	—	200	3	33	1	6	1.50	—	—	—
35008	Cranberry walnut relish	¾	cup(s)	210	—	350	3	75	3	5	0.00	—	—	—
34980	Creamed spinach	¾	cup(s)	181	—	260	9	11	2	20	13.00	—	—	—
34981	Glazed carrots	¾	cup(s)	153	—	280	1	35	4	15	3.00	—	—	—
34983	Green bean casserole	¾	cup(s)	170	—	80	1	9	2	5	1.50	—	—	—
34982	Green beans	¾	cup(s)	85	—	70	1	6	2	4	0.50	—	—	—
34967	Half chicken, w/skin	1	item(s)	277	—	590	70	4	0	33	10.00	—	—	—
34984	Homestyle mashed potatoes	¾	cup(s)	173	—	210	4	30	2	9	5.00	—	—	—
34985	Homestyle mashed potatoes & gravy	1	cup(s)	201	—	230	4	32	3	9	5.00	—	—	—
34969	Honey glazed ham	5	ounce(s)	142	—	210	24	10	0	8	3.00	—	—	—
34988	Hot cinnamon apples	¾	cup(s)	181	—	250	0	56	3	5	0.50	—	—	—
34989	Macaroni & cheese	¾	cup(s)	192	—	280	13	33	1	11	6.00	—	—	—
34970	Meatloaf	5	ounce(s)	142	—	282	20	15	1	17	7.28	—	—	—
35012	Old-fashioned potato salad	¾	cup(s)	150	—	200	3	22	2	12	2.00	—	—	—

PAGE KEY: A–4 = Breads/Baked Goods   A–8 = Cereal/Rice/Pasta   A–12 = Fruit   A–16 = Vegetables/Legumes   A–26 = Nuts/Seeds   A–28 = Vegetarian   A–30 = Dairy   A–36 = Eggs   A–36 = Seafood   A–38 = Meats   A–42 = Poultry   A–42 = Processed meats   A–44 = Beverages   A–48 = Fats/Oils   A–50 = Sweets   A–52 = Spices/Condiments/Sauces   A–54 = Mixed Foods/Soups/Sandwiches   A–60 = Fast food   A–76 = Convenience   A–78 = Baby foods

Chol (mg)	Calc (mg)	Iron (mg)	Magn (mg)	Pota (mg)	Sodi (mg)	Zinc (mg)	Vit A (RAE) (µg)	Thia (mg)	Vit E (mg α)	Ribo (mg)	Niac (mg)	Vit B$_6$ (mg)	Fola (µg)	Vit C (mg)	Vit B$_{12}$ (µg)	Sele (µg)
0	0	0.00	—	—	386	—	0	—	—	—	—	—	—	0	—	—
90	100	3.60	—	—	1240	—	0	—	—	—	—	—	—	1	—	—
5	60	1.80	—	—	1290	—	0	—	—	—	—	—	—	15	—	—
45	500	0.72	—	—	370	—	38	—	—	—	—	—	—	2	—	—
0	0	2.70	—	—	1540	—	0	—	—	—	—	—	—	21	—	—
0	0	1.80	—	—	990	—	0	—	—	—	—	—	—	15	—	—
43	90	3.78	—	450	864	—	10	0.35	—	0.31	5.60	—	—	1	—	—
90	150	2.70	—	—	1450	—	20	—	—	—	—	—	—	1	—	—
0	0	1.80	—	—	1070	—	0	—	—	—	—	—	—	30	—	—
0	0	1.08	—	—	710	—	0	—	—	—	—	—	—	21	—	—
0	0	0.72	—	—	570	—	0	—	—	—	—	—	—	15	—	—
120	250	2.70	—	—	2440	—	—	—	—	—	—	—	—	2	—	—
80	200	1.80	—	—	920	—	—	—	—	—	—	—	—	42	—	—
42	130	4.50	—	392	826	—	40	0.27	—	0.49	8.40	—	—	8	—	—
110	60	5.40	—	—	1440	—	0	—	—	—	—	—	—	0	—	—
70	60	2.70	—	—	740	—	0	—	—	—	—	—	—	0	—	—
85	60	3.60	—	—	950	—	0	—	—	—	—	—	—	0	—	—
85	80	3.60	—	—	1130	—	40	—	—	—	—	—	—	1	—	—
130	300	4.50	—	—	2230	—	40	—	—	—	—	—	—	4	—	—
40	100	2.70	—	—	1010	—	—	—	—	—	—	—	—	2	—	—
40	80	1.80	—	—	980	—	—	—	—	—	—	—	—	1	—	—
15	350	0.36	—	—	340	—	36	—	—	—	—	—	—	1	—	—
100	200	0.36	—	—	2170	—	—	—	—	—	—	—	—	2	—	—
45	500	1.08	—	—	360	—	39	—	—	—	—	—	—	2	—	—
10	100	0.00	—	—	510	—	—	—	—	—	—	—	—	0	—	—
0	20	1.98	—	—	410	—	0	—	—	—	—	—	—	0	—	—
25	30	2.34	—	—	430	—	—	—	—	—	—	—	—	0	—	—
0	0	0.00	—	—	180	—	—	—	—	—	—	—	—	0	—	—
0	30	2.34	—	—	900	—	0	—	—	—	—	—	—	0	—	—
10	30	2.16	—	—	930	—	—	—	—	—	—	—	—	0	—	—
10	80	1.80	—	—	780	—	—	—	—	—	—	—	—	1	—	—
30	60	1.80	—	—	660	—	—	—	—	—	—	—	—	1	—	—
0	20	2.88	—	—	840	—	0	—	—	—	—	—	—	0	—	—
15	20	2.70	—	—	860	—	—	—	—	—	—	—	—	0	—	—
0	30	1.98	—	—	920	—	—	—	—	—	—	—	—	0	—	—
10	40	2.16	—	—	930	—	—	—	—	—	—	—	—	0	—	—
40	0	0.00	—	—	120	—	0	—	—	—	—	—	—	0	—	—
0	30	1.98	—	—	1100	—	0	—	—	—	—	—	—	0	—	—
10	30	2.34	—	—	1120	—	—	—	—	—	—	—	—	0	—	—
0	100	3.60	—	—	540	—	42	—	—	—	—	—	—	6	—	—
0	40	1.80	—	—	1050	—	0	—	—	—	—	—	—	4	—	—
20	80	1.08	—	—	560	—	1150	—	—	—	—	—	—	30	—	—
15	100	0.72	—	—	690	—	—	—	—	—	—	—	—	9	—	—
0	0	0.00	—	—	180	—	0	—	—	—	—	—	—	0	—	—
110	40	4.50	—	—	1530	—	—	—	—	—	—	—	—	1	—	—
20	60	0.72	—	—	540	—	108	—	—	—	—	—	—	36	—	—
25	0	1.08	—	—	390	—	0	—	—	—	—	—	—	0	—	—
0	0	5.40	—	—	0	—	0	—	—	—	—	—	—	0	—	—
55	250	2.70	—	—	740	—	—	—	—	—	—	—	—	9	—	—
0	40	1.08	—	—	80	—	1000	—	—	—	—	—	—	1	—	—
5	20	0.72	—	—	670	—	—	—	—	—	—	—	—	2	—	—
0	40	0.36	—	—	250	—	30	—	—	—	—	—	—	5	—	—
290	0	2.70	—	—	1010	—	0	—	—	—	—	—	—	0	—	—
25	40	0.36	—	—	590	—	53	—	—	—	—	—	—	15	—	—
25	60	0.36	—	—	780	—	—	—	—	—	—	—	—	15	—	—
75	0	1.08	—	—	1460	—	0	—	—	—	—	—	—	0	—	—
0	20	0.36	—	—	45	—	—	—	—	—	—	—	—	0	—	—
30	300	1.44	—	—	890	—	—	—	—	—	—	—	—	0	—	—
68	91	2.46	—	—	592	—	—	—	—	—	—	—	—	1	—	—
15	60	1.08	—	—	450	—	0	—	—	—	—	—	—	6	—	—

TABLE A–1
Food Composition

(Computer code number is for Wadsworth Diet Analysis program)   (For purposes of calculations, use "0" for t, <1, <.1, <.01, etc.)

DA + Code	Food Description	Quantity	Measure	Wt (g)	H₂O (g)	Ener (kcal)	Prot (g)	Carb (g)	Dietary Fiber (g)	Fat (g)	Sat	Mono	Poly	Trans
	FAST FOOD—Continued													
34965	Quarter chicken, dark meat, no skin	1	item(s)	95	—	190	22	1	0	10	3.00	—	—	—
34966	Quarter chicken, dark meat, w/skin	1	item(s)	125	—	320	30	2	0	21	6.00	—	—	—
34963	Quarter chicken, white meat, no skin or wing	1	item(s)	140	—	170	33	2	0	4	1.00	—	—	—
34964	Quarter chicken, white meat, w/skin & wing	1	item(s)	152	—	280	40	2	0	12	3.50	—	—	—
34993	Rice pilaf	1	cup(s)	137	—	140	2	24	1	4	0.50	—	—	—
34968	Rotisserie turkey breast, skinless	5	ounce(s)	142	—	170	36	3	0	1	0.00	—	—	—
34998	Savory stuffing	1	cup(s)	132	—	190	4	27	2	8	1.50	—	—	—
34999	Squash casserole	¾	cup(s)	187	—	330	7	20	3	24	13.00	—	—	—
35003	Steamed vegetables	1	cup(s)	102	—	30	2	6	2	0	0.00	—	—	0
35004	Sweet potato casserole	¾	cup(s)	181	—	280	3	39	2	13	4.50	—	—	—
35005	Whole kernel corn	¾	cup(s)	146	—	180	5	30	2	4	0.50	—	—	—
	Burger King													
29731	Biscuit with sausage, egg, & cheese	1	item(s)	189	—	650	20	38	1	46	14.00	—	—	1
3739	BK Broiler chicken sandwich	1	item(s)	258	—	550	30	52	3	25	5.00	—	—	—
14249	Cheeseburger	1	item(s)	133	—	360	19	31	2	17	8.00	—	—	0.50
14251	Chicken sandwich	1	item(s)	224	—	660	25	53	3	39	8.00	—	—	2.20
3808	Chicken Tenders, 8 pieces	1	serving(s)	123	—	340	22	20	1	19	5.00	—	—	3.50
14259	Chocolate shake, small	1	item(s)	333	—	620	12	72	2	32	21.00	—	—	0
29732	Croissanwich w/sausage & cheese	1	item(s)	107	—	420	14	23	1	31	11.00	—	—	2
14261	Croissanwich w/sausage, egg, & cheese	1	item(s)	157	—	520	19	24	1	39	14.00	—	—	1.93
3809	Double cheeseburger	1	item(s)	189	—	540	32	32	2	31	15.00	—	—	1.50
14244	Double Whopper	1	item(s)	374	—	980	52	52	4	62	22.00	—	—	2
14245	Double Whopper w/cheese	1	item(s)	399	—	1070	57	53	4	70	27.00	—	—	2.50
14250	Fish Fillet sandwich	1	item(s)	185	—	520	18	44	2	30	8.00	—	—	1.12
14255	French fries, medium, salted	1	item(s)	117	—	360	4	46	4	18	5.00	—	—	4.50
14262	French toast sticks	1	serving(s)	112	—	390	6	46	2	20	4.50	—	—	4.50
14248	Hamburger	1	item(s)	121	—	310	17	31	2	13	5.00	—	—	0.50
14263	Hash brown rounds, small	1	serving(s)	75	—	230	2	23	2	15	4.00	—	—	5.0
14256	Onion rings, medium	1	serving(s)	91	—	320	4	40	3	16	4.00	—	—	3.50
39000	Tendercrisp chicken sandwich	1	item(s)	310	—	810	28	72	6	47	8.00	—	—	4.28
14258	Vanilla shake, small	1	item(s)	305	—	560	11	56	1	32	21.00	—	—	0
1736	Whopper	1	item(s)	291	—	710	31	52	4	43	13.00	—	—	1
14243	Whopper w/cheese	1	item(s)	316	—	800	36	53	4	50	18.00	—	—	2
	Carl's Jr													
10801	Carl's Catch fish sandwich	1	item(s)	201	—	530	18	55	2	28	7.00	—	1.89	—
10862	Carl's Famous Star hamburger	1	item(s)	254	—	590	24	50	3	32	9.00	—	—	—
10866	Charboiled chicken salad-to-go	1	item(s)	350	—	200	25	12	4	7	3.00	—	1.02	—
10855	Charboiled Sante Fe chicken sandwich	1	item(s)	220	—	540	28	37	2	31	8.00	—	—	—
10790	Chicken stars (6 pieces)	6	item(s)	90	—	260	13	14	1	16	4.50	—	1.71	—
34864	Chocolate shake, small	1	item(s)	595	—	530	14	96	0	10	7.00	—	—	—
10797	Crisscut fries	1	serving(s)	139	—	410	5	43	4	24	5.00	—	—	—
10799	Double western bacon cheeseburger	1	item(s)	308	—	920	51	65	3	50	21.00	—	6.55	—
34855	Famous bacon cheeseburger	1	item(s)	279	—	700	31	51	3	41	13.00	—	—	—
14238	French fries, small	1	serving(s)	92	—	290	5	37	3	14	3.00	—	—	—
10798	French toast dips w/o syrup	1	serving(s)	105	—	370	6	42	1	20	2.50	—	1.35	—
34856	Hamburger	1	item(s)	119	—	280	14	36	1	9	3.50	—	—	—
10802	Onion rings	1	serving(s)	127	—	430	7	53	3	22	5.00	—	0.84	—
38925	Six Dollar burger	1	item(s)	539	—	1000	39	72	6	82	25.00	—	—	—
34858	Spicy chicken sandwich	1	item(s)	198	—	480	14	47	2	26	5.00	—	—	—
34867	Strawberry shake, small	1	item(s)	595	—	510	14	91	0	10	7.00	—	—	—
10865	Super Star hamburger	1	item(s)	345	—	790	41	51	3	47	15.00	—	—	—
10818	Vanilla shake, small	1	item(s)	595	—	470	15	78	0	11	7.00	—	—	—
10770	Western bacon cheeseburger	1	item(s)	225	—	660	31	64	3	30	12.00	—	4.85	—
	Chick Fil-A													
38746	Biscuit w/bacon, egg, & cheese	1	item(s)	155	—	430	16	38	1	24	9.00	—	—	2.85
38747	Biscuit w/egg	1	item(s)	135	—	340	11	38	1	16	4.50	—	—	3
38748	Biscuit w/egg & cheese	1	item(s)	148	—	390	13	38	1	21	7.00	—	—	2.98
38753	Biscuit w/gravy	1	item(s)	191	—	310	5	44	1	13	3.50	—	—	3.98
38752	Biscuit w/sausage, egg, & cheese	1	item(s)	189	—	540	18	43	1	33	13.00	—	—	2.67
38741	Biscuit, plain	1	item(s)	78	—	260	4	38	1	11	2.50	—	—	2.97
38771	Carrot & raisin salad	1	item(s)	91	—	130	1	22	2	5	1.00	—	—	0
38761	Chargrilled chicken cool wrap	1	item(s)	245	—	380	29	54	3	6	3.00	—	—	0
38766	Chargrilled chicken garden salad	1	item(s)	275	—	180	22	9	3	6	3.00	—	—	0
38758	Chargrilled chicken sandwich	1	item(s)	157	—	280	26	30	1	7	1.50	—	—	0

PAGE KEY: A–4 = Breads/Baked Goods   A–8 = Cereal/Rice/Pasta   A–12 = Fruit   A–16 = Vegetables/Legumes   A–26 = Nuts/Seeds   A–28 = Vegetarian
A–30 = Dairy   A–36 = Eggs   A–36 = Seafood   A–38 = Meats   A–42 = Poultry   A–42 = Processed meats   A–44 = Beverages   A–48 = Fats/Oils
A–50 = Sweets   A–52 = Spices/Condiments/Sauces   A–54 = Mixed Foods/Soups/Sandwiches   A–60 = Fast food   A–76 = Convenience   A–78 = Baby foods

Chol (mg)	Calc (mg)	Iron (mg)	Magn (mg)	Pota (mg)	Sodi (mg)	Zinc (mg)	Vit A (RAE) (µg)	Thia (mg)	Vit E (mg α)	Ribo (mg)	Niac (mg)	Vit B$_6$ (mg)	Fola (µg)	Vit C (mg)	Vit B$_{12}$ (µg)	Sele (µg)
115	0	1.08	—	—	440	—	0	—	—	—	—	—	—	0	—	—
155	0	1.80	—	—	500	—	0	—	—	—	—	—	—	0	—	—
85	0	0.72	—	—	480	—	0	—	—	—	—	—	—	0	—	—
135	0	1.08	—	—	510	—	0	—	—	—	—	—	—	0	—	—
0	20	1.08	—	—	520	—	—	—	—	—	—	—	—	4	—	—
100	20	1.80	—	—	850	—	0	—	—	—	—	—	—	0	—	—
5	40	1.44	—	—	620	—	—	—	—	—	—	—	—	2	—	—
70	200	0.72	—	—	1110	—	—	—	—	—	—	—	—	5	—	—
0	40	0.35	—	—	135	—	389	—	—	—	—	—	—	18	—	—
10	40	1.08	—	—	190	—	—	—	—	—	—	—	—	9	—	—
0	0	0.36	—	—	170	—	20	—	—	—	—	—	—	5	—	—
190	150	2.70	—	—	1600	—	90	—	—	—	—	—	—	0	—	—
105	60	3.60	—	—	1110	—	—	0.46	—	0.23	10.50	—	—	6	—	—
50	150	3.60	—	—	790	—	63	0.25	—	0.32	4.18	—	—	1	—	—
70	80	2.70	—	—	1330	—	—	0.47	—	0.30	9.59	—	—	0	—	—
50	20	0.72	—	—	840	—	—	0.14	—	0.12	10.93	—	—	0	—	—
95	350	1.08	—	—	310	—	42	0.11	—	0.56	0.24	—	—	0	—	—
45	100	3.60	—	—	840	—	—	—	—	—	—	—	—	0	—	—
210	300	4.50	—	—	1090	—	140	0.36	—	0.42	4.35	—	—	0	—	—
100	250	4.50	—	—	1050	—	100	0.26	—	0.45	6.37	—	—	1	—	—
160	150	9.00	—	—	1070	—	—	0.40	—	0.60	11.08	—	—	9	—	—
185	300	9.00	—	—	1500	—	—	0.40	—	0.67	11.07	—	—	9	—	—
55	150	2.70	—	—	840	—	14	—	—	—	—	—	—	1	—	—
0	20	0.72	—	—	640	—	0	0.16	—	0.48	2.32	—	—	9	—	—
0	60	1.80	—	—	440	—	0	0.19	—	0.22	2.86	—	—	0	—	—
40	76	3.60	—	—	580	—	9	0.25	—	0.29	4.26	—	—	1	—	—
0	0	0.36	—	—	450	—	0	0.11	—	0.07	2.11	—	—	1	—	—
0	97	0.00	—	—	460	—	0	0.14	—	0.09	2.33	—	—	1	—	—
60	80	4.50	—	—	1800	—	—	—	—	—	—	—	—	9	—	—
95	300	0.36	—	—	220	—	39	0.11	—	0.64	0.22	—	—	0	—	—
85	150	6.30	—	—	980	—	52	0.39	—	0.44	7.33	—	—	9	—	—
110	250	6.30	—	—	1420	—	157	0.39	—	0.51	7.31	—	—	9	—	—
80	150	1.80	—	—	1030	—	60	—	—	—	—	—	—	2	—	—
70	100	4.50	—	—	910	—	—	—	—	—	—	—	—	6	—	—
75	150	1.80	—	—	440	—	—	—	—	—	—	—	—	5	—	—
95	200	2.70	—	—	1210	—	—	—	—	—	—	—	—	6	—	—
40	20	1.08	—	—	480	—	0	—	—	—	—	—	—	0	—	—
45	600	1.08	—	—	350	—	0	—	—	—	—	—	—	0	—	—
0	20	1.80	—	—	950	—	0	—	—	—	—	—	—	12	—	—
155	300	7.20	—	—	1770	—	—	—	—	—	—	—	—	1	—	—
95	200	5.40	—	—	1310	—	102	—	—	—	—	—	—	6	—	—
0	0	1.08	—	—	180	—	0	—	—	—	—	—	—	21	—	—
0	40	1.08	—	—	430	—	0	0.26	—	0.24	2.00	—	—	0	—	—
35	80	2.70	—	—	480	—	0	—	—	—	—	—	—	1	—	—
0	20	0.72	—	—	700	—	0	—	—	—	—	—	—	4	—	—
135	350	5.40	—	—	1690	—	—	—	—	—	—	—	—	21	—	—
40	100	2.70	—	—	1220	—	—	—	—	—	—	—	—	6	—	—
45	600	0.00	—	—	330	—	0	—	—	—	—	—	—	0	—	—
130	100	7.20	—	—	980	—	—	—	—	—	—	—	—	9	—	—
50	600	0.00	—	—	350	—	0	—	—	—	—	—	—	0	—	—
85	200	5.40	—	—	1410	—	40	—	—	—	—	—	—	1	—	—
265	150	3.60	—	—	1070	—	—	—	—	—	—	—	—	0	—	—
245	80	2.70	—	—	740	—	—	—	—	—	—	—	—	0	—	—
260	150	2.70	—	—	960	—	—	—	—	—	—	—	—	0	—	—
5	60	1.80	—	—	930	—	0	—	—	—	—	—	—	0	—	—
280	150	3.60	—	—	1030	—	—	—	—	—	—	—	—	0	—	—
0	60	1.80	—	—	670	—	0	—	—	—	—	—	—	0	—	—
0	20	0.36	—	—	90	—	—	—	—	—	—	—	—	4	—	—
70	200	2.70	—	—	1060	—	—	—	—	—	—	—	—	6	—	—
70	150	0.72	—	—	660	—	—	—	—	—	—	—	—	30	—	—
70	80	1.80	—	—	980	—	0	—	—	—	—	—	—	2	—	—

TABLE A–1
Food Composition

(Computer code number is for Wadsworth Diet Analysis program)   (For purposes of calculations, use "0" for t, <1, <.1, <.01, etc.)

DA + Code	Food Description	Quantity	Measure	Wt (g)	H₂O (g)	Ener (kcal)	Prot (g)	Carb (g)	Dietary Fiber (g)	Fat (g)	Sat	Mono	Poly	Trans
	FAST FOOD—Continued													
38759	Chargrilled deluxe chicken sandwich	1	item(s)	195	—	290	27	31	2	7	1.50	—	—	0
38742	Chicken biscuit	1	item(s)	137	—	400	16	43	2	18	4.50	—	—	2.83
38743	Chicken biscuit w/cheese	1	item(s)	151	—	450	19	43	2	23	7.00	—	—	2.85
38762	Chicken caesar wrap	1	item(s)	227	—	460	36	52	2	10	6.00	—	—	0
38757	Chicken deluxe sandwich	1	item(s)	208	—	420	28	39	2	16	3.50	—	—	0
38764	Chicken salad sandwich	1	item(s)	153	—	350	20	32	5	15	3.00	—	—	0
38756	Chicken sandwich	1	item(s)	170	—	410	28	38	1	15	3.50	—	—	0
38768	Chick-n-Strip salad	1	item(s)	331	—	390	34	22	4	18	5.00	—	—	0
38763	Chick-n-Strips	4	item(s)	127	—	290	29	14	1	13	2.50	—	—	0
38770	Coleslaw	1	item(s)	105	—	210	1	14	2	17	2.50	—	—	0
38755	Hash browns	1	serving(s)	84	—	170	2	20	2	9	4.50	—	—	1
38765	Hearty breast of soup	1	cup(s)	241	—	140	8	18	1	4	1.00	—	—	0
38778	Icedream, small cone	1	item(s)	135	—	160	4	28	0	4	2.00	—	—	0
38774	Icedream, small cup	1	serving(s)	213	—	230	5	38	0	6	3.50	—	—	0
38775	Lemonade	1	cup(s)	255	—	170	0	41	0	1	0.00	—	—	0
38776	Lemonade, diet	1	cup(s)	255	—	25	0	5	0	0	0.00	0.00	0.00	0
38777	Nuggets	8	item(s)	113	—	260	26	12	1	12	2.50	—	—	0
38769	Side salad	1	item(s)	108	—	60	3	4	2	3	1.50	—	—	0
38767	Southwest chargrilled salad	1	item(s)	303	—	240	22	17	5	8	3.50	—	—	0
38772	Waffle potato fries, small, salted	1	serving(s)	85	—	280	3	37	5	14	5.00	—	—	1.50
	Cinnabon													
39569	Caramel Pecanbon	1	item(s)	272	—	1100	16	141	8	56	10.00	—	—	5
39572	Caramellata Chill w/whipped cream	16	fluid ounce(s)	480	—	406	10	61	0	14	8.00	—	—	—
39571	Cinnapoppers	1	serving(s)	74	—	368	4	41	2	21	11.00	—	—	1
39567	Classic roll	1	item(s)	221	—	813	15	117	4	32	8.00	—	—	5
39568	Minibon	1	item(s)	92	—	339	6	49	2	13	3.00	—	—	2
39573	Mochalatta chill w/whipped cream	16	fluid ounce(s)	480	—	362	9	55	0	13	8.00	—	—	—
39570	Stix	5	item(s)	85	—	379	6	41	1	21	6.00	—	—	4
	Dairy Queen													
1466	Banana split	1	item(s)	369	—	510	8	96	3	12	8.00	3.00	0.50	0
38552	Brownie Earthquake	1	serving(s)	304	—	740	10	112	0	27	16.00	—	—	3
38561	Chocolate chip cookie dough blizzard, small	1	item(s)	319	—	720	12	105	0	28	14.00	—	—	2.50
1464	Chocolate malt, small	1	item(s)	418	—	650	15	111	0	16	10.00	—	—	0.50
38541	Chocolate shake, small	1	item(s)	397	—	560	13	93	1	15	10.00	—	—	0.50
17257	Chocolate soft serve	½	cup(s)	94	—	150	4	22	0	5	3.50	—	—	0
1463	Chocolate sundae, small	1	item(s)	163	—	280	5	49	0	7	4.50	1.00	1.00	0
1462	Dipped cone, small	1	item(s)	156	—	340	6	42	1	17	9.00	4.00	3.00	1
38555	Oreo cookies blizzard, small	1	item(s)	283	—	570	11	83	1	21	10.00	—	—	2.50
38547	Royal Treats Peanut Buster parfait	1	item(s)	305	—	730	16	99	2	31	17.00	—	—	0
17256	Vanilla soft serve	½	cup(s)	94	—	140	3	22	0	5	3.00	—	—	0
	Domino's													
31606	Barbeque wings	1	item(s)	25	—	50	6	2	<1	2	0.65	—	—	—
31604	Breadsticks	1	item(s)	37	—	116	3	18	1	4	0.79	—	—	—
37551	Buffalo chicken kickers	1	item(s)	24	14	47	4	3	<1	2	0.39	—	—	—
37548	Cinnastix	1	item(s)	32	8	122	2	15	1	6	1.15	—	—	—
	Classic hand tossed pizza													
31573	America's favorite feast, 12"	2	slice(s)	205	99	508	22	57	4	22	9.20	—	—	—
31574	America's favorite feast, 14"	2	slice(s)	283	138	697	30	79	5	30	12.70	—	—	—
37543	Bacon cheeseburger feast, 12"	2	slice(s)	198	60	549	25	55	3	26	11.62	—	—	—
37545	Bacon cheeseburger feast, 14"	2	slice(s)	275	121	762	35	75	4	36	16.10	—	—	—
37546	Barbeque feast, 12"	2	slice(s)	192	85	506	22	62	3	20	9.08	—	—	—
37547	Barbeque feast, 14"	2	slice(s)	262	115	691	30	85	4	27	12.24	—	—	—
31569	Cheese, 12"	2	slice(s)	159	—	375	15	55	3	11	4.81	—	—	—
31570	Cheese, 14"	2	slice(s)	219	—	516	21	75	4	15	6.72	—	—	—
37538	Deluxe feast, 12"	2	slice(s)	201	102	465	20	57	3	18	7.66	—	—	—
37540	Deluxe feast, 14"	2	slice(s)	273	138	627	26	78	5	24	10.20	—	—	—
31685	Deluxe, 12"	2	slice(s)	213	—	465	20	57	3	18	7.65	—	—	—
31694	Deluxe, 14"	2	slice(s)	273	—	627	26	78	5	24	10.20	—	—	—
31686	Extravaganzza, 12"	2	slice(s)	245	127	576	27	59	4	27	11.56	—	—	—
31695	Extravaganzza, 14"	2	slice(s)	329	171	773	36	88	5	36	15.42	—	—	—
31575	Hawaiian feast, 12"	2	slice(s)	204	105	450	21	58	3	16	7.20	—	—	—
31576	Hawaiian feast, 14"	2	slice(s)	283	147	623	29	80	5	22	10.09	—	—	—
31687	Meatzza, 12"	2	slice(s)	213	—	560	26	57	3	26	11.40	—	—	—
31696	Meatzza, 14"	2	slice(s)	293	139	753	35	78	5	34	15.24	—	—	—

PAGE KEY: A–4 = Breads/Baked Goods   A–8 = Cereal/Rice/Pasta   A–12 = Fruit   A–16 = Vegetables/Legumes   A–26 = Nuts/Seeds   A–28 = Vegetarian
A–30 = Dairy   A–36 = Eggs   A–36 = Seafood   A–38 = Meats   A–42 = Poultry   A–42 = Processed meats   A–44 = Beverages   A–48 = Fats/Oils
A–50 = Sweets   A–52 = Spices/Condiments/Sauces   A–54 = Mixed Foods/Soups/Sandwiches   A–60 = Fast food   A–76 = Convenience   A–78 = Baby foods

Chol (mg)	Calc (mg)	Iron (mg)	Magn (mg)	Pota (mg)	Sodi (mg)	Zinc (mg)	Vit A (RAE) (µg)	Thia (mg)	Vit E (mg α)	Ribo (mg)	Niac (mg)	Vit B$_6$ (mg)	Fola (µg)	Vit C (mg)	Vit B$_{12}$ (µg)	Sele (µg)
70	80	1.80	—	—	990	—	—	—	—	—	—	—	—	5	—	—
30	60	2.70	—	—	1200	—	0	—	—	—	—	—	—	0	—	—
45	150	2.70	—	—	1430	—	—	—	—	—	—	—	—	0	—	—
80	500	2.70	—	—	1390	—	—	—	—	—	—	—	—	1	—	—
60	100	2.70	—	—	1300	—	—	—	—	—	—	—	—	2	—	—
65	150	1.80	—	—	880	—	—	—	—	—	—	—	—	0	—	—
60	100	2.70	—	—	1300	—	—	—	—	—	—	—	—	0	—	—
80	200	0.36	—	—	860	—	—	—	—	—	—	—	—	30	—	—
65	20	0.36	—	—	730	—	—	—	—	—	—	—	—	1	—	—
20	40	0.36	—	—	180	—	—	—	—	—	—	—	—	27	—	—
10	0	0.72	—	—	350	—	—	—	—	—	—	—	—	0	—	—
25	40	1.08	—	—	900	—	—	—	—	—	—	—	—	0	—	—
15	100	0.36	—	—	80	—	—	—	—	—	—	—	—	0	—	—
25	150	0.00	—	—	100	—	—	—	—	—	—	—	—	0	—	—
0	0	0.36	—	—	10	—	0	—	—	—	—	—	—	15	—	—
0	0	0.36	—	—	5	—	0	—	—	—	—	—	—	15	—	—
70	40	1.08	—	—	1090	—	0	—	—	—	—	—	—	0	—	—
10	100	0.00	—	—	75	—	—	—	—	—	—	—	—	15	—	—
60	200	1.08	—	—	770	—	—	—	—	—	—	—	—	24	—	—
15	20	0.00	—	—	105	—	0	—	—	—	—	—	—	21	—	—
63	—	—	—	—	600	—	—	—	—	—	—	—	—	—	—	—
46	—	—	—	—	187	—	—	—	—	—	—	—	—	—	—	—
62	—	—	—	—	104	—	—	—	—	—	—	—	—	—	—	—
67	—	—	—	—	801	—	—	—	—	—	—	—	—	—	—	—
27	—	—	—	—	337	—	—	—	—	—	—	—	—	—	—	—
46	100	0.00	—	—	252	—	—	—	—	—	—	—	—	0	—	—
16	—	—	—	—	413	—	—	—	—	—	—	—	—	—	—	—
30	250	1.80	—	860	180	—	—	0.15	—	0.60	0.20	—	—	15	—	—
50	250	1.80	—	—	350	—	—	—	—	—	—	—	—	0	—	—
50	350	2.70	—	—	370	—	—	—	—	—	—	—	—	1	—	—
55	450	1.80	—	—	370	—	—	—	—	—	—	—	—	2	—	—
50	450	1.44	—	—	280	—	—	0.12	—	—	—	—	—	2	—	—
15	100	0.72	—	—	75	—	—	—	—	—	—	—	—	0	—	—
20	200	1.08	—	278	140	—	—	0.06	—	0.24	0.20	—	—	0	—	—
20	200	1.08	—	290	130	—	—	0.06	—	0.26	0.20	—	—	1	—	—
40	350	2.70	—	—	430	—	—	—	—	—	—	—	—	1	—	—
35	300	1.80	—	—	400	—	—	—	—	—	—	—	—	1	—	—
15	150	0.72	—	—	70	—	150	—	—	—	—	—	—	0	—	—
26	6	0.32	—	—	175	—	—	—	—	—	—	—	—	<.1	—	—
0	<.1	0.87	—	—	152	—	—	—	—	—	—	—	—	6	—	—
9	3	0.00	—	—	163	—	—	—	—	—	—	—	—	0	—	—
0	6	0.70	—	—	110	—	—	—	—	—	—	—	—	<.1	—	—
49	202	3.70	—	—	1221	—	—	—	—	—	—	—	—	1	—	—
68	281	5.10	—	—	1685	—	—	—	—	—	—	—	—	1	—	—
60	293	3.56	—	—	1274	—	—	—	—	—	—	—	—	0	—	—
84	395	4.96	—	—	1809	—	—	—	—	—	—	—	—	0	—	—
46	—	—	—	—	1206	—	—	—	—	—	—	—	—	—	—	—
63	393	4.42	—	—	1672	—	—	—	—	—	—	—	—	2	—	—
23	187	2.99	—	—	776	—	131	—	—	—	—	—	—	0	—	—
32	261	4.13	—	—	1080	—	184	—	—	—	—	—	—	0	—	—
40	199	3.56	—	—	1063	—	—	—	—	—	—	—	—	1	—	—
53	276	4.84	—	—	1432	—	—	—	—	—	—	—	—	2	—	—
40	199	3.56	—	—	1063	—	—	—	—	—	—	—	—	1	—	—
53	276	4.85	—	—	1432	—	—	—	—	—	—	—	—	2	—	—
60	290	4.08	—	—	1348	—	—	—	—	—	—	—	—	1	—	—
89	403	5.48	—	—	1780	—	—	—	—	—	—	—	—	2	—	—
41	274	3.30	—	—	1102	—	—	—	—	—	—	—	—	2	—	—
57	384	4.57	—	—	1544	—	—	—	—	—	—	—	—	3	—	—
344	282	3.71	—	—	1463	—	—	—	—	—	—	—	—	<1	—	—
85	393	5.04	—	—	1947	—	—	—	—	—	—	—	—	<1	—	—

TABLE A–1
Food Composition

(Computer code number is for Wadsworth Diet Analysis program)   (For purposes of calculations, use "0" for t, <1, <.1, <.01, etc.)

DA + Code	Food Description	Quantity	Measure	Wt (g)	H₂O (g)	Ener (kcal)	Prot (g)	Carb (g)	Dietary Fiber (g)	Fat (g)	Sat	Mono	Poly	Trans
	FAST FOOD—Continued													
31571	Pepperoni feast, extra pepperoni & cheese, 12"	2	slice(s)	196	87	534	24	56	3	25	10.92	—	—	—
31572	Pepperoni feast, extra pepperoni & cheese, 14"	2	slice(s)	270	121	732	33	77	4	34	15.00	—	—	—
31577	Vegi feast, 12"	2	slice(s)	203	107	439	19	57	4	16	7.09	—	—	—
31578	Vegi feast, 14"	2	slice(s)	278	147	304	27	78	5	22	9.89	—	—	—
37549	Dot cinnamon	1	item(s)	28	8	99	2	15	1	4	0.68	—	—	—
31605	Double cheesy bread	1	item(s)	35	11	123	4	13	1	6	2.06	—	—	—
31607	Hot wings	1	item(s)	25	—	45	5	1	<1	2	0.65	—	—	—
	Thin crust pizza													
31583	America's favorite, 12"	¼	item(s)	159	—	408	19	34	2	23	9.77	—	—	—
31584	America's favorite, 14"	¼	item(s)	202	—	557	26	47	3	31	13.19	—	—	—
31579	Cheese, 12"	¼	item(s)	106	—	273	12	31	2	12	9.37	—	—	—
31580	Cheese, 14"	¼	item(s)	148	—	382	17	43	2	17	6.72	—	—	—
31688	Deluxe, 12"	¼	item(s)	159	—	363	16	34	2	19	7.64	—	—	—
31697	Deluxe, 14"	¼	item(s)	202	—	494	22	47	3	25	10.20	—	—	—
31689	Extravaganzza, 12"	¼	item(s)	159	—	425	20	34	3	24	9.41	—	—	—
31698	Extravaganzza, 14"	¼	item(s)	202	—	571	27	48	4	31	12.44	—	—	—
31585	Hawaiian, 12"	¼	item(s)	159	—	349	18	35	2	16	7.20	—	—	—
31586	Hawaiian, 14"	¼	item(s)	202	—	489	25	48	3	23	10.09	—	—	—
31690	Meatzza, 12"	¼	item(s)	159	—	458	23	33	2	27	11.39	—	—	—
31699	Meatzza, 14"	¼	item(s)	202	—	619	31	46	3	36	15.24	—	—	—
31581	Pepperoni, extra pepperoni & cheese 12"	¼	item(s)	159	—	420	20	32	2	24	10.46	—	—	—
31582	Pepperoni, extra pepperoni & cheese 14"	¼	item(s)	202	—	586	28	45	3	34	14.55	—	—	—
31587	Vegi, 12"	¼	item(s)	159	—	338	16	34	3	17	7.08	—	—	—
31588	Vegi, 14"	¼	item(s)	202	—	471	22	47	3	23	9.89	—	—	—
	Ultimate deep dish pizza													
31596	America's favorite, 12"	2	slice(s)	235	—	617	26	59	4	33	12.88	—	—	—
31702	America's favorite, 14"	2	slice(s)	311	—	851	36	84	5	44	17.35	—	—	—
31590	Cheese, 12"	2	slice(s)	181	—	482	19	56	3	22	7.91	—	—	—
31591	Cheese, 14"	2	slice(s)	257	—	677	26	80	5	30	10.88	—	—	—
31589	Cheese, 6"	1	item(s)	215	—	598	23	68	4	28	9.94	—	—	—
31691	Deluxe, 12"	2	slice(s)	235	—	527	23	59	4	29	10.75	—	—	—
31700	Deluxe, 14"	2	slice(s)	311	—	788	31	84	5	38	14.36	—	—	—
31692	Extravaganzza, 12"	2	slice(s)	235	—	635	27	59	4	34	12.52	—	—	—
31701	Extravaganzza, 14"	2	slice(s)	311	—	866	36	85	6	45	16.60	—	—	—
31599	Hawaiian, 12"	2	slice(s)	235	—	558	24	60	4	26	10.31	—	—	—
31600	Hawaiian, 14"	2	slice(s)	311	—	784	35	85	5	36	14.25	—	—	—
31693	Meatzza, 12"	2	slice(s)	235	—	667	30	58	4	37	14.50	—	—	—
31703	Meatzza, 14"	2	slice(s)	311	—	914	40	83	5	49	19.40	—	—	—
31593	Pepperoni, extra pepperoni & cheese 12"	2	slice(s)	235	—	629	26	57	4	34	13.57	—	—	—
31594	Pepperoni, extra pepperoni & cheese 14"	2	slice(s)	311	—	880	37	82	5	47	18.71	—	—	—
31602	Vegi, 12"	2	slice(s)	235	—	547	22	59	4	26	10.19	—	—	—
31603	Vegi, 14"	2	slice(s)	311	—	765	32	84	6	36	14.05	—	—	—
31598	With ham & pineapple tidbits, 6"	1	item(s)	430	—	619	25	70	4	28	10.19	—	—	—
31595	With Italian sausage, 6"	1	item(s)	430	—	642	25	70	4	31	11.33	—	—	—
31592	With pepperoni, 6"	1	item(s)	430	—	647	25	69	4	32	11.70	—	—	—
31601	With vegetables, 6"	1	item(s)	430	—	619	23	71	5	29	10.11	—	—	—
	In-n-Out Burger													
34374	Cheeseburger	1	item(s)	268	—	480	22	39	3	27	10.00	—	—	—
34391	Cheesburger w/mustard & ketchup	1	item(s)	268	—	400	22	41	3	18	9.00	—	—	—
34390	Cheeseburger, lettuce leaves instead of buns	1	item(s)	300	—	330	18	11	2	25	9.00	—	—	—
34377	Chocolate shake	1	item(s)	425	—	690	9	83	0	36	24.00	—	—	—
34375	Double-Double cheeseburger	1	item(s)	328	—	670	37	40	3	41	18.00	—	—	—
34393	Double-Double cheeseburger w/mustard & ketchup	1	item(s)	328	—	590	37	42	3	32	17.00	—	—	—
34392	Double-Double cheeseburger, lettuce leaves instead of buns	1	item(s)	361	—	520	33	11	2	39	17.00	—	—	—
34376	French fries	1	item(s)	125	—	400	7	54	2	18	5.00	—	—	—
34373	Hamburger	1	item(s)	243	—	390	16	39	3	19	5.00	—	—	—
34389	Hamburger w/mustard & ketchup	1	item(s)	243	—	310	16	41	3	10	4.00	—	—	—
34388	Hamburger, lettuce leaves instead of buns	1	item(s)	275	—	240	12	10	2	17	4.50	—	—	—

PAGE KEY: A–4 = Breads/Baked Goods A–8 = Cereal/Rice/Pasta A–12 = Fruit A–16 = Vegetables/Legumes A–26 = Nuts/Seeds A–28 = Vegetarian A–30 = Dairy A–36 = Eggs A–36 = Seafood A–38 = Meats A–42 = Poultry A–42 = Processed meats A–44 = Beverages A–48 = Fats/Oils A–50 = Sweets A–52 = Spices/Condiments/Sauces A–54 = Mixed Foods/Soups/Sandwiches A–60 = Fast food A–76 = Convenience A–78 = Baby foods

A

Chol (mg)	Calc (mg)	Iron (mg)	Magn (mg)	Pota (mg)	Sodi (mg)	Zinc (mg)	Vit A (RAE) (µg)	Thia (mg)	Vit E (mg α)	Ribo (mg)	Niac (mg)	Vit B6 (mg)	Fola (µg)	Vit C (mg)	Vit B12 (µg)	Sele (µg)
57	279	3.36	—	—	1349	—	155	—	—	—	—	—	—	<1	—	—
78	390	4.66	—	—	1855	—	233	—	—	—	—	—	—	<1	—	—
34	279	3.44	—	—	987	—	—	—	—	—	—	—	—	1	—	—
47	389	4.71	—	—	1369	—	—	—	—	—	—	—	—	2	—	—
0	6	0.59	—	—	86	—	—	—	—	—	—	—	—	<.1	—	—
6	47	0.66	—	—	164	—	—	—	—	—	—	—	—	<1	—	—
26	5	0.30	—	—	354	—	—	—	—	—	—	—	—	1	—	—
51	318	1.52	—	—	1285	—	—	—	—	—	—	—	—	<1	—	—
69	444	2.07	—	—	1751	—	—	—	—	—	—	—	—	1	—	—
23	225	0.97	—	—	835	—	125	—	—	—	—	—	—	0	—	—
32	315	1.36	—	—	1172	—	175	—	—	—	—	—	—	0	—	—
40	237	1.54	—	—	1123	—	—	—	—	—	—	—	—	1	—	—
53	330	2.08	—	—	1523	—	—	—	—	—	—	—	—	2	—	—
53	245	1.95	—	—	1408	—	—	—	—	—	—	—	—	1	—	—
69	340	2.59	—	—	1871	—	—	—	—	—	—	—	—	2	—	—
41	312	1.28	—	—	1162	—	—	—	—	—	—	—	—	2	—	—
57	437	1.80	—	—	1635	—	—	—	—	—	—	—	—	3	—	—
64	320	1.69	—	—	1523	—	—	—	—	—	—	—	—	<1	—	—
454	446	2.27	—	—	2039	—	—	—	—	—	—	—	—	<1	—	—
54	316	1.34	—	—	1362	—	162	—	—	—	—	—	—	<1	—	—
76	442	1.87	—	—	1900	—	227	—	—	—	—	—	—	<1	—	—
34	317	1.42	—	—	1047	—	—	—	—	—	—	—	—	1	—	—
47	442	1.94	—	—	1460	—	—	—	—	—	—	—	—	2	—	—
58	334	4.43	—	—	1573	—	—	—	—	—	—	—	—	1	—	—
78	464	6.24	—	—	2155	—	—	—	—	—	—	—	—	1	—	—
30	241	3.88	—	—	1123	—	151	—	—	—	—	—	—	<1	—	—
41	335	5.53	—	—	1575	—	210	—	—	—	—	—	—	1	—	—
36	295	4.67	—	—	1341	—	174	—	—	—	—	—	—	1	—	—
47	253	4.45	—	—	1410	—	—	—	—	—	—	—	—	2	—	—
62	349	6.25	—	—	1927	—	—	—	—	—	—	—	—	2	—	—
60	261	4.86	—	—	1696	—	—	—	—	—	—	—	—	2	—	—
78	359	6.76	—	—	2275	—	—	—	—	—	—	—	—	2	—	—
48	328	4.19	—	—	1449	—	—	—	—	—	—	—	—	2	—	—
67	457	5.97	—	—	2039	—	—	—	—	—	—	—	—	3	—	—
379	336	4.60	—	—	1810	—	—	—	—	—	—	—	—	1	—	—
501	466	6.44	—	—	2443	—	—	—	—	—	—	—	—	1	—	—
61	332	4.25	—	—	1650	—	187	—	—	—	—	—	—	1	—	—
85	462	6.04	—	—	2304	—	260	—	—	—	—	—	—	1	—	—
41	333	4.33	—	—	1334	—	—	—	—	—	—	—	—	2	—	—
57	462	6.11	—	—	1864	—	—	—	—	—	—	—	—	2	—	—
43	298	4.84	—	—	1498	—	—	—	—	—	—	—	—	1	—	—
45	302	4.89	—	—	1478	—	—	—	—	—	—	—	—	1	—	—
47	299	4.81	—	—	1524	—	168	—	—	—	—	—	—	1	—	—
36	307	5.10	—	—	1472	—	—	—	—	—	—	—	—	5	—	—
60	200	3.60	—	—	1000	—	188	—	—	—	—	—	—	15	—	—
55	200	3.60	—	—	1080	—	182	—	—	—	—	—	—	15	—	—
60	200	1.08	—	—	720	—	—	—	—	—	—	—	—	18	—	—
95	300	0.72	—	—	350	—	143	—	—	—	—	—	—	0	—	—
120	350	5.40	—	—	1430	—	184	—	—	—	—	—	—	15	—	—
115	350	5.40	—	—	1510	—	229	—	—	—	—	—	—	15	—	—
120	350	1.08	—	—	1160	—	275	—	—	—	—	—	—	18	—	—
0	20	1.80	—	—	245	—	0	—	—	—	—	—	—	0	—	—
40	40	3.60	—	—	640	—	50	—	—	—	—	—	—	15	—	—
35	40	3.60	—	—	720	—	75	—	—	—	—	—	—	15	—	—
40	40	1.08	—	—	370	—	—	—	—	—	—	—	—	18	—	—

TABLE A–1
Food Composition

(Computer code number is for Wadsworth Diet Analysis program)   (For purposes of calculations, use "0" for t, <1, <.1, <.01, etc.)

DA + Code	Food Description	Quantity	Measure	Wt (g)	H₂O (g)	Ener (kcal)	Prot (g)	Carb (g)	Dietary Fiber (g)	Fat (g)	Fat Breakdown (g)			
											Sat	Mono	Poly	Trans
	FAST FOOD—Continued													
34379	Strawberry shake	1	item(s)	425	—	690	8	91	2	33	22.00	—	—	—
34378	Vanilla shake	1	item(s)	425	—	680	9	78	2	37	25.00	—	—	—
	Jack in the Box													
30392	Bacon ultimate cheeseburger	1	item(s)	353	—	1120	52	59	2	55	28.00	—	—	3.13
1740	Breakfast Jack	1	item(s)	133	—	310	14	34	1	14	5.00	—	—	0
14074	Cheeseburger	1	item(s)	116	—	300	14	31	1	13	6.00	—	—	0.89
14106	Chicken breast pieces	5	piece(s)	150	—	360	27	24	1	17	3.00	—	—	4.48
37241	Chicken club salad	1	item(s)	535	—	310	28	15	5	16	6.00	—	—	0
14111	Chocolate ice cream shake	1	item(s)	315	—	660	11	89	1	29	18.00	—	—	1
14075	Double cheeseburger	1	item(s)	155	—	410	20	32	1	22	11.00	—	—	—
14098	French fries, jumbo	1	serving(s)	142	—	410	4	55	4	20	4.50	—	—	5.34
14099	French fries, super scoop	1	serving(s)	198	—	580	6	77	6	28	6.00	—	—	7.07
14073	Hamburger	1	item(s)	104	—	250	12	30	2	9	3.50	—	—	0.88
14090	Hash browns	1	serving(s)	57	—	150	1	13	2	10	2.50	—	—	3
14072	Jack's Spicy Chicken sandwich	1	item(s)	253	—	580	24	53	3	31	6.00	—	—	2.81
1468	Jumbo Jack hamburger	1	item(s)	269	—	600	22	58	3	31	11.00	—	—	1.55
1469	Jumbo Jack hamburger w/cheese	1	item(s)	294	—	690	26	60	3	38	16.00	—	—	1.55
1470	Onion rings	1	serving(s)	119	—	500	6	51	3	30	5.00	—	—	10
33141	Sausage, egg, & cheese biscuit	1	item(s)	223	—	760	25	33	2	60	20.00	—	—	5.72
14095	Seasoned curly fries	1	serving(s)	125	—	400	6	45	5	23	5.00	—	—	7
14077	Sourdough Jack	1	item(s)	244	—	700	30	36	3	49	16.00	—	—	2.98
37249	Southwest chicken salad	1	serving(s)	598	—	340	28	31	9	13	6.00	—	—	0
14112	Strawberry ice cream shake	1	item(s)	313	—	640	10	84	0	28	18.00	—	—	1
14078	Ultimate cheeseburger	1	item(s)	328	—	990	41	59	2	66	28.00	—	—	3.05
14110	Vanilla ice cream shake	1	item(s)	285	—	570	12	65	0	29	18.00	—	—	1
	Jamba Juice													
31646	Banana berry smoothie	24	fluid ounce(s)	719	—	470	5	112	5	2	0.50	—	—	—
31647	Caribbean passion smoothie	24	fluid ounce(s)	730	—	440	4	102	4	2	1.00	—	—	—
38422	Carrot juice	16	fluid ounce(s)	472	—	100	3	23	0	1	0.00	—	—	—
31648	Chocolate mood smoothie	24	fluid ounce(s)	612	—	690	16	142	2	8	4.50	—	—	—
31649	Citrus squeeze smoothie	24	fluid ounce(s)	729	—	450	4	105	5	2	1.00	—	—	—
31650	Coffee mood smoothie	24	fluid ounce(s)	560	—	596	13	121	1	6	4.00	—	—	—
31651	Coldbuster smoothie	24	fluid ounce(s)	724	—	430	5	100	5	3	1.00	—	—	—
31652	Cranberry craze smoothie	24	fluid ounce(s)	731	—	420	6	97	4	2	1.00	—	—	—
31654	Jamba powerboost smoothie	24	fluid ounce(s)	730	—	440	6	103	7	2	0.00	—	—	—
38423	Lemonade	16	fluid ounce(s)	483	—	300	1	75	0	0	0.00	0.00	0.00	0
31656	Lime sublime smoothie	24	fluid ounce(s)	721	—	450	3	104	6	2	1.00	—	—	—
31657	Mango-a-go-go smoothie	24	fluid ounce(s)	739	—	500	4	117	4	2	1.00	—	—	—
38424	Orange juice, freshly squeezed	16	fluid ounce(s)	496	—	220	3	52	1	1	0.00	—	—	—
38426	Orange/carrot juice	16	fluid ounce(s)	484	—	160	3	37	0	1	0.00	—	—	—
31660	Orange-a-peel smoothie	24	fluid ounce(s)	726	—	440	9	102	5	1	0.00	—	—	—
31665	Protein berry pizzaz smoothie	24	fluid ounce(s)	710	—	440	20	92	6	2	0.00	—	—	—
31667	Raspberry refresher smoothie	24	fluid ounce(s)	636	—	442	3	101	8	3	0.90	—	—	—
31668	Razzmatazz smoothie	24	fluid ounce(s)	730	—	480	3	112	4	2	1.00	—	—	—
31669	Strawberries wild smoothie	24	fluid ounce(s)	725	—	450	6	105	4	0	0.00	—	—	—
38421	Strawberry tsunami smoothie	24	fluid ounce(s)	740	—	530	4	128	4	2	1.00	—	—	—
38427	Vibrant C juice	16	fluid ounce(s)	448	—	210	2	50	1	0	0.00	0.00	0.00	0
38428	Wheatgrass juice, freshly squeezed	1	ounce(s)	32	—	5	1	1	0	0	0.00	0.00	0.00	0
	Kentucky Fried Chicken (KFC)													
31850	BBQ baked beans	1	serving(s)	156	—	190	6	33	6	3	1.00	—	—	0.29
31853	Biscuit	1	item(s)	56	—	180	4	20	1	10	2.50	—	—	3.44
31851	Coleslaw	1	serving(s)	142	—	232	2	26	3	14	2.00	—	—	0.27
31842	Colonel's Crispy Strips	3	item(s)	150	—	340	28	20	0	16	4.50	—	—	4.47
31849	Corn on the cob	1	item(s)	162	—	150	5	35	2	2	0.00	—	—	0
3761	Extra Crispy chicken, breast	1	item(s)	162	—	470	34	19	0	28	8.00	—	—	4.50
3762	Extra Crispy chicken, drumstick	1	item(s)	60	—	160	12	5	0	10	2.50	—	—	1.50
3763	Extra Crispy chicken, thigh	1	item(s)	114	—	370	21	12	0	26	7.00	—	—	3
3764	Extra Crispy chicken, whole wing	1	item(s)	52	—	190	10	10	0	12	3.50	—	—	2
31833	Honey BBQ wing pieces	6	item(s)	189	—	607	33	33	1	38	10.00	—	—	5.42
10810	Hot & spicy chicken, breast	1	item(s)	179	—	450	33	20	0	27	8.00	—	—	0
10813	Hot & spicy chicken, drumstick	1	item(s)	60	—	140	13	4	0	9	2.50	—	—	0
10811	Hot & spicy chicken, thigh	1	item(s)	128	—	390	22	14	0	28	8.00	—	—	0
10812	Hot & spicy chicken, whole wing	1	item(s)	55	—	180	11	9	0	11	3.00	—	—	0
10859	Hot wings pieces	6	piece(s)	135	—	471	27	18	2	33	8.00	—	—	4.03
31848	Macaroni & cheese	1	serving(s)	153	—	180	7	21	2	8	3.00	—	—	2.81

PAGE KEY: A–4 = Breads/Baked Goods A–8 = Cereal/Rice/Pasta A–12 = Fruit A–16 = Vegetables/Legumes A–26 = Nuts/Seeds A–28 = Vegetarian A–30 = Dairy A–36 = Eggs A–36 = Seafood A–38 = Meats A–42 = Poultry A–42 = Processed meats A–44 = Beverages A–48 = Fats/Oils A–50 = Sweets A–52 = Spices/Condiments/Sauces A–54 = Mixed Foods/Soups/Sandwiches A–60 = Fast food A–76 = Convenience A–78 = Baby foods

A

Chol (mg)	Calc (mg)	Iron (mg)	Magn (mg)	Pota (mg)	Sodi (mg)	Zinc (mg)	Vit A (RAE) (µg)	Thia (mg)	Vit E (mg α)	Ribo (mg)	Niac (mg)	Vit B$_6$ (mg)	Fola (µg)	Vit C (mg)	Vit B$_{12}$ (µg)	Sele (µg)
85	250	0.00	—	—	280	—	134	—	—	—	—	—	—	0	—	—
90	300	0.00	—	—	390	—	145	—	—	—	—	—	—	0	—	—
160	300	7.20	—	600	2260	—	—	—	—	—	—	—	—	1	—	—
210	150	3.60	—	210	770	—	—	—	—	—	—	—	—	4	—	—
40	150	3.60	—	180	840	—	40	—	—	—	—	—	—	0	—	—
80	20	1.80	—	430	970	—	—	—	—	—	—	—	—	1	—	—
65	300	3.60	—	1010	890	—	—	—	—	—	—	—	—	54	—	—
110	350	0.36	—	720	270	—	215	—	—	—	—	—	—	0	—	—
70	250	4.50	—	280	920	—	—	—	—	—	—	—	—	1	—	—
0	20	1.08	—	550	690	—	0	—	—	—	—	—	—	6	—	—
0	20	1.44	—	770	960	—	0	—	—	—	—	—	—	9	—	—
30	100	3.60	—	155	610	—	0	—	—	—	—	—	—	0	—	—
0	10	0.18	—	190	230	—	0	—	—	—	—	—	—	0	—	—
60	150	1.80	—	470	950	—	—	—	—	—	—	—	—	9	—	—
45	164	4.92	—	390	980	—	—	—	—	—	—	—	—	10	—	—
75	250	4.50	—	420	1360	—	—	—	—	—	—	—	—	9	—	—
0	40	2.70	—	140	420	—	40	—	—	—	—	—	—	18	—	—
280	100	2.70	—	240	1390	—	—	—	—	—	—	—	—	0	—	—
0	40	1.80	—	580	890	—	—	—	—	—	—	—	—	0	—	—
80	200	4.50	—	450	1220	—	—	—	—	—	—	—	—	9	—	—
60	300	4.50	—	1020	920	—	—	—	—	—	—	—	—	48	—	—
110	350	0.00	—	610	220	—	202	—	—	—	—	—	—	0	—	—
130	300	7.20	—	480	1670	—	—	—	—	—	—	—	—	1	—	—
115	400	0.00	—	630	220	—	218	—	—	—	—	—	—	0	—	—
5	200	1.08	32	1000	85	0.30	—	0.06	0.32	0.26	1.20	0.40	33	15	0	0
5	100	1.80	24	810	60	0.30	—	0.09	0.64	0.26	5.00	0.50	100	78	0	1
0	150	2.70	80	1030	250	0.90	0	0.53	—	0.26	5.00	0.70	80	18	0	6
25	500	1.08	32	760	280	0.60	0	0.09	0.00	0.85	0.40	0.08	9	6	1	4
5	150	1.80	60	1150	50	0.30	—	0.30	0.40	0.26	1.90	0.40	100	168	0	1
28	455	0.30	49	634	429	1.50	—	0.10	0.16	0.60	0.30	0.10	18	7	1	3
5	100	1.08	60	1240	35	15.00	—	0.38	17.71	0.34	3.00	0.40	122	1302	0	1
5	250	1.44	16	500	90	0.30	—	0.03	0.64	0.26	5.00	0.50	100	54	0	1
0	1100	1.44	480	1110	40	15.00	—	5.25	17.71	5.78	66.00	6.80	640	294	10	70
0	20	0.00	8	200	10	0.00	0	0.03	0.00	0.17	14.00	1.80	320	36	0	0
5	150	1.80	32	660	75	0.60	—	0.12	0.32	0.26	7.00	0.80	160	66	<1	1
5	100	1.08	24	800	60	0.30	—	0.15	1.61	0.26	5.00	0.70	120	72	0	1
0	60	1.08	60	990	0	0.30	—	0.45	—	0.14	2.00	0.20	160	246	0	0
0	100	1.80	60	1010	125	0.60	0	0.45	—	0.26	3.00	0.50	120	132	0	3
0	250	1.80	60	1350	100	0.30	—	0.38	0.64	0.43	3.00	0.40	140	240	0	1
0	1100	2.62	39	650	240	0.58	0	0.09	0.31	0.10	1.55	0.40	58	60	0	4
3	104	2.20	56	806	47	0.80	—	0.10	0.40	0.30	1.60	0.40	43	35	<1	1
5	150	1.80	32	790	70	0.60	—	0.09	0.32	0.26	6.00	0.90	160	60	0	1
0	250	1.80	32	1020	115	0.30	—	0.03	0.32	0.34	1.20	0.20	32	60	0	1
5	100	1.08	24	480	10	0.30	0	0.06	—	0.34	14.00	1.80	320	90	0	1
0	20	1.08	40	720	0	0.30	0	0.30	—	0.10	1.60	0.40	80	678	0	0
0	0	1.80	8	80	0	0.00	0	0.03	—	0.03	0.40	0.04	16	4	0	3
5	80	1.80	—	—	760	—	—	—	—	—	—	—	—	1	—	—
0	20	1.08	—	—	560	—	—	—	—	—	—	—	—	1	—	—
8	30	0.18	—	—	284	—	65	—	—	—	—	—	—	34	—	—
70	10	0.72	—	—	1140	—	—	—	—	—	—	—	—	1	—	—
0	10	0.18	—	—	20	—	10	—	—	—	—	—	—	4	—	—
135	19	1.44	—	—	1230	—	—	—	—	—	—	—	—	1	—	—
70	9	0.65	—	—	415	—	—	—	—	—	—	—	—	1	—	—
120	19	1.04	—	—	710	—	—	—	—	—	—	—	—	1	—	—
55	9	0.34	—	—	390	—	—	—	—	—	—	—	—	1	—	—
193	40	1.44	—	—	1145	—	—	—	—	—	—	—	—	5	—	—
130	10	1.07	—	—	1450	—	—	—	—	—	—	—	—	1	—	—
65	20	0.68	—	—	380	—	—	—	—	—	—	—	—	1	—	—
125	10	1.44	—	—	1240	—	—	—	—	—	—	—	—	1	—	—
60	10	0.72	—	—	420	—	—	—	—	—	—	—	—	1	—	—
150	40	1.44	—	—	1230	—	—	—	—	—	—	—	—	1	—	—
10	150	0.18	—	—	860	—	350	—	—	—	—	—	—	1	—	—

TABLE A–1
Food Composition

(Computer code number is for Wadsworth Diet Analysis program)   (For purposes of calculations, use "0" for t, <1, <.1, <.01, etc.)

DA + Code	Food Description	Quantity	Measure	Wt (g)	H₂O (g)	Ener (kcal)	Prot (g)	Carb (g)	Dietary Fiber (g)	Fat (g)	Sat	Mono	Poly	Trans
	FAST FOOD—Continued													
31847	Mashed potatoes with gravy	1	serving(s)	136	—	120	1	17	2	6	1.00	—	—	0.50
10825	Original Recipe chicken, breast	1	item(s)	161	—	370	40	11	0	19	6.00	—	—	2.50
10826	Original Recipe chicken, drumstick	1	item(s)	59	—	140	14	4	0	8	2.00	—	—	1
10827	Original Recipe chicken, thigh	1	item(s)	126	—	360	22	12	0	25	7.00	—	—	1.50
10828	Original Recipe chicken, whole wing	1	item(s)	47	—	145	11	5	0	9	2.50	—	—	1
3760	Original Recipe chicken sandwich w/sauce	1	item(s)	200	—	450	29	33	2	22	5.00	—	—	—
31834	Original Recipe chicken sandwich w/o sauce	1	item(s)	187	—	360	29	21	1	13	3.50	—	—	—
31852	Potato salad	1	serving(s)	160	—	230	4	23	3	14	2.00	—	—	0.31
10845	Potato wedges	1	serving(s)	156	—	376	6	53	5	15	4.20	—	—	6.12
10853	Rotisserie Gold chicken, breast & wing w/skin	4	ounce(s)	114	—	218	26	1	0	12	3.51	—	—	—
10851	Rotisserie Gold chicken, thigh & leg w/skin	4	ounce(s)	114	—	260	23	1	0	18	5.15	—	—	—
10852	Rotisserie Gold chicken, thigh & leg w/o skin	4	ounce(s)	117	—	217	27	0	0	12	3.50	—	—	—
31843	Spicy Crispy Strips	3	item(s)	115	—	335	25	23	1	15	4.00	—	—	—
10854	Tender Roast chicken, breast w/o skin	1	item(s)	118	—	169	31	1	0	4	1.20	—	—	—
	Long John Silver													
39392	Baked cod	1	serving(s)	101	—	120	22	1	0	5	1.00	—	—	—
3777	Batter dipped fish sandwich	1	item(s)	177	—	440	17	48	3	20	5.00	—	—	—
37568	Battered fish	1	item(s)	92	—	230	11	16	0	13	4.00	—	—	—
37569	Breaded clams	1	serving(s)	85	—	240	8	22	1	13	2.00	—	—	—
39404	Clam chowder	1	item(s)	227	—	220	9	23	0	10	4.00	—	—	—
39398	Cocktail sauce	1	ounce(s)	28	—	25	0	6	0	0	0.00	0.00	0.00	0
3770	Coleslaw	1	serving(s)	113	—	200	1	15	4	15	2.50	1.76	4.10	—
39394	Crunchy shrimp basket	21	item(s)	114	—	340	12	32	2	19	5.00	—	—	—
39400	French fries, large	1	item(s)	142	—	390	4	56	5	17	4.00	—	—	—
3774	Fries regular	1	serving(s)	85	—	230	3	34	3	10	2.50	7.40	5.10	—
3779	Hushpuppy	1	piece(s)	23	—	60	1	9	1	3	0.50	—	—	—
3781	Shrimp batter-dipped	1	piece(s)	14	—	45	2	3	0	3	1.00	—	—	—
39399	Tartar sauce	1	ounce(s)	28	—	100	0	4	0	9	1.50	—	—	—
39395	Ultimate fish sandwich	1	item(s)	199	—	500	20	48	3	25	8.00	—	—	—
	McDonald's													
2247	Barbecue sauce	1	serving(s)	28	—	45	0	10	0	0	0.00	0.00	0.00	0
737	Big Mac hamburger	1	item(s)	216	—	590	24	47	3	34	11.00	—	—	1.48
738	Cheeseburger	1	item(s)	121	—	330	15	36	2	14	6.00	—	—	1.02
29775	Chicken McGrill sandwich	1	item(s)	213	—	400	25	37	2	17	3.00	—	—	0
3792	Chicken McNuggets	4	item(s)	72	—	210	10	12	1	13	2.50	—	—	1.13
1873	Chicken McNuggets	6	item(s)	108	—	310	15	18	2	20	4.00	—	—	1.69
73	Chocolate milkshake	8	fluid ounce(s)	227	164	270	7	48	1	6	3.81	1.77	0.23	—
29774	Crispy chicken sandwich	1	item(s)	219	—	500	22	46	2	26	4.50	—	—	1.50
743	Egg McMuffin	1	item(s)	138	—	300	18	29	2	12	4.50	—	—	0.42
742	Filet-o-fish sandwich	1	item(s)	156	—	470	15	45	1	26	5.00	—	—	1.11
2257	French fries, large	1	serving(s)	176	—	540	8	68	6	26	4.50	—	—	6.18
1872	French fries, small	1	serving(s)	68	—	210	3	26	2	10	1.50	—	—	2.30
2244	French fries, super size	1	serving(s)	198	—	610	9	77	7	29	5.00	—	—	—
33822	Fruit n' yogurt parfait	1	item(s)	338	—	380	10	76	2	5	2.00	—	—	0.18
2251	Garden salad	1	item(s)	177	—	35	2	7	3	0	0.00	0.00	0.00	0
739	Hamburger	1	item(s)	107	—	280	12	35	2	10	4.00	—	—	0.51
2003	Hash browns	1	item(s)	53	—	130	1	14	1	8	1.50	—	—	2
2249	Honey sauce	1	item(s)	14	—	45	0	12	0	0	0.00	0.00	0.00	—
33816	McSalad Shaker chef salad	1	item(s)	206	—	150	17	5	2	8	3.50	—	—	—
33817	McSalad Shaker garden salad	1	item(s)	149	—	100	7	4	2	6	3.00	—	—	—
33818	McSalad Shaker grilled chicken caesar salad	1	item(s)	163	—	100	17	3	2	3	1.50	—	—	—
38396	Newman's Own cobb salad dressing	1	item(s)	59	—	120	1	9	0	9	1.50	—	—	0.01
38397	Newman's Own creamy caesar salad dressing	1	item(s)	59	—	190	2	4	0	18	3.50	—	—	0.29
38398	Newman's Own low fat balsamic vinaigrette salad dressing	1	item(s)	44	—	40	0	4	0	3	0.00	—	—	0.01
38399	Newman's Own ranch salad dressing	1	item(s)	59	—	290	1	4	0	30	4.50	—	—	0.22
1874	Plain hotcakes w/syrup & margarine	3	item(s)	228	—	600	9	104	0	17	3.00	—	—	4
740	Quarter Pounder hamburger	1	item(s)	172	—	430	23	37	2	21	8.00	—	—	1.01
741	Quarter Pounder hamburger w/cheese	1	item(s)	200	—	530	28	38	2	30	13.00	—	—	1.51
2005	Sausage McMuffin w/egg	1	item(s)	164	—	450	20	29	2	28	10.00	—	—	0.59

PAGE KEY: A–4 = Breads/Baked Goods   A–8 = Cereal/Rice/Pasta   A–12 = Fruit   A–16 = Vegetables/Legumes   A–26 = Nuts/Seeds   A–28 = Vegetarian
A–30 = Dairy   A–36 = Eggs   A–36 = Seafood   A–38 = Meats   A–42 = Poultry   A–42 = Processed meats   A–44 = Beverages   A–48 = Fats/Oils
A–50 = Sweets   A–52 = Spices/Condiments/Sauces   A–54 = Mixed Foods/Soups/Sandwiches   A–60 = Fast food   A–76 = Convenience   A–78 = Baby foods

Chol (mg)	Calc (mg)	Iron (mg)	Magn (mg)	Pota (mg)	Sodi (mg)	Zinc (mg)	Vit A (RAE) (µg)	Thia (mg)	Vit E (mg α)	Ribo (mg)	Niac (mg)	Vit B$_6$ (mg)	Fola (µg)	Vit C (mg)	Vit B$_{12}$ (µg)	Sele (µg)
1	10	0.36	—	—	440	—	—	—	—	—	—	—	—	1	—	—
145	20	1.14	—	—	1145	—	—	—	—	—	—	—	—	1	—	—
75	10	0.70	—	—	440	—	—	—	—	—	—	—	—	1	—	—
165	10	1.00	—	—	1060	—	—	—	—	—	—	—	—	1	—	—
60	10	0.36	—	—	370	—	—	—	—	—	—	—	—	1	—	—
70	40	1.80	—	—	940	—	—	—	—	—	—	—	—	1	—	—
60	40	1.80	—	—	890	—	—	—	—	—	—	—	—	1	—	—
15	20	2.70	—	—	540	—	100	—	—	—	—	—	—	1	—	—
4	36	1.55	—	—	1323	—	—	—	—	—	—	—	—	8	—	—
102	7	0.12	—	—	718	—	—	—	—	—	—	—	—	1	—	—
127	8	0.14	—	—	764	—	—	—	—	—	—	—	—	1	—	—
128	10	0.18	—	—	772	—	—	—	—	—	—	—	—	1	—	—
70	20	0.90	—	—	1140	—	—	—	—	—	—	—	—	1	—	—
112	10	0.18	—	—	797	—	—	—	—	—	—	—	—	1	—	—
90	20	0.72	—	—	240	—	—	—	—	—	—	—	—	0	—	—
35	60	3.60	—	—	1120	—	—	—	—	—	—	—	—	9	—	—
30	20	1.80	—	—	700	—	—	—	—	—	—	—	—	5	—	—
10	20	1.08	—	—	1110	—	—	—	—	—	—	—	—	0	—	—
25	150	0.72	—	—	810	—	—	—	—	—	—	—	—	0	—	—
0	0	0.00	—	—	250	—	—	—	—	—	—	—	—	0	—	—
20	40	0.36	—	223	340	0.70	34	0.07	—	0.08	2.35	—	—	18	—	—
105	500	1.80	—	—	720	—	—	—	—	—	—	—	—	1	—	—
0	0	0.00	—	—	580	—	—	—	—	—	—	—	—	24	—	—
0	0	0.00	—	370	350	0.30	—	0.09	—	0.02	1.60	—	—	15	—	—
0	20	0.36	—	—	200	—	—	—	—	—	—	—	—	0	—	—
15	0	0.00	—	—	125	—	—	—	—	—	—	—	—	1	—	—
15	0	0.00	—	—	250	—	—	—	—	—	—	—	—	0	—	—
50	150	3.60	—	—	1310	—	—	—	—	—	—	—	—	9	—	—
0	10	0.18	—	45	250	—	3	—	—	—	—	—	—	4	—	—
85	300	4.50	—	430	1090	—	60	—	—	—	—	—	—	4	—	—
45	250	2.70	—	250	830	—	60	—	—	—	—	—	—	2	—	—
60	200	2.70	—	440	890	—	—	—	—	—	—	—	—	6	—	—
35	20	0.72	—	180	460	—	—	—	—	—	—	—	—	1	—	—
50	20	0.72	—	260	680	—	—	—	—	—	—	—	—	1	—	—
25	299	0.70	36	508	252	1.09	41	0.11	0.11	0.50	0.28	0.06	11	0	1	4
50	200	2.70	—	400	1100	—	—	—	—	—	—	—	—	6	—	—
235	300	2.70	—	210	830	—	—	—	0.72	—	—	—	—	1	—	—
50	200	1.80	—	280	890	—	40	—	—	—	—	—	—	1	—	—
0	20	1.44	—	1210	350	—	—	—	—	—	—	—	—	21	—	—
0	10	0.36	—	470	135	—	—	—	—	—	—	—	—	9	—	—
0	20	1.44	—	1370	390	—	—	—	—	—	—	—	—	24	—	—
15	300	1.80	—	550	240	—	—	—	—	—	—	—	—	24	—	—
0	40	1.09	—	410	20	—	—	—	—	—	—	—	—	24	—	—
30	200	2.70	—	230	590	—	5	—	—	—	—	—	—	2	—	—
0	10	0.36	—	210	330	—	—	—	—	—	—	—	—	2	—	—
0	10	0.18	—	7	0	—	—	—	—	—	—	—	—	1	—	—
95	150	1.44	—	360	740	—	323	—	—	—	—	—	—	15	—	—
75	150	1.08	—	290	120	—	273	—	—	—	—	—	—	15	—	—
40	100	1.08	—	420	240	—	—	—	—	—	—	—	—	12	—	—
10	40	0.18	—	13	440	—	—	—	0.00	—	—	—	—	1	—	—
20	60	0.18	—	16	500	—	—	—	15.40	—	—	—	—	1	—	—
0	10	0.18	—	9	730	—	—	—	0.00	—	—	—	—	2	—	—
20	40	0.18	—	64	530	—	—	—	—	—	—	—	—	1	—	—
20	100	4.50	—	280	770	—	—	—	—	—	—	—	—	1	—	—
70	200	4.50	—	370	840	—	10	—	—	—	—	—	—	2	—	—
95	350	4.50	—	420	1310	—	100	—	—	—	—	—	—	2	—	—
255	300	2.70	—	260	930	—	115	—	0.72	—	—	—	—	1	—	—

TABLE A–1
Food Composition

(Computer code number is for Wadsworth Diet Analysis program) (For purposes of calculations, use "0" for t, <1, <.1, <.01, etc.)

DA + Code	Food Description	Quantity	Measure	Wt (g)	H₂O (g)	Ener (kcal)	Prot (g)	Carb (g)	Dietary Fiber (g)	Fat (g)	Fat Breakdown (g)			
											Sat	Mono	Poly	Trans
	FAST FOOD—Continued													
3163	Strawberry milkshake	8	fluid ounce(s)	226	168	256	8	43	1	6	3.93	—	—	—
74	Vanilla milkshake	8	fluid ounce(s)	227	169	254	9	40	0	7	4.28	1.98	0.26	—
	Pizza Hut													
39009	Hot chicken wings	2	item(s)	57	—	110	11	1	0	6	2.00	—	—	0.25
14025	Meat Lovers hand tossed pizza	1	slice(s)	125	—	320	16	30	2	15	7.00	—	—	0.53
14026	Meat Lovers pan pizza	1	slice(s)	130	—	360	16	29	2	20	7.00	—	—	0.53
31009	Meat Lovers stuffed crust pizza	1	slice(s)	188	—	500	25	44	3	25	11.00	—	—	1.11
14024	Meat Lovers thin 'n crispy pizza	1	slice(s)	112	—	310	15	22	2	18	8.00	—	—	0.57
14031	Pepperoni Lovers hand tossed pizza	1	slice(s)	114	—	300	15	30	2	14	7.00	—	—	0.50
14032	Pepperoni Lovers pan pizza	1	slice(s)	119	—	350	15	29	2	19	8.00	—	—	0.50
31011	Pepperoni Lovers stuffed crust pizza	1	slice(s)	171	—	480	23	44	3	24	11.00	—	—	1.05
14030	Pepperoni Lovers thin 'n crispy pizza	1	slice(s)	94	—	270	13	22	2	14	7.00	—	—	0.51
10834	Personal Pan pepperoni pizza	1	slice(s)	59	—	150	7	18	—	6	2.50	—	—	0.97
10842	Personal Pan supreme pizza	1	slice(s)	73	—	170	8	19	1	7	3.00	—	—	0.95
39013	Personal Pan Veggie Lovers pizza	1	slice(s)	69	—	150	6	19	1	6	2.00	—	—	0.50
14028	Veggie Lovers hand tossed pizza	1	slice(s)	120	—	220	10	31	2	6	3.00	—	—	0.25
14029	Veggie Lovers pan pizza	1	slice(s)	125	—	260	10	31	2	12	4.00	—	—	0.26
31010	Veggie Lovers stuffed crust pizza	1	slice(s)	181	—	370	17	45	3	14	7.00	—	—	0.53
14027	Veggie Lovers thin 'n crispy pizza	1	slice(s)	110	—	190	8	23	2	7	3.00	—	—	0.54
39012	Wing blue cheese dipping sauce	1	item(s)	43	—	230	2	2	0	24	5.00	—	—	1
39011	Wing ranch dipping sauce	1	item(s)	43	—	210	1	4	0	22	3.50	—	—	0.50
	Starbucks													
38042	Apple cider, tall steamed	12	fluid ounce(s)	360	—	180	0	45	0	0	0.00	0.00	0.00	0
38052	Cappuccino, tall	12	fluid ounce(s)	360	—	120	7	10	0	6	4.00	—	—	—
38053	Cappuccino, tall nonfat	12	fluid ounce(s)	360	—	80	7	11	0	0	0.00	0.00	0.00	0
38054	Cappuccino, tall soy milk	12	fluid ounce(s)	360	—	100	5	13	1	3	0.00	—	—	—
38059	Cinnamon spice mocha, tall nonfat w/o whipped cream	12	fluid ounce(s)	360	—	170	11	32	0	0	0.50	0.00	0.00	0
38057	Cinnamon spice mocha, tall w/whipped cream	12	fluid ounce(s)	360	—	320	10	31	0	17	11.00	—	—	—
38051	Espresso, single shot	1	fluid ounce(s)	30	—	5	0	1	0	0	0.00	0.00	0.00	—
38088	Flavored syrup, 1 pump	1	serving(s)	10	—	20	0	5	0	0	0.00	0.00	0.00	—
32562	Frappuccino coffee drink, lite mocha	9½	fluid ounce(s)	281	—	100	7	12	3	3	2.00	—	—	0
38079	Frappuccino, grande chocolate malt	16	fluid ounce(s)	480	—	470	15	87	2	10	3.50	—	—	—
38075	Frappuccino, grande mocha malt	12	fluid ounce(s)	360	—	430	14	91	1	7	4.00	—	—	—
32561	Frappuccino low fat coffee drink, all flavors	9½	fluid ounce(s)	281	—	190	6	39	0	3	2.00	—	—	—
38067	Frappuccino, tall caramel	12	fluid ounce(s)	360	—	210	4	43	0	3	1.50	—	—	—
38078	Frappuccino, tall chocolate	12	fluid ounce(s)	360	—	290	13	52	1	5	1.00	—	—	—
38069	Frappuccino, tall chocolate brownie	12	fluid ounce(s)	360	—	270	5	51	1	7	4.50	—	—	—
38070	Frappuccino, tall coffee	12	fluid ounce(s)	360	—	190	4	38	0	3	1.50	—	—	—
38071	Frappuccino, tall espresso	12	fluid ounce(s)	360	—	160	4	33	0	2	1.50	—	—	—
38073	Frappuccino, mocha	12	fluid ounce(s)	360	—	220	5	44	0	3	1.50	—	—	—
38072	Frappuccino, tall mocha coconut	12	fluid ounce(s)	360	—	300	5	58	2	7	5.00	—	—	—
38080	Frappuccino, tall vanilla	12	fluid ounce(s)	360	—	260	11	47	0	4	1.00	—	—	—
38074	Frappuccino, tall white chocolate	12	fluid ounce(s)	360	—	240	5	48	0	4	2.50	—	—	—
33111	Latte, tall w/nonfat milk	12	fluid ounce(s)	360	335	123	12	17	0	1	0.40	0.16	0.02	0
33112	Latte, tall w/whole milk	12	fluid ounce(s)	360	325	212	11	17	0	11	6.90	3.24	0.42	—
33109	Macchiato, tall caramel w/nonfat milk	12	fluid ounce(s)	360	—	140	7	27	0	1	0.40	—	—	—
33110	Macchiato, tall caramel w/whole milk	12	fluid ounce(s)	360	—	190	6	27	0	7	4.00	—	—	—
33107	Mocha coffee drink, tall nonfat, w/o whipped cream	12	fluid ounce(s)	360	—	180	12	33	1	2	1.50	0.68	0.08	—
38089	Mocha syrup	1	serving(s)	17	—	25	1	6	0	1	0.00	—	—	—
33108	Mocha, tall w/whole milk	12	fluid ounce(s)	360	—	340	12	33	1	20	12.00	3.48	0.44	—
38084	Tazo chai black tea, tall	12	fluid ounce(s)	360	—	210	6	36	0	5	3.50	—	—	—
38083	Tazo chai black tea, tall nonfat	12	fluid ounce(s)	360	—	170	6	37	0	0	0.00	0.00	0.00	0
38087	Tazo chai black tea, tall soy milk	12	fluid ounce(s)	360	—	190	4	39	1	2	0.00	—	—	—
38063	Tazo chai creme frappuccino, tall	12	fluid ounce(s)	360	—	280	11	51	0	4	1.00	—	—	—
38076	Tazo iced tea, tall	12	fluid ounce(s)	360	—	60	0	16	0	0	0.00	0.00	0.00	0
38077	Tazo tea, grande lemonade	16	fluid ounce(s)	480	—	120	0	31	0	0	0.00	0.00	0.00	0
38065	Tazoberry creme frappuccino, tall	12	fluid ounce(s)	360	—	240	4	54	1	1	0.00	—	—	—
38066	Tazoberry frappuccino, tall	12	fluid ounce(s)	360	—	140	1	36	1	0	0.00	0.00	0.00	0
38045	Vanilla creme steamed nonfat milk, tall w/whipped cream	12	fluid ounce(s)	360	—	180	12	32	0	0	0.00	0.00	0.00	—
38046	Vanilla creme steamed soy milk, tall w/whipped cream	12	fluid ounce(s)	360	—	300	8	37	1	12	6.00	—	—	—

PAGE KEY: A–4 = Breads/Baked Goods A–8 = Cereal/Rice/Pasta A–12 = Fruit A–16 = Vegetables/Legumes A–26 = Nuts/Seeds A–28 = Vegetarian A–30 = Dairy A–36 = Eggs A–36 = Seafood A–38 = Meats A–42 = Poultry A–42 = Processed meats A–44 = Beverages A–48 = Fats/Oils A–50 = Sweets A–52 = Spices/Condiments/Sauces A–54 = Mixed Foods/Soups/Sandwiches A–60 = Fast food A–76 = Convenience A–78 = Baby foods

A

Chol (mg)	Calc (mg)	Iron (mg)	Magn (mg)	Pota (mg)	Sodi (mg)	Zinc (mg)	Vit A (RAE) (µg)	Thia (mg)	Vit E (mg α)	Ribo (mg)	Niac (mg)	Vit B$_6$ (mg)	Fola (µg)	Vit C (mg)	Vit B$_{12}$ (µg)	Sele (µg)
25	256	0.25	29	412	188	0.82	59	0.10	—	0.44	0.40	0.10	7	2	1	5
27	331	0.23	27	415	215	0.88	57	0.07	0.11	0.44	0.33	0.10	16	0	1	5
70	0	0.36	—	—	450	—	—	—	—	—	—	—	—	0	—	—
40	150	1.80	—	—	830	—	—	—	—	—	—	—	—	6	—	—
40	150	2.70	—	—	810	—	—	—	—	—	—	—	—	6	—	—
65	250	2.70	—	—	1450	—	—	—	—	—	—	—	—	9	—	—
45	150	1.80	—	—	880	—	—	—	—	—	—	—	—	9	—	—
40	200	1.80	—	—	730	—	58	—	—	—	—	—	—	2	—	—
40	200	2.70	—	—	710	—	58	—	—	—	—	—	—	2	—	—
65	300	2.70	—	—	1300	—	—	—	—	—	—	—	—	4	—	—
40	200	1.44	—	—	700	—	58	—	—	—	—	—	—	2	—	—
15	80	1.44	—	—	340	—	38	—	—	—	—	—	—	1	—	—
15	80	1.86	—	—	400	—	—	—	—	—	—	—	—	4	—	—
10	80	1.80	—	—	280	—	—	—	—	—	—	—	—	4	—	—
15	150	1.80	—	—	490	—	—	—	—	—	—	—	—	9	—	—
15	150	2.70	—	—	470	—	—	—	—	—	—	—	—	9	—	—
35	250	2.70	—	—	980	—	—	—	—	—	—	—	—	12	—	—
15	150	1.44	—	—	480	—	—	—	—	—	—	—	—	12	—	—
25	20	0.00	—	—	550	—	0	—	—	—	—	—	—	0	—	—
10	0	0.00	—	—	340	—	0	—	—	—	—	—	—	0	—	—
0	0	1.08	—	—	15	—	0	—	—	—	—	—	—	0	0	—
25	250	0.00	—	—	95	—	0	—	—	—	—	—	—	1	0	—
3	200	0.00	—	—	100	—	0	—	—	—	—	—	—	0	0	—
0	250	0.72	—	—	75	—	0	—	—	—	—	—	—	0	0	—
5	300	0.72	—	—	150	—	0	—	—	—	—	—	—	0	0	—
70	350	1.08	—	—	140	—	0	—	—	—	—	—	—	2	0	—
0	0	0.00	—	—	0	—	0	—	—	—	—	—	—	0	0	—
0	0	0.00	—	—	0	—	0	—	—	—	—	—	—	0	0	—
13	200	1.08	—	—	80	—	—	—	—	—	—	—	—	0	—	—
15	250	2.70	—	—	420	—	0	—	—	—	—	—	—	12	0	—
20	250	1.08	—	—	390	—	0	—	—	—	—	—	—	0	0	—
12	220	0.00	—	—	110	—	—	—	—	—	—	—	—	0	—	—
10	150	0.00	—	—	180	—	0	—	—	—	—	—	—	0	0	—
3	400	1.80	—	—	300	—	0	—	—	—	—	—	—	5	0	—
10	150	1.44	—	—	220	—	0	—	—	—	—	—	—	0	0	—
10	150	0.00	—	—	180	—	0	—	—	—	—	—	—	0	0	—
10	100	0.00	—	—	160	—	0	—	—	—	—	—	—	0	0	—
10	150	0.72	—	—	180	—	0	—	—	—	—	—	—	0	0	—
10	150	1.08	—	—	220	—	0	—	—	—	—	—	—	0	0	—
3	400	0.00	—	—	280	—	0	—	—	—	—	—	—	4	0	—
10	150	0.00	—	—	210	—	0	—	—	—	—	—	—	0	0	—
6	420	0.18	40	—	174	1.35	—	0.12	—	0.47	0.36	0.14	18	4	1	—
46	400	0.18	47	254	165	1.28	—	0.13	—	0.54	0.35	0.14	17	3	1	—
25	250	0.36	—	—	110	—	—	—	—	—	—	—	—	2	—	—
25	200	0.36	—	—	105	—	—	—	—	—	—	—	—	1	—	—
5	350	2.70	—	—	150	—	—	—	—	—	—	—	—	2	—	—
0	0	0.72	—	—	0	—	0	—	—	—	—	—	—	0	0	—
47	300	0.18	—	—	169	—	—	—	—	—	—	—	—	2	—	—
20	200	0.36	—	—	85	—	0	—	—	—	—	—	—	1	0	—
5	200	0.36	—	—	95	—	0	—	—	—	—	—	—	0	0	—
0	200	0.72	—	—	70	—	0	—	—	—	—	—	—	0	0	—
3	400	0.00	—	—	280	—	0	—	—	—	—	—	—	4	0	—
0	0	0.00	—	—	0	—	0	—	—	—	—	—	—	0	0	—
0	0	0.00	—	—	15	—	0	—	—	—	—	—	—	5	0	—
0	150	0.00	—	—	125	—	0	—	—	—	—	—	—	1	0	—
0	0	0.00	—	—	30	—	0	—	—	—	—	—	—	0	0	—
5	350	0.00	—	—	170	—	0	—	—	—	—	—	—	0	0	—
30	400	1.44	—	—	130	—	0	—	—	—	—	—	—	0	0	—

TABLE A–1
Food Composition (Computer code number is for Wadsworth Diet Analysis program) (For purposes of calculations, use "0" for t, <1, <.1, <.01, etc.)

DA + Code	Food Description	Quantity	Measure	Wt (g)	H₂O (g)	Ener (kcal)	Prot (g)	Carb (g)	Dietary Fiber (g)	Fat (g)	Sat	Mono	Poly	Trans
	FAST FOOD—Continued													
38044	Vanilla creme steamed whole milk, tall w/whipped cream	12	fluid ounce(s)	360	—	340	10	31	0	18	12.00	—	—	—
38090	Whipped cream	1	serving(s)	27	—	100	0	2	0	9	6.00	—	—	—
38062	White chocolate mocha, tall nonfat w/o whipped cream	12	fluid ounce(s)	360	—	260	12	45	0	4	3.00	—	—	—
38061	White chocolate mocha, tall w/whipped cream	12	fluid ounce(s)	360	—	410	11	44	0	20	13.00	—	—	—
38048	White hot chocolate, tall w/o whipped cream	12	fluid ounce(s)	360	—	300	15	51	0	5	3.50	—	—	—
38047	White hot chocolate, tall w/whipped cream	12	fluid ounce(s)	360	—	460	13	50	0	22	15.00	—	—	—
38050	White hot chocolate soy milk, tall w/whipped cream	12	fluid ounce(s)	360	—	420	11	56	1	16	9.00	—	—	—
	Subway													
34023	Asiago caesar chicken wrap	1	item(s)	244	—	413	22	47	2	15	3.00	—	—	0
38622	Atkins-friendly chicken bacon ranch wrap	1	item(s)	213	—	480	40	19	11	27	9.00	—	—	0
38623	Atkins-friendly turkey bacon melt wrap	1	item(s)	199	—	430	32	22	12	25	9.00	—	—	0
34029	Bacon & egg breakfast sandwich	1	item(s)	127	—	302	14	29	1	15	4.00	—	—	0
32045	Chocolate chip cookie	1	item(s)	48	—	209	3	29	1	10	3.50	—	—	1.07
32048	Chocolate chip M&M cookie	1	item(s)	48	—	210	2	29	1	10	3.00	—	—	2.67
32049	Chocolate chunk cookie	1	item(s)	48	—	210	2	30	1	10	3.00	—	—	2.67
4024	Classic Italian B.M.T. sandwich, 6", white bread	1	item(s)	250	—	453	21	40	3	24	8.00	—	—	0
16397	Club salad	1	item(s)	323	—	145	17	12	3	4	1.00	—	—	0
3422	Club sandwich, 6", white bread	1	item(s)	253	—	294	22	40	3	5	1.50	—	—	0
4030	Cold cut trio sandwich, 6", white bread	1	item(s)	254	—	415	19	40	3	20	7.00	—	—	0
34030	Ham & egg breakfast sandwich	1	item(s)	147	—	291	15	30	1	12	3.00	—	—	0
3885	Ham sandwich, 6", white bread	1	item(s)	219	—	261	17	39	3	5	1.50	—	—	0
34026	Honey mustard melt sandwich, 6", Italian bread	1	item(s)	258	—	373	23	47	3	11	5.00	—	—	—
34027	Horseradish roast beef sandwich, 6", Italian bread	1	item(s)	230	—	401	18	42	3	17	3.00	—	—	—
4651	Meatball sandwich, 6", white bread	1	item(s)	284	—	501	23	46	4	25	10.00	—	—	0.75
15839	Melt sandwich, 6", white bread	1	item(s)	256	—	380	23	41	3	15	5.00	—	—	0
32046	Oatmeal raisin cookie	1	item(s)	48	—	197	3	29	1	8	2.00	—	—	2.67
32047	Peanut butter cookie	1	item(s)	48	—	220	3	26	1	12	3.00	—	—	1.07
3957	Roast beef sandwich, 6", white bread	1	item(s)	220	—	264	18	39	3	5	1.00	—	—	0
16403	Roasted chicken breast salad	1	item(s)	304	—	137	16	12	3	3	0.50	—	—	—
16378	Roasted chicken breast sandwich, 6", white bread	1	item(s)	234	—	311	25	40	3	6	1.50	—	—	0
34028	Southwest steak & cheese sandwich, 6", Italian bread	1	item(s)	255	—	412	23	42	4	18	6.00	—	—	—
4032	Spicy italian sandwich, 6", white bread	1	item(s)	213	—	458	19	42	2	24	9.00	—	—	0
4031	Steak & cheese sandwich, 6", white bread	1	item(s)	253	—	362	23	41	4	13	4.50	—	—	0
34024	Steak & cheese wrap	1	item(s)	245	—	353	22	46	3	9	4.00	—	—	—
32050	Sugar cookie	1	item(s)	48	—	222	2	28	1	12	3.00	—	—	3.73
16402	Tuna salad	1	item(s)	314	—	238	13	11	3	16	4.00	—	—	—
15844	Tuna sandwich, 6", white bread	1	item(s)	252	—	419	18	39	3	21	5.00	—	—	—
15834	Turkey breast & ham sandwich, 6", white bread	1	item(s)	229	—	267	18	40	3	5	1.00	—	—	0
34025	Turkey breast & bacon wrap	1	item(s)	228	—	318	19	45	2	7	2.50	—	—	—
16376	Turkey breast sandwich, 6", white bread	1	item(s)	220	—	254	16	39	3	4	1.00	—	—	0
16375	Veggie delite, 6", white bread	1	item(s)	163	—	200	7	37	3	3	0.50	—	—	0
32051	White macadamia nut cookie	1	item(s)	48	—	221	2	27	1	12	3.00	—	—	1.07
	Taco Bell													
29906	7-layer burrito	1	item(s)	283	—	530	18	67	10	22	8.00	—	—	3
744	Bean burrito	1	item(s)	198	—	370	14	55	8	10	3.50	—	—	2
749	Beef burrito supreme	1	item(s)	248	—	440	18	51	7	18	8.00	—	—	2
33417	Beef chalupa supreme	1	item(s)	153	—	390	14	31	3	24	10.00	—	—	3
29910	Beef gordita supreme	1	item(s)	153	—	310	14	30	3	16	7.00	—	—	0.50
2014	Beef soft taco	1	item(s)	99	—	210	10	21	2	10	4.50	—	—	1
10860	Beef soft taco supreme	1	item(s)	134	—	260	11	22	3	14	7.00	—	—	1
2018	Big beef burrito supreme	1	item(s)	291	—	510	23	52	11	23	9.00	6.55	1.61	—
14467	Big chicken burrito supreme	1	item(s)	255	—	460	27	50	3	17	6.00	—	—	—
34472	Chicken burrito supreme	1	item(s)	248	—	410	21	50	5	14	6.00	—	—	2
33418	Chicken chalupa supreme	1	item(s)	153	—	370	17	30	1	20	8.00	—	—	3

PAGE KEY: A–4 = Breads/Baked Goods   A–8 = Cereal/Rice/Pasta   A–12 = Fruit   A–16 = Vegetables/Legumes   A–26 = Nuts/Seeds   A–28 = Vegetarian
A–30 = Dairy   A–36 = Eggs   A–36 = Seafood   A–38 = Meats   A–42 = Poultry   A–42 = Processed meats   A–44 = Beverages   A–48 = Fats/Oils
A–50 = Sweets   A–52 = Spices/Condiments/Sauces   A–54 = Mixed Foods/Soups/Sandwiches   A–60 = Fast food   A–76 = Convenience   A–78 = Baby foods

Chol (mg)	Calc (mg)	Iron (mg)	Magn (mg)	Pota (mg)	Sodi (mg)	Zinc (mg)	Vit A (RAE) (µg)	Thia (mg)	Vit E (mg α)	Ribo (mg)	Niac (mg)	Vit B$_6$ (mg)	Fola (µg)	Vit C (mg)	Vit B$_{12}$ (µg)	Sele (µg)
75	40	0.00	—	—	160	—	0	—	—	—	—	—	—	2	0	—
40	0	0.00	—	—	10	—	0	—	—	—	—	—	—	0	0	—
5	400	0.00	—	—	210	—	0	—	—	—	—	—	—	0	0	—
70	400	0.00	—	—	210	—	0	—	—	—	—	—	—	2	0	—
10	450	0.00	—	—	250	—	0	—	—	—	—	—	—	0	0	—
75	500	0.00	—	—	250	—	0	—	—	—	—	—	—	4	0	—
35	500	1.44	—	—	210	—	0	—	—	—	—	—	—	0	0	—
46	40	2.70	—	—	1320	—	—	—	—	—	—	—	—	15	—	—
90	350	2.70	—	—	1340	—	—	—	—	—	—	—	—	7	—	—
65	300	2.70	—	—	1650	—	—	—	—	—	—	—	—	5	—	—
185	60	1.80	—	—	480	—	—	—	—	—	—	—	—	15	—	—
12	0	1.00	—	—	135	—	0	—	—	—	—	—	—	0	—	—
13	0	1.00	—	—	135	—	0	—	—	—	—	—	—	0	—	—
12	0	1.00	—	—	150	—	0	—	—	—	—	—	—	0	—	—
56	100	2.70	—	—	1740	—	—	—	—	—	—	—	—	24	—	—
30	40	1.80	—	—	1070	—	—	—	—	—	—	—	—	30	—	—
30	40	3.60	—	—	1250	—	60	—	—	—	—	—	—	24	—	—
57	150	3.60	—	—	1670	—	100	—	—	—	—	—	—	24	—	—
189	60	2.70	—	—	700	—	67	—	—	—	—	—	—	15	—	—
25	40	2.70	—	—	1260	—	—	—	—	—	—	—	—	24	—	—
41	100	2.70	—	—	1570	—	—	—	—	—	—	—	—	24	—	—
27	40	3.60	—	—	880	—	—	—	—	—	—	—	—	24	—	—
56	100	3.60	—	—	1350	—	—	—	—	—	—	—	—	24	—	—
41	100	2.70	—	—	1690	—	—	—	—	—	—	—	—	24	—	—
14	0	1.00	—	—	180	—	0	—	—	—	—	—	—	0	—	—
0	0	1.00	—	—	200	—	0	—	—	—	—	—	—	0	—	—
20	40	3.60	—	—	840	—	60	—	—	—	—	—	—	24	—	—
36	40	1.08	—	—	730	—	—	—	—	—	—	—	—	30	—	—
48	60	3.60	—	—	880	—	—	—	—	—	—	—	—	24	—	—
44	100	6.30	—	—	1120	—	—	—	—	—	—	—	—	24	—	—
57	30	3.00	—	—	1498	—	—	—	—	—	—	—	—	13	—	—
37	100	6.30	—	—	1200	—	—	—	—	—	—	—	—	24	—	—
37	150	7.20	—	—	1400	—	—	—	—	—	—	—	—	15	—	—
18	0	1.00	—	—	170	—	0	—	—	—	—	—	—	0	—	—
42	100	1.08	—	—	880	—	177	—	—	—	—	—	—	30	—	—
42	100	2.70	—	—	1180	—	100	—	—	—	—	—	—	24	—	—
23	40	2.70	—	—	1210	—	—	—	—	—	—	—	—	24	—	—
24	60	2.70	—	—	1490	—	—	—	—	—	—	—	—	15	—	—
15	40	2.70	—	—	1000	—	—	—	—	—	—	—	—	24	—	—
0	40	1.80	—	—	500	—	—	—	—	—	—	—	—	24	—	—
13	0	1.00	—	—	140	—	0	—	—	—	—	—	—	0	—	—
25	300	3.59	—	—	1360	—	—	—	—	—	—	—	—	5	—	—
10	200	2.69	—	—	1200	—	53	—	—	—	—	—	—	5	—	—
40	200	2.70	—	—	1330	—	351	—	—	—	—	—	—	9	—	—
40	150	1.80	—	—	600	—	—	—	—	—	—	—	—	5	—	—
35	150	2.70	—	—	590	—	—	—	—	—	—	—	—	5	—	—
25	100	1.80	—	—	620	—	44	—	—	—	—	—	—	2	—	—
40	150	1.80	—	—	630	—	73	—	—	—	—	—	—	5	—	—
60	150	2.70	—	493	1500	—	877	—	—	0.07	—	—	—	5	—	—
70	101	1.46	—	—	1200	—	—	—	—	—	—	—	—	2	—	—
45	200	2.70	—	—	1270	—	—	—	—	—	—	—	—	9	—	—
45	100	1.08	—	—	530	—	—	—	—	—	—	—	—	5	—	—

TABLE A–1
Food Composition

(Computer code number is for Wadsworth Diet Analysis program)   (For purposes of calculations, use "0" for t, <1, <.1, <.01, etc.)

DA + Code	Food Description	Quantity	Measure	Wt (g)	H₂O (g)	Ener (kcal)	Prot (g)	Carb (g)	Dietary Fiber (g)	Fat (g)	Fat Breakdown (g)			
											Sat	Mono	Poly	Trans
	FAST FOOD—Continued													
29900	Chicken fajita wrap supreme	1	item(s)	255	—	510	20	53	3	24	7.76	—	—	—
29895	Choco taco ice cream dessert	1	item(s)	113	—	310	3	37	1	17	10.00	—	—	—
10794	Cinnamon twists	1	serving(s)	35	—	160	0	28	0	5	1.00	—	—	1.50
14465	Grilled chicken burrito	1	item(s)	198	—	390	19	49	3	13	4.00	—	—	—
29911	Grilled chicken gordita supreme	1	item(s)	153	—	290	17	28	2	12	5.00	—	—	0
14463	Grilled chicken soft taco	1	item(s)	99	—	190	14	19	0	6	2.50	—	—	—
29912	Grilled steak gordita supreme	1	item(s)	153	—	290	16	28	2	13	6.00	—	—	0.50
29904	Grilled steak soft taco	1	item(s)	127	—	280	12	21	1	17	4.50	—	—	1
29905	Grilled steak soft taco supreme	1	item(s)	135	—	240	15	20	2	11	5.00	—	—	—
2021	Mexican pizza	1	serving(s)	216	—	550	21	46	7	31	11.00	—	—	5
2011	Nachos	1	serving(s)	99	—	320	5	33	2	19	4.50	—	—	5
2012	Nachos bellgrande	1	serving(s)	308	—	780	20	80	12	43	13.00	—	—	10
34473	Steak burrito supreme	1	item(s)	248	—	420	19	50	6	16	7.00	—	—	2
33419	Steak chalupa supreme	1	item(s)	153	—	370	15	29	2	22	8.00	—	—	3
29899	Steak fajita wrap supreme	1	item(s)	255	—	510	21	52	3	25	8.00	—	—	—
747	Taco	1	item(s)	78	—	170	8	13	3	10	4.00	—	—	0.50
2015	Taco salad w/salsa, with shell	1	serving(s)	533	—	790	31	73	13	42	15.00	—	—	8.75
14459	Taco supreme	1	item(s)	113	—	220	9	14	3	14	7.00	—	—	1
748	Tostada	1	item(s)	170	—	250	11	29	7	10	4.00	—	—	1.50
29901	Veggie fajita wrap supreme	1	item(s)	255	—	470	11	55	3	22	7.00	—	—	—
	CONVENIENCE MEALS													
	Banquet													
29961	Barbeque chicken meal	1	item(s)	281	—	330	16	37	2	13	3.00	—	—	—
14788	Boneless white fried chicken meal	1	item(s)	234	—	490	14	49	2	27	7.00	—	—	—
29960	Fish sticks meal	1	item(s)	187	—	270	13	31	3	10	3.00	—	—	—
29957	Lasagna with meat sauce meal	1	item(s)	312	—	320	15	46	7	9	4.00	—	—	—
14777	Macaroni & cheese meal	1	item(s)	340	—	420	15	57	5	14	8.00	—	—	—
1741	Meatloaf meal	1	item(s)	269	—	240	14	20	4	11	4.00	—	—	—
39418	Pepperoni pizza meal	1	item(s)	191	—	480	11	56	5	23	8.00	—	—	—
33759	Roasted white turkey meal	1	item(s)	255	—	230	14	30	5	6	2.00	—	—	—
1743	Salisbury steak meal	1	item(s)	269	197	380	12	28	3	24	12.00	—	—	—
	Budget Gourmet													
1914	Cheese manicotti w/meat sauce	1	item(s)	284	194	420	18	38	4	22	11.00	6.00	1.34	—
1915	Chicken w/fettucini	1	item(s)	284	—	380	20	33	3	19	10.00	—	—	—
3986	Light beef stroganoff	1	item(s)	248	177	290	20	32	3	7	4.00	—	—	—
3996	Light sirloin of beef in herb sauce	1	item(s)	269	214	260	19	30	5	7	4.00	2.30	0.31	—
3987	Light vegetable lasagna	1	item(s)	298	227	290	15	36	5	9	1.79	0.89	0.60	—
	Healthy Choice													
36979	Bowls chicken teriyaki with rice	1	item(s)	298	—	330	19	50	5	6	2.00	2.00	2.00	—
9425	Cheese French bread pizza	1	item(s)	170	—	360	20	57	5	5	1.50	—	—	—
9306	Chicken enchilada suprema meal	1	item(s)	320	252	360	13	59	8	7	3.00	2.00	2.00	—
9316	Lemon pepper fish meal	1	item(s)	303	—	280	11	49	5	5	2.00	1.00	2.00	—
9322	Traditional salisbury steak meal	1	item(s)	354	250	360	23	45	5	9	3.50	4.00	1.00	—
9359	Traditional turkey breasts meal	1	item(s)	298	—	330	21	50	4	5	2.00	1.50	1.50	—
9451	Zucchini lasagna	1	item(s)	383	—	280	13	47	5	4	2.50	—	—	—
	Stouffers													
2363	Cheese enchiladas with mexican rice	1	serving(s)	276	—	370	12	48	5	14	5.00	—	—	—
2313	Cheese French bread pizza	1	serving(s)	294	—	370	14	43	3	16	6.00	—	—	—
11138	Cheese manicotti w/tomato sauce	1	item(s)	255	—	330	17	35	3	13	8.00	—	—	—
2366	Chicken pot pie	1	item(s)	284	—	740	23	56	4	47	18.00	12.41	10.48	—
11116	Homestyle baked chicken breast w/mashed potatoes & gravy	1	item(s)	252	—	260	19	21	1	11	3.00	—	—	—
11146	Homestyle beef pot roast & potatoes	1	item(s)	252	—	270	16	25	3	12	4.50	—	—	—
11152	Homestyle roast turkey breast w/stuffing & mashed potatoes	1	item(s)	273	—	300	16	34	2	11	3.00	—	—	—
11043	Lean Cuisine Cafe Classics baked chicken & whipped potatoes w/stuffing	1	item(s)	227	—	240	17	33	3	5	1.50	1.50	1.00	0
11046	Lean Cuisine Cafe Classics honey mustard chicken	1	item(s)	213	—	260	18	37	1	4	1.50	1.00	1.00	0
360	Lean Cuisine Everyday Favorites chicken chow mein w/rice	1	item(s)	255	—	210	12	33	2	3	1.00	1.00	0.50	0
9467	Lean Cuisine Everyday Favorites fettucini alfredo	1	item(s)	262	—	280	13	40	2	7	3.50	2.00	1.00	0
11055	Lean Cuisine Everyday Favorites lasagna w/meat sauce	1	item(s)	291	—	300	19	41	3	8	4.00	2.00	0.50	0
9479	Lean Cuisine French bread deluxe pizza	1	item(s)	174	—	330	18	44	3	9	3.50	1.50	1.00	0

PAGE KEY: A–4 = Breads/Baked Goods A–8 = Cereal/Rice/Pasta A–12 = Fruit A–16 = Vegetables/Legumes A–26 = Nuts/Seeds A–28 = Vegetarian
A–30 = Dairy A–36 = Eggs A–36 = Seafood A–38 = Meats A–42 = Poultry A–42 = Processed meats A–44 = Beverages A–48 = Fats/Oils
A–50 = Sweets A–52 = Spices/Condiments/Sauces A–54 = Mixed Foods/Soups/Sandwiches A–60 = Fast food A–76 = Convenience A–78 = Baby foods

A

Chol (mg)	Calc (mg)	Iron (mg)	Magn (mg)	Pota (mg)	Sodi (mg)	Zinc (mg)	Vit A (RAE) (µg)	Thia (mg)	Vit E (mg α)	Ribo (mg)	Niac (mg)	Vit B$_6$ (mg)	Fola (µg)	Vit C (mg)	Vit B$_{12}$ (µg)	Sele (µg)
57	165	1.52	—	—	1182	—	—	—	—	—	—	—	—	7	—	—
20	60	0.72	—	—	100	—	—	—	—	—	—	—	—	0	—	—
0	0	0.37	—	—	150	—	0	—	—	—	—	—	—	0	—	—
40	151	1.44	—	—	1240	—	—	—	—	—	—	—	—	2	—	—
45	100	1.80	—	—	530	—	—	—	—	—	—	—	—	5	—	—
30	100	1.08	—	—	550	—	15	—	—	—	—	—	—	1	—	—
35	100	2.70	—	—	520	—	—	—	—	—	—	—	—	4	—	—
30	100	1.44	—	—	650	—	29	—	—	—	—	—	—	4	—	—
35	100	1.08	—	—	510	—	29	—	—	—	—	—	—	4	—	—
45	350	3.60	—	—	1030	—	—	—	—	—	—	—	—	6	—	—
4	80	0.72	—	—	530	—	0	—	—	—	—	—	—	0	—	—
35	200	2.70	—	—	1300	—	162	—	—	—	—	—	—	6	—	—
35	200	2.70	—	—	1260	—	789	—	—	—	—	—	—	9	—	—
35	100	1.44	—	—	520	—	—	—	—	—	—	—	—	4	—	—
50	150	1.80	—	—	1200	—	—	—	—	—	—	—	—	6	—	—
25	60	1.08	—	—	350	—	44	—	—	—	—	—	—	2	—	—
65	400	6.23	—	—	1670	—	—	—	—	—	—	—	—	21	—	—
40	80	1.44	—	—	360	—	73	—	—	—	—	—	—	5	—	—
15	150	1.44	—	—	710	—	281	—	—	—	—	—	—	5	—	—
30	150	1.44	—	—	990	—	—	—	—	—	—	—	—	6	—	—
50	40	1.08	—	—	1210	—	0	—	—	—	—	—	—	5	—	—
65	60	1.08	—	—	1150	—	—	—	—	—	—	—	—	0	—	—
30	60	1.44	—	—	690	—	—	—	—	—	—	—	—	2	—	—
20	100	2.70	—	—	1170	—	—	—	—	—	—	—	—	0	—	—
20	150	1.44	—	—	1330	—	0	—	—	—	—	—	—	0	—	—
30	0	1.80	—	—	1040	—	0	—	—	—	—	—	—	0	—	—
35	150	1.80	—	—	870	—	0	—	—	—	—	—	—	0	—	—
25	60	1.80	—	—	1070	—	—	—	—	—	—	—	—	4	—	—
60	40	1.44	—	—	1140	—	0	—	—	—	—	—	—	3	—	—
85	300	2.70	45	484	810	2.29	—	0.45	—	0.51	4.00	0.23	31	0	1	—
85	100	2.70	—	—	810	—	—	0.15	—	0.43	6.00	—	—	0	—	—
35	40	1.80	39	280	580	4.71	—	0.17	—	0.37	4.28	0.27	19	2	3	—
30	40	1.80	58	540	850	4.81	—	0.16	—	0.29	5.53	0.37	38	6	2	—
15	283	3.03	79	420	780	1.39	—	0.22	—	0.45	3.13	0.32	75	59	<1	—
40	20	0.72	—	—	600	—	—	—	—	—	—	—	—	15	—	—
10	350	3.60	—	—	600	—	—	—	—	—	—	—	—	12	—	—
30	40	1.44	—	—	580	—	—	—	—	—	—	—	—	4	—	—
30	40	0.36	—	—	580	—	—	—	—	—	—	—	—	30	—	—
45	80	2.70	—	—	580	—	—	—	—	—	—	—	—	21	—	—
35	40	1.44	—	—	600	—	—	—	—	—	—	—	—	0	—	—
10	200	1.80	—	—	310	—	—	—	—	—	—	—	—	0	—	—
25	200	1.44	—	360	890	—	—	—	—	—	—	—	—	12	—	—
15	200	1.80	—	240	880	—	—	—	—	—	—	—	—	0	—	—
40	350	1.08	—	430	810	—	—	—	—	—	—	—	—	1	—	—
65	150	2.70	—	—	1170	—	—	—	—	—	—	—	—	2	—	—
50	20	0.72	—	500	760	—	0	—	—	—	—	—	—	0	—	—
35	20	1.80	—	790	820	—	—	—	—	—	—	—	—	6	—	—
35	40	0.72	—	450	1190	—	0	—	—	—	—	—	—	0	—	—
30	80	0.72	—	480	690	—	—	—	—	—	—	—	—	0	—	—
35	60	0.36	—	370	640	—	—	—	—	—	—	—	—	0	—	—
30	20	0.36	—	310	620	—	—	—	—	—	—	—	—	0	—	—
20	200	0.36	—	260	670	—	0	—	—	—	—	—	—	0	—	—
30	200	1.08	—	590	650	—	—	—	—	—	—	—	—	5	—	—
20	100	1.80	—	390	630	—	—	—	—	—	—	—	—	9	—	—

TABLE A–1

Food Composition   (Computer code number is for Wadsworth Diet Analysis program)   (For purposes of calculations, use "0" for t, <1, <.1, <.01, etc.)

DA + Code	Food Description	Quantity	Measure	Wt (g)	H₂O (g)	Ener (kcal)	Prot (g)	Carb (g)	Dietary Fiber (g)	Fat (g)	Sat	Mono	Poly	Trans
	CONVENIENCE MEALS—Continued													
	Weight Watchers													
11164	Smart Ones chicken enchiladas suiza entree	1	serving(s)	255	—	270	15	33	2	9	3.50	—	—	—
11155	Smart Ones garden lasagna entree	1	item(s)	312	—	270	14	36	5	7	3.50	—	—	—
11187	Smart Ones pepperoni pizza	1	item(s)	158	—	390	23	46	4	12	4.00	—	—	—
31514	Smart Ones spicy penne pasta & ricotta	1	item(s)	289	—	280	11	45	4	6	2.00	—	—	—
31512	Smart Ones spicy szechuan style vegetables & chicken	1	item(s)	255	—	220	11	39	3	2	0.50	—	—	—
	BABY FOODS													
787	Apple juice	4	fluid ounce(s)	127	112	60	0	15	<1	<1	0.02	0.00	0.04	—
778	Applesauce, strained	4	tablespoon(s)	64	55	31	<1	8	1	<1	0.02	0.01	0.04	—
779	Bananas w/tapioca, strained	4	tablespoon(s)	60	50	34	<1	9	1	<.1	0.02	0.01	0.01	—
604	Carrots, strained	4	tablespoon(s)	56	52	15	<1	3	1	<.1	0.01	0.00	0.03	—
770	Chicken noodle dinner, strained	4	tablespoon(s)	64	55	42	2	6	1	1	0.38	0.55	0.30	—
801	Green beans, strained	4	tablespoon(s)	60	0.05	15	0.77	3.53	1.13	0.05	0.01	0	0.03	—
910	Human milk, mature	2	fluid ounce(s)	62	54	43	1	4	0	3	1.24	1.02	0.31	—
760	Mixed cereal, prepared w/whole milk	4	ounce(s)	114	85	128	5	18	1	4	2.19	1.25	0.43	—
772	Mixed vegetable dinner, strained	2	ounce(s)	57	50	23	1	5	1	<.1	0.00	0.00	0.06	—
762	Rice cereal, prepared w/whole milk	4	ounce(s)	114	85	131	4	19	<1	4	2.64	1.02	0.16	—
758	Teething biscuits	1	item(s)	11	1	43	1	8	<1	<1	0.17	0.16	0.09	—

PAGE KEY: A–4 = Breads/Baked Goods   A–8 = Cereal/Rice/Pasta   A–12 = Fruit   A–16 = Vegetables/Legumes   A–26 = Nuts/Seeds   A–28 = Vegetarian
A–30 = Dairy   A–36 = Eggs   A–36 = Seafood   A–38 = Meats   A–42 = Poultry   A–42 = Processed meats   A–44 = Beverages   A–48 = Fats/Oils
A–50 = Sweets   A–52 = Spices/Condiments/Sauces   A–54 = Mixed Foods/Soups/Sandwiches   A–60 = Fast food   A–76 = Convenience   A–78 = Baby foods

Chol (mg)	Calc (mg)	Iron (mg)	Magn (mg)	Pota (mg)	Sodi (mg)	Zinc (mg)	Vit A (RAE) (µg)	Thia (mg)	Vit E (mg α)	Ribo (mg)	Niac (mg)	Vit B$_6$ (mg)	Fola (µg)	Vit C (mg)	Vit B$_{12}$ (µg)	Sele (µg)
50	250	1.08	—	—	660	—	—	—	—	—	—	—	—	4	—	—
30	350	1.80	—	—	610	—	—	—	—	—	—	—	—	6	—	—
45	450	1.80	—	320	650	—	55	—	—	—	—	—	—	5	—	—
5	150	2.70	—	250	400	—	—	—	—	—	—	—	—	6	—	—
10	150	1.80	—	—	730	—	—	—	—	—	—	—	—	2	—	—
0	5	0.72	4	115	4	0.04	1	0.01	0.76	0.02	0.11	0.04	0	73	0	<1
0	3	0.14	2	45	1	0.01	1	0.01	0.38	0.02	0.04	0.02	1	25	0	<1
0	3	0.12	6	53	5	0.04	1	0.01	0.36	0.02	0.11	0.07	4	10	0	<1
0	12	0.21	5	110	21	0.08	321	0.01	0.29	0.02	0.26	0.04	8	3	0	<1
10	17	0.41	9	89	15	0.35	70	0.03	0.13	0.04	0.46	0.04	7	<.1	<.1	2
0	23.39	0.44	14.39	94.8	1.2	0.12	27	0.01	0.31	0.05	0.2	0.02	21	3.11	0	0.18
9	20	0.02	2	31	10	0.10	38	0.01	0.05	0.02	0.11	0.01	3	3	<.1	1
12	250	11.85	31	226	53	0.81	28	0.49	—	0.66	6.56	0.07	12	1	<.1	—
0	12	0.19	6	69	5	0.09	77	0.01	—	0.02	0.29	0.04	5	2	0	<1
12	272	13.85	51	216	52	0.73	25	0.53	—	0.57	5.91	0.13	9	1	<.1	4
0	29	0.39	4	36	40	0.10	3	0.03	0.03	0.06	0.48	0.01	5	1	<.1	3

The FDA lists caffeine as a multipurpose GRAS substance that may be added to foods and beverages. Drug manufacturers use caffeine in many kinds of drugs: stimulants, pain relievers, cold remedies, diuretics, and weight-loss aids.

Caffeine Content of Beverages, Foods, and Over-the-Counter Drugs

Beverages and Foods	Average (mg)	Range (mg)	Drugs[a]	Average (mg)
Coffee (5-oz cup)			Cold remedies (standard dose)	
Brewed, drip method	130	110–150	Dristan	0
Brewed, percolator	94	64–124	Coryban-D, Triaminicin	30
Instant	74	40–108	Diuretics (standard dose)	
Decaffeinated, brewed or instant	3	1–5	Aqua-ban, Permathene H2Off	200
Tea (5-oz cup)			Pre-Mens Forte	100
Brewed, major U.S. brand	40	20–90	Pain relievers (standard dose)	
Brewed, imported brands	60	25–110	Excedrin	130
Instant	30	25–50	Midol, Anacin	65
Iced (12-oz can)	70	67–76	Aspirin, plain (any brand)	0
Soft drinks (12-oz can)			Stimulants	
Dr. Pepper	40		Caffedrin, NoDoz, Vivarin	200
Colas and cherry cola			Weight-control aids (daily dose)	
Regular		30–46	Prolamine	280
Diet		2–58	Dexatrim, Dietac	200
Caffeine-free		0–trace		
Jolt	72			
Mountain Dew, Mello Yello	52			
Fresca, Hires Root Beer, 7-Up, Sprite, Squirt, Sunkist Orange	0			
Cocoa beverage (5-oz cup)	4	2–20		
Chocolate milk beverage (8 oz)	5	2–7		
Milk chocolate candy (1 oz)	6	1–15		
Dark chocolate, semisweet (1 oz)	20	5–35		
Baker's chocolate (1 oz)	26			
Chocolate flavored syrup (1 oz)	4			

Note: A pharmacologically active dose of caffeine is defined as 200 milligrams.
[a]Because products change, contact the manufacturer for an update on products you use regularly.

This appendix first presents nutrition recommendations from the World Health Organization (WHO) and then provides details for Canadians on Canada's *Food Guide to Healthy Eating* and meal planning system.

Nutrition Recommendations from WHO

The World Health Organization (WHO) has assessed the relationships between diet and the development of chronic diseases. Its recommendations include:

- Total energy: sufficient to support normal growth, physical activity, and healthy body weight (body mass index = 20 to 22).
- Total fat: 15 to 30 percent of total energy.
- Saturated fat: less than 10 percent of total energy.
- Total carbohydrate: 55 to 75 percent of total energy.
- Added sugars: less than 10 percent of total energy.
- Protein: 10 to 15 percent of total energy.
- Salt: less than 5 grams/day, preferably iodized.
- Fruit and vegetables: at least 400 grams (almost 1 pound) daily.
- Physical activity: one hour per day of moderate intensity on most days of the week.

Canada's *Food Guide to Healthy Eating*

Figure B-1 presents the 1992 Canada's *Food Guide to Healthy Eating*, which interprets Canada's *Guidelines for Healthy Eating* (see Table 1-5 on p. 16) for consumers and recommends a range of servings to consume daily from each of the four food groups. The following publications, which are available from Health Canada, through its website, explain how to use the *Guide: Using the Food Guide; Food Guide Facts: Background for Educators and Communicators; Canada's Food Guide to Healthy Eating—Focus on Preschoolers: Background for Educators and Communicators;* and *Canada's Food Guide to Healthy Eating—Focus on Children Six to Twelve Years: Background for Educators and Communicators.* Figure B-2 presents Canada's Physical Activity Guide.

Canada's *Guidelines for Healthy Eating* and *Canada's Food Guide to Healthy Eating* are being reviewed for consistency with the Dietary Reference Intakes. Check the website for the Health Canada Office of Nutrition Policy and Promotion, **www.hc-sc.gc.ca/ahc-asc/branch-dirgen/hpfb-dgpsa/onpp-bppn/index_e.html**, for the status of the review.

Canada's Meal Planning for Healthy Eating

Beyond the Basics: Meal Planning for Healthy Eating, Diabetes Prevention and Management is Canada's system of meal planning.[1] Similar to the U.S. exchange system, *Beyond the Basics* sorts foods into groups and defines portion sizes to help people manage their blood glucose and maintain a healthy weight. Because foods that contain carbohydrate raise blood glucose, the food groups are organized into two sections—those that contain carbohydrate (presented in Table B-1) and those that contain little or no carbohydrate (shown in Table B-2). One portion from any of the food groups listed in Table B-1 provides about 15 grams of available carbohydrate (total carbohydrate minus fiber) and counts as one carbohydrate choice. Within each group, foods are identified as those to "choose more often" (generally higher in vitamins, minerals, and fiber) and those to "choose less often" (generally higher in sugar, saturated fat, or *trans* fat).

CONTENTS

[1]The tables for the Canadian meal planning system are adapted from *Beyond the Basics: Meal Planning for Healthy Eating, Diabetes Prevention and Management*, copyright 2005, with permission of the Canadian Diabetes Association. Additional information is available from **www.diabetes.ca**.

FIGURE B-1 Canada's *Food Guide to Healthy Eating*

Healthy Canada

Health and Welfare Canada

Santé et Bien-être social Canada

CANADA'S Food Guide TO HEALTHY EATING

Enjoy a variety of foods from each group every day.

Choose lower-fat foods more often.

Grain Products
Choose whole grain and enriched products more often.

Vegetables & Fruit
Choose dark green and orange vegetables and orange fruit more often.

Milk Products
Choose lower-fat milk products more often.

Meat & Alternatives
Choose leaner meats, poultry and fish, as well as dried peas, beans and lentils more often.

CANADA'S

Food Guide

TO HEALTHY EATING

FOR PEOPLE FOUR YEARS AND OVER

Different People Need Different Amounts of Food

The amount of food you need every day from the 4 food groups and other foods depends on your age, body size, activity level, whether you are male or female and if you are pregnant or breast-feeding. That's why the Food Guide gives a lower and higher number of servings for each food group. For example, young children can choose the lower number of servings, while male teenagers can go to the higher number. Most other people can choose servings somewhere in between.

Grain Products

5–12
SERVINGS PER DAY

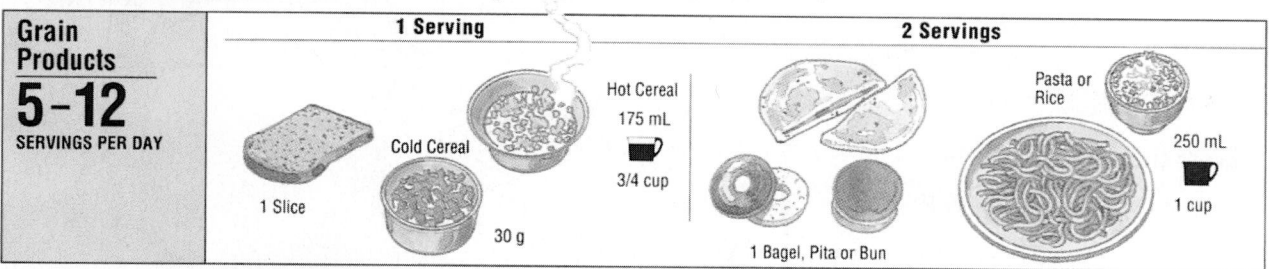

1 Serving — 1 Slice — Cold Cereal 30 g — Hot Cereal 175 mL 3/4 cup

2 Servings — 1 Bagel, Pita or Bun — Pasta or Rice 250 mL 1 cup

Vegetables & Fruit

5–10
SERVINGS PER DAY

1 Serving — 1 Medium Size Vegetable or Fruit — Fresh, Frozen or Canned Vegetables or Fruit 125 mL 1/2 cup — Salad 250 mL 1 cup — Juice 125 mL 1/2 cup

Milk Products

SERVINGS PER DAY
Children 4–9 years: 2–3
Youth 10–16 years: 3–4
Adults: 2–4
Pregnant & Breast-feeding
Women: 3–4

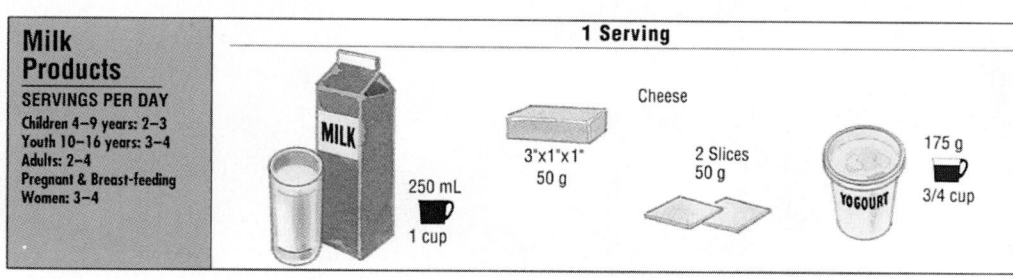

1 Serving — MILK 250 mL 1 cup — Cheese 3"x1"x1" 50 g — 2 Slices 50 g — YOGOURT 175 g 3/4 cup

Other Foods

Taste and enjoyment can also come from other foods and beverages that are not part of the 4 food groups. Some of these foods are higher in fat or Calories, so use these foods in moderation.

Meat & Alternatives

2–3
SERVINGS PER DAY

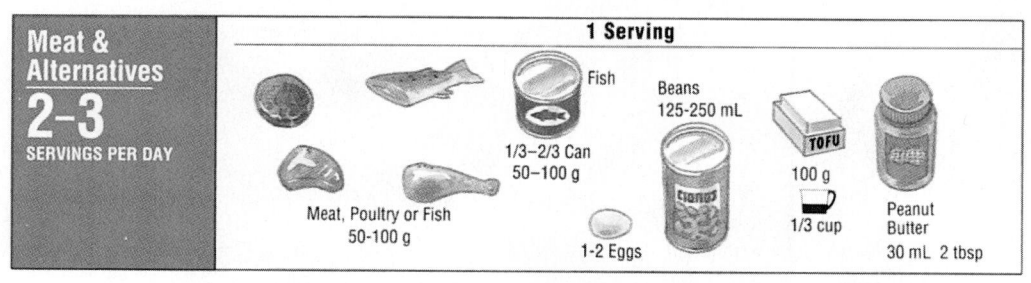

1 Serving — Meat, Poultry or Fish 50–100 g — 1-2 Eggs — Fish 1/3–2/3 Can 50–100 g — Beans 125–250 mL 1/3 cup — TOFU 100 g — Peanut Butter 30 mL 2 tbsp

Enjoy eating well, being active and feeling good about yourself. That's VITALIT

© Minister of Supply and Services Canada 1992   Cat. No. H39-252/1992E   No changes permitted. Reprint permission not required.
ISBN 0-662-19648-1

Choose a variety of activities from these three groups:

Endurance

4-7 days a week
Continuous activities for your heart, lungs and circulatory system.

Flexibility

4-7 days a week
Gentle reaching, bending and stretching activities to keep your muscles relaxed and joints mobile.

Strength

2-4 days a week
Activities against resistance to strengthen muscles and bones and improve posture.

Starting slowly is very safe for most people. Not sure? Consult your health professional.

For a copy of the *Guide Handbook* and more information: **1-888-334-9769**, or **www.paguide.com**

Eating well is also important. Follow *Canada's Food Guide to Healthy Eating* to make wise food choices.

Get Active Your Way, Every Day – For Life!

Scientists say accumulate 60 minutes of physical activity every day to stay healthy or improve your health. As you progress to moderate activities you can cut down to 30 minutes, 4 days a week. Add-up your activities in periods of at least 10 minutes each. Start slowly... and build up.

Time needed depends on effort

Very Light Effort	Light Effort *60 minutes*	Moderate Effort *30-60 minutes*	Vigorous Effort *20-30 minutes*	Maximum Effort
• Strolling	• Light walking	• Brisk walking	• Aerobics	• Sprinting
• Dusting	• Volleyball	• Biking	• Jogging	• Racing
	• Easy gardening	• Raking leaves	• Hockey	
	• Stretching	• Swimming	• Basketball	
		• Dancing	• Fast swimming	
		• Water aerobics	• Fast dancing	

Range needed to stay healthy

You Can Do It – Getting started is easier than you think

Physical activity doesn't have to be very hard. Build physical activities into your daily routine.

- Walk whenever you can – get off the bus early, use the stairs instead of the elevator.
- Reduce inactivity for long periods, like watching TV.
- Get up from the couch and stretch and bend for a few minutes every hour.
- Play actively with your kids.
- Choose to walk, wheel or cycle for short trips.

- Start with a 10 minute walk – gradually increase the time.
- Find out about walking and cycling paths nearby and use them.
- Observe a physical activity class to see if you want to try it.
- Try one class to start, you don't have to make a long-term commitment.
- Do the activities you are doing now, more often.

Benefits of regular activity:

- better health
- improved fitness
- better posture and balance
- better self-esteem
- weight control
- stronger muscles and bones
- feeling more energetic
- relaxation and reduced stress
- continued independent living in later life

Health risks of inactivity:

- premature death
- heart disease
- obesity
- high blood pressure
- adult-onset diabetes
- osteoporosis
- stroke
- depression
- colon cancer

ACTIVE LIVING

No changes permitted. Permission to photocopy this document in its entirety not required.
Cat. No. H39-429/1998-1E ISBN 0-662-86627-7

CANADA'S *Physical Activity Guide* to Healthy Active Living

• = choose more often

• = choose less often

TABLE B-1	Food Groups That Contain Carbohydrate

1 serving = 15 g carbohydrate or 1 carbohydrate choice

Food	Measure
Grains and starches: 15 g carbohydrate, 2 g protein, 0 g fat, 286 kJ (68 kcal)	
• Bagel, large	$1/4$
• Bagel, small	$1/2$
• Bannock, fried	$1.5'' \times 2.5''$
• Bannock, whole grain baked	$1.5'' \times 2.5''$
• Barley, cooked	125 mL ($1/2$ c)
• Bread, white	30 g (1 oz)
• Bread, whole grain	30 g (1 oz)
• Bulgur, cooked	125 mL ($1/2$ c)
• Bun, hamburger or hot dog	$1/2$
• Cereal, flaked unsweetened	125 mL ($1/2$ c)
• Cereal, hot	$3/4$ c
• Chapati, whole wheat (6")	1
• Corn	125 mL ($1/2$ c)
• Couscous, cooked	125 mL ($1/2$ c)
• Crackers, soda type	7
• Croutons	$2/3$ c
• English muffin, whole grain	$1/2$
• French fries	10
• Millet, cooked	$1/3$ c
• Naan bread (6")	$1/4$
• Pancake (4")	1
• Pasta, cooked	125 mL ($1/2$ c)
• Pita bread, white (6")	1
• Pita bread, whole wheat (6")	1
• Pizza crust (12")	$1/12$
• Plantain, mashed	$1/3$ c
• Potatoes, boiled or baked	$1/2$ medium
• Rice, cooked	$1/3$ c
• Roti, whole wheat (6")	1
• Soup, thick type	250 mL (1 c)
• Sweet potato, mashed	$1/3$ c
• Taco shells (5")	2
• Tortilla, whole wheat (6")	1
• Waffle (4")	1
Fruits: 15 g carbohydrate, 1 g protein, 0 g fat, 269 kJ (64 kcal)	
• Apple	1 medium
• Apple sauce, unsweetened	125 mL ($1/2$ c)
• Banana	1 small
• Blackberries	500 mL (2 c)
• Cherries	15

TABLE B-1   Food Groups That Contain Carbohydrate—continued

Food	Measure
Fruits: 15 g carbohydrate, 1 g protein, 0 g fat, 269 kJ (64 kcal)—continued	
• Fruit, canned in juice	125 mL (1/2 c)
• Fruit, dried	50 mL (1/4 c)
• Grapefruit	1 small
• Grapes	15
• Kiwi	2 medium
• Juice	125 mL (1/2 c)
• Mango	1/2 medium
• Melon	250 mL (1 c)
• Orange	1 medium
• Other berries	250 mL (1 c)
• Pear	1 medium
• Pineapple	3/4 c
• Plum	2 medium
• Raspberries	500 mL (2 c)
• Strawberries	500 mL (2 c)
Milk and alternatives: 15 g carbohydrate, 8 g protein, variable fat, 386–651 kJ (92–155 kcal)	
• Chocolate milk, 1%	125 mL (1/2 c)
• Evaporated milk, canned	125 mL (1/2 c)
• Milk, fluid	250 mL (1 c)
• Milk powder, skim	30 mL (2 tbs)
• Soy beverage, flavored	125 mL (1/2 c)
• Soy beverage, plain	250 mL (1 c)
• Soy yogurt, flavored	1/3 c
• Yogurt, nonfat, plain	3/4 c
• Yogurt, skim, artificially sweetened	250 mL (1 c)
Other choices (sweet foods and snacks): 15 g carbohydrate, variable protein and fat	
• Brownies, unfrosted	2″ × 2″
• Cake, unfrosted	2″ × 2″
• Cookies, arrowroot or gingersnap	3–4
• Jam, jelly, marmalade	15 mL (1 tbs)
• Milk pudding, skim, no sugar added	125 mL (1/2 c)
• Muffin	1 small (2″)
• Oatmeal granola bar	1 (28 g)
• Popcorn, low fat	750 mL (3 c)
• Pretzels, low fat, large	7
• Pretzels, low fat, sticks	30
• Sugar, white	15 mL (3 tsp or packets)

TABLE B-2   Food Groups That Contain Little or No Carbohydrate

Food	Measure
Vegetables: To encourage consumption, most vegetables are considered "free"	
• Asparagus	
• Beans, yellow or green	
• Bean sprouts	
• Beets	
• Broccoli	
• Cabbage	
• Carrots	
• Cauliflower	
• Celery	
• Cucumber	
• Eggplant	
• Greens	
• Leeks	
• Mushrooms	
• Okra	
• Parsnips[a]	
• Peas[a]	
• Peppers	
• Rutabagas (turnips)[a]	
• Salad vegetables	
• Snow peas	
• Squash, winter[a]	
• Tomatoes	
Meat and alternatives: 0 g carbohydrate, 7 g protein, 3–5 g fat, 307 kJ (73 kcal)	
• Cheese, skim (<7% milk fat)	30 g (1 oz)
• Cheese, light (<17% milk fat)	30 g (1 oz)
• Cheese, regular (17–33% milk fat)	30 g (1 oz)
• Cottage cheese (1–2% milk fat)	50 mL ($^{1}/_{4}$ c)
• Egg	1 large
• Fish, canned in oil	50 mL ($^{1}/_{4}$ c)
• Fish, canned in water	50 mL ($^{1}/_{4}$ c)
• Fish, fresh, cooked	30 g (1 oz)
• Hummus[b]	$^{1}/_{3}$ c
• Legumes, cooked[b]	125 mL ($^{1}/_{2}$ c)
• Meat, game, cooked	30 g (1 oz)

[a]These vegetables provide significant carbohydrate when more than 125 mL ($^{1}/_{2}$ c) is eaten.

[b]Legumes contain 15 g carbohydrate in a 125 mL ($^{1}/_{2}$ c) serving.

TABLE B-2	Food Groups That Contain Little or No Carbohydrate—continued

Food	Measure
Meat and alternatives: 0 g carbohydrate, 7 g protein, 3–5 g fat, 307 kJ (73 kcal)	
• Meat, ground, lean, cooked	30 g (1 oz)
• Meat, ground, medium-regular, cooked	30 g (1 oz)
• Meat, lean, cooked	30 g (1 oz)
• Meat, organ or tripe, cooked	30 g (1 oz)
• Meat, prepared, low fat	30 g (1 oz)
• Meat, prepared, regular fat	30 g (1 oz)
• Meat, regular, cooked	30 g (1 oz)
• Peameal/back bacon, cooked	30 g (1 oz)
• Poultry, ground, lean, cooked	30 g (1 oz)
• Poultry, skinless, cooked	30 g (1 oz)
• Poultry/wings, skin on, cooked	30 g (1 oz)
• Shellfish, cooked	30 g (1 oz)
• Tofu (soybean)	1/2 block (100 g)
• Vegetarian meat alternatives	30 g (1 oz)
Fats: 0 g carbohydrate, 0 g protein, 5 g fat, 189 kJ (45 kcal)	
• Avocado	1/6
• Bacon	30 g (1 oz)
• Butter	5 mL (1 tsp)
• Cheese, spreadable	15 mL (1 tbs)
• Margarine, non-hydrogenated	5 mL (1 tsp)
• Mayonnaise, light	30 mL (2 tbs)
• Nuts	15 mL (1 tbs)
• Oil, canola or olive	5 mL (1 tsp)
• Salad dressing, regular	15 mL (1 tbs)
• Seeds	15 mL (1 tbs)
• Tahini	7.5 mL (1/2 tbs)
Extras: <5 g carbohydrate, 84 kJ (20 kcal)	
Broth	
Coffee	
Herbs and spices	
Ketchup	
Mustard	
Sugar-free soft drinks	
Sugar-free gelatin	
Tea	

United States: Exchange Lists

CONTENTS

The Exchange Groups
and Lists

Chapter 20 introduced the exchange system, and this appendix provides details from the 2003 edition. Appendix B presents Canada's choice system for meal planning.

The Exchange Groups and Lists

The exchange system sorts foods into three main groups by their proportions of carbohydrate, fat, and protein. These three groups—the carbohydrate group, the fat group, and the meat and meat substitutes group (protein)—organize foods into several exchange lists (see Table C-1). Then any food on a list can be "exchanged" for any other on that same list. The carbohydrate group covers these exchange lists:

- Starch (cereals, grains, pasta, breads, crackers, snacks, starchy vegetables, and dried beans, peas, and lentils).
- Fruit.
- Milk (fat-free, reduced fat, and whole).
- Other carbohydrates (desserts and snacks with added sugars and fats).
- Vegetables.

TABLE C-1	The Exchange Groups and Lists				
Group/Lists	Typical Item/Portion Size	Carbohydrate (g)	Protein (g)	Fat (g)	Energy[a] (kcal)
Carbohydrate Group					
Starch[b]	1 slice bread	15	3	0–1	80
Fruit	1 small apple	15	—	—	60
Milk					
Fat-free, low-fat	1 c fat-free milk	12	8	0–3	90
Reduced-fat	1 c reduced-fat milk	12	8	5	120
Whole	1 c whole milk	12	8	8	150
Other carbohydrates[c]	2 small cookies	15	varies	varies	varies
Vegetable (nonstarchy)	1/2 c cooked carrots	5	2	—	25
Meat and Meat Substitute Group[d]					
Meat					
Very lean	1 oz chicken (white meat, no skin)	—	7	0–1	35
Lean	1 oz lean beef	—	7	3	55
Medium-fat	1 oz ground beef	—	7	5	75
High-fat	1 oz pork sausage	—	7	8	100
Fat Group					
Fat	1 tsp butter	—	—	5	45

[a]The energy value for each exchange list represents an approximate average for the group and does not reflect the precise number of grams of carbohydrate, protein, and fat. For example, a slice of bread contains 15 grams of carbohydrate (that's 60 kcalories), 3 grams protein (that's another 12 kcalories), and a little fat—rounded to 80 kcalories for ease in calculating. A half-cup of vegetables (not including starchy vegetables) contains 5 grams carbohydrate (20 kcalories) and 2 grams protein (8 more), which has been rounded down to 25 kcalories.

[b]The starch list includes cereals, grains, breads, crackers, snacks, starchy vegetables (such as corn, peas, and potatoes), and legumes (dried beans, peas, and lentils).

[c]The other carbohydrates list includes foods that contain added sugars and fats such as cakes, cookies, doughnuts, ice cream, potato chips, pudding, syrup, and frozen yogurt.

[d]The meat and meat substitutes list includes legumes, cheeses, and peanut butter.

The fat group covers this exchange list:

- Fats.

The meat and meat substitutes group (protein) covers these exchange lists:

- Meat and meat substitutes (very lean, lean, medium-fat, and high-fat).

See Tables C-2 through C-10 for the exchange lists.

Portion Sizes The exchange system helps people control their energy intakes by paying close attention to portion sizes. The portion sizes have been carefully adjusted and defined so that a portion of any food on a given list provides roughly the same amount of carbohydrate, fat, and protein and, therefore, total kcalories. Any food on a list can then be exchanged, or traded, for any other food on that same list without significantly affecting the diet's balance or total kcalories. For example, a person may select either 17 small grapes or ¹/₂ large grapefruit as one fruit portion, and either choice would provide roughly 60 kcalories. A whole grapefruit, however, would count as 2 portions.

Both the exchange system and the USDA Food Guide list meats in single ounces; that is, one *portion* (or *exchange*) of meat is 1 ounce. Calculating meat by the ounce encourages a person to keep close track of the exact amounts eaten. This in turn helps control energy and fat intakes. Be aware, though, that most people do not serve foods in carefully measured portions, nor do the amounts reflect the exchange system or USDA Food Guide amounts.

To apply the system successfully, users must become familiar with portion sizes. A convenient way to remember the portion sizes and energy values is to keep in mind a typical item from each list (review Table C-1).

The Foods on the Lists Foods do not always appear on the exchange list where you might first expect to find them. They are grouped according to their energy-nutrient contents rather than by their source (such as milks), their outward appearance, or their vitamin and mineral contents. Notice, for example, that cheeses are grouped with meats (not milk) because, like meats, cheeses contribute energy from protein and fat but provide negligible carbohydrate. Similarly, starchy vegetables such as potatoes are found on the starch list with breads and cereals, not with the vegetables, and bacon is with the fats and oils, not with the meats.

Users of the exchange lists learn to view mixtures of foods, such as casseroles and soups, as combinations of foods from different exchange lists. They also learn to interpret food labels with the exchange system in mind (see Figure C-1).

Controlling Energy and Fat By assigning items like bacon to the fat list, the exchange system alerts consumers to foods that are unexpectedly high in fat. Even the starch list specifies which grain products contain added fat (such as biscuits, muffins, and waffles). In addition, the exchange system encourages users to think of fat-free milk as milk and of whole milk as milk with added fat, and to think of very lean meats as meats and of lean, medium-fat, and high-fat meats as meats with added fat. To that end, foods on the milk and meat lists are separated into categories based on their fat contents. The milk group is classed as fat-free, reduced-fat, and whole; the meat group as very lean, lean, medium-fat, and high-fat.

Control of food energy and fat intake can be highly successful with the exchange system. Exchange plans do not, however, guarantee adequate intakes of vitamins and minerals. Food group plans work better from that standpoint because the food groupings are based on similarities in vitamin-mineral content. In the exchange system, for example, meats are grouped with cheeses, yet the meats are iron-rich and calcium-poor, whereas the cheeses are iron-poor and calcium-rich. To take advantage of the strengths of both food group plans and exchange patterns, and to compensate for their weaknesses, diet planners often combine these two diet-planning tools.

Nutrition Facts

Serving size 10¹/₂ oz (298 g)
Servings per Package 1

Amount per serving

Calories 361   Calories from Fat 117

	% Daily Value
Total Fat 13 g	20%
Saturated Fat 8 g	40%
Cholesterol 87 mg	29%
Sodium 860 mg	36%
Total Carbohydrate 37 g	12%
Dietary fiber 0 g	
Sugars 8 g	
Protein 26 g	

Can you "see" these exchanges in the label above?

Exchange	Carbohydrate	Protein	Fat
2 starches	30 g	6 g	—
1 vegetable	5 g	2 g	—
3 medium-fat meats	—	21 g	15 g
Exchange totals	35	29	15
Label totals	37	26	13

FIGURE C-1 Seeing Exchanges on a Food Label

Knowing that foods on the starch list provide 15 grams of carbohydrate and those on the vegetable list provide 5, you can count a lasagna dinner that provides 37 grams of carbohydrate as "2 starches and 1 vegetable"; knowing that foods on the meat list provide 7 grams of protein, you might count it as "3 meats"; the grams of fat suggest that the meat (and cheese) is probably medium-fat.

1 starch exchange = 15 g carbohydrate, 3 g protein, 0–1 g fat, and 80 kcal

Note: In general, one starch exchange is $^1/_2$ c cooked cereal, grain, or starchy vegetable; $^1/_2$ c cooked rice or pasta; 1 oz of bread; $^3/_4$ to 1 oz snack food.

Serving Size	Food
Bread	
$^1/_4$ (1 oz)	Bagel, 4 oz
2 slices (1$^1/_2$ oz)	Bread, reduced-kcalorie
1 slice (1 oz)	Bread, white (including French and Italian), whole-wheat, pumpernickel, rye
4 ($^2/_3$ oz)	Bread sticks, crisp, 4″ × $^1/_2$″
$^1/_2$	English muffin
$^1/_2$ (1 oz)	Hot dog or hamburger bun
$^1/_4$	Naan, 8″ × 2″
1	Pancake, 4″ across, $^1/_4$″ thick
$^1/_2$	Pita, 6″ across
1 (1 oz)	Plain roll, small
1 slice (1 oz)	Raisin bread, unfrosted
1	Tortilla, corn, 6″ across
1	Tortilla, flour, 6″ across
$^1/_3$	Tortilla, flour, 10″ across
1	Waffle, 4″ square or across, reduced-fat
Cereals and Grains	
$^1/_2$ c	Bran cereals
$^1/_2$ c	Bulgur, cooked
$^1/_2$ c	Cereals, cooked
$^3/_4$ c	Cereals, unsweetened, ready-to-eat
3 tbs	Cornmeal (dry)
$^1/_3$ c	Couscous
3 tbs	Flour (dry)
$^1/_4$ c	Granola, low-fat
$^1/_4$ c	Grape nuts
$^1/_2$ c	Grits, cooked
$^1/_2$ c	Kasha
$^1/_3$ c	Millet
$^1/_4$ c	Muesli
$^1/_2$ c	Oats
$^1/_3$ c	Pasta, cooked
1$^1/_2$ c	Puffed cereals
$^1/_3$ c	Rice, white or brown, cooked
$^1/_2$ c	Shredded wheat
$^1/_2$ c	Sugar-frosted cereal
3 tbs	Wheat germ
Starchy Vegetables	
$^1/_3$ c	Baked beans
$^1/_2$ c	Corn
$^1/_2$ cob (5 oz)	Corn on cob, large
1 c	Mixed vegetables with corn, peas, or pasta
$^1/_2$ c	Peas, green

Serving Size	Food
$^1/_2$ c	Plantains
$^1/_2$ medium (3 oz) or $^1/_2$ c	Potato, boiled
$^1/_4$ large (3 oz)	Potato, baked with skin
$^1/_2$ c	Potatoes, mashed
1 c	Squash, winter (acorn, butternut, pumpkin)
$^1/_2$ c	Yams, sweet potatoes, plain
Crackers and Snacks	
8	Animal crackers
3	Graham crackers, 2$^1/_2$″ square
$^3/_4$ oz	Matzoh
4 slices	Melba toast
24	Oyster crackers
3 c	Popcorn (popped, no fat added or low-fat microwave)
$^3/_4$ oz	Pretzels
2	Rice cakes, 4″ across
6	Saltine-type crackers
15–20 ($^3/_4$ oz)	Snack chips, fat-free or baked (tortilla, potato)
2–5 ($^3/_4$ oz)	Whole-wheat crackers, no fat added
Beans, Peas, and Lentils (count as 1 starch + 1 very lean meat)	
$^1/_2$ c	Beans and peas, cooked (garbanzo, lentils, pinto, kidney, white, split, black-eyed)
$^2/_3$ c	Lima beans
3 tbs	Miso 🖉
Starchy Foods Prepared with Fat (count as 1 starch + 1 fat)	
1	Biscuit, 2$^1/_2$″ across
$^1/_2$ c	Chow mein noodles
1 (2 oz)	Cornbread, 2″ cube
6	Crackers, round butter type
1 c	Croutons
1 c (2 oz)	French-fried potatoes (oven baked)
$^1/_4$ c	Granola
$^1/_3$ c	Hummus
$^1/_5$ (1 oz)	Muffin, 5 oz
3 c	Popcorn, microwave
3	Sandwich crackers, cheese or peanut butter filling
9–13 ($^3/_4$ oz)	Snack chips (potato, tortilla)
$^1/_3$ c	Stuffing, bread (prepared)
2	Taco shells, 6″ across
1	Waffle, 4$^1/_2$″ square or across
4–6 (1 oz)	Whole-wheat crackers, fat added

🖉 = 400 mg or more of sodium per serving.

TABLE C-3 U.S. Exchange System: Fruit List

1 fruit exchange = 15 g carbohydrate and 60 kcal
Note: In general, one fruit exchange is 1 small fresh fruit; $^1/_2$ c canned or fresh fruit or unsweetened fruit juice; $^1/_4$ c dried fruit.

Serving Size	Food	Serving Size	Food
1 (4 oz)	Apple, unpeeled, small	1 (4 oz)	Peach, medium, fresh
$^1/_2$ c	Applesauce, unsweetened	$^1/_2$ c	Peaches, canned
4 rings	Apples, dried	$^1/_2$ (4 oz)	Pear, large, fresh
4 whole (5$^1/_2$ oz)	Apricots, fresh	$^1/_2$ c	Pears, canned
8 halves	Apricots, dried	$^3/_4$ c	Pineapple, fresh
$^1/_2$ c	Apricots, canned	$^1/_2$ c	Pineapple, canned
1 (4 oz)	Banana, small	2 (5 oz)	Plums, small
$^3/_4$ c	Blackberries	$^1/_2$ c	Plums, canned
$^3/_4$ c	Blueberries	3	Plums, dried (prunes)
$^1/_3$ melon (11 oz) or 1 c cubes	Cantaloupe, small	2 tbs	Raisins
12 (3 oz)	Cherries, sweet, fresh	1 c	Raspberries
$^1/_2$ c	Cherries, sweet, canned	1$^1/_4$ c whole berries	Strawberries
3	Dates	2 (8 oz)	Tangerines, small
1$^1/_2$ large or 2 medium (3$^1/_2$ oz)	Figs, fresh	1 slice (13$^1/_2$ oz) or 1$^1/_4$ c cubes	Watermelon
1$^1/_2$	Figs, dried	**Fruit Juice, Unsweetened**	
$^1/_2$ c	Fruit cocktail		
$^1/_2$ (11 oz)	Grapefruit, large	$^1/_2$ c	Apple juice/cider
$^3/_4$ c	Grapefruit sections, canned	$^1/_3$ c	Cranberry juice cocktail
17 (3 oz)	Grapes, small	1 c	Cranberry juice cocktail, reduced-kcalorie
1 slice (10 oz) or 1 c cubes	Honeydew melon	$^1/_3$ c	Fruit juice blends, 100% juice
1 (3$^1/_2$ oz)	Kiwi	$^1/_3$ c	Grape juice
$^3/_4$ c	Mandarin oranges, canned	$^1/_2$ c	Grapefruit juice
$^1/_2$ (5$^1/_2$ oz) or $^1/_2$ c	Mango, small	$^1/_2$ c	Orange juice
1 (5 oz)	Nectarine, small	$^1/_2$ c	Pineapple juice
1 (6$^1/_2$ oz)	Orange, small	$^1/_3$ c	Prune juice
$^1/_2$ (8 oz) or 1 c cubes	Papaya		

TABLE C-4 U.S. Exchange System: Milk List

Note: In general, one milk exchange is 1 c milk or yogurt.

Serving Size	Food	Serving Size	Food
Fat-Free and Low-Fat Milk		**Reduced-Fat Milk**	
1 fat-free/low-fat milk exchange = 12 g carbohydrate, 8 g protein, 0–3 g fat, 90 kcal		1 reduced-fat milk exchange = 12 g carbohydrate, 8 g protein, 5 g fat, 120 kcal	
1 c	Fat-free milk	1 c	2% milk
1 c	$^1/_2$% milk	1 c	Soy milk
1 c	1% milk	1 c	Sweet acidophilus milk
1 c	Fat-free or low-fat buttermilk	$^3/_4$ c	Yogurt, plain low-fat
$^1/_2$ c	Evaporated fat-free milk	**Whole Milk**	
$^1/_3$ c dry	Fat-free dry milk		
1 c	Soy milk, low-fat or fat-free	1 whole milk exchange = 12 g carbohydrate, 8 g protein, 8 g fat, 150 kcal	
$^2/_3$ c (6 oz)	Yogurt, fat-free or low-fat, flavored, sweetened with nonnutritive sweetener and fructose	1 c	Whole milk
$^2/_3$ c (6 oz)	Yogurt, plain fat-free	$^1/_2$ c	Evaporated whole milk
		1 c	Goat's milk
		1 c	Kefir
		$^3/_4$ c	Yogurt, plain (made from whole milk)

TABLE C-5   U.S. Exchange System: Sweets, Desserts, and Other Carbohydrates List

1 other carbohydrate exchange = 15 g carbohydrate, or 1 starch, or 1 fruit, or 1 milk exchange

Food	Serving Size	Exchanges per Serving
Angel food cake, unfrosted	1/12 cake (2 oz)	2 carbohydrates
Brownies, small, unfrosted	2″ square (1 oz)	1 carbohydrate, 1 fat
Cake, unfrosted	2″ square (1 oz)	1 carbohydrate, 1 fat
Cake, frosted	2″ square (2 oz)	2 carbohydrates, 1 fat
Cookies or sandwich cookies with creme filling	2 small (2/3 oz)	1 carbohydrate, 1 fat
Cookies, sugar-free	3 small or 1 large (3/4–1 oz)	1 carbohydrate, 1–2 fats
Cranberry sauce, jellied	1/4 c	1 1/2 carbohydrates
Cupcake, frosted	1 small (2 oz)	2 carbohydrates, 1 fat
Doughnut, plain cake	1 medium (1 1/2 oz)	1 1/2 carbohydrates, 2 fats
Doughnut, glazed	3 3/4″ across (2 oz)	2 carbohydrates, 2 fats
Energy, sport, or breakfast bar	1 bar (1 1/3 oz)	1 1/2 carbohydrates, 0–1 fat
Energy, sport, or breakfast bar	1 bar (2 oz)	2 carbohydrates, 1 fat
Fruit cobbler	1/2 c (3 1/2 oz)	3 carbohydrates, 1 fat
Fruit juice bar, frozen, 100% juice	1 bar (3 oz)	1 carbohydrate
Fruit snacks, chewy (pureed fruit concentrate)	1 roll (3/4 oz)	1 carbohydrate
Fruit spreads, 100% fruit	1 1/2 tbs	1 carbohydrate
Gelatin, regular	1/2 c	1 carbohydrate
Gingersnaps	3	1 carbohydrate
Granola or snack bar, regular or low-fat	1 bar (1 oz)	1 1/2 carbohydrates
Honey	1 tbs	1 carbohydrate
Ice cream	1/2 c	1 carbohydrate, 2 fats
Ice cream, light	1/2 c	1 carbohydrate, 1 fat
Ice cream, low-fat	1/2 c	1 1/2 carbohydrates
Ice cream, fat-free, no sugar added	1/2 c	1 carbohydrate
Jam or jelly, regular	1 tbs	1 carbohydrate
Milk, chocolate, whole	1 c	2 carbohydrates, 1 fat
Pie, fruit, 2 crusts	1/6 of 8″ commercially prepared pie	3 carbohydrates, 2 fats
Pie, pumpkin or custard	1/8 of 8″ commercially prepared pie	2 carbohydrates, 2 fats
Pudding, regular (made with reduced-fat milk)	1/2 c	2 carbohydrates

TABLE C-5 **U.S. Exchange System: Sweets, Desserts, and Other Carbohydrates List—continued**

Food	Serving Size	Exchanges per Serving
Pudding, sugar-free (made with fat-free milk)	$1/2$ c	1 carbohydrate
Reduced-calorie meal replacement (shake)	1 can (10–11 oz)	$1^1/2$ carbohydrates, 0–1 fats
Rice milk, low-fat or fat-free, plain	1 c	1 carbohydrate
Rice milk, low-fat, flavored	1 c	$1^1/2$ carbohydrates
Salad dressing, fat-free 🖉	$1/4$ c	1 carbohydrate
Sherbet, sorbet	$1/2$ c	2 carbohydrates
Spaghetti or pasta sauce, canned 🖉	$1/2$ c	1 carbohydrate, 1 fat
Sports drinks	8 oz (1 c)	1 carbohydrate
Sugar	1 tbs	1 carbohydrate
Sweet roll or danish	1 ($2^1/2$ oz)	$2^1/2$ carbohydrates, 2 fats
Syrup, light	2 tbs	1 carbohydrate
Syrup, regular	1 tbs	1 carbohydrate
Syrup, regular	$1/4$ c	4 carbohydrates
Vanilla wafers	5	1 carbohydrate, 1 fat
Yogurt, frozen	$1/2$ c	1 carbohydrate, 0–1 fat
Yogurt, frozen, fat-free	$1/3$ c	1 carbohydrate
Yogurt, low-fat with fruit	1 c	3 carbohydrates, 0–1 fat

🖉 = 400 mg or more of sodium per serving.

TABLE C-6 U.S. Exchange System: Nonstarchy Vegetable List

1 vegetable exchange = 5 g carbohydrate, 2 g protein, 0 g fat, and 25 kcal
Note: In general, one vegetable exchange is $^1/_2$ c cooked vegetables or vegetable juice; 1 c raw vegetables. Starchy vegetables such as corn, peas, and potatoes are on the starch list (Table C-2).

Artichokes	Mushrooms
Artichoke hearts	Okra
Asparagus	Onions
Beans (green, wax, Italian)	Pea pods
Bean sprouts	Peppers (all varieties)
Beets	Radishes
Broccoli	Salad greens (endive, escarole, lettuce, romaine, spinach)
Brussels sprouts	Sauerkraut 🖊
Cabbage	Spinach
Carrots	Summer squash (crookneck)
Cauliflower	Tomatoes
Celery	Tomatoes, canned
Cucumbers	Tomato sauce 🖊
Eggplant	Tomato/vegetable juice 🖊
Green onions or scallions	Turnips
Greens (collard, kale, mustard, turnip)	Water chestnuts
Kohlrabi	Watercress
Leeks	Zucchini
Mixed vegetables (without corn, peas, or pasta)	

🖊 = 400 mg or more of sodium per serving.

TABLE C-7 — U.S. Exchange System: Meat and Meat Substitutes List

Note: In general, a meat exchange is 1 oz meat, poultry, or cheese; 1/2 c dried beans (weigh meat and poultry and measure beans after cooking).

Serving Size	Food

Very Lean Meat and Substitutes

1 very lean meat exchange = 7 g protein, 0–1 g fat, 35 kcal

Serving Size	Food
1 oz	Poultry: Chicken or turkey (white meat, no skin), Cornish hen (no skin)
1 oz	Fish: Fresh or frozen cod, flounder, haddock, halibut, trout, lox (smoked salmon 🖊); tuna, fresh or canned in water
1 oz	Shellfish: Clams, crab, lobster, scallops, shrimp, imitation shellfish
1 oz	Game: Duck or pheasant (no skin), venison, buffalo, ostrich
	Cheese with ≤1g fat/oz:
1/4 c	Fat-free or low-fat cottage cheese
1 oz	Fat-free cheese
1 oz	Processed sandwich meats with ≤1 g fat/oz (such as deli thin, shaved meats, chipped beef 🖊, turkey ham)
2	Egg whites
1/4 c	Egg substitutes, plain
1 oz	Hot dogs with ≤1 g fat/oz 🖊
1 oz	Kidney (high in cholesterol)
1 oz	Sausage with ≤1 g fat/oz

Count as 1 very lean meat + 1 starch exchange:

Serving Size	Food
1/2 c	Beans, peas, lentils (cooked)

Lean Meat and Substitutes

1 lean meat exchange = 7 g protein, 3 g fat, 55 kcal

Serving Size	Food
1 oz	Beef: USDA Select or Choice grades of lean beef trimmed of fat (round, sirloin, and flank steak); tenderloin; roast (rib, chuck, rump); steak (T-bone, porterhouse, cubed), ground round
1 oz	Pork: Lean pork (fresh ham); canned, cured, or boiled ham; Canadian bacon 🖊; tenderloin, center loin chop
1 oz	Lamb: Roast, chop, leg
1 oz	Veal: Lean chop, roast
1 oz	Poultry: Chicken, turkey (dark meat, no skin), chicken (white meat, with skin), domestic duck or goose (well drained of fat, no skin)
	Fish:
1 oz	Herring (uncreamed or smoked)
6 medium	Oysters
1 oz	Salmon (fresh or canned), catfish
2 medium	Sardines (canned)
1 oz	Tuna (canned in oil, drained)
1 oz	Game: Goose (no skin), rabbit

Serving Size	Food
	Cheese:
1/4 c	4.5%-fat cottage cheese
2 tbs	Grated Parmesan
1 oz	Cheeses with ≤3 g fat/oz
1 1/2 oz	Hot dogs with ≤3 g fat/oz 🖊
1 oz	Processed sandwich meat with ≤3 g fat/oz (turkey pastrami or kielbasa)
1 oz	Liver, heart (high in cholesterol)

Medium-Fat Meat and Substitutes

1 medium-fat meat exchange = 7 g protein, 5 g fat, and 75 kcal

Serving Size	Food
1 oz	Beef: Most beef products (ground beef, meat loaf, corned beef, short ribs, prime grades of meat trimmed of fat, such as prime rib)
1 oz	Pork: Top loin, chop, Boston butt, cutlet
1 oz	Lamb: Rib roast, ground
1 oz	Veal: Cutlet (ground or cubed, unbreaded)
1 oz	Poultry: Chicken (dark meat, with skin), ground turkey or ground chicken, fried chicken (with skin)
1 oz	Fish: Any fried fish product
	Cheese with ≤5 g fat/oz:
1 oz	Feta
1 oz	Mozzarella
1/4 c (2 oz)	Ricotta
1	Egg (high in cholesterol, limit to 3/week)
1 oz	Sausage with ≤5 g fat/oz
1 c	Soy milk
1/4 c	Tempeh
4 oz or 1/2 c	Tofu

High-Fat Meat and Substitutes

1 high-fat meat exchange = 7 g protein, 8 g fat, 100 kcal

Serving Size	Food
1 oz	Pork: Spareribs, ground pork, pork sausage
1 oz	Cheese: All regular cheeses (American 🖊, cheddar, Monterey Jack, swiss)
1 oz	Processed sandwich meats with ≤8 g fat/oz (bologna, pimento loaf, salami)
1 oz	Sausage (bratwurst, Italian, knockwurst, Polish, smoked)
1 (10/lb)	Hot dog (turkey or chicken) 🖊
3 slices (20 slices/lb)	Bacon
1 tbs	Peanut butter (contains unsaturated fat)

Count as 1 high-fat meat + 1 fat exchange:

Serving Size	Food
1 (10/lb)	Hot dog (beef, pork, or combination) 🖊

🖊 = 400 mg or more of sodium per serving.

TABLE C-8 U.S. Exchange System: Fat List

1 fat exchange = 5 g fat and 45 kcal
Note: In general, one fat exchange is 1 tsp regular butter, margarine, or vegetable oil; 1 tbs regular salad dressing. Many fat-free and reduced-fat foods are on the Free Foods List (Table C-9).

Serving Size	Food
Monounsaturated Fats	
2 tbs (1 oz)	Avocado
1 tsp	Oil (canola, olive, peanut)
8 large	Olives, ripe (black)
10 large	Olives, green, stuffed
6 nuts	Almonds, cashews
6 nuts	Mixed nuts (50% peanuts)
10 nuts	Peanuts
4 halves	Pecans
½ tbs	Peanut butter, smooth or crunchy
1 tbs	Sesame seeds
2 tsp	Tahini or sesame paste
Polyunsaturated Fats	
4 halves	English walnuts
1 tsp	Margarine, stick, tub, or squeeze
1 tbs	Margarine, lower-fat spread (30% to 50% vegetable oil)
1 tsp	Mayonnaise, regular
1 tbs	Mayonnaise, reduced-fat
1 tsp	Oil (corn, safflower, soybean)
1 tbs	Salad dressing, regular
2 tbs	Salad dressing, reduced-fat
2 tsp	Mayonnaise type salad dressing, regular
1 tbs	Mayonnaise type salad dressing, reduced-fat
1 tbs	Seeds (pumpkin, sunflower)
Saturated Fats*	
1 slice (20 slices/lb)	Bacon, cooked
1 tsp	Bacon, grease
1 tsp	Butter, stick
2 tsp	Butter, whipped
1 tbs	Butter, reduced-fat
2 tbs (½ oz)	Chitterlings, boiled
2 tbs	Coconut, sweetened, shredded
1 tbs	Coconut milk
2 tbs	Cream, half and half
1 tbs (½ oz)	Cream cheese, regular
1½ tbs (¾ oz)	Cream cheese, reduced-fat
	Fatback or salt pork†
1 tsp	Shortening or lard
2 tbs	Sour cream, regular
3 tbs	Sour cream, reduced-fat

 = 400 mg or more of sodium per serving.

*Saturated fats can raise blood cholesterol levels.

†Use a piece 1″ × 1″ × ¼″ if you plan to eat the fatback cooked with vegetables. Use a piece 2″ × 1″ × ½″ when eating only the vegetables with the fatback removed.

TABLE C-9 U.S. Exchange System: Free Foods List

Note: A serving of free food contains less than 20 kcalories or no more than 5 grams of carbohydrate; those with serving sizes should be limited to 3 servings a day whereas those without serving sizes can be eaten freely.

Serving Size	Food
Fat-Free or Reduced-Fat Foods	
1 tbs (1/2 oz)	Cream cheese, fat-free
1 tbs	Creamers, nondairy, liquid
2 tsp	Creamers, nondairy, powdered
4 tbs	Margarine spread, fat-free
1 tsp	Margarine spread, reduced-fat
1 tbs	Mayonnaise, fat-free
1 tsp	Mayonnaise, reduced-fat
1 tbs	Mayonnaise type salad dressing, fat-free
1 tsp	Mayonnaise type salad dressing, reduced-fat Nonstick cooking spray
1 tbs	Salad dressing, fat-free or low-fat
2 tbs	Salad dressing, fat-free, Italian
1 tbs	Sour cream, fat-free, reduced-fat
1 tbs	Whipped topping, regular
2 tbs	Whipped topping, light or fat-free
Sugar-Free Foods	
1 piece	Candy, hard, sugar-free Gelatin dessert, sugar-free Gelatin, unflavored Gum, sugar-free
2 tsp	Jam or jelly, light Sugar substitutes
2 tbs	Syrup, sugar-free
Drinks	
	Bouillon, broth, consommé 🖋
	Bouillon or broth, low-sodium
	Carbonated or mineral water
	Club soda
1 tbs	Cocoa powder, unsweetened

Serving Size	Food
	Coffee
	Diet soft drinks, sugar-free
	Drink mixes, sugar-free
	Tea
	Tonic water, sugar-free
Condiments	
1 tbs	Catsup Horseradish Lemon juice Lime juice Mustard
1 tbs	Pickle relish
1 1/2 medium	Pickles, dill 🖋
2 slices	Pickles, sweet (bread and butter)
3/4 oz	Pickles, sweet (gherkin)
1/4 c	Salsa
1 tbs	Soy sauce, regular or light 🖋
1 tbs	Taco sauce Vinegar
2 tbs	Yogurt
Seasonings	
	Flavoring extracts
	Garlic
	Herbs, fresh or dried
	Hot pepper sauces
	Pimento
	Spices
	Wine, used in cooking
	Worcestershire sauce

🖋 = 400 mg or more of sodium per serving.

TABLE C-10   U.S. Exchange System: Combination Foods List

Food	Serving Size	Exchanges per Serving
Entrées		
Tuna noodle casserole, lasagna, spaghetti with meatballs, chili with beans, macaroni and cheese ✐	1 c (8 oz)	2 carbohydrates, 2 medium-fat meats
Chow mein (without noodles or rice)	2 c (16 oz)	1 carbohydrate, 2 lean meats
Tuna or chicken salad	½ c (3½ oz)	½ carbohydrate, 2 lean meats, 1 fat
Frozen Entrées and Meals		
Dinner-type meal ✐	Generally 14–17 oz	3 carbohydrates, 3 medium-fat meats, 3 fats
Entrée or meal with <340 kcal ✐	About 8–11 oz	2–3 carbohydrates, 1–2 lean meats
Meatless burger, soy based	3 oz	½ carbohydrate, 2 lean meats
Meatless burger, vegetable and starch based	3 oz	1 carbohydrate, 1 lean meat
Pizza, cheese, thin crust ✐	¼ of 12″ (6 oz)	2 carbohydrates, 2 medium-fat meats, 1 fat
Pizza, meat topping, thin crust ✐	¼ of 12″ (6 oz)	2 carbohydrates, 2 medium-fat meats, 2 fats
Pot pie ✐	1 (7 oz)	2½ carbohydrates, 1 medium-fat meat, 3 fats
Soups		
Bean ✐	1 c	1 carbohydrate, 1 very lean meat
Cream (made with water) ✐	1 c (8 oz)	1 carbohydrate, 1 fat
Instant ✐	6 oz prepared	1 carbohydrate
Instant with beans/lentils ✐	8 oz prepared	2½ carbohydrates, 1 very lean meat
Split pea (made with water) ✐	½ c (4 oz)	1 carbohydrate
Tomato (made with water)	1 c (8 oz)	1 carbohydrate
Vegetable beef, chicken noodle, or other broth-type ✐	1 c (8 oz)	1 carbohydrate
Fast Foods		
Burrito with beef ✐	1 (5–7 oz)	3 carbohydrates, 1 medium-fat meat, 1 fat
Chicken nuggets ✐	6	1 carbohydrate, 2 medium-fat meats, 1 fat
Chicken breast and wing, breaded and fried ✐	1 each	1 carbohydrate, 4 medium-fat meats, 2 fats
Chicken sandwich, grilled ✐	1	2 carbohydrates, 3 very lean meats
Chicken wings, hot	6 (5 oz)	1 carbohydrate, 3 medium-fat meats, 4 fats
Fish sandwich/tartar sauce ✐	1	3 carbohydrates, 1 medium-fat meat, 3 fats
French fries ✐	1 medium serving (5 oz)	4 carbohydrates, 4 fats
Hamburger, regular	1	2 carbohydrates, 2 medium-fat meats
Hamburger, large ✐	1	2 carbohydrates, 3 medium-fat meats, 1 fat
Hot dog with bun ✐	1	1 carbohydrate, 1 high-fat meat, 1 fat
Individual pan pizza ✐	1	5 carbohydrates, 3 medium-fat meats, 3 fats
Pizza, cheese, thin crust ✐	¼ of 12″ (about 6 oz)	2½ carbohydrates, 2 medium-fat meats
Pizza, meat, thin crust ✐	¼ of 12″ (about 6 oz)	2½ carbohydrates, 2 medium-fat meats, 1 fat
Soft serve cone	1 small (5 oz)	2½ carbohydrates, 1 fat
Submarine sandwich ✐	1 sub (6″)	3 carbohydrates, 1 vegetable, 2 medium-fat meats, 1 fat
Submarine sandwich (<6 g fat) ✐	1 sub (6″)	2½ carbohydrates, 2 lean meats
Taco, hard or soft shell	1 (3–3½ oz)	1 carbohydrate, 1 medium-fat meat, 1 fat

✐ = 400 mg or more of sodium per serving.

Chapter 6 described how to calculate your estimated energy requirements by using an equation that accounts for your gender, age, weight, height, and physical activity level. This appendix first helps you determine the correct physical activity factor to use in the equation, either by calculating your physical activity level or by guesstimating it. Then the appendix presents tables that provide a shortcut to estimating total energy expenditure.*

Calculating Your Physical Activity Level

To calculate your physical activity level, record all of your activities for a typical 24-hour day, noting the type of activity, the level of intensity, and the duration. Then, using a copy of Table D-1, find your activity in the first column (or an activity that is reasonably similar) and multiply the number of minutes spent on that activity by the factor in the third column. Put your answer in the last column and total the accumulated values for the day. Now add the subtotal of the last column to 1.1 (to account for basal energy and the thermic effect of food) as shown. This score indicates your level of physical activity. Using Table D-2, find the PA (physical activity) factor for your gender that correlates with your physical activity level score and use it in the energy equation presented on p. 139 of Chapter 6.

Guesstimating Your Physical Activity Level

As an alternative to recording your activities for a day, you can use the first two columns of Table D-3 to decide if your daily activity is sedentary, low active, active, or very active. Find the PA factor for your gender that correlates with your typical physical activity level and use it in the energy equation presented on p. 139.

Using a Shortcut to Estimate Total Energy Expenditure

The DRI committee has developed estimates of total energy expenditure based on the equations presented in Chapter 6. These estimates are presented in Table D-4 for women and Table D-5 for men. You can use these tables to estimate your energy requirement—that is, the number of kcalories needed to maintain your current body weight. On the table appropriate for your gender, find your height in meters (or inches) in the left-hand column. Then follow the row across to find your weight in kilograms (or pounds). (If you can't find your exact height and weight, choose a value between the two closest ones.) Look down the column to find the number of kcalories that corresponds to your activity level.

Importantly, the values given in the tables are for 30-year-old people. Women 19 to 29 should add 7 kcalories per day for each year below age 30; older women should subtract 7 kcalories per day for each year above age 30. Similarly, men 19 to 29 should add 10 kcalories per day for each year below age 30; older men should subtract 10 kcalories per day for each year above age 30.

*This appendix, including the tables, is adapted from Committee on Dietary Reference Intakes, Food and Nutrition Board, Institute of Medicine, *Dietary Reference Intakes for Energy, Carbohydrate, Fiber, Fat, Fatty Acids, Cholesterol, Protein, and Amino Acids* (Washington, D.C.: National Academies Press, 2002).

TABLE D-1   Physical Activities and Their Scores

If your activity was equivalent to this . . .	Then list the number of minutes here and . . .	Multiply by this factor . . .	Add this column to get your physical activity level score:
Activities of Daily Living			
Gardening (no lifting)		0.0032	
Household tasks (moderate effort)		0.0024	
Lifting items continuously		0.0029	
Loading/unloading car		0.0019	
Lying quietly		0.0000	
Mopping		0.0024	
Mowing lawn (power mower)		0.0033	
Raking lawn		0.0029	
Riding in a vehicle		0.0000	
Sitting (idle)		0.0000	
Sitting (doing light activity)		0.0005	
Taking out trash		0.0019	
Vacuuming		0.0024	
Walking the dog		0.0019	
Walking from house to car or bus		0.0014	
Watering plants		0.0014	
Additional Activities			
Billiards		0.0013	
Calisthenics (no weight)		0.0029	
Canoeing (leisurely)		0.0014	
Chopping wood		0.0037	
Climbing hills (carrying 11 lb load)		0.0061	
Climbing hills (no load)		0.0056	
Cycling (leisurely)		0.0024	
Cycling (moderately)		0.0045	
Dancing (aerobic or ballet)		0.0048	
Dancing (ballroom, leisurely)		0.0018	

(continued)

TABLE D-1   Physical Activities and Their Scores—continued

If your activity was equivalent to this . . .	Then list the number of minutes here and . . .	Multiply by this factor . . .	Add this column to get your physical activity level score:
Dancing (fast ballroom or square)		0.0043	
Golf (with cart)		0.0014	
Golf (without cart)		0.0032	
Horseback riding (walking)		0.0012	
Horseback riding (trotting)		0.0053	
Jogging (6 mph)		0.0088	
Music (playing accordion)		0.0008	
Music (playing cello)		0.0012	
Music (playing flute)		0.0010	
Music (playing piano)		0.0012	
Music (playing violin)		0.0014	
Rope skipping		0.0105	
Skating (ice)		0.0043	
Skating (roller)		0.0052	
Skiing (water or downhill)		0.0055	
Squash		0.0106	
Surfing		0.0048	
Swimming (slow)		0.0033	
Swimming (fast)		0.0057	
Tennis (doubles)		0.0038	
Tennis (singles)		0.0057	
Volleyball (noncompetitive)		0.0018	
Walking (2 mph)		0.0014	
Walking (3 mph)		0.0022	
Walking (4 mph)		0.0033	
Walking (5 mph)		0.0067	
Subtotal			
Factor for basal energy and the thermic effect of food		1.1	
Your physical activity level score			

TABLE D-2   Physical Activity Level Scores and Their PA Factors

Physical Activity Level Score	Description	Men: PA Factor	Women: PA Factor
1.0 to 1.39	Sedentary	1.0	1.0
1.4 to 1.59	Low active	1.11	1.12
1.6 to 1.89	Active	1.25	1.27
1.9 and above	Very active	1.48	1.45

TABLE D-3   Physical Activity Equivalents and Their PA Factors

Description	Physical Activity Equivalents	Men: PA Factor	Women: PA Factor
Sedentary	Only those physical activities required for normal independent living	1.0	1.0
	Activities equivalent to walking at a pace of 2–4 mph for the following distances:		
Low active	1.5 to 3.0 miles/day	1.11	1.12
Active	3 to 10 miles/day	1.25	1.27
Very active	10 or more miles/day	1.48	1.45

TABLE D-4	Total Energy Expenditure (TEE in kCalories per Day) for Women 30 Years of Age[a] at Various Levels of Activity and Various Heights and Weights

Height m (in)	Physical Activity Level	Weight[b] kg (lb)					
1.45 (57)		38.9 (86)	45.2 (100)	52.6 (116)	63.1 (139)	73.6 (162)	84.1 (185)
				kCalories			
	Sedentary	1564	1623	1698	1813	1927	2042
	Low active	1734	1800	1912	2043	2174	2304
	Active	1946	2021	2112	2257	2403	2548
	Very active	2201	2287	2387	2553	2719	2886
1.50 (59)		41.6 (92)	48.4 (107)	56.3 (124)	67.5 (149)	78.8 (174)	90.0 (198)
				kCalories			
	Sedentary	1625	1689	1771	1894	2017	2139
	Low active	1803	1874	1996	2136	2276	2415
	Active	2025	2105	2205	2360	2516	2672
	Very active	2291	2382	2493	2671	2849	3027
1.55 (61)		44.4 (98)	51.7 (114)	60.1 (132)	72.1 (159)	84.1 (185)	96.1 (212)
				kCalories			
	Sedentary	1688	1756	1846	1977	2108	2239
	Low active	1873	1949	2081	2230	2380	2529
	Active	2104	2190	2299	2466	2632	2798
	Very active	2382	2480	2601	2791	2981	3171
1.60 (63)		47.4 (104)	55.0 (121)	64.0 (141)	76.8 (169)	89.6 (197)	102.4 (226)
				kCalories			
	Sedentary	1752	1824	1922	2061	2201	2340
	Low active	1944	2025	2168	2327	2486	2645
	Active	2185	2276	2396	2573	2750	2927
	Very active	2474	2578	2712	2914	3116	3318
1.65 (65)		50.4 (111)	58.5 (129)	68.1 (150)	81.7 (180)	95.3 (210)	108.9 (240)
				kCalories			
	Sedentary	1816	1893	1999	2148	2296	2444
	Low active	2016	2102	2556	2425	2594	2763
	Active	2267	2364	2494	2682	2871	3059
	Very active	2567	2678	2824	3039	3254	3469
1.70 (67)		53.5 (118)	62.1 (137)	72.3 (159)	86.7 (191)	101.2 (223)	115.6 (255)
				kCalories			
	Sedentary	1881	1963	2078	2235	2393	2550
	Low active	2090	2180	2345	2525	2705	2884
	Active	2350	2453	2594	2794	2994	3194
	Very active	2662	2780	2938	3166	3395	3623

(continued)

[a]For each year below 30, add 7 kcalories/day to TEE. For each year above 30, subtract 7 kcalories/day from TEE.

[b]These columns represent a BMI of 18.5, 22.5, 25, 30, 35, and 40, respectively.

TABLE D-4 Total Energy Expenditure (TEE in kCalories per Day) for Women 30 Years of Age[a] at Various Levels of Activity and Various Heights and Weights—continued

Height m (in)	Physical Activity Level	Weight[b] kg (lb)					
1.75 (69)		56.7 (125)	65.8 (145)	76.6 (169)	91.9 (202)	107.2 (236)	122.5 (270)
				kCalories			
	Sedentary	1948	2034	2158	2325	2492	2659
	Low active	2164	2260	2437	2627	2817	3007
	Active	2434	2543	2695	2907	3119	3331
	Very active	2758	2883	3054	3296	3538	3780
1.80 (71)		59.9 (132)	69.7 (154)	81.0 (178)	97.2 (214)	113.4 (250)	129.6 (285)
				kCalories			
	Sedentary	2015	2106	2239	2416	2593	2769
	Low active	2239	2341	2529	2731	2932	3133
	Active	2519	2634	2799	3023	3247	3472
	Very active	2855	2987	3172	3428	3684	3940
1.85 (73)		63.3 (139)	73.6 (162)	85.6 (189)	102.7 (226)	119.8 (264)	136.9 (302)
				kCalories			
	Sedentary	2083	2179	2322	2509	2695	2882
	Low active	2315	2422	2624	2836	3049	3262
	Active	2605	2727	2904	3141	3378	3615
	Very active	2954	3093	3292	3562	3833	4103
1.90 (75)		66.8 (147)	77.6 (171)	90.3 (199)	108.3 (239)	126.4 (278)	144.4 (318)
				kCalories			
	Sedentary	2151	2253	2406	2603	2800	2996
	Low active	2392	2505	2720	2944	3168	3393
	Active	2693	2821	3011	3261	3511	3760
	Very active	3053	3200	3414	3699	3984	4270
1.95 (77)		70.3 (155)	81.8 (180)	95.1 (209)	114.1 (251)	133.1 (293)	152.1 (335)
				kCalories			
	Sedentary	2221	2328	2492	2699	2906	3113
	Low active	2470	2589	2817	3053	3290	3526
	Active	2781	2917	3119	3383	3646	3909
	Very active	3154	3309	3538	3838	4139	4439

[a]For each year below 30, add 7 kcalories/day to TEE. For each year above 30, subtract 7 kcalories/day from TEE.

[b]These columns represent a BMI of 18.5, 22.5, 25, 30, 35, and 40, respectively.

TABLE D-5	Total Energy Expenditure (TEE in kCalories per Day) for Men 30 Years of Age[a] at Various Levels of Activity and Various Heights and Weights

Height m (in)	Physical Activity Level	Weight[b] kg (lb)					
1.45 (57)		38.9 (86)	47.3 (100)	52.6 (116)	63.1 (139)	73.6 (163)	84.1 (185)
		kCalories					
	Sedentary	1777	1911	2048	2198	2347	2496
	Low active	1931	2080	2225	2393	2560	2727
	Active	2127	2295	2447	2636	2826	3015
	Very active	2450	2648	2845	3075	3305	3535
1.50 (59)		41.6 (92)	50.6 (107)	56.3 (124)	67.5 (149)	78.8 (174)	90.0 (198)
		kCalories					
	Sedentary	1848	1991	2126	2286	2445	2605
	Low active	2009	2168	2312	2491	2670	2849
	Active	2215	2394	2545	2748	2951	3154
	Very active	2554	2766	2965	3211	3457	3703
1.55 (61)		44.4 (98)	54.1 (114)	60.1 (132)	72.1 (159)	84.1 (185)	96.1 (212)
		kCalories					
	Sedentary	1919	2072	2205	2376	2546	2717
	Low active	2089	2259	2401	2592	2783	2974
	Active	2305	2496	2646	2862	3079	3296
	Very active	2660	2887	3087	3349	3612	3875
1.60 (63)		47.4 (104)	57.6 (121)	64.0 (141)	76.8 (169)	89.6 (197)	102.4 (226)
		kCalories					
	Sedentary	1993	2156	2286	2468	2650	2831
	Low active	2171	2351	2492	2695	2899	3102
	Active	2397	2601	2749	2980	3210	3441
	Very active	2769	3010	3211	3491	3771	4051
1.65 (65)		50.4 (111)	61.3 (129)	68.1 (150)	81.7 (180)	95.3 (210)	108.9 (240)
		kCalories					
	Sedentary	2068	2241	2369	2562	2756	2949
	Low active	2254	2446	2585	2801	3017	3234
	Active	2490	2707	2854	3099	3345	3590
	Very active	2880	3136	3339	3637	3934	4232
1.70 (67)		53.5 (118)	65.0 (137)	72.3 (159)	86.7 (191)	101.2 (223)	115.6 (255)
		kCalories					
	Sedentary	2144	2328	2454	2659	2864	3069
	Low active	2338	2542	2679	2909	3139	3369
	Active	2586	2816	2961	3222	3483	3743
	Very active	2992	3265	3469	3785	4101	4417

(continued)

[a]For each year below 30, add 10 kcalories/day to TEE. For each year above 30, subtract 10 kcalories/day from TEE.

[b]These columns represent a BMI of 18.5, 22.5, 25, 30, 35, and 40, respectively.

TABLE D-5	Total Energy Expenditure (TEE in kCalories per Day) for Men 30 Years of Age[a] at Various Levels of Activity and Various Heights and Weights—continued

Height m (in)	Physical Activity Level	Weight[b] kg (lb)					
1.75 (69)		56.7 (125)	68.9 (145)	76.6 (169)	91.9 (202)	107.2 (236)	122.5 (270)
		kCalories					
	Sedentary	2222	2416	2540	2757	2975	3192
	Low active	2425	2641	2776	3020	3263	3507
	Active	2683	2927	3071	3347	3623	3900
	Very active	3108	3396	3602	3937	4272	4607
1.80 (71)		59.9 (132)	72.9 (154)	81.0 (178)	97.2 (214)	113.4 (250)	129.6 (285)
		kCalories					
	Sedentary	2301	2507	2628	2858	3088	3318
	Low active	2513	2741	2875	3132	3390	3648
	Active	2782	3040	3183	3475	3767	4060
	Very active	3225	3530	3738	4092	4447	4801
1.85 (73)		63.3 (139)	77.0 (162)	85.6 (189)	102.7 (226)	119.8 (264)	136.9 (302)
		kCalories					
	Sedentary	2382	2599	2718	2961	3204	3447
	Low active	2602	2844	2976	3248	3520	3792
	Active	2883	3155	3297	3606	3915	4223
	Very active	3344	3667	3877	4251	4625	4999
1.90 (75)		66.8 (147)	81.2 (171)	90.3 (199)	108.3 (239)	126.4 (278)	144.4 (318)
		kCalories					
	Sedentary	2464	2693	2810	3066	3322	3579
	Low active	2693	2948	3078	3365	3652	3939
	Active	2986	3273	3414	3739	4065	4390
	Very active	3466	3806	4018	4413	4807	5202
1.95 (77)		70.3 (155)	85.6 (180)	95.1 (209)	114.1 (251)	133.1 (293)	152.1 (335)
		kCalories					
	Sedentary	2547	2789	2903	3173	3443	3713
	Low active	2786	3055	3183	3485	3788	4090
	Active	3090	3393	3533	3875	4218	4561
	Very active	3590	3948	4162	4578	4993	5409

[a]For each year below 30, add 10 kcalories/day to TEE. For each year above 30, subtract 10 kcalories/day from TEE.

[b]These columns represent a BMI of 18.5, 22.5, 25, 30, 35, and 40, respectively.

Several chapters in this book described data from nutrition assessments that help health professionals evaluate patients' nutrition status and nutrient needs. This appendix provides additional information that may be useful for complete assessments.

Weight Gain during Pregnancy

Chapter 10 described desirable weight-gain patterns during pregnancy. Figure E-1 shows prenatal weight-gain grids, used to plot the rate of weight gain during pregnancy.

Growth Charts

Health professionals generally evaluate physical development by monitoring the growth rate of a child and comparing this rate with standard charts. Standard charts compare length or height to age, weight to age, and body mass index to age; ideally, height and weight are in roughly the same percentile. Although individual growth patterns may vary, a child's growth curve will generally stay at about the same percentile throughout childhood. In children whose growth has been retarded, nutrition rehabilitation will ideally induce height and weight to increase to higher percentiles. In overweight children, the goal is for weight to remain stable as height increases, until weight becomes appropriate for height.

To evaluate growth in infants, an assessor uses charts such as those in Figures E-2 (A and B) through E-5 (A and B). The assessor follows these steps to plot a weight measurement on a percentile graph:

- Select the appropriate chart based on age and gender.
- Locate the child's age along the horizontal axis on the bottom or top of the chart.
- Locate the child's weight in pounds or kilograms along the vertical axis.
- Mark the chart where the age and weight lines intersect, and read off the percentile.

To assess length, height, or head circumference, the assessor follows the same procedure, using the appropriate chart. (When length is measured, use the chart for birth to 36 months; when height is measured, use the chart for 2 to 20 years.) Head circumference percentile should be similar to the child's height and weight percentiles. With height,

CONTENTS

Reminder: The *body mass index (BMI)* is an index of a person's weight in relation to height, determined by dividing the weight in kilograms by the square of the height in meters:

$$BMI = \frac{Weight\ (kg)}{Height\ (m)^2}.$$

Additional growth charts are available at **www.cdc.gov/growthcharts**.

FIGURE E-1 Recommended Prenatal Weight Gain Based on Prepregnancy Weight

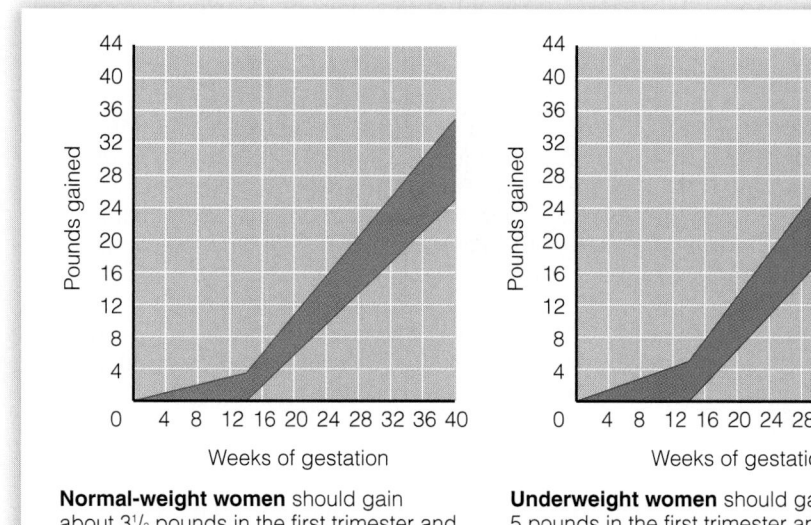
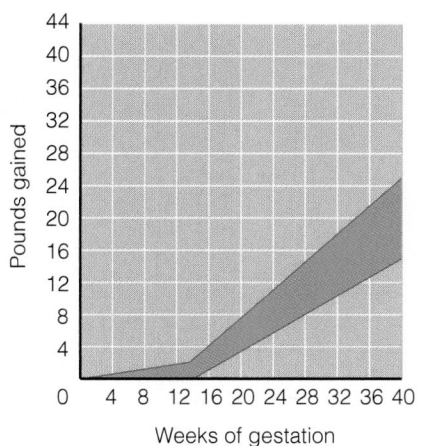

Normal-weight women should gain about 3½ pounds in the first trimester and just under 1 pound/week thereafter, achieving a total gain of 25 to 35 pounds by term.

Underweight women should gain about 5 pounds in the first trimester and just over 1 pound/week thereafter, achieving a total gain of 28 to 40 pounds by term.

Overweight women should gain about 2 pounds in the first trimester and ⅔ pound/week thereafter, achieving a total gain of 15 to 25 pounds.

weight, and head circumference measures plotted on growth percentile charts, a skilled clinician can begin to interpret the data.

Percentile charts divide the measures of a population into 100 equal divisions. Thus half of the population falls above the 50th percentile, and half falls below. The use of percentile measures allows for comparisons among people of the same age and gender. For example, a six-month-old female infant whose weight is at the 75th percentile weighs more than 75 percent of the female infants her age.

Head circumference is generally measured in children under two years of age. Since the brain grows rapidly before birth and during early infancy, extreme and chronic malnutrition during these times can impair brain development, curtailing the number of brain cells and the size of head circumference. Nonnutritional factors, such as certain disorders and genetic variation, can also influence head circumference.

Measures of Body Fat and Lean Tissue

Significant weight changes in both children and adults can reflect overnutrition or undernutrition with respect to energy and protein. To estimate the degree to which fat stores or lean tissues are affected by overnutrition or malnutrition, several anthropometric measurements are useful.

Common sites for skinfold measures:

- Triceps.
- Biceps.
- Subscapular (below shoulder blade).
- Suprailiac (above hip bone).
- Abdomen.
- Upper thigh.

Skinfold Measures Skinfold measures provide a good estimate of total body fat and a fair assessment of the fat's location. Approximately half the fat in the body lies directly beneath the skin, and the thickness of this subcutaneous fat reflects total body fat. In some parts of the body, such as the back and the back of the arm over the triceps muscle, this fat is loosely attached; a person can pull it up between the thumb and forefinger to obtain a measure of skinfold thickness. To measure the skinfold, a skilled assessor follows a standard procedure using reliable calipers (illustrated in Figure E-6) and then compares the measurement with standards. Triceps skinfold measures greater than 15 millimeters in men or 25 millimeters in women suggest excessive body fat.

Skinfold measurements correlate directly with the risk of heart disease. They assess central obesity and its associated risks better than do weight measures alone. If a person gains body fat, the skinfold increases proportionately; if the person loses fat, it decreases. Measurements taken from central-body sites (around the abdomen) better reflect changes in fatness than those taken from upper sites (arm and back). A major limitation of the skinfold test is that fat may be thicker under the skin in one area than in another. A pinch at the side of the waistline may not yield the same measurement as a pinch on the back of the arm. This limitation can be overcome by taking skinfold measurements at several (often three) different places on the body (including upper-, central-, and lower-body sites) and comparing each measurement with standards for that site. Multiple measures are not always practical in clinical settings, however, and most often, the triceps skinfold measurement alone is used because it is easily accessible.

Waist Circumference Chapter 6 described how fat distribution correlates with health risks and mentioned that the waist circumference is a valuable indicator of fat distribution. To measure waist circumference, the assessor places a nonstretchable tape around the person's body, crossing just above the upper hip bones and making sure that the tape remains on a level horizontal plane on all sides (see Figure E-7). The tape is tightened slightly, but without compressing the skin.

To calculate the waist-to-hip ratio, divide the waistline measurement by the hip measurement. For example, a woman with a 28-inch waist and 38-inch hips would have a ratio of 28 ÷ 38 = 0.74.

Waist-to-Hip Ratio Alternatively, some clinicians measure both the waist and the hips. The waist-to-hip ratio also assesses abdominal obesity, but provides no more information than using the waist circumference alone. In general, women with a waist-to-hip ratio of 0.80 or greater and men with a waist-to-hip ratio of 0.90 or greater have a high risk of health problems.

Hydrodensitometry To estimate body density using hydrodensitometry, the person is weighed twice—first on land and then again when submerged under water. Underwater weighing usually generates a good estimate of body fat and is useful in research, although the technique has drawbacks: it requires bulky, expensive, and nonportable equipment. Furthermore, submerging some people (especially those who are very young, very old, ill, or fearful) under water is not always practical.

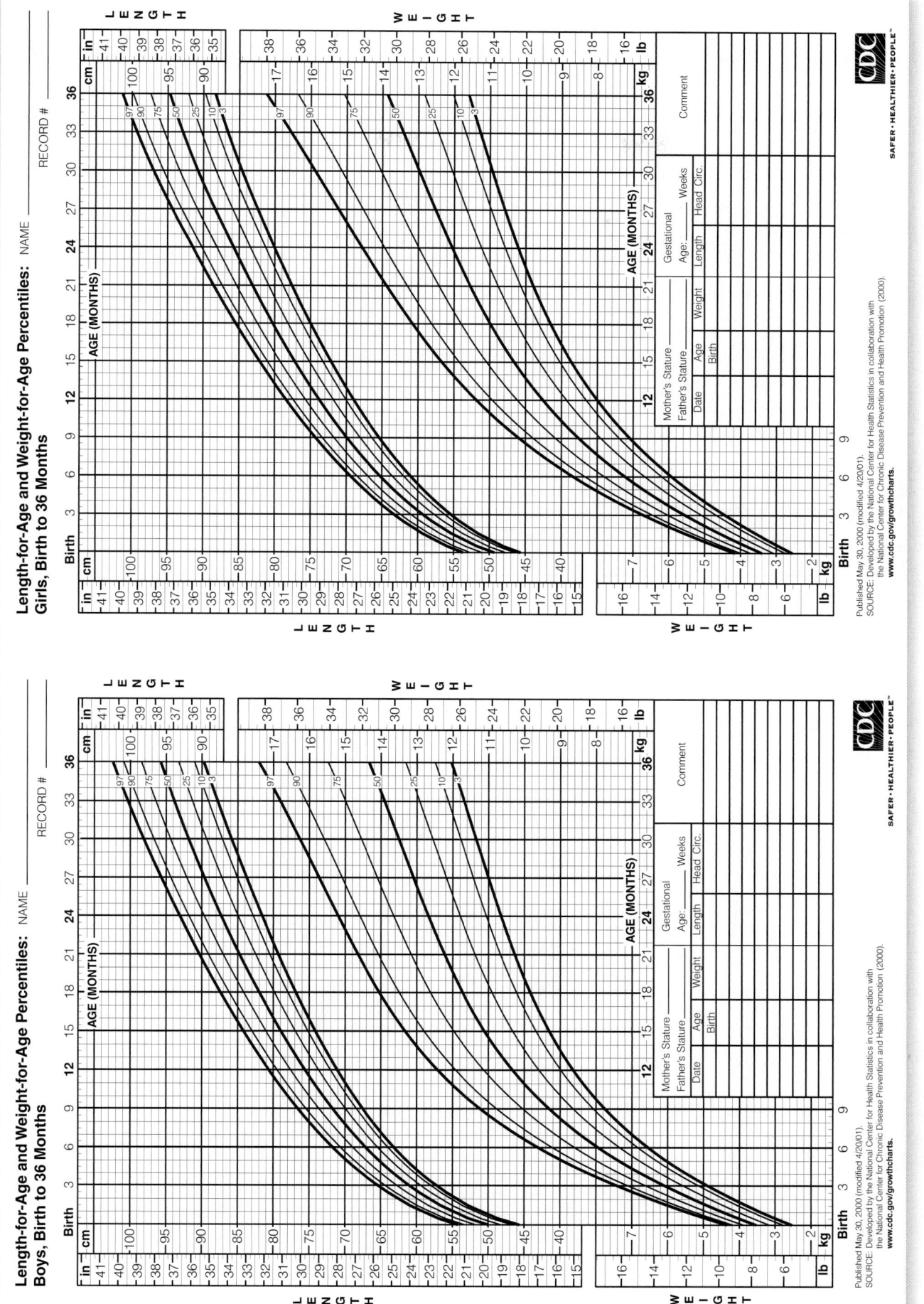

FIGURE E-2A Length-for-Age and Weight-for-Age Percentiles:
Boys, Birth to 36 Months
http://www.cdc.gov/nchs/data/nhanes/growthcharts/set1clinical/cj41l017.pdf

FIGURE E-2B Length-for-Age and Weight-for-Age Percentiles:
Girls, Birth to 36 Months
http://www.cdc.gov/nchs/data/nhanes/growthcharts/set1clinical/cj41l018.pdf

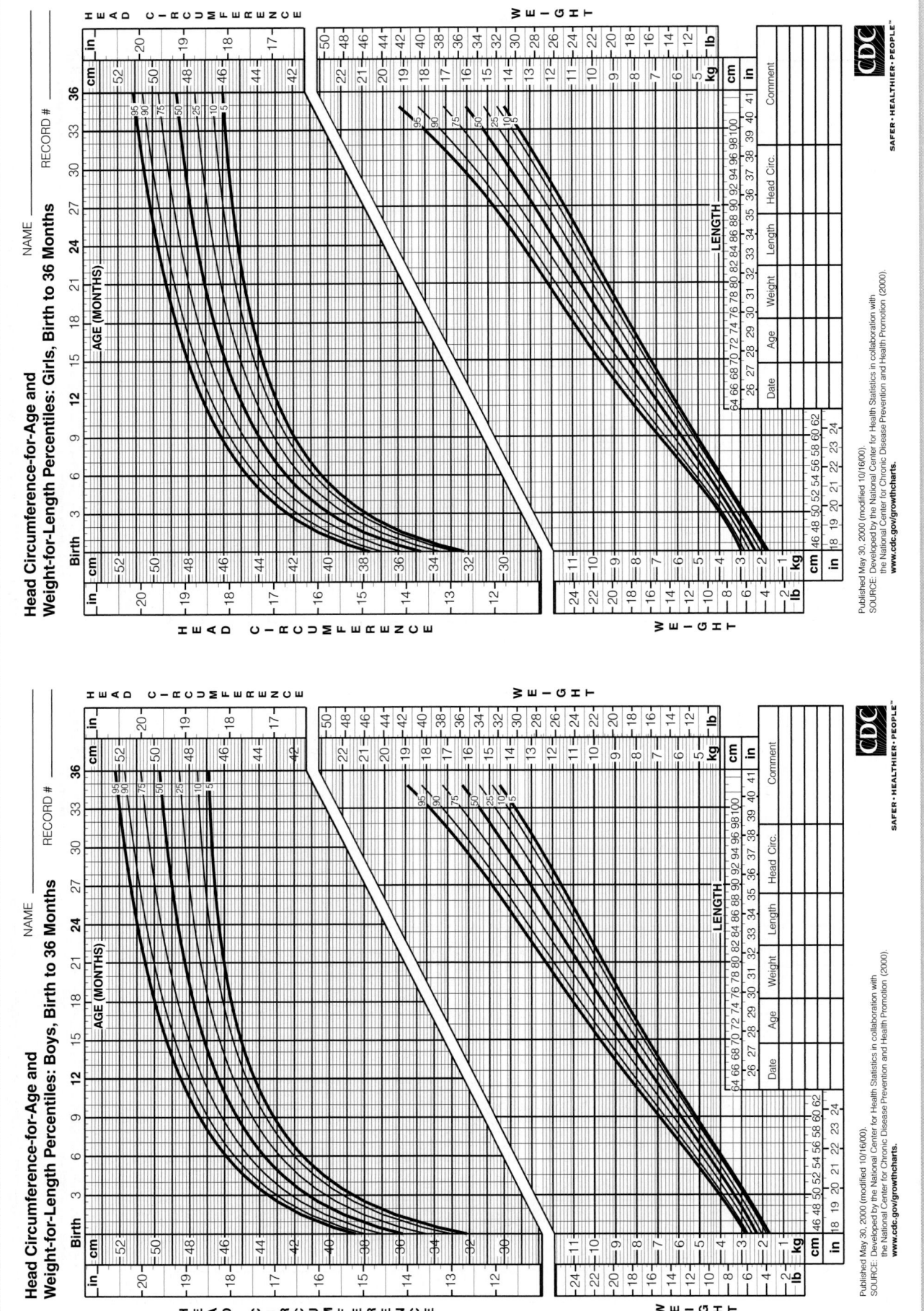

Published May 30, 2000 (modified 10/16/00).
SOURCE: Developed by the National Center for Health Statistics in collaboration with
the National Center for Chronic Disease Prevention and Health Promotion (2000).
www.cdc.gov/growthcharts.

FIGURE E-3B Head Circumference-for-Age and Weight-for-Length
Percentiles: Girls, Birth to 36 Months
http://www.cdc.gov/nchs/data/nhanes/growthcharts/set1clinical/cj41l020.pdf

Published May 30, 2000 (modified 10/16/00).
SOURCE: Developed by the National Center for Health Statistics in collaboration with
the National Center for Chronic Disease Prevention and Health Promotion (2000).
www.cdc.gov/growthcharts.

FIGURE E-3A Head Circumference-for-Age and Weight-for-Length
Percentiles: Boys, Birth to 36 Months
http://www.cdc.gov/nchs/data/nhanes/growthcharts/set1clinical/cj41l019.pdf

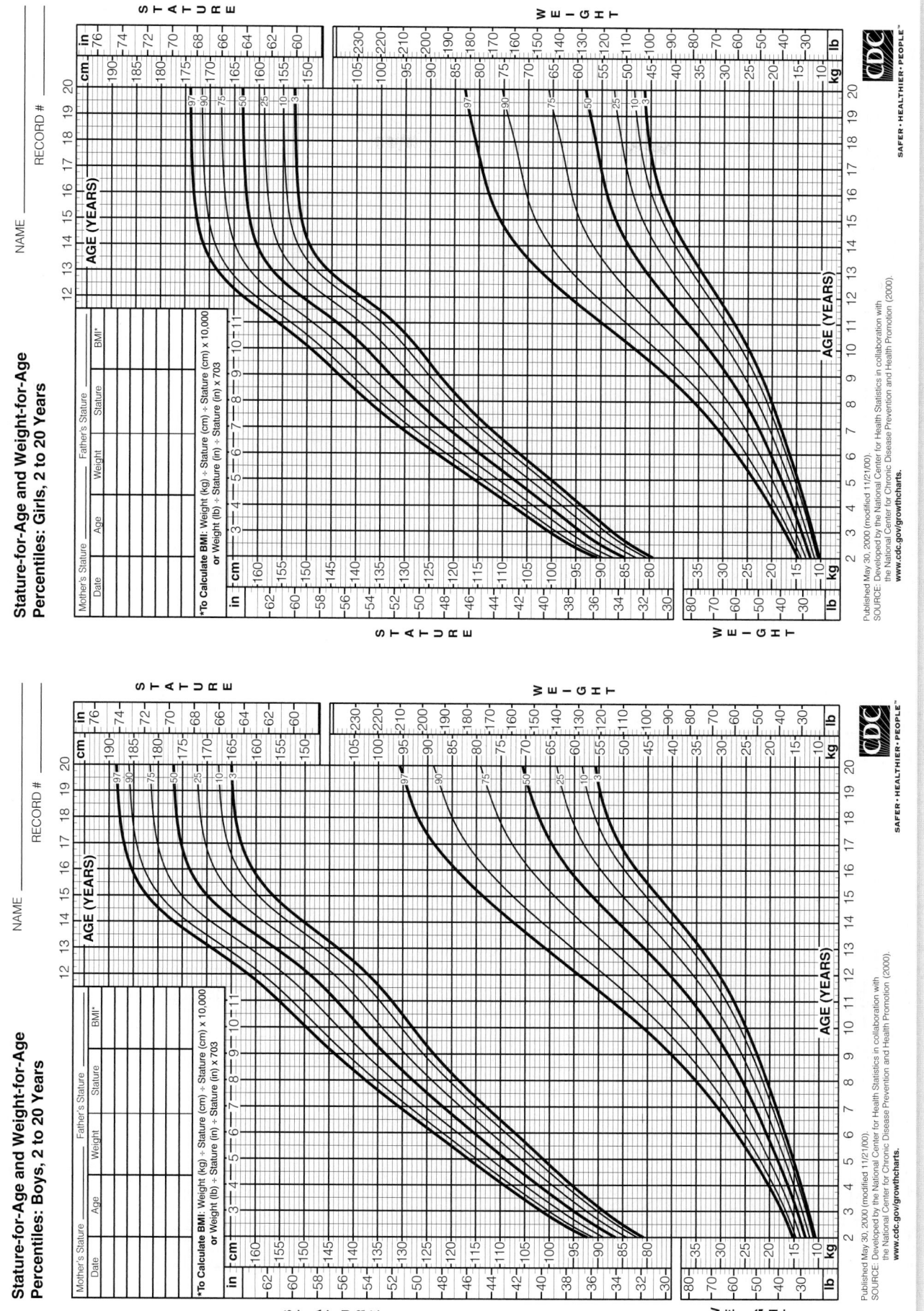

FIGURE E-4A Stature-for-Age and Weight-for-Age Percentiles:
Boys, 2 to 20 Years

http://www.cdc.gov/nchs/data/nhanes/growthcharts/set1clinical/cj41l021.pdf

FIGURE E-4B Stature-for-Age and Weight-for-Age Percentiles:
Girls, 2 to 20 Years

http://www.cdc.gov/nchs/data/nhanes/growthcharts/set1clinical/cj41l022.pdf

E

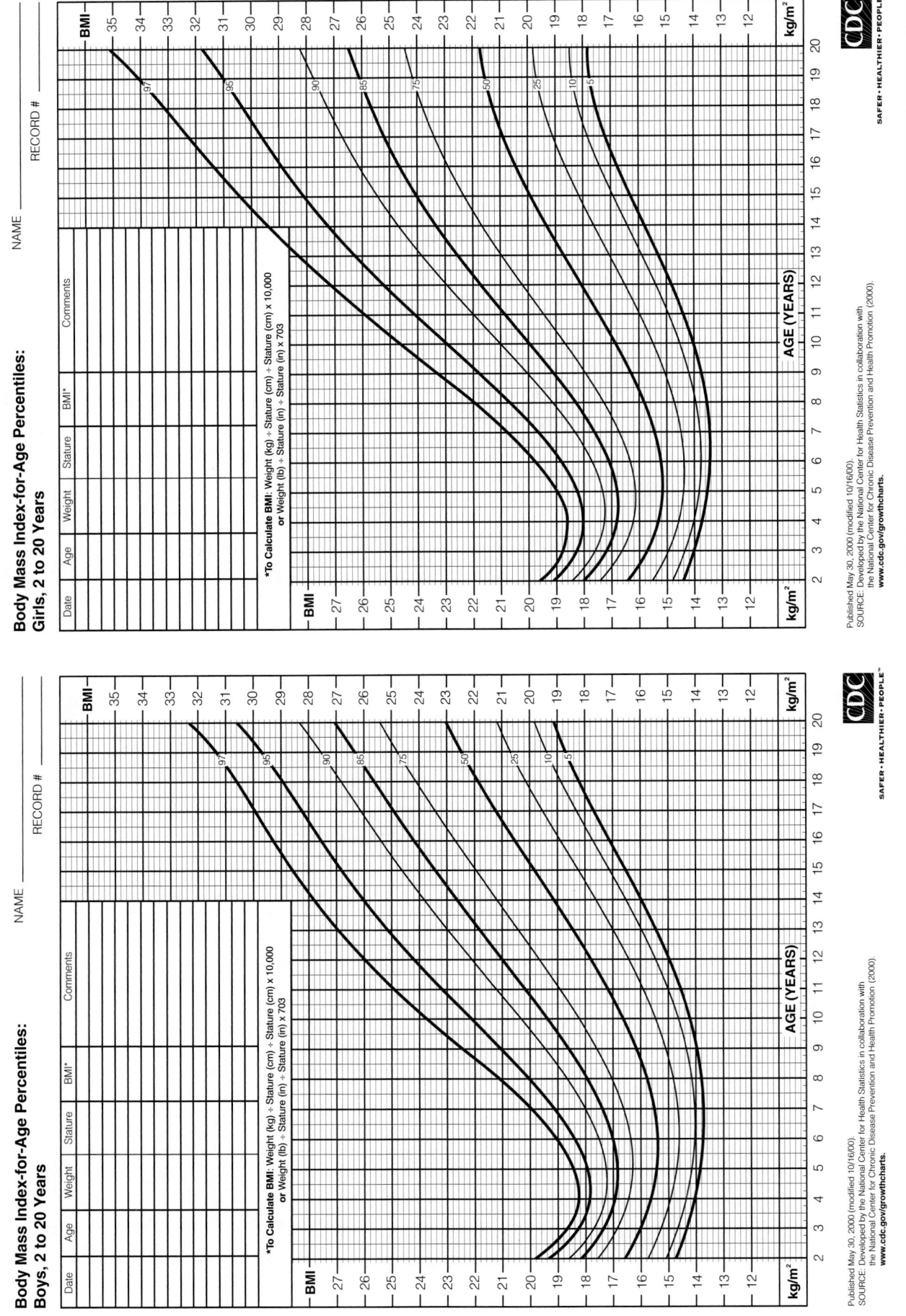

Body Mass Index-for-Age Percentiles: Boys, 2 to 20 Years

NAME _____ RECORD # _____

Date	Age	Weight	Stature	BMI*	Comments

*To Calculate BMI: Weight (kg) ÷ Stature (cm) ÷ Stature (cm) × 10,000
or Weight (lb) ÷ Stature (in) ÷ Stature (in) × 703

Published May 30, 2000 (modified 10/16/00).
SOURCE: Developed by the National Center for Health Statistics in collaboration with
the National Center for Chronic Disease Prevention and Health Promotion (2000).
www.cdc.gov/growthcharts.

FIGURE E-5A Body Mass Index-for-Age Percentiles: Boys, 2 to 20 Years
http://www.cdc.gov/nchs/data/nhanes/growthcharts/set1clinical/cj41l023.pdf

Body Mass Index-for-Age Percentiles: Girls, 2 to 20 Years

NAME _____ RECORD # _____

Date	Age	Weight	Stature	BMI*	Comments

*To Calculate BMI: Weight (kg) ÷ Stature (cm) ÷ Stature (cm) × 10,000
or Weight (lb) ÷ Stature (in) ÷ Stature (in) × 703

Published May 30, 2000 (modified 10/16/00).
SOURCE: Developed by the National Center for Health Statistics in collaboration with
the National Center for Chronic Disease Prevention and Health Promotion (2000).
www.cdc.gov/growthcharts.

FIGURE E-5B Body Mass Index-for-Age Percentiles: Girls, 2 to 20 Years
http://www.cdc.gov/nchs/data/nhanes/growthcharts/set1clinical/cj41l024.pdf

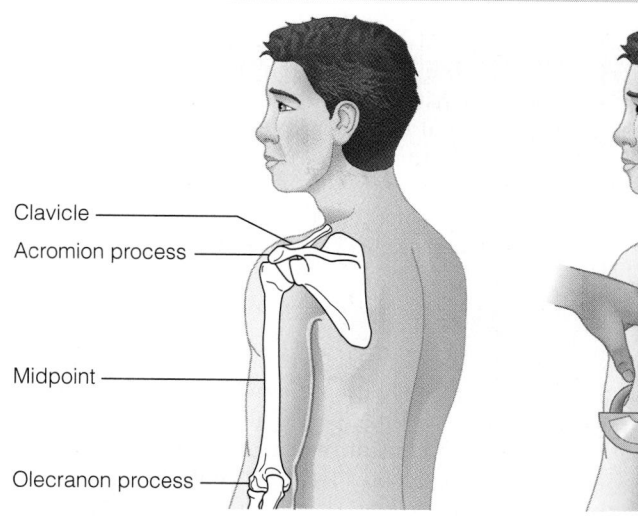

Clavicle

Acromion process

Midpoint

Olecranon process

A. Find the midpoint of the arm:
1. Ask the subject to bend his or her arm at the elbow and lay the hand across the stomach. (If he or she is right-handed, measure the left arm, and vice versa.)
2. Feel the shoulder to locate the acromion process. It helps to slide your fingers along the clavicle to find the acromion process. The olecranon process is the tip of the elbow.
3. Place a measuring tape from the acromion process to the tip of the elbow.

Divide this measurement by 2 and mark the midpoint of the arm with a pen.
B. Measure the skinfold:
1. Ask the subject to let his or her arm hang loosely to the side.
2. Grasp a fold of skin and subcutaneous fat between the thumb and forefinger slightly above the midpoint mark. Gently pull the skin away from the underlying muscle. (This step takes a lot of practice. If you want to be sure you don't have muscle as well as fat, ask the subject to

contract and relax the muscle. You should be able to feel if you are pinching muscle.)
3. Place the calipers over the skinfold at the midpoint mark, and read the measurement to the nearest 1.0 millimeter in two to three seconds. (If using plastic calipers, align pressure lines, and read the measurement to the nearest 1.0 millimeter in two to three seconds.)
4. Repeat steps 2 and 3 twice more. Add the three readings, and then divide by 3 to find the average.

FIGURE E-6 How to Measure the Triceps Skinfold

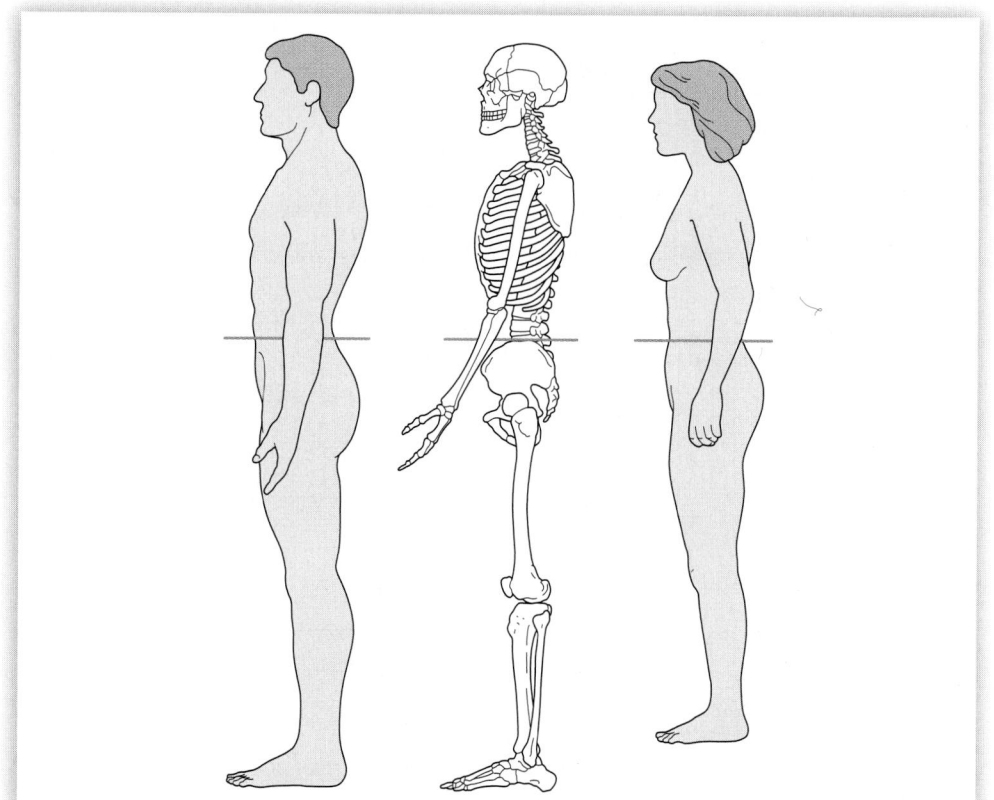

FIGURE E-7 How to Measure Waist Circumference
Place the measuring tape around the waist just above the bony crest of the hip. The tape runs parallel to the floor and is snug (but does not compress the skin). The measurement is taken at the end of normal expiration.

Source: National Institutes of Health Obesity Education Initiative, *Clinical Guidelines on the Identification, Evaluation, and Treatment of Overweight and Obesity in Adults* (Washington, D.C.: U.S. Department of Health and Human Services, 1998), p. 59.

Bioelectrical Impedance To measure body fat using the bioelectrical impedance technique, a very-low-intensity electrical current is briefly sent through the body by way of electrodes placed on the wrist and ankle. As is true of other anthropometric techniques, bioelectrical impedance requires standardized procedures and calibrated instruments to provide reliable results. Recent food intake and hydration status, for example, influence results.

Clinicians use many other methods to estimate body fat and its distribution. Each has its advantages and disadvantages, as Table E-1 summarizes.

Nutritional Anemias

Anemia, a symptom of a wide variety of nutrition- and nonnutrition-related disorders, is characterized by a reduced number of red blood cells. Iron, folate, and vitamin B_{12} deficiencies caused by inadequate intake, poor absorption, or abnormal metabolism of these nutrients are the most common nutritional anemias. Some nonnutrition-related causes of anemia include massive blood loss, infections, hereditary blood disorders such as sickle-cell anemia, and chronic liver or kidney disease.

ASSESSMENT OF IRON-DEFICIENCY ANEMIA

Iron deficiency, a common mineral deficiency, develops in stages. Chapter 9 describes iron deficiency in detail. This section describes tests used to uncover iron deficiency as it progresses. Table E-2 shows which laboratory tests detect various nutrition-related anemias, and Table E-3 provides values used for assessing iron status. Although other tests are more specific in detecting early deficiencies, hemoglobin and hematocrit are the commonly available tests.

TABLE E-1	Methods of Estimating Body Fat and Its Distribution			
Method	Cost	Ease of Use	Accuracy	Measures Fat Distribution
Height and weight	Low	Easy	High	No
Skinfolds	Low	Easy	Low	Yes
Circumferences	Low	Easy	Moderate	Yes
Ultrasound	Moderate	Moderate	Moderate	Yes
Hydrodensitometry	Low	Moderate	High	No
Heavy water tritiated	Moderate	Moderate	High	No
Deuterium oxide, or heavy oxygen	High	Moderate	High	No
Potassium isotope (^{40}K)	Very high	Difficult	High	No
Total body electrical conductivity (TOBEC)	High	Moderate	High	No
Bioelectrical impedance (BIA)	Moderate	Easy	High	No
Dual energy X-ray absorptiometry (DEXA)	High	Easy	High	No
Computed tomography (CT)	Very high	Difficult	High	Yes
Magnetic resonance imaging (MRI)	Very high	Difficult	High	Yes

Source: Adapted with permission from G. A. Bray, a handout presented at the North American Association for the Study of Obesity and Emory University School of Medicine Conference on Obesity Update: Pathophysiology, Clinical Consequences, and Therapeutic Options, Atlanta, Georgia, August 31–September 2, 1992.

Hemoglobin Iron forms an integral part of the hemoglobin molecule that transports oxygen to the cells. In iron deficiency, the body cannot synthesize hemoglobin. Low hemoglobin values signal depleted iron stores. Table E-3 provides hemoglobin values used in nutrition assessment. Hemoglobin's usefulness in evaluating iron status is limited, however, because hemoglobin concentrations drop fairly late in the development of iron deficiency, and other nutrient deficiencies and medical conditions can also alter hemoglobin concentrations.

Hematocrit Hematocrit is commonly used to diagnose iron deficiency, even though it is an inconclusive measure of iron status. To measure the hematocrit, a clinician spins a volume of blood in a centrifuge to separate the red blood cells from the plasma. The hematocrit is the percentage of red blood cells in the total blood volume. Table E-3 includes values used to assess hematocrit status. Low values indicate incomplete hemoglobin formation, which is manifested by microcytic (abnormally small-celled), hypochromic (abnormally lacking in color) red blood cells.

Stages of iron deficiency:

1. Iron stores diminish.
2. Transport iron decreases.
3. Hemoglobin production falls.

TABLE E-2	Laboratory Tests Useful in Evaluating Nutrition-Related Anemias
Test or Test Result	**What It Reflects**
For Anemia (general)	
Hemoglobin (Hg)	Total amount of hemoglobin in the red blood cells (RBC)
Hematocrit (Hct)	Percentage of RBC in the total blood volume
Red blood cell (RBC) count	Number of RBC
Mean corpuscular volume (MCV)	RBC size; helps to determine if anemia is microcytic (iron deficiency) or macrocytic (folate or vitamin B_{12} deficiency)
Mean corpuscular hemoglobin concentration (MCHC)	Hemoglobin concentration within the average RBC; helps to determine if anemia is hypochromic (iron deficiency) or normochromic (folate or vitamin B_{12} deficiency)
Bone marrow aspiration	The manufacture of blood cells in different developmental states
For Iron-Deficiency Anemia	
↓ Serum ferritin	Early deficiency state with depleted iron stores
↓ Transferrin saturation	Progressing deficiency state with diminished transport iron
↑ Erythrocyte protoporphyrin	Later deficiency state with limited hemoglobin production
For Folate-Deficiency Anemia	
↓ Serum folate	Progressing deficiency state
↓ RBC folate	Later deficiency state
For Vitamin B_{12}-Deficiency Anemia	
↓ Serum vitamin B_{12}	Progressing deficiency state
Schilling test	Absorption of vitamin B_{12}

TABLE E-3	Criteria for Assessing Iron Status		
Test	Age (yr)	Gender	Deficiency Value
Hemoglobin (g/dL)	0.5–10	M–F	<11
	11–15	M	<12
		F	<11.5
	>15	M	<13
		F	<12
	Pregnancy		<11
Hematocrit (%)	0.5–4	M–F	<32
	5–10	M–F	<33
	11–15	M	<35
		F	<34
	>15	M	<40
		F	<36
Serum ferritin (μg/L)	0.5–15	M–F	<10
	>15	M–F	<12
Total iron-binding capacity (μg/dL)	>15	M–F	>400
Serum iron (μg/dL)	>15	M–F	<60
Transferrin saturation (%)	0.5–4	M–F	<12
	5–10	M–F	<14
	>10	M–F	<16
Erythrocyte protoporphyrin (μg/dL RBC)	0.5–4	M–F	>80
	>4	M–F	>70

Low hemoglobin and hematocrit values alert the assessor to the possibility of iron deficiency. However, many nutrients and other conditions can affect hemoglobin and hematocrit. The other tests of iron status help pinpoint true iron deficiency.

Serum Ferritin In the first stage of iron deficiency, iron stores diminish. Serum ferritin measures provide a noninvasive estimate of iron stores. Such information is most valuable to iron assessment. Table E-3 shows serum ferritin cutoff values that indicate iron store depletion in children and adults. Serum ferritin is not reliable for diagnosing iron deficiency in infants because normal serum ferritin values are often present in conjunction with iron-responsive anemia.

A decrease in transport iron characterizes the second stage of iron deficiency. This is revealed by an increase in the iron-binding capacity of the protein transferrin and a decrease in serum iron. These changes are reflected by the transferrin saturation, which is calculated from the ratio of the other two values as described in the following paragraphs.

Total Iron-Binding Capacity (TIBC) Iron travels through the blood bound to the protein transferrin. TIBC is a measure of the total amount of iron that transferrin can carry. Lab technicians measure iron-binding capacity directly. Table E-3 includes the value cutoff for TIBC.

Serum Iron Lab technicians can also measure serum iron directly. Elevated values indicate iron overload; reduced values indicate iron deficiency. Table E-3 shows the deficient value for serum iron.

Transferrin Saturation The percentage of transferrin that is saturated with iron is an indirect measure that is derived from the serum iron and total iron-binding capacity measures as follows:

$$\%\text{Transferrin} = \frac{\text{serum iron}}{\text{total iron-binding capacity}} \times 100.$$

Table E-3 shows deficient transferrin saturation values for various age groups.

The third stage of iron deficiency occurs when the supply of transport iron diminishes to the point that it limits hemoglobin production. It is characterized by increases in erythrocyte protoporphyrin, a decrease in mean corpuscular volume, and decreased hemoglobin and hematocrit.

Erythrocyte Protoporphyrin The iron-containing portion of the hemoglobin molecule is heme. Heme is a combination of iron and protoporphyrin. Protoporphyrin accumulates in the blood when iron supplies are inadequate for the formation of heme. Lab technicians can measure erythrocyte protoporphyrin directly in a blood sample. The cutoffs for abnormal values of erythrocyte protoporphyrin are shown in Table E-3.

Mean Corpuscular Volume (MCV) A direct or calculated measure of the mean corpuscular volume (MCV) determines the average size of a red blood cell. Such a measure helps to classify the type of nutrient anemia. In iron deficiency, the red blood cells are smaller than average.

ASSESSMENT OF FOLATE AND VITAMIN B$_{12}$ ANEMIAS

Folate deficiency and vitamin B$_{12}$ deficiency present a similar clinical picture—an anemia characterized by abnormally large red blood cell precursors (megaloblasts) in the bone marrow and abnormally large, mature red blood cells (macrocytic cells) in the blood. Distinguishing between these two deficiencies is particularly important because their treatments differ. Giving folate to a person with vitamin B$_{12}$ deficiency improves many of the lab test results indicative of vitamin B$_{12}$ deficiency, but this is a dangerous error because vitamin B$_{12}$ deficiency causes nerve damage that folate cannot correct. Thus inappropriate folate administration masks vitamin B$_{12}$-deficiency anemia, and nerve damage worsens. For this reason, it is critical to determine whether the anemia results from a folate deficiency or from a vitamin B$_{12}$ deficiency. The following biochemical assessment techniques help to make this distinction.

Mean Corpuscular Volume (MCV) As previously mentioned, the MCV is a measure of red blood cell size. In folate and vitamin B$_{12}$ deficiencies, the red blood cells are larger than average (macrocytic). Additional tests must be performed to differentiate folate from vitamin B$_{12}$ deficiency.

Folate Levels Serum folate levels fluctuate with changes in folate intake and metabolism. Thus serum folate concentrations reflect current status, but provide little information about folate stores. As folate deficiency progresses and low serum levels persist, folate stores decline, resulting in folate depletion. Folate depletion is characterized by a fall in the folate concentrations of red blood cells (erythrocytes). As erythrocyte folate levels diminish, folate-deficiency anemia develops. Because low erythrocyte folate concentrations also occur with vitamin B$_{12}$ deficiency, serum vitamin B$_{12}$ concentrations must also be measured. Table E-4 shows standards for folate assessment.

Vitamin B₁₂ Levels Serum and urinary methylmalonic acid are elevated in vitamin B_{12} deficiency, but not in folate deficiency. Thus this measure is useful in distinguishing between the two. Vitamin B_{12} deficiency usually arises from malabsorption. To determine whether malabsorption is the cause, a small oral dose of vitamin B_{12} is given, and urinary excretion is measured. This procedure measures vitamin B_{12} absorption and is called a Schilling test.

Early stages of vitamin B_{12} deficiency can be detected by a low percentage saturation of its transport protein, a measure similar to iron's transferrin saturation. As the deficiency progresses, serum vitamin B_{12} concentrations fall. Table E-4 shows standards for vitamin B_{12} assessment.

TABLE E-4	Criteria for Assessing Folate and Vitamin B₁₂		
	Deficient	Borderline	Acceptable
Serum folate (ng/mL)[a]	<3.0	3.0–5.9	>6.0
Erythrocyte folate (ng/mL)[a]	<140	140–159	>160
Serum vitamin B₁₂ (pg/mL)	<150	150–200	≥201
Serum methylmalonic acid (nmol/L)	<376	—	—

Note: A nanogram (ng) is one-billionth of a gram; a picogram (pg) is one-trillionth of a gram.

[a]To convert folate values (ng/mL) to international standard units (nmol/L), multiply by 2.266.

Many mathematical problems have been worked out in the "How to" sections of the text. These pages provide additional help and examples.

Conversion Factors

A *conversion factor* is a fraction that converts a measurement expressed in one unit to another unit; for example, the fraction may be used to convert pounds to kilograms or feet to inches. To create a conversion factor, an equality (such as *1 kilogram = 2.2 pounds*) is expressed as a fraction:

$$\frac{1 \text{ kg}}{2.2 \text{ lb}} \text{ and } \frac{2.2 \text{ lb}}{1 \text{ kg}}.$$

Because a conversion factor has a value of *1* (the value in the numerator is equal to the value in the denominator), it can be used as a multiplier to change the *unit* of measure without changing the *value* of the measurement. To convert the units of a measurement, the fraction must have the desired unit in the numerator, as the unit in the denominator will cancel out the original unit.

Example 1 Convert the weight of 130 pounds to kilograms.

• Multiply 130 pounds by a conversion factor (fraction) that includes both pounds and kilograms and is arranged so that the desired unit (kilograms) is in the numerator:

$$130 \text{ lb} \times \frac{1 \text{ kg}}{2.2 \text{ lb}} = \frac{130 \text{ kg}}{2.2} = 59 \text{ kg}.$$

As this example shows, the unit for *pounds* cancels out and the unit for *kilograms* remains in the solution.

Example 2 The food label on a bottle of apple juice shows the contents in both fluid ounces and liters. How many liters are contained in a bottle that holds 64 fluid ounces?

• Multiply 64 ounces by a conversion factor that includes both fluid ounces and liters, with the desired unit (liters) in the numerator:

$$64 \text{ fl oz} \times \frac{1 \text{ L}}{33.8 \text{ fl oz}} = \frac{64 \text{ L}}{33.8} = 1.89 \text{ L}.$$

Percentages

A percentage expresses a fraction that has 100 in the denominator; for example, the term *50 percent* is equivalent to the fraction 50/100. Similar to other fractions, percentages are used to express a *proportion* of the *whole;* therefore, the units in the numerator and denominator must be similar. Any fraction can be expressed in hundredths and converted to a per-

centage by dividing the numerator by the denominator and multiplying by 100:

$$^1/_4 = 0.25.$$
$$0.25 \times 100 = 25\%.$$

Example 3 Suppose your energy intake for the day is 2000 kcalories (kcal) and your recommended energy intake is 2400 kcalories. What percent of the recommended energy intake did you consume?

• Divide your intake by the recommended intake:

$$2000 \text{ kcal (your intake)} \div$$
$$2400 \text{ kcal (recommended intake)} = 0.83.$$

• Multiply by 100 to express the decimal as a percentage:

$$0.83 \times 100 = 83\%.$$

Example 4 A percentage can also be more than 100. Suppose your intake of vitamin C is 120 milligrams and your RDA (male) is 90 milligrams. What percent of the RDA for vitamin C did you consume?

$$120 \text{ mg (your intake)} \div 90 \text{ mg (RDA)} = 1.33.$$
$$1.33 \times 100 = 133\%.$$

Weights and Measures

LENGTH

1 meter (m) = 39 in.
1 centimeter (cm) = 0.4 in.
1 inch (in) = 2.5 cm.
1 foot (ft) = 30 cm.

TEMPERATURE

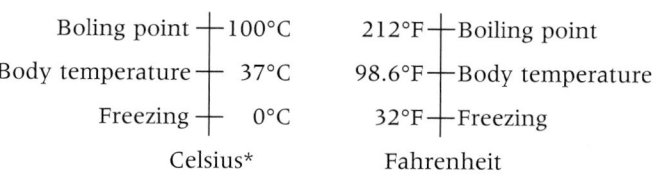

	Celsius*			Fahrenheit
Boling point	100°C		212°F	Boiling point
Body temperature	37°C		98.6°F	Body temperature
Freezing	0°C		32°F	Freezing

• To find degrees Fahrenheit (°F) when you know degrees Celsius (°C), multiply by 9/5 and then add 32.
• To find degrees Celsius (°C) when you know degrees Fahrenheit (°F), subtract 32 and then multiply by 5/9.

VOLUME

1 liter (L) = 1000 mL, 0.26 gal, 1.06 qt, 2.1 pt, or 33.8 fl. oz.
1 milliliter (mL) = 1/1000 L or 0.03 fluid oz.
1 gallon (gal) = 128 oz, 8 c, or 3.8 L.
1 quart (qt) = 32 oz, 4 c, or 0.95 L.
1 pint (pt) = 16 oz, 2 c, or 0.47 L.
1 cup (c) = 8 oz, 16 tbs, about 250 mL, or 0.25 L.
1 ounce (oz) = 30 mL.
1 tablespoon (tbs) = 3 tsp or 15 mL.
1 teaspoon (tsp) = 5 mL.

*Also known as *centigrade.*

WEIGHT

1 kilogram (kg) = 1000 g or 2.2 lb.
1 gram (g) = 1/1000 kg, 1000 mg, or 0.035 oz.
1 milligram (mg) = 1/1000 g or 1000 μg.
1 microgram (μg) = 1/1000 mg.
1 pound (lb) = 16 oz, 454 g, or 0.45 kg.
1 ounce (oz) = about 28 g.

ENERGY

1 kilojoule (kJ) = 0.24 kcal.
1 millijoule (mJ) = 240 kcal.
1 kcalorie (kcal) = 4.2 kJ.
1 g carbohydrate = 4 kcal = 17 kJ.
1 g fat = 9 kcal = 37 kJ.
1 g protein = 4 kcal = 17 kJ.
1 g alcohol = 7 kcal = 29 kJ.

The large number of enteral formulas available allows health care professionals to meet a variety of their patients' medical needs, but also complicates the process of selecting an appropriate formula. The first step in narrowing the choice of formulas is to determine the patient's ability to digest and absorb nutrients. Table G-1 on pp. G-2 through G-3 lists examples of standard formulas for patients who can adequately digest and absorb nutrients, and Table G-2 on p. G-4 provides examples of hydrolyzed formulas for patients with limited ability to digest or absorb nutrients. Products promoted to the general public and intended primarily as oral supplements, such as Carnation Instant Breakfast® (Nestlé), Boost® (Mead John-

son), and Ensure (Ross), are not included as examples. Each formula is listed only once, although a formula may have more than one use. A high-protein formula, for example, may also be a fiber-containing formula. Tables G-3 through G-5 on p. G-5 list modular formulas.

The information listed in this appendix reflects the literature provided by manufacturers and does not suggest endorsement by the authors. Manufacturers frequently add new formulas, discontinue old ones, and change formula composition. Consult the manufacturers' literature and websites for updates and additional examples of enteral formulas. The following products are listed in this appendix:

- Mead Johnson Nutritionals:[a]
  Kindercal® TF

- Nestlé Nutrition:[b]
  Crucial®
  Nutren® 1.0
  Nutren® 1.0 Fiber
  Nutren® 1.5
  Nutren® 2.0
  Nutren® Glytrol
  Nutren® Junior
  Nutren® Probalance
  Nutren® Pulmonary
  Nutren® Renal
  Nutren® Replete
  NutriHep®
  Peptamen®
  Peptamen Junior®

- Novartis Medical Nutrition:[c]
  Compleat® Pediatric
  Impact®
  Impact® 1.5
  Impact® Glutamine
  Isocal®
  Isocal® HN
  Isosource® Standard
  Isosource® VHN
  MCT Oil
  Microlipid®
  Novasource® Pulmonary
  Novasource® Renal
  Resource® Beneprotein®
    Instant Protein Powder
  Resource® Diabetic TF
  Vivonex® Pediatric
  Vivonex® Plus
  Vivonex® T.E.N.

- Ross Medical Nutritionals:[d]
  Glucerna®
  Jevity® 1 Cal
  Nepro®
  Optimental®
  Osmolite®
  Oxepa®
  Perative®
  Polycose Liquid®
  Polycose Powder®
  ProMod®
  Promote®
  Promote® with Fiber
  Pulmocare®

[a]Mead Johnson Nutritionals, **www.meadjohnson.com**, visited March 5, 2006.
[b]Nestlé Nutrition, **www.nestleclinicalnutrition.com**, visited March 5, 2006.
[c]Novartis Medical Nutrition, **www.novartis.com**, visited March 5, 2006.
[d]Ross Medical Nutritionals, **www.ross.com**, visited March 5, 2006.

TABLE G-1 Standard Formulas

Product[a]	Volume to Meet 100% RDI[b] (mL)	Energy (kcal/mL)	Protein or Amino Acids (g/L)	Carbohydrate (g/L)	Fat (g/L)	Osmolality[c] (mOsm/kg)	Notes
Lactose-Free, Standard Formulas							
Isocal®	1890	1.06	34	135	44	270	20% fat from MCT
Isosource® Standard	1165	1.20	43	170	39	490	50% fat from MCT
Nutren® 1.0	1500	1.00	40	127	38	315	25% fat from MCT
Osmolite®	1887	1.06	37	151	35	300	20% fat from MCT
Lactose-Free, Fiber-Containing Formulas							
Jevity® 1 Cal	1321	1.06	44	155	35	300	14 g fiber/L
Nutren® 1.0 Fiber	1500	1.00	40	127	38	330	14 g fiber/L
Nutren® ProBalance	1000	1.20	54	156	41	350	10 g fiber/L
Promote® with Fiber	1000	1.00	63	138	28	380	14 g fiber/L
Lactose-Free, High-kCalorie Formulas							
Nutren® 1.5	1000	1.50	60	169	68	430	50% fat from MCT
Nutren® 2.0	750	2.00	80	196	104	745	75% fat from MCT
Lactose-Free, High-Protein Formulas							
Isocal® HN	1180	1.06	44	124	45	270	Low residue
Promote®	1000	1.00	63	130	26	340	20% fat from MCT, low residue
Special-Use Formulas: Pediatric (1 to 10 years)							
Compleat® Pediatric	Varies[d]	1.00	38	130	39	380	Blenderized formula, 6.8 g fiber/L
Kindercal® TF	Varies[d]	1.06	30	135	44	345	12% fat from MCT
Nutren® Junior	Varies[d]	1.00	30	110	50	350	21% fat from MCT
Special-Use Formulas: Glucose Intolerance							
Glucerna®	1420	1.00	42	96	54	355	14 g fiber/L
Nutren® Glytrol	1400	1.00	45	100	48	280	15 g fiber/L; 20% fat from MCT
Resource® Diabetic TF	1180	1.06	58	100	50	400	15 g fiber/L

Note: MCT = Medium-chain triglycerides.

[a]Formulas come in ready-to-use (liquid) form unless specified under "Notes."

[b]RDI = Reference Daily Intakes, which are labeling standards for vitamins, minerals, and protein. Consuming 100 percent of the RDI will meet the nutrient needs of most people using the product.

[c]Osmolality may vary, depending on the flavorings added to a product.

[d]Depends on age of child.

TABLE G-1 Standard Formulas—continued

Product[a]	Volume to Meet 100% RDI[b] (mL)	Energy (kcal/mL)	Protein or Amino Acids (g/L)	Carbohydrate (g/L)	Fat (g/L)	Osmolality[c] (mOsm/kg)	Notes
Special-Use Formulas: Immune System Support							
Impact®	1500	1.00	56	130	28	375	Enriched with arginine, nucleic acids, and omega-3 fatty acids
Impact® 1.5	1250	1.50	84	140	69	550	Same as above
Impact® Glutamine	1000	1.30	78	150	43	630	Same as above and enriched with glutamine; 10 g fiber/L
Special-Use Formulas: Renal Failure							
Nepro®	947	2.00	70	223	96	665	High-calcium, low-phosphorus; intended for use once dialysis has been instituted
Novasource® Renal	1000	2.00	74	200	100	700	Low in electrolytes; intended for use once dialysis has been instituted
Nutren® Renal	750	2.00	70	205	104	650	50% fat from MCT; enriched with vitamins C and B_6, folate, zinc, and selenium; intended for use once dialysis has been instituted
Special-Use Formulas: Respiratory Insufficiency							
Novasource® Pulmonary	933	1.50	75	150	68	650	8 g fiber/L
Nutren® Pulmonary	1000	1.50	68	100	95	330	55% kcal from fat, 40% fat from MCT
Oxepa	947	1.50	63	106	94	493	55% kcal from fat, enriched with antioxidant nutrients
Pulmocare®	947	1.50	63	106	93	475	55% kcal from fat, 20% fat from MCT, enriched with antioxidant nutrients
Special-Use Formulas: Wound Healing							
Isosource® VHN	1250	1.00	62	130	29	300	10 g fiber/L, enriched with vitamins A and C and zinc
Nutren® Replete	1000	1.00	62	113	34	300	Enriched with vitamins A and C and zinc; 25% fat from MCT

TABLE G-2   Hydrolyzed Protein Formulas

Product	Volume to Meet 100% RDI[a] (mL)	Energy (kcal/mL)	Protein or Amino Acids (g/L)	Carbohydrate (g/L)	Fat (g/L)	Osmolality[b] (mOsm/kg)	Notes
Special-Use Hydrolyzed Formulas: Hepatic Insufficiency							
NutriHep®	1000	1.50	40	290	21	790	Free amino acids, high in branched-chain amino acids, low in aromatic amino acids
Special-Use Hydrolyzed Formulas: Immune System Support							
Crucial®	1000	1.50	94	134	68	490	Enriched with arginine, antioxidant nutrients, and zinc
Perative®	1155	1.30	67	180	37	460	Enriched with arginine and beta-carotene
Vivonex® Plus	1800	1.00	45	190	7	650	Powder form; 100% free amino acids, enriched with glutamine, arginine, and branched-chain amino acids
Special-Use Hydrolyzed Formulas: Malabsorption							
Optimental®	1422	1.00	51	139	28	540	Contains MCT and arginine; enriched with vitamins C and E and beta-carotene
Peptamen®	1500	1.00	40	127	39	270	70% fat from MCT
Vivonex® T.E.N.	2000	1.00	38	210	3	630	Powder form; 100% free amino acids, enriched with glutamine
Special-Use Hydrolyzed Formulas: Pediatric (1 to 10 years)							
Peptamen Junior®	Varies[c]	1.0	30	138	39	260	60% fat from MCT
Vivonex® Pediatric	Varies[c]	0.8	24	130	24	360	Powder form; 100% free amino acids

[a]RDI = Reference Daily Intakes, which are labeling standards for vitamins, minerals, and protein. Consuming 100 percent of the RDI will meet the nutrient needs of most people using the product.

[b]Osmolality may vary depending on the flavorings added to a product.

[c]Depends on age of child.

TABLE G-3 Protein Modules

Product	Form	Major Protein Source	Energy (kcal/g)	Protein (g/100 g)
ProMod®	Powder	Whey protein	4.2	75
Resource® Beneprotein® Instant Protein Powder	Powder	Whey protein	3.6	86

TABLE G-4 Carbohydrate Modules

Product	Form	Major Carbohydrate Source	Energy (kcal/mL or g)
Polycose Liquid®	Liquid	Hydrolyzed cornstarch	2.0 kcal/mL
Polycose Powder®	Powder	Hydrolyzed cornstarch	3.8 kcal/g

TABLE G-5 Fat Modules

Product	Form	Major Fat Source	Energy (kcal/mL)	Fat (g/100 mL)
MCT Oil	Liquid	Coconut oil	7.7	93
Microlipid®	Liquid	Safflower oil	4.5	50

Answers to Self Check Questions

CHAPTER 1
1. d, 2. d, 3. c, 4. d, 5. c, 6. b, 7. c, 8. b, 9. c, 10. c

CHAPTER 2
1. b, 2. a, 3. d, 4. a, 5. c, 6. a, 7. d, 8. c, 9. d, 10. d

CHAPTER 3
1. b, 2. b, 3. c, 4. a, 5. a, 6. a, 7. b, 8. a, 9. d, 10. c

CHAPTER 4
1. b, 2. c, 3. a, 4. d, 5. a, 6. a, 7. d, 8. d, 9. b, 10. a

CHAPTER 5
1. c, 2. b, 3. d, 4. d, 5. a, 6. d, 7. c, 8. b, 9. c, 10. d

CHAPTER 6
1. b, 2. c, 3. d, 4. d, 5. b, 6. a, 7. a, 8. b, 9. c, 10. c

CHAPTER 7
1. b, 2. a, 3. a, 4. c, 5. b, 6. c, 7. d, 8. b, 9. c, 10. b

CHAPTER 8
1. d, 2. d, 3. c, 4. a, 5. d, 6. d, 7. b, 8. c, 9. c, 10. a

CHAPTER 9
1. d, 2. b, 3. a, 4. d, 5. c, 6. d, 7. a, 8. d, 9. d, 10. b

CHAPTER 10
1. a, 2. d, 3. c, 4. c, 5. b, 6. d, 7. c, 8. d, 9. d, 10. c

CHAPTER 11
1. b, 2. a, 3. b, 4. d, 5. c, 6. d, 7. c, 8. b, 9. c, 10. c

CHAPTER 12
1. d, 2. c, 3. d, 4. d, 5. d, 6. d, 7. b, 8. b, 9. a, 10. b

CHAPTER 13
1. b, 2. d, 3. d, 4. a, 5. c, 6. d, 7. c, 8. a, 9. d, 10. a

CHAPTER 14
1. c, 2. b, 3. a, 4. b, 5. d, 6. c, 7. b, 8. d, 9. c, 10. d

CHAPTER 15
1. d, 2. c, 3. a, 4. b, 5. c, 6. d, 7. b, 8. a, 9. a, 10. c

CHAPTER 16
1. d, 2. a, 3. b, 4. c, 5. d, 6. b, 7. c, 8. c, 9. d, 10. a

CHAPTER 17
1. c, 2. b, 3. a, 4. d, 5. a, 6. c, 7. d, 8. b, 9. b, 10. c

CHAPTER 18
1. c, 2. d, 3. b, 4. b, 5. c, 6. a, 7. c, 8. b, 9. d, 10. c

CHAPTER 19
1. d, 2. a, 3. c, 4. b, 5. a, 6. b, 7. d, 8. d, 9. c, 10. a

CHAPTER 20
1. b, 2. d, 3. d, 4. c, 5. a, 6. c, 7. d, 8. c, 9. a, 10. b

CHAPTER 21
1. a, 2. d, 3. c, 4. a, 5. d, 6. c, 7. d, 8. b, 9. c, 10. d

CHAPTER 22
1. a, 2. c, 3. b, 4. d, 5. c, 6. a, 7. b, 8. b, 9. c, 10. d

CHAPTER 23
1. d, 2. b, 3. c, 4. b, 5. d, 6. b, 7. c, 8. a, 9. a, 10. d

Glossary

24-hour recall a record of foods consumed in the previous 24 hours; sometimes modified to include foods consumed in a typical day.

2-in-1 solution a parenteral solution that contains dextrose and amino acids but excludes lipids.

abscesses (AB-sess-es) accumulated pus that is surrounded by inflamed tissue.

Acceptable Daily Intake (**ADI**) the amount of an artificial sweetener that individuals can safely consume each day over the course of a lifetime without adverse effect. It includes a 100-fold safety factor.

Acceptable Macronutrient Distribution Ranges (**AMDR**) ranges of intakes for the energy-yielding nutrients that provide adequate energy and nutrients and reduce the risk of chronic disease.

acetone breath distinctive fruity odor on the breath of a person with ketosis.

achalasia (ack-ah-LAY-zhah) an esophageal disorder characterized by weakened peristalsis and impaired relaxation of the lower esophageal sphincter.

achlorhydria (AY-clor-HIGH-dree-ah) absence of gastric acid secretion.

acid-base balance the balance maintained between acid and base concentrations in the blood and body fluids.

acidosis acid accumulation in the blood and body fluids; depresses the central nervous system and can lead to disorientation and, eventually, coma.

acids compounds that release hydrogen ions in a solution.

acquired immune deficiency syndrome (**AIDS**) the late stage of illness caused by infection with the human immunodeficiency virus (HIV); characterized by severe damage to immune function.

acute erosive gastritis a condition in which the gastric mucosa is acutely injured, often by the toxic effects of chemical substances or radiation treatment. Damage may include hemorrhaging, tissue erosion, and ulcers.

acute renal failure abrupt loss of kidney function over a period of hours or days.

acute respiratory distress syndrome (**ARDS**) respiratory failure triggered by acute lung injury; a medical emergency that causes dyspnea and pulmonary edema and usually requires assisted (mechanical) ventilation.

acute-phase response changes in body chemistry resulting from infection, inflammation, or injury; characterized by alterations in plasma proteins.

added sugars sugars and syrups added to a food for any purpose, such as to add sweetness or bulk or to aid in browning (baked foods). Also called carbohydrate sweeteners, they include glucose, fructose, corn syrup, concentrated fruit juice, and other sweet carbohydrates.

adequacy the characteristic of a diet that provides all the essential nutrients, fiber, and energy necessary to maintain health and body weight.

Adequate Intakes (**AI**) a set of values that are used as guides for nutrient intakes when scientific evidence is insufficient to determine an RDA.

adipose tissue the body's fat, which consists of masses of fat-storing cells called adipose cells.

adolescence the period of growth from the beginning of puberty until full maturity. Timing of adolescence varies from person to person.

AIDS-defining illnesses diseases and complications associated with the later stages of an HIV infection; include wasting, recurrent bacterial pneumonia, opportunistic infections, and certain cancers.

alcohol-related birth defects (**ARBD**) a condition caused by prenatal alcohol exposure. ARBD is diagnosed when there is a history of substantial, regular maternal alcohol intake or heavy episodic drinking and birth defects known to be associated with alcohol exposure.

alcohol-related neurodevelopmental disorder (**ARND**) a condition caused by prenatal alcohol exposure. ARND is diagnosed when there is a confirmed history of substantial, regular maternal alcohol intake or heavy episodic drinking and behavioral, cognitive, or central nervous system abnormalities known to be associated with alcohol exposure.

aldosterone (al-DOS-ter-own) a hormone secreted by the adrenal glands that stimulates the reabsorption of sodium by the kidneys; also regulates chloride and potassium concentrations.

alkalosis excessive base in the blood and body fluids.

alpha-lactalbumin (lackt-AL-byoo-min) the chief protein in human breast milk, as casein (*CAY-seen*) is the chief protein in cow's milk.

alveoli (al-VEE-oh-lie) air sacs in the lungs. One sac is an *alveolus*.

Alzheimer's disease a progressive, degenerative disease that attacks the brain and impairs thinking, behavior, and memory.

amino (a-MEEN-oh) **acids** building blocks of protein. Each has a hydrogen atom, an amino group, and an acid group attached to a central carbon, which also carries a distinctive side chain.

amniotic (am-nee-OTT-ic) **sac** the "bag of waters" in the uterus in which the fetus floats.

anaphylactic (AN-ah-feh-LAC-tic) **shock** a life-threatening whole-body allergic reaction to an offending substance.

anencephaly (AN-en-SEF-a-lee) an uncommon and always fatal type of neural tube defect; characterized by the absence of a brain.

anthropometric (AN-throw-poe-MEH-trik) related to physical measurements of the human body, such as height, weight, body circumferences, and percentage of body fat.

antibodies large proteins of the blood and body fluids, produced in response to invasion of the body by unfamiliar molecules (mostly proteins) called *antigens*. Antibodies inactivate the invaders and so protect the body.

antidiuretic hormone (**ADH**) a hormone released by the pituitary gland in response to high salt concentrations in the blood. The kidneys respond by reabsorbing water.

antioxidant (anti-OX-ih-dant) a compound that protects other compounds from oxygen by itself reacting with oxygen. Oxidation is a potentially damaging effect of normal cell chemistry involving oxygen.

appetite the psychological desire to eat; a learned motivation that is experienced as a pleasant sensation that accompanies the sight, smell, or thought of appealing foods.

artery a vessel that carries blood away from the heart.

arthritis inflammation of a joint, usually accompanied by pain, swelling, and structural changes.

artificial fats zero-energy fat replacers that are chemically synthesized to mimic the sensory and cooking qualities of naturally occurring fats but are totally or partially resistant to digestion.

artificial sweeteners noncarbohydrate, nonkcaloric synthetic sweetening agents; sometimes called *nonnutritive sweeteners*.

ascites (ah-SIGH-teez) the abnormal accumulation of fluid in the abdominal cavity.

ascorbic acid one of the two active forms of vitamin C. Many people refer to vitamin C by this name.

aspiration drawing in by suction or breathing; a common complication of enteral feedings in which foreign material enters the lungs, often from reflux of stomach contents.

atherosclerosis (ath-er-oh-scler-OH-sis) a type of artery disease characterized by accumulations of lipid-containing material on the inner walls of the arteries.

atrophic gastritis (a-TRO-fik gas-TRI-tis) a condition characterized by chronic inflammation of the stomach accompanied by a diminished size and functioning of the mucosa and glands.

autoimmune an immune response directed against the body's own tissues.

bacterial overgrowth excessive bacterial colonization of the stomach and small intestine; may be caused by low gastric acidity, altered gastrointestinal motility, mucosal damage, or contamination; interferes with normal digestion and absorption.

balance the dietary characteristic of providing foods of a number of types in proportion to each other, such that foods rich in some nutrients do not crowd out foods that are rich in other nutrients.

bariatric (BAH-ree-AH-trik) **surgery** surgery that treats severe obesity.

Barrett's esophagus a condition in which esophageal cells damaged by chronic exposure to stomach acid are replaced by cells that resemble those in the stomach or small intestine, sometimes becoming cancerous.

basal metabolic rate (**BMR**) the rate of energy use for metabolism under specified conditions after a 12-hour fast and restful sleep, without any physical activity or emotional excitement, and in a comfortable setting. It is usually expressed as kcalories per kilogram of body weight per hour.

basal metabolism the energy needed to maintain life when a person is at complete digestive, physical, and emotional rest. Basal metabolism is normally the largest part of a person's daily energy expenditure.

bases compounds that accept hydrogen ions in a solution.

behavior modification the changing of behavior by the manipulation of antecedents (cues or environmental factors that trigger behavior), the behavior itself, and consequences (the penalties or rewards attached to behavior).

beriberi the thiamin-deficiency disease; characterized by loss of sensation in the hands and feet, muscular weakness, advancing paralysis, and abnormal heart action.

beta-carotene a vitamin A precursor made by plants and stored in human fat tissue; an orange pigment.

BHA, BHT preservatives commonly used to slow the development of "off" flavors, odors, and color changes caused by oxidation.

bifidus (BIFF-id-us, by-FEED-us) **factors** factors in colostrum and breast milk that favor the growth of the "friendly" bacterium *Lactobacillus* (lack-toe-ba-SILL-us) *bifidus* in the infant's intestinal tract.

bioavailability the rate and extent to which a nutrient is absorbed and used.

blenderized formulas enteral formulas that are made by blenderizing whole foods.

body mass index (**BMI**) an index of a person's weight in relation to height; determined by dividing the weight (in kilograms) by the square of the height (in meters).

bolus (BOH-lus) the portion of food swallowed at one time.

bolus (BOH-lus) **feeding** delivery of about 250 to 500 mL of formula in less than 15 minutes.

bone marrow transplants procedures that replace bone marrow that has been destroyed by cancer treatments; also used to treat certain types of cancers and blood disorders.

bronchi, bronchioles the main airways of the lungs. The singular form of bronchi is *bronchus*.

buffalo hump the accumulation of fatty tissue at the base of the neck.

buffers compounds that can reversibly combine with hydrogen ions to help keep a solution's acidity or alkalinity constant.

calcium rigor hardness or stiffness of the muscles caused by high blood calcium.

calcium tetany intermittent spasms of the extremities due to nervous and muscular excitability caused by low blood calcium.

calories units in which energy is measured. Food energy is measured in kilocalories (1000 calories equal 1 kilocalorie), abbreviated kcalories or kcal. One kcalorie is the amount of heat necessary to raise the temperature of 1 kilogram (kg) of water 1°C. The scientific use of the term kcalorie is the same as the popular use of the term *calorie.*

cancer cachexia (ka-KEK-see-ah) a wasting syndrome, associated with cancer, that is characterized by anorexia, muscle wasting, weight loss, and fatigue.

cancers malignant growths or tumors that result from abnormal and uncontrolled cell division.

capillaries small vessels that branch from an artery. Capillaries connect arteries to veins. Oxygen, nutrients, and waste materials are exchanged across capillary walls.

carbohydrates energy nutrients composed of monosaccharides.

carbohydrate-to-insulin ratio the amount of carbohydrate that can be handled per unit of insulin. On average, every 15 g of carbohydrate requires about 1 unit of rapid- or short-acting insulin.

carcinogenesis (CAR-sin-oh-JEN-eh-sis) the process of cancer development.

carcinogens (CAR-sin-oh-jenz or car-SIN-oh-jenz) substances that can cause cancer (the adjective is *carcinogenic*).

cardiac cachexia the severe muscle wasting and weight loss that accompany congestive heart failure

cardiac output the volume of blood pumped by the heart within a specified period of time.

cardiovascular disease (**CVD**) a general term for all diseases of the heart and blood vessels.

cataracts thickenings of the eye lenses that impair vision and can lead to blindness.

cathartics (ca-THART-ics) strong laxatives.

catheter a thin tube placed within a narrow lumen (such as a blood vessel) or body cavity; can be used to infuse or withdraw fluids or to keep a passage open.

celiac (SEE-lee-ack) **disease** a condition characterized by an abnormal immune reaction to wheat gluten that causes severe intestinal damage and nutrient malabsorption; also called *gluten-sensitive enteropathy* or *celiac sprue.*

cellulite (SELL-you-light or SELL-you-leet) supposedly, a lumpy form of fat; actually, a fraud. Fatty areas of the body may appear lumpy when the strands of connective tissue that attach the skin to underlying muscles pull tight where the fat is thick.

central obesity excess fat around the trunk of the body; also called *abdominal fat* or *upper-body fat.*

central veins large-diameter veins located close to the heart.

cerebral cortex the outer surface of the cerebrum, which is the largest part of the brain.

Certified Diabetes Educator (**CDE**) a health care professional who specializes in diabetes management education. Certification is obtained from the National Certification Board for Diabetes Educators.

chemotherapy the use of drugs to arrest or destroy cancer cells. Such drugs are called *antineoplastic agents.*

choline a nonessential nutrient that can be made in the body from an amino acid.

chronic bronchitis (bron-KYE-tis) persistent inflammation of the mucous membranes lining the main airways of the lungs. Chronic inflammation leads to narrower airways and difficulty with breathing.

chronic diseases degenerative diseases characterized by deterioration of the body organs. Examples include heart disease, cancer, and diabetes.

chronic obstructive pulmonary disease (**COPD**) a group of lung diseases characterized by persistent obstructed airflow through the lungs and airways; include chronic bronchitis and emphysema.

chronological age a person's age in years from his or her date of birth.

chylomicrons (kye-lo-MY-crons) the lipoproteins that transport lipids from the intestinal cells into the body. The cells of the body remove the lipids they need from the chylomicrons, leaving chylomicron remnants to be picked up by the liver cells.

chyme (KIME) the semiliquid mass of partly digested food expelled by the stomach into the duodenum (the top portion of the small intestine).

cirrhosis (sih-ROE-sis) an advanced stage of liver disease in which extensive scarring replaces healthy liver tissue, causing impaired liver function and liver failure.

claudication (CLAW-dih-KAY-shun) pain in the legs while walking; usually due to an inadequate supply of blood to muscles.

clear liquid diet a diet that consists of foods that are liquid at room temperature and leave almost no residue (undigested material) in the intestines after digestion and absorption.

clinically severe obesity a BMI of 40 or greater or a BMI of 35 or greater and one or more serious conditions such as hypertension. Another term used to describe the same condition is *morbid obesity.*

closed feeding systems delivery systems in which formula comes prepackaged in containers that are ready to be attached to feeding tubes for administration.

club drug any of a wide variety of drugs used by young adults at all-night dance parties such as "raves."

coenzyme (co-EN-zime) a small molecule that works with an enzyme to promote the enzyme's activity. Many coenzymes have B vitamins as part of their structure.

cofactor a mineral element that, like a coenzyme, works with an enzyme to facilitate a chemical reaction.

colectomy removal of a portion or all of the colon.

collagen the characteristic protein of connective tissue.

collaterals blood vessels that enlarge in order to allow an alternative pathway for diverted blood.

colostomy (co-LAHS-toe-me) a surgical procedure that creates a stoma using a section of the colon.

colostrum (co-LAHS-trum) a milklike secretion from the breasts that is rich in protective factors. Colostrum is present during the first day or so after delivery, before milk appears.

complement a group of plasma proteins that assist the activities of antibodies.

complementary proteins two or more proteins whose amino acid assortments complement each other in such a way that the essential amino acids missing from one are supplied by the other.

conditionally essential amino acid an amino acid that is normally nonessential but must be supplied by the diet in special circumstances when the need for it becomes greater than the body's ability to produce it.

congestive heart failure (**CHF**) a condition in which the heart is unable to pump adequate blood, resulting in fluid congestion in tissues and in the veins leading to the heart.

continuous feedings slow delivery of formula at a constant rate over an 8- to 24-hour period.

continuous parenteral nutrition continuous administration of parenteral solutions over a 24-hour period.

cornea (KOR-nee-uh) the hard, transparent membrane covering the outside of the eye.

C-reactive protein an acute-phase protein released from the liver during acute inflammation or stress.

cretinism (CREE-tin-ism) an iodine-deficiency disease characterized by mental and physical retardation.

critical pathways interdisciplinary care plans for specific diagnoses, treatments, or procedures that merge the medical and nursing plans with those of other disciplines, such as physical therapy, nutrition, and mental health.

critical period a finite period during development in which certain events occur that will have irreversible effects on later developmental stages; usually a period of rapid cell division.

Crohn's disease an inflammatory bowel disease that usually occurs in the lower portion of the small intestine and the colon. Inflammation may pervade the entire intestinal wall.

cryptosporidiosis (KRIP-toe-spo-rid-ee-OH-sis) a food-borne illness caused by the parasite *Cryptosporidium parvum.*

cyanosis (sigh-ah-NOH-sis) a bluish cast in skin due to the color of deoxygenated hemoglobin. Cyanosis is most evident in individuals with lighter, thinner skin; it is mostly seen on lips, cheeks, and ears and under nails.

cyclic parenteral nutrition administration of a parenteral solution over a 10- to 16-hour period.

cystic fibrosis an inherited disease characterized by the production of abnormally viscous exocrine secretions; often leads to respiratory illness and pancreatic insufficiency.

cystinuria (SIS-tin-NOO-ree-ah) an inherited disorder characterized by elevated urinary excretion of several amino acids, including cystine.

cytokines (SIGH-toe-kynes) proteins produced by white blood cells that regulate immune cell development and immune responses.

Daily Values reference values developed by the FDA specifically for use on food labels.

dawn phenomenon morning hyperglycemia that is caused by the early morning release of growth hormone, which counteracts insulin's glucose-lowering effects.

debridement the surgical removal of dead, damaged, or contaminated tissue resulting from burns or wounds; helps to prevent infection and hasten healing.

deficient in regard to nutrient intake, the amount below which almost all healthy people can be expected, over time, to experience deficiency symptoms.

dehydration the loss of water from the body that occurs when water output exceeds water input. The symptoms progress rapidly from thirst, to weakness, to exhaustion and delirium and end in death if not corrected.

denaturation (dee-nay-cher-AY-shun) the change in a protein's shape brought about by heat, acid, or other agents. Past a certain point, denaturation is irreversible.

dental caries the gradual decay and disintegration of a tooth.

dermatitis herpetiformis (DERM-ah-TYE-tis HER-peh-tih-FOR-mis) a gluten-sensitive disorder characterized by a severe skin rash. Gastrointestinal symptoms may be mild or absent.

diabetes (DYE-ah-BEE-teez) **mellitus** a group of metabolic disorders characterized by hyperglycemia and disordered insulin metabolism.

diabetic coma a coma that occurs in uncontrolled diabetes; may be due to diabetic ketoacidosis, the hyperosmolar hyperglycemic state, or excessive doses of insulin or certain antidiabetic drugs.

dialysate (dye-AL-ih-sate) the solution used in dialysis to draw wastes and fluids from the blood.

dialysis (dye-AH-lih-sis) a procedure for removing wastes and excess fluid from the blood after the kidneys have stopped functioning. The two main types are hemodialysis and peritoneal dialysis.

dialyzer (DYE-ah-LYE-zer) a machine used for hemodialysis; also called an artificial kidney.

diet history a comprehensive record of a person's food intake and dietary practices.

diet manual a resource that specifies the foods allowed and restricted on modified diets and provides sample menus.

diet orders specific instructions concerning dietary management; also called *diet prescriptions.*

diet progression a change in diet as a patient's tolerances permit.

dietary fibers a general term denoting in plant foods the polysaccharides cellulose, hemicellulose, pectins, gums, and mucilages, as well as the nonpolysaccharide lignins, that are not digested by human digestive enzymes, although some are digested by GI bacteria.

dietary folate equivalents (**DFE**) the amount of folate available to the body from naturally occurring sources, fortified foods, and supplements, accounting for differences in bioavailability from each source.

Dietary Reference Intakes (**DRI**) a set of values for the dietary nutrient intakes of healthy people in the United States and Canada. These values are used for planning and assessing diets.

differentiation the development of specific functions different from those of the original.

digestion the process by which complex food particles are broken down to smaller absorbable particles.

digestive system all the organs and glands associated with the ingestion and digestion of food.

dipeptide two amino acids bonded together.

disaccharides (dye-SACK-uh-rides) pairs of sugar units bonded together.

discretionary kcalorie allowance the kcalories remaining in a person's energy allowance after consuming enough nutrient dense foods to meet all nutrient needs for a day.

disease-specific formulas enteral formulas designed to meet the nutrient needs of patients with specific illnesses.

diuresis (DYE-uh-REE-sis) increased urine production.

diuretics (dye-yoo-RET-ics) medications that promote the excretion of water through the kidneys.

diverticulitis (DYE-ver-tic-you-LYE-tis) an inflammation or infection that involves diverticula.

diverticulosis (DYE-ver-tic-you-LOH-sis) a disorder characterized by the presence of small outpockets (called diverticula) in the intestinal wall, which often develop by middle age.

dumping syndrome symptoms that result from the rapid emptying of an osmotic load from the stomach into the small intestine. Early symptoms include nausea, abdominal cramps, weakness, and diarrhea; later symptoms are those of hypoglycemia.

duodenal ulcers peptic ulcers that develop in the duodenum. Approximately 85% of ulcers occur in the duodenum.

dysentery (DIS-en-terry) an infection of the gastrointestinal tract caused by an amoeba or bacterium that gives rise to severe diarrhea.

dyspepsia a feeling of pain, bloating, or discomfort in the upper abdominal area, often called "indigestion"; a symptom of illness rather than a disease itself.

dysphagia difficulty swallowing.

dyspnea (DISP-nee-ah) shortness of breath.

eclampsia a severe complication during pregnancy in which convulsions occur.

ecstasy a street or slang term used for the drug methylene-dioxymethamphetamine (MDMA). Ecstasy is chemically similar to amphetamine (speed) and mescaline (a hallucinogen).

edema (eh-DEEM-uh) the swelling of body tissue caused by leakage of fluid from the blood vessels and accumulation of the fluid in the interstitial spaces.

eicosanoids (eye-KO-sah-noyds) 20-carbon molecules derived from dietary fatty acids that help to regulate blood pressure, blood clotting, and other body functions.

electrolyte a salt that dissolves in water and dissociates into charged particles called *ions*.

electrolyte solutions solutions that can conduct electricity.

embryo (EM-bree-oh) the developing infant from two to eight weeks after conception.

emphysema (EM-fih-ZEE-mah) disease characterized by progressive damage to the alveoli (air sacs) in the lungs; causes difficulty with breathing.

emulsifier a substance that mixes with both fat and water and that disperses the fat in the water, forming an emulsion.

end-stage renal disease (**ESRD**) an advanced stage of chronic renal failure in which dialysis or a kidney transplant is necessary to sustain life.

energy density a measure of the energy a food provides relative to the amount of food (kcalories per gram).

energy-yielding nutrients the nutrients that break down to yield energy the body can use. The three energy-yielding nutrients are carbohydrate, protein, and fat.

enrichment the addition to a food of nutrients to meet a specified standard. In the case of refined bread or cereal, five nutrients have been added: thiamin, riboflavin, niacin, and folate in amounts approximately equivalent to, or higher than, those originally present and iron in amounts to alleviate the prevalence of iron-deficiency anemia.

enteral (EN-ter-al) **nutrition** the provision of nutrients using the GI tract, including the use of tube feedings and oral diets.

enteric-coated refers to medications or enzyme preparations that can withstand gastric acidity and dissolve only at a higher pH.

environmental tobacco smoke (**ETS**) the combination of exhaled smoke (mainstream smoke) and smoke from lighted cigarettes, pipes, or cigars (sidestream smoke) that enters the air and may be inhaled by other people.

enzymes protein catalysts. A *catalyst* is a compound that facilitates chemical reactions without itself being changed in the process.

EPA, DHA omega-3 fatty acids made from linolenic acid. The full name for EPA is *eicosapentaenoic* (EYE-cosa-PENTA-ee-NO-ick) *acid*. The full name for DHA is *docosahexaenoic* (DOE-cosa-HEXA-ee-NO-ick) *acid*.

epinephrine one of the stress hormones secreted whenever emergency action is needed; prescribed therapeutically to relax the bronchioles during allergy or asthma attacks.

epithelial (ep-i-THEE-lee-ul) **cells** cells on the surface of the skin and mucous membranes.

epithelial tissue tissue composing the layers of the body that serve as selective barriers between the body's interior and the environment (examples are the cornea, the skin, the respiratory lining, and the lining of the digestive tract).

erythrocyte (er-RITH-ro-cite) **hemolysis** (he-MOLL-uh-sis) rupture of the red blood cells, caused by vitamin E deficiency.

erythrocyte protoporphyrin (PRO-toe-PORE-fe-rin) a precursor to hemoglobin

erythropoietin (eh-RITH-ro-POY-eh-tin) a hormone made by the kidneys that stimulates red blood cell production.

esophageal dysphagia an inability to move food through the esophagus; usually caused by an obstruction or a motility disorder.

essential amino acids amino acids that the body cannot synthesize in amounts sufficient to meet physiological need; also called *indispensable amino acids*. Nine amino acids are known to be essential for human adults.

essential fatty acids fatty acids that the body requires but cannot make in amounts sufficient to meet its physiological needs.

essential nutrients nutrients a person must obtain from food because the body cannot make them for itself in sufficient quantities to meet physiological needs.

Estimated Average Requirements (EAR) the average daily nutrient intake levels estimated to meet the requirements of half of the healthy individuals in a given age and gender group; used in nutrition research and policymaking and the basis on which RDA values are set.

Estimated Energy Requirement (EER) the dietary energy intake level that is predicted to maintain energy-yielding balance in a healthy adult of a defined age, gender, weight, and physical activity level consistent with good health.

ethnic diets foodways and cuisines typical of national origins, races, cultural heritages, or geographic locations.

expected outcomes patient-oriented goals that are derived from nursing diagnoses.

extracellular fluid fluid residing outside the cells; includes the fluid between the cells (interstitial fluid), plasma, and the water of structures such as the skin and bones. Extracellular fluid accounts for about one-third of the body's water.

fat replacers ingredients that replace some or all of the functions of fat in foods and may or may not provide energy.

fats lipids that are solid at room temperature.

fatty acids organic compounds composed of a chain of carbon atoms with hydrogens attached and an acid group at one end.

fatty liver an accumulation of fat in the liver.

fertility the capacity of a woman to produce a normal ovum periodically and of a man to produce normal sperm; the ability to reproduce.

fetal alcohol spectrum disorders (FASD) a spectrum of physical, behavioral, and cognitive disabilities caused by prenatal alcohol exposure.

fetal alcohol syndrome (FAS) the cluster of symptoms seen in an infant or child whose mother consumed excessive alcohol during her pregnancy. FAS includes, but is not limited to, brain damage, growth retardation, mental retardation, and facial abnormalities.

fetus (FEET-us) the developing infant from eight weeks after conception until its birth.

fistulas (FIST-you-luz) abnormal passages between body tissues; may lead from one hollow organ to another or to an organ surface.

flatulence the condition of having excessive intestinal gas, which causes abdominal discomfort.

fluid and electrolyte balance maintenance of the necessary amounts and types of fluid and minerals in each compartment of the body fluids.

fluorapatite (floor-APP-uh-tite) the stabilized form of bone and tooth crystal, in which fluoride has replaced the hydroxy portion of hydroxyapatite.

fluorosis (floor-OH-sis) mottling of the tooth enamel from ingestion of too much fluoride during tooth development.

follicle (FOLL-i-cul) a group of cells in the skin from which a hair grows.

food allergies adverse reactions to foods that involve an immune response; also called *food-hypersensitivity reactions*.

food aversions strong desires to avoid particular foods.

food cravings deep longings for particular foods.

food frequency questionnaire a survey of foods routinely consumed. Some questionnaires ask about the types of food eaten and yield only qualitative information, whereas others include questions about portions consumed and yield semiquantitative data as well.

food group plan a diet-planning tool that sorts foods into groups based on nutrient content and then specifies that people should eat certain amounts of food from each group.

food intolerance an adverse response to a food or food additive that does not involve the immune system.

food record a detailed log of food eaten during a specified time period, usually several days. A food record may also include information regarding disease symptoms, physical activity, and medication use; also called a *food diary*.

foodborne illness illness transmitted to human beings through food and water, caused either by an infectious agent (foodborne infection) or a poisonous substance (food intoxication); commonly known as *food poisoning*.

fortification the addition to a food of nutrients that were either not originally present or present in insignificant amounts.

free radicals highly reactive chemical forms that can cause destructive changes in nearby compounds, sometimes setting up a chain reaction.

French units units of measure used to indicate the size of a feeding tube's outer diameter. One French unit equals 1/3 millimeter.

fructose a monosaccharide; sometimes known as *fruit sugar*. It is abundant in fruits, honey, and saps.

functional foods foods that may provide health benefits beyond their nutrient contributions. Functional foods may include whole foods, fortified foods, and modified foods.

galactose a monosaccharide; part of the disaccharide lactose.

gangrene death of tissue due to a deficient blood supply and/or infection.

gastrectomy (gah-STREK-ta-mee) the surgical removal of part of the stomach (partial gastrectomy) or the entire stomach (total gastrectomy).

gastric banding a surgical means of producing weight loss by restricting stomach size with a constricting band; used in people whose severe obesity brings extreme health risks.

gastric bypass surgery that restricts stomach size and reroutes food from the stomach to the lower part of the small intestine; creates a chronic, lifelong state of malabsorption by preventing normal digestion and absorption of nutrients.

gastric decompression the removal of the stomach contents (including swallowed saliva, stomach secretions, and gas) of patients who have motility disorders or obstructions that prevent stomach emptying.

gastric residual volume the volume of formula remaining in the stomach from a previous feeding.

gastric ulcers peptic ulcers that develop in stomach tissue. Approximately 15% of ulcers occur in the stomach.

gastritis inflammation of stomach tissue.

gastrointestinal motility: spontaneous motion in the digestive tract accomplished by involuntary muscular contractions.

gastrointestinal (GI) tract: the digestive tract. The principal organs are the stomach and intestines.

gastroparesis (GAS-troe-pah-REE-sis) delayed stomach emptying.

gatekeeper with respect to nutrition, a key person who controls other people's access to foods and thereby exerts a profound impact on their nutrition. Examples are the spouse who buys and cooks the food, the parent who feeds the children, and the caregiver in a day-care center.

gestation the period of about 40 weeks (three trimesters) from conception to birth; the term of a pregnancy.

gestational diabetes the presence of abnormal glucose tolerance during pregnancy.

ghrelin (GRELL-in) a hormone produced primarily by the stomach cells. It signals the hypothalamus of the brain to stimulate appetite and food intake.

glands single cells or groups of cells that secrete materials for special uses in the body. Glands may be *exocrine glands,* secreting their materials "out" (into the digestive tract or onto the surface of the skin), or *endocrine glands,* secreting their materials "in" (into the blood).

glomerular filtration rate (GFR) the rate at which filtrate is formed within the kidneys, normally approximately 125 mL/min.

glomerulus (gloh-MEHR-yoo-lus) a tuft of capillaries within the nephron that filters water and solutes from blood as urine production begins (plural *glomeruli*).

glucagon (GLOO-ka-gon) a hormone that is secreted by special cells in the pancreas in response to low blood glucose concentration and elicits release of glucose from storage.

glucose a monosaccharide, the sugar common to all disaccharides and polysaccharides; also called *blood sugar* or *dextrose.*

glycated hemoglobin (HbA$_{1c}$) hemoglobin molecules to which glucose is attached. The percentage of such molecules is used to evaluate long-term glycemic control. Also called *glycosylated hemoglobin.*

glycemic (gly-SEE-mic) pertaining to blood glucose.

glycemic effect a measure of the extent to which a food raises the blood glucose concentration and elicits an insulin response, as compared with pure glucose.

glycerol (GLISS-er-ol) an organic compound, three carbons long, that can form the backbone of triglycerides and phospholipids.

glycogen (GLY-co-gen) a polysaccharide composed of glucose, made and stored by liver and muscle tissues of human beings and animals as a storage form of glucose.

glycosuria (GLY-koh-SOOR-ee-ah) an abnormal amount of glucose in urine.

goiter (GOY-ter) an enlargement of the thyroid gland due to an iodine deficiency, malfunction of the gland, or overconsumption of a thyroid antagonist. Goiter caused by iodine deficiency is *simple goiter.*

gout (GOWT) a metabolic disorder characterized by elevated uric acid in the blood and urine and the deposition of uric acid in and around the joints, causing acute joint inflammation.

Harris-Benedict equation an equation used to estimate basal energy expenditure.

health a range of states with physical, mental, emotional, spiritual, and social components. At a minimum, health means freedom from physical disease, mental disturbances, emotional distress, spiritual discontent, social maladjustment, and other negative states. At a maximum, health means "wellness."

health claims statements that characterize the relationship between a nutrient or other substance in food and a disease or health-related condition.

Helicobacter pylori a type of bacterium that colonizes gastric mucosa; a major cause of gastritis and peptic ulcer disease.

helper T cells lymphocytes that have a specific protein called CD4 on their surfaces and therefore are also known as CD4$^+$ T cells; the cells most affected in HIV infection.

hematocrit (hee-MAT-oh-crit) measurement of the volume of the red blood cells packed by centrifuge in a given volume of blood.

hematuria (HE-mah-TOO-ree-ah) blood in the urine.

hemochromatosis (heem-oh-crome-a-TOE-sis) iron overload characterized by deposits of iron-containing pigment in many tissues, with tissue damage. Hemochromatosis is usually caused by a hereditary defect in iron metabolism.

hemodialysis (HE-moh-dye-AL-ih-sis) removal of fluids and wastes from blood by passing the blood through a dialyzer.

hemoglobin the oxygen-carrying protein of the red blood cells.

hemorrhage severe bleeding; a copious flow of blood from blood vessels.

hemorrhagic (hem-oh-RAJ-ik) **disease** the vitamin K–deficiency disease in which blood fails to clot.

hemorrhagic strokes strokes that result from bleeding within the brain, which destroys or compresses brain tissue.

hepatic coma loss of consciousness resulting from severe liver disease.

hepatic encephalopathy (en-SEF-ah-LOP-ah-thee) a condition in advanced liver disease that is characterized by altered neurological functioning, including personality changes, reduced mental abilities, and disturbances in motor function.

hepatitis (hep-ah-TYE-tis) inflammation of the liver.

hepatomegaly (HEP-ah-toe-MEG-ah-lee) enlargement of the liver.

herpes simplex virus a common virus that can cause blisterlike lesions on the lips and in the mouth.

hiatal hernia a condition in which the upper portion of the stomach protrudes above the diaphragm. Most cases are asymptomatic.

high-density lipoproteins (**HDL**) the type of lipoproteins that transport cholesterol back to the liver from peripheral cells; composed primarily of protein.

high-fructose corn syrup (**HFCS**) the predominant sweetener used in processed foods today. HFCS is mostly fructose; glucose makes up the balance.

high-quality proteins dietary proteins containing all the essential amino acids in relatively the same amounts that human beings require. They may also contain nonessential amino acids.

histamine-2 receptor blockers a class of drugs that suppresses acid secretion by inhibiting receptors on acid-producing cells (commonly called *H2 blockers*).

HIV (human immunodeficiency virus) the virus that causes acquired immune deficiency syndrome (AIDS). HIV destroys immune cells and progressively impedes the body's ability to fight infections and certain cancers.

HIV-lipodystrophy (LIP-oh-DIS-tro-fee) **syndrome** a collection of abnormalities in fat and glucose metabolism that result from drug treatments for HIV; includes body fat redistribution, abnormal lipid levels, and insulin resistance.

homeostasis (HOME-ee-oh-STAY-sis) the maintenance of constant internal conditions (such as chemistry, temperature, and blood pressure) by the body's control system.

hormones chemical messengers. Hormones are secreted by a variety of glands in the body in response to altered conditions. Each travels to one or more target tissues or organs and elicits specific responses to restore normal conditions.

hunger the physiological need to eat, experienced as a drive to obtain food; an unpleasant sensation that demands relief.

hydrogenation (high-dro-gen-AY-shun) a chemical process by which hydrogens are added to monounsaturated or polyunsaturated fats to reduce the number of double bonds, making the fats more saturated (solid) and more resistant to oxidation.

hydrolyzed formulas enteral formulas that contain macronutrients that have been partially or fully hydrolyzed; also called *monomeric,* or *elemental formulas.*

hyperactivity inattentive and impulsive behavior that is more frequent and severe than is typical of others of a similar age; professionally called *attention deficit/ hyperactivity disorder* (*ADHD*).

hypercalcemia (HIGH-per-kal-SEE-me-ah) elevated serum calcium levels.

hypercalciuria (HIGH-per-kal-see-YOO-ree-ah) elevated urinary calcium levels.

hypercapnia (high-per-CAP-nee-ah) excessive carbon dioxide in the blood.

hyperglycemia elevated blood glucose concentrations.

hyperkalemia (HIGH-per-ka-LEE-me-ah) elevated serum potassium levels.

hypermetabolism a higher-than-normal metabolic rate.

hyperosmolar hyperglycemic state extreme hyperglycemia that is associated with hyperosmolar blood, dehydration, and altered mental status; formerly called *hyperglycemic hyperosmolar nonketotic coma.*

hyperoxaluria (HIGH-per-ox-ah-LOO-ree-ah) elevated urinary oxalate levels.

hyperphosphatemia (HIGH-per-fos-fa-TEE-me-ah) elevated serum phosphate levels.

hypertension high blood pressure.

hypertonic formula a formula with an osmolality greater than that of blood serum.

hypoallergenic formulas clinically tested infant formulas that do not provoke reactions in 90% of infants or children with confirmed cow's milk allergy.

hypochlorhydria (HIGH-poe-clor-HIGH-dree-ah) a reduction in gastric acid secretion.

hypoglycemia abnormally low concentrations of blood glucose.

hypokalemia (HIGH-po-ka-LEE-me-ah) low serum potassium levels.

hypothalamus (high-poh-THALL-uh-mus) a part of the brain that helps regulate many body balances, including fluid balance.

hypoxemia (high-pox-SEE-me-ah) a low level of oxygen in the blood.

hypoxia (high-POX-see-ah) a low amount of oxygen in body tissues.

ileostomy (ill-ee-OS-toe-me) a surgical procedure that creates a stoma using the ileum.

implantation the stage of development in which the zygote embeds itself in the wall of the uterus and begins to develop; occurs during the first two weeks after conception.

inflammatory response the metabolic responses of the immune system to infection or injury.

inorganic not containing carbon or pertaining to living things.

insoluble fibers the tough, fibrous structures of fruits, vegetables, and grains; indigestible food components that do not dissolve in water.

insulin a hormone secreted by the pancreas in response to high blood glucose. It promotes cellular glucose uptake for use or storage.

insulin resistance reduced sensitivity to insulin in muscle, adipose, and liver cells.

intermittent feedings delivery of about 250 to 400 mL of formula over 20 to 40 minutes.

intestinal adaptation after resection, the process of intestinal recovery that leads to improved absorptive capacity.

intestinal flora the bacterial inhabitants of the GI tract.

intra-abdominal fat fat stored within the abdominal cavity in association with the internal abdominal organs, as opposed to fat stored directly under the skin (subcutaneous fat); also called *visceral fat.*

intractable vomiting vomiting that is not easily managed or controlled.

intradialytic parenteral nutrition the infusion of nutrients during hemodialysis, often providing amino acids, dextrose, lipids, and some trace minerals.

intravenous feedings the provision of nutrients through a vein, bypassing the intestine. The intravenous provision of nutrients is called *parenteral nutrition.*

intrinsic factor inside the system. Anemia that reflects a vitamin B$_{12}$ deficiency caused by lack of intrinsic factor is known as *pernicious anemia.*

iron deficiency having depleted iron stores.

iron overload toxicity from excess iron.

iron-deficiency anemia a blood iron deficiency characterized by small, pale red blood cells; also called microcytic hypochromic anemia.

irritable bowel syndrome an intestinal disorder of unknown cause that affects the functioning of the lower bowel. Symptoms include abdominal pain, flatulence, diarrhea, and constipation.

ischemic strokes strokes that result from the obstruction of blood flow to brain tissue.

isotonic formula a formula with an osmolality similar to that of blood serum (300 mOsm/kg).

jaundice (JAWN-dis) yellow discoloration of the skin and mucous membranes due to an accumulation of bilirubin, a breakdown product of hemoglobin that exits the body via bile secretions.

Joint Commission on Accreditation of Healthcare Organizations (JCAHO) a nonprofit organization that sets standards for health care performance and safety and confers accreditation on health care organizations that meet those standards.

Kaposi's (cap-OH-seez) **sarcoma** a common cancer in HIV-infected persons that is characterized by lesions on the skin, in the lungs, and in the GI tract.

kcalorie (energy) control management of food energy intake.

kcalorie counts the determination of food energy (and often, protein) consumed by patients during one or more days.

keratin (KERR-uh-tin) a water-insoluble protein; the normal protein of hair and nails. Keratin-producing cells may replace mucus-producing cells in vitamin A deficiency.

ketoacidosis (KEY-to-ah-sih-DOE-sis) lowering of pH in blood and tissues due to excessive production of ketone bodies.

ketones (KEY-tones) acidic, fat-related compounds formed from the incomplete breakdown of fat when carbohydrate is not available; technically known as *ketone bodies.*

kidney stones crystalline masses that form in the urinary tract; also called *renal calculi* and *nephrolithiasis.*

kwashiorkor (kwash-ee-OR-core or kwash-ee-or-CORE) a severe form of PEM that occurs more frequently after 18 months of age. Kwashiorkor is characterized by failure to

grow and develop, changes in the pigmentation of the hair and skin, edema, and fatty liver.

lactadherin (lack-tad-HAIR-in) a protein in breast milk that attacks diarrhea-causing viruses.

lactation production and secretion of breast milk for the purpose of nourishing an infant.

lactoferrin (lack-toe-FERR-in) a protein in breast milk that binds iron and keeps it from supporting the growth of the infant's intestinal bacteria.

lactose a disaccharide composed of glucose and galactose; commonly known as milk sugar.

laparoscopic weight-loss surgery a procedure in which surgeons gain access to the abdomen via several small incisions. A tiny video camera is inserted through one of the incisions and surgical instruments through the others.

latent the period in the course of a disease when the condition is present but the symptoms have not begun to appear.

lecithins one type of phospholipid.

legumes (lay-GYOOMS, LEG-yooms) plants of the bean and pea family, with seeds that are rich in protein compared with other plant-derived foods.

leptin a hormone produced by fat cells under the direction of the *ob* gene. It decreases appetite and increases energy expenditure.

life expectancy the average number of years lived by people in a given society.

life span the maximum number of years of life attainable by a member of a species.

limiting amino acid an essential amino acid that is present in dietary protein in the shortest supply relative to the amount needed for protein synthesis in the body.

linoleic acid, linolenic acid polyunsaturated fatty acids that are essential for human beings.

lipids a family of compounds that includes triglycerides (fats and oils), phospholipids, and sterols.

lipomas (lih-POE-muz) benign tumors composed of fatty tissue.

lipoprotein a cluster of lipids associated with proteins that serves as a transport vehicle for lipids in the lymph and blood.

lipoprotein lipase (LPL) an enzyme mounted on the surface of fat cells (and other cells). It hydrolyzes triglycerides in the blood into fatty acids and glycerol for absorption into the cells. There they are metabolized or reassembled for storage.

listeriosis a serious foodborne infection that can cause severe brain infection or death in a fetus or newborn; caused by the bacterium *Listeria monocytogenes,* which is found in soil and water.

longevity long duration of life.

low birthweight (LBW) a birthweight less than 5^{1}/$_{2}$ lb (2500 g); indicates probable poor health in the newborn and poor nutrition status of the mother during pregnancy. Normal birthweight for a full-term infant is 6^{1}/$_{2}$ to 9 lb (about 3000 to 4000 g).

low-density lipoproteins (LDL) the type of lipoproteins derived from VLDL as cells remove triglycerides from them.

LDL carry cholesterol and triglycerides from the liver to the cells of the body and are composed primarily of cholesterol.

lymph (LIMF) the body fluid found in lymphatic vessels. Lymph consists of all the constituents of blood except red blood cells.

lymphatic system a loosely organized system of vessels and ducts that conveys the lipid products of digestion toward the heart.

macrosomia (MAK-roh-SOH-mee-ah) the condition of having an abnormally large body. In infants, refers to birthweights of 4000 g (8 lb 13 oz) and above.

macular (MACK-you-lar) **degeneration** deterioration of the macular area of the eye that can lead to loss of central vision and eventual blindness.

malignant (ma-LIG-nant) describes a cancerous cell or tumor, which can injure healthy tissue and spread cancer to other regions of the body.

malnutrition any condition caused by deficient or excess energy or nutrient intake or by an imbalance of nutrients.

maltose a disaccharide composed of two glucose units; sometimes known as *malt sugar.*

marasmus (ma-RAZZ-mus) the most common form of severe PEM before one year of age. Marasmus is characterized by generalized muscle wasting associated with extreme deprivation, or impaired absorption, of energy, protein, vitamins, and minerals.

mast cells cells within connective tissue (close to blood vessels) that produce and release histamine.

medical nutrition therapy nutrition care provided by a registered dietitian; includes diagnosing nutrition problems, prescribing diet plans, and providing dietary counseling.

medium-chain triglycerides (MCT) triglycerides that contain fatty acids that are 8 to 10 carbons in length. MCT do not require digestion and can be absorbed in the absence of lipase or bile.

metabolic stress a disruption in the body's chemical environment due to the effects of disease or injury. Metabolic stress is characterized by changes in metabolic rate, heart rate, blood pressure, hormonal status, and nutrient metabolism.

metastasize (meh-TAS-tah-size) the spread of cancer cells from one part of the body to another.

microalbuminuria the presence of small amounts of albumin (protein) in the urine, a sign of diabetic nephropathy.

microvilli (MY-cro-VILL-ee or MY-cro-VILL-eye) tiny, hair-like projections on each cell of every villus that can trap nutrient particles and transport them into the cells. The singular form is *microvillus.*

moderation providing enough, but not too much, of a substance.

modified diet a diet that is adjusted in consistency, in energy or nutrient content, or by the inclusion or elimination of certain foods; sometimes called a *therapeutic diet.*

modular formulas enteral formulas that contain only one or two macronutrients; used to enhance other formulas, meet specific nutrient needs, or create individualized formulas for people with unique needs.

molybdenum (mo-LIB-duh-num) a trace element.

monosaccharides (mon-oh-SACK-uh-rides) single sugar units.

monounsaturated fatty acid (MUFA) a fatty acid that has one point of unsaturation; for example, the oleic acid found in olive oil.

mucous membrane membrane composed of mucus-secreting cells that lines the surfaces of body tissues. *Mucus* is the smooth, slippery substance secreted by these cells.

muscular dystrophy (DIS-tro-fee) a hereditary disease in which the muscles gradually weaken, with the most debilitating effects occurring in the lungs.

myoglobin the oxygen-carrying protein of the muscle cells.

naturally occurring sugars sugars that are not added to a food but are present as its original constituents, such as the sugars of fruit or milk.

nephron (NEF-ron) the functional unit of the kidneys, consisting of a glomerulus and tubules.

nephrotic (neh-FROT-ik) **syndrome** a kidney disorder characterized by urinary protein losses exceeding 3.5 g per day. Accompanying symptoms often include low serum albumin, elevated blood lipids, and edema.

neural tube the embryonic tissue that later forms the brain and spinal cord.

neural tube defects (NTD) malformations of the brain, spinal cord, or both during embryonic development.

neurons nerve cells; the structural and functional units of the nervous system. Neurons initiate and conduct nerve transmissions.

niacin equivalents (NE) the amount of niacin present in food, including the niacin that can theoretically be made from tryptophan, its precursor, present in the food.

night blindness the slow recovery of vision after exposure to flashes of bright light at night; an early symptom of vitamin A deficiency.

nitrogen balance the amount of nitrogen consumed (N in) as compared with the amount of nitrogen excreted (N out) in a given period of time.

nonessential amino acids amino acids that the body can synthesize.

nursing bottle tooth decay extensive tooth decay due to prolonged tooth contact with formula, milk, fruit juice, or other carbohydrate-rich liquid offered to an infant in a bottle.

nursing diagnoses clinical judgments about actual or potential health problems that provide the basis for selecting appropriate nursing interventions.

nutrient claims statements that characterize the quantity of a nutrient in a food.

nutrient density a measure of the nutrients a food provides relative to the energy it provides. The more nutrients and the fewer kcalories, the higher the nutrient density.

nutrients substances obtained from food and used in the body to provide energy and structural materials and to serve as regulating agents to promote growth, maintenance, and repair.

nutrition the science of foods and the nutrients and other substances they contain, and of their ingestion, digestion, absorption, transport, metabolism, interaction, storage, and excretion. A broader definition includes the study of the environment and of human behavior as it relates to these processes.

nutrition care plans strategies for meeting an individual's nutritional needs.

nutrition care process a problem-solving method that dietetics professionals use to evaluate nutrition-related problems and provide safe and effective nutrition care.

nutrition screening an examination process that identifies patients who require intervention for existing or potential nutritional problems.

nutrition support the delivery of formulated nutrients by feeding tube or intravenous infusion.

nutrition support teams health care professionals responsible for the provision of nutrients by tube feeding or intravenous infusion.

nutritive sweeteners sweeteners that yield energy, including both the sugars and the sugar alcohols.

obesity overfatness with adverse health effects, as determined by reliable measures and interpreted with good medical judgment. Obesity is officially defined as a body mass index (BMI) of 30 or higher.

oils lipids that are liquid at room temperature.

olestra a synthetic fat made from sucrose and fatty acids that provides zero kcalories per gram; also known as *sucrose polyester.*

oliguria (OL-lih-GOO-ree-ah) an abnormally low amount of urine, often less than 400 mL per day.

omega-3 fatty acids polyunsaturated fatty acids in which the endmost double bond is three carbons back from the end of the carbon chain; *linolenic acid* is an example.

omega-6 fatty acid a polyunsaturated fatty acid with its endmost double bond six carbons back from the end of its carbon chain; *linoleic acid* is an example.

open feeding systems delivery systems that require formula to be transferred from the original packaging to feeding containers before being administered through feeding tubes.

opportunistic infections infections from microorganisms that normally do not cause disease in healthy people but are damaging to persons with compromised immune function.

oral glucose tolerance test a test that evaluates a person's ability to tolerate a glucose load. A common protocol for diabetes diagnosis is ingestion of a 75 g glucose load followed by measurement of plasma glucose after a two-hour interval.

organic carbon containing. The four organic nutrients are carbohydrate, fat, protein, and vitamins.

oropharyngeal dysphagia (or-oh-fah-ren-JEE-al diss-FAY-jee-ah) an inability to transfer food from the mouth and pharynx to the esophagus; usually caused by a neurological or muscular disorder.

osmolality (OZ-moe-LAL-ih-tee) the osmotic property of a solution, based on its concentrations of molecules and ionic particles. Osmolality is expressed as milliosmoles (mOsm) per kilogram.

osmolarity the concentration of osmotically active particles in a solution, expressed as milliosmoles per liter (mOsm/L).

osteoarthritis a painful, chronic disease of the joints that occurs when the cushioning cartilage in a joint breaks down; joint structure is usually altered, with loss of function; also called *degenerative arthritis.*

osteomalacia (os-tee-oh-mal-AY-shuh) a bone disease characterized by softening of the bones. Symptoms include bending of the spine and bowing of the legs. The disease occurs most often in adults with renal failure or malabsorption disorders.

osteoporosis (os-tee-oh-por-OH-sis) literally, porous bones; reduced density of the bones, also known as adult bone loss.

overnutrition overconsumption of food energy or nutrients sufficient to cause disease or increased susceptibility to disease; a form of malnutrition.

overt out in the open, full-blown.

overweight overfatness of a moderate degree; defined as a body mass index (BMI) of 25.0 through 29.9.

ovum (OH-vum) the female reproductive cell, capable of developing into a new organism upon fertilization; commonly referred to as an egg.

oxidation (OKS-ee-day-shun) the process of a substance combining with oxygen.

paracentesis (pahr-ah-sen-TEE-sis) a surgical puncture of a body cavity with an aspirator to draw out excess fluid.

parenteral (par-EN-ter-al) **nutrition** the intravenous provision of nutrients that bypasses the GI tract.

pellagra (pell-AY-gra) the niacin-deficiency disease. Symptoms include the "4 Ds" diarrhea, dermatitis, dementia, and, ultimately, death.

peptic ulcers ulcers in the gastrointestinal mucosa resulting from exposure to gastric secretions; may develop in the esophagus, stomach, or duodenum.

peripheral parenteral nutrition (PPN) a type of nutrition support in which intravenous feedings are delivered into peripheral veins.

peripheral resistance the resistance to pumped blood in the small arterial branches (arterioles) that carry blood to tissues.

peripheral veins small-diameter veins that carry blood from the arms and legs.

peristalsis (peri-STALL-sis) successive waves of involuntary muscular contractions passing along the walls of the GI tract that push the contents along.

peritoneal (PEH-rih-toe-NEE-al) **dialysis** removal of fluids and wastes by using the peritoneal membrane to filter blood.

peritoneovenous (PEH-rih-toe-nee-oh-VEE-nus) **shunt** a surgical passage created between the peritoneum and the jugular vein to divert fluid and relieve ascites. The peritoneum is the membrane that surrounds the abdominal cavity.

pH the concentration of hydrogen ions. The lower the pH, the stronger the acid. Thus pH 2 is a strong acid; pH 6 is a weak acid; pH 7 is neutral; and a pH above 7 is alkaline.

phagocytes (FAG-oh-sites) white blood cells (neutrophils and macrophages) that have the ability to engulf and destroy antigens.

phospholipids one of the three main classes of lipids; compounds similar to triglycerides but with choline (or another compound) and a phosphorus-containing acid in place of one of the fatty acids.

physiological age a person's age as estimated from her or his body's health and probable life expectancy.

phytates nonnutrient components of grains, legumes, and seeds. Phytates can bind minerals such as iron, zinc, calcium, and magnesium in insoluble complexes in the intestine, and the body excretes them unused.

phytochemicals: nonnutrient compounds in plant-derived foods that have biological activity in the body.

pica (PIE-ka) a craving for nonfood substances; also known as *geophagia* (jee-oh-FAY-jee-uh) when referring to clay-eating behavior.

pituitary (pit-TOO-ih-tary) **gland** in the brain, the "king gland" that regulates the operation of many other glands.

placenta (pla-SEN-tuh) an organ that develops inside the uterus early in pregnancy, in which maternal and fetal blood circulate in close proximity and exchange materials.

polydipsia (pah-lee-DIP-see-ah) excessive thirst.

polypeptide ten or more amino acids bonded together. An intermediate strand of between four and ten amino acids is an oligopeptide.

polyphagia (pah-lee-FAY-jee-ah) excessive appetite or eating.

polysaccharides long chains of monosaccharide units arranged as starch, glycogen, or fiber.

polyunsaturated fatty acids (PUFA) fatty acids with two or more points of unsaturation. For example, linoleic acid has two such points, and linolenic acid has three.

polyuria (pah-lee-YOOR-ee-ah) excessive urine secretion.

portal hypertension elevated blood pressure in the portal vein; may be caused by obstructed blood flow through the liver.

potassium iodide a medication approved by the FDA as safe and effective for the prevention of thyroid cancer caused by radioactive iodine known to be released during radiation emergencies.

precursors compounds that can be converted into other compounds; with regard to vitamins, compounds that can be converted into active vitamins; also known as *provitamins.*

prediabetes the condition in which blood glucose levels are higher than normal but not high enough to be diagnosed as diabetes; considered a major risk factor for future diabetes and cardiovascular diseases. Formerly called *impaired glucose tolerance.*

preeclampsia a condition characterized by hypertension, fluid retention, and protein in the urine.

preformed vitamin A vitamin A in its active form.

protein digestibility a measure of the amount of amino acids absorbed from a given protein intake.

protein turnover the continuous breakdown and synthesis of body proteins involving the recycling of amino acids.

protein-energy malnutrition (PEM) a deficiency of protein and food energy; the world's most widespread malnutrition problem, including both marasmus and kwashiorkor.

proteins compounds made from strands of amino acids composed of carbon, hydrogen, oxygen, and nitrogen atoms. Some amino acids also contain sulfur atoms.

proteinuria (PRO-teen-NEW-ree-ah) loss of protein, especially albumin, in the urine; also known as *albuminuria.*

proton-pump inhibitors a class of drugs that inhibits the enzyme that pumps hydrogen ions (protons) into the stomach. Example include omeprazole (Prilosec) and lansoprazole (Prevacid).

puberty the period in life in which a person becomes physically capable of reproduction.

purine (PYOO-reen) a product of nucleotide metabolism that degrades to uric acid.

pyloroplasty (pye-LORE-oh-PLAS-tee) surgery that enlarges the pyloric sphincter.

quality of life a person's perceived physical and mental well-being.

radiation enteritis inflammation of intestinal tissue caused by exposure to radiation.

radiation therapy the use of X-rays, gamma rays, or atomic particles to destroy cancer cells.

rancid the term used to describe fats when they have deteriorated, usually by oxidation. Rancid fats often have an "off" odor.

rebound hyperglycemia hyperglycemia that results from the release of counterregulatory hormones following nighttime hypoglycemia; also called the *Somogyi phenomenon.*

Recommended Dietary Allowances (RDA) a set of values reflecting the average daily amounts of nutrients considered adequate to meet the known nutrient needs of practically all healthy people in a particular life stage and gender group; a goal for dietary intake by individuals.

refeeding syndrome a condition that sometimes develops when a severely malnourished person is aggressively fed; characterized by electrolyte and fluid imbalances and hyperglycemia.

refined grain a product from which the bran, germ, and husk have been removed, leaving only the endosperm.

reflux esophagitis inflammation in the esophagus related to the reflux of acidic stomach contents.

renal colic the intense pain that occurs when a kidney stone passes through the ureter.

renal osteodystrophy a bone disorder that develops in patients with chronic renal failure as a consequence of increased secretion of parathyroid hormone (which stimulates calcium release from bone), reduced serum calcium, acidosis, and impaired vitamin D activation in the kidneys.

renal threshold the point at which the blood concentration of a substance exceeds the kidneys' capacity for reabsorption, causing the substance to spill into the urine.

requirement the lowest continuing intake of a nutrient that will maintain a specified criterion of adequacy.

resection the surgical removal of part of an organ or body structure.

residue material left in the intestine after digestion; includes mostly dietary fiber and undigested starches and proteins.

resistant starches starches that escape digestion and absorption in the small intestine of healthy people.

respiratory stress inadequate gas exchange between the air and blood, resulting in lower-than-normal oxygen levels and higher-than-normal carbon dioxide levels.

resting metabolic rate (RMR) a measure of the energy use of a person at rest in a comfortable setting; similar to the BMR but with less stringent criteria for recent food intake and physical activity. Consequently, the RMR is slightly higher than the BMR.

retina (RET-in-uh) the layer of light-sensitive nerve cells lining the back of the inside of the eye; consists of rods and cones.

retinol activity equivalents (RAE) a measure of vitamin A activity; the amount of retinol that the body will derive from a food containing preformed retinol or its precursor beta-carotene.

retinol-binding protein (RBP) the specific protein responsible for transporting retinol.

rheumatoid arthritis a disease of the immune system involving painful inflammation of the joints and related structures.

rickets the vitamin D–deficiency disease in children.

salts compounds composed of charged particles (ions). An example of a salt is potassium chloride (K^+Cl^-).

salt-sensitive an individual who responds to a high salt intake with an increase in blood pressure or to a low salt intake with a decrease in blood pressure.

sarcopenia (SAR-koh-PEE-nee-ah) loss of skeletal muscle mass, strength, and quality.

saturated fatty acid a fatty acid carrying the maximum possible number of hydrogen atoms (having no points of unsaturation).

scurvy the vitamin C–deficiency disease.

segmentation a periodic squeezing or partitioning of the intestine by its circular muscles that both mixes and slowly pushes the contents along.

self-monitoring of blood glucose home monitoring of blood glucose levels using a glucose meter.

senile dementia the loss of brain function beyond the normal loss of physical adeptness and memory that occurs with aging.

sepsis an acute inflammatory response caused by infection; characterized by symptoms similar to those of the systemic inflammatory response syndrome (SIRS).

set-point theory the theory that proposes that the body tends to maintain a certain weight by means of its own internal controls.

shock a dangerous physiological response to injury, bleeding, or infection resulting from an insufficient blood supply; associated with reduced blood pressure, raised heart and respiratory rates, and muscle weakness.

short-bowel syndrome a malabsorption syndrome that follows resection of the small intestine which causes insufficient absorptive capacity in the remaining intestine.

simple sugars the monosaccharides (glucose, fructose, and galactose) and the disaccharides (sucrose, lactose, and maltose).

sinusoids the small capillary-like passages that carry blood through liver tissue.

Sjögren's (SHOW-grenz) **syndrome** an autoimmune disease characterized by the destruction of secretory glands, resulting in dry mouth and dry eyes.

skinfold measure a clinical estimate of total body fatness in which the thickness of a fold of skin on the back of the arm (over the triceps muscle), below the shoulder blade (subscapular), or in other places is measured with a caliper.

soaps chemical compounds formed between positively charged minerals and fatty acids.

soluble fibers indigestible food components that readily dissolve in water and often impart gummy or gel-like characteristics to foods. An example is pectin from fruit, which is used to thicken jellies.

Special Supplemental Food Program for Women, Infants, and Children (WIC) a high-quality, cost-effective health care and nutrition services program administered by the U.S. Department of Agriculture for low-income women, infants, and children who are nutritionally at risk.

sphincter (SFINK-ter) a circular muscle surrounding, and able to close, a body opening.

spina (SPY-nah) **bifida** (BIFF-ih-dah) one of the most common types of neural tube defects; characterized by the incomplete closure of the spinal cord and its bony encasement.

standard diet a diet that includes all foods and meets the nutrient needs of healthy people; sometimes called a regular diet.

standard formulas general-purpose enteral formulas that contain mostly intact proteins and polysaccharides; also called *polymeric formulas.*

starch a plant polysaccharide composed of glucose and digestible by human beings.

steatorrhea (stee-AH-tor-REE-ah) excessive fat in the stools resulting from fat malabsorption; characterized by stools that are loose, frothy, and foul smelling due to a high fat content.

sterile free of microorganisms, such as bacteria.

steroids (STARE-oids) medications used to reduce tissue inflammation, to suppress the immune response, or to replace certain steroid hormones in people who cannot synthesize them.

sterol esters compounds, derived from plants, that belong to the sterol family of lipids and have been shown experimentally to reduce blood cholesterol when consumed in place of other fats in a low–saturated fat diet.

sterols one of the three main classes of lipids; include cholesterol, vitamin D, and the sex hormones (such as testosterone).

stoma (STOE-ma) a surgical opening made in the abdominal wall.

stress response the chemical and physical changes that occur within the body during stress.

stricture abnormal narrowing of a passageway due to inflammation, scarring, or other structural changes.

structure-function claims statements that describe how a product may affect a structure or function of the body; for example, "calcium builds strong bones." Structure-function claims do not require FDA authorization.

struvite (STROO-vite) crystals of magnesium ammonium phosphate.

sucrose a disaccharide composed of glucose and fructose; commonly known as table sugar, beet sugar, or cane sugar.

sugar alcohols sugarlike compounds. Like sugars, they are sweet to taste but yield 2 to 3 kcal per gram, slightly less than sucrose. Examples are maltitol, mannitol, sorbitol, isomalt, lactitol, and xylitol.

systemic (sih-STEM-ic) relating to the entire body.

systemic inflammatory response syndrome (SIRS) a whole-body response to acute inflammation; characterized by raised heart and respiratory rates, abnormal white blood cell counts, and altered body temperature.

tannins compounds in tea (especially black tea) and coffee that bind iron.

teratogenic (ter-AT-oh-jen-ik) causing abnormal fetal development and birth defects.

thermic effect of food an estimation of the energy required to process food (digest, absorb, transport, metabolize, and store ingested nutrients).

thrush a fungal infection of the mouth and throat, most often caused by *Candida albicans*.

tissue rejection destruction of donor tissue by the recipient's immune system, which recognizes the donor cells as foreign.

tocopherol (tuh-KOFF-er-ol) a general term for several chemically related compounds, one of which has vitamin E activity.

Tolerable Upper Intake Levels (UL) A set of values reflecting the highest average daily nutrient intake levels that are likely to pose no risk of toxicity to almost all healthy individuals in a particular life stage and gender group. As intake increases above the UL, the potential risk of adverse health effects increases.

total nutrient admixture (TNA) a parenteral solution that contains dextrose, amino acids, and lipids; also called a *3-in-1* or an *all-in-one* solution.

total parenteral nutrition (TPN) a type of nutrition support in which intravenous feedings are delivered into a central vein.

trans-**fatty acids** fatty acids in which the hydrogens next to the double bond are on opposite sides of the carbon chain.

transferrin (trans-FERR-in) the body's iron-carrying protein.

transient ischemic attacks brief ischemic episodes that cause short-term neurological symptoms, such as blurred vision, slurred speech, numbness, paralysis, or difficulty speaking.

triglycerides (try-GLISS-er-rides) one of the main classes of lipids; the chief form of fat in foods and the major storage form of fat in the body; composed of glycerol with three fatty acids attached.

trimester a period representing one-third of the term of gestation. A trimester is about 13 to 14 weeks.

tripeptide three amino acids bonded together.

tube feedings liquid formulas delivered through a tube placed in the stomach or intestine.

tubules tubelike structures of the nephron that process filtrate during urine production. The tubules are surrounded by capillaries that reabsorb substances retained by tubule cells.

tumor an abnormal tissue mass that has no physiological function; also called a *neoplasm* (NEE-oh-plazm).

type 1 diabetes the type of diabetes that accounts for 5 to 10% of diabetes cases and usually results from autoimmune destruction of pancreatic beta cells. In this type of diabetes, the pancreas produces little or no insulin.

type 2 diabetes the type of diabetes that accounts for 90 to 95% of diabetes cases and usually results from insulin resistance coupled with insufficient insulin secretion.

ulcerative colitis (ko-LYE-tis) an inflammatory bowel disease that involves the colon. Inflammation affects the mucosa and submucosa.

umbilical (um-BIL-ih-cul) **cord** the ropelike structure through which the fetus's veins and arteries reach the placenta; the route of nourishment and oxygen into the fetus and the route of waste disposal from the fetus.

undernutrition underconsumption of food energy or nutrients severe enough to cause disease or increased susceptibility to disease; a form of malnutrition.

unsaturated fatty acid a fatty acid with one or more points of unsaturation where hydrogens are missing (includes monounsaturated and polyunsaturated fatty acids).

uremia (you-REE-me-ah) abnormal accumulation of nitrogen-containing substances, especially urea, in the blood; also called *azotemia* (AZE-oh-TEE-me-ah).

uremic syndrome the cluster of symptoms associated with a GFR below 15 mL/min, including uremia, anemia, bone disease, hormonal imbalances, bleeding impairment, increased cardiovascular disease risk, and reduced immunity.

USDA Food Guide the USDA's food group plan for ensuring dietary adequacy that assigns foods to five major food groups.

uterus (YOO-ter-us) the womb, the muscular organ within which the infant develops before birth.

vagotomy (vay-GOT-oh-mee) surgery that severs the vagus nerve in order to suppress gastric acid secretion. This surgery may require a follow-up pyloroplasty procedure to allow stomach drainage.

vagus nerve the cranial nerve that regulates hydrochloric acid secretion and peristalsis. Effects elsewhere in the body include regulation of heart rate and bronchiole constriction.

varices (VAH-rih-seez) abnormally dilated blood vessels (singular *varix*).

variety (dietary) eating a wide selection of foods within and among the major food groups (the opposite of monotony).

vein a vessel that carries blood back to the heart.

very-low-density lipoproteins (**VLDL**) the type of lipoproteins made primarily by liver cells to transport lipids to various tissues in the body; composed primarily of triglycerides.

villi (VILL-ee or VILL-eye) fingerlike projections from the folds of the small intestine. The singular form is *villus.*

viscous a gel-like consistency.

vitamin A a fat-soluble vitamin. Its three chemical forms are *retinol* (the alcohol form), *retinal* (the aldehyde form), and *retinoic acid* (the acid form).

vitamins essential, nonkcaloric, organic nutrients needed in tiny amounts in the diet.

voluntary activities the component of a person's daily energy expenditure that involves conscious and deliberate muscular work—walking, lifting, climbing, and other physical activities. Voluntary activities normally require less energy in a day than basal metabolism does.

waist circumference a measurement used to assess a person's abdominal fat.

wasting the breakdown of lean tissue that results from disease or malnutrition.

water balance the balance between water intake and water excretion that keeps the body's water content constant.

water intoxication the rare condition in which body water contents are too high. The symptoms may include confusion, convulsion, coma, and even death in extreme cases.

weight training the use of free weights or weight machines to provide resistance for developing muscle strength and endurance; also called *resistance training.*

wellness maximum well-being; the top range of health states; the goal of the person who strives toward realizing his or her full potential physically, mentally, emotionally, spiritually, and socially.

wheat gluten (GLU-ten) a family of water-insoluble proteins in wheat. The protein in wheat gluten that has toxic effects in celiac disease is called *gliadin* (GLY-ah-din).

xerostomia (ZEE-roh-STOE-me-ah) dry mouth caused by reduced salivary flow.

zygote (ZY-goat) the product of the union of ovum and sperm; so-called for the first two weeks after fertilization.

Photo Credits

CHAPTER 1

1: center, © Rosenfeld/Corbis 3: top left, © PhotoDisc, Inc. 4: bottom center, © Michael Newman/PhotoEdit 4: center left, © PhotoDisc, Inc. 4: center left, © PhotoDisc, Inc. 4: top left, © Becky Luigart-Stayner/Corbis 4: center left, © Bonnie Kamin/PhotoEdit 16: center, © David Hanover Photography 16: center left, © PhotoDisc, Inc. 16: center right, © David Hanover Photography 17: top right, © Warren Morgan/Corbis 18: top left, © Polara Studios, Inc. 18: center left, © Polara Studios, Inc. 18: bottom left, © Polara Studios, Inc. 19: center left, © Matthew Farruggio 19: bottom left, © Matthew Farruggio 19: top left, © Polara Studios, Inc. 19: center left, © Polara Studios, Inc. 20: top left, © Matthew Farruggio 22: top left, Polara Studios, Inc.

CHAPTER 2

38: center, © Envision/Corbis 57: top left, © Charles Gupton/PictureQuest 59: top right, © AP Wide World Photos 59: top left, © Skjold/The Image Works

CHAPTER 3

60: center, © photocuisine/Corbis 61: top right © Polara Studios, Inc. 64: top left, Brian Leatart/Foodpix/Jupiter Images 65: center right, © Polara Studios, Inc. 68: top left, Matthew Farruggio 69: bottom left, © Craig Moore 77: top left, © Polara Studios, Inc. 77: center left, © Polara Studios, Inc. 77: center left, © Polara Studios, Inc. 77: bottom left, © Polara Studios, Inc.

CHAPTER 4

84: center, © age fotostock/SuperStock 85: top right, © Stock Image/SuperStock 93: center right, © FoodCollection/StockFood America 99: bottom left, © Polara Studios, Inc 99: bottom right © Polara Studios, Inc. 99: top left, © Polara Studios, Inc. 99: top center, © Polara Studios, Inc. 99: top right © Polara Studios, Inc. 99: bottom right © Polara Studios, Inc. 99: bottom center, © Polara Studios, Inc. 105: top right Matthew Farruggio 106: top right © www.comstock.com 106: top left, Matthew Farruggio

CHAPTER 5

110: center, © PhotoEdit 116: top right ©Ariel Skelley/Corbis 119: bottom right FAO/13670/G. Kent 119: bottom left, FAO/7752/F. Bolts 122: bottom left, © Polara Studios, Inc 127: center left, © Polara Studios, Inc. 127: center, © Polara Studios, Inc.

CHAPTER 6

130: center, © Envision/Corbis 136: bottom left, Image Source/Getty Images 143: bottom right, © Rick Schaff 149: bottom left, © Geri Engberg 150: top right, Matthew Farruggio 153: top right, © PhotoDisc, Inc.

CHAPTER 7

156: center, © photocuisine/Corbis 158: top left, Courtesy Amgen, Inc. Photo John Shoitis/The Rockefeller University 160: bottom left, Stockbyte/Jupiter Images 165: bottom right, Lori Adamski Peek/Getty Images 168: top left, Matthew Farruggio 168: top center, Matthew Farrugio 168: top right, Matthew Farrugio 169: top right © Joe Sampson, Courtesy of Jennifer Portnick 176: top right, Matthew Farruggio 181: top right © Tony Freeman/PhotoEdit 183: top left, © Michael Newman/PhotoEdit

CHAPTER 8

186: center, © Corbis 192: top left, © 2002 Massachusetts Medical Society 193: top right © Julie Houck/Corbis 194: top left, Biophoto Associates/Photo Researchers, Inc. 197: top right © Polara Studios, Inc. 198: top left © Polara Studios, Inc. 201: top right © Polara Studios, Inc. 202: top left, © Polara Studios, Inc. 202: center left © Polara Studios, Inc. 204: top left, © Polara Studios, Inc. 219: top left, Brassica Protection Products 219: bottom center, Courtesy of Flax Council of Canada 219: top right, EyeWire, Inc 219: center, EyeWire, Inc 219: bottom left, Matthew Farruggio 219: center, PhotoDisc, Inc/Getty Images 219: top, PhotoDisc, Inc/Getty Images 219: center, PhotoDisc, Inc/Getty Images 219: center right PhotoDisc, Inc/Getty Images 219: bottom right, PhotoDisc, Inc/Getty Images 219: top, PhotoDisc, Inc/Getty Images 219: center left, PhotoDisc/Getty Images 219: bottom center, PhotoDisc/Getty Images 221: top left, © Craig M. Moore

CHAPTER 9

223: center, © Corbis 224: top left © David Young Wolff/PhotoEdit 227: top right © Craig M. Moore 229: top left, Matt Farruggio 229: top, Matt Farruggio 229: top center, Matt Farruggio 229: top, Matt Farruggio 229: top right Matt Farruggio 229: bottom left, Matt Farruggio 229: bottom, Matt Farruggio 229: bottom center. Matt Farruggio 229: bottom, Matt Farruggio 229: bottom right Matt Farruggio 231: top right © Polara Studios, Inc. 231: bottom right Courtesy of Gjon Mili 240: top left, © L. V. Bergman & Associates/Corbis 240: center left, © L. V. Bergman & Associates/Corbis 243: top right, © Benjamin F. Fink Jr./Getty Images 245: bottom right © Bob Daemmrich/The Image Works 247: bottom right © Dr. P Marrazi/Science Photo Library/Photo Researchers, Inc. 252: top left, © Tom Carter/ PhotoEdit

CHAPTER 10

256: center, Betsie Van Der Meer/Getty Images 257: top right © Masterfile 260: top left, Petit Format/Nestle/Photo Researchers 260: bottom left, Petit Format/Nestle/Photo Researchers 260: top right Petit Format/Nestle/Photo Researchers 260: bottom right, Randy Lincks/Masterfile 264: top left, Biophoto/Photo Researchers, Inc. 269: top center, Tracy Frankel/lmage Bank/Getty Images 274: bottom left, © James W. Hanson, M.D./NICHD 276: top left, © Ariel Skelley/Corbis 279: bottom right Mark Scott/Getty Images 283: bottom right, © Vic Bider/PhotoEdit 284: center left. Courtesy of Pamela R. Erickson 284: bottom left, © E.H. Gill/Custom Medical Stock Photo 285: bottom right © Polara Studios, Inc. 287: bottom right, © Felicia Martinez/PhotoEdit

Index

This index lists primarily topics that received significant mention in the text. Inclusive pages (for example, 53–56) indicate major discussions; pages in **bold-face** refer to defined terms; pages in *italics* refer to figures, diagrams, or chemical structures; pages followed by a "t" refer to tables; pages followed by an "n" refer to notes; letter-number combinations (such as C-24) refer to appendixes.

Race. *See* Ethnicity/race
Radiation emergencies, potassium iodide in, 246
Radiation enteritis, **597**
Radiation therapy, **596–597,** 597t, 599
RAE (retinol activity equivalents), **191**
Rancid fats, **87**
Ratios, F–2
RBC (red blood cell) count, 367t, E–9 to E–10
RBP (retinol-binding protein), **188,** 366, 368
RDs (registered dietitians), **36–37,** 355
"Reach for It" program, 350
Rebound hyperglycemia, **527**
Recall interviews, 360–361, 360t
Recommended Dietary Allowances (RDA), **9,** *10*
  copper, 246
  folate, 204, 263, 339
  iodine, 246
  iron, 241, 266, 314, 339
  magnesium, 236, 265
  niacin, 202
  phosphorus, 236, 265
  protein, 121, 261–262, 575
  riboflavin, 201
  thiamin, 201
  vitamin A, 191
  vitamin B$_6$, 204
  vitamin B$_{12,}$ 206, 264, 338
  vitamin C, 209
  zinc, 244, 266, 339
Rectum, *39, 40,* 42
Red blood cell (RBC) count, 367t, E–9 to E–10
Refeeding syndrome, **416,** 431
Refined grains, **200**
Reflexology, 611t, **615**
Reflux esophagitis, **451**
Registered dietetic technicans, 355
Registered dietitians (RDs), **36–37,** 355
Rehydration solutions, 472
Religions, diet and, 3
Renal colic, **582**
Renal diets, *396,* 400
Renal diseases, 566–586. *See also* Kidneys
  acute renal failure, **568–572,** 570t
  chronic renal failure, 572–580, 573t, 575t
  diet-drug interactions in, 579t
  enteral formulas for, G–3
  kidney stones, **580–583,** *580,* 581t, 582t
  kidney transplants, 577–580, 579t, 580t
  nephrotic syndrome, **566–568,** *568*
Renal osteodystrophy, **573**
Renal reserve, 572
Renal threshold, **514**
Renin, 224, 566
Renin-angiotensin mechanism, 224
Requirements, nutrient, **9**
Resections
  gastric, **457,** *458*
  intestinal, 482–485, *484*
Residue, **383**
Resistance training, in HIV infections, 605, *605*
Resistant starches, **64**
Resistin, 535, **537**
Respiratory failure, 438–439

Respiratory muscle strength, 369
Respiratory stress, 435–439
  chronic obstructive pulmonary disease, **435–438,** *435, 436*
  defined, **428**
  enteral formulas for, G–3
  respiratory failure, 438–439
Respiratory system, 435, *435, 436*
Resting metabolic rate (RMR), **135**
Resveratrol, 216n
Retina, **189,** *189,* 518, 528
Retinal, 188. *See also* Vitamin A
Retinoic acid, 188–189. *See also* Vitamin A
Retinol, 188. *See also* Vitamin A
Retinol activity equivalents (RAE), **191**
Retinol-binding protein (RBP), **188,** 366, 368
Retinopathy, diabetes and, 518, 528
Rheumatoid arthritis, **334**
Riboflavin, 128, 201–202, *202,* 211t. *See also* B vitamins
Rice, white, 200
Rickets, **193,** 281, 567, *568*
Right to die, 424–426
Risk factors
  cancer, 591–593, 592t, 593t, 595t
  cardiovascular disease, 543t
  chronic diseases, *349*
  coronary heart disease, 541–548, 542t, 543t, 544t
  HIV infection, 602
  malnutrition, 342, 356–357, 356t, 357t
  multiple organ failure, 444, 444t
RMR (resting metabolic rate), **135**
Rooting reflex, 295, **296**
Rotavirus, 282n
Roux-en-Y gastric bypass, 460–461, *460*

Saccharin, **70,** 71t
Safety. *See* Food safety
Saliva, **45**
  dental caries and, 82
  in digestion, 39, *40–41,* 46
Salivary glands, **45,** 46
Salt, **226.** *See also* Sodium
  in cystic fibrosis, 478
  in fluid and electrolyte balance, 226
  high blood pressure and, 229, 350, 552
  iodized, 246
  reducing intake of, 230
  sodium amount in, 228
Salt-sensitive individuals, **229**
Saponins, 545
Sarcomas, 591
Sarcopenia, **336–337,** *337*
Saturated fats
  blood cholesterol and, 91, 106–107, *107,* 544
  intake recommendations, 94–95, 104, 106–107, 544–545, 544t
  replacing in diet, 107–108, 108t
  sources of, *99,* 108t
Saturated fatty acids
  stearic acid, 91n
  structure of, **86–87,** *87*
Saturation, 86–87, *87*
Schiavo, Terri, 425
Schilling test, E–12

Schizophrenia, malnutrition and, **493**
School Breakfast Program, 311
Schools
  hunger and performance in, 302
  nutrition at, 311–313, 312t
  recess followed by lunch in, 309
Scurvy, **207,** 208–209
Secondary amenorrhea, **185**
Secondhand smoke, 272
Secretory diarrhea, 472
Segmentation, 43, *43,* 44, **44**
Selective menus, 395, **397**
Selenium, 245, 248t
Self-esteem, physical activity and, 170
Self-monitoring of blood glucose, **519**
Semipermeable membranes, 587, *587,* **589**
Semiselective menus, 395, **397**
Semivegetarians, **129**
Senile cataracts, 333
Senile dementia, **334–335**
Senior Farmers Market Nutrition Program, 343
Sepsis, 430, **431,** 443–444, *443*
Serotonin, 334
Serum, blood, 366
Serum ferritin, E–10 to E–11
Serum iron, E–10
Serum phosphate, 570
Serving sizes
  in carbohydrate counting, 523t
  in the exchange system, C–1
  on food labels, 24–25, *25*
  portion control, 22
  in weight-gain diets, 174
  in weight-loss diets, 167
Set-point theory, **159–160**
SGA (small for gestational age), 257
Shock, **431**
Short-bowel syndrome, 482–485, **483,** *484*
Short-term memory loss, 334, 335t
Sibutramine, 164
Sickle-cell anemia, 209, 374
SIDS (sudden infant death syndrome), 272
Silicon, 249
Simple carbohydrates, 61n. *See also* Monosaccharides
Simple goiter, **245,** *245*
Simple sugars, **61**
Single-nucleotide polymorphisms, 374, **376**
Sinusoids, **500**
SIRS (systemic inflammatory response syndrome), 430–**431,** 443–444, *443*
Sjögren's syndrome, **447**
Skin color
  beta-carotene excess and, 191, *192*
  vitamin D synthesis and, 194
Skinfold measures, **142,** E–2, E–7
Skip lesions, 480
Sleeping, 286, 419
Small for gestational age (SGA), 257
Small intestine, **39**
  absorption in, 48, *49*
  digestion in, *40–41,* 42, 44, 46–47
  HIV infection and changes in, 603
  peristalsis in, *43*
Smokeless tobacco, 317–318

Daily Values for Food Labels

The Daily Values are standard values developed by the Food and Drug Administration (FDA) for use on food labels. Daily Values for protein, vitamins, and minerals reflect average allowances based on the RDA. Daily Values for nutrients and food components, such as fat and fiber, that do not have an established RDA but do have important relationships with health are based on recommended calculation factors as noted.

Nutrient	Amount
Protein[a]	50 g
Thiamin	1.5 mg
Riboflavin	1.7 mg
Niacin	20 mg NE
Biotin	300 µg
Pantothenic acid	10 mg
Vitamin B_6	2 mg
Folate	400 µg
Vitamin B_{12}	6 µg
Vitamin C	60 mg
Vitamin A	5000 IU
Vitamin D	400 IU
Vitamin E	30 IU
Vitamin K	80 µg
Calcium	1000 mg
Iron	18 mg
Zinc	15 mg
Iodine	150 µg
Copper	2 mg
Chromium	120 µg
Selenium	70 µg
Molybdenum	75 µg
Manganese	2 mg
Chloride	3400 mg
Magnesium	400 mg
Phosphorus	1000 mg

[a]The Daily Values for protein vary for different groups of people: pregnant women, 60 g; nursing mothers, 65 g; infants under 1 year, 14 g; children 1 to 4 years, 16 g.

Food Component	Amount	Calculation Factors
Fat	65 g	30% of kcalories
Saturated fat	20 g	10% of kcalories
Cholesterol	300 mg	Same regardless of kcalories
Carbohydrate (total)	300 g	60% of kcalories
Fiber	25 g	11.5 g per 1000 kcalories
Protein	50 g	10% of kcalories
Sodium	2400 mg	Same regardless of kcalories
Potassium	3500 mg	Same regardless of kcalories

Note: Daily Values were established for adults and children over 4 years old. The values for energy-yielding nutrients are based on 2000 kcalories a day.

Glossary of Nutrient Measures

kcal: kcalories; a unit by which energy is measured (Chapter 1 provides more details).

g: grams; a unit of weight equivalent to about 0.03 ounces.

mg: milligrams; one-thousandth of a gram.

µg: micrograms; one-millionth of a gram.

IU: international units; an old measure of vitamin activity determined by biological methods (as opposed to new measures that are determined by direct chemical analyses). Many fortified foods and supplements use IU on their labels.

- For vitamin A, 1 IU = 0.3 µg retinol, 3.6 µg β-carotene, or 7.2 µg other vitamin A carotenoids.
- For vitamin D, 1 IU = 0.025 µg cholecalciferol.
- For vitamin E, 1 IU = 0.67 natural α-tocopherol (other conversion factors are used for different forms of vitamin E).

mg NE: milligrams niacin equivalents; a measure of niacin activity (Chapter 8 provides more details).

- 1 NE = 1 mg niacin.
  - = 60 mg tryptophan (an amino acid).

µg DFE: micrograms dietary folate equivalents; a measure of folate activity (Chapter 8 provides more details).

- 1 µg DFE = 1 µg food folate.
  - = 0.6 µg fortified food or supplement folate.
  - = 0.5 µg supplement folate taken on an empty stomach.

µg RAE: micrograms retinol activity equivalents; a measure of vitamin A activity (Chapter 8 provides more details).

- 1 µg RE = 1 µg retinol.
  - = 12 µg β-carotene.
  - = 24 µg other vitamin A carotenoids.

Body Mass Index (BMI)

Height	18	19	20	21	22	23	24	25	26	27	28	29	30	31	32	33	34	35	36	37	38	39	40
									Body Weight (pounds)														
4'10"	86	91	96	100	105	110	115	119	124	129	134	138	143	148	153	158	162	167	172	177	181	186	191
4'11"	89	94	99	104	109	114	119	124	128	133	138	143	148	153	158	163	168	173	178	183	188	193	198
5'0"	92	97	102	107	112	118	123	128	133	138	143	148	153	158	163	168	174	179	184	189	194	199	204
5'1"	95	100	106	111	116	122	127	132	137	143	148	153	158	164	169	174	180	185	190	195	201	206	211
5'2"	98	104	109	115	120	126	131	136	142	147	153	158	164	169	175	180	186	191	196	202	207	213	218
5'3"	102	107	113	118	124	130	135	141	146	152	158	163	169	175	180	186	191	197	203	208	214	220	225
5'4"	105	110	116	122	128	134	140	145	151	157	163	169	174	180	186	192	197	204	209	215	221	227	232
5'5"	108	114	120	126	132	138	144	150	156	162	168	174	180	186	192	198	204	210	216	222	228	234	240
5'6"	112	118	124	130	136	142	148	155	161	167	173	179	186	192	198	204	210	216	223	229	235	241	247
5'7"	115	121	127	134	140	146	153	159	166	172	178	185	191	198	204	211	217	223	230	236	242	249	255
5'8"	118	125	131	138	144	151	158	164	171	177	184	190	197	203	210	216	223	230	236	243	249	256	262
5'9"	122	128	135	142	149	155	162	169	176	182	189	196	203	209	216	223	230	236	243	250	257	263	270
5'10"	126	132	139	146	153	160	167	174	181	188	195	202	209	216	222	229	236	243	250	257	264	271	278
5'11"	129	136	143	150	157	165	172	179	186	193	200	208	215	222	229	236	243	250	257	265	272	279	286
6'0"	132	140	147	154	162	169	177	184	191	199	206	213	221	228	235	242	250	258	265	272	279	287	294
6'1"	136	144	151	159	166	174	182	189	197	204	212	219	227	235	242	250	257	265	272	280	288	295	302
6'2"	141	148	155	163	171	179	186	194	202	210	218	225	233	241	249	256	264	272	280	287	295	303	311
6'3"	144	152	160	168	176	184	192	200	208	216	224	232	240	248	256	264	272	279	287	295	303	311	319
6'4"	148	156	164	172	180	189	197	205	213	221	230	238	246	254	263	271	279	287	295	304	312	320	328
6'5"	151	160	168	176	185	193	202	210	218	227	235	244	252	261	269	277	286	294	303	311	319	328	336
6'6"	155	164	172	181	190	198	207	216	224	233	241	250	259	267	276	284	293	302	310	319	328	336	345

Under-weight (<18.5) **Healthy Weight** (18.5–24.9) **Overweight** (25–29.9) **Obese** (≥30)

Find your height along the left-hand column and look across the row until you find the number that is closest to your weight. The number at the top of that column identifies your BMI. Chapter 6 describes how BMI correlates with disease risks and defines obesity. The area shaded in green represents healthy weight ranges.

Take Assessment: Course Final Exam

Name	Course Final Exam
Instructions	Once your open the test, you have 4 hours to complete it.
Multiple Attempts	Not allowed. This Test can only be taken once.
Force Completion	This Test can be saved and resumed later.

▼ **Question Completion Status:**

Question 1　　　　　　　　　　　　　　　　　　　　　**1 points**　Save

An infant's appetite declines markedly around the first birthday.

○ True
○ False

Question 2　　　　　　　　　　　　　　　　　　　　　**1 points**　Save

Children who eat nutritious breakfasts function better than their peers who do not.

○ True
○ False

Question 3　　　　　　　　　　　　　　　　　　　　　**1 points**　Save

The more often a food is presented to a young child, the less likely the child will like that food.

○ True
○ False

Question 4　　　　　　　　　　　　　　　　　　　　　**1 points**　Save

No clear evidence currently supports a benefit of restricting sodium to lower blood pressure in children and adolescents.

○ True
○ False

Question 5　　　　　　　　　　　　　　　　　　　　　**1 points**　Save

Recommendations limiting fat and cholesterol are not intended for infants or children under two years of age.

○ True
○ False

Question 6   **1 points**   Save

Children who spend more than one or two hours watching televion (or using other sedentary media) can become obese even while consuming fewer kcalories than more active children.

- ○ True
- ● False

Question 7   **1 points**   Save

How many ounces from the meat and beans group are needed daily to meet nutrient needs for a child who needs 1000 kcal/day?

- ● 2 ounces
- ○ 4 ounces
- ○ 6 ounces
- ○ about 1/2 pound.

Question 8   **1 points**   Save

To prevent iron deficiency, a child needs _____ milligrams of iron per day.

- ○ 2-3
- ○ 3-5
- ● 7-10
- ○ 9-12

Question 9   **1 points**   Save

The adolescent growth spurt:

- ● begins and ends earlier in girls than in boys.
- ○ affects every organ except the brain.
- ○ decreases total nutrient needs.
- ○ causes a greater weight gain in girls.

Question 10   **1 points**   Save

204

Research shows that the requirement for this nutrient is higher for smokers than for non-smokers.

- ○ beta carotene.
- ○ folate.
- ● vitamin C.
- ○ fiber.

Question 11   **1 points**   Save

Studies suggest that a diet providing ample carotenoids, vitamin C, and vitamin E may help to prevent the early onset of cataracts.

- ⬤ True
- ◯ False

Question 12   **1 points**   Save

The protein needs of older adults appear to be higher than those of younger people.

- ◯ True
- ⬤ False

Question 13   **1 points**   Save

Zinc intake in the elderly is commonly higher than needed.

- ◯ True
- ⬤ False

Question 14   **1 points**   Save

Older adults spend more money per person on foods to eat at home than other age groups.

- ⬤ True
- ◯ False

Question 15   **1 points**   Save

Which of the following vitamins is NOT involved in cognition in older adults?

- ◯ folate
- ◯ vitamin B12
- ⬤ vitamin A
- ◯ vitamin B6

Question 16   **1 points**   Save

Severe fat restriction among older adults may lead to:

- ⬤ nutrient deficiencies.
- ◯ weight gain.
- ◯ greater risk of cancer.
- ◯ dehydration.

Question 17   **1 points**   Save

The DRI Committee found that _____ percent of adults aged 51 years or older lose

the ability to absorb vitamin B12.

- ○ 5 to 10
- ● 10 to 30
- ○ 25 to 50
- ○ 50 to 69

Question 18   1 points   Save

One nutrient that is underconsumed by many elderly people is:

- ○ vitamin A.
- ○ protein.
- ○ vitamin C.
- ● vitamin D.

Question 19   1 points   Save

Which of the following groups of older adults is most likely to be poorly nourished?

- ○ those living as intact couples in retirement community that has communal dining.
- ○ those with the college degrees or higher
- ● men who live alone

Question 20   1 points   Save

In the absence of smoking or drinking excessively, what factor has the most influence on your long-term health prospects?

- ● physical activity
- ○ genetics
- ○ dietary choices
- ○ occupation

Question 21   1 points   Save

All inborn errors cause severe illness or death.

- ○ True
- ● False

Question 22   1 points   Save

Phenylketonuria is a metabolic disorder that affects lipid metabolism.

- ○ True
- ● False

Question 23

1 points   Save

Repeated bouts of pneumonia may be associated with:

- ◉ dysphagia.
- ○ irritable bowel syndrome.
- ○ peptic ulcers.
- ○ diverticular disease.

Question 24

1 points   Save

Clients with reflux esophagitis should be instructed to:

- ○ drink liquids with meals.
- ◉ eat small, frequent meals.
- ○ lie down soon after eating.
- ○ consume caffeine containing beverages with meals.

Question 25

1 points   Save

General discomfort in the upper abdominal area with complaints of stomach pain, heartburn, fullness, nausea, and bloating are descriptive of:

- ○ Sjogren's Syndrome.
- ◉ dyspepsia.
- ○ hiatal hernia.
- ○ peptic ulcer disease.

Question 26

1 points   Save

One of the most COMMON causes of acute gastritis is:

- ○ radiation therapy.
- ◉ viral infections.
- ○ food allergies.
- ○ repeated use of aspirin.

Question 27

1 points   Save

Which of the following plays the PRIMARY role in treatment of ulcers?

- ○ surgery
- ○ relaxation therapy
- ◉ drug therapy
- ○ phlebotomy

Question 28 **1 points** Save

Liquids are restricted following a gastrectomy because they:

- ○ are likely to cause aspiration.
- ● can speed up the emptying rate of the stomach.
- ○ add unwanted kcalories.
- ○ slow down the rate of gastric emptying.

Question 29 **1 points** Save

Frequent complications of gastric surgery include all of the following except:

- ○ bone disease.
- ○ weight loss.
- ● protein malabsorption.
- ○ iron-deficiency anemia.

Question 30 **1 points** Save

Dietary treatment of PKU includes:

- ○ restriction of L-caritine.
- ● restriction of phenylalanine
- ○ restriction of Nutrasweet/ aspartame.
- ○ Answers 2 and three are correct.

Question 31 **1 points** Save

Daily exercise can help treat constipation by stimulating peristalsis.

- ● True
- ○ False

Question 32 **1 points** Save

MCT (medium chain triglyceride) oil requires bile for digestion and absorption.

- ○ True
- ● False

Question 33 **1 points** Save

Celiac disease is a genetic disease characterized by an abnormal immune response to certain fragments of gluten proteins. The answer to this question is true. I'm not sure it's well elaborated in your text. So, this is a gift. Select "true."

- ● True

○ False

Question 34   1 points   Save

Irritable bowel syndrome occurs more frequently in men than women and episodes of acute symptoms often increase with age.

○ True

◉ False

Question 35   1 points   Save

The client with a colostomy or ileostomy should have a diet that is restricted in fluids.

◉ True

○ False

Question 36   1 points   Save

Which of the following foods will the nurse INCLUDE in the diet plan for a patient with celiac disease?

○ oatmeal

◉ scrambled eggs

○ whole wheat toast

○ rye crackers

○ barley soup

Question 37   1 points   Save

Shawna is a 21-year-old college student who have been diagnosed with Crohn's disease. She is underweight and has a poor appetite. Realizing that Shawna is at nutritional risk, the nurse advises that Shawna:

◉ consume a liquid nutrition supplement two to three times a day.

○ increase the fiber in her diet.

○ Drink at least four cups of milk each day.

○ Eat more fast foods and sodas.

Question 38   1 points   Save

The nurse is instructing a client with diverticular diease on appropriate foods to include in his diet in order to prevent progression of the disease, She would tell him to consume more of all of the following except:

○ whole grain breads.

◉ enriched rice or pasta.

○ legumes.

○ fresh vegetables.

Question 39

1 points   Save

A patient with an ileostomy complains about excessive odor. What information would cause the nurse to suspect a cause for this odor?

○ The patient tells the nurse he eats yogurt every day for breakfast.

○ The patient smokes.

● The patient eats eggs each day for breakfast.

⊘ The patient routinely consumes unsweetened cranberry juice.

Question 40

1 points   Save

People with malnutrition frequently experience fatty liver.

● True

○ False

Question 41

1 points   Save

A patient with cirrhosis may present with no symptoms.

● True

○ False

Question 42

1 points   Save

Hepatic encephalopathy may be caused by an increased ratio of aromatic amino acids to branched amino acids in brain tissue.

● True

○ False

Question 43

1 points   Save

Diuretics are often prescribed to control ascites.

● True

⊘ False

Question 44

1 points   Save

High-fat foods are not recommended for patients with cirrhosis who need to increase their kcalorie intake.

○ True

● False

Question 45

1 points   Save

Many patients with cirrhosis require medication or insulin to control hyperglycemia.

- ◉ True
- ○ False

Question 46

1 points   Save

When triglycerides accumulate in the liver, the liver:

- ○ becomes smaller.
- ◉ enlarges.
- ○ turns to stone.
- ○ functions less effectively.
- ◉ Answers 2 and 4 are correct.

Question 47

1 points   Save

Laboratory findings associated with fatty liver include all of the following except:

- ◉ elevated HDL.
- ○ elevated ALT and AST.
- ○ elevated glucose.
- ○ elevated triglcyerides.

Question 48

1 points   Save

A client with cirrhosis who develops changes in personality and motor function may be exhibiting early symptoms of:

- ○ fatigue.
- ○ hepatic regression.
- ○ hepatitis.
- ○ malnutrition.
- ◉ hepatic coma.

Question 49

1 points   Save

The number of deaths from CVD is greater among men than women.

- ○ True
- ◉ False

Question 50

1 points   Save

The most common cause of an aneurysm is high blood pressure.

○ True

● False

Question 51   **1 points**   [Save]

People with Metabolic Syndrome may benefit from a slightly higher fat intake in order to reduce their carbohydrate intake.

○ True

● False

Question 52   **1 points**   [Save]

The DASH diet is useful for treating hypertension and also reduces the risk of CHD.

● True

○ False

Question 53   **1 points**   [Save]

Patients who have had a stroke may develop nutritional problems due their inability to communicate their food preferences.

● True

○ False

Question 54   **1 points**   [Save]

Which of the following is NOT TRUE regardings trans fatty acids?

○ They raise LDL levels.

○ They are found in foods made with partially hydrogenated oils.

○ Their intake should be kept as low as possible.

● They raise HDL levels.

Question 55   **1 points**   [Save]

Weight loss is associated with all of the following EXCEPT:

○ a decline in blood pressure.

○ a decrease in LDL cholesterol.

● an increase in triglycerides.

○ a decrease in insulin resistance.

Question 56   **1 points**   [Save]

Chronic protein-energy malnutrition that develops as a consequence of heat failure is called _____.

- CHF
- cardiomegaly
- cardiac cachexia.
- tachycardia.

Question 57

1 points   Save

The nurse is counseling a patient who recently had a heart attack. Which of the following dietary recommendations is LEAST appropriate?

- Eat two or three servings of fatty fish per week.
- When having grain foods, choice whole grains, not refined grains.
- Include fresh fruits and vegetables at most meals.
- Drink diet sodas.

Question 58

1 points   Save

The medical nutrition therapy for nephrotic syndrome helps to prevent protein-energy malnutrition and alleviate edema.

- ● True
- ○ False

Question 59

1 points   Save

Late in the course of acute renal failure, urine volume may increase significantly and large amounts of fluid may have to be provided.

- ○ True
- ● False

Question 60

1 points   Save

Individuals on peritoneal dialysis have lower protein needs than those on hemodialysis.

- ○ True
- ● False

Question 61

1 points   Save

Dialysis removes excess fluids and wastes from the blood by employing the principles of diffusion, osmosis, and ultrafiltration.

- ● True
- ○ False

Question 62

1 points   Save

Manifestations of the nephrotic syndrome include:

- ● low serum albumin.
- ○ uremia.
- ○ low blood lipids.
- ○ dehydration.

Question 63

1 points Save

Which of the following is a characteristic of an enteral formula designed for a renal patient?

- ● more kcalorically dense
- ○ higher in protein
- ○ more free water
- ○ higher in branched chain amino acids

Question 64

1 points Save

Chronic renal failure is typically diagnosed once _____ percent of kidney function has been lost.

- ○ 10
- ○ 25
- ○ 50
- ● 75
- ○ 95

Question 65

1 points Save

A client in acute renal failure has a urine output of 500 milliliters. The nurse determines his fluid need to be:

- ○ 500 milliliters
- ○ 800 milliliters
- ● 1000 milliliters
- ○ 1200 milliliters

Question 66

1 points Save

Following kidney transplantation, dietary restrictions are generally liberalized; however, if fluid retention and hypertension are present, the nurse understands that a restriction of _____ may be necessary.

- ○ protein
- ○ potassium
- ● sodium

○

○ calcium

○ magnesium

Question 67　　　　　　　　　　　　　　　　　　　　　　　**1 points**　　Save

Most often, cancer is caused by interactions between a person's genes and environmental agents.

● True

○ False

Question 68　　　　　　　　　　　　　　　　　　　　　　　**1 points**　　Save

Nutrition-related factors may increase or decrease the risk of developing cancer.

● True

○ False

Question 69　　　　　　　　　　　　　　　　　　　　　　　**1 points**　　Save

Some people with cancer fail to regain lean body mass even when they are receiving adequate nutrients and energy.

● True

○ False

Question 70　　　　　　　　　　　　　　　　　　　　　　　**1 points**　　Save

Recurrent bacterial pneumonia is classified as an AIDS-defining illness.

● True

○ False

Question 71　　　　　　　　　　　　　　　　　　　　　　　**1 points**　　Save

Fat accumulation is a common side effect of drug treatments for HIV infection.

● True

○ False

Question 72　　　　　　　　　　　　　　　　　　　　　　　**1 points**　　Save

An example of a patient who may require long-term tube feeding is a patient:

○ with breast cancer.

● with neck cancer.

○ with bone cancer.

○ with skin cancer.

Question 73

1 points   Save

The nurse is teaching a class on cancer prevention. Someone in the audience asks if one's diet affects his or her risk of developing cancer. The nurse makes the following statements. Which one is INCORRECT?

○ "Citrus fruits may decrease the risk of lung cancer."

○ "Cruciferous vegetables, like broccoli and cauliflower, may reduce the risk of several types of cancer."

○ "Foods rich in folate seem to prevent cancer."

● "Dairy foods have been implicated in many types of cancers."

Question 74

1 points   Save

The nurse understands that methotrexate is a chemotherapeutic agent used in the treatment of cancer and that it can cause a patient to become deficient in:

● folate.

○ essential fatty acids.

○ vitamin B12.

○ calcium.

Question 75

1 points   Save

The nurse is working with a client who is in the early stages of HIV. The nurse recognizes all of the following as early symptoms of HIV EXCEPT:

○ malaise.

● infections of the GI tract.

○ nausea.

○ diarrhea.

Save   Submit